HUDSON BAY
GREENLAND SEA
NORWEGIAN SEA
ATLANTIC
OCEAN
IFIC
OCEAN
EQUATOR
SCOTIA SEA
BELLINGHAUSEN SEA
WEDDELL SEA

THE

DEEP SEA DRILLING PROJECT:

A DECADE OF PROGRESS

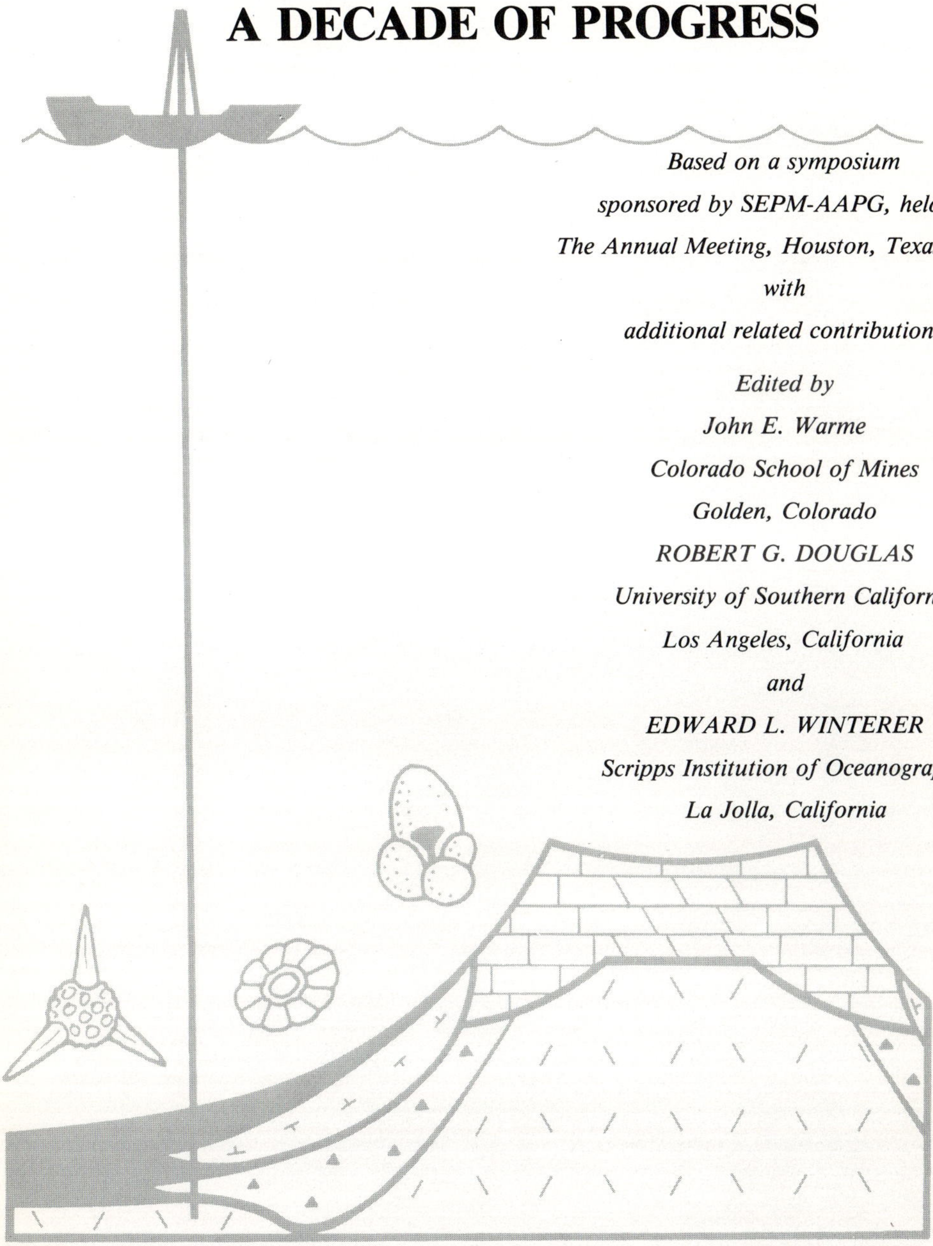

Based on a symposium

sponsored by SEPM-AAPG, held at

The Annual Meeting, Houston, Texas, 1979,

with

additional related contributions

Edited by

John E. Warme

Colorado School of Mines

Golden, Colorado

ROBERT G. DOUGLAS

University of Southern California

Los Angeles, California

and

EDWARD L. WINTERER

Scripps Institution of Oceanography

La Jolla, California

SOCIETY OF ECONOMIC PALEONTOLOGISTS AND MINERALOGISTS

Special Publication No. 32

Tulsa, Oklahoma, U.S.A. October, 1981

PREFACE

At the present the *Glomar Challenger* has drilled over 500 holes over the world ocean, involving hundreds of scientists from dozens of countries. This volume is intended as a review of some of the important results from the most comprehensive, ambitious and successful earth-bound geologic project ever undertaken, rivalled only by the space probes and lunar sampling program. A log of the *Glomar Challenger,* as represented by the Initial Reports volumes for Legs published over the first twelve years of program, is presented in Appendix One. These volumes represent the raw data and initial analyses of those data as they have been collected. Special Publication 32 is a synthesis of much of that information, largely cutting across geographic and time boundaries.

The symposium upon which this volume originated was generated through the support and activities of the Deep Sea Drilling Project (DSDP) Paleoceanography Panel, the SEPM Research Committee, the AAPG Marine Geology Committee, and the 1979 AAPG-SEPM Program Committee. It was held April 4, 1979 at the Annual Meeting in Houston. Additional papers were solicated to develop a more complete published overview, and a few topics that were presented at the symposium subsequently were not submitted as manuscripts or are published elsewhere.

No comprehensive synthesis of all aspects of the DSDP has appeared, and the topic coverage in this volume is by no means complete, being biased towards the sediments and fossils, and their significance for certain aspects of earth history—paleo-geography, -bathymetry, -climatology, -oceanography, -ecology, -environments—all in keeping with our anticipated audience of sedimentary geologists.

Organization of this volume is somewhat arbitrary because most contributions cover too many topics to be simply categorized. Therefore, we encourage the reader to become familiar with the papers in each section, to avoid missing a discussion that may be of special interest to you. A history of the birth and first-decade development of the Deep Sea Drilling Project is given in the opening chapter (also see Nierenberg, 1978). It is hoped that within fifteen years of commencement of DSDP a newly-converted *Glomar Explorer* will begin drilling even deeper holes at critical sites with "riser" (circulating drilling mud) capability and the blowout protection it affords, and with the cooperation and financial support of the petroleum industry. This phase of scientific ocean drilling will include penetration of thicker stratigraphic sections and thicker segments of oceanic crystalline basement.

Problems continue to be generated and attacked by Deep-Sea Drilling; exciting results stemming from careful planning and execution have become its hallmark; we expect more of the same in the current second decade of scientific ocean drilling.

A companion volume of photos from DSDP cores is in preparation by Arthur, Kelts and Scholle, to be published by AAPG.

Editors' Acknowledgements

We wish to thank all the authors and manuscript reviewers for their splendid and prompt cooperation, and the SEPM Publications Committee who backed the volume from its inception. Edwards Brothers, Inc., the publishers, provided exemplary professional service and advice, insuring timely publication. We thank DSDP, SEPM and AAPG, our respective departments, and all the individuals and institutions that helped support the compilation of this volume. Editorial assistance was provided by Kata McCarville, Mark and Elane Hudson, Lise Brinton, and, especially, John Ziegler, all of the Colorado School of Mines, who carefully checked bibliographies and provided other editorial services. Darwin Spearing kindly designed the cover from a sketch by J. Warme, which was inspired by Figure 13 in Kelts and Arthur, herein.

John E. Warme
Robert G. Douglas
Edward L. Winterer
Editors
June, 1981

REFERENCE CITED

NIERENBERG, W. A., 1978, The Deep Sea Drilling Project after Ten Years: Amer. Scientist, v. 66, p. 20–29.

CHALLENGER

THE DEEP SEA DRILLING PROJECT:
A DECADE OF PROGRESS
Edited by John E. Warme, Robert G. Douglas, and Edward L. Winterer

CONTENTS

PART 1
HISTORY AND OVERVIEW

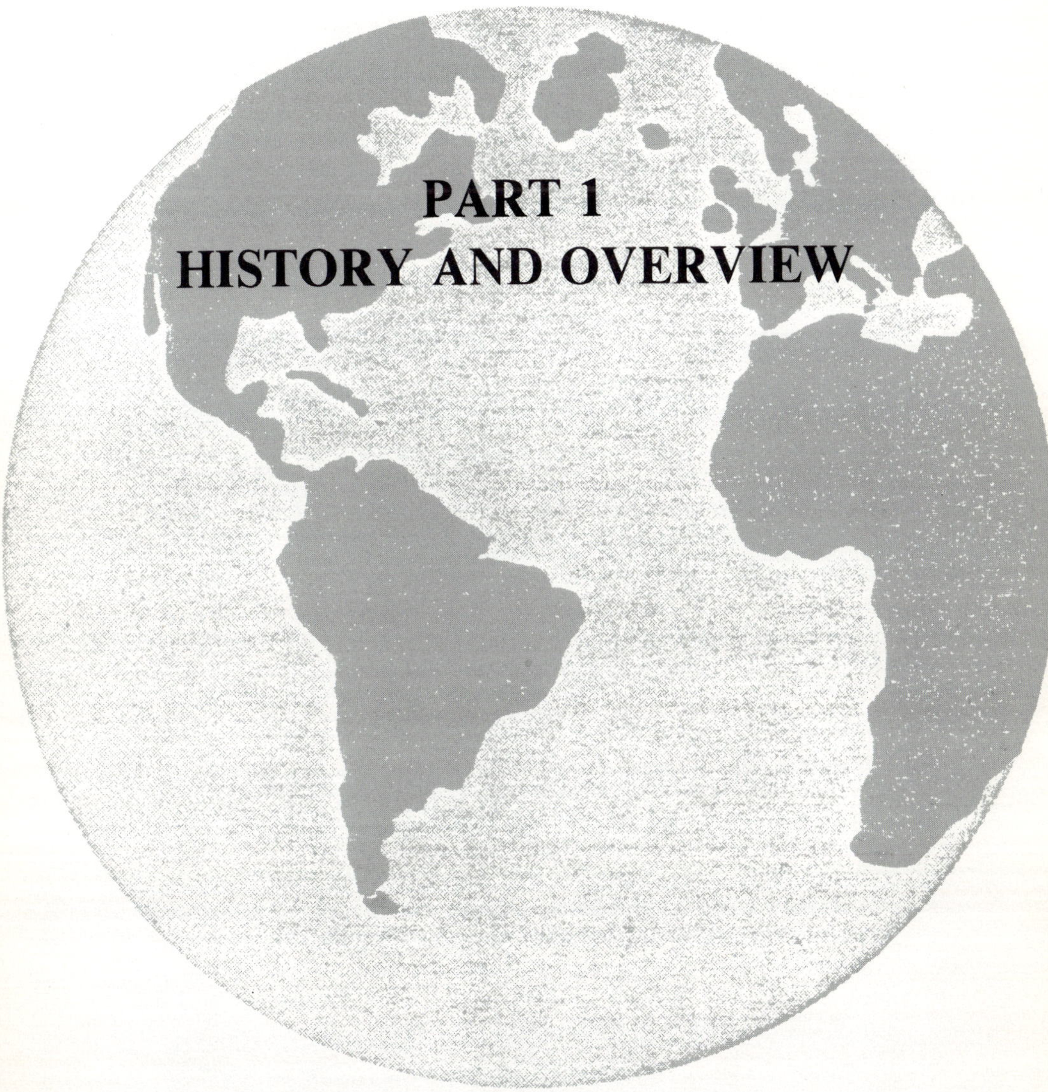

SEPM Special Publication No. 32, p. 1–4, October 1981

THE PAST AND FUTURE OF OCEAN DRILLING*

ROGER REVELLE
Science and Public Policy, University of California, San Diego, California 92093 U.S.A.

**This address was presented on the occasion of the tenth anniversary of scientific ocean drilling, October 30, 1978, Washington, D. C.*

I am talking to you tonight mainly because I am a survivor. Of the small group who formed the informal association of oceanographic laboratories for ocean drilling, and thought up the name JOIDES, Maurice Ewing, Fritz Koczy, and Columbus Iselin are no longer with us, nor is Harry Hess, who inspired us all.

One of the values of being a survivor is that I can remember the ideas about the earth we had when we were young. This can give some perspective to the revolution in understanding that has come about during the last 15 years.

The fundamental doctrine of American geology in the late 1940's was that the continents and ocean basins had been permanent features of the earth's surface, almost since the beginning of geologic time. The crust of the earth beneath the sea was unimaginably old, and the continents had always been about where they are today, though they had probably grown in size over several billion years. The differences in elevation between the sea floor and the continents had perhaps slowly become greater over the geologic ages, as the volume of sea water increased. When Harry Hess first discovered the flat-topped sea mounts of the Central Pacific, he thought they must have been slowly submerged over the last two billion years. It was believed that sediments had gradually but steadily accumulated on the deep-sea floor throughout all this time. The most learned and mathematically able geophysicists published conclusive demonstrations that continental drift was impossible, because of the known physical properties of rocks.

These doctrines received an abrupt shock in 1950 when shallow-water reef corals only about a hundred million years old were found on the flat-topped sea mounts of the Mid-Pacific Mountains at depths of two kilometers, and when Russell Raitt's seismic refraction studies indicated that the deep-sea sediments over vast areas are only one or two hundred meters thick. But I am reasonably certain Maurice Ewing went to his grave believing that somewhere in the deep sea there was a place where sediments of all geologic ages, back to one or two billion years, had been deposited one on top of the other, and that all these sediments were still there, in what earth scientists call a complete geologic column—in effect a complete book of records of most of the geologic history of the earth.

Now, 15 years later, we know that none of these doctrines even faintly resembled the truth. Far from being as old as the earth, the oldest rocks on the ocean floor are less than 4% of the earth's age. One of our planet's major oceans, the Atlantic, did not even exist 200 million years ago. The six continents we know today were born and attained their present location by the fragmenting of a giant single continent, PANGEA, and the subsequent gradual but steady motion of the fragments in different directions over and through the ocean floor.

Ten years ago the new ideas that we now subsume under the rubric of plate tectonics were a reasonable but startling hypothesis, which offered an integrated explanation of many kinds of observations of the earth including analyses of seismic waves, measurements of gravity, magnetism, and heat-flow, delineations of under-sea topography, and the fitting together of the continental margins on the two sides of the Atlantic. But the direct proof of this hypothesis from actual samples of rocks and sediments has come about very largely from the research we are celebrating tonight, the ten years of probing into the sea floor by the Deep Sea Drilling Project. This is surely one of the great achievements in the entire history of the earth sciences.

Let us list some of the principal results of the drilling:

1. Demonstration of the extreme youth, in geologic terms, of the earth's crust beneath the sea.

The oldest rocks found in any of the 460 bore holes during ten years of drilling are between 150 and 200 million years old.

2. Parts of the ocean crust have moved horizontally for great distances, away from the zones of crust formation along mid-ocean ridges.

This is shown by several lines of evidence in the cores of sediments and rocks obtained by the drilling. Three of these are related to the distribution of organic remains, magnetic dip, and times of deposition, respectively. First, in the topmost layers of sediments, a high concentration of organic remains occurs in a narrow band right along the equator, owing to the high organic productivity of the subsurface equatorial waters. In the Pacific, this narrow band is found at progressively greater distances north of the equator with increasing depth in the cores, that is, increasing age of the sediments. These older sediments must have been deposited underneath the equator and afterward moved northward to their present position; the older the sediments, the farther they have moved from the equator. Second, the remanent magnetic dip of older sediments and the underlying rocks in many cores is greater or less, depending on the north-south component of their horizontal displacement since they were deposited, than the dip corresponding to their present position relative to the earth's magnetic pole. Third, sediments of progressively greater age directly overlie the basaltic basement at increasing distances from the zone of formation of the basalt along the mid-ocean ridges.

These and other lines of evidence indicate that the horizontal velocities of the crustal motion have been between 1 and 13 centimeters per year, and that parts of the sea floor have moved several thousand kilometers during the past 200 million years.

3. As it has moved away from the mid-ocean ridge, the sea floor has subsided relative to the ocean surface.

This is shown by older sediments in the drill cores, which were laid down in much shallower water than the depths at which they are found. The older the sediments, the shallower the site of deposition, up to the original depth of the mid-ocean ridge. The rate of subsidence has been 60 to 100 meters per million years. Presumably this subsidence resulted from contraction due to cooling of the sub-oceanic rocks, as they moved away from the zone of magmatic activity along the mid-ocean ridge. It is possible also that portions of the continental margin in the Atlantic and elsewhere have foundered as they were torn apart by rifting.

4. The existence of salt domes, which may have acted as traps for hydrocarbons, in the bottom of the Gulf of Mexico.

The salt was presumably laid down as an evaporite in the early stages of subsidence of this basin, and was afterwards forced upward through the overlying sediments in giant plugs.

5. The drying up and later refilling of the Mediterranean Sea.

Massive salt deposits, laid down between 5 and 12 million years ago, are found in cores from the Mediterranean. Evidently, the entrances to the Mediterranean were blocked during this period and the water in the basin evaporated, forming a huge, mile-deep depression, five times as far below sea level as the present Dead Sea,—the deepest spot on earth today—and as large in area as the Indian subcontinent.

6. The growth of the Atlantic Ocean from a series of narrow stagnant basins into the broad ocean of today.

The oldest sediments in the Atlantic are black shales, salt deposits, and cherts, occurring in limited areas on or near the continental margins.

7. Past climatic and ecological changes in the ocean.

Studies of the sequences of marine fossils found in the cores have contributed greatly to the construction of a geological and evolutionary timetable for all parts of the world ocean, to precise correlation of different types of fossils representing different marine ecosystems and to an understanding of the paleoclimatology of the ocean.

8. Discovery of resource-rich deposits.

New oceanic crust formed along plate margins is hydrothermally altered by circulating sea water, resulting in some areas in the formation of deposits with high concentrations of valuable metals. Along both active and passive continental margins, marine sediments and their contained organic matter are apparently metamorphosed by heat and pressure, with possible formation of fluid hydrocarbons.

9. The age of the Antarctic icecap.

Sediment cores from high latitudes in the Southern Ocean indicate that Antarctica was covered by ice 20 million years ago, and that the ice cap was much more extensive 5 million years ago than at present. At that time worldwide sea levels were about 15 meters lower than today.

10. The fixed location of a "hot spot" in the mantle.

Paleontological and magnetic evidence indicates that the Emperor sea mounts in the northwest Pacific are a continuation of the Hawaiian Island chain, and that all these elevated portions of the sea floor were formed by volcanic activity near the present position of the island of Hawaii. The northern-most of the Emperor sea mounts is the oldest, formed about 70 million years ago, and the ages became progressively younger towards the south and southeast. The first motion of the sea floor away from Hawaii was to the north, and later the direction of motion shifted toward the northwest.

To an old fashioned oceanographer like myself, the technical and engineering accomplishments of the Deep Sea Drilling Project seem as remarkable as the scientific ones. The drilling ship, *Glomar Challenger,* is not fastened to the sea floor in any way, but simply floats freely at the surface. The only means of communication between the ocean bottom and the people on the ship is by underwater sound and the behavior of the drill string. Yet with a sonar transponder planted on the sea floor, a series of sonar receivers at different points on the hull, and powerful thrusters which can move the ship in any direction, all in a computer-linked system, it is possible to maintain the ship for months at a time within about a hundred feet of a fixed geographical position.

When the project began, the drilling bit and drill stem could not be put back into the same hole once they had been pulled out. This meant that when the bit wore out, as it does pretty quickly in rock drilling, the hole had to be abandoned. Now it is possible to re-enter the same hole as many times as desired. The difficulty of doing this can be imagined when one realizes that a string of drill pipe 15 to 20 thousand feet long is exactly that—a string which bends and twists as if it were made of cotton rather than steel, subject to every current and vagary of motion all the way from the surface to the bottom.

During a recent cruise, the drill string was pulled out after a hole that had been well started, and the ship sailed to a port more than 1,000 miles away, for repairs. After the repairs were accomplished, the ship returned to within a few hundred feet of the hole, located the bottom-mounted sonar transponder, re-entered the hole, and continued drilling.

The Deep Sea Drilling Project is an example of Big Science with a capital B. But in terms of results obtained and numbers of scientists involved it has been fairly inexpensive.

Thirty miles of cores have been recovered from a total length of holes drilled of nearly 100 miles. Ninety thousand core samples have been distributed to over 800 separate investigators, coming from 253 academic, industrial, and government organizations. About half of these investigators are Americans and the other half are citizens of other countries. Four hundred and eighty scientists have participated in the work at sea; 360 others have been members of JOIDES advisory panels; an additional 600 have contributed to the Initial Reports of the project and 809 have made subsequent requests for samples. In all, at least 2,000 scientist-years have been devoted to the project. Quantitatively, the most impressive products have been printed words. Fifty volumes of Initial Reports have been or are about to be published, averaging over 1,000 pages per volume and about 1,000 words per page, in addition to some 500 pages each, or a total of around 5 million words. My own estimate, subject to a large margin of error, is that a total of 55,535,373.7 words have been printed! (The 0.7 is the result of a typographical error in which three letters were left out of a ten-letter word!)

The cost of the project for the twelve years from FY 1967 to FY 1980 will be 137 million dollars. This corresponds to $2.50 per published scientific word and less than $60,000 per scientist-year.

What about the future? In spite of its impressive accomplishments, the Deep Sea Drilling Project has almost literally gone little farther than to scratch the surface of the ocean floor. Holes have been sunk at 460 locations to an average depth of 300 meters, roughly 1,000 feet. This is one hole for every 800,000 square kilometers of the sea floor. 800,000 square kilometers represents an area about equal to that of California, Oregon, and Washington, combined. How much could we expect to know about these three states if we only had core samples from one hole drilled to a depth of 1,000 feet? Our conclusions would be vastly different if the hole were drilled in the Sierra Nevada, the San Joaquin Valley, or the Coast Ranges.

The deepest penetration into the ocean floor was about 5,700 feet, a little over a mile. This is less than one-fourth of the thickness of the oceanic crust. The greatest depth of water was 24,000 feet. Yet some of the most interesting places in the ocean—the great deep sea trenches—are around 35,000 feet deep.

Though much remains to be done within the limited capability of *Glomar Challenger,* the drilling technology has now been developed to the point where a larger ship, capable of drilling in deeper water, penetrating further beneath the sea bottom, and suspending much heavier loads from the drilling platform, could be effectively used. These greater drilling capabilities are needed to take advantage of the scientific opportunities revealed by the work already done.

To date, not much drilling has been accomplished in the continental margins, including the continental slope and the rise to seaward of the slope in the passive margins of the Atlantic, and the landward sides of trenches in the active margins that encircle the Pacific. Yet these are the regions that may contain many of the undiscovered keys to the dynamics of plate tectonics. The sediments of the continental margins are thick and complex, and deep penetration by drilling is necessary to understand their relationships. At the same time, these sediments may contain liquid and gaseous hydrocarbons, and the danger of pollution requires that heavy auxiliary equipment be used to prevent blowouts.

Several practical reasons exist for drilling in the continental margins. The active margins are the loci of many destructive earthquakes and volcanic eruptions. Measurements of stress and strain at depth in bore holes may be important in the development of forecasting methods for these catastrophic phenomena. Enormous volumes of sediments have accumulated on the passive margins. Hydrocarbons may have been formed and trapped in these sediments on a large scale. We need to know the nature of the chemical and physical alterations that occurred in these sediments and their contained organic matter, as they were buried under later accumulations.

From a scientific point of view, we need to remind ourselves that only 4% of earth history is explained by our present ideas of continental drift, sea-floor spreading and plate tectonics. What happened 200 million years ago to cause the great single continent of Pangea to break up into fragments? Prior to that time were there repeated cycles of continental masses coming together and splitting up, then coming together again? Some geologists believe that evidence exists in the continents for such a series of events.

But an alternative is at least possible. Could only one continent have existed throughout a major part of the earth's lifetime, gradually growing in size by accretion of ocean crust to the edges, until finally it grew so large that it became unstable and could no longer remain in one piece? If this was the case, the last 200 million years have been unique in the history of the earth.

The coast on North America now intersects a zone of active crustal formation. Was this same zone formerly located at a distance from shore and subsequently moved laterally, relative to the continent, to its present position? Or has a new axis of spreading formed and extended northward into the land? A similar question can be asked about the spreading axis that now lies along the Mid-Atlantic ridge. If this existed prior to 200 million years ago, it was in the center of Pangea. Did it form there, and if so how? Or did the split occur as an extension of an active spreading axis in the ocean which crept under the continent?

If questions about events prior to splitting of Pangea can be answered, it will only be by a deeper understanding of the geologic history of the continents, for only the continents contain a record in the rocks of the early history of the earth. But to understand this record we need to know a great deal more about the processes that are now occurring, and have occurred in the relatively recent geologic past, in the continental margins.

Insofar as possible, the objective of future drilling must be to answer well-formulated, significant questions. Consequently, drilling of each hole should be preceded and accompanied by extensive geophysical studies, using all the new tools that have become available during the last few years—multi-channel seismic arrays, bottom-towed side-scanning instruments, and the rest. But as our drill bits penetrate deeper into the earth, we should also expect the unexpected. In this sense, future drilling must be thought of as exploration into the unknown as well as process of question-answering. Many of the most remarkable results of the last ten years of deep sea drilling were unexpected surprises. Consider for example, the vast salt deposits underlying the Mediterranean, the black shales of Cretaceous age in both the Atlantic and the Pacific, the extensive layers of chert in the Paleogene sediments, the hydrothermal mounds of the Galapagos spreading center, and the unexpectedly great age of the Antarctic Ice Cap.

Alfred North Whitehead once said "the universe is not only strange, but it is stranger than we can imagine". A long lifetime in the earth sciences has convinced me that this wise statement applies to the earth beneath our feet as well as to the starry universe above us. The earth is full of wonders. We may not be able to imagine these until we see them, but by careful observation and clear-headed reasoning it should be possible for us to understand them. This is the underlying justification for a continuing program of exploration of the earth beneath the sea, with the powerful drilling tools we have learned to use so well.

PART 2
STRUCTURE AND SEDIMENTS

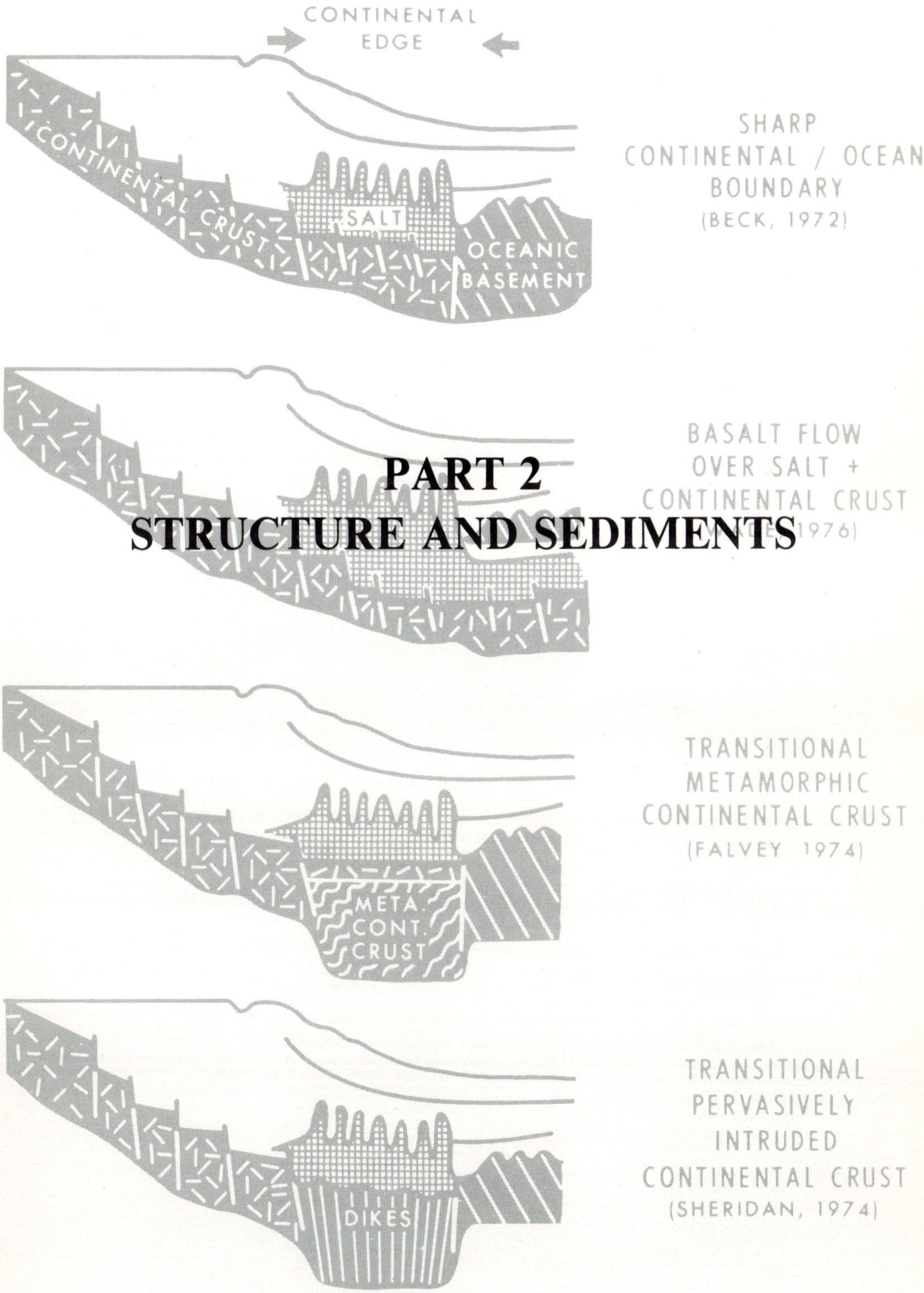

SEPM SPECIAL PUBLICATION NO. 32, P. 7–22, OCTOBER 1981

LINEAR ISLAND AND SEAMOUNT CHAINS, ASEISMIC RIDGES AND INTRAPLATE VOLCANISM: RESULTS FROM DSDP

DAVID A. CLAGUE
U.S. Geological Survey, 345 Middlefield Road, Menlo Park, Calif. 94025 U.S.A.

ABSTRACT

The Deep Sea Drilling Project drilled a substantial number of sites that bear on the origin of linear island and seamount chains, aseismic ridges and other more regional expressions of intraplate volcanism. Drilling in the Emperor Seamounts during Leg 55 was particularly successful. Results from this leg include: 1) the volcanoes of the Hawaiian-Emperor chain continue to increase in age away from Kilauea as predicted. 2) Suiko Seamount formed at a paleolatitide of 26.9±3.5°N, 7° north of present-day Hawaii, but far south of its present latitude of 44.8°N. 3) the volcanic rock types recovered include hawaiite, mugearite, alkalic basalt and tholeiitic basalt in the sequence and relative volume expected for Hawaiian volcanoes. 4) the tholeiitic and alkalic basalts recovered are geochemically similar to those in the Hawaiian Islands, only the ratio of $^{87}Sr/^{86}Sr$ appears to change through time. All the lavas appear to be derived from a source that has small-scale heterogeneities, but is homogeneous on a large scale. 4) The Emperor Seamounts were once volcanic islands that underwent subaerial and shallow marine erosion, and deposition of shallow-water biogenic carbonate sediments that capped all or most of each volcano.

Drilling in other regions has yielded less conclusive results. For example, it is uncertain if the Line Islands are an age progressive chain (hot-spot trace) or result from some other type of intraplate volcanism. The mid-Pacific Mountains also show evidence of originating from a regional episode of volcanism in the mid-Cretaceous. Drilling in the Nauru Basin encountered a voluminous mid-Cretaceous volcanic flow-sill complex that overlies Jurassic magnetic anomalies, yet is composed of depleted tholeiite. In the Indian Ocean, drilling on the Ninety-East Ridge established that it 1) is volcanic in origin; 2) is older to the north; 3) formed in shallow water, and 4) formed further south and has moved northward. It appears that the Ninety-East Ridge, like the Hawaiian-Emperor chain, is a hot spot trace. In the Atlantic Ocean, drilling on the Iceland-Faeroe Ridge and the Rio Grande Rise-Walvis Ridge suggests that all these aseismic ridges are hot spot traces generated by the Iceland and Tristan de Cunha hot-spots.

INTRODUCTION

The Deep Sea Drilling Project has drilled a number of sites that bear on the nature of mid-plate volcanism, aseismic ridges and linear island chains. In this paper I summarize the results of drilling at sites that are located on or near aseismic ridges and linear island and seamount chains as well as a few sites where evidence for widespread non-edifice building intraplate volcanic complexes have been found. A number of holes have also been drilled on oceanic plateaus such as the Magellan Rise, Hess Rise, and a number of others; these are probably formed at or near spreading ridges and will not be discussed in this paper. Drilling in the Hawaiian-Emperor chain, the Line Islands and the mid-Pacific Mountains in the Pacific Ocean will be discussed first. Subsequent sections discuss the results from drilling on the Ninety-East Ridge in the Indian Ocean and the Iceland-Faeroe and Rio Grande-Walvis Ridges in the Atlantic Ocean. A large part of the paper is devoted to the Hawaiian-Emperor chain, because of the number of drilled sites and the success of the Emperor Seamount Leg (Leg 55) in achieving all its major objectives.

In this paper I have used the terms "linear island and seamount chain" and "aseismic ridge" to describe similar geomorphic features that appear to be formed away from active spreading ridges and on active spreading ridges, respectively. Neither term implies a formative mechanism, although many of these features represent the volcanic traces of hot spots. In addition to hot-spot traces, there appear to be other types of intraplate volcanism, particularly in the Central Pacific. Finally, a third, and new, type of intraplate volcanism has been discovered by DSDP at several locations. This voluminous volcanism is non-edifice building (unlike the first two types) and occurs far from active spreading centers.

Until recently, basalts formed at mid-ocean ridges were thought to be quite uniform tholeiitic basalts depleted in the large radius lithophile elements (including the light rare earth elements) and having ratios of $^{87}Sr/^{86}Sr$ of about 0.7026. Basalts erupted on oceanic islands, such as Iceland and Hawaii, were thought to be mainly tholeiitic basalts that were not depleted in the large radius lithopile elements and that had higher $^{87}Sr/^{86}Sr$ ratios. These island tholeiitic basalts are commonly associated with younger alkalic basalts, and, in the case of Hawaii, even younger strongly alkalic basalts erupted after a prolonged period of erosion (the so-called post-erosional basalts). Recent work on basalts both dredged and drilled in the oceans and collected on islands has complicated this simple picture. In particular, the compositional variation of basalts drilled in the North Atlantic during Leg 49 has led to the view that source regions beneath oceanic spreading ridges are complex with both regions of depleted mantle and *variably* undepleted mantle (Wood and others, 1979; Tarney and others, 1979; Cann and others, 1979). These results argue strongly against simple mixing models with only two main types of mantle source—the depleted low velocity zone and the undepleted mantle plume (Schilling, 1973). The petrologic portions of the following discussion will attempt to place the drilled basalts in terms of these types of mantle source regions.

PACIFIC OCEAN

The Pacific Ocean Basin contains examples of intra-plate volcanism. The most spectacular examples are the volcanic islands such as Hawaii and Tahiti, but the basin is littered with thousands of other smaller volcanic edifices. The principal goals of drilling related to intraplate volcanism in the Pacific Basin have concentrated on the age relations and origin of linear volcanic chains (possible hotspot traces). Morgan (1972) proposed that the Hawaiian-Emperor, Line-Tuumotu Sala y Gomez, and Marshall-Gilbert-Cook-Austral chains were the traces of three hot spots that had remained nearly stationary with respect to each other and to the earth's spin axis as the Pacific plate rotated over them. The following sections will summarize the results of drilling along the Hawaiian-Emperor chain, the Line Island chain, in the mid-Pacific mountains, and in the Nauru Basin where there is evidence for widespread non-edifice building intraplate volcanism.

Hawaiian-Emperor Chain

DSDP Leg 55 was the first leg to be entirely devoted to a single major objective (Jackson, Koizumi and others, 1978; 1980): to test the hot spot hypothesis for the origin of the linear Hawaiian-Emperor island-seamount chain. The four principal goals were to determine: 1) whether the known increase in the age of the volcanoes on the Hawaiian chain with distance from Kilauea continues northward along the Emperor chain; 2) whether the lavas of the Emperor volcanoes are of the same chemical composition and were erupted in the same sequence as those of Hawaiian volcanoes; 3) the latitude of formation of Suiko Seamount as a test of hot spot fixity; and, 4) whether the Emperor Seamounts were once islands and, if so, to determine their post-volcanic and subsidence history.

Prior to leg 55, Site 192 was drilled on Meiji Seamount during Leg 19 (Scholl, Creager, 1973), Sites 308 and 309 were drilled on Kōkō Guyot during Leg 32, and Site 311 was drilled to the west of Midway Atoll also during Leg 32 (Larson, Moberly and others, 1975a). The site on Meiji Seamount at the north end of the Emperor Seamounts reached volcanic basement; the basalt recovered is highly altered island (undepleted) tholeiitic basalt, based on interpretation of the immobile elements Y, Nb, Zr, and TiO_2 (Dalrymple, Lanphere and Natland, 1980). The oldest sediment above basement is Lower Maestrichtian based on nannoplankton. The two sites on Kōkō Guyot did not reach basement, so the paleo-latitude of the guyot could not be determined. However, the sites did confirm that Kōkō Guyot was Eocene or older in age. The data from the drill sites also show that corals were not a major component of the shallow-water carbonate bank that capped the guyot, thereby suggesting formation in relatively cool waters (Schlanger and Konishi, 1975). The paleontologic age of the recovered oldest sediment was several million years older than the age determined by K-Ar dating of dredged volcanic rocks (Clague and Dalrymple, 1973), but the difference probably lies within the accuracy of the paleontologic zone boundaries and the K-Ar age. Site 311 was drilled primarily to date the young portion of the M-anomaly sequence, but recovered turbidites shed from the western Hawaiian Ridge about 27–30 m.y. ago. This age suggested that the previously determined age for Midway Atoll (Dalrymple and others, 1974) of 17.9 ± 0.6 was too young. Subsequent studies of additional Midway samples show that Midway is 27.7 ± 0.6 m.y. old (Dalrymple and others, 1977; Dalrymple, Lanphere and Clague, 1980) and that the seamount nearest Site 311 is about 28.0 ± 0.4 m.y. old (Clague and others, 1975; Dalrymple, Lanphere and Clague, 1980).

During Leg 55, eleven holes were drilled at four sites on four seamounts in the Emperor Seamount chain, including a multiple re-entry hole on Suiko Seamount. The locations of these sites are shown in Figure 1, together with the three sites drilled during Leg 32 and the site drilled during leg 19.

Site 430 was drilled in 1485 meters of water on Ōjin Seamount in what appeared from seismic records to be a lagoonal sediment pond. Hole 430A penetrated 118.0 meters of volcaniclastic sand and ash and five underlying basalt flows. The upper four flows are hawaiite, and the lower flow is a tholeiitic basalt similar in composition to those that make up the Hawaiian shield volcanoes. The sediment above the basalt is late Paleocene in age and contains benthonic foraminifers, ostracodes, bryozoans, echinoderms, bivalves, and calcareous algae typical of a shallow-water carbonate bank environment.

Site 431 was drilled in 1714 meters of water on Yōmei Seamount in what appeared from seismic

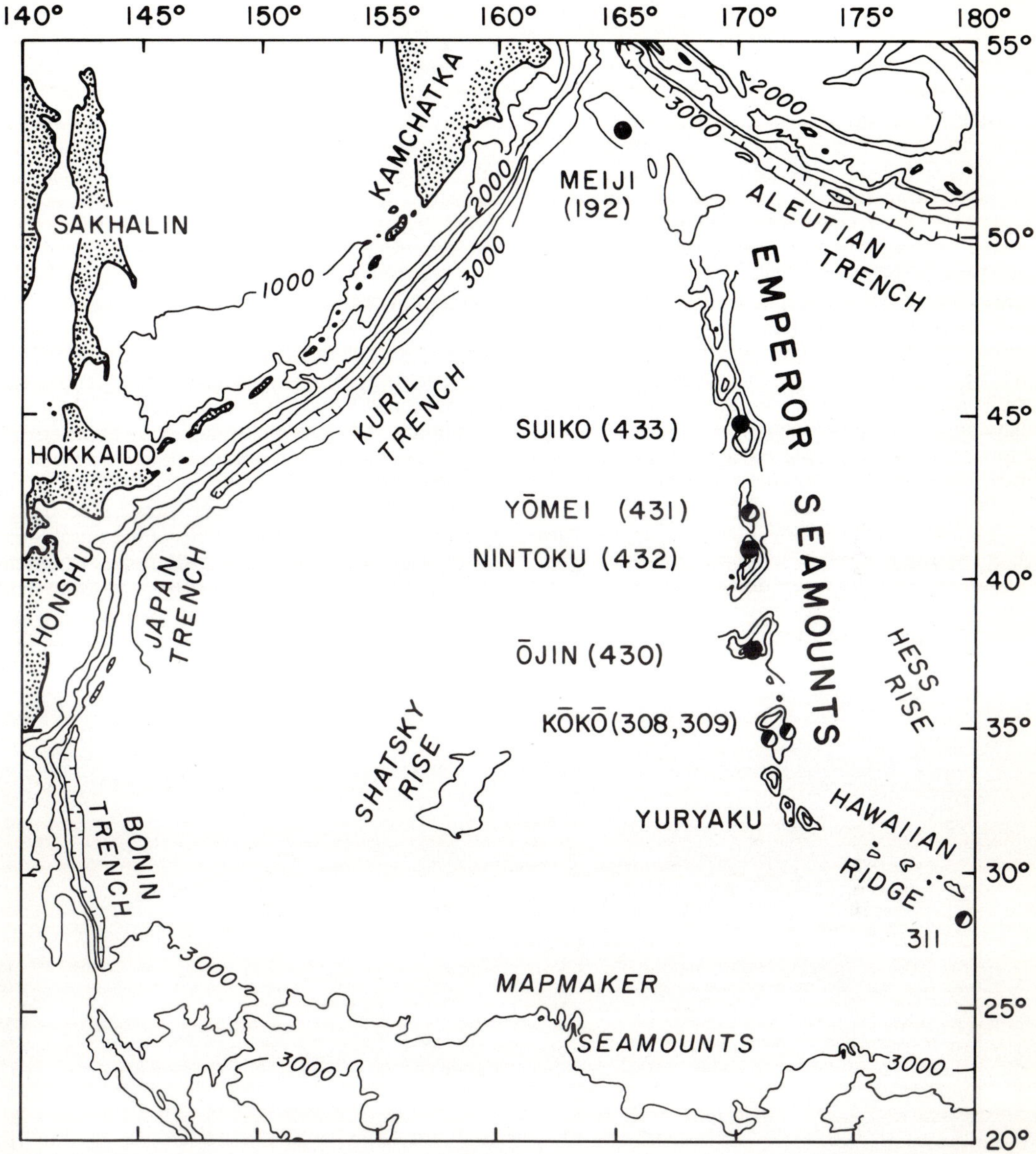

FIG. 1.—Location of DSDP sites in the Hawaiian-Emperor chain. Sites where basement was recovered are shown as filled dots; sites that did not recover basement, as half-filled dots. Diagram is modified from Jackson, Koizumi, and others (1978).

records to be a faulted terrace. Neither hole reached basement. The oldest sediments are upper Middle Eocene and are directly overlain by iron-manganese crust and Pleistocene ice-rafted pebbly sand.

Site 432 was drilled in 1320 meters of water on Nintoku Seamount in what appeared from seismic data to be perched lagoonal deposits. At Hole 432A, well consolidated sandstone and conglomerate underlain by a thin red clay above basalt, were recovered.

The sandstone contains Paleocene large foraminifers, planktonic foraminifers, and calcareous benthonic foraminifers. The sediment was deposited in a shallow, warm water, high energy environment near an organic reef or bank. The three basalt flows are all similar to basalts of the Hawaiian alkalic eruptive stage and probably formed during a post-caldera stage of Nintoku volcano. The red clay and the vesicular tops and bottoms of the flows show that the flows were erupted subaerially.

Four holes, including a multiple re-entry hole, were drilled at Site 433 on Suiko Seamount. The site was drilled in 1874 meters of water in what appeared from seismic data to be ponded pelagic sediments. The recovered section consists of Pleistocene to Early Miocene mainly pelagic sediments overlying a Paleocene shallow water bank assemblage of carbonate sand and sandy mud containing algal nodules overlying basalt. The top three flows are alkalic basalt and are underlain by more than 100 flows of tholeiitic basalt. All the basalt flows were erupted subaerially indicating that Suiko volcano once stood as a high island similar to Hawaii. More than 300 onboard paleomagnetic measurements indicate that Suiko Seamount formed far south of its present location.

The principal results which correspond to objectives 1–4 listed above are summarized briefly below.

Ages of Volcanoes and Progression of Volcanism. — Reliable K-Ar ages have been determined for Ōjin, Nintoku, and Suiko, the three seamounts where volcanic basement was reached. The best ages are 55.2 ± 0.7, 56.2 ± 0.6, and 64.7 ± 1.1 m.y. for Ojin, Nintoku, and Suiko, respectively (Dalrymple, Lanphere and Clague, 1980). These ages are plotted as a function of distance from Kilauea in Figure 2 with preexisting data from the Hawaiian-Emperor chain. The new ages confirm that the Emperor chain increases in age to the north as predicted by the hot-spot hypothesis and that the Emperor Seamount chain is a continuation of the Hawaiian volcanic chain.

Hawaiian Hot Spot Fixity. — The primary purpose of the re-entry hole on Suiko Seamount was to obtain a sufficient number of samples for paleomagnetic study so that a reliable paleolatitude could be determined for a volcano in the central Emperor Seamounts. All 450+ paleomagnetic samples from Ojin, Nintoku and Suiko Seamounts have reversed polarity and magnetic inclinations indicative of paleolatitudes far south of their present locations. The lavas are magnetically stable with natural remnant magnetizations essentially those acquired when the rocks cooled (Kono, 1980a, b; Kono and others, 1980). In addition to the stability of magnetization, a key question in paleomagnetic studies

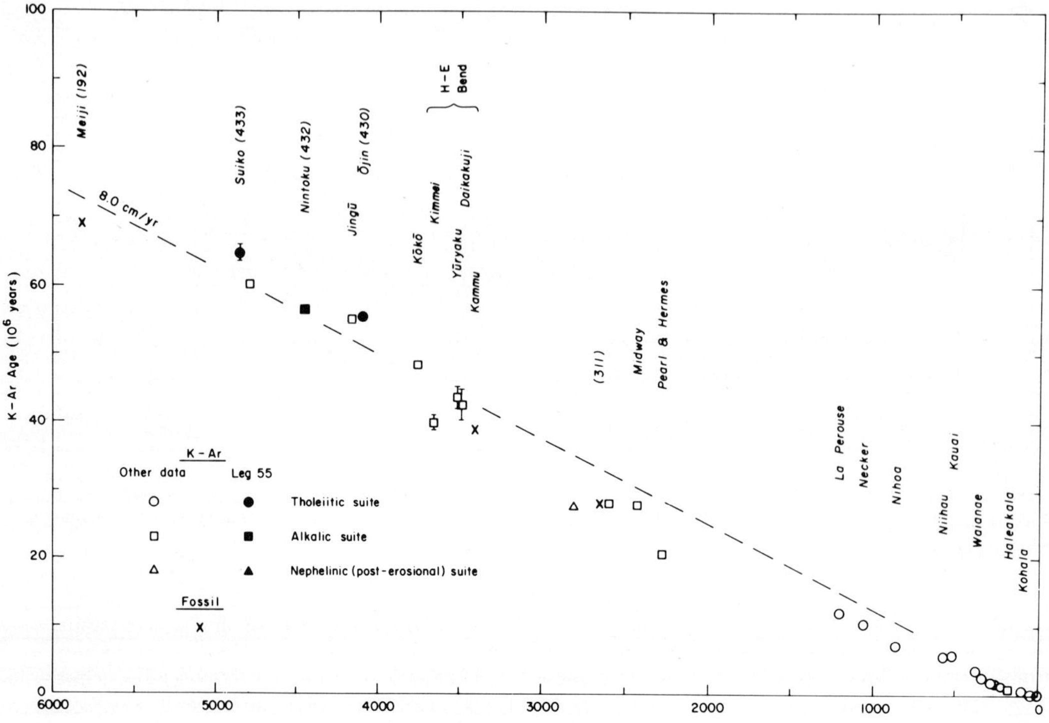

FIG. 2. — Plot of age versus distance to Kilauea along the Hawaiian-Emperor chain. Diagram is from Dalrymple, Lanphere, and Clague (1980). The DSDP Leg 55 results are consistent with the hot-spot hypothesis in that the seamounts increase in age away from Hawaii.

is whether sampling is sufficient to average out paleosecular variation (PSV) so that the measured paleomagnetic inclination represents the paleo-inclination of the main dipole field at the site. The number of flows from Ojin and Nintoku are clearly too small to average out the PSV. At Suiko Seamount, however, there was sufficient sampling so that PSV has probably been averaged out (Kono, 1980b).

The mean inclination for the Suiko lava flows is $45.7 \pm 4.8°$ (2σ) which corresponds to a paleolatitude of $26.9 \pm 3.5°$ (2σ). This is significantly different from both the present latitude of Suiko Seamount at 44.8°N and that of Kilauea (19.5°N).

The paleomagnetic data are consistent with the hypothesis that the Hawaiian hot spot has been at relatively low latitudes for the last 70 m.y. and that the volcanoes are at their present locations largely due to plate motion relative to the Earth's spin axis. The data for Suiko, however, indicate that the hot spot has not been exactly fixed, but has moved southward about 7° (Kono, 1980b) since Suiko formed 65 m.y.BP (Dalrymple, Lanphere, Clague, 1980).

The biofacies represented by the Paleogene shallow water carbonate sediments above the basalts substantiate the Suiko paleolatitude. The bioclastic sediments consist primarily of coraline algae and bryozoa with ostracodes, foraminifers, and shell fragments typical of a shallow water high-energy environment (Butt, 1980). Schlanger and Kinoshi (1975) have pointed out that the bryozoan-algal facies of carbonate buildups occurs in cooler waters than the coral-algal facies. They locate the boundary between these facies at about 25°N. An analysis of paleotemperature distribution in the Pacific (Green and others, 1978) suggests that at the time Suiko Seamount formed, the local temperatures would be below 22°C (appropriate for the bryozoan-algal facies) if Suiko formed at 27°N.

Although the paleomagnetic and biofacies data collected on Leg 55 suggest that the Hawaiian hot spot was at 27°N, 65 m.y. ago, this does not necessarily mean that the Hawaiian hot spot has moved relative to the hot spot framework. Kono (1980) has proposed that the entire hot spot frame, the deep mantle and the crust shifted relative to the spin axis (true polar wander). He proposed that such a shift might be caused by a change in the surface distribution of mass of the Earth, which results in a change in the moment of inertia.

Petrology, Geochemistry and Volcanic Lithostratigraphy.—The rock types recovered from Ōjin, Nintoku and Suiko Seamounts include hawaiite, mugearite, alkalic basalt and tholeiitic basalt in the sequence and relative volumes expected for Hawaiian volcanoes. That is, relatively small volumes of alkalic basalts overlying voluminous tholeiitic basalt. All have mineralogy and whole rock major and trace element chemistry well within the range of rock types found in the Hawaiian Islands and western Hawaiian Ridge (Kirkpatrick and others, 1980; Bence and others, 1980; Clague and Frey, 1980; Cambon and others, 1980). Variations in major and trace element chemistry along the chain are small and are within the range of compositions found within single volcanoes and between adjacent volcanoes (Clague and Frey, 1979).

It is significant for the verification of the hot spot hypothesis that tholeiitic basalts, similar to Hawaiian tholeiites but distinct from mid-ocean ridge basalts, have erupted along the entire chain. The major element chemistry of the tholeiitic basalts does not appear to change with distance from Kilauea in any systematic way, although the lavas erupted from 45–7 m.y. ago (Daikakuji to Nihoa) have distinctly lower Al_2O_3 and higher TiO_2 than tholeiitic basalts from the principal Hawaiian Islands or from Suiko Seamount (Kirkpatrick and others, 1980). This chemical variation appears to be related in a rough way to seamount volume, with larger volcanoes being built of undepleted tholeiitic basalts having higher Al_2O_3 and lower TiO_2 than those of smaller volcanoes. This relationship suggests that the depth of magma segregation varies as a function of volume with smaller volumes of magma equilibrated at greater depth.

Hawaiian tholeiites have distinctive rare earth element (REE) patterns, with little or no light REE fractionation and more extensive mid- to heavy-REE fractionation, and commonly have chondrite normalized La $\leq$ Ce. The REE patterns of the tholeiitic basalts drilled on Suiko and Ōjin during Leg 55 and recovered from Kōkō, Nihoa, Necker and Midway (Clague and Frey, 1980) have these same features (see Fig. 3) and resemble Mauna Loa tholeiites (Leeman and others, 1980). It is apparent that the source region for Hawaiian-Emperor volcanism has had very similar REE characteristics for the last 65 m.y. Computer modeling shows that these REE patterns can be generated by small degrees of partial melting of a previously depleted source region, by larger percentages of partial melting of an undepleted source leaving residual garnet, or by partial melting of an enriched source with no garnet in the residuum (Leeman and others, 1980; Clague and Frey, 1980). Trace element ratios such as La/Ta, Ta/Th and Nb/Th indicate that all the lavas recovered during Leg 55 are derived from an undepleted mantle source (Cambon and others, 1980). The inferred source composition for the Emperor Seamount lavas is similar to that inferred for basalts erupted along the northern Mid-Atlantic Ridge (Wood and others, 1979; Bougault and others, 1979; Tarney and others, 1979), particularly those from 63°N.

The Sr isotopic data (Lanphere and others, 1980)

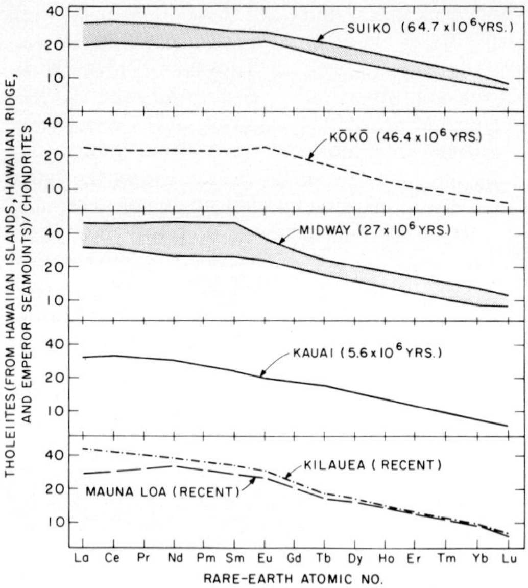

FIG. 3.—Chondrite-normalized rare earth element (REE) patterns for Suiko Seamount (Site 433) tholeiitic basalts compared to other tholeiites from the Hawaiian-Emperor chain. The patterns are unusual in having nearly unfractionated light REE and more strongly fractionated mid- to heavy-REE. The similarity in REE distribution through time suggests that the mantle source is fairly homogeneous through time. Diagram from Clague and Frey (1980).

provide the only conclusive evidence that the source region for Hawaiian-Emperor tholeiites has changed systematically through time. The $^{87}Sr/^{86}Sr$ ratios of the tholeiites increases from about 0.7033 at Suiko Seamount to 0.7037 at the Hawaiian-Emperor bend (Figure 4). For the past 43 m.y., the $^{87}Sr/^{86}Sr$ ratio has remained essentially constant in the tholeiites, although there is a range of values. The change in Sr isotopic ratio appears to correlate with the difference in age between the underlying crust and the individual seamounts. Seamounts built on younger crust have lower Sr isotopic ratios than those built on older crust.

The trace element data suggest that the tholeiitic basalts and the alkalic basalts from along the entire Hawaiian-Emperor chain are derived from source regions that are broadly similar but differ in detail, and that these sources have remained relatively constant in composition through time (Clague and Frey, 1979; Kirkpatrick, 1980).

Post-volcanic Subsidence History.—The post-volcanic history of the drilled seamounts can be divided into three main stages: subaerial, shallow water, and deep water or bathyal.

The evidence that Ōjin, Nintoku, and Suiko Seamounts were once volcanic islands is both abundant and decisive. The titanomagnetite in most of the flows has undergone high temperature oxidation (Kono and others, 1980), a feature common in subaerial flows but rare in submarine flows. Many of the flows have vesicular flow tops that are commonly oxidized bright red to brown, a phenomenon that does not occur in the submarine environment. At Suiko Seamount, a layer of oxidized pumice lapilli was recovered that appears to be an airfall deposit. Parts of oxidized soil zones were recovered at Ōjin and Nintoku Seamounts. Many of the Suiko basalt flows contain carbonate veins that, based on oxygen isotopic data, began depositing while the basalts were above sea level (McKenzie and others, 1980). Other evidence suggestive of a subaerial or near sea-level history include carbonate sands and wave-rounded volcanic pebbles that overlie basalt flows at Ōjin and Nintoku, and a possible beach sand between flows 1 and 2 at Suiko Seamount.

After cessation of volcanism, the seamounts underwent a period of subaerial and shallow water marine erosion and deposition of shallow water biogenic carbonate sediments that eventually capped most or all of each volcano. The biogenic carbonate sediments found on Ōjin, Nintoku and Suiko clearly represent the bryozoan-algal facies of Schlanger and Konishi (1975), indicating deposition in water cooler than presently found at the latitude of Hawaii. A fairly detailed subsidence history may be reconstructed from the microfossil assemblages (see Butt, 1980; Hägn and others, 1980). The shallow-water conditions favorable to the growth of organisms of the bryozoan-algal facies persisted for

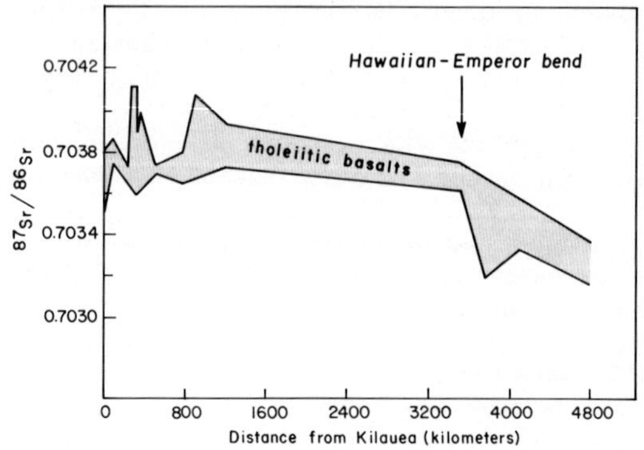

FIG. 4.—Plot of $^{87}Sr/^{86}Sr$ for tholeiitic basalts from the Hawaiian-Emperor chain plotted against distance to Kilauea along the chain. Note that the basalts drilled in the northern Emperor seamounts have lower Sr isotopic ratios than those from the Hawaiian Islands or Ridge. These lower ratios trend towards mid-ocean ridge basalt values as the age of the crust, at the time the seamount was built, decreases. Diagram is from Lanphere and others (1980).

only a few million years at each of the seamounts. Carbonate bank growth probably terminated due to subsidence rather than northward movement of the Pacific plate or global cooling (see discussion in Greene and others, 1978). The temperatures of the surface waters at Suiko Seamount probably had cooled by only about a degree due to northward motion by Middle to Late Eocene. The same is true for Nintoki and Ōjin. In addition, the drastic worldwide cooling of the oceans in Late Eocene to Early Oligocene occurred after the shallow-water carbonate buildup had ceased. It appears likely that the organisms of the bryozoan-algal facies, which are not very productive, simply could not keep pace with subsidence.

At Ōjin, Yomei, Nintoku and Suiko, a hiatus representing a significant period of non-accumulation of sediment occurs above the shallow-water carbonate section. At Ōjin, Yomei and Nintoku, the Eocene to Pliocene section is missing. At Suiko only the Eocene and Oligocene are absent. Apparently, after the tops of the seamounts had subsided to depths below 100 m or so, where the shallow-water organisms could not survive, pelagic sediment did not accumulate because of strong currents across the tops of the seamounts. The exception was at Suiko where local faulting formed small basins in which sediment accumulated. Butt (1980) has developed a detailed subsidence history for Suiko Seamount based on benthonic foraminifers, indicating that subsidence was relatively slow during the Paleogene and more rapid during the Neogene.

Line Islands

Drilling in the Line-Tuamotu chain during Leg 33 (Schlanger, Jackson, and others, 1976) marked the first time that a major portion of a DSDP leg was devoted to seamount studies. The objective of that leg was to test Morgan's (1972) hot spot hypothesis by drilling a series of holes along the Line Island chain and Tuamotu chain to complement Site 165 drilled in the vicinity of the Line Island chain during Leg 17 (Winterer, Ewing and others, 1973). A major objective was to determine if the volcanism that produced this bathymetrically complex chain was age progressive, as had been documented in the Hawaiian Islands (McDougall, 1964) and along the Hawaiian Ridge and southern Emperor Seamounts (Dalrymple and others, 1974; Clague and Dalrymple, 1973), and thereby compatible with a hot-spot origin for the chain. From north to south, the sites drilled are 314, 165, 315, and 316 in the Line Island chain, and Site 318 in the Tuamotu chain (see Fig. 5). All the sites were drilled on the flanks of seamounts or in deep-water troughs between seamounts.

Site 314 was drilled in a trough southeast of Johnson Island. Only 45 m of penetration was achieved, and drilling terminated in Middle to Late Eocene brown mud containing rare, reworked, Upper Cretaceous nannoplankton. The age of the basement is unknown, but is at least Cretaceous (>65 m.y.) in age.

Site 165 was drilled about 300 km northwest of Kingman Reef. The oldest sediments are Middle to Early Campanian in age and overlie basalt that is strongly alkalic. Bass and others (1973) described the basaltic basement rocks as hawaiites and mugearites, although the chemistry and mineralogy suggest that they had a basanite or nephelinite parent similar to rocks dredged from the same region (Natland, 1976). The strongly alkalic composition of these basalts suggests that they may be equivalent to Hawaiian post-erosional basalts and that true basement may not have been reached. Estimates of the age of the sediments directly overlying basalt are highly speculative. Lanphere and Dalrymple (1976) argue that ages between 88 and 102 are possible although Jackson and Schlanger (1976) propose that the sediment overlying basalt is 81–85 m.y. old, but that the underlying shield volcano could be as old as 106 m.y.

Site 315 was drilled on the east slope of Fanning Island and penetrated the entire sedimentary section and 38 meters of basaltic basement. The basalts are transitional between island tholeiites and alkalic basalts and are 93.5 ± 2.7 m.y.BP (Lanphere and Dalrymple, 1976). The basement is overlain by a thick sequence dominated by turbidites that contain redeposited sediment and volcanogenic material. The oldest sediments are Santonian or older (>82–86 m.y.BP), consistent with the K/Ar age of the basement. Paleomagnetic studies of the sediments indicate 13° of northward motion instead of the approximately 30° predicated by the hot spot model (Cockerham and Jarrard, 1976).

Site 316 was drilled near the southern end of the Line Island chain, about 200 km south of Christmas Island. The oldest sediment recovered is Early Campanian (79–82 m.y.BP). The hole did not reach basement, so extrapolation to obtain an estimate of the basement age is highly speculative. It is probable that basement was within 75 meters of the bottom of the hole, as the dominant sediment was volcanogenic turbidite similar to those recovered at Site 165 above basement. Jackson and Schlanger (1976) estimate the basement age as 79–85 m.y.BP.

Site 318 was drilled near the northwest end of the Tuamotu Islands. The oldest sediment recovered is Early Eocene (50–51 m.y.BP). The hole did not reach basement, but clasts of basalt, hawaiite, mugearite, trachyte, and quartz syenite are present in a thin volcaniclastic unit near the bottom of the hole. These rock types are typical alkalic island basalts similar to those on Hawaii or Tahiti.

The results from drilling in the Line-Tuamotu Island chains may be summarized as follows: 1) The

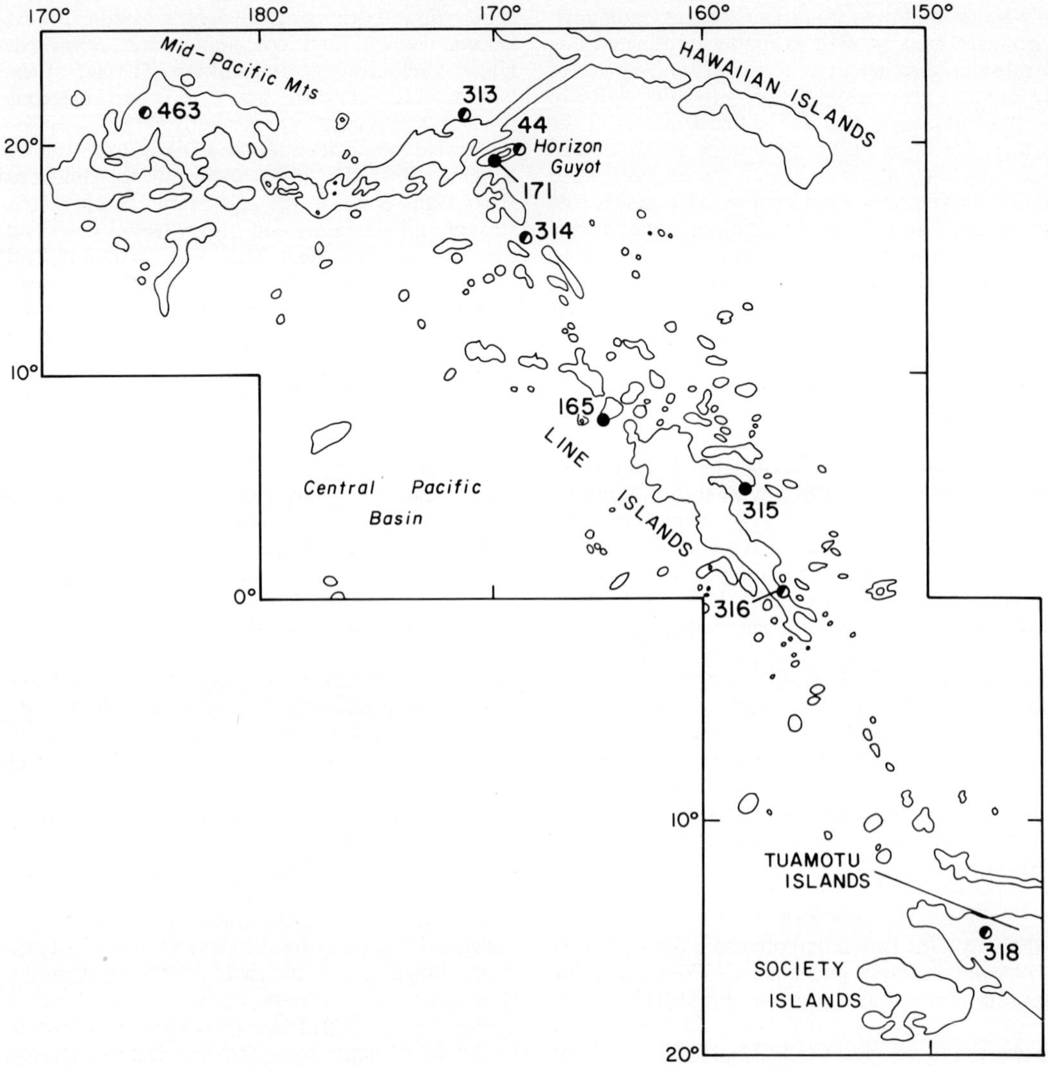

FIG. 5.—Location of DSDP sites in the Line-Tuamotu Islands and the mid-Pacific Mountains in the Pacific Ocean. Symbols as in Figure 1. Chart is from Chase and others (1971).

age data obtained do not allow for an unambiguous test of the hot spot model for the origin of this volcanic feature. It is possible that the chain is age-progressive from north to south as Morgan (1972) proposed, but it is equally possible that the entire chain formed almost simultaneously about 80–90 m.y.BP. The range of tectonic models proposed (see Winterer, 1976; Jackson and Schlanger, 1976) is in part due to the ambiguity of the data. 2) The presence of strongly alkalic basalts suggest that some edifices, and perhaps the entire region, underwent one or more episodes of post-erosional volcanism. The carbonate banks on top of the volcanic edifices are coeval (~70–80 m.y.BP) along the entire Line Island chain. 3) The Line Island chain is older than the Emperor Seamount chain, and the Tuamotu chain is older than the oldest seamounts in the Hawaiian Ridge. These data demonstrate that the Line-Tuamotu bend is not coeval with the Emperor-Hawaiian bend as proposed by Morgan (1972). Jackson and Schlanger (1976) proposed that the entire region underwent epeirogenic uplift 80–85 m.y.BP which caused synchronous volcanism of highly alkalic post-erosional basalts and uplifted the pre-existing shields into shallow water. The islands were capped by carbonate banks about 70–80 m.y.BP during subsidence following the epeirogenic pulse.

The origin of the Line Island chain appears to be more complex than that of the Hawaiian-Emperor chain. It is still unclear whether the major volcanic edifices of the Line Islands were constructed sequentially. However, the entire chain, and the Central Pacific region, apparently underwent a period of widespread volcanic activity from about 80–85 m.y.BP. This period of volcanism was apparently unrelated to either oceanic spreading centers or hot spots and resulted in eruption of SiO_2-undersaturated lavas in the Wake Islands and mid-Pacific Mountains (Natland, 1975) as well as in the Line Islands.

Mid-Pacific Mountains

Site 44 (Leg 6) and Site 171 (Leg 17), Site 313 (Leg 32) and Site 463 (Leg 62) were drilled on Horizon Guyot, in the northeast mid-Pacific Mountains, and in the western mid-Pacific Mountains, respectively (see Fig. 5). Drilling during Leg 6 on Horizon Guyot ended with the drillstring twisted off in Eocene chert (Fischer, Heezen and others, 1971). Site 171 reached basement in a saddle between two summit platforms and recovered sediments as old as the Cenomanian-Turonian boundary (~92 m.y.BP) (Winterer, Ewing and others, 1973) but none as old as the Albian fossils dredged near the summit (Lonsdale and others, 1972). Horizon Guyot resided in shallow water for a period of time as shallow water limestone and lagoonal deposits are present. Volcanic activity persisted at least into the Late Cenomanian and was accompanied by periods of uplift. Site 171 basalt is island or undepleted tholeiite; some flows have affinities with mid-ocean ridge tholeiite. Horizon Guyot apparently formed with several summits projecting as islands in close proximity to a mid-ocean spreading ridge similar to the setting of the present day Galapagos Islands.

Site 313 in the northeastern mid-Pacific Mountains recovered Early Campanian sediment overlying nephelinite (Marshall, 1975; Larson, Moberly and others, 1975b). Volcanic activity persisted at least into the Coniacian.

Drilling at Site 169 in the Central Basin recovered a sill of alkalic basalt intruding Cenomanian to Turonian sediments overlying mid-ocean ridge basalt pillow lavas (Bass and others, 1973). The sill is probably related to widespread alkalic volcanism in this region in the Cretaceous.

Site 463 in the western mid-Pacific mountains did not reach basement but recovered ash beds of Aptian age overlying sediments as old as Barremian (Vallier and Thiede and others, 1979). On the basis of the cores recovered, it appears that volcanic activity in the region occurred over a span of 40–50 m.y. This is in contrast to well-dated volcanism in the Hawaiian Islands that rarely lasts longer than 3 m.y. It is not clear whether the mid-Pacific Mountains are a hot-spot trace or some other type of anomalous volcanism. The similarity in age and in basalt geochemistry between basalts from the mid-Pacific Mountains and the Line Islands suggests that volcanism in both areas is related, perhaps to a regional uplift as suggested by Jackson and Schlanger (1976).

Nauru Basin

Site 462, drilled on Leg 61 in the Nauru Basin, encountered a mid-Cretaceous volcanic flow-sill complex greater than 500 meters thick that is thought to overlie crust of Late Jurassic age (Larson, Schlanger and others, 1978). The complex apparently covers the entire basin without obscuring the Jurassic magnetic lineations and is an expression of non-edifice building mid-plate volcanism. The lavas are geochemically similar to depleted mid-ocean ridge basalt (Batiza, 1979). The Nauru Basin volcanic complex marks the discovery of a new volumetrically significant type of oceanic volcanism, perhaps similar to the volcanic complex described from the Caribbean (Donnelly and others, 1973) during Leg 15.

It is tempting to try to relate the Nauru Basin volcanism to that in the Line Islands, Wake Seamounts and mid-Pacific Mountains; however, the differences in age and basalt chemistry suggest that the Nauru Basin basaltic volcanism is not directly related to the widespread younger alkalic volcanism in and around the Central Basin.

Aleutian Abyssal Plain

Drilling at Site 183 (Stewart and others, 1973) recovered alkalic basalt similar to Hawaiian alkalic basalt. The age of the oldest sediment is Middle Eocene, about 12 m.y. younger than the crustal age inferred from the magnetic anomaly pattern in the region. Stewart and others (1973) suggest that the recovered basalt is a seamount-derived flow from nearby Dirikson or Sirius Seamounts.

INDIAN OCEAN

The *D/V Glomar Challenger* drilled 72 holes at 64 sites during 7 cruises in the Indian Ocean. The vast majority of the sites are located in the deep basins, but some also were drilled on several of the shallow aseismic ridges and platforms. The most prominent aseismic ridges, the Chagos-Laccadive (Site 219) and the Ninetyeast (Sites 214, 215, 216, 253, and 254), were drilled as were the Mascarene Plateau (Site 237), the Madagascar Ridge (Site 246), the Mozambique Ridge (Site 249) and Broken Ridge (Site 255). Figure 6 shows the general physiographic features and the locations of these drill sites.

The discussion that follows will be based on results from the Ninetyeast Ridge where the large number of sites and the recovery of basement rocks at four of the five sites make the interpretation less

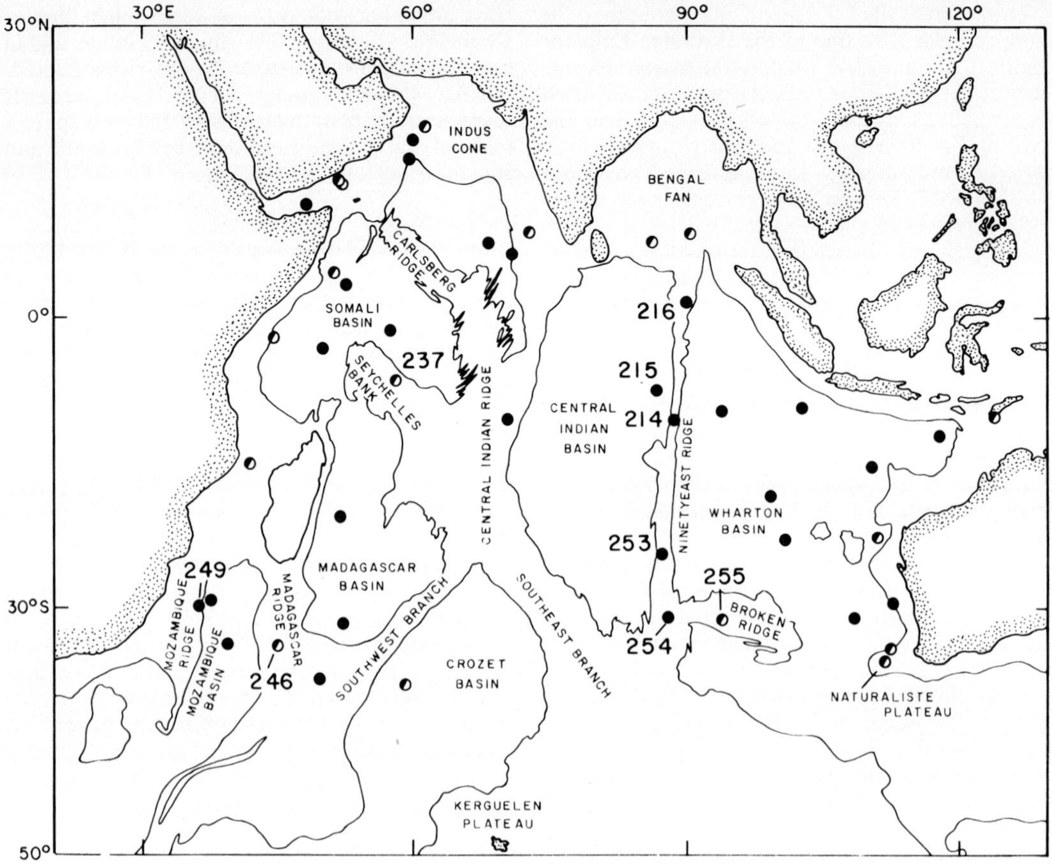

FIG. 6.—Location of DSDP sites on the Ninetyeast Ridge in the central Indian Ocean. Symbols as in Figure 1. The diagram is modified from Luyendyk (1977). DSDP Site 215 is also discussed in the text.

speculative than for the other drilled aseismic ridges. The primary data are presented in von der Borch, Sclater and others (1974) and Davies, Luyendyk and others (1974).

Ninetyeast Ridge

The Ninetyeast Ridge is a major N-S feature in the central Indian Ocean that extends from about 32°S to 10°N where it becomes buried beneath sediments of the Bengal Fan (Moore and others, 1974). The ridge trends slightly east of north and essentially follows the 90°E meridian. Summit depths range from nearly 3000 m in the north to less than 2000 m in the south where the ridge intersects the Broken Ridge. The Ninetyeast Ridge can be divided into three distinct morphologic sections that Luyendyk (1977) believed related to the tectonic origin of the ridge. The three provinces are: 1) en echelon lobate structures north of 7°S that are similar to Pacific seamount chains, 2) a narrow (100 km) symmetric ridge with straight steep sides from 7°S to 14°S, and 3) a wider (200 km) shallow asymmetrical ridge with a very steep eastern side from 14°S to 30°S. Prior to deep drilling on the Ninetyeast Ridge a variety of hypotheses had been proposed to explain its origin including 1) it was a continental fragment or 2) horst (Francis and Raitt, 1967); 3) it was uplifted in response to subduction of the Wharton plate (McKenzie and Sclater, 1971); 4) it was an abandoned spreading center (Veevers and others, 1971) or 5) it formed by a combination of spreading and subduction (Falvey, 1972). All the early hypotheses were abandoned and two additional hypotheses were added following drilling: 6) it formed from a hot spot or mantle plume (Curray and Moore, 1974), or 7) it formed by a combination of a hot spot and a leaky fault (e.g., Sclater and Fisher, 1974).

Luyendyk (1977) summarized the principal results of drilling on the Ninetyeast Ridge. These results include: 1) The ridge is volcanic in origin, and the extrusive basement rocks recovered have distinctive petrochemistry. 2) The ridge is older to the north with ages of at least Campanian at 9°N to Eocene-Oligocene at 31°S. This is the same age gradient as the Indian Basin to the West and indicates that the Ridge is part of the Indian plate. 3) The ridge formed in shallow water (in some cases, subaerially) and subsided along known age-depth curves. 4) The ridge formed in more southerly latitudes and moved northward.

These observations led Luyendyk (1977) to propose that the Ninetyeast Ridge was a hot-spot trace recorded on the Indian plate as it moved northward during Late Cretaceous and Early Tertiary time over the Amsterdam-St. Paul and the Kerguelen (?) hot spots.

Frey and Sung (1974) and Frey and others (1977) studied the altered basement rocks recovered at Sites 214, 216, 253, and 254. The rocks from Sites 214, 216, and 254 include light rare earth enriched tholeiitic basalts, ferro-basalts and oceanic andesite that appear to be derived by crystal fractionation from similar parental magmas. This suite of rocks is petrographically and geochemically similar to tholeiitic island sequences found on Iceland, the Galapagos Islands, the Faeroes, and the Amsterdam-St. Paul Islands in the Indian Ocean.

Basement rock recovered at Site 253 is light rare earth depleted tholeiite similar to mid-ocean ridge basalt (Frey and others, 1977). Its occurrence on the Ninetyeast Ridge is anomalous and unexplained (Frey and others, 1972) although similar occurrences of depleted tholeiite are known from Iceland (Wood, 1978; O'Nions and others, 1976). The presence of both light rare earth enriched and depleted tholeiite on Iceland, in the *FAMOUS* area, and from the Ninetyeast Ridge, suggest that the two-source model proposed by Schilling (1973) for the formation of mid-ocean ridge and island basalts is inadequate and that the melting process in the mantle is quite complicated (Langmuir and others, 1977; 1978).

Frey and others (1977) proposed that the Ninetyeast Ridge formed as a result of hot-spot activity presently centered under Amsterdam and St. Paul Islands. It is important to note that the fractionation sequence tholeiitic basalt → ferro-basalt → oceanic andesite is common on tholeiitic islands located on or very near active mid-ocean ridges but is quite rare on tholeiitic islands located in plate interiors (*e.g.*, Hawaii). This observation suggests that the Amsterdam-St. Paul hot spot has remained in close proximity to the active South-east Indian Ridge throughout the formation of the Ninetyeast Ridge. The reconstruction presented by Luyendyk (1977), based on different data, shows exactly this relationship. The Ninetyeast Ridge is an aseismic ridge in this model rather than a linear island and seamount chain.

Site 215 in the Central Indian Basin

Site 215 rocks may show evidence for non-edifice building off-ridge volcanism similar to that recently discovered in the Nauru Basin in the western Pacific (see section on the Pacific). Site 215 is located 240 km west of the Ninetyeast Ridge at about 8°S and was drilled in greater than 5000 m of water. The recovered basement rocks are tholeiitic basalts enriched in K_2O, P_2O_5, and the light rare earth elements (Frey and others, 1977). The younger-than-expected basal sediment age (61–62 m.y.BP) and the possibility that the recovered volcanic rock is a sill rather than a flow (Frey and others, 1977) suggest that Site 215 may have sampled the uppermost section of an off-ridge volcanic/intrusive complex similar to that encountered at Site 462 in the Nauru Basin. The possible significance of this type of intra-plate volcanism will be discussed in a later section.

ATLANTIC OCEAN

Only six sites in the Atlantic Ocean have direct bearing on intraplate volcanism or hot spot traces. These include Sites 336 and 352 on the Iceland-Faeroe Ridge, Site 357 on the Rio Grande Rise, Sites 359 and 363 on the Walvis Ridge, and Site 368 near the Cape Verde Islands. Figure 7 shows the generalized physiography of the Atlantic Ocean and the location of these sites.

Drilling at Site 368 recovered tholeiite, similar in composition to mid-ocean ridge tholeiite (depleted), as a sill of Miocene age (Natland, 1978). This sill is apparently related to initiation of alkalic to strongly alkalic volcanism on the Cape Verde Islands. Natland (1978) suggested that these mid-ocean ridge tholeiitic magmas may be the first magmas erupted to form the Cape Verde Islands. This suggests that intraplate volcanism might begin with eruption of depleted tholeiite followed by eruption of undepleted tholeiitic and alkalic lavas.

Iceland-Faeroe Ridge

The results from Sites 336 and 352 (Talwani, Udintsev and others, 1976) on the Iceland-Faeroe Ridge are consistent with an origin of the ridge as a hot spot trace formed on the Eurasian plate by the Iceland hot spot as proposed by Morgan (1972). Site 352 did not reach basement, but the oldest sediments present are Middle Oligocene. Further from Iceland at Site 336 the basement is 41–44 m.y. old and consists of quartz tholeiite that has a relatively low Mg number ($\sim$48) and low K_2O and TiO_2. The tholeiite is similar to those that form the upper series Plateau basalts on the Faeroes. Schilling (1976) reported the presence of both light rare earth enriched and depleted basalts from this site with the upper younger flows being the more depleted. Both light rare earth enriched and depleted basalts are

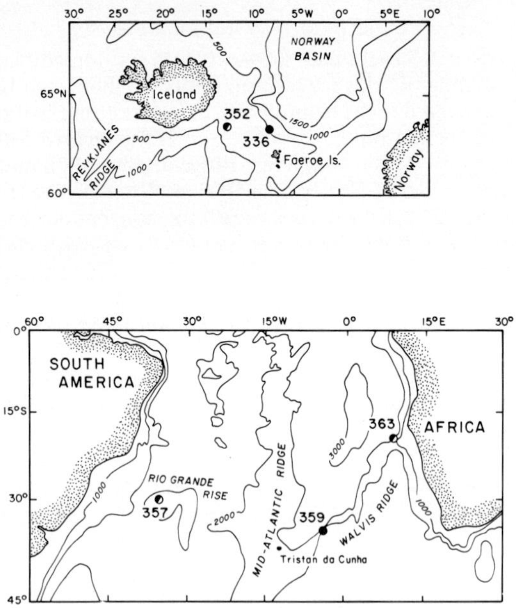

FIG. 7.—Location of DSDP sites on the Iceland-Faeroe Ridge, the Rio Grande Rise, and the Walvis Ridge in the Atlantic Ocean. Symbols as in Figure 1. The diagrams are modified from Talwani, Udintsev and others, 1976, and from Davies, Luyendyk and others, 1974.

also present on the Faeroes (Schilling and Noe-Nygaard, 1974) and on Iceland (O'Nions and others, 1976; Wood, 1978). The limited data are consistent with an origin for the Iceland-Faeroe Ridge as an aseismic ridge marking the trace of the Iceland hot spot, but are certainly not compelling, particularly in view of the widespread occurrence of variably undepleted mantle sources in the North Atlantic (Wood and others, 1979; Tarney and others, 1979).

Rio Grande Rise and Walvis Ridge

Morgan (1972) proposed that the Rio Grande Rise and the Walvis Ridge were the traces of the Tristan de Cunha hot spot recorded on the American and African plates, respectively. In order to generate the roughly symmetrical ridges, the Tristan de Cunha hot spot must remain beneath the crest of the Mid-Atlantic Ridge during the opening of the south Atlantic. The model predicts that the basement ages will increase away from the active spreading center along both the Walvis Ridge and the Rio Grande Rise, that the basement ages will be the same or only slightly younger than the adjacent sea-floor, that the ridges each formed in shallow water and subsided with the adjacent crust along known age-depth curves, and that the basement of both ridges should be basaltic and have chemical affinities to the basalts that make up Tristan de Cunha. Site 357 on the Rio Grande Rise and Site 359 on the Walvis Ridge were drilled during Leg 39, in part, to test these predictions (Supko, Perch-Nielson and others, 1977). These sites were followed by Site 363 on the Walvis Ridge during Leg 40 (Bolli, Ryan and others, 1978). The drilling was not totally successful in testing the hot spot models because basement rock was reached only at Site 359.

Site 357 on the Rio Grande Rise was located close to Sites 21 and 22 drilled during Leg 3. The oldest sediment recovered at Site 357 was Lower Santonian in age (~87 m.y.BP) and was deposited in shallow water. The section contains altered Eocene hyaloclastite fragments that were probably derived from alkalic basalts (Fodor and Thiede, 1977) based on microprobe analyses of fresh clinopyroxene. The recovery of alkalic volcanic rock is in agreement with the results of dredging in the same area which also recovered alkalic basalts (McDowell and others, 1977). Fodor and Thiede (1977) suggest that the Eocene hyaloclastite fragments were derived from nearby oceanic islands located near the Mid-Atlantic ridge crest.

At Site 359 on the Walvis Ridge, a subaerially erupted trachytic tuff was recovered that is Late Eocene in age (Fodor and others, 1977) and compositionally similar to trachyte on nearby Gough and Tristan de Cunha Islands. Site 363 did not reach basement, but the basal sediment recovered was deposited in shallow water and is Aptian in age, suggesting that the age of the Walvis Ridge increases towards the African continent as predicted by the hot spot model.

The Rio Grande and Walvis Ridges are aseismic ridges that apparently mark the paths of the American and African plates relative to the Tristan de Cunha hot spot. The Mid-Atlantic Ridge has remained centered over this hot spot for at least 85 m.y.

SUMMARY AND CONCLUSIONS

DSDP drilling on linear island chains has provided a rigorous test of the hot spot hypothesis (Morgan, 1972). The Leg 55 results demonstrate that drilling on top of the volcanic edifices provides less ambiguous data than drilling along the flanks adjacent to the seamount chain as was done on Leg 33. The operational difficulties of drilling in shallow water into rather thin sediment cover are more than offset by the value of the data obtained.

In addition, the Leg 55 drilling in the Emperor Seamounts provided a wealth of data that could not have been obtained by standard oceanographic techniques. In particular, the recovery of vertical cores enabled the determination of the paleolatitude of Suiko Seamount and a valuable test of hot spot fixity. In addition, drilling was the only way to 1) test whether the lavas on the seamounts were erupted in the same sequence as those in the well-studied Hawaiian Islands; 2) recover lavas for geochemical and petrological study, and radiometric age dating because the central and northern Emperor Seamounts are blanketed with ice-rafted glacial pebbles and sediment; 3) determine the detailed post-volcanic history of the seamounts through the study of stratigraphic sections.

The simple picture of oceanic volcanism presented in the Introduction is more complicated following routine recovery of basement volcanic rocks by drilling. It is now known that mid-ocean ridges erupt not only depleted tholeiite but also undepleted tholeiite. Most of the locations where undepleted tholeiite occurs are associated with high-elevation sections of the ridge, such as near Iceland or the Azores, but undepleted tholeiite is also found along "normal" sections of ridgecrest (Wood and others, 1979; Tarney and others, 1979; Bougault and others, 1979).

The recovery of depleted tholeiite, similar to mid-ocean ridge basalt, from Site 253 on the Ninetyeast Ridge demonstrates that these basalts are not unique to mid-ocean ridges. A larger-scale example is the volcanic sill-flow complex recovered in the Nauru Basin. The basalt recovered was erupted far from a mid-ocean ridge, yet has many of the chemical characteristics once attributed only to those basalts erupted at mid-ocean ridges.

Off-ridge non-edifice building volcanic complexes such as that recovered in the Nauru Basin, or sills (*e.g.*, Site 169, Leg 17, and Site 215, Leg 22) mark the discovery of a new type of intraplate oceanic volcanism. These offridge volcanic complexes are geochemically variable, the volumetrically large Nauru Basin complex is depleted tholeiite while the smaller sills recovered are undepleted basalt.

The picture of island volcanism is also complex, although drilling in the Hawaiian-Emperor chain confirmed the well known sequence of volcanism known from the Hawaiian Islands. Elsewhere, as in the Line Islands, it is not clear whether the presence of strongly alkalic basalts indicates post-erosional volcanism (as in Hawaii) because we do not know whether these volcanoes are similar to Hawaiian volcanoes or whether they undergo a different sequence of eruptions. The period of quiescence is clearly longer in the Line Islands than in the Hawaiian Islands where post-erosional eruptions generally follow the shield building and alkalic stages by only a few million years.

Several key questions remain unanswered: Are all oceanic volcanoes built mainly of tholeiitic basalt (depleted or undepleted?) and are only capped by alkalic or strongly alkalic basalts? What caused the outbreak of widespread volcanic activity in the Central Pacific during the Cretaceous? Are compositionally discrete mantle source regions small enough that single volcanoes could tap more than one source region during their growth? Could these small scale heterogeneities account for observed compositional differences between adjacent volcanoes such as Kilauea and Mauna Loa (Leeman and others, 1979)? Are the Line Islands an age-progressive chain (a hot-spot trace) with widespread superimposed younger volcanics or was the entire chain formed by some non-hot spot type of intraplate volcanism? Future drilling may provide answers to these key questions.

ACKNOWLEDGMENTS

T. Vallier, G. B. Dalrymple, J. H. Natland, and an anonymous referee reviewed the manuscript and suggested changes that significantly improved the final version.

REFERENCES CITED

NOTE: All references to Deep Sea Drilling Project Initial Reports have been abbreviated; for complete citations see *Appendix I* at the end of this volume, or the individual Initial Reports as published.

Bass, M. N., Moberly, R., Rhodes, M. J., and Shih, C., 1973, Volcanic rocks cored in the central Pacific, Leg 17, Deep Sea Drilling Project: Init. Repts. DSDP, v. 17, p. 429–503.

Batiza, R., 1979, Petrology and chemistry of volcanic rocks drilled during DSDP Leg 61: a possible off-ridge flood basalt in the western equatorial Pacific (abs): Hawaii Symposium on Intraplate Volcanism and Submarine Volcanism, p. 6.

Bence, A. E., Taylor, S. R., and Fisk, M., 1980, Major and trace element geochemistry of basalts from Ojin, Nintoku, and Suiko Seamounts of the Emperor Seamount chain: DSDP-IPOD Leg 55: Init. Repts. DSDP, v. 55, p. 599–605.

Bolli, H. M., Ryan, W. B. F., and others, 1978, Init. Repts. DSDP, v. 40.

Bougault, H., Joron, J. L., and Treuil, M., 1979, Alteration, fractional crystallization, partial melting; mantle properties from trace elements in basalts recovered from the North Atlantic, *in* Talwani, M., Harrison, C. G., and Hayes, D. E., eds., Deep Sea Drilling Results in the Atlantic Ocean: Ocean Crust: Washington D.C., American Geophys. Union, p. 352–368.

Butt, A., 1980, Biostratigraphic and paleoenvironmental analyses of the sediments at the Emperor Seamounts, DSDP Leg 55 NW Pacific: Foraminifera: Init. Repts. DSDP, v. 55, p. 289–325.

CAMBON, P., JORON, J. L., BOUGAULT, H., AND TREUIL, M., 1980, Leg 55: Emperor Seamounts; trace elements in transitional tholeiites, alkalic basalts, and hawaiites; mantle homogeneity or heterogeneity and magmatic processes: Init. Repts. DSDP, v. 55, p. 585–597.

CANN, J. R., TARNEY, J., VARET, J., AND WOOD, D. A., 1979, Mantle heterogeneity in the North Atlantic: Evidence from Leg 49 geochemistry: Init. Repts. DSDP, v. 49, p. 841–850.

CHASE, T. E., MENARD, H. W., AND MAMMERICK, J., 1971, Topography of the North Pacific: IMR Tech. Rept. Ser. TR 17, Scripps Inst. Oceanography, La Jolla, Calif.

CLAGUE, D. A., AND DALRYMPLE, G. B., 1973, Age of Kōkō Seamount, Emperor Seamount chain: Earth Planet. Sci. Letters, v. 17, p. 411–415.

———, AND FREY, F. A., 1979, Petrology of the Hawaiian Emperor volcanic chain (abs): Hawaii Symposium on Intraplate Volcanism and Submarine Volcanism, p. 118.

———, AND ———, 1980, Trace element geochemistry of tholeiitic basalts from Site 433C, Suiko Seamount: Init. Repts. DSDP, v. 55, p. 559–569.

———, DALRYMPLE, G. B., AND MOBERLY, R., 1975, Petrography and K-Ar ages of dredged volcanic rocks from the western Hawaiian Ridge and southern Emperor Seamount chain: Geol. Soc. America Bull., v. 86, p. 991–998.

COCKERHAM, R. S., AND JARRARD, R. D., 1976, Paleomagnetism of some Leg 33 sediments and basalts: Init. Repts. DSDP, v. 33, p. 631–647.

CURRAY, J. R., AND MOORE, D. G., 1974, Sedimentary and tectonic processes in the Bengal Deep Sea Fan and geosyncline: *in* Burke, C. A., and Drake, C. L., *eds.,* The Geology of Continent Margins: New York, Springer-Verlag, p. 617–627.

DALRYMPLE, G. B., CLAGUE, D. A., AND LANPHERE, M. A., 1977, Revised age for Midway volcano, Hawaiian Volcanic chain: Earth Planet. Sci. Letters, v. 37, p. 107–116.

———, LANPHERE, M. A., AND JACKSON, E. D., 1974, Contributions to the petrography and geochronology of volcanic rocks from the leeward Hawaiian Islands: Geol. Soc. America Bull., v. 85, p. 727–738.

———, LANPHERE, M. A., AND CLAGUE, D. A., 1980, Conventional and $^{40}Ar/^{39}Ar$ K-Ar ages of volcanic rocks from Ojin (Site 430), Nintoku (Site 432), and Suiko (Site 433) Seamounts and the chronology of volcanic propagation along the Hawaiian-Emperor chain: Init. Repts. DSDP, v. 55, p. 659–676.

———, ———, AND NATLAND, J. H., 1980, K-Ar minimum age for Meiji Guyot, Emperor Seamount chain: Init. Repts. DSDP, v. 55, p. 677–683.

DAVIES, T. A., LUYENDYK, B. P., AND OTHERS, 1974, Init. Repts. DSDP, v. 26.

DONNELLY, T. W., MELSON, W., KAY, R., AND ROGERS, J. J. W., 1973, Basalts and dolerites of late Cretaceous age from the central Caribbean: Init. Repts. DSDP, v. 15, p. 989–1011.

FALVEY, D. A., 1972, Sea-floor spreading in the Wharton Basin (northeast Indian Ocean) and the breakup of eastern Gondwanaland: Aust. Petroleum Expl. Assoc. Jour., v. 12, p. 86–88.

FISCHER, A. G., HEEZEN, B. C., 1971, Init. Repts. DSDP, v. 6, 1329 p.

FODOR, R. V., AND THIEDE, J., 1977, Volcanic breccias from DSDP Site 357: Implications for the composition and origin of the Rio Grande Rise: Init. Repts. DSDP, v. 39, p. 537–544.

———, KEIL, K., HUSLER, J. W., AND MCKEE, E. M., 1977, Petrology and K-Ar age of volcanic tuff and ash from the Walvis Seamount Province, DSDP Site 359, Leg 39: Init. Repts. DSDP, v. 39, p. 525–536.

FRANCIS, T. J. G., AND RAITT, R. W., 1967, Seismic refraction measurements in the southern Indian Ocean: Jour. Geophys. Res., v. 72, p. 3015–3041.

FREY, F. A., AND SUNG, C. M., 1974, Geochemical results for basalts from Sites 253 and 254: Init. Repts. DSDP, v. 26, p. 657–572.

———, DICKEY, J. S., THOMPSON, G., AND BRYAN, W. B., 1977, Eastern Indian Ocean DSDP Sites: correlations between petrography, geochemistry and tectonic setting, *in* Heirtzler, J. R., and others, eds., Indian Ocean Geology and Biostratigraphy: Washington, D.C., Amer. Geophys. Union, p. 189–257.

GREENE, H. G., DALRYMPLE, G. B., AND CLAGUE, D. A., 1978, Evidence for northward movement of the Emperor Seamounts: Geology, v. 6, p. 70–74.

HAGN, H., BUTT, A., AND MALZ, H., 1980, Paleocene shallow water facies at Emperor Seamounts: DSDP Leg 55, NW Pacific: Init. Repts. DSDP, v. 55, p. 327–347.

JACKSON, E. D., KOIZUMI, I., AND OTHERS, 1978, Drilling confirms hot-spot origins: Geotimes, v. 23, p. 23–26.

———, AND SCHLANGER, S. O., 1976, Regional syntheses, Line Island Chain, Tuamotu Island Chain, and Manikiki Plateau, Central Pacific Ocean: Init. Repts. DSDP, v. 33, p. 915–928.

———, KOIZUMI, I., DALRYMPLE, G. B., CLAGUE, D. A., KIRKPATRICK, R. J., AND GREENE, H. G., 1980, Introduction and summary of results from DSDP Leg 55: the Hawaiian-Emperor hot-spot experiment: Init. Repts. DSDP, v. 55, p. 5–31.

KIRKPATRICK, R. J., CLAGUE, D. A., AND FRIESEN, W., 1980, Petrology and Geochemistry of volcanic rocks, DSDP Leg 55: Emperor Seamount Chain: Init. Repts. DSDP, v. 55, p. 509–557.

KONO, M., 1980a, Magnetic properties of DSDP Leg 55 basalts: Init. Repts. DSDP, v. 55, p. 723–736.

———, 1980b, Paleomagnetism of DSDP Leg 55 basalts and implications to the tectonics of the Pacific plate: Init. Repts. DSDP, v. 55, p. 737–752.

———, CLAGUE, D. A., AND LARSON, E. E., 1980, Fe-Ti oxide mineralogy of DSDP Leg 55 basalts: Init. Repts. DSDP, v. 55, p. 639–652.

LANGMUIR, C. H., BENDER, J. F., BENCE, A. E., AND HANSON, G. N., 1977, Petrogenesis of basalts from FAMOUS area: mid-Atlantic Ridge: Earth Planet. Sci. Letters, v. 36, p. 133–156.

———, Voche, R. D., Hanson, G. N., and Hart, S. R., 1978, A general mixing equation with applications to Icelandic basalts: Earth Planet. Sci. Letters, v. 37, p. 380–392.

Lanphere, M. A., and Dalrymple, G. B., 1976, K-Ar ages of basalts from DSDP Leg 33: Sites 315 (Line Islands) and 317 (Manikiki Plateau): Init. Repts. DSDP, v. 33, p. 649–654.

———, ———, and Clague, D. A., 1980, Rb-Sr systematics of basalts from the Hawaiian-Emperor Volcanic chain: Init. Repts. DSDP, v. 55, p. 695–706.

Larson, R. L., Moberly, Ralph, and others, 1975a, Site 311: Hawaiian Magnetic Lineations: Init. Repts. DSDP, v. 32, p. 295–309.

———, ———, ———, 1975b, Site 313: Mid-Pacific Mountains: Init. Repts. DSDP, v. 32, p. 313–390.

Larson, R. L., Schlanger, S., and others, 1978, Leg 61 drills in the central Nauru Basin: Geotimes, v. 23, p. 21–24.

Leeman, W. P., Budahn, J. R., Gerlach, D. C., Smith, D. R., and Powell, B. N., 1980, Origin of Hawaiian tholeiites: trace element constraints: Amer. Jour. Sci. (Jackson volume), v. 280A, p. 794–819.

Lonsdale, P., Normark, W., and Newman, W., 1972, Sedimentation and erosion on Horizon Guyot: Geol. Soc. America Bull., v. 83, p. 289–316.

Luyendyk, B. P., 1977, Deep Sea Drilling on the Ninetyeast Ridge: Synthesis and a Tectonic Model: *in* Indian Ocean Geology and Biostratigraphy, Heirtzler, J. R., and others, eds.: Washington D.C., Amer. Geophy. Union.

Marshall, M., 1975, Petrology and chemical composition of basaltic rocks recovered on Leg 32, Deep Sea Drilling Project: Init. Repts. DSDP, v. 32, p. 563–570.

McDougall, I., 1964, Potassium-argon ages from lavas of the Hawaiian Islands: Geol. Soc. America Bull., v. 75, p. 107–128.

McDowell, S., Kumar, N., Jacobi, R. D., Johnson, D. A., and Bance, E. T., 1977, Regional setting of Site 357, North flank of Rio Grande Rise: Init. Repts. DSDP, v. 39, p. 955–970.

McKenzie, D. P., and Sclater, J. G., 1971, The evolution of the Indian Ocean since the late Cretaceous: Royal Astron. Soc. Geophys. Jour., v. 25, p. 437–528.

McKenzie, J., Bernoulli, D., and Schlanger, S. O., 1980, Shallow water carbonate sediments from the Emperor Sea-mounts: Their diagenesis and paleogeographic significance: Init. Repts. DSDP, v. 55, p. 415–455.

Morgan, W. J., 1972, Plate motions and deep mantle convection: *in* Shagam, R., ed., Studies in Earth and Space Sciences, (Hess volume): Geol. Soc. America Mem., v. 32, p. 265–282.

Moore, D. G., Curray, J. R., Russell, W. R., and Frans, J. E., 1974, Stratigraphic-seismic section correlations and implications to Bengal Fan history: Init. Repts. DSDP, v. 22, p. 403–412.

Natland, J. H., 1975, Petrologic Studies of Linear Island Chains, Part 2: The Line Islands: unpublished Ph.D. Dissertation, University of California, San Diego, 384 p.

———, 1976, Petrology of volcanic rocks dredged from seamounts in the Line Islands: Init. Repts. DSDP, v. 33, p. 749–778.

———, 1978, Composition of basaltic rocks recovered at Sites 367 and 368, Deep Sea Drilling Project, near the Cape Verde Islands: Init. Repts. DSDP, v. 41, p. 1107–1112.

O'Nions, R. K., Pankurst, R. J., and Gronvold, K., 1976, Nature and development of basalt magma sources beneath Iceland and the Reykjanes Ridge: Jour. Petrology, v. 17, p. 315–338.

Schilling, J. R., 1973, Iceland mantle plume: geochemical evidence along Reykjanes Ridge: Nature, v. 242, p. 565–571.

———, 1976, Rare earth, Sc, Cr, Fe, Co and Na abundances in DSDP Leg 38 basement basalts: some additional evidence on the evolution of the Thulean volcanic province: Init. Repts. DSDP, v. 38, p. 741–750.

———, and Noe-Nygaard, A., 1974, Faeroe-Iceland plume: rare earth evidence: Earth Planet. Sci. Letters., v. 24, p. 1–14.

Schlanger, S. O., and Konishi, K., 1975, The geographic boundary between the Coral-Algal and the Bryozoan-Algal limestone facies: a paleolatitude indicator: IX International Congress of Sedimentology, Nice, Theme 1, Sedimentologic Indicators, p. 187–190.

———, Jackson, E. D., and others, 1976, Init. Repts. DSDP, v. 33.

Scholl, D. W., and Creager, J. S., 1973, Geologic synthesis of Leg 19 (DSDP) results: Far north Pacific, Aleutian Ridge, and Bering Sea: Init. Repts. DSDP, v. 19, p. 897–913.

Sclater, J. G., and Fisher, R. L., 1974, Evolution of the east central Indian Ocean, with emphasis on the tectonic setting of the Ninetyeast Ridge: Geol. Soc. America Bull., v. 85, p. 683–702.

Stewart, R. J., Natland, J. H., and Glassley, W. R., 1973, Petrology of volcanic rocks recovered on DSDP Leg 19 from the North Pacific Ocean and the Bering Sea: Init. Repts. DSDP, v. 19, p. 615–627.

Supko, P. R., Perch-Nielson, K., and others, 1977, Init. Repts. DSDP, v. 39.

Talwani, M., Udintsev, G., and others, 1976, Init. Repts. DSDP, v. 38.

Tarney, J., Wood, D. A., Varet, J., Saunders, A. D., and Cann, J. R., 1979, Nature of mantle heterogeneity in the North Atlantic: Evidence from Leg 41 basalts, *in* Talwani, M., Harrison, C. G., and Hayes, D. E., eds., Deep Drilling Results in the Atlantic Ocean: Ocean Crust: Washington D.C., Amer. Geophys. Union, p. 285–301.

Vallier, T., Thiede, J., and others, 1979, Leg 62 probes the paleoenvironments: Geotimes, v. 24, p. 24–26.

Veevers, J. J., Jones, J. G., and Talent, J. A., 1971, Indo-Australian stratigraphy and the configuration and dispersal of Gondwanaland: Nature, v. 229, p. 383–388.

von der Borch, C. C., Sclater, J. G., and others, 1974, Init. Repts. DSDP, v. 22.

Winterer, E. L., 1976, Bathymetry and regional tectonic setting of the Line Island chain: Init. Repts. DSDP, v. 33, p. 731–748.

———, Ewing, J. I., and others, 1973, Init. Repts. DSDP, v. 17.

Wood, D. A., 1978, Major and trace element variations in the Tertiary lavas of eastern Iceland with respect to the Iceland geochemical anomaly: Jour. Petrology, v. 19, p. 393–436.

———, Tarney, J., Varet, J., Saunders, A. D., Bougault, J., Joron, J. L., Treuil, M., and Cann, J. R., 1979, Geochemistry of basalts drilled in the North Atlantic by IPOD, Leg 49: Implications for mantle heterogeneity: Earth Planet. Sci. Letters, v. 42, p. 77–97.

SEPM SPECIAL PUBLICATION NO. 32, P. 23–37, OCTOBER 1981

GEOLOGIC SIGNIFICANCE OF SEISMIC REFLECTORS IN THE DEEP WESTERN NORTH ATLANTIC BASIN*

BRIAN E. TUCHOLKE
Woods Hole Oceanographic Institution, Woods Hole, Massachusetts 02543 U.S.A.

ABSTRACT

Several major seismic horizons in the deep basin of the western North Atlantic Ocean have been calibrated according to age and physical and lithologic character by JOIDES drilling. Three horizons are discussed that are representative of the spectrum of stratigraphic relations in the basin. Horizon β, which ranges from Hauterivian to Barremian in age, correlates with a sharp upward transition from limestones to black clays coincident with a rise in the calcite compensation depth. In Eocene time, sediments enriched in biogenic silica were widely deposited, and subsequent diagenesis formed chert beds; the upper Lower to lower Middle Eocene surface of these cherts correlates with Horizon A^c which is one of the strongest and most widespread reflectors in the basin. Widely distributed borehole data indicate that both Horizon β and Horizon A^c normally occur within continuously deposited sedimentary sections. Beneath the continental rise, Horizon A^u correlates with a major unconformity eroded between Late Eocene and Early Miocene time by southward-flowing abyssal currents. The reflector may have limited chronostratigraphic significance because of lateral migration of the abyssal current system with time. Several simple models are used to assess potential age significance of the seismic and lithologic boundaries and are compared to borehole data.

INTRODUCTION

Geologic interpretation of seismic reflection stratigraphy in the ocean basins has received increasing attention in recent years. Two methods of stratigraphic interpretation commonly have been used. One, based primarily on studies of shallow continental-margin seismic data, relates major seismic reflectors to time-synchronous unconformities and their correlative conformities, and it correlates the reflection boundaries with sea-level changes (Vail and others, 1977, 1980). A second method, applied to the study of deep-basin seismic data, relates major reflectors to impedance contrasts that are commonly, but not consistently, coincident with significant lithologic boundaries, both conformable and unconformable. Such reflecting horizons result from a variety of oceanographic events including sea-level changes, and they need not be time-synchronous (Tucholke and Mountain, 1977, 1979). Each method has been used successfully over its primary depth range in predicting stratigraphic relations, but distinct differences in interpretational methods become apparent in stratigraphic studies at intermediate depths (about 2000–4000 m) along continental margins.

The purpose of this paper is to summarize observational data on three prominent seismic horizons in the deep western North Atlantic basin and to relate these data to the lithostratigraphic and chronostratigraphic frameworks determined at DSDP drill sites. The mechanics of correlation between reflectors and borehole results have been dealt with elsewhere (*e.g.,* Ewing and others, 1970; Tucholke, Vogt and others, 1979) and will not be repeated here. However, it should be noted that the strength and position of reflectors can be a composite result of signal interference patterns from a series of adjacent beds of varying impedance (*e.g.,* Woods, 1956; Widess, 1973; Mayer, 1979), and the physical reality of an impedance or lithologic change at the level of a given reflector therefore must be evaluated. The major reflecting horizons discussed in this paper are laterally coherent, but they do cross regions of changing bed thickness and geometry and therefore exhibit amplitude variations and slight changes in apparent position. However, the reflectors also show consistent correlations with physical and lithologic changes in boreholes from a wide variety of depositional and seismic provinces, and therefore they appear to represent real impedance and lithologic changes rather than signal-interference artifacts. Throughout this discussion, the seismic stratigraphic horizons will be treated separately from chronostratigraphic and lithostratigraphic horizons, although there are notable coincidences among the three.

The three seismic horizons discussed here each represent different stratigraphic relations. Horizon β generally correlates with a conformable contact between Neocomian limestones and Middle Cretaceous black clays. Horizon A^c matches the top of a Lower to Middle Eocene diagenetic chert sequence, and Horizon A^u corresponds to a major erosional unconformity developed along the continental

*Contribution No. 4459 of Woods Hole Oceanographic Institution
Contribution No. 3160 of Lamont-Doherty Geological Observatory

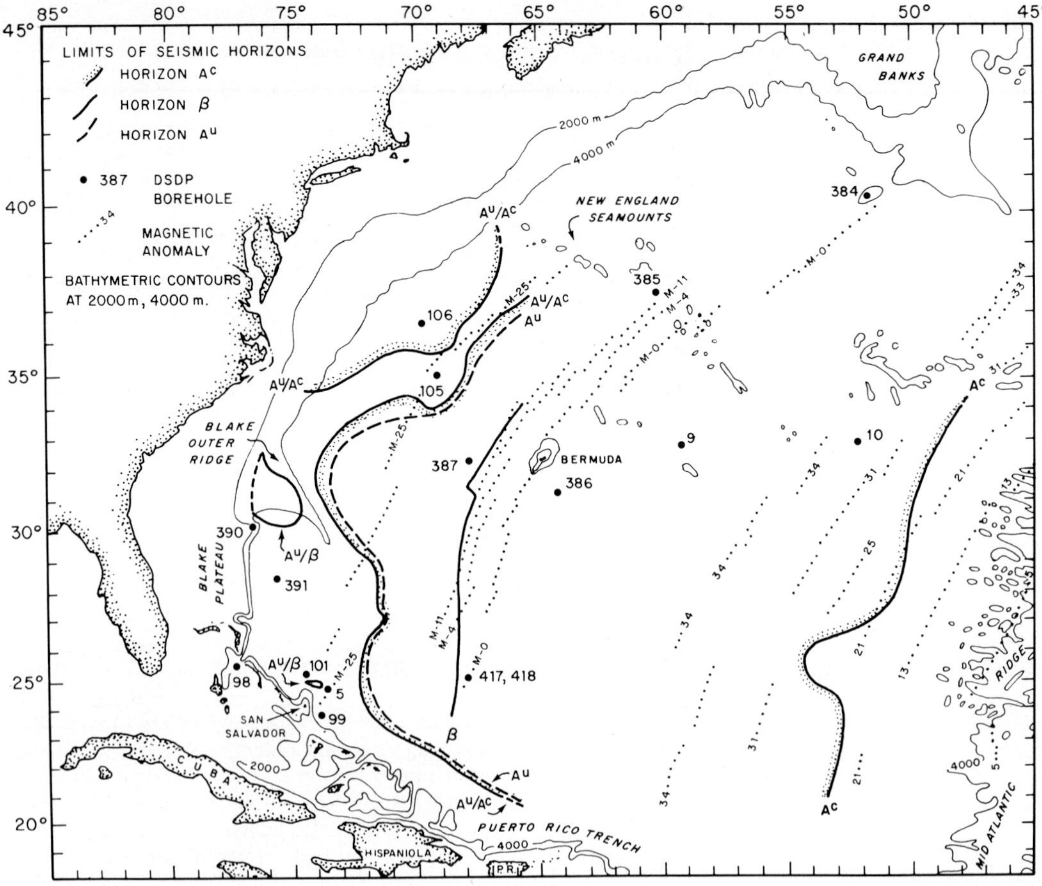

FIG. 1.—Distribution in western North Atlantic of seismic horizons discussed in this paper (from Tucholke and Mountain, 1979). Horizons A^c and β are truncated along the margin by Horizon A^u as shown; a long "channel" was eroded through A^c in the vicinity of DSDP Site 105. The western limit of Horizon A^u probably lies near the present 2000m isobath.

margin between Late Eocene and Early Miocene time (Ewing and Hollister, 1972; Tucholke, 1979). This paper summarizes the lithologic and chronologic significance of these seismic horizons, the possible causes for their age variations, the evidence for sediment accumulation or erosion across the horizons, and the relation of these ages and accumulation patterns to changes in sea-level and other oceanographic processes. Several simple models are used to illustrate how seismic stratigraphic and lithostratigraphic horizons can deviate from chronostratigraphic boundaries. The discussion is not definitive, but it serves to illustrate the application of some basic stratigraphic principles to seismic data, to point out several limitations and uncertainties, and to stimulate more critical evaluation of seismic reflection data in ocean basins.

HORIZON β

Horizon β is a deep, well-defined seismic horizon overlying ocean crust older than about Barremian age (Figs. 1, 2). The reflector has been drilled in five DSDP holes (5A, 101A, 105, 387, 391C) where correlation between the reflector and the lithology is possible (Tucholke, 1979). The impedance contrast correlating with the reflector at each hole coincides with or lies within a few meters of the lithologic boundary between limestones or chalks and overlying dark green-gray to black claystones (Fig. 3). In some cases the reflector may not be exactly coincident with the lithologic boundary; for example, at Site 387 at least the upper 9 meters of the calcareous sequence are chalk with an impedance similar to that of the overlying black clays and significantly lower than the immediately underlying limestones. At certain other sites (*e.g.,* Site 391C) thin chalk and limestone lenses (<30cm) within the black clays overlie Horizon β (Benson, Sheridan and others, 1978), but reflection amplitudes of these lenses probably are insignificant compared to that of the underlying massive limestone. Thus, available data show that Horizon β closely approximates

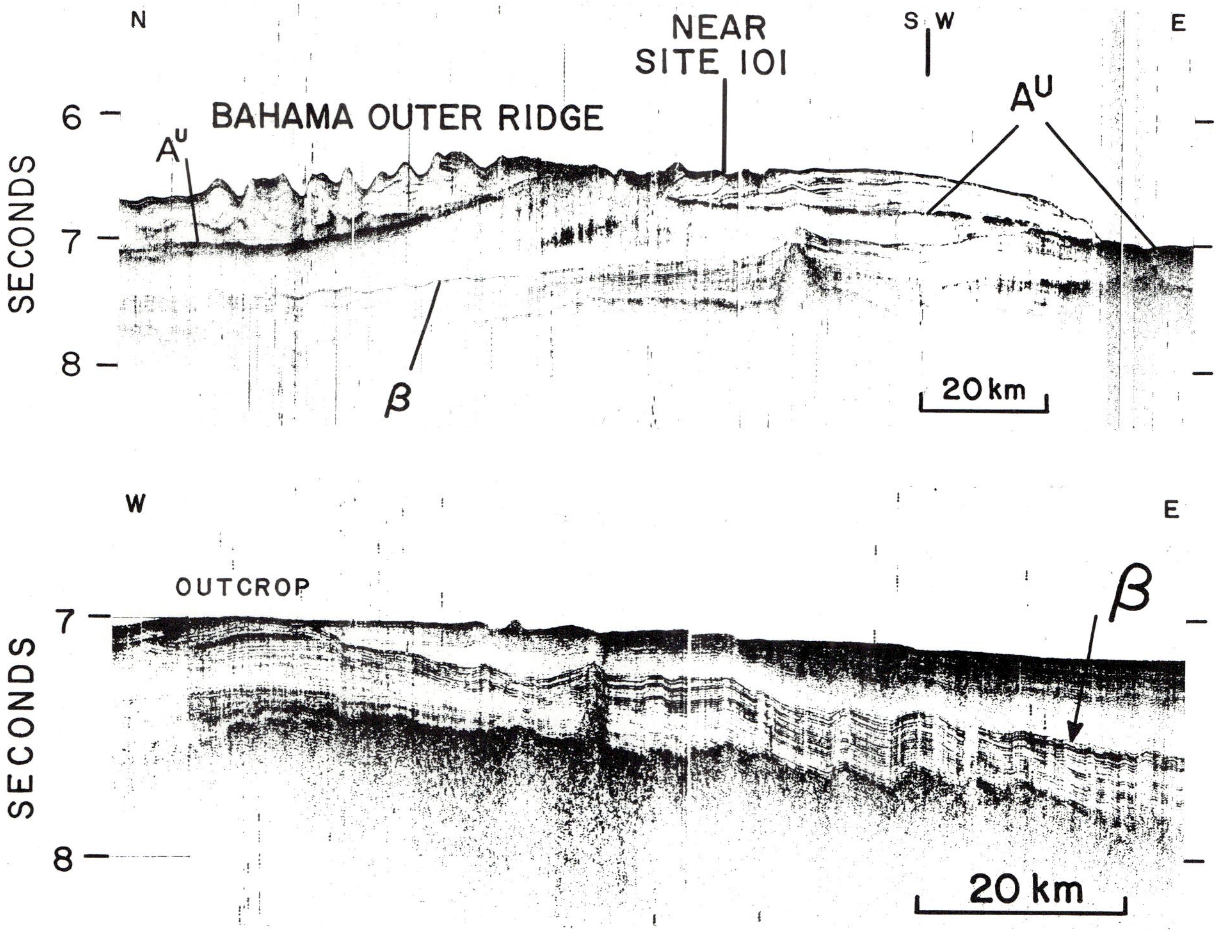

FIG. 2.—Upper profile: single-channel seismic profile extending south across the southern Bahama Outer Ridge and DSDP Site 101, then east to the Horizon A—Horizon β outcrop area (Ewing and others, 1966; Windisch and others, 1968) north of San Salvador. Horizons β and A^u are identified. Lower profile: seismic profile extending east from Horizon β outcrop area.

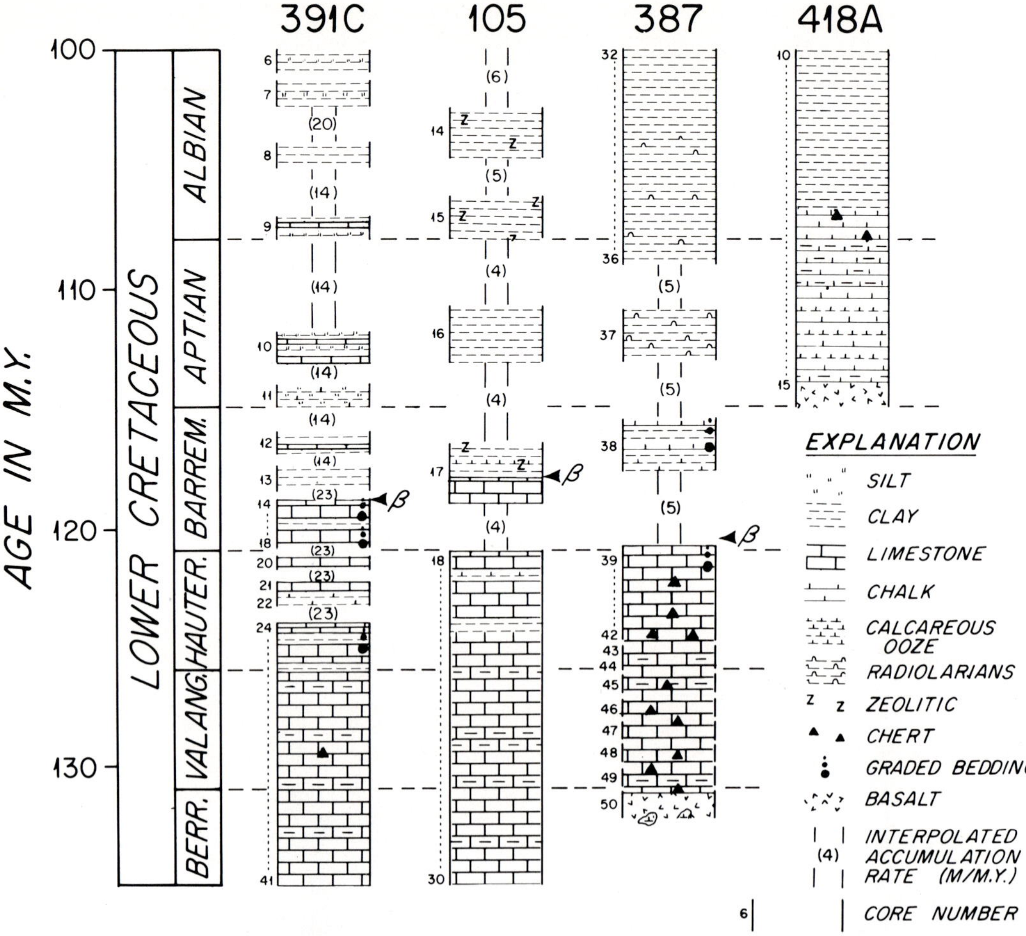

FIG. 3.—Summary of age and lithology of sediments cored at three DSDP boreholes that penetrated Horizon β and at Hole 418A which penetrated the same lithologic transition but in a position seaward of the limit where Horizon β is resolved seismically. Borehole locations in Figure 1.

the boundary between a dominantly calcareous and a dominantly non-calcareous facies.

To evaluate the age significance (as determined by biostratigraphy) of the lithologic boundary and thus of the seismic horizon, we must consider temporal changes in the oceanographic factors that created the lithologic transition. The sediments document a change from deposition of calcareous sediments on a well-oxygenated seafloor to deposition of essentially carbonate-free clays and silts under alternating low-oxygen and anoxic conditions (Lancelot and others, 1972; Tucholke and Vogt, 1979). As shown in Figure 4, the change was coincident with a rise in the calcite compensation depth (CCD; defined as the level in the water column below which the rate of solution of calcium carbonate exceeds the rate of its deposition). The simple model in Figure 5 predicts the age distribution for the lithologic boundary along a midocean ridge flank, assuming a linear rise in the CCD with time. One significant aspect of this model is that the lateral shift of the lithologic boundary per unit time is much greater over deeper, gently sloping seafloor than over the more steeply sloping seafloor of the upper ridge flank. Thus for a linear CCD rise the lithologic boundary is more nearly time synchronous in the deep basin, and it departs most markedly from synchronism along the ridge flank.

Although the model is simplified by assuming a linear CCD rise, it reasonably agrees with data from DSDP sites. The lithologic boundary is first observed nearly simultaneously in several widely spaced holes (101A, 105, 387, 391C) that were within the deep basin in Barremian time (Figs. 1, 3, 4), but it did not develop at shallow ridge-crest positions (Sites 417, 418) until the late Aptian, about

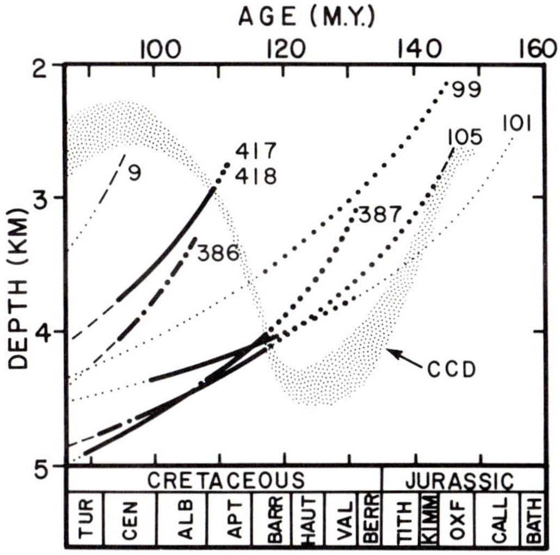

FIG. 4. — Reconstructed seafloor age-versus-depth curves for western North Atlantic DSDP boreholes, showing the Early Cretaceous rise in the CCD (shaded; from Tucholke & Vogt, 1979). Along the curves large solid dots represent calcareous sediments, heavy black lines are black clays, thin dashed lines are multicolored clays, and thin dotted lines represent uncored intervals and hiatuses.

6–8 m.y. later (Figs. 3, 4). The diachronism suggested for the lithologic boundary also could be expected to apply to the seismic reflector, Horizon β, because of its close relation to lithologic type. In practice, however, Horizon β is extremely difficult to trace seismically onto crust younger than Hauterivian age for a variety of reasons. First, the sedimentary section below the reflector is very thin above younger crust and therefore often is not resolvable in seismic data; for example, the basal calcareous sediments at Sites 417 and 418 are only a few meters thick (Donnelly and others, 1979). In addition, side echoes from the irregular basement in surface-ship reflection profiles often mask the thin bed; recently published seismic reflection profiles obtained near Sites 417 and 418 with a deep-towed hydrophone receiver and a surface airgun clearly show a deep reflector, probably Horizon β, that is masked by hyperbolae in surface profiles (Purdy and others, 1980). Finally, the carbonate beds overlying this younger crust locally may be unlithified chalks without sufficient impedance contrast with the overlying claystones to generate a significant reflector.

Assuming such factors to be equally important throughout the basin, some diachronism of Horizon β is indicated by the location of the eastward pinchout of the reflector on basaltic basement (Fig. 1). In the central part of the basin Horizon β can be traced onto crust of Valanginian/Hauterivian age (magnetic anomalies M-10, M-11), but farther south the seismic pinchout is observed on younger crust (Barremian, anomalies M-2 to M-4). If this pattern also reflects the distribution of the lithologic boundary between limestones or chalks and the black clays, there are two explanations for the

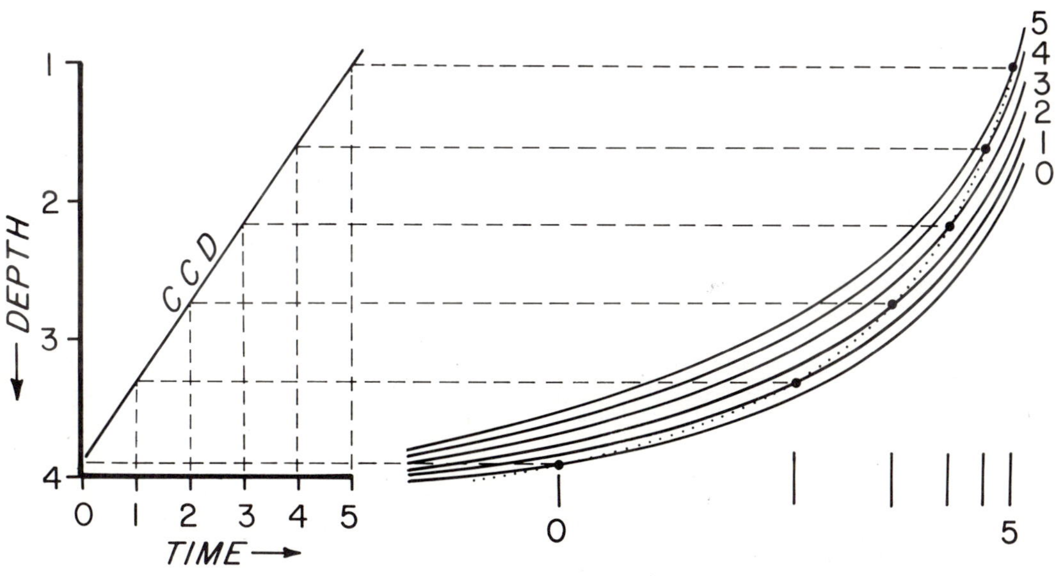

FIG. 5. — Model showing time-transgressive nature of facies change from calcareous to clayey sediments (dotted line), assuming linear CCD rise and seafloor depth along a midocean ridge flank that is an exponential function of crustal age. Note increased diachronism of boundary per unit distance over steeper seafloor slopes.

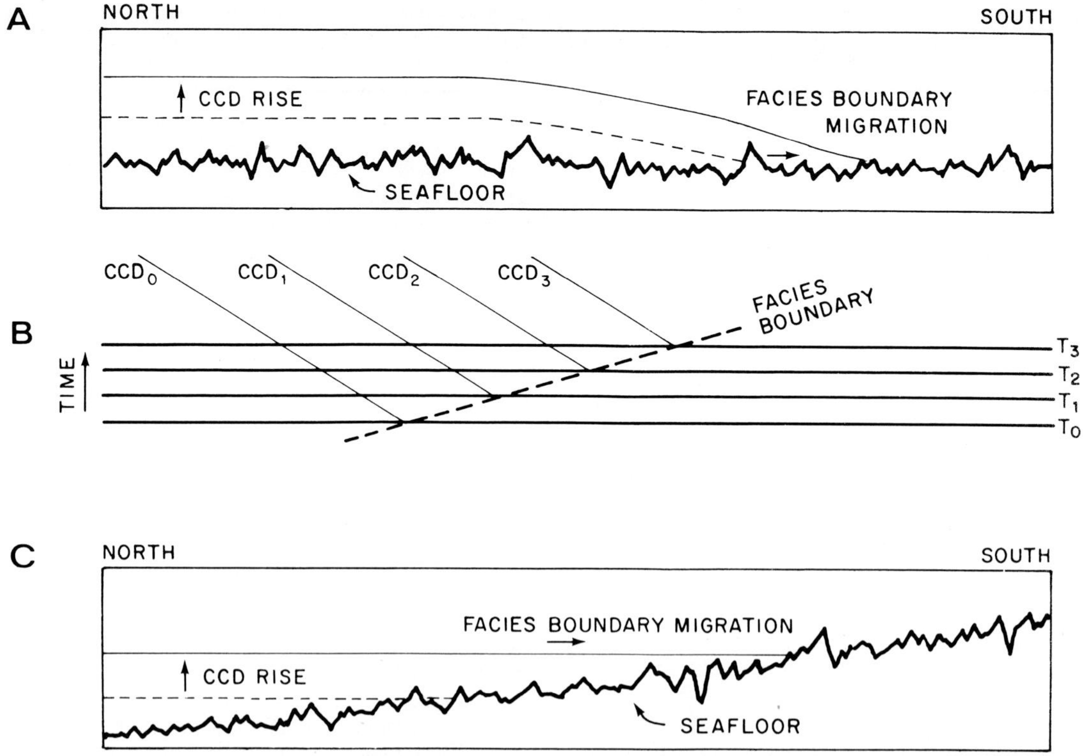

FIG. 6.—A, B. Simple model of equatorward-sloping CCD in the North Atlantic overlying crust of uniform depth along a seafloor isochron. The rise of the sloping CCD with time (T_0 to T_3) produces a time-transgressive facies boundary. C. Model as in A and B, but with a level CCD and southward-shoaling crust along a seafloor isochron. Both models produce a carbonate/clay facies boundary that becomes younger to the south, similar to the apparent diachronous pinchout of Horizon β in Figure 1.

distribution. First, if the CCD during the Early Cretaceous deepened toward the equator rather than being horizontal, then a CCD rise with time would result in younger ages of the lithologic boundary toward the south (Fig. 6A, B). In the Pacific Ocean, the existence of an equatorward-sloping CCD has been documented for most of the Cenozoic; it results from the large carbonate flux from the productive surface waters of the equatorial current system (van Andel, 1975). In the Early Cretaceous Atlantic Ocean, a circumglobal "equatorial" flow probably also was present (Berggren and Hollister, 1974), and a similar depression of the CCD may have occurred beneath associated productive surface waters in the southern North Atlantic.

A second possible explanation is that sea-floor depth along a given isochron may have shoaled equatorward beneath a horizontal CCD, producing the same diachronous distribution of the lithologic boundary as the CCD raised with time (Fig. 6C). This explanation seems less likely because most present midocean ridges deepen toward the equator (Menard and Dorman, 1976; Rona, 1977). However the latitudinal slopes of mid-ocean ridges may have been much different in the Early Cretaceous. In either case, the available lithologic data, seismic reflection data, and simple predictive models all indicate that the CCD-determined lithologic boundary is diachronous and that the diachronism is likely to be more pronounced over younger, more steeply sloping seafloor.

The continuity of deposition across Horizon β can be examined both on the basis of deepsea borehole data and on the basis of seismic stratigraphy. Three boreholes that not only penetrated Horizon β but also were cored well enough to recover a fairly complete sedimentary record across the reflector are shown in Figure 7. Although biostratigraphic age resolution generally is no better than to the stage level, it has produced no evidence that significant unconformities interrupt the sedimentary sequence at or near Horizon β. This is not surprising in view of the fact that Horizon β occurs at the transition to stagnant-basin conditions and thus at a time when any bottom currents should have had a

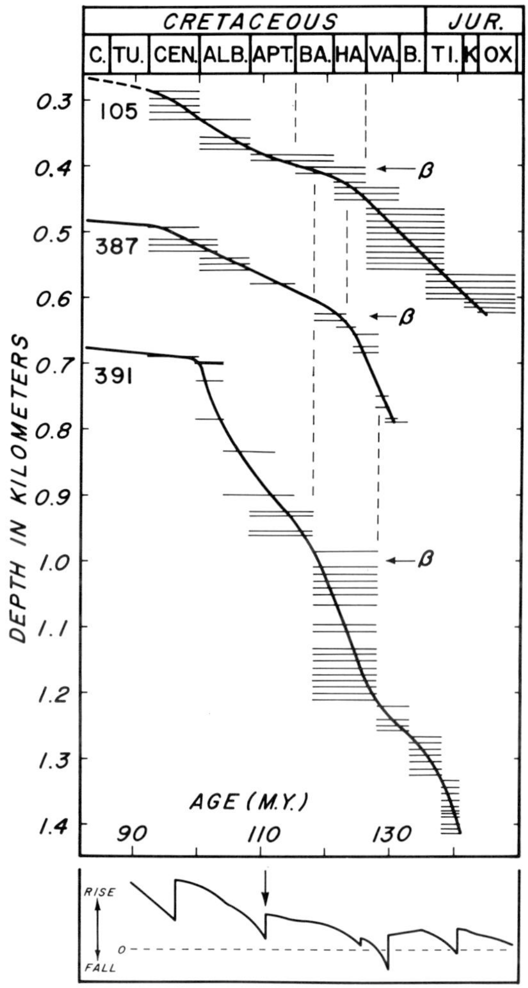

FIG. 7.—Age versus depth-in-borehole at three DSDP sites that penetrated Horizon β. Horizontal lines give biostratigraphic age determinations in cores. Relative sealevel at bottom from Vail and others (1977). Note that Horizon β (age constraints indicated by vertical dashed lines at each site) predates the mid-Aptian sealevel lowstand.

minimal effect on the continuity of the sedimentary record. The record at Site 105 may be especially important in evaluating potential effects of bottom currents because this borehole was drilled on a prominent basement swell, where abyssal currents typically would cause erosional effects. The lack of any significant unconformity at this site therefore suggests that abyssal circulation was very limited at the time that Horizon β was deposited.

In the seismic stratigraphic record it is possible to define seismic discordances that may or may not indicate true unconformities in the sedimentary record; these seismic discordances can appear

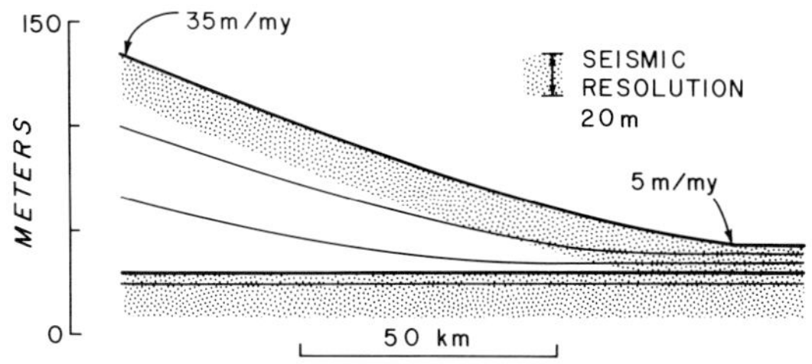

FIG. 8.—Model of terrigenous influx forming a seaward-encroaching onlap sequence in an abyssal pelagic basin. Although continuous accumulation occurs throughout the sequence, a seismic discordance may be identified because of limited seismic resolution.

simply because of the limited resolution of seismic methods. A possible example is shown in a study by Shipley and Watkins (1978) who demonstrated the presence of a flat-lying Horizon β locally overlying hummocky reflectors in a small area beneath the lower continental rise off the eastern United States. They interpreted this as truncation of the underlying irregular topography by bottom currents, but it also is possible that the reflector represents onlap of flat-lying turbidites onto hummocky pelagic sediments, possibly with minor erosion by turbidity currents. Such progradation of turbidites with mixed clastic and pelagic components almost certainly was an important element in the construction of the continental rise along this passive margin. In fact, deep-basin turbidite sedimentation of this kind is documented immediately below Horizon β at Hole 391C in the Blake-Bahama Basin (Fig. 3; Benson, Sheridan and others, 1978). Although the limited seismic resolution does not allow verification of either interpretation, Horizon β over much larger regions of the basin rarely shows seismic discordances, suggesting that deposition generally was not interrupted at the boundary. This suggestion is supported by the fact that no unconformity has been detected at Horizon β in DSDP boreholes within the basin.

The problem of seismically determining true unconformities is most severe along continental margins where high sedimentation-rate and low sedimentation-rate environments are juxtaposed. In such areas seismically determined "downlap" sequences can define a seismic discordance that does not exist as a true unconformity in the sedimentary record (Fig. 8). The sevenfold lateral change in accumulation rates illustrated in Figure 8 is approximately the actual variation observed for the black clays above Horizon β over a 300km distance in the southwestern North Atlantic (Tucholke and Mountain, 1979). Because of the typically long pulse-length of the acoustic signal and signal interference between the more closely spaced beds, the best seismic resolution that can be obtained may only approach 20 meters; therefore a physically continuous sequence can be misinterpreted as an unconformity in seismic reflection profiles.

Vail and others (1980) have related Horizon β beneath the Blake Outer Ridge to a seismic discordance (inferred unconformity) defined by downlapping sediments and correlative with a mid-Aptian relative sea-level lowstand. This correlation appears unlikely in that Horizon β is no younger than Barremian (Fig. 7). However at prevailing accumulation rates of 3–5m/MY for the black clays (Tucholke and Vogt, 1979) only 20–40 meters of sediment (13–27ms reflection time at 3.0km/s) would have accumulated above Horizon β before a mid-Aptian onlap sequence began to accumulate. Because this thickness is at the limits of seismic resolution, it is possible that Horizon β could be miscorrelated with an overlying mid-Aptian seismic discordance. These kinds of uncertainties in the use of seismic stratigraphy serve to illustrate the necessity of integrating other data from the lithologic and biostratigraphic record to make accurate interpretations of the geological history.

HORIZON A^c

One of the most prominent groups of sedimentary reflectors in the western North Atlantic occurs within Paleogene sediments overlying Horizon β. This group of reflectors collectively was termed Horizon A for many years (Ewing and Ewing, 1963), but detailed study has shown that several widespread, prominent reflectors are present within the group and that each has a distinct lithologic correlation (Tucholke and Mountain, 1979). For this reason the reflector group is termed the "Horizon-A complex". One of the reflectors, Horizon A^c, correlates with the top of a diagenetic boundary consisting of upper Lower to lower Middle Eocene cherts.

Horizon A^c is nearly ubiquitous above ocean crust older than Lower Eocene, although it is only patchily distributed in the mostly thin and ponded sediments east of Bermuda (Fig. 1). The reflector also is absent along much of the continental margin because it was truncated by later bottom-current erosion (Horizon A^u, see later discussion). Horizon A^c normally is the most reflective layer in the Horizon-A complex, and its attitude generally is conformable to both underlying and overlying reflectors in the basin.

Horizon A^c has been drilled at ten DSDP sites in the North American basin, and at all sites but one it correlates with the top of porcelaneous cherts (Fig. 9). The exception is Site 386 on the central Bermuda

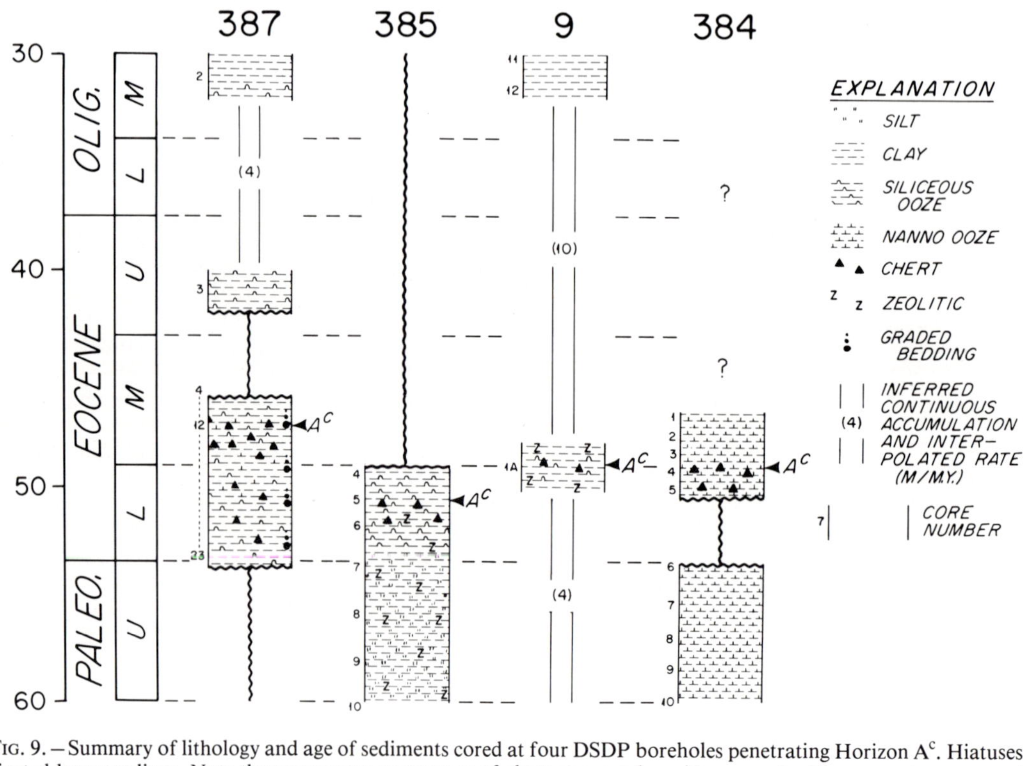

FIG. 9.—Summary of lithology and age of sediments cored at four DSDP boreholes penetrating Horizon A^c. Hiatuses indicated by wavy lines. Note the uppermost occurrence of chert at a nearly uniform age in siliceous, calcareous and clayey sediments. Borehole locations in Figure 1.

Rise where overlying high-impedance chalks seismically mask the chert (Tucholke 1979). Horizon A^c thus represents another instance where a seismic-stratigraphic marker quite consistently matches a lithostratigraphic boundary.

The cherts occur in a variety of lithofacies (Fig. 9), but in each facies biogenic silica is a significant component at the level of chert development. The Lower to Middle Eocene CCD in this area probably was between 4 and 4.5 km depth (Tucholke and Vogt, 1979), and at shallower paleodepths the cherts therefore are developed in largely calcareous sediments (*e.g.,* Site 384). On the eastern Bermuda Rise cherts occur in mostly clayey sediments that were deposited below the CCD (Site 9). From the central Bermuda Rise west to the continental margin cherts are present in siliceous clays and in fine- to coarse-grained turbidites containing calcareous, siliceous and terrigenous debris. The terrigenous debris and some of the pelagic material in the turbidites probably was eroded from the continental margin and redeposited in the area of the present Bermuda Rise before uplift formed the Rise in Middle to Late Eocene time (Tucholke, 1979). Cherty calcareous sediments of Early to Middle Eocene age also occur on the Blake Nose (Hole 390A) and in the Bahamas (Northeast Providence Channel—Site 98) although the seismic reflector Horizon A^c has not been traced to these areas.

The age of the upper boundary of the cherts is remarkably uniform considering that the surface is a diagenetic horizon. The maximum age variation probably is less than 4 m.y., ranging from late Early to early Middle Eocene (Fig. 9). The widespread development of these cherts in highly variable lithofacies and over great ranges of paleodepth indicates significant production of biogenic silica during the Eocene. This production possibly is related to enhanced upwelling (Berggren and Hollister, 1974), increased supply of silica and nutrient phosphorous from volcanism (Gibson and Towe, 1971; Mattson and Pessagno, 1971), and/or increased silica preservation on the seafloor due to silica enrichment of bottom waters by submarine volcanism (Herman, 1972). The restricted age range of the upper chert surface, even within thick siliceous sequences (*e.g.,* Site 387, Fig. 9), further suggests that a distinct episode of biogenic silica production or preservation imposed a geochemical stress that has controlled chert diagenesis in the sediments. However the exact nature of the controls on the overall cycle of silica production, and the cause of the chronologic restriction of the chert boundary remain largely unresolved.

The continuity of sediment accumulation across Horizon A^c at many DSDP drill sites is not well determined because of discontinuous coring or incomplete core recovery (*e.g.,* Site 9, Fig. 10). At those sites with continuous or closely-spaced cores

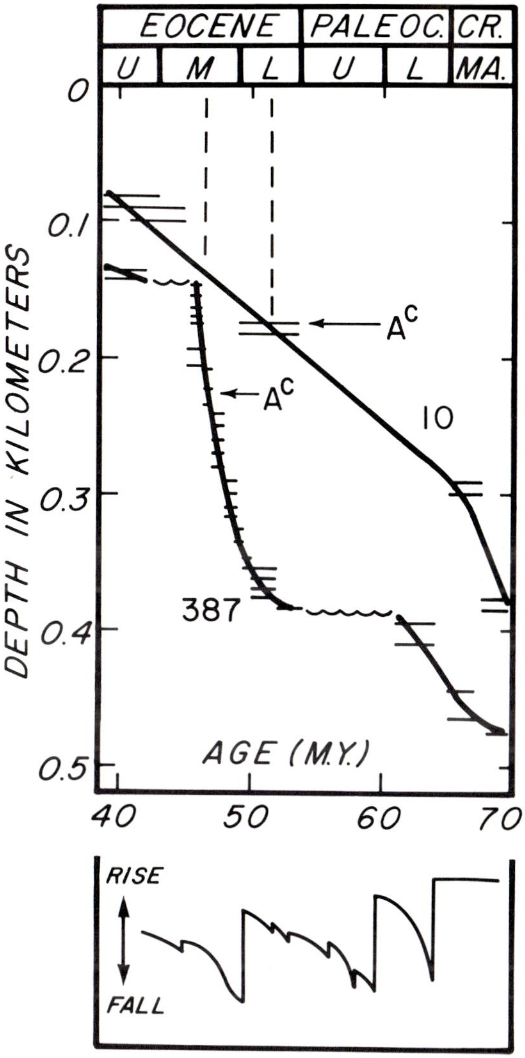

FIG. 10.—Age versus depth-in-borehole for two sites penetrating Horizon A^c. Horizontal lines show biostratigraphic age ranges for cores. Wavy lines show hiatuses. Relative sealevel at bottom from Vail and others (1977). Note the lack of coherent response between sealevel lowstands and sediment accumulation rate at Site 387.

biostratigraphic data indicate that deposition was continuous across Horizon A^c (Site 387, Fig. 10), but at some sites a significant unconformity lies within a few tens of meters of the reflector. At Site 385, the ?Middle Eocene to ?Upper Oligocene section is missing above Horizon A^c, and at Site 384 the lower Lower Eocene section is absent below Horizon A^c. However, the lack of a consistent relationship between Horizon A^c and the position or presence of an unconformity indicates that the development of the cherts was independent of for-

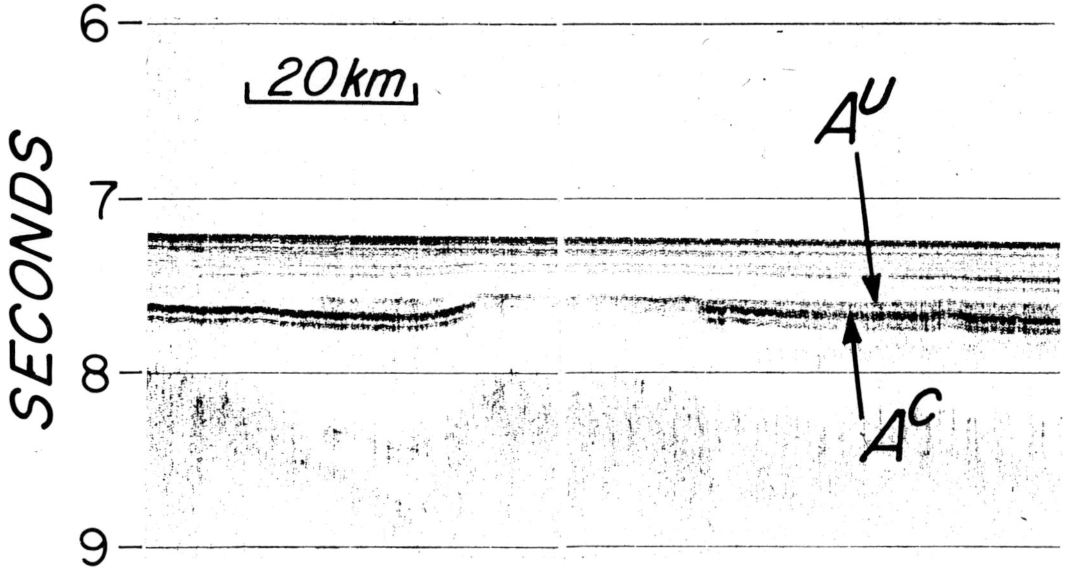

FIG. 11.—Seismic profile from the southern Hatteras Abyssal Plain (27°N, 71°30′W) near the eastern limit of the Horizon A^u unconformity. Note the truncation of Horizon A^c by the unconformity at center.

mation of the apparently local unconformities.

In the deep basin there appears to be no consistent relation between variations in sediment accumulation rate and the timing and magnitude of sealevel fluctuations. For example at Site 387 a significant increase in accumulation rates of turbiditic sediments correlates closely with a major lowstand about 49 m.y.BP, but unconformities are developed at the time of a preceeding major lowstand and a subsequent minor lowstand (Fig. 10). Similar variability at other drillsites suggests that the sealevel "signal" in the deep basin commonly is subsidiary to or modified by the effects of other processes (*e.g.*, abyssal currents and surface productivity patterns) that may be independent of changes in sea level.

HORIZON A^U

Beneath the continental rise off the eastern U.S. the reflectors of the Horizon-A complex and some deeper reflectors are truncated by a major Late Paleogene erosional surface (Fig. 1). This surface of unconformity has been termed Horizon A^u (Tucholke, 1979). Horizon A^u can be traced in reflection profiles eastward beneath the Hatteras Abyssal Plain where it truncates Horizon A^c and overlying Upper Eocene turbidites (Fig. 11). Erosion of sediments at the level of Horizon A^c must have occurred before the time of chert formation, but the amount of time necessary for silica diagenesis to have formed these erosion-resistant strata is uncertain. The exact eastern limit of the Horizon A^u unconformity is also difficult to resolve because of the very low angle cross-cutting relations and because similar impedances of sediments above and below the unconformity may preclude formation of a reflecting interface. The depth of excavation along the unconformity generally increases from east to west beneath the continental margin. Horizon A^u cuts as deeply as Horizon β beneath the landward end of the Blake Outer Ridge and in the Horizon β outcrop area north of San Salvador (Fig. 1; Tucholke and Mountain, 1979). The restricted location and increased westward development of the unconformity along the margin provide strong evidence that the unconformity was eroded by a southward-flowing, westward-intensified abyssal boundary current, possibly a precursor to the modern Western Boundary Undercurrent.

The oldest sediments known to overlie the unconformity are Lower to Middle Miocene hemipelagic silty clays at Sites 105, 106B and 388A on the lower continental rise off New Jersey. Thus the timing of the erosional event is restricted to the period between Late Eocene and Early Miocene. Vail and others (1980) have suggested that periods of erosion by abyssal currents correlate with sea-level lowstands and that, in particular, the erosion that formed Horizon A^u correlates with a major sealevel lowstand at the beginning of the Late Oligocene (29 m.y.BP). Such a general correlation would be extremely useful, but the sheer number of lowstands allows approximate correlation in time to virtually any observed unconformity in the deep sea, and the possible correlation thus may be fortuitous. If

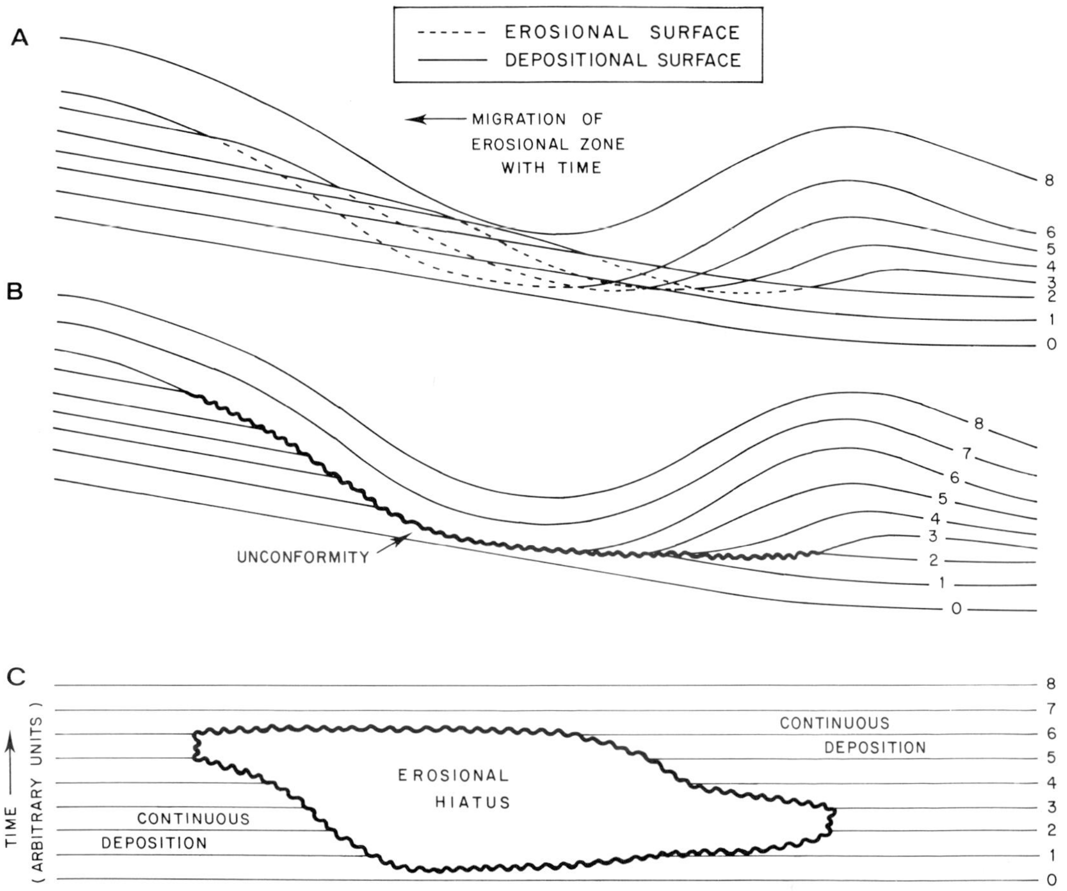

FIG. 12.—Model of erosion and deposition beneath a laterally migrating current system. A. Seafloor surfaces at arbitrary times T_0 to T_8. B. Resulting depositional sequence. C. Distribution in time and space of resulting hiatus. Note that the hiatus is diachronous, being bounded below and above by T_2 and T_3 at the right, and by T_5 and T_6 at the left.

either 1) the correlation could be clearly documented or 2) a reasonable mechanism that causally related low stands and increased abyssal circulation could be demonstrated, then the use of the correlation would be better justified. While the erosion of Horizon A^u and the 29 m.y.BP sealevel lowstand could very well coincide, it is unclear whether such a correlation would result from a causal relation or whether both events were the product of some other oceanographic process.

It is thought that the source of bottom water in the abyssal current system that eroded the Horizon A^u unconformity was in the Labrador and Norwegian-Greenland seas (Ewing and Hollister, 1972; Berggren and Hollister, 1974; Tucholke and Vogt, 1979), but the details of the tectonic and climatic controls on this source are poorly known. At least during the latter part of the Oligocene the Iceland-Faroe Ridge appears to have blocked access of Norwegian-Greenland Sea water to the North Atlantic (Talwani, Udintsev and others, 1976), and any bottom-water sources thus may have been restricted to the Labrador Sea. Detailed stratigraphic study of the entire northern North Atlantic region will be necessary to determine the specific controls on Tertiary bottom-water sources.

The chronostratigraphic value of deep-sea unconformities like Horizon A^u that are eroded by abyssal currents merits careful consideration. Unconformities eroded in shallow water during sea-level lowstands may have chronostratigraphic value because the associated hiatus not only separates strata of different ages, but it appears also to enclose a geologic time boundary extending laterally to a correlative conformity (Vail and others, 1977). However, this may not hold true for hiatuses resulting from erosion and non-deposition controlled by abyssal currents. It is possible that an abyssal current system and thus the boundaries between the main erosional (or non-depositional) zone and the

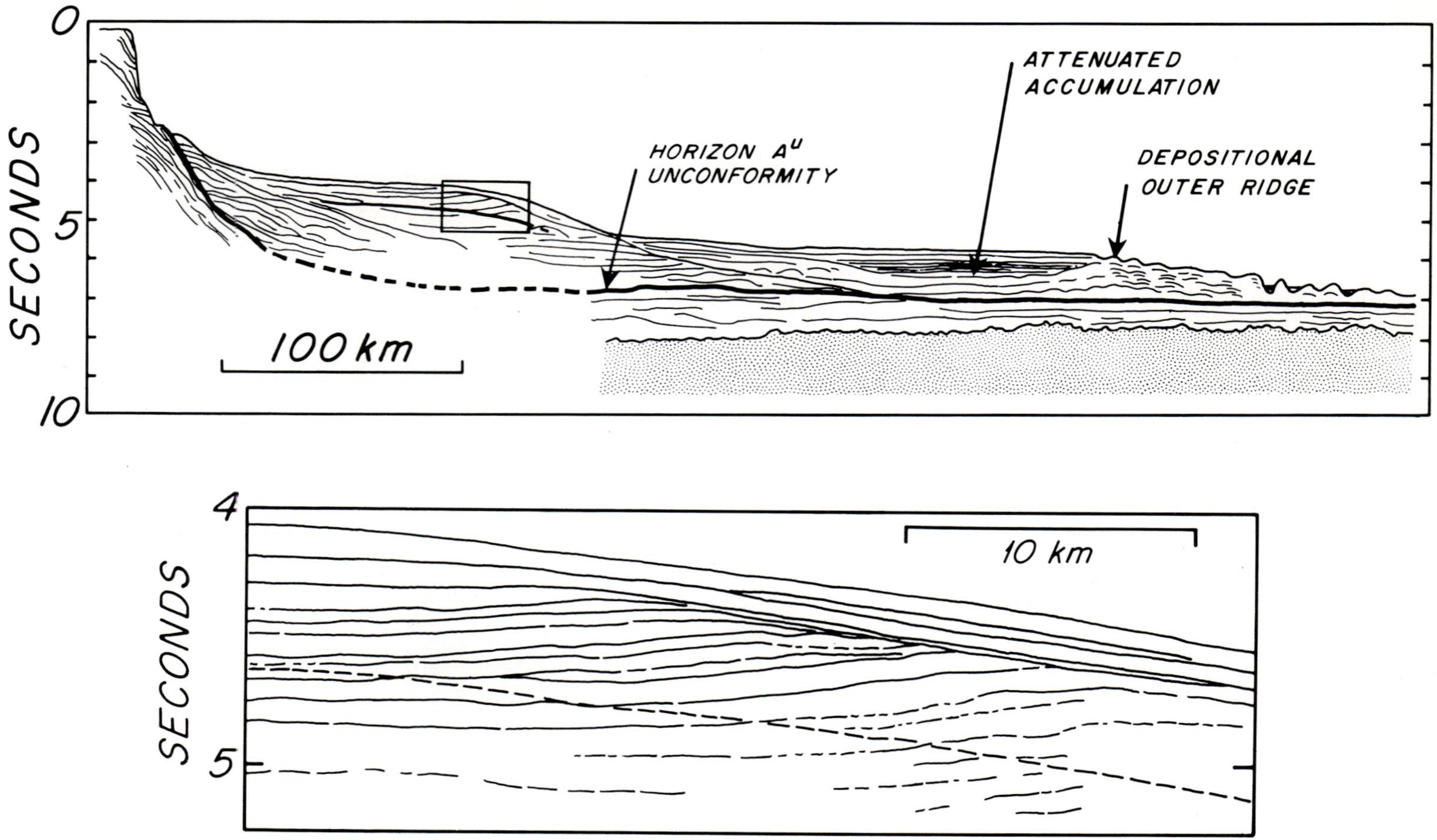

FIG. 13.—Top: Simplified tracing of CHAIN 73 seismic profile extending from the shelf break off New Jersey southeast to near DSDP Site 105. Note that the widespread zone of erosion/non-deposition that formed Horizon A^u subsequently contracted and migrated to a central position; this zone then was flanked by depositional loci (outer ridge and upper rise) that formed beneath the boundaries of the current. Bottom: Enlarged tracing of profile outlined by box above. Dashed reflector probably is base of gas-hydrate zone (Tucholke and others, 1977). Note similarity to model produced by migrating current system in Figure 12B.

adjacent depositional zones can migrate with time (Fig. 12). In such instances a hiatus can form that contains no common chronostratigraphic surface. Borehole data are not available to test details of this concept, but seismic profiles suggest that the critical boundaries of abyssal currents do migrate with time. For example, off the eastern U.S. (Fig. 13) abyssal currents were eroding sediments to form Horizon A^u over a broad area of the continental rise at some time between the Late Eocene and Early Miocene. However, by Early to Middle Miocene time a depositional outer ridge was being developed above the seaward edge of this unconformity, and the zone of erosion or non-deposition was restricted to the region of the present central continental rise, west of the outer ridge (Tucholke and Laine, in prep., for AAPG Hedberg Symp.). Such migration in abyssal current patterns might be expected whenever there was a significant change in the temperature/salinity characteristics of the bottom-water sources. Thus, if we consider the changing climatic regimes, and consequently the changing locales of bottom-water source areas during the Mesozoic and Cenozoic history of the Atlantic (Berggren and Hollister, 1974; Tucholke and Vogt, 1979), few deepsea unconformities eroded by abyssal currents may have regional chronostratigraphic significance.

CONCLUSIONS

Three major seismic horizons in the western North Atlantic have been reviewed in terms of their lithostratigraphic and chronostratigraphic significance; they are typical of the spectrum of stratigraphic variations observed within the deep basin.

Where tested by drilling, seismic Horizon β appears to correlate with a conformable boundary between high-impedance limestones and overlying lower-inpedance black claystones or thin local chalks. The seismic and lithologic boundary is related to a CCD rise during the Early Cretaceous and is no younger than Barremian age. However, the lithologic boundary and thin underlying limestones or chalks probably extend eastward immediately above basaltic crust beyond the limits where Horizon β can be resolved seismically. The boundary over this younger crust may be diachronous because of the time-dependent lateral shift of the CCD-seafloor contact along the sloping mid-ocean ridge flank.

Within a few million years, and thus a few tens of meters above Horizon β at prevailing sedimentation rates, a mid-Aptian sea-level lowstand may have facilitated seaward dispersal of sediment to create seaward-onlapping sedimentary sequences along the western margin of the basin. Because of the limited resolution of seismic-reflection techniques, the base of this sequence seismically can have the appearance of an unconformity; however, available borehole data suggest that there is no unconformity in the lithologic record, and the seismic discordance probably both overlies and postdates Horizon β.

Horizon A^c is one of the most prominent sedimentary reflectors in the western North Atlantic, and it systematically correlates with the top of Eocene chert layers developed at virtually all paleodepths and in all lithofacies in the basin. Considering that it is a diagenetic boundary, this seismic and lithologic horizon is remarkably close to being a time boundary (late Early to early Middle Eocene). The boundary has no known genetic relationship to overlying and underlying, seemingly randomly distributed, unconformities.

Horizon A^u is a major unconformity that was eroded by southward-flowing abyssal currents along the eastern U.S. continental rise at some period between the Late Eocene and Early Miocene. The depth of excavation generally increased from east to west, so that even Neocomian limestones underlying Horizon β were exposed beneath the western end of the Blake Outer Ridge and adjacent to the western reaches of the Bahama Banks. Hemipelagic silty clays at least as old as Lower to Middle Miocene overlie the unconformity. Except in the broadest terms, Horizon A^u may have little chronostratigraphic significance because of probable lateral migration with time of erosional and depositional zones associated with the abyssal current system.

In general, seismic stratigraphic markers in the deep basin correlate rather well with lithostratigraphic boundaries, and both commonly have chronostratigraphic significance, although there are clear exceptions. Depth-dependent lithologic boundaries (*e.g.*, CCD-dependent) are likely to be diachronous in seafloor regions with significant paleoslope.

Intuitively, we expect that increased sediment influx during sealevel lowstands would develop seaward onlapping sedimentary sequences along the margin of the deep basin. At the base of these sequences seismic discordances may be identified, but because of the increased sediment influx it is unlikely that true unconformities are present. The significance of this influx probably decreases markedly in a basinward direction where other parameters that control sediment deposition become more important in determining depositional sequences and boundaries. These deep-basin controls include surface and bottom circulation, productivity patterns, and the position and attitude of the CCD. A consistent relationship between sealevel lowstands and intensified abyssal circulation (and unconformities) as suggested by Vail and others (1980) has not been observed in the deep basin.

Because sea-level effects on the sedimentary record are maximized in shallow, proximal margin environments but play a subsidiary role to other controls in the off-margin, pelagic realm, the bulk of the continental margin probably contains a heterogeneous sedimentary record dominated alternately by sealevel and deep-ocean events. Satisfactory understanding of the geologic evolution of these critical areas of the ocean floor therefore will depend on understanding the relative significance of these events and on careful analysis of the limitations and the relations of each of the seismic-stratigraphic, lithostratigraphic, and chronostratigraphic frameworks.

ACKNOWLEDGEMENTS

This research was supported by NSF Grant OCE 76-02038 and ONR Contract N00014-75-C-0210 during my residence at Lamont-Doherty Geological Observatory, Columbia University, and by ONR Contract N00014-79-C-0071, NR083-004 at Woods Hole Oceanographic Institution. I thank J. I. Ewing, E. T. Bunce, and H. Schouten for discussions and review of the manuscript, and R. T. Buffler and P. R. Vail for valuable dialogue on seismic-stratigraphic concepts.

REFERENCES CITED

NOTE: All references to Deep Sea Drilling Project Initial Reports have been abbreviated; for complete citations see *Appendix I* at the end of this volume, or the individual Initial Reports as published.

Benson, W., Sheridan, R. E., and others, 1978, Init. Repts. DSDP, v. 44, 1005 p.

Berggren, W. A., and Hollister, C. D., 1974, Paleogeography, paleobiogeography, and the history of circulation in the Atlantic Ocean: *in* Hay, W., ed., Studies in paleo-oceanography: Soc. Econ. Paleontologists Mineralogists, Spec. Pub. No. 20, p. 126–186.

Donnelly, T., Francheteau, J., Bryan, W., Robinson, P., Flower, M., Salisbury, M., and others, 1979, Init. Repts. DSDP, v. 51, 52, 53, 1613 p.

Ewing, J. I., and Hollister, C. D., 1972, Regional aspects of deep-sea drilling in the western North Atlantic: Init. Repts. DSDP, v. 11, p. 951–973.

———, Windisch, C. C., and Ewing, M., 1970, Correlation of Horizon A with JOIDES borehole results: Jour. Geophys. Res., v. 75, p. 5645–5653.

———, Worzel, J. L., Ewing, M., and Windisch, C. C., 1966, Ages of Horizon A and the oldest Atlantic sediments: Science, v. 154, p. 1125–1132.

Ewing, M., and Ewing, J. I., 1963, Sediments at proposed LOCO drilling sites: Jour. Geophys. Res., v. 68, p. 251–256.

Gibson, T. G., and Towe, K. M., 1971, Eocene volcanism and the origin of Horizon A: Science, v. 172, p. 152–154.

Herman, Y., 1972, Origin of deep-sea cherts in the North Atlantic: Nature, v. 238, p. 392–393.

Lancelot, Y., Hathaway, J. C., and Hollister, C. D., 1972, Lithology of sediments from the western North Atlantic Leg 11 Deep-Sea Drilling Project: Init. Repts. DSDP, v. 11, p. 901–949.

Mattson, P. H., and Pessagno, E. A., Jr., 1971, Carribbean Eocene volcanism and the extent of Horizon A: Science, v. 174, 138–139.

Mayer, L., 1979, The origin of fine scale acoustic stratigraphy in deep-sea carbonates: Jour. Geophys. Res., v. 84, p. 6177–6184.

Menard, H. W., and Dorman, L. M., 1976, On the variation of ocean depth with latitude: EOS Trans. American Geophysical Union, v. 57, p. 933.

Purdy, G. M., Ewing, J. I., and Bryan, G. M., 1980, A deep-towed hydrophone seismic reflection survey around IPOD Sites 417 and 418: Marine Geology, v. 35, p. 1–19.

Rona, P. A., 1977, Paleobathymetry of the crest of spreading ridges related to the age of ocean basins—comment: Earth Planet. Sci. Letters, v. 35, p. 454–456.

Shipley, T. H., and Watkins, J. S., 1978, Fine-scale seismic stratigraphy in the western North Atlantic: Geology, v. 6, p. 635–639.

Talwani, M., Udintsev, G., and others, 1976, Init. Repts. DSDP, v. 38, 1256 p.

Tucholke, B. E., 1979, Relationships between acoustic stratigraphy and lithostratigraphy in the western North Atlantic basin: Init. Repts. DSDP, v. 43, p. 827–846.

———, Bryan, G. M., and Ewing, J. I., 1977, Gas-hydrate horizons detected in seismic-profiler data from the western North Atlantic: Amer. Assoc. Petroleum Geologists Bull., v. 61, p. 698–707.

———, and Mountain, G. S., 1977, The Horizon-A complex: lithostratigraphic correlation and paleoceanographic significance of reflectors in the western North Atlantic: EOS Trans. American Geophysical Union, v. 58, p. 406

———, and ———, 1979, Seismic stratigraphy, lithostratigraphy and paleosedimentation patterns in the North American basin: *in* Talwani, M., Hay, W., and Ryan, W. B. F., eds., Deep drilling results in the Atlantic Ocean: continental margins and paleoenvironment: Amer. Geophys. Union, M. Ewing Series, v. 3, p. 58–85.

———, and Vogt, P. R., 1979, Western North Atlantic: Sedimentary evolution and aspects of tectonic history: Init. Repts. DSDP, v. 43, p. 791–825.

———, and ———, and others, 1979: Init. Repts. DSDP, v. 43, 1115 p.

Vail, P. R., Todd, R. G., Widmier, J. M., Thompson, S., III, Sangree, B., Bubb, J. N., and Hatlelid, W. G., 1977, Seismic stratigraphy and global changes of sea level: *in* Payton, C. E., ed., Seismic stratigraphy—applications to hydrocarbon exploration: Amer. Assoc. Petroleum Geologists, Mem. 26, p. 49–212.

———, Mitchum, R. M., Jr., Shipley, T. H., and Buffler, R. T., 1980, Unconformities of the North Atlantic: Phil. Trans. Royal Soc. London, v. A294, p. 137–155.

van Andel, T. H., 1975, Mesozoic/Cenozoic calcite compensation depth and the global distribution of calcareous sediments: Earth Planet. Sci. Letters, v. 26, p. 187–194.

Widess, M. B., 1973, How thin is a thin bed?: Geophysics, v. 38, p. 1176–1180.

Windisch, C. C., Leyden, R. J., Worzel, J. L., Saito, T., and Ewing, J. I., 1968, Investigation of Horizon β: Science, v. 162, p. 1473–1479.

Woods, J. P., 1956, The composition of reflections: Geophysics, v. 21, p. 261–276.

SEPM SPECIAL PUBLICATION NO. 32, P. 39–55, OCTOBER 1981

RECENT RESEARCH ON PASSIVE CONTINENTAL MARGINS

ROBERT E. SHERIDAN
Department of Geology, University of Delaware, Newark, Delaware 19711 U.S.A.
for
The JOIDES Passive Margin Panel

ABSTRACT

Recently available multichannel-seismic data have provided a detailed look at many Atlantic passive margins. DSDP holes and COST (Continental Offshore Stratigraphic Test) wells have provided geologic calibration. Reefal-carbonate bank underpinnings were involved in continental slope migration seaward of the original continental edge, especially in the Jurassic-Early Cretaceous. Tertiary erosion has caused a large landward retreat of the continental slope. These erosional events are nearly coeval on both sides of the Atlantic Ocean, indicating that they were caused by some basin-wide oceanographic change.

Deeper crustal layers are identified on the modern reflection and refraction data. Intermediate seismic velocity layers, 7.1 km/sec, near the continental edge on both sides of the Atlantic might be characteristic of transitional-type crusts, or merely continuations of Layer 3 under the slope and shelf.

Deep, listric normal faults are observed where the soles of the faults merge into a lower crustal layer (6.3 km/sec velocity). Thinning with listric faulting of apparent continental crust has brought the mantle (8.2 km/sec) to within 14 km depths of the surface. Viscous creep in the lower continental crust appears necessary to account for the measured crustal thinning.

Detailed analysis of the multichannel reflection data permits sequence identification within the thick margin sediments. Sealevel cycles can be identified, and correlations reveal the configuration of genetically related stratigraphic units. Such analyses define the subsidence history and paleobathymetry of the margins. Some passive margins start with an uplift and rifting phase, whereas others are rifted through previous deep basins without uplift or volcanism. Other margins are dominated by volcanism in the early stages, and outer ridge structures have formed. Others have been involved in ridge-jump early in the seafloor spreading history which isolated transitional crusts of dispersed volcanism and continental slivers.

INTRODUCTION

In the study of passive margins drilling for the recovery of cores, logs, and down-hole measurements at selected sites is considered only one of several techniques, albeit a very vital one, for solving passive margin problems. In addition, the past, present and proposed JOIDES drilling with *Glomar Challenger* is limited because of engineering and safety considerations, so that the *Challenger* contributions to research on passive margins, while extremely valuable, will not be complete. Accordingly, a rational plan has been devised to use the *Challenger* capabilities to methodically complement other information gained from geophysics, both marine and on land, from commercial stratigraphic and exploration drilling on the shelves and slopes, from drilling on land, from conceptual modeling, from studies of ancient passive margins now uplifted and exposed on the continents, and from marine geological sampling with cores, dredges and submersibles. While *Challenger* drilling data must be integrated with these other investigations for the maximum understanding of passive margins, *there are phenomena which will never be understood or proven without Challenger or more sophisticated drilling*. This statement must be emphasized lest the impression be given that *Challenger* drilling is not necessary for a complete understanding of passive margins. For example, major contributions have been made by the past *Challenger* drilling in documenting the precise times of rifting and precise marginal subsidence rates for several particular regions. Times of erosion and rates of deposition by margin—following contour currents, and massive deposition by gravity flows off the continental margins have been documented.

In this paper I will review some recent geological and geophysical studies of passive continental margins that have resulted in the identification and interpretation of several features and phenomena common to many parts of the world. The commonality stems from the similar evolution that passive margins undergo, as plate tectonic theory indicates, with the initial rifting of continental crust and subsequent seafloor spreading of the new ocean forming in the lee of the drifting continents. I will attempt to collect and classify these common phenomena and relate them to the processes of formation of passive continental margins.

The two important stages in the process of forming passive continental margins can be divided by the point of breakup and onset of formation of true ocean crust. Before breakup the *rift stage* is accomplished by several different occurrences of

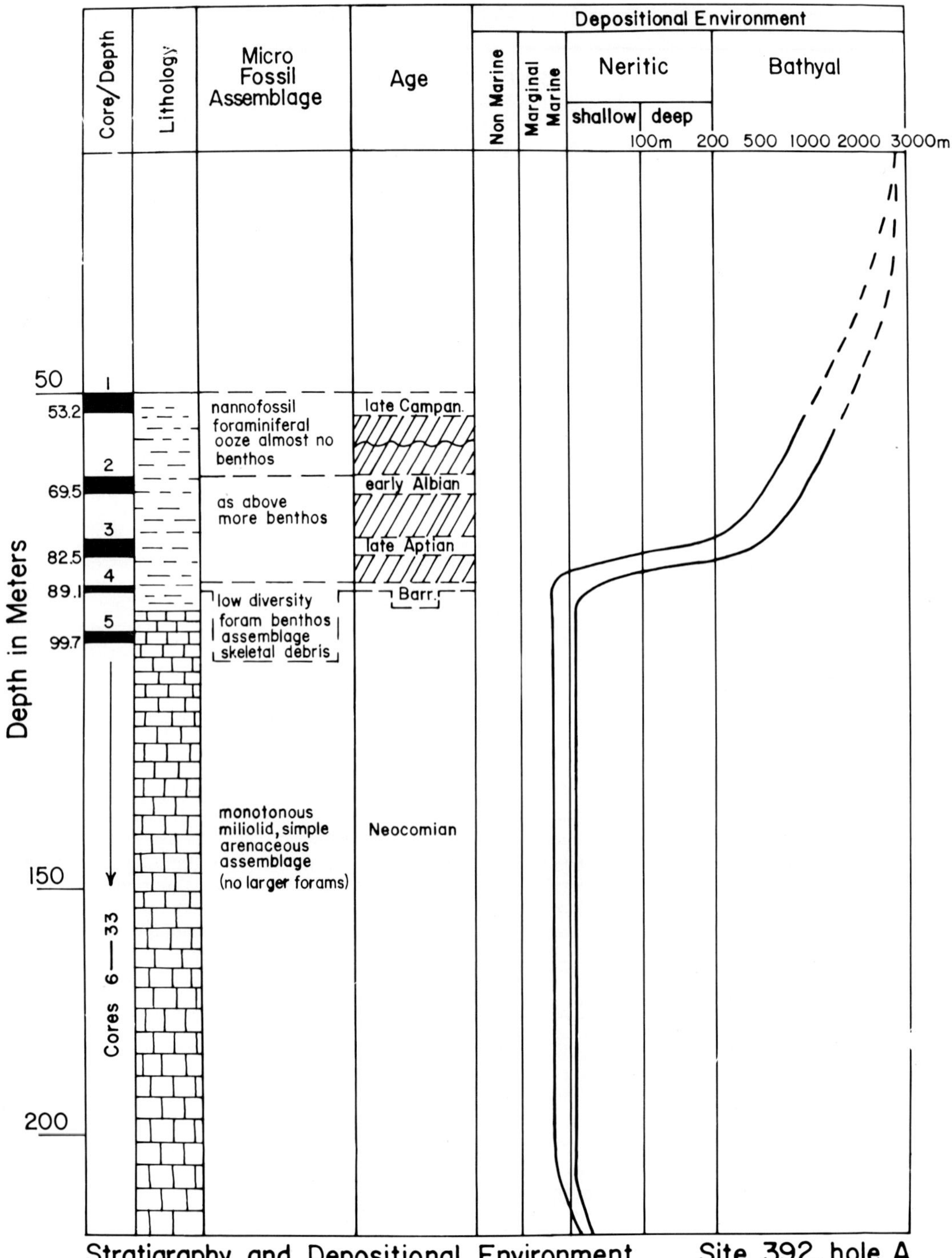

FIG. 1.—Paleobathymetry at DSDP Site 392 (Benson, Sheridan and others, 1978a).

structural, igneous, and sedimentary phenomena, the combined effect of which results in several different basement styles along the rifted continental edge. After breakup, the *seafloor spreading stage* involves the creation of the oceanic crust at the accreting plate boundary, and consequent subsidence with distance away from the boundary of ocean crust and the attached passive margin. Sedimentation accompanying this subsidence occurs with several common variations found in many places in the world. These variations have profound effects on the post-rift evolution of passive margins. Related to these two important genetic stages, the common phenomena and features of passive continental margins can be grouped loosely into those created in the rift stage (*basement structure variations*), and those created during the sea-floor spreading stage (*intrasedimentary features*).

PASSIVE MARGIN STUDIES

Paleobathymetry and Subsidence History

One of the important things achieved by JOIDES drilling on passive margins has been the precise determination of paleobathymetry. An example of this is at DSDP Site 392 on the Blake Nose, a topographic spur of the Blake Escarpment (Fig. 1) (Benson, Sheridan and others, 1978a). At this site on the North American Atlantic margin the water depth increased abruptly after the Barremian and bathyal water depths were present in Aptian and younger times. Using the core data available at Site 392 the water depth is seen to have increased approximately 500 m in 10 million years.

Such an abrupt deepening could be explained by abrupt vertical movements by faulting or slumping in this area. However, recent seismic reflection profiles across the Blake Nose and Blake Escarpment, that were tied into the DSDP Sites 392 and 390 (Fig. 2), reveal that faulting is not involved in the structure of the nose (Sheridan and others, 1979a). Instead, the correlation of the seismic profiles and the carbonate petrology at Sites 390 and 392 indicate that a carbonate bank along the Blake Nose was rapidly submerged when shallow-water

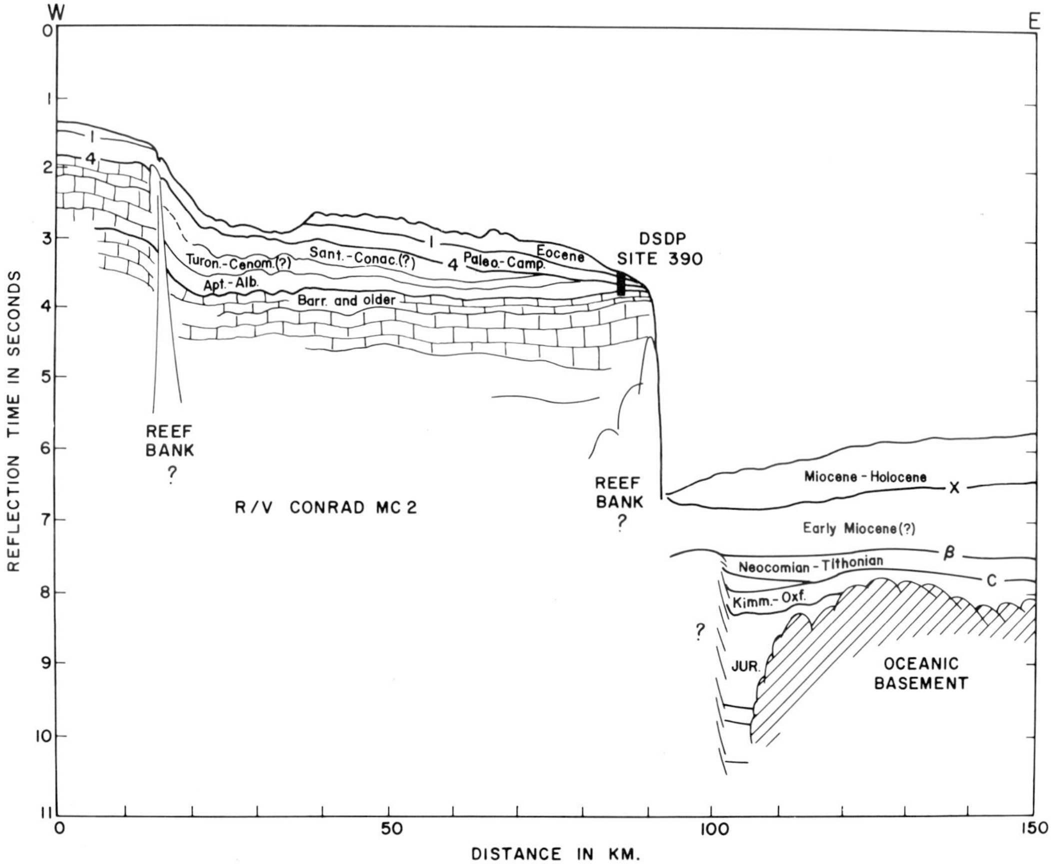

FIG. 2.—Geologic structure of the Blake Escarpment based on seismic reflection profiles (Sheridan and others, 1979a) and correlation with Deep Sea Drilling Project Site 390 (Sheridan and Enos, 1979).

carbonate deposition failed to keep up with subsidence and eustatic sealevel rise (Sheridan and Enos, 1979). The seismic data indicate that the shallow-water carbonate bank-edge shifted landward after the Barremian and maintained a steep bank-margin or reef structure along the Blake Escarpment. This bank edge persisted at least through the Albian, after which sometime in the Late Cretaceous the broader area of the Blake Plateau deepened and the shallow-water carbonate bank environment ceased.

Projection of the drilling data west of Sites 390 and 392 to the seismic layers under the Blake Plateau can give an estimation of Blake Plateau subsidence since the Barremian (Fig. 3) (Sheridan and Enos, 1979). By comparison with the subsidence of the same datum, the Barremian for the B2 well on the New Jersey continental shelf, the similarity in subsidence vs. age curves for the two areas implies that the Blake Plateau subsided along with the overall regional subsidence of the entire Atlantic margin. It was not abruptly downfaulted as a separate structural block.

The apparent abrupt deepening of the Blake Nose is not unlike the regional subsidence rate for the entire margin, being about 5 cm/1000 yrs. So that these relatively high rates of subsidence in the Early Cretaceous, coupled with some possible eustatic sealevel rise, must have caused the deepening of the Blake Nose and Plateau when shallow-water limestone deposition failed to keep up with sealevel rise. No local faulting needs to be involved to explain the deepening.

Control of Margin Structure by Carbonate Banks

Calibration of the seismic character of the documented carbonate bank margin by the Blake Nose drilling can be used to infer the presence of a similar bank margin buried under the continental slope off New Jersey and Delaware (Fig. 4) (Grow and others, 1979). High-amplitude hyperbolic reflectors at the top of a reflectorless mass into which the stratified reflectors terminate from the landward side are the acoustic characteristics of a carbonate bank margin. On the seaward side the draping of the off-bank reflectors indicate the initial dips of bank-margin debris deposits inclined into the adjacent basin. High seismic-interval velocities in the reflectorless mass is caused by the cemented limestones, similar to those drilled at Site 392 with velocities of 5.0–5.5 km/sec. (Benson, Sheridan and others, 1978a).

Based on the seismic reflection data, the buried carbonate bank off New Jersey has migrated seaward during its buildup (Fig. 5) (Grow and others,

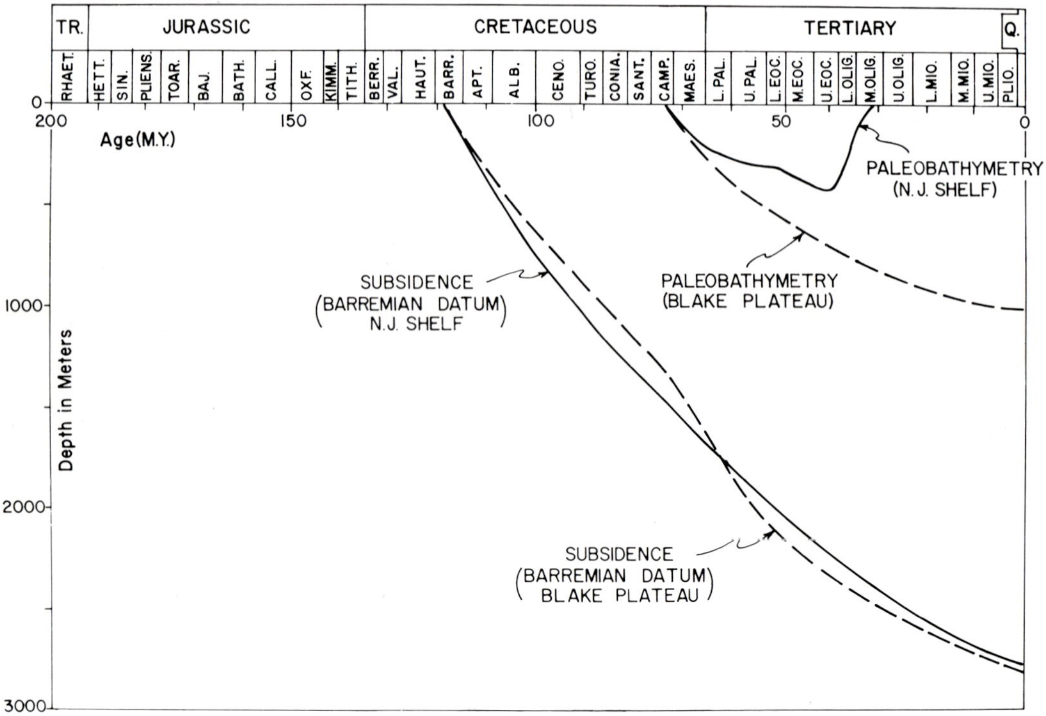

FIG. 3.—Geohistory diagram of the subsidence of the Barremian datum and paleobathymetry for the Blake Plateau and New Jersey shelf (Sheridan and Enos, 1979).

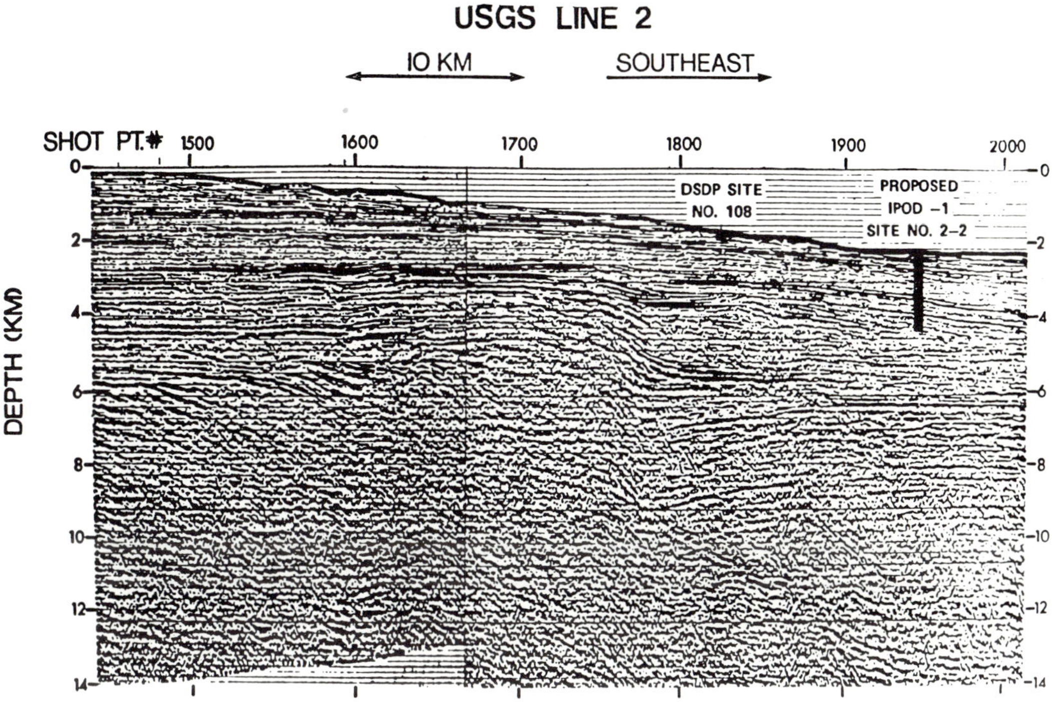

FIG. 4.—Seismic depth section of multichannel Line 2 of the U.S.G.S. across the New Jersey margin (Grow and others, 1979). Also shown are DSDP Site 108 and possible future drilling site considered for the IPOD program.

1979). Consequently the bank-margin formed a shelf edge in the Jurassic and Early Cretaceous which was several tens of kilometers seaward of the initial structural edge of the continental crust. The overlying Late Cretaceous marine shales drape the bank-margin structure, which forms the structural foundation for the present continental slope.

Seismic refraction profiles on the continental shelf edge and slope off New Jersey indicate that a seismic velocity layer of 7.1 km/sec. continues from the oceanic crust to just landward of the present shelf edge (Fig. 5) (Sheridan and others, 1979b). One possible interpretation of these refraction velocities is that the oceanic layer 3b with 7.1 km/sec. velocity comes in under the edge of the New Jersey shelf. If this is true, the carbonate bank has migrated well seaward of the continent/ocean boundary.

This interpretation agrees with the magnetic data which shows that the prominent East Coast magnetic anomaly is landward of the shelf edge off New Jersey, and west of the landward extent of the 7.1 km/sec. refracting layer. The East Coast Magnetic anomaly is taken to be the boundary between continental and oceanic crust (Fig. 5) (Klitgord and Behrendt, 1979; Grow and others, 1979).

In the Tertiary the draping shale and hemipelagic ooze that covered the continental slope was eroded back, probably by slope-following currents. This produced a major unconformity along the Atlantic continental slope (Fig. 5). Eocene beds are nearly exposed close to DSDP Site 108 (Fig. 4), and Miocene and Late Oligocene are drilled nearby in the COST No. B2 well (Fig. 8), indicating the erosion which formed this unconformity occurred in the Oligocene. This was the time of a regression on the Atlantic margin, and a relationship is therefore indicated by these data between regressions and intensification of slope-water currents.

Deeper Crustal Structure of Continental Edge

While the sharp ocean/continent transition under the East Coast magnetic anomaly (Fig. 5) is a plausible interpretation of the geophysical data, the true structure of the continental edge has not been clearly detected with seismic reflection data. Such an interpretation that the oceanic crust comes under the shelf of New Jersey is based on the refraction data (Sheridan and others, 1979b) and the magnetics and gravity (Grow and others, 1979). The thick high-velocity carbonate bank-margin structure obscures the continental edge structure on most profiles. Consequently, other geological interpretations are possible.

Below the obscuring intrasedimentary structures at the continental slope, which off New Jersey is a

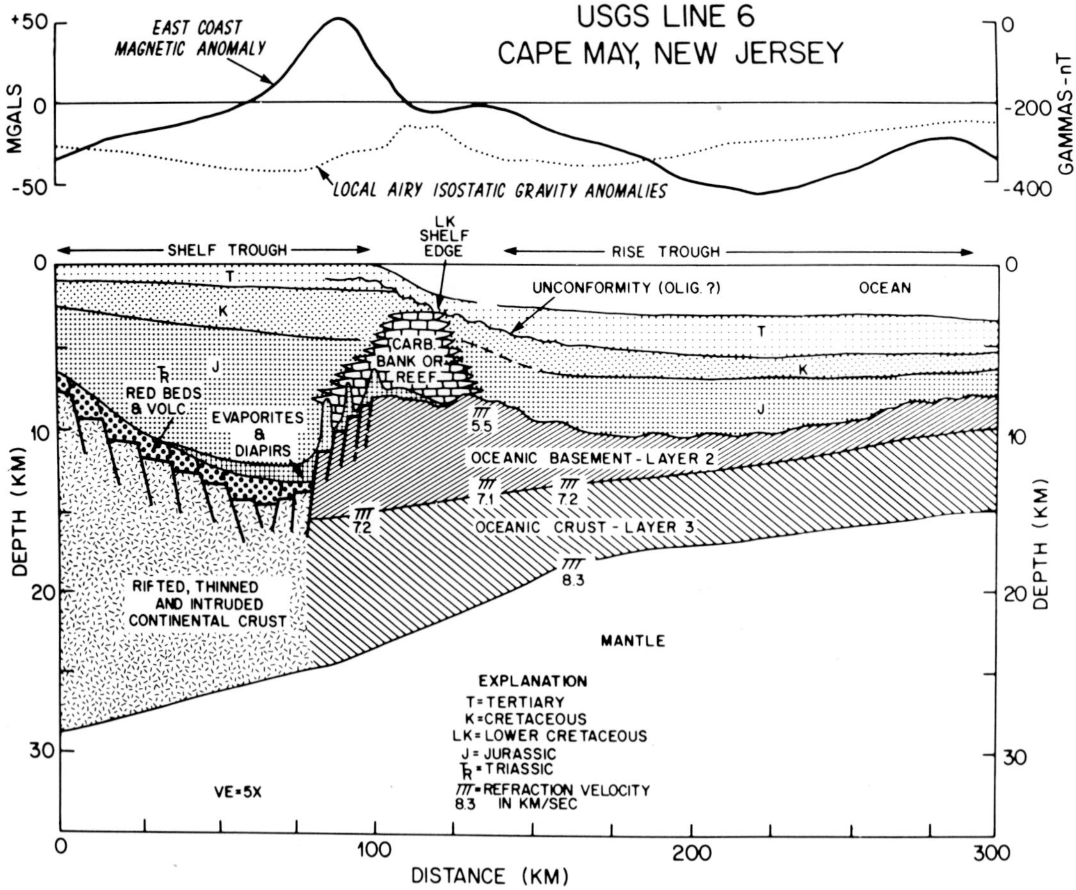

FIG. 5.—Diagrammatic geologic cross-section across New Jersey based on multichannel seismic Line 6 (U.S.G.S.) magnetic, gravity, and seismic-refraction data (Grow and others, 1979; Sheridan and others, 1979b).

reef-bank but off Cape Hatteras and Nova Scotia (Fig. 6) also includes a diapiric salt ridge, there are several possible alternate models for the existence of the 7.1 km/sec refracting crustal layer obscured at the continental edge (Fig. 7). Such an intermediate velocity, between the 6.2–6.5 km/sec velocities of continental crust and the 8.0–8.2 km/sec velocities of the oceanic mantle, might represent a transitional crust produced by the sandwiching of basic and ultrabasic dikes between septa of granitic crust along a splintered continental edge (Sheridan, 1974). On the other hand such an intermediate 7.1 km/sec velocity could correspond to that of an amphibolite crust, formed by thermal metamorphism during rifting of the greenschist facies of the continental crust (Falvey, 1974). More geophysical research on the nature of the continental edge and control by drilling is needed to resolve these alternative models.

Age of Continental Separation

The age of formation of the continental edge along the eastern North American margin is presently based on extrapolation of Deep Sea Drilling Project results well seaward of the margin (Fig. 8). Seismic reflection data indicate the presence of as yet undrilled sediments in a structural trough at the base of the continental slope and base of the Blake Escarpment (Unit 1, Fig. 9) (Sheridan and others, 1978). These sediments are bounded on the west by the basement structures of the East Coast magnetic anomaly (Fig. 5) and the Blake Escarpment acoustic basement block (of questionable origin) (Fig. 2), while on the seaward side they pinch out near the basement escarpment associated with the Blake Spur magnetic anomaly. The age of the structural trough between the East Coast and magnetic anomaly and the Blake Spur anomaly is dated by these sediments. Extrapolation of sedimentation rates from drilled sections such as DSDP Site 391 (Fig. 9) (Benson, Sheridan and others, 1978b) and extrapolation of magnetic anomaly seafloor spreading rates based on DSDP Sites 386, 387, and 105 which reached basement (Fig. 8), indicates an age of 195–165 m.y. (van Hinte, 1976) for the formation of the trough.

Apparently in this time interval an oceanic rift

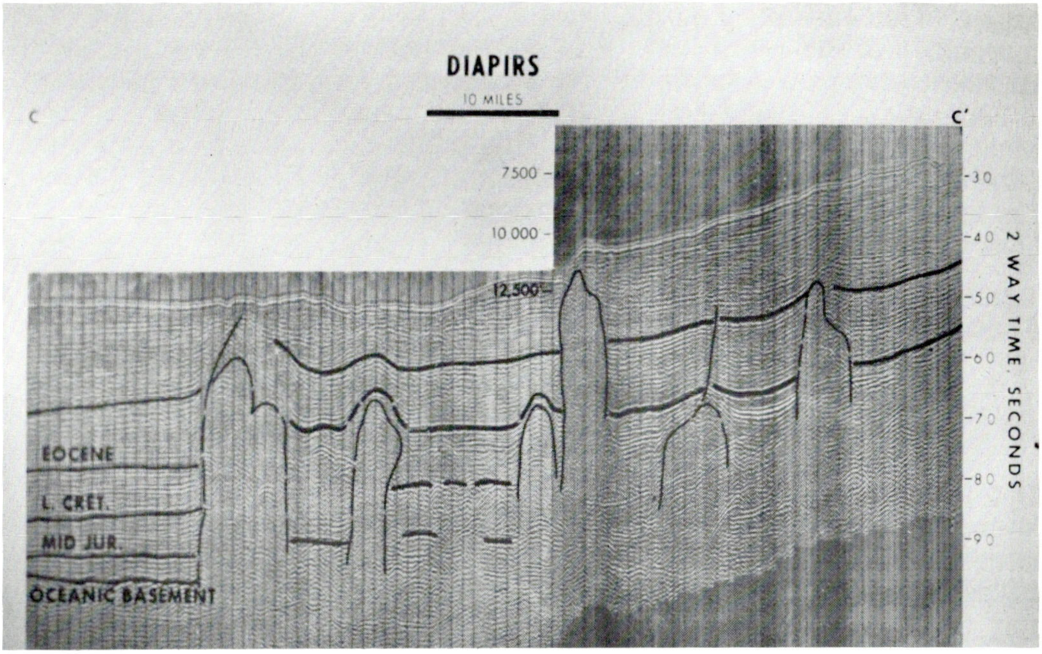

FIG. 6.—Photograph of seismic reflection profile across the continental slope off Nova Scotia showing diapiric salt structures (Parsons, 1975).

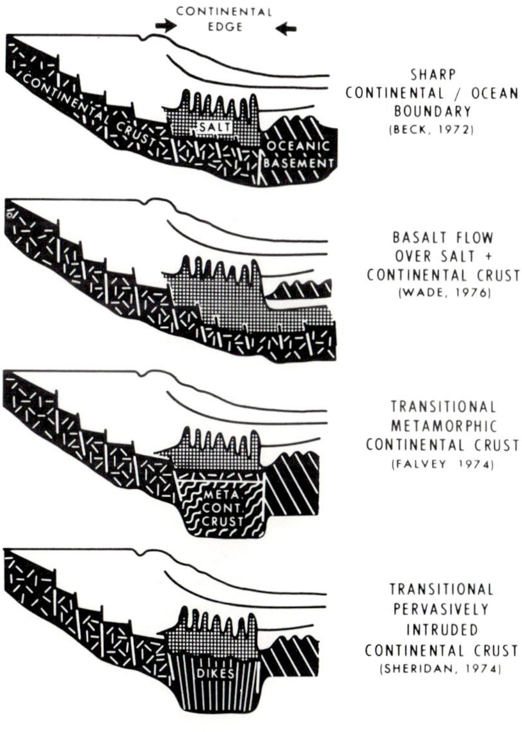

FIG. 7.—Alternative continental edge structures which might exist beneath the obscurring effects of salt-ridge structures (Beck, 1972; Wade, 1978; Falvey, 1974; Sheridan, 1974).

existed in the East Coast-Blake Spur anomaly corridor north of the Blake Plateau (Klitgord and Behrendt, 1979). Under the Blake Plateau, no single rift existed, but instead a stretched and pervasively dike-intruded continental crust existed as a transitional intermediate-density crustal structure. At the time of the Blake Spur anomaly, about 165 m.y., there was a ridge-jump and the new oceanic rift which became the modern Atlantic began along this anomaly. This explains why there is no equivalent crustal corridor as the East Coast-Blake Spur trough along the African margin (Klitgord and Behrendt, 1979).

A Model of Crustal Thinning

Another important achievement of JOIDES drilling on passive margins was the drilling at DSDP Site 400 on the Biscay Margin off France (Fig. 10) (Montadert and others, 1979). Penetration of the post-rift sediments above clearly defined listric normal faults of the rifted basement indicates that the rifting took place before the Aptian and that the margin had already subsided more than 2000 m by that time (Fig. 11). Thus a portion of the subsidence of the Biscay margin was accomplished in the rift stage as the listric normal faults slipped and caused a thinning of the brittle upper crust. The good seismic reflectivity along this margin permits the observation of the soles of the listric normal faults, which appear to merge into a reflector above a 6.3 km/sec lower crustal layer only 4 km thick (Fig. 12). This lower crustal layer is thought to be

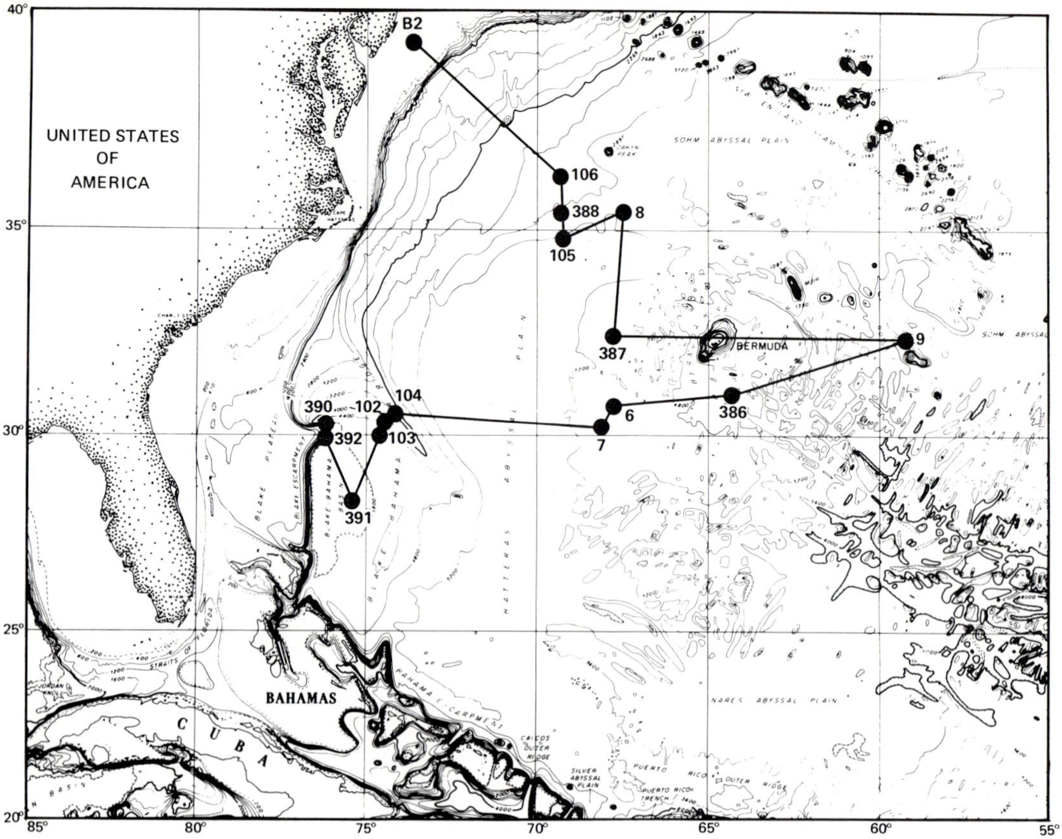

FIG. 8.—Location map of stratigraphic cross-sections across the eastern North American margin based on Deep Sea Drilling Project Data (Sheridan and others, 1978).

the ductile lower crust which by viscous creep thinned to allow the sub-moho rocks to rise and isostatically balance the thinned and subsided upper brittle crust.

Correlation of Prominent Unconformities

Drilling at Site 400 off France also revealed a marked hiatus caused by deepsea current erosion between the Campanian chalks and Aptian-Albian black shales (Montadert and others, 1979). This same hiatus is found on the North American margin at Sites 390 and 392 (Fig. 2). Apparently there was a basin-wide sculpting event which eroded the continental slopes at approximately 1000–1500 m paleodepths during the Middle Cretaceous.

Again, as in the Oligocene event mentioned earlier, there is a correlation of the sculpting of the continental slope in the Middle Cretaceous with the mid-Cenomanian global sealevel lowering estimated by Vail and others (1977) (Fig. 13). Based on good seismic reflection profiles of passive margins around the world (Fig. 14), the onlap, downlap, and termination of reflections reveal key unconformities bounding stratigraphic sequences. Measurements by Vail and others (1977) of sequence thicknesses corrected for tectonic tilt yields an estimate of sealevel rise and fall. The seismic profiles of Todd and Mitchum (1977) off West Africa document the mid-Cenomanian and the Oligocene regressions to be major sealevel drops. The profiles (Fig. 14) also show that the African continental slope was eroded to great depths in the Oligocene leaving a slope unconformity very similar to that off the United States. This Oligocene erosion appears to be basin wide as well.

BASEMENT STRUCTURES

Past geological and geophysical studies of passive margins, including drilling, have documented many basement features. Basement structures of various kinds are "seen" seismically and magnetically, and have been drilled on some youthful passive margins. Some basement structures have been inferred to exist on modern margins because they are seen exposed in uplifted ancient margins and would fit the computed geophysical anomalies. From these

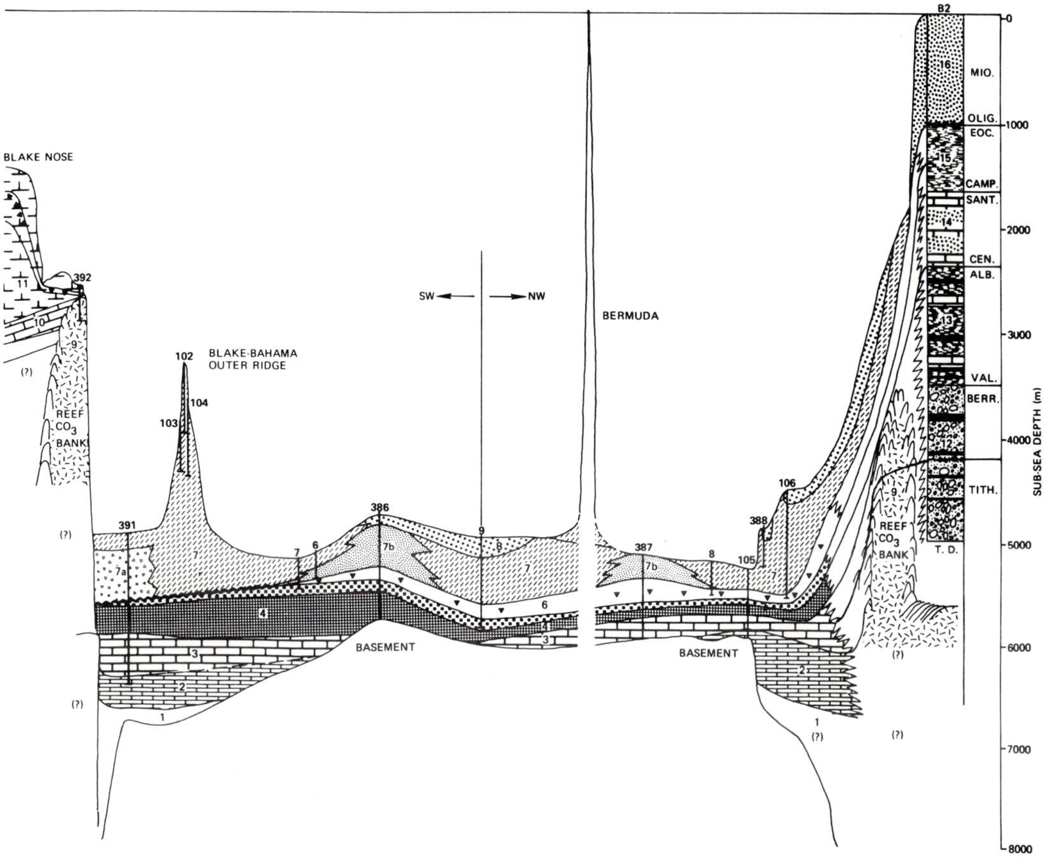

FIG. 9.—Stratigraphic cross-sections across the eastern North American margin based on Deep Sea Drilling Project data (Sheridan and others, 1978).

studies of the passive margin basement, several distinct variations have been identified (Fig. 15).

Apparently just prior to and during rifting the continental crust may undergo two different vertical motions, either uplift and doming or subsidence into an intracratonic basin. The uplifts might be domal or subcircular as the patterns over hot spots, with linear rifts emanating from the triple point junction over the center of uplift. In contrast there are cases where the rifts apparently cut through already deepened epeiric and epicontinental seaways of pre-existing intracratonic basins. The uplift and dome caused by the rise of the partially molten mantle of the hot spot, followed by extensional rifting of the stretched dome crust, is a logical site where basaltic magmas would intrude the shallow crust and extrude on the surface.

This igneous activity apparently can occupy a narrow zone of intrusion and extrusion, creating a volcanic ridge in the central area of the rift basin. This volcanic ridge in some cases formed a positive structure shedding detritus into the rift and a structural barrier to onlapping sediments separated into two sub-basins. This situation contrasts with that where igneous intrusions are dispersed over a wide zone and cover hundreds of kilometers of crust. Many pervasively intruded dikes of basalt are interspersed with septa of continental crust yielding a crust of effectively intermediate density and seismic velocity. Rather than an extrusive buildup of a volcanic ridge, the igneous activity could take the form of numerous sills intercalated with rapidly deposited rift sediments. Such a combination of sills and sediments, and dikes and continental septa, forms a broad zone of heterogeneous crust sometimes referred to as "rift" or "transitional" crust. The ocean/continent boundary in the pervasive dike situation would be transitional over a broad zone, whereas the boundary would be narrow in the case of the initial volcanic ridge.

With post-rift subsidence in the seafloor spreading stage the volcanic ridge would be buried to become an outer-ridge structure typical of those observed on many passive margins. This might bear a strong magnetic anomaly because of the strong magnetization of the extrusive rocks. On the other hand, the

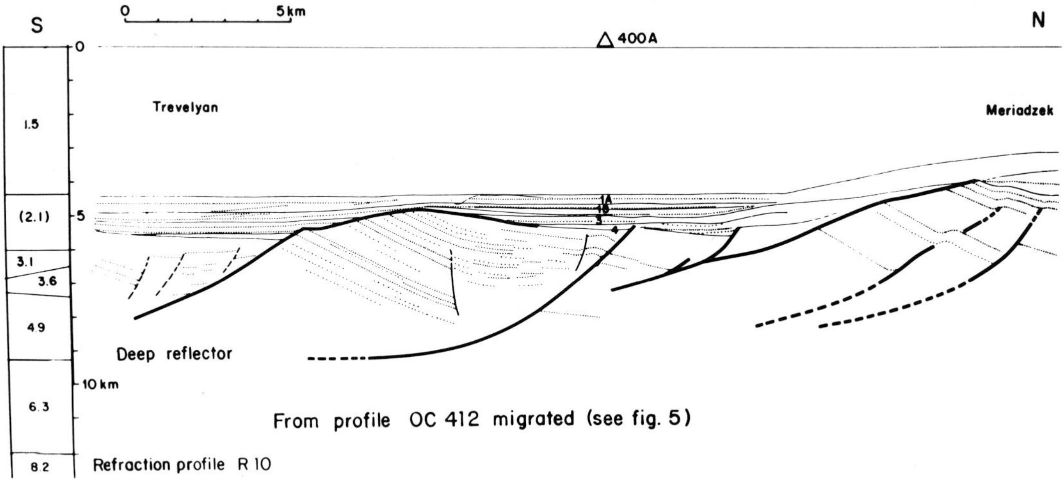

FIG. 10.—Geologic cross-section across the Biscay continental margin off France through DSDP Site 400 (Montadert and others, 1979).

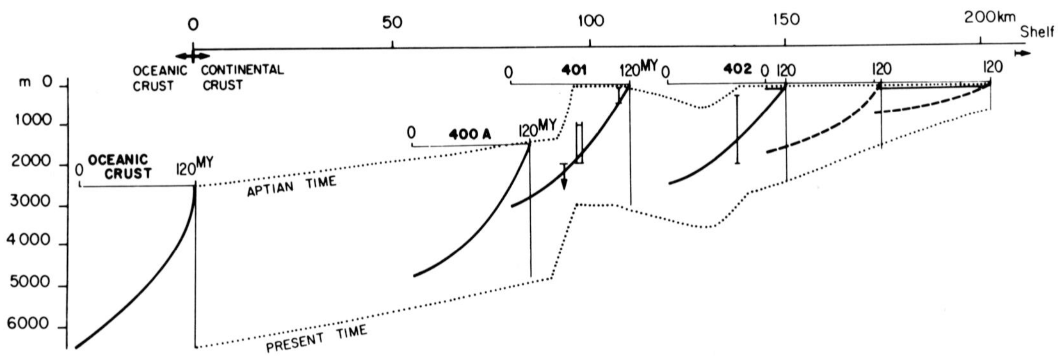

FIG. 11.—Subsidence history of the Biscay continental margin plotted at various places across the margin based on DSDP Sites 400, 401, and 402 (Montadert and others, 1979). The dotted lines represent the sea-floor profile in the past and at present.

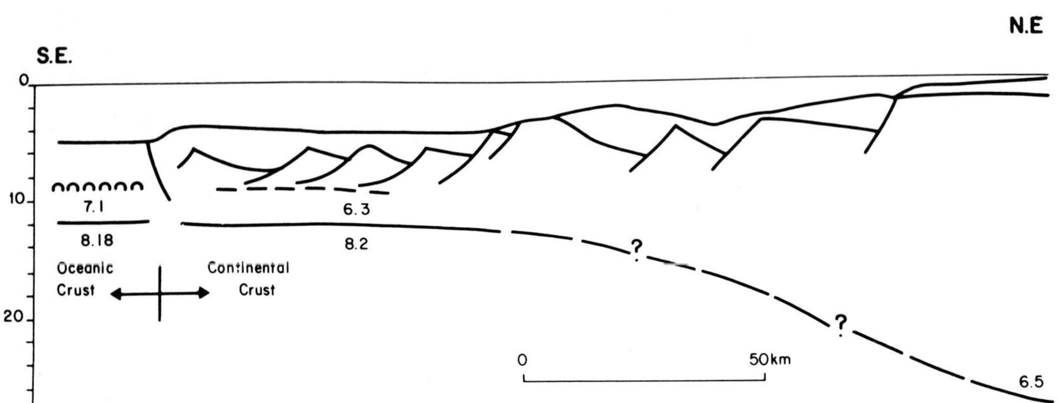

FIG. 12.—Diagrammatic crustal cross-section across the Biscay continental margin showing the deep crustal layer, 6.3 km/sec, and the relatively shallow mantle under the listric normal-faulted basement (Montadert and others, 1979).

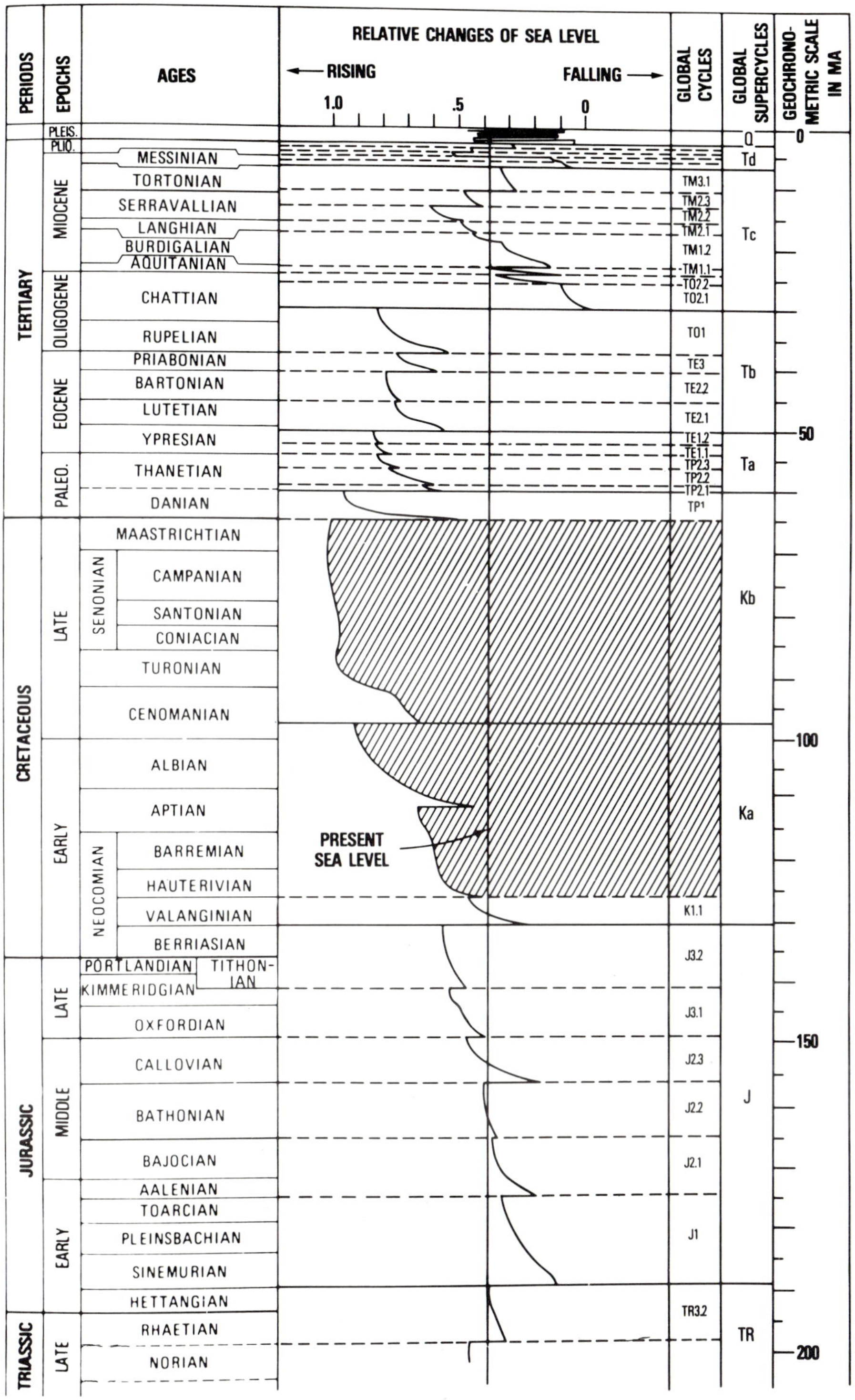

FIG. 13.—Global sealevel fluctuation curve showing sealevel changes vs. geologic time (Todd and Mitchum, 1977).

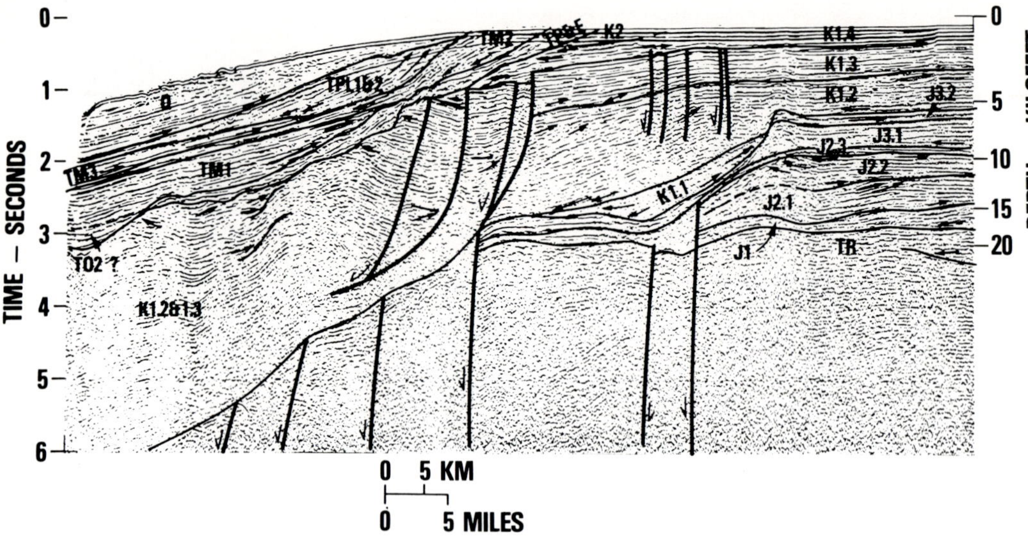

FIG. 14.—Interpreted seismic reflection profile across the west African continental margin showing seismic sequence identification based on onlap, downlap and unconformities (Vail and others, 1977).

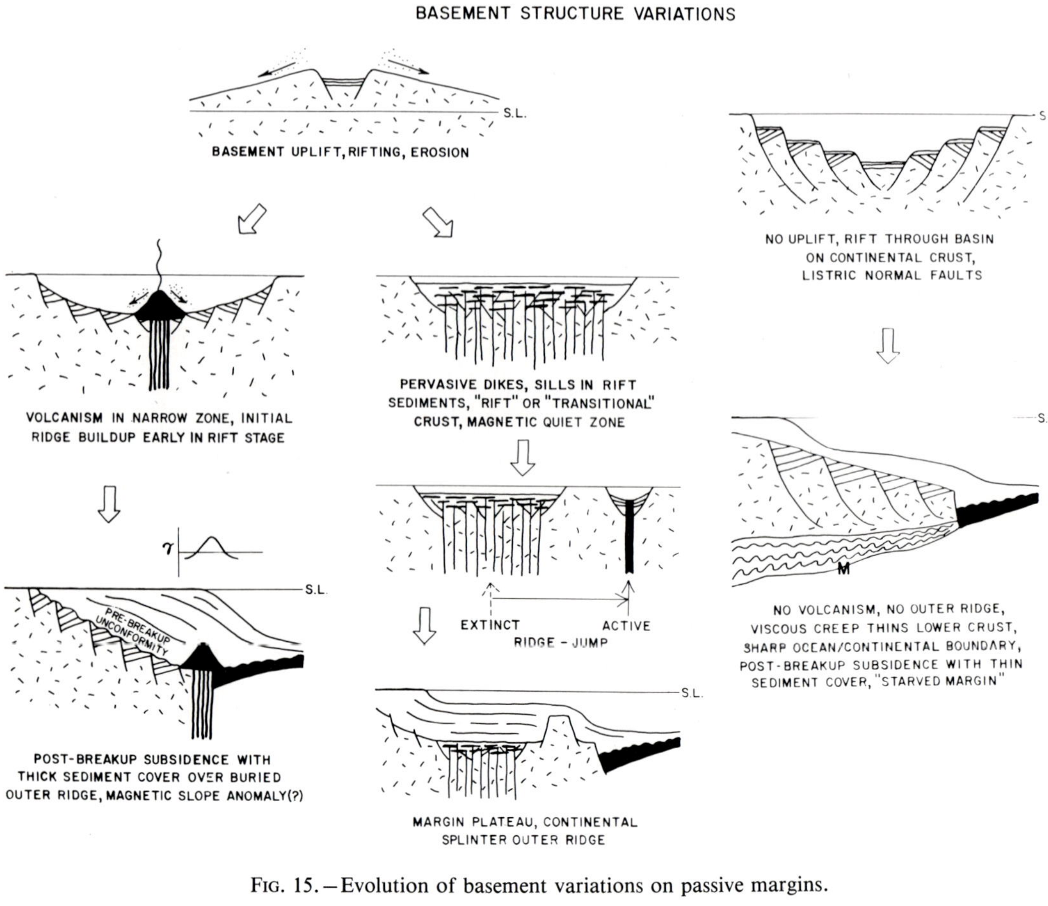

FIG. 15.—Evolution of basement variations on passive margins.

transitional crust with its largely intrusive basaltic components allowed to cool slowly in a thermal blanket of thick sediments might acquire a weak magnetization. This could explain the observations of magnetic quiet zones near rifted margins.

In several regions where there are rift-type transitional crusts ridge-jumps occur. Apparently in the process of forming the transitional crust, the central rift is not fully developed, and when the central rift does finally develop, it shifts to one side of the broader rifted zone. Such a ridge-jump could isolate a sliver or splinter of continental crust as an outlier separated from the larger continent by a broad zone of transitional crust. The intermediate density transitional crust would subside less than normal ocean crust but more than the adjacent continent and continental splinter, which would tend to buoy up the transitional crust. As a result, intermediate depth marginal plateaus are created.

Ridge-jumps commonly occur within about 10–20 m.y. after break-up. During this early stage of sea-floor spreading the continental salients and recesses are still likely to be interacting with shearing along transform margins. Once this continental interference is passed, the divergent plate boundary reorganizes, and the spreading direction often changes once the continents are clear of one another. In a case where a divergent plate boundary is itself moving away relative to one of the rifted continents, the ridge jump apparently shifts the spreading center toward that continent away from which the plate boundary is moving. Thus the transitional crust isolated adjacent to a trailing edge margin should be attached to the continent with the most active drift in an absolute sense.

A different set of circumstances influences the basement structure and evolution in the case where the rift cuts through a pre-existing intracratonic basin. With a rift occurring in an already deepwater situation there is no chance for supra-crustal erosion, as occurs with doming and diversion of drainage. So no appeal can be made for surficial erosion of many km of crust to explain crustal thinning so commonly observed on passive continental margins. Also, when deepwater rifting occurs without doming it seems to indicate that the hot spot is not located at that particular point on the incipient margin. Without the hot spot's presence there apparently is not enough heat to cause extensive igneous intrusion and extrusion. This might be thought of as a cold rift margin.

Without supra-crustal thinning by surficial erosion, other mechanisms of crustal thinning must be considered. Where listric normal faults occur the stretched brittle upper crust is rifted and the faults grow from the upper surface downward to where they sole out along a horizontal surface at the transition to the lower ductile crust. With the listric faults in place early in the rifting the upper brittle crust can only be thinned by rotation on these faults. Observations of 30° rotations are common on passive margins, accounting for only 3–4 km of effective crustal thinning. Thinning, as required by the observed subsidence documented by drilling and dredging (5 km of subsidence requires about 15 km of crustal thinning of shoaling of the M discontinuity), must be accounted for by other means. In a situation of no igneous activity, magmatic stoping cannot be called on to raise the M discontinuity. Thus, viscous creep in the metamorphic rocks of the lower continental crust must permit the thinning of the continental crust by some 11 km, and subsequent shoaling of the Moho. Observations indicate that such thinning is very effective in reducing continental crustal thickness to proportions not unlike the thickness of oceanic crust. In some cases the depth to the M discontinuity is nearly the same under a thinned continental crust, which has subsided yet been starved of sediments, as under a well-sedimented oceanic crustal section. Thus depth to the M discontinuity is not a good criterion for the origin of the crust.

These variations in basement structure occur in many cases along the same continental margin. They appear to be separated from one another by the impinging transform faults of the adjacent oceanic crust which cut the passive continental margin into segments. During the rift stage the initial segments are defined possibly along old lines of structural weakness in the continental crust. Transform and rift segments radiate from hot-spot uplifts. Those segments near the initial hot-spot might have a strong volcanic and intrusive igneous activity, while other segments away from hot-spots might cut previous intracratonic basins. Narrow zones of volcanism occur along one segment, while across the transform fault boundary of that segment with another, the more pervasive widespread intrusions might occur. Some circumstantial evidence suggests that the pervasive intrusions occur near the hot-spots while the narrower volcanic zones are farther afield. Patterns of this type have yet to be well-documented.

INTRASEDIMENTARY FEATURES

Past geological and geophysical investigations of passive continental margins, including drilling, have led to the documentation of many structures whose origin solely involves the sedimentary section (Fig. 16). Also, documentation of unconformities in the sedimentary section has led to inferences about the relationship of the hiatuses to sealevel rise and fall, and to the generation of deepsea currents. While there is a variety of these intrasedimentary features, a number of significant ones occur commonly on the modern passive margins in many parts of the world (Fig. 16). These common features probably occurred on past continental margins, but

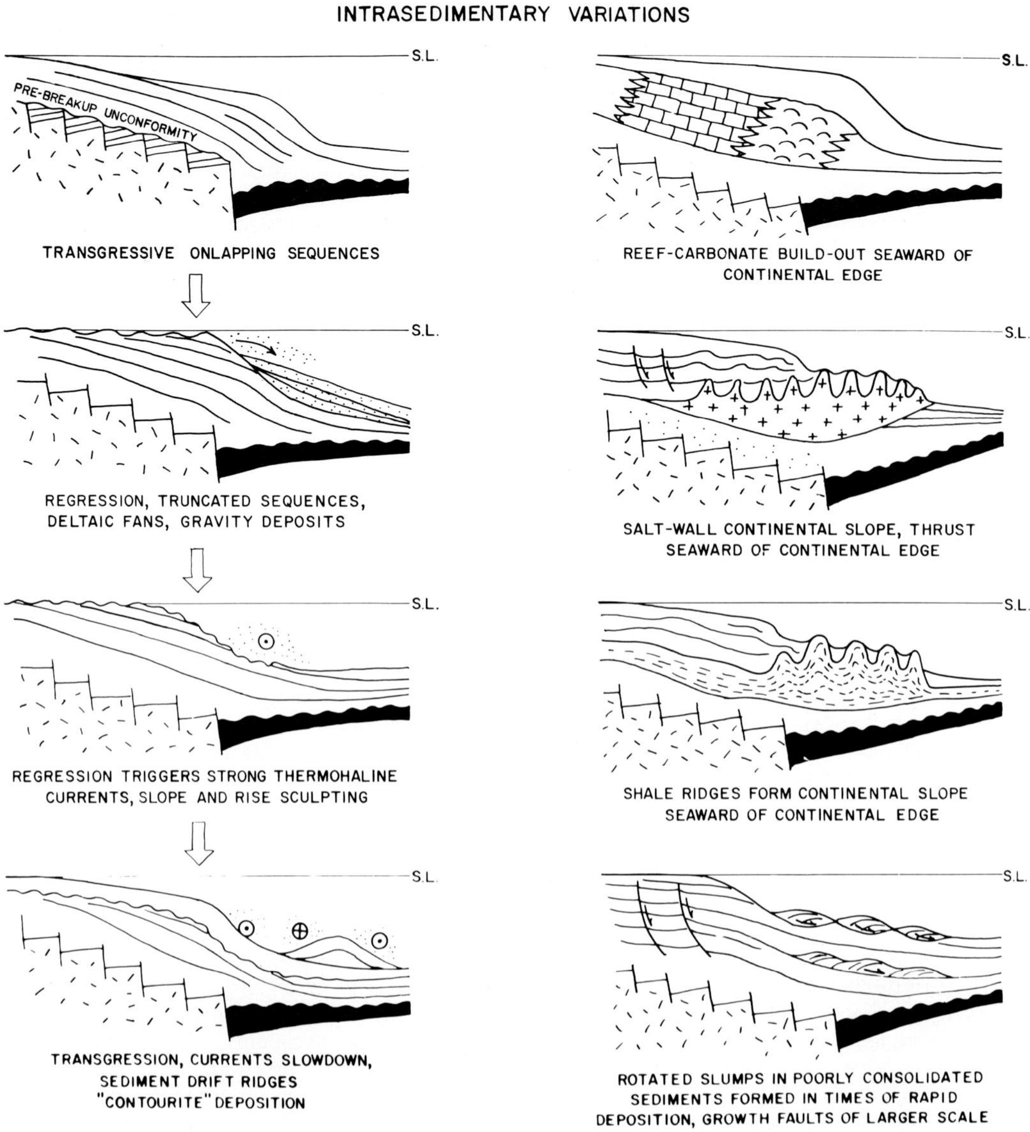

FIG. 16.—Some intrasedimentary features common to many passive margins.

their existence is more easily obscured by subsequent tectonism than are original basement features, so some are not easy to document by the study of ancient margins. The more brittle structures such as massive reef and carbonate bank margins are retained, for example, in the Appalachians, Rocky Mountains, and Alps, but the more ductile structures such as salt and shale diapirs are so extensively remobilized in mountain building that their original passive-margin configuration is hard to ascertain from the study of uplifted equivalents. It can be inferred that they occurred in similar configurations to the modern margins, however.

On modern passive margins geophysical studies, mainly of the seismic type, have revealed that the intrasedimentary features control the first-order topography of the continental slope, the position of which has migrated tens of kilometers in response to these sedimentary phenomena. For example, internal reflections in carbonate-bank complexes indicate that in some cases they have "grown" seaward, building out the continental shelf edge well beyond the original continent/ocean boundary. In other areas a thick wedge of salt has been thrust seaward and upward over the horizontal oceanic sedimentary sequences. The rising salt diapirs have formed a wall along the base of the newly formed continental slope, which has been thrust well seaward of the original continent/ocean boundary. Shale diapirism has been documented to create similar continental-slope structures.

The question arises as to why in the first place some passive margins have thick salt deposits while others have shale. This indicates that these sedimentary formations, which produce these interesting slope structures, also pose interesting paleogeographic questions. Salt requires restricted shallow seas, in most cases, for its formation. Where evidence exists that it is thrust seaward over oceanic crust, its shallow-water origin is not a problem. But in some cases along passive margins salt is observed over oceanic crust as if it were deposited there initially. Documentation of the paleogeographic origin of such salt is important.

On a smaller scale, topography on the continental rise of modern passive margins commonly has a ridge-and-swale nature which in some cases is caused by rotated slump blocks. Local gravity gliding in poorly consolidated sediments has caused the slumps. Detailed seismic reflection studies have documented past events of rotational blocks caused by slumping on the continental rise, which are now buried by subsequent deposition. Apparently periods of rapid sedimentation on the continental slope and rise lead to accumulations of poorly consolidated muds that fail and slump under the load of gravity. In the Gulf of Mexico active slumping has caused collapse of offshore drill rigs, and potential slump terrain of the Atlantic continental slope off the eastern U.S. has been deemed too environmentally hazardous to permit leasing of a few tracts, although many other east coast slope tracts have been put up for lease. Understanding slumping phenomena is important for more than academic reasons.

Studies of the sequences of sedimentary deposits of passive continental margins indicate that the boundary unconformities on the continental shelves are related to global rise and fall of sealevel. These profound changes in sealevel cause characteristic onlap of transgressive shelf sequences, and characteristic wedges of slope and rise deposits laid down during regressions. During regressions the initial change in the sedimentary environment is the increased availability of unconsolidated shelf sediments which can be washed as gravity deposits (turbidites and debris flows) into the deep marginal basin. With regressions the shoreline is near the shelf edge, and sediment laden rivers deliver their loads directly to the top of the continental slope. Here deepsea fan deposits form as the distal members of deltaic lobes.

It is thought that after some time regressions exposing perhaps 10% more land area on the earth's surface, with the higher albedo of the land, will cause a cooling of the earth and an intensification of the latitudinal thermal gradients. In response, stronger deepwater thermohaline circulation would occur. Strong contour-following currents would then scour the slope and rise causing, perhaps by undercutting, mass movements and slumping. For example, such erosion has caused several tens of kilometers retreat of the continental slope on both sides of the Atlantic.

With subsequent transgressions and warming, the thermohaline currents would decelerate and drop their suspended sediment load. Then deposition of "contourite" deposits would form the deepsea sediment drifts to shape large spit-like outer ridges. Recurrence of this cycle of transgression and regression, scour and deposition, creates the sequences recorded in the passive margin sediments. Thus, the unravelling of these cycles of sedimentation, and from them the global sealevel and scour effects, is fundamental to understanding sedimentary processes on passive continental margins.

After the thick sediments of passive margins are deposited, interesting diagenetic processes occur. One involves the development of gas-hydrate ice at water depths of 3 to 5 km, because of the very high pressures and the cold pore-water temperatures in the presence of large amounts of biogenic gas (methane). The inversion phase-change of ice to gas yields a very strong bottom-simulating reflector (BSR) observed in several places. Geochemical theory supports the gas-hydrate explanation for some of the BSR, but it has never been tested by DSDP drilling. Other causes of BSR phenomena are possible, such as the suggestion that organic accumulations of opal were being converted to chert at some pressure and temperature determined by depth of burial below the seafloor. More drilling is needed to document this phenomenon.

GLOMAR CHALLENGER PLAN FOR 1979–81

With these important basement and sedimentary features and phenomena identified above, the JOIDES Passive Margin Panel devised a drilling program to meet specific scientific objectives. As stated previously, the lack of drill-hole control with blowout prevention and the present 6800 m drillstring limit on the *Challenger* restricts the application of deepsea drilling on passive margins. Many of the basement-related phenomena cannot be drilled directly, but must be attacked indirectly by drilling into the sedimentary section above geophysically identified basement features. With these limitations in mind, the primary scientific objectives for the 1979–81 drilling are to examine:

1. The subsidence history of ocean crust and continental margins in relation to rifting and seafloor spreading;

2. The mechanisms and rates of crustal thinning during continental rifting;
3. The nature of the first Atlantic oceanic crust and paleo-environment;
4. Pre- and syn-rift sedimentation, including tectonic influence;
5. The relationship of transgressions and regressions to deep-marine current erosion;
6. The nature of gas hydrates and bottom-simulating reflectors on continental margins.

To accomplish these objectives the 1979–81 *Challenger* drilling focuses on candidate sites on the eastern North American continental margin, and off Europe, west Africa, and in the Caribbean. The strategy is to develop transects, in a broad sense, where the particular continental-margin features are well-developed and documented geophysically. Also, we are attempting to position the transects to take maximum advantage of offshore commercial and COST type stratigraphic drilling as well as geological and geophysical studies on land.

With the expectation that more sophisticated drilling techniques will be available in the future, the transects are positioned where presently inaccessible deeper targets are identified so that the *Glomar Challenger* drilling will also serve as preparation for possible *Glomar Explorer*-mode drilling. Stratigraphic information on the lateral variations in the shallower section and engineering data necessary for riser and casing programs will be vital for *Explorer* drilling. The transects are constrained very loosely into corridors several hundred kilometers wide to maximize the selection of the best specific site, where more complete composite sections might be obtained in "windows" opened by canyon erosion, for example.

The approach that is favored by the Passive Margin Panel is to *test and document models of important common passive margin phenomena by a concentration on these transects in well-studied areas, rather than by continued reconnaissance of all the passive margins of the world.* Presently transects have been identified off the Gulf Coast, the Blake Plateau, the mid-Atlantic United States, off eastern Canada, Biscay, and the Rockall-Faroes-Norway region, off Morocco and Mauritania, and in the Caribbean. Extensive geophysical studies and previous JOIDES and IPOD drilling exist in these

TABLE 1.—IMPORTANT AND COMMON PASSIVE CONTINENTAL MARGIN PHENOMENA

Basement Phenomena

- Sharp Continent/Ocean Boundary[1,2]
  - Narrow Volcanic Zone[1,2]
  - Volcanic Outer Ridge[2,4]
- Broad Transitional Crust[4]
  - Pervasive Dike Intrusion[4]
- Ridge-Jump Isolation
  - Extinct Rift Crust[4]
  - Exotic Continental Splinter[4]
  - Marginal Plateau[3,4]
- Thinned Continental Crust[2,4]
  - Listric Normal Faults[2,4]
  - Viscous Creep of Lower Crust[2,4]
- Variants in Segments Along One Margin[1,2]
  - Transform Bounded Segments[2,3]
- Syn-Rift Subsidence[2,4]
- Post-Rift Subsidence[1,2]
- Magnetic Slope Anomalies[4]
- Magnetic Quiet Zones[1,3,4]

Combined Phenomena

- Plate Tectonic Reconstructions[1,2,3,4]
- Age of Rifting[1,2,3,4]
- Sea Floor Spreading Rates[1,2,3,4]

Sedimentary Phenomena

- Depositional Processes[1,2,3]
- Gravity Deposits[1,2,3]
  - Contour Current Deposits[1,2,3]
  - Sediment Drift Ridges[1,2,3]
- Hiatuses[1,2,3]
  - Breakup Unconformity[4]
  - Global Sea-Level Change[4]
  - Slope and Rise Sculpting[1,2,3]
- Diapirs[4]
  - Salt[4]
  - Shale[4]
- Mass Sediment Movements[1,4]
  - Rotational Slumps[1,4]
  - Growth Faults[4]
  - Gravity Gliding[1,4]
- Diagenesis[1,2,4]
  - Gas Hydrates[1,2,3,4]
  - Chertification[1,2,3]
- Paleoenvironment[1,2,3,4]
  - Syn-Rift, Restricted[2,4]
- Early Sea Floor Spreading, Restricted[2,4]
- Late Sea Floor Spreading, Open[1,2,3]
- Carbonate Banks[1,4]
  - Continental Slope Migration[4]
  - Sediment Damming[4]
- Paleobathymetry[1,3,4]

[1] Addressed Partially in Past JOIDES & IPOD Drilling
[2] To be Addressed 1979–81
[3] Possible 1981–83 Objective
[4] Possible *Explorer* Objective

areas so that additional drilling will methodically build on previous knowledge to complete the documentation of passive-margin problems.

Listed in Table 1 are the identified phenomena associated with passive margins. Noted with a footnote are those features and phenomena already addressed by JOIDES and IPOD drilling and those to be studied through 1981. From this table those remaining subjects of passive margin research can be identified, some of which could be attacked in a proposed 1981–83 IPOD I extension. Others will have to await *Glomar Explorer* capabilities.

REFERENCES CITED

NOTE: All references to Deep Sea Drilling Project Initial Reports have been abbreviated; for complete citations see *Appendix I* at the end of this volume, or the individual Initial Reports as published.

BECK, R. H., 1972, The oceans, the new frontier in exploration: Australian Petroleum Exploration Assoc. Jour., v. 12, p. 1-21.

BENSON, W. E., SHERIDAN, R. E., AND OTHERS, 1978a, Site 392: South rim of the Blake Nose: Init. Repts. DSDP, v. 44, p. 337-393.

———, ———, AND ———, 1978b, Sites 391: Blake-Bahama Basins: Init. Repts. DSDP, v. 44, p. 153-336.

FALVEY, D. A., 1974, The development of continental margins in plate tectonic theory: Australian Petroleum Exploration Assoc. Jour., v. 14, p. 95–106.

GROW, J. A., MATTICK, R. E., SCHLEE, J. S., 1979, Multichannel Seismic depth sections and interval velocities of outer continental shelf and upper continental slope between Cape Hatteras and Cape Cod: Am. Assoc. Petroleum Geologists, Mem. 29, p. 65–83.

KLITGORD, K. D., AND BEHRENDT, J. C., 1979, Basin structure of the U.S. Atlantic margin: Am. Assoc. Petroleum Geologists, Mem. 29, p. 85–112.

MONTADERT, L., DE CHARPAL, O., ROBERTS, D. G., GEUNNOC, P., AND SIBUET, J. C., 1979, Northeast Atlantic Passive Continental Margins: Rifting and Subsidence Processes: Amer. Geophys. Union, Maurice Ewing Series, v. 3, p. 154–186.

PARSON, M. G., 1975, The geology of the Laurentian Fan and Scotia Rise: Canadian Soc. Petroleum Geologists, Mem. 4, p. 155–167.

SHERIDAN, R. E., 1974, Atlantic continental margin off North America, *in* Burk, C. A., and Drake, C. L., eds., The Geology of Continental Margins: New York, Springer Verlag, p. 391–407.

SHERIDAN, R. E., AND ENOS, P., 1979, Stratigraphic evolution of the Blake Plateau after a decade of scientific drilling: Amer. Geophys. Union, Maurice Ewing Series, v. 3, p. 109–122.

———, ———, GRADSTEIN, F. M., AND BENSON, W. E., 1978, Mesozoic and Cenozoic sedimentary environments in the western North Atlantic: Init. Repts. DSDP, v. 44, p. 971–979.

———, WINDISCH, C. C., EWING, J. I., AND STOFFA, P. L., 1979a, Structure and stratigraphy of the Blake Escarpment based on seismic reflection profiles: Am. Assoc. Petroleum Geologists, Mem. 29, p. 177–186.

———, GROW, J. A., BEHRENDT, J. C., AND BAYER, K. C., 1979b, Seismic refraction study of the continental edge of the eastern United States: Tectonophysics, v. 59, p. 1–26.

TODD, R. G., AND MITCHUM, R. M., JR., 1977, Seismic stratigraphy and global changes in sea-level, part 8: Identification of Upper Triassic, Jurassic and Lower Cretaceous seismic sequences in Gulf of Mexico and offshore west Africa: Am. Assoc. Petroleum Geologist, Mem. 26, p. 145–163.

VAIL, P. R., MITCHUM, R. M., JR., AND THOMPSON, S., III, 1977, Seismic stratigraphy and global changes of sea-level: part 4: Global cycles of relative changes of sea-level: Am. Assoc. Petroleum Geologists, Mem. 26, p. 83–97.

VAN HINTE, J. E., 1976, A Jurassic time scale: Am. Assoc. Petroleum Geologist Bull., v. 60, p. 489–497.

WADE, J., 1978, Mesozoic-Cenozoic history of northeast margin of North America: Proc. Offshore Tech. Conference, Houston, Paper 3266, p. 1849–1858.

SEPM Special Publication No. 32, p. 57–66, October 1981

REVIEW OF EARLY RESULTS FROM DRILLING OF THE IPOD-1 ACTIVE MARGIN TRANSECTS ACROSS THE JAPAN, MARIANA, AND MIDDLE-AMERICA CONVERGENT MARGINS

ROLAND VON HUENE
U.S. Geological Survey, Menlo Park, California 94025 U.S.A.

ABSTRACT

IPOD transects across convergent margins have provided time stratigraphic and environmental information to support the general concepts of spreading during the opening of backarc basins and tectonic accretion at the front of the underthrust plate. Three convergent margins are without large accretionary complexes, and other less frequently applied concepts are necessary to interpret them. Much sediment is subducted beneath the four margins investigated; along all of the margins there are periods when the continental framework was truncated and somehow removed by tectonic erosion. Vertical motions have been recorded during active arc magmatism (and, by implication, subduction) including not only uplift but also equally rapid subsidence. Off Japan the seaward part of the margin subsided from above sea level to 6 km depth. The scope of tectonic processes along convergent margins is much greater than was anticipated when the program was first planned.

INTRODUCTION

Transects across the Japan, Mariana, and middle-America convergent margins were drilled by the *Glomar Challenger* between 1977 and 1979. This completed the first phase of IPOD-1 studies that include not only drilling but also a wide variety of other geophysical and geological work. The results of earlier studies supporting the hypothesis that the Philippine Sea backarc basins originated by seafloor spreading were confirmed. Tectonic accretion of sediment to the front of a margin, here termed subduction accretion*, was also supported by paleontological and structural evidence. Generally unanticipated was evidence that much of the ocean basin sediment carried into the trench by plate convergence is not scraped off at the front of the continental framework but is subducted, that large parts of the continental framework are missing, and that hemipelagic rather than deep-sea pelagic sediment is widely distributed hundreds of kilometers seaward of the trenches. Most of these results are published only in preliminary form (Langseth, Okada, and others, 1978; von Huene, Nasu and others, 1978; Klein, Kobayashi, and others, 1978; Kroenke and Scott, 1978; Hussong, Uyeda and others, 1978; Moore, Watkins, and others, 1979a, b; von Huene, Aubouin, and others, 1980). This is a general summary of these preliminary results with emphasis on tectonic processes at convergent margins and not a comprehensive summary of the original data which at this time is premature and beyond the scope of this paper. For the predrilling geophysical work and the drill sampling data from each leg, I refer the reader to the Initial Reports of the Deep Sea Drilling Project for Legs 56–60, 66 and 67, which began appearing in 1980. In this narrative, I comment briefly on the planning phase during which IPOD-1 objectives were defined (Uyeda, 1978), the tectonic processes discovered along the three margins transected, the subsequent change in emphasis on objectives after IPOD-1 drilling, and the proposing follow-on programs considered by the Active Margins Panel.

*Tectonic accretion is made more specific by using the term subduction accretion, which refers to the concept of active tectonism at the front of the margin where thrusts and folds deform deep ocean basin, trench, and slope sediment into a subduction complex (Dickinson and Seely, 1979; Scholl and others, 1980).

PLANNING THE IPOD-1 ACTIVE MARGIN PROGRAM

When IPOD began in 1974, about 10 DSDP holes had been drilled at the front of active margins and about twice that number in backarc basins. The results of Leg 18, the first to drill in the active margin environment, indicated that the thick sediment of Astoria Fan and in the Aleutian Trench were no older than Late Pliocene and Late Pleistocene respectively, and that deformed sediment of similar age and lithology occur on the adjacent trench lower slopes (Kulm, von Huene, and others, 1973). However, no distinctly oceanic sediment was recovered from the trench slope to clearly substantiate subduction accretion. Much of the rock beneath the trench slope, whose refraction velocities are typical of sedimentary rock, was assumed to be highly deformed because reflection seismic records were largely masked by diffractions from numerous point reflectors and contained little evidence of coherent bedding. On Leg 19, two holes were drilled in the central Aleutian forearc area.

The Leg 19 staff emphasized the Late Miocene age of material on the trench lower slope (Creager, Scholl, and others, 1973). Later, on Leg 31, the core from a site near the Nankai Trough off southern Honshu Island showed overturned beds and fracturing which was interpreted to come from a recumbent fold at the front of the subduction zone (Karig, Ingle, and others, 1975; Moore and Karig, 1976). The Leg 31 staff concluded that in both the West Philippine and Shikoku-Parece Vela backarc basins of the Philippine Sea, crustal spreading was the most likely mode of origin. It was also proposed that the active volcanic arc had migrated eastward since the Eocene (Karig, 1975).

Just prior to Leg 31, Seely and his colleagues showed that seismic records across convergent margins could be improved greatly by using multichannel seismic-reflection techniques, and some structure could be resolved beneath the trench landward slope (Seely and others, 1974). Subsequent multichannel records of trenches acquired off Peru and Ecuador were interpreted by Kulm and others (1975, 1977) and Hussong and others (1976) as showing an abrupt juncture of the Paleozoic and Mesozoic continental framework and an accreted subduction complex forming the base of the trench landward slope. These were some of the principal new data when active-margin planning for the IPOD program began. The Active Margin Panel decided that efforts should be concentrated on only a few convergent margins along transects beginning seaward of the trench and continuing through the backarc basin. This investigation needed to begin with study of multichannel seismic-reflection records. It was assumed that during a follow-on program to IPOD-1, a ship capable of drilling in trenches deeper than 6000 m would become available, so certain transects could be partially drilled by *Glomar Challenger* during IPOD-1, and then be completed during a subsequent phase of drilling.

With the six legs allotted for Active Margins work, three transects across different types of arcs were accepted by the Planning Committee. The Japan Trench transect, drilled on Legs 56 and 57, crosses a detached intra-oceanic arc-trench system (Dickinson and Seely, 1979) where continental crust is separated from the Asian mainland by a backarc basin. The Mariana Trench transect, drilled on Legs 59 and 60, crosses a migratory intra-oceanic arc-trench system with a broad backarc basin. The middle-America Trench transects, drilled on Legs 66 and 67, cross a continental margin arc-trench system. These were chosen instead of the Peru-Chile trench transects because of logistic considerations. The two short transects drilled off middle America were off Oaxaca, Mexico, where the continental framework of Paleozoic and older rock appeared to be truncated, and off Guatemala, where steady-state subduction accretion was postulated for most of the Tertiary (Seely, 1979). Prior to drilling, each transect area was studied with multichannel seismic reflection and other geophysical techniques. In addition, the northern region of the Philippine Sea was drilled on Leg 58 to test its origin by backarc spreading.

THE DEVELOPMENT OF BACKARC BASINS BY SEAFLOOR SPREADING

Based on studies of surface samples, bathymetry, and geophysics, Karig (1971) proposed that backarc basins in the Philippine Sea (for example, the Mariana Trough) formed by spreading of ocean crust. The results of *Glomar Challenger* drilling on Leg 31 are generally consistant with the spreading hypothesis, but it was not clear how backarc spreading was related to plate convergence and subduction (Karig, 1975). A second question concerned the timing and origin of features in the Diato Ridge-and-Basin province of the northern West Philippine Basin; here metamorphic, plutonic, and shallow water rocks had been recovered from the seafloor (Mizuno and others, 1975; Shiki and others, 1974).

In the complex Diato Ridge-and-basin province of the northern Philippine Sea, backarc spreading simultaneous with arc magmatism was confirmed by the studies of the scientific staff on Leg 58 (Klein, Kobayashi, and others, 1978). The ridges are subsided remnant arcs; the basins contain volcanic rock formed during arc and post-arc volcanism. The complex includes about 35° migration north and 55° rotation of a backarc basin and subsequent subduction of parts of the basin. The remaining questions concern the symmetry of spreading in the west Philippine basin and the age of its oldest crust.

The staff of Leg 59 concluded, from a basement age determined on the arc and ages inferred from magnetic anomalies in the basin, that backarc spreading in the south part of the west Philippine Basin may have been simultaneous with volcanism along the Palau-Kyushu Ridge. This initial spreading and volcanism was followed by subsequent episodes of spreading and arc volcanism to the east during which the Mariana Ridge and Arc was active. The staffs of Legs 59 and 60 documented more thoroughly than before that new volcanic arcs developed in an eastward progression leaving behind a sequence of remnant volcanic ridges and associated backarc basins. Problems now emphasized appear to concern petrochemical and evolutionary models that relate the spreading process with volcanism and convergence. Although development of the extension forming the backarc basin appears to occur simultaneously with island-arc volcanism, there are periods when one process might operate independent of the other. The mechanisms for generating tensional forces behind a zone of presumed convergence between lithospheric plates are still not clear.

MAGMATIC ROCKS IN THE FOREARC REGION

Along the Japan and Mariana transects, arc-related volcanic and plutonic rocks were recovered at sites near the trench slope, far seaward of the presently active arc (Fig. 1). Off Japan, the composition of the recovered rocks is interpreted as the product of a briefly active island arc (Fujioka, 1980). Refraction data (Nagumo and others, 1980) indicate that another igneous body occurs nearby, and despite a low magnetic susceptibility measured in the rocks recovered, other plutons are suggested by low-amplitude magnetic anomalies (Oshima and others, 1975). The crystallization age of these rocks is 22–24 m.y.BP and about the same age as the initial volcanism along the present arc (Yanagasawa and others, 1980). Where these rocks occur, the lithosphere is shallow, and the present temperatures from downhole thermometry projected to the base of the upper plate are about one quarter of those needed to produce normal island arc magma (Fig. 1).

In the Mariana transect, arc-related volcanic and plutonic rock has been recovered by dredging from as far seaward as the trench lower slope (Dietrich and others, 1978). The dredged and drilled rocks date from the earliest history of the Mariana arc (Eocene-Oligocene) (Hussong, Uyeda, and others, 1978). Their altered state and low melting temperatures indicate that they were formed in the presence of abundant water. Perhaps this water was derived from the dehydration of the lower plate during subduction (Anderson and others, 1976; J. Natland, written communication).

The position of ancient arc-related rocks, much closer to the trench than the presently active arc, suggests that the upper plate was once thicker and may formerly have extended farther seaward unless forearc magmas are produced by some unrecognized process of island-arc development.

SUBDUCTION ACCRETION

Prior to drilling, multichannel seismic reflection records were interpreted as showing subduction accretion off Japan and middle America. The clearest evidence was across the Middle-America Trench transect off Oaxaca (Shipley and others, 1980). Here, a sequence of strong landward-dipping reflections is covered by a thin veneer of sediment with reflections paralleling the seafloor (Fig. 2). Drilling at three sites on Leg 66 indicates an age progression on the top of the landward-dipping reflections, from Pleistocene at the foot of the trench slope to Upper Miocene near the middle of the slope. This age progression can be interpreted as evidence for the progressive stacking of thrust slices of off-scraped trench deposits (Moore and others, 1979). However, despite the general increase of small-scale deformation with depth in every hole, no major faults were recognized from repeated ages or from the physical character of cored sediment (Moore, Watkins, and others, 1979). Penetration (maximum

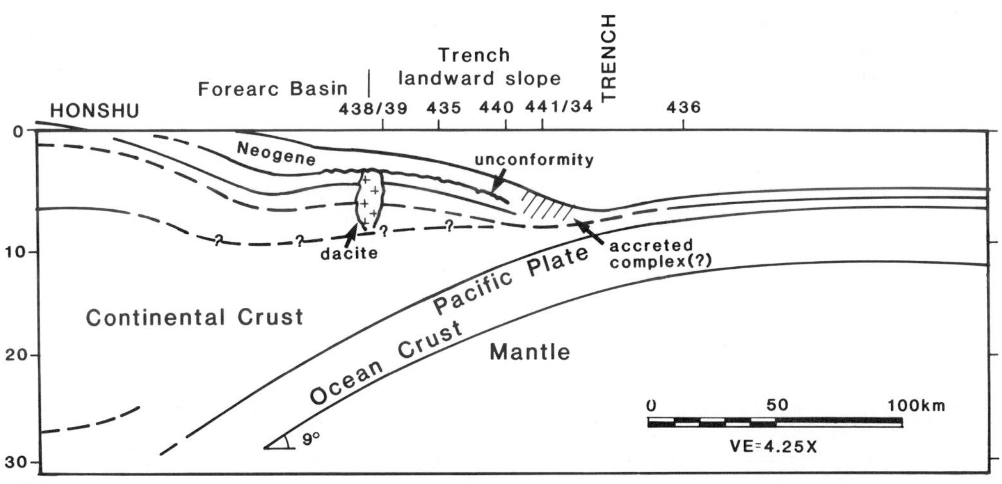

FIG. 1.—Composite section along the Japan Trench transect based on seismic-reflection data, with landward part after Ishiwada and Ogawa (1976). Lower layers shown in heavy lines are from refraction data summarized by Murauchi and Ludwig (1980). Seismic-refraction velocities are noted; the angular unconformity formed by subaerial erosion in Late Paleogene and Early Neogene time is indicated by the wavey line. An accreted complex is inferred from seismic data; only overlying slope sediment was drilled. Dacite and associated island-arc rock was recovered near the bottom of Hole 439.

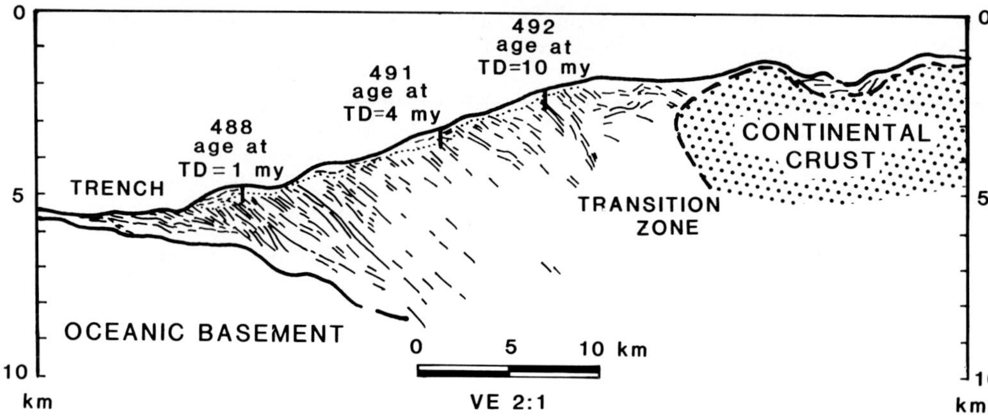

FIG. 2.—Composite section along the middle-America Trench transect off Oaxaca, Mexico, showing landward-dipping reflections and the truncated continental framework. At the bottom of Hold 492, the sediment is about 12 m.y. old, whereas at the bottom of Hole 488 it is of Quaternary age (from Moore, Watkins, and others, 1979).

542 m) was less than hoped for, and perhaps deeper drilling would have penetrated the thrust faults interpreted from the seismic records and dating of cores.

Off Japan and Guatemala small-scale deformation increases rapidly with depth in slope sediment, but no accreted subduction complex or major thrust faults were recognized in the core; however, drilling probably did not penetrate well into the section inferred to be an accreted sequence from seismic data. On the Japan Trench transect, an interval of repeated ages at the bottom of Hole 434 could signify thrust faulting, slumping, or possibly environmental changes recorded by the microfossil flora. Thus an imbricate thrust sequence might have been penetrated, but this interpretation is not unique for the data available.

Sediment recovered from the trench slopes of all transects is mainly hemipelagic mud and interbedded thin volcanic ash. No pelagic sediment characteristic of the deep ocean basin was recovered. This observation does not in itself rule out the possibility that offscraped oceanic material was recovered because hemipelagic sediment was recovered in the upper parts of three holes as much as 20 km seaward, and 2000 m above the trench floor (middle-America and Japan). If only the top of the deep ocean section and trench floor deposits are accreted (Moore, 1975), they would be difficult to distinguish from sections deposited on the trench landward slope. Nonetheless, benthic foraminiferal fauna on the Japan and Guatemala landward trench slopes indicate that only slope deposits were penetrated there.

Deformation of shallow sediment on the landward trench slope may occur as microfracturing, which was commonly observed at a depth of 200 m to more than 600 m. This may be a dominant mode of deformation in the fine-grained sediment recovered (Japan, Mariana, Guatemala). There is evidence from the logging in holes off Japan for geopressured sections beneath the foot of the slope (von Huene, Nasu, and others, 1978; Arthur and others, 1980b). This leads me to speculate that, 1) perhaps the most intense dislocations within some underthrust margins are concentrated along narrow zones that are saturated with overpressured water, 2) that the lack of teleseismic earthquakes at the seaward end of the Benioff Zone is explained by water-lubricated faults and abundant microfracturing, and 3) that a small frictional resistance along lubricated faults is an explanation for the less than chaotic or tight structure in terrains beneath which thousands of kilometers of oceanic crust have been subducted.

SEDIMENT SUBDUCTION

Soft sediment entering a subduction zone is commonly presumed to be scraped off the more competent ocean crust and to be mechanically attached to the margin's existing framework. Some authors have argued that sediment subduction can occur possibly as a result of sediment entrained in graben of the ocean plate that is thrust beneath the trench wall (Lister, 1971; Hilde and Sharman, 1978; Schweller and Kulm, 1978). Accretion of oceanic sediment was studied along the active margin transects by comparing the potential input from sections drilled on the ocean crust and global plate motion convergence rates, with the corresponding sediment volume of the accreted body at the front of the subduction zone. In all cases, the volume of accreted sediment was inferred from interpretation of seismic records (Fig. 3). Despite the inability of the *Glomar Challenger* to drill completely through the landward-dipping sediment sections and recover

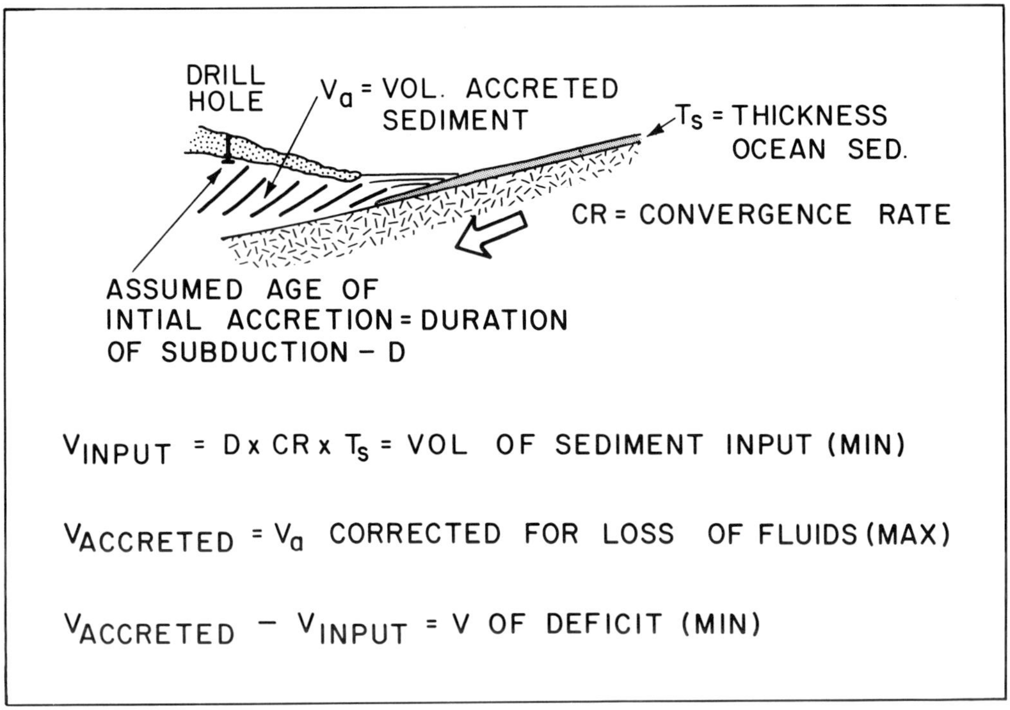

FIG. 3.—Diagram showing how the volume of subducted sediment was derived from drill and geophysical data. Assumptions have been made at the limits that result in a minimum deficit of sediment. Drilling above the accreted complex has generally not penetrated into, or very far into, the accretion complex, which is commonly inferred from landward-dipping reflections in seismic data.

the oldest sediment in a subduction complex, the oldest rocks penetrated at sites on the trench lower slope can be used as the minimum age of the accreted subduction complex (Fig. 3). This would give the largest possible volume of sediment scraped off the ocean crust per unit time, compared with the smallest volume of sediment input from the oceanic plate; it results in an estimate of the minimum amount of sediment that has not been accreted. Such an estimate on the Japan Trench transect indicates that during the past 6 m.y. about 20% of the sediment input is accreted and the rest is presumed to be subducted. Off Guatemala there is virtually no sediment younger than Cretaceous at the foot of the trench slope, thus all sediment is presently being subducted. Even for the Oaxaca margin, where subduction accretion is best demonstrated, a maximum of 66% and a minimum of 33% of the sediment is estimated to be accreted. Along the Mariana arc most of the rocks are igneous and sediment is rare. The imprecision of data in these estimates of subducted sediment is far less than the large amount of sediment not accreted.

SUBDUCTION EROSION

Tectonic erosion, which is the retreat or subsidence of a margin through removal of material from a trench landward slope, can be caused either by strike-slip faulting or by subduction erosion.* Subduction erosion has been suggested infrequently in the last few decades (i.e., Hussong and others, 1976; Kulm and others, 1977; Scholl and others, 1980). However, evidence for older arc magmatism in the present forearc area, evidence of massive subsidence during subduction in the forearc and trench slope areas, and abrupt truncation of the arc or continental framework on the trench landward slope all suggest that the trench slope has retreated. Evidence of forearc magmatism within a few tens of kilometers of the trench axis has been found on the Japan and Mariana transects as previously mentioned.

Subsidence of an ancient landmass by as much as 6 km dominates Neogene tectonic history in the Japan forearc and trench area, and net subsidence

*Subduction erosion is a term equivalent to consumption (Kulm and others, 1977), to some uses of tectonic erosion (Karig and others, 1978), or just simply to erosion (Hussong and others, 1976). Subduction erosion can occur either at the leading edge of a margin, leaving a truncated margin, or along the subsurface base of the upper plate, causing subsidence as the less dense crust is thinned.

rather than uplift dominates Neogene vertical movement in the Mariana forearc area. Off Oaxaca, the shelf first subsided 2 to 3 km, and then rose to the present depths of 1 to 1.5 km during the episode of subduction accretion. Subsidence since Early Miocene time was indicated by the study of benthic fauna in the Neogene slope sediment off Guatemala, and either subsidence or sealevel fluctuations are reported from the Guatemalan shelf (Seely, 1979). Subsidence of the margin within some bounds of isostacy would seem to require thinning of low density continental crust and replacement with higher density oceanic crust. An alternative mechanism suggested by Langseth (personal communication, 1979) accomplishes subsidence by the thermal effect of ridge subduction or change in rates of subduction which provides cooler rock of greater density. Such an exchange of less dense for more dense crust would probably occur at the crustal contact formed by the present subduction zone.

The subsided paleolandmass off Japan had a Miocene Pacific shoreline that is now beneath the trench landward slope within 20 km of the present trench axis. Reconstruction of the margin with such a shoreline suggests that the trench axis was located 30 to 50 km further east than at present and that during the Neogene it retreated toward Honshu. Off Oaxaca, subduction accretion began 12 m.y.BP against a truncated margin of Paleozoic and older cratonic rock. Here truncation may have resulted from strike-slip or convergent tectonism. Truncation by subduction erosion is perhaps the most attractive of three explanations for the Cretaceous section recovered by drilling at the foot of the slope off Guatemala. Truncated margins are not uncommon but the type of tectonic erosion that has caused truncation is not always clear. There appears to be considerable evidence for subduction erosion especially off Japan (von Huene, Nasu, and others, 1978; Murauchi and Ludwig, 1980; Langseth and others, 1981).

Perhaps the concept of subduction erosion has been applied infrequently by geoscientists because it requires processes that seem unlikely, such as stuffing young rock beneath older rock of greater density or large-scale abrasion of the continental framework. The concept of structural truncation by lateral faulting has been more frequently invoked to explain truncated margins, and indeed it is much less difficult to imagine (for instance, Karig, 1974). However, if the plate convergence vector at margins truncated in the past 10 m.y. has not changed appreciably, then the lateral component of fault motion, and hence the lateral transport of crustal fragments, should be small. The Neogene crustal fragments missing from the foot of the margin have not been located anywhere along strike near the Japan and Guatemala transects. Therefore, subduction erosion deserves consideration as one cause of structural truncation at underthrust margins despite the conceptual problems.

VERTICAL TECTONIC MOVEMENTS IN THE FOREARC

In the previous discussions, vertical uplift and subsidence are mentioned in connection with subduction at convergent margins. Subduction accretion, where sediment is tectonically attached to a continental or island arc framework, is commonly inferred to cause uplift of the landward trench slope. From cores recovered along the middle-America transect off Oaxaca, the staff of Leg 66 inferred a period of uplift coincident with accretion. Similar uplift is reported for the Guatemala shelf (Seely, 1979). Despite periods of uplift, a period of subsidence is also indicated by microfossil assemblages in the forearc areas on all of the IPOD-1 transects. The 3 to 6 km subsidence of a Neogene subaerial erosion surface on the Japan Trench transect is perhaps the most dramatic example. Here the rate of subsidence was 1.7 km/m.y. maximum, and this is about the rate of steady uplift measured off Oaxaca (Moore, Watkins, and others, 1979b). If uplift has raised the seafloor to levels of erosion, then subsequent subsidence and sedimentation will bury the erosion surface. Off Japan, where a deeply buried erosion surface was penetrated, it is overlain by a 50-m-thick conglomerate bed and a 100-m-thick unit of medium-grained sand derived from a nearby surf zone. An erosion surface was also drilled off Oaxaca (Moore, Watkins, and others, 1979a).

The presence of erosion surfaces and evidence for large amounts of subsidence along convergent margins suggests possibilities for the development of hydrocarbon reservoirs. For instance, if the Neogene section penetrated at Site 438 and 439 off Japan were resting on source beds, the area would be prospective (Fig. 4). Depression of the section below the unconformity to levels of increased temperature might cause hydrocarbon maturation and migration. Perhaps the greatest deterrent to developing commercial hydrocarbon accumulations is the generally low heat flow observed in many convergent margin settings.

SEDIMENT FACIES DISTRIBUTION

Most of the sediment recovered by drilling in deep forearc areas and trench slopes, especially off continents, is hemipelagic silt and claystone rich in volcanic ash (including sites on Legs 18, 19, and 31). Discrete beds of sand are a relatively minor constituent in the core material; however, it is more difficult to recover than mudstone. On the middle-America transect off Oaxaca, caving sand terminated drilling in the trench and on the trench lower

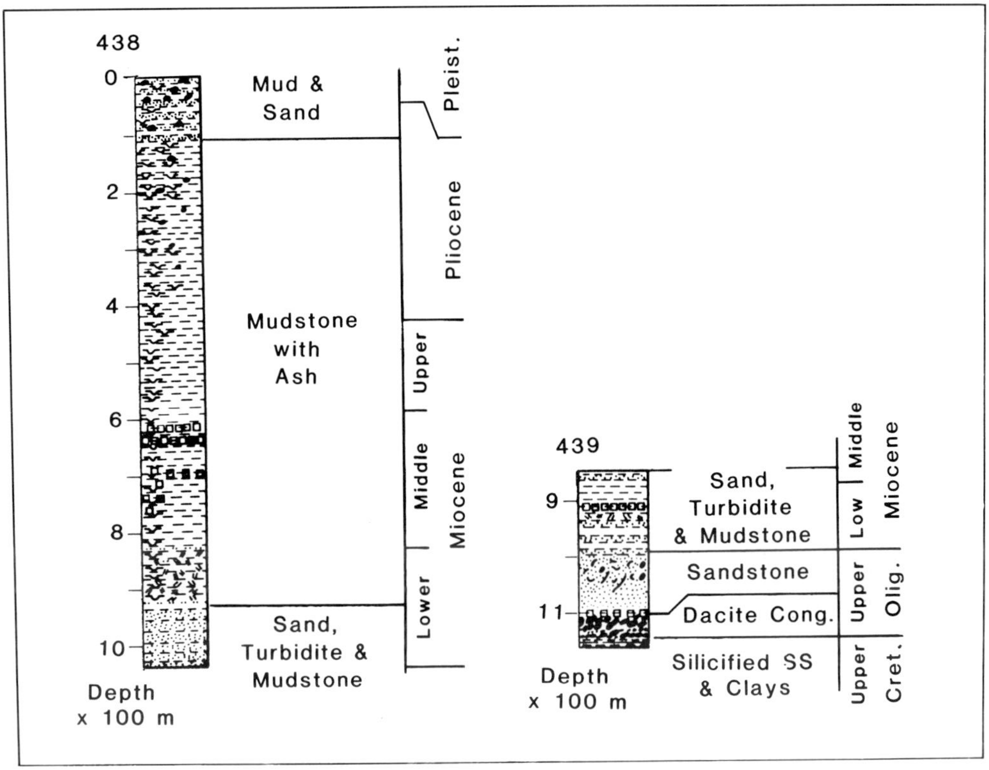

FIG. 4.—Lithologic section penetrated at sites 438 and 439 (from von Huene, Nasu, and others, 1978).

slope. On the other hand, off Guatemala the trench fill (sampled in nine holes) is predominently mudstone, as was also found in the Aleutian Trench off Kodiak (Kulm, von Huene, and others, 1973). Thus, although sand is under-represented in recovered material, the trench and trench lower slope drilling penetrated mainly hemipelagic mudstone, a conclusion confirmed where the holes were logged.

Sediment transport and deposition along a convergent margin are highly influenced by tectonism. Off Japan, sediment is transported across the forearc in major channels 40 to 60 km apart, and coarse as well as fine fractions commonly bypass the continental shelf to be deposited in trench-slope basins and ultimately the trench floor (Arthur and others, 1980a). Measured rates of accumulation in the trench-slope basin were greater than in the forearc basin off Honshu.

The hemipelagic sediment lithology across the transects of continental arc-trench systems extends beyond the trench to the deep ocean basin. The hemipelagic sections seaward of the trenches are of Late Miocene age, although near the Aleutian Trench (Site 178) and Kamchatka Trench (Site 192) the deepest hemipelagic material is of Early or perhaps Middle Miocene age. When backtracked to their position in Miocene time, by applying global plate motion history, these sites were apparently influenced by terrigenous sediment when they were several hundred kilometers out to sea. One consequence of the thick hemipelagic sediment on the ocean plate is that considerably more hemipelagic sediment is available to be accreted than if the whole oceanic section were typical pelagic ooze.

The great uniformity of sediment facies recovered by IPOD-1 drilling is probably influenced by the choice of sites on structurally high areas where the drill can reach deep targets. Thus much sampling on the traverses was done in interchannel areas. However, the sediment sampled in the interchannel areas has a uniform hemipelagic character and there is no distinct lithologic imprint of the shelf, slope, trench, and perhaps the trench seaward slope interchannel environments.

POST IPOD-1 PLANNING

Most earth scientists are familiar with the concept of subduction accretion and, in fact, much of the IPOD-1 active margin program was designed to test this concept. However, the results of IPOD-1 provided direct evidence for extensive subduction accretion only along the middle-America transect off Oaxaca. Unanticipated evidence for sediment subduction, subduction erosion, and for large vertical motion that includes kilometers of subsidence was uncovered. Rather than only building upward and seaward by accretion, some convergent margins may undergo episodes of subsidence and retreat with a net loss of volume at the seaward edge of the forearc. Apparently there are large volumes of sediment conveyed beneath the margin, and just how far down the subduction zone such material is taken is unknown.

The results of IPOD-1 drilling on active margins indicate that of all the tectonic elements of convergent margins, the forearc areas are the least understood. Development of western Pacific backarc basins by seafloor spreading is now well established, although the relation of the spreading processes and subduction is not known. Perhaps a more precise relation between explosive volcanism and backarc spreading can be developed in a backarc basin with thick sediment like the Japan Sea. The hypothesis for development of backarc basins by entrapment of pre-arc ocean crust (Bering Sea) is untested. The early developmental sequence of arc magmas is poorly known. This is a complex problem of economic as well as scientific importance that can be approached in part by studies of insular exposures, but drilling in the forearc area seems essential to its resolution.

The Active Margins Panel was asked to recommend programs to follow IPOD-1. The original objectives of IPOD-1 were reviewed and prioritized. The panel felt that because least is known about the forearc area, it should have a high priority. It also seemed necessary to explore the scope of tectonic processes in forearc areas which will require investigation of different convergent margin types. The panel grouped convergent margins into four categories for this purpose: 1) intraoceanic (Mariana), 2) continental (middle-America, Peru-Chile), 3) intermediate or transitional (Japan, Aleutian), and 4) collision (Banda-Sunda, Oman). The panel prioritized key objectives to be addressed: 1) structural evolution of the forearc, 2) sedimentary sequences in modern trench and forearc basins as models for the interpretation of ancient fold belts, 3) tectonic processes during collision, 4) evolution of island arc magmatism, 5) origin of backarc basins.

To achieve the objectives of highest priority requires drilling in great water depths, deeper than is possible with present *Glomar Challenger* capability. A capability to penetrate 1.5 km in 7–8 km of water would be ideal. The program envisioned by the panel would also benefit greatly from better deep-water seismic resolution, such as could be achieved, for example, with deep-towed sensors. To achieve all the backarc objectives will require deep-water riser capability, e.g., fluid circulation during drilling.

The present models of tectonic processes operating at modern convergent margins are based more on hypothetical considerations than on kinematic geologic sections established with modern geophysics and rock information. Thus, future multidisciplinary investigations along active margins like those of IPOD-1 hold promise for making significant scientific advances which relate directly to an assessment of their economic potential. The possibility that large amounts of sediment are subducted has geochemical implications related to the genesis of mineral deposits along Pacific coastal belts. The large uplift and subsidence that sometimes affects the forearc area has significance for the development of hydrocarbon reservoirs and must affect thermal history. Another significant reason to mount a costly investigation of modern convergent ocean margins is the potential advance in predictive capability for earth scientists working in ancient and modern convergent margins.

ACKNOWLEDGMENTS

Much of what is contained in this paper was discussed at length in meetings of the IPOD Active Margins Panel. I am very grateful to my colleagues on the original panel and on the present one. Many ideas were also discussed with Michael Arthur, Marcus Langseth and Bobb Carson. Nori Nasu, cochief of Leg 57 and chairman of the Japanese IPOD panel, was instrumental in acquiring the geophysical data off Japan that have been so unusually informative. This manuscript benefited greatly from review by David Scholl, David McCulloch, William Dickinson, and William Ludwig.

REFERENCES CITED

NOTE: All references to Deep Sea Drilling Project Initial Reports have been abbreviated; for complete citations see *Appendix I* at the end of this volume, or the individual Initial Reports as published.

ANDERSON, R. N., UYEDA, S., AND MIYASHIRO, A., 1976, Geophysical and geochemical constraints at converging plate boundaries—Part I: Dehydration in the downgoing slab: Geophysical Jour. Royal Astronomical Soc., v. 44, p. 333–357.

ARTHUR, M. A., VON HUENE, R., AND ADELSECK, C., 1980a, Sedimentary evolution of the Japan forearc region off northern Honshu (DSDP Legs 56 and 57): Init. Repts. DSDP, v. 57, p. 521–568.

ARTHUR, M. A., CARSON, B., VON HUENE, R., 1980b, Initial tectonic deformation of hemipelagic sediment at the leading edge of the Japan convergent margin: Init. Repts. DSDP, v. 57, p. 569–614.

Creager, J. S., Scholl, D. W., and others, 1973: Init. Repts. DSDP, v. 19, 913 p.
Dietrich, V., Ermann, R., Oberhansli, R., Purchelt, H., 1978, Geochemistry of basaltic and gabbroic rocks from the west Mariana Basin and the Mariana Trench: Earth Planet. Sci. Letters, v. 39, p. 127–144.
Dickinson, W. R., and Seely, D. R., 1979, Structure and stratigraphy of forearc regions: Am Assoc. Petroleum Geologists Bull., v. 63, p. 2–31.
Fujioka, K., 1980, Conglomerates of volcanic rocks of Deep Sea Drilling Project, Site 439: Init. Repts. DSDP, v. 57, p. 1075–1082.
Hilde, T. W. C. and Sharman, G. F., 1978, Fault structure of the descending plate and its influence of the subduction process (abs.): Proc. Amer. Geophys. Union Meeting, Dec., 1978.
Hussong, D. M., Edwards, P. B., Johnson, S. H., Campbell, J. R., and Sutton, G. H., 1976, Crustal structure of the Peru-Chile Trench; 8–12°S. latitude, *in* Sutton, G. H., Manghnani, M. H., Moberly, R., and McAfee, E. U., eds., The geophysics of the Pacific Ocean Basin and its margin, Geophysical Monograph 19: Washington, D.C., Amer. Geophys. Union, p. 71–85.
———, Uyeda, S., and others, 1978, Leg 60 ends in Guam: Geotimes, v. 23, no. 10, p. 19–22.
Ishiwada, Y. and Ogawa, K., 1976, Petroleum geology of offshore areas around the Japanese islands: UN-ESCAP, CCOP Tech. Bull., v. 10, p. 23–34.
Karig, D. E., 1971, Origin and development of marginal basins in the western Pacific: Jour. Geophys. Res., v. 87, p. 1259–1268.
———, 1974, Tectonic erosion at trenches: Earth Planet. Sci. Letters, v. 21, p. 209–212.
———, Ingle, J. C. Jr., and others, 1975: Init. Repts. DSDP, v. 31, 927 p.
———, ———, and ———, 1975, Basin genesis in the Philippine Sea: Init. Repts. DSDP, v. 31, p. 857–879.
Klein, G. DeV., Kobayashi, K., and others, 1978, Philippine Sea drilled (Leg 58): Geotimes, v. 23, no. 5, p. 23–25.
Kroenke, L. and Scott, R., 1978, Old questions answered—and new ones asked (Leg 59): Geotimes, v. 23, no. 7, p. 20–23.
Kulm, L. D., von Huene, R., and others, 1973: Init. Repts. DSDP, v. 18, 1077 p.
———, Prince, R. A., French, W., Masais, A., and Johnson, S., 1975, Evidence of imbricate thrusting in the Peru Trench and continental slope from multi-fold seismic reflection data: EOS Trans. Amer. Geophys. Union, v. 56, p. 442–443.
Kulm, L. D., Schweller, W. J., and Masias, A., 1977, A preliminary analysis of the subduction processes along the Andean continental margin, 6° to 45°S: *in* Talwani, M., and Pitman, W. C., III (eds.), Island arcs, deep-sea trenches, and backarc basins: Washington, D.C., Amer. Geophys. Union, p. 285–302.
Langseth, M., Okada, H., and others, 1978, Near the Japan Trench; transects begun (Leg 56): Geotimes, v. 23, no. 3, p. 22–26.
Langseth, M., von Huene, R., Nasu, N., and Okada, H., 1981, Subsidence of the Japan Trench forearc region of Northern Honshu: Oceanologica Acta, in press.
Lister, C. R. B., 1971, Tectonic movement in the Chile Trench: Science, v. 173, p. 719–722.
Mizuno, A., and others, 1975, Marine geology and geological history of the Daito Ridge area, northwestern Philippine Sea: Marine Science, v. 7, p. 484–491, 543–548.
Moore, J. C., 1975, Selective subduction: Geology, v. 3, no. 9, p. 530–532.
———, and Karig, D. E., 1976, Sedimentology, structural geology and tectonics of the Shikoku subduction zone: Geol. Soc. America Bull., v. 87, p. 1259–1268.
———, and others, 1979, Progressive accretion in the Middle America Trench, southern Mexico: Nature, v. 281, p. 633–642.
———, Watkins, J., and others, 1979, The Middle America Trench off Mexico (Leg 66): Geotimes, v. 24, no. 9, p. 20–22.
Murauchi, S. and Ludwig, W. J., 1980, Crustal structure of the Japan Trench: The effect of subduction of ocean crust: Init. Repts. DSDP, p. 463–470.
Nagumo, S., Kasahara, J., and Sadayuki, K., 1980, OBS airgun seismic refraction survey near Sites 441 and 434 (J-1A), 438 and 439 (J-12), and proposed site J-2B: Legs 56 and 57, Deep Sea Drilling Project: Init. Repts. DSDP, v. 57, p. 459–462.
Oshima, S., Kondu, T., Tsukamoto, T. and Onodera, K., 1975, Magnetic anomalies at sea around the northern part of Japan, Report of Hydrographic Researches, no. 10: Tokyo, Hydrographic Office of Japan.
Scholl, D. W., von Huene, R., Vallier, T. L., and Howell, D. G., 1980, Sedimentary masses and concepts about tectonic processes at underthrust ocean margins: Geology, v. 8, p. 564–568.
Schweller, W. J., and Kulm, L. D., 1978, Extensional rupture of oceanic crust in the Chile Trench: Marine Geology, v. 28, p. 271–291.
Seely, D. R., 1979, The evolution of structural highs bordering major forearc basins, *in* Watkins, J. S. and Montadert, L., eds., Geological and geophysical investigation of continental margins: Tulsa, Oklahoma, Am. Assoc. Petroleum Geologists, Memoir 29, p. 245–260.
Seely, D. R., Vail, P. R., and Walton, G. G., 1974, Trench slope model, *in* Burk, C. A., and Drake, C. L., eds., Geology of continental margins: New York, Springer-Verlag.
Shiki, T., Aoki, H., Suzuki, H., Masashino, M., and Okuda, Y., 1974, Geological and petrographical results of the GDP-8 cruise in the Philippine Sea: Marine Science, v. 6, p. 51–55.
Shipley, T. H., McMillen, K. J., Watkins, J. S., Moore, J. C., Sandoval-Ochoa, J. H., and Worzel, J. L., 1980, Continental margin and lower slope structures of the Middle America Trench near Acapulco (Mexico): Marine Geology, v. 35, no. 1/3, p. 65–82.

Uyeda, S., 1978, Active margins actively considered: Geotimes, v. 23, no. 2, p. 27–29.
von Huene, R., Nasu, N., and others, 1978, Japan Trench transected on Leg 57: Geotimes, v. 23, no. 4, p. 16–21.
———, and others, 1980, Leg 67: The DSDP Mid-America Trench transect off Guatemala: Geo. Soc. America Bull., v. 91, p. 421–432.
Yanagisawa, M., Takigami, Y., Ozima, M., and Kaneoka, I., 1980, $^{40}Ar/^{39}Ar$ Ages of boulders drilled at Site 439, Leg 57, Deep Sea Drilling Project: Init. Repts. DSDP, v. 56, 57, pt. 2, p. 1281–1284.

SEPM SPECIAL PUBLICATION NO. 32, P. 67–90, OCTOBER 1981

THE DISTRIBUTION OF MAJOR PELAGIC SEDIMENT COMPONENTS IN THE MESOZOIC AND CENOZOIC NORTH ATLANTIC OCEAN

JÖRN THIEDE, JAN-ERIK STRAND* AND TORLEIV AGDESTEIN*
Department of Geology, University of Oslo, P.O. Box 1047, Blindern, Oslo 3, Norway

ABSTRACT

Ten years of deep-ocean drilling have helped to assemble an enormous and largely untapped body of new data about the evolution of the Mesozoic and Cenozoic oceans. The rapid physiographic-tectonic evolution of the ocean basins and the fluctuations of the earth's climate during the Late Mesozoic and Cenozoic, including the climatic deterioration of the past 30 m.y., led to very important changes in the depositional regimes in the deep oceans. These changes are related to the initiation of a vigorous polar bottom-water formation and the generation of steep zonal hydrographic gradients in the surface water-masses. The effects of these changes on pelagic sedimentation cannot be separated easily from each other, but sedimentation rates of pelagic deposits have been used here to quantify the sediment flux into the Late Mesozoic and Cenozoic North Atlantic Ocean. Phases of high ($>10-$ >50 m/10^{-6}y) sediment input at 140–110 m.y.BP, 55–45 m.y.BP and during the past 25 m.y. alternated with intervals of much lower sediment input. Hiatuses which interrupted the sediment record are most frequent at 2–3 km and 4–5 km paleodepth. Processes such as dissolution or downslope displacement generated these hiatuses. They produced a distinctive depth-related accumulation pattern of the North Atlantic deep-sea deposits during the past 140 m.y. Concentrations of pelagic sediment components (calcareous, siliceous, organic and terrigenous detrital material) have been used to describe the temporal and spatial distribution of North Atlantic deep-sea deposits.

Biogenic components outweighed by far the terrigenous detrital flux throughout the past 140 m.y. Carbonate-secreting organisms made quantitatively the most important contribution to the Late Mesozoic and Cenozoic North Atlantic deep-sea sediments. Two phases of sedimentation of carbonate-rich deposits were separated by an interval between 100 and 80 m.y.BP when $CaCO_3$ particles were deluted by chiefly terrigenous detrital material. Before 110 m.y.BP ago the highest concentrations of calcareous matter were confined to the deepest part of the then 4.0–4.5 km-deep North Atlantic Ocean. Since 80 m.y.BP sediments with high concentrations of calcareous matter were deposited above 3 km paleodepth and also between 4 and 5.5 km paleodepth during the past 25 m.y.BP. The deeper occurrence is associated with indications of downslope displacement of calcareous material into the deep abyssal plains in the North Atlantic.

Siliceous particles are the second most important biogenic components (> 20%), but they were preserved only during two relatively brief intervals, 120–100 and 50–40 m.y.BP. Organic carbon concentrations were high (> 1%) 130–100 m.y.BP, coincident with the older maximum of opaline material, and during the past 5 m.y. when they were confined to the part of the basin deeper than 5 km. The North Atlantic did not receive a very high input of terrigenous material during its early history, but about 100–90 m.y.BP sediments with high concentrations of terrigenous material were shed into the basin. During the past 80 m.y. terrigenous material was essentially confined to the deepest part of the basins.

INTRODUCTION

After more than a decade of deep-sea drilling we are able to unravel the history of most ocean basins with hitherto unknown detail and with quantitative methods which have never been available before. In this study we try to demonstrate how the tectonic evolution of an ocean basin and the global paleoclimate interacted and controlled the abyssal depositional paleoenvironment of a characteristic subbasin of the world ocean during the past 140 m.y. The effects of both processes are not easily separated from each other. The matter is further complicated by the evolution of the main producers of pelagic sediment components, namely planktonic foraminifers, coccolithophorids and diatoms which have only existed since mid-Mesozoic times (Tappan & Loeblich, 1973). Together with radiolarians they produce the bulk of the dominantly biogenic calcareous and siliceous pelagic sediments in the Atlantic Ocean today as they have for the past 140 m.y. As we know from many deep-sea drill sites, the sediment facies changed character frequently suggesting that the flux of sediment particles to the abyss, their mode of preservation in the water column or in the upper sediment layers and the processes displacing sediments in the deep-sea have fluctuated in intensity, time and space.

The reasons for choosing North Atlantic data for this study are several-fold, but the overriding attraction is the good drill-site coverage (Fig. 1, Appendix). The North Atlantic has been visited by a considerable number of cruises of the *Glomar Challenger,* and many of the later sites have excellent coring records. Furthermore it is surrounded almost entirely by passive ocean margins such that

*Present address: Norsk Hydro, Kjørbokollen, 1300 Sandvika, Norway.

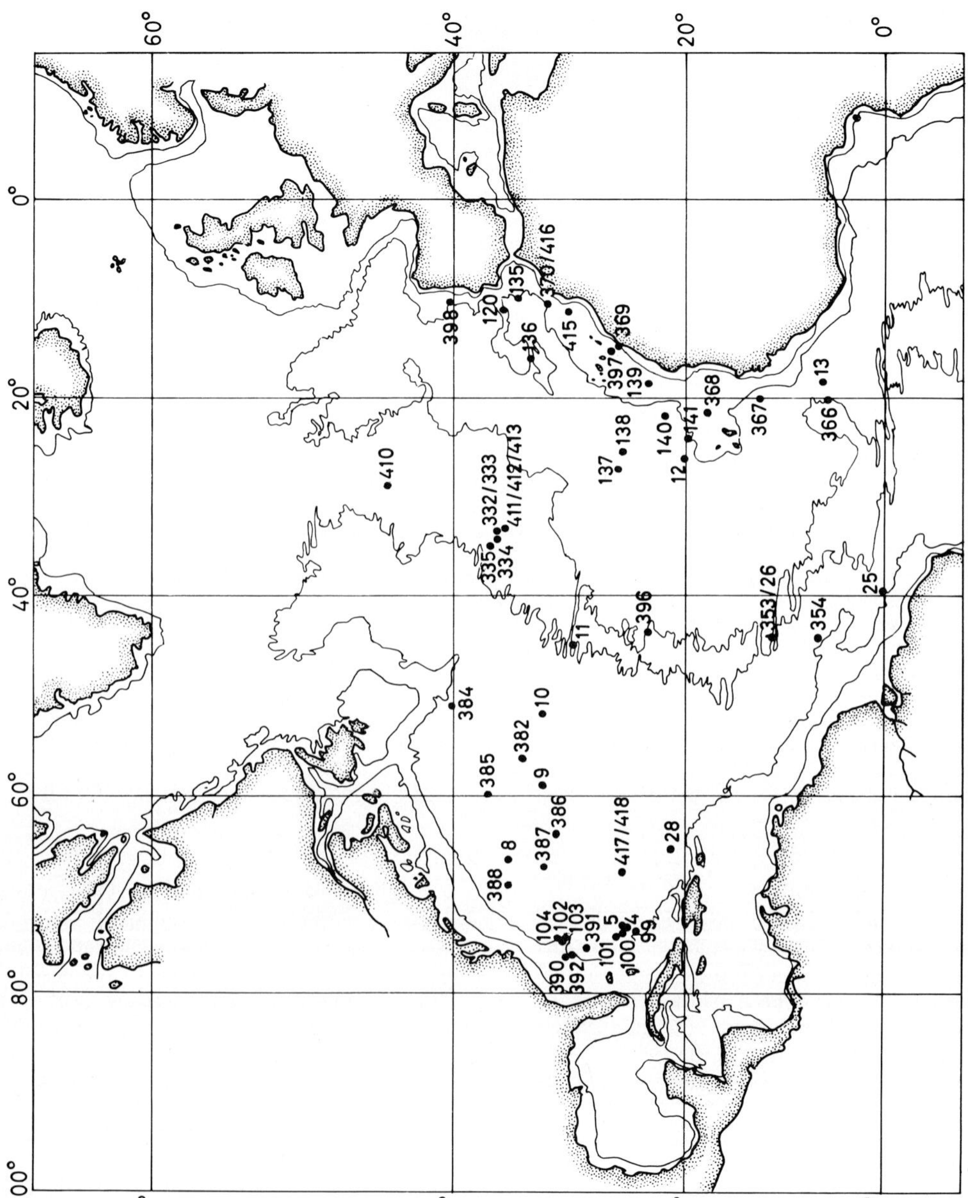

FIG. 1.—Distribution of deep-sea drill sites in the main North Atlantic Ocean.

most of its history has been preserved. The wide range of paleodepths that can be reconstructed at these sites allow us to describe the distribution of pelagic sediment facies and their depositional environment in the main North Atlantic basin during the last 140 m.y. and for the entire water column below 1.5 km paleodepth. The age structure of the oceanic crust under the North Atlantic is well known, and its paleogeographic (Thiede, 1979) and paleobathymetric (Sclater and others, 1977b) evolution have been described in some detail. The sediment data can therefore be placed in a regional context.

Physiographic Evolution of the Upper Mesozoic and Cenozoic North Atlantic Ocean

For more than a hundred years, similarities in the shape of the coastline and of the geological structures on both sides of the Atlantic Ocean suggested that it was a relatively young geological feature that did not exist during the Paleozoic (Snidèr, 1858; Wegener, 1929; Boucot and Gray, 1979). It comprises a complex assemblage of basins with oceanic crust which started to open in the Jurassic (Hayes and Rabinowitz, 1975; Vogt and others, 1971; Vogt and Einwich, 1979), and whose continental margins follow belts of weakness along parts of the Caledonian, Variscan and Alpine orogens. In the North Atlantic a tectonic regime composed of a variable number of plates and spreading axes cut the early Mesozoic Laurasian continent into a complex jigsaw puzzle (Dietz and Holden, 1970). It has therefore proved very difficult to reconstruct and date the original rifting episode through Laurasia (Pitman and Talwani, 1972; Beurlen, 1974; Russel and Smythe, 1978) and to reassemble its various pieces of continental crust.

Linear magnetic anomalies that have been generated parallel to the midocean ridges due to reversals of the earth's magnetic fields (Pitman and Talwani, 1972; Laughton, 1975) permit a reconstruction of the horizontal component of growth in the North Atlantic for the parts of the basin which are underlain by oceanic crust. The oldest magnetic anomalies have been found along the continental margins off northwest Africa and off southeast North America, and are of late Middle or early Late Jurassic age (Hayes and Rabinowitz, 1975). The subsidence of oceanic crust with time and increasing distance from the midocean ridge as a consequence of its thermal history provides the vertical component of the physiographic evolution of an ocean basin (Sclater and others, 1971). These processes explain the position of the Atlantic Ocean's deep basins along the continental rises where the oldest crust has been preserved; under ideal conditions (*e.g.*, in the South Atlantic) they would lead to very simple morphology of the ocean floor. However, in the North Atlantic, especially in the vicinity of the Iceland-Faeroe Ridge, the spreading center has jumped laterally several times (Roberts, 1975; Talwani and Eldholm, 1977) cutting through older oceanic or even continental crust and disturbing its regular subsidence. The Iceland-Faeroe Ridge covers the most important portion of a structural high which separates the Cenozoic Norwegian-Greenland Seas from the deep and old main North Atlantic basin. It was a subaerial ridge little more than 10 m.y.BP (Thiede, 1979).

The evolution of the Mesozoic and Cenozoic North Atlantic from Jurassic to Recent times has gone through a number of specific settings that controlled its hydrographic regime. Evaporites in the Gulf of Mexico, along the northwest African and southeastern North American continental margins (Lehner and de Ruiter, 1977), seem to represent the remains of the first isolated narrow marine troughs that developed along the early rift and into which saltwater from the Mesozoic Pacific and Tethys oceans spilled (Burke, 1975). The oldest part of the main North Atlantic basin opened between Late Jurassic and Early Cretaceous times. The sediments which were deposited during this early stage of the North Atlantic resemble Tethyan Mesozoic lithofacies (Bernoulli and Jenkyns, 1974). It is therefore believed that the Mesozoic North Atlantic basin was connected both in the southwest to the Pacific and in the northeast to the ancient Tethys Ocean. After the connection to the South Atlantic broke through during mid-Cretaceous times (van Andel and others, 1977; Ponte and Asmus, 1978), the North Atlantic became part of a north-south trending ocean basin which was virtually closed to the north, maybe with the exception of a narrow and shallow epicontinental seaway through the Norwegian-Greenland Sea. The Norwegian-Greenland Seas opened in Late Paleocene times (Talwani and Eldholm, 1977). The final subsidence of the Iceland-Faeroe Ridge, and the disconnection from the circumequatorial current system during Miocene or possibly Pliocene times (Keigwin, 1978), resulted in a meridional ocean basin that connects the cold polar water-masses of both the northern and the southern hemispheres. The North Atlantic is an important part of this meridional ocean basin. The hydrography of the dominant portion of the water masses in the entire Atlantic Ocean is therefore to a high degree controlled by climatic conditions in the polar realms.

The opening of the North Atlantic also had profound consequences for the evolution of the surrounding land masses of the Eurafrican and American continents. The large basins which today drain the dominant portion of all continental areas directly or indirectly into the Atlantic cannot have existed prior to the formation of this ocean. Therefore the drainage patterns observed on these land masses may well be directly or indirectly the result of the extensional plate-tectonic movements along the rifting and spreading areas (dominantly meridional structural elements) or of compressional plate tectonic movements along the ancient mobile Tethys belt (dominantly zonal structural elements). Even passive ocean margins may undergo vertical tectonic movements during their evolution from a juvenile rift to the boundary of a mature ocean basin (Curray, 1980). Although it is clear from the establishment of the present drainage pattern around the Atlantic that its continental margins have undergone considerable subsidence, a generally applicable model which would enable us to describe this subsidence quantitatively has yet to be found (Watts and Ryan, 1976; Beaumont, 1978; Pitman, 1978). While marginal seas were of minor importance around the South Atlantic, they have at times covered wide regions of the continents around the North Atlantic and Arctic Oceans. The most important events of continental flooding around this part of the world ocean are the evolution of a circum-Fennoscandian sea in the Late Jurassic, the development of a trans-North American sea during mid-Cretaceous, of a trans-Saharan sea during Late Cretaceous and very Early Tertiary, and of a trans-European sea during Middle Tertiary times (Thiede, 1979).

The above discussion of physiographic-tectonic events illustrate the development of the North Atlantic Ocean from a narrow zonal tropical sea as part of the Tethys Ocean during the Mesozoic, to a wide meridional basin which as part of the Atlantic Ocean connects both polar oceans, and whose water masses carry therefore important hydrographic characteristics from the Arctic and Antarctic realms. With respect to surface water circulation, these events are the establishment of: 1) a marine connection from the northwestern European epicontinental seas-Tethys realm to the Gulf of Mexico-Caribbean area and through that to the Pacific during Early Jurassic, 2) a surface-water pathway to the South Atlantic during mid-Cretaceous, and 3) a seaway through the Norwegian-Greenland Sea to the Arctic Ocean during Early Tertiary, although a narrow and shallow epicontinental sea may have already existed during Mesozoic times. The passage through the Caribbean realm to the Pacific and the gap between North Africa and Southern Europe closed during the Late Miocene and Pliocene. To understand the sediment distribution in this growing ocean basin, it is probably even more important to consider pathways for the deep-water circulation. Deep connections to the Pacific and Tethys oceans probably existed from Late Jurassic/Early Cretaceous to Late Tertiary times. A deep basin connected the North and South Atlantic oceans since the Late Cretaceous, and the deep Norwegian-Greenland Seas broke through to the Arctic Ocean only since Miocene times. It is unclear and probably not resolvable when and where deep and narrow channels through fracture zones allowed an exchange of bottom water masses between the deep eastern and western basins of the North Atlantic Ocean.

Approach to Study the Paleoceanography of a Growing Ocean Basin

The paleoceanography of the Upper Mesozoic and Cenozoic oceans was not only a function of the plate tectonic developments of the past 140 m.y., but it responded as well to the fluctuations of the paleoclimate during the same time span (Berggren and Hollister, 1977; Arthur, 1979), and in particular to the climatic deterioration during the past 30–40 m.y. (Savin, 1977). In reality, it was probably an interaction of both processes which controlled the paleoenvironment. In our attempt to understand the depositional paleoenvironment of the North Atlantic we relied in part on the information stored in the sedimentary sequences which have been drilled and in part on its tectonic history. We also wanted to consider the paleoenvironment of the entire main North Atlantic basin that allowed us only to describe gross changes in depositional regime. This limitation required considerable simplification because the North Atlantic is subdivided into several sub-basins which all had their own history of sedimentation and erosion (Rona, 1973) as well as preservation of important biogenous pelagic sediment components (Berger, 1968). Because of the development of eastern and western boundary current regimes we expected major zonal and meridional asymmetries of sediment distributions (Berger and von Rad, 1972; Thiede, 1979). In an ocean basin whose water column probably was stratified throughout most of its history it seemed to us important to try to understand the distribution of major sediment components in a depth profile rather than in maps; we have therefore plotted most of the sediment data in diagrams displaying age versus paleodepth.

There are several very distinct elements of the depositional environment which have to be considered separately in this attempt to describe the flux of pelagic sediments and their various components to the abyss. Rates at which sediments are accumulating on the ocean floor are the results of a multitude of processes. They are controlled by the flux of sediment particles through the oceanic water column, by the preservation of dissolvable components and by the stability of the sediments once they are resting on the sea floor. Many of these processes operate in certain water depths and in certain regions only, and they can therefore be studied through a relatively simple analysis of the spatial and temporal variations of bulk sedimentation rates in the deep ocean. Long hiatuses interrupt the continuity of the stratigraphic sequences in many areas of the North Atlantic (Rona, 1973; Moore and others, 1978). We have integrated across hiatuses which resulted in very low linear sedimentation rates in certain depths and age intervals. Their temporal and spatial distribution is therefore an important variable in a study of the sediment flux in the Upper Mesozoic and Cenozoic North Atlantic.

Pelagic sediments which cover most of the world ocean floor consist in many areas of dominantly biogenic calcareous and siliceous particles that have been secreted by planktonic living algae (diatoms, coccolithophorids) and protozoans (radiolarians, foraminifers). With exception of the radiolarians, these organisms are believed to have populated the oceanic surface water masses since mid-Mesozoic time (Tappan and Loeblich, 1973). The development of these plankters in a paleoenvironment which had previously lacked carbonate secreting plankton and which had supported relatively poor siliceous faunas, resulted in many areas not only in a drastic change of the composition of deep-sea sediments, but also to have altered the chemistry of the oceanic water masses (Broecker, 1974). We have therefore plotted the age versus paleodepth distributions of calcareous and opaline fossil components in the North Atlantic during the past 140 m.y., and we have tried to analyze these data for the effects of the primary input which have survived the alterations due to differential preservation and to downslope displacement. The concentrations of opaline fossils allowed us to describe the fertility of the surface water masses in a very gross way (Broecker, 1974; Heath, 1974). Sediments with relatively high organic carbon concentrations required a paleoenvironment favorable for their preservation (Tissot and others, 1979). In a pelagic environment processes controlling the primary inputs of sediment particles are expected to be traceable across the entire water column. Hydrographic properties of the different parts of the oceanic water column control preservation and displacement and generate horizontal zonations.

Almost all detrital terrigenous material has to enter the oceans across the continental margins, either as fine-grained airborne dust or as suspensions from the drainage basins and from the coastal areas of the surrounding land masses. The Sahara is the only extensive desert bordering the North Atlantic, but it did not develop before Late Miocene time (van Zinderen Bakker, 1978). Probably most of the detrital terrigenous components therefore entered the North Atlantic in aqueous suspensions except during the past few million years. As outlined above, the size of the drainage basins together with many other factors is also a measure of the age of the passive ocean margins which separate the continents from the ocean and which in the case of the main North Atlantic basin are probably of Jurassic age. The amount of erosional products that are delivered from the continental surfaces are believed to be directly dependent on the size of the area available for the removal of this material (Hay and Southam, 1977). Age and depth distribution of concentrations of terrigenous detrital material in pelagic sediments enable us to at least indirectly make an estimate of when the most intensive periods of basin formation occurred around the North Atlantic.

METHODS AND DATA

The information used in this study has been extracted from the Initial Reports of the Deep-Sea Drilling Project, from the Geotimes-reports of legs 1–4, 11, 13–14, 37, 39, 41, 43–45, 47, and 49–52, and from the DSDP data bank at Scripps Institution of Oceanography. We have tried to select only sites whose paleodepth of deposition could be described with reasonable certainty (Fig. 1, Appendix). The reconstruction of this important variable has been carried out following the backtracking method (Berger, 1972; Berger and von Rad, 1972; van Andel and others, 1975) which required the depth and age of oceanic basement, as well as thickness and age of the overlying sediments, to be known. In selecting the drill sites we tried to obtain a reasonably even regional coverage of the eastern and western North Atlantic basins, though this proved difficult because of the bias of the much denser drill-site coverage of the western North Atlantic. The comparison of the western and eastern basins of the North Atlantic also revealed major asymmetries in the distribution of their sedimentary facies due to the differences in surface and bottom water circulation (Thiede, 1979). These asymmetries, and the occurrence of hiatuses that last much longer in the western than in the eastern North Atlantic (Rona, 1973), contribute to the fact that only very broad trends of the sediment distributions can be discussed here.

Information about age and composition of the sediments has also been extracted from the site reports of the various deep-sea drill sites. Because the time-scales and biostratigraphies used can vary from author to author, we have redrawn all data following the time-scales of Berggren (1972) and van Hinte (1976a, b). Subsidence curves for all drill sites were then plotted in diagrams of age (in m.y.BP) versus depth (paleodepth of deposition, in km).

The following data have been used here to describe the Upper Mesozoic and Cenozoic paleoenvironment of the North Atlantic Ocean (Table 1):

TABLE 1.—Availability of compositional data from deep-sea drill sites in the main North Atlantic basin (x = data available).

Site No.	$CaCO_3$(%)	Org.C. (%)	Biog. Opal (%)	Terr.(%)
4,4A		x		
5,5A	x	x	x	x
8,8A	x	x		x
9,9A	x	x	x	x
10	x	x	x	x
11,11A		x	x	x
12	x		x	
13	x			
25,25A	x	x	x	x
26A	x	x		
28		x		
99A	x	x		
100	x	x		
101A	x	x		
102	x			
103	x	x		
104	x	x	x	
120	x			
135	x			
136	x		x	
137	x		x	
138	x			
139	x	x	x	
140	x		x	
141	x		x	
332A	x	x	x	x
333	x	x	x	x
334		x	x	x
335		x	x	x
353	x	x		
354	x	x	x	x
366	x		x	
367	x			
368	x			
369	x			
370	x			
382	x	x	x	x
384		x	x	x
385		x	x	x
386	x	x	x	x
387	x	x	x	x
388A	x	x		
390,390A	x	x		
391A,391C	x	x		
392,392A	x	x		
396	x			
397,397A	x		x	
398,398D	x	x	x	
410			x	x
411A		x	x	
412		x	x	
413	x			
415,415A	x	x	x	
416A	x	x	x	
417A	x	x	x	x
417D	x	x		x
418A,418B	x	x	x	x

linear sedimentation rates and hiatuses, concentrations of calcium carbonate, organic carbon, siliceous microfossils, and terrigenous detrital components. These variables have been measured or estimated by a wide variety of methods, most of which are documented in detail in the Initial Reports of the Deep-Sea Drilling Project of the respective DSDP legs. Inconsistencies in the data coverage due to changing procedures between early and late DSDP legs meant that the distributional patterns of some of the sediment components could only be described for the western main basin of the North Atlantic.

To estimate the rate at which sediments accumulated the measured length of section was divided by the time required for its accumulation. This value has then been termed the linear sedimentation rate (either in cm 10^{-3} yr or in m 10^{-6} yr). Because the lower portions of sedimentary sections have undergone compaction to one-half or one-third of their initial thickness it seemed problematic to compare linear sedimentation rates of near-surface and deeply-buried sections (van Andel and others, 1975). But it has recently been shown (Worsley and Davies, 1979) that gross distribution patterns of sedimentation rates uncorrected for compaction, and rates which expressed the accumulation of sediments in terms of mass per unit area and per unit time are reasonably compared as long as the range of linear sedimentation rates exceeded the effect of loss of porosity with depth. Sedimentation rates of all sites which have been calculated for each 2.5 m.y. interval, and which appear in rows of data points along the individual subsidence curves (Fig. 2), were then plotted into a composite diagram.

Hiatuses have interrupted the continuity of the sedimentary record of the North Atlantic in certain age (Moore and Heath, 1977; Moore and others, 1978) and depth intervals (Thiede and others, 1980a) such that the temporal and spatial data coverage remain fragmentary. Because linear sedimentation rates are also interpolated across hiatuses, we have made a special effort to analyze their spatial and temporal distribution in the western North Atlantic (Fig. 3) where they were best developed (Rona, 1973). Because we wanted to study the gross sedimentation rate patterns we have only considered hiatuses which exceeded 5 m.y. in length (Table 2), and we have tried to locate the intensity of processes generating hiatuses with respect to their frequency, in both time and space. We have counted how many of the sample points displayed in Figure 2 fell into a hiatus, and we have calculated the proportion of the total number of sample points per depth (200 m paleodepth intervals) and time interval (Fig. 3, Table 3) to assess how large a proportion of the North Atlantic sediment record was able to survive (Moore and Heath, 1977).

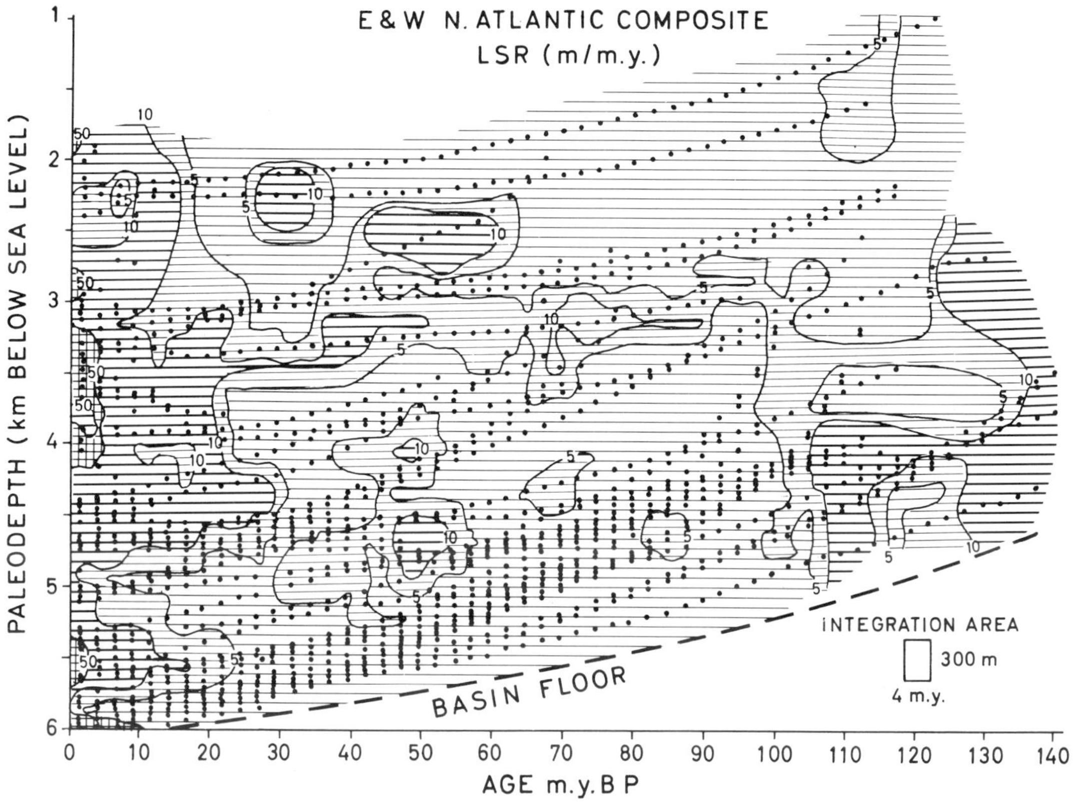

FIG. 2.—Distribution (age versus paleodepth) of linear sedimentation rates (LSR, in meters per million years = m 10^{-6} yr) in the Upper Mesozoic and Cenozoic North Atlantic Ocean during the past 140 m.y.

The measurements of the calcium carbonate concentrations (%) have been carried out by a variety of methods which allow us to determine the carbonate concentrations of the bulk sediment with only limited precision. However, for the purpose of this study the accuracy of the estimates and their comparability was considered to be sufficient. Most of the measurements have been carried out by means of the LECO WR-12 analyzer or the carbonate bomb (Müller and Gastner, 1971). From many cores numerous determinations of the carbonate concentrations are available. In these instances we used their average. However, there are also many depth intervals where the data coverage was not as good. In general, we have tried to obtain at least one measurement per section (corresponding to 1.5 m length) of each core (corresponding to 9.5 m of the penetrated sedimentary column) of the sites. Organic carbon concentrations (%) have been determined by means of the LECO WR-12 analyzer after acid treatment of the bulk sediment to remove the carbonate.

The weakest data used in this study were probably the visual estimates of concentrations of biogenic opal, and of terrigenous detrital components which have been taken from the smear-slide descriptions. Being aware of the variable and questionable quality of these data we have tried to contour at widely-spaced intervals. The most important observation in favor of these data was the fact that although originating from separate legs (hence studied by different colleagues) the data from different drill sites showed good agreement, in age as well as in paleodepth. Biogenous opal refers to the composite contents of radiolarians, diatoms, sponge spicules, silicoflagellates and chert in the smear slides. Terrigenous components resulted from summing the concentrations of all detrital minerals (including unspecified clay minerals).

The sedimentation rates and the compositional data have all been plotted into age versus paleodepth diagrams (Figs. 2, 4–7). Where available the data points fell on the curve describing the paleodepth of deposition of the single sites. Because of the numerous data points, the large scatter of values and the fact that we wanted to elucidate the broad sedimentation rate patterns rather than regional details, we smoothed the data over 4 m.y.

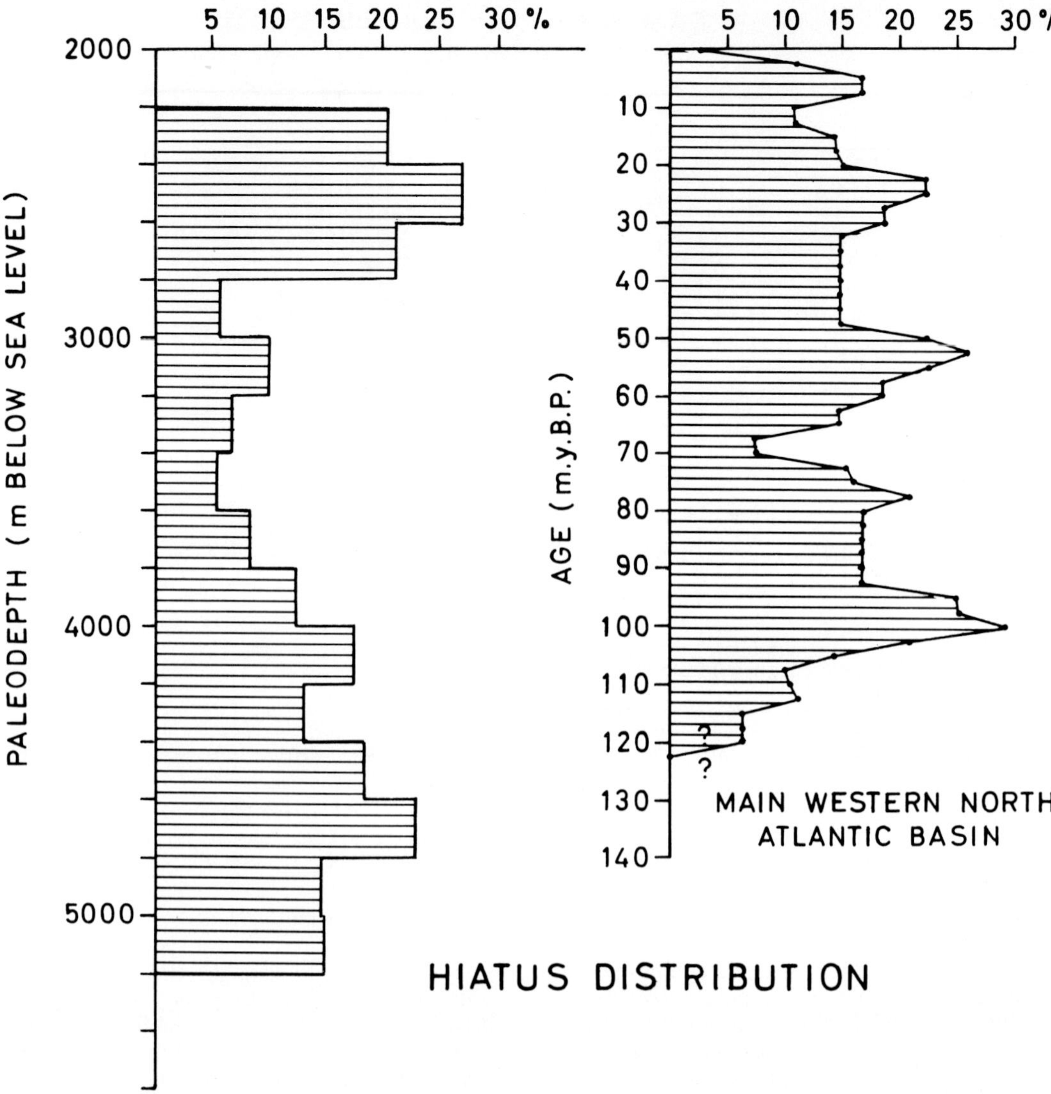

FIG. 3.—Temporal and spatial distribution of hiatuses in the western North Atlantic during the past 140 m.y.

and 300 m depth intervals after which the diagrams have been contoured.

TEMPORAL AND SPATIAL DISTRIBUTION OF LINEAR SEDIMENTATION RATES AND HIATUSES

Oceanic sedimentation rates exceeding 5-10 m 10^{-6} yr were an exception during the Cenozoic (Davies and others, 1977), and probably also during the preceding part of the late Mesozoic that could be documentated with pelagic sediments. Linear sedimentation rates in the North Atlantic were high ($\geq$ 10 m 10^{-6} yr) during three intervals in the past 140 m.y.: 140–105, 60–40 and less than 30 m.y.BP. The dramatic increase of sediment input in the Pacific, Indian and Atlantic oceans during the past 15 m.y. (Davies and others, 1977; Worsley and Davies, 1979) is reflected in these data (Fig. 2) because sedimentation rates exceeded 10 m 10^{-6} yr over most of the North Atlantic Ocean from 15–25 m.y.BP; but during the past 3–4 m.y. they rose in many areas above 50 m 10^{-6} yr.

The distribution of linear sedimentation rates in the North Atlantic Ocean during the past 140 m.y. was very inhomogenous, with respect to both time and paleodepth of deposition. Besides a number of

TABLE 2.—Distribution and length of major hiatuses (>5 m.y. duration) in the main western basin of the North Atlantic Ocean.

Site No.	Time interval of hiatuses (m.y.BP)
4,4A	95–100
5,5A	
8,8A	
9,9A	
10	4–21.5, 22–30
11,11A	
25,25A	
26A	
28	
99A	6–120
100	
101A	13–103
102	
103	1–8
104	1–8
333	
334	
335	
353	
354	
382	
384	49.5–56
385	
386	22–26, 94–100
387	45–50
388	
390,390A	0–45, 51–56, 57–62, 70.5–108
391C	1.8–5, 22.5–66
392,392A	71–104
395	
410	
417A	49–66
417D	99–106
418A	109–113
418B	

TABLE 3.—Paleodepth distribution of hiatuses in the western basin of North Atlantic Ocean

Paleodepth (km)	% Hiatus
2.2–2.4	20.5
2.4–2.6	26.9
2.6–2.8	21.1
2.8–3.0	5.6
3.0–3.2	10.0
3.2–3.4	6.8
3.4–3.6	5.4
3.6–3.8	8.2
3.8–4.0	12.1
4.0–4.2	17.4
4.2–4.4	13.0
4.4–4.6	18.4
4.6–4.8	22.9
4.8–5.0	14.7
5.0–5.2	15.0
5.2–5.4	0.0
5.4–5.6	0.0

temporal fluctuations which were described above, the most important feature was a clear quasi-horizontal stratification. A belt of high sedimentation rates could be traced at 3–4 km paleodepth almost through the entire history of the North Atlantic basin, with exception of the first and the last 30 m.y. when sedimentation rates decreased towards shallower as well as towards deeper waters to values < 5 m 10^{-6} yr. During the first 30 m.y. high sedimentation rates of an almost purely calcareous material (Fig. 4) were confined to the part of the basin deeper than 3 km paleodepth. During the past 30 m.y. sedimentation rates were high throughout the upper 4.8 km of the basin, but they were reduced at 4.8–5.2 and 5.7–5.9 km paleodepths. The latter two intervals over- and underlied a region of high sedimentation rates of probably largely displaced calcareous deposits (Thiede and others, 1980b). This pattern of sedimentation rates was considerably different from South Atlantic Cenozoic distributions, which suggested a relatively simple increase and deepening of the isopleths as the crust subsided and moved away from the ridge crest (Berger, 1972), probably because the depth interval that could be studied with the South Atlantic samples was much narrower. The early phase of high sedimentation-rates coincided with the deposition of high concentrations of calcareous and siliceous biogenic components as well as of organic matter; the second one coincided with high concentrations of calcareous and siliceous components. They were not correlated with the influx of terrigenous particles, which remained modestly small until 105–110 m.y.BP. During the past 80 m.y. important concentrations of terrigenous material were confined to the deep North Atlantic basins.

As we know from global compilations of the record of hiatuses in the ocean basins (Moore and others, 1978), the intensity of these processes fluctuated with time on a global scale, and on ocean-basin-wide and local scales. In the western North Atlantic (Fig. 3, Table 2), where hiatuses lasted longer and where they were more widely distributed than in the eastern basin (Rona, 1973), erosional processes were most active 90–100 m.y.BP ($\geq 30\%$), around 50 m.y.BP (25–30%), 20–30 m.y.BP (20–25%) and 5–7 m.y.BP (15–20%). Phases when hiatuses were rare have been found during the past few million years, around 70 m.y.BP, and during the first 10 m.y. of western

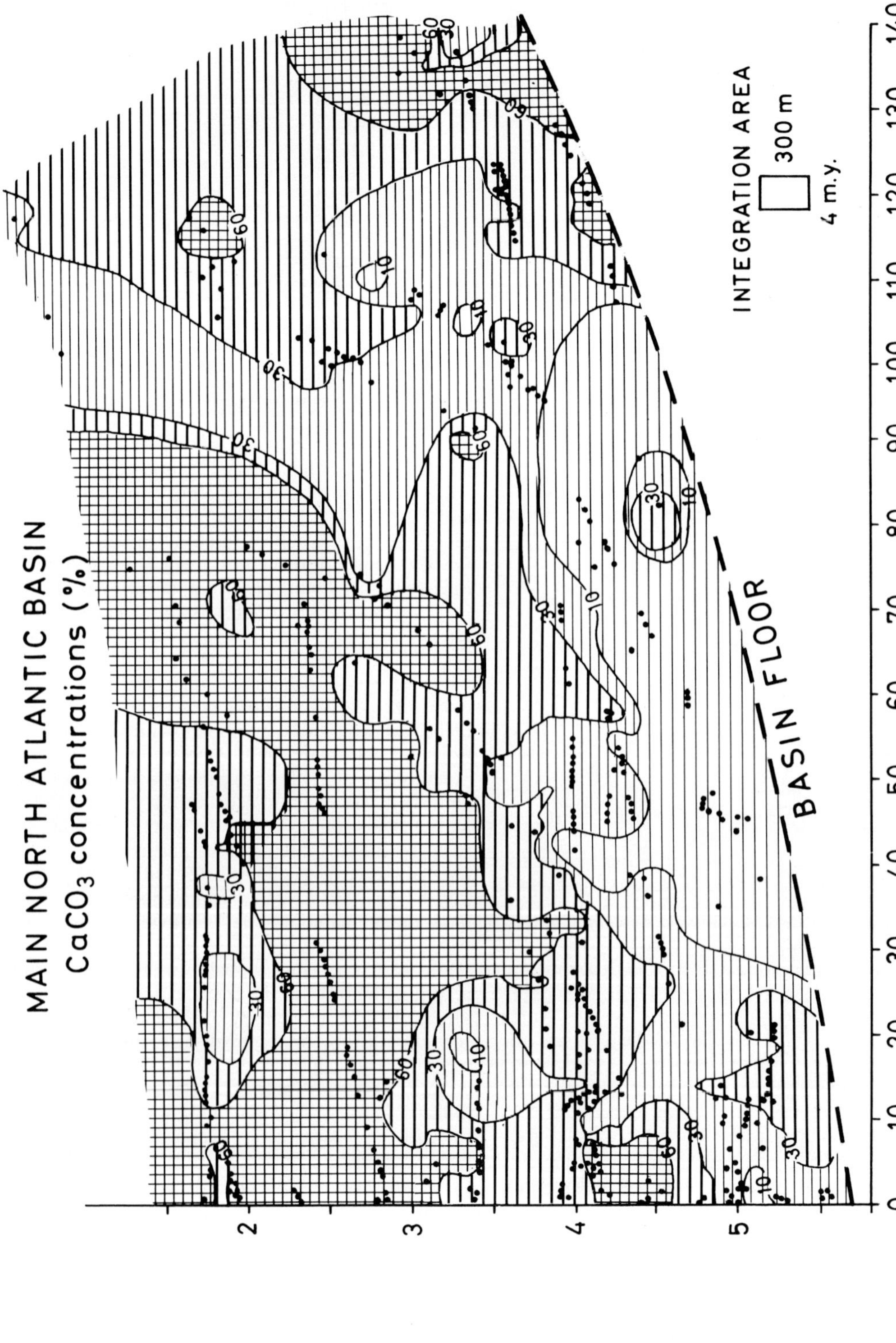

FIG. 4.—Distribution (age versus paleodepth) of calcium carbonate (dominantly calcite) concentrations (in weight percent of the bulk sediment) in the main North Atlantic basin.

North Atlantic history, although it seems doubtful that the data from the latter interval were representative because of the few data points available. These former two minima coincide well with South Atlantic data (van Andel and others, 1977).

To characterize the spatial distribution of hiatuses we have chosen to reconstruct their frequency versus paleodepth rather than to try to map their regional extent (Moore and others, 1978), because water columns of subpolar to tropical oceans (which the North Atlantic was during the past 120 m.y.) tend to be more-or-less intensively quasi-horizontally stratified. The region shallower than 2 km paleodepth and deeper than 5.6 km paleodepth could not be analyzed because of insufficient data points (Fig. 3). Processes generating hiatuses were most intensively at work at 2.2–2.8 km paleodepth (25–30%) and at 3.8–5.2 km paleodepth (25–30%), while both intervals enclose a well-defined minimum where the hiatus frequency decreased to approximately 5%. Again, there is no depth interval where no part of the sediment record has been removed, maybe with the exception of the lowermost few hundred meters of the basin during the past 5 m.y., which received a very high input of displaced calcareous sediment and where the bulk of the terrigenous material was accumulating (Thiede and others, 1980a, 1980b). There was virtually no phase of the past 120 m.y. of North Atlantic history that was free of hiatuses. On the contrary, this ocean basin received continuously a certain input of sediment, either as suspended material from the surrounding land masses and adjacent basins, or as biogenic particles which represented the transformation of chemical components from a dissolved (= nutrients) to a discrete form (mainly skeletal particles composed of silica or calcium carbonate). But chemical (Berger and Winterer, 1974) and mechanical (Seibold, 1978) processes which redistributed this sediment seemed continuously at work. An average of 16% of the past 120 m.y. of North Atlantic history is not represented by a sediment record.

TEMPORAL AND SPATIAL DISTRIBUTION OF MAJOR SEDIMENT COMPONENTS

Calcareous Components

Calcium carbonate particles were the single most important sediment component in the Upper Mesozoic and Cenozoic North Atlantic Ocean. However, their temporal and spatial distribution was much more complicated than originally anticipated. During this time span the North Atlantic grew from a relatively narrow zonal ocean into a broad, elongate latitudinal ocean basin (Sclater and others, 1977a; Thiede, 1979) whose maximum depth increased from about 4 km to more than 5.5 km paleodepth, as schematically indicated in Figure 4. Two phases of high carbonate concentrations can be observed during the past 140 m.y., the first one lasting until approximately 100 m.y.BP, the second one beginning somewhere between 70 and 80 m.y.BP and lasting until modern times.

The highest carbonate concentrations during the early phase of deposition of carbonate-rich sediments were confined to the deepest part of the basin, although the sediments deposited at higher paleodepth levels had high calcium carbonate contents too. The distributional pattern during the later phase was radically different from the above mentioned early phase, because the deepest part of the ocean was covered with sediments relatively poor in $CaCO_3$ and carbonate-rich sediments were confined to paleodepths shallower than 3.5 km. This trend is considerably more complicated during the Neogene when sedimentation rates in the ocean basins rose dramatically (Davies and others, 1977). During the last 20 m.y., carbonate concentrations have remained high in the upper 3.5 km of the basins, but they were depressed at depth levels around 3.5–4 km and 5 km paleodepths. Carbonate-rich deposits that occur below 4 km paleo-water depths are commonly confined to parts of the sedimentary columns which are characterized by indications of redeposition.

Biogenic Opal

The opal in pelagic sediments has originated dominantly from concentrations of siliceous microfossils that were either preserved in the sediments or that have been dissolved and whose silica has been precipitated to produce chert. Except under very specific oceanographic features of the surface water-current regimes which generated nutrient-rich surface waters with a high productivity (Lisitzyn and others, 1967), the number of siliceous skeletal components which survived dissolution on their way to the sea floor was so small that most areas of the world ocean were covered by opal-free or opal-poor sediments. It has been estimated that only 4% of the original input reached the sea floor, and that only 2% escaped postdepositional dissolution (Heath, 1974).

The Atlantic surface sediments have been depleted in silica (Broecker, 1974) because the Atlantic has lost silica-enriched deep-water to the other oceans (lagoonal model of water exchange; see Berger, 1970). It was therefore surprising to note that during the past 120 m.y. the North Atlantic has twice gone through phases of high input and preservation of siliceous microfossils which resulted in high concentrations of opal in its pelagic sediments (Fig. 5). The early phase lasted approximately from >120–100 m.y.BP, the later from 40–50 m.y.BP. Sediments with high opal contents during these intervals have been found distributed over a wide

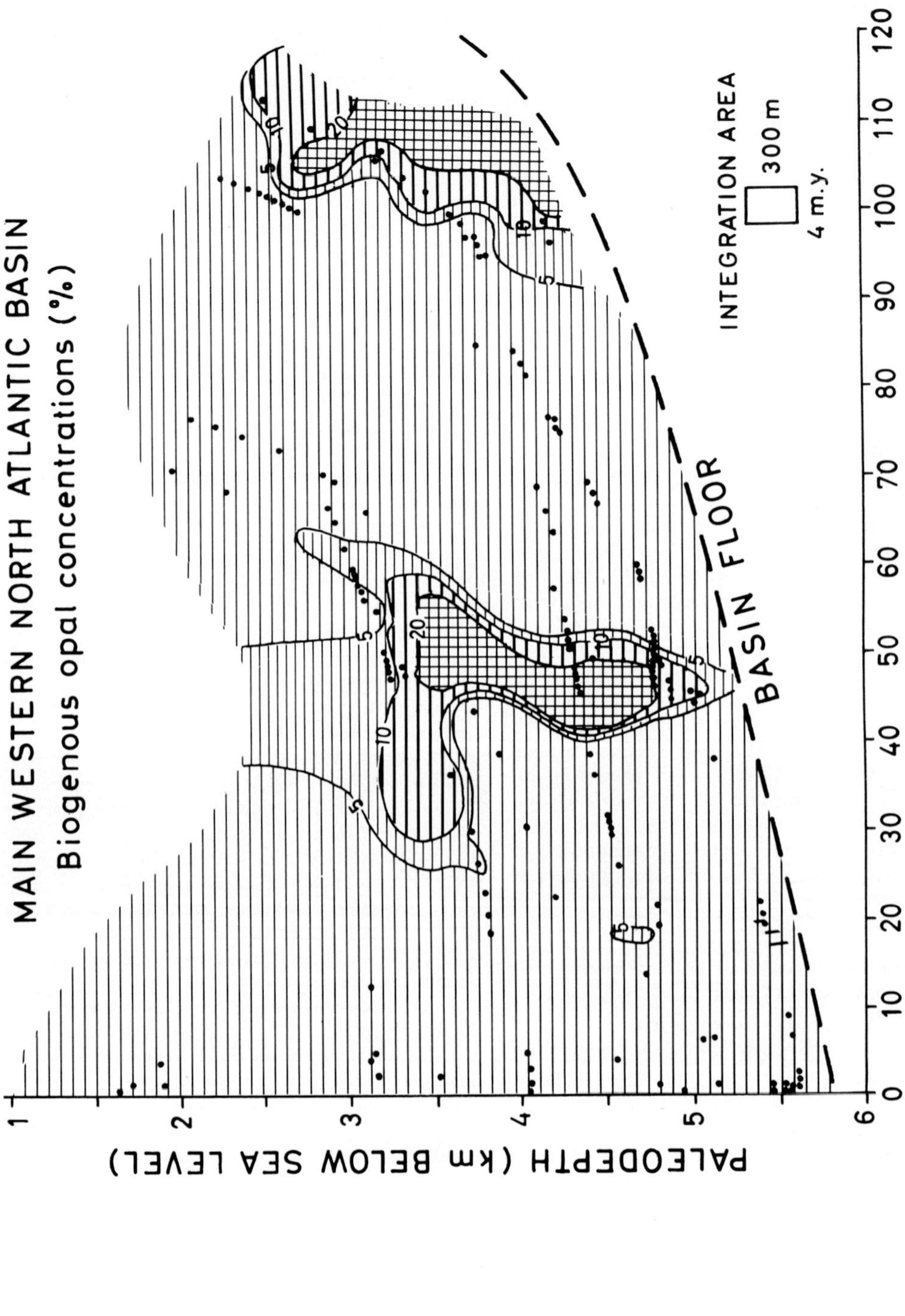

FIG. 5.—Distribution (age versus paleodepth) of biogenous opal concentrations (in weight percent of the bulk sediment) in the main western North Atlantic basin.

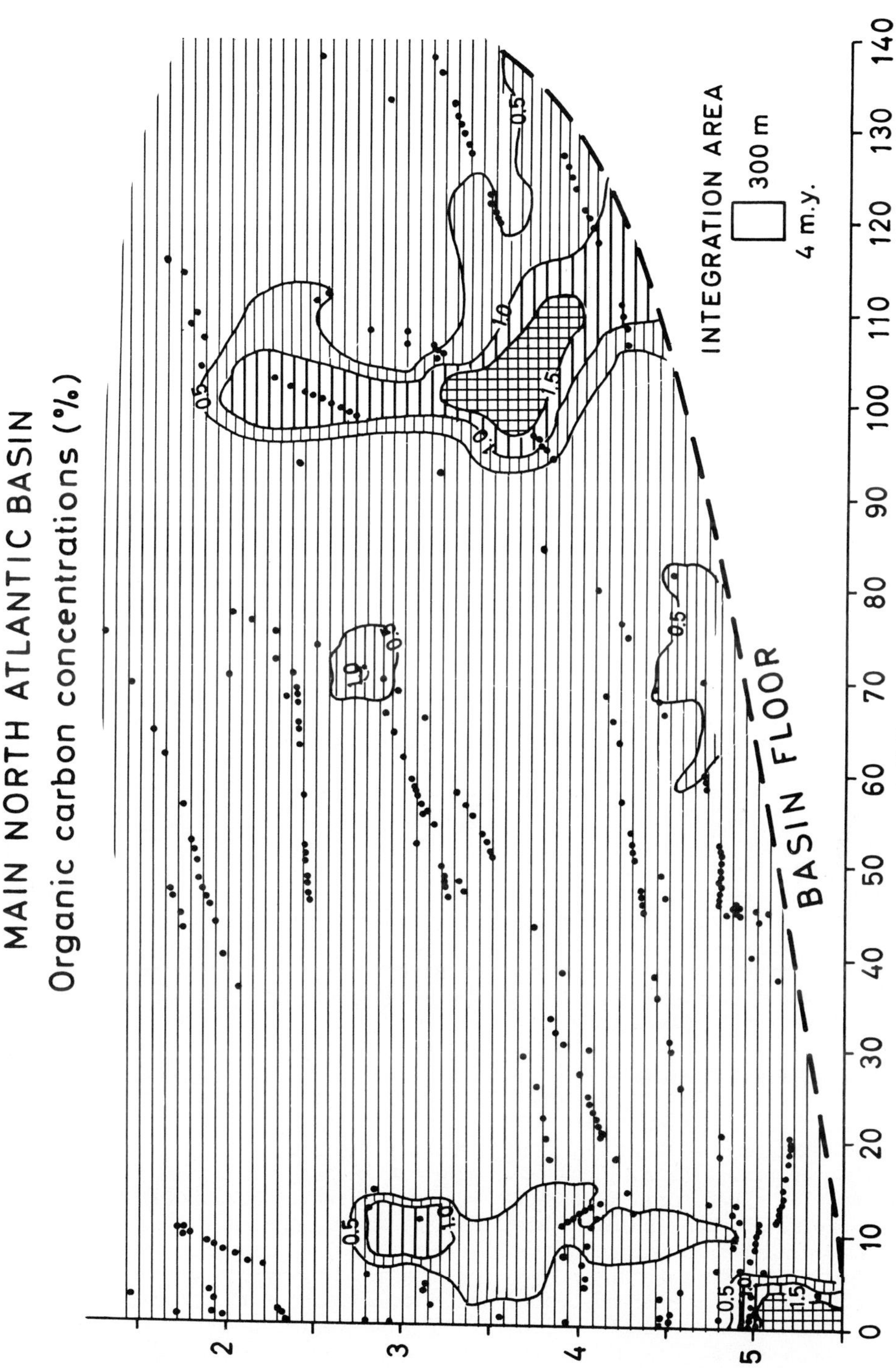

FIG. 6.—Distribution (age versus paleodepth) of organic carbon concentrations (in weight percent of the bulk sediment) in the main North Atlantic basin.

range of water depths, as one could expect from truly pelagic sediment components, but the concentrations seem to decrease in the upper 2.3 km of the water column (Fig. 5). Both intervals of opal deposition coincided with high linear sedimentation rates in the North Atlantic (Fig. 2); the younger one also coincided with unusually high opal accumulation-rates in the tropical Pacific (Leinen, 1975). While the latter example was believed to be due to an intensification of the Pacific equatorial current regime (van Andel and others, 1975; Leinen, 1975; Leinen and Stakes, 1979), resulting in nutrient-rich surface water masses with a high productivity, this explanation could not easily be applied to a meridional basin such as the Atlantic Ocean.

Organic Carbon

Most of the organic carbon in the oceans is produced in or near the photic zone of the surface water masses, but it never reaches the sea floor because it is decomposed biologically or chemically within the water column. Organic carbon concentrations in pelagic sediments are therefore in general <0.5% of the bulk sediment (Heath and others, 1977a). Considerably higher concentrations of organic carbon are today preserved only in a few regions where part of a high input survived oxidation because of the development of intensive midwater oxygen minima. Good examples for such organic carbon-rich lithofacies in the oceans have been found along the continental margins around the northern Indian Ocean (Stackelberg, 1972), in the Gulf of California (van Andel, 1964), or under regions of intensive coastal upwelling along the western continental margins (Seibold, 1979).

The North Atlantic deep-sea sediments which have been deposited during the past 140 m.y. were therefore also depleted of organic carbon except during an interval of about 10 m.y. length in mid-Cretaceous time (approximately 100 m.y.BP) and during the past 4.5 m.y. (Fig. 6). Both occurrences are of widely different character because the early one can be traced as a broad band across the deepest 3 km of the North Atlantic basin while the latter one is confined to water depths >5 km in the deepest part of the abyssal plains where no or little carbonate is deposited (Fig. 4). We know from detailed core descriptions that the mid-Cretaceous depositional phase of organic carbon-rich sediments was composed of many relatively short events of oxygen deficient paleoenvironments in the deep North Atlantic (Jansa and others, 1978; McCave, 1979). The penecontemporaneous deposition of similar lithofacies in the mid-Cretaceous South Atlantic, Indian and Pacific Oceans (Schlanger and Jenkyns, 1976; Thiede and van Andel, 1977; Arthur and Natland, 1979; Vallier and others, 1979; Dean and others, *in press*) called for a global and probably paleoclimatic cause for the preservation of the high organic-carbon concentrations of these sediments, regardless of whether this happened from the influence of a midwater oxygen minimum creating euxinic conditions, or from locally oxygen-deficient bottom waters (Thiede and others, *in press*). The high organic-carbon concentrations deposited in the abyssal plains during the past 3–5 m.y. coincided with an interval of very high sedimentation rates (Fig. 2) of dominantly fine-grained terrigenous muds (Fig. 7).

Terrigenous Components

Detrital terrigenous grains or particle assemblages in pelagic sediments bear some important information about the nature and paleoclimate of the source area, about the transport agent from the source area to the ocean and about processes distributing them within the ocean. When plotting the concentrations of terrigenous detrital material in the Upper Mesozoic and Cenozoic North Atlantic into an age versus paleodepth diagram (Fig. 7) their great spatial and temporal variability came as a surprise. Their distributional pattern with time and depth could be subdivided into three different intervals: 1) 100–120 m.y.BP (a relatively poorly documented interval), when almost no terrigenous grains reached the North Atlantic, particularly not the bottom of the then only 3.0–4.5 km-deep ocean basin; 2) 70–100 m.y.BP, when sediments with high concentrations of terrigenous detrital material were deposited throughout the entire water column; and 3) the past 70 m.y., when high concentrations of detrital terrigenous components were confined to the deepest part of the ocean basin, although the interface between dominantly biogenic sediments fluctuated between 4.5 and 5.5 km paleodepth.

It was interesting to compare these distributions with those of the other major pelagic sediment components. High concentrations of terrigenous detrital material did not coincide with high linear sedimentation rates (Fig. 2), neither paleodepth- nor age-wise, but more detrital material seemed to have reached the ocean during periods of intensive bottom erosion (Fig. 3). The detrital material entering the North Atlantic 70–100 m.y.BP diluted the calcareous components so that the depositional history of the pelagic carbonates divided into two phases. High concentrations of detrital components coincide with a high input of neither opaline components (Fig. 5) nor organic carbon (the exception of the area >5 km water depth during the past 5 m.y.; see discussion above). Also, during the past 50 m.y. the Norwegian-Greenland Sea received a dominantly terrigenous input (Thiede, 1979), while this component in most areas of the North Atlantic was masked by biogenic and dominantly calcareous particles.

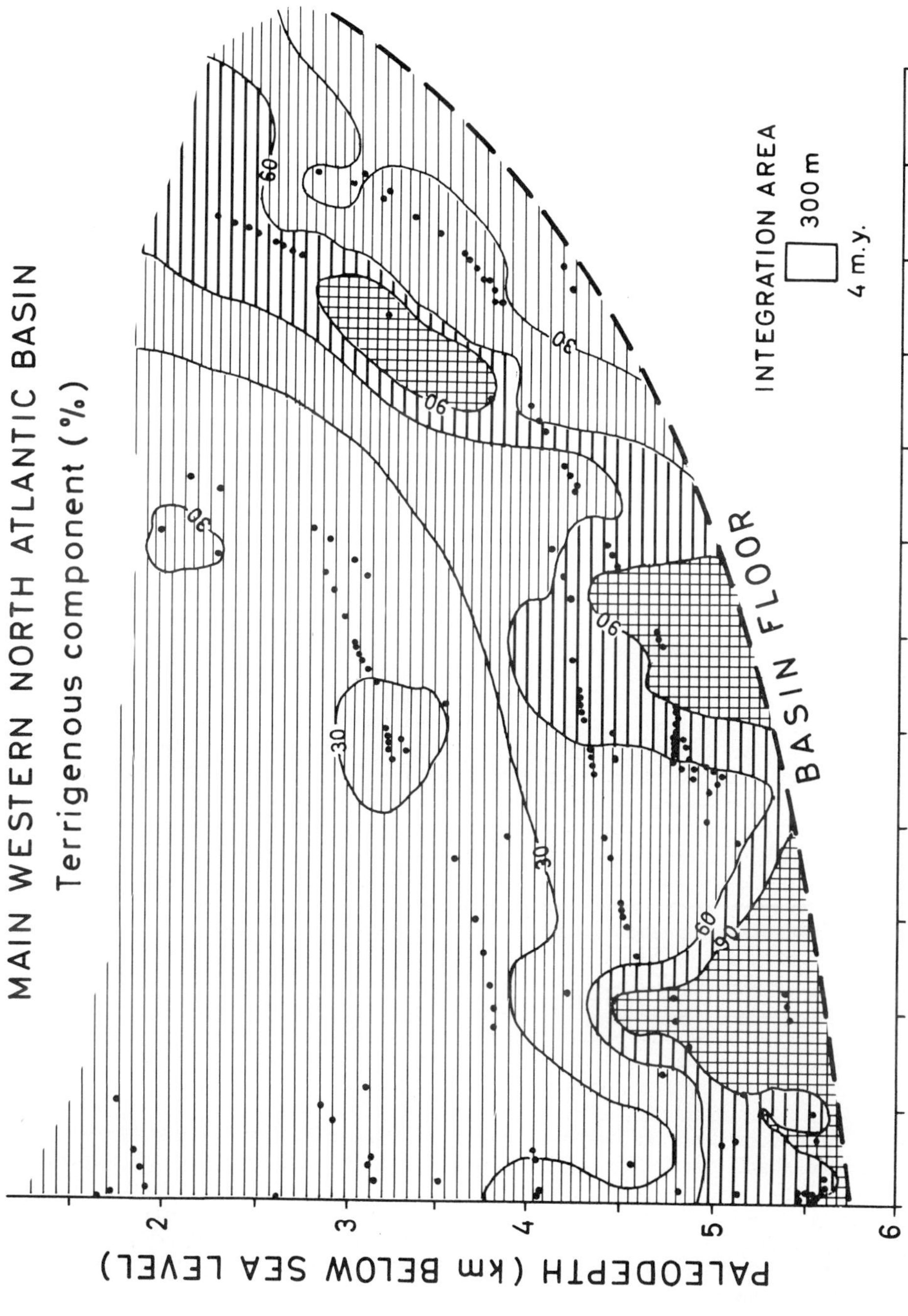

FIG. 7.—Distribution (age versus paleodepth) of terrigenous components (in weight percent of the bulk sediment) in the main western North Atlantic basin during that past 120 m.y.

ASPECTS OF THE NORTH ATLANTIC PALEO-OCEANOGRAPHY DURING THE PAST 140 M.Y.

The data described above allow us to quantify the sediment flux and its composition to the upper Mesozoic and Cenozoic North Atlantic Ocean. Using these data we have tried to discuss some of the processes which controlled the sediment flux and preservation of the depositional record of a growing ocean that was surrounded by passive ocean margins (Curray, 1980). The North Atlantic developed from a relatively narrow and immature ocean, as part of the zonal Mesozoic circum-equatorial ocean, into a meridional basin as part of the Atlantic Ocean connecting the polar regions of both the northern and the southern hemispheres. The depositional regime of the North Atlantic responded to a complicated and changing network of boundary conditions of the global paleoclimate (Savin, 1977; Shackleton, 1978), and resulted in the sedimentation rates and the distributional patterns of the major deep-sea sediment components described above, and the paleogeographic and paleobathymetric evolution of this ocean (Thiede, 1979). The paleodepth and age distribution of the variables used in this study allow us to describe qualitatively properties of the surface, intermediate and bottom water masses of the main North Atlantic basin during the past 140 m.y.

Sedimentation rates determined for fossil deep-sea deposits result from many constructive and destructive depositional and postdepositional processes. The two Cenozoic phases of high sedimentation rates in the North Atlantic (Fig. 2) largely reflected the flux of biogenic calcareous and siliceous skeletal particles to the ocean floor. They coincided with simultaneous sedimentation-rate increases in the South Atlantic (van Andel and others, 1977), the equatorial Pacific (van Andel and others, 1975) and in the world ocean in general (Davies and others, 1977). The sedimentation-rate increases in the Upper Mesozoic North Atlantic were also caused by biogenic particles. We were surprised to discover that pelagic sediments in an ocean basin such as the North Atlantic, which is almost entirely surrounded by continents and which received the dominant share of the continental fresh-water runoff from the vast and probably old drainage basins on the adjacent land masses, were dominated by calcareous and siliceous biogenic sediment components rather than by terrigenous detrital material during the past 140 m.y.

The major temporal changes in the sedimentation rates can be best explained as the result of fluctuations in the input of the dominant sediment components. However, the linear sedimentation rates (Fig. 3) displayed also a stratification according to paleodepth, because they were found to be in general <5 m 10^{-6} yr in water depths <2 km, >10 m 10^{-6} yr in 3–4 km paleodepth, but to decrease below that depth. We therefore had to discuss processes redistributing the primary input which produced either hiatuses or accumulations of displaced material. The temporal and paleodepth distribution of hiatuses in the North Atlantic during the past 120 m.y. (Fig. 3) was highly variable. It implies major changes of the intensity of the intermediate and deep-water current regimes. Hiatuses at 2–3 km paleodepth pointed to revolutions of the intermediate water current regimes and to the injection of intermediate water masses from marginal basins such as today from the Mediterranean and from the Norwegian-Greenland Sea. Hiatuses affecting the basin floor below 4 km paleodepth gained in importance only after approximately 95 m.y.BP after the development of oxygen-deficient bottom water masses had ceased.

By far the most important biogenic components of the Upper Mesozoic and Cenozoic North Atlantic deep-sea sediments consist of calcareous material. Although the major producers of calcareous pelagic sediment components (mostly calcitic skeletal material produced by planktonic foraminifers and coccoliths) did not evolve before Late Triassic or Early Jurassic (Tappan and Loeblich, 1973), deep-sea sediments covering the basin floor of the Upper Jurassic North Atlantic were dominantly calcareous. They have also remained quantitatively the most important components throughout the past 140 m.y., but we were able to discern two phases of depositional regimes which resulted in high carbonate concentrations (Fig. 4) of North Atlantic deep-sea sediments. One reason for this twofold subdivision has to be sought in the influx of terrigenous material into the North Atlantic 80–100 m.y.BP, which did not reach this ocean in large quantities before, and which was later confined to the deepest part of its eastern and western basins (Fig. 7).

As discussed above, the primary input of biogenous calcitic particles must have gone through a number of major fluctuations during the past 140 m.y. They seem to have resulted in vertical movements of the calcite compensation depth (CCD) over 1,000–2,000 m. However, the results described in this study were also difficult to reconcile with the concept of a CCD which could be applied to the entire time span of the last 140 m.y. of North Atlantic history. If considering the 10%-isopleth of carbonate concentrations (Fig. 4), which is between the values of 20% used by van Andel (1975) and 0% or only very few percent used by other authors, a CCD could only be observed during part of this interval.

For the first 30–40 m.y. portion of North Atlantic history, when carbonate-rich sediments covered the entire basin floor, three possible scenarios are considered: 1) The basin depth was shallower than the global CCD, which is at variance with previously-published information from the Atlantic and from the Indian Ocean, but not from the Pacific where the CCD seemed to have been located at 4 km paleodepth or below during Early and Late Cretaceous (van Andel, 1975; Thierstein, 1979). 2) The production of $CaCO_3$ particles during Early and mid-Cretaceous was unusually high such that the CCD was depressed into water depths considerably deeper than in the other ocean basins and well below the North Atlantic basin floor. However, the sedimentation rates at this time rarely exceeded 5–10 m 10^{-6} yr, but remained over large parts of the basin, even in its shallow parts, below 5 m 10^{-6} yr (Fig. 2), which would not suggest a very high productivity of the surface waters on an ocean-wide scale. 3) The paleoceanography of the Early and mid-Cretaceous North Atlantic was radically different from any model which can be applied to the Upper Cretaceous and Cenozoic North Atlantic record. The widespread occurrence of anoxic sediments during Early and mid-Cretaceous times in the North Atlantic (Arthur and Natland, 1979) suggest a very sluggish bottom water renewal at times when sea surface temperatures were elevated (Savin, 1977) and when temperatures of the bottom waters seemed to have been as much as 15° warmer than they are today (Shackleton, 1978). The shape of the carbonate profile for the first 40 million years of the North Atlantic history as described here also implies the possibility of intermediate water-masses which could dissolve more carbonate than the bottom water-masses.

A well-defined interface separating carbonate-poor ($\leq$ 10% $CaCO_3$) from carbonate-rich ($\geq$ 10% $CaCO_3$) sediments could only be observed 100–25 m.y.BP (Fig. 4). This interface rose during late Cretaceous to approximately 4 km paleodepth, but subsided slowly afterwards to 4.5 km paleodepth by Paleocene time (approximately 60 m.y.BP and below 5.5 km paleodepth about 25 m.y.BP. This agrees reasonably well with previously published information (van Andel, 1975), although the association of this drop with the initiation of southern hemisphere glaciation was not substantiated because the CCD's drop is lagging 10–15 m.y. behind (Kennett and others, 1975; Kennett, 1978).

The distributional patterns of calcareous matter in the North Atlantic during the last 25 m.y. can be described with great accuracy because of the many data points available (Fig. 3). The interface between the carbonate-poor and the carbonate-rich sediments appeared to have intersected the basin floor approximately 25 m.y.BP at 5.5 km paleodepth, and it did not rise above this level again until modern times. It is interesting to note that an abrupt rise of the CCD during Late Miocene and a subsequent fall cannot be observed, although it has been described in previous studies from the Atlantic (Berger and von Rad, 1972), and Pacific (Heath and others, 1977b) and the Indian Ocean (Sclater and others, 1977b). The picture obtained for the past 20–25 m.y. is rather confused (Fig. 4) and did not allow a clear definition of a CCD. There were several fields of values indicating the occurrence of sediments with high carbonate contents in paleodepths between 4.0 and 6.0 km. This descrepancy with the previously published result was obviously caused by the massive downslope displacement of carbonate-rich sediments into water depths below 5 km paleodepth. The carbonate-rich portions of the sedimentary columns corresponding to these depth intervals during the past 25 m.y. are characterized by sedimentary structures (layering, turbidites, slumps) indicative of bottom-water transport. Although it is not quite clear where precisely the upper boundary of the dominantly displaced carbonates was situated, this limit could be as shallow as 3.5–4.0 km paleodepth. This time interval coincided with the widespread generation of hiatuses (Moore and others, 1978) which were particularly well developed in the North Atlantic in intermediate water depths at 2–3 km paleodepth (Fig. 2). The pelagic origin of the calcareous components of the displaced material suggests a source region distant from the continental margins, and above the CCD. These observations imply that the displacement of these carbonates was closely related to changes of the intermediate water-mass structure in the North Atlantic, either due to major changes of the intermediate water-current regimes or due to the injection of intermediate water masses from marginal basins. These past 25 m.y. coincided with a time when sedimentation rates in the Atlantic increased by almost an order of magnitude (Davies and others, 1977), when the other major subbasins of the world ocean received a much higher sediment supply than at any earlier time, when thick sediment sequences were deposited in the regions close to the continental margins around the North Atlantic Ocean (Ewing and Hollister, 1972), and when the temperate to subtropical regions of the northern hemisphere continents were affected by major climatic shifts (Sarnthein and Diester Haass, 1977; Lancelot and Seibold, 1978). The fact that the accumulation rates of carbonate and of the bulk sediment are similar and fluctuate simultaneously in all oceans, even in the Pacific which was protected from a major influx of suspended terrigenous matter, linked the deposition of the carbonate during the time span discussed here to the flux of dissolved nutrients from the surrounding land masses and epicontinental seas and/or to the fertility of the oceanic surface-water masses (Davies and others, 1977; Bender and Graham, 1978), henceforth to an interaction of the tectonic and climatic evolution of the surface environment of our globe (Fischer and Arthur, 1977).

Today the Atlantic surface sediments are silica-starved (Broecker, 1974), and relatively few opaline microfossils survive dissolution. However, during the past 140 m.y. the North Atlantic has experienced two phases when considerable amounts of silica (up to >20% of the bulk sediment) have been preserved, approximately 110–120 and 40–50 m.y.BP (Fig. 5). The younger of these two intervals, and probably both, coincided with periods of elevated bulk and carbonate sedimentation rates in all three major ocean basins (Davies and others, 1977). Comparing silica and organic carbon concentrations (Fig. 6) as far as they could be related to primary input, but not to displaced material, the major discrepancy between the mid-Cretaceous and the Eocene events seemed to have resulted from important differences in the depositional rather than the producing paleoenvironment. The high organic-carbon concentrations which have been preserved in the North Atlantic as well as in other oceans during mid-Cretaceous time, points to oxygen-deficient bottom water masses. The pattern of organic carbon concentrations (Fig. 6) suggest oxygen deficiency of the entire bottom-water column in the North Atlantic Ocean, but the techniques applied in this study did not allow us to analyze in detail the nature of each of the frequent and probably short-lived events leading to the deposition of lithofacies indicative of an anoxic paleoenvironment (McCave, 1979). These events might have been rather different from each other, both with respect to nature and source of the primary input of the organic carbon contents and to the mode of its preservation. The low organic-carbon concentrations of the Eocene/Lower Oligocene sediments suggest a well-oxygenated paleoenvironment in the deep North Atlantic which required a very important, major shift of the mode of bottom-water formation and circulation since mid-Cretaceous times. The large amounts of organic carbon that accumulated during the past 5 m.y. below 5 km paleodepth (Fig. 6) have probably reached the ocean floor in conjunction with fine-grained terrigenous input. Since the bottom water-masses have been well-oxygenated during the past few million years due to their glacial mode of renewal, the high organic-carbon concentrations must have been preserved because of rapid burial (Suess and Müller, 1979). The influx of detrital terrigenous material did not occur simultaneously with high overall sedimentation rates or with high concentrations of opal and/or organic carbon. Therefore, the high detrital terrigenous input did not coincide with either a high input of dissolved nutrients into the ocean or a high rate of nutrient regeneration. The response time of chemical, life-sustaining substances for these plankters due to processes of nutrient regeneration is on the order of 10^3–10^5 yr (Broecker, 1974), but the fluctuations discussed here have occurred on the order of 10^7 yr. Therefore, the chemical and detrital load of water masses entering the Atlantic from the drainage basins on the surrounding land masses or from their epicontinental seas could have been highly variable depending on the size of the drained area or epicontinental seas and on the paleoclimatic conditions over the adjacent continents.

Terrigenous detrital components have been supplied in pulses to the North Atlantic Ocean during the past 140 m.y. Peaks of high concentrations (Fig. 7) of detrital material after its initial injection into the basin 80–100 m.y.BP have been found in the intervals 50–65 and 15–25 m.y.BP. They coincided with hiatus maxima (Fig. 3) and indications of rapidly rising sea levels on a global scale (Vail and others, 1977). Why only minor quantities of detrital terrigenous material reached the North Atlantic prior to 100 m.y.BP remains a mystery. It is tempting to speculate that the size of the drainage basins after formation of the early North Atlantic rift was insufficient to supply large volumes of detrital material and that its margins acted as sediment traps.

The variables used in this study could be used to discuss a variety of aspects of the paleo-oceanography of the Upper Mesozoic and Cenozoic North Atlantic. These aspects concerned mainly the contribution of nutrients to the ocean and/or the fertility of its surface water-masses, the depositional paleoenvironment under the intermediate and bottom water-masses, and the interaction of deep and shallow paleoenvironments versus sealevel fluctuations. In many studies it has been argued that the CCD was closely related to eustatic sea level fluctuations (Berger and Von Rad, 1972), which during the Late Mesozoic and Cenozoic were believed to have been caused by climatic (Fischer and Arthur, 1977) and plate tectonic (Frerichs, 1970; Pitman, 1978) processes. This correlation was not very obvious when considering the data presented here because the general trend of sea level which was believed to have dropped since Late Mesozoic by approximately 300–500 m, was modulated by numerous short-time eustatic sealevel changes (Vail and others, 1977). Therefore, no correlation could be made between the Mesozoic and Cenozoic carbonate distributions in the North Atlantic and the rapid eustatic sealevel fluctuations which had such profound consequences for the distribution of sedimentary facies along the continental margins (Vail and others, 1977). However, there appeared to be a good correlation with the calculated sea level curves (Hays and Pitman, 1973; Pitman, 1978) whose fluctuations were based on volume changes of the mid-ocean ridge system.

If the assumption was correct that the chemical

input into the sea could not have significantly fluctuated with time (Hay and Southam, 1977), but that it was rather due to changes in the size of the shelf seas, one should have expected a close correlation of high sedimentation rates with times of low sealevels (Worsley and Davies, 1979) and vice versa. There is indeed some agreement to the first order cycles of relative changes of sealevel (Vail and others, 1977), because sedimentation rates were low during the interval 100–60 m.y.BP when sealevel was particularly high, and high before and after that time. However, a consistent synchroneity of sealevel fluctuations of short duration which modulate this general trend (Vail and others, 1977), and of sedimentation rate changes, could not be observed.

Major changes of the fertility of the Upper Mesozoic and Cenozoic North Atlantic surface water masses were not only suggested by the carbonate distributions (Fig. 4) and by the temporal fluctuations of the linear sedimentation rates (Fig. 2), but also by the silica (Fig. 5) and organic carbon (Fig. 6) distributions. The global synchroneity of Cenozoic phases of high and low pelagic sedimentation rates of biogenic components suggested large changes in the chemistry of the ocean resulting in important fluctuations of surface water-mass productivity (Broecker, 1974). By analogy, it is also concluded that the Upper Mesozoic high sedimentation-rate phase in the North Atlantic was caused by fertility increases of the surface water-masses where the skeletal material was secreted by planktonic living organisms, either due to higher inputs of dissolved nutrients from the continents into the oceans or to higher rates of nutrient recycling. Both possibilities required changes of the paleoclimate as a driving mechanism.

The paleoenvironment of the intermediate and bottom water-masses could be addressed by means of organic-carbon concentrations and hiatus distributions. As outlined above, the preservation of high organic-carbon contents required either oxygen-deficient bottom water-masses or rapid burial after sedimentation. The spatial and temporal variations of the hiatuses and the distributions of sediment components, especially silica, indicate large changes in the structure of the Upper Mesozoic and Cenozoic North Atlantic water column. These observations also suggest major changes in the mode of circulation and nature of the North Atlantic bottom water-masses because silica particles had to be produced in surface waters, but they also had to be preserved until they were protected against dissolution by sedimentation. Therefore, the North Atlantic might well have changed several times from a lagoonal-type basin (Berger, 1970) with a deep-water outflow (as today) to an estuarine type with a deepwater inflow and vice versa, which would have resulted in a mode of fractionation of biogenous deep-sea sediments between the North Atlantic and adjacent ocean basins which was different from today.

ACKNOWLEDGMENTS

This study is based on data extracted from site descriptions published in the Initial Reports of the Deep-Sea Drilling Project and in the preliminary "Geotimes" reports of the North Atlantic legs of *Glomar Challenger* as well as from the Scripps Institution of Oceanography data bank. These data have been generated by shipboard scientists, and the authors are grateful to numerous colleagues and to the data-handling group of the Deep-Sea Drilling Project at Scripps. This investigation was supported by the Norwegian Petroleum Directorate. The manuscript has benefitted from the comments of D. Gorsline (Los Angeles) and B. Dale (Oslo).

REFERENCES CITED

NOTE: All references to Deep Sea Drilling Project Initial Reports have been abbreviated; for complete citations see *Appendix I* at the end of this volume, or the individual Initial Reports as published.

Arthur, M. A., 1979, Paleoceanographic events—recognition, resolution, and reconsideration: Rev. Geophys. Space Physics, v. 17, p. 1474–1494.

———, and Natland, J. H., 1979, Carbonaceous sediments in the North and South Atlantic: the role of salinity in stable stratification of early Cretaceous basins: Maurice-Ewing-Ser., Amer. Geophys. Union, v. 3, p. 375–401.

Beaumont, C., 1978, The evolution of sedimentary basins on a viscoelastic lithosphere: theory and examples: Roy. Astron. Soc. Geophys. Jour., v. 55, p. 471–497.

Bender, M. L., and Graham, D. W., 1978, Long term constraints on the global marine carbonate system: Jour. Marine Res., v. 36, p. 551–567.

Berger, W. H., 1968, Planktonic foraminifera: selective solution and paleo-climatic interpretation: Deep-sea Res., v. 15, p. 31–43.

———, 1970, Biogenous deep-sea sediments: fractionation by deep-sea circulation: Geol. Soc. America Bull., v. 81, p. 1385–1402.

———, 1972, Deep sea carbonates: dissolution facies and age-depth constancy: Nature, v. 236, p. 392–395.

———, and Rad., U., von, 1972, Cretaceous and Cenozoic sediments from the Atlantic Ocean: Init. Repts. DSDP, v. 14, p. 787–954.

———, and Winterer, E. L., 1974, Plate stratigraphy and the fluctuating carbonate line: Internat. Assoc. Sedimentol. Spec. Pub., v. 1, p. 11–48.

BERGGREN, W. A., 1972, A Cenozoic time scale—some implications for regional geology and paleobiogeography: Lethaia, v. 5, p. 195–215.
———, AND HOLLISTER, C. D., 1977, Plate tectonics and paleocirculation-commotion in the ocean: Tectonophysics, v. 38, p. 11–48.
BERNOULLI, D., AND JENKYNS, H. C., 1974, Alpine, Mediterranean, and central Atlantic Mesozoic facies in relation to the early evolution of the Tethys: Soc. Econ. Paleontologists Mineralogists Spec. Publ., v. 19, p. 129–160.
BEURLEN, K., 1974, Die geologische Entwicklung des Atlantischen Ozeans: Geotekt. Forsch., v. 46, p. 1–69.
BOUCOT, A. J., AND GRAY, J., 1979, Epilogue: a Paleozoic Pangaea, *in* Boucot, A. J., and Gray, J., eds., Historical Biogeography, Plate Tectonics and the Changing Environment: Corvallis, Oregon State Univ. Press, p. 465–482.
BROECKER, W. S., 1974, Chemical Oceanography: New York, Harcourt, Brace, Jovancovich, Inc., 214 pp.
BURKE, K., 1975, Atlantic evaporites formed by evaporation of water spilled from Pacific, Tethyan and Southern oceans: Geology, v. 3, p. 613–616.
CURRAY, J. R., 1980, The IPOD programme on passive continental margins, *in* Kent, P., Laughton, A. S., Robert, D. G., and Jones, E. J. W., eds., The evolution of passive continental margins in the light of recent deep drilling results: Phil. Trans. Royal Soc. London, A 294, p. 17–33.
DAVIES, T. A., HAY, W. W., SOUTHAM, J. R., AND WORSLEY, T. R., 1977, Estimates of Cenozoic oceanic sedimentation rates: Science, v. 197, p. 53–55.
DEAN, W. E., CLAYPOOL, G. E., AND THIEDE, J., *in press,* Origin of organic carbon-rich mid-Cretaceous limestones, Mid-Pacific Mountains and southern Hess Rise: Init. Repts. DSDP, v. 62.
DIETZ, R. S., AND HOLDEN, J. C., 1970, Reconstruction of Pangaea: breakup and dispersion of continents, Permian to present: Jour. Geophys. Res., v. 75, p. 4939–4956.
EWING, J., AND HOLLISTER, C. H., 1972, Regional aspects of deep-sea drilling in the western North Atlantic: Init. Rept. DSDP, v. 11, p. 951–973.
EWING, M., CARPENTER, G., WINDISCH, C., AND EWING, J., 1973, Sediment distribution in the oceans: Geol. Soc. America Bull., v. 84, p. 71–88.
FISCHER, A. G., AND ARTHUR, M. A., 1977, Secular variations in the pelagic realm: Soc. Econ. Paleontologists Mineralogists Spec. Pub., v. 25, p. 19–50.
FRERICHS, W. E., 1970, Paleobathymetry, paleotemperature, and tectonism: Geol. Soc. America Bull., v. 81, p. 3445–3452.
HAY, W. W., AND SOUTHAM, J. R., 1977, Modulation of marine sedimentation by the continental shelves, *in* Andersen, N. R., and Malahoff, A., eds., The fate of fossil fuel CO_2 in the oceans: New York, Plenum Press, p. 569–604.
HAYES, D. E., AND RABINOWITZ, P. D., 1975, Mesozoic magnetic lineations and the magnetic quiet zone off northwest Africa: Earth Planet. Sci. Letters, v. 28, p. 105–115.
HAYS, J. D., AND PITMAN, W. C., 1973, Lithospheric plate motion, sea level changes, and climatic and ecological consequences: Nature, v. 246, p. 18–22.
HEATH, G. R., 1974, Dissolved silica and deep-sea sediments: Soc. Econ. Paleontologists Mineralogists Spec. Pub., v. 20, p. 77–93.
———, MOORE, T. C., AND DAUPHIN, J. P., 1977a, Organic carbon in deep-sea sediments, *in* Andersen, N. R., and Malahoff, A., eds., The fate of fossil fuel CO_2 in the oceans: New York, Plenum Press, p. 605–625.
———, ———, AND VAN ANDEL, T. H., 1977b, Carbonate accumulation and dissolution in the equatorial Pacific during the past 45 million years, *in* Andersen, N. R., and Malahoff, A., eds., The fate of fossil fuel CO_2 in the oceans: New York, Plenum Press, p. 627–639.
JANSA, L., GARDNER, J. V., AND DEAN, W. E., 1978, Mesozoic sequences of the central North Atlantic: Init. Repts. DSDP, v. 41, p. 991–1031.
KEIGWIN, L. D., 1978, Pliocene closing of the isthmus of Panama, based on biostratigraphic evidence from nearby Pacific Ocean and Caribbean Sea cores: Geology, v. 6, p. 630–634.
KENNETT, J. P., 1978, Cenozoic evolution of circumantarctic palaeoceanography, *in* van Zinderen Bakker, E. M., ed., Antarctic glacial history and paleoenvironments: Rotterdam, A. A. Balkema, p. 41–56.
———, HOUTZ, R. E., AND OTHERS, 1975, Cenozoic paleoceanography in the southwest Pacific Ocean, Antarctic glaciation, and the development of the circum-Antarctic current: Init. Repts. DSDP, v. 29, p. 1155–1169.
LANCELOT, Y., AND SEIBOLD, E., 1978, The evolution of the central northeastern Atlantic—summary of results of DSDP Leg 41: Init. Repts. DSDP, v. 41, p. 1215–1245.
LAUGHTON, A. S., 1975, Tectonic evolution of the northeast Atlantic Ocean; a review: Norg. Geol. Unders., v. 316, p. 169–193.
LEHNER, P., AND DE RUITER, P. A. C., 1977, Structural history of Atlantic margin of Africa: Amer. Assoc. Petroleum Geologists Bull., v. 61, p. 961–981.
LEINEN, M., 1975, Biogenic silica sedimentation in the central equatorial Pacific during the Cenozoic: MS-thesis, Corvallis, School of Oceanography, Oregon State University, 136 p.
———, AND STAKES, D., 1979, Metal accumulation rates in the central equatorial Pacific during Cenozoic time: Geol. Soc. America Bull., v. 90, p. 357–375.
LISITZYN, A. P., BELAYAYEV, Y. I., BOGDANOV, Y. A., AND BOGOYAVLENSKIY, A. N., 1967, Distribution relationships and forms of silicon suspended in waters of the world ocean: Intern. Geol. Rev., v. 9, p. 604–623.
MCCAVE, I. N., 1979, Depositional features of organic-rich black and green mudstones at DSDP Sites 386 and 387, western North Atlantic: Init. Repts. DSDP, v. 43, p. 411–416.
MOORE, T. C., AND HEATH, G. R., 1977, Survival of deep-sea sedimentary sections: Earth Planet. Sci. Letters, v. 37, p. 71–80.

———, VAN ANDEL, T. H., SANCETTA, C., AND PISIAS, N. G., 1978, Cenozoic hiatuses in pelagic sediments: Micropaleontology, v. 24, p. 113–138.

MÜLLER, G., AND GASTNER, M., 1971, The "Karbonat-Bombe", a simple device for the determination of the carbonate content in sediments, soils and other materials: N. Jb. Miner. Mh., v. 10, p. 466–469.

PITMAN, W. C., 1978, Relationship between eustacy and stratigraphic sequences at passive margins: Geol. Soc. America Bull., v. 89, p. 1389–1403.

———, AND TALWANI, M., 1972, Sea-floor spreading in the North Atlantic: Geol. Soc. America Bull., v. 83, p. 619–646.

PONTE, F. C., AND ASMUS, H. E., 1978, Geological framework of the Brazilian continental margin: Geol. Rundschau., v. 67, p. 201–235.

ROBERTS, D. G., 1975, Marine geology of the Rockall Plateau and Trough: Phil. Trans. Royal Soc. London, v. A278, p. 447–509.

RONA, P. A., 1973, Worldwide unconformities in marine sediments related to eustatic changes of sea level: Nature Phys. Sci., v. 244, p. 25–26.

RUSSELL, M. J., AND SMYTHE, D. K., 1978, Evidence for an early Permian oceanic rift in the northern North Atlantic, *in* Neumann, E. R., and Ramberg, I. B., eds., Petrology and geochemistry of continental rifts: Dordrecht, D. Reidel Publ. Co., p. 173–179.

SARNTHEIN, M., AND DIESTER-HAASS, L., 1977, Eolian-sand turbidites: Jour. Sed. Petrology, v. 47, p. 868–890.

SAVIN, S., 1977, The history of the earth's surface temperature during the past 100 million years: Ann. Rev. Earth Planet. Sci., v. 5, p. 319–355.

SCHLANGER, S. O., AND JENKYNS, H. C., 1976, Cretaceous oceanic anoxic events: causes and consequences: Geol. en Mijnbouw, v. 55, p. 179–184.

SCLATER, J. G., ANDERSON, R. N., AND BELL, M. L., 1971, The elevation of ridges and evolution of the central eastern Pacific: Jour. Geophys. Res., v. 76, p. 7888–7915.

———, HELLINGER, S., AND TAPSCOTT, C., 1977a, The paleobathymetry of the Atlantic Ocean from the Jurassic to the Present: Jour. Geology, v. 85, p. 509–552.

———, ABBOTT, D., AND THIEDE, J., 1977b, Paleobathymetry and sediments of the Indian Ocean, *in* Heirtzler, J. and others, eds., Indian Ocean geology and biostratigraphy: Washington D.C., Amer. Geophys. Union, p. 25–29.

SEIBOLD, E., 1978, Mechanical processes influencing the distribution of pelagic sediments: Micropaleontology, v. 24, p. 407–421.

———, 1979, Sediments in upwelling areas, particularly off northwest Africa: Rapp. Proc.-verb. ICES (Symp. Canary Current: Upwelling and living resources).

SHACKLETON, N. J., 1978, Evolution of the earth's climate during the Tertiary era, *in* Proc. Coll. Intern. "Evolution des atmosphères planetaires et climatologie de la terre" (Nice, 16–20 Oct. 1978): Toulouse, Centre Nation. Etud. Spat., p. 49–58.

SNIDÈR, A., 1858, La Création et Ses Mystères Devoicès: Paris, A. Frank & E. Dentu.

SUESS, E., AND MÜLLER, P. J., 1979, Productivity, sedimentation rate and sedimentary organic matter in the oceans. II.—Elemental Fractionation: Abstr. Coll. Intern. C. N. R. S. "Biogéochimie de la Matière Organique a l'Interface Eau-Sediment Marin" (Marseille, Avr. 24–27, 1979).

TALWANI, M., AND ELDHOLM, O., 1977, Evolution of the Norwegian-Greenland Sea: Geol. Soc. America Bull., v. 88, p. 969–999.

TAPPAN, H., AND LOEBLICH, A. R., 1973, Evolution of the oceanic plankton: Earth Science Rev., v. 9, p. 207–240.

THIEDE, J., 1979, History of the North Atlantic Ocean: evolution of an asymmetric zonal paleoenvironment in a latitudinal ocean basin: Maurice-Ewing-Ser., Amer. Geophys. Union, v. 3, p. 275–296.

———, AGDESTEIN, T., AND STRAND, J. E., 1980a, Temporal and spatial variations of the upper Mesozoic and Cenozoic sediment flux to the deep North Atlantic Ocean: Marine Geology, v. 36, p. M11–M19.

———, ———, AND ———, 1980b, Depth distribution of calcareous sediments in the Mesozoic and Cenozoic North Atlantic Ocean: Earth Planet. Sci. Letters, v. 47, p. 416–422.

———, AND VAN ANDEL, T. H., 1977, The paleoenvironment of anaerobic sediments in the late Mesozoic South Atlantic Ocean: Earth Planet. Sci. Letters, v. 33, p. 301–309.

———, DEAN, W. E., AND CLAYPOOL, G. E., *in press,* Oxygen deficient depositional paleoenvironments in the mid-Cretaceous tropical and subtropical central Pacific Ocean: Cretaceous Research.

THIERSTEIN, H. R., 1979, Paleoceanographic implication of organic carbon and carbonate distribution in Mesozoic deep-sea sediment: Maurice-Ewing-Ser., Amer. Geophys. Union, v. 3, p. 249–274.

TISSOT, B., DEROO, G., AND HERBIN, J. P., 1979, Organic matter in Cretaceous sediments of the North Atlantic: contribution to sedimentology and paleo-geography: Maurice-Ewing-Ser., Amer. Geophys. Union, v. 3, p. 362–374.

VAIL, P. R., MITCHUM, R. M., AND THOMPSON, S., 1977, Seismic stratigraphy and global changes of sea level. Parts 3 and 4: Amer. Assoc. Petroleum Geologists Mem., v. 26, p. 63–97.

VALLIER, T., AND OTHERS, 1979, Leg 62 probes the paleoenvironments of the Mid-Pacific: Geotimes, v. 24, p. 24–26.

VAN ANDEL, T. H., 1964, Recent marine sediments in the Gulf of California: Amer. Assoc. Petroleum Geologists Mem., v. 3, p. 216–310.

———, 1975, Mesozoic-Cenozoic calcite compensation depth and global distribution of calcareous sediments: Earth Planet. Sci. Letters, v. 26, p. 187–194.

———, HEATH, G. R., AND MOORE, T. C., 1975, Cenozoic history and paleoceanography of the central equatorial Pacific Ocean: Geol. Soc. America Mem., v. 143, p. 1–134.

———, THIEDE, J., SCLATER, J. G., AND HAY, W. W., 1977, Depositional history of the South Atlantic Ocean during

the last 125 million years: Jour. Geology, v. 85, p. 651–698.
VAN HINTE, J. E., 1976a, A Jurassic time scale: Amer. Assoc. Petroleum Geologists Bull., v. 60, p. 489–497.
———, 1976b, A Cretaceous time scale: Amer. Assoc. Petroleum Geologists Bull., v. 60, p. 498–516.
VAN ZINDEREN BAKKER, E. M., 1978, Late Mesozoic and Tertiary paleoenvironments of the Sahara region, *in* van Zinderen Bakker, E. M., ed., Antarctic glacial history and world paleoenvironments: Rotterdam, A. A. Balkema, p. 129–135.
VOGT, P. R., ANDERSON, C. N., AND BRACEY, D. R., 1971, Mesozoic magnetic anomalies, sea-floor spreading and geomagnetic reversals in the southwestern North Atlantic: Jour. Geophys. Res., v. 76, p. 4796–4823.
———, AND EINWICH, A. M., 1979, Magnetic anomalies and sea-floor spreading in the western North Atlantic, and a revised calibration of the Keathley (M) geomagnetic reversal chronology: Init. Repts. DSDP, v. 43, p. 857–876.
VON STACKELBERG, U., 1972, Faziesverteilung in Sedimenten des indisch-pakistanischen Kontinentalrandes (Arabisches Meer): "Meteor" Forsch.-Ergebn., v. C9, p. 1–73.
WATTS, A. B., AND RYAN, W. B. F., 1976, Flexure of the lithosphere and continental margin basins: Tectonophysics, v. 36, p. 25–44.
WEGENER, A., 1929, Die Entstehung der Kontinente und Ozeane: Braunschweig, F. Vieweg & Sohn, 231 p.
WORSLEY, T. R., AND DAVIES, T. A., 1979, Sea-level fluctuations and deep-sea sedimentation: Science, v. 203, p. 455–456.

APPENDIX.—LOCATION DATA OF DEEP-SEA DRILL SITES IN THE MAIN NORTH ATLANTIC BASIN

Leg. no	Site no	Latitude °N	Longitude °W or °E	Water-depth (m)	Basem. depth below sea bottom (m)	Basem. Age m.y.BP	Source of data	
1	4	24.4856	−73.7978	5319	S 259	A 145.0	S: IR	A: AM
1	4A	24.4856	−73.7978	5319	S 259	A 145.0	S: IR	A: AM
1	5	24.7331	−73.6461	5361	S 278	A 145.0	S: IR	A: AM
1	5A	24.7331	−73.6461	5361	S 278	A 145.0	S: IR	A: AM
2	8	35.3833	−67.5556	5169	S 1300	A 145.0	S: Ewing and others, 1973	A: AM
2	8A	35.3833	−67.5556	5169	S 1300	A 145.0	S: Ewing and others, 1973	A: AM
2	9	32.7778	−59.2028	4981	D 836	A 120.0	D: IR	A: AM
2	9A	32.7778	−59.2028	4981	D 838	A 120.0	D: IR	A: AM
2	10	32.8603	−52.2256	4612	D 457	A 77.0	D: IR	A: AM
2	11	29.9494	−44.7556	3571	D 282	A 19.0	D: IR	A: AM
2	12	19.7036	−26.0008	4557	S 700	A 103.0	S: IR	A: AM
2	12A	19.7031	−26.0008	4557	S 700	A 103.0	S: IR	A: AM
2	12B	19.7036	−26.0008	4557	S 700	A 103.0	S: IR	A: AM
2	12C	19.7036	−26.0008	4557	S 700	A 103.0	S: IR	A: AM
2	12D	19.7036	−26.0008	4557	S 700	A 103.0	S: IR	A: AM
3	13	6.0444	−18.2364	4585	S 515	A 82.0	S: van Andel and others, 1977	A: van Andel and others, 1977
3	13A	6.0444	−18.2364	4585	S 515	A 82.0	S: van Andel and others, 1977	A: van Andel and others, 1977
4	25	-0.5167	−39.2444	1916	S 840	A 75.0	S: van Andel and others, 1977	A: van Andel and others, 1977
4	25A	-0.5167	−39.2444	1916	S 840	A 75.0	S: van Andel and others, 1977	A: van Andel and others, 1977
4	26	10.8986	−44.0492	5169	S 225	A 3.0	S: IR	A: AM
4	26A	10.8986	−44.0492	5169	S 225	A 3.0	S: IR (Site 353)	A: AM
4	28	20.5886	−65.6258	5521	S 700	A 100.0	S: IR	A: AM
11	99	23.6872	−73.8608	4914	S 300	A 143.0	S: IR	A: AM
11	99A	23.6872	−73.8608	4914	S 300	A 143.0	S: IR	A: AM
11	100	24.6911	−73.8097	5325	D 317	Z 160.0	D: IR	Z: IR
11	101	25.2092	−74.4419	4868	S 1400	A 145.0	S: Ewing and others, 1973	A: AM
11	101A	25.2092	−74.4419	4868	S 1400	A 145.0	S: Ewing and others, 1973	A: AM
11	102	30.7425	−74.4539	3426	S 3200	A 180.0	S: Ewing and others, 1973	A: AM
11	103	30.4522	−74.5942	3964	S 2500	A 180.0	S: Ewing and others, 1973	A: AM

APPENDIX.—LOCATION DATA OF DEEP-SEA DRILL SITES IN THE MAIN NORTH ATLANTIC BASIN (CONTINUED)

Leg. no	Site no	Latitude °N	Longitude °W or °E	Water-depth (m)	Basem. depth below sea bottom (m)	Basem. Age m.y.BP	Source of data	
11	104	30.8347	−74.3344	3811	S 3000	A 180.0	S: Ewing and others, 1973	A: AM
14	135	35.3556	−10.4294	4152	S 1039	A 150.0	S: IR	A: AM
14	136	34.1703	−16.3053	4169	D 308	Z 110.0	D: IR	Z: IR
14	137	25.9314	−27.0678	5361	D 397	Z 101.0	D: IR	Z: IR
14	138	25.9269	−25.5719	5288	D 437	A 103.0	D: IR	A: AM
14	139	23.5206	−18.7072	3047	S 2000	A 168.0	S: Ewing and others, 1973	A: AM
14	140	21.7603	−21.7978	4483	S 1800	A 148.0	S: Ewing and others, 1973	A: AM
14	140A	21.7603	−21.7978	4483	S 1800	A 148.0	S: Ewing and others, 1973	A: AM
14	141	19.4211	−24.0086	4148	D 295	A 125.0	D: IR	A: AM
37	332	36.8867	−33.6461	1818	D 104	A 3.5	D: IR	A: AM + spreading rate
37	332A	36.8867	−33.6461	1818	D 104	A 3.5	D: IR	A: AM + spreading rate
37	332B	36.8878	−33.6492	1806	D 148	A 3.5	D: IR	A: AM + spreading rate
37	333	36.8458	−33.6681	1666	D 221	A 3.5	D: IR	A: AM + spreading rate
37	333A	36.8458	−33.6681	1666	D 219	A 3.5	D: IR	A: AM + spreading rate
37	334	37.0369	−34.4242	2632	D 259	A 8.9	D: IR	A: AM
37	335	37.3039	−35.2089	3198	D 454	A 16.5	D: IR	A: AM
39	353	10.9250	−44.0403	5165	S 656	A 3.0	S: IR	A: AM + spreading rate
39	353A	10.9275	−44.0392	5165	S 224	A 3.0	S: IR	A: AM + spreading rate
39	353B	10.9303	−44.0414	5165	S 224	A 3.0	S: IR	A: AM + spreading rate
39	354	5.9097	−44.2050	4045	D 886	Z 72.0	D: IR	Z: IR
41	366	5.6856	−19.8522	2853	S 960	A 75.0	S: van Andel and others, 1977	A: van Andel and others, 1977
41	366A	5.6861	−19.8528	2853	S 960	A 75.0	S: van Andel and others, 1977	A: van Andel and others, 1977
41	367	12.4892	−20.0564	4748	D 1146	Z 150.0	D: IR	Z: IR
41	368	17.5119	−21.3564	3366	S 1000	A 135.0	S: IR	A: AM
41	369	26.5986	−15.0089	1752	S 1400	A 180.0	S: Ewing and others, 1973	A: AM
41	369A	26.5981	−15.0100	1752	S 1400	A 180.0	S: Ewing and others, 1973	A: AM
41	370	32.8403	−10.7822	4214	S 3280	A 170.0	S: IR	A: AM
43	382	34.4178	−56.5403	5526	S 1000	A 100.0	S: Ewing and others, 1973	A: AM
43	384	40.3681	−52.6722	3909	D 325	A 112.0	D: IR	A: AM
43	385	37.3714	−60.1625	4936	S 1200	A 125.0	S: van Andel and others, 1977	A: van Andel and others, 1977
43	386	31.1892	−64.2594	4792	D 966	Z 105.0	D: IR	Z: IR
43	387	32.3222	−67.6667	5117	D 792	Z 130.0	D: IR	Z: IR
44	388	35.5258	−69.4044	4919	S 1300	A 160.0	S: Ewing and others, 1973	A: AM
44	388A	35.5258	−69.4044	4919	S 1300	A 160.0	S: Ewing and others, 1973	A: AM
44	390	30.1483	−76.1206	2670	S 1300	A 182.0	S: Ewing and others, 1973	A: AM
44	390A	30.1483	−76.1206	2670	S 1300	A 182.0	S: Ewing and others, 1973	A: AM

APPENDIX.—LOCATION DATA OF DEEP-SEA DRILL SITES IN THE MAIN NORTH ATLANTIC BASIN (CONTINUED)

Leg. no	Site no	Latitude °N	Longitude °W or °E	Water-depth (m)	Basem. depth below sea bottom (m)	Basem. Age m.y.BP	Source of data	
44	392	29.9175	−76.1856	2607	S 1150	A 182.0	S: Ewing and others, 1973	A: AM
44	392A	29.9175	−76.1856	2607	S 1150	A 182.0	S: Ewing and others, 1973	A: AM
45	396	22.9911	−43.5264	4450	D 125	A 9.0	D: IR	A: AM
?	396A	?	−43.5250	4459	D 120	A 9.0	D: IR	A: AM
46	396B	22.9870	−43.5250	4459	D 151	A 9.0	D: IR	A: AM
47	397	26.8528	−15.1889	2900	S 1600	A 170.0	S: Ewing and others, 1973	A: AM
47	397A	26.8528	−15.1889	2900	S 1600	A 170.0	S: Ewing and others, 1973	A: AM
47	398	40.9667	−10.7194	3890	S 1858	A 180.0	S: Ewing and others, 1973	A: AM
47	398A	40.9667	−10.7194	3890	S 1850	A 180.0	S: Ewing and others, 1973	A: AM
47	398B	40.9667	−10.7194	3890	S 1850	A 180.0	S: Ewing and others, 1973	A: AM
47	398C	40.9667	−10.7194	3890	S 1850	A 180.0	S: Ewing and others, 1973	A: AM
47	398D	40.9667	−10.7194	3890	S 1850	A 180.0	S: Ewing and others, 1973	A: AM
49	410	45.5142	−29.4822	2975	? 330	? 10.0	D: IR	A: IR
49	410A	45.5147	−29.4822	2977	D 330	A 10.0	D: IR	A: IR
49	411	36.7769	−33.3917	1925	D 74	A 1.0	D: IR	A: GEOTIMES
49	411A	36.7769	−33.3917	1935	D 74	A 1.0	D: IR	A: GEOTIMES
49	412	36.5706	−33.1767	2609	D 160	A 1.6	D: IR	A: GEOTIMES
49	412A	36.5706	−33.1767	2609	D 160	A 1.6	D: IR	A: GEOTIMES
49	413	36.5497	−33.1806	2598	D 110	A 3.5	D: IR	A: GEOTIMES
50	415	31.0367	−11.6531	2794	S 6800	A 175.0	S: IR	A: AM
50	415A	31.0367	−11.6531	2794	S 6800	A 175.0	S: IR	A: AM
50	415B	31.0367	−11.6531	2794	S 6800	A 175.0	S: IR	A: AM
50	416	32.8383	−10.8017	4191	S 7500	A 170.0	S: IR	A: AM
50	416A	32.8383	−10.8017	4101	S 3280	A 170.0	S: IR	A: AM
51	417	25.1197	−68.0492	5468	D 211	A 108.0	D: IR	A: AM
51	417A	25.1175	−68.0467	5468	D 211	A 108.0	D: IR	A: AM
51	417B	25.1181	−68.0550	5479	S 250	A 108.0	S: IR	A: AM
51	417C	25.1156	−68.0508	5479	S 250	A 108.0	S: IR	A: AM
51	417D	25.1192	−68.0561	5479	D 343	A 108.0	D: IR	A: AM
52	417D	25.1192	−68.0561	5472	D 343	A 108.0	D: IR	A: AM
52	418	25.0356	−68.0625	5511	D 344	Z 113.0	D: IR	Z: IR
52	418A	25.0356	−68.0625	5511	D 324	Z 113.0	D: IR	Z: IR
53	418A	25.0361	−68.0622	5511	D 324	Z 113.0	D: IR	Z: IR
53	418A	25.0356	−68.0625	5514	D 324	Z 113.0	D: IR	Z: IR

APPENDIX 1.—Location data of deep-sea drill sites in the main North Atlantic basin. D = Drilled, S = Seismic, A = Magnetic anomaly, Z = Sediment age, IR = Initial Report Deep Sea Drilling Project, AM = Magnetic anomaly map (cf. Thiede, 1979).

SEPM SPECIAL PUBLICATION NO. 32, P. 91–127, OCTOBER 1981

TURBIDITES AFTER TEN YEARS OF DEEP-SEA DRILLING—WRINGING OUT THE MOP?

KERRY KELTS AND MICHAEL A. ARTHUR
Geological Institute, ETH-Z, CH8092 Zurich, Switzerland
U.S. Geological Survey, Denver, Colorado 80225 U.S.A.

ABSTRACT

Deep-sea drilling has provided data on the distribution and changes in composition of redeposited sediment in deep-sea settings through time. The collection of such data by drilling, however, has not been very systematic or purposeful, and has contributed relatively little to the direct understanding of mechanisms of sediment redeposition. A review of reported occurrences of redeposited sediment in DSDP initial reports through Leg 54 shows some temporal patterns: Pleistocene continental margin sequences are thick and dominated by redeposited, relatively coarse-grained terrigenous material, whereas Middle to Upper Eocene and Albian-Cenomanian sequences are characterized by radiolarian turbidites. Moreover, DSDP has recovered an inordinate number of redeposited volcanogenic sequences in the Upper Cretaceous (centered on the Campanian) of all ocean basins and to a lesser extent in the Eocene. These commonly drape the flanks of aseismic rises, ridges, and plateaus and imply episodic volcanism along these features.

Deep Sea Drilling Project cores have also contributed greatly to increasing recognition of the importance of redeposition of large volumes of mud both in fan and non-fan turbidite settings along active and passive margins. Pelagic environments are not devoid of redeposited material; so-called "pelagic turbidites" are common and comprise redeposited pelagic calcareous and siliceous biogenic components derived from the flanks of submarine highs. Redeposited sediments found in DSDP cores have also been extremely important in reconstruction of the tectonic history of numerous features, such as the drowning and demise of Jurassic-Lower Cretaceous carbonate platforms on young passive margins bordering the Atlantic, the destruction and upbuilding of the continental rise along Atlantic margins in the mid-Cenozoic, the history of uplift and volcanism on linear island chains in the Pacific, and the formation and destruction of slope basins along active margins.

With the advent of the Hydraulic Piston Coring system, the future holds promise for more complete documentation of deep-sea fans, a closer examination of controls on timing and frequency of turbidite events, and detailed studies of turbidite physical properties and early diagenesis.

INTRODUCTION

The Role of the Turbidite Concept in DSDP

In a delightful paper entitled "Mopping up the Turbidite Mess", Walker (1973) summarized knowledge on turbidite sedimentology. He pointed out that modern thinking about turbidites stems from the classic paper by Keunen and Migliorini (1950), and that we may "claim that now, in 1971, the 'turbidite' is twenty-one years old and has finally come of age." At that time the Deep Sea Drilling Project (DSDP) was only three years old, and Walker suggested that the full importance of JOIDES drilling for turbidite studies could not be evaluated. He correctly pointed out that the rotary coring procedure eliminates most sedimentary structures in relatively unconsolidated sediment in the upper parts of the cored section, and that the drilling was therefore not likely to elucidate the relationship between turbidite sedimentary structures and specific depositional environments. Our paper is such an evaluation of the highlights of the impact of DSDP on concepts of sediment redeposition in the ocean basins, with emphasis on the regional implications of turbidite occurrences and on the remaining problems of turbidite sedimenlogy.

DSDP has never directly attacked the problem of the theory of turbidity current initiation and flow. Nor has it attempted to systematically study the distribution of turbidites in ocean basins or the anatomy of deep-sea fans. Early drilling was designed to test the sea-floor spreading concept, and during the first 35 Legs or so there was relatively little continuous coring. Thick sediment sections, which would be expected to contain turbidite sequences, were generally avoided because of time limitations and because maximum penetration into older sedimentary horizons or to "basement" was desired. This eliminated the opportunity for sequential analysis of turbidites. Berger (1979) discussed the adverse impact that these philosophies have had on paleoceanographic studies using DSDP cores. The same reasoning applies to turbidite sedimentology. Thus, it is not surprising that when asked what contribution DSDP has made to turbidite sedimentation, most of our colleagues would say "none", because they are unaware of such contributions.

The above problems notwithstanding, DSDP has

provided valuable data on some major problems of sediment redeposition to the ocean basins. In fact, as we propose later in this paper, with new drilling/coring technology available, DSDP can now contribute much to the study of continuous sequences of sedimentary environments of Neogene age; particularly important will be the systematic study of sedimentary facies on a modern or submodern deep-sea fan. So by "wringing out the mop," this sketch is not meant to provide a final summary of turbidite sedimentology, but rather to set the stage for another step forward. An up-to-date summary of turbidity sedimentology may be found in Rupke (1978), and recent volumes edited by Stanley and Kelling (1978) and Doyle and Pilkey (1979) contain a series of papers in part devoted to these topics.

A preliminary compilation of the reported occurrences of redeposited sediments (mainly turbidites) in Deep Sea Drilling Project sites (through Leg 54) is given in Figure 1. Some aspects of our compilation may be misleading due to several biases. Drilling to date naturally has favored Quaternary deposits due to their immense thickness and their widespread distribution on the continental rises. In spite of this there is a "gap" in our knowledge of younger (mainly Neogene and Quaternary) deposits because of the deformation associated with drilling in relatively unconsolidated sediment. Furthermore older deposits are usually cored from mid-plate or ridge provinces because Mesozoic marginal sequences are generally buried below the reach of the drill-string. The compilation shows that downslope redeposition of sediment is a major process and that it occurs in both continental margin and pelagic environments. In addition it illustrates some distinct sedimentation patterns through time. Terrigenous clastic turbidites are most common in the Quaternary, the Eocene, and in association with Cretaceous continental margin sequences. Mass flow events, often consisting of carbonate sediment, are common in the Lower to Middle Miocene of many basins. Volcanogenic turbidites are most common in the Late Cretaceous, and radiolarian or biosiliceous turbidites in the Eocene. These and other problems are reviewed below, following a section on the lithologic features of redeposited sediment types found in DSDP cores.

Why Study Turbidites?

The answer to this question is multifold and probably rather obvious to most sedimentologists. Yet we have seen during 10 years of DSDP a general reluctance to study turbidite sequences in detail when recovered, and an even greater reluctance to drill them in the first place.

Redeposited sediments represent local or regional "noise" in the often-desired complete sequences of pelagic sediment. However, the study of such mass flows in marine sequences can provide evidence for structural relief, for the structural evolution of edifices such as volcanic pedestals or continental slopes, and for the composition of exposed source-terranes long after other positive evidence has disappeared. In geochemical studies and sampling—*i.e.,* those of chemical fluxes or mass balances or of sources of organic matter in marine sediments—it is important to recognize redeposited components. In "hemipelagic" or continental-margin mud sequences it may be difficult to distinguish individual redeposited layers, yet it is crucial to the organic geochemist, for example, to sample and analyze material of both pelagic and redeposited origin and to recognize differences between the two. Biostratigraphers can both benefit by, and help in recognition of, redeposited sediment. In most deep basins, the only way to date sediment is by calcareous fossils preserved in or just below turbidites sedimented beneath the carbonate compensation depth (CCD). Modern studies of the relative importance of various sediment-transport processes along continental margins can utilize the perspective gained from study of ancient sequences in order to compare and contrast various stages in the evolution of those margins. In other words, turbidites are major deep-water reservoir rocks, and they provide clues to basin formation, evolution, sediment by-passing, and passing sediment sources. Finally, we can develop predictive models of sedimentary facies distribution, hydrocarbon source and reservoir relationships, and trends in diagenesis with age and depth of burial for petroleum exploration. This can best be done with information, much of which is still to come, from DSDP drilling.

"Just Another Turbidite"— A Chronology of DSDP Milestones

The recognitition of turbidites in DSDP cores parallels developments in the study of sequences on land. When the *Glomar Challenger* set sail, Mutti and Ricci Lucchi were still mapping turbidite facies and developing their submarine fan model in the Apennines, and the oil crunch was five years away. Sedimentologists were just beginning to interpret flysch sequences in terms of global tectonics, expecting to find modern equivalents along continental rises and on trench walls. Turbidite facies concepts were dominated by Bouma cycles (*e.g.,* Bouma, 1962) and proximal-distal transitions (Walker, 1967).

DSDP Leg 1 had already documented impressive evidence for long-distance transport and redeposition of sands. Shallow-water miliolid carbonates were recovered in the Gulf of Mexico, but were displaced over 500 km from their presumed origin along the Campeche escarpment (Ewing, Worzel

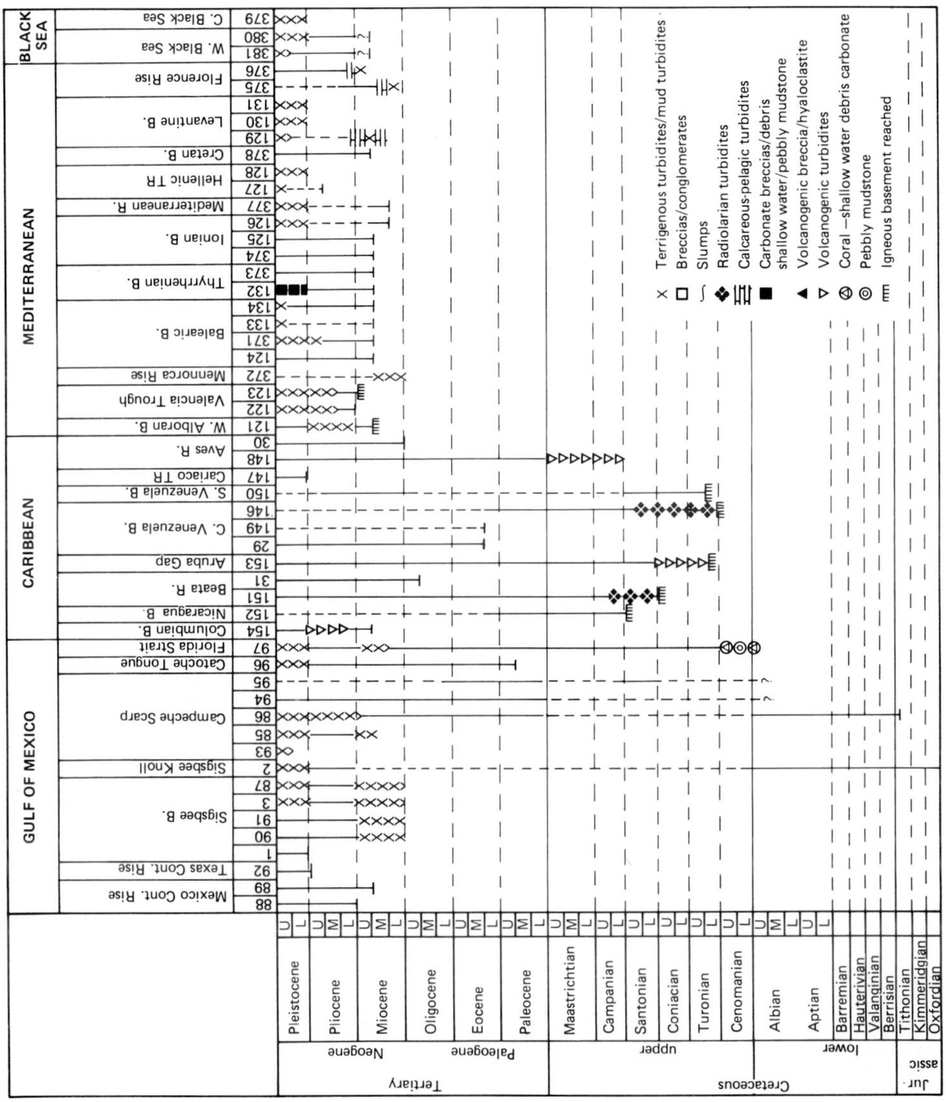

FIG. 1.—(Parts A to E): Chart showing the age distribution of different types of redeposited sediment recovered and described in DSDP Sites through Leg 54. Symbols are explained on key. The biostratigraphic base is modified after Bolli (1980, and *in press*) and Bolli and Saunders (1977). Compilation is based largely on information published in DSDP Initial Reports volumes. (Fig. 1A).

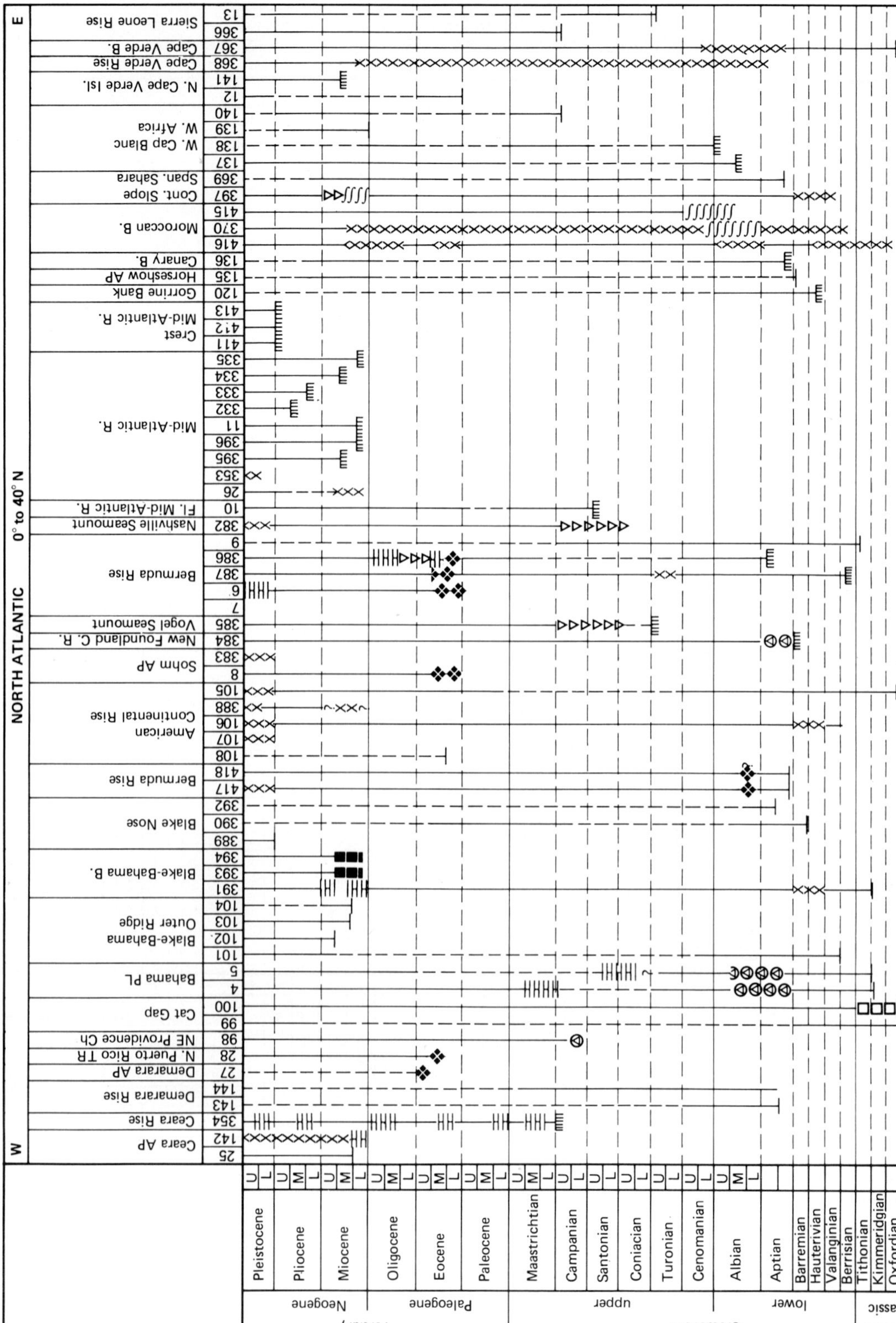
W
NORTH ATLANTIC 0° to 40° N
E
Ceara AP
Ceara Rise
Demarara Rise
Demarara AP
N. Puerto Rico TR
NE Providence Ch
Cat Gap
Bahama PL
Blake-Bahama
Outer Ridge
Blake-Bahama B.
Blake Nose
Bermuda Rise
American
Continental Rise
Sohm AP
New Foundland C. R.
Vogel Seamount
Bermuda Rise
Nashville Seamount
Fl. Mid-Atlantic R.
Mid-Atlantic R.
Mid-Atlantic R.
Crest
Gorrine Bank
Horseshow AP
Canary B.
Moroccan B.
Cont. Slope
Span. Sahara
W. Cap Blanc
W. Africa
N. Cape Verde Isl.
Cape Verde Rise
Cape Verde B.
Sierra Leone Rise
Tertiary
Neogene
Paleogene
Cretaceous
upper
lower
Jur-
assic
Pleistocene
Pliocene
Miocene
Oligocene
Eocene
Paleocene
Maastrichtian
Campanian
Santonian
Coniacian
Turonian
Cenomanian
Albian
Aptian
Barremian
Hauterivian
Valanginian
Berrisian
Tithonian
Kimmeridgian
Oxfordian

Fig. 1B

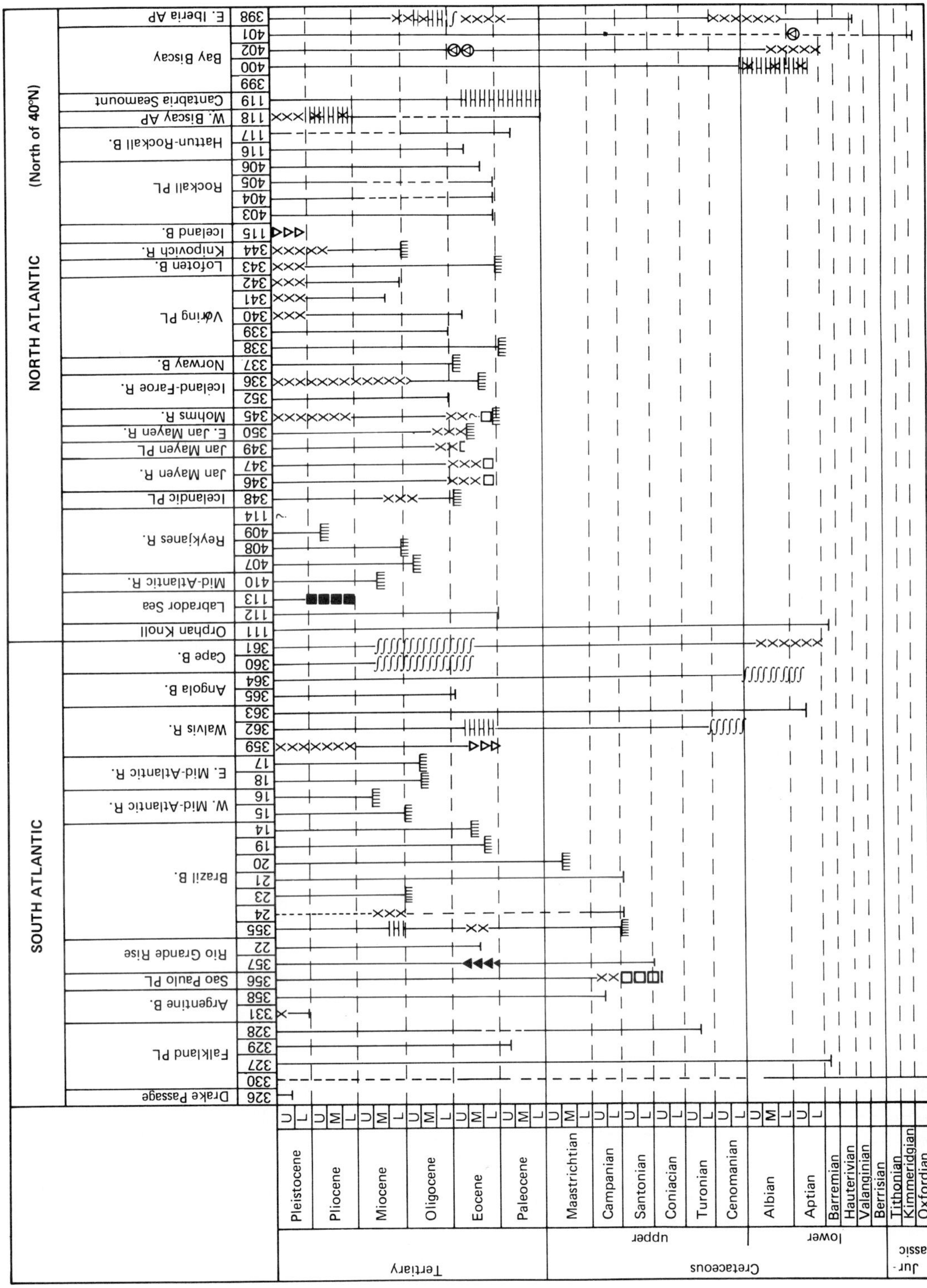
NORTH ATLANTIC
(North of 40°N)
SOUTH ATLANTIC
E. Iberia AP
398
Bay Biscay
401
402
400
399
Cantabria Seamount
119
W. Biscay AP
118
Hattun-Rockall B.
117
116
Rockall PL
406
405
404
403
Iceland B.
115
Knipovich R.
344
Lofoten B.
343
Vøring PL
342
341
340
339
338
Norway B.
337
Iceland-Faroe R.
336
352
Mohms R.
345
E. Jan Mayen R.
350
Jan Mayen PL
349
Jan Mayen R.
347
346
Icelandic PL
348
Reykjanes R.
114
409
408
407
Mid-Atlantic R.
410
Labrador Sea
113
112
Orphan Knoll
111
Cape B.
361
360
Angola B.
364
365
Walvis R.
363
362
359
E. Mid-Atlantic R.
17
18
W. Mid-Atlantic R.
16
15
Brazil B.
14
19
20
21
23
24
355
Rio Grande Rise
22
357
Sao Paulo PL
356
Argentine B.
358
331
Falkland PL
328
329
327
330
Drake Passage
326
Pleistocene
Pliocene
Miocene
Oligocene
Eocene
Paleocene
Maastrichtian
Campanian
Santonian
Coniacian
Turonian
Cenomanian
Albian
Aptian
Barremian
Hauterivian
Valanginian
Berrisian
Tithonian
Kimmeridgian
Oxfordian
upper
lower
Tertiary
Cretaceous
Jur-assic

Fig. 1C

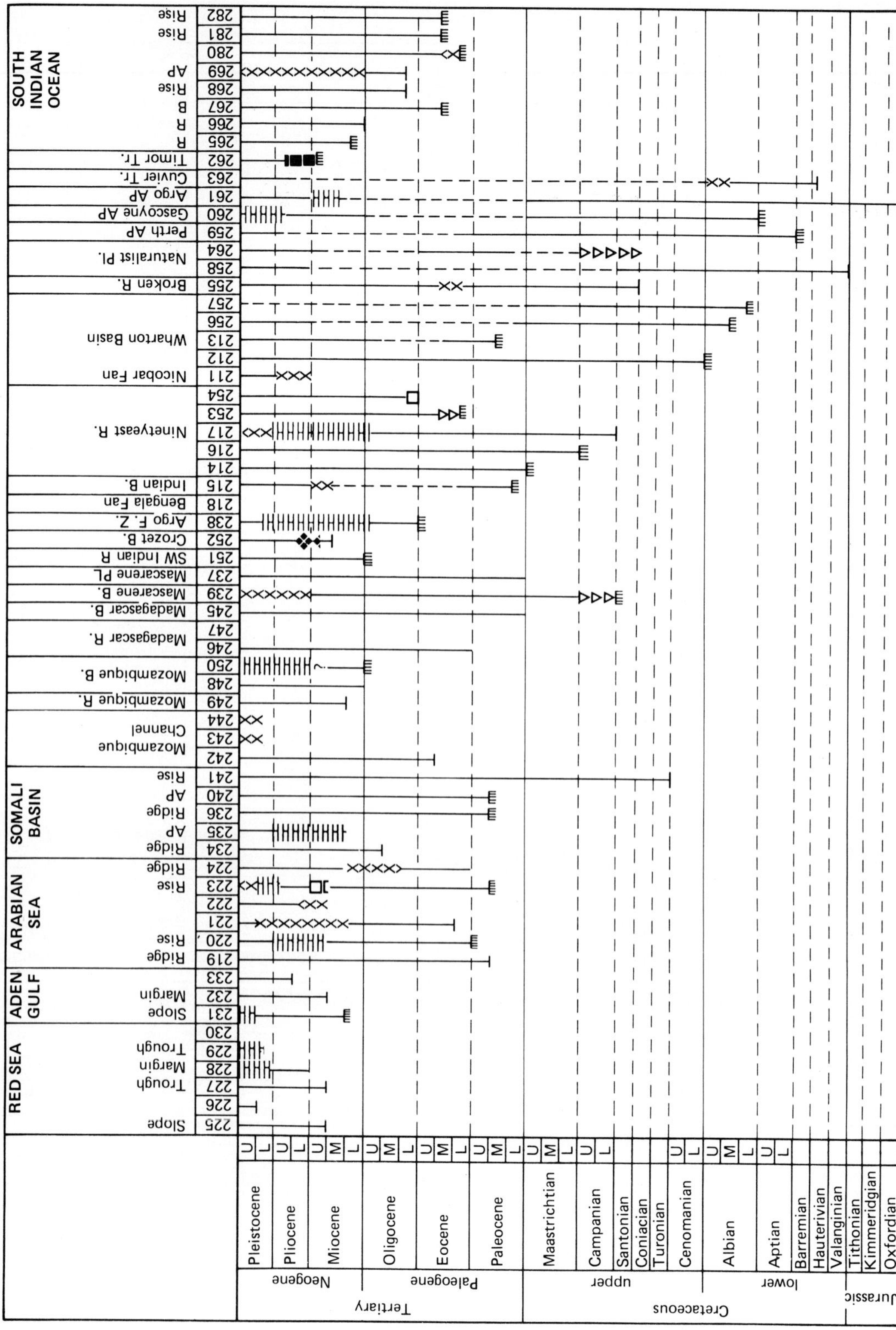

RED SEA
ADEN GULF
ARABIAN SEA
SOMALI BASIN
SOUTH INDIAN OCEAN
Pleistocene
Pliocene
Miocene
Oligocene
Eocene
Paleocene
Maastrichtian
Campanian
Santonian
Coniacian
Turonian
Cenomanian
Albian
Aptian
Barremian
Hauterivian
Valanginian
Tithonian
Kimmeridgian
Oxfordian
Neogene
Paleogene
Tertiary
upper
lower
Cretaceous
Jurassic

Fig. 1D

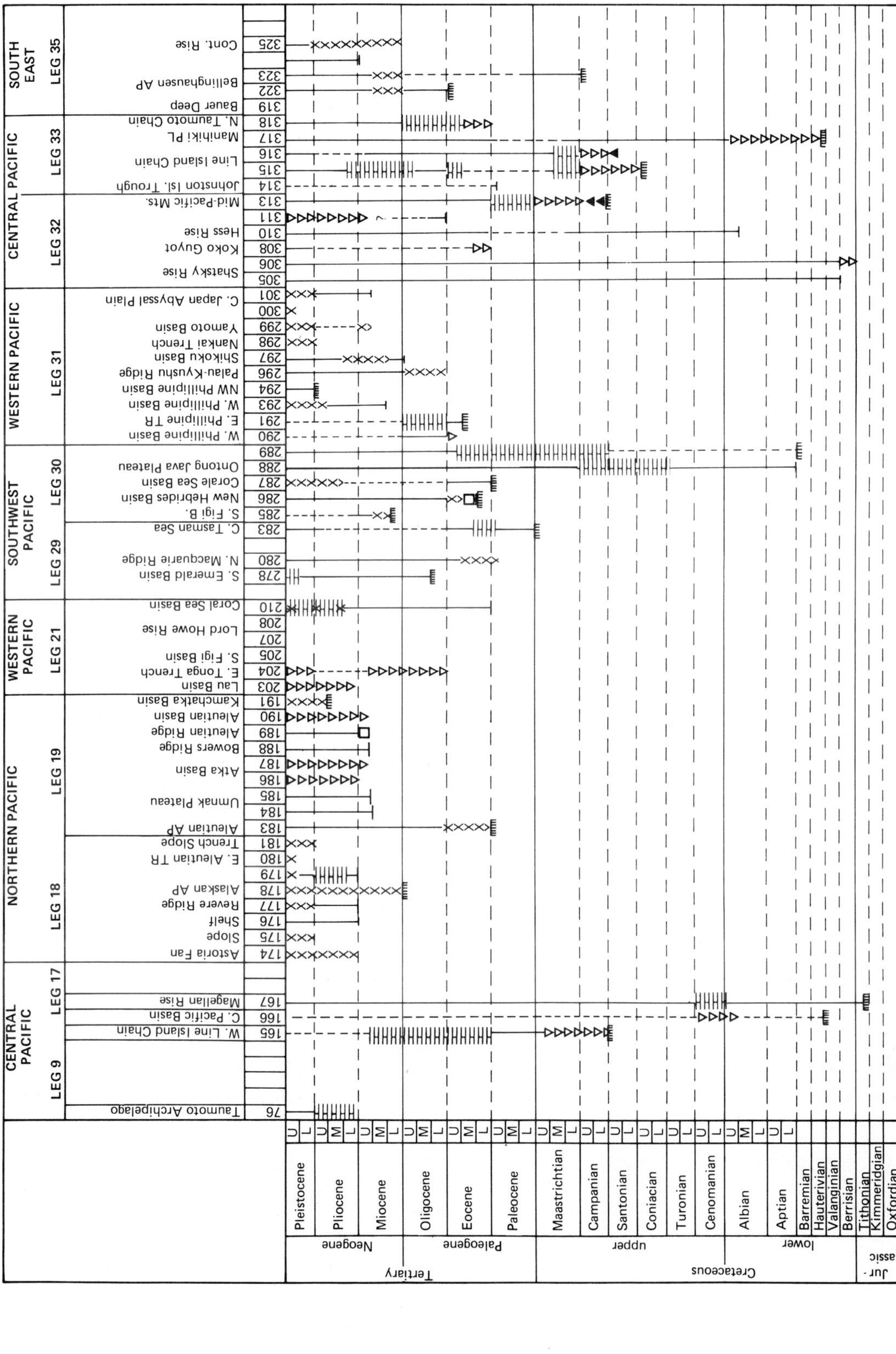
CENTRAL PACIFIC
LEG 9
LEG 17
NORTHERN PACIFIC
LEG 18
LEG 19
WESTERN PACIFIC
LEG 21
SOUTHWEST PACIFIC
LEG 29
LEG 30
WESTERN PACIFIC
LEG 31
CENTRAL PACIFIC
LEG 32
LEG 33
SOUTH EAST
LEG 35
76 Taumoto Archipelago
165 W. Line Island Chain
166 C. Pacific Basin
167 Magellan Rise
174 Astoria Fan
175 Slope
176 Shelf
177 Revere Ridge
178 Alaskan AP
179
180 E. Aleutian TR
181 Trench Slope
183 Aleutian AP
184
185 Umnak Plateau
186
187 Atka Basin
188 Bowers Ridge
189 Aleutian Ridge
190 Aleutian Basin
191 Kamchatka Basin
203 Lau Basin
204 E. Tonga Trench
205 S. Figi Basin
207
208 Lord Howe Rise
210 Coral Sea Basin
278 S. Emerald Basin
280 N. Macquarie Ridge
283 C. Tasman Sea
285 S. Figi B.
286 New Hebrides Basin
287 Corale Sea Basin
288 Ontong Java Plateau
289
290 W. Philipine Basin
291 E. Philipine TR
293 W. Philipine Basin
294 NW Philipine Basin
296 Palau-Kyushu Ridge
297 Shikoku Basin
298 Nankai Trench
299 Yamoto Basin
300
301 C. Japan Abyssal Plain
305
306 Shatsky Rise
308 Koko Guyot
310 Hess Rise
311
313 Mid-Pacific Mts.
314 Johnston Isl. Trough
315
316 Line Island Chain
317 Manihiki PL
318 N. Taumoto Chain
319 Bauer Deep
322
323 Bellinghausen AP
325 Cont. Rise
Tertiary
Neogene
Paleogene
Pleistocene
Pliocene
Miocene
Oligocene
Eocene
Paleocene
Cretaceous
upper
lower
Maastrichtian
Campanian
Santonian
Coniacian
Turonian
Cenomanian
Albian
Aptian
Barremian
Hauterivian
Valanginian
Berrisian
Jur-assic
Tithonian
Kimmeridgian
Oxfordian

Fig. 1E

and others, 1969; Beall and Fischer, 1969). Scientists on Leg 1 recognized and described many of the types of redeposited sediments which only much later in the program found widespread application: they identified thick graded sands as turbidites from the Mississippi Delta; thought diatom- and spicule-rich clays were redeposited; recognized Cretaceous "pelagic" turbidites consisting of calcarenites grading to nannofossil oozes; and described Albian pebbly mudstones with limestone exotic clasts. Leg 4, Hole 26 in the Vema Fracture Zone contained a thick section of dark sands, some graded, containing abundant plant remains. These were interpreted as evidence for highly competent turbidity currents which crossed the Demerara Abyssal Plain carrying sediment from the Amazon delta (Bader, Gerard and others, 1970). On Leg 10, Beall and others (1973) used changes in the character of turbidites to follow subsidence history and eustatic sealevel changes in the Gulf of Mexico. Cores from the center of the Sigsbee plain indicated to them a vertical evolution from distal to medial to proximal turbidite facies. Decimeter to 2 m thick graded beds occur with sand contents around 10 percent. Beall and others (1973) also enlarged the turbidite concept to include many poorly sorted, finer-grained silty clays which might now be termed mud turbidites.

After Phase I of DSDP, it was clear that flysch-type sands from high-concentration turbidity flows were relatively rare in deep ocean sequences, but that redeposition processes were immensely important for deep-sea facies development on the whole. Enormous volumes of redeposited fine-grained muds are delivered to the ocean basins (*e.g.*, Piper, 1978). In addition mid-ocean ridges, rises, plateaus, and linear seamount chains were recognized as important source regions for turbidity current deposits (Heath and Moberly, 1971; Cook and others, 1976; Winterer, 1973). Improved drill bits increased the depth of penetration. Sedimentary structures became more obvious on the cut surfaces of indurated sediment cores with the advent of the "super saw" at about Leg 17. Lancelot and others (1972) examined cores from the passive Atlantic margin and described graded beds in Oxfordian to Kimmeridgian red clayey limestones attributed to pelagic turbidites, and graded, volcaniclastic Upper Jurassic breccias above basalt at Site 105. A significant impetus arose as land-based sedimentologists began to compare their sedimentary sequences with similar ones in DSDP cores (*e.g.*, Bernoulli, 1972; see Hsü and Jenkyns, 1974; Bernoulli and Jenkyns, 1974). Characteristics of an overall sequence of redeposited beds became more important than single redeposition events as a mirror of depositional environment changes. Hesse (1975) drew parallels between mud-turbidite occurrences in DSDP sites and his own observations in the Cretaceous Flysch belt of the Eastern Alps and thereby devised a set of criteria for their recognition in ancient rocks. In the meantime, Legs 7 and 17 recovered massive volcanic turbidites and set the groundwork for testing the hot-spot theory using principles of turbidite evolution in archipelagic settings (Heath and Moberly, 1971; Winterer, 1973).

Publications by Hampton (1972), Middleton and Hampton (1973), Walker (1973), Mutti and Ricci Lucchi (1972), and others developed complex three-dimensional models for redepositional facies and environments, particularly for deep-sea fans. These had a corresponding echo in the number of core descriptions of slumps, debris flows, and high- and low-concentration turbidites. Some studies attempted to classify redeposited sediment sequences by sedimentary structures and composition, in a single DSDP hole, into one of the model subfacies (*e.g.*, Bouma, 1975a; Klein, 1975), but the widespread use of numerous facies designations was inhibited by cumbersome new terminologies.

In the core descriptions for the 69 legs completed to 1980, thousands of references, most of them cursory, have been made to the intimate details of graded beds. Since the IPOD (International Program of Ocean Drilling) phase began continuous coring, sequential analysis has become an established tool in deciphering the evolution of passive and active margins and young marginal ocean basins (*e.g.*, Arthur and von Rad, 1979; deGraciansky and Chenet, 1979; Lancelot and Winterer, 1980; Schlager, 1980; Curray, Moore, and others, 1980; von Huene and others, 1980; Arthur and others, 1980b). In the following section we review definitions and descriptions of the basic types of redeposited sediments as seen in DSDP cores in preparation for later discussion of depositional environment and trends in resedimentation through time as revealed in DSDP drillholes.

REDEPOSITED SEDIMENT TYPES

Classic Clastics

Detailed descriptions and photographs illustrating varieties of sedimentary structures have previously been compiled for some Pacific marginal basins (*e.g.*, Bouma, 1975a; Klein, 1975).

Sandy turbidites derived from terrigenous sources which show typical Bouma-type structures have been cored at several sites, but are best illustrated by Sites 416, 415, 370, and 397 along the Moroccan continental margin. These turbidites are the classic types that are easily recognized by sedimentologists familiar with flysch sequences on land (Figs. 2–4). For example, Site 416, Leg 50 (and Site 370, Leg 41) is dominated by various types of redeposited terrigeneous deposits. The Tithonian to

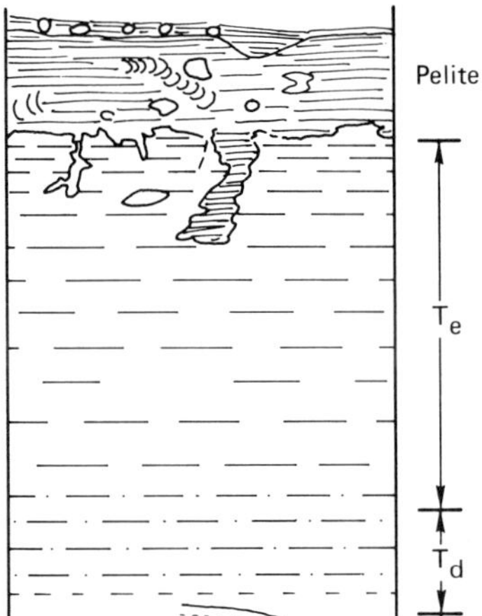

FIG. 2.—Bouma turbidite cycle. T_e: Pelite interval. No primary sedimentary structure, commonly burrowed. T_d: Upper interval of parallel lamination. Usually indistinct, overprinted grading from fine sand to mudstone. T_c: Interval of current ripples of less than 5 cm amplitude. Commonly wispy foreset bedding, in rare cases convolute. T_b: Lower interval of parallel laminations. Clay-silt and coarse sand laminations. Grading may continue. Gradational contacts. T_a: Graded interval. Sandy base with more or less distinct grading, vague when components are well-sorted. Basal scour common.

Hauterivian section of Site 416 is over 700 m thick and consists almost entirely of repeated, massive terrigeneous turbidite beds with some interbedded calcareous turbidites. This sequence is thought to represent part of a fan environment (Lancelot and

FIG. 3.—The basal portion of an indurated, clastic sandstone bed showing the classic Bouma-type sequence of grading (T_a) followed upward and overprinted by parallel laminations (T_b) and some crossbedding (T_c). Such beds are common from Albian sediments drilled off the Moroccan continental margin (sample 50-415A-45-1, 18-35cm). Core section is 6 cm wide.

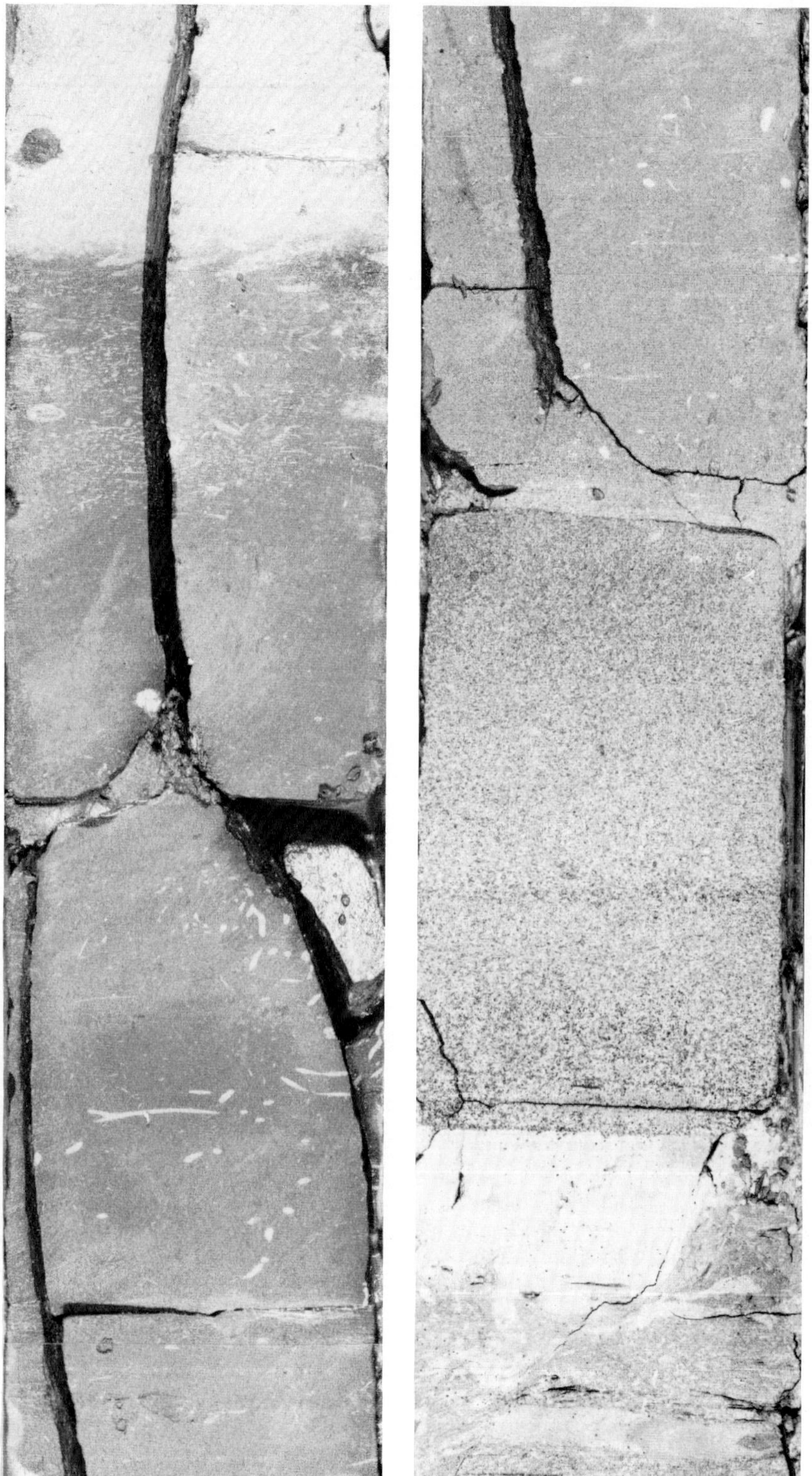

Winterer, 1980; Gardner and others, 1978). The turbidite sequences show normal grading upward from medium- to fine-grained sandstone to siltstone to mudstone. The sands are moderately mature, rounded, and composed of both quartzose-feldspathic and calcareous grains. All of the Bouma cycle intervals have been recognized (Fig. 2). Core 416A-9 is a typical example. It contains 72 turbidite beds within 835 cm. Most of the cycles begin with the T_b or T_c interval but 10–20 cm graded beds (T_a) are also common. From Core 416A-9 to 18 the average thickness of beds remains about 8.5 to 12 cm/bed, although the sand layer percentages fluctuate from 15–50% of the core. A typical graded cycle 5–20 cm thick contains from top to bottom:

- P pelite, grayish green mudstone, intensively burrowed
- T_e grayish olive to olive green calcareous mudstone to marlstone, some burrows of *Chondrites*
- T_d medium gray (N 6) siltstone to silty mudstone, laminated, friable
- T_c medium to light gray sandstone; ripple-cross laminae; mainly fine grained, well-cemented; scattered plant debris and mica flakes; rare convolute layers
- T_b medium to light gray parallel-laminated sandstone

The composition of coarse-grained terrigenous turbidite sands has been used to trace provenance in the Indus Fan (Jipa and Kidd, 1974) and the Nicobar and Bengal fans (Ingersoll and Suczek, 1979). Natland (1978) studied the composition of coarse-grained turbidites from a 900 m thick sequence of middle-Cretaceous clastics from Site 361 in the Cape Basin. Using bulk chemical analysis he linked Unit F (Aptian) sands to the Cape granites. These sands are clean, well-sorted, with calcareous cement. Components include quartz, microcline and plagioclase with accessory biotite muscovite, magnetite, tourmaline and zircon. In the same interval mud-rich graded sandstones occur which have high organic-carbon contents, ranging from 0.7–14% (avg. 3.5%) including abundant wood chips and fiber. Sands of slightly younger age in the same sequence apparently had a different source, linked to the Orange River, and chemically resemble Karoo Basalt.

FIG. 5.—Thin-bedded quartzose turbidites (or contourites?) with parallel bedding and flaser-like cross bedding in very dark claystone (sample 47B- 398D- 117-5, 84–105cm; Upper Barremian).

FIG. 4.—Thick sandy-silt turbidite beds which are more common along the outer continental slope. They generally show only basal grading, a massive midsection, and a heavily burrowed (*Chondrities*) top (sample 50-416A3-3, 55–80 cm; Upper Eocene to Lower Oligocene). Top of core, upper left; bottom, lower right.

Thin-bedded Turbidites vs. Contourites

Thin-bedded turbidites often exhibit the same suite of sedimentary structures as the thicker units discussed above; more commonly, only the T_{cde} or T_{de} intervals of the Bouma sequence occur, and small-scale crosslamination, ripple drift, starved ripples and sand-silt lenses are typical (see Fig. 5). These beds have been defined as being generally less than 5 cm thick, and are typically composed of silt to fine-grained sand-sized particles in the lower part of varying thickness (Fig. 6) grading upward to clay. They are "distal" turbidites in the sense of Walker (1967), but can occur in "proximal"environments such as interchannel areas on submarine fans (*e.g.*, Nelson and others, 1978). In the extreme, thin clastic turbidites may consist largely of mud, as discussed in the next section (*e.g.*, Piper, 1978).

Thin-bedded turbidites are ubiquitous in DSDP Sites, especially in continental rise and abyssal-plain settings. Piper and Brisco (1975) have illustrated some typical examples from the Antarctic continental rise on Leg 28 (Site 268), and Piper has discussed thin silt turbidites in the Aleutian Trench from Site 180 (Leg 18) deposited at rates in excess of 1800 m/m.y.

A major problem is to distinguish between some fine-grained, crossbedded, lenticular or laminated thin-bedded units deposited by dilute-suspension turbidity currents and those winnowed by bottom-currents, often termed "contourites" (*e.g.*, Heezen and others, 1966; Bouma and Hollister, 1973). Stow and Lovell (1979) have recently discussed the difficulties in distinguishing contour-current deposits from turbidites in distal settings, especially where the two types are interbedded. The original criteria used (*e.g.*, Hollister and Heezen, 1972; Bouma and Hollister, 1973) differentiated between thick-bedded, coarser-grained, poorly-sorted units as turbidites and thin-bedded, crosslaminated, better-sorted units having sharp tops as contourites. Although these criteria are often applied, they do not distinguish satisfactorily between thin-bedded turbidite silt sequences and possible current-winnowed or current-transported deposits (*e.g.*, Piper and Brisco, 1975; Hollister, Craddock and others, 1976; Piper, 1978; Rupke and Stanley, 1974; Stow, 1979; Stow and Lovell, 1979; Kolla and others, 1976).

Many of the identifications of contourites are by inference using knowledge of present patterns and velocities of circulation, on the North Atlantic mar-

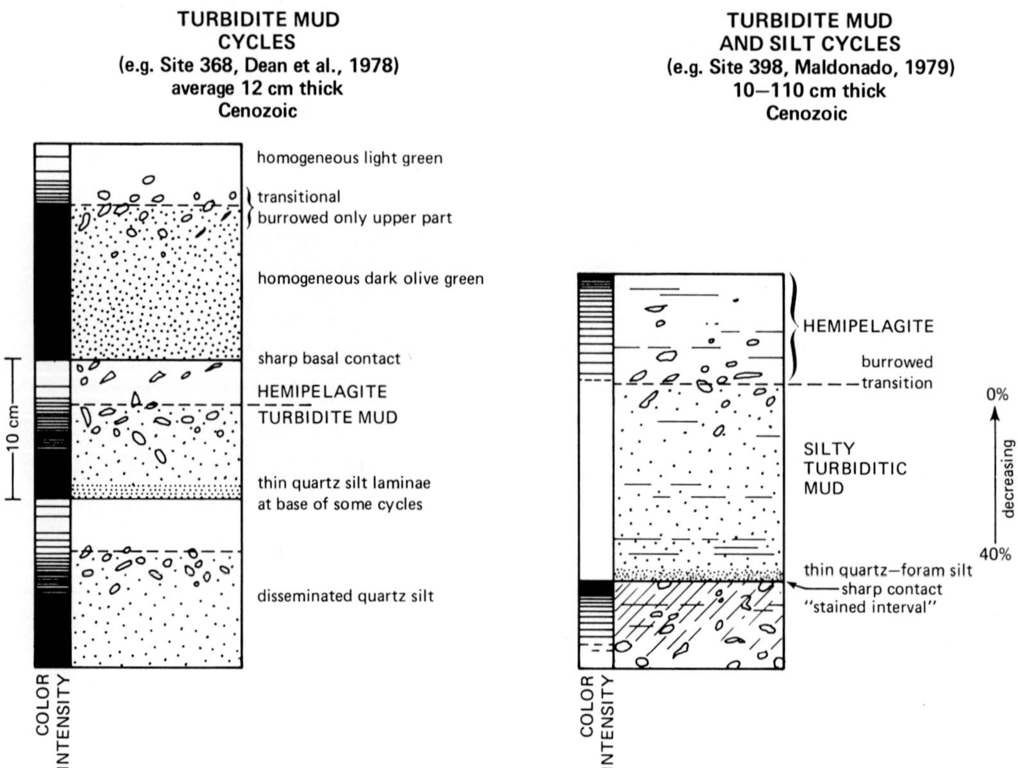

FIG. 6.—Diagram of typical turbidite mud (and silt) cycles from DSDP Sites in the North Atlantic. Mud turbidites at Site 368 are Eocene in age, while those from Site 398 are Oligocene-Miocene (see text). Cycles are about 10–20 cm thick.

gin, for example. In DSDP sites it is not always possible to infer paleocirculation patterns. It is, however, important to recognize contourite facies. Documentation of contourites in DSDP cores is sparse, but suspected examples have been described from Antarctic continental rise and abyssal plain areas on Legs 28 (*e.g.,* Piper and Brisco, 1975) and 35 (*e.g.,* Tucholke and others, 1976). Other suggested contourite laminae occur in sediment drifts drilled on Leg 49 (*e.g.,* Site 408: Shor and Poore, 1978).

Detailed studies of suspected contourite deposits in DSDP cores are needed. Sequence, composition, fauna, grain-size and texture, and changes in grain orientation (anisotropy) are necessary to distinguish turbidites from contourites, because visual identification is nearly impossible (see Stow and Lovell, 1979, for description of these criteria).

Turbidite Muds

Deep-sea terrigenous sequences are commonly dominated by muds, much of which may have been deposited from turbidity currents. Piper (1978) has, for example, estimated that at least 50 percent of the mud deposited in rise and abyssal plain settings was of turbidite origin. Hesse (1975) attributed some 80 percent of fine-grained sediments in the Cretaceous Flysch of the Eastern Alps to turbidity-current deposition. Turbidite muds have not been widely recognized in sediment sequences from DSDP Sites; this is undoubtedly because they are often difficult to discern in intervals disturbed by drilling and because many sequences of hemipelagic mud are monotonous, of little interest to most shipboard scientists, and commonly described in the shipboard core descriptions as "S.O.S." (Same Old Stuff). However, under close scrutiny, many mud sequences may divulge their depositional origins using the following criteria outlined by Rupke and Stanley (1974), Rupke (1975), Hesse (1975), Piper (1978), McCave (1979), Gardner and others (1978), and Stow (1979). The reader should consult these papers for more detail.

Primary Sedimentary Structures and Textures. — Turbidite muds may be finely laminated, faintly graded, or homogeneous depending on the grain-size distribution and sorting. Grain-size analysis by Coulter-counter is also useful to detect textural grading in otherwise homogeneous mud sequences as employed by McCave (1979) in his study of Eocene turbidite muds from Site 386 in the Western North Atlantic. Piper (1978) has constructed a sequential threefold division of turbidite muds (Van der Lingen, 1969; Hesse, 1975), analogous to the Bouma sequence, as follows from top to bottom:

F - hemipelagic sediment
E3 - ungraded mud
E2 - graded mud
E1 - laminated mud
D - laminated sand and silt

As with the familiar Bouma sequences, not all of these divisions may be found in each bed. Turbidite muds often contain substantial silt-sized material (<50%), either as discrete laminae or as disseminated particles. The muds commonly overlie and are in continuity with sand or silt turbidites. There are many examples of this relationship in DSDP sequences (Piper, 1978). Turbidite muds also occur interbedded with so-called "hemipelagic" muds having relatively few coarse-grained turbiditic layers (*e.g.,* Rupke, 1975). Rupke (1975) suggests that these two mud types may be distinguished, in part, by silt content—the hemipelagic intervals generally containing lower amounts of silt (<15%) than the turbiditic muds. Perhaps the best example of relatively homogeneous interbedded turbiditic and non-turbiditic mud in a DSDP site comes from the Upper Eocene through Lower Miocene of Site 368 on the Cape Verde Rise (Dean and others, 1978). This sequence consists of simple couplets of light and dark olive-green claystone with an average thickness of about 12 cm. However, the dark olive green muds, which contain slightly more organic carbon, have sharp basal contacts with underlying light green mud, and transitional upper contacts which are obscured by bioturbation (*e.g.,* Fig. 6). The lower parts of the darker layers are unburrowed and in some cases have very thin detrital silt layers at the base. These features are characteristic of turbidite mud sequences. Similar but thicker, more "proximal" turbidite muds from the Upper Aptian of Site 400A in the Bay of Biscay were described by deGraciansky and others (1979). The individual turbidite mud units are up to one meter thick with sharp basal contacts. Laminated silty claystone occurs at the base of each unit grading upward to homogeneous claystone, and only the upper 20–30 cm are burrowed.

Compositional distinctions. — Aside from the differences in total silt content between turbiditic muds and hemipelagic muds in the same sequence (Rupke, 1975) mentioned above, there may be other compositional distinctions. In many cases, inferred turbiditic muds in DSDP sites contain both more carbonate and organic carbon than interbedded hemipelagic muds. Commonly some shallow-water fossils are included. Maldonado (1979) showed the pronounced increase in carbonate content and preservation of calcareous microfossils within redeposited silty muds interbedded with carbonate-depleted hemipelagic muds deposited at or below the CCD in the Tertiary section at Site 398 off Portugal. In the examples described by Dean and others (1978) and deGraciansky and others (1979) outlined above, the turbidite muds contain substantially more organic carbon than the interbedded more slowly-deposited hemipelagic intervals. For example, at Site 400A, the Upper Aptian turbiditic muds contain up to 2.4 percent organic carbon, with an average of 0.6 percent (largely

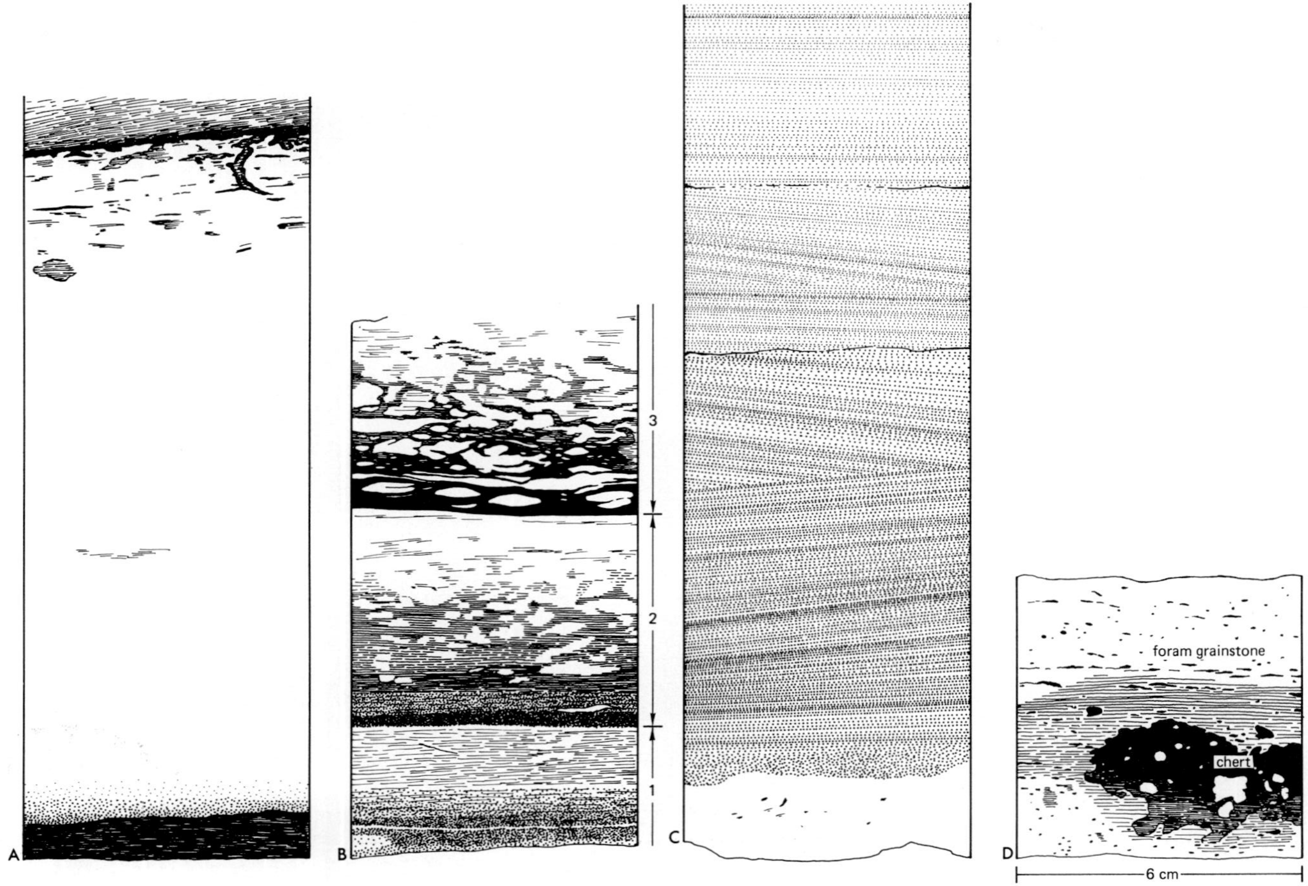

FIG. 7.—Sketches of carbonate "pelagic" turbidites and redeposited volcanogenic material in sequences from near the Line Islands, Pacific (Leg 33; see Cook and others, 1976). A) 33-314S-24-6 (30–45 cm): Calcareous ooze "pelagic" turbidite of Mid-Maestrichtian age. Faintly graded base; homogeneous mid-section; increase in burrowing near top. B) 33-215A-20-5 (111-123 cm): Sequence of three graded, mixed volcanogenic turbidites (Lower Maestrichtian). Amount of calcareous material (lighter) increaes toward top of each graded interval. Note downward decrease in burrowing of each, which practically obliterates bed number 3. C) 33-315A-23-4 (115–150 cm): Light to dark greenish-gray graded volcanogenic sand showing sequence of structures (scour, grading, cross-lamination, and parallel laminae) typical of turbidites (Bouma sequence). D) Diagenetic silicification front in a foraminiferal grainstone—example of preferred diagenesis along the coarser basal portion of a graded bed.

plant debris), while interbedded pelagic calcareous mudstones contain less than 0.35 percent organic carbon. Similar examples can be seen in the Lower Cretaceous of Site 398 (*e.g.*, Arthur, 1979; deGraciansky and Chenet, 1979).

Turbiditic muds are, therefore, recognizable and probably quite common in terrigenous sequences cored by DSDP. They are believed to originate as dilute suspension currents (Piper, 1978; see Rupke, 1978, for discussion). Further examples can be found in marginal areas of deep-sea fans or cones such as the Nile Cone (Site Report 131) where mud layers up to 3–5 m thick are common, and on the Indus Cone (Jipa and Kidd, 1974; Thiede, 1974). Although no turbidite muds were identified as such in the Black Sea, depositional rates of mud sequences of over 800 m/m.y. in the Quaternary at Site 380 imply rapid emplacement of muddy sediment, probably by turbidity currents, reflecting a change in the course of the Danube Delta (Hsü, 1978).

Pelagic Turbidites

Ocean ridges are regions of considerable topographic relief, tectonic activity, episodic currents and pelagic sedimentation. Ponding in low areas is common, suggesting injections of low-density turbidity currents from spreading flanks (van Andel and Komar, 1969). The role of Neogene turbidite sedimentation in these areas is imperfectly known, partly because of drilling disturbances. But DSDP cores have shown that redeposition of calcareous and siliceous biogenic oozes has been a major process around mid-ocean rises, ridges, seamounts, and other submarine highs. Of key importance in interpreting pelagic turbidites is consideration of the distribution of components in the source regions and the hydraulic equivalence of various biogenic components. If the source is a fine coccolith ooze, with scattered foraminifers and/or radiolarians, it is not surprising the grading in such beds would be imperfect. The hydraulic equivalence of the biogenic components is difficult to surmise, partly because of lack of knowledge of the original particle size. Tests may be filled or empty, individual, flocculated or grouped as fecal pellets, or as whole tests or platelets. Threshold erosion curves for pelagic components are lacking, but Miller and Komar (1977) estimated that currents of 5–10 cm/sec will erode foraminifers or radiolarians, whereas coccolith muds require 15–30 cm/sec (Southard and others, 1971). Here we discuss the various types of pelagic biogenic turbidites, and suggest some possibilities for their origins.

Calcareous biogenic turbidites.—Redeposited calcareous oozes are most easily recognized when the sediments have been transported to depths below the CCD or lysocline. Pelagic turbidites are generally inferred to be the result of low-concentration, sluggish turbidity currents, which have limited erosive capabilities. Currents are estimated at 5–30 cm/sec. The base is generally sharp, seldom with any evidence of cross lamination or erosion. There may be a concentration of foraminiferal sand, or a few scattered volcanogenic grains (Figs. 7, 8) at the base. The body is usually massive, white to gray, and highly calcareous, with a hint of grading shown by color gradations. The upper section is commonly

FIG. 8.—Thin, graded foraminiferal grainstone with shallow-water carbonate debris and benthic forams along with dark green palagonite clasts. The basal contact with reddish-orange burrowed limestone deposited near the CCD is sharp, but the upper contact is diffusely burrowed. Such turbidites document the emergent periods of nearby mid-ocean volcanic rises or seamounts (sample 33-315A-19-2, 52–50cm; Middle Maestrichtian from near the Pacific Line Island Chain.)

delineated by an exponentially-downward-diminishing series of dark-filled burrow structures (*e.g.*, Cook and others, 1976). The host sediment usually contains less carbonate than do the pelagic turbidites which are commonly initiated on slopes above the CCD. In areas with the CCD depressed, such as the mid-Pacific equatorial bulge, the recognition of redeposited calcareous oozes is more problematic. The basal enrichment of foraminifers may still occur, and some faint burrowing at the top may be visible. Redeposited oozes commonly have a slightly more purplish hue, sometimes gradational, caused by the presence of slightly more organic matter or disseminated pyrite than in the host sediment.

Burns, Andrews and others (1974) report aragonite preserved in pelagic sediment redeposited below the CCD at Site 210. A typical example of a redeposited biogenic calcareous sequence is taken from the Lower Oligocene to Lower Miocene section of Site 398. Beds 1-5 cm thick, which are foraminifer- and nannofossil-rich and contain scattered pyrite and glauconite grains, are assembled into rhythms 10–50 cm thick. The base (3–5 cm) comprises a thin turbiditic silt which is either finely laminated or faintly graded. This interval is overlain by a pelitic fraction (T_e) richer in carbonate than the overlying part which is a true pelagic interval (E_p) which is correspondingly lighter in color (see Fig. 6).

Calcareous turbidites deposited below the CCD may contain the only datable fossils. However, the fossils may be coeval or significantly older than the time of redeposition. Using ichthyoliths found in brown clays, for example, Kaneps and others (1980) have shown that an 80-m-thick unit of Middle Eocene chalks sandwiched between brown clay in Site 212 (Wharton Basin, Indian Ocean) must have been redeposited, because the brown clays below are of Oligocene age. Therefore, caution must be exercised in interpreting such redeposited sequences.

Diagenesis may occur earlier in calcareous pelagic turbidites than in the host sediments partly because the turbidites may contain relatively unstable carbonate minerals. At several DSDP sites in the Pacific, foraminiferal grainstone layers are preferential sites of silicification (Keene, 1975; Kelts, 1976; Fig. 7). This process may be due to concentration of siliceous tests in these layers, increased porosities for fluid movement, or trapped geochemical gradients.

Diatomaceous turbidites.—In the Gulf of California, DSDP Leg 64 cored turbidites (Sites 477, 478, 481) composed almost entirely of diatom frustules, with an admixture of silty clay (Curray, Moore and others, 1980). Pyrite and organic matter (up to 4% organic carbon) are abundant. Individual beds, which are 30–80 cm thick, dark olive gray, and homogeneous, show little visible evidence of grading. Because the Guaymas Basin floor is marginally anoxic, the tops of these turbidites are not heavily burrowed. Two characteristics are noteworthy: 1) along the base there is commonly a thin laminae of benthic foraminifera with some bivalve hash, and 2) the tops of each unit are commonly marked by a 2–5 cm layer of pale olive, pure diatom ooze. The color changes, burrowing relationships, and differentiation of constituents into certain parts of the beds suggest that these are turbidites (Fig. 9). The small amount of flocculated clay and fine-grained silt in these turbidity currents apparently settles out of suspension faster than some lacy diatom frustules common in upwelling regions. The question of hydraulic equivalence of diatoms is unresearched in the marine environment. Clearly, such turbidites may cause size- as well as species-sorting (*e.g.*, Lerman and others, 1974).

The Guaymas Basin sites penetrated up to 400 meters of uppermost Pleistocene turbidite fill, about half of which are diatom turbidite layers. The accumulation rates are enormous, probably more than 1,500 m/m.y. The components and sedimentary structures suggest that these diatomaceous turbidites were redeposited by low-density turbidity currents initiated from bathyal depths of the outer hemipelagic slopes. Episodic movements in this tectonically active region might easily dislodge sediments from the unstable slopes. Such diatomaceous turbidites are prone to early diagenetic changes, such as the recrystallization of the opaline frustules to porcellanite, which might soon mask the turbidite origin of the beds (Einsele and Kelts, 1980).

Redeposited siliceous biogenic deposits.—Paleogene (Middle Eocene to Lower Oligocene) sediment rich in radiolarians, diatoms, siliceous sponge spicules, and other siliceous microfossils are widespread in the North Atlantic (*e.g.*, Tucholke and Vogt, 1979). The perplexing widespread occurrence of these deposits (*e.g.*, Pisciotto, this volume) has led to a number of hypotheses to account for their origin. These include enhanced productivity due to the greatly increased silica content in sea water caused by increased volcanic activity during the Eocene (Gibson and Towe, 1971; Mattson and Pessagno, 1971), or to enhanced productivity due to upwelling of nutrient-rich, oxygen-depleted deep water and enhanced preservation in silica-saturated deep waters (*e.g.*, Fischer and Arthur, 1977).

Whatever the ultimate cause of the apparent increased productivity of siliceous organisms in the North Atlantic, their widespread distribution and concentration of siliceous microfossils (and chert) in discrete beds has not been satisfactorily explained. A number of sites where these deposits occur were relatively distant from presumed areas of upwelling along the continental margin or equatorial zone. Biogenic siliceous beds at many of these basinal

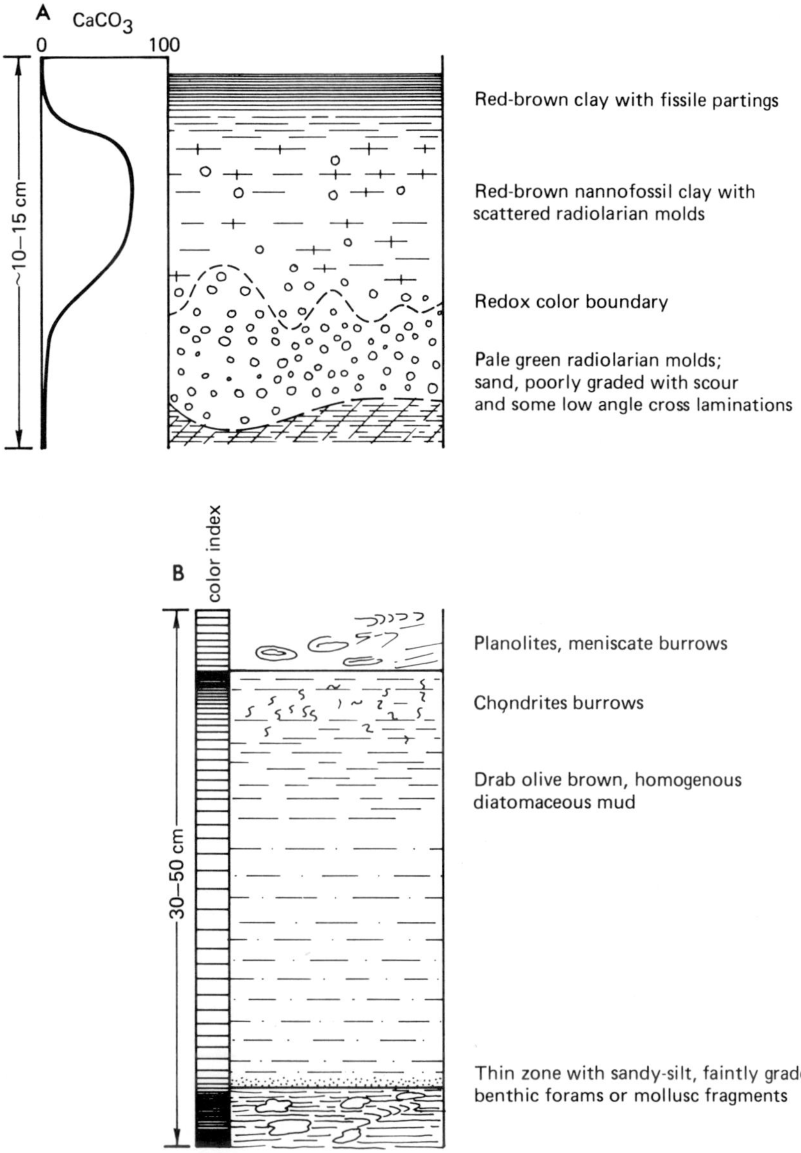

FIG. 9.—Top. Paradoxical example from the Albian of the North Atlantic of a radiolarian-rich bed deposited below the CCD, which could be a turbidite, a winnowing cycle, or one of many radiolarian "blooms" (see text). This sequence shows concentration of radiolarians near base of unit, with more carbonate-rich material in mid-bed, capped by red clay. The pore waters in the basal thin bed may have become anoxic after deposition of the unit due to initial high organic content, and therefore the green color of the basal zone may be a diagenetic reduction effect possibly wrought by a slightly higher organic concentration in the initial radiolarian accumulation. Bottom. Diatomaceous mud turbidite from the Gulf of California (Leg 64). In these homogeneous muds the repeated occurrence of a zone of *chondrites* burrows, overlain by a mixed population of trace fossils, is used as one main criterion for subdividing beds. *Chondrites* appear to show a tolerance to lower oxygen levels than other burrowers.

deeper water sites exhibit sedimentary structures characteristic of turbidite sedimentation, including textural grading, lamination, and cross-lamination (*e.g.*, Sites 6, 7, 8, 387, 27, 28). These characteristics were pointed out originally for some Middle Eocene sediments of Sites 6 and 7 by Beall and Fischer (1969) in their study of Leg 1 materials. Peterson, Edgar, and others (1970) also noted cherty siltstone and turbidites containing glauconite and biogenic silica in Middle Eocene strata at Site 8 (Leg 2). More recently, McCave (1979), Kagami (1978), and Riech and von Rad (1979) have analyzed and described Eocene and Oligocene turbiditic units rich in siliceous biogenic debris in the North Atlantic DSDP sites. Kagami (1978) presented evidence that bottom currents and turbidity currents in the Bay of Biscay and along Rockall Bank were responsible for concentrating biosiliceous material in discrete beds, especially Oligocene and Lower Miocene beds containing abundant siliceous sponge spicules. McCave (1979) used a Coulter-counter to perform textural analyses of cryptically graded siliceous muds of Eocene-Oligocene age on the Bermuda Rise, and showed that these massive, relatively homogeneous and dark-colored units at Sites 386 and 387 were probably redeposited.

These observations suggest that much of the Eocene and Lower Oligocene biosiliceous sequences in the deep North Atlantic basin were redeposited and derived from the continental margins. The CCD was elevated during the Eocene (Tucholke and Vogt, 1979), sea level was relatively high (Vail and others, 1977), and rates of movement of clastic material across the shelves into the deep-sea were probably low. Drilling on the western North Atlantic shelf of the United States (Hathaway and others, 1979; Scholle, 1977, 1978) has demonstrated thick sequences of Eocene pelagic carbonate rocks which were deposited along the entire margin. Coastal upwelling along the margins of a highly fertile North Atlantic basin and high rates of sedimentation of biogenic deposits on the outer shelf and slope may have led to frequent slumping and generation of turbidity currents along the margin. Undercutting of the slope by deep contour currents during the Oligocene time may also have triggered slumping and generation of turbidity currents. High rates of carbonate dissolution at the base of slope may have induced slumping from submarine highs and along slopes by progressively undermining slope deposits (*e.g.*, Berger and Johnson, 1976). The turbidites contain abundant quartz, glauconite, and shallow-water fossil debris (typically outer shelf-slope faunas) which suggest derivation from the margin.

Bedded radiolarian cherts of Jurassic age showing redepositional features, graded beds, laminations, and the like, have been described from Tethyan areas partially interpreted as mid-ocean ridge settings (see Nisbet and Price, 1974; Imoto and Fukutomi, 1975; Kalin and others, 1979). Equivalent deposits have not been recovered in DSDP cores, although two time-stratigraphic units, the Middle Eocene and the Albian of the North Atlantic, show prolific radiolarian productivity. Albian radiolarian-rich layers from Sites 386, 417, 367 and 136 show faint similarity to bedded cherts, including features common to turbidites. A typical bed (Fig. 9) is reddish-brown, in some cases cross- or parallel-laminated with a basal zone of radiolarian sand. This portion of a bed may also display some reduced greenish patches. The middle section is a radiolarian-nannofossil marl, and the top comprises a few centimeters of claystone, with partings. Silicified radiolarians and molds are common. Beds rarely exceed 10–15 cm, and contacts are sharp. The best developed cycles of interbedded turbidites and clay occur in sites drilled near the crest of the Late Aptian to Albian Mid-Atlantic Ridge, and it has been suggested that these deposits represent oxidized pelagic turbidites, wafted from nearby highs, and redeposited in lows (Tucholke and Vogt, 1979). After his grain-size studies, McCave (1979) concluded that cycles from site 387 and 386 were the result of "long" radiolarian blooms resulting from turnover in the oxygen-deficient Mid-Cretaceous ocean and subsequent bottom-current winnowing. Support for a non-turbidite origin is also given by the distribution of radiolarian-rich beds. The thin, oxic Upper Aptian or Albian sequence containing such cycles can be correlated along 2000 km of the presumed ridge trace, with almost identical sedimentary features. It seems unlikely that low-density turbidity currents could travel these distances along rugged ridge topography, and that local turbidites should not occur simultaneously along larger ridge segments. In the basinal site 367, radiolarian beds also have similar features and thicknesses. This problem remains unresolved.

Volcanogenic Turbidites

Turbidites composed mostly of volcanogenic constituents are transported by the same mechanisms as their terrigenous counterparts, but their unique occurrences in mid-ocean, arc-trench and archipelagic settings warrant some special treatment. Redeposited volcanic debris is commonly used as an indicator of nearby active volcanic phases (*e.g.*, Vallier, 1974; Heath and Moberly, 1971). For example, Sites 315 and 316 of Leg 33 were drilled along the eastern margin of the Line Islands. Redeposited archipelagic sediments comprise up to 50 percent of some stratigraphic intervals there (Cook and others, 1976). Fine-grained beds exhibit well-defined primary sedimentary structures such as normal size-grading, parallel and low angle cross-laminations, and selective sorting of various grain components. Coarse-grained beds tend to show poor sorting and massive bedding. Three main types of beds are recognized (Cook and others, 1976): 1) Graded units composed solely of pelagic nannofossils and foraminifers with a smattering of palagonite grains that are common but difficult to recognize. These thick (60–250 cm) beds are characterized by a slight color change, from white at the base to yellowish brown at the top, presumably related to clay-mineral content. Burrow mottling of the top is common. 2) Mixed layers with pelagic and reefal carbonate and moderately abundant volcanogenic constituents are very common. Grading is obvious from the presence of a thin basal sand laminae and a peppering of green palagonitic grains and large foraminifera. Primary sedimentary structures are clearly defined as is common basal scour. Smectite-rich lutite tops have conspicuous burrowing. The amount of downward burrowing can be used as a relative indicator of the time elapsed between subsequent turbidite events (see Fig. 7). Turbidite beds may be thin, from 5–10 cm, or up to 60–80 cm thick; 3) Resedimented layers comprising only volcanogenic components occur either as massive breccias or as dark greenish-gray turbidite sands to silts with textural grading, cross- and parallel-laminations, and burrowed lutite tops. The breccias are commonly grain-supported.

The clasts in volcanogenic turbidites may include vitric glass shards, vesicular hyaloclastic fragments, fresh basaltic pebbles, rounded and weathered volcanic grains, clay clasts, euhedral feldspar or mafic minerals, and feldspathic crystal aggregates. A notable feature is that when cored, the composition and texture rarely match the original redeposited condition because of relatively rapid diagenesis upon burial. Unstable glass quickly devitrifies, forming soft palagonite, smectite clay minerals, and zeolites. Halmyrolosis (submarine weathering) decomposes mafic grains. Volcanogenic turbidites may be sites of preferred early diagenesis and cementation, or can act as conduits for hydrothermal fluids from later stages of edifice volcanism which forms new mineral assemblages including chlorite, authigenic feldspars, zeolites, carbonates, and sulfides (*e.g.*, Kelts and McKenzie, 1976; Jenkyns, 1976; Natland, 1976). Thus, conventional size analysis after diagenesis has little application to understanding transport mechanisms. Volcanogenic turbidite sequences are particularly prone to post-depositional creep, slumping, and stretching of burrows.

Settings where volcanogenic turbidites occur provide examples of mass-wasting and transport in complex non-fan environments, although the mechanics of emplacement for hyaloclastic breccias from submarine eruptions remain poorly understood. Tectonic movements and mineral changes lead to the collapse of unstable mounds and redeposition in the cold state by debris flow avalanches and turbidites. These may occur both from shallow-water sources and from the steep flanks of volcanic pedestals.

Other examples were cored on Leg 21 in the Lau Basin at Site 203 (Pleistocene volcanogenic sands), and on the outer slopes of the Tonga Trench (Site 204) where 409 m of young volcaniclastic breccia and sand were recovered (Burns, Andrews, and others, 1974).

Slumps and Other Mass Flow Deposits

Slump deposits, pebbly mudstones (Fig. 10), and sedimentary breccias, products of downslope mass movement, have been encountered in DSDP Sites, especially in those from along the continental rise or slope. The attributes of such deposits are fairly obvious, but are sometimes difficult to recognize in narrow DSDP cores, especially when individual slump masses or debris-flow units are tens of meters thick, core recovery is spotty, or drilling-induced deformation is extensive. If individual clasts in pebbly mudstones, conglomerates, or breccias are much larger than 5 cm in diameter it may be difficult to distinguish them from interbeds between units.

Sheared, stretched, or highly microfaulted beds, folded or overturned layers, and frequent reversals in dip are all good indicators of probable slumped beds in DSDP cores. Good examples are illustrated in Site Chapters for DSDP Sites 360 through 365 (Leg 40; Ryan, Bolli and others, 1978) and Site 398 (Leg 47).

Debris flow and breccia deposits, which may be part of a continuum between slide or slump deposits and turbidites (*e.g.*, Hampton, 1972; Middleton and Hampton, 1973; Embley, 1976), are best illustrated by a 554-m-thick sequence of interbedded slump units, debris flows and turbidites recovered in Site 397 on the upper rise off northwest Africa (Arthur and von Rad, 1979). Pebbles of various more indurated lithologies and plastic muds (often smeared-out or elongated in presumed flow directions) are enclosed in a dark, organic carbon-rich silty mud matrix containing benthic foraminifers typically dwelling at shelf and slope depths. Another example of a sequence of mass flow deposits was cored in Site 391 (Benson, Sheridan, and others, 1978) in the Blake-Bahama Basin. Over 500 meters of redeposited intraclastic chalk breccia of Early to Late Miocene age and interbedded indigenous radiolarian mudstones, deposited below the CCD, were recovered. The intraclastic chalks were probably derived from the adjacent Blake Plateau. These mass flows, in addition to clasts of chalk, contain radiolarian mudstone clasts, also of Miocene age, eroded from the seabed over which they traveled. The mass flows apparently were trapped in a basin bounded by the "ancestral" Blake-Bahama outer Ridge. Such mass flow deposits sometimes exhibit sedimentary structures of the classic "Bouma sequence" such as textural grading and cross lamination.

FACIES ASSOCIATIONS OF REDEPOSITED SEDIMENT AND TECTONIC-PALEOENVIRONMENTAL RECONSTRUCTIONS

Aside from the descriptive aspects of redeposited sediments documented in DSDP cores, as discussed above, the DSDP has provided data and core material on the reconstruction of facies associations and facies organization in a number of geologic settings. Many of the facies characteristics of modern and submodern submarine sedimentary environments elucidated by DSDP drilling have been invaluable in interpreting depositional environments of ancient sedimentary sequences. As drilling continues and large amounts of existing data are collated we suspect that the results of DSDP will have an even greater impact on sedimentology. In this section we review some of the marine depositional settings and associations of redeposited sediment sampled by DSDP. The main theme is to demonstrate how the study of redeposited sediment in DSDP cores has provided clues to the tectonic evolution of active and passive margins and linear island chains, to the history of development of deep-sea fans and their internal facies relationships, and to the imprint of sealevel and climatic and oceanographic processes on the sedimentary record. We have selected examples from various DSDP legs, but by no means have we exhaustively reviewed all aspects of redeposited sediments exhibited in DSDP cores or in the burgeoning literature on these topics.

Drilling on Submarine Fans

Since the late 1960's the emphasis of many marine geological studies has shifted to studies of sediment dispersal along modern continental margins, mainly through investigations of submarine fans and abyssal cones. However, these studies have been performed primarily using seismic reflection, side-scan sonar or other remote-sensing methods (*e.g.,* Normark, 1970; Normark and Hess, 1980). As a result, we now know the surface morphology of a number of fans, and, using results from conventional piston cores, we have some idea of the surficial distribution of sediment types (Nelson and Kulm, 1973; Nelson, 1976; Wilde and others, 1978). High-resolution seismic reflection records have also elucidated the basic internal geometry of a few fans and cones. Yet, we know very little of the actual internal distribution of sediment types, their depositional environment, sedimentary structures, sand contents, or reservoir properties. Fan depositional models and internal geometries are based almost entirely on studies of several well-exposed ancient deep-sea fans on land (*e.g.,* Mutti and Ricci Lucchi, 1972; Walker and Mutti, 1973; Walker, 1975, 1978). These models have been extrapolated to modern fans. The models are useful, but it is desirable to have information on at least one fan where detailed sedimentological studies could interface with high-resolution seismic profiling in order to construct a full three-dimensional geometry of the last several million years of fan deposition. Such knowledge would be useful for petroleum exploration in older and more deeply buried parts of fans, particularly because we might have a better understanding of the relationship between total deposition on fans, the thickness and geometry of fan sand-bodies, and changes in fan depositional regimes related to climate change, oceanographic events, sealevel fluctuations, basin type, and tectonism.

DSDP drilling has contributed only marginally to knowledge of sedimentary facies in submarine fans, although a number of holes have been drilled on marginal or outer parts of active fans (shown schematically in Fig. 11). The simple presence of turbidites does not constitute evidence of a submarine fan configuration. It is difficult, in 6-cm-wide cores, to adequately define fan facies because this requires knowledge of shapes of sedimentary bodies and lateral changes in sedimentary structures, as well as of the vertical sequence exhibited in cores. These relationships require data from more than one hole and adequate seismic resolution.

On the basis of the vertical sequences of sedimentary structures in Site 299 drilled in the Yamato Basin, Bouma (1975b) interpreted sediments, though most were highly deformed in drilling, as part of a submarine fan, mainly as levee or proximal overbank deposits with one possible channel deposit. The upper 76 meters were considered to be more proximal outer levee deposits. Bouma argued that the deposits of predominantly silty clay were dissimilar to previously described distal or thin-bedded turbidites, and that they must represent a sand-poor fan system or confinement of sand to a narrow channel. This is an example of over interpretation of possible fan facies based on DSDP cores, but one which is not unreasonable considering that the Site was drilled in an area having submarine fan morphology connected to the shelf by a major submarine channel (the Toyama Trough). Interpre-

←

FIG. 10. — Left. Part of a 3.5 m-thick slumped interval from the flanks of the Vigo Seamount which consists of Upper Aptian medium gray to bluish white marly limestone and shows some incipient shear planes (sample 47 B-398D-105-1,28–57cm; Aptian). Right. A matrix-supported, graded, pebbly mudstone from the flank of the Vigo Seamount on the continental margin of western Iberia. Pebbles consist of white oolitic limestone, deformed white to gray-brown chalks, black claystone, green claystone and some graded siltstone clasts (sample 47B-398D-124-4, 6–29cm; Lower Aptian).

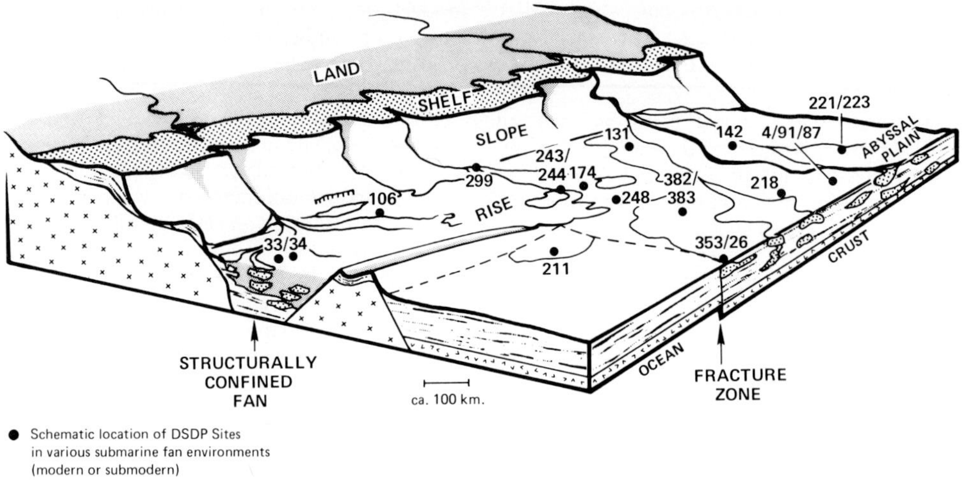

FIG. 11.—Diagram showing schematic location of various DSDP Sites on a hypothetical modern submarine fan environment. Relatively few have penetrated proximal fan facies.

tation becomes more difficult in even older turbidite sequences such as those of Tithonian-Hauterivian and possibly younger age cored in the Moroccan Basin in Sites 416 (*e.g.,* Lancelot and Winterer, 1980) and 370 (*e.g.,* Gardner and others, 1978), which are suspected of being part of a large fan complex, and the distal part of an Aptian to Upper Cretaceous possibly structurally confined fan at Site 361 in the Orange Basin (*e.g.,* Natland, 1978; Kagami, 1979).

Other DSDP sites have penetrated marginal parts of more modern fan systems. Site 174 was drilled on the lower part of the Astoria Fan in an interchannel area (Kulm, von Huene, and others, 1973). The fan deposits are over 280 m thick there and include sand beds up to 8 m thick. The major part of fan growth has apparently occurred since Middle to Late Pleistocene time. Fan deposits apparently prograded over a Plio-Pleistocene abyssal plain sequence containing thin-bedded turbidites. The overall fan section coarsens upward. Beds of poorly- to well-sorted lithic arkoses, classic turbidites, from 50 to 100 cm thick, are common (62% of the fan sequence is sand). Diagenesis has had little affect on the sand beds.

Site 218 (Leg 22) was drilled on the distal portion of the Bengal Fan, some 900 km or more from the source. Over 770 m of sediment of Middle Miocene through Holocene age were penetrated. Interbedded and interlaminated fine-grained sandstone, siltstone, and silty clay were recovered; about 50 percent of the sequence consisted of pelagic sediment (clayey nannofossil ooze), but relatively coarse-grained terrigenous clastics apparently were deposited in at least four pulses: in the upper Middle Miocene, middle Upper Miocene, uppermost Miocene-Pliocene, and Upper Pleistocene. Graded sand and silt layers from 20–60 cm thick are common. A nannofossil ooze of Holocene age caps the sequence. These pulses could be related either to channel avulsion on the lower part of the fan (*e.g.,* Rupke, 1978), to increases in orogenic activity in the Himalayas (Moore and others, 1974), to changes in climate or sea level, or to a combination of the above. The amount of coarse detritus at that distance from the apex of the Bengal Fan is staggering. Site 211 (von der Borch and others, 1974) recovered some silty sand and silty clay, probably thin-bedded turbidites, of Pliocene age from the distal part of the now extinct Nicobar Fan, from which the sediment supply was cut off by the northward movement of the Ninety-East Ridge. Ingersoll and Suczek (1979) and Thompson (1974) showed the clastics on the Bengal and Nicobar Fans to have been derived from uplifted crystalline basement and older sediment sources based on petrographic studies of DSDP core samples. Sites 221 and 222 penetrated terrigenous clastic sediment which was part of the Indus Cone sequence (Whitmarsh and others, 1974). Carbonate-rich detrital clay with interbedded sand and silt beds accumulated at Site 222 at rates of 600 m/m.y. in the Late Miocene, 135–350 m/m.y. in the Pliocene and less than 50 m/m.y. in the Quaternary. The sequence at Site 221 in the Arabian Basin shows that prograding Indus Cone sediment reached the site in the Miocene; at that time sedimentation rates began increasing to a maximum of 170 m/m.y. in detrital clays with interbedded graded sandy-silty turbidites having sharp lower boundaries; the sequence overlies Paleogene pelagic limestones.

In the Madagascar Basin, Sites 243/244 (Leg 25)

could not be spudded-in because of thick, very coarse-grained quartz sand in the Zambesi Canyon, over 400 km from the shelf (Simpson, Schlich, and others, 1974). Site 248, another 400 km down the Zambesi Fan, recovered clayey silt, silty clay, sand and silty sand, and very coarse-grained quartz-sand layers of probable Mid-Miocene and younger age overlying Eocene brown clay and volcanic silty clay.

These examples illustrate the importance of single canyon-fan and abyssal cone settings in transporting terrigenous clastic detritus to the deep-sea. Many of these fans have built out substantially since Miocene time, and especially during Pleistocene time, in response to increased rates of tectonism and uplift of source terranes such as in the case of the Himalayas (Bengal Fan and Indus Cone), and in response to drastically lowered sea levels and to changing climates.

Continental-Rise Sedimentation on Passive Margins

Young Cretaceous Margins. —Terrigenous sand-silt turbidites are common in Early to Middle Cretaceous sediment recovered in DSDP sites from the North and South Atlantic (Fig. 1). The mineralogy of these turbidites suggests sources largely of granitic composition with admixed shallow water and pelagic carbonate. Early Cretaceous shelves were relatively narrow, and clastic sediment derived from uplifted terranes along parts of the rifted margins, especially in the northern North Atlantic and the South Atlantic, was swept across the shelves and into adjoining deep basins. Many of the accumulations of terrigenous turbidites may have been related to the development of deep-sea fans seaward of large delta systems. In some areas continental margin reef growth may have prevented bypassing of sediment into the deep sea. By about Turonian time, however, the sources of most of the clastic detritus to the deep sea were cut off, possibly by a series of major sealevel rises and flooding of the shelves. Over most of the North Atlantic only slowly deposited multicolored or red pelagic clays accumulated during the Late Cretaceous.

In the northern North Atlantic, times of peak accumulation of terrigenous detritus appear to be from Mid-Aptian to Late Albian (deGraciansky and others, 1979) along the northern margin of the Bay of Biscay (Sites 400, 402) and from Barremian to Late Albian along the continental margin of Portugal (Site 398: deGraciansky and Chenet, 1979; Arthur, 1979). The terrigenous influx followed rifting and opening of the Bay of Biscay and the North Atlantic north of the Newfoundland Fracture Zone respectively (*e.g.*, Roberts, Montadert, and others, 1978). Redeposited shallow-water carbonate detritus and reworked pelagic biogenic components are common in these Lower Cretaceous sequences. Turbiditic mud comprises a large proportion of the redeposited terrigenous material, and organic contents are relatively high; much of the organic matter is reworked terrestrial higher plant detritus.

In the eastern and western North Atlantic basins, relatively few terrigenous turbidites are found interbedded with pelagic limestones and carbonaceous claystones of Early Cretaceous age. The proportion of terrigenous clastic detritus is low in the western basin, possibly because of the Cretaceous reefal buildups along the outer shelf which apparently blocked much of the terrigenous flux (*e.g.*, Lancelot, Seibold, and others, 1978; Tucholke and Vogt, 1979; Benson, Sheridan, and others, 1978; Jansa and others, 1979). These reefs and banks may have existed on both sides of the Atlantic from Late Jurassic through Middle to Late Cretaceous time. Even so, Tucholke and Mountain (1979) showed pronounced increases in thickness of the Early Cretaceous black clay facies at local points along the continental margin, perhaps near point sources or breeches in the reef trend. Some terrigenous sand (quartz, oolites, and other shelf-derived material) occurs in thin-bedded, cross-laminated and graded limestone of Neocomian age at Site 391C (Freeman and Enos, 1978) nearest the margin, but at the other sites at which the Early Cretaceous sediments were cored, only carbonate slump units (Site 105) or "pelagic" turbidites were found interbedded in the pelagic limestone or claystone sequences (Lancelot and others, 1972). Lithologic trends in the eastern North Atlantic basin during the Early to Middle Cretaceous were similar. Some thin terrigenous silt turbidites were deposited during the Aptian-Turonian at Site 367 in the Cape Verde Basin (Dean and others, 1978), and are interbedded with black and green claystones. At Sites 370 and 416 in the deep Moroccan basin, however, a thick Lower Neocomian to Albian and Cenomanian terrigenous sequence was penetrated (Lancelot, Seibold, and others, 1978; Lancelot, Winterer, and others, 1980). Thin- to thick-bedded graded to massive muddy sands and silts containing quartz, feldspar, and a variety of shallow water and pelagic calcareous components (*e.g.*, Meyer, 1978) are interpreted as having been deposited in a deepwater fan setting. These are interbedded with thick, dark-colored claystones. The flood of clastic detritus may have originated in response to uplift of the Western Atlas region during the first Late Jurassic-Early Cretaceous tectonic phases there (*e.g.*, Lancelot, Seibold, and others, 1978). Late Cretaceous nannofossil claystones overlie the presumed fan facies.

In the South Atlantic, DSDP sites have penetrated strata no older than Late Aptian, with the exception of those sites on the Falkland Plateau. Most sites are located on submarine highs, and those in the deep basins (such as Site 355 in the

Brazil Basin) did not penetrate strata older than the Late Cretaceous (Supko, Perch-Nielsen, and others, 1977). Site 361, however, was drilled in the deep Cape Basin (Ryan, Bolli, and others, 1978) and recovered a Cretaceous clastic sequence about 900 m thick. Coarse, massive to graded sandstone and siltstone layers (classic turbidites) and pebbly and sandy mudstone units occur interbedded with carbonaceous mudstones in the lower 300 meters or so of the sequence, which is of Aptian age. These give way to interbedded thin- and cross-bedded, finely-laminated, well-sorted, quartzose siltstones and shale of Albian to Maestrictian age. Kagami (1978) has suggested that these deposits represent an outer deepsea fan environment. Natland (1978) postulated that the upper thin-bedded siltstones were formed by winnowing bottom-traction currents. Sites from the remainder of the South Atlantic show relatively little terrigenous influx to deeper basins during the Late Cretaceous.

Terrigenous mud also dominates the Lower Cretaceous sequences drilled on the rifted margin of Western Australia, especially those of Aptian-Albian Ages, as exemplified by the section at Site 263 (*e.g.* Davies and Kidd, 1977; Cook, 1977). By Cenomanian time, however, the rapid deposition of clastics had ceased and brown pelagic clays or nannofossil oozes accumulated (above the CCD) at most sites. Similarities in sedimentation over broad areas of the world ocean suggest that the Mid-Cretaceous rise in sea level flooded shelf and interior basins, thereby trapping much terrigenous sediment on the continents and shallow margins and preventing much bypassing to the deeper basins. This pattern continued through most of the Late Cretaceous.

Late Cenozoic Continental Margin Destruction and Rise Upbuilding

Seismic studies and DSDP sites have confirmed a widespread erosional event that occurred along the continental margins of the western North Atlantic, probably during Oligocene-Early Miocene time (*e.g.* Tucholke and Mountain, 1979; Tucholke and Vogt, 1979). A similar event took place in the eastern North Atlantic (*e.g.* Lancelot and Siebold, 1978; Arthur and others, 1979; Lancelot and Winterer, 1980). Material was eroded from the deepsea floor and along the continental rise and slope as well, probably by intensified geostrophic currents. On the western margin, some of the eroded material was accumulated as sediment drifts, such as the Blake outer ridge, but on the eastern side of the Atlantic there is as yet no known sediment drift resulting from a margin-destructive event (Arthur and others, 1979). In fact there seems to be a paucity of sediment of Oligocene age in the North Atlantic as a whole.

Off northwest Africa, a sequence drilled in Site 397 (von Rad, Ryan, and others, 1979) shows the sequential upbuilding of the upper continental rise (Fig. 12) following the erosional event which removed much of the upper rise and lower slope prism (Arthur and von Rad, 1979). Slope and shelf-derived material initially cascaded down the slope as slumps and debris flows and came to rest on the upper rise during Early Miocene time. As gradients stabilized, the frequency of redeposition events declined and debris-flow deposits gave way to occasional mud or thin-bedded sand turbidites. Finally, from the Late Miocene to the present a thick calcareous hemipelagic mud drape has covered the base-of-slope locality. Interestingly, much of the redeposited material was rich in organic carbon, probably derived from upper slope settings under a well-developed oxygen-minimum zone. This mechanism provides potential hydrocarbon source beds in deeper-water environments (see Arthur and others, 1979). Sloughing off of carbonate detritus from the Blake Plateau escarpment probably occurred at about the same time as the slumping on the African margin, as shown by drilling at Site 391 (Benson, Sheridan, and others, 1978).

The rapid upbuilding of the continental rise in the western North Atlantic basin during Miocene to Holocene time is illustrated by DSDP sites which have penetrated thick sequences of hemipelagic mud and interbedded biosiliceous and clastic turbidites deposited at rates often in excess of 150 m/m.y. (see compilation in Jansa and others, 1979). These are apparently common facies associations for continental rise settings. Nearer the margin, they are interbedded with slump and slide material. The seismic expression of this rise prism, as much as 2 to 3 km thick, varies depending on the number and thickness of interbedded sand and silt layers (*e.g.*, Tucholke and Mountain, 1979); acoustically laminated sediments at Site 382 are, for example, interbedded turbidite sands and muds derived from the Laurentian Channel (Tucholke, Vogt, and others, 1979). The upper part of the sequence at Site 106 is also highly reflective, and contains ponded turbidites of Pleistocene age (*e.g.*, Tucholke and Mountain, 1979). The reflections can be quite chaotic in slumped terranes (*e.g.*, Schlee and others, 1979). In most DSDP sites there are real increases in sand and silt content in the Pleistocene. These increases are attributed, in part, to the effects of the Pleistocene low stands of sealevel, which allowed bypassing of coarse-grained sediment across the shelf to deepsea basins, and also to the increased detrital supply resulting from glacial erosion. The Holocene rapid rise has apparently led to trapping of sediment on shelves (*e.g.*, Hay and Southam, 1977), and to inactivity of many fans and fan channels (*e.g.*, Nelson and others, 1978). This

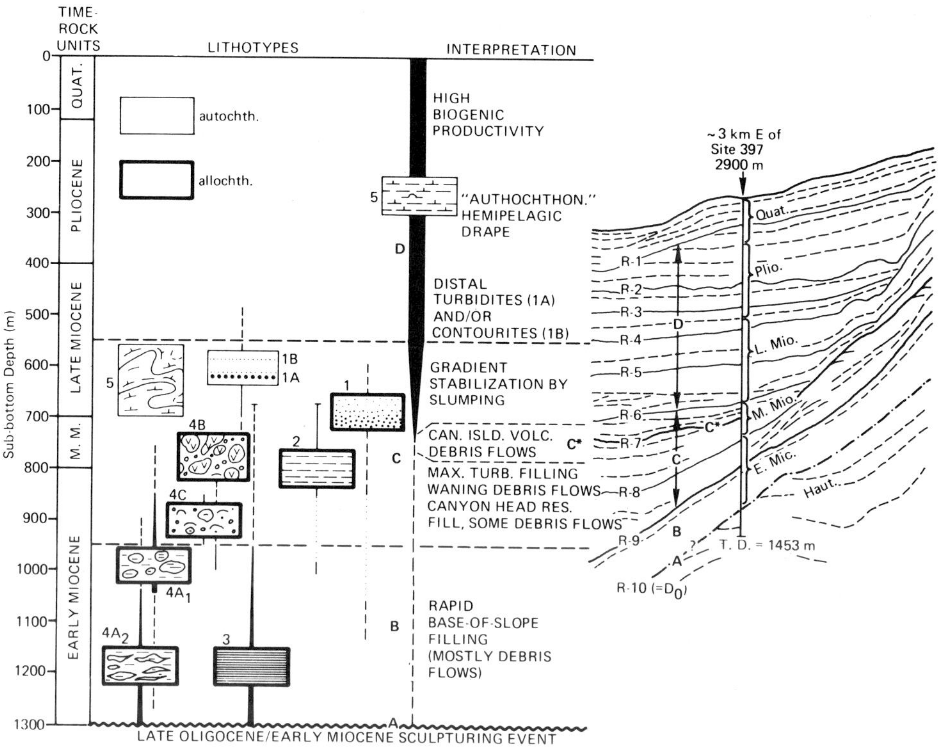

FIG. 12.—Sequence of redeposited sediments cored at Site 397 on the upper continental rise off NW Africa (Arthur and von Rad, 1979). Rapid sedimentation followed pre-Late Oligocene erosion of part of the rise-slope prism.

underscores the possible influence of sea-level changes on redeposition of terrigenous clastic material to the deep sea (see also Worsley and Davies, 1979).

Turbidites and Tectonic Arguments

Linear Island Chains.—Certainly one of the most original uses of turbidite sedimentation has been to follow the evolution of mid-ocean seamount edifice construction as a test of the Hot Spot theory (Schlanger, Jackson, and others, 1976; Clague, this volume). Earlier, Leg 7, Site 64 recovered thin (approx. 10 cm) foraminiferal grainstone layers sporadically interbedded with calcareous oozes. The grainstones show crude grading and truncated low-angle crossbedding. Heath and Moberly (1971) used these as evidence of heightened tectonism along the Marshall islands around 35–45 m.y.BP, and postulated a submarine ridge outside the Marianas Arc as the likely source. Similarly, Leg 9, Site 76 found evidence of Early Eocene volcanism near the Taumoto Archipelago by examining a Miocene mass-gravity flow containing clasts of Eocene reefal debris and benthic foraminifera along with weathered volcanics. Winterer (1973) suggested the potential of using the sequence of debris shed off seamount chains to trace their tectonic history. Sites 165, 315 and 316 were drilled into the thick archipelagic drape of the Line Islands Seamount Chain. In each hole, partially graded volcanogenic breccias record Late Cretaceous subaqueous mass flows of hyaloclastics during the constructional phase (Winterer, 1973; Schlanger, Jackson, and others, 1976: see Fig. 13). As the top of the edifice approached sealevel during the emergent phase, shallow-water benthic foraminifers appear mixed with hyaloclastics in graded beds showing features of Bouma cycles. Beds up to 2–4 m thick are not uncommon. Finally in the subsidence phase, the thick archipelagic drape is subjected to frequent sloughing, forming sequences of interbedded pelagic calcareous ooze and pelagic turbidites. Later evolutionary stages show quiescence and normal pelagic sedimentation. The timing of the basal sequence of beds indicates that edifice construction is rapid, reaching the emergent stage within a few million years. The application of these principles has shown that the timing of sedimentation at the three sites mentioned above match closely. The constructional phase is bracketed at between 85–94 m.y. (Schlan-

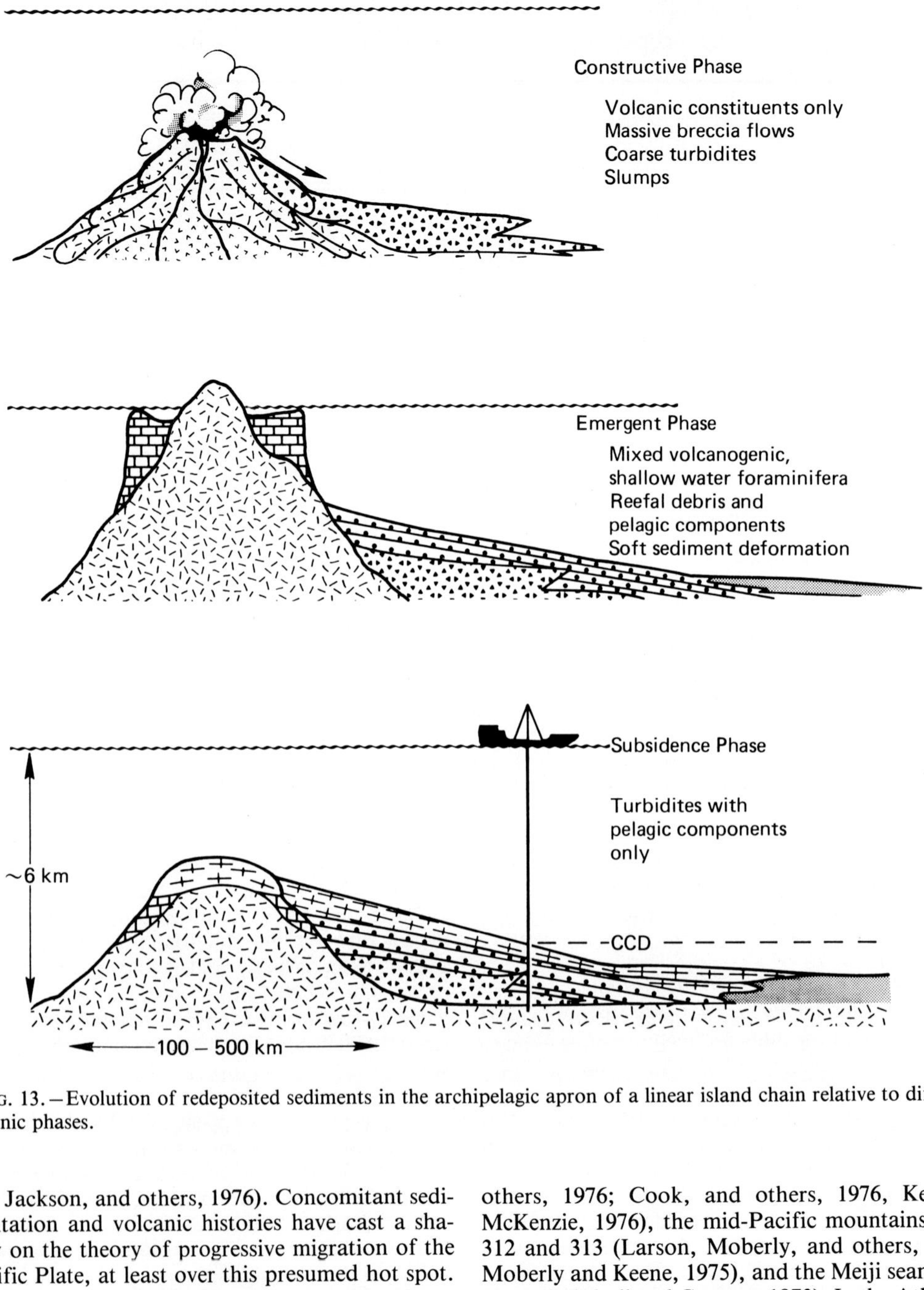

FIG. 13.—Evolution of redeposited sediments in the archipelagic apron of a linear island chain relative to different tectonic phases.

ger, Jackson, and others, 1976). Concomitant sedimentation and volcanic histories have cast a shadow on the theory of progressive migration of the Pacific Plate, at least over this presumed hot spot.

Synchronous (?) Volcanogenic Events.—Upper Cretaceous, particularly Campanian-Turonian, volcanogenic turbidites have been cored at numerous sites in the world ocean (Fig. 1). Volcaniclastic turbidite sequences of this age are described from the Caribbean Sites 146/149 (Edgar, Saunders and others, 1973), the Line island Sites 165 (Winterer, 1973), 315 and 316 (Schlanger, Jackson, and others, 1976; Cook, and others, 1976, Kelts & McKenzie, 1976), the mid-Pacific mountains Sites 312 and 313 (Larson, Moberly, and others, 1975; Moberly and Keene, 1975), and the Meiji seamount Site 192 (Scholl and Creager, 1973). In the Atlantic, both Sites 382 on the Nashville seamount and 385 on the Vogel seamount bottomed in thick (520 m) Upper Cretaceous volcanogenic breccias, presumedly Campanian. The Sierrra Leone Rise and Ceara Rise (Leg 39, Site 354) are both thought to have formed around 80 m.y.BP (Supko, Perch-Nielsen, and others, 1977) on the basis of sedimentary sequences associated with these features. In the Indian ocean, several sites cored thick ash flows characterized by grading and cross- and parallel-laminations. These are Mid- to Late Cretaceous in age. A thick volcaniclastic wedge of similar age occurs on the southern margin of the Naturaliste Plateau and in Site 264 (Veevers, Hiertzler, and others, 1974). Several sites in the Northeastern, western and southeastern Indian Ocean regions have cored Upper Cretaceous volcanogenic sediment (Vallier, 1974) which Vallier and Kidd (1977) attribute to the initial collison of India and Asia. A widespread Campanian volcanogenic event may correspond to global readjustments in plate motion, perhaps initiated by this collison, which caused widespread mid-plate fissuring. Natland (1976) attributed an epic of Upper Cretaceous mid-plate alkalic basalts in the Pacific to an overprint of a tectonic cross-trend at this time. On the other hand, there is a possibility that volcanic events are periodic, perhaps related to poorly understood processes in the earth's interior, as suggested by Vogt (1979). Santonian-Campanian, Eocene, and Mid-Miocene, for example, are apparently times of increased mid-plate volcanism.

The construction of several large plateaus in the Pacific also seems related to active volcanic periods documented by thick volcaniclastic debris flows and turbidites. The basal sequences on Ontong Java (Sites 288 and 281) and Manihiki (Site 317) plateaus both have thick units of Neocomian graded, subaqueous mass-gravity volcaniclastic flows (Klein, 1975; Jenkyns, 1976). Eocene turbidites with abundant volcanogenic components have also been encountered at numerous DSDP sites and the Eocene in general has been suggested as an interval of heightened volcanic activity and biosiliceous productivity (*e.g.*, Vallier and Kidd, 1977; Moberly and Keene, 1975). Graded volcanic breccias with clasts up to 3 cm in diameter and with shallow-water carbonates were cored in the Middle Eocene sections from Site 357 on the Rio Grande Rise (Supko, Perch-Nielsen, and others, 1977).

Perched or Isolated Turbidites.—Beall and Fischer (1969) used the occurrence of fine-grained Eocene turbidites from Site 6 to infer an old abyssal plain running up the Bermuda Rise. From this, they surmised an Oligocene age for the mid-plate uplift phase of the Rise. This was later confirmed (*e.g.*, Donnelly, and others, 1979) and linked to concomitant volcanic activity along the Rise (*e.g.*, Tucholke, Vogt and others, 1979). Galehouse (1979) traced the provenance of Lower to Middle Oligocene volcaniclastic turbidite sands from Site 386, nearly 200 km southeast of Bermuda Island. He showed that the sediment source persisted for about 15 m.y. Most of these volcaniclastic turbidites, although up to 30–50 cm thick, exhibit only textural grading (McCave, 1979). The 160 + m-thick sequence forms a reflector (Horizon Av) recognizable over hundreds of square kilometers (Tucholke and Mountain, 1979).

There are several other interesting examples: a pulse of turbidites in the Mid-Miocene sediments of Site 397 was used to bracket mid-plate volcanism and the rise of Fuerteventura and the Canary Islands (*e.g.*, Arthur and von Rad, 1979). Fifteen coarse-grained Pleistocene hyaloclastic turbidites cored at Site 115 off Rockall Banks show cross stratification and grading (Davies and Laughton, 1972). They are thought to have derived from subglacial or submarine eruptions around Iceland, which generated highly competent turbidity currents that crossed the Maury Midocean Canyon.

Pleistocene-age beds, comprising faintly graded, quartzose, black terrigenous mud cycles with a thin sandy base, occur interbedded with pelagic marl-oozes upon the Levantine-eastern Mediterranean Ridge at Site 130 (Ryan, Hsü, and others, 1973). The source area for these perched mud turbidites is the Nile Cone. They are considered key evidence of continued compressional Quaternary folding for the origin of the Eastern Mediterranean ridge.

Paleocene-Eocene pelagic turbidites consisting of gray calcareous marls to soft calcareous arenites sit atop the Cantabria Seamount in the Bay of Biscay (Site 119); these are interlayered with host red clays and overlain by normal pelagic carbonates. They are evidence of the tectonic uplift of this structure coeval with the Pyrenean orogeny (Davies and Laughton, 1972). Turbidites can, however, apparently flow uphill for some distance, probably depending on the velocity and thickness of the flow. Damuth and Embley (1979) have suggested this mechanism for turbidites found "perched" about 400 m high on Ceara Rise in the western Equatorial Atlantic.

Using turbidite occurrences in a different vein, Mueller and others (1978) argued that the lack of turbidites before the Plio-Pleistocene in the Eastern Mediterranean was evidence that the basin was relatively shallow at the time of the Messinian Salinity Crisis (Ryan, Hsu, and others, 1973). Conversely, scattered graded beds in Lower Miocene green mudstones and pelagic calcareous oozes from the Balearic Basin (Site 372) and Florence Rise (Site 375) are construed as evidence for a deep basin prior to evaporite deposition (Hsu, Montadert, and others, 1978).

At Site 183 on the northern edge of the Aleutian Abyssal Plain, 250 meters of Lower Oligocene to Lower Eocene olive-gray clayey-silt to silty-sand turbidites are sandwiched between pelagic sediments. Graded intervals include up to 1-m-thick sands. These were interpreted as a fossil abyssal plain and were used to limit the absolute amount of northward movement of the Pacific Plate to about 500 km (Scholl and Creager, 1973; Stewart, 1976). They argued that this figure was considerably less than that predicted by previous models of plate motion.

In another trench setting, turbidites were used to document the Quaternary tilting of the basin plain of the Hellenic Trench (Site 127; Ryan, Hsu and others, 1973; Bartolini and others, 1975).

Redeposited Sediments Along Active Margins

Deep-Sea Drilling has begun to help elucidate patterns of sedimentation in different morphotectonic environments within active margin settings (Fig. 14). Drilling has not, until recently, been systematic in terms of recovering sediment sequences from a number of structural settings on one margin (e.g., von Heune, this volume), nor have offset holes been drilled in order to examine the lateral facies changes within one particular environment. Therefore DSDP has had an important, yet inconclusive, influence on studies of active margin sedimentation. The major problems which can be attacked by DSDP drilling are: 1) to understand dispersal patterns and the distribution of coarse-grained sediment to the outer part of the forearc region such as in slope basins and in the trench—large channels may be most important in this regard; 2) to know enough about variability in sedimentary facies and compositions to use them in interpreting the structural evolution of active margins; and 3) to tie the results from DSDP drilling to detailed mapping in exposed but structurally complex parts of ancient active margins in order to aid in interpretation of depositional environments and structural evolution there. Schweller and Kulm (1978) and Underwood and others (1980) have recently reviewed many aspects of trench and trench-slope sedimentation, while Dickinson and Seely (1979) have described tectono-sedimentary aspects of active-margin evolution in general. The following discussion will point out several salient examples of the contribution (or failings) of DSDP studies of resedimentation in active-margin settings. Locations of various key sites are shown schematically with respect to morphotectonic environment in Figure 14.

Trench Sedimentation.—Drilling in the eastern Aleutian Trench (Leg 18, Site 180) gave the first details of the nature of trench sedimentation (*e.g.*, Piper and others, 1973; Kulm, von Huene, and others, 1973). This site drilled into the trench wedge sequence and recovered about 470 m of graded silt beds and interbedded muds interpreted as turbidites. The turbidite sequence, which is horizontal, overlies and laps onto a slightly landward-tilted thick pelagic-plate facies which has accumulated since Miocene time on the incoming oceanic plate (hemipelagic mud and thin silt turbidites in Site 174). The trench sediments were deposited at rates in excess of 1500 m/m.y. at least 10 times that of the subjacent terrigenous abyssal plain facies. Piper and others (1973) underscored the importance of channels in transporting large volumes of coarse detritus to and within the trench. Sediment fans of various sizes may build into the trench at point sources such as canyons leading down the slope. The role of such slope canyons has been discussed more recently by Underwood and others (1980) using the Middle America Trench as an example.

The facies model for trench sedimentation developed by Piper and others (1973) on the basis of drilling on DSDP Leg 18 also involved axial transport as turbidity currents in large confined chan-

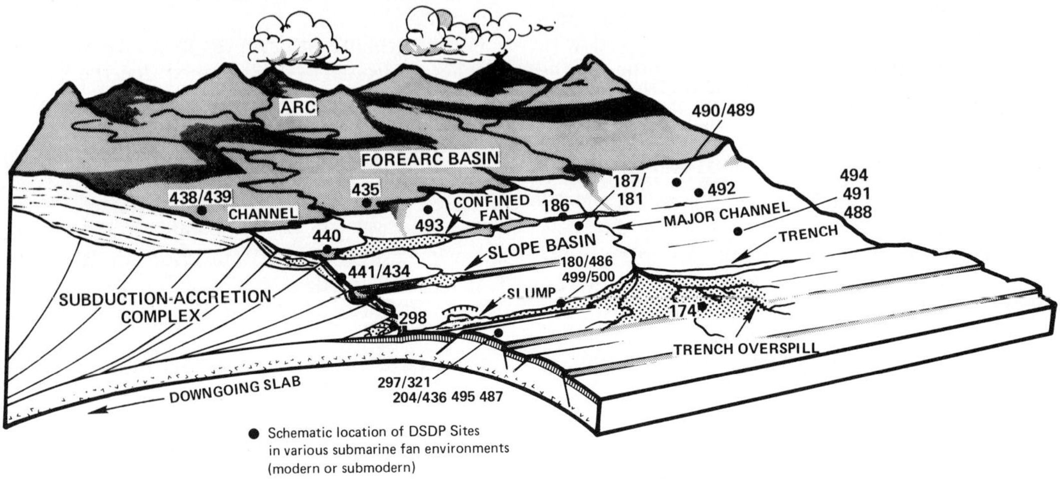

FIG. 14.—Diagram schematically illustrating tectonic-sedimentary environments drilled in various DSDP Sites along active continental margins (marginal or back-arc basin sites not shown).

nels, since emphasized in review by Schweller and Kulm (1978). The thickness of the trench wedge, and the coarseness and thickness of individual turbidites, of course, depend on distance from the point sources and on the convergence rates which control the rates of accretion or subduction of trench sediments. Site 194 was drilled into the outer part of the Astoria Fan off the Oregon margin, which is an example of fan buildout by extreme rates of sediment feed to the margin from the Columbia River and overfilling of the trench; the trench has no topographic expression in the region. The converging rates along this margin are also relatively slow. Another example of this process, but on a rapidly converging margin, is found off Japan. A large trough or channel (the Hidaka Channel) apparently carries large volumes of sediment across a broad forearc region to the slope. A pronounced "toe" to the slope has been built out seaward of the channel, and has pushed the trench axis over 20 km seaward (Arthur and others, 1980a; Nasu and others, 1980). The toe is presumed to consist of coarse-grained turbidite sequences, mud, and material slumped from the slope. The toe is an anomalous feature, because it appears that elsewhere relatively little sediment reaches the trench over much of its length. Drilling during Legs 56–57 established that tectonic accretion of sediment to the overriding plate has been unexpectedly low (von Huene and others, 1980). The Japan Trench is largely empty of sediment for most of its length (over 600 km) except in the vicinity of Tokyo Bay, where sediment is again funneled downslope via a large channel. Most piston cores from the trench recovered hemipelagic mud (Inouchi and others, 1977).

A number of DSDP sites have documented the onset of increased terrigenous sedimentation on the incoming plate. The transition from deposition of clay and pelagic carbonate to hemipelagic mud occurred at Site 321 at about 5 m.y.BP which took place when the Site was over 750 km west of the axis of the Peru Trench (Yeats, Hart, and others, 1976). Site 436 east of the Japan Trench began accumulating hemipelagic mud about 15 m.y.BP, while the Site was nearly 1,200 km from the present trench axis (Arthur and others, 1980a). The onset of hemipelagic sedimentation at Site 495 off the Middle America Trench took place at 10 m.y.BP when the Site may have been more than 900 km away (von Huene, Auboin and others, 1980). No turbidites were recovered at any of these sites, so alternative depositional mechanism are required. The trench must generally act as a barrier to coarse sediment transport and most of the fine-grained muddy sediment seaward of the trench is derived from eolian transport, nepheloid-layer transport and/or dilute-suspension overflows topping the topographic trench. An example of the use of this concept is from Site 297 in the Shikoku Basin, east of the present Nankai Trough. There 240 m of Lower Pliocene turbidite sands and silts containing displaced shallow-water faunas were recovered beneath 330 m of claystone, nannofossil ooze, *and diatom and ash-rich clay* (Karig, Ingle, and others, 1975). Shipboard scientists inferred from this that the Nankai Trough came into existence sometime during the Early Pliocene, marked by the cessation of turbidite deposition.

Site 298 of Leg 31 (Moore, J. C., and Karig, 1976) penetrated a presumed uplifted and deformed trench wedge sequence on the landward wall of the Nankai Trough. The folded and possibly partly overturned Early Pleistocene to Holocene sequence is 791 m thick and consists of interbedded turbidite sands, silts, and "hemipelagic" muds. This sequence coarsens upward and is consistent with facies models depicting the transition from distal to proximal trench sedimentation as any one point on the converging ocean plate approaches the base of the inner slope (Moore, J. C., and Karig, 1976; Piper and others, 1973). Sites 486 (Moore, Watkins, and others, 1979) and 499–500 (von Huene, Auboin, and others, 1980) also recovered coarse sand of Quaternary age in the Middle America Trench.

Trench Slopes.—Only in more recent IPOD transects (*e.g.,* Legs 56–57, 59–60, 66–67) has much attention been devoted to trench inner slopes, slope basin sequences, or to slope drapes. However, Site 175 on the Oregon continental slope and Site 181 on the lower continental slope in the Gulf of Alaska (Kulm, von Huene, and others, 1973) recovered primarily hemipelagic mud with relatively few silt layers, probably of turbiditic origin. Site 182, upslope from Site 181, recovered apparently massive sand (which may have been homogenized by drilling). Scholl and Creager (1973) drilled in the Atka Basin (Site 186) on the Aleutian Trench Slope. Again, a thick sequence (926 m) of diatomaceous hemipelagic mud was recovered, although the section was sparsely cored. Sand layers were found interbedded in the sequence at several levels. The individual beds were as much as 4.5 m thick, medium- to coarse-grained, and consist largely of redeposited volcaniclastic material. Many of the beds exhibit sedimentary structures characteristic of turbidites and include benthic foraminifers which are displaced shallow-water species.

Drilling on the Middle America Trench slope during Leg 66 (Moore, Watkins, and others, 1979) showed that hemipelagic muds with thin interbedded silt and sand layers, deposited on the slope, overlie coarse-grained sediments presumed to be uplifted trench wedge deposits. Sites on the Japan Trench inner slope (Sites 434, 441, 435) also recovered hemipelagic mud sequences with few interbedded sand or silt layers (*e.g.,* Arthur and others, 1980b; von Huene and others, 1980). Site

440, drilled in a midslope basin, did recover coarse-grained turbiditic silt and sand layers interbedded with mud in the Pleistocene section. Sedimentation rates of 270 m/m.y. were over twice that of the underlying sequence of Upper Miocene through Pliocene mudstones. This and other structural evidence suggest that the slope basin was formed abruptly in Pleistocene time.

Other mass-flow deposits such as slumps and debris flows also occur on trench slopes and in trenches and have been cored in DSDP Sites. A large exotic block of Middle Miocene age was cored in a section of Pliocene mud at Site 186 in the Atka trench slope basin (Scholl and Creager, 1973). Piper and others (1973) documented slumps at the base of the inner slope of the Aleutian Trench near Site 180. Contorted slumped units of the same age and lithology as interbedded mudstones were found interbedded in Pliocene mudstones in Site 440 in the midslope basin off northeast Japan (*e.g.*, Arthur and others, 1980a). Several redeposited sedimentary breccia units consisting of older indurated clasts or chips of vitric and diatomaceous mudstone set in a diatomaceous mud matrix of younger age were also recovered at Sites 434 and 441 on the lower slope. Based on Leg 56–57 drilling results, Arthur and others (1980b) suggest that fracturing and rapid dewatering of the slope sediment mass, in response to the stress of convergence, coupled with rapid sedimentation of a relatively impermeable mud cap on the slope, causes a mildly overpressured zone at depths of a few hundred meters. This overpressuring in fractured rock reduces shear stress and can lead to widespread slumping on the slope, analogous to that occurring on slopes of the Gulf of Mexico (*e.g.*, Booth, 1979).

It is likely that much of the mud drape on trench inner slopes is deposited from dilute fine-grained turbidity currents which spill over the slope-edge or from slope basins. The dominance of mud, except in slope basins, suggests the important role of slope basins, and the forearc basin itself, in trapping coarse-grained detritus (*e.g.*, Scholl and Marlow, 1974; Moore, G. F., and Karig, 1976; Barnard, 1978; Moore and others, 1980; Arthur and others, 1980b; Underwood and others, 1980). So far, however, DSDP drilling has done little to establish the details of sedimentary facies in individual basins within the forearc region. Future drilling should attempt several holes within a slope basin, in order to examine lateral facies changes in one basin. Most productive work of this nature will likely come from mapping in land-based, presumed ancient slope-basin complexes (*e.g.*, Bachman, 1978; Moore and others, 1980).

Marginal Basins. — Drilling has been extensive in so-called marginal or "backarc" basins, and the origin and evolution of many of these basins are complex and still controversial (*e.g.*, Karig 1971; 1974). Redeposited sediments, mainly volcanogenic turbidites and breccias, are common in many DSDP Sites from these regions. The DSDP sites here have given information primarily on redeposited sediment composition and the evolution of sediment in relation to tectonics in marginal basins.

Sites 190 in the Aleutian Basin and 191 in the Kamchatka Basin penetrated thick sequences of redeposited sediment and interbedded diatomaceous mud (*e.g.*, Scholl and Creager, 1973). Site 191 was located at least 150 km from a presumed source (the Kamchatka Peninsula) for the mainly volcanogenic turbidites (*e.g.*, Stewart, 1977). Graded sand layers up to 1 m thick are common in the upper 300 meters of the sequences recovered; these coarse layers are mainly of Pleistocene age. The entire Miocene through Pleistocene sequence over 900 meters thick, which apparently overlies a regionally-extensive basalt layer of Oligocene age (Scholl and Creager, 1973), seems to represent a gradually coarsening-upward sequence deposited on an abyssal plain. Site 190, in a more distal position from the presumed source on the Bering Sea Shelf also recovered size-graded terrigenous turbidites of Pleistocene age, but these are fine-grained and generally thinner-bedded than those at Site 191. Many silty mud layers at Site 190 are visually unrecognizable as redeposited sediment, but careful grain-size analysis shows them to be texturally graded (Fullam and others, 1973). The turbidite facies is mappable in seismic records over the Bering Abyssal Plain as a highly reflective sequence (*e.g.*, Ludwig and others, 1971).

Several factors may influence the coarsening of layers in the Pleistocene part of the sequence at both sites, but are difficult to assess. These include gradual progradation of coarse-grained facies from the shelf margin to the deeper basin, and Pleistocene climatic and sealevel changes. The Pleistocene lowering of sealevel is regarded as a major factor (*e.g.*, Fullam and others, 1973).

Harrold and Moore (1975) used compositions of turbidite sand layers from the West Philippine Sea (Site 293), and the Shikoku (Sites 297, 298) and Japan Sea basins (Site 301), to define the relative influence of differing source areas on sediment composition. From these primarily lithic wackes they were able to differentiate tectonic setting, and to estimate proximality and polarity to the volcanic arc on the basis of relative proportions of various lithic fragments and on quartz-feldspar ratios. This study was done at a time when such analyses were rare. Dickinson and Suzcek (1979) have recently summarized and strengths and weaknesses of this technique and its application in tectonic reconstructions (partly using DSDP data).

Sites drilled in marginal basins near the volcanic arc typically encounter thick seismically opaque volcaniclastic aprons (*e.g.*, Karig, 1975; Scholl and

Creager, 1973). Karig (1975) differentiated between some aprons associated with volcanic activity (such as that at Site 189 at the base of the Aleutian Ridge) and those cored in Site 293, which represent redeposited volcaniclastic material associated with the cessation of volcanism and the Quaternary uplift and denudation of Luzon. Rates of deposition and average grain-size both increase with time in the latter case. Pulses in redeposition of volcaniclastic material may also be used generally to date pulses in arc volcanism.

Turbidite Frequency—Rise and Abyssal Plains

Study of the frequency of turbidity-current events in the deep sea, based on DSDP cores, is hampered by the general lack of continuously cored sequences, by deformation due to drilling, and by poor age control in older sequences or those deposited below the CCD. Nevertheless, a brief survey of clastic turbiditic sediments of various ages from Atlantic DSDP sites suggests that during many intervals a turbidite was deposited from a maximum of one every 1,000 years to a minimum of one every 20,000 years at any one point on the continental rise or basin plain. These turbidites are typical "distal" units (*e.g.*, Walker, 1967) and consist mainly of thin-bedded (a few mm to 30 cm), homogeneous to graded, muddy silt to fine- or medium-grained sand. Cyclic sequences of homogeneous mud-turbidites are also common. Good examples of these can be seen in the sediments of Late Eocene through Early Miocene age of Site 368 cored on the Cape Verde Rise. Dean and others (1978) have described these turbidite sequences as cyclic and estimated a frequency of 6000–8000 years in the lower portion of the sequence and about 12,000 years in the upper part. Lower to Middle Eocene siliceous mud-turbidite units in the western North Atlantic basin (McCave, 1979) are an average of 30–40 cm thick, and were deposited at a rate of about one every 5,000 to 10,000 years. At Site 142 on the Ceara Abyssal Plain, graded terrigenous sand and silt turbidites were deposited about one every 1,000–5,000 years during much of the Pleistocene (Berger and von Rad, 1972). Calcareous turbidites, consisting largely of pelagic components, were deposited with a frequency of about one every 50,000 to 100,000 years during the Miocene at the same site. Similar but thicker calcareous turbidites, with mixed pelagic and terrigenous components, were deposited at a frequency of about 18,000 to 50,000 years at Site 398 off Galicia Bank (Maldonado, 1979).

At Site 210 in the Coral Sea Basin (Burns, Andrews, and others, 1974) Plio-Pleistocene cycles of graded silt and clay interbedded with nannofossil ooze were recovered. Over 1,900 of these turbidite cycles, averaging 14 cm thick, represent an average of one event every 4,000–5,000 years for about 5 m.y. or more. These turbidites contain quartz, volcanic glass, mica and plagioclase, and were probably derived via Moresby Submarine Canyon from New Guinea.

It is not clear what controls the frequency of turbidite events, nor is it clear that such events have a constant periodicity. Obvious controlling factors are changes in climate and runoff (delivery of terrigenous detritus to the oceans), fluctuations in carbonate dissolution parameters (periodic slumping events and "pelagic" turbidites induced by increased dissolution at the base of slopes or oceanic rises), and/or periodic earthquake-generated mass flows in tectonically active areas. The possible existence of fundamental frequencies is certainly a topic for further study using DSDP cores.

CONCLUSION: TURBIDITE SEDIMENTOLOGY—NOW AND THE FUTURE

As outlined above, DSDP has contributed little to the knowledge of the origin and dynamics of turbidite sedimentation. The basic framework and theory of turbidite sedimentation existed prior to the advent of deep-sea drilling, as discussed by Walker (1973) in "Mopping up the Turbidite Mess." He eloquently pointed out that the basic paradigm has been challenged and accepted by the geologic community, and that the major future effort should be towards applying models of turbidite dispersal to the ancient record in order to understand the tectonic and sedimentary evolution of ancient continental margins.

The program of exploration during the first 10 years of the Deep Sea Drilling Project has generally not addressed problems of redeposited sediments, including turbidites, directly. Most of the examples presented above came about as a byproduct of other scientific objectives, lack of knowledge of the sedimentary sequence to be penetrated, or by target of opportunity. Also, redeposited sediments provide undesirable noise in sequences with biostratigraphic or paleoenvironmental importance. Thick sedimentary sequences are also avoided for safety reasons such as the possibility of encountering hydrocarbon accumulations.

Only with the advent of the IPOD drilling program have we attempted to examine continental margin sequences dominated by redeposited sediments. As a result we are just beginning to unravel the relationship of deep-sea sedimentation to changes in climate and sealevel or to the tectonic evolution of active and passive margins as touched on in this paper.

The Deep Sea Drilling Project now has a tool which provides a potential for carrying out detailed study of turbidite facies. The Storms-Serocki Hydraulic Piston Coring Device (HPC), which has now been successfully deployed on several Legs (*e.g.*, Leg 64; see Moore, Curray, and others,

1979), might enable us to drill a number of 200–300-m-deep holes on a modern deepsea fan, minimizing the effects of drilling deformation which previously prohibited such a study. In the event that it is proven that the HPC is effective and can recover thick sand units, a program of drilling a number of holes and sampling different environments on a well-known deepsea fan should be designed and carried out. In our opinion, the first drilling should be attempted on a well-studied and relatively small continental margin deepsea fan such as the Monterrey Fan (*e.g.,* Normark, 1970; Wilde and others, 1978). Later programs could be designed to study abyssal cones or larger fans that have somewhat different morphologies and internal geometries.

The Deep Sea Drilling Project will continue to contribute to turbidite sedimentology in the future: indirectly, we will continue to collect information on the distribution of redeposited sediments in time and space as further drilling is completed along continental margins, and directly if one or more deepsea fans are cored in detail.

ACKNOWLEDGEMENTS

We thank E. L. Winterer for inviting us to contribute to this volume. We also sincerely thank Chiye Wenkam for her patience and thoughtful suggestions for improvement of this paper. Michael Field and Roger G. Walker critically reviewed the manuscript and provided many helpful comments. Gerhard Einsele, Lee Krystinik, and Helmut Weissert read all or part of an earlier manuscript; we appreciate their freely-given advice and ideas. We are indebted to Kathy Sanderson and Lisa Peters for the final artwork, and to DSDP La Jolla for furnishing space and access to DSDP data files. Earlier drafts of this manuscript were kindly typed and reprocessed by Pat Krupa, Jo Lefebre, and Helen Colburn (USGS). Inspiration was provided by Django Rheinhardt, Larry Coryell, Philip Catherine and Pat Metheny.

REFERENCES CITED

NOTE: All references to Deep Sea Drilling Project Initial Reports have been abbreviated; for complete citations see *Appendix I* at the end of this volume, or the individual Initial Reports as published.

ARTHUR, M. A., 1979, North Atlantic Cretaceous black shales: The record at Site 398 and a brief comparison with other occurrences: Init. Repts. DSDP, v. 47, p. 719–751.

———, CARSON, B., AND VON HUENE, R., 1980a, Initial tectonic dewatering of hemipelagic sediment at the leading edge of the Japan convergent margin: Init. Repts. DSDP, v. 56–57/1, p. 569–614.

———, VON HUENE, R., AND ADELSECK, C. G., 1980b, Sedimentary evolution of the Japan forearc region off Northern Honshu, Legs 56 and 57 Deep Sea Drilling Project: Init. Repts. DSDP, v. 56–57/1, p. 521–568.

———, AND VON RAD, U., 1979, Early Neogene base-of-slope sediment at Site 397, DSDP Leg 47A: sequential evolution of gravitative mass transport processes and redeposition along the northwest African passive margin: Init. Repts. DSDP, v. 47/1, p. 603–640.

———, AND ———, CORNFORD, C., MCCOY, F., AND SARNTHIEN, M., 1979, Evolution and sedimentary history of the Cape Bojador continental margin, Northwestern Africa: Init. Repts. DSDP, v. 47/2, p. 773–815.

BACHMAN, S. B., 1978, A Cretaceous and Early Tertiary subduction complex, Mendocino Coast, northern California, *in* Howell, D. G., and McDougall, K. A., eds., Mesozoic Paleogeography of the Western U.S.: Soc. Econ. Paleontologists Mineralogists, Pacific Section, Pacific Coast Paleogeography Symposium 2, p. 419–430.

BADER, R. G., GERARD, R. D., AND OTHERS, 1970: Init. Repts. DSDP, v. 4.

BARNARD, W. B., 1978, The Washington continental slope: Quaternary tectonics and sedimentation: Marine Geology, v. 27, p. 70–114.

BARTOLINI, C., MALESANI, P. G., MANGETTI, P., AND WEZEL, F. C., 1975, Sedimentology and petrology of Quaternary sediments from the Hellenic Trench, Mediterranean Ridge and the Nile Cone from the DSDP, Leg 13 cores: Sedimentology, v. 22, p. 205–236.

BEALL, A. O., JR., AND FISCHER, A. G., 1969, Sedimentology: Init. Repts. DSDP, v. 1, p. 521–593.

———, LAURY, R., DICKINSON, K., AND PUSEY, W. C., III, 1973, Sedimentology: Init. Repts. DSDP, v. 10, p. 699–730.

BENSON, W. E., SHERIDAN, R. E. AND OTHERS, 1978: Init. Repts. DSDP, v. 44.

BERGER, W. H., 1979, The impact of deep sea drilling on paleoceanography, *in* Talwani, M., Hay, W. W., and Ryan, W. B. F., eds., Deep drilling results in the Atlantic Ocean: continental margins and paleoenvironments, Maurice Ewing Series 3, p. 297–314.

———, AND JOHNSON, T. C., 1976, Deep-sea carbonates: dissolution and mass-wasting on Ontong-Java Plateau: Science, v. 192, p. 785–787.

———, AND VON RAD, U., 1972, Cretaceous and Cenozoic sediments from the Atlantic Ocean: Init. Repts. DSDP, v. 14, p. 787–954.

BERNOULLI, D., 1972, North Atlantic and Mediterranean Mesozoic facies: a comparison: Init. Repts. DSDP, v. 11, p. 801–876.

———, and Jenkyns, H. C., 1974, Alpine, Mediterranean, and Central Atlantic Mesozoic facies in relation to the early evolution of the Tethys: Soc. Econ. Paleontologists Mineralogists Spec. Pub. 19, p. 129–160.

Bolli, H. M., 1980, The age of sediments recovered from DSDP Legs 5–9, 16–35, Pacific Ocean: Init. Repts. DSDP, v. 54.

———, and Saunders, J. B., 1977, Introduction to stratigraphy and paleontology, *in* Heirtzler, J. R., and others, eds., Indian Ocean Geology and Biostratigraphy: Amer. Geophys. Union, p. 331–324.

Booth, J. S., 1979, Recent history of mass-wasting on the upper continental slope, northern Gulf of Mexico, as interpreted from the consolidation states of the sediment: Soc. Econ. Paleontologists and Mineralogists Spec. Pub. 27, p. 153–164.

Bouma, A. H., 1962, Sedimentology of some flysch deposits: A graphic approach to facies interpretation: Amsterdam, Elsevier, 168 p.

———, 1975a, Sedimentary structures of Philippines Sea and Sea of Japan sediments, DSDP Leg 31: Init. Repts. DSDP, v. 31, p. 471–488.

———, 1975b, Deep-sea fan deposits from Toyama Trough, Sea of Japan: Init. Repts. DSDP, v. 31, p. 489–495.

———, and Hollister, C. D., 1973, Deep ocean basin sedimentation, *in* Middleton, G. V., and Bouma, A. H., eds., Turbidites and deep-water sedimentation: Soc. Econ. Paleontologists Mineralogists Pacific Section, 79–118.

Burns, R. E., Andrews, J. E., and others, 1974: Init. Repts. DSDP, v. 21.

Cook, H. E., Jenkyns, H. C., and Kelts, E., 1976, Redeposited sediments along the Line Islands, Equatorial Pacific: Init. Repts. DSDP, v. 33, p. 837–848.

Cook, P. J., 1977, Mesozoic-Cenozoic sediments of the Eastern Indian Ocean, *in* Heirtzler, J. R., and others, eds., Indian Ocean Geology and Biostratigraphy: Amer. Geophys. Union, p. 119–150.

Curray, J. R., Moore, D. G., and others, *in press,* Drilling in the Gulf of California, DSDP Leg 64: Nature.

Damuth, J. E., and Embley, R. W., 1979, Upslope flow of turbidity currents on the northwest flank of the Ceara Rise: western Equatorial Atlantic: Sedimentology, v. 26, p. 825–834.

Davies, T. A., and Kidd, R. B., 1977, Sedimentation in the Indian Ocean through time, *in* Heirtzler, J. R., and others, eds., Indian Ocean Geology and Biostratigraphy: Amer. Geophys. Union, p. 61–86.

———, and Laughton, A. S., 1972, Sedimentary processes in the North Atlantic: Init. Repts. DSDP, v. 12, p. 905–934.

Dean, W. E., Gardner, J. V., Jansa, L. F., Cepek, P., and Seibold, E., 1978, Cyclic sedimentation along the continental margin of North Africa: Init. Repts. DSDP, v. 41, p. 965–990.

deGraciansky, P. C., and Chenet, P. Y., 1979, Sedimentological study of cores 138 to 56 (Upper Hauterivian to Middle Cenomanian): an attempt at reconstruction of paleoenvironments: Init. Repts. DSDP, v. 47/2, p. 403–418.

———, Auffret, G. A., Dupeuble, P., Montadert, L., and Muller, C., 1979, Interpretation of depositional environments of the Aptian/Albian black shales on the north margin of the Bay of Biscay: Init. Repts. DSDP, v. 48, p. 877–908.

Dickinson, W. R., and Seely, D. R., 1979, Structure and stratigraphy of forearc regions: Amer. Assoc. Petroleum Geologists Bull., v. 63, p. 1–31.

———, and Suczek, C. A., 1979, Plate tectonics and sandstone composition: Amer. Assoc. Petroleum Geologists Bull., v. 63, p. 2164–2182.

Donnelly, W., 1979, Chemistry of sediments of the western Atlantic: Site 417 compared with Sites 9, 105, 386, and 387: Init. Repts. DSDP, v. 51–53/2, p. 1515–1523.

Doyle, L., and Pilkey, O., eds., 1979, Geology of Continental Slopes: Soc. Econ. Paleontologists Mineralogists Spec. Pub. 27, 374 p.

Edgar, N. T., Saunders, J. B., and others, 1973: Init. Repts. DSDP, v. 15.

Einsele, G., and Kelts, K., *in press,* Diatomaceous mud turbidites from the Guaymas Basin, DSDP Leg 64: Init. Repts. DSDP, v. 64.

Embley, R. W., 1976, New evidence for occurrence of debris flow deposits in the deep sea: Geology, v. 4, p. 371–374.

Ewing, M., Worzel, J. L., and others, 1969: Init. Repts. DSDP, v. 1.

Fischer, A. G., and Arthur, M. A., 1977, Secular variations in the pelagic realm: Soc. Econ. Paleontologists Mineralogists Spec. Pub. 25, p. 19–50.

Freeman, T., and Enos, P., 1978, Petrology of Upper Jurassic-Lower Cretaceous limestones, Site 391: Init. Repts. DSDP, v. 44, p. 463–475.

Fullam, T. J., Supko, P. R., Boyce, R. E., and Stewart, R. W., 1973, Some aspects of Late Cenozoic sedimentation in the Bering Sea and North Pacific Ocean: Init. Repts. DSDP, v. 19, p. 887–896.

Galehouse, J. S., 1979, Heavy mineralogy and provenance of volcaniclastic turbidites of Site 386: Init. Repts. DSDP, v. 43, p. 407–409.

Gardner, J. V., Dean, W. E., and Jansa, L. F., 1978, Sediments recovered from the Northwest African continental margin, Leg 41, DSDP: Init. Repts. DSDP, v. 41, p. 1121–1134.

Gibson, T. G., and Towe, K. M., 1971, Eocene volcanism and the origin of Horizon A: Science, v. 172, p. 152.

Hampton, M. A., 1972, The role of subaqueous debris flow in generating turbidity currents: Jour. Sed. Petrology, v. 42, p. 775–793.

Hathaway, J. C., Poag, C. W., Valentine, P. C., Miller, R. E., Schultz, D. M., Manheim, F. T., Kohout, F. A., Bothner, M. H., and Sangrey, D. A., 1979, U.S. Geological Survey core drilling on the Atlantic Shelf: Science, v. 206, p. 515–527.

Harold, P. J., and Moore, J. C., 1975, Composition of deep-sea sands from marginal basins of the northwestern Pacific: Init. Repts. DSDP, v. 31, p. 507–514.

HAY, W. W., AND SOUTHAM, J. R., 1977, Modulation of marine sedimentation by the continental shelves, *in* Anderson, N. R., and Malahoff, A., eds., The Fate of Fossil Fuel CO_2 in the Oceans: New York, Plenum Press, p. 569–604.
HEATH, G. R., AND MOBERLY, R., 1971, Deep-sea turbidites from the western Pacific: Init. Repts. DSDP, v. 7, p. 1009–1011.
HEEZEN, B. C., HOLLISTER, C. D., AND RUDDIMAN, W. R., 1966, Shaping of the continental rise by deep geostrophic contour currents: Science, v. 152, p. 502–508.
HESSE, R., 1975, Turbiditic and non-turbiditic mudstone of Cretaceous flysch sections of the East Alps and other basins: Sedimentology, v. 22, p. 387–416.
HOLLISTER, C. D., CRADDOCK, C., AND OTHERS, 1976: Init. Repts. DSDP, v. 35.
———, AND HEEZEN, B. C., 1972, Geologic effects of ocean-bottom currents: western North Atlantic, *in* Gordon and Breach, eds., Studies in Physical Oceanography: v. 2, p. 37–66.
HSÜ, K. J., 1978, Stratigraphy of lacustrine sedimentation in the Black Sea: Init. Repts. DSDP, v. 42/2, p. 489–498.
———, AND JENKYNS, H., eds., 1974, Pelagic sediments—on land and under the sea: International Assoc. Sedimentologists Spec. Pub. 1, 447 p.
———, MONTADERT, L., AND OTHERS, 1978: Init. Repts. DSDP, v. 42/1.
IMOTO, N., AND FUKUTOMI, M., 1975, Genesis of bedded cherts in the Tamba Belt, southwest Japan: Assoc. Geol. Collabor. Japan, Monogr., v. 19, p. 35–42.
INGERSOLL, R. V., AND SUCZEK, C. A., 1979, Petrology and provenance of Neogene sand from Nicobar and Bengal fans, DSDP Sites 211 and 218: Jour. Sed. Petrology, v. 49, p. 1217–1228.
INOUCHI, Y., YUASA, M., AND ONODERA, K., 1977, Cored Materials (VIII), *in* Honza, E., ed., Geological Investigation of Japan and southern Kurile Trench and slope areas: Japan, Geol Surv. GH76-2 Cruise Rept. 7, p. 78–80.
JANSA, L. F., ENOS, P., TUCHOLKE, B. E., GRADSTEIN, F. M., AND SHERIDAN, R. E., 1979, Mesozoic-Cenozoic sedimentary formations of the North American basin: western North Atlantic, *in* Talwani, M., Hay, W. W., and Ryan, W. B. F., eds., Deep drilling results in the Atlantic Ocean: continental margins and paleoenvironments, Maurice Ewing Series 3, p. 1–57.
JENKYNS, H. C., 1976, Sediments and sedimentary history, Manihiki Plateau, South Pacific Ocean: Init. Repts. DSDP, v. 33, p. 873–890.
JIPA, D., AND KIDD, R. B., 1974, Sedimentation of coarser grained interbeds in the Arabian Sea, and sedimentation processes of the Indus Cone: Init. Repts. DSDP, v. 23, p. 471–495.
KALIN, O., PATACCA, E., AND RENE, O., 1979, Jurassic pelagic deposits from southeastern Tuscany: aspects of sedimentation and new biostratigraphic data: Eclogae Geol. Helv., v. 72/3, p. 715–762.
KAGAMI, H., 1978, Sedimentary features of Capa Basin and Angola Basin sediments, DSDP Leg 40: Init. Repts. DSDP, v. 40, p. 525–540.
———, 1979, Transformation of opaline silica in sediments from Bay of Biscay and Rockall Bank: Init. Repts. DSDP, v. 48, p. 757–764.
KANEPS, A., DOYLE, P. S., AND RIEDEL, W. R., *in press,* Further icthyolith age determinations of otherwise unfossiliferous deep sea drilling cores: Micropaleontology.
KARIG, D. E., 1971, Origin and development of marginal basins in the western Pacific: Jour. Geophys. Res., v. 76, p. 2542–2571.
———, 1974, Evolution of arc systems in the western Pacific: Annual Review Earth Planet. Sci., v. 2, p. 51–76.
———, 1975, Basin genesis in the Philippines Sea: Init. Repts. DSDP, v. 31, p. 857–879.
———, INGLE, J. C., AND OTHERS, 1975: Init. Repts. DSDP, v. 31.
KEENE, J. B., 1975, Cherts and porcellanites from the north Pacific, DSDP, Leg 32: Init. Repts. DSDP, v. 32, p. 429–508.
KELTS, K., 1976, Summary of chert occurrences from Line Islands Sites 314, 315, 316, DSDP Leg 33: Init. Repts. DSDP, v. 33, p. 855–832.
———, AND MCKENZIE, J. A., 1976, Cretaceous volcanogenic sediments from the Line Islands Chain: diagenesis and formation of K-feldspar. DSDP, Leg 33, Hole 315A and Site 316: Init. Repts. DSDP, v. 33, p. 789–832.
KEUNEN, PH. H., AND MIGLIORINI, C. I., 1950, Turbidity currents as a cause of graded bedding: Jour. Geology, v. 58, p. 91–127.
KLEIN, G. DEVRIES, 1975, Depositional facies of Leg 30 Deep Sea Drilling Project sediment cores: Init. Repts. DSDP, v. 30, p. 423–443.
KOLLA, V. MOORE, D. G., AND WRAY, J. R., 1976, Recent bottom current activity in the deep western Bay of Bengal: Marine Geology, v. 21, p. 255–270.
KULM, L. D., VON HUENE, R., AND OTHERS, 1973: Init. Repts. DSDP, v. 18.
LANCELOT, Y., HATHAWAY, J. C., AND HOLLISTER, C. D., 1972, Lithology of sediments from the western North Atlantic, Leg 11, DSDP: Init. Repts. DSDP, v. 11, p. 901–949.
———, SEIBOLD, E., AND OTHERS, 1978: Init. Repts. DSDP, v. 41.
———, WINTERER, E. L., AND OTHERS, 1980: Init. Repts. DSDP, v. 50.
———, AND ———, 1980, Evolution of the Moroccan oceanic basin and adjacent continental margin—a synthesis: Init. Repts. DSDP, v. 50, p. 801–821.
LARSON, R. L., MOBERLY, R., AND OTHERS, 1975: Init. Repts. DSDP, v. 32.
LERMAN, A., LAL, D., AND DACEY, M. F., 1974, Stokes settling and chemical reactivity of suspended particles in natural waters, *in* Gibbs, R. J., ed., Suspended Solids in Water: New York, Plenum Press, p. 17–47.
LUDWIG, W. J., HOUTZ, R. E., AND EWING, M., 1971, Sediment distribution in the Bering Sea: Bowers Ridge, Shirshov

Ridge, and enclosed basins: Jour. Geophys. Res., v. 76, p. 6367.

MALDONADO, A., 1979, Upper Cretaceous and Cenozoic depositional processes and facies in the distal North Atlantic continental margin of Portugal, DSDP Site 398: Init. Repts. DSDP, v. 47/2, p. 373–401.

MATTSON, P. H., AND PESSAGNO, E. A., 1971, Caribbean Eocene volcanism and the extent of Horizon A: Science, v. 174, p. 138–139.

MCCAVE, I. N., 1979, Diagnosis of turbidites at Sites 386 and 387 by particle-counter size analysis of the silt (2–4 μm) fraction: Init. Repts. DSDP, v. 43, p. 395–405.

MEYER, D. E., 1978, Microfacies and microfabrics of Early to Middle Cretaceous sediments selected from Site 370, DSDP Leg 41 (deep basin off Morocco): Init. Repts. DSDP, v. 41 supplement, p. 961–981.

MIDDLETON, G. V., AND HAMPTON, M. A., 1973, Sediment gravity flows: mechanics of flow and deposition *in* Middleton, G. V. and Bouma, A. H., eds., Turbidites and deep-water sedimentation, Soc. Econ. Paleontologists Mineralogists Pacific Section, Short Course Notes, p. 1–38.

MILLER, M. C., AND KOMAR, P. D., 1977, The development of sediment threshold curves for unusual environments (Mars) and for inadequately studied materials (foram sands): Sedimentology, v. 24, p. 709–721.

MOBERLY, R., AND KEENE, J. B., 1975, Origin and diagenesis of volcanic-rich sediments from North Pacific seamounts, DSDP, Leg 32: Init. Repts. DSDP, v. 32, p. 537–546.

MOORE, D. G., CURRAY, J. R., RAITT, R. W., AND EMMEL, F. J., 1974, Stratigraphic-seismic section correlations and implications to Bengal Fan history: Init. Repts. DSDP, v. 22, p. 403–412.

MOORE, G. F., AND KARIG, D. E., 1976, Development of sedimentary basins on the lower trench slope: Geology, v. 4, p. 693–697.

———, BILLMAN, H. G., HEHANUSSA, P. E., AND KARIG, D. E., 1980, Sedimentology and paleobathymetry of Neogene trench-slope deposits, Nias Island, Indonesia: Jour. Geology, v. 88, p. 161–180.

MOORE, J. C., AND KARIG, D. E., 1976, Sedimentology, structural geology, and tectonics of the Shikoku subduction zone, southwestern Japan: Geol. Soc. America Bull., v. 87, p. 1259–1268.

MOORE, J. C., WATKINS, J., AND OTHERS, 1979, Progressive accretion in the middle America trench, southern Mexico: Nature, v. 281, p. 638–642.

MUELLER, J., HEIKE, W., AND FABRICIUS, F., 1978, Turbidites at Site 374: their composition, provenance and paleobathymetric significance: Init. Repts. DSDP, p. 397–400.

MUTTI, E., AND RICCI LUCCHI, R., 1972, Le torbiditi del "Appennino settentrionale:" introduzione all'analisi de facies: Soc. Geol. Italiana Mem., v. 11, p. 161–199.

NASU, N., VON HUENE, R., ISHIWADA, Y., LANGSETH, M., BRUNS, J., AND HONZA, E., 1980, Interpretation of multichannel seismic reflection data, Legs 56 and 57, Japan Trench transect, Deep Sea Drilling Project: Init. Repts. DSDP, v. 56–57/1, p. 489–504.

NATLAND, J. H., 1976, Petrology of volcanic rocks dredged from seamounts in the Line Islands: Init. Repts. DSDP, v. 33, p. 788.

———, 1978, Composition, provenance, and diagenesis of Cretaceous clastic sediments drilled on the Atlantic continental rise off southern Africa, DSDP Site 361—implications for the early circulation of the South Atlantic: Init. Repts. DSDP, v. 40, p. 1025–1061.

NELSON, C. H., 1976, Late Pleistocene and Holocene depositional trends, processes, and history of Astoria deep-sea fan, northeast Pacific: Marine Geology, v. 20, p. 129–173.

———, AND KULM, V., 1973, Submarine fans and channels, *in* Middleton, G. V., and Bouma, A. H., eds., Turbidites and deep water sedimentation: Soc. Econ. Paleontologists Mineralogists Pacific Section, p. 39–78.

———, NORMARK, W. R., BOUMA, A. H., AND CARLSON, P. R., 1978, Thin-bedded turbidites in modern submarine canyons and fans, *in* Stanley, D. J., and Kelling, G., eds., Sedimentation in Submarine Canyons, Fans, and Trenches: Stroudsburg, Pa., Dowden, Hutchinson and Ross, p. 177–190.

NISBET, E. G., AND PRICE, I., 1974, Silicified turbidites: graded cherts as redeposited ocean-ridge derived sediments, *in* Hsü, K. J., and Jenkyns, H. C., eds., Pelagic Sediments: On Land and Under the Sea: International Assoc. Sedimentologists Spec. Pub. 1, p. 351–366.

NORMARK, W. R., 1970, Growth patterns of deep-sea fans: Amer. Assoc. Petroleum Geologists, v. 54, p. 2170–2195.

———, AND HESS, G. R., 1980, Quaternary growth-patterns of California submarine fans, *in* Field, M. E., Bouma, A. H., and Colburn, I. P., eds., Quaternary Depositional Environments on the Pacific Continental Margin: Soc. Econ. Paleontologists Mineralogists, Pacific Section Symposium Volume.

PETERSON, M. N. A., EDGAR, N. T., AND OTHERS, 1970: Init. Repts. DSDP, v. 2.

PIPER, D. J. W., 1973, The sedimentology of silt turbidites from the Gulf of Alaska: Init. Repts. DSDP, v. 18, p. 847–868.

———, 1978, Turbidite muds and silts on deep sea fans and abyssal plains, *in* Stanley, D. J., and Kelling, G., eds., Sedimentation in Submarine Canyons, Fans, and Trenches: Stroudsburg, Pa., Dowden, Hutchinson and Ross, p. 163–176.

———, AND BRISCO, C. D., 1975, Deep water continental margin sedimentation, DSDP Leg 28, Antarctica: Init. Repts. DSDP, v. 28, p. 727–755.

———, VON HUENE, R., AND DUNCAN, J. R., 1973, Late Quaternary sedimentation in the active eastern Aleutian Trench: Geology, v. 1, p. 19–22.

RIECH, V., AND VON RAD, U., 1979, Silica diagenesis in the Atlantic Ocean: diagenetic potential and transformations, *in* Talwani, M., Hay, W. W., and Ryan, W. B. F., eds., Deep drilling results in the Atlantic Ocean: continental margins and paleoenvironments, Maurice Ewing Series 3, p. 315–341.

ROBERTS, D. G., MONTADERT, L., AND OTHERS, 1978: Init. Repts. DSDP, v. 48.
RUPKE, N. A., 1975, Deposition of fine-grained sediments in the abyssal environment of the Algero-Balearic Basin, western Mediterranean Sea: Sedimentology, v. 22, p. 95–109.
———, 1978, Deep clastic seas, *in* Reading, H. G., ed., Sedimentary Environments and Facies: New York, Elsevier, p. 372–415.
———, AND STANLEY, D. J., 1974, Distinctive properties of turbidite and hemipelagic mud layers in the Algero-Balearic Basin, western Mediterranean Sea: Smithsonian Contrib. Earth Science, v. 13, 40 p.
RYAN, W. B. F., BOLLI, H. M., AND OTHERS, 1978: Init. Repts. DSDP, v. 40.
RYAN, W. B. F., HSÜ, K. J., AND OTHERS, 1973: Init. Repts. DSDP, v. 13.
SARNTHEIN, M., 1978, Neogene sand layers off Northwest Africa: composition and source environment: Init. Repts. DSDP, v. 41 supplement, p. 939–959.
SCHLAGER, W., 1980, Mesozoic calciturbidites in Deep Sea Drilling Project Hole 416A: recognition of a drowned carbonate platform: Init. Repts. DSDP, v. 50, p. 733–749.
SCHLANGER, S. O., JACKSON, E. D., AND OTHERS, 1976: Init. Repts. DSDP, v. 33, p. 966.
SCHLEE, J. S., DILLON, W. P., AND GROW, J. A., 1979, Structure of the continental slope off the eastern United States, *in* Doyle, L. J., and Pilkey, O. H., Jr., eds., Geology of Continental Slopes: Soc. Econ. Paleontologists Mineralogists Spec. Pub. 27, p. 119–130.
SCHOLL, D. W., AND CREAGER, J. S., 1973, Geologic synthesis of Leg 19 (DSDP) results; for North Pacific, Aleutian Ridge, and Bering Sea: Init. Repts. DSDP, v. 19, p. 897–913.
———, MARLOW, M. S., AND COOPER, A. K., 1977, Sediment subduction and offscraping at Pacific margins: in island arcs, deep sea trenches and back-arc basins, Maurice Ewing Series 1, p. 199–209.
SCHOLLE, P. A., ed., 1977, Geological studies on the COST No. B-2 well, United States Mid-Atlantic outer continental shelf: U.S. Geological Survey Circ. 750, p. 1–71.
———, 1978, Geological studies of the COST No. GE-1 well, United States South Atlantic outer continental shelf area: U.S. Geological Survey Circ. 800, 92 p.
SCHWELLER, W. J., AND KULM, L. D., 1978, Depositional patterns and channelized sedimentation in active eastern Pacific trenches, *in* Stanley, D. J., and Kelling, G., eds., Sedimentation in Submarine Canyons, Fans, and Trenches: Stroudsburg, Pa., Dowden, Huchinson and Ross, p. 311–324.
SHOR, A. N., AND POORE, R. Z., 1978, Bottom currents and ice rafting in the North Atlantic: interpretation of Neogene depositional environments of Leg 49 cores: Init. Repts. DSDP, v. 49, p. 859–872.
SIMPSON, E., SCHLICH, J., AND OTHERS, 1974: Init. Repts. DSDP, v. 25, p. 872.
SOUTHARD, J. B., YOUNG, R. A., AND HOLLISTER, C. D., 1971, Experimental erosion of calcareous ooze: Jour. Geophys. Res., v. 76, p. 5903–5909.
STANLEY, D. J., AND KELLING, G., eds., 1978, Sedimentation in Submarine Canyons, Fans and Trenches: Stroudsburg, Pa., Dowden, Hutchinson and Ross, 395 p.
STEWART, R. J., 1976, Turbidites of the Aleutian abyssal plain: mineralogy, provenance, and constraints for Cenozoic motion of the Pacific Plate: Geol. Soc. America Bull., v. 87, p. 793–808.
———, 1977, Neogene turbidite sedimentation in Komandorskig Basin, western Bering Sea: Amer. Assoc. Petroleum Geologists Bull., v. 61, p. 192–206.
STOW, D. A. V., 1979, Method for silt and sand fabric analysis in deep-sea cores: Jour. Sed. Petrology, v. 49, p. 627–630.
———, AND LOVELL, T. P. B., 1979, Contourites: their recognition in modern and ancient sediments: Earth Science Review, v. 14, p. 251–291.
SUPKO, P., PERCH-NIELSEN, K., AND OTHERS, 1977: Init. Repts. DSDP, v. 39.
THIEDE, J., 1974, Sediment coarse fractions from the western Indian Ocean and the Gulf of Aden (DSDP Leg 24): Init. Repts. DSDP, v. 24, p. 651.
THOMPSON, R. W., 1974, Mineralogy of sands from the Bengal and Nicobar fans, Sites 218 and 211, eastern Indian Ocean: Init. Repts. DSDP, v. 22, p. 711–713.
TUCHOLKE, B. E., HOLLISTER, C. D., WEAVER, F. M., AND VENNUM, W., 1976, Continental rise and abyssal plain sedimentation in the southeast Pacific Basin—Leg 35, Deep Sea Drilling Project: Init. Repts. DSDP, v. 35, p. 359–400.
———, AND MOUNTAIN, G., 1979, Seismic stratigraphy, lithostratigraphy, and paleosedimentation patterns in the western North Atlantic, *in* Talwani, M., Hay, W. W., Ryan, W. B. F., eds., Deep drilling results in the Atlantic Ocean: continental margins and paleoenvironments, Maurice Ewing Series 3, p. 58–86.
———, AND VOGT, P. R., 1979, Western North Atlantic—sedimentary evolution and aspects of tectonic history: Init. Repts. DSDP, v. 43, p. 791–826.
———, ———, AND OTHERS, 1979: Init. Repts. DSDP, v. 43.
UNDERWOOD, M. B., BACHMAN, S. B., AND SCHWELLER, W. J., 1980, Sedimentary processes and facies associations within trench and trench-slope settings, Field, M. E., Bouma, A. H. and Colburn, I. P., eds., Quaternary Depositional Environments on the Pacific Continental Margin: Soc. Econ. Paleontologists Mineralogists, Pacific Section Symposium Volume, p. 211–228.
VAIL, P. R., MITCHUM, R. M., JR., AND THOMPSON, S., 1977, Seismic stratigraphy and global changes of sea level, part 3: Relative changes of sea level from coastal onlap: Amer. Assoc. Petroleum Geologists Mem. 26, p. 63–98.
———, 1974, Volcanogenic sediments and their relation to landmass volcanism and sea floor-continent movements, Western Indian Ocean Leg 25, Deep Sea Drilling Project: Init. Repts. DSDP, v. 25, p. 515–542.

———, AND KIDD, R. B., 1977, Volcanogenic sediments in the Indian Ocean, Heirtzler, J. R., and others, eds., Indian Ocean Geology and Biostratigraphy: Amer. Geophys. Union, , p. 87–118.

VAN ANDEL, T. H., AND KOMAR, P. D., 1969, Ponded sediments of the mid-Atlantic ridge between 22° and 23° North latitude: Geol. Soc. America Bull., v. 80, p. 1163–1190.

VAN DER LINGEN, G. J., 1969, The turbidite problem: New Zealand Jour. Geol. Geophys., v. 12, p. 7–50.

VEEVERS, J. J., HEIRTZLER, J. R., AND OTHERS, 1974: Init. Repts. DSDP, v. 27.

VOGT, P. R., 1979, Global magmatic episodes: new evidence and implications for the steady-state midoceanic ridge: Geology, v. 7, p. 93–98.

VON DER BORCH, C. C., SCLATER, J. G., AND OTHERS, 1974, Initial Reports of the Deep Sea Drilling Project, v. 22: Washington, U.S. Govt. Printing Office, 881 p.

VON HUENE, R., AUBOIN, J., AND OTHERS, 1980, Drilling on the Middle America Trench off Guatemala: Geol. Soc. America Bull., v. 91, p. 421–432.

VON HUENE, R. LANGSETH, M., NASU, N., AND OKADA, H., 1980, Summary, Japan Trench transect: Init. Repts. DSDP, v. 56-57/1, p. 473–489.

VON RAD, U., RYAN, W. B. F., AND OTHERS, 1979: Init. Repts. DSDP, v. 47.

WALKER, R. G., 1967, Turbidite sedimentary structures and their relationship to proximal and distal depositional environments: Jour. Sed. Petrology, v. 37, p. 24–43.

———, 1973, Mopping up the turbidite mess, *in* Ginsburg, R. N., ed., Evolving Concepts in Sedimentology: Baltimore, Johns Hopkins University Press, p. 1–37.

———, 1975, Generalized facies models for resedimented conglomerates of turbidite association: Geol. Soc. America Bull., v. 86, p. 737–748.

———, 1978, Deep-water sandstone facies and ancient submarine fans—models for exploration for stratigraphic traps: Amer. Assoc. Petroleum Geologists Bull., v. 62, p. 932–966.

———, AND MUTTI, E., 1973, Turbidite facies and facies associations, *in* Middleton, G. V., and Bouma, A. H., eds., Turbidites and Deep Water Sedimentation: Soc. Econ. Paleontologists Mineralogists Pacific Section, p. 119–157.

WHITMARSH, R. B., ROSS, D., WESER, O. E., AND OTHERS, 1974: Init. Repts. DSDP, v. 23.

WILDE, P., NORMARK, W. R., AND CHASE, T. E., 1978, Channel sands and petroleum potential of Monterey deep-sea fan: Amer. Assoc. Petroleum Geologists Bull., v. 62, p. 967–983.

WINTERER, E. L., 1973, Regional synthesis: Init. Repts. DSDP, v. 17, p. 911–922.

WORSLEY, T., AND DAVIES, T., 1979, Sea-level fluctuations and deep-sea sedimentation rates: Science, v. 203, p. 455–456.

YEATS, R. S., HART, S. R., AND OTHERS, 1976: Init. Repts. DSDP, v. 34.

SEPM Special Publication No. 32, p. 129–147, October 1981

DISTRIBUTION, THERMAL HISTORIES, ISOTOPIC COMPOSITIONS, AND REFLECTION CHARACTERISTICS OF SILICEOUS ROCKS RECOVERED BY THE DEEP SEA DRILLING PROJECT

KENNETH A. PISCIOTTO
Deep Sea Drilling Project, Scripps Institution of Oceanography, La Jolla, California 92093 U.S.A.

ABSTRACT

A synthesis of deep sea drilling results through Leg 70 indicates that siliceous rocks have been recovered in a variety of lithofacies from almost all major ocean basins. Reconstructed distributions of these rocks delineate original sites of deposition. During the Neogene, these areas included the equatorial Pacific, the Southern Ocean, the southern Bering Sea, the Red Sea, and local centers off Japan, California and northwest Africa. Paleogene occurrences describe a wide equatorial belt in the Pacific, Caribbean, Atlantic, and Indian Oceans, and several smaller regions in the North Atlantic, South Atlantic, and southern Indian Ocean. Cretaceous areas of deposition also included the equatorial Pacific and Atlantic with a few sites off western Australia. By analogy with present circulation patterns and areas of opal accumulation, reconstructed distributions of DSDP siliceous rocks delineate areas of past oceanic divergence. The rate of accumulation of silica, when compared with better known accumulation rates of opal, is a crude index to past productivity in these areas.

Opal-CT and quartz are the principal authigenic silica minerals in siliceous rocks recovered from the deep sea; both are derived predominantly from biogenic silica comprising sponge spicules and the tests of radiolarians and diatoms. Time and temperature are important, though not the only, controls of the transformation rates of silica. Plots of age versus present subsurface temperature of opal-CT and quartz in DSDP siliceous rocks generally match the time and temperature stability fields of these two minerals determined from experiments. Using the experimental results, the transformation history of silica can be estimated. High thermal gradients and rates of burial apparently favor rapid transformations.

Oxygen isotopic compositions of opal-CT and quartz depend upon the temperature of formation and the isotopic composition of the equilibrating fluid. Comparisons of present subsurface temperatures with temperatures computed from isotopic values, in conjunction with isotopic data on interstitial water squeezed from deep sea sediments, suggest the $\delta^{18}O$ values of opal-CT and quartz in deep sea siliceous rocks reflect a restricted range of formation temperatures, generally less than 50°C, and equilibration with interstitial water having a $\delta^{18}O$ range of 0‰ to −3‰.

Siliceous rocks are important seismic reflectors in the western North Atlantic (Horizon A^c), in the southern Bering Sea (Bottom Simulating Reflector), and in the northwest Pacific. The strength and continuity of these reflectors are complex functions of degree of silicification and bedding characteristics. Any seismic analysis must treat siliceous rocks as both sedimentary and diagenetic facies because stratigraphic and diagenetic boundaries do not always coincide.

INTRODUCTION

Chert and silicified rocks were recovered in the deep sea at four sites in the northwestern Atlantic Ocean on the first cruise of the *Glomar Challenger* (Ewing, Worzel and others, 1969). To date over forty percent of the holes drilled in the seafloor by the Deep Sea Drilling Project (DSDP) encountered and recovered siliceous rocks (Fig. 1). Detailed studies of these rocks (Table 1) meshed with complementary experimental studies (*e.g.*, Mizutani, 1966, 1967, 1970; Ernst and Calvert, 1969; Kastner and others, 1977) and with investigations of siliceous rocks on land (*e.g.*, Murata and Larson, 1975; Murata and others, 1977) form the basis for interpretations of siliceous rocks.

In light of published compilations (*e.g.*, Ramsay, 1973; Davies and Supko, 1973; Greenwood, 1973; Wise and Weaver, 1974; Calvert, 1977; Riech and von Rad, 1979b; Kastner, *in press*), this synthesis of DSDP results focuses on four aspects of siliceous rocks: 1) past and present distributions in the oceans, 2) transformation histories of silica minerals as functions of time and temperature, 3) oxygen isotopic compositions, and 4) seismic reflection characteristics.

DATA BASE

Site chapters and individual contributions in the Initial Reports of the Deep Sea Drilling Project constitute the principal storehouse of data summarized in the illustrations of this paper. The DSDP prime data bank facilitated retrieval of some of these data, specifically information on lithology, mineralogy, and stratigraphic age. I used the Initial Core Descriptions, shipboard hole summaries, and current site summaries through Leg 70 to add the latest drilling information. Additional data came from syntheses and reports outside the Initial Reports. I used the paleoceanic reconstructions of Firstbrook and others (1979), tied to the 1964 Geological Society of London Time Scale, to restore DSDP sites to their past geographic positions, and

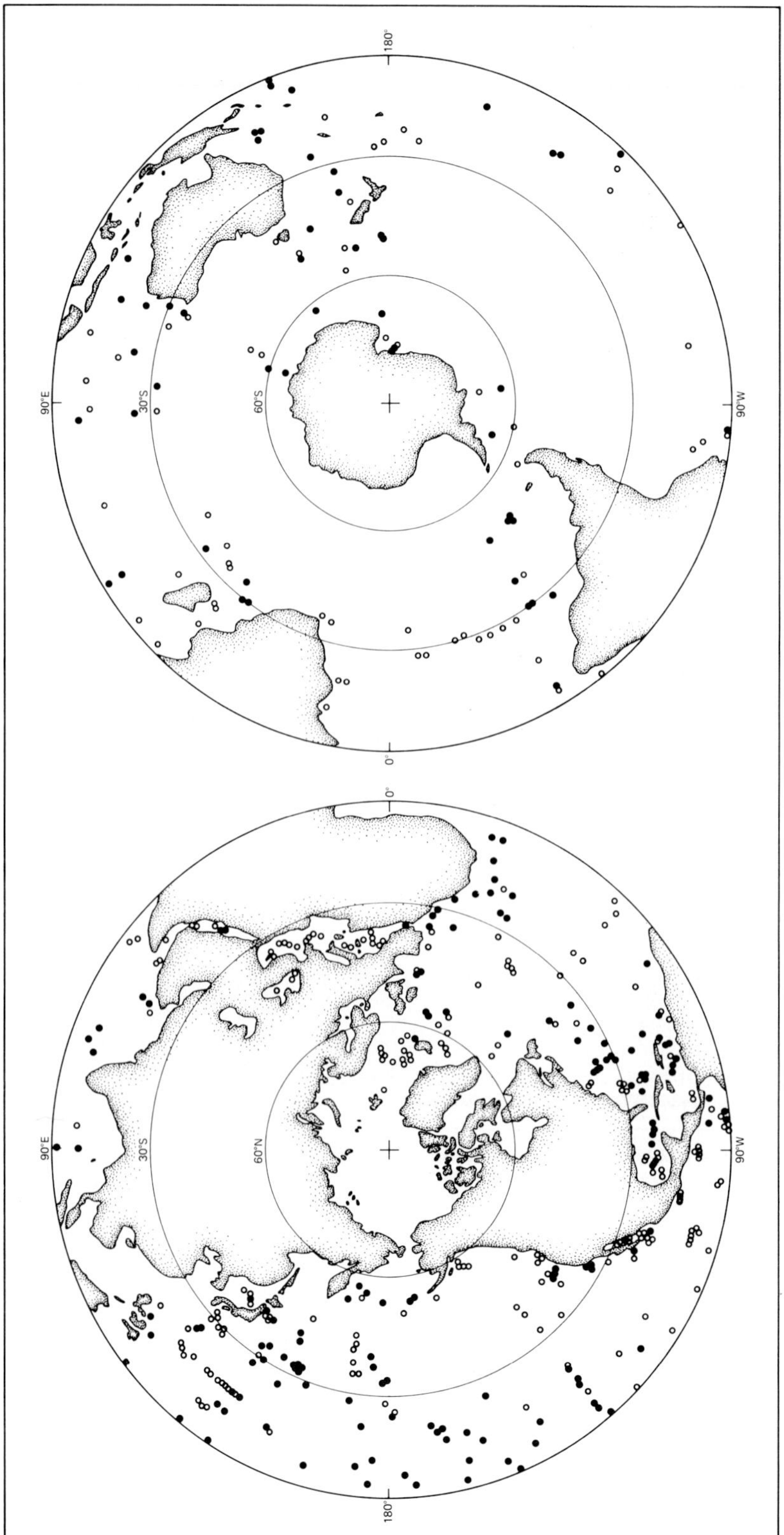

FIG. 1.—Summary of DSDP drill sites through Leg 70. Filled symbols denote sites where authigenic silica was recovered; open symbols = no authigenic silica.

TABLE 1.—REFERENCES TO DETAILED STUDIES OF DSDP SILICEOUS ROCKS.

Atlantic Region

Beall and Fischer (1969), Beall and others (1973), Berger and von Rad (1972), Bernoulli (1972), Calvert (1971), Greenwood (1973), Kagami (1979), Laughton and others (1972), Peterson Edgar and others (1970), Riech (1979, 1980), Riech and von Rad (1979a, 1979b), von Rad and Rösch (1972, 1974, 1977).

Pacific Region

Anderson and Lawrence (1976), Garrison and others (1975), Greenwood (1973), Heath (1973), Heath and Moberly (1971), Hein and others (1978), Iijima and others (1980), Kastner (1976), Keene (1975, 1976), Kelts (1976), Knauth and Epstein (1975), Kolodny and Epstein (1976), Lancelot (1973), Lawrence and others (1976), Levitan and others (1975), Packham and van der Lingen (1973), Pimm and others (1971), Pisciotto (1980), van der Lingen and Packham (1975), von der Borch and others (1971).

Indian Ocean

Leclaire (1974), Robinson and others (1974), Vallier (1974).

Caribbean Area

Donnelly and Nalli (1973), Knauth and Epstein (1975), Levitan and others (1975).

Red Sea

Stoffers and Kuhn (1974).

$5° \times 5°$ compilations of thermal conductivity and heat flow (Langseth and von Herzen, 1971) to compute present thermal gradients and subsurface temperatures.

Figures 1 and 2 provide some measure of the limitations of the data. Areal drilling coverage is generally good, although the results may be slightly biased in favor of the more intensively sampled North Pacific, North Atlantic, and Caribbean areas (Fig. 1). Successful recovery of material within cored siliceous intervals is extremely variable (Fig. 2). The cluster of values along the x-axis in this figure illustrates that siliceous lithologies often comprise less than 10% of the recovered material at many sites. Recovery was best in those intervals containing little silica, typically weakly silicified chalks, and poor when nothing but siliceous lithologies, usually chert and porcelanite in radiolarian mudstone, were recovered.

TERMINOLOGY

I follow the terminology of Jones and Segnit (1971) for the three varieties of opaline silica: opal-A = high disordered, x-ray-amorphous opal, opal-CT = "disordered" α-cristobalite and α-tridymite, and, opal-C = "well-ordered" α-cristobalite. Although a variety of pathways are possible, most commonly in nature biogenic silica (opal-A) transforms to quartz through an intermediate cristobalitic stage (opal-CT) with increasing diagenetic grade (*e.g.*, Riech and von Rad, 1979b; Kastner, *in press*). Opal-C is generally associated with volcanic rocks and with sediments containing volcanic debris.

DSDP usage of the terms chert and porcelanite to denote siliceous sedimentary rocks composed mainly of quartz and opal-CT, respectively, follows Calvert (1971). In practice, however, it is often difficult to verify the specific silica mineralogy without x-ray diffraction (XRD). These terms enjoy greater flexibility if used as textural designations, adding the dominant silica mineral, when known, as a modifier. In this report I follow the usage of Taliaferro (1934), Bramlette (1946), and Keene (1975): chert is a hard, conchoidally-fracturing, cryptocrystalline, varicolored sedimentary rock with semivitreous, vitreous or waxy luster, consisting dominantly of silica; porcelanite refers to a siliceous sedimentary rock with a dull or matte luster resembling that of unglazed porcelain. Figure 3 summarizes the relationship between lithology and silica mineralogy.

DISTRIBUTION OF DSDP SILICEOUS ROCKS

Siliceous sediments accumulate at present in the productive equatorial zones of the Pacific and Indian Oceans (mainly radiolaria and diatoms), at high latitudes in the Pacific and Southern Oceans (diatoms) and in areas of coastal upwelling controlled by eastern boundary currents and local coastal and bathymetric configurations (Riedel, 1959, 1971; Hays, 1970; Lisitzin, 1971; Orr, 1972; Ramsay, 1973; Calvert, 1974). The rate of accumulation of opal in these sediments is an index to fluctuations of surface productivity (Leinen, 1979; Brewster, 1980).

The environmental significance of DSDP cherts, porcelanites and siliceous rocks lies in their descendence principally from biogenic silica. To outline areas of past surface productivity and opal accumulation, it is necessary to reconstruct the original geographic positions of these siliceous rocks. Previous published reconstructions demonstrate the following patterns:

1) Occurrences of Neogene siliceous sediments and their diagenetic derivatives, when restored to their original sites of deposition, match modern distributions (Riedel, 1971; Ramsay, 1973; Kastner, *in press*).

2) Paleogene reconstructions define an equatorial zone of siliceous deposits that, unlike the Neogene, includes a continuous high-fertility belt across both the Pacific and Atlantic joined by a middle America seaway (Riedel, 1971; Ramsay, 1973; Keene, 1976; Riech and von Rad, 1979b). Siliceous

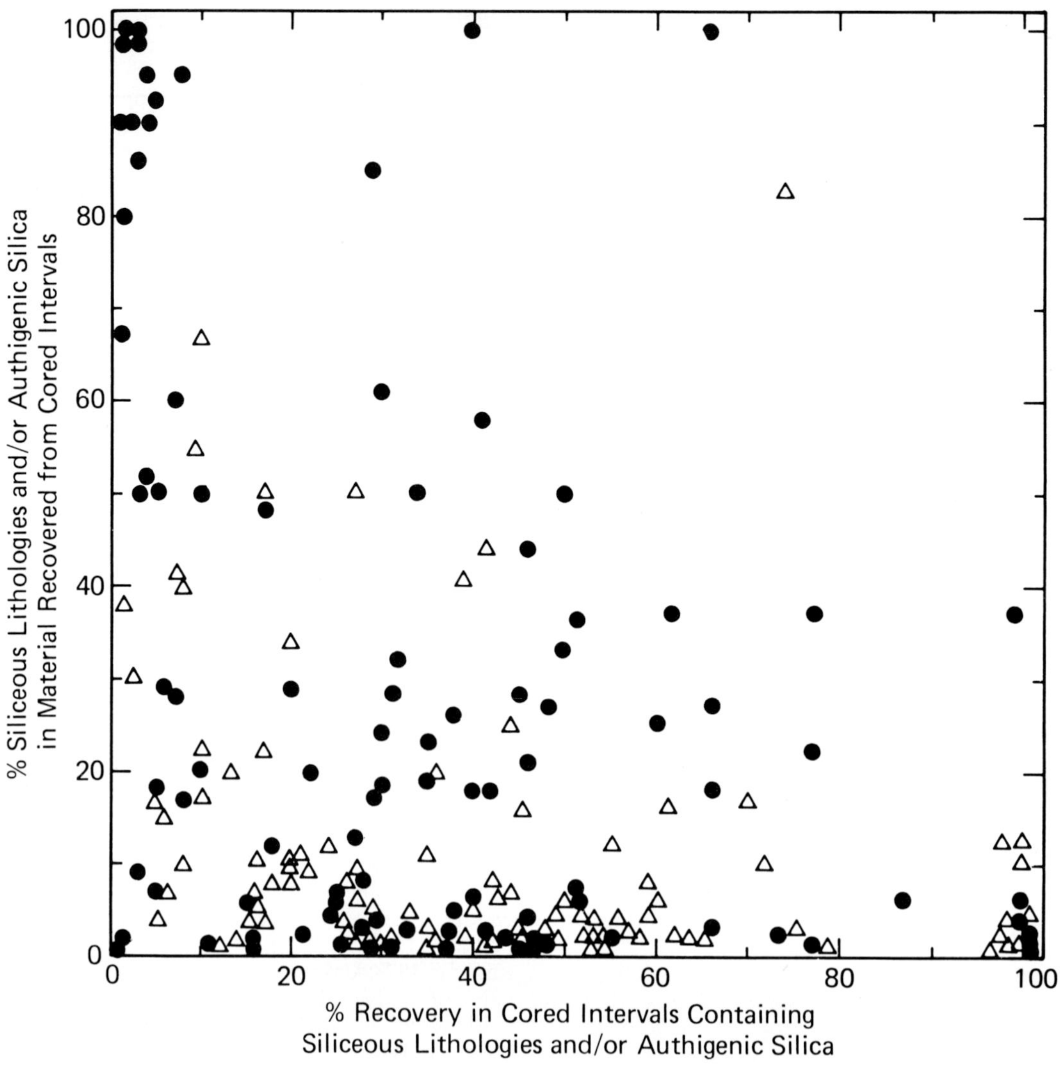

FIG. 2.—Summary of DSDP core recovery data in siliceous intervals. Filled symbols indicate carbonate-poor host rock; open symbols denote carbonate host rock. Data from the Initial Reports of DSDP.

sediments also accumulated locally in the eastern North Atlantic and along the margins of the Argentine Basin in the South Atlantic during the Paleogene (Riech and von Rad, 1979b; Ramsay, 1973).

3) Cretaceous reconstructions show a broad equatorial zone of siliceous deposits and part of a southern belt, much like the Paleogene distribution (Ramsay, 1973).

4) Jurassic and older siliceous rocks are too few to permit delineation of past distributions (Ramsay, 1973).

5) Estimates of the relative abundances of chert and porcelanite for different periods of geologic time differ with the method of compilation (Kastner, *in press*). If computed as the total number of occurrences for each geologic time interval, siliceous rocks are apparently most abundant in Eocene and Upper Cretaceous sedimentary sequences (Davies and Supko, 1973). In contrast, no intervals of peak abundance occur, at least for Pacific data, if occurrences of chert and porcelanite are tallied as per cent of time stratigraphic units at each geologic age (Levitan and others, 1975; Keene, 1976).

With continued *Challenger* drilling, many new discoveries of siliceous rocks have been made since the publication of these syntheses. These new data and recently published paleoceanic reconstructions of DSDP sites (Firstbrook and others, 1979)

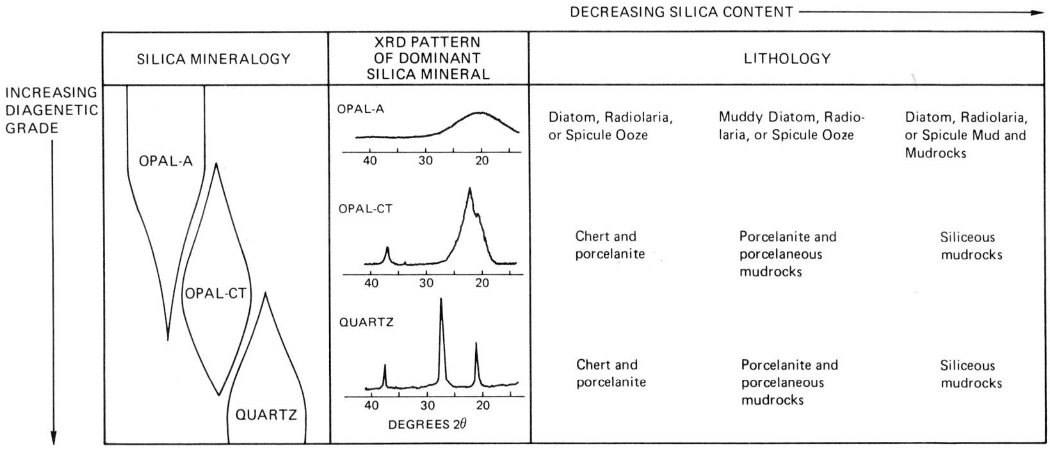

FIG. 3.—Generalized relationship between silica mineralogy and rock types.

prompted the updated compilations illustrated in Figure 4. In addition, I list approximate accumulation rates of silica and compare them with published accumulation rates of opal (Leinen, 1979; Brewster, 1980) to estimate surface productivities for specific intervals of time at key DSDP sites (Table 2). I computed these rates from bulk accumulation rates of the sediments (van Andel and others, 1975) and from recovery data illustrated in Figure 2. The results, summarized below, generally support the earlier compilations of Ramsay (1973) and others with several significant additions.

Neogene (Fig. 4)—Much of the biogenic opal deposited in the oceans during the Neogene has not been coverted to opal-CT or quartz. Areas where the conversions have occurred are shown in the first frame of Figure 4. The youngest siliceous rocks that contain opal-CT or quartz accumulated as rapidly deposited diatomaceous sediments in the Southern Ocean, southern Bering Sea, and in areas of local divergence off Japan, California, northwest Africa and in the Red Sea. Accumulation rates of silica at Sites 184 and 185 in the southern Bering Sea were extremely high, 900 to 2000 g/cm^2/m.y. For comparison, accumulation rates of opal at Site 266 in the Antarctic and at Sites 77 through 83 in the eastern equatorial Pacific were about 100 g/cm^2/m.y. (Brewster, 1980) and 50 to 1000 g/cm^2/m.y. (Leinen, 1979), respectively, during the same period. At Site 315 in the equatorial Pacific, the Late Oligocene accumulation rate of silica in carbonate ooze (40 g/cm^2/m.y.; Table 2) closely matched the accumulation rate of opal (50–100 g/cm^2/m.y.; Leinen, 1979) at nearby Site 69 for the same period.

Paleogene (Fig. 4)—Paleogene siliceous rocks were deposited in a broad equatorial belt including the Pacific, Caribbean, and Atlantic Oceans. They occur in a variety of lithofacies. In the Pacific, silica commonly forms nodules, lenses and stringers of chert and porcelanite or occurs as dispersed cement (opal-CT and quartz) in carbonates and radiolarian mudstones. At Site 167 in the equatorial Pacific, the accumulation rate of silica during the Early Paleocene through Middle Eocene was only about 3 g/cm^2/m.y., several orders of magnitude lower than Middle Eocene and younger accumulation rates of opal in an area just to the east (Leinen, 1979). In the Caribbean and Gulf of Mexico siliceous rocks are mainly nodular chert and porcelanite in carbonate ooze and chalk. However, at Site 146 in the Venezuela Basin, graded, quartzose, cemented radiolarian sands are interbedded with limestones and layers of graded basaltic ash (Edgar and others, 1973, p. 25). Along the western margin of the North Atlantic, areally extensive turbidites occur that are composed of terrigenous silt and fine sand, sponge spicules, radiolaria, ostracodes, and fish debris and are cemented by opal-CT and quartz (Beall and Fischer, 1969; Riech and von Rad, 1979b). Further east, in the west central North Atlantic, radiolarian mudstones containing chert and porcelanite or sparse opal-CT and quartz cement are common (Calvert, 1971).

Sites at higher latitudes record variable levels of productivity during the Paleogene. For example, at Site 405 in the North Atlantic (Rockall Plateau) the accumulation rate of silica during the Early Eocene was 630 to 945 g/cm^2/m.y. (Table 2), similar to high Pliocene through Recent accumulation rates of opal in the Antarctic at Site 266 (Brewster, 1980) and to the Late Miocene peak at Sites 82 and 83 in the eastern equatorial Pacific (Leinen, 1979). The high accumulation rate at Site 405 may reflect increased productivity associated with the entrance of cold arctic bottom water from the newly opened Norwegian Sea during the Early Eocene (Riech and

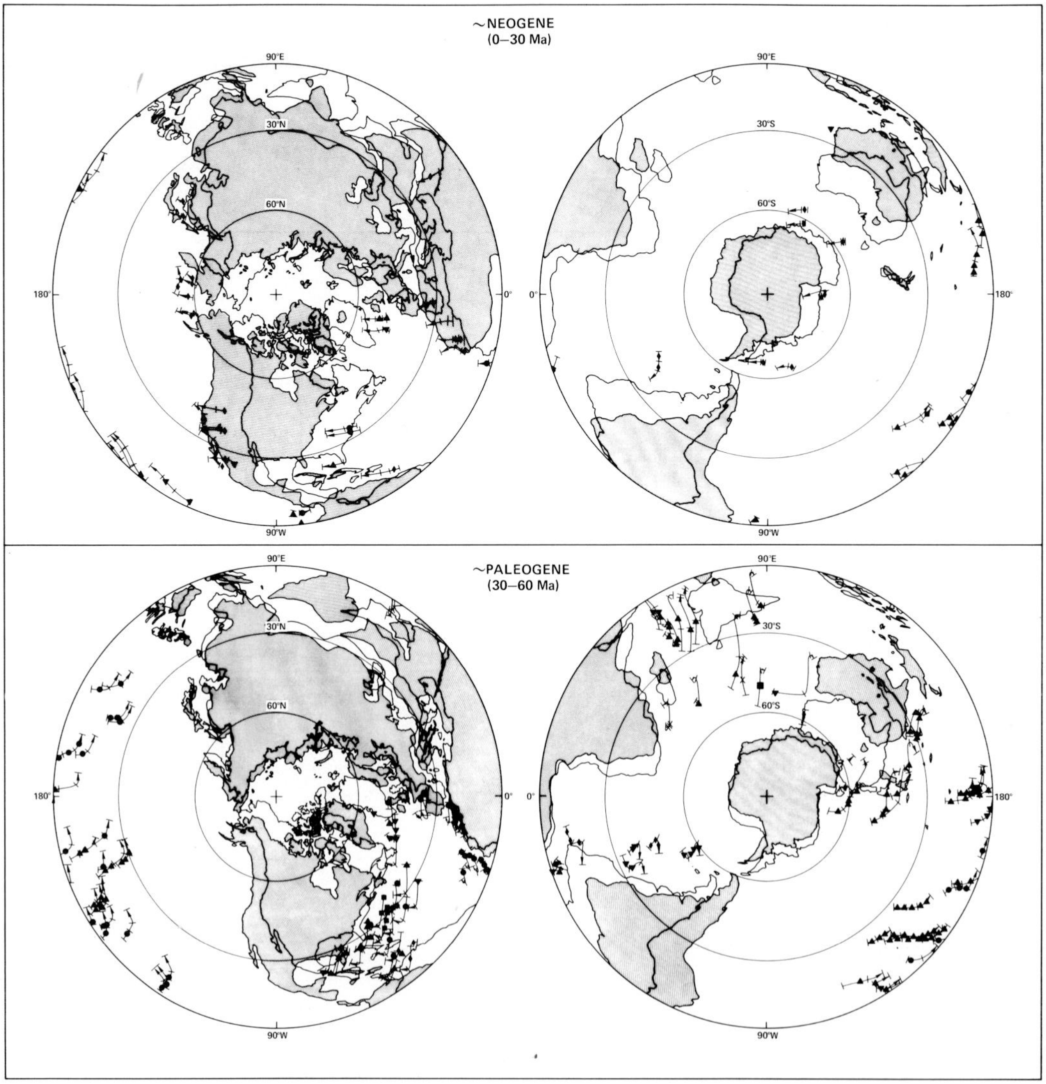

FIG. 4.—Distribution of DSDP siliceous rocks (authigenic silica only) for selected time periods. Key to symbols in Figure 5 with the difference that all symbols are filled regardless of silica mineralogy. Curved lines through symbols track the movement of sites within individual time periods; head of arrow corresponds to youngest age of time frame,

von Rad, 1979b). Similarly, the occurrences of Lower Eocene chert and porcelanite along the margins of the Argentine Basin in the South Atlantic perhaps record increased productivity sparked by the influx of nutrients provided by cold, subantarctic bottom water moving across the subsiding Falkland Plateau and Rio Grande Rise (Riech and von Rad, 1979b).

In the southern Indian Ocean off southwest Africa (Site 245) nodular chert occurs in carbonate ooze and chalk. The Late Paleocene through Early Eocene accumulation rate of silica here was 132 $g/cm^2/m.y.$ (Table 2), about twice the accumulation rate of opal for the same time period at Site 277 in the Antarctic (Brewster, 1980). During this period India occupied a central position in the Indian Ocean. Oceanic circulation and productivity west of the landmass at Site 245 was apparently influenced by an open Tethyan connection to the north and a southern passage to the South Atlantic, whereas waters east of the continent felt the effects of the South Pacific equatorial current through a seaway north of Australia (Luyendyk and Davies, 1974; Vincent, 1974). In contrast, the Late Paleo-

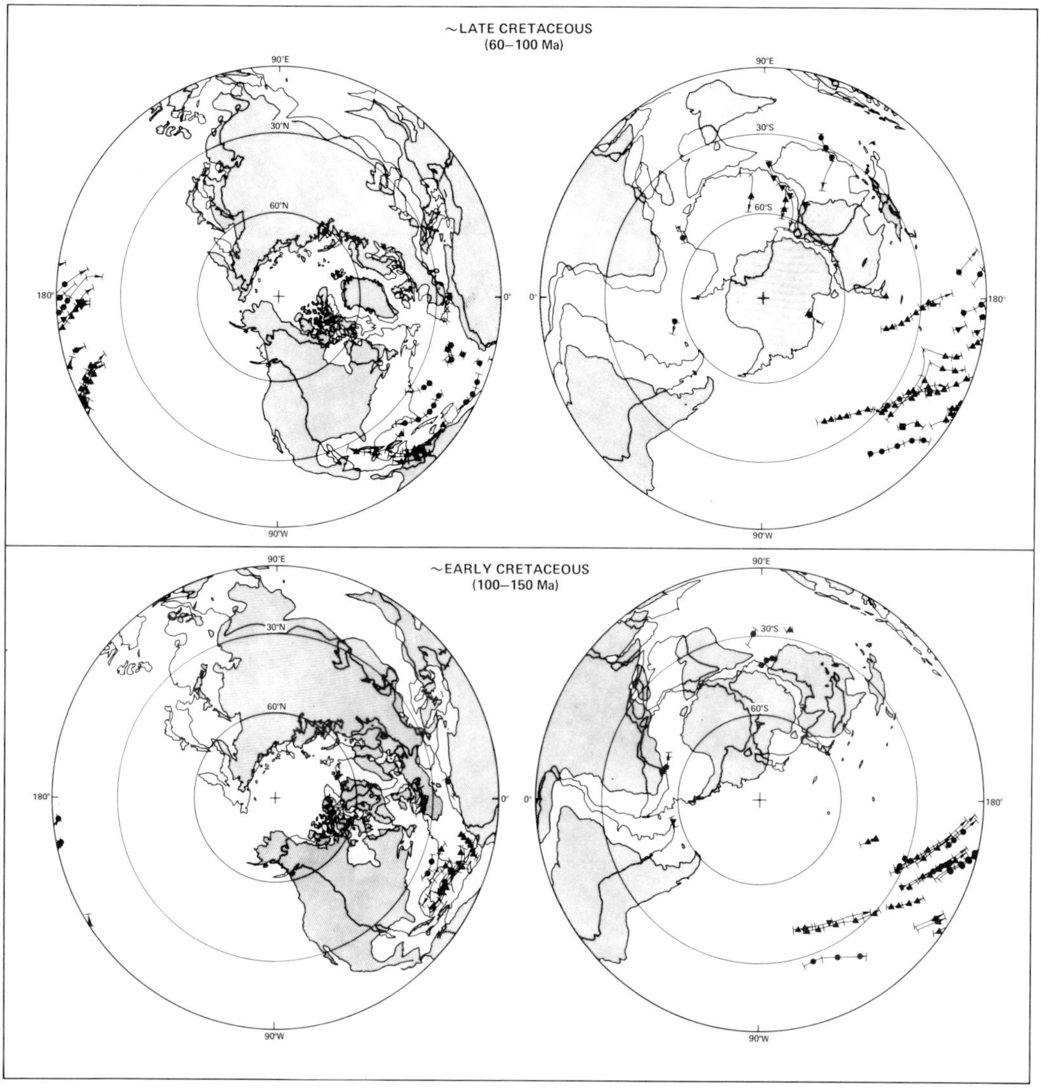

tick marks are successively older ages in 10 m.y. increments. Shaded areas mark the positions of continents (present coast lines) corresponding to the youngest age in the time frame; open or outlined areas refer to continental positions (present coast lines) at the greater age shown. Paleoceanic reconstructions after Firstbrook and others (1979).

cene through Middle Eocene accumulation rate of silica at Site 283 in the South Tasman Sea (Table 2) is similar to accumulation rates of opal in coeval sediments at Site 277 in the Antarctic; neither site reflects any dramatic circulation or climatic changes during this period (Kennett, 1977; Brewster, 1980).

Cretaceous (Fig. 4)—As during the Paleogene, silica accumulated in an equatorial belt in both the Pacific and Atlantic during the Cretaceous. In the Pacific, the apparent asymmetric distribution of siliceous rocks about the equator during the Early Cretaceous illustrates that much of the sediment of that age that was deposited north of the equator has been obliterated by subduction. In both oceans, the siliceous rocks occur almost exclusively in carbonates and in radiolarian mudstones; turbidites, similar to the Eocene turbidites containing abundant opal-CT and quartz that occur in the western North Atlantic, either do not occur or have not been drilled and recovered. Accumulation rates of silica at Sites 167, 303, and 305 in the equatorial Pacific were 5 to 30 g/cm^2/m.y. These rates match the lower range of accumulation rates of opal in the

TABLE 2.—APPROXIMATE SILICA ACCUMULATION RATES FOR SELECTED DSDP SITES COMPUTED FROM DATA IN THE INITIAL REPORTS.

Site	Time Interval (m.y.BP)	Present Location (Lat./Long.)	Approximate Paleolatitudes	Accumulation Rate of Sediments ($g/cm^2/m.y.$)	Accumulation Rate of Silica ($g/cm^2/m.y.$)
		NEOGENE			
184	L. Miocene (7.5–10)	S. Bering Sea (53.7°N/170.9°W)	57°N @ 7.5 m.y.BP 59°N @ 10 m.y.BP	7231–10948	1323–2003
185	L. Miocene (7–8)	S. Bering Sea (54.4°N/169.3°W)	58°N @ 7 m.y.BP 58°N @ 8 m.y.BP	7565–11454	953–1443
315	L. Oligocene (23–30)	Central Pacific (4.2°N/158.5°W)	0° @ 23 m.y.BP 2°S @ 20 m.y.BP	2272	40
		PALEOGENE			
167	M. Eocene-E. Paleocene (46–65)	Central Pacific (7.1°N/176.8°W)	4°S @ 46 m.y.BP 13°S @ 65 m.y.BP	492–536	3
405	E. Eocene (50–52)	N. Atlantic (55.3°N/22.1°W)	47°N @ 50 m.y.BP 47°N @ 52 m.y.BP	8734–13130	630–945
245	E. Eocene-L. Paleocene (51–56)	S. Indian Ocean (31.5°S/52.3°E)	49°S @ 51 m.y.BP 48°S @ 56 m.y.BP	4403	132
283	M. Eocene-E. Paleocene (46–55)	SW Pacific (43.9°S/154.3°E)	53°S @ 46 m.y.BP 58°S @ 55 m.y.BP	303	22
		CRETACEOUS			
167	Maestrichtian-Tithonian (65–146)	Central Pacific (7.1°N/176.8°W)	13°S @ 65 m.y.BP 41°S @ 146 m.y.BP	869–1100	5–6
258	Coniacian-Cenomanian (82–100)	SE Indian Ocean (33.8°S/112.5°E)	53°S @ 82 m.y.BP 41°S @ 100 m.y.BP	913	165
258	L.-M. Albian (100–103)	SE Indian Ocean (33.8°S/112.5°E)	41°S @ 100 m.y.BP 41°S @ 103 m.y.BP	5523	1014
303	L. Albian-Valanginian (100–124)	NW Pacific (40.8°N/154.5°E)	2°S @ 100 m.y.BP 8°S @ 124 m.y.BP	537–808	11–17
305	M. Albian-Valanginian (103–124)	NW Pacific (32.0°N/157.9°E)	7°S @ 103 m.y.BP 17°S @ 124 m.y.BP	1390–1534	28–30

equatorial Pacific during the Cenozoic (Leinen, 1979) and may indicate that surface productivity of siliceous plankton was consistently low in this area during the Cretaceous.

In contrast, at Site 258, off western Australia (Naturaliste Plateau), a dramatic decrease in the accumulation rate of silica at the beginning of the Cenomanian (Table 2) coincides with an abrupt lithologic transition from siliceous black shales (Albian) to Upper Cretaceous carbonates. The high Albian rate (1014 $g/cm^2/m.y.$) may reflect intense upwelling in the restricted Wharton Basin during that period (Luyendyk and Davies, 1974); this rate is similar to the Late Miocene accumulation rate of opal in the eastern equatorial Pacific (Leinen, 1979) and to the Pliocene through Recent rate in the Antarctic (Brewster, 1980). The change to much lower accumulation rates of silica in the Cenomanian may record the adjustments of oceanic circulation that accompanied the separation of India from Australia and Antarctica (Luyendyk and Davies, 1974).

To summarize, the correlation between areas of opal accumulation and zones of divergence in the present ocean suggests that a similar correlation can be made between reconstructed distributions of DSDP siliceous rocks and areas of past oceanic divergence. Accumulation rates of silica provide an estimate of past surface productivity in the regions when compared with better-known opal accumulation rates.

THERMAL HISTORY

Of the many factors that control rates of transformation of diagenetic silica (*e.g.*, Kastner and others, 1977; and references therein) time and temperature play crucial roles. Experimental studies demonstrate that transformation rates accelerate as

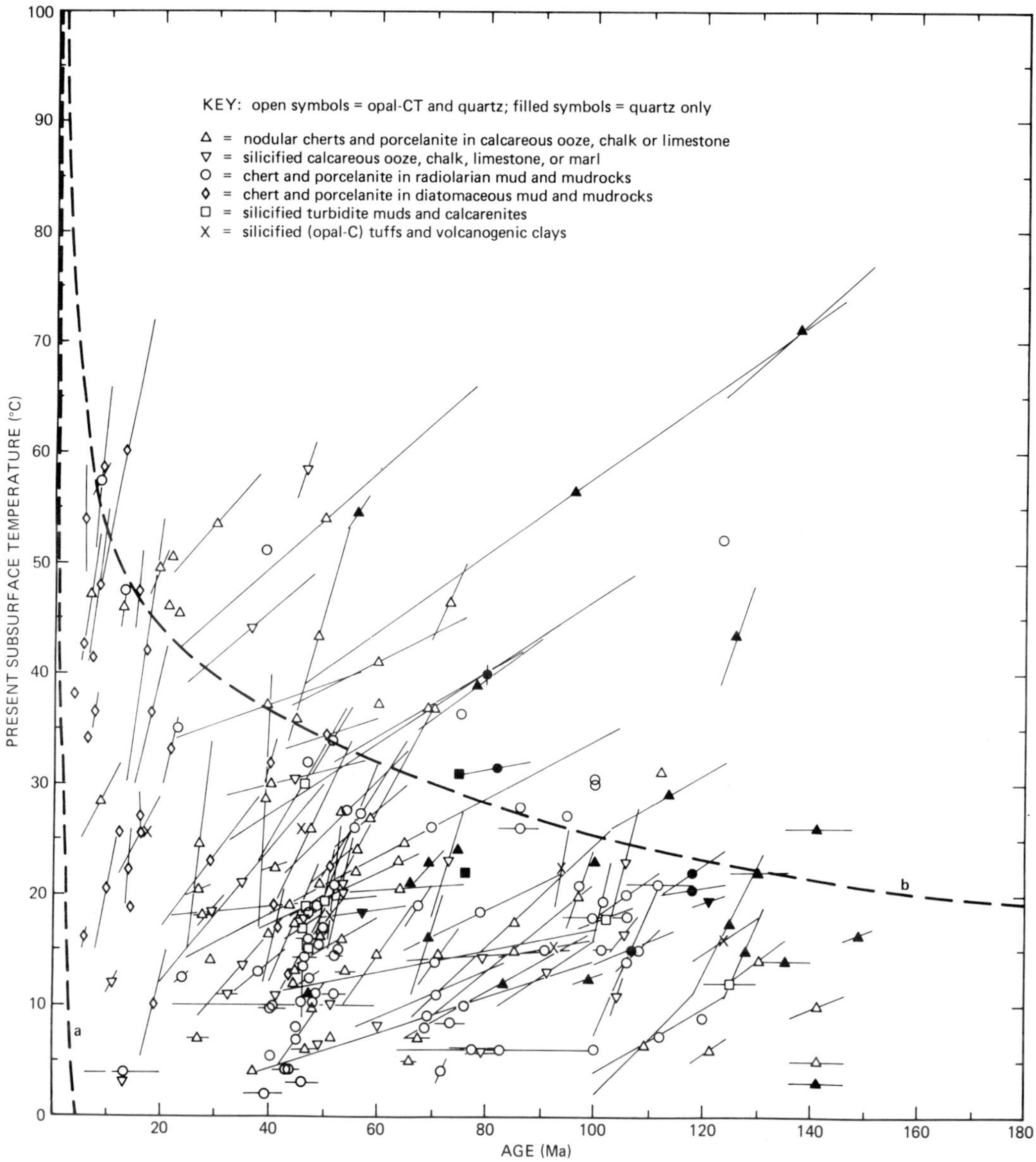

FIG. 5.—Relationship between age and present subsurface temperature of DSDP siliceous rocks and sediments containing opal-CT and/or quartz. Bars on symbols indicate ranges in age and temperature for siliceous rocks at individual sites. Data from the Initial Reports of DSDP. Curves a and b define the lower and upper boundaries, respectively, of the time and temperature stability field for opal-CT based on the experimental work of Mizutani (1966, 1967, 1970).

temperature increases (*e.g.*, Mizutani, 1966, 1967; Ernst and Calvert, 1969). The ranges of temperatures at which the silica transformations occur may depend largely on a unique thermal history at individual localities; higher temperature ranges for the transformations characterize young, rapidly buried rocks in areas of high heat flow (Mizutani, 1970).

Thermal and burial histories of DSDP siliceous rocks are often better documented than those of similar rocks exposed on land. As a first approximation, a plot of age versus present subsurface temperature for DSDP siliceous rocks provides a measure of the natural stability fields of opal-CT and quartz (Fig. 5). Opal-CT commonly falls within

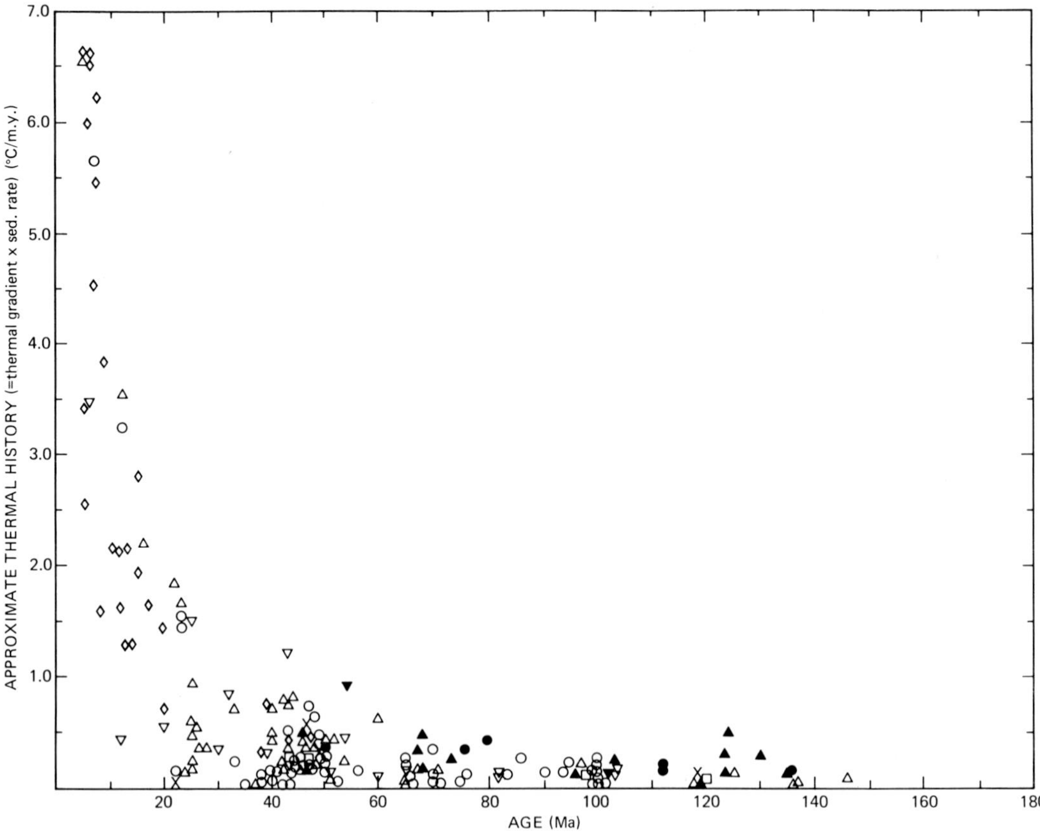

FIG. 6.—Relation between approximate thermal history and age for DSDP siliceous rocks. See Figure 5 for key to symbols. Open symbols denote the shallowest occurrence of opal-CT at each site; filled symbols indicate the deepest occurrence of opal-CT only at sites where underlying quartzose rocks can be identified. Data from the Initial Reports of DSDP.

the limiting time and temperature fields predicted by Mizutani (1970) from experiments. Less supportive are the occurrences of siliceous rocks that contain quartz as the only silica mineral (Fig. 5).

Consider next the rates at which opal-CT would form and transform by examining approximate thermal histories of siliceous rocks at DSDP sites. The product of the present thermal gradient and average sedimentation rate for the period following deposition of the silica provides an estimate of thermal history (Mizutani, 1970). Figure 6 shows data for the shallowest and deepest stratigraphic occurrences of opal-CT at each site; opal-A is present above the shallowest occurrences of opal-CT. The results illustrate that rates of silica transformations are rapid if thermal gradients and sedimentation rates are high. Thermal history values greater than 1.0 generally correspond to rapidly deposited Neogene diatomaceous sediments along continental margins or in small ocean basins having moderately high heat flow. Values less than 1.0 are typically from Oligocene and older rocks deposited in the pelagic realm away from non-biogenic sediment supplies, a distinction made earlier by Hein and others (1978).

Two case histories offer further refinement of this approach (Fig. 7). During the past 120 m.y. Sites 303 and 305, presently located in the northwest Pacific, tracked similar northwesterly paths across the Pacific behind the ancient Japanese Spreading Center (Lancelot and Larson, 1975). The theoretical heat flow model of Sclater and others (1971) predicts that both sites probably experienced similar exponential decay in heat flow as they migrated away from the spreading center. Sedimentary histories, however, were significantly different. At Site 305, on the Shatsky Rise and above the CCD throughout its entire sedimentary history, a thick pile of carbonate sediments accumulated, whereas Site 303, on the abyssal sea floor, rapidly subsided below the CCD and received

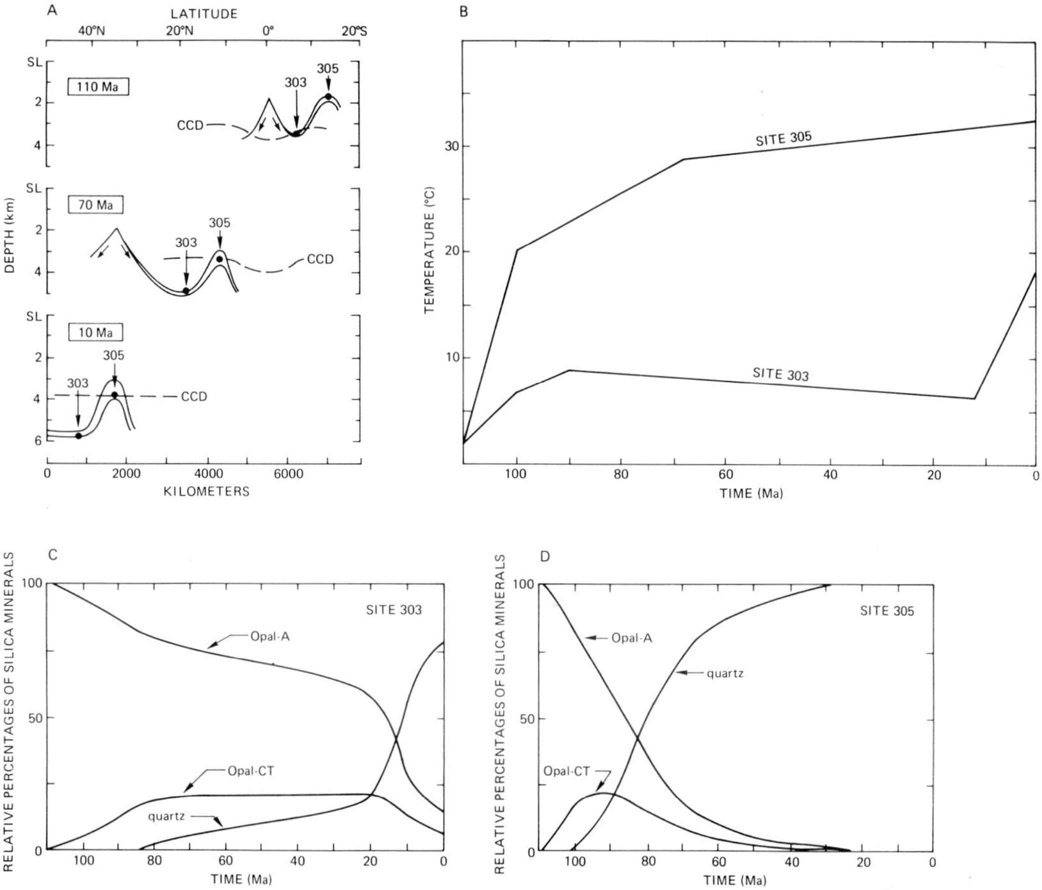

FIG. 7.—Oceanographic, thermal, and diagenetic histories of siliceous sediments deposited 110 Ma at Sites 303 and 305 in the Pacific. A. Oceanographic and tectonic framework and the burial histories at these sites (after Lancelot and Larson, 1975). Dots mark positions of siliceous rocks whose thermal and diagenetic histories are plotted in B, C, and D. B. Possible thermal histories of these siliceous rocks using the burial histories in A (Lancelot and Larson, 1975) and theoretical heat flow from Sclater and others (1971). C, D. Transformation histories of silica deposited 110 Ma at Sites 303 and 305 using thermal histories in B and the diagenetic model of Mizutani (1970). See text for further discussion.

much less sediment (Fig. 7A). Decreasing heat flow and a suppressed sedimentation rate may have actually led to a slight decrease in temperature over a 78 m.y. period at Site 303 (Fig. 7B). Using Mizutani's (1970) model for silica transformations and the unique thermal histories at Sites 303 and 305 (Fig. 7B), quite different diagenetic histories unfold. Figures 7C and 7D illustrate these histories for biogenic silica that was initially deposited 110 Ma as radiolarian remains in the productive equatorial region. If heat flow decayed similarly at both sites and if host lithology [*i.e.*, carbonate vs. clay (Kastner and others, 1977)] is not a factor, the differences in the diagenetic histories mainly reflect differences in thermal histories directly related to varying rates of sediment accumulation. Continued rapid accumulation of carbonate sediments at Site 305 dictated faster rates of transformation of silica through burial heating than at Site 303. The transformation rates of silica may have been even faster in the carbonate-rich sediments at Site 305 if the effect of host lithology was important (Kastner and others, 1977).

Summarizing, plots of age and present subsurface temperature generally describe the time and temperature stability field for opal-CT determined from experiments. More importantly, a unique thermal history, specific for each occurrence, strongly influences the diagenetic history of siliceous rocks at different localities.

OXYGEN ISOTOPIC COMPOSITIONS

Assuming an infinite oxygen reservoir, $^{18}O/^{16}O$ ratios of diagenetic silica in chert, porcelanite, and other siliceous rocks depend directly upon the temperature of formation and the isotopic composition of the equilibrating fluid. Since most DSDP siliceous rocks are products of burial diagenesis, their oxygen isotopic compositions record the thermal and chemical conditions of their formation in the subsurface rather than surface or bottom-water paleotemperatures.

Most isotopically analyzed DSDP cherts and porcelanites come from Eocene and Cretaceous sediments deposited at low latitudes in the Pacific (Knauth and Epstein, 1975; Levitan and others, 1975; Kolodny and Epstein, 1976). Additional analyses from other areas include nodular opal-CT and quartz chert and porcelanite in carbonate rocks from Site 366 (Early to Middle Eocene) and Site 367 (Oxfordian to Hauterivian) off northwestern Africa in the Atlantic (Riech and von Rad, 1979b), similar rocks (Early to Middle Eocene) from Site 149 in the Caribbean (Levitan and others, 1975; Knauth and Epstein, 1975), and an opal-CT porcelanite in Lower Miocene diatomaceous sediments from Site 323 in the Southern Ocean (Anderson and Lawrence, 1976). Oxygen isotopic compositions of DSDP siliceous rocks are some of the highest and most variable of all natural silicates: opal-CT = 31.0–41.8‰; microcrystalline quartz = 30.1–38.9‰. The considerable overlap in values demonstrates that unique diagenetic histories at different localities can produce opal-CT and quartz with identical oxygen isotopic compositions. Where opal-CT and quartz coexist, quartz is always depleted by 0.5–2‰ independent of depth of burial (Knauth and Epstein, 1975; Kolodny and Epstein, 1976). Mixtures of these two minerals yield intermediate $\delta^{18}O$ values, but because the transformation of opal-CT to quartz most likely involves solution and precipitation (Stein and Kirkpatrick, 1976) $\delta^{18}O$ values for the mixtures are not useful for estimating temperatures of formation or isotopic compositions of fluids associated with formation unless the two minerals are separated and analyzed individually.

The general correlation of decreasing $\delta^{18}O$ values of siliceous rocks from Pacific DSDP sites with increasing age has been variously interpreted to reflect the isotopic evolution of the ocean (Levitan and others, 1975) or an overall cooling of Pacific bottom water during the last 150 m.y. (Knauth and Epstein, 1975, Kolodny and Epstein, 1976). The second interpretation relies heavily upon the correlation of the $\delta^{18}O$ versus time plot for silica, translated 40m.y. towards the present to account for the gestation period of the silica transformations, with bottom-water temperatures derived from isotopic measurements of well-preserved benthic foraminifers at low latitude Pacific sites (Douglas and Savin, 1975). It also hinges upon the assumption that silica transformations, particularly opal-A to opal-CT, occur at shallow burial depths and attendant low temperatures that correspond roughly to bottom-water temperatures. Kolodny and Epstein (1976) point out some of the problems with these assumptions; the preceding discussion of thermal histories adds fuel to this fire. Nevertheless, this interpretation may apply in a general way; interstitial waters inherit their starting isotopic compositions from the original $\delta^{18}O$ values of bottom water.

A second way to interpret $\delta^{18}O$ trends in DSDP siliceous rocks is from a diagenetic viewpoint. Unlike the well-constrained carbonate paleothermometer, quartz-water and opal-A-water fractionation expressions are less certain and the opal-CT-water fractionation is virtually unknown (Friedman and O'Neil, 1977; Mikkelsen and others, 1978). Absolute temperatures of formation computed from isotopic data therefore rely heavily on poorly-known fractionation expressions, the isotopic composition of the equilibrating fluid, and attainment of isotopic equilibrium.

Figure 8 shows present subsurface temperature, computed from depth of burial, present bottom water temperature, heat flow, and conductivity data, versus isotopic temperature. Isotopic temperature ranges come from the extrapolated quartz-water fractionation expressions of Knauth and Epstein (1975, 1976) and Clayton and others (1972) that are used to estimate minimum and maximum temperatures, respectively. I chose standard mean ocean water (SMOW, $\delta^{18}O$ = 0‰) as the equilibrating fluid. Since opal-CT and quartz retain the isotopic compositions acquired at the time of formation until dissolved (Murata and others, 1977), points above the line x = y in Figure 8 indicate samples that have undergone continued burial following the formation of opal-CT or quartz or that have equilibrated with water having $\delta^{18}O$ greater than 0‰. Values below this line correspond to samples that either were once at higher temperatures but have since cooled, perhaps the result of erosion or nondeposition (*e.g.*, Site 303, Fig. 7B) or of higher bottom-water temperatures when formed, or they reflect equilibrium during formation with a fluid having $\delta^{18}O$ less than 0‰.

To ascribe the deviations from the x = y line in Figure 8 solely to equilibration with fluids of variable oxygen isotopic compositions requires a range in $\delta^{18}O$ values between +10‰ and −9‰. Whereas this range is well within the spread quoted for meteoric and formation waters (Craig, 1961; Fleischer and others, 1977), the range of observed values for interstitial waters extracted from DSDP materials is more restricted (Fig. 9). At a few sites there are isotopic data for both the silica and the interstitial water, and these data indicate that for newly-formed opal-CT or quartz present subsurface temperatures fall within the isotopic temperature range (Table 3). Although further measurements are needed, these results suggest that equi-

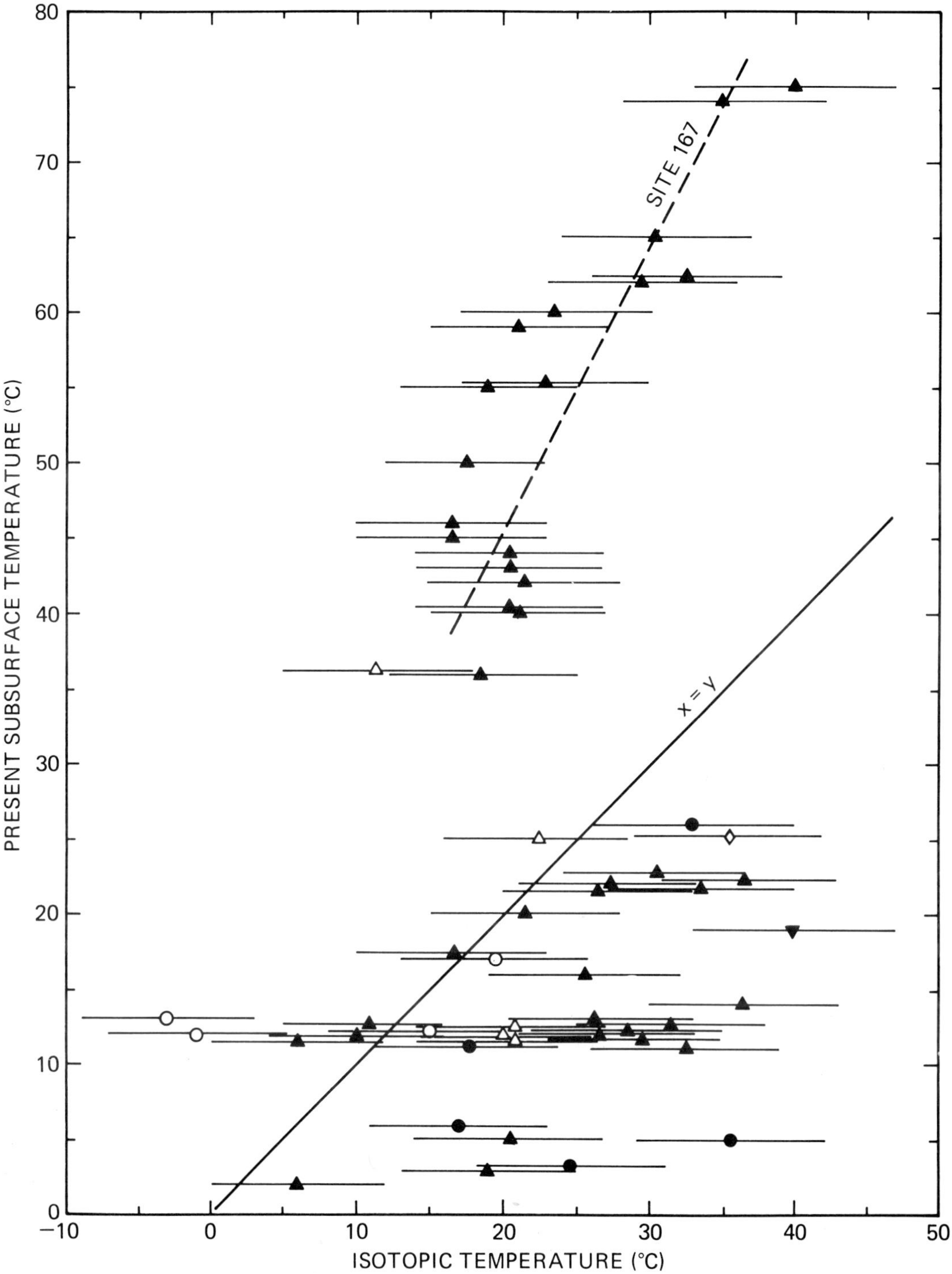

FIG. 8.—Relation between isotopic temperatures and present subsurface temperatures for DSDP cherts and porcelanites. Data from Knauth and Epstein (1975), Levitan and others (1975), Anderson and Lawrence (1976), and Kolodny and Epstein (1976). Minimum and maximum isotopic temperatures computed from Knauth and Epstein (1975, 1976) and Clayton and others (1972), respectively. Site 167 is the only DSDP site where a complete vertical sequence of cherts (all quartz in this case) has been isotopically analyzed (Kolodny and Epstein, 1976); the dashed line shows the general trend through these data. Key to symbols in Figure 5.

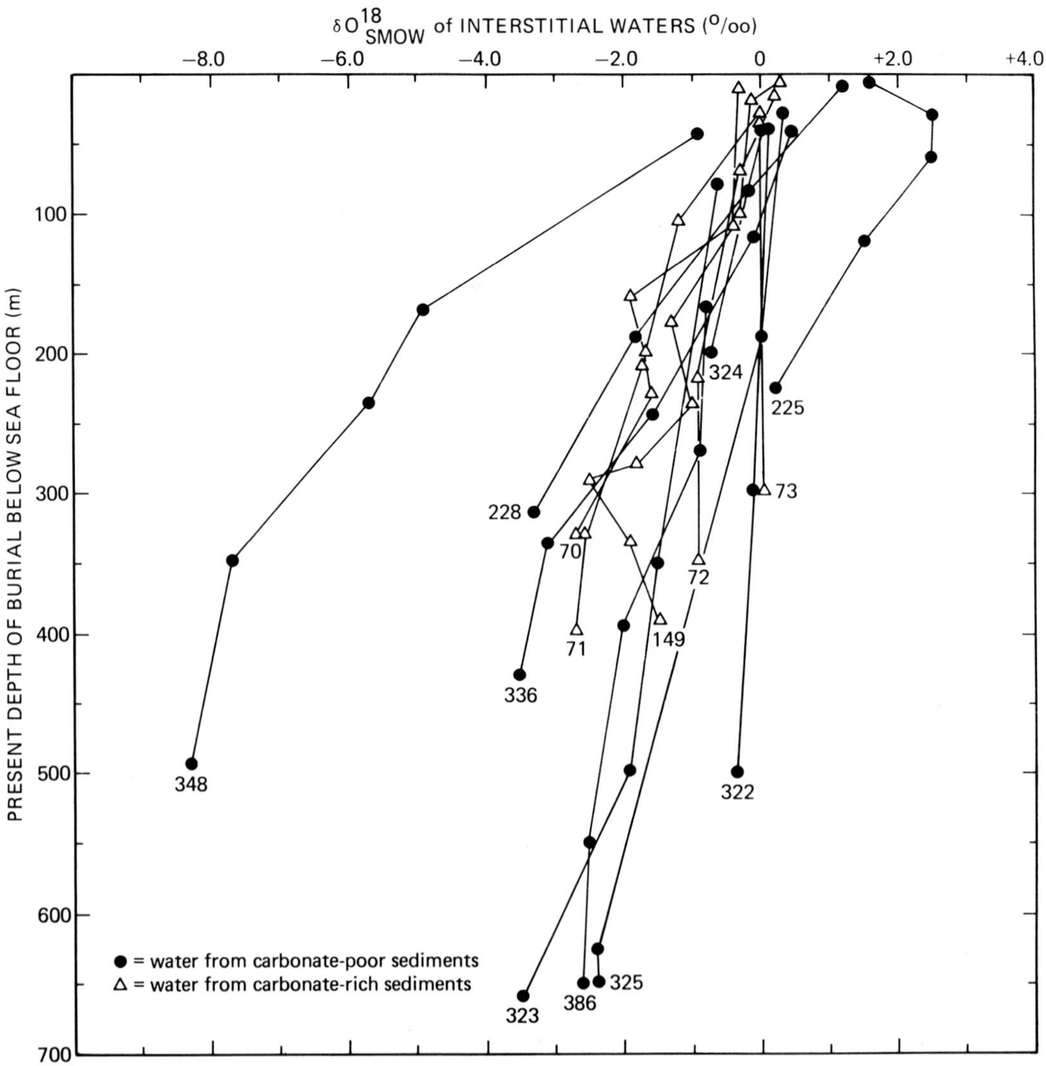

FIG. 9.—$\delta^{18}O$ profiles of interstitial waters at DSDP sites. Data from Lawrence (1973, 1974), Anderson and Lawrence (1976), Lawrence and others (1976), Gieskes and others (1978), and Miller and others (1979). $\delta^{18}O$ profiles at Site 225 in the Red Sea and at Site 348 near the Jan Mayen Ridge north of Iceland diverge from the general trend; see Lawrence (1974) and Gieskes and others (1978), respectively, for possible explanations.

libration of authigenic silica with slightly depleted interstitial waters may explain isotopic temperatures of opal-CT and quartz that exceed present subsurface temperatures. For those samples with isotopic temperatures below ambient temperatures, depleted interstitial waters may still have contributed, but the effect is masked by elevation of the silica above formation temperature by continued burial (*e.g.*, Site 167 trend in Fig. 8). In any event, the isotopic data suggest a restricted range of formation temperatures for opal-CT and quartz in DSDP siliceous rocks, generally less than 50°C.

This diagenetic interpretation of oxygen isotopic compositions does not exclude the climatic factor. It does, however, suggest that $\delta^{18}O$ values of diagenetic opal-CT and quartz may reflect somewhat elevated temperatures concomitant with burial and equilibration with slightly depleted interstitial waters. Furthermore, since diagenetic histories of siliceous rocks are unique at each site, caution is essential when comparing isotopic data from different localities.

REFLECTION CHARACTERISTICS

The scenario of Horizon A in the western North Atlantic illustrates the importance and complexity

TABLE 3.—COMPARISON OF ISOTOPIC TEMPERATURES OF OPAL-CT AND QUARTZ WITH PRESENT SUBSURFACE TEMPERATURES AT SELECTED DSDP SITES.

Site	Depth (m)	Stratigraphic Age of Enclosing Sediment	Silica Mineral	$\delta^{18}O$ Silica (‰ SMOW)	$\delta^{18}O$ Water (‰ SMOW)	Isotopic Temperatures[1] Min. (°C)	Max. (°C)	Present Subsurface Temperature (°C)	References for Isotopic Data
70	330–383	M. Eocene	opal-CT	34.9	−2.7	4	16	10–12	Knauth & Epstein (1975) Lawrence and others (1975)
70	330–383	M. Eocene	quartz	32.6	−2.7	12	24	10–12	Knauth & Epstein (1975) Lawrence and others (1975)
149	285	M. Eocene	opal-CT	34.0	−2.2	9	21	19	Levitan and others (1975) Lawrence (1973)
149	390	M. Eocene	opal-CT	34.3	−1.5	10	22	25	Knauth & Epstein (1975) Lawrence (1973)
323	350–440	M.-E. Miocene	opal-CT	31.0	−1.5	23	36	22–25	Anderson & Lawrence (1976) Lawrence and others (1976)

[1] Minimum isotopic temperatures computed from Knauth and Epstein (1975, 1976). Maximum isotopic temperatures calculated from Clayton and others (1972).

of silicified rocks as seismic reflectors. First thought to correspond to a fossil abyssal plain (Ewing and Ewing, 1962), then to Maestrichtian turbidites (Ewing and others, 1966), subsequent deep sea drilling and closer scrutiny of seismic profiles have shown Horizon A to consist of several reflectors only one of which (Horizon A^c) correlates with siliceous rocks (Tucholke and Mountain, 1979, and references therein). Off the east coast of North America Horizon A^c is continuous and corresponds to areally extensive turbidites that are cemented by opal-CT and quartz. To the east this reflector is patchy and correlates with chert and porcelanite associated with radiolarian mudstone in sedimentary sequences overlying basement depressions (Tucholke and Mountain, 1979).

In the Pacific, some seismic reflectors also correlate with siliceous rocks. Many of these in the northwest Pacific correspond to chert and porcelanite developed from biogenic silica that was deposited in the equatorial region (Lancelot and Larson, 1975). In the southern Bering Sea, a strong reflector—the so-called Bottom Simulating Reflector or BSR—marks the time-transgressive, bottom-parallel diagenetic front between diatomaceous sediment (opal-A) and siliceous claystone (opal-CT)

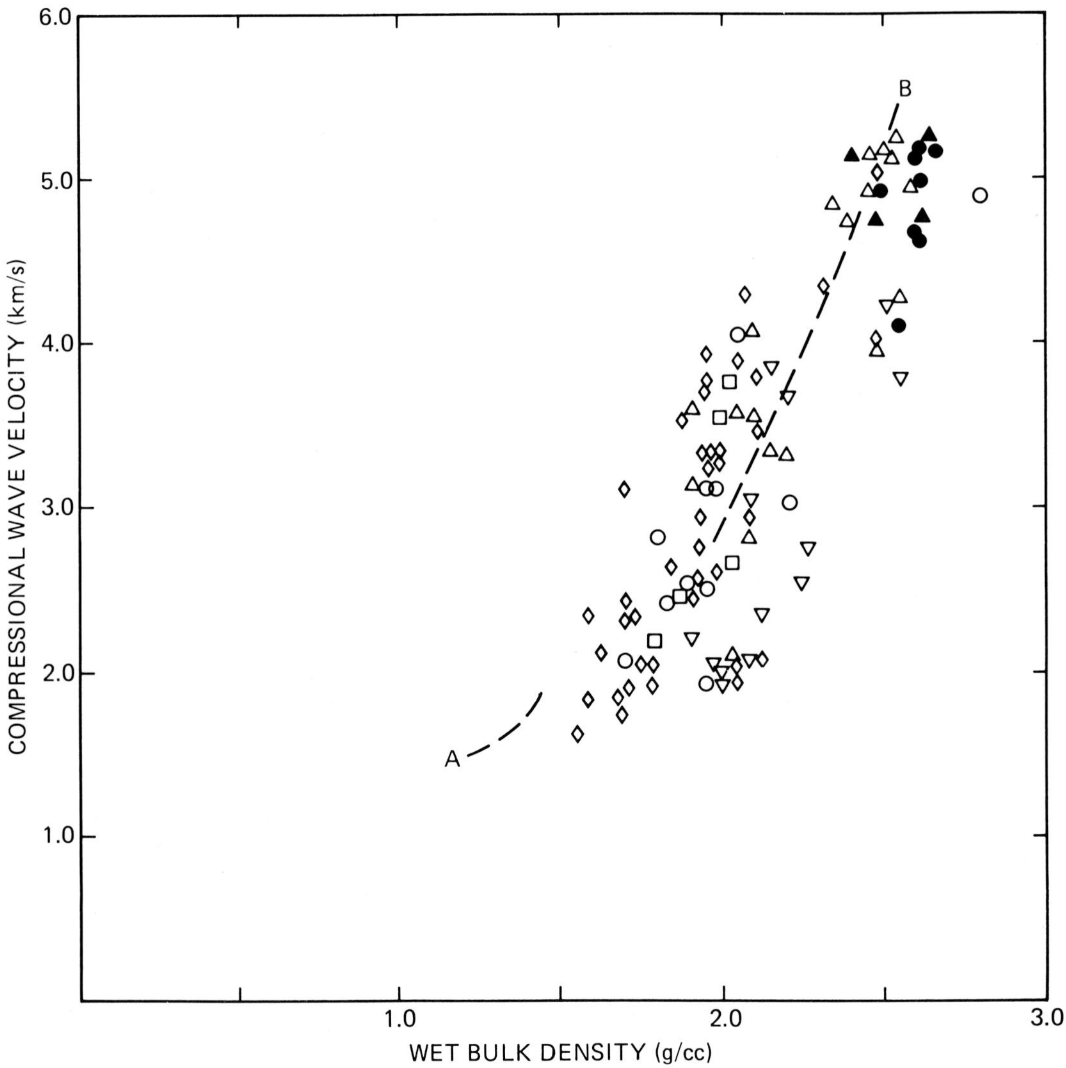

FIG. 10.—Wet bulk densities and compressional wave velocities of DSDP siliceous rocks. Data from DSDP data bank and from Boyce (1976, 1980). Dashed curves A and B mark trends for diatomite and for chert and porcelanite, respectively, reported by Hamilton (1978). Key to symbols in Figure 5.

(Scholl and Creager, 1973; Hein and others, 1978). In a similar occurrence at Site 471 west of central Baja California, a strong reflector that can be traced over a distance of about 80km marks the position of the opal-A-opal-CT diagenetic front (Haq and others, 1979). Not all BSRs, however, correspond to fronts of diagenetic silica; some mark the basal zones of gas hydrates (Shipley and others, 1979).

Densities and velocities of DSDP siliceous rocks vary considerably and more or less directly with degree of silicification (Fig. 10). This variation translates to a wide range in reflection coefficients for boundaries between these rocks and enclosing or overlying sediments (Table 4). Bedding characteristics exert further control over whether siliceous rocks will produce a seismic reflector. For example, along a transect between abyssal Sites 303 and 304 and Sites 305 and 306 on the Shatsky Rise in the northwest Pacific, Lancelot and Larson (1975) recognized a deterioration in strength and continuity of reflectors corresponding to chert and porcelanite. These siliceous rocks occur as discrete, discontinuous nodules in the thick carbonate sequence capping the rise; at the abyssal sites the thin sequence of chert and porcelanite occurs in radiolarian mudstone. Even though reflection coefficients computed from discrete samples are considerable (0.67–0.74 for Sites 303 and 304; 0.42 for Site 306), the scattered occurrences of nodules at the rise sites produce weak, patchy reflections.

The northwest Pacific is fertile ground for further detailed seismic investigations. Numerous DSDP sites offer the opportunity to compare tectonic and sedimentologic histories with seismic data. The recognition of siliceous rocks as both sedimentary and diagenetic facies should constitute an important element of any synthesis.

TABLE 4.—REFLECTION COEFFICIENTS FOR CHERTS, PORCELANITES AND SILICEOUS ROCKS.

Lithologic Combinations	Reflection Coefficients[1] Low	High	Mean
1. Nodular chert and porcelanite in calcareous ooze, chalk and limestone	0.16	0.78	0.44
2. Weakly silicified carbonate sediments and rocks	0.06	0.47	0.26
3. Chert, porcelanite and siliceous rocks in radiolarian muds and mudrocks	0.07	0.74	0.46
4. Chert, porcelanite and siliceous rocks in diatomaceous muds and mudrocks	0.01	0.57	0.29
5. Silicified turbidites	0.09	0.47	0.24

[1] Data from Figure 10 and references therein.

ACKNOWLEDGMENTS

I am grateful to the Data Services group at the Deep Sea Drilling Project, especially Tom Birtley, for retrieval of some of the data used in this paper. Special thanks to Kathleen Sanderson for preparation of figures and to Lola Boyce for typing the manuscript. Miriam Kastner, E. L. Winterer, and J. R. Hein read the manuscript and offered helpful suggestions.

REFERENCES CITED

NOTE: All references to Deep Sea Drilling Project Initial Reports have been abbreviated; for complete citations see *Appendix I* at the end of this volume, or the individual Initial Reports as published.

ANDERSON, T. F., AND LAWRENCE, J. R., 1976, Stable isotope investigations of sediments, basalts, and authigenic phases from Leg 35 cores: Init. Repts. DSDP, v. 35, p. 507–512.

BEALL, A. O., JR., AND FISCHER, A. G., 1969, Sedimentology: Init. Repts. DSDP, v. 1, p. 521–593.

———, LAURY, R., DICKINSON, K., AND PUSEY, W. C., III, 1973, Sedimentology: Init. Repts. DSDP, v. 10, p. 699–730.

BERGER, W. H., AND VON RAD, U., 1972, Cretaceous and Cenozoic sediments from the Atlantic Ocean: Init. Repts. DSDP, v. 14, p. 787–954.

BERNOULLI, D., 1972, North Atlantic and Mediterranean Mesozoic facies: Init. Repts. DSDP, v. 11, p. 801–872.

BOYCE, R. E., 1976, Sound velocity-density parameters of sediment and rock from DSDP from DSDP Sites 315–318 on the Line Islands Chain, Manihiki Plateau, and Tuamotu Ridge in the Pacific Ocean: Init. Repts. DSDP, v. 33, p. 695–728.

———, 1980, Determination of the relationships of electrical resistivity, sound velocity, and density/porosity of sediment and rock by laboratory techniques and well logs from DSDP Sites 415 and 416 off the coast of Morocco: Init. Repts. DSDP, v. 50, p. 305–318.

BRAMLETTE, M. N., 1946, The Monterey Formation of California and the origin of its siliceous rocks: U.S. Geological Survey Prof. Paper 212, 57 p.

BREWSTER, N. A., 1980, Cenozoic biogenic silica sedimentation in the Antarctic Ocean: Geol. Soc. America Bull., v. 91, p. 337–347.

CALVERT, S. E., 1971, Composition and origin of North Atlantic deep sea cherts: Contrib. Mineral. Petrology, v. 33, p. 273–288.

———, 1974, Deposition and diagenesis of silica in marine sediments, *in* Hsü, K. J. and Jenkyns, H. C. eds., Pelagic sediments: on land and under the sea: Internat. Assoc. Sedimentol. Spec. Pub. 1, p. 273–299.
———, 1977, Mineralogy of silica phases in deep-sea cherts and porcelanites: Phil. Trans. Royal Soc. London A, v. 286, p. 239–252.
CLAYTON, R. N., O'NEIL, J. R., AND MAYEDA, T. K., 1972, Oxygen isotope exchange between quartz and water: Jour. Geophys. Res., v. 77, p. 3057–3067.
CRAIG, H., 1961, Isotopic variations in meteoric waters: Science, v. 133, p. 1702–1703.
DAVIES, T. A., AND SUPKO, P. R., 1973, Oceanic sediments and their diagenesis: some examples from deep-sea drilling: Jour. Sed. Petrology, v. 43, p. 381–390.
DONNELLY, T. W., AND NALLI, G., 1973, Mineralogy and chemistry of Caribbean sediments: Init. Repts. DSDP, v. 15, p. 929–962.
DOUGLAS, R. G., AND SAVIN, S. M., 1975, Oxygen and carbon isotope analysis of Tertiary and Cretaceous microfossils from Shatsky Rise and other sites in the North Pacific Ocean: Init. Repts. DSDP, v. 32, p. 509–520.
EDGAR, N. T., SAUNDERS, J. B. AND OTHERS, 1973: Init. Repts. DSDP, v. 15, 1137 p.
ERNST, W. G. AND CALVERT, S. E., 1969, An experimental study of the recrystallization of porcelanite and its bearing on the origin of some bedded cherts: Amer. Jour. Sci., v. 267-A, p. 114–133.
EWING, J. I., AND EWING, M., 1962, Reflection profiling in and around the Puerto Rico Trench: Jour. Geophys. Res., v. 67, p. 4762–4739.
———, WORZEL, J. L., EWING, M., AND WINDISCH, C. C., 1966, Ages of Horizon A and the oldest Atlantic sediments: Science, v. 154, p. 1125–1132.
EWING, M., WORZEL, J. L. AND OTHERS, 1969: Init. Repts. DSDP, v. 1, 672 p.
FIRSTBROOK, P. L., FUNNELL, B. M., HURLEY, A. M., AND SMITH, A. G., 1979, Paleoceanic reconstructions—160-0 Ma: National Science Foundation, 41 p.
FLEISCHER, E., GOLDBERG, M., GAT, J. R., AND MARGARITZ, M., 1977, Isotopic composition of formation waters from deep drillings in southern Israel: Geochim. et Cosmochim. Acta, v. 41, p. 411–425.
FRIEDMAN, I., AND O'NEILL, J. R., 1977, Compilation of stable isotopic fractionation factors of geochemical interest: U.S. Geological Survey Prof. Paper 440K, 12 p.
GARRISON, R. E., ROWLAND, S. M., HORAN, L. J., AND MOORE, J. C., 1975, Petrology of siliceous rocks recovered from marginal seas of the western Pacific, Leg 31, Deep Sea Drilling Project: Init. Repts. DSDP, v. 31, p. 519–529.
GIESKES, J. M., LAWRENCE, J. R., AND GALLEISKY, G., 1978, Interstitial water studies, Leg 38: Init. Repts. DSDP, v. 38–41 Supplement, p. 121–133.
GREENWOOD, R., 1973, Cristobalite: its relationship to chert formation in selected samples from the Deep Sea Drilling Project: Jour. Sed. Petrology, v. 43, p. 700–708.
HAQ, B. U., YEATS, R. S. AND OTHERS, 1979, Eastern Pacific boundary currents: Geotimes (April), p. 30–31.
HAMILTON, E. L., 1978, Sound velocity-density relations in sea-floor sediments and rocks: Jour. Acoust. Soc. America, v. 63, p. 365–377.
HAYS, J. D., 1970, Stratigraphy and evolutionary trends of radiolaria in North Pacific deep-sea sediments: Geol. Soc. America Mem. 126, p. 185–218.
HEATH, G. R., 1973, Cherts from the eastern Pacific, Leg 16, Deep Sea Drilling Project: Init. Repts. DSDP, v. 16, p. 609–614.
———, AND MOBERLY, R., JR., 1971, Cherts from the western Pacific, Leg 7, Deep Sea Drilling Project: Init. Repts. DSDP, v. 7, p. 991–1008.
HEIN, J. R., SCHOLL, D. W., BARRON, J. A., JONES, M. G., AND MILLER, J., 1978, Diagenesis of late Cenozoic diatomaceous deposits and formation of the bottom simulating reflector in the southern Bering Sea: Sedimentology, v. 25, p. 155–181.
IIJIMA, A., MATSUMOTO, R., AND TADA, R., 1980, Zeolitic and silica diagenesis and sandstone petrography at Sites 438 and 439, DSDP/IPOD Leg 57 off Sanriku, Northwest Pacific: Init. Repts. DSDP, v. 56/57, p. 1143–1158.
JONES, J. B., AND SEGNIT, E. R., 1971, The nature of opal. I. Nomenclature and constituent phases: Jour. Geol. Soc. Australia, v. 18, p. 57–68.
KAGAMI, H., 1979, Transformation of opaline silica in sediments from Bay of Biscay and Rockall Bank: Init. Repts. DSDP, v. 48, p. 757–764.
KASTNER, M., 1976, Diagenesis of basal sediments and basalts of Sites 322 and 323, Leg 35, Bellinghausen abyssal plain: Init. Repts. DSDP, v. 35, p. 513–528.
KASTNER, M., *in press,* Authigenic silicates in deep-sea sediments: formation and diagenesis, *in* Emiliani, C. (ed.), The Sea: New York, Wiley-Interscience, v. 7.
———, KEENE, J. B., AND GIESKES, J. M., 1977, Diagenesis of siliceous oozes: I. Chemical controls on the rate of opal-A diagenesis—an experimental study: Geochim. et Cosmochim. Acta., v. 41, p. 1041–1059.
KEENE, J. B., 1975, Cherts and porcelanites from the North Pacific, DSDP Leg 32: Init. Repts. DSDP, v. 32, p. 429–507.
———, 1976, Petrography of biogenic and authigenic silica from the Pacific Basin: PhD. thesis, Scripps Inst. Oceanography, Univ. Calif., San Diego, California, 264 p.
KELTS, K. R., 1976, Summary of chert occurrences from Line Islands Sites 314, 315, 316 DSDP Leg 33: Init. Repts. DSDP, v. 33, p. 855–866.
KENNETT, J. P., 1977, Cenozoic evolution of Antarctic glaciation, the circum-Antarctic Ocean, and their impact on

global paleoceanography: Jour. Geophys. Res., v. 82, p. 3843-3860.

Knauth, L. P., and Epstein, S., 1975, Hydrogen and oxygen isotope ratios in silica from JOIDES Deep Sea Drilling Project: Earth Planet. Sci. Letters, v. 25, p. 1-10.

———, and Epstein, S., 1976, Hydrogen and oxygen isotope ratios in nodular and bedded cherts: Geochim. et Cosmochim. Acta., v. 40, p. 1095-1108.

Kolodny, Y., and Epstein, S., 1976, Stable isotope geochemistry of deep sea cherts: Geochim. et Cosmochim. Acta., v. 34, p. 3-24.

Lancelot, Y., 1973, Chert and silica diagenesis in sediments from the central Pacific: Init. Repts. DSDP, v. 17, p. 377-406.

———, and Larson, R. L., 1975, Sedimentary and tectonic evolution of the Northwestern Pacific: Init. Repts. DSDP, v. 32, p. 925-939.

Langseth, M. G., and von Herzen, R. P., 1971, Heat flow through the floor of the world oceans, *in* Maxwell, A. E. ed., The Sea: New York, Wiley-Interscience, v. 4, pt. 1, p. 229-352.

Laughton, A. S., Berggren, W. A., and others, 1972: Init. Repts. DSDP, v. 12, p. 411-414.ımmmmmmmmmm

Lawrence, J. R., 1973, Interstitial water studies, Leg 15—stable oxygen and carbon isotope variations in water, carbonates, and silicates from the Venezuela Basin (Site 149) and the Aves Rise (Site 148): Init. Repts. DSDP, v. 20, p. 891-900.

———, 1974, Stable oxygen and carbon isotope variations in pore waters, carbonates, and silicates, Sites 225 and 228, Red Sea: Init. Repts. DSDP, v. 23, p. 939-942.

———, Gieskes, J., and Anderson, T. F., 1976, Oxygen isotope material balance calculations, Leg 35: Init. Repts. DSDP, v. 35, p. 507-512.

———, ———, and Broecker, W. S., 1975, Oxygen isotope and cation composition of DSDP pore waters and the alteration of Layer II basalts: Earth Planet. Sci. Letters, v. 27, p. 1-10.

Leclaire, L., 1974, Late Cretaceous and Cenozoic pelagic deposits—paleoenvironment and oceanography of the central western Indian Ocean: Init. Repts. DSDP, v. 25, p. 481-512.

Leinen, M., 1979, Biogenic silica accumulation in the central Equatorial Pacific and its implications for Cenozoic paleoceanography: Geol. Soc. America Bull., v. 90, p. 1310-1376.

Levitan, M. A., Contsova, E. I., Lisitzin, A. P., and Bogdanov, Y. A., 1975, The origin of chert in sediments of the Pacific Ocean from data of oxygen isotopic analysis and a study of the distribution of chert: Geochem. Internat., v. 12, p. 95-104.

Lisitzin, A. P., 1971, Distribution of siliceous microfossils in suspension and bottom sediments, *in* Funnel, B. M. and Riedel, W. R. eds., Micropaleontology of Oceans: Cambridge University Press, p. 173-195.

Luyendyk, B. P., and Davies, T. A., 1974, Results of DSDP Leg 26 and the geologic history of the southern Indian Ocean: Init. Repts. DSDP, v. 24, p. 909-944.

Mikkelsen, N., Labeyrie, L., and Berger, W. H., 1978, Silica oxygen isotopes in diatoms: a 20,000 year record in deep-sea sediments: Nature, v. 271, p. 536-538.

Miller, R. S., Lawrence, J. R., and Gieskes, J. M., 1979, Interstitial water studies, Sites 386 and 387, Leg 43: Init. Repts. DSDP, v. 43, p. 667-674.

Mizutani, S., 1966, Transformation of silica under hydrothermal conditions: Nagoya Univ.: Jour. Earth Sci., v. 14, p. 56-88.

———, 1967, Kinematic aspects of diagenesis of silica in sediments: Nagoya Univ.: Jour. Earth Sci., v. 15, p. 99-111.

———, 1970, Silica minerals in the early stage of diagenesis: Sedimentology, v. 15, p. 419-436.

Murata, K. J., Friedman, I., and Gleason, J. D., 1977, Oxygen isotope relations between diagenetic silica minerals in Monterey Shale, Temblor Range, California: Amer. Jour. Science, v. 277, p. 259-272.

———, and Larson, R. R., 1975, Diagenesis of Miocene siliceous shales, Temblor Range, California: Jour. Research U.S. Geol. Survey, v. 3, p. 553-566.

Orr, W. N., 1972, Pacific Northwest siliceous phytoplankton: Paleogeography, Palaeoclimatology, Palaeoecology, v. 12, p. 95-114.

Packham, G. H., and van der Lingen, G. J., 1973, Progressive carbonate diagenesis at Deep Sea Drilling Sites 206, 207, 208, and 210 in the Southwest Pacific and its relationship to sediment physical properties and seismic reflections: Init. Repts. DSDP, v. 21, p. 495-522.

Peterson, M. N. A., Edgar, N. T., and others, 1970: Init. Repts. DSDP, v. 2, 491 p.

Pimm, A. G., Garrison, R. E., and Boyce, R. E., 1971, Sedimentology synthesis: Lithology, chemistry and physical properties of sediments in the Northwestern Pacific Ocean: Init. Repts. DSDP, v. 6, p. 1132-1252.

Pisciotto, K. A., 1980, Chert and porcelanite at Site 436, Northwest Pacific, Deep Sea Drilling Project: Init. Repts. DSDP, v. 56/57, p. 1133-1142.

Ramsay, A. T. S., 1973, A history of organic siliceous sediments in oceans, *in* Hughes, N. F. ed., Organisms and continents through geologic time: Spec. Pap. Palaeont., v. 12, p. 199-234.

Riech, V., 1979, Diagenesis of silica, zeolites, and phyllosilicates at Sites 397 and 398: Init. Repts. DSDP, v. 47, p. 741-759.

———, 1980, Diagenesis of siliceous sediments, porcelanites, and cherts of the Moroccan Basin (Sites 370, 415 and 416): Init. Repts. DSDP, v. 50, p. 725-732.

———, and von Rad, U., 1979a, Eocene porcelanites and early Cretaceous cherts from the western North Atlantic Basin: Init. Repts. DSDP, v. 43, p. 437-456.

———, and ———, 1979b, Silica diagenesis in the Atlantic Ocean: diagenetic potential and transformations, *in*

Talwani, M., Hay, W., Ryan, W. B. F., eds., Implications of deep drilling results in the Atlantic Ocean: Amer. Geophys. Union, M. Ewing Series, v. 2, p. 315–341.

RIEDEL, W. R., 1959, Siliceous organic remains in pelagic sediments: Soc. Econ. Paleontologists Mineralogists, Spec. Pub. 7, p. 80–91.

———, 1971, The occurrence of pre-Quaternary radiolaria in deep-sea sediments, *in* Funnel, B. M. and Riedel, W. R. eds., Micropaleontology of Oceans: Cambridge Univ. Press, p. 567–594.

ROBINSON, P. T., THAYER, P. A., COOK, P. J., AND MCKNIGHT, B. M., 1974, Lithology of Mesozoic and Cenozoic sediments of the eastern Indian Ocean, Leg 27, Deep Sea Drilling Project: Init. Repts. DSDP, v. 27, p. 1001–1048.

SCHOLL, D. W., AND CREAGER, J. S., 1973, Geologic synthesis of Leg 19 (DSDP) results: Far North Pacific, Aleutian Ridge and Bering Sea: Init. Repts. DSDP, v. 19, p. 897–913.

SCLATER, J. G., ANDERSON, R. N., AND BELL, M. L., 1971, Elevation of ridges and evolution of the central Eastern Pacific: Jour. Geophys. Res., v. 76, p. 7888–7915.

SHIPLEY, T. H., HOUSTON, M. H., BUFFLER, R. T., SHAUB, F. J., MCMILLEN, K. J., LADD, J. W., AND WORZEL, J. L., 1979, Seismic reflection evidence for the widespread occurrence of possible gas-hydrate horizons on continental slopes and rises: Amer. Assoc. Petroleum Geologists Bull., v. 63, p. 2204–2221.

STEIN, C. L., AND KIRKPATRICK, R. J., 1976, Experimental porcelanite recrystallization kinetics: A nucleation and growth model: Jour. Sed. Petrology, v. 46, p. 430–435.

STOFFERS, P., AND KUHN, R., 1974, Red Sea evaporites: a petrographic and geochemical study: Init. Repts. DSDP, v. 23, p. 821–847.

TALIAFERRO, N. L., 1934, Contraction phenomena in cherts: Geol. Soc. America Bull., v. 45, p. 189–232.

TUCHOLKE, B. E., AND MOUNTAIN, G. S., 1979, Seismic stratigraphy and paleosedimentation patterns in the North American Basin, *in* Talwani, M., Hay, W. and Ryan, W. B. F., eds., Deep drilling results in the Atlantic Ocean: Continental margins and paleoenvironment: Amer. Geophys. Union, Maurice Ewing Series, v. 3, p. 58–86.

VALLIER, T. L., 1974, Volcanogenic sediments and their relation to landmass volcanism and sea floor-continent movements, western Indian Ocean, Leg 25, Deep Sea Drilling Project: Init. Repts. DSDP, v. 25, p. 515–542.

VAN ANDEL, T. H., HEATH, G., AND MOORE, T. C., JR., 1975, Cenozoic history and paleoceanography of the central Equatorial Pacific Ocean: Geol. Soc. America Mem., v. 143, 134 p.

VAN DER LINGEN, G. J., AND PACKHAM, G. H., 1975, Relationships between diagenesis and physical properties of biogenic sediments of the Ontong-Java Plateau (Sites 288 and 289, Deep Sea Drilling Project): Init. Repts. DSDP, v. 30, p. 443–482.

VINCENT, E., 1974, Cenozoic planktonic biostratigraphy and paleoceanography of the tropical western Indian Ocean: Init. Repts. DSDP, v. 24, p. 1111–1150.

VON DER BORCH, C. C., GALEHOUSE, J., AND NESTEROFF, W. D., 1971, Silicified limestone-chert sequences cored during Leg 8 of the Deep Sea Drilling Project: a petrologic study: Init. Repts. DSDP, v. 8, p. 819–822.

VON RAD, U., AND RÖSCH, H., 1972, Mineralogy and origin of clay minerals, silica and authigenic silicates in Leg 14 sediments: Init. Repts. DSDP, v. 14, p. 727–751.

———, AND ———, 1974, Petrography and diagenesis of deep-sea cherts from the central Atlantic, *in* Hsü, K. J. and Jenkyns, H. C. eds., Pelagic sediments: on land and under the sea: Internat. Assoc. Sedimentol. Spec. Pub. 1, p. 327–347.

———, AND ———, 1977, Silica diagenesis in continental margin sediments off northwest Africa: Init. Repts. DSDP, v. 41, p. 879–905.

WISE, S. W., JR., AND WEAVER, F. M., 1974, Chertification of oceanic sediments, *in* Hsü, K. J. and Jenkyns, H. C. eds., Pelagic sediments: on land and under the sea: Internat. Assoc. Sedimentol. Spec. Pub. 1, p. 301–326.

SEPM SPECIAL PUBLICATION NO. 32, P. 149–167, OCTOBER 1981

DEEP-SEA DRILLING INTERSTITIAL WATER STUDIES: IMPLICATIONS FOR CHEMICAL ALTERATION OF THE OCEANIC CRUST, LAYERS I AND II

JORIS M. GIESKES
Scripps Institution of Oceanography, La Jolla, California 92093 U.S.A.

ABSTRACT

This paper represents a review of observations made on the composition of interstitial waters of sediments recovered during the Deep Sea Drilling Project.

In pelagic sediments with relatively slow rates of deposition (<50 m/10^6 yr) increases commonly are measured in dissolved calcium and decreases in dissolved magnesium and potassium as well as $H_2{}^{18}O$. These changes, to a large extent, can be understood in terms of alteration reactions occurring in the basalts of Layer II of the oceanic crust, and, to a lesser extent, in terms of alteration of volcanic matter dispersed in the sediment column (Layer I).

Dissolved silica concentrations are generally enhanced in sediments containing opaline biogenic silica. Such enhanced silica activities in turn provide an environment in which reactions involving the alteration of volcanic matter in the sediments may occur. These reactions are particularly evident in zones of silicification, in which opal-A is transformed to opal-CT and quartz.

Dissolved strontium concentrations in carbonate sediments are affected by carbonate recrystallization processes, which release strontium to the interstitial waters. These increases imply that Sr/Ca distribution coefficients in calcites must be smaller than generally accepted. In carbonate-free sediments increased strontium is often the result of reactions involving volcanic matter in the sediments and/or the underlying basalts. This relationship is particularly evident from observations on the $^{87}Sr/^{86}Sr$ ratio of dissolved strontium.

Dissolved manganese and lithium appear to be released by biogenic silica. For Mn such sediments contain relatively more reactive organic carbon, which appears to act as the reducing agent in marine sediments.

In rapidly-deposited hemipelagic sediments complex concentration-depth gradients result, particularly as a result of processes involving sulfate reduction and methane gas generation. High alkalinities (bicarbonate) result, as well as high dissolved ammonia values. Carbonate precipitation (calcite, ankerite, dolomite) can be important in such sediments.

Freshwater acquifers near salt deposits can result in saline brines being advected into the sediment column in areas on the continental shelves and slopes.

INTRODUCTION

From the initiation of the Deep Sea Drilling Project (DSDP) a program of collection and analysis of interstitial waters has been carried out under the sponsorship of the JOIDES Inorganic Geochemistry Panel. Particularly active in the scientific program have been F. T. Manheim and F. L. Sayles of the Woods Hole Oceanographic Institution, as well as I. R. Kaplan (University of California at Los Angeles) and B. J. Presley (Texas A&M University). These investigators analyzed the various major and minor dissolved constituents of the interstitial waters obtained during the first 23 legs of *Glomar Challenger*. Subsequently, many samples of interstitial waters collected from Leg 24 to the present have been analyzed in my laboratory at Scripps Institution of Oceanography. In addition, from Leg 36 onwards, a shipboard analytical program for the analysis of some key parameters (pH, alkalinity, chloride, salinity, calcium and magnesium) has been implemented successfully.

The major philosophy underlying the program has been that changes in the chemical composition of the interstitial waters may indicate that reactions are taking place in the sediments that might go undetected from the study of the solid phases alone (Sayles and Manheim, 1975). This might be so because for many of the interstitial water components, the quantity in the solid phases of the sediments exceeds those in the interstitial waters by a large extent (Table 1; McDuff, 1978). A requisite, of course, is that the reactions taking place in the sediments involve a net uptake or release of dissolved constituents from or to the interstitial waters. Special geochemical efforts during Legs 15, 25 and 35 were designed to check these concepts and to define more closely the reactions that occur. In addition to the routine interstitial-water program, these efforts included studies of the oxygen and hydrogen isotopic composition of the interstitial waters, studies of the chemistry and mineralogy of the bulk sediments and clay mineral fractions, as well as studies of the oxygen isotopic composition of the solid phases.

In this paper I emphasize some of the major accomplishments of the Inorganic Geochemistry Program, which has centered around the study of the chemistry of interstitial waters. I will briefly describe the methodology used to retrieve and analyze interstitial waters, followed by a discussion of observations in open ocean pelagic sediments and in rapidly

TABLE 1.—ELEMENTAL COMPOSITION OF SEDIMENTS AT 50% WATER CONTENT (McDUFF, 1978).

Element	Solid Phases (moles)	Aqueous Phase (moles)	Ratio
O	3.1×10^{1}	5.5×10^{1}	5.6×10^{-1}
Na	5.2×10^{-1}	4.7×10^{-1}	1.1
Mg	8.8×10^{-1}	5.5×10^{-2}	1.6×10^{1}
K	7.2×10^{-1}	1.0×10^{-2}	7.2×10^{1}
Ca	9.8×10^{-1}	1.0×10^{-2}	9.8×10^{1}
C	9.8×10^{-1}	2.3×10^{-3}	4.3×10^{2}
Li	$3\ \times10^{-3}$	2.7×10^{-7}	1.1×10^{4}
Si	1.1×10^{1}	$1\ \times10^{-4}$	1.1×10^{5}
Al	3.3	$2\ \times10^{-7}$	1.7×10^{7}

deposited hemi-pelagic sediments. This distinction is mostly made because of complications introduced in the latter sediment types as a result of sulfate reduction processes and resulting complex distribution profiles in interstitial-water constituents.

METHODOLOGY

The routine method of obtaining interstitial waters from DSDP cores has not been changed since the origin of the project. The method, using stainless steel squeezing-apparatus, has been described in detail by Manheim and Sayles (1974). During the early part of the project, however, some concern arose about the possible artifacts that might be introduced as a result of squeezing a sediment at room temperature rather than at the in situ temperature (Mangelsdorf and others, 1969; Bischoff and others, 1970; Fanning and Pilson, 1971). For these reasons, a detailed program of study of the effects of temperature on the sediment-squeezing process was carried out. Temperature effects were generally small with respect to overall changes in concentration, with the exception of potassium and dissolved silica (Sayles and Manheim, 1975). These results were confirmed during Leg 25 (Gieskes, 1974). The usually small temperature-induced artifacts, coupled with the knowledge that over a sediment column of 500 m, temperatures may vary from 2°C to 25°C or higher, has led to the abandonment of a routine program involving squeezing at different temperatures. Concentration-depth gradients will generally be affected by less than 5%.

The recent development of the in situ sampler by R. Barnes (Barnes and others, 1979) has shed additional light on the validity of the routine methods described above. Good agreement with routine shipboard data was obtained during its first deployment during Leg 47B (Barnes and others, 1979). Further investigations during Leg 57 (Moore and Gieskes, 1980) and Leg 64 (Gieskes, in preparation) again showed an excellent comparison for most dissolved constituents. The clear advantage of the in situ sampler, of course, is that for the first time reliable samples of dissolved gases can be obtained, which is a major step forward (Barnes and Will, in press). Thus shipboard squeezing procedures yield interstitial waters that closely resemble, in composition, the interstitial waters obtained in situ; therefore, conclusions based on these studies are not invalidated by sampling artifacts.

Of importance is evidence that even interstitial waters obtained from cores stored for as long as two years in the DSDP core lockers yield interstitial water compositions close to those obtained on nearby samples collected by shipboard squeezing (McDuff, 1978; Gieskes and Johnson, *in press*). This evidence is especially useful when uncertainties exist concerning particular concentration-depth profiles.

Chemical methods for the analysis of major and minor dissolved constituents of interstitial waters have been described by Presley (1971), Manheim and Sayles (1974), Gieskes (1974), and Sayles and Manheim (1975). Accuracy is generally more than adequate for the establishment of concentration gradients, and the various analytical methods compare favorably (Hammond, 1973).

PELAGIC DEEP-SEA SEDIMENTS

Major Cations and Oxygen Isotopes

Much of the published interpretation of the chemistry of the interstitial waters of DSDP cores has centered around concentration-depth profiles of calcium, magnesium and—to a lesser extent—potassium and sodium. These elements are of particular interest to geochemists because of uncertainties in their geochemical cycle (Garrels and Perry, 1974; Drever, 1974; MacKenzie, 1975). An excellent example of such an uncertainty is the so-called magnesium problem, for which the exchange between the ocean and the underlying oceanic crust (sediments of Layer I as well as Layer II) is still debated (Broecker, 1971; Drever, 1974; Edmond and others, 1979). In order to solve this exchange problem, one needs a more accurate knowledge of the concentration gradients in surface sediments (Sayles, 1979), of exchange reactions taking place as a result of hydrothermal interaction of sea water with basalts at ridge crests (Corliss and others, 1979; Edmond and others, 1979), as well as of fluxes which result from reactions taking place in deeply-buried sediments and/or the underlying rocks of Layer II of the oceanic crust. In this paper the latter fluxes and the reactions leading to the observed concentration changes are the subject of discussion.

In pelagic sediments, especially in those with higher sedimentation rates (>5 m/10^6 years) such as biogenic oozes or sediments with high detrital inputs, often concentration increases in dissolved calcium and decreases in dissolved magnesium are observed. Interpretations of these changes have varied. As an example, a few concentration-depth profiles obtained in the Mascarene Basin, Site 239 (Gieskes, 1974) are presented in Figure 1.

Sayles and Manheim (1975) explained the above observations in terms of reactions involving carbonate diagenesis. This interpretation was based on the often noticed, rather close one-to-one exchange of calcium for magnesium, which in principle could be explained in terms of the formation of high-magnesium calcites or dolomites. In some cases, however, these authors did not rule out the possibility of silicate diagenesis as an additional contributor to these observations (see also Manheim, 1976). Other explanations of observed decreases in dissolved magnesium and increases in dissolved calcium have included the formation of sepiolite (Site 149; Wollast, 1974) by precipitation from solution upon silica dissolution, as well as the adsorption of magnesium ions on opaline silica (Donnelly and Merrill, 1977).

Lawrence and others (1975) noted the rather good correlation between profiles of dissolved calcium and of the changes in $\delta^{18}O$ of the interstitial waters. Subsequently this relationship has been established for a large number of sites (Lawrence and Gieskes, in preparation). Lawrence and others (1975) argued that the good correlation between calcium concentration increases and decreases in $\delta^{18}O$ of the pore fluids has a causal origin, in that the reactions leading to a $\delta^{18}O$ depletion would also lead to a release of calcium. On the basis of oxygen isotope mass-balance calculations, these authors concluded that neither carbonate nor silica diagenesis could account for the relatively large depletions in $\delta^{18}O$ of the interstitial waters. The main reactions that are likely to contribute to such large $\delta^{18}O$ depletions are alteration reactions involving volcanic matter in the sediments or basaltic material in Layer II. Studies of interstitial waters and sediments from Site 149 (Perry and others, 1976) and from Site 323 (Lawrence and others, 1979) have supported these conclusions. Additional evidence from other DSDP sites (252, 285, 315, 336 and 386) also agrees with this interpretation (Gieskes and Lawrence, *in press*).

Although concentration-depth gradients in dissolved calcium and magnesium are often observed, many of these gradients are nonlinear in nature (see, for instance, Fig. 1). Thus investigation into the nature of the gradients is necessary in order to establish whether they are the results of reactions within the sediment column over which the gradient

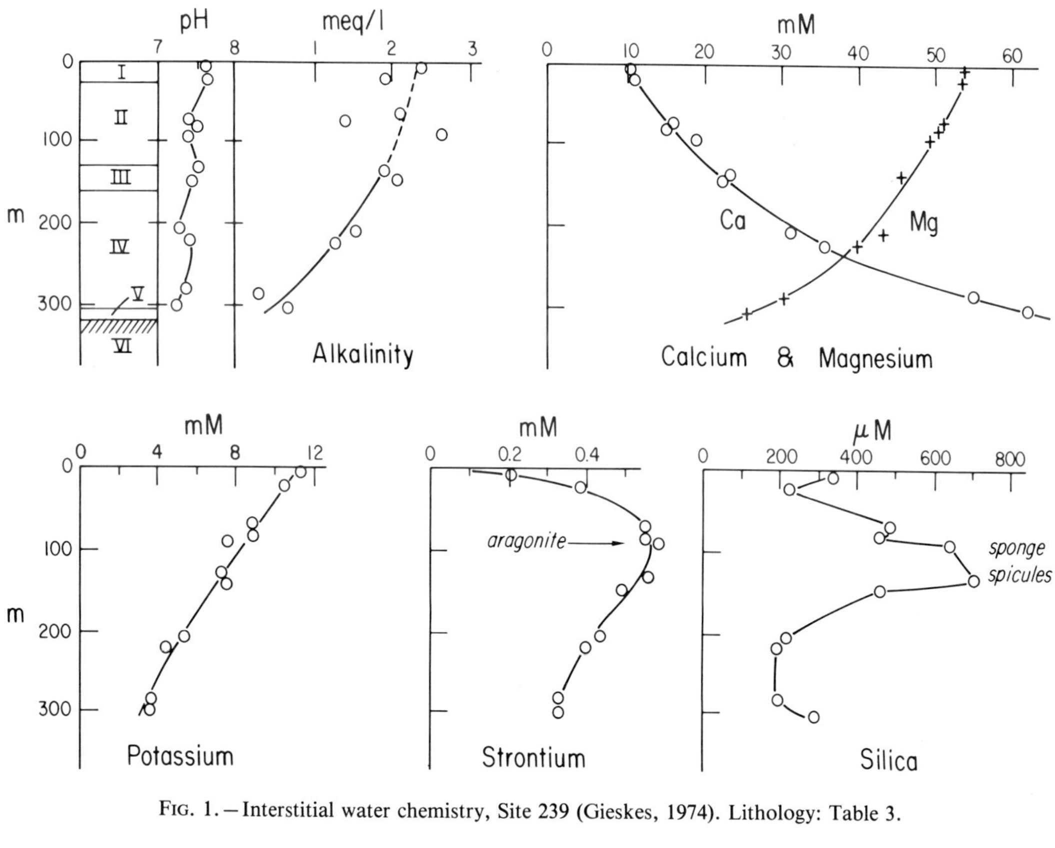

FIG. 1.—Interstitial water chemistry, Site 239 (Gieskes, 1974). Lithology: Table 3.

is observed, or whether they are due to reactions occurring in deeper unsampled sediments and/or underlying rocks of Layer II. Lerman (1975, 1977) examined a number of calcium and magnesium profiles in DSDP holes and treated these with simple advection-diffusion-reaction models with a constant diffusion coefficient in preference to models with variable diffusion coefficients. Such models imply that reactions occur in all cases where nonlinear concentration-depth profiles occur. This problem was investigated by McDuff and Gieskes (1976) and by McDuff (1978) who used electrical conductivity measurements (Manheim and Waterman, 1974) to estimate diffusion coefficients as a function of depth in a sediment column. Using this information, these investigators examined a number of concentration-depth gradients for their possible conservative nature, *i.e.,* whether only transport processes could account for the gradients. Especially in cases where changes in dissolved calcium and magnesium were linearly correlated, concentration gradients were found to be conservative (McDuff and Gieskes, 1976; McDuff, 1978). Increases in calcium and decreases in magnesium were found to be due to release or uptake of these elements by the underlying basalts. This was shown to be the case for the calcium and magnesium profiles of Site 239 (Fig. 1). In many cases, however, reactions in the sediments cause additional complications in the concentration-depth profiles, as was demonstrated by Kastner and Gieskes (1976) in Site 323 in the Bellingshausen Abyssal Plain. There is strong evidence that, in the latter case, reactions involving volcanic matter dispersed in the sediments are responsible for these observations (Kastner and Gieskes, 1976; Lawrence and others, 1979).

In Figure 2 a summary of observed average calcium gradients in pelagic DSDP sites is presented. In agreement with McDuff (1978, *in press*), it is apparent that gradients generally are not found in areas near ridge crests, whereas at greater distances from the mid ocean ridges the number of sites with gradients increases as do their magnitudes. McDuff interpreted these observations, particularly in those sites with linearly correlated calcium and magnesium gradients, in terms of continued alteration of Layer II long after burial beneath a thick sediment blanket. This confirms the observations of Hart (1973), Muehlenbachs and Clayton (1972; 1976) and Lawrence and others (1975).

The concentration gradients shown in Figure 2, when averaged over the entire ocean floor (Lawrence and Gieskes, in preparation), lead to an average gradient of dissolved calcium in pelagic sediments of 3.5 mM/100 m (where mM is millimoles per liter), which supports a flux of dissolved calcium to the ocean of about 1% of the river flux. The associated flux of magnesium from the ocean into the oceanic crust also would amount to less than 2% of the river flux. Clearly fluxes in surface sediments (Sayles, 1979) and those associated with hydrothermal reactions at oceanic spreading centers (Edmond and others, 1979) are far more important with respect to the geochemical balance of magnesium and calcium. Lawrence and Gieskes (in preparation) estimated from the average correlation between the change in $\delta^{18}O$ of the pore waters and the change in dissolved calcium (-0.1 ‰/mM), that a flux of 2×10^{15} moles^{18}O per million years into the oceanic crust results from these suggested low temperature alteration reactions in Layer II. This flux is nontrivial and implies the alteration of 5–20% of the upper 1000 meters of Layer II during its average 60 million years transit from ocean ridges to subduction zones.

Dissolved potassium generally shows decreases with depth. These decreases are due to reactions in the sediments and the basalts of Layer II (Gieskes and Lawrence, *in press;* see also Fig. 1). The average gradient for dissolved potassium is less than -2mM/100m, which would make the flux of potassium about 10% of the river flux and thus of relatively minor importance. Fluxes into the near surface sediments (Sayles, 1979) and fluxes into the ocean from hydrothermal systems (Edmond and others, 1979) appear to be of much greater importance. Studies of low temperature basalt alteration in Layer II suggest that substantial uptake of potassium occurs during the stage of relatively free circulation of sea water through the ocean crust, before burial by a sedimentary cover (Hart, 1969, 1970; Seidemann, 1977; Donnelly and others, 1979).

Dissolved Silica in Pelagic Sediments

From data collected during DSDP, a large body of information is now available on dissolved silica in deep sea sediments. Notwithstanding so-called "temperature of squeezing" artifacts (Fanning and Pilson, 1971; Gieskes, 1973, 1974; Sayles and Manheim, 1975), some general trends can be clearly discerned. In sites with relatively large contributions of biogenic silica (opal-A), one often observes a rather good correlation between the concentration of dissolved silica and the Si/Al ratio of the bulk solids (Fig. 3). The latter ratio is a rather crude measure of the contribution of opal-A to the mineral assemblage, though more refined methods to estimate this contribution are available (Leinen, 1977). Reasons for the apparent linear dependence of dissolved silica concentrations on the Si/Al ratio are not clear, and indeed the data in Figure 3 suggest a nonlinear dependence on the opal-A contents. Some authors have proposed that the above observations can be understood in terms of a dynamic equilibrium state between the release of silica by dissolution of opal-A and the uptake of silica by clay minerals, presumably by reaction rather

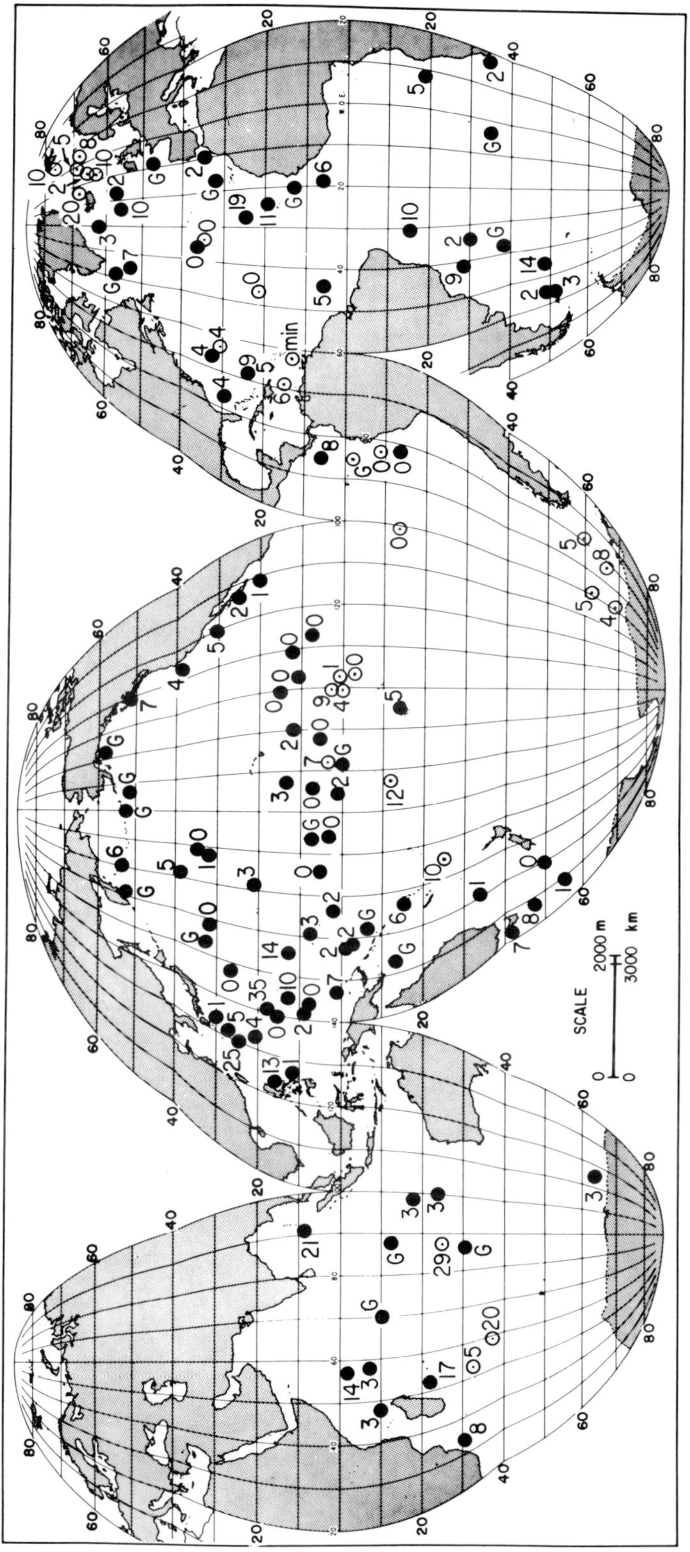

FIG. 2. — Average calcium concentration-depth gradients in pelagic DSDP cores (mM/100 meters). Points with circles have also $\delta^{18}O$ data (Lawrence and Gieskes, in preparation). G = gradient positive but ill-defined.

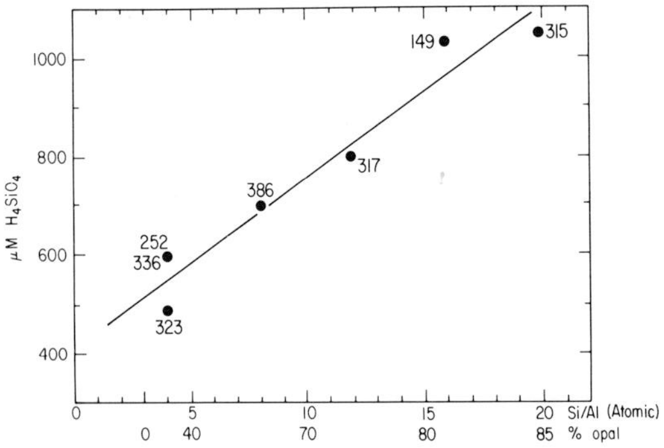

FIG. 3.—Correlation between dissolved silica concentrations and Si/Al (atomic) ratios of solids in pelagic DSDP cores. Data sources: Site 149—Perry and others, 1976; Site 323—Kastner and Gieskes, 1976; Donnelly and Wallace, 1976; Sites 252, 315, 317, 336 and 386—Gieskes and Lawrence, *in press*; T. W. Donnelly, personal communication. Also given is estimated opal-A (based on silicate Si/Al = 3.0).

than by absorption (Bischoff and Ku, 1971; Sholkovitz, 1973). There is little doubt that dissolved silica concentrations in most DSDP sites are highly variable with depth and are highly dependent on the local lithology. This is clearly demonstrated in the various dissolved silica profiles presented herein.

The presence of high dissolved-silica values does not per se imply the presence of biogenic silica (opal-A). For instance, high dissolved silica values persist in DSDP sites in which all opal-A has undergone transformation to opal-CT and quartz (Site 245, Gieskes and others, 1975; Site 317, Gieskes, 1976). Only when opal-CT has converted completely to quartz do the values of dissolved silica drop significantly. This, of course, is in accord with the different solubilities of various silica phases (*e.g.*, Kastner and others, 1977). Also, in sites with large contributions of volcanic glass, high dissolved silica values often are observed despite the absence of biogenic silica.

Elevated concentrations of dissolved silica, whether caused by the presence of biogenic silica or volcanic glass, appear to be of great importance as a requisite for reactions in the sediment column. For instance, alteration of igneous material in Site 323 in the Bellingshausen Abyssal Plain is associated with a zone of silicification (Kastner and Gieskes, 1976). Similar observations have been made in Site 474 in the Gulf of California (Fig. 10). In other sites, *e.g.*, Site 149 (Perry and others, 1976) and Sites 252, 315 and 317 (Gieskes and Lawrence, in preparation), again alteration of volcanic matter has been shown to be associated with enhanced dissolved silica activities as a result of the presence of biogenic opal-A. Mobilization of silica to the interstitial waters in these reaction zones will be enhanced because of its rapid removal into new silica phases (opal-CT and/or quartz) and associated minerals, such as clinoptilolite, smectite, palygorskite, sepiolite, K-feldspar (see, *e.g.*, Kastner and Gieskes, 1976).

Minor Elements in Pelagic Sediments

Strontium.—Values of dissolved strontium are usually observed to increase in DSDP cores. This is particularly true in sites with large contributions of calcium carbonate to the solid phases. Generally these increases have been explained in terms of recrystallization of calcium carbonate (Sayles and Manheim, 1975). In many carbonate sites a rough correlation exists between dissolved strontium increases and sedimentation rates (Sayles and Manheim, 1975; Gieskes, 1976), and steady state values of dissolved strontium or maximum concentrations of this component are often observed (Gieskes, 1976). This is demonstrated for two carbonate sites in the Southwest Pacific Ocean in Figure 4 (Gieskes, unpublished data). The position of the maxima or the depth horizon in which steady state concentrations are reached is a function of the sedimentation rate, much more so than the maximum value of the concentration increase (Table 2). This can be understood best in terms of the so-called communication length for diffusion (Gieskes, 1975) or the diffusion scale length (Lerman, 1975; 1977). To a depth that is in diffusive communication with the overlying ocean a strontium gradient will exist, below which either steady state concentrations of

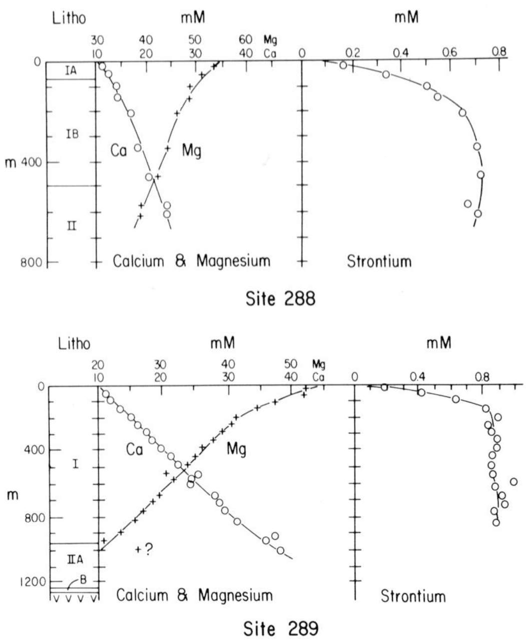

FIG. 4.—Dissolved strontium (and calcium and magnesium) in Sites 288 and 289, Leg 30 (Gieskes, unpublished data).

TABLE 2. — EXTREMES IN STRONTIUM CONCENTRATION AS FUNCTIONS OF SEDIMENTATION RATES.

Site	u cm/1000y	Z_c meters	$\delta_{Sr_{max}}$ m moles/1	$Z_{Sr_{max}}$ meters
288	1.5	420	0.65	450
289	3.0	210	0.79	200
315	2.3	275	1.10	250
317	0.9	700	0.55	>500

u = sedimentation rate (cm/1000 years)

Z_c = communication length (Gieskes, 1975) = pD_b/u

pD_b = diffusion coefficient; $2 \times 10^{-6} cm^2/sec$ (McDuff, 1978)

$\delta_{Sr_{max}}$ = maximum increase in Sr^{++}

$Z_{Sr_{max}}$ = position Sr^{++} maximum

dissolved strontium are reached or values decrease as a result of deeper-lying sink for dissolved strontium (*e.g.,* Sites 315 and 317; Gieskes, 1976).

The increases in the dissolved strontium concentration must be related to carbonate recrystallization reactions, but these observations pose a dilemma. The data in Figure 4 show a steady increase in the Sr/Ca ratio with depth, with increases as high as three times the seawater ratio. If the distribution coefficients for calcite as chosen by Gieskes (1978) were correct, no strontium concentration increase should be observed in the interstitial waters as a result of calcite recrystallization. If, however, the distribution coefficients of Katz and others (1972) were valid no such problem would exist. Indeed, large increases in dissolved strontium could occur, and the observed decreases in the Sr/Ca ratio of the carbonates could be expected (Matter and others, 1976). Recent experiments by Paul Baker (personal communication) have supported the lower values of the distribution coefficients as determined by Katz and others (1972). The Sr/Ca ratio of biogenic calcite ($\sim 20.10^{-4}$ atomic) appears not to be an equilibrium ratio, at least not with sea water. Thus strontium can be released to interstitial waters upon carbonate recrystallization. This problem is at present a subject of more detailed investigation.

In sediments with small carbonate contributions or from which calcium carbonate is essentially absent, small but distinct increases in dissolved strontium often have been observed (Gieskes and Lawrence, 1976; Gieskes and others, 1978). Studies of the $^{87}Sr/^{86}Sr$ ratio of the dissolved strontium in such sites have shown substantial contributions of volcanic material in the sediments and underlying basalts as contributors to observed changes of major cations in the interstitial waters (Hawkesworth and Elderfield, 1978).

Manganese. — The problem of manganese migration in marine sediments has been the subject of intensive research during the past few years, especially because of the often observed enrichment of manganese oxide in surface sediments (Li and others, 1969; Bonatti and others, 1971). Hydrothermal inputs near ocean ridges are considerable (Corliss and others, 1979, and references therein) leading to manganese-enriched sediments (Bostrom and Peterson, 1969). Bender (1971) showed that manganese enrichments of ridge-crest sediments could not be accounted for by manganese migration over distances of more than a few meters. Similarly, concentration gradients of dissolved manganese in DSDP sites are generally small and less than +0.1 mM/100 m, which would result in an upward flux of manganese of considerably less than 5% of the hydrothermal flux (Corliss and others, 1979).

Observed concentration-depth profiles of dissolved manganese, however, are often rather complex in DSDP sites; for these reasons, a brief inquiry into the processes that may cause these observations follows. Previous work on the distribution of dissolved manganese in marine sediments has indicated that reducing conditions are required to mobilize manganese by reduction of manganese oxide phases (Lynn and Bonatti, 1965; Li and others, 1969; Elderfield, 1976). Several workers have suggested that the reduction of manganese oxide may follow a biogenic pathway involving the oxidation of organic matter (Bostrom, 1967; Li and others, 1969; Elderfield, 1976; Emerson and others, 1980). Upon reduction, dissolved manganese values reach a maximum, which can be understood in terms of the precipitation of manganese carbonate, generally as a mixed carbonate (Hartmann, 1964). Elderfield (1976) discussed the types of concentration-depth profiles that may be encountered in sedimentary pore fluids. Occurrences of secondary maxima in DSDP cores were explained in terms of the occurrence of more than one oxidation-reduction boundary in the sediment column. This, however, is not likely to occur because reducing conditions, especially anoxia, can be expected to persist throughout such sites. This will be so notwithstanding the apparent "oxidizing" nature of the solid phases. Here the frequently observed parallelism of dissolved manganese and dissolved silica profiles is of importance (Fig. 5). This phenomenon can be understood best in terms of the mobilization of dissolved manganese in the siliceous sections, which may be enriched in both labile organic carbon and in manganese-oxide phases adsorbed on siliceous material. Thus in order to explain secondary maxima in dissolved manganese profiles, it is not necessary to invoke the presence of distinct oxidation-reduction boundaries, but rather of zones within which manganese can be mobilized because of the presence of the necessary reducing agents, *i.e.,* organic carbon.

Lithium. — Recent investigations of hydrothermal inputs at oceanic spreading centers (Corliss and others, 1979; Edmond and others, 1979) have shown that high-temperature interaction between seawater and basalts can lead to significant increases in dis-

TABLE 3.—LITHOLOGY OF SITES CONSIDERED IN THIS PAPER (SEE FIGURES).

Site	Lithology
Site 148. 13°25′N 63°43′W	I: marls and clays with minor ash above 63 m; II: volcanic sands and clays.
Site 239. 21°18′S 51°41′E	I: silty clay; II: nanno ooze and nanno clay; III: green and gray silty clay and nanno ooze; IV: brown silty clay and clayey silt; V: nanno clay; VI: basalt.
Site 288. 5°58′S 161°50′E	IA: pyrite bearing ash-rich foram-nanno ooze; IB: bioturbated nanno-ooze; cherts below 250 m; IIA: nanno-foram chalk and nanno ooze to chalk with cherts.
Site 289. 0°30′S 158°31′E	I: nanno-foram ooze and chalk (below 300 m); IIA: radiolarian limestone; siliceous limestone; nanno-foram chalks; cherts; IIB: limestone and tuff.
Site 336. 63°21′N 7°47′W	I: gray mud, silty mud; II: olive and dusky yellow mud, claystones; mudstones; III: red, black-red clay, volcanic rubble; IV: basalt.
Site 440. 39°44′N 143°56′E	I: clayey sand, gravel, diatomaceous clay; II: diatomaceous clay and claystone (below 168 m); III: interbedded claystone and diatomaceous claystone.
Site 445. 25°31′N 133°13′E	I: clayey foram nanno ooze; II: foram nanno ooze; III: clayey nanno chalk; IV: clayey nanno chalk and calcareous siliceous mudstone; V: clayey nanno chalk and calcareous mudstone; VI: clayey nanno chalk and siliceous nanno chalk; VII: siliceous chalk and mudstone; VIII: chert-radiolarite; IX: mudstones and conglomerates.
Site 446. 25°42′N 132°47′E	I: brown terriginous mud, clay; II: pelagic clay and ash; siliceous; III: calcareous claystones, mudstones; IV: basaltic sills and intrusions.
Site 459. 17°52′N 147°18′E	I: siliceous mud, calcareous ooze; II: vitric mud, sandy ash; III: turbidites; vitric marly nanno chalk; vitric tuff; IV: turbidites; claystone, mudstone, vitric tuff; V: vitric chalk and limestone.
Site 474. 22°58′N 108°59′W	I: diatomaceous muds, turbidites; II: silty claystones and clayey siltstones; III: dolerite sills.

solved lithium in the hydrothermal fluids. Similar dissolved lithium increases were observed by Vidal and others (1978) in hydrothermal waters from Punta Banda, Baja California, which is associated with hydrothermal activity within the Aqua Blanca Fault, and by myself in the hydrothermal fluids of Site 477 in the Guaymas Basin in the Gulf of California. Estimated hydrothermal inputs of lithium into the ocean exceed the river input by a factor of 8–10 (Edmond and others, 1979). Thus a renewed examination of the geochemical cycle of lithium appears warranted.

Studies of dissolved lithium by Kaplan, Presley and co-workers, as well as recent work in my own laboratory, have indicated that the concentration of dissolved lithium generally increases with depth in DSDP cores. Concentration gradients are well below 15 μM/100 m (where μM is micromoles per liter), implying a flux out of the sediments into the ocean of less than 1.5×10^9 moles/year, which is less than 1% of the estimated hydrothermal flux (Corliss and others, 1979; Edmond and others, 1979). Therefore the study of DSDP interstitial-water profiles of dissolved lithium is not important with respect to its geochemical balance. However, an understanding of the processes leading to its distribution is useful with regard to its geochemical behavior in the ocean.

Dissolved lithium is transported to the ocean by rivers, via pore fluids and mostly by the hydrothermal flux at oceanic spreading centers. Two major sinks must therefore be considered for this element: (1) uptake into basalts during low-temperature alteration, probably closely parallel to the processes leading to the uptake of potassium (Humphris and Thompson, 1978), and (2) uptake into sediments. The former process has received little attention but is probably very significant. Uptake in sediments does occur, and particularly deep-sea clays and zeolites appear enriched in lithium (Goldberg and Arrhenius, 1958; Heier and Billings, 1970). Another sink for lithium appears in manganese oxide phases (J. D. Isaacs and W. Fenical, personal communication).

In Figure 6 a few typical concentration-depth profiles of lithium in DSDP drill holes are presented. These profiles suggest that lithium released to interstitial waters is associated with reactions involving biogenic silica. Siliceous sediments contain between 20–26 ppm Li (J. D. Isaacs and W. Fenical, personal communication). In Site 459 a sink for lithium occurs in the lowest sediment layers or in the underlying basalts. No such sink is apparent in Site 446 (Fig. 6), though profiles of dissolved calcium, manganese, and potassium indicate sources and sinks for these elements in Layer II. Indeed, the interstitial water profiles of lithium and potassium indicate very different geochemistries of these alkali metals.

HEMIPELAGIC SEDIMENTS

In this section I wish to discuss those DSDP sites that are characterized by high sedimentation rates (>50 m/10^6 years) and that are generally located around the periphery of the ocean on continental shelves or on the slopes or bases thereof. Sedimentation is characterized by large terriginous inputs, though biogenic contributions, both in the form of carbonate and silica are often large. In addition, contents of organic matter are higher than in pelagic sediments, mainly as a result of high primary production and the rapid burial and incomplete biochemical utilization of this material in the surface sediments.

Many interstitial-water profiles are complex in

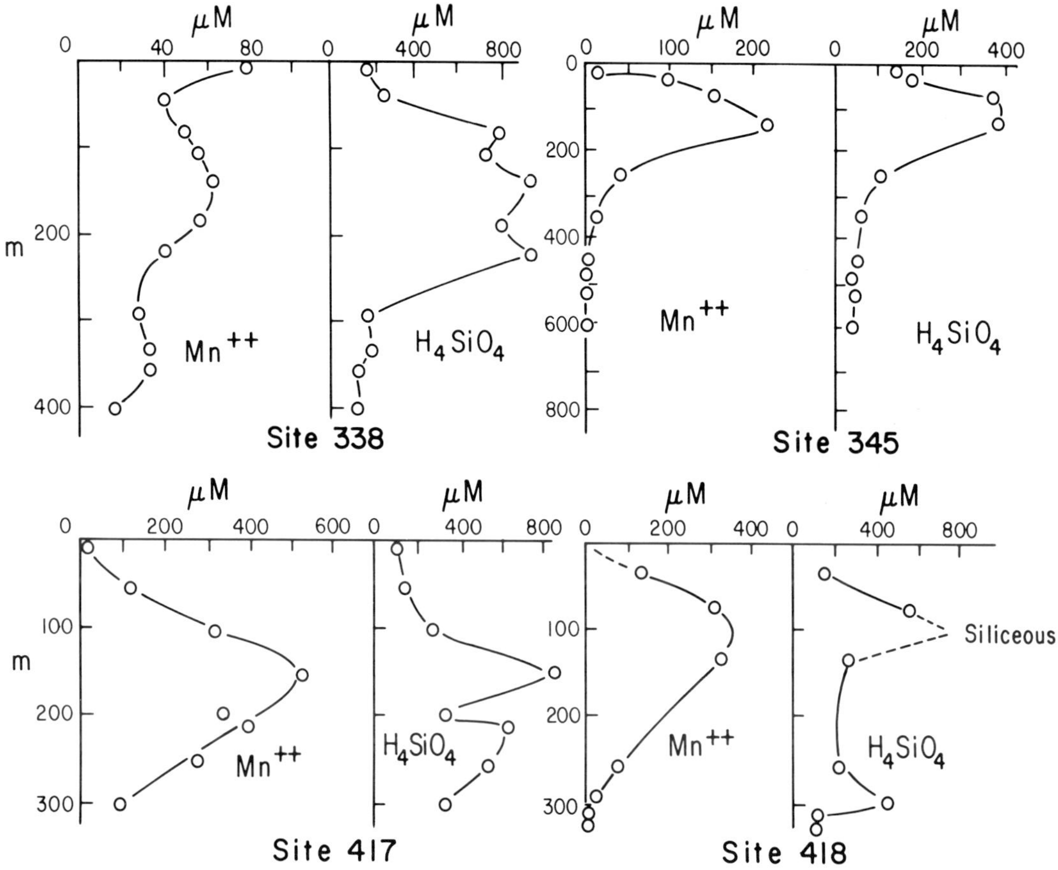

FIG. 5.—Dissolved manganese and dissolved silica in selected DSDP cores. Data sources: Sites 338 and 345—Gieskes and others, 1978, Sites 417 and 418—Gieskes and Reese (1980).

hemi-pelagic sites, especially because of the importance of reduction-oxidation processes involving interstitial-water components with reactive organic carbon. The complexity of the profiles is enhanced as a result of rapid sedimentation rates, which limits diffusive communication with the overlying ocean to relatively shallow depths (Gieskes, 1975). In Figures 7 through 10, typical interstitial water profiles are presented, covering sites of different sediment types and sedimentation rates.

In all these sites, a common feature is that decomposition of organic carbon occurs to great depths. The main processes responsible for the oxidation of the reactive organic carbon are sulfate reduction and methane formation, especially because other common oxidants of organic matter, *e.g.*, dissolved oxygen and nitrate, are usually depleted at the very surface of sediments of this nature (Sholkovitz, 1973; Gieskes, 1975; Emerson and others, 1980).

The oxidation of organic matter by sulfate-reducing bacteria can be written as follows (Gieskes, 1974):

$$(CH_2O)_x\,(NH_3)_y\,(H_3PO_4)_z + \frac{x}{2}\,SO_4^{=} =$$

$$x\,CO_2 + \frac{x}{2}\,S^{=} + y\,NH_3 + z\,H_3\,PO_4 + x\,H_20$$

in which the exact stoichiometry depends on the composition of the buried organic matter (Gieskes, 1973; Sholkovitz, 1973; Hartmann and others, 1976; Berner, 1977). Using the stoichiometric composition of oceanic organic matter ($x = 106$; $y = 16$; $z = 1$), Emerson and others (1980) predict maximum increases of alkalinity and dissolved ammonia as a result of sulfate reduction of 67 meq/l and 8.8 mM respectively. Increases in alkalinity and ammonia in excess of these values must be due to methane production processes in zones below the sulfate reduction zone (Reeburgh, 1976; Martens and Berner, 1977; Barnes and Will, *in press*). The frequently observed absence of methane in the sulfate reduction zone has been explained by these authors in terms of utilization of methane by sulfate reducers.

The products of these oxidation processes of organic matter, especially dissolved sulfide, bicarbon-

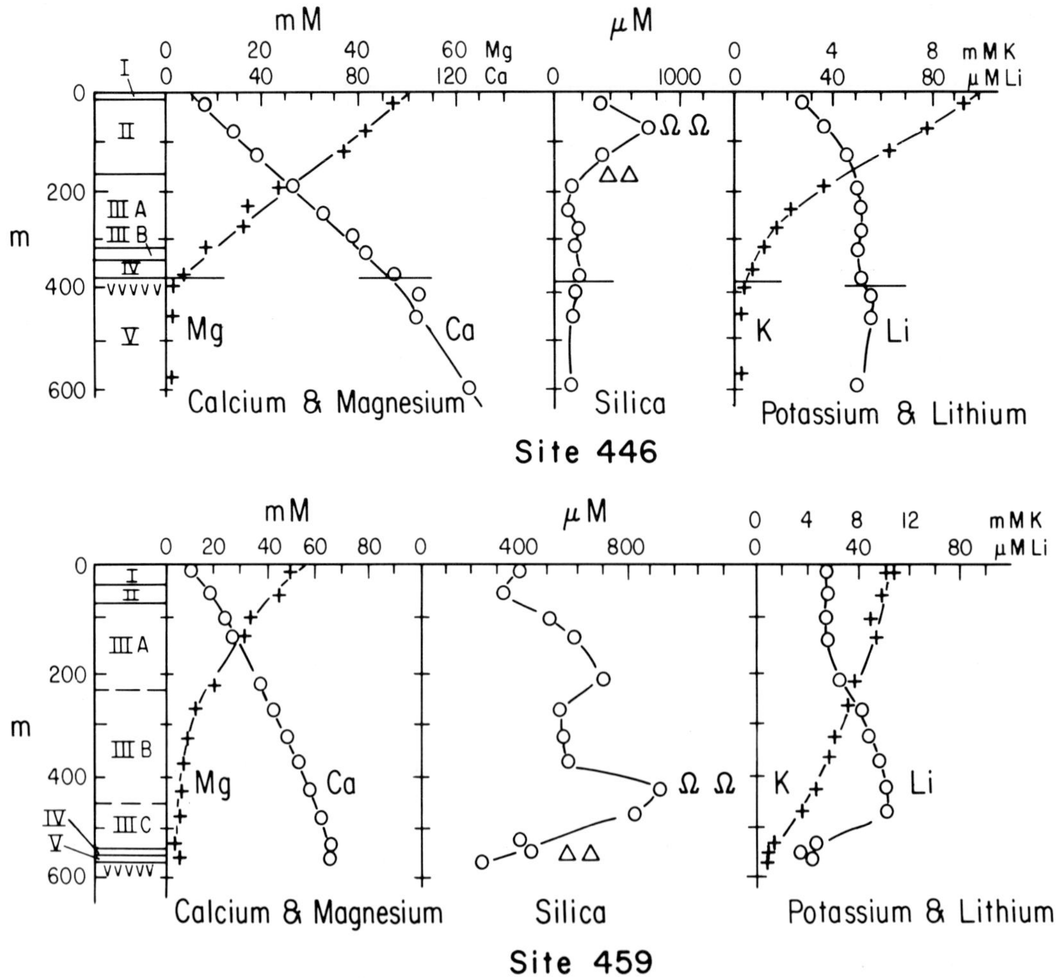

FIG. 6.—Dissolved calcium, magnesium, silica, potassium and lithium in Sites 446 and 459. Lithologies: Table 3. Data sources: Site 446 (Leg 58)—Gieskes and Johnson, *in press;* Site 459 (Leg 60)—Gieskes and Johnson, *in press.*

ate (alkalinity) and ammonia can cause diagenetic processes in these sediments. Dissolved sulfide generally is removed rapidly from the interstitial waters to form solid sulfides, in particular FeS and FeS_2 (Berner, 1970; Goldhaber and Kaplan, 1974), and rarely is dissolved sulfide detected at depths exceeding 20–50 meters in DSDP sites. Bicarbonate causes supersaturation of the interstitial waters with respect to calcite and leads to carbonate precipitation, usually as calcite, and results typically in the often observed minimum in dissolved calcium (Figs. 7, 9, 10). This process has been confirmed by carbon-isotope studies of sedimentary calcites (Lawrence, 1973; Anderson and Lawrence, 1976). In addition to this authigenic calcite, other carbonates can be precipitated, *e.g.*, rhodochrosite, siderite, ankerite, and dolomite. More often, however, less well-defined, mixed carbonates are formed. Additional decreases in dissolved bicarbonate can result from processes involving silicate diagenesis or alteration of igneous material in the sediments, processes usually responsible for magnesium removal. Finally, ammonia produced during the diagenesis of organic matter can interact with clay minerals by ion-exchange for other cations into exchangeable and non-exchangeable positions in clay lattices. This process explains the almost invariably observed decreases in dissolved ammonia with depth in deeper sediment sections. Diagenetic modeling of the interstitial water data is complicated by these processes and can only be achieved in a meaningful fashion if reasonably precise information is available on the solid phases. Lerman (1977) modeled some of the observed DSDP dissolved ammonia profiles and indicated that rates of ammonia production generally are much smaller than rates observed in near-

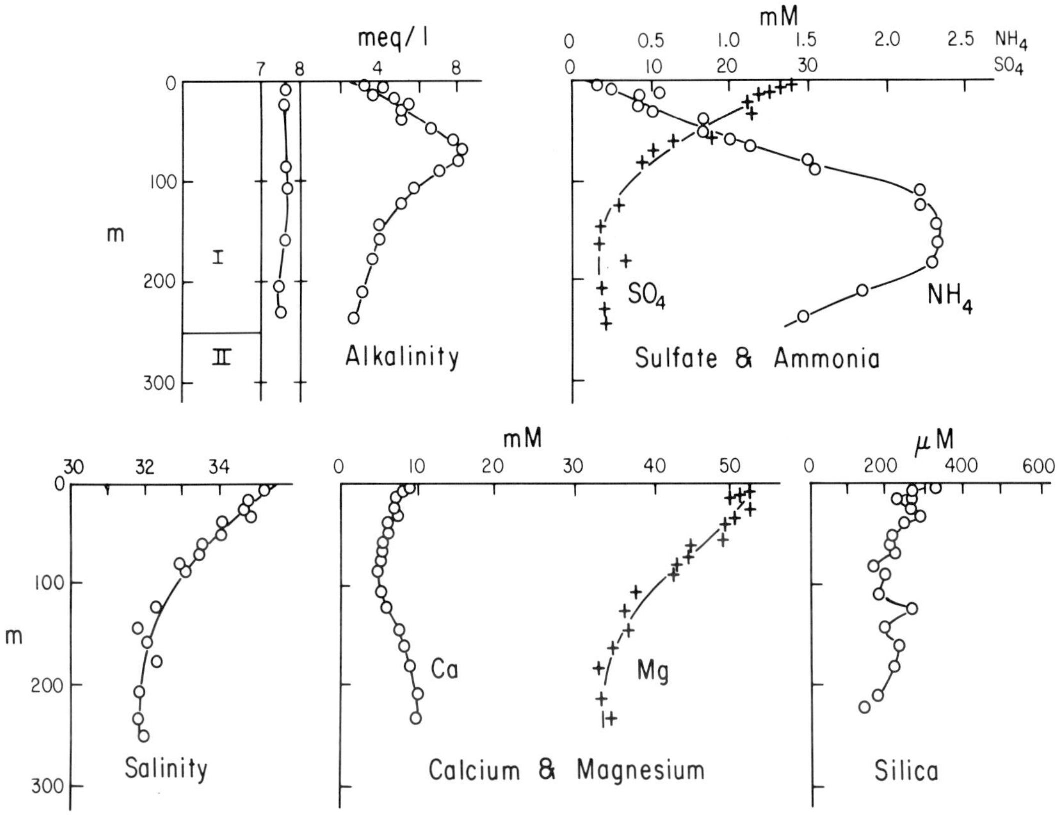

FIG. 7.—Interstitial water chemistry, Site 148 (Gieskes, 1973; Sayles and others, 1973). Lithology: Table 3.

shore sediments. Although this undoubtedly is true, Lerman's estimates must be considered minimum estimates because of the neglected uptake of ammonia in the solid phases. Thus, because of the complexity of the processes involved and because of the dependence of these reactions on the host sediments, each DSDP site presents its own unique case history and generalizations are not readily made. For these reasons I will only describe some salient features common to many hemi-pelagic sites in order to emphasize that, notwithstanding the complexity of the interstitial-water concentration-depth profiles, the information gained from interstitial water studies is extremely useful in reconstructing the diagenetic history of the sediments. To achieve this, I will briefly discuss the interstitial water profiles presented in Figures 7–10.

Site 148 (Figure 7)

Site 148, situated on the Aves Rise, is characterized by sedimentation rates of about 80 m/10^6 years. Sulfate reduction is an important phenomenon in this site, but no complete depletion of dissolved sulfate is observed. Alkalinity increases are considerably less than expected from the dissolved sulfate depletions, partly as a result of calcium carbonate precipitation (Lawrence, 1973) and also because of magnesium removal processes. Dissolved ammonia values reach a maximum of about 2.4 mM, less than expected from the sulfate depletion, presumably due to uptake processes in the sediments by clay minerals. Salinity values drop to as low as 31.9‰, which reflects the removal of dissolved sulfate from the interstitial waters. Gieskes (1974) pointed out the diagnostic use of salinity as an indicator of sulfate reduction in DSDP cores. Only if salinity values drop below 31.5‰ can there be valid reasons to suspect fresh water input at greater depths (Gieskes, 1974). Dissolved silica values are relatively low, indicating the minor importance of biogenic silica in these sediments characterized by terriginous inputs with volcanic admixtures as a result of Caribbean volcanism.

In addition to the formation of authigenic calcite (Lawrence, 1973), siderite has been noticed as a contributor to the carbonate fraction between 100m and 200m (Fan and others, 1973). The absence of siderite above 100m may be partly due to the continuing reduction of sulfate and the concomitant production of sulfide. This would cause iron sulfides to precipitate in preference to siderite (Garrels and Christ, 1965). Below 100m, where sulfate reduction has almost ceased, bicarbonate concentrations are still elevated enough to exceed the siderite solubility product.

Below the calcium minimum, a gradual increase in dissolved calcium occurs. This increase and the simultaneous decreases in magnesium and $\delta^{18}O$ of the interstitial waters indicate that alteration reactions involving volcanic matter again contribute to these concentration changes, in a manner similar to those observed in pelagic sediments.

Site 336 (Figure 8)

This site, located on the northern side of the Iceland-Faroe Ridge, is characterized by a rapidly-deposited Pleistocene section (50–60 m/10^6 years) underlain by sediments that have accumulated at much lower sedimentation rates.

The rapidly deposited upper-section, which is also enriched in biogenic silica, presumably contains more reactive organic matter than the underlying sediments. As a result, a minimum in dissolved sulfate occurs around 80m depth. Clearly the dissolved sulfate profile represents non steady-state conditions (Gieskes and others, 1978). Such observations have been made previously in Sites 241 and 242 in the Indian Ocean (Gieskes, 1974), in Site 343 in the Norwegian Sea (Gieskes and others, 1978), and in Site 400 of the Bay of Biscaye (Ellis and others, 1979). Alkalinity values in Site 336 show only a weak maximum, and similarly dissolved ammonia values remain relatively low. Interestingly, dissolved calcium data show no minimum. This does not mean that no precipitation of calcium carbonate occurs in these sediments, but rather that upward transport of dissolved calcium from the lower strata prevents the establishment of a calcium minimum. This has also been observed in sites drilled during Leg 41 off North Africa (Couture and others, 1978), where upward diffusion of calcium from salt-enriched zones prevents the establishment of such a minimum. Dissolved magnesium values decrease throughout the hole, and large depletions in $\delta^{18}O$ of the interstitial waters have also been observed (Gieskes and others, 1978; Gieskes and Lawrence, *in press*). The sediments contain a large volcanic component, and the observations on the calcium, magnesium and $\delta^{18}O$ can be understood in terms of alteration of this volcanic material and of the underlying basalts (Gieskes and Lawrence, *in press*). This interpretation is supported by the strontium isotope studies of Hawkesworth and Elderfield (1978), who found values of $^{87}Sr/^{86}Sr$ as low as 0.70719 for dissolved strontium, much lower than contemporaneous Eocene seawater.

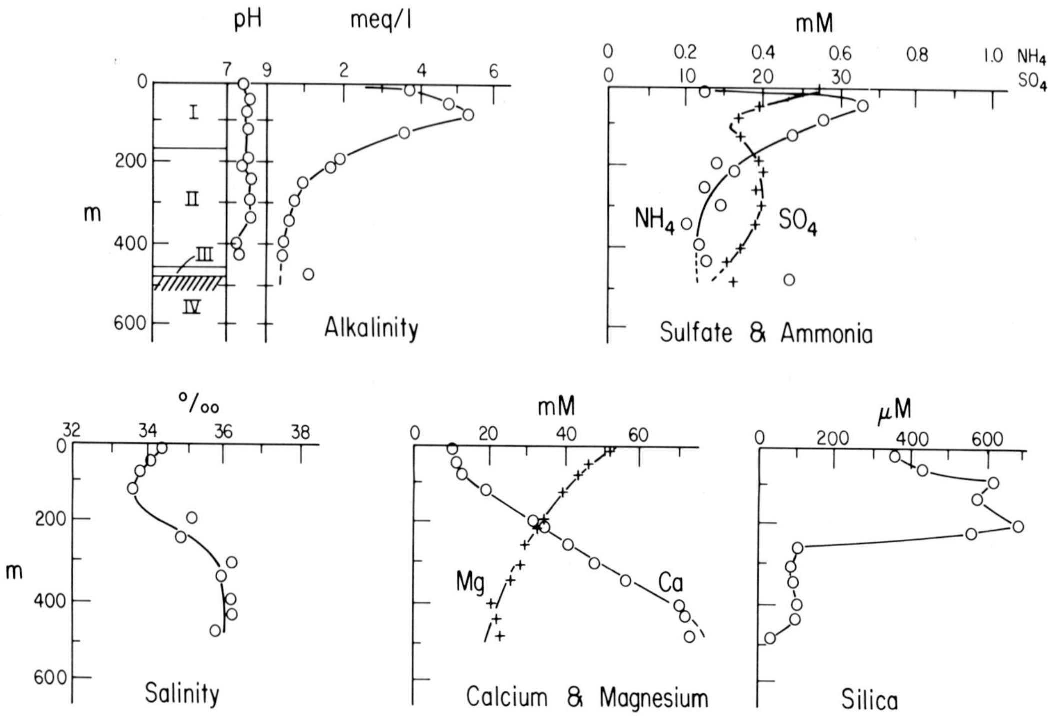

FIG. 8.—Interstitial water chemistry, Site 336 (Gieskes and others, 1978). Lithology: Table 3.

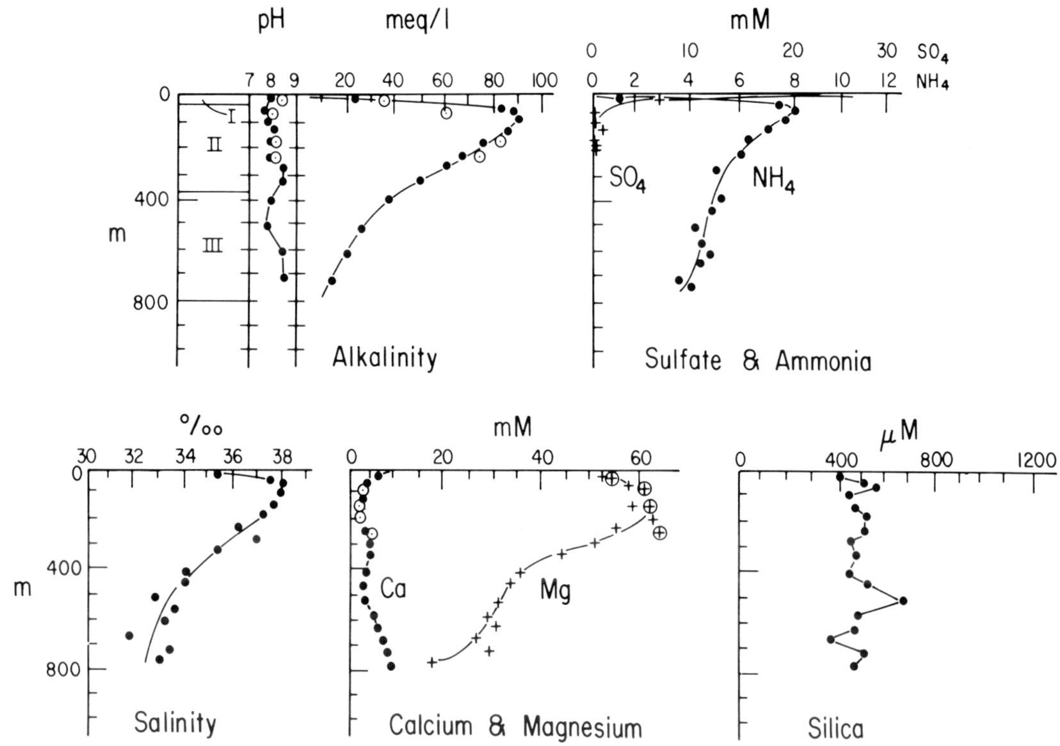

FIG. 9.—Interstitial water chemistry, Site 440 (Moore and Gieskes, 1980). Lithology: Table 3. Data with open circles are in situ values (Moore and Gieskes, 1980).

Site 440 (Figure 9)

This site is located on the westward side of the Japan Trench and is characterized by terriginous siliceous sediments accumulated at rates of about 200 m/10^6 years.

As a result of the high organic-carbon contents and the high accumulation rates, sulfate is completely depleted below 100 meters (or less), where methane production becomes dominant. As a result of the latter process, alkalinity values increase to maximum values of 92 milliequivalents per liter, and also dissolved ammonia concentrations are above 8 mM. Moore and Gieskes (1980) discussed the consequences of this phenomenon and observed that, as a result of the high ammonia values, important cation exchange has occurred with the clay mineral phases; this has led to the observed maximum in dissolved magnesium, as shown in Figure 9. Similar observations were made in Site 262 (Timor Trough) by Cook (1974) and in various DSDP sites drilled during Leg 64 (Gieskes, in preparation) in the Gulf of California. The increase in alkalinity, magnesium and, presumably, potassium and sodium (Moore and Gieskes, 1980) has led to a salinity maximum at 100m depth. Below this, magnesium concentrations decrease rapidly, and again a slight increase occurs in dissolved calcium. Although some of the magnesium is consumed by clay minerals, another sink for both magnesium and iron may be located in the calcite and ankerite in these sediments (Moore and Gieskes, 1980). The elevated concentrations of bicarbonate and magnesium, and the absence of bisulfide below relatively shallow depths may promote the formation of ankerite and the co-precipitation of magnesium in this mineral.

Site 474 (Figure 10)

This site is located at the mouth of the Gulf of California, and is characterized by Quaternary sedimentation rates of 240 m/10^6 years. Again, complete removal of dissolved sulfate occurs at shallow depths, and high values of alkalinity and ammonia are observed. These maxima are reflected in a maximum in salinity, in magnesium, and in potassium (not shown), similar to observations in Sites 262 (Cook, 1974) and in Site 440. Dissolved calcium again shows a minimum, with increases to above in values to greater than those for seawater below this. At about 300m, dissolved silica values drop rapidly, and sediments become indurated as a result of silicification. Concentration-depth profiles of dissolved calcium and magnesium below 200m are similar to

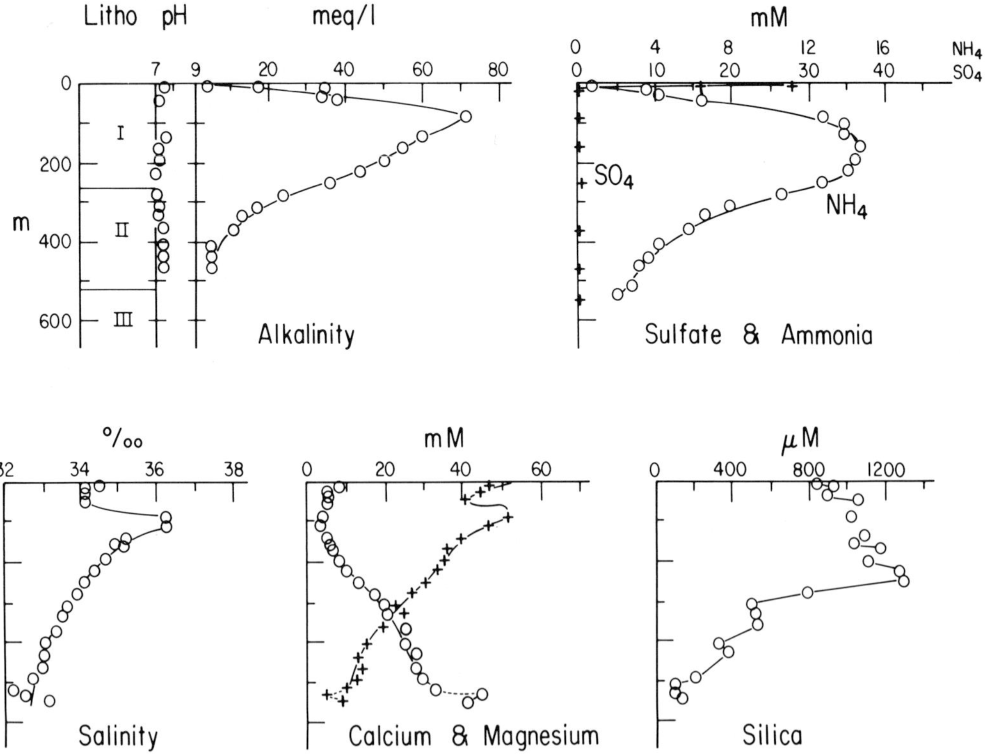

FIG. 10.—Interstitial water chemistry, Site 474 (Gieskes, in preparation). Lithology: Table 3.

those observed in Site 323 (Kastner and Gieskes, 1976), and presumably diagenetic processes occurring in Site 474 are much like those in Site 323. A basaltic imprint is evident, especially because of the rapid concentration changes below the doleritic sills.

The DSDP sites discussed above demonstrate the complexity of interstitial-water concentration-depth profiles in hemipelagic sediments. Notwithstanding these complexities, however, there is little doubt that these studies are most useful for obtaining a complete understanding of diagenetic processes that affect the sediments, whether they may have occurred in the past or whether they are still continuing. Close sampling is necessary, as is well demonstrated in Figures 9 and 10, especially in rapidly depositing hemipelagic sediments in which diffusion lengths are relatively short (Gieskes, 1975; Lerman, 1975, 1977). The lower the sedimentation rate, the smoother the concentration profiles will be, especially when reaction rates are relatively low. A notable exception to this is dissolved silica, concentrations of which (both in pelagic and hemipelagic sediments) are highly dependent on the lithology of the immediate environment from which the sample is retrieved. This must be the result of the high reactivity of silica, as was discussed above. Lerman (1977) estimated reaction rates for silica in some selected pelagic DSDP sites and found reaction rates involving this substance to be much larger than those estimated for other interstitial water components.

Fresh and Saline Water Intrusions in Continental Margin Sediments

Often in hemipelagic sites, decreases in chlorinity have been observed to occur (Presley and others, 1973; Manheim and Sayles, 1974; Gieskes, 1974; Mann and Gieskes, 1975; Gieskes and others, 1978; Moore and Gieskes, 1980). Most of the sites where such freshwater penetrations have been observed characteristically have layers of sediments consisting of permeable sandstones, *e.g.,* Site 241 off Kenya (Gieskes, 1974), Site 270 in the Ross Sea (Mann and Gieskes, 1975), and Site 339 on the Voring Plateau (Gieskes and others, 1978). Penetration of continentally-derived fresh waters through aquifers can occur to great depths and over long distances from land (at least 200 miles in Site 241). Although in most cases freshwater aquifers that are presumably continuous to the continent have been invoked, alternate explanations of freshwater occurrences are possible. For instance, Moore and Gieskes (1980) suggest a fossil freshwater remnant to explain the chloride decrease in Site 438 off the east coast of Japan. This could be due to a recent rapid subsidence of a freshwater-charged structural block of continental origin. Usually no sections through aquifers have been drilled, with the exception of the JOIDES drill holes off southern Florida (Manheim and Sayles, 1974). For these reasons, no attempts have been made to relate changes in interstitial water chemistry to diagenetic reactions, as has been done on land (Plummer, 1977).

Of special interest, however, are the sites drilled during Legs 14, 41, 47, and 50 off North Africa. Many of these sites show the presence of large increases in chloride concentrations with depth. These have usually been interpreted in terms of diffusion from inferred underlying evaporite beds. However, sites drilled during Leg 40 in the Angola Basin (Sotelo and Gieskes, 1978) and Leg 41 (Couture and others, 1978) suggested the penetration of saline brines through aquifers from evaporite strata near the coast. This interpretation was confirmed during Leg 50 (Gieskes and others, 1980) by clear maxima of chloride in Sites 415 and 416 (Fig. 11). The saline brines are derived from continental waters that flow past salt domes in this area, as is evident from decreased D/H values of the interstitial waters ($\delta D = -12.8‰$ [SMOW] at chloride maxima in Site 415 (J. R. Lawrence, personal communication). Associated with the chloride maxima are maxima in dissolved magnesium (Figure 11). This phenomenon may be of great importance for the continued formation of palygorskite and sepiolite in these sediments (Couture, 1977; Couture and others, 1978; Gieskes and others, 1980; Riech, 1980), especially

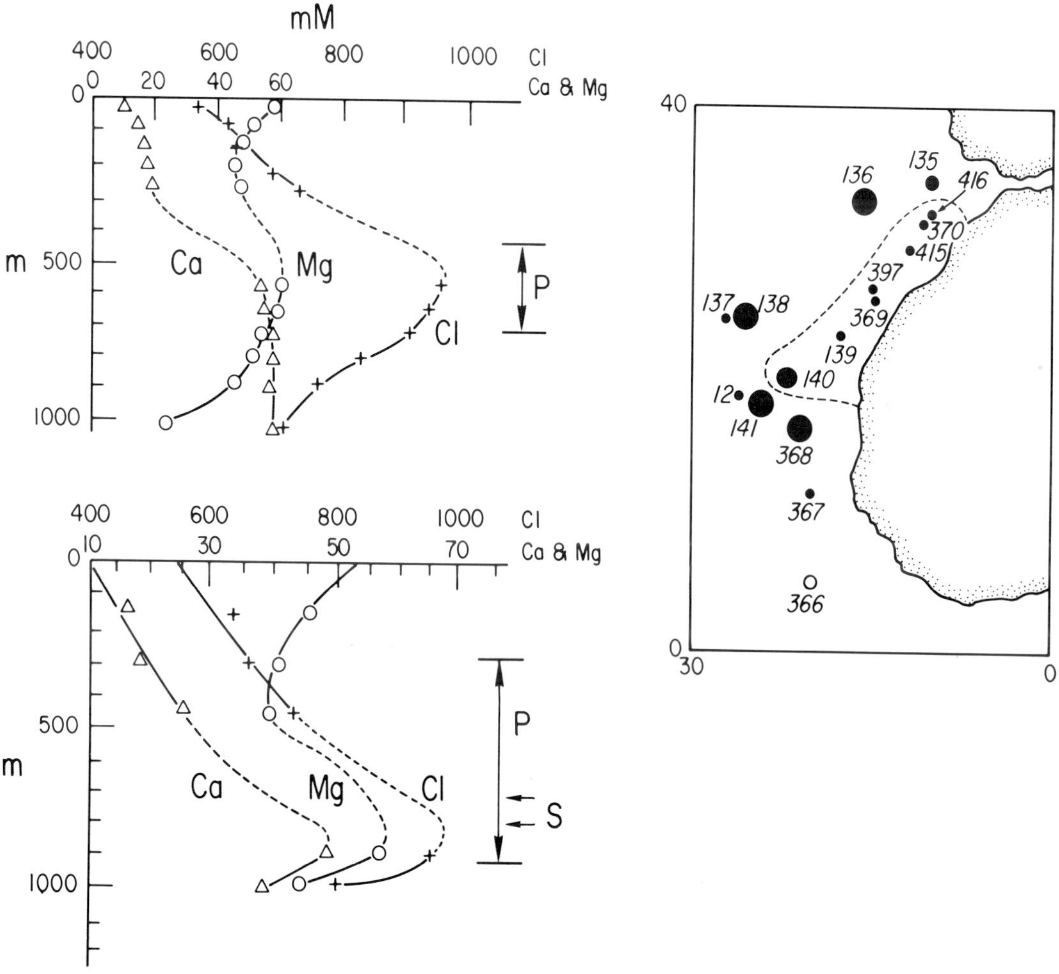

FIG. 11.—Interstitial water chemistry, Sites 415 and 416 (Gieskes and others, 1980); map indicates area of enhanced salinities off northwest Africa. P = palygorskite; S = sepiolite.

in cases where the brines penetrate layers rich in biogenic silica in a stage of transformation to more stable phases (Couture, 1977). Typically, palygorskite and sepiolite show maxima in the zones of maximum chlorinity in Sites 415 and 416. It should be noted, however, that the layers containing these minerals exist of sandstones or turbidities, which could serve as acquifers. The problem of whether the palygorskite and sepiolite are caused by in situ formation in these sediments is presently under further investigation in our laboratory.

SUMMARY AND CONCLUSIONS

Interstitial-water studies carried out during the Deep Sea Drilling Project over the last decade have served as an important aide to understanding the geochemical processes occurring in the sediments and/or underlying basalts of Layer II of the oceanic crust, and vice versa. There is now a large body of information on the interstitial-water chemistry of almost every DSDP hole drilled. In addition we have intensive, though limited, information on the oxygen-isotopic composition of the interstitial waters, as well as on the chemistry and mineralogy of the solid phases. This has led to the deduction of some general trends in the pathways of the alteration of Layer I (sediments) and Layer II (basalts) of the oceanic crust. Some of these deductions can be summarized as follows:

1) Alteration of oceanic crust Layer II leads to a release of calcium to the pore fluids of the sediments and to the overlying ocean. This calcium increase is accompanied by decreases in magnesium, potassium, and sodium. In many pelagic DSDP sites, concentration gradients (particularly those of calcium and magnesium) can be understood in terms of transport processes between the ocean and the basalts of Layer II through the sediments. Concentration gradients increase in number and magnitude as a function of age, with changes in calcium concentrations eventually exceeding those of magnesium.

2) In many DSDP sites, particularly those characterized by high volcanic inputs to the sediments, alteration of volcanic matter leads to increases in dissolved calcium and decreases in dissolved magnesium and/or potassium. These are superimposed on concentration changes, discussed above, as a result of alteration of Layer II.

3) Concentration-depth gradients of dissolved calcium and magnesium in DSDP drill holes, when averaged over the entire ocean, support fluxes between the ocean and the oceanic crust (Layers I and II) amounting to less than 2% of the river flux. For these elements exchange in near-surface sediments and alteration reactions during seawater circulation through Layer II on midocean ridges appear more significant with regard to their fluxes. Data on DSDP potassium gradients support a flux of about 10% of the river flux into Layers I and II. Again, exchange appears more important in near surface sediments and by direct interaction of seawater with basalts before burial by the sediments. On the other hand, ^{18}O fluxes from the ocean into the oceanic crust, associated with observed DSDP calcium gradients, are significant in the ^{18}O geochemical budget. The estimated ^{18}O flux implies as much as 20% alteration of the upper 1000m of Layer II of the oceanic crust during its transit between ocean ridges and subduction zones.

4) As a result of the presence of biogenic opal-A in the sediments, reactions in sediments are enhanced by the presence of high dissolved-silica concentrations in the interstitial waters. Where enhanced transport of silica through the interstitial waters occurs, particularly in zones of silicification, silica is utilized to form the more stable silica phases of opal-CT and/or quartz; it also is utilized to form such authigenic minerals as smectite, palygorskite, sepiolite, and/or K-feldspar.

5) Dissolved strontium increases in carbonate sediments often indicate carbonate recrystallization; in non-carbonate sediments, however, contributions due to alteration reactions involving volcanic material are also noticeable. The latter processes are particularly evident from strontium isotope studies of the interstitial waters.

6) In hemipelagic DSDP sites, diagenesis of organic matter in rapidly-depositing sediments leads to substantial or complete depletion of dissolved sulfate as a result of sulfate-reduction processes. When sulfate reduction is complete, methane formation occurs. As a result of these processes, large increases in bicarbonate (alkalinity) and dissolved ammonia may result. This in turn leads to more complicated depth profiles, both in calcium and magnesium, as well as in other constituents. High alkalinities result in calcium carbonate precipitation, yielding typical minima in concentration-depth profiles of this constituent. Magnesium is rapidly decreased as a result of co-precipitation with carbonates (calcite, ankerite, dolomite) as well as by enhanced silicate diagenesis. High ammonia values can lead to complex profiles in magnesium and potassium because of ion exchange of ammonia with clay minerals. In general, detailed interstitial-water profiles in rapidly accumulating sediments can be of great value in unraveling the diagenetic history of DSDP sites.

7) Freshwater input through aquifers from land can often be observed in continental-margin sediments. Transport can occur over large distances from the continents. In sediments off northwest Africa, these meteoric waters flow past salt deposits and show increased chloride, calcium and magnesium concentrations.

ACKNOWLEDGMENTS

I wish to thank my colleagues of the JOIDES Inorganic Geochemistry Advisory Panel and my direct collaborators for the many discussions and information exchanges that have made the DSDP interstitial-water program a success. Special thanks go to Drs. Frank Manheim, Fred Sayles, Ian Kaplan and Bob Presley for initiating this program, and to Dr. Robert Garrels for talking me into becoming a participant. The group of geochemists whose direct collaboration has led to a deeper understanding of the interstitial water chemistry of DSDP cores include Drs. James R. Lawrence, Miriam Kastner, James I. Drever, the late Edward Perry, Henry Elderfield, Russell E. McDuff, Thomas W. Donnelly, and many others. It is gratifying to know that a reasonably consistent picture has developed from the vast information collected during the duration of the Deep Sea Drilling Project. The diligent participation of all the DSDP chemical technicians should be especially acknowledged. I also thank my wife Barbara for letting me go to sea on *Glomar Challenger* during several Christmas-New Year and summer periods.

This research was generously supported by the National Science Foundation through Grants GA-33229; DES 72-01410; GA-40005; OCE 76-20151; and OCE 77-24102 to the author, as well as by NSF funding of the Deep Sea Drilling Project.

REFERENCES CITED

NOTE: All references to Deep Sea Drilling Project Initial Reports have been abbreviated; for complete citations see *Appendix I* at the end of this volume, or the individual Initial Reports as published.

Anderson, T. F., and Lawrence, J. R., 1976, Stable isotope investigations of sediments, basalts and authigenic phases from Leg 35 cores: Init. Repts. DSDP, v. 35, p. 497–505.

Barnes, R. O., Gieskes, J. M., Horvath, J., and Akiyama, W., 1979, Interstitial water studies, Leg 47a, b: Init. Repts. DSDP, v. 47, p. 577–582.

———, and Will, B. R., *in press,* Physical and diagenetic controls on the concentrations of dissolved Ar, N_2 and CH_4 in pore fluids from DSDP Legs 47 to 57: Geochim. Cosmochim. Acta.

Bender, M. L., 1971, Does upward diffusion supply the excess manganese in pelagic sediments?: Jour. Geophys. Res., v. 76, p. 4212–4215.

Berner, R. A., 1970, Sedimentary pyrite formation: Amer. Jour. Sci., v. 268, p. 1–23.

———, 1977, Stoichiometric models for nutrient regeneration in anoxic sediments: Limnology Oceanography, v. 22, p. 781–786.

Bischoff, J. L., and Ku, T., 1971, Pore fluids of recent marine sediments: II. Anoxic sediments of 35° to 45°N Gibraltar to mid-Atlantic Ridge: Jour. Sed. Petrology, v. 41, p. 1008–1017.

———, Greer, R. E., and Luistro, A. O., 1970, Composition of interstitial waters of marine sediments; temperature of squeezing effect: Science, v. 167, p. 1245–1246.

Bonatti, E., Fisher, D. E., Joensuu, O., and Rydell, H. S., 1971, Post-depositional mobility of some transition elements, phosphorus, uranium and thorium in deep sea sediments: Geochim. Cosmochim. Acta, v. 35, p. 189–201.

Bostrom, K., 1967, The problem of excess manganese in pelagic sediments, *in* Abelson, P. H., ed., Researches in Geochemistry, v. 2: New York, Wiley, v. 2, p. 421–452.

———, and Peterson, M. N. A., 1969, Origin of aluminum poor ferro-manganoan sediments in areas of high heat flow on the East Pacific Rise: Marine Geology, v. 7, p. 427–447.

Broecker, W. S., 1971, A kinetic model for the chemical composition of seawater: Quaternary Res., v. 1, p. 188–207.

Cook, P. J., 1974, Geochemistry and diagenesis of interstitial fluids and associated calcareous oozes, DSDP, Leg 27, Site 262, Timor Trough: Init. Repts. DSDP, v. 27, p. 463–480.

Corliss, J. B., Dymond, J., Gordon, L. I., Edmond, J. M., von Herzen, R. P., Ballard, R. D., Green, K., Williams, D., Bainbridge, A., Crane, K., and van Andel, T. H., 1979, Submarine thermal springs on the Galapagos Rift: Science, v. 203, p. 1073–1083.

Couture, R. A., 1977, Synthesis of some clay minerals at 25°C; palygorskite and sepiolite in the oceans (Ph.D. thesis): La Jolla, California, University of California, San Diego, 237 p.

———, Miller, R. S., and Gieskes, J. M., 1978, Interstitial water and mineralogical studies, Leg 41: Init. Repts. DSDP, v. 41, p. 907–914.

Donnelly, T. W., and Merrill, L., 1977, The scavenging of magnesium and other chemical species by biogenic opal in deep sea sediments: Chem. Geology, v. 19, p. 167–186.

———, and Wallace, J., 1976, Major and minor element chemistry of Antarctic clay rich sediments: Init. Repts. DSDP, v. 35, p. 427–446.

———, Pritchard, R. A., Emmermann, R., and Puchelt, H., 1979, The aging of oceanic crust: mineralogic and chemical synthesis of the results of legs 51–53: Init. Repts. DSDP, v. 51–53, p. 1563–1577.

Drever, J. I., 1974, The magnesium problem, *in* Goldberg, E. D., ed., The Sea, v. 5: New York, Wiley-Interscience, p. 337–357.

Edmond, J. M., Measures, C., McDuff, R. E., Chan, L. H., Grant, B., Gordon, L. I., and Corliss, J. B., 1979,

Ridgecrest hydrothermal activity and the balances of the major and minor elements in the ocean: the Galapagos data: Earth Planet. Sci. Letters, v. 46, p. 1–18.

ELDERFIELD, H., 1976, Manganese fluxes to the oceans: Marine Chemistry, v. 4, p. 103–132.

ELLIS, R., PINE, J., AND GIESKES, J. M., 1979, Interstitial water studies, Leg 48: Init. Repts. DSDP, v. 48, p. 297–303.

EMERSON, S., JAHNKE, R., BENDER, M., FROELICH, P., KLINKHAMMER, G., BOWSER, C., AND SETLOCK, G., *in press,* Early diagenesis in sediments from the eastern equatorial Pacific: I. Pore water nutrient and carbonate results: Earth Planet Sci. Letters, v. 49, p. 57–80.

FAN, P. F., REX, R. W., COOK, H. E., AND ZEMMELS, I., 1973, X-ray mineralogy of the Caribbean Sea, Leg 15: Init. Repts. DSDP, v. 15, p. 847–921.

FANNING, K. A., AND PILSON, M. E. Q., 1971, Interstitial silica and pH in marine sediment: some effects of sampling procedures: Science, v. 173, p. 1225–1231.

GARRELS, R. M., AND CHRIST, C., 1965, Solutions, Minerals and Equilibria: New York, Harper and Row.

———, AND PERRY, E. A., 1974, Cycling of carbon, sulfur and oxygen through geologic time, *in* Goldberg, E. D., ed., The Sea, v. 5: New York, Wiley-Interscience, p. 303–336.

GIESKES, J. M., 1973, Interstitial water studies, Leg 15—alkalinity, pH, Mg, Ca, Si, PO_4 and NH_4: Init. Repts. DSDP, v. 20, p. 813–829.

———, 1974, Interstitial water studies, Leg 25: Init. Repts. DSDP, v. 25, p. 361–394.

———, 1975, Chemistry of interstitial waters of marine sediments: Ann. Rev. Earth Planet. Sci., v. 3, p. 433–453.

———, 1976, Interstitial water studies, Leg 33: Init. Repts. DSDP, v. 33, p. 563–570.

———, 1978, Reactions controlling strontium abundance in natural waters, *in* Handbook of Geochemistry, v. II/5, p. 38H1–38H5.

———, AND JOHNSON, J., *in press,* Interstitial water studies, Leg 58: Init. Repts. DSDP, v. 61.

———, AND ———, *in press,* Interstitial water studies, Leg 60: Init. Repts. DSDP, v. 60.

———, AND LAWRENCE, J. R., 1976, Interstitial water studies, Leg 35: Init. Repts. DSDP, v. 35, p. 407–423.

———, AND ———, *in press,* Alteration of volcanic matter in deep sea sediments: evidence from the chemical composition of interstitial waters from Deep Sea Drilling cores: Geochim. Cosmochim. Acta.

———, AND REESE, H., 1980, Interstitial water studies, Legs 51–53: Init. Repts. DSDP, v. 51–53, p. 747–751.

———, GRAHAM, D., AND ELLIS, R., 1980, Interstitial water studies, DSDP sites 415 and 416: Init. Repts. DSDP, v. 50, p. 691–693.

———, KASTNER, M., AND WARNER, 1975, Evidence for extensive diagenesis, Madagascar Basin, Deep Sea Drilling Site 245: Geochim. Cosmochim. Acta, v. 39, p. 1385–1393.

———, LAWRENCE, J. R., AND GALLEISKY, G., 1978, Interstitial water studies, Leg 38: Init. Repts. DSDP, Suppl. v. 38-39-40-41, p. 121–133.

GOLDHABER, M., AND KAPLAN, I. R., 1974, The sulfur cycle, *in* Goldberg, E. D., ed., The Sea, v. 5: New York, Wiley-Interscience, p. 569–655.

GOLDBERG, E. D., AND ARRHENIUS, G., 1958, Chemistry of Pacific pelagic sediments: Geochim. Cosmochim. Acta, v. 13, p. 153–212.

HAMMOND, D. E., 1973, Interstitial water studies, Leg 15—a comparison of the major element and carbonate chemistry data from Sites 147, 148 and 149: Init. Repts. DSDP, v. 20, p. 831–850.

HART, S. R., 1969, K, Rb, Cs contents and K/Rb, K/Cs ratios of fresh and altered submarine basalts: Earth Planet. Sci. Letters, v. 6, p. 295–303.

HART, R. A., 1970, Chemical exchange between seawater and deep ocean basalts: Earth Planet. Sci. Letters, v. 9, p. 269–279.

———, 1973, A model for chemical exchange in the basalt-seawater system of oceanic layer II: Canadian Jour. Earth Sci., v. 10, p. 799–816.

HARTMANN, M., 1964, Zur Geochemie von Mangan und Eisen in der Ostsee: Meyniana, v. 14, p. 3–20.

———, MUELLER, P. J., SUESS, E., VAN DER WEIJDEN, D. H., 1976, Chemistry of late Quaternary sediments and their interstitial waters from the N.W. African continental margin: *Meteor* Forsch. Ergebn., v. C 24, p. 1–67.

HAWKESWORTH, C. J., AND ELDERFIELD, H., 1978, The strontium isotope composition of interstitial waters from Sites 245 and 336 of the Deep Sea Drilling Project: Earth Planet. Sci. Letters, v. 40, p. 423–432.

HEIER, K. S., AND BILLINGS, G. K., 1970, Lithium. 3-K. Abundance in common sediments and sedimentary rock types, *in* Wedepohl, K. H., ed., Handbook of Geochemistry, v. II/2, p. 3K1–3K3.

HUMPHRIS, S. E., AND THOMPSON, G., 1978, Trace element mobility during hydrothermal alteration of oceanic basalts: Geochim. Cosmochim. Acta, v. 42, p. 127–136.

KASTNER, M., AND GIESKES, J. M., 1976, Interstitial water profiles and sites of diagenetic reactions, Leg 35, DSDP, Bellingshausen Abyssal Plain: Earth Planet. Sci. Letters, v. 33, p. 11–20.

———, KEENE, J. B., AND GIESKES, J. M., 1977, Diagenesis of siliceous oozes—I. Chemical controls on the rate of opal-A to opal-CT transformation—an experimental study: Geochim. Cosmochim. Acta, v. 41, p. 1041–1059.

KATZ, A., SASS, E., STARINSKY, A., AND HOLLAND, H. D., 1972, Strontium behavior in the aragonite-calcite transformation: an experimental study at 40–98°C: Geochim. Cosmochim. Acta, v. 36, p. 481–496.

LAWRENCE, J. R., 1973, Interstitial water studies, Leg 15—stable oxygen and carbon isotope variations in water, carbonates, and silicates from the Venezuela Basin (Site 149) and the Aves Ridge (Site 148): Init. Repts. DSDP, v. 20, p. 891–899.

———, AND GIESKES, J. M., in preparation, Constraints on water transport and alteration in the oceanic crust from the isotopic composition of pore water: Jour. Geophys. Res.

———, ———, AND BROECKER, W. F., 1975, Oxygen isotope and cation composition of DSDP pore waters and the alteration of layer II basalts: Earth Planet. Sci. Letters, v. 27, p. 1–10.

———, Drever, J. I., Anderson, T. F., and Brueckner, H. K., 1979, Importance of alteration of volcanic material in the sediments of Deep Sea Drilling Site 323: Chemistry, $^{18}O/^{16}O$ and $^{87}Sr/^{86}Sr$: Geochim. Cosmochim. Acta, v. 43, p. 573-588.

Leinen, M., 1977, A normative calculation technique for determining opal in deep sea sediments: Geochim. Cosmochim. Acta, v. 41, p. 671-675.

Lerman, A., 1975, Maintenance of steady state in oceanic sediments: Amer. Jour. Sci., v. 275, p. 609-635.

———, 1977, Migrational processes and chemical reactions in interstitial waters, *in* Goldberg, E. D., McCave, I. N., O'Brian, J. J., and Steele, J. H., eds., The Sea, v. 6: New York, Wiley-Interscience, p. 695-738.

Li, Y.-H., Bischoff, J., and Mathieu, 1969, The migration of manganese in Arctic Basin sediment: Earth Planet Sci. Letters, v. 7, p. 265-270.

Lynn, D. C., and Bonatti, E., 1965, Mobility of manganese in diagenesis of deep-sea sediments: Marine Geology, v. 3, p. 457-474.

MacKenzie, F. T., 1975, Sedimentary cycling and the evolution of seawater, *in* Riley, J. P., and Skirrow, G., eds., Chemical Oceanography, v. 2, second edition: New York, Academic Press, p. 309-364.

Mangelsdorf, P. C., Wilson, T. R. S., and Daniel, E., 1969, Potassium enrichments in interstitial waters of recent sediments: Science, v. 165, p. 171-174.

Manheim, F. T., 1976, Interstitial waters of marine sediments, *in* Riley, J. P., and Chester, R., eds., Chemical Oceanography, v. 7, second edition: New York, Academic Press, p. 115-186.

———, and Sayles, F. L., 1974, Composition and origin of interstitial waters of marine sediments based on deep sea drill cores, *in* Goldberg, E. D., ed., The Sea, v. 5: New York, Wiley-Interscience, p. 527-568.

———, and Waterman, L. S., 1974, Diffusimetry (diffusion coefficient estimation) on sediment cores by resistivity probe: Init. Repts. DSDP, v. 22, p. 663-670.

Mann, R., and Gieskes, J. M., 1975, Interstitial water studies, Leg 28: Init. Repts. DSDP, v. 28, p. 805-814.

Martens, C. S., and Berner, R. A., 1977, Interstitial water chemistry of anoxic Long Island Sound sediments. I. Dissolved gases: Limnology Oceanography, v. 22, p. 10-25.

Matter, A., Douglas, R. G., Perch-Nielsen, K., 1974, Fossil preservation, geochemistry, and diagenesis of pelagic carbonates from the Shatsky Rise, Northwest Pacific: Init. Repts. DSDP, v. 32, p. 891-921.

McDuff, R. E., 1978, Conservative behavior of calcium and magnesium in the interstitial waters of marine sediments: identification and interpretation (Ph.D. thesis): La Jolla, California, University of California, San Diego, 183 p.

———, *in press,* Major cation gradients in DSDP interstitial waters: the role of diffusive exchange between seawater and upper oceanic crust: Geochim. Cosmochim. Acta.

———, and Gieskes, J. M., 1976, Calcium and magnesium in DSDP interstitial waters: diffusion or reaction?: Earth Planet. Sci. Letters, v. 33, p. 1-10.

Moore, G. W., and Gieskes, J. M., 1980, Interaction between sediment and interstitial water near the Japan Trench, DSDP Leg 57: Init. Repts. DSDP, v. 57, p. 1269-1275.

Muehlenbachs, K., and Clayton, R. N., 1972, Oxygen isotope studies of fresh and weathered submarine basalts: Canadian Jour. Earth Sci., v. 9, p. 172-184.

———, and ———, 1976, Oxygen isotope composition of the oceanic crust and its bearing on seawater: Jour. Geophys. Res., v. 81, p. 4365-4369.

Perry, E. A., Gieskes, J. M., and Lawrence, J. R., 1976, Mg, Ca, and $^{18}O/^{16}O$ exchange in the sediment-pore water system, Hole 149, DSDP: Geochim. Cosmochim. Acta, v. 40, p. 413-423.

Plummer, L. N., 1977, Defining reactions and mass transfer in part of the Floridan Aquifer: Water Resources Res., v. 13, p. 801-812.

Presley, B. J., 1971, Techniques for analyzing interstitial water samples. I. Determination of selected minor and major inorganic constituents: Init. Repts. DSDP, v. 7, p. 1749-1755.

———, Culp, J., Petrowski, C., and Kaplan, I. R., 1973, Interstitial water studies, Leg 15—major ions, Br, Mn, NH_3, Li, B, Si, and $\delta^{13}C$: Init. Repts. DSDP, v. 20, p. 805-809.

Reeburgh, W. S., 1976, Methane consumption in Cariaco Trench waters and sediments: Earth Planet. Sci. Letters, v. 28, p. 337-344.

Riech, V., 1980, Diagenesis of siliceous sediments, porcellanites, and cherts of the Moroccan Basin, DSDP Sites 370, 415, and 416: Init. Repts. DSDP, v. 50, p. 725-731.

Sayles, F. L., 1979, The composition and diagenesis of interstitial solutions. I: fluxes across the seawater-sediment interface in the Atlantic Ocean: Geochim. Cosmochim. Acta, v. 43, p. 527-545.

———, and Manheim, F. T., 1975, Interstitial solutions and diagenesis in deeply buried marine sediments: results from the Deep Sea Drilling Project: Geochim. Cosmochim. Acta, v. 39, p. 103-127.

———, ———, and Waterman, L. S., 1973, Interstitial water studies on small core samples, Leg 15: Init. Repts. DSDP, v. 20, p. 783-804.

Seidemann, D. E., 1977, Effects of submarine alteration on K-Ar dating of deep-sea igneous rocks: Geol. Soc. America Bull., v. 88, p. 1660-1666.

Sholkovitz, E., 1973, Interstitial water chemistry of the Santa Barbara Basin sediments: Geochim. Cosmochim. Acta, v. 37, p. 2043-2073.

Sotelo, V., and Gieskes, J. M., 1978, Interstitial water studies, Leg 40: shipboard results: Init. Repts. DSDP, v. 40, p. 549-551.

Vidal, V. M. V., Vidal, F. V., Isaacs, J. D., and Young, D. R., 1978, Coastal submarine hydrothermal activity off northern Baja California: J. Geophys. Res., v. 83, p. 1757-1774.

Wollast, R., 1974, The Silica Problem, *in* Goldberg, E. D., ed., The Sea, v. 5: New York, Wiley-Interscience, p. 359-392.

SEPM Special Publication No. 32, p. 169–179, October 1981

PALEOENVIRONMENTAL IMPLICATIONS OF OCEANIC CARBONATE SEDIMENTATION RATES

THOMAS A. DAVIES*
Department of Geology, Middlebury College, Middlebury, Vermont 05753 U.S.A.
THOMAS R. WORSLEY
Department of Geology, Ohio University, Athens, Ohio 45701 U.S.A.

ABSTRACT

Global land-sea carbonate flux for the past 60 m.y. averages 10.3–12.5 × 10^{14} g/y, surprisingly close to 12.2 × 10^{14} g/y calculated using data from today's rivers. However, oceanic carbonate accumulation rates vary between 7.8 × 10^{14} g/y and 28.6 × 10^{14} g/y, a factor of four. Furthermore carbonate accumulation oscillates between periods of high (0–6, 22–30, 45–53 m.y.) and low deposition. Prior to 30 m.y.BP all oceans behaved in concert, but since then significant partitioning between the Pacific and Atlantic-Indian oceans has complicated the picture. Prior to 15 m.y.BP the Pacific consumed two-thirds of the total pelagic carbonate, but since that time has never consumed more than 50 percent of the total, and for the past 3 m.y. only 38 percent. This trend is related to hypsometry and changes in carbonate dissolution rates as well as to changes in relative size of the oceans resulting from seafloor spreading.

Global carbonate flux through time appears to be simply related to changing land-sea ratios as calculated from sea level curves. In this system, maximum exposed continental area correlates with high pelagic carbonate flux. Deviations from this simple relationship are attributable to changes in carbonate production-dissolution ratios, latitudinal hypsometric differences, global climatic changes, and biases introduced by the simple averaging techniques used in the calculations.

INTRODUCTION

A major outcome of deepsea drilling over the past decade has been the exciting possibility of developing a truly global view of the Earth's history over the past 150 million years (m.y.). Paleoceanography, the study of ancient oceans, has emerged as a major component of this developing world view. Quite appropriately, at this early stage in its development, the main thrust of paleoceanography to date has been to describe, on a regional or topical basis, conditions within the oceans at various times in the past. A number of excellent regional studies (*e.g.*, van Andel and others, 1975, 1977; Kidd and Davies, 1978; McCoy and Zimmerman, 1977) have recently been published, while other workers (*e.g.*, Shackleton and Kennett, 1974; Moore and others, 1978) have concentrated on particular aspects of the geologic record of the ocean floor which indicate significant trends or changes in global oceanic conditions (principally temperature and circulation). In this paper we present some quantitative estimates of oceanic carbonate sedimentation over the past 60 million years and discuss the implications of these estimates concerning the Earth's environment in the past.

Inherent in the fundamental concept of the rock cycle is the idea of mass balance. If we consider the Earth to be a closed system in quasi-steady state, then inputs and outputs in any part of the system should, over geologically significant periods of time (greater than 10^6 y.), be balanced. Because the deep oceans are the ultimate repository for all materials eroded from the continents, it follows that the sedimentary record of the ocean floor contains information not only about conditions within the oceans but also about erosion rates, and thus conditions upon the continents at various times in the past.

The erosional products that comprise marine sediments consist primarily of physically transported detritus, derived from physical weathering, and solution-transported chemicals (principally carbonate and silica). The detrital component (except for a small volume of atmospheric dust) is always closely associated with a discrete source, *i.e.*, river, volcano, sea cliff, etc. Therefore, although detrital material may locally dominate the sedimentation regime, it is generally restricted to areas near continental margins or to the sub-aerial portions of oceanic volcanos and mid-ocean ridges. The paleoenvironmental signal it contains is equally localized and therefore difficult to use in inferring global trends.

Biogenous precipitates, on the other hand, have the chemicals dissolved in seawater itself as their source and need no physical transport to be deposited. Their abundance is determined by chemical erosion rates, and their distribution is clearly the

*Present address: JOI, Inc. (Joint Oceanographic Institutions, Inc.) 2600 Virginia Ave., N.W., Suite 512, Washington, D.C., 20037

product of climate and circulation. Thus biogenous precipitates provide an accurate signal with which to interpret global paleoenvironmental history.

We have chosen to discuss only the carbonate system for this study, mainly because the easily measured carbonate content of deepsea sediments has been more thoroughly analyzed than the opaline silica content, which is more difficult to measure.

THE NATURE OF THE OCEANIC SEDIMENTARY RECORD

As Moore and others (1978) have pointed out, the results of deepsea drilling have shattered once and for all the notion of the deep oceans as a vast and quiet abyss. Accumulation rates of oceanic sediments have been shown to vary widely in space and time, and the sedimentary record at any one location on the ocean floor is commonly far from continuous.

Whether or not material accumulates at a given location is a function of the balance between supply of material and those processes which operate to remove it. In a few cases, hiatuses in the sedimentary record are clearly the result of uplift and erosion or of tectonic events which cut off the supply of sediment. In most cases, however, hiatuses are a result of sedimentary bypassing, erosion by deep ocean-bottom currents and/or dissolution of sedimentary materials in deep water-masses. This interpretation is supported by the observations of several authors (*e.g.,* Kennett and others, 1972; Davies and others, 1975; Moore and others, 1978) that the distribution in space and time of hiatuses in the deep ocean sedimentary record can be interpreted in terms of the deep circulation history of the oceans.

Of critical importance in any quantitative estimates of oceanic sedimentation is the quality of the available data. Moore (1972) and Whitman and Davies (1979) have examined this aspect of the Deep Sea Drilling Project (DSDP) core material and documented a number of biases resulting from both technical problems of core recovery and from problems of site selection. Most significant of these, from our point of view, are the facts that the site locations are biased in favor of the northern hemisphere (Fig. 1A) and shallower water depths (Fig. 1B). The bias towards the northern hemisphere will become less significant in older parts of the record since the floors of both the Pacific and Indian oceans have been moving northward with time. The bias in favor of water depths in the range 1 to 3 kilometers is significant, however, since sediment accumulation rates, especially of carbonate sediments, are strongly depth dependent. Furthermore, this bias will become progressively worse in older samples, because ocean floor sites have subsided through time.

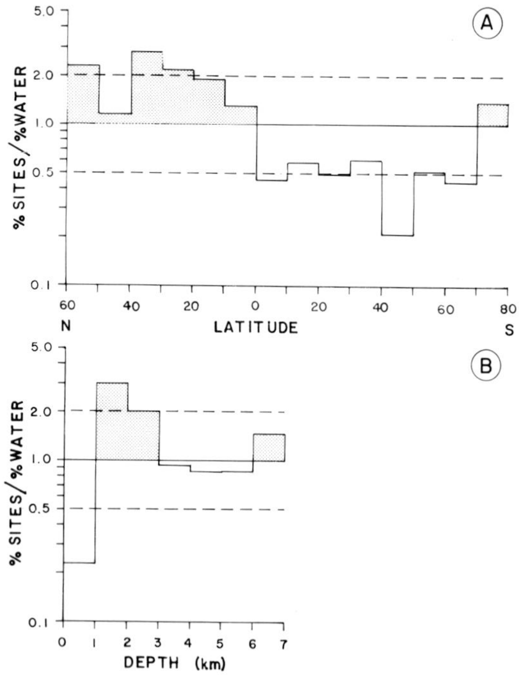

FIG. 1.—A. Distribution of drill sites compared to distribution of ocean floor with respect to latitude. B. Distribution of drill sites compared to distribution of ocean floor with respect to water depth. (From Whitman and Davies, 1979.)

Other biases in the DSDP core collection result from the fact that younger sediments tend to have been sampled in locations where the sedimentary record is thick to improve resolution, whereas older sediments have been sampled at localities where overall Cenozoic sediment accumulation rates are low and unconformities are often present, in order to bring the older parts of the record within reach of the drill.

Nevertheless, Whitman and Davies (1979) reached the conclusion that, provided appropriate allowances are made, the DSDP core material can be treated as a reasonably representative sample of the geologic record of the ocean floor and can permit meaningful conclusions to be drawn.

A MODEL FOR OCEANIC CARBONATE SEDIMENTATION

Figure 2 depicts the essential components of the oceanic sedimentation system, from the perspective of mass balance. The land areas are predominantly areas of erosion. Conversely, the oceanic areas are regions of net accumulation. The overwhelming majority of calcium carbonate in the system is chemically eroded from the land and enters the ocean in the dissolved state. It leaves the ocean via chemical (*i.e.,* inorganic) or biochemical precipitation. Shelf carbonate sediments include a large,

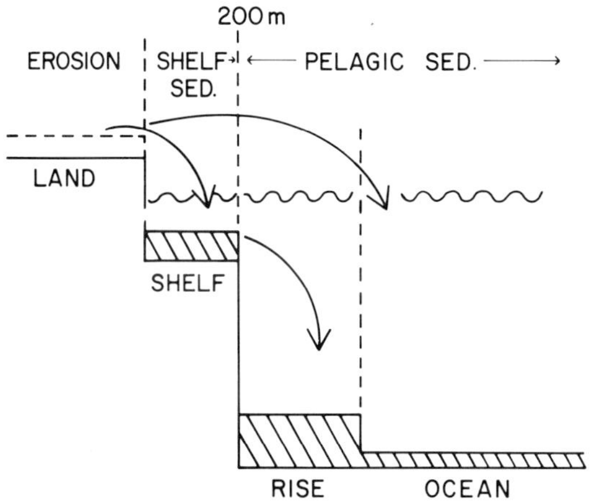

FIG. 2.—Principal elements of the oceanic sedimentation system.

usually dominant, benthonic component (corals, algae, molluscs, etc.) along with pelagic remains, whereas open ocean sediments consist almost entirely of the remains of pelagic organisms. Rise sedimentation consists of a pelagic component mixed with material physically eroded from the shelves. The arrows in Figure 2 represent a complex of processes operating to weather, erode, transport and distribute material from the continents. In this paper we are concerned less about the details of these processes than about their ultimate effects.

The two major sinks that compete for carbonate washed into the ocean by rivers are the submerged continental shelves and the deep ocean basins and continental rises. The controls governing these two sinks are very different.

Shelf accumulation requires marine submergence and is therefore dependent on relative sea level. Accumulation progresses discontinuously, usually occurring rapidly over short, *i.e.,* less than 10^4 y., intervals interspersed with long intervals of erosion or non-deposition. Estimates of average shelf accumulation rates are, consequently, strongly dependent on the length of the interval over which they are averaged. Deepsea carbonate accumulation consists of material by-passed by the shelf-rise system. Its mode of accumulation is less episodic than that of shelf sediments, being governed by the difference between the rate at which pelagic organisms produce biogenic carbonate and the rate of preburial dissolution. These two rates are influenced by both river input and shelf consumption rate, but their relative independence from local effects renders them less sensitive to the length of time over which they are averaged.

Calculated Holocene shelf carbonate deposition rates offer a good example of how the global carbonate system works and point out the inherent difference between short and long term accumulation rates. Holocene shelf rates (35 $g/cm^2/10^3$ y. according to Milliman, 1974), if they could be sustained, would be high enough to fill a 100 m deep shelf to sea level with 250 m of sediment in approximately 0.6 m.y. (The excess 150 m of sediment would be accomodated by isostatic subsidence of the shelf.) However, Hay and Southam (1977) have shown that riverine carbonate input alone cannot sustain such rates, so that a fraction of the carbonate must be obtained from the dissolved carbonate reservoir of the ocean. In the extreme case, if all Holocene shelf carbonate were supplied from the ocean reservoir, and could be precipitated at typical Holocene shelf rates, the ocean's supply would be completely depleted in approximately 250 years. However, slow ocean mixing rates preclude the possibility of such high accumulation rates being sustained from the ocean reservoir alone. Thus, until the depositional interface approaches sea level, the rate-governing step in our hypothetical aggrading shelf is river input. Global river input of dissolved carbonate, if evenly distributed over the submerged shelves of the world would yield an accumulation rate of approximately 5 $g/cm^2/10^3$ y. However, continued carbonate accumulation in the deep sea suggests that the river-controlled shelf rate must be even lower than this. As the depositional interface builds up to sea level two* possibilities arise: 1) carbonate production continues at the previous rate with excess particulate carbonate being removed to the continental rise via marine transport agencies at a rate proportional to shelf precipitation minus shelf subsidence; or 2) shelf precipitation slows to a rate commensurate with shelf subsidence with no material being eroded and added to the rise and the excess dissolved carbonate becoming directly available to calcareous plankton for precipitation to the deep sea floor. Either of the above scenarios will have the effect of increasing deepwater carbonate accumulation, but the first will sharply increase rates on the rise whereas the second will increase overall pelagic rates. The ultimate control of shelf carbonate accumulation, then, is thermal subsidence of the continental margin (Sleep, 1971; Berger and Winterer, 1974; Hay and Southam, 1977; Pitman, 1978).

In light of the above, deepsea carbonate accumulation can be thought of as that part of the riverine input not consumed by the shelf-rise system. Following the principle of mass balance, total carbon-

*A third possibility is that the total quantity of dissolved carbonate in the ocean reservoir simply increases. This seems implausible in view of the fact that the surface waters of the ocean are supersaturated with respect to carbonate (Berner, 1974). Any further increase in dissolved carbonate would surely lead to increased precipitation more or less instantaneously.

ate eroded from land should equal total shelf, rise and ocean sedimentation.

This simple model of oceanic sedimentation is complicated by the effects of pre- and post-burial recycling of carbonate in the deep sea. Approximately 80 percent of the carbonate fixed in the surface waters by organisms each year redissolves before burial (Broecker, 1974). Moore and Heath (1977), applying the concept of differential recycling suggested by Garrels and MacKenzie (1971), have suggested that at least half of the carbonate sediment buried on the ocean floor is subsequently redissolved over the following 20 m.y. This model implies that less than 10 percent of the carbonate removed from the oceans annually by biochemical precipitation in the surface waters survives to become part of the "permanent" geologic record. Material that redissolves before burial can be thought of as instantaneously recycled and can be ignored in our discussion. Removal of material after burial must be taken into account, however. If material is removed from one location it must sooner or later be deposited somewhere else in order to maintain balance. This additional deposition introduces a two-way bias into the sedimentary record because the recycled material is mixed with younger sediments thereby reducing the apparent accumulation rate of older material and increasing the sedimentation rate of younger material. If redeposition is penecontemporaneous with initial deposition (*i.e.,* within 10^6 y.), it may be considered geologically instantaneous and will have no effect on the overall long-term picture. However, the greater the time lag between burial and redeposition, the more it will influence any quantitative assessment of the oceanic sedimentary record. In short, total carbonate deposition equals total carbonate eroded from land plus carbonate recycled from the ocean floor. As deep ocean waters are undersaturated with respect to carbonate (Berner, 1974), it remains unclear to what extent variations in dissolution are a simple consequence of factors controlling the rate of dissolution of material on the seafloor, *e.g.,* deep ocean circulation rates, and to what extent they may be a more subtle response to "demands" in excess of riverine input by the shelf-rise system.

Two separate sets of data are available for testing global carbonate budgets: 1) measurements of current carbonate erosion rates calculated from sampling the world's rivers, and 2) calculations based on measurements of marine carbonate accumulation rates through time. The best available estimate of global river input to the oceans is that of Garrels and MacKenzie (1971), who calculate it to be 12.2 $\times 10^{14}$ g/y. Current estimates of long-term global carbonate accumulation are given by Hay and Southam (1977) as 17.8 $\times 10^{14}$ g/y, based on Milliman (1974) and 18.0 $\times 10^{14}$ g/y., based on Broecker (1974). As Hay and Southam (1977) point out, the estimated long-term carbonate accumulation rates exceed by 50 percent the calculated riverine input rates. They suggest that either (1) there is a source which has not been taken into account, or (2) that the long term carbonate flux to the deep sea has been overestimated because calculations were based upon high sedimentation rate, Pleistocene piston cores. However, our long-term carbonate accumulation rates based on DSDP cores and on Moore and Heath's (1977) estimate that in 20 m.y. approximately half of the carbonate deposited on the deep ocean floor is subsequently recycled, reconcile this discrepancy. We will show below that long-term net removal of carbonate from the oceans is approximately equal to riverine input.

OCEANIC CARBONATE SEDIMENTATION THROUGH TIME

Examination of the variations in the supply of material and partitioning between the various parts of the system through time require a more detailed data base. As a first step in this direction, we have calculated carbonate accumulation for each of the major ocean basins as a function of time over the past 60 m.y.

Data from DSDP Sites 1 through 335, with the exception of the Mediterranean, were selected for study. The general scheme for processing the data have been discussed in some detail in Worsley and Davies (1979b). By combining the lithologic and paleontologic data for each site we can calculate rates of accumulation of material at each location for successive increments of time. In this study we used a standard time increment of 3 million years. Combining data from all sites which penetrated a given time increment, it is possible to calculate oceanwide average accumulation rates. In calculating such averages we have excluded sites located on the shelf-influenced continental rises or archipelagic aprons in the hope of obtaining a better estimate of oceanic (pelagic) sedimentation rates. The results are shown in Figure 3.

Most studies covering large regions of oceanic sedimentation (*e.g.,* van Andel and others, 1975; Davies and others, 1977; Worsley and Davies, 1979a) show high oceanic sediment accumulation rates for the past few million years that steadily decrease back through time. Upon this unmistakeable trend are superimposed distinct fluctuations. As discussed earlier, Moore and Heath (1977) suggested that the apparent decrease in accumulation rate in older materials is a consequence of post-depositional destruction of material, and they developed a simple section survival model in which the most recently deposited sediments have suffered the least amount of destruction, and the oldest the most. Using their model, it is possible to estimate that fraction of the material originally deposited

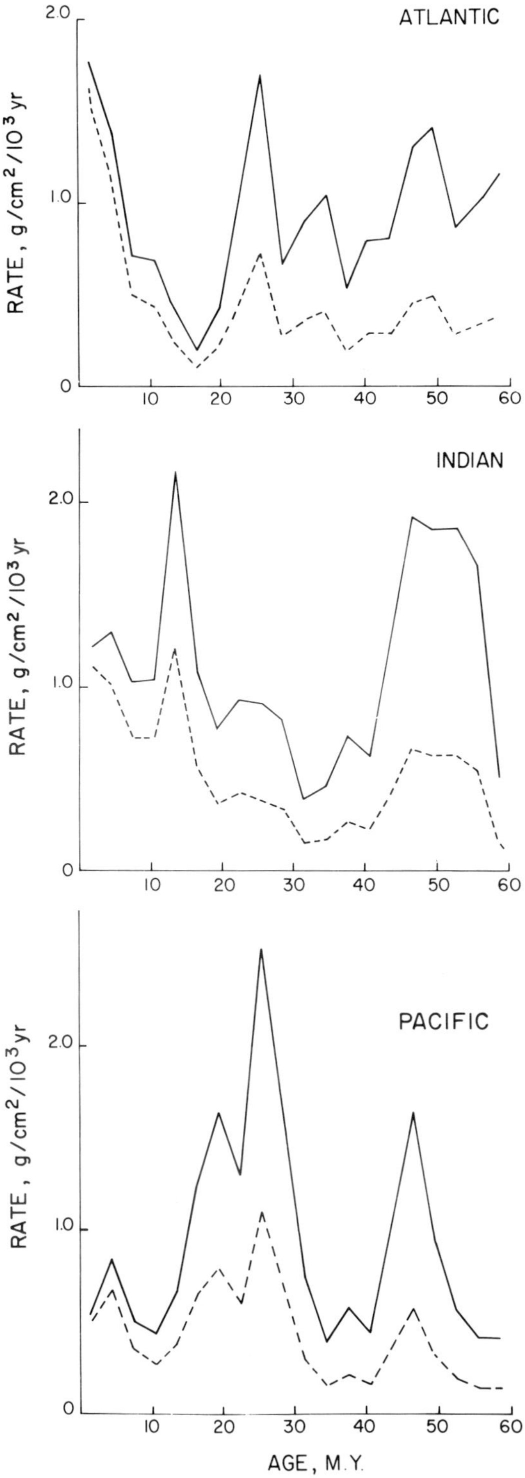

FIG. 3.—Average carbonate accumulation rates, uncorrected (broken line) and corrected (solid line) according to the Moore and Heath (1977) section survival model.

which now survives in the record, and thus to correct the observed accumulation rates to give what we believe to be the best currently available estimates of rates of deposition of shallow-water (less than 3 km) pelagic carbonates through time.

Ultimately, the best estimates of total oceanic carbonate sedimentation will come from putting all the available accumulation-rate data on paleogeographic maps for successive time increments and contouring the data. The contoured maps can then be used to estimate areas over which different accumulation rates were operative and thus estimate total masses of sediment. Such an approach will, however, require a more extensive geographical coverage of DSDP sites than is presently available. In the meantime, if we assume that the average calculated deposition rates shown in Figure 3 are applicable oceanwide, we can make a preliminary assessment of oceanic carbonate deposition through time.

Some further simplifying assumptions are necessary. First, we know that the oceans have changed in size and shape through time (Fig. 4). To allow for this we will assume that the increasing area of the Atlantic has been compensated by a corresponding decrease in the area of the Pacific. Similarly, the growth of the Indian Ocean has been accommodated by closing of the Tethys Sea. Secondly, we have assumed that the present cycle of seafloor spreading began 150 m.y.BP and has proceeded at a constant rate ever since. This means that over the past 50 m.y. all the change in the area of the individual ocean basins has occurred in water depths greater than 3 km. The dimensions used in this calculation are shown in Table 1.

As noted earlier, if we compare the distribution of drill sites with the distribution of ocean floor with respect to water depth (Figure 1B), we find that water depths less than 3 km are over represented in the drilling data by a factor of 2. This means that the average accumulation rates calculated earlier are biased towards shallower water depths, especially since most locations have subsided with time. This is probably not significant for the non-carbonate portion of the sediments but is critical for the carbonate material because carbonate dissolves at increasing water depths. To allow for this we have assumed that the corrected average deposition rates calculated from DSDP results are a good estimate of the accumulation rates for water depths of less than 3 km and that the rate for depths greater than 3 km is one third the rate calculated. The basis for this assumption is illustrated in Figure 5. The net accumulation rate is assumed to decrease linearly from a value of R $g/cm^2/10^3$ y. at 3 km depth to zero at 5 km depth. Taking into account areas of present day ocean floor between different elevations, this gives a net accumulation of 85.5 R $g/cm^2/10^3$ y. for depths greater than 3 km. The

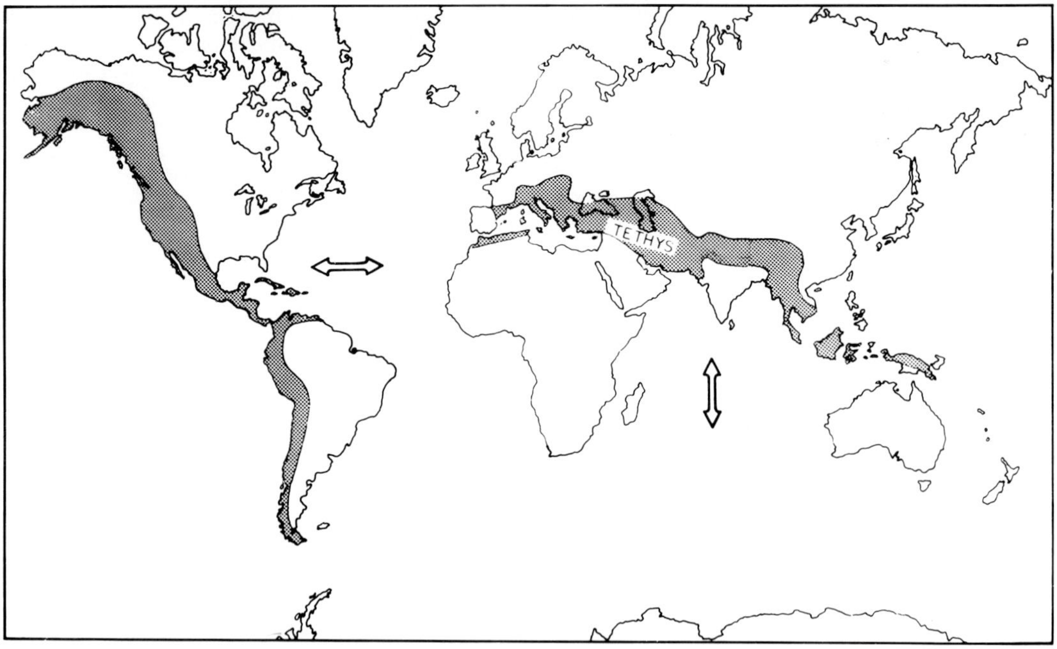

FIG. 4.—Elements of the simplified Cenozoic plate tectonic model. East-west expansion of the Atlantic is accommodated by closing of the Pacific. Opening of the Indian Ocean is balanced by closing of the Tethys.

total area of the ocean floor at depths greater than 3 km is 272.2×10^6 km^2 (Menard and Smith, 1966), yielding an average accumulation rate of 0.32 R g/cm^2/10^3 y.

Further assumptions necessary in order to estimate oceanic carbonate budgets are that 1) the addition of hydrothermal Ca^{++} ions from mid-ocean ridges is negligible, and 2) marine productivity changes, while clearly influencing the distribution of carbonate sediment upon the ocean floor, have no influence upon total global carbonate fluxes.

Having made all of the above assumptions, we can now estimate the total mass of carbonate sediment deposited on the ocean floor for successive 3 m.y. time increments over the past 60 m.y. by multiplying appropriate rates by areas. A sample calculation, for the Atlantic Ocean for 27–30 m.y.BP, follows. The present oceanic area (Table 1) is 85.49 $\times 10^6$ km^2. If seafloor spreading has proceeded at a constant rate, beginning at 150 m.y.BP, the oceanic area at 28.5 m.y.BP would be given by

$$85.49 \times \frac{121.5}{150} \times 10^6 = 69.25 \times 10^6 \text{ km}^2.$$

If all the change in area has taken place in water depths greater than 3 km, the oceanic area at 28.5 m.y.BP would have been distributed thus:

Area of Ocean Floor (10^6km^2)	Depth (km)	Accumulation Rate (gm/cm^2/10^3y)	Total Accumulation (gm/10^3y)
	3	R	
75.8			56.85 R
	4	R/2	
114.7			28.67R
	5	0	
81.7			0
Total 272.2	>3	0.32R	85.5R

FIG. 5.—Rationale for estimating deep ocean (>3 km) carbonate accumulation rates to be one third shallow (<3 km) rates (see discussion in text).

TABLE 1.—PRESENT OCEAN AREAS IN MILLIONS OF SQUARE KILOMETERS.*

Ocean	Shelf	Ocean	Ocean <3km	Ocean >3km
Atlantic	8.56	85.49	18.80	66.69
Indian	3.19	70.72	11.88	58.84
Pacific	10.61	170.22	25.31	144.91
Arctic	6.04	6.18	4.33	1.84
Total	28.40	332.61		

*Based upon data from Menard and Smith (1966) and Gross (1977)

TABLE 2.—TOTAL CARBONATE DEPOSITION IN UNITS OF 10^{14} G/Y.

Time Interval (my)	Atlantic	Indian	Pacific	Total
0–3	6.81	3.77	4.02	14.60
3–6	6.73	4.01	6.20	15.94
6–9	2.82	3.17	3.74	9.74
9–12	2.69	3.52	3.33	9.55
12–15	1.58	6.38	5.07	13.03
15–18	0.74	3.11	9.54	13.40
18–21	1.57	2.19	12.59	16.35
21–24	3.62	2.60	10.15	16.37
24–27	6.22	2.51	19.85	28.58
27–30	2.36	2.18	13.00	17.50
30–33	3.15	1.04	5.82	10.01
33–36	3.59	1.19	3.14	7.92
36–39	1.76	1.88	4.60	8.23
39–42	2.64	1.55	3.61	7.80
42–45	2.65	2.85	8.35	13.85
45–48	4.19	4.65	13.52	22.36
48–51	4.47	4.39	7.78	16.64
51–54	2.63	4.33	4.67	11.63
54–57	3.02	3.75	3.47	10.24
57–60	3.43	1.14	3.52	8.09

$$<3 \text{ km depth} = 18.80 \times 10^6 \text{ km}^2$$

$$>3 \text{ km depth} = 50.45 \times 10^6 \text{ km}^2$$

Average accumulation rate for the Atlantic for 27–30 m.y.BP, based on DSDP data, is 0.277 g/cm²/10³ y. The correction factor (from Moore and Heath, 1978) is 2.39, hence the actual average accumulation rate is 0.662 g/cm²/10³ y. This is taken to be the best estimate for water depths less than 3 km. For water depths greater than 3 km we use one third of this value, *i.e.*, 0.221 g/cm²/10³ y. (see above). Hence the total carbonate accumulation in the Atlantic Ocean for the period 27–30 m.y.BP is given by

$$\frac{(18.80 \times 10^{16} \times 0.662) + (50.45 \times 10^{16} \times 0.221)}{10^3} \text{ g/y.}$$

This simplfies to 2.36×10^{14} g/y.

The results of these calculations for all the ocean basins except the Arctic, where we have assumed no carbonate accumulation, are shown in Table 2 and Figure 6. Clearly, total carbonate deposition in each of the major ocean basins, and overall, has been far from constant. The variations presumably reflect changes in sediment supply and shelf sedimentation, as well as processes operating within the oceans.

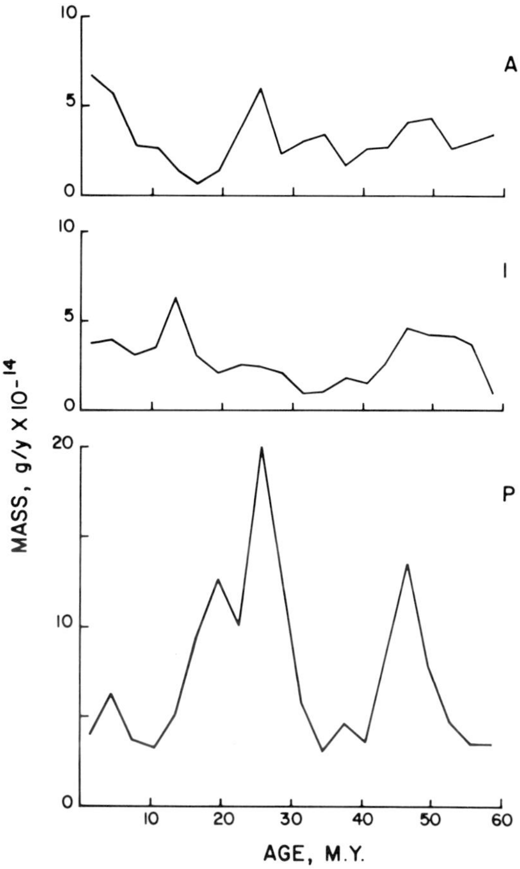

FIG. 6.—Pelagic carbonate deposition in each of the major oceans through time. A: Atlantic; I: Indian; P: Pacific.

DISCUSSION

Figure 6 suggests that oceanic carbonate deposition has oscillated between periods of high deposition (0–6, 22–30, 45–53 m.y.BP) and periods of low deposition. These large-scale fluctuations in total carbonate accumulation will be discussed below.

Superimposed upon the major variations are more subtle inter-ocean effects, principally partitioning between the Pacific and the other oceans. Prior to 30 m.y.BP all the oceans behaved in concert. The dramatic increase in carbonate deposition at 25–30 m.y.BP, almost double that of any other interval, is predominantly in the Pacific, while at 12–15 m.y.BP, although there is little change in total carbonate deposition, there is a marked increase in the Atlantic and Indian oceans corresponding to a decrease in the Pacific (Fig. 7).

The 12–15 m.y. interval corresponds to closing of the Tethys and consequent closure of low-latitude communication between the ocean basins. Incursion of cold water from the Norwegian Sea into the Atlantic (Thiede, 1979) and extensive glaciation of Antarctica (Kennett, 1977) also occurred at this time. All of these events lead to the establishment of the present circulation regime in the oceans. Berger (1970) has shown that the present circulation

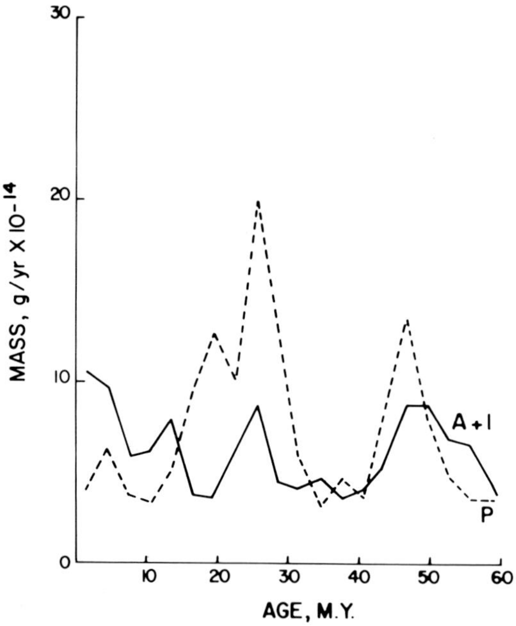

FIG. 7.—Pacific carbonate deposition through time compared to Atlantic plus Indian Oceans.

regime leads to a situation in which the Atlantic tends to collect carbonate sediments at the expense of the Pacific. Prior to 15 m.y.BP approximately two thirds of the total carbonate accumulated in the Pacific. This was a reflection of the dominantly latitudinal circulation of the time, coupled with the presence of many relatively shallow oceanic rises to provide suitable sites for carbonate accumulation. (A comparison of the hypsometry of the ocean basins (Menard and Smith, 1966) suggests that, in the absence of circulation effects, the proportion of carbonate accumulating in the Pacific would be highly sensitive to fluctuations in sea level. During periods of low sea level, such as the Late Oligocene-Early Miocene, the Pacific would gain at the expense of the other oceans.)

Looking at the overall picture, we calculate from Table 2 that the long-term average total oceanic carbonate deposition rate is 13.0×10^{14} g/y. This is comparable to the figure of 11.8×10^{14} g/y. (Milliman, 1974) derived from studies of piston cores, and when combined with current estimates of the long-term shelf slope-rise and epicontinental sea deposition rate (Hay and Southam, 1977), gives a global total carbonate deposition rate of 19.0×10^{14} g/y. (In estimating shelf and slope-rise rates Hay and Southam made use of Milliman's estimates but excluded the Red and Black seas and the Sunda Shelf.) If we accept Moore and Heath's (1977) contention that only between one half and one third of the carbonate deposited on the deep seafloor survives to be incorporated into the permanent geologic record, then the long-term rate of removal of carbonate from the oceans to the deepsea floor, based upon DSDP cores, is $4.3 - 6.5 \times 10^{14}$ g/y. Added to estimates of long-term shelf and slope-rise rates this yields a figure of $10.3 - 12.5 \times 10^{14}$ g/y., comparable to the estimated riverine input of 12.2×10^{14} g/y. (Garrels and MacKenzie, 1971).

Excursions of the system away from the long-term averages, following the model proposed earlier, must principally reflect changes in erosion from land and changes in shelf sedimentation. As a first approximation, following Garrels and MacKenzie (1971), we may assume that global chemical erosion rates are proportional to continental area and are relatively independent of climate and elevation. Thus changes in carbonate supply (and therefore deposition rate) are directly related to changes in land area, with the shelf region being either a source or a sink depending upon global sea level.

Both land area and shelf areas may be estimated from sea-level changes, although the assumption that eustatic sea-level changes can be directly related to transgressive-regressive cycles is a gross oversimplification (Pitman, 1978). This oversimplification arises because the area of transgression or regression is a function of isostatic adjustment, sediment supply, and thermal subsidence of the ocean margin as well as the interaction between hypsography and changing sea level. Because of these effects, the widely publicized sea-level curve of Vail and others (1977), although globally applicable, is now recognized as indicative of changes in coastal onlap rather than of eustatic sea level (Vail and Hardenbol, 1979). Nevertheless, since topographic gradients near the margins of the continental blocks are near linear, the curve of Vail and others provides a useful tool for estimating changes in the proportion of submerged shelf area through time, which is our main interest. If the simple model proposed earlier is correct, increases in pelagic carbonate sedimentation should be positively correlated with increases in land area (increased supply) and decreases in shelf area. This relationship is implied in Figure 8. It is important to realize that the modulating effect of the continental shelves on oceanic sedimentation includes an element of positive feedback, so that shelf sedimentation operates to enhance the effects of changing supply resulting from changes in land area.

In Figure 9 we show a simple scatter plot of pelagic carbonate accumulation against land area and shelf area. The correlation is not as clear cut as would be expected from our extremely simple model, but the general trend is unmistakable. That the fit is as good as it is suggests that the model is a good first approximation. Deviations from the predicted values must reflect the effects of other fac-

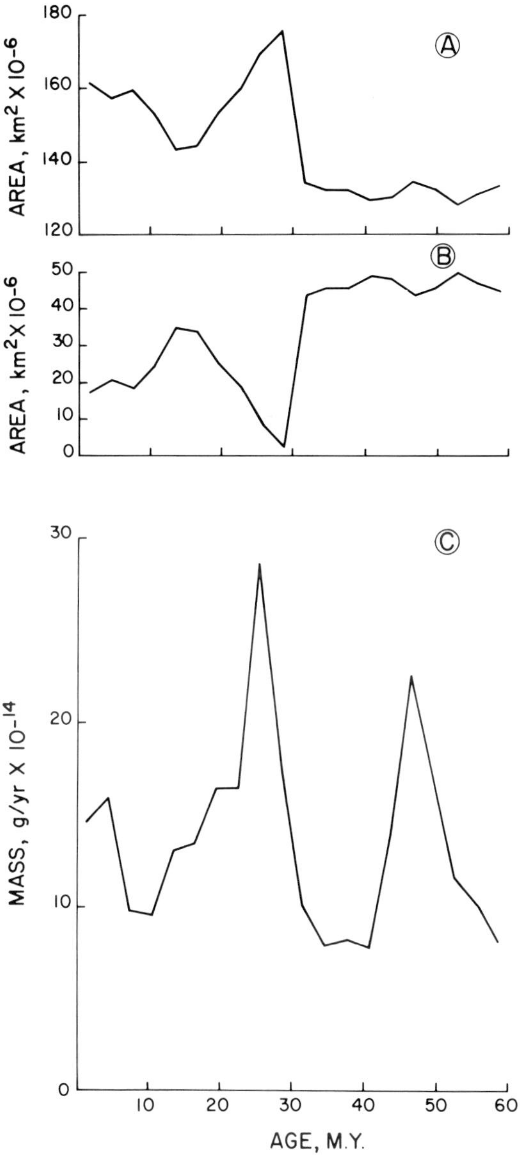

FIG. 8.—Total carbonate accumulation compared to changes in shelf area and land area estimated from the sea level curve of Vail and others (1977). A: Continental area; B: continental shelf area; C: total pelagic carbonate accumulation.

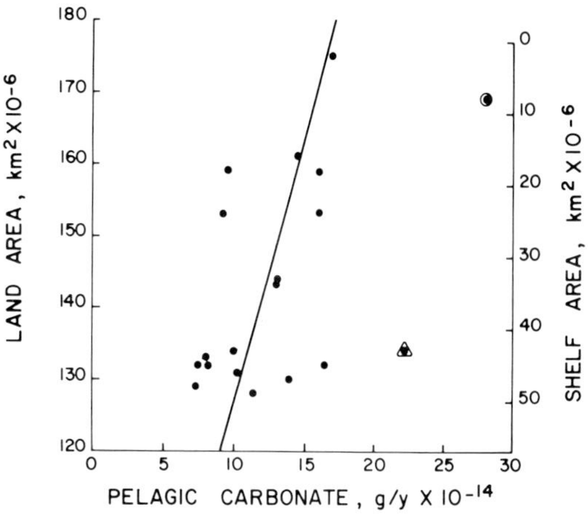

FIG. 9.—Total pelagic carbonate plotted against land area and shelf area. The solid line is a best fit regression line representing the equation: Carbonate = 0.137 × land area − 7.45. The two intervals of abnormally high accumulation, 24–27 m.y.BP (circled) and 45–48 m.y.BP (triangle), were excluded from the calculation.

tors that influence the supply of carbonate to the deep sea: latitudinal hypsographic variations, the presence of isolated carbonate sinks, climate (rates of chemical weathering), regional geologic differences, changing areas of internal drainage, etc. From Figure 10, it is clear that there were three times in particular during the past 60 m.y. when oceanic carbonate sedimentation has departed significantly from the values predicted by the model: 6–12 m.y.BP (less than expected), 24–27 m.y.BP, and 42–51 m.y.BP (both cases much more than expected). In each case, the environment was markedly different from the present. At 6–12 m.y.BP: extensive glaciation of Antarctica, intense deep ocean circulation, isolation of the Mediterranean and of other marginal seas (*e.g.*, Caribbean). At 24–27 m.y.BP: extreme low stand of sea level, exposure of carbonate-covered shelves in low-latitude areas. At 42–51 m.y.BP: high sea levels, warm equable climate, low-latitude connections between the major oceans. To our thinking, these three periods represent sharp excursions from what otherwise appears to be a well-buffered system in quasi-equilibrium. It is possible that our having broken geologic time into uniform 3 m.y. increments might have biased the data from the natural flow of shorter-period oscillations and that refinement of our (or another) model will help reconcile the discrepancies. At this time we prefer not to speculate as to the significance of these excursions other than to suggest that they clearly offer the potential for much fruitful further study.

The oscillations in pelagic carbonate deposition suggested by Figure 8 can be correlated with secular variations documented for a whole range of phenomena by Fischer and Arthur (1977). These authors also attribute their observations to the effects of changes in relative sea level but consider consequent changes in ocean circulation and climate to be primary factors. It seems ever more clear that sea level, climate, ocean circulation, sedimentation and

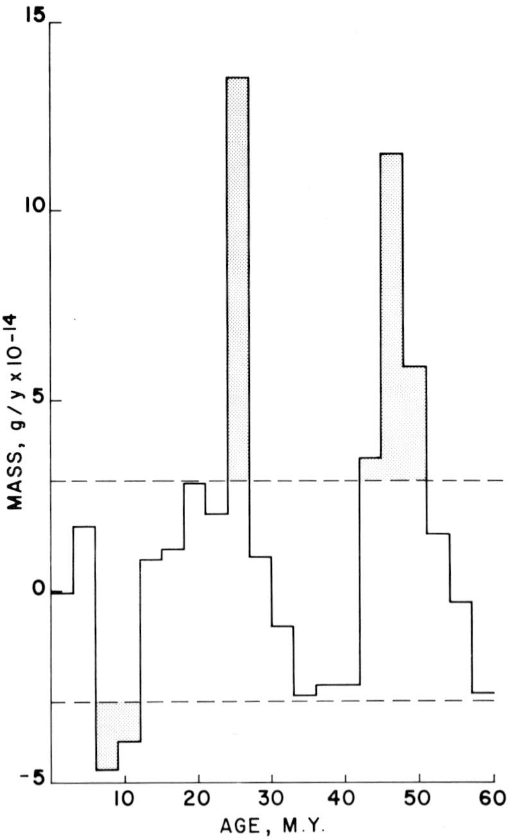

FIG. 10.—Differences between observed values of pelagic carbonate accumulation and values predicted from the regression line in Figure 9. The broken lines represent one standard deviation either side of the predicted value.

the biological realm are linked in dynamic balance, and that fluctuations in sea level provide the primary driving mechanism. Because of the apparently simply relationship between land area and carbonate supply, the deep sea carbonate record provides a delicate indicator of sea level changes. The obvious next step is twofold: (1) to examine the sedimentary history of the ocean margins and the Tethys for evidence to add further details to our steadily developing picture of the Earth's evolving environment, and (2) to extend that picture backward in time to the Mesozoic.

ACKNOWLEDGEMENTS

W. W. Hay, through various provocative discussions, provided much of the initial stimulus for this work, while T. C. Moore and W. H. Berger, by their comments and suggestions, significantly influenced the development of our ideas.

P. B. Woodbury and R. G. Blank were instrumental in helping us to develop the methodology. Programs to process the lithologic data and generate accumulation rates for each site were developed for general use by P. B. Woodbury at DSDP. C. Suchland wrote the programs to process the paleontologic data. R. G. Blank oversaw the building of the paleontologic data base. B. Baldwin, W. E. Glassley and J. B. Moody reviewed the manuscript. Their thoughtful comments greatly improved it. Linda Keble patiently prepared the illustrations.

This study was supported by grants OCE-75-04066, OCE-76-81934, and OCE-76-82514 from the Oceanography Section of the National Science Foundation. The Deep Sea Drilling Project is supported by the National Science Foundation through contract C-482 with the University of California.

REFERENCES CITED

NOTE: All references to Deep Sea Drilling Project Initial Reports have been abbreviated; for complete citations see *Appendix I* at the end of this volume, or the individual Initial Reports as published.

BERGER, W. H., 1970, Biogenous deep-sea sediments, Fractionation by deep-sea circulation: Geol. Soc. America Bull., v. 81, p. 1385–1402.

———, AND WINTERER, E. L., 1974, Plate stratigraphy and the fluctuating carbonate line, *in* Hsü, K. J., and Jenkyns, H. C., eds., Pelagic sediments: on land and under the sea: Spec. Pub. 1, Int. Assoc. Sedimentol., p. 11–48.

BERNER, R. A., 1974, Physical chemistry of carbonates in the oceans, *in* Hay, W. W., ed., Studies in paleo-oceanography: Soc. Econ. Paleontologists Mineralogists Spec. Pub. 20, p. 37–43.

BROECKER, W. S., 1974, Chemical oceanography: New York, Harcourt Brace Jovanovich, Inc., 214 p.

DAVIES, T. A., LUYENDYK, B. P., KIDD, R. B., AND WESER, O. E., 1975, Unconformities in the sediments of the Indian Ocean: Nature, v. 253, p. 15–19.

———, HAY, W. W., SOUTHAM, J. R., AND WORSLEY, T. R., 1977, Estimates of Cenozoic oceanic sedimentation rates: Science, v. 197, p. 53–55.

FISCHER, A. G., AND ARTHUR, M. A., 1977, Secular variations in the pelagic realm, *in* Cook, H. E., and Enos, P., eds., Deep water carbonate environments: Soc. Econ. Paleontologists Mineralogists Spec. Pub. 25, p. 19–50.

GARRELS, R. M., AND MACKENZIE, F. T., 1971, Evolution of sedimentary rocks: New York, W. W. Norton & Co., 397 p.

GROSS, M. G., 1977, Oceanography—a view of the earth: Englewood Cliffs, Prentice Hall, 497 p.

HAY, W. W., AND SOUTHAM, J. R., 1977, Modulation of marine sedimentation by the continental shelves, *in* Anderson, N. R., and Malahoff, A., eds., The fate of fossil fuel CO_2 in the oceans: New York, Plenum Publishing Corp., p. 569–604.

Kennett, J. P., 1977, Cenozoic evolution of Antarctic glaciation, the circum-Antarctic Ocean, and their impact on global paleoceanography: Jour. Geophys. Res., v. 82, p. 3843–3860.

———, Burns, R. E., Andrews, J. E., Churkin, M., Davies, T. A., Dumitrica, P., Edwards, A. R., Galehouse, J. S., Packham, G. H., and Van der Lingen, G. J., 1972, Australian-Antarctic continental drift, Paleo-circulation changes and Oligocene deep-sea erosion: Nature, v. 239, p. 51–55.

Kidd, R. B., and Davies, T. A., 1978, Indian Ocean sediment distribution since the late Jurassic: Marine Geology, v. 26, p. 49–70.

McCoy, F. W., and Zimmerman, H. B., 1977, A history of sediment lithofacies in the South Atlantic Ocean: Init. Repts. DSDP, v. 39, p. 1047–1080.

Menard, H. W., and Smith, S. M., 1966, Hypsometry of ocean basin provinces: Jour. Geophys. Res., v. 71, p. 4305–4325.

Milliman, J. D., 1974, Marine carbonates, *in* Milliman, J. D., Mueller, G., and Foerstuer, U., eds., Recent sedimentary carbonates: New York, Springer-Verlag, 375 p.

Moore, T. C., 1972, DSDP: successes, failures, proposals: Geotimes, v. 17, p. 27–31.

———, and Heath, G. R., 1977, Survival of deep-sea sedimentary sections: Earth Planet. Sci. Letters, v. 37, p. 71–80.

———, van Andel, Tj. H., Sancetta, C., and Pisias, N., 1978, Cenozoic hiatuses in pelagic sediments: Micropaleontology, v. 24, p. 113–138.

Pitman, W. C., 1978, Relationship between eustacy and stratigraphic sequences of passive margins: Geol. Soc. America Bull., v. 89, p. 1389–1403.

Shackleton, N. J., and Kennett, J. P., 1974, Paleotemperature history of the Cenozoic and the initiation of Antarctic glaciations; oxygen and carbon isotope analyses at DSDP sites 77, 279, and 281: Init. Repts. DSDP, v. 29, p. 743–755.

Sleep, N. H., 1971, Thermal effects of the formation of Atlantic continental margins by continental breakup: Geophys. Jour. Royal Astronom. Soc., v. 24, p. 325–350.

Thiede, Jörn, 1979, History of the North Atlantic Ocean: evolution of an asymmetric zonal paleo-environment in a latitudinal basin, *in* Talwani, Manik, Hay, W. W., and Ryan, W. B. F., eds., Deep drilling results in the Atlantic Ocean: continental margins and paleoenvironment: Washington, D.C., Amer. Geophys. Union, p. 275–296.

Vail, P. R., and Hardenbol, Jan, 1979, Sea-level changes during the Tertiary: Oceanus, v. 22, p. 71–79.

———, Mitchum, R. M., Todd, R. G., Widmier, J. M., Thompson, S., Sangree, J. B., Bubb, J. N., and Hatlelid, W. G., 1977, Seismic stratigraphy and global changes of sea level, *in* C. E. Payton, ed., Seismic stratigraphy—applications to hydrocarbon exploration: Amer. Assoc. Petroleum Geologists Mem. 26, p. 49–212.

van Andel, Tj. H., Heath, G. R., and Moore, T. C., 1975, Cenozoic history and paleoceanography of the central equatorial Pacific Ocean: Geol. Soc. America Mem. 143, 134 p.

———, Thiede, Jörn, Sclater, J. G., and Hay, W. W., 1977, Depositional history of the South Atlantic Ocean during the last 125 million years: Jour. Geology, v. 85, p. 651–698.

Whitman, J. M., and Davies, T. A., 1979, Cenozoic oceanic sedimentation rates: how good are the data?: Marine Geology, v. 30, p. 269–284.

Worsley, T. R., and Davies, T. A., 1979a, Sea level fluctuations and deep-sea sedimentation rates: Science, v. 203, p. 455–456.

———, and ———, 1979b, Cenozoic sedimentation in the Pacific Ocean: steps toward a quantitative evaluation: Jour. Sed. Petrology, v. 49, p. 1131–1146.

SEPM Special Publication No. 32, p. 181–207, October 1981

DIAGENESIS OF OCEANIC CARBONATE SEDIMENTS: A REVIEW OF THE DSDP PERSPECTIVE

ROBERT E. GARRISON
Earth Sciences Board, University of California, Santa Cruz, California 95064 U.S.A.

ABSTRACT

DSDP cores at selected sites provide a continuous carbonate sediment record characterized by comparative compositional simplicity, vertical homogeneity, and uniformity of diagenetic environment. These circumstances have allowed detailed reconstruction of diagenetic events, most notably those during burial. Burial diagenesis converts carbonate oozes to chalks which in turn may become transformed into dense limestones. The most important variables affecting these first-order changes are burial depth and pore water chemistry, but burial lithification is not related to burial depth in a simple way. Instead, differences in original sediment composition appear to predetermine the depths at which the transformations occur. Such differences in composition, caused by major and minor ocean events, may also be important in developing acoustic reflectors in the sediment pile. Some cementation occurs early during diagenesis; but this is comparatively rare in DSDP cores and apparently took place in chemically reactive sediments, such as those with abundant braarudosphaerid pentaliths. Carbonate sediments have a "diagenetic potential" which results from their preburial history and which in part determines subsequent diagenetic events during burial. Gravitational compaction is significant in the upper 50 to 200 meters of sediment; the dominant processes below these depths are pressure solution and reprecipitation of calcite cement, both of which are highly taxa- and site- specific. This lithification proceeds without external sources of calcium carbonate, thus cementation is calcite conservative. Sr^{++} and ^{18}O in the carbonate tend to decrease with increasing burial depth, a consequence of solution-reprecipitation. But variations in the composition of the original sediment can cause deviations from these trends.

Lithification of carbonate sediments associated with oceanic basalts appears to be a consequence of the submarine weathering of basalt. Shallow-water carbonate sediments have been infrequently encountered in the DSDP; but recognition of diagenetic features in them has allowed reconstruction of important geologic events such as intervals of subaerial exposure and of extremely rapid subsidence. Dolomite and other authigenic carbonate minerals in oceanic carbonate sediments apparently have formed in quite diverse ways. Some pelagic oozes were dolomitized by saline fluids emanating from underlying evaporites, some by Mg-rich pore waters derived from subaqueously weathered basalt; other dolomites formed as a by-product of the diagenesis of organic matter in sediments. The origin of other dolomite as well as siderite, ankerite, and other authigenic minerals in pelagic carbonates has not been systematically investigated.

INTRODUCTION

The special insights on carbonate diagenesis resulting from the Deep Sea Drilling Project derive from several circumstances. Foremost among these is the comparative simplicity of composition and diagenetic environments of pelagic carbonate sediments in the ocean basins. Most are rather homogenous nannofossil-foraminiferal oozes composed of low magnesium calcite. Most have not been touched by subaerial diagenesis and meteoric waters. They have not been exposed during sea level lowerings or tectonic uplifts, nor have they been subjected to tectonic deformation that would alter their petrology and physical properties. This means we must contend with fewer variables and less noise in reconstructing their post-depositional history.

Further advantages accrue from the method of sampling. Continuous coring of pure, thick pelagic carbonate sections, particularly on submerged plateaus, provides a uniform and continuous record that cannot be duplicated in on-land sequences which are scarred by tectonic-eustatic events. Moreover, the oceanic sections can be dated precisely by pelagic microfossils, allowing us to calculate rates of processes and to backtrack to paleooceanographic and paleogeothermal conditions which previously affected the sediment. These thick oceanic sections also commonly preserve transitional diagenetic phases that are lost from the on-land record.

The comparative simplicity and uniformity of composition and diagenetic environments has allowed attention to focus on many subtle yet quite important attributes which affect carbonate diagenesis. An example is the recognition that seemingly slight variations in grain type and grain size in the original carbonate sediment predetermine the subsequent course of diagenesis. Thus a carbonate sediment has a "diagenetic potential" as the consequence of "conditioning" events that affected it before it began the burial progression (Schlanger and Douglas, 1974; Berger and Mayer, 1978).

Finally, the study of carbonate diagenesis in DSDP cores has fostered unusually close co-operative work among sedimentologists, paleontologists, geochemists, and geophysicists. The scale and breadth of this eclectic approach is unprecedented in the study of sedimentary materials. It may serve as a model for what can be achieved, an accomplishment which could be one of the enduring contributions of the Deep Sea Drilling Project.

As in any complicated technical enterprise, however, a number of uncertainties and limitations exist, and some of the more important of these should be noted. Foremost among the limitations is the fact that DSDP coring frequently alters the physical properties of the sediments and rocks recovered. Thus it has proved difficult in some instances to separate the effects of drilling disturbances from diagenetic changes, particularly in investigations of progressive lithification of carbonate oozes during burial diagenesis. In addition, lack of continuous coring at most sites has meant that interpretations of diagenetic changes with burial depth have frequently involved extrapolations through uncored intervals. Finally, over the course of DSDP investigations, inconsistencies in terminology and variability in the methods and intensity of laboratory investigations have invariably occurred, sometimes making it difficult to achieve a satisfactory comparative analysis.

Regarding carbonate diagenesis, we have thus far learned most about the processes and products of burial diagenesis. Therefore the review which follows focuses chiefly on this and allied topics such as acoustistratigraphy. Other topics, such as alteration of shallow-water carbonate sediments and carbonate diagenesis in non-carbonate sequences, receive correspondingly less attention; they are included chiefly so that they may direct attention to the possibilities for future work.

EARLY DIAGENESIS OF PELAGIC CARBONATE SEDIMENTS

Relatively little evidence for very early diagenesis in carbonate sediments has emerged from DSDP cores. Perhaps this is because sea floor cementation and hardground genesis are most prominent at atypical oceanic locations (*e.g.*, current-swept submarine highs), whereas DSDP drilling sites are predominately in deep-water basinal regions characterized by more or less continuous sedimentation.

Schneidermann (1973), however, has described crusts of limestone, partly silicified marls and limestone breccias at sites 146 and 153 in the central Caribbean which suggest sea-floor cementation at the Cretaceous-Tertiary boundary. These rocks have anomalously heavy oxygen isotope compositions ($\delta^{18}O_{PDB} = -0.9$ to -0.6) which are consistent with cementation at or near the sea floor. Similarly, anomalously heavy oxygen isotopes in Jurassic limestones above basalt at site 367 in the North Atlantic also suggest cementation at low temperatures and shallow burial depths (Brenneke, 1977).

The best documented example of early diagenesis is the lithification of the Oligocene *Braarudosphaera* chalk in the South Atlantic (Wise and Hsü, 1971; Wise and Kelts, 1972; Wise, 1977). This chalk, which averages 40% porosity, occurs as thin layers within soft carbonate ooze at burial depths of 13 to 133 meters. Observations with the scanning electron microscope (SEM) indicate the chalk is composed primarily of braarudosphaerid pentalith segments that have been overgrown by secondary low magnesium calcite cement; opal-CT lepispheres form an additional, minor cementing element. The oxygen isotope composition confirms that the cementation occurred near the sea floor at ambient bottom water temperatures of about 3° to 6°C (Lloyd and Hsü, 1972). Textural studies by Wise and Kelts (1972) suggest this sediment experienced simultaneous dissolution and reprecipitation; small coccoliths dissolved and reprecipitated calcite cement crystals nucleated preferentially on the relatively large braarudosphaerid pentalith segments which acted as "seed crystals."

Cementation was concentrated in these braarudosphaerid-rich layers because of the abundance of the seed crystals. Burns (1972) has found similar overgrowth on discoasters in near-surface Pacific sediments; but these may have been reworked and redeposited, and the question remains whether overgrowths can form at the sea floor (Roth and Berger, 1975). Nonetheless, it is evident that the primary composition of pelagic carbonate sediments influences their diagenesis (cf. Schlanger and Douglas, 1974), a matter pursued in the following section.

BURIAL DIAGENESIS OF PELAGIC CARBONATE OOZES

Carbonate petrologists for the past two decades have speculated on how fine-grained carbonate sediments become lithified (see review in Van der Lingen, 1977, p. 16–18). The central questions have been: (1) What is the major source of cement in the micritic limestones in the geologic record, (2) how much compaction do these sediments experience, and (3) what are the major processes causing lithification?

Bathurst (1975, p. 504) articulated the problem most clearly:

> "The central difficulty is tantalizingly familiar: how to cement a carbonate ooze, while it is still largely uncompacted with initially a primary porosity of 50% or more. Slightness of compaction is indicated by the absence of crushing of delicate tests, the scarcity of drag or penetration effects among rigid allochems in a micritic matrix, and the general similarity between the intergranular matrix of micrite and the intragranular micrite occupying body chambers."

Information gathered during the Deep Sea Drilling Project has shed considerable light on these questions insofar as burial diagenesis is concerned, though perplexing uncertainties remain. Numerous DSDP sites have yielded data on burial diagenesis of carbonates, but studies in seven regions have been especially useful:

1) Site 167, Magellan Rise in the central Pacific (Schlanger and others, 1973; Schlanger and Douglas, 1974).

2) Sites 206–208 and 210, Southwest Pacific (New Caledonia basin, Lord Howe Rise, Coral Sea Abyssal Plain) (Packham and van der Lingen, 1973).

3) Sites 220 and 223, Arabian Sea (Matter, 1974).

4) Sites 288 and 289, Ontong-Java Plateau in the southwest Pacific (van der Lingen and Packham, 1975).

5) Sites 305 and 306, Shatsky Rise in the northwest Pacific (Matter and others, 1975).

6) Site 317, Manihiki Plateau, South Pacific (Schlanger, Jackson and others, 1976; Jenkyns, 1976).

7) Site 366, Sierra Leone Rise, northwest African continental margin (Gardner and others, 1977).

All of these areas have thick carbonate sections which record progressive changes in diagenesis with depth. Many of these sections are unusually pure, especially those from the Pacific plateaus. Continuous or nearly continuous coring of these sections has provided the data base upon which much of the following discussion is formulated.

Lithification and Age/Burial Depth Relationships

Coring at numerous sites has revealed progressive changes in lithification with depth, from soft ooze to friable chalk to limestone (Figs. 1, 7). Although a survey of Initial Report volumes reveals some inconsistencies in terminology, there appears fairly widespread acceptance of the following descriptive categories (*cf.* Schlanger and Douglas, 1974):

> *Ooze* has little strength and readily deforms under the finger or the broad blade of a spatula. Some workers have recognized an upper zone of soft or creamy ooze and a lower zone of stiff ooze (Packham and van der Lingen, 1973; Matter, 1974; van der Lingen and Packham, 1975).
>
> *Chalk* is partly indurated ooze; it is friable limestone that is readily deformed under the fingernail or the edge of a spatula blade (some early reports refer to this category as "semilithified ooze").
>
> *Limestone* is hard carbonate rock that is not readily deformed under the fingernail or the edge of a spatula blade (pelagic limestones recovered in DSDP cores vary from porous, somewhat chalky rocks to low porosity, dense rocks with a conchoidal fracture).

Since the composition of the sediment showing these changes with depth is broadly similar (*i.e.*, nannofossil-foraminiferal ooze), the lithologic categories may be regarded as approximate zones of burial diagenesis. It is important, however, to emphasize that the boundaries between the lithologic zones in most cases are not sharp and in many instances are markedly gradational (*cf.* Fig. 1). For example, at Site 220 the ooze and chalk zones are separated by a 60-meter thick interval of alternating ooze and chalk layers; at the top of the interval the chalk layers are thin (1–5 cm) and infrequent but become thicker and more numerous until all sediment is chalk (Matter, 1974).

In addition, most sections have short-interval reversals in the degree of lithification, *e.g.*, from ooze to chalk and back to ooze or from chalk to limestone and back to chalk (*e.g.*, Rezak and Henry, 1975). A specific instance is at Site 289 where a thin interval of soft ooze of Maestrictian age lies several hundred meters below the top of the chalk zone; soft ooze of the same age lies within the limestone zone at Site 289 (van der Lingen and Packham, 1975). These reversals seem related to differences in original sediment compositions, as discussed later.

The interpretation of these changes is complicated, however, by the coring disturbances previously mentioned. Thus, the physical properties of ooze can be markedly altered during coring; and chalk and even soft limestone can, on occasion, be ground up into a pasty sediment that resembles ooze but in fact is a drilling artifact. The seven studies previously mentioned, however, are noteworthy in that the investigators were aware of this problem and appear to have made an attempt to separate these artifacts from variations caused by drilling processes.

Table 1 shows the burial depth and age relationships of chalk and limestone recorded in these seven studies. The depths shown for the tops of the chalk and limestone zones are those levels at which thick sections of the respective lithologies are first encountered, even though short-interval reversals in lithification may occur lower in the section. The variation recorded in Table 1 makes it evident that the diagenetic transformations which convert ooze to chalk and chalk to limestone are not related to age and burial depth in a simple way.

Packham and van der Lingen (1973) noted that at four Sites in the southwest Pacific (206–208, 210), different degrees of lithification showed no systematic relationships to either burial depth or to the position of a regional unconformity. They found

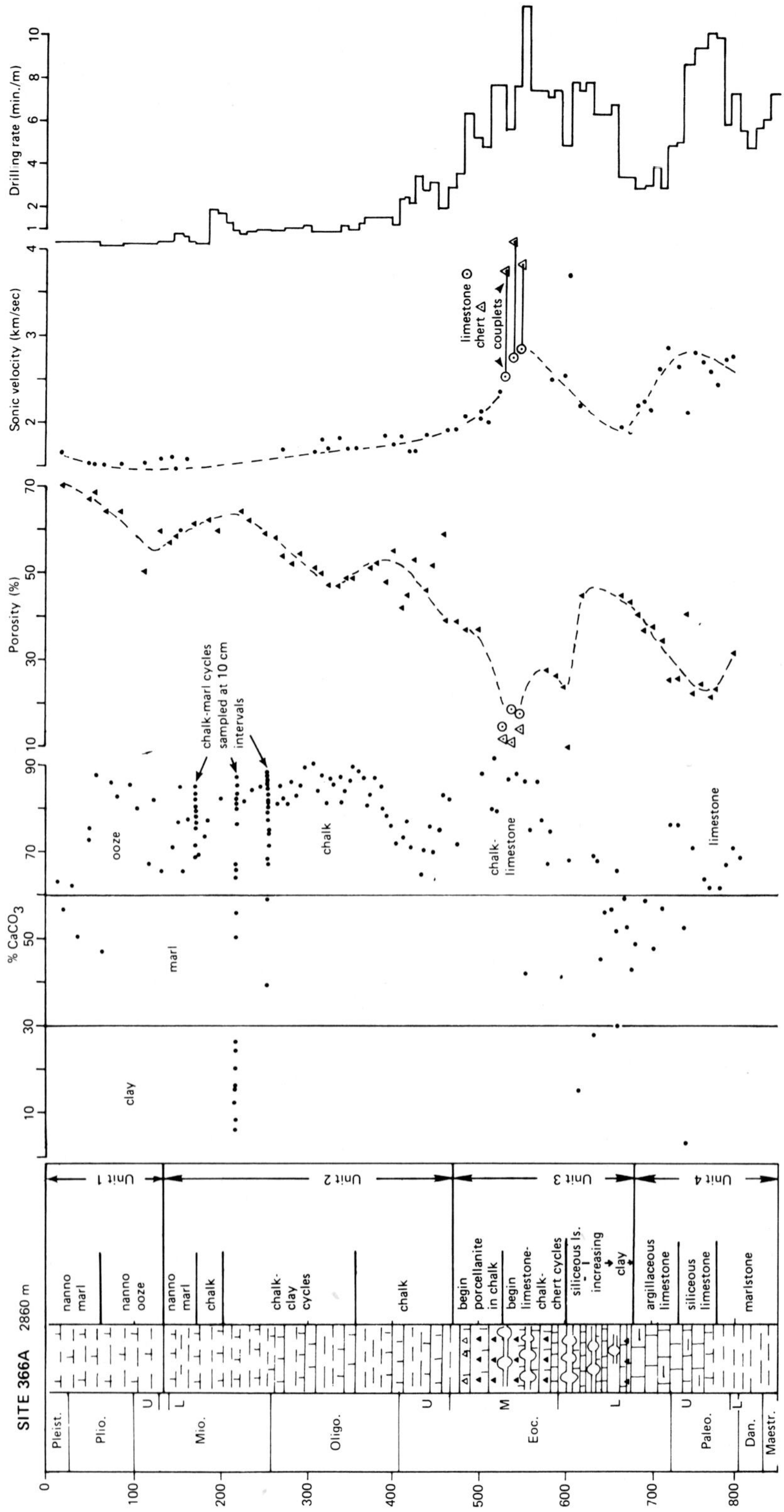

FIG. 1.—Lithology, stratigraphy and physical properties of sediments at Site 366, Sierra Leone Rise, eastern Atlantic (from Gardner and others, 1977).

TABLE 1.—AGE-BURIAL DEPTH RELATIONSHIPS OF LITHIFICATION ZONES AT DSDP SELECTED SITES.

	(NR = not recovered) Site Numbers											
	167	220	233	206	207	208	210	288	289	305	317	366
Depth to top of chalk zone (m)	200	155	230	410	225	325	440	150	240	145	304	165
Age at top of chalk zone	Early Miocene	Early Oligocene	Late Miocene	Early Miocene	Middle Eocene	Early Miocene	Middle Miocene	Middle Miocene	Late Miocene	Late Cretaceous (Maestr.)	Early Eocene	Early Miocene
Depth to top of limestone zone (m)	580	NR	610	NR	Silty claystone	NR	NR	745	1005	540	647	530
Age at top of limestone zone	Late Creta-ceous	NR	Early Eocene	NR	Silty claystone	NR	NR	Late Creta-ceous (Coniacian)	Late Eocene (Aptian)	Early Creta-ceous (Aptian)	Early Creta-ceous	Middle Eocene

Sources of Data
Site 167: Magellan Rise, Schlanger and Douglas, 1974
Sites 220 and 223, Arabian Sea, Matter, 1974
Sites 206, 207, 208, and 210, Southwest Pacific, Packham and van der Lingen, 1973
Sites 288 and 289, Ontong-Java Plateau, Pacific, van der Lingen and Packham, 1975
Site 305, Shatsky Rise, Pacific, Matter, Douglas and Perch-Nielsen, 1975
Site 317, Manihiki Plateau, Pacific, Schlanger, Jackson and others, 1976
Site 366, Sierra Leone Rise, Atlantic, Gardiner and others, 1977

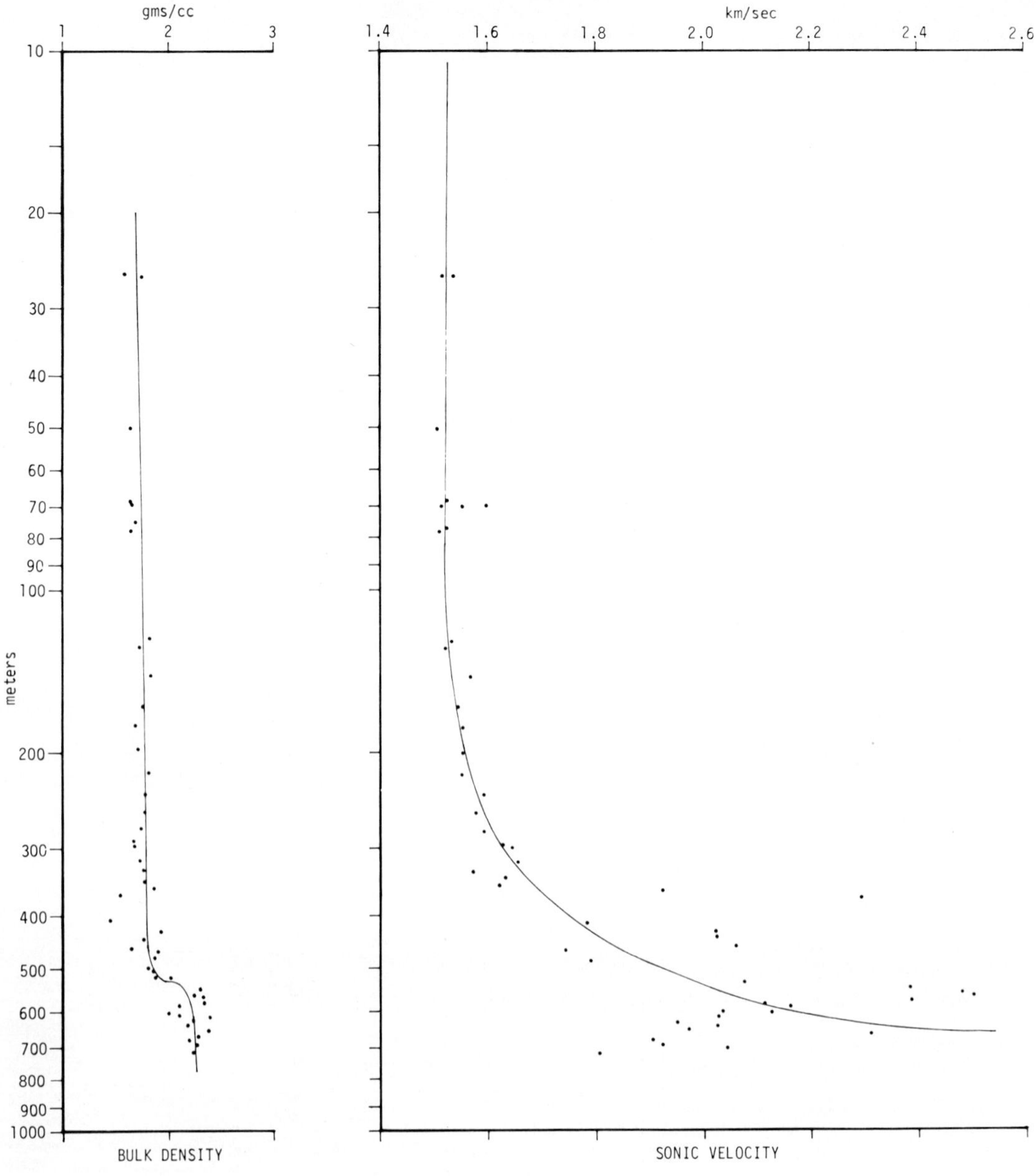

FIG. 2.—Plot of bulk density and sonic velocity versus depth, Site 210, Coral Sea Abyssal Plain, Southwest Pacific (from Packham and van der Lingen, 1973).

that the transitions in lithification (e.g., from ooze to chalk) are shallowest where sedimentation rate was lowest, deepest where the rate was highest, but data from other regions do not corroborate this relationship (Table 1).

Changes in Physical Properties

Paralleling the increase in degree of lithification with depth over long stratigraphic intervals are long-interval decreases in porosity and increases in density and sonic velocity, as recorded by shipboard measurements on sediment cores (Fig. 1). Van der Lingen and Packham (1975) have published the most complete data, and Table 2 summarizes these for the different lithologic zones at two sites. Other data on changes in physical properties of pelagic carbonate sediments with burial depth are in Schlanger and others (1973), Schlanger and Douglas (1974), Packham and van der Lingen (1973), Matter and others (1975), and Gardner and others (1977).

Porosities reported in the above sources decrease from values of 60 to 80% in oozes at the surface to values as low as 13% in silicified limestones (site 289) at depth of about 1100 meters. A fairly rapid porosity decrease in the upper 200 meters is succeeded by more modest decreases with depth. Numerous porosity reversals occur over short intervals, therefore the correlation between burial depth and porosity is inexact.

Bulk densities decrease from about 1.3 to 1.5 g/cm³ in surface oozes to values as high as 2.6 g/cm³ in limestone. As with porosity, many reversals in this trend occur over short intervals.

Packham and van der Lingen (1973) and van der Lingen and Packham (1975) recorded systematic downhole density and sonic velocity measurements at Sites 206–208, 210, 288 and 289. At all these sites the sonic velocities show small increases for the first few hundred meters followed by more rapid increases (Fig. 2). The increases occur within the ooze at Sites 206–208 and 210 and in the chalk zone at Sites 288 and 289. Bulk densities at Sites 206–208 and 210 show small increases with depth until there is a fairly rapid increase of 1.9 to 2.2 g/cm³, after which densities again slowly increase with depth (Fig. 2). Interestingly, the rapid increase in density occurs well within the chalk zone and coincides with neither an obvious lithologic change nor a change in sonic velocity; as seen in Figure 2, the elbow in the velocity-depth curve occurs well above the inflection point in the density-depth curve.

Physical properties in the limestone zone show a great deal of variability (*cf.* Table 2), probably because of local silicification.

Schlanger and Douglas (1974) noted that sonic-velocity plots of long intervals of pelagic carbonate sediments deviate markedly from the smooth curve derived through compactional experiments. This is because pelagic carbonates undergo cementation in addition to compaction during burial.

TABLE 2.—PHYSICAL PROPERTIES IN LITHIFICATION ZONES (FROM VAN DER LINGEN AND PACKHAM, 1975).

		Site 288	Site 289
Ooze Zone	Porosity	60–65	60–70
	Bulk Density	1.5–1.7	1.5–1.7
	Sonic Velocity	1.4–1.6	1.3–1.4
Chalk Zone	Porosity	40–60	40–62
	Bulk Density	1.6–2.1	1.6–2.0
	Sonic Velocity	1.4–2.2	1.4–2.4
Limestone Zone	Porosity	20–42	13–41
	Bulk Density	1.7–2.4	2.0–2.6
	Sonic Velocity	1.9–3.6	1.8–4.4

Porosity in percent
Bulk density in g/cm³
Sonic velocity is horizontal component in Km/sec

Textural Studies and Inferred Diagenetic Processes

Based on petrologic and paleontologic studies, Schlanger and Douglas (1974) deduced that the two major processes in burial diagenesis are dissolution of foraminiferal tests and very small coccolith elements, and reprecipitation of secondary calcite on larger crystals, particularly those comprising discoasters and large coccoliths. Numerous subsequent investigations have confirmed this suggestion.

Schlanger and Douglas found that planktonic foraminiferal tests appear especially susceptible to dissolution during burial thus yielding progressive down-section decreases in their abundance and degree of preservation at Site 167. They noted that "dissolution and collapse of weakened (foraminiferal) tests can account for most of the porosity loss within the sedimentary column. The amount of calcite dissolved from the tests almost equals the amount of new calcite formed as interstitial cement, calcite overgrowths and infillings of remaining tests" (Schlanger and Douglas, 1974, p. 124). Estimated loss of foraminiferal tests is nearly 50% in the transition from ooze to chalk and almost 100% from ooze to limestone. Among the nannofossils they found that small coccoliths tend to disaggregate along element or segment sutures due to dissolution, thus producing micron-sized crystals which in turn may be prone to solution. Matter (1974) has coined the term "micarb" (for microcrystalline carbonate) for these grains (Fig. 3). In contrast, discoasters and larger coccoliths acquire secondary overgrowths and persist into the later stages of diagenesis. These broad outlines of diagenetic changes were earlier confirmed in specific detail through hydrothermal experiments of Adelseck and others (1973).

A number of excellent studies using SEM have documented the ultramicroscopic details of dissolution and overgrowth fabrics formed during burial diagenesis (Figs. 3, 4) (Packham and van der Lingen, 1973; van der Lingen and Packham, 1975; Matter, 1974; Matter and others, 1975; Wise, 1973, 1977). A synthesis of this information suggests the following progression with depth:

1. Accretionary overgrowth of calcite on discoasters begins in the early stages of burial; as this continues it creates difficulties in species identification.

2. Significant calcite precipitation on other fossils commences only after compaction and expulsion of water have created a grain-supporting framework and have triggered the onset of substan-

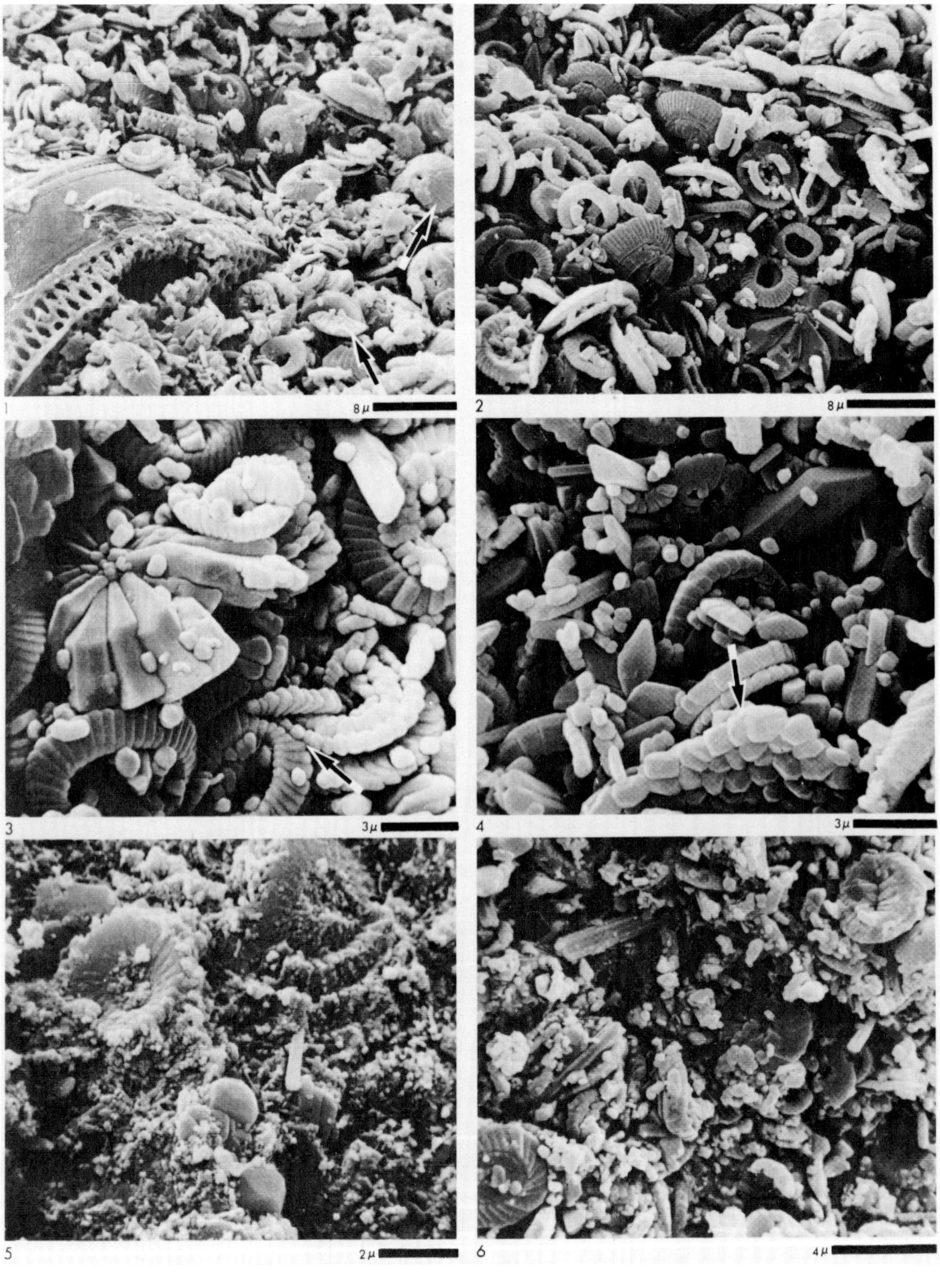
1
8 μ
2
8 μ
3
3 μ
4
3 μ
5
2 μ
6
4 μ

FIG. 4.—Diagenetic textures of foraminiferal nannofossil limestones, Middle Eocene, Site 289, Ontong-Java Plateau, Southwestern Pacific (from van der Lingen and Packham, 1975). Scanning electron micrographs, scale bars equal 5 μ. *Left:* Advanced welding of nannofossils through calcite overgrowths (burial depth = 1027.1 m). *Right:* Euhedral calcite crystals lining walls of foraminiferal chambers (burial depth = 1027.1 m).

tial pressure solution of foraminiferal tests and solution-prone nannofossils. At most sites this seems to occur within the upper 200 meters of sediment. Accretionary calcite then begins to precipitate syntaxially on micarb particles and on the edges and in the central parts of coccoliths.

3. Next, secondary calcite begins to connect the gap between the proximal and distal shields of placoliths (Fig. 5), and void-filling euhedral calcite crystals begin to grow within foraminiferal chambers (Fig. 4).

4. Continuation of the above processes creates limestones with as little as 25% porosity, the rest having been largely filled with granular calcite cement (Fig. 4).

Several studies, particularly that of Matter and others (1975) have shown that accretionary overgrowths are highly site selective; for example, calcite cement preferentially overgrows discoasters, braarudosphaerid plates, and the proximal sides of the distal shields of the more solution-resistant coccoliths (Figs. 3–5). Moreover, overgrowth is species selective as well, in a manner analogous to selective dissolution of planktic skeletal components.

Isotope Studies

Several studies have reported a general decrease of the ^{18}O content in pelagic carbonate sediments with increasing age and depth of burial (Figs. 6, 7). Different workers have interpreted this relationship in diverse ways. Coplen and Schlanger (1973), for example, suggested a global change in the condition of the ocean, such as an increase in temperature or a depletion of ^{18}O, to explain the isotopic shift

FIG. 3.—Microfacies and diagenetic textures of carbonate sediments at site 305, Shatsky Rise, Northwestern Pacific (from Matter and others, 1975). Scanning electron micrographs of fracture surfaces. Arranged in order of increasing burial depth, which is given in parentheses. *1:* Soft siliceous foram-rich nanno ooze with many broken shields of *Coccolithus leptoporous* (Murray and Blackman) (arrows), abundant micritic carbonate particles (micarb) formed by disaggregation of nannofossils, and moderately preserved diatoms (39.0 m). *2* and *3:* Soft foraminiferal nannofossil ooze. No. *2* shows many isolated distal coccolith shields with dissolved central structures. Heavy accretionary overgrowths on discoasters are visible. Pressure dissolution between coccoliths is indicated by arrow in no. *3*. Note much denser packing in no. *3* than in no. *2*. (no. *2* = 86.3 m; no. *3* = 115.9 m). *4:* Foram-rich nannofossil chalk. Abundant small micarb particles and heavy overgrowths on nannofossils, some showing euhedral facies. Arrow points to a remnant of a foraminiferal wall (212.2 m). *5:* Silicified radiolarian nannofossil limestone showing a few dissolution-resistant coccoliths or their replicas or molds in a dense silicified ground mass (439.3 m). *6:* Weakly silicified radiolarian nannofossil limestone showing many tightly packed and overgrown dissolution-resistant *Watanaueria barnesae* (Black) and small micarb particles (599.0 m).

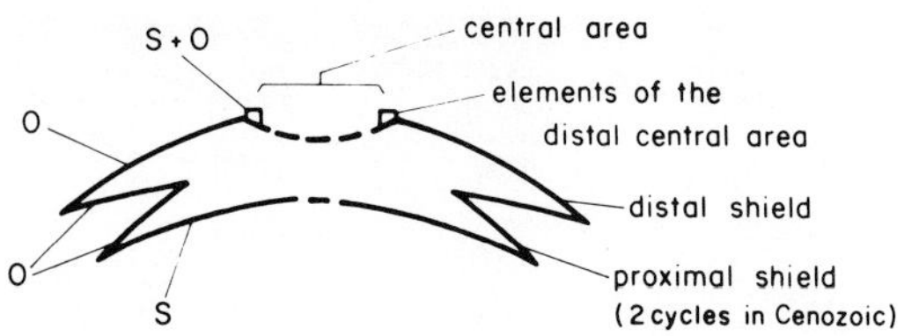

FIG. 5.—Cross section of a placolith similar to *Watznaeria* (Cretaceous), *Ericsonia* and *Coccolithus* (Tertiary) showing major structural features and sites of preferential deposition of overgrowth cement (O) or of dissolution (S). From Matter and others (1975).

noted in Upper Cretaceous sediments on the Magellan Rise. Anderson and Schneidermann (1973) found a similar depletion in Upper Cretaceous carbonates from the Caribbean and linked this to elevated temperatures resulting from regional volcanic activity. Létolle and others (1978) recorded parallel trends in both $\delta^{13}C$ and $\delta^{18}O$ from contemporaneous Lower Cretaceous to Upper Jurassic carbonates at Site 391 in the Atlantic and in an onland section in the Alps. The Alpine Mesozoic carbonates were apparently never buried as deeply as the correlative carbonates at Site 391. $\delta^{18}O$ at both localities decreases with age, but there are increases in Berriasian and Tithonian carbonates, leading the authors to suggest that the overall trends were determined by similar depositional environments in both areas.

Lawrence (1973), in a study of Cenozoic sequences at Sites 148 and 147 in the Caribbean, was able to measure isotopic ratios in both carbonate sediments and associated pore waters; he could thus demonstrate that the more deeply buried carbonates were nearly in isotopic equilibrium with their pore waters, a result he attributed to recrystallization or isotopic exchange during burial diagenesis.

Eade and Anderson (1975) found that variations in lithification of pelagic carbonates at Site 287 in the Coral Sea produces variations in $\delta^{18}O$, which is lowered by solution and recrystallization. Matter and others (1975) measured the $\delta^{18}O$ values of nannofossils in cores from Sites 305 and 306, reasoning that this might more precisely record diagenetic contributions because of the propensity of these fossils to become overgrown with secondary calcite. They likewise found a consistent negative shift in the $\delta^{18}O$ ratio with age and depth of burial (Fig. 7), a trend attributed by them to progressive lithification during burial diagenesis under a normal geothermal gradient. Douglas and Savin (1975) noted that the isotopic values of nannofossils, especially discoasters, at these sites are significantly altered by 200 meters of burial, probably by overgrow cement. McKenzie and others (1978) obtained comparable results and interpretations from a study of Neogene marly sediments at Site 372 in the Balearic Basin, western Mediterranean (Fig. 6).

Other investigations, however, have found less systematic changes in ^{18}O content. Brenneke (1977) measured $\delta^{18}O$ values in Upper Jurassic-Lower Cretaceous carbonates at Sites 105 and 367 in the North Atlantic. The carbonates at both Sites are of the same composition and age, but sediments at Site 367 are buried some 500 to 600 meters deeper than contemporaneous deposits at Site 105. This probably accounts for the fact that the oxygen isotopic compositions at Site 367 are considerably more negative than at Site 105. But at both sites there are rather wide fluctuations in $\delta^{18}O$, possibly the result of variations in original sediment composition. A number of studies (Perry and others, 1976; Eade and Anderson, 1975; Anderson and Lawrence, 1976; Anderson and others, 1976; Lawrence and others, 1979) have shown that the alteration of volcanic debris to authigenic silicates can produce ^{18}O-depleted pore waters from which isotopically light secondary carbonates can be subsequently precipitated. Therefore variations in the amount of volcanic ash in the original carbonate sediments will significantly affect the isotopic composition of secondary carbonates.

Brenneke's (1977) data also demonstrate that the amount of clay in carbonate sediments has an important effect on their diagenesis and oxygen isotope composition. In deeply buried carbonates at Site 367, clay-rich sediments tend to have higher $\delta^{18}O$ values, whereas clay-poor, carbonate-rich layers are relatively depleted in ^{18}O by as much as 3‰. Brenneke suggested this may be due in part to the fact that clay-rich sediments tend to be zones of intense carbonate dissolution, and the carbonate thus dissolved commonly precipitates in immediately adjoining carbonate-rich layers. Because this secondary carbonate precipitates at the higher temperatures of deep burial diagenesis, it will be depleted in ^{18}O relative to the original carbonate sediment. Clay-rich zones, in contrast, retain a more positive ^{18}O signature closer to that of the original carbonate.

In summary, $\delta^{18}O$ values in buried carbonate oozes may reflect in part the composition of the original sediment, in part the diagenetic environment. A progressive decrease of $\delta^{18}O$ with burial depth is found in many sequences and appears to reflect precipitation of cement at progressively higher temperatures. In contrast, the ^{13}C content in the same sequences apparently is little altered and more nearly reflects the composition in the original sediment.

FIG. 6.—Stratigraphic section of Site 372, Balearic Basin, Western Mediterranean, showing induration, water content, percent calcium carbonate, and stable isotope data for the sediments as a function of depth. From McKenzie and others (1978).

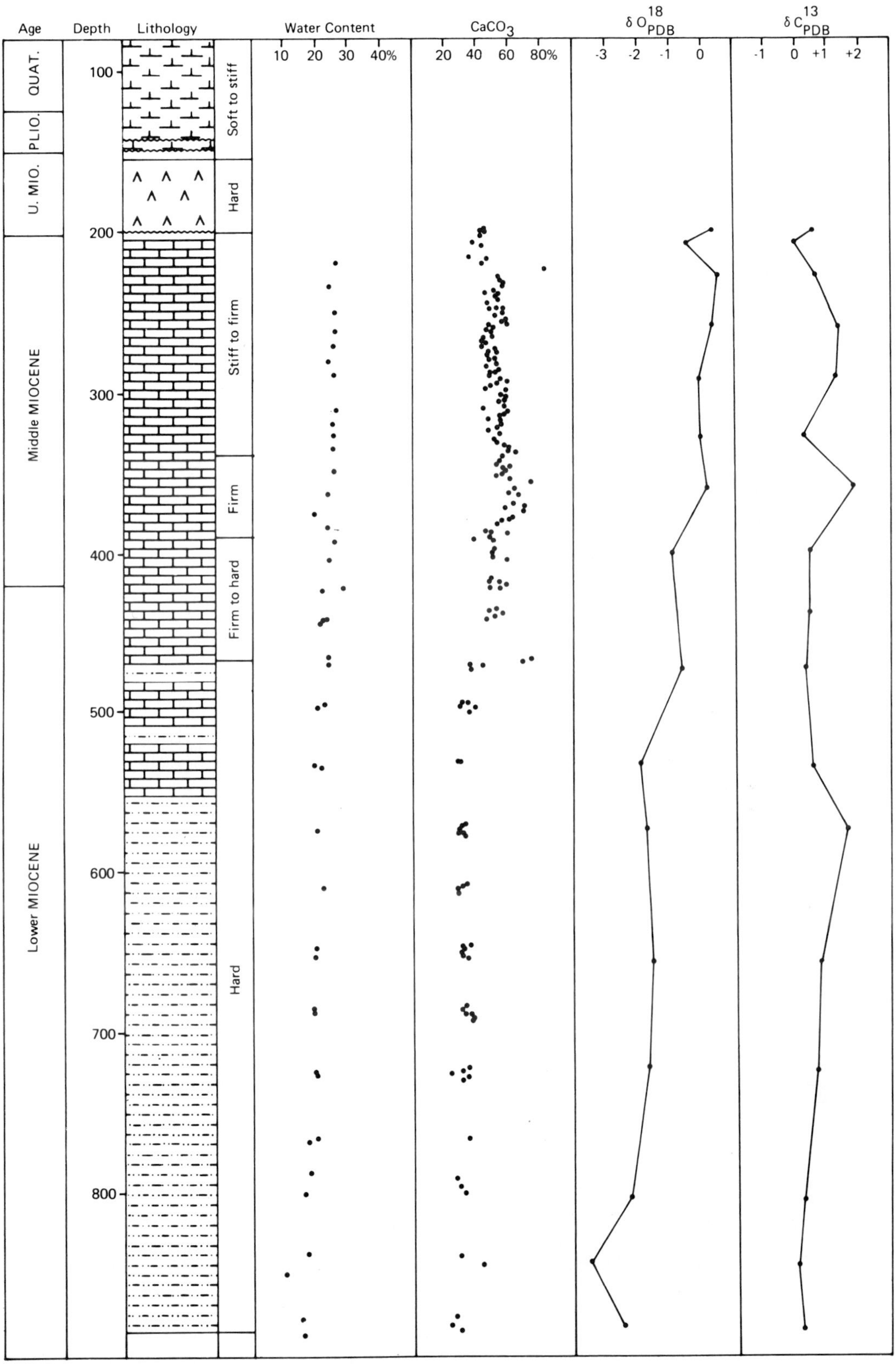
Age
Depth
Lithology
Water Content
$CaCO_3$
δO^{18}_{PDB}
δC^{13}_{PDB}
QUAT.
PLIO.
U. MIO.
Middle MIOCENE
Lower MIOCENE
100
200
300
400
500
600
700
800
Soft to stiff
Hard
Stiff to firm
Firm
Firm to hard
Hard
10 20 30 40%
20 40 60 80%
-3 -2 -1 0
-1 0 +1 +2

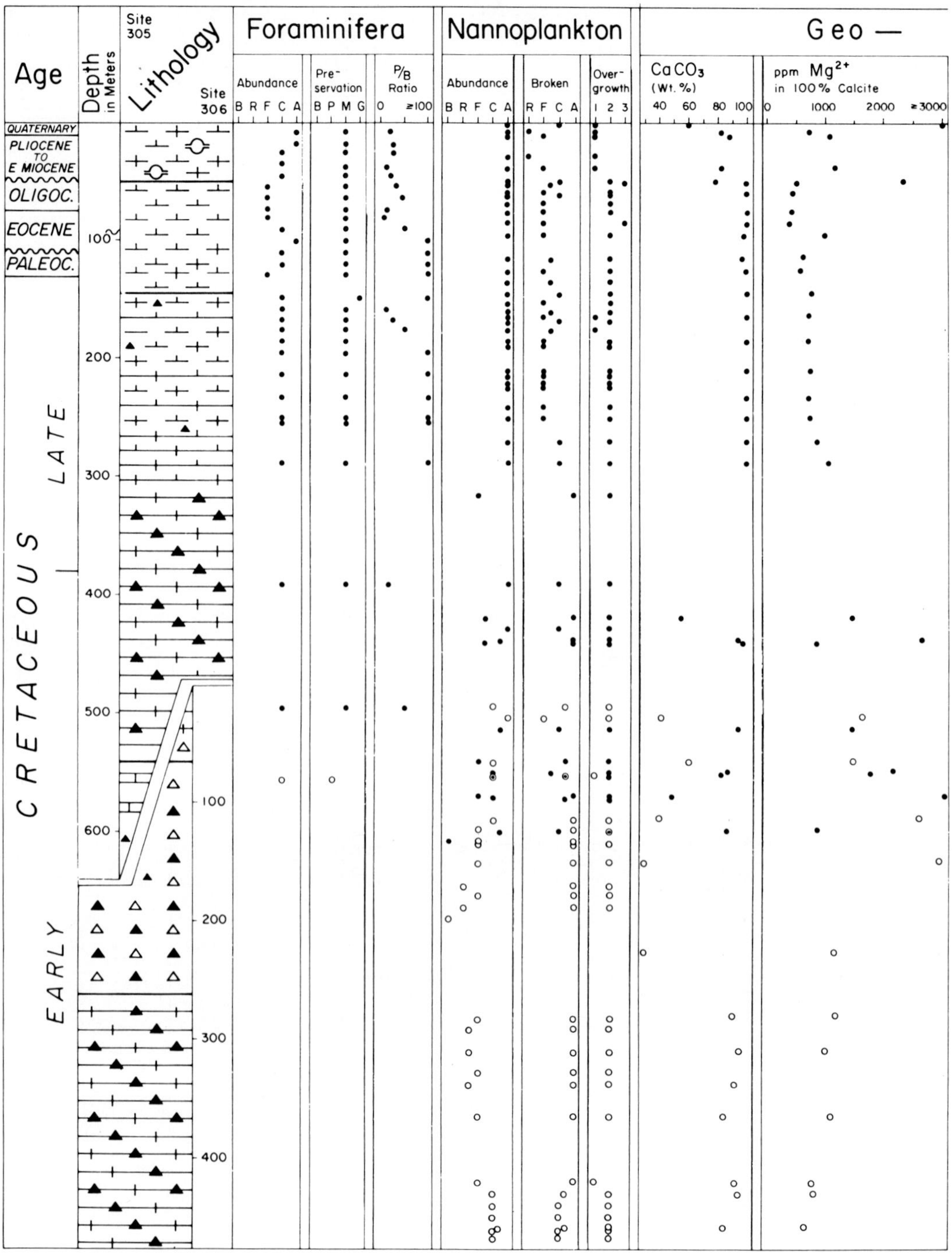

FIG. 7.—Composite stratigraphic sections of sites 305 and 306, Shatsky Rise, Northwestern Pacific, showing foraminifera-nannofossil data, geochemical data, porosities, and sedimentation rates. Site 305 is indicated by closed circles, Site 306 by open circles. From Matter and others (1975).

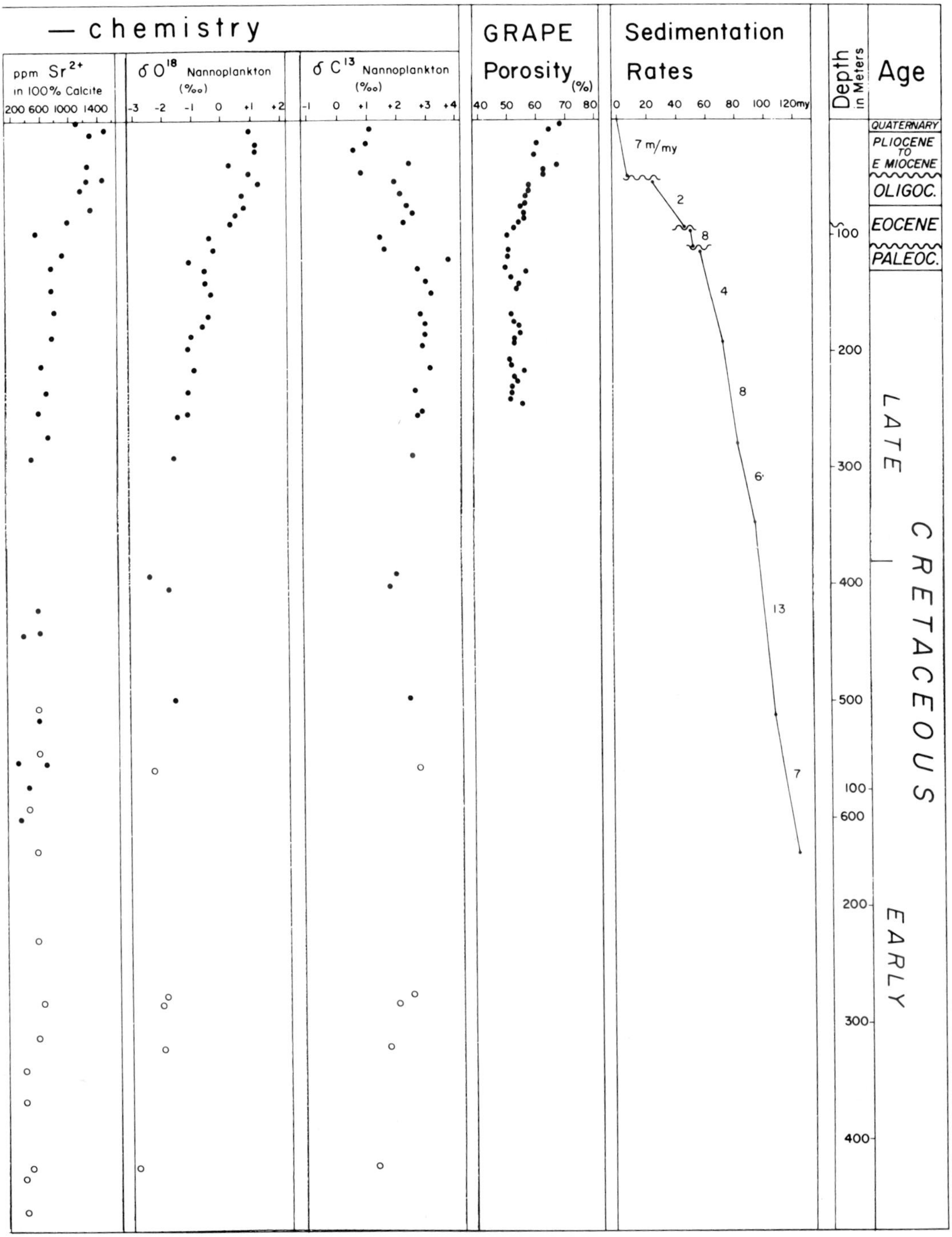
— chemistry
ppm Sr^{2+} in 100% Calcite
200 600 1000 1400
δO^{18} Nannoplankton (‰)
-3 -2 -1 0 +1 +2
δC^{13} Nannoplankton (‰)
-1 0 +1 +2 +3 +4
GRAPE Porosity (%)
40 50 60 70 80
Sedimentation Rates
0 20 40 60 80 100 120my
7 m/my
2
8
4
8
6
13
7
Depth in Meters
100
200
300
400
500
100
600
200
300
400
Age
QUATERNARY
PLIOCENE TO E MIOCENE
OLIGOC.
EOCENE
PALEOC.
LATE
CRETACEOUS
EARLY

Models of Burial Diagenesis in Pelagic Carbonate Sediments

To account for the observed changes in physical and textural properties of pelagic carbonate sediments with burial, Schlanger and Douglas (1974) proposed a model whereby about 3 cm³ of foraminiferal-nannofossil ooze with a porosity of 80% becomes reduced to 1 cm³ of limestone with a porosity of 40% as a result of the transformation (through pressure solution-reprecipitation) ooze → chalk → limestone (Fig. 8).

Schlanger and Douglas conceived the notion of diagenetic realms to encompass the major processes affecting the sediment (Table 3). Realms I through IV precede burial of the sediment and partly predetermine the course of burial diagenesis by influencing the composition of the sediment. For example, a sediment which experiences considerable dissolution of planktic foraminiferal tests prior to burial would be less susceptible to diagenesis during burial since one of the more reactive components had been reduced. Out of this evolved the concept of "diagenetic potential," defined by Schlanger and Douglas (1974, p. 133) "as the length of the diagenetic pathway left for the original dispersed foraminiferal-nannoplankton assemblage to traverse before it reaches the very low free-energy of a crystalline mosaic".

Diagenetic potential according to Schlanger and Douglas would explain observed local deviations in progressive diagenesis, e.g., short-interval lithification reversals from chalk back to ooze or from limestone back to ooze or chalk. These reversals would result when layers of differing diagenetic potential were buried sequentially and thus experienced varying degrees of diagenesis during burial. The *Braarudosphaea* chalk described by Wise and his colleagues (Wise and Hsü, 1971; Wise and Kelts, 1972) is an excellent example of a sediment that had high diagenetic potential. The Schlanger-Douglas view of the important parameters affecting the diagenetic potential (DP) of carbonate oozes is expressed in the formula:

$$DP = f\,(\text{water depth, sedimentation rate, Temperature (surface), Productivity (surface), foraminifer + coccolith : discoaster ratio; } Size_{(max)} : Size_{(min)} \text{ ratio, predation rate} \ldots)$$

Schlanger and Douglas surmised that variations of the diagenetic potential are related to original variations in basin depth, the calcite compensation depth, and the calcareous plankton productivity of the upper-water layers.

Consensus exists that pressure solution provides $CaCo_3$ for all of the cement precipitated during burial diagenesis of carbonate ooze (Lancelot and others, 1972; Schlanger and Douglas, 1974; Matter,

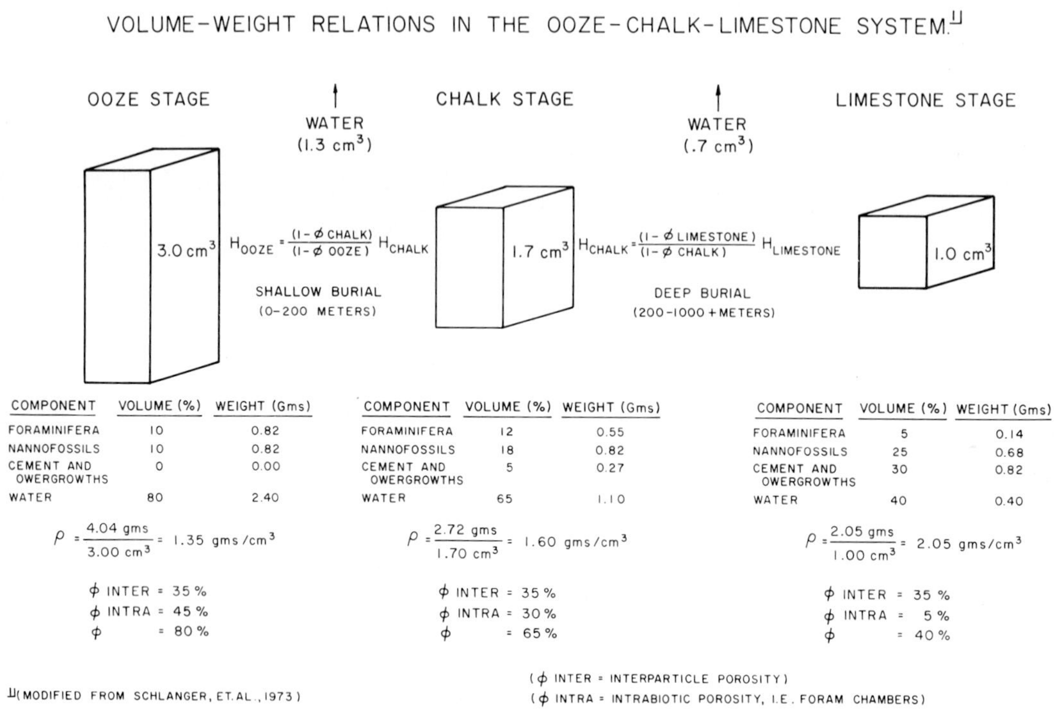

FIG. 8.—Volume-weight relationships in the transformation ooze ¨chalk ¨limestone (from Schlanger and Douglas, 1974).

TABLE 3.—DIAGENETIC REALMS (FROM SCHLANGER AND DOUGLAS, 1974)

Depth	Realm	Residence Time	Petrography	Porosity (ϕ) % Velocity (V_c) km/s	Diagenetic potential O ←——→ ∞
0–200 m (surface water)	I Initial production	Weeks	Highly dispersed calcite-sea water system; 10–10[2] forams/m^3, 10^4–10^6 nannoplankton/m^3 (Lisitzin, 1971; Berger, 1971)		
200 m to sea floor	II Settling	Days to weeks for forams; months to years for coccoliths depending on pelletization (Smayda, 1971)	Pelletized coccoliths, ratio of broken to whole nannoplankton increases downward, ratio of living to empty foram tests decreasing during settling (Lisitzin, 1971; Berger, 1971)		
3000–5000 m (see Diagenetic Potential)	III Deposition	Inversely proportional to sedimentation rate and dissolution rate	'Honeycombed' structure (Tschebotarioff, 1952). Large foram tests supported by chains of coccolith discs. This surface is actually part of Realm IV	$\phi \simeq$ 80%+ $V_c \simeq$ 1·45–1·50 (Nafe & Drake, 1963)	← slope at 3000 m ←Slope at 5000 m
0–1 m sub-bottom)	IV Bioturbation	50,000 years (at 20 m/10^6 years sedimentation rate)	Remoulded 'honeycomb', slight compaction, burrowing, destruction by ingestion and solution	$\phi \simeq$ 75–80% $V_c \simeq$ 1·45–1·6	
1 200 m (sub-bottom)	V Shallow-burial	10 × 10^6 years (at 20 m/10^6 years sedimentation rate)	Ooze affected by gravitational composition, establishment of firm grain contacts; dissolution of fossils and initiation of overgrowths	$\phi \simeq$ 75–60% $V_c \simeq$ 1·6–1·8	
200–1000 m + (sub-bottom)	VI Deep-burial	Up to $\simeq$ 120 × 10^6 years (by then either subducted or uplifted)	Chalk with strong development of interstitial cement and overgrowths; transition down to limestone with dissolution of forams, pervasion by cement and overgrowths—grain interpenetration, welding and 'ameboid mosaics' (Fischer *et al.*, 1967)	$\phi \simeq$ 60% down to 35–40% $V_c \simeq$ 1·8 increasing to 3·3 km/s	
1–10 km sub-surface	VII Metamorphic	10^6–10^7 years	Recrystallization trending to 'pavement mosaic' (Fischer *et al.*, 1967) of completely interlocking crystals	$\phi \simeq$ 40% down to < 5% $V_c \simeq$ 3+up to 6 km/s	

1974; van der Lingen and Packham, 1975). Matter (1974) called this process autholithification and suggested that it is not related to the age of the sediment and is less controlled by burial depth than by primary content. The higher the carbonate content the more nuclei the sediment contains and the more thoroughly it will become cemented, *i.e.,* layers with higher primary carbonate content have higher diagenetic potential. In addition, a carbonate-rich layer may accrete dissolved calcium carbonate from adjacent, less carbonate-rich layers.

Matter contrasted the burial diagenesis of pelitic and carbonate sediments. The lithification of pelitic muds results largely from compaction under increasing overburden pressure. Pelagic carbonates, in contrast, become lithified at much shallower depths due to solution-reprecipitation.

Van der Lingen and Packham (1975) raised the question of whether silica diagenesis does not also importantly influence carbonate diagenesis (see also van der Lingen, 1977, p.73), noting that when carbonate is replaced by chert, this process must result in surplus $CaCO_3$ which can be precipitated in pore spaces of the adjacent sediment. They point out, for example, that carbonate lithification zones at Site 288 are shallower than those at Site 289 (Table 1). At Site 288 chert is present at a subbottom depth of about 270 meters in sediments as young as Late Oligocene, whereas the first chert encountered at Site 289 is of Eocene age at 1030 meters. This suggests a correlation between the presence of chert and the lithification of carbonate sediment, and thus whether chert forms early or late during burial diagenesis may determine the path of carbonate lithification.

A further complicating factor discussed by van der Lingen and Packham is variations in heat flow. High heat flow, as near an oceanic ridge, might promote inter- and intragranular precipitation of calcite cement at comparatively shallow burial depths; but this effect would decrease with time as the sea floor moved away from the ridge and as the sediment pile thickened.

Matter and others (1975) have combined detailed paleontologic, textural and geochemical observations from sites 305 and 306 to construct a comprehensive model of burial diagenesis. They recognize three diagenetic zones:

> *Zone 1,* 0 to 50 meters, consists of soft oozes whose grains show scant evidence of diagenesis. Ca^{++} concentrations in pore waters are identical to sea water and tend to confirm that little or no dissolution has occurred. Gravitational compaction, causing pore water expulsion and a porosity reduction of about 10%, is the dominant process.
>
> *Zone 2,* 50 to 140 meters, comprised of stiff oozes in which there is strong evidence of dissolution and fragmentation of nannofossils and planktic foraminiferal tests. Conversely, some nannofossils, particularly discoasters, show pronounced secondary overgrowths. Accompanying these textural changes are large decreases in strontium and magnesium in the solid phase and significant shifts to lower $\delta^{18}O$ and higher $\delta^{13}C$ values. These are in part due to diagenetic solution-reprecipitation, in part a reflection of the original composition and bottom conditions (*e.g.,* relatively lowered bottom temperatures) at the time of deposition.
>
> *Zone 3,* below 140 meters, encompasses the transition first into chalk and subsequently into radiolarian limestones and porcelanites. Dissolution of planktic foraminifera and nannofossils is particularly severe below 250 meters, and only the most robust nannofossil forms survive in the lower parts of the zone. Silicification is locally widespread, and silica replaces many coccoliths and micarb grains in the carbonates. Strontium content of the solid phase decreases progressively with depth of burial (Fig. 7), and this is taken by the authors to reflect progressive lithification and stabilization of the pelagic carbonates. Renard and others (1978) found a similar inverse correlation between strontium content of the solid phase and burial depth in Cretaceous-Jurassic carbonates at Sites 390 and 391.

Sayles and Manheim (1975) found that pore waters in DSDP biogenic sediments deposited at rates of a few $cm/10^3$ yr become progressively enriched in calcium and strontium and depleted in magnesium. They interpreted this as reflecting the dissolution of biogenic calcite and reprecipitation of higher magnesium, lower strontium calcite in which Mg^{++} took the place of some Ca^{++}. Their supposition about strontium is supported by the data from Sites 305 and 306 (Fig. 7). The interpretation of the data on magnesium from these sites (Fig. 7), however, is equivocal because of the presence of dolomite and because of leaching of magnesium from clay minerals during sample preparation.

Matter and others (1975) attempted to estimate how much original sediment was dissolved and how much cement was reprecipitated by using measurements of strontium concentrations in pore waters and in the bulk samples. The results shown in Table 4 assume 1350 ppm strontium in the unaltered sediment and a partition coefficient of 0.05; they also assume that the cement was precipitated in equilibrium with the pore water present today. Thus strontium emerges as a potentially excellent measure of the degree of diagenesis in pelagic carbonates.

TABLE 4.—ESTIMATES OF THE AMOUNTS OF BIOGENIC CALCITE PARTICLES AND CALCITE CEMENT PRESENT AT DIFFERENT BURIAL DEPTHS (FROM MATTER AND OTHERS, 1975).

Burial Depth (m)	Composition of Bulk Carbonate: Biogenic Grains (%)	Composition of Bulk Carbonate: Cement (%)	Sr^{++} in Cement (ppm)
53	87	13	510
100	37	64	480
240	24	76	440

Assumptions:
1. Average concentration of Sr^{++} in original unaltered sediment = 1350 ppm
2. Partition coefficient of Sr^{++} = 0.05
3. Cement precipitated in equilibrium with pore waters now present

Correlation of Lithology, Texture and Physical Properties

Although long-interval decreases in porosity and increases in bulk density and sonic velocity parallel the transformations ooze → chalk → limestone, over short intervals a lack of correlation commonly exists between lithology and these properties. Schlanger and Douglas (1974) noted, for example, that ooze-chalk-ooze-chalk sequences can occur in sections within which there is no discernable change in porosity.

Correlations may, however, exist on a much more subtle level. At sites 206 and 210 Packham and van der Lingen (1973) noted that the change in the gradient of the depth-sonic velocity curve (Fig. 2) corresponds with a change from plastic stiff ooze to crumbly stiff ooze; they speculated this could be the point where compaction produces significant grain contacts and thus more efficient transmission of sound through the sediment.

Similarly, as noted before, the inflection in the depth-density curve (Fig. 2) does not correlate with an obvious lithologic change but occurs instead well within the chalk zone. Detailed textural observations, however, show that this change coincides with the beginning of extensive void-filling cementation within the chalk zone. Packham and van der Lingen postulated that this abrupt increase in density may occur where the process of grain breakage with some solution at grain contacts is replaced by rapid solution and reprecipitation of cement.

At sites 220 and 223 in the Arabian Sea, Matter (1974) correlated texture with lithology by means of exceptionally detailed SEM observations. He noted that mechanical compaction largely accounts for the transformation of soft to stiff ooze within the upper 100 meters of sediment. Further lithification, reduction of porosity and increase in density below 100 meters results from cementation which increases with depth. The top of the chalk zone coincides with stabilization of the grain framework through cementation at point contacts; significant selective dissolution of some fossils and secondary overgrowths on others begins below this level.

A number of studies have shown that chalks and limestones recovered in DSDP cores exhibit significant sound velocity anisotropy, with horizontal velocities parallel to bedding being higher than vertical velocity (van der Lingen and Packham, 1975; Boyce, 1976; Tucholke and others, 1976; Carlson and Christensen, 1979). Carlson and Christensen (1979) investigated velocity anisotropy at in situ pressures for chalks and limestones from Sites 354, 356 and 357. Their studies showed that velocity anisotropy is closely related to burial depths. It is well developed at burial depths less than 400 meters and increases markedly with depth; between 400 and 800 meters, anisotropy in the samples they investigated increased at a rate of about 19% per kilometer, reaching a value near 16% at the maximum burial depths they sampled (750 meters).

Although the cause for the velocity anisotropy is not firmly established, Carlson and Christensen suggested it may result from preferred orientation of calcite due to mechanical or chemical processes. They noted that the compressional wave velocity for propagation parallel to the *c* crystallographic axis of calcite is about 24% less than that in the plane normal to the *c* axis; therefore a preferred orientation of calcite crystals with *c* axes perpendicular to bedding could produce the observed anisotropies. Such a preferred orientation in a carbonate ooze could be partially attained by gravitational compaction of nannofossils such as discoasters, braarudosphaerids, and many coccoliths whose crystal elements have *c* axes perpendicular to the long dimension of the fossil. This effect would be enhanced during burial diagenesis and accretionary overgrowth. Bukry (1971) observed that nannofossil forms most resistant to dissolution (and thus most likely to survive and become overgrown during burial) are those with *c* axes normal to their long dimensions. Bain (1940, quoted in Adelseck and others, 1973) has shown that syntaxial overgrowths on crinoid ossicles is fastest parallel to the *c* axis. Matter and others (1975) noted that this probably explains why the intershield space of placoliths is bridged by accretionary calcite relatively early in diagenesis. Thus it is likely that a preferred orientation of secondary calcite crystals would become progressively better developed during burial diagenesis, a process that could help to explain the increase in velocity anisotropy with burial depth (Carlson and Christensen, 1979).

Studies of On-Land Chalks

The study of on-land Late Mesozoic and Cenozoic chalks of epicontinental seas has both benefited from and contributed to knowledge of the diagenesis of deep sea carbonate oozes. Neugebauer's

(1973, 1974) studies of European and North American Cretaceous chalks, for example, have yielded especially useful insights. His theoretical calculations (1973) suggest chalk should become lithified into limestone by pressure solution at burial depths of a few hundred meters, yet friable chalk sequences as thick as 1500 meters exist. In Neugebauer's view this anomaly results from two properties: (1) chalk of largely low magnesium calcite, and (2) pore fluids with Mg/Ca ratios similar to that of sea water. Pore fluids with the latter composition will be supersaturated with respect to low magnesium calcite, thus little pressure solution will occur until threshold pressures of 250 to 1000 atmospheres have been reached, irrespective of burial depth. These pressures are attained very locally at point contacts within the first 100 meters of burial, but significant cementation and lithification as a consequence of wholesale pressure solution should require an overburden of 1000 meters or more.

In 1974 Neugebauer was able to collate DSDP observations with his theoretical calculations to produce an outline of chalk diagenesis. He noted that pressure solution could convert chalk to dense limestone at relatively shallow depths if magnesium in the pore fluids became significantly reduced. Pore fluid analyses of DSDP cores (e.g., Manheim and Sayles, 1971; Sayles and others, 1974; Sayles and Manheim, 1975) suggest, however, that pore-water concentrations of magnesium are not appreciably lower in pelagic carbonates until burial depths of 500 to 700 meters. Neugebauer postulates that below these depths magnesium reduction in pore waters results from precipitation of magnesium-bearing calcite cement, a view also held by Manheim and Sayles (1971; see also Sayles and others, 1974; Sayles and Manheim, 1975); the progressive depletion of magnesium would in turn promote further pressure solution so that the chalk → limestone transition should be expected at subbottom depths of 100–1600 meters.

DSDP results show that chalks commonly alter to limestones at shallower depths (Table 1), but these limestones are usually chert- or porcelanite-bearing whereas Neugebauer dealt with theoretically pure chalks. As noted earlier, van der Lingen (in van der Lingen and Packham, 1975) has suggested that chert formation enhances carbonate lithification by supplying excess $CaCO_3$ for cement. Later (1977, p. 216) he called attention to the possible complementary role of magnesium in both chert genesis and carbonate cementation. The experimental work of Kastner and others (1977) suggested that the transformation of biogenic opal-A to opal-CT is promoted through the precipitation of nuclei with magnesium hydroxide as an important component. Following Neugebauer's hypothesis the resulting reduced Mg^{++} content in pore waters would enhance pressure solution of carbonates. The formation of dolomite (as in the limestones at Site 306) would also serve to reduce Mg in contiguous pore waters. Another possible magnesium-depleting reaction is magnesium uptake by clay minerals, especially during the alteration of volcanic material (Lawrence and others, 1979).

Mapstone (1975) and Scholle (1974, 1977) have investigated aspects of diagenesis in Cretaceous chalks of Britain and adjacent areas. Scholle's work in particular has emphasized the role of compositional variations in determining bed-to-bed differences in diagenetic response during burial. His studies have confirmed and extended important conclusions reached by several of the earlier studies of DSDP materials, especially the importance of pressure solution and autolithification during chalk diagenesis and the progressive depletion of ^{18}O in cements with increasing burial depth.

A Summary of Current Views and Uncertainties

Out of this welter of information, the following consensus emerges:

1) Burial diagenesis converts carbonate oozes to chalks and chalks to limestones. Decreases in porosity and increases in bulk density and sonic velocity accompany these first order changes.

2) Gravitational compaction is significant only in the first 50 to 200 meters of burial.

3) Below these depths, the dominant processes are pressure solution and reprecipitation which may act simultaneously. Solution removal of the more soluble components becomes an important means of compaction.

4) Solution and reprecipitation are highly site- and taxa-selective. Solution affects certain taxa selectively and certain parts of skeletons selectively; reprecipitation of cement occurs preferentially on specific taxa and on specific skeletal sites.

5) The rates of burial dissolution and cementation will thus depend on the original composition of the carbonate components (its diagenetic potential).

6) The presence of non-carbonate components (siliceous microfossils, volcanic ash, clay minerals) affects burial diagenesis in ways that are not precisely understood. Conversion of biogenic silica to chert seems to enhance carbonate cementation in the surrounding sediment, whereas clay minerals appear to promote burial dissolution and solution-transfer.

7) For the above reasons (nos. 5 and 6), burial lithification does not proceed in a smoothly progressive manner and both bed-by-bed variations and short interval reversals in degree of lithification are common. Physical properties vary accordingly with depth over short intervals.

8) Physical properties may not correlate precisely with first order changes in lithology and textures; the relationships may be very subtle ones that

involve ultramicroscopic details of fabrics.

9) These variations not withstanding, two main factors, burial depth and pore-water chemisty, appear to control the first order changes ooze → chalk → limestone.

10) The above transformation in oceanic oozes is calcite conservative; there is no external source of calcium carbonate.

11) The Mg^{++}/Ca^{++} ratio in pore waters may significantly affect burial diagenesis of carbonates; high ratios retard solution-reprecipitation of low magnesium calcite, thus any mechanisms (*e.g.*, clay mineral diagenesis, dolomitization) which remove Mg^{++} will enhance cementation.

12) Progressive burial lithification is accompanied by a loss of strontium through dissolution of strontium-richer skeletal calcite and reprecipitation of strontium-poorer calcite cement.

13) There is a comparable depletion of ^{18}O during burial lithification due to precipitation of calcite cement in progressively deeper and warmer zones. Variations in the trend toward more negative $\delta^{18}O$ values with burial depth may result from differences in both the carbonate and non-carbonate composition of the original sediment. $\delta^{13}C$ remains comparatively unaffected.

The questions about lithification of micritic limestones posed at the beginning of this section can therefore be answered. The source of the cement is pressure dissolution of biogenic calcite grains; no outside source is needed. Gravitational compaction operates only in the upper part of the sediment pile; below this, solution-compaction occurs which tends to remove grains; those that remain become secondarily overgrown by cement and thus progressively more insulated from crushing. The major processes causing lithification are therefore solution and reprecipitation.

Major remaining uncertainties are mainly geochemical and include precise delineation of the effects of non-carbonate components, the relationships between silica and carbonate diagenesis, the role of pore fluid chemistry, the effects of pore water circulation, and the effects of variable heat flow. Non-carbonate components, especially clay minerals, may be particularly significant in affecting diagenesis and acoustic properties, but the direction as well as the extent of their role remains largely unknown. Whereas studies of on-land chalks have suggested that abundant clays enhance burial dissolution (Garrison and Kennedy, 1977; Scholle, 1977), experimental work by Baker and others (1980) implies that clays and other silicates retard both dissolution and precipitation of fine-grained calcite.

Among other additional studies needed are further attempts to quantify lithification processes along the line suggested by Matter and others (1975; *cf.* Table 4) and to determine whether the magnesium content of calcite cements increases with burial depth, as predicted by Neugebauer (1974) and Sayles and Manheim (1975). But additional petrologic studies will be productive also, especially of cements in the deeper levels of lithification. Van der Lingen and Packham (1975, p. 458) and Matter and others (1975, p. 905), for example, noted evidence that some cements had been later partially dissolved. This matter seems not to have been pursued in subsequent work, yet it has important consequences for our understanding of the effects of diagenetic textures on sonic properties in order to advance the field of acoustistratigraphy.

CARBONATE DIAGENESIS AND ACOUSTISTRATIGRAPHY

Schlanger and Douglas (1974) called attention to possible relationships between carbonate diagenesis and acoustistratigraphy. They noted that closely spaced reflections are a characteristic of pure pelagic carbonate sections in the deep oceans. These result from discontinuities, sharp gradients and reversals in acoustic impedance values which, at Site 167, can be related to changes in the degree of lithification evident in the cores recovered. Schlanger and Douglas argued that the degree of cementation of a pelagic carbonate sediment is largely correlative with the diagenetic potential it had when it began the burial process; therefore acoustistratigraphy of pelagic carbonates is related to the degree of cementation which in turn is controlled by the diagenetic potential. The latter in turn may reflect more universal events such as glacio-eustatic changes which affect plankton productivity and CCD levels. For example, higher carbonate productivity due to increased upwelling during glacial intervals would yield a higher rate of carbonate deposition, higher diagenetic potential, and a reflecting horizon.

Studies in the southwest Pacific and on the Ontong-Java Plateau (Packham and van der Lingen, 1973; van der Lingen and Packham, 1975), however, emphasize that reflecting horizons may be due to quite subtle changes in sediment properties. For example at Site 288, a Middle Miocene reflector at about 172 meters may be related to erosion or slumping but is otherwise not lithologically explainable, whereas no reflector is associated with the ooze-chalk transition at about 150 meters. Van der Lingen and Packham therefore caution that a precise correlation between seismic reflection and diagenetic modification may be difficult due to the complications of carbonate and silica diagenesis.

Insight on subtle changes in sediment properties that may determine reflecting horizons has come from further work on the Ontong-Java plateau by Johnson and others (1978) and Berger and Mayer (1978). According to the latter authors the origin of reflectors in carbonates, although not completely understood, must be related to lithologic changes that produce impedance contrasts. They suggest such lithologic changes are a consequence of sediment or acoustic "conditioning" (*i.e.,* changes in type and rate of calcareous sediment accumulation) which is controlled by productivity, dissolution and winnowing. This "conditioning" thus determines the diagenetic potential of the sediment during burial. Focusing on the effects of dissolution, Johnson and others (1978) showed that carbonate oozes deposited below the lysocline have anomalously low acoustic velocities in the upper 200 meters of sediment but normal velocities at greater burial depths. They interpret this to mean that carbonate oozes deposited below the lysocline have a lower diagenetic potential (due to solution of reactive components) and lithify more slowly during burial than those deposited above the lysocline.

Berger and Mayer (1978) recorded a strong depth dependence of the number and spacing of near surface acoustic reflectors on the Ontong-Java Plateau. They took this as an indication that carbonate dissolution (*e.g.,* below the lysocline) is one of the major processes controlling the development of reflecting intervals. Dissolution strongly influences the grain size which in turn affects acoustic impedance. Shifts in dissolution boundaries (lysocline and CCD) through time may thus condition contemporaneously deposited sediment for creating reflections during burial. Because dissolution events are cyclic, they might help to explain the evenly spaced reflections characteristic of thick pelagic carbonate sections. More recently Mayer (1979) has extended this work in a study of a small area in the eastern equatorial Pacific. There, variations in acoustic impedance are controlled by changes in bulk density which in turn are highly correlated with the percent of calcium carbonate. Variation in the latter resulted from climatic changes; warm periods were times of enhanced dissolution which resulted in low $CaCO_3$ content, relatively low density, and low impedance; the opposite was true for cold periods.

These investigations raise the possibility that if precise relationships between sediment conditioning and acoustistratigraphy can be established, major oceanic events of the past could be recognized from high-resolution profiling. This underscores the importance of further work to refine our understanding of relationships between diagenesis and sonic properties, in particular the effects of non-carbonate components. Abundance of the latter may be a function of primary sedimentological events (*e.g.,* changes in terrigenous input or patterns of organic productivity or dissolution), and these components very likely affect the diagenetic potential (and hence the acoustistratigraphy) in ways not presently understood.

DIAGENESIS IN CARBONATE SEDIMENTS ASSOCIATED WITH OCEANIC BASALTS

Pelagic limestones spatially associated with ophiolites and other mafic igneous rocks are common in many mobile belts on land. These limestones, however, are ordinarily so intensely deformed and metamorphosed that little can be determined about their diagenetic history. The recovery of closely associated carbonates and basalts in DSDP cores is therefore ideal for reconstructing this history and for judging the extent of interactions between carbonate sediments and igneous rocks.

Whether thermal metamorphism has affected carbonate sediments in contact with oceanic rocks is a critical question for geologic interpretations. Anderson and Schneidermann (1973) and Schneidermann (1973) interpreted low $\delta^{18}O$ values at sites 146 and 153 in the central Caribbean as reflecting thermal alteration associated with emplacement of basalts and dolerites. Many oceanic limestones lying above basalt, interbedded with basalt or occurring in interpillow areas show lithification and alteration which is commonly attributed to contact thermal metamorphism. At several sites, for example, a zone of lithified limestone or dolomite, a few centimeters to several meters thick and often with a reddish color, occurs on top of basalt (*e.g.,* site 53, Pimm and others, 1971).

One of the surprising results of DSDP studies, however, is that isotope analyses show the alteration frequently occurred at relatively low temperatures (*e.g.,* Site 14, Lloyd and Hsü, 1972; Sites 105 and 367, Brenneke, 1977; Site 373, Bernoulli and others, 1978; Site 417, McKenzie and Kelts, 1979). Similarly, oxygen isotopic compositions of calcite veins in basalts at several sites indicate a low temperature origin (Anderson and Lawrence, 1976; Brenneke, 1977; Muehlenbachs and Hodges, 1978).

Bernoulli and others (1978) and McKenzie and Kelts (1979) carried out detailed petrographic-geochemical studies of limestone associated with basaltic breccias and pillow lavas in the western Mediterranean (Site 373) and North Atlantic (Site 417), respectively. Neogene basaltic breccias at Site 373 contain complex patterns of sediment infiltration into interclast space and of cementation (Fig. 9). Four generations of internally resedimented interclast limestones and four generations of void-filling calcite cement are present. Two generations of the sediment were pelagic carbonate oozes that infiltrated from the sea floor; the other two generations apparently originated as diagenetic sediment which was precipitated and deposited within breccia interstices. Interspersed with these diagenetic events were repeated episodes of brecciation and fracturing. Oxygen and carbon isotopic compositions indicate that cementation of the breccias and lithification of the sediments occurred at low temperaatures characteristic of bottom waters. R. L. Folk (personal communication, 1978) has suggested that the void-filling coarse sparry calcite cements in these rocks (Fig. 9) resembles those precipitated from low-Mg meteoric waters (*cf.* Folk, 1974). This cement, however, is magnesium calcite with 4 to 7 mole % $MgCo_3$, and the submarine origin is further suggested by its isotopic composition and the close association with coccolith-bearing limestones.

In contrast to the infiltrated limestones of Site 373, most of the Upper Cretaceous interpillow limestones at Site 417 apparently formed when the pillows intruded soft pelagic ooze (McKenzie and Kelts, 1979). Lithification of the ooze yielded a micritic limestone which is extensively recrystallized. Its ^{18}O content ($\delta^{18}O_{PDB}$ = −0.14 to −0.78 per mil) indicates the recrystallization process occurred at Late Cretaceous bottom-water temperatures of 10 to 15°C. Lithification apparently took

FIG. 9.—Thin-section photograph of basaltic breccia with a variety of interclast sediment infillings and cements, Site 373, Tyrrhenian Basin, Western Mediterranean. Scale bar at lower right equals 2 mm. The letter *P* indicates areas of pelagic limestone that formed by infiltration of nannofossil-foraminiferal ooze from the sea floor; this limestone includes clasts of basalt. The upper part of the photograph is dominated by several generations of void-filling sparry calcite cement (white) and diagenetic limestone (*D*); the latter formed within the breccia and was redeposited to form geopetal floors. The letter *V* at upper right indicates a large unfilled void. Note also numerous shelters or "umbrellas" below dark volcanic clasts, now filled with spar (especially in lower right quadrant). From Bernoulli and others (1978).

place early during diagenesis and near the seawater-rock interface. This acted to form a seal or shield around some of the basalt so that fresh basaltic glass of Middle Cretaceous age remains preserved. Some pillows contain coarse authigenic calcite that fills voids or replaces basalts. It has rather negative ^{18}O contents (mostly in the range of $\delta^{18}O_{PDB}$ = -4.01 to -4.92). McKenzie and Kelts interpret this as the product of late stage alteration of basalt which produces isotopically light interstitial water (*cf.* Anderson and Lawrence, 1976; see discussion below).

Like the interpillow pelagic limestones at Site 417, limestone immediately above basaltic basement at Site 367 in the North Atlantic apparently became lithified very early during diagenesis at relatively shallow burial depths (Brenneke, 1977). This limestone lies at burial depths of 1120 meters and should have more negative $\delta^{18}O$ values than the overlying sediments; instead it is significantly heavier and must have been lithified at lower temperatures and with a thinner overburden.

In Cenozoic shallow water limestones and carbonate sediments lying on seamounts in the northwestern Pacific, McKenzie and others (1980) have found late stage calcite and zeolitic cements. They believed these were precipitated from pore fluids whose composition was influenced by low-temperature alterations of the underlying basalts.

The processes responsible for the kind of low temperature alteration described in previous paragraphs are not precisely understood, but the evidence suggests that it may be a by-product of the submarine weathering of basalt to authigenic clays or zeolites. Numerous studies have shown this increases the concentration of calcium and the alkalinity in interstitial waters, circumstances favorable to the precipitation of calcite (see discussions in Garrison and others, 1973, and Sayles and Manheim, 1975). Whether this could also promote the kind of extensive low temperature recrystallization evident at site 417 (McKenzie and Kelts, 1979) remains unknown.

A further consequence of the subaqueous weathering of basalt is the production of interstial waters which become progressively more depleted in ^{18}O. Anderson and Lawrence (1976) found that the oxygen isotopic composition of low temperature calcite veins in basalt correlates inversely with that of basalt adjacent to the vein. They interpreted this to mean that the veins were precipitated from solutions which were isotopically depleted by interaction (*i.e.,* weathering) with basalt in closed systems. Thus late-stage low-temperature alteration of basalts will produce isotopically light interstitial water and hence isotopically light calcite that might be misinterpreted as hydrothermal.

DIAGENESIS OF SHALLOW-WATER CARBONATE SEDIMENTS

Because shallow-water carbonate sediments have been so infrequently cored, relatively little new insight has emerged regarding their diagenesis as the result of DSDP operations. Instead, concepts developed from study of modern and ancient on-land shallow-water carbonates have been utilized to interpret DSDP materials, sometimes in quite imaginative ways.

An example is at Sites 390 and 392 on the outer edge of the Blake Plateau, where Enos and Freeman (1978, 1979) have investigated shallow-water skeletal and oolitic limestones. These rocks exhibit extensive textural evidence of subaerial diagenesis, including large solution caverns with dimensions up to 10 meters, stalactitic intergranular cement, meniscus sediments and cements, and fenestral fabrics. Three limestone units are present and form part of a single regressive sequence: a lower skelmoldic limestone, a middle oolitic unit, and an upper fenestral limestone. Porosity in the skelmoldic limestone is largely secondary, and much of it is preserved. This contrasts with the two overlying units in which porosity was predominantly primary but was later nearly completely occluded by cement and diagenetic sediment. These rocks, formed in shallow water and subjected to intermittent subaerial exposure, are abruptly overlain by goethitic laminated crusts and ooids and then by Early Cretaceous to Eocene pelagic oozes. They now lie at water depths in excess of 2600 meters, and consequently must have experienced quite rapid subsidence.

Several investigations have focused on diagenesis of shallow-water carbonates recovered from the tops of seamounts in the western Pacific. Hesse (1973) reconstructed the diagenetic history of a pre-Eocene oolitic limestone cored from the top of Ita Matae seamount at Site 202 in the western Pacific. The sediment was originally a mixture of high-magnesium calcite and aragonite but is now mostly low-magnesium calcite with very minor aragonite. Among the important diagenetic processes inferred from detailed textural study are replacement of aragonitic oolites by calcite, dissolution of skeletal aragonite to form cavities, conversion of high magnesium skeletal grains to low magnesium calcite, syntaxial overgrowths on echinoderm fragments, and internal cementation and crystallization of external sparry rim cements. The limestones apparently formed in shallow to moderately shallow water but now occur at water depths in excess of 1600 meters. All of their diagenesis apparently occurred subaqueously during subsidence.

Matter and Gardner (1975) made somewhat

similar observations on Eocene shallow-water biogenous volcanic sandstone cored at Site 304 on Koko Guyot in the Emperor Seamount chain, northwestern Pacific. The aragonitic and high-magnesium calcite skeletal elements of these rocks have been replaced by low-magnesium calcite. The cements are also relatively low in magnesium (about 3 mol % $MgCO_3$); but this calcite may have replaced an earlier high-magnesium calcite because the cement fabrics resemble modern high-magnesium calcite cements of submarine origin. No indications of vadose diagenesis are present, and the authors conclude that this diagenesis occurred subaqueously.

If Hesse (1973) and Matter and Gardner (1975) are correct in judging that the diagenesis they describe is entirely subaqueous, then it is evident that mineralogical stabilization to low Mg calcite and development of common textures such as solution vugs, syntaxial overgrowths and sparry rim cements can occur in the submarine environment.

McKenzie and others (1980) investigated diagenesis in lithified and unlithified shallow-water carbonates cored from the top of three other seamounts in the Emperor Seamount chain. These materials are now completely low-magnesium and low-strontium calcites. Textural evidence suggests the diagenesis occurred in two stages. Early-stage diagenesis commenced with dissolution of aragonite and precipitation of autocthonous calcite rim cement in the phreatic environment to produce cemented limestones. Unlithified to weakly lithified sediments on one of the seamounts also experienced complete transformation to low magnesium calcite but with only minor cementation by micrite and intraparticle sparite. Isotopic studies of these latter sediments suggests this diagenesis occurred subaqueously in cool intermediate waters during a 5 to 10 million-year interval of subsidence.

A second stage of diagenesis involved the precipitation of "dog-tooth" calcite simultaneously with the zeolites phillipsite and clinoptilolite. McKenzie and others (1980) infer that both calcite and zeolite cements were precipitated from solutions derived from low-temperature alteration of the underlying basalt.

DOLOMITE AND OTHER AUTHIGENIC MINERALS

Although dolomite commonly occurs in generally small amounts in pelagic carbonate oozes and other sediments recovered in DSDP cores, few systematic investigations have been made of it. Moreover, it has proved difficult in many instances to differentiate between authigenic dolomite formed in situ and redeposited detrital dolomite (see discussion in Lloyd and Hsü, 1972, and Davies and Supko, 1973). For example, Berger and von Rad (1972) have described a variety of dolomitic occurrences in Atlantic pelagic sediment. Silt-size dolomite rhombs are present in a number of pelagic facies, and the authors suggest they form authigenically wherever calcium carbonate and magnesium (derived perhaps from alteration of ash) are available. Dolomite becomes enriched near the boundary of carbonate and non-carbonate facies due to dissolution. These authors also describe cyclic sequences of black shale and dolomite silts, the latter redeposited from shallower areas.

A thin dolomite layer lying above basaltic basement at Site 373 in the eastern Mediterranean appears to be dolomitized pelagic carbonate ooze. As noted, Bernoulli and others (1978) believed this replacement occurred as a consequence of low-temperature submarine alteration of the basalt.

The dolomite occurring within evaporitic sequences in the Mediterranean and elsewhere and in restricted basins like the Black Sea are beyond the scope of this report. It is pertinent to mention, however, the Pliocene dolomitized nannofossil oozes, chalks and limestone that lie above the Miocene evaporite sections in the Red Sea and Mediterranean. Supko and others (1974) studied the former and interpreted them as in part early diagenetic authigenic precipitates in muds and replacements of biogenic oozes. Detailed SEM observations of the Mediterranean dolomite at Site 374 by Bernoulli and Mélières (1978) confirmed this origin. This dolomite occurs as platy pseudohexagonal crystals which replaced coccoliths and foraminiferal tests and grew in inter- and intra-particle voids. It is calcium-rich dolomite with 53 to 55% $CaCO_3$. Bernoulli and Mélières concluded that the dissolution of calcite and dolomitization in these Early Pliocene carbonates is related to saline interstitial waters rich in Mg^{++} and with a Mg/Ca ratio of about 7; the unusual crystal morphology of the platy dolomite is attributed to the inhibiting effect on crystal growth perpendicular to (0001) of ions such as Cl^- in the highly saline interstitial brine. This brine was probably derived from the underlying Miocene evaporites.

Cook (1974) suggests a similar source in evaporites for fluids which partly dolomitized Pliocene shallow water carbonates at Site 262 in the Timor Trough.

Matsumoto and others (1978) recorded dolomite in Cretaceous carbonates cored on the Walvis Ridge (Site 363) and on the continental margin of the Angola Basin (Site 364). Based on chemical compositions and occurrences, the authors recognized two types of dolomites. One is dolostone from the margin of the Angola basin; this has a low iron and high sodium content, and the authors postulated that the dolomitizing fluids were derived from underlying evaporites. The other occurs in argillaceous dolomitic limestones both in the Angola Basin and the Walvis Ridge; it is iron-rich (average of 5 mol % $FeCO_3$), and Matsumoto and others suggested it formed a reducing environment with excess carbonate ion derived from decomposition of organic materials in the sediment.

Lancelot and Ewing (1972) correlated gas zonation and carbonate diagenesis at several sites in the northwestern Atlantic. Authigenic calcite and dolomite occur in the clathrate zone, whereas scattered siderite crystals characterize an underlying transition zone which contains both clathrates and free gas. Below this is the free-gas zone which contains nodules and lenses of siderite and, locally, ankerite. Chemical properties of the interstitial waters parallel the mineralogical changes, leading the authors to suggest that diagenesis of organic material determines both the gas zonation and (by controlling Eh-pH values) the type of authigenic carbonate formed.

Berger and von Rad (1972) recorded a variety of authigenic carbonate minerals, in addition to dolomite, in Atlantic pelagic deposits. These include siderite, ankerite, and rhodochrosite. Most are associated with organic-rich sediments. Lancelot and others (1972) described dolomitization and sideritization of Atlantic limestones at Site 101 and common siderite in black Lower Cretaceous clays at Sites 101 and 105.

Dean and Schreiber (1978) found authigenic barite and calcitized barite in limestones and marls at Sites 369 and 370 along the margin of northwest Africa. This replacement apparently occurred during pauses in sedimentation, and the main source of barium was from oxidation of organic matter.

Varied authigenic carbonate minerals also occur in noncarbonate sequences cored by DSDP. For example, Robinson and others (1974) have reported dolomite and barite nodules as well as secondary calcite in Mesozoic claystones of the eastern Indian Ocean. Dean and Schreiber (1978), in a tabulation of occurrences of authigenic barite in DSDP materials, have shown that fine-grained barite is common in organic-rich sediments of nearly every type and age.

Hein and others (1979) have extensively investigated authigenic carbonates occurring in Cenozoic diatomaceous sediments and mudstones cored during Leg 19 in the northwest Pacific and the southern Bering Sea. The carbonates include low and high magnesium calcite (0.0 to 24.0 mol % $MgCO_3$), protodolomite (53 to 58 mol % $CaCO_3$), siderite, rhodochrosite and manganosiderite. The carbonates occur as cements and replacement zones. The average $\delta^{13}C_{PDB}$ value for the carbonates is -17.2 per mil, which led the authors to conclude that the bicarbonate supplied to the diagenetic carbonates was produced by bacterial sulfate reduction and/or methane generation through bacterial degradation of organic debris. Ca^{++}, Mg^{++}, Fe^{++} and Mn^{++} in the carbonates are believed to have been derived mainly from the alteration of volcanic ash. Hein and others concluded that there were no calcitic precursors of biogenic origin for these carbonates which they postulated formed relatively late in diagenesis at burial depths of 440 to 550 meters and some 3 to 4 m.y. follow deposition of the host sediment.

Pisciotto and Mahoney (*in press*) reported authigenic dolomite and minor calcite in organic-rich Cenozoic sediments cored at six sites off Southern California and Baja California during Leg 63. These carbonates have highly variable carbon isotopes (-30 to $+16$‰ that reflect the isotopic reservoirs in which they formed. Authigenic dolomites, for example, apparently formed simultaneously at several different burial depths or at different times within the same stratigraphic interval. Oxidation of organic matter by bacterial reduction of sulfate at shallow burial depths produced light-carbon carbonates whereas heavy-carbon carbonates formed at deeper levels where $H^{12}CO_3^-$ was preferentially removed by reduction of CO_2 to methane during bacterial fermentation.

In summary, one type of dolomite found in DSDP cores can be related to dolomitizing fluids derived from underlying evaporites. Another type of dolomite, along with other authigenic carbonates and barite, appears to form in connection with diagenesis of organic matter in sediments. A third type of dolomite may be formed as a consequence of submarine weathering of basalt. Doubtless other types of dolomite and authigenic carbonates also are present in DSDP cores but have not yet been systematically analyzed.

CONCLUSION

The DSDP contribution to knowledge of carbonate diagenesis is mainly in the burial realm. From cores in storage and those yet to be recovered, much remains to be learned about burial diagenesis as well as about other forms of carbonate alteration in oceanic regions. But one major result to date that bears reiteration is the recognition that seemingly inconspicuous variations in sediment composition may reflect major oceanic events; they may also predetermine the levels of diagenetic changes which form prominent markers in seismic profiles. An improved knowledge of carbonate diagenesis will thus be an important link to more comprehensive interpretations of the acoustistratigraphic record.

ACKNOWLEDGEMENTS

I am grateful to many colleagues for sharing reprints and ideas, most especially to Seymour O. Schlanger, Judith McKenzie and Kenneth A. Pisciotto. Partial financial support for preparation of this manuscript was provided by grants from the Faculty Research Committee of the University of California, Santa Cruz. Acknowledgement is also made to the donors of the Petroleum Research Fund (PRF #10758-AC2), administered by the American Chemical Society, for partial support of this research.

REFERENCES CITED

NOTE: All references to Deep Sea Drilling Project Initial Reports have been abbreviated; for complete citations see *Appendix I* at the end of this volume, or the individual Initial Reports as published.

Adelseck, C. G., Jr., Geehman, G. W., and Roth, P. H., 1973, Experimental evidence for the selective dissolution and overgrowth of calcareous nannofossils during diagenesis: Geol. Soc. America Bull., v. 84, p. 2755–2762.

Anderson, T. F., Donnelly, T. W., Drever, J. I., Eslinger, E., Gieskes, J. M., Kastner, M., Lawrence, J. R., and Perry, E. A., 1976, Geochemistry and diagenesis of deep-sea sediments from Leg 35 of the Deep Sea Drilling Project: Nature, v. 261, p. 473–476.

———, and Lawrence, J. R., 1976, Stable isotope investigations of sediments, basalts and authigenic phases from Leg 35 cores: Init. Repts. DSDP, v. 35, p. 497–505.

———, and Schneidermann, Nahum, 1973, Stable isotope relationships in pelagic limestones from the central Caribbean: Leg 15, Deep Sea Drilling Project: Init. Repts. DSDP v. 15, p. 795–803.

Bain, G. W., 1940, Geological, chemical, and physical problems in the marble industry: Amer. Inst. Mining and Metall. Engineers, Tech. Pub. no. 1261, 16 p.

Baker, P. A., Kastner, M., Byerlee, J. D., and Lockner, D. A., 1980, Pressure solution and hydrothermal recrystallization of carbonate sediments—an experimental study: Marine Geology, v. 38, p. 185–204.

Bathurst, R. G. C., 1975, Carbonate sediments and their diagenesis: Amsterdam, Elsevier Scientific Publishing Co., 658 p.

Berger, W. H., and Mayer, L. A., 1978, Deep-sea carbonates: acoustic reflectors and lysocline fluctuations: Geology, v. 6, p. 11–15.

———, and von Rad, U., 1972, Cretaceous and Cenozoic sediments from the Atlantic Ocean: Init. Repts. DSDP, v. 14, p. 787–954.

Bernoulli, D., Garrison, R. E., and McKenzie, J., 1978, Petrology, isotope geochemistry, and origin of limestone and dolomite associated with basaltic breccia, Hole 373A, Tyrrhenian Basin: Init. Repts. DSDP, v. 42, p. 541–558.

———, and Mélières, F., 1978, Dolomitization in early Pliocene pelagic limestone, Site 374, Ionian Abyssal Plain: Init. Repts. DSDP, v. 42, p. 1079–1094.

Boyce, R. E., 1976, Sound velocity-density parameters and rocks from DSDP Sites 315–318 on the Line Islands, Manihiki Plateau, and Tuamotu Ridge in the Pacific Ocean: Init. Repts. DSDP, v. 33, p. 695–728.

Brenneke, J. C., 1977, A comparison of the stable oxygen and carbon isotope composition of Early Cretaceous and Late Jurassic carbonates from DSDP Sites 105 and 367: Init. Repts. DSDP, v. 41, p. 937–956.

Burns, D. A., 1972, Discoasters in Holocene sediments: Southwest Pacific Ocean: Marine Geology, v. 12, p. 301–306.

Bukry, D., 1971, Cenozoic calcareous nannofossils from the Pacific Ocean: Trans. San Diego Soc. Natural History, v. 16, p. 303–327.

Carlson, R. L., and Christensen, N. I., 1979, Velocity anisotropy in semi-indurated calcareous deep sea sediments: Jour. Geophys. Res., v. 84, p. 205–211.

Cook, P. J., 1974, Geochemistry and diagenesis of interstitial fluids and associated calcareous oozes, Deep Sea Drilling Project, Leg 27, Site 262, Timor Trough: Init. Repts. DSDP, v. 27, p. 463–480.

Coplan, T. B., and Schlanger, S. O., 1973, Oxygen and carbon isotope studies of carbonate sediments from Site 167, Magellan Rise, Leg 17: Init. Repts. DSDP, v. 17, p. 505–509.

Davies, T. A., and Supko, P. R., 1973, Oceanic sediments and their diagenesis: some examples from deep-sea drilling: Jour. Sed. Petrology, v. 43, p. 381–390.

Dean, W. E., and Schreiber, B. C., 1978, Authigenic barite, Leg 41, Deep Sea Drilling Project: Init. Repts. DSDP, v. 41, p. 915–932.

Douglas, R. G., and Savin, S. M., 1975, Oxygen and carbon isotope analyses of Tertiary and Cretaceous microfossils from Shatsky Rise and other sites in the North Pacific Ocean: Init. Repts. DSDP, v. 32, p. 509–520.

Eade, J. V., and Anderson, T. F., 1975, Oxygen and carbon isotope composition and diagenesis of Eocene clay nannochalk at DSDP Site 287, Coral Sea Basin, Leg 30: Init. Repts. DSDP, v. 30, p. 419–422.

Enos, P., and Freeman, T., 1978, Shallow water limestone from the Blake Nose, Sites 340 and 392. Init. Repts. DSDP, v. 44, p. 413–461.

———, and ———, 1979, Lower Cretaceous peritidal limestones at 2,700 m depth, Blake Nose, Atlantic Ocean: Geology, v. 7, p. 83–87.

Folk, R. L., 1974, The natural history of crystalline calcium carbonate: effect of magnesium content and salinity: Jour. Sed. Petrology, v. 44, p. 40–53.

Gardner, J. V., Dean, W. E., and Jansa, L., 1977, Sediments recovered from the northwest African continental margin, Leg 41, Deep Sea Drilling Project: Init. Repts. DSDP, v. 41, p. 1121–1134.

Garrison, R. E., Hein, J. R., and Anderson, T. F., 1973, Lithified carbonate sediment and zeolitic tuff in basalts, Mid-Atlantic Ridge: Sedimentology, v. 20, p. 399–410.

———, and Kennedy, W. J., 1977, Origin of solution seams and flaser structure in Upper Cretaceous chalks of southern England: Sed. Geology, v. 19, p. 107–137.

Hein, J. R., O'Neil, J. R., and Jones, M. G., 1979, Origin of authigenic carbonates in sediment from the deep Bering Sea: Sedimentology, v. 26, p. 681–705.

Hesse, R., 1973, Diagenesis of a seamount oolite from the West Pacific, Leg 20, DSDP: Init. Repts. DSDP, v. 20, p. 363–387.

Jenkyns, H. C., 1976, Sediments and sedimentary history of the Manihiki Plateau, South Pacific Ocean: Init. Repts. DSDP, v. 33, p. 873–890.

Johnson, T. C., Hamilton, E. L., Bachman, R. T., and Berger, W. H., 1978, Sound velocities in calcareous oozes and chalks from sonobuoy data: Ontong Java Plateau, Western Equatorial Pacific: Jour. Geophys. Res., v. 83, p. 283–288.

Kastner, M., Keene, J. B., and Gieskes, J. M., 1977, Diagenesis of siliceous oozes: I. Chemical controls on the rate of opal-A diagenesis—an experimental study: Geochim. Cosmochim. Acta, v. 41, p. 1041–1059.

Lancelot, Y., and Ewing, J. I., 1972, Correlation of natural gas zonation and carbonate diagenesis in Tertiary sediments from the north-west Atlantic: Init. Repts. DSDP, v. 11, p. 791–799.

———, Hathaway, J. C., and Hollister, C. D., 1972, Lithology of sediments from the western North Atlantic, Leg 11, Deep Sea Drilling Project: Init. Repts. DSDP, v. 11, p. 901–949.

Lawrence, J. R., 1973, Interstitial water studies, Leg 15—stable oxygen and carbon isotope variations in water, carbonates, and silicates from the Venezuela Basin (Site 149) and the Aves Rise (Site 148): Init. Repts. DSDP, v. 20, p. 891–899.

———, Dreyer, J. I., Anderson, T. F., and Brueckner, H. K., 1979, Importance of alteration of volcanic material in the sediments of Deep Sea Drilling site 323: Chemistry, $^{18}O/^{16}O$ and $^{87}Sr/^{86}Sr$: Geochim. Cosmochim. Acta, v. 47, p. 573–588.

Létolle, R., Renard, M., Bourbom, M., and Filly, A., 1978, O^{18} and C^{13} isotopes in Leg 44 carbonates: A comparison with the Alpine series: Init. Repts. DSDP, v. 44, p. 567–573.

Lloyd, R. M., and Hsü, K. J., 1972, Stable-isotope investigations of sediments from DSDP III cruise to South Atlantic: Sedimentology, v. 19, p. 45–58.

Manheim, F. T., and Sayles, F. L., 1971, Interstitial water studies on small core samples, Deep Sea Drilling Project, Leg. 8: Init. Repts. DSDP, v. 8, p. 857–869.

Mapstone, N. B., 1975, Diagenetic history of a North Sea chalk: Sedimentology, v. 22, p. 601–614.

Matsumoto, R., Utada, M., and Kagami, H., 1978, Sedimentary petrology of DSDP cores from Sites 362 and 363, the Walvis Ridge, and Site 364, the Angola Basin, drilled on Leg 40: Init. Repts. DSDP, v. 40, p. 469–486.

Matter, A., 1974, Burial diagenesis of pelitic and carbonate deepsea sediments from the Arabian Sea: Init. Repts. DSDP, v. 23, p. 421–470.

———, Douglas, R. G., and Perch-Nielsen, K., 1975, Fossil preservation, geochemistry and diagenesis of pelagic carbonates from Shatsky Rise, northwest Pacific: Init. Repts. DSDP, v. 32, p. 891–922.

———, and Gardner, J. V., 1975, Carbonate diagenesis at Site 308, Koko Guyot: Init. Repts. DSDP, v. 32, p. 521–530.

Mayer, L. A., 1979, Deep sea carbonates: Acoustic, physical, and stratigraphic properties: Jour. Sed. Petrology, v. 49, p. 819–836.

McKenzie, J., Bernoulli, D., and Garrison, R. E., 1978, Lithification of pelagic-hemipelagic sediments at DSDP Site 372: Oxygen isotope alteration with diagenesis: Init. Repts. DSDP, v. 41, p. 473–478.

———, ———, and Schlanger, S.D., 1980, Shallow-water carbonate sediments from the Emperor Seamounts: Their diagenesis and paleogeographic significance: Init. REpts. DSDP, v. 55, p. 415–455.

———, and Kelts, K. R., 1979, A study of interpillow limestones from the M-zero anomaly, DSDP Leg 51, Site 417D: Init. Repts. DSDP, v. 51–53, part 2, p. 753–769.

Muehlenbachs, K., and Hodges, F. N., 1978, Oxygen isotope geochemistry of rocks from DSDP Leg 46: Init. Repts. DSDP, v. 46, p. 257–258.

Neugebauer, J., 1973, The diagenetic problem of chalk: the role of pressure solution and pore fluid: Neues Jahrb. Geol. Palaont, Abh., v. 143, p. 223–245.

———, 1974, Some aspects of cementation in chalk: *in* Hsü, K. J. and Jenkyns, H. C. (eds.), Pelagic sediments: on land and under the sea: Int. Assoc. Sed. Spec. Pub. No. 1, p. 149–176.

Packham, G. H., and van der Lingen, G. J., 1973, Progressive carbonate diagenesis at Deep Sea Drilling Sites 206, 207, 208, and 210 in the southwest Pacific and its relationship to sediment physical properties and seismic reflectors: Init. Repts. DSDP, v. 21, p. 495–521.

Perry, E. A., Jr., Gieskes, J. M., and Lawrence, J. R., 1976, Mg, Ca, and O^{18}/O^{16} exchange in the sediment-pore water system, Hole 149, DSDP: Geochim. Cosmochim. Acta, v. 40, p. 413–423.

Pimm, A. C., Garrison, R. E., and Boyce, R. E., 1971, Sedimentology synthesis: Lithology, chemistry and physical properties of sediments in the northwestern Pacific Ocean: Init. Repts. DSDP, v. 6, p. 1131–1252.

Pisciotto, K. A., and Mahoney, J. J., *in press,* Isotopic survey of diagenetic carbonates, DSDP Leg 63: Init. Repts. DSDP, v. 63.

RENARD, M., **Létolle**, R., BOURBON, M., AND RICHEBOIS, G., 1978, Some trace elements in the carbonate samples recovered from Holes 390, 390A, 391C, and 392A of DSDP Leg 44: Init. Repts. DSDP, v. 44, p. 557–566.
REZAK, R., AND HENRY, M. E., 1975, Diagenesis of calcareous nannofossils: Init. Repts. DSDP, v. 3, p. 531–533.
ROBINSON, P. T., THAYER, P. A., COOK, P. J., AND MCKNIGHT, B. K., 1974, Lithology of Mesozoic and Cenozoic sediments of the eastern Indian Ocean, Leg 27, Deep Sea Drilling Project: Init. Repts. DSDP, v. 24, p. 1001–1047.
ROTH, P. A., AND BERGER, W. A., 1975, Distribution and dissolution of coccoliths in the South and Central Pacific: Spec. Pub. No. 13, Cushman Foundation for Foraminiferal Research, p. 87–113.
SAYLES, F. L., AND MANHEIM, F. T., 1975, Interstitial solutions and diagenesis in deeply buried marine sediments: results from the Deep Sea Drilling Project: Geochim. Cosmochim. Acta, v. 39, p. 103–127.
———, ———, AND WATERMAN, L. S., 1974, Interstitial water studies on small core samples, Leg 15: Init. Repts. DSDP, v. 20, p. 783–804.
SCHLANGER, S. O., AND DOUGLAS, R. G., 1974, The pelagic ooze-chalk-limestone transition and its implications for marine stratigraphy, *in* Hsü, K. J. and Jenkyns, H. C. (eds.), Pelagic sediments: On land and under the sea: Int. Assoc. Sed. Spec. Pub. No. 1, p. 117–148.
———, ———, LANCELOT, Y., MOORE, T. C., JR., AND ROTH, P. A., 1973, Fossil preservation and diagenesis of pelagic carbonates from the Magellan Rise, Central North Pacific Ocean: Init. Repts. DSDP, v. 17, p. 407–477.
———, AND JACKSON, E. D., AND OTHERS, 1976, Init. Repts. DSDP, v. 33, 973 p.
SCHOLLE, P. A., 1974, Diagenesis of Upper Cretaceous chalks from England, Northern Ireland, and the North Sea: *in* Hsü, K. J. and Jenkyns, H. C. (eds.), Pelagic sediments: On land and under the sea: Int. Assoc. Sed. Spec. Public. No. 1, p. 177–210.
———, 1977, Chalk diagenesis and its relation to petroleum exploration: Oil from chalks, a modern miracle?: Amer. Assoc. Petroleum Geologists Bull., v. 61, p. 982–1009.
SCHNEIDERMANN, NAHUM. 1973, Pelagic limestones of the central Caribbean, Leg 15: Init. Repts. DSDP, v. 15, p. 773–793.
SUPKO, P. R., STOFFERS, P., AND COPLEN, T. B., 1974, Petrography and geochemistry of Red Sea dolomite: Init. Repts. DSDP, v. 23, p. 867–878.
TUCHOLKE, B. E., EDGAR, N. T., AND BOYCE, R. E., 1976, Physical properties of sediments and correlations with acoustic stratigraphy: Init. Repts. DSDP, v. 35, p. 229–249.
VAN DER LINGEN, G. J., (Ed.), 1977, Diagenesis of deep-sea biogenic sediments: Stroudsburg, Penn., Dowden, Hutchinson & Ross, Inc., 385 p.
———, AND PACKHAM, G. H., 1975, Relationships between diagenesis and physical properties of biogenic sediments of the Ontong-Java Plateau (Sites 288 and 289, Deep Sea Drilling Project): Init. Repts. DSDP, v. 30, p. 443–481.
WISE, S. W., JR., 1973, Calcareous nannofossils from cores recovered during Leg 18, Deep Sea Drilling Project: Biostratigraphy and observations of diagenesis: Init. Repts. DSDP, v. 18, p. 569–615.
———, 1977, Chalk formation: Early diagenesis, *in* Anderson, N. R., and Malahoff, A. (eds), The fate of fossil CO_2 in the oceans: New York and London, Plenum Press, p. 717–739.
———, AND HSÜ, K. J., 1971, Genesis and lithification of a deep sea chalk: Ecolog. Geol. Helv., v. 64, p. 273–278.
———, AND KELTS, K. R., 1972, Inferred diagenetic history of a weakly silicified deep sea chalk: Trans. Gulf. Coast Assoc. Geol. Soc., v. 22, p. 177–203.

SEPM SPECIAL PUBLICATION NO. 32, P. 209–226, OCTOBER 1981

SHALLOW-WATER LIMESTONES IN OCEANIC BASINS AS TECTONIC AND PALEOCEANOGRAPHIC INDICATORS[1]

SEYMOUR O. SCHLANGER
Hawaii Institute of Geophysics, University of Hawaii, Honolulu, Hawaii 96822 U.S.A.

ABSTRACT

Limestones whose constituents were deposited in shallow water are useful indicators of post-depositional uplift, subsidence and latitudinal changes through time. A review of such limestones, drilled at several dozen DSDP sites in all of the major ocean basins, demonstrates their utility in deciphering the geologic history of major structural features and in contributing to our knowledge of paleoceanographic conditions.

Drilling in the northeast Providence Channel of the Bahamas, on the Blake Nose and off the Grand Banks of Newfoundland has extended our knowledge of the Cretaceous reef trends and carbonate shelf margins in the Atlantic passive margin province and has demonstrated subsidence of several kilometers there since Cretaceous time. The Rio Grande Rise and the Walvis Ridge both had Late Cretaceous and Eocene shallow-water histories related to their subsidence as paired aseismic ridges. On Orphan Knoll and Rockall Bank, Jurassic to Paleogene shallow-water sediments containing rich Paleocene bryozoan faunas confirm plate reconstructions for the early opening of the Atlantic. On the aseismic Ninetyeast Ridge in the Indian Ocean shallow-water sediments, including lignite deposits, directly above the basalt basement indicate a progressive subsidence history for the ridge similar to that of an island-seamount chain. In the Pacific Ocean the widespread occurrence of Campanian-Maestrichtian reef faunas from the Line Islands to the Nauru Basin-Marshall Islands leads to an interpretation of central Pacific bathymetric history reminiscent of the Darwin Rise of Menard in that mid-plate volcanism resulted in regional uplift in the central Pacific between ~115–110 and 70 m.y.BP. This uplift contributed to the major Cretaceous transgression during that period. Invocation of a spreading rate increase along mid-ocean ridges as the sole explanation for this transgression may not be necessary. The large benthic foraminifers of Late Cretaceous age recovered along the Line Islands and in the Nauru Basin have affinities to those of the Caribbean, indicating that the mid-Pacific atoll scene was not dominated by an Indo-Pacific fauna until Cenozoic time. Further, the Caribbean affinities of the central Pacific Late Cretaceous reef faunas lends credence to earlier suggestions that the Caribbean plate is a relic of the Farallon plate. Analysis of vertical tectonics in the Pacific shows that estimates of paleo-CCD levels should take into account periods of uplift which interrupted normal plate subsidence. Cenozoic reef debris found in turbidite units in deep-water fan deposits adjacent to the Line and Marshall islands chains record global sealevel changes. Drilling on the Emperor Seamounts has revealed Paleogene bryozoan-algal limestones whose latitude of deposition confirms paleomagnetic data from Suiko Seamount which indicate that this seamount did not form at the latitude of the present Hawaiian hot spot.

INTRODUCTION

Limestones whose original constituents were deposited in shallow water have proven useful as indicators of post-depositional uplift, subsidence and paleolatitude. In particular, limestones made up largely of calcitic or aragonitic skeletal debris of animals or plants whose communal association indicates deposition in sunlit waters—approximately the upper 200 m of the sea depending on latitude and water transparency—are excellent "benchmarks" for determining both magnitude and rate of uplift or subsidence where the post-depositional vertical motion has been large relative to the errors involved in determining the depth of original deposition. A cautionary note should be sounded here on interpreting assemblages of skeletal debris made up of fragments of calcareous algae, echinoids, bryozoans, barnacles, corals, molluscs, etc. in terms of water depth, temperature and latitude. Ideally, interpretations should be based on complete taxonomic descriptions and with the possibility of resedimentation in mind. In working from the descriptions available, I have had to make some secondary interpretations. Papers calling attention to possible pitfalls include those by Teichert (1958), Chave (1967), Schlanger and Konishi (1975) and Lees (1975).

Prior to the Deep Sea Drilling Project, dredge hauls in the oceans and drilling operations on isolated atolls and limestone islands had recovered reefal limestones whose present depth indicated subsidence on the order of kilometers in the ocean basins. Along the Blake Escarpment, limestones of Cretaceous age from reef and associated facies were dredged from depths as great as 4800m (Heezen and Sheridan, 1966; Sheridan and others, 1971). Modeling of the evolution of passive margins has been made easier because of the information these limestones contain on the extent of subsidence associated with the opening of the Atlantic. Drilling on the Bahama Platform (reviewed by Paulus, 1972) revealed reef and associated limestones of Cretace-

[1] Hawaii Institute of Geophysics Contribution No. 1066

ous age lying 5200m below sea level. In the Pacific, drilling at Funa-Tuti, Enewetak, Bikini and Midway (see Schlanger, 1963; Ladd and others, 1970 for review) showed post-Eocene subsidence in the region of up to 1400m. This subsidence history was recently used by Detrick and Crough (1978) and Crough (1978) to develop an important tectonic model relating uplift to lithospheric thinning caused by passage of the Pacific plate over hot spots. Hamilton in his classic 1956 paper "Sunken Islands of the Mid-Pacific Mountains" interpreted dredged reef limestone to be indicative of major subsidence which confirmed Darwin's theory. Since the advent of the Deep Sea Drilling Project, cores containing shallow-water limestones have been recovered aboard the *Glomar Challenger* from several dozen sites in all of the major ocean basins of the world (Appendix; Fig. 1).

In a review paper such as this I believe that a discussion of the geologic significance of shallow-water limestone discoveries in the oceanic basins is facilitated by integration of the evidence from DSDP sites with other geological evidence by using a regional approach. I have also included, where appropriate, remarks on the reasons for the selection of the drilling sites. To this end I have reviewed DSDP volumes available to October 1979, journal articles that used DSDP data, and Geotimes articles on individual legs, and have included data soon to be published in the DSDP Initial Reports on Legs 55 and 61. Pushcharovskiy (1978) also has reviewed evidence for shallow-water conditions in oceanic basins and applied it to tectonic problems.

I have not attempted to discuss every site where shallow-water limestones have been recovered but have used the example method and selected what I consider to be significant sites or groups of sites where the presence of shallow-water limestones or skeletal debris has advanced our knowledge of the geologic evolution and paleogeography of a region. Because of space limitations I have not included sites in marginal seas such as the Mediterranean and Caribbean. Significance, like beauty, surely lies in the eye of the beholder. If I have neglected some geologists's favorite site, I apologize in advance.

ATLANTIC OCEAN

Bahama Platform and Blake Plateau

Drilling and surface studies around the perimeter of the Gulf of Mexico had revealed the presence of a Cretaceous barrier-reef system stretching from Mexico to the Bahamas; dredging (Heezen and Sheridan, 1966) and seismic profiling on the Blake Plateau indicated an extension of the reef trend to the east. Paulus (1972) wrote a detailed review on the subject of the Bahama Platform and the numerous individual references will not be cited here. Deep, land based drilling on Andros Island and New Providence Island had by 1970 (Paulus, 1972) reinforced the argument that the major Bahamian platforms were the result of Cretaceous to Recent reef growth. It remained to be determined, however, what geological processes had been responsible for the development of the deep Bahama channels that separate the shallow banks and islands (Fig. 2). "Site 98 was selected by the Atlantic Advisory Panel for the purpose of determining whether the Bahama Channels are primarily the product of: (1) differential rates of deposition, (2) tectonism (block faulting), or (3) erosional processes, or possibly a combination of these phenomena" (Paulus, 1972, p. 877). The Site 98 results (Fig. 2) confirmed the presence in the northeast Providence Channel of deepwater facies of Late Cretaceous to Pleistocene age underlain by turbidites containing fragments of peri-reef limestones. Paulus (1972) summarized the contribution of Site 98 to our knowledge of the Bahama Platform evolution and pointed out that evidently the higher bathymetric relief in the region is the result of differing depositional rates on the reef platforms and in the channels where pelagic sedimentation has prevailed; submarine erosion has accentuated this difference by removing some of the pelagic interplatform deposits. Newell's (1955) argument that the Bahama banks are of atoll reef origin is supported by the Site 98 data.

Reef limestone recovered from dredge hauls (Heezen and Sheridan, 1966) and inferred from seismic profiles (Sheridan and others, 1971; Sheridan, 1974) led to the selection of DSDP Sites 390 and 392 on the Blake Nose, the objectives of which were "...to determine the nature of the lower Cretaceous shelf edge that has been traced from Mexico around the U.S. Gulf Coast, and extrapolated into Cuba, the Bahamas, and up the Atlantic Coast..." (Enos and Freeman, 1978, p. 413). Cores recovered at these sites were studied by Enos and Freeman (1978, 1979) who concluded that:

1) The Lower Cretaceous limestones consist of three lithologic units: limestone with moldic porosity, oolitic limestone and fenestral limestone.

2) The moldic limestones containing pelletoids, green algae and small benthic foraminifers were deposited as a low-relief blanket near the shelf edge. The oolitic unit displays incipient hard grounds. The fenestral limestone containing pelletoids, miliolid foraminifers, gastropods, ostracodes and algal filaments was deposited on tidal flats.

3) These units were subjected to extensive subaerial diagenesis prior to the deposition of overlying Barremian pelagic calcareous oozes.

4) The present sub-sealevel depth of 2700m serves to "...dramatize the apparent degree of Barremian and later subsidence of this part of the Atlantic outer Continental shelf" (Enos and Freeman, 1979, p. 83).

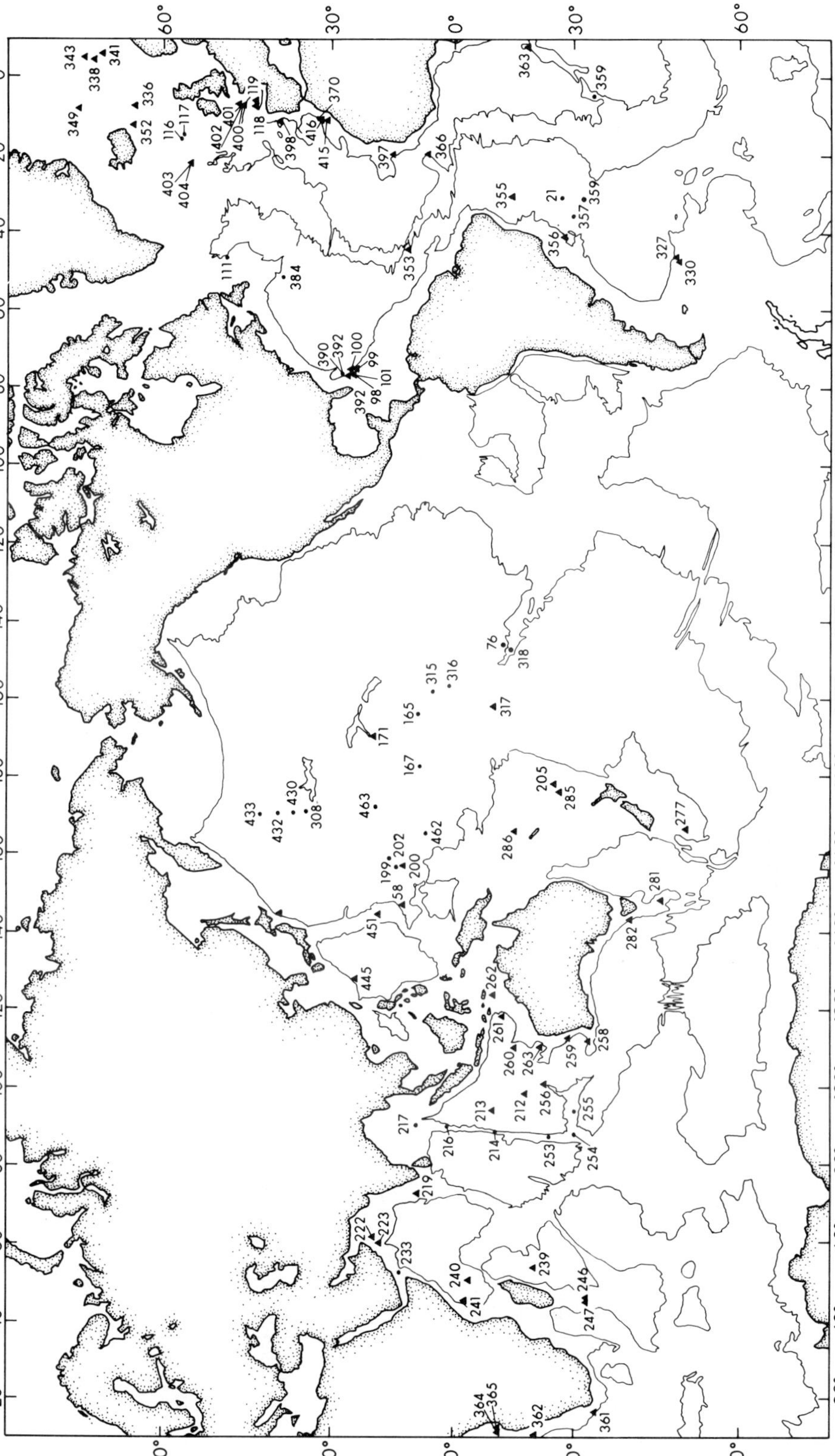

FIG. 1.—Locations of DSDP drill sites from which significant shallow-water limestones have been reported. Those sites marked by dots are discussed in the text.

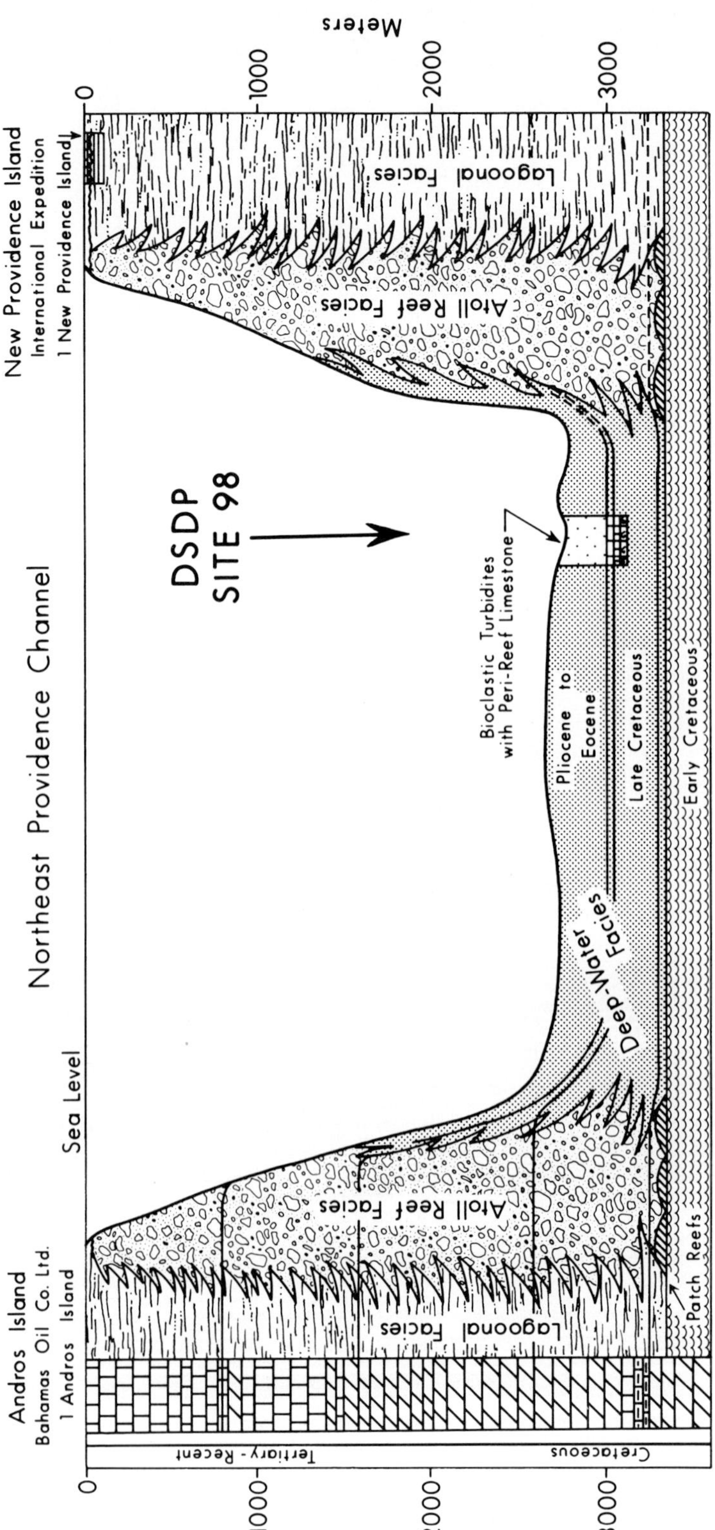

FIG. 2.—Geologic section from Andros Island through DSDP Site 98 to New Providence Island (modified from Paulus, 1972).

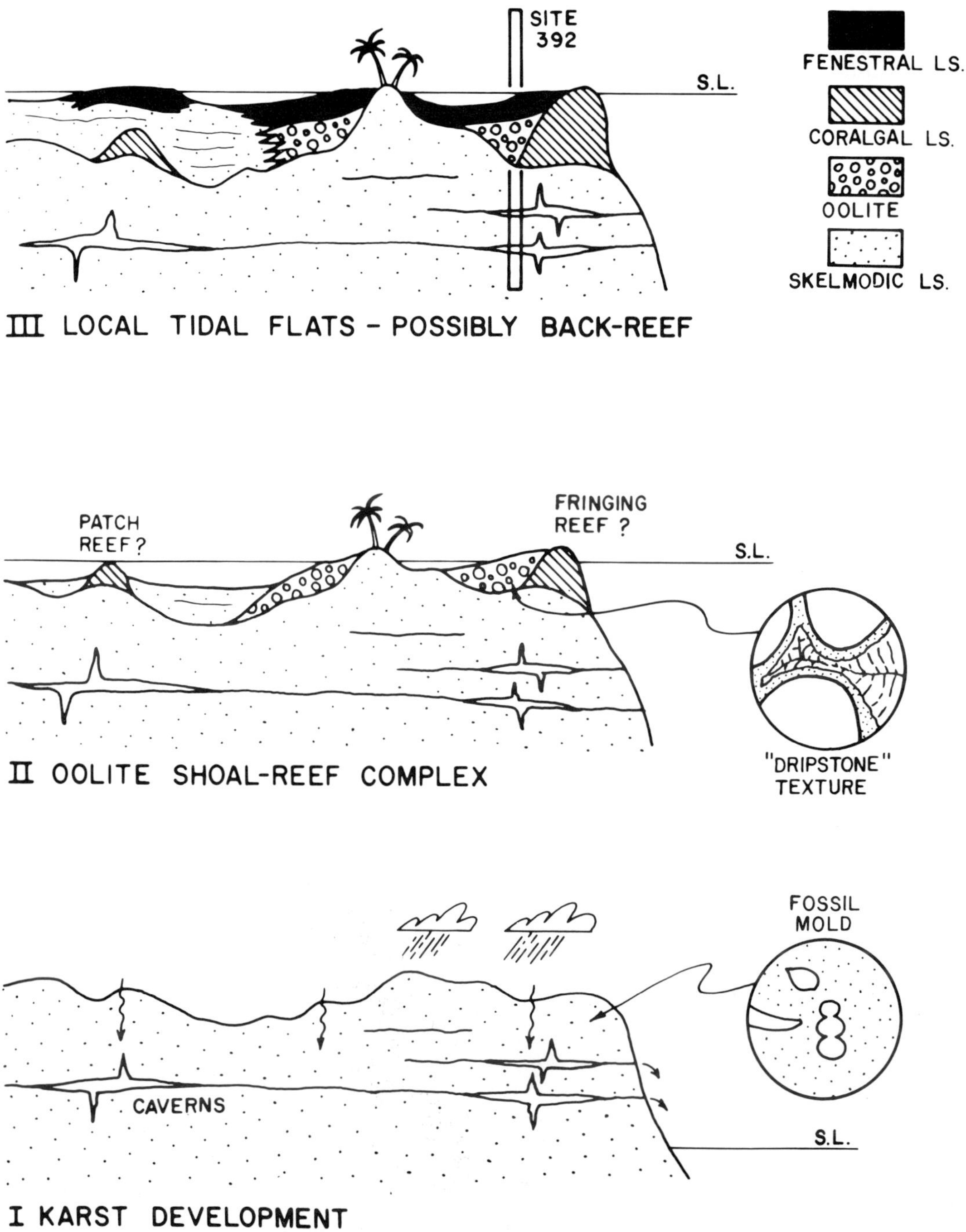

FIG. 3. — Sequence of depositional events at Site 392 which produced units of identifiable paleoenvironments of Early Cretaceous age (from Sheridan and Enos, 1979).

Figure 3 (from Sheridan and Enos, 1979) serves to illustrate the paleogeographic detail derived from such petrographic studies.

On Leg 43, (Tucholke, Vogt and others, 1979) limestones of Late Barremian to Early Albian age containing rudists, corals, gastropods and orbitoline foraminifers of back-reef origin were recovered at Site 384 from the J-Anomaly Ridge just south of the Grand Banks. This reef was exposed to subaerial diagenesis and dolomitized (Rothe, 1979). This discovery extends the Early Cretaceous carbonate shelf margin to 40°N latitude. The basalt basement, which lies directly below 123 m of this shallow water limestone, is now at a depth of 4235 m below sealevel indicating subsidence of more than 4 km since Early Cretaceous time (Tucholke, Vogt and others, 1979).

The circum-Caribbean Cretaceous reef and associated facies are excellent reservoirs and contribute to the prolific petroleum production in the region. Arthur and Schlanger (1979) have argued that the simultaneous buildup of shelf-edge reefs along passive margins and the deposition of organic carbon-rich basinal and outer slope black shales in Middle Cretaceous time, during the "Oceanic Anoxic Events" of Schlanger and Jenkyns (1976), may have been responsible for the concentration of oil in reservoirs of this type and age. Therefore, studies of the lithology and distribution of Cretaceous reef facies from DSDP sites such as 98, 390, 392 and 384 have contributed to our knowledge concerning oil possibilities in the areas discussed.

Rio Grande Rise and Walvis Ridge

Thiede (1977, p. 929) considered the bathymetric history of aseismic ridges and rises in the world ocean and came to the conclusion that ". . . the best indicator for the subsidence history of the rises is the distribution of shallow-water sediments along their flanks and peaks." He reviewed the evidence from Site 21 of Leg 3 and Site 357 of Leg 39 (Maxwell and others, 1970; Supko, Perch-Nielsen and others, 1977) and dredge hauls. At Site 21 a sorted coquina, cemented by sparry calcite, of Campanian or pre-Campanian age, was cored at a sub-sealevel depth of 2250 m. The limestone contains bivalve fragments, echinoderms and red algae, which Thiede (1977) interpreted as indicating deposition in water less than 100 m deep. At Site 357 an Eocene volcanic breccia containing red algae, larger benthic foraminifers and bivalve and gastropod shell fragments was cored. This fossil evidence allowed Thiede to conclude that the Rio Grande Rise has been present as a 2-km high volcanic island since at least Santonian time; its highest peaks subsided below sea level during Oligocene time. The southeastern extremity of the Walvis Ridge where Site 359 was drilled mirrors the Rio Grande Rise in terms of position and distance from the Mid-Atlantic Ridge (Supko, Perch-Nielsen and others, 1977). There, evidence for Eocene shallow-water environments of less than 1-km depths includes the presence of oyster-like bivalves, bryozoans, barnacles, brachiopods and a solitary coral. Such an assemblage could be interpreted as a cold and deep fauna (Teichet, 1958), and the shipboard party con-

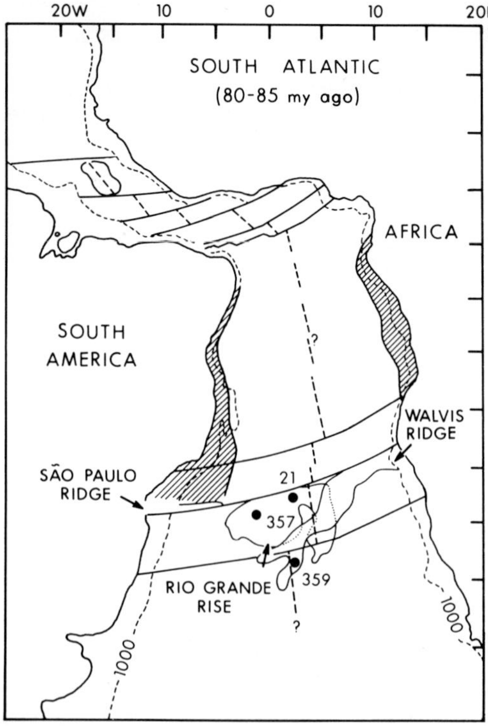

FIG. 4. – Reconstruction of the South Atlantic approximately 80–85 m.y.BP showing relationship of the Rio Grande Rise to the Walvis Ridge and locations of DSDP Sites 21, 357 and 359 (modified from Kumar, 1979).

siders deposition to have taken place between 300 and 900 m. The geographic juxtaposition of the Rio Grande Rise and the Walvis Ridge, the data for an early shallow-water history, and considerations of the volcanic geology led Kumar (1979) to propose a model for the evolution of these two features (Fig. 4) in which "excessive" volcanism along the Mid-Atlantic Ridge created them between 100 and 80 m.y.BP. This period of volcanism in the Atlantic is of interest because similar data were used from DSDP drill sites in the central Pacific to account for regional shallowing and reef buildups which accompanied widespread mid-plate volcanism during the same time period (see section below on the Pacific Basin).

Orphan Knoll and Rockall Bank

On Leg 12 *Glomar Challenger* drilled a North Atlantic transect that included sites on Orphan Knoll, in the Labrador Sea, on the Reykjanes Ridge, on Rockall Bank and in the Biscay Abyssal Plain (Laughton, Berggren and others, 1972). At Site 111 on Orphan Knoll and Sites 116 and 117 on Rockall Bank the discovery of shallow-water sediments added dramatically to our knowledge of the early opening of the North Atlantic. In a water depth of 1780 m, drilling at Site 111 penetrated 250 m into coarse sands of Bajocian age derived from Late Paleozoic strata. High-rank anthracite coal fragments found there are geochemically similar to the South Wales anthracite. This evidence led to the conclusion that these Jurassic sands were derived from the *east*! Orphan Knoll appears to be a piece of Europe abandoned during early opening of the Atlantic. An angular unconformity separates these sands from overlying carbonates of Albian, Cenomanian and Maestrichtian age. The hard sandy limestones of Albian age contain abundant foraminifers and ostracodes indicative of depths of deposition of less than 100 meters in cool marine waters. The Cenomanian strata contain oysters, echinoderms, gastropods and miliolid foraminifers indicative of deposition on a shelf less than 100 meters deep. These deposits mark "...the time of development of the first marine separation between Europe and America" (Laughton, Berggren and others, 1972, p. 77). The Maestrichtian strata are typical shelf chalks rather than oceanic pelagic sediments. Orphan Knoll has subsided 2050 m since Bajocian time; the major subsidence occurred in Paleocene time.

Geophysical evidence (see Laughton, Berggren and others, 1972, for a review) indicated that Rockall Bank was a continental fragment; Sites 116 and 117 were drilled to gain data on its tectonic history. Site 116 was drilled to the northwest of the Bank in the Rockall-Hatton Basin in 1150 m of water; Site 117 was drilled higher up on the flank of the Bank in 1038 m of water where seismic stratigraphy showed a better chance of penetrating to the basement with extant drilling capabilities. Basalt was reached at 311 m subbottom depth. Sediments directly on the basalt are characterized by shallow-water fossils including the alga *Lithothamnium,* fragments of thick-shelled molluscs, gastropods, bryozoans and foraminifers (including *Operculina*). Fractures in the basalt are also filled with fossiliferous sediments. These are of Late Paleocene age. Because of their abundance the ostracodes and bryozoans were subjects of intensive analyses by R. H. Benson and A. H. Cheetham, and E. Håkansson, respectively, in the Site report (Laughton, Berggren and others, 1972, p. 422–441). Paleodepth data on these two groups are displayed on Figure 5. In addition to paleodepth information the composition of the rich bryozoan fauna (94 species from 68 genera were identified) indicated a close biogeographic connection with both Europe and North America during Paleocene time and closer connections with

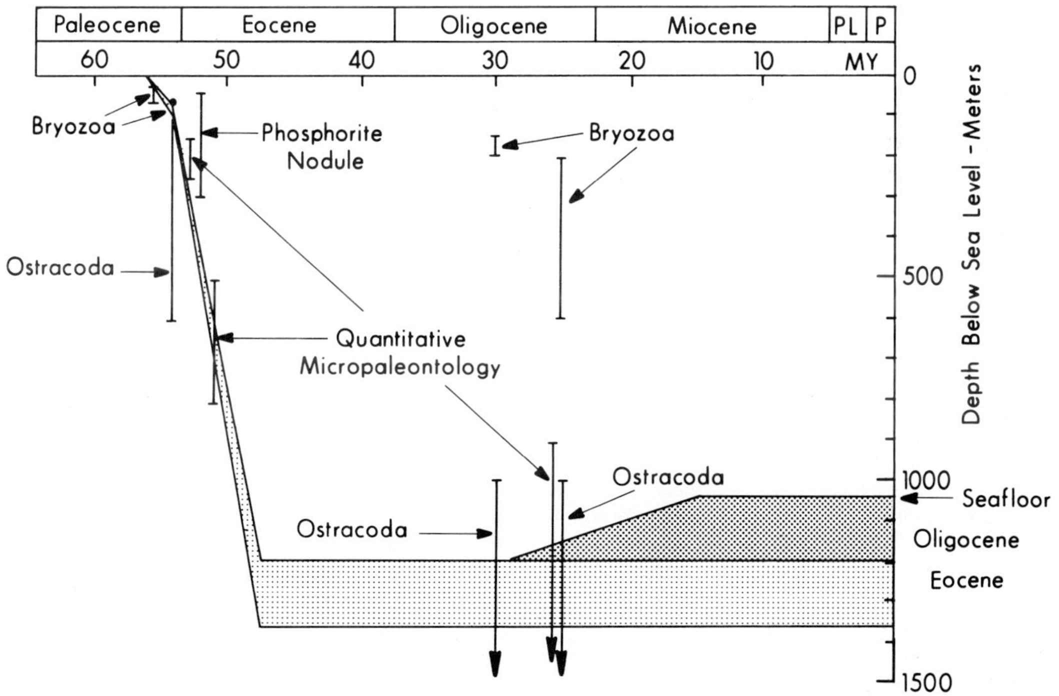

FIG. 5.—Crustal subsidence at DSDP Site 117 based on estimates of paleoenvironments indicated by fossils. The area shaded and dotted beneath the curve indicates sediment accumulation in the Rockall-Hatton Basin (from Laughton, Berggren and others, 1972).

Europe than North America in Oligocene time. This Paleocene bryozoan fauna is similar in abundance, species diversity and general aspect to that from the Late Paleocene Vincentown Formation in New Jersey. The Vincentown Formation is a typical shallow-shelf deposit, rich in glauconite, quartz sand and bryozoans of the northern Atlantic Coastal Plain (Schlanger, 1954). As Cheetham points out (Laughton, Berggren and others, 1972, p. 433), "Changes in the assemblages are consistent with a progressive isolation of the Rockall Plateau and subsidence of the seafloor from slightly less than 60 m to slightly more than 200 m below sea level from Late Paleocene to Oligocene time."

The biogeographic data on European and North American affinities and the paleodepth interpretations derived from the Site 117 material contributed to the paleogeographic reconstructions of Laughton (1972, figs. 16–19). Shortly prior to 150 m.y.BP the Orphan Knoll area was receiving its Bajocian continentally derived sands and coal fragments from South Wales, and the sedimentary record at Rockall Bank had not begun. By Paleocene time Rockall Bank was separated from both Greenland and the Irish shelf and began accumulating its oldest shallow benthic skeletal debris.

INDIAN OCEAN

The Indian Ocean was the scene of intensive drilling efforts during DSDP Legs 22 through 29. The results of this campaign have been integrated with geophysical studies and synthesized in two recent volumes by Heirtzler and others (1977) and von der Borch (1978). At a number of sites shallow-water sediments were recovered, which was to be expected because many sites were located on shelves and basins near the African, Australian and Indian continental masses. However, one of the Indian Ocean's more remarkable geologic features, the history of which has been elucidated by analysis of shallow-water sediments, is the Ninetyeast Ridge, which in this paper will serve as the Indian Ocean example. Pimm and others (1974), Detrick and others (1977), Sclater and others (1977) and Luyendyk (1977) have all written on the history of the Ninetyeast Ridge using DSDP data from Sites 254, 253, 214, 216 and 217. At Site 254 a Middle or Upper Eocene to Lower Oligocene fauna of bryozoans, benthic foraminifers, ostracodes, pelecypods and gastropods was found in volcanically derived sediment. This unit represents a lagoonal and littoral facies. The volcanogenic sediments indicate subaerial erosion of highly weathered basalt terrain nearby. At Site 253 a shallow-bank facies overlies the basement. At Site 214 lignite is associated with basal volcanogenic units above the basalt basement; Paleocene glauconitic silts and sands containing gastropods, bivalves and echinoderms underlie early Eocene to Pleistocene pelagic carbonate oozes. The lignite is taken as an indicator of a subaerial history. At Site 216 a shallow-water facies of Maestrichtian age was cored above the basement. At Site 217 algal and reef structures of Campanian age are described as dolomitized and as representative of lagoonal and littoral facies. This brief summary emphasizes not only the presence of shallow-water facies at all of the Ninetyeast Ridge sites but also the fact that these facies are progressively older from south to north. These data are graphically summarized in Figure 6. Evidently the Ninetyeast Ridge formed progressively; "The geology of the ridge is more comparable to that of an island-seamount chain such as the Hawaiian-Emperor, than of any other oceanic feature" (Luyendyk, 1977, p. 183). As discussed in the section below on the Pacific Basin, shallow-water limestones also helped greatly in deciphering the history of the Hawaiian-Emperor chain.

On Broken Ridge, which abuts the south end of Ninetyeast Ridge, drilling at Site 255 recovered coarse-grained, sandy, chert gravel with abundant bryozoans and Eocene large benthic foraminifers such as *Asterocyclina.* This sediment is taken as evidence of reworking during Middle Eocene time and shows that Broken Ridge as well as parts of the Ninetyeast Ridge may have been emergent or at near sealevel at that time.

PACIFIC BASIN

Late Cretaceous Reef Fossils in Deepwater Strata

Displaced reef fossils of Campanian-Maestrichtian age were recovered at DSDP drill sites in the Line Islands and in the Nauru Basin just west of the Marshall Islands (Fig. 7). During Leg 17, Site 165 was drilled on a large, deep submarine fan northwest of Kingman Reef in the northern Line Islands. At this site an integrated reef fauna of large benthic foraminifers—including *Pseudorbitoides*—and broken and abraded skeletal fragments of molluscs, echinoids and bryozoans were found associated with volcaniclastic strata of Campanian-Maestrichtian age. These were interpreted to be turbidites containing redeposited skeletal and lithic debris eroded from reefs and edifices of Late Cretaceous age (Winterer, Ewing and others, 1973). On Leg 33, Sites 315 and 316 were drilled adjacent to the Line Islands in geologic settings similar to Site 165 (Schlanger, Jackson and others, 1976). At both sites Cretaceous faunas virtually identical to those found at Site 165 were encountered in similar volcaniclastic stratigraphic units. The Leg 33 sites yielded a diverse fauna that included *Pseudorbitoides, Asterorbis* and *Sulcoperculina;* coralline algae and rudist fragments were also noted (Beckmann, 1976). These occurrences of coeval Campan-

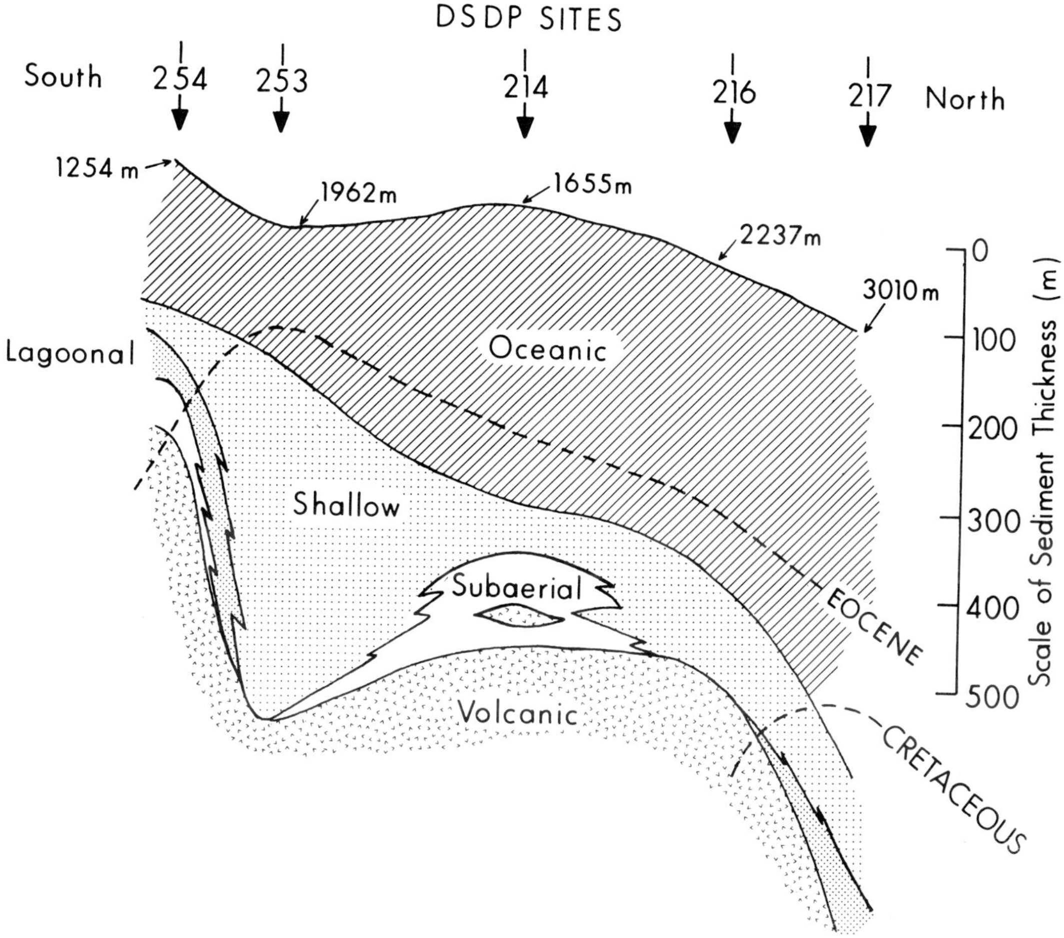

FIG. 6.—Diachronous facies on the Ninetyeast Ridge showing areal distribution and age of shallow-water facies (from Pimm and others, 1974; Luyendyk, 1977).

ian-Maestrichtian reefal faunas in identical volcaniclastic sections along the Line Islands from Site 165 to 316 record volcanic activity that ended ~70 m.y.BP after creating edifices that had grown up into the euphotic zone, thereby allowing Late Cretaceous reef growth simultaneously along the entire length of the central Line Islands (Jackson and Schlanger, 1976).

At Site 462, drilled in the Nauru Basin during Leg 61 (Larson, Schlanger and others, *in press*), Campanian-Maestrichtian reef faunas were again found in a stratigraphic context identical to that found along the Line Islands (Premoli-Silva and Brusa, *in press*), indicating that Late Cretaceous reefs grew simultaneously along both the Line and Marshall islands—the latter being the presumed source of the Nauru Basin reefal debris (Fig. 7). Heretofore, the Marshall edifices were thought to be Eocene in age, on the basis of drilling results from Enewetak (Schlanger, 1963). The presence at Site 462 of a thick section of Cretaceous sills and extrusive basalt units apparently younger than the underlying plate dated at 155 m.y. (Larson and Hilde, 1975) suggested that edifice-building, permitting reef growth on a Cretaceous Marshall chain, may have accompanied Cretaceous deeper water volcanism (Schlanger and Premoli-Silva, *in press*).

Rea and Thiede (*in press*), in a study of mass accumulation rates of volcaniclastic debris, point out a marked increase in accumulation rate over a wide area of the Central Pacific from the Caroline Islands to the Line Islands between Late Coniacian and Early Maestrichtian time—85 to 68 m.y.BP. Within this time interval there was an especially marked peak during Campanian and early Mid-Maestrichtian time.

The striking similarity of the volcaniclastic sec-

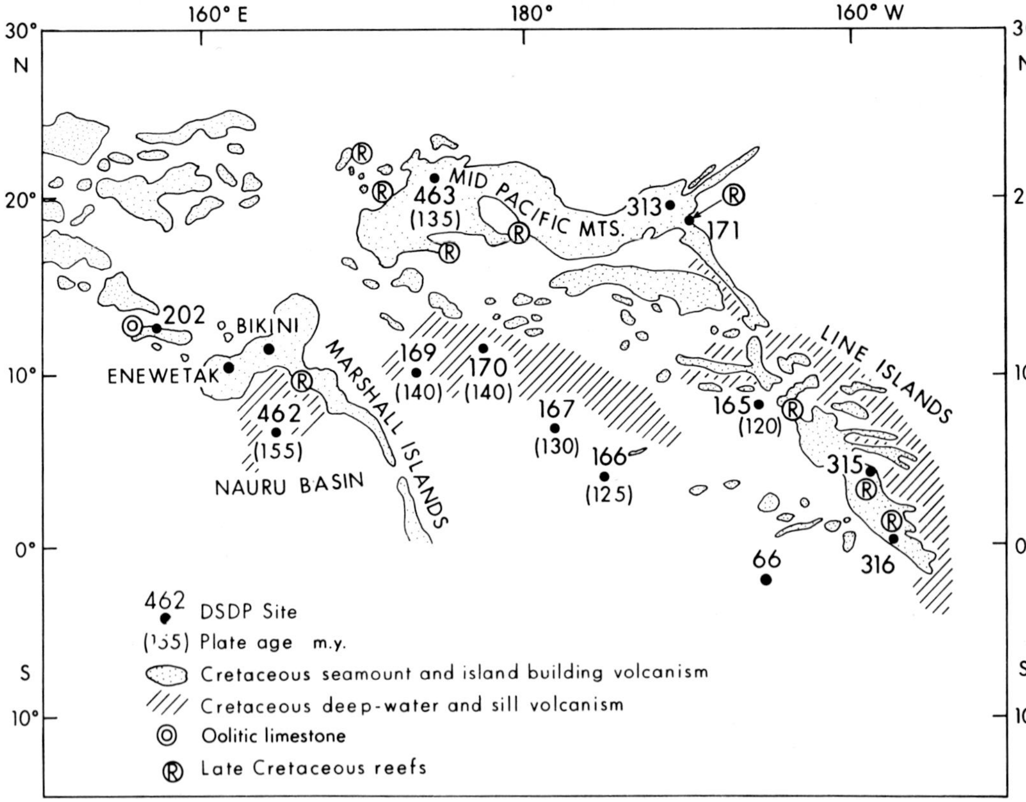

FIG. 7.—Central Pacific Basin showing distribution of Cretaceous reefs, deepwater and seamount and island-building volcanism, plate ages, and DSDP drill sites (modified from Winterer, 1976).

tions with their included fossils at Site 462 and at Sites 165, 315 and 316 along the Line Islands argues for the fact that the Line Islands and the Marshall Islands may have had very similar Late Cretaceous histories. The Campanian-Maestrichtian volcaniclastic section at Site 313 (Larson, Moberly and others, 1975) is also similar to the Line Islands and Site 462 sections; however, shallow-water reefal debris was not found there.

These Late Cretaceous fossiliferous volcaniclastic sections represent the culminating phase of a period of mid-plate volcanism that seems to have begun in the central pacific ~115–110 m.y.BP (Figs. 7, 8). Matthews and others (1974) reviewed the distribution of rudist reefs in the Mid-Pacific Mountains region (see also Ladd and others, 1974) and showed that reef growth in these mountains started in Aptian-Albian time. Dean and others (1979) based on the presence of land-derived, humic-rich organic matter in carbonaceous limestone of Early Aptian age associated with volcanic ash at Site 463, propose that numerous volcanic islands existed in the Mid-Pacific Mountains in Aptian time.

Data from DSDP Sites 169 and 170 (Winterer, 1976) showed Albian mid-plate volcanism in the form of deepwater sill intrusions. Winterer (1976) proposed that regional uplift accompanied this mid-plate volcanism. At Site 462 (Larson, Schlanger and others, *in press*) a sill and extrusive submarine flow complex records volcanic activity there during Mid-Cretaceous time long after the oceanic crust formed ~155 m.y.BP (Larson and Hilde, 1975). Terrestrial plant debris in Aptian sediments in the Nauru Basin indicates that the Cretaceous Marshall Islands included vegetation-covered volcanic islands (Jenkyns and Schlanger, *in press*). Schlanger and Premoli-Silva (*in press*) applied the thermal uplift model of Detrick and Crough (1978) and Crough (1978) to the Nauru Basin in an attempt to estimate the magnitude of uplift in the region and proposed that from ~115–110 to 70 m.y.BP the area was uplifted, departing from the "normal" subsidence path determined by Parsons and Sclater (1977), and then subsided ~1.5 km as have Bikini, Enewetak and Site 202 (Fig. 8) where the oolitic limestone (see below) serves as a measure of subsidence. This tectonic-volcanic history is reminiscent of that proposed for the Darwin Rise by Menard (1964).

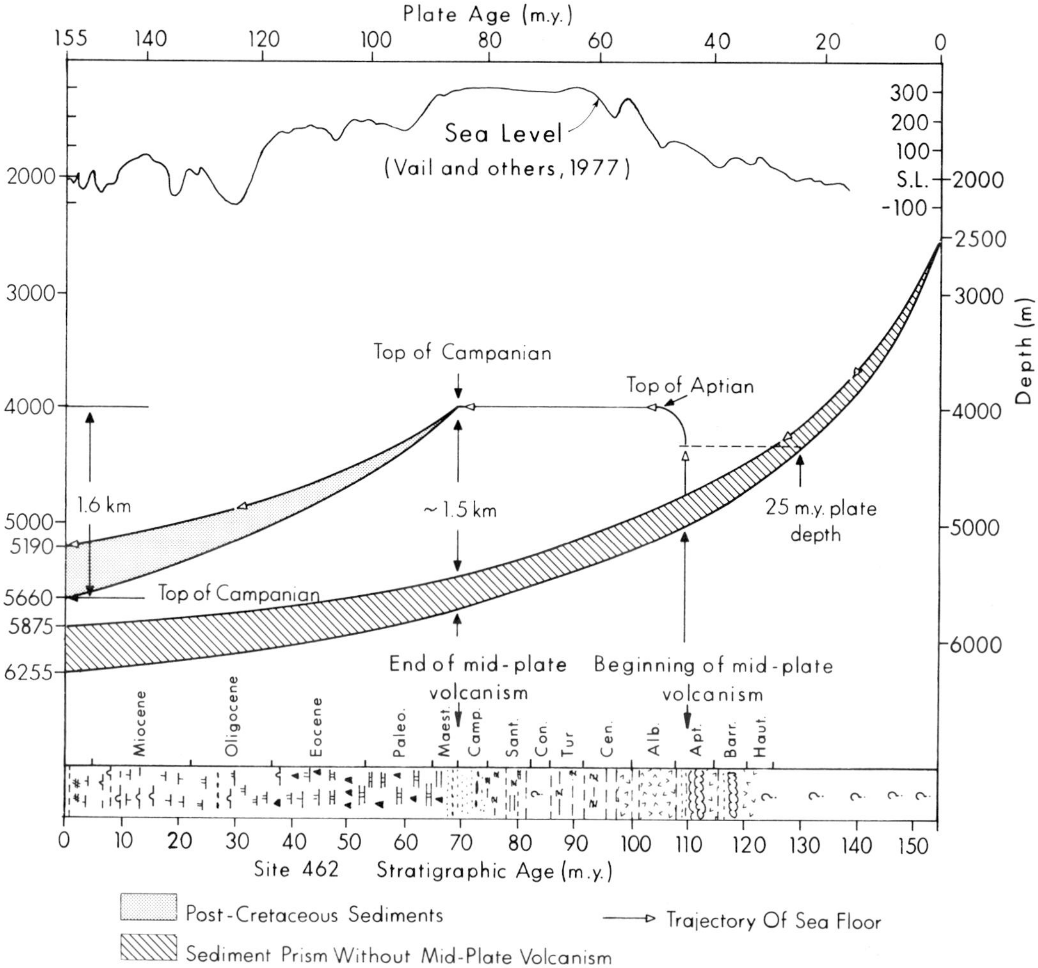

FIG. 8.—Uplift and subsidence history of the Nauru Basin (modified from Schlanger and Premoli-Silva (*in press*) and Schlanger and others (1981). The bathymetric history of Site 462, had there not been Cretaceous mid-plate volcanism, is shown by the diagonally shaded prism. Starting at a ridge-crest elevation 155 m.y.BP, sediments would have accumulated as the lithospheric plate sank so that the depth of the Nauru Basin would have been 5875 meters and a sediment cover of ~300 m would have overlain the top of the plate at 6255 m (data from Parsons and Sclater, 1977, table 4). Due to mid-plate volcanism beginning ~115–110 m.y.BP the lithospheric plate was thermally reset to a 25 m.y.BP depth (Detrick and Crough, 1978). The area stayed high until the cessation of mid-plate volcanism ~70 m.y.BP. The Campanian sediments then being deposited now lie 5660 m below sea level (Larson, Schlanger and others, *in press*). A basin fill of Cenozoic turbidites reaches to 5190 m below sea level—the present depth of the Nauru Basin. The post-Cretaceous subsidence of ~1.6 km is approximately equal to the subsidence of Enewetak, Bikini and Site 202.

The uplift of a large area in the central Pacific during Cretaceous time, due to a Cretaceous mid-plate volcanic-thermal event followed by volcanism extending to Campanian-Maestrichtian time, ~70 m.y.BP, may have contributed to the marked Cretaceous sea level rise beginning about 110 m.y.BP (Fig. 8), which Vail and others (1977) have described as a worldwide event. Schlanger and others (1981) and Schlanger and Premoli-Silva (*in press*) estimated the maximum uplift to be ~2 km. However, a recent analysis of the vertical bounds on Cretaceous uplift by Larson and Schlanger (*in press*) shows that the maximum uplift may have been closer to 1.5 km (Fig. 8). Increases in spreading rates along rise crests between 110 and 85 m.y.BP (Larson and Pitman, 1972; Hays and Pitman, 1973; see Pitman, 1978, for a review) may, therefore, not have been the only major factor in producing this great Cretaceous transgression; epeirogenic, thermally-induced regional uplift also contributed to some extent.

The Nauru Basin evidently had a history similar

to that of the Caribbean whose shallow depth and thick crust have been attributed to the intrusion of sills approximately 80 m.y.BP (Saunders and others, 1972; Burke and others, 1978). A number of workers (*e.g.,* Mattson, 1969; Edgar and others, 1971; Malfait and Dinkelman, 1972) have argued that the present Caribbean plate is a detached relic of the Farallon plate. Beckman (1976) pointed out that the large benthic foraminifers found at Site 165 have Caribbean affinities. The Campanian-Maestrichtian reef faunas discussed above, now known from Sites 165, 315, 316 and 462 are all related to Caribbean faunas rather than to Pacific faunas (Premoli-Silva and Brusa, *in press*). These faunal relations lend credence to the argument that the Caribbean plate is a relic of the Farallon plate. The recent work of Kumar (1979) in the Atlantic shows that "excessive volcanism" 100 to 80 m.y.BP formed the Ceara-Sierra Leone and Rio Grande-Walvis Ridge paired aseismic ridges (see section on the Atlantic Basin). These volcanic events in the Atlantic and Caribbean, accompanied by regional uplift, could also have contributed to the major Cretaceous sea level rise (Schlanger and others, 1981).

Late Cretaceous CCD Excursions

Thierstein (1979) reviewed carbonate contents of Late Mesozoic sediments in the Pacific that were cored on several DSDP legs and concluded (p. 261) that "Increased carbonate contents in the early Campanian sediments at Sites 167 and 170 indicate a CCD excursion into the deep Pacific, that extended down to 6 km paleodepth in the late Campanian to early Maestrichtian *Tetralithus trifidus* Zone at Site 199". Consideration of possible shallow-water indicators and subsidence history at Sites 202 and 167, however, shows that the CCD may well have been shallower than 6 km in Late Cretaceous time in the central Pacific. At Site 167 "Recrystallized benthonic foraminifera similar to outer neritic and upper bathyal species in California, occur in the reddish marls of Cores 56 and 60" (Winterer, Ewing and others, 1973, p. 148). These cores span Late Turonian to Early Campanian time, 75 to 87 m.y. Thus, the top of the Magellan Rise (Site 167) could have been shallower than its presumed depth obtained by backtracking it along a Parsons-Sclater curve. The increase in carbonate content noted by Thierstein (1979) need not be ascribed to a deepening of the CCD. Site 199 (Heezen, MacGregor and others, 1973) was drilled in 6090 m of water at the east margin of the Caroline Abyssal Plain 70 km north and 80 km west of Site 202, which was drilled in 1505 m of water atop Ita Maitai Guyot (Fig. 7). Oolitic limestone and underlying lagoonal carbonates of probable Paleocene age cored atop the basalt edifice show that Ita Maitai Guyot was at sealevel at that time (Hesse, 1973). Ita Maitai Guyot evidently built upon a volcanic edifice, with eruptions culminating in Campanian-Maestrichtian time, and then subsided. The oolitic limestone has subsided 1600 m since Paleocene time, an amount remarkably similar to the subsidence at Enewetak. Jones (1973) used sonobuoy data obtained over Ita Maitai Guyot to show that the limestone cap there could be 525 m thick and suggested post-volcanic subsidence of 2090 m. The Campanian chalks at Site 199 now lie at a sub-sea level depth of 6500 m. If Site 199 sank in the same manner as did Site 202 then the maximum depth of deposition of the Campanian chalks could have been as shallow as 4400 m and there is no need to invoke a CCD deepening to 6 km to account for the Campanian chalks.

Estimation of paleo-CCD depths should take into account the possibility that the seafloor in the area under consideration was uplifted above a relatively stable CCD level after it had once passed below the CCD, *i.e.,* the seafloor did not follow an uninterrupted subsidence path (Detrick and Crough, 1978; Crough, 1978). Winterer (1976) recognized this possibility when he pointed out that Site 170 had apparently risen above the CCD for a short period during Campanian time.

Cenozoic Reef Debris in Deepwater Strata

Skeletal debris characteristic of eroded reefs have been recorded from a number of deepwater DSDP sites in the central Pacific; such material is particularly abundant from Sites 165, 315, 316 and 462 drilled along the flanks of the Line and Marshall islands chains. Premoli-Silva and Brusa (*in press*) have undertaken extensive studies of the taxonomy and stratigraphic distribution of these shallow-water fossils; this section is derived largely from their work. These fossils are found in carbonate turbidite sequences that make up much of the archipelagic aprons around these major atoll chains. The composition of the skeletal debris varies from site to site, but typically it is represented by assemblages of abraded and broken fragments of coral, bryozoans, molluscs, red and green algae, echinoids and large benthic foraminifers such as *Discocyclina, Asterocyclina, Miogypsinoides, Spiroclypeus, Assilina, Operculina* and *Lepidocyclina.* The fragments are commonly chalky, and many are coated by calcite crystals. In addition to individual fossil fragments, calcite-cemented reef-rock fragments have been identified. Also, clusters of coarse calcite crystals occur in the debris-rich sections. The presence of the calcite cements and the limestone fragments suggests that lithified reef-rock as well as loose reef sediments were being eroded and then transported by turbidity currents. Present-day reefs are contributing to deepwater deposits particularly where high-velocity currents through reef passes carry lagoonal material past the reef front. Also typhoons cause large amounts of reef debris to move into deeper waters. However, the reef debris found along the Line Islands and in the Nauru Basin—where it was probably derived from the Marshall Atolls—is concentrated in deepwater turbidites deposited during Early and Late Oligocene and Middle Miocene time. The reef debris is interpreted as the product of erosion of reefs during sealevel lowerings. As pointed out by Schlanger and Douglas (1974) Oligocene and Middle Miocene time were periods when atolls such as Enewetak and Bikini were emergent during low sealevel stands. Savin and others (1975) documented a severe drop in sea surface temperature beginning in Late Eocene time and culminating in Late Oligocene time. This corresponds to a major phase of glaciation in the southern hemisphere which could account for a major sealevel drop and consequent erosion of Eocene and Oligocene reefs over much of the Pacific. Vail and Hardenbohl (1979) showed marked sealevel falls in both Middle to Late Oligocene and Middle to Late Miocene time after a relative high during Eocene time. The Late Oligocene sea level drop of approximately 400m suggested by Vail and Hardenbohl (1979) is of particular interest inasmuch as the turbidites of Oligocene age from Sites 462 and 315 contain not only Cenozoic but also Cretaceous reef debris attesting to the severity of the Oligocene erosional period in the Pacific. Thus these displaced reef fossils convey information on sealevel changes and may be valuable in calibrating sealevel curves.

At Sites 76 (Hays and others, 1972) and 318 (Schlanger, Jackson and others, 1976) drilled on the northeast flank of the Tuamoto Ridge, and on the ridge itself, reefal debris of Early to Late Eocene age was found (Premoli-Silva and Brusa, *in press*). The Tuamotu reefs have therefore contributed to deep-water sedimentation since Early Eocene time. These fossils can be interpreted as indicative of a possible date for cessation of volcanism in the Late Cretaceous to Early Eocene for at least the northwestern part of the Tuamotu chain.

Paleolatitude of the Emperor Seamounts

Leg 55, along the Emperor Seamounts (Fig. 9), was designed to test the hot-spot hypothesis for the origin of the Hawaiian-Emperor chain (Jackson, Koizumi and others, 1980). Data and interpretations derived from Leg 55 are discussed in detail by Jackson and others (1980), and only a brief summary is presented here. Prior to Leg 55 shallow-water carbonates from Kōko Guyot were described from dredge hauls (Davies and others, 1972) and DSDP cores recovered at Site 308 (Matter and Gardner, 1975; Larson, Moberly and others, 1975). Also, seismic profiles (Davies and others, 1972) showed apparent reef structures on Suiko, Nintoku, Jingu, Ojin and Kōko seamounts. On the basis of an interpretation of these carbonates as

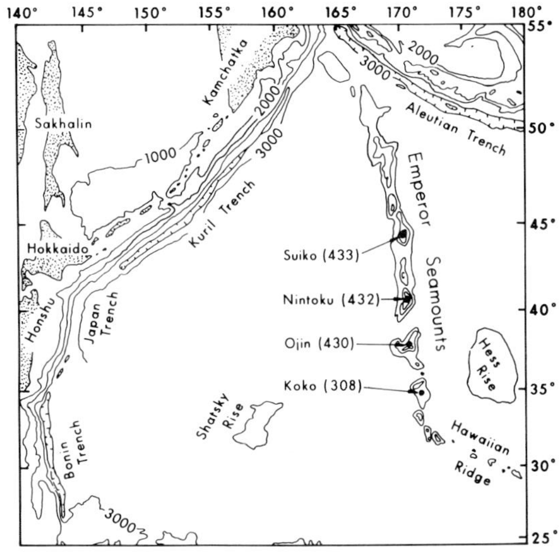

FIG. 9.—Emperor Seamounts drilled on DSDP Legs 32 and 55.

true tropical reefs, Greene and others (1978, p. 70) concluded that these reefs "...indicate northward movement of the Emperor Seamounts as predicted by the hot-spot hypothesis." However, the Paleogene carbonate sections drilled on Suiko, Nintoku and Ojin seamounts are rich in skeletal debris of bryozoans, echinoids, molluscs and red algae; red algal and bryozoan encrustations are pervasive. Corals are present but rare. Larson, Moberly and others (1975) pointed out that the main difference between these sediments at Kōko and modern Hawaiian sublittoral deposits is the higher bryozoan content of the former.

Study of these Paleogene skeletal bioclastic limestones from the Emperor Seamounts by McKenzie and others (1980) show them to be representative of the Pacific Bryozoan-Algal facies rather than the Coral-Algal facies. On the basis of a study of Eocene to Recent shallow-water carbonate facies in the western Pacific, Schlanger and Konishi (1966, 1975) developed a regional facies model (Fig. 10) that explained the bipolar distribution of bryozoan-rich limestones in the region in terms of water temperature and solar insolation, both of which vary systematically with latitude (see also Chave, 1967; Lees, 1975). They pointed out that limestones of the Coral-Algal facies, dominated by hermatypic corals and calcareous algae, including green algae such as *Halimeda,* are largely restricted to latitudes lower than ~23°. In higher latitudes, bryozoans, red algae and locally even brachiopods are abundant, and true coral reefs are rare; such deposits belong to the Bryozoan-Algal facies.

Paleomagnetic data (Jackson and others, 1980) indicate a paleolatitude of formation for Suiko of

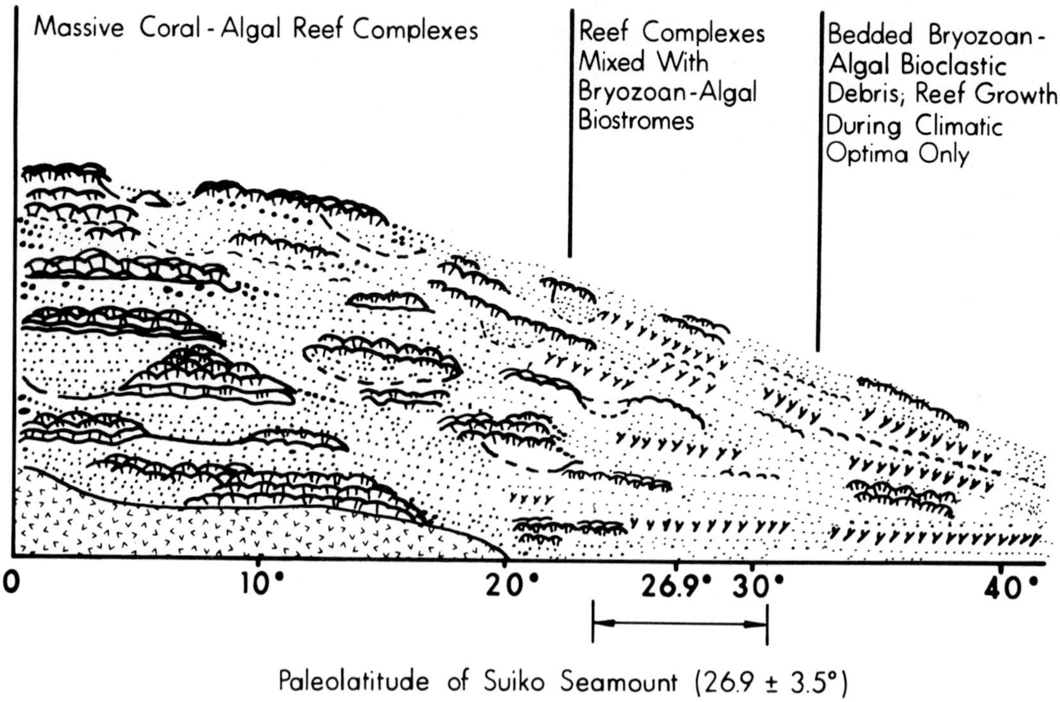

FIG. 10.—Coral-Algal and Bryozoan-Algal facies model of Schlanger and Konishi (1975) showing the latitudinal distribution of these two facies in the western Pacific. The paleolatitude of formation of Suiko Seamount (Jackson and others, 1980) is shown in its relation to the facies model.

26.9 ± 3.5°N latitude. This latitude is significantly different from the present latitude of Suiko (Site 433) at 44.8°N and of Kilauea, the present location of the hot spot postulated to be responsible for the generation of the Hawaiian-Emperor chain, at 19.5°N. As shown on Figure 10 the 26.9 ± 3.5°N paleolatitude of Suiko fits precisely into the facies scheme of Schlanger and Konishi. Thus the evidence from these shallow-water limestones substantiates the 7° difference between the paleolatitude of formation of Suiko and the present Hawaiian hot spot.

CONCLUSIONS

Limestones of shallow-water origin make up only a vanishingly small percentage of the rocks recovered during DSDP and IPOD drilling. However, analysis and interpretation of this material have added to our knowledge of the geological evolution of major geologic features such as the Rio Grande–Walvis Ridge paired aseismic ridges, the Bahama Platform, Orphan Knoll, Rockall Bank and the Ninetyeast Ridge. Drilling on the Blake Nose and off Newfoundland revealed Cretaceous reefs and carbonate shelf margins similar to those found in the circum-Caribbean region; because these reefs formed at sealevel, precise measurements of post-Cretaceous subsidence can be obtained. In the central Pacific the discovery of Campanian-Maestrichtian reef faunas has resulted in an interpretation of the region's paleobathymetric history reminiscent of that for the Darwin Rise by Menard. This interpretation suggests that the major Cretaceous transgression of 110 to 70 m.y.BP may be related to regional uplift caused by mid-plate volcanism as well as any increase in spreading rates along the mid-ocean ridge system. The Pacific Campanian-Maestrichtian reef faunas are of Caribbean affinities and therefore reinforce older arguments that the Caribbean plate is a relic of the Farallon plate. Estimates of paleo-CCD excursions can be refined by consideration of regional uplift and subsidence deduced from present depths of shallow-water limestones. The stratigraphic distribution of redeposited shallow-water reef debris in submarine fans and aprons around the Line and Marshall islands records periods of atoll emergence due to Cenozoic sealevel fluctuations. On the Emperor Seamounts the presence of Paleogene bryozoan-algal limestones on the basalt edifices confirms paleomagnetic data from Suiko Seamount indicating that it formed ~7° north of the present Hawaiian hot spot.

The paleoenvironmental information recovered from shallow-water limestones has been a somewhat unexpected but important product of 10 years of drilling by that worthy successor to H.M.S. *Challenger,* the *Glomar Challenger.*

ACKNOWLEDGMENTS

I thank my shipmates on DSDP Legs 17, 33 and 61 for their aid and advice while we traced the extent of Late Cretaceous reefal faunas from the Line Islands to the Nauru Basin and worked out the geological implications of these fossils. The late Dale Jackson invited me to work on Leg 55 material from the Emperor Seamounts and Brent Dalrymple kindly gave permission to use data from that leg in this paper. Isabella Premoli-Silva permitted use of our data on Cretaceous and Tertiary rocks and mid-plate volcanism obtained on Leg 61. Ralph Moberly read drafts on the manuscript and contributed patient discussions. Paul Enos and Michael Arthur kindly reviewed the manuscript and made valuable suggestions. Janet Haggerty helped in the literature search through the DSDP Initial Report "Blue Books" and Geotimes articles. To Josh Tracey, who 30 years ago on Guam taught me how to study reefs and reef limestones, and to Harry Ladd, who encouraged me to study the cores from Enewetak, I owe a special debt.

REFERENCES CITED

NOTE: All references to Deep Sea Drilling Project Initial Reports have been abbreviated; for complete citations see *Appendix I* at the end of this volume, or the individual Initial Reports as published.

Arthur, M. A. and Schlanger, S. O., 1979, Cretaceous "oceanic anoxic events" as causal factors in development of reef-reservoired giant oil fields: Am. Assoc. Petroleum Geologists Bull., v. 63, p. 870–885.

Beckmann, J. P., 1976, Shallow water foraminifers and associated microfossils from Sites 315, 316 and 318, DSDP Leg 33: Init. Repts. DSDP, v. 33, p. 467–490.

Burke, K., Fox, P. J. and Sengör, A. M. C., 1978, Buoyant ocean floor and the evolution of the Caribbean: Jour. Geophys. Research, v. 83, p. 3949–3954.

Chave, K. E., 1967, Recent carbonate sediments—an unconventional view: Jour. Geological Education, v. 15, p. 200–204.

Crough, S. T., 1978, Thermal origin of mid-plate hot-spot swells: Geophys. Jour. Royal Astro. Society, v. 55, p. 451–469.

Davies, T. A., Wilde, P., and Clague, D. A., 1972, Kōko Seamount: A major guyot at the southern end of the Emperor Seamount: Mar. Geol., v. 13, p. 311–321.

Dean, W. E., Thiede, J. and Claypool, G. E., 1979, Origin of organic-rich limestones of mid-Cretaceous age, mid-Pacific Mountains and Southern Hess Rise, central north Pacific Ocean: abstracts with programs, Geol. Soc. Amer. Annual Meeting, v. 11, p. 411.

Detrick, R. S. and Crough, S. T., 1978, Island subsidence, hot spots, and lithospheric thinning: Jour. Geophys. Research, v. 83, p. 1236–1244.

Detrick, R. S., Sclater, J. G. and Thiede, J., 1977, The subsidence of aseismic ridges: Earth and Planetary Sci. Letters, v. 34, p. 185–196.

Edgar, N. T., Ewing, J. I. and Hennion, J., 1971, Seismic refraction and reflection in the Caribbean Sea: Am. Assoc. Pet. Geologists Bull., v. 55, p. 833–870.

Enos, P. and Freeman, T., 1978, Shallow-water limestones from the Blake Nose, Sites 390 and 392: Init. Repts. DSDP, v. 44, p. 413–461.

Enos, P. and Freeman, T., 1979, Lower Cretaceous peritidal limestones at 2700 m depth, Blake Nose, Atlantic Ocean: Geology, v. 7, p. 83–87.

Greene, H. G., Dalrymple, C. B., Clague, D. A., 1978, Evidence for northward movement of the Emperor Seamounts: Geology, v. 6, p. 70–74.

Hamilton, E. L., 1956, Sunken islands of the Mid-Pacific Mountains: Geol. Soc. Amer. Mem. 64, 97 p.

Hays, J. D. and Pitman, W. C. III, 1973, Lithospheric plate motion, sea level changes and climatic and ecological consequences: Nature, v. 246, no. 5427, p. 18–22.

Hays, J. D. and others, 1972, Init. Repts. DSDP, v. 9, p. 21–42.

Heezen, B. C. and Sheridan, R. E., 1966, Lower Cretaceous rocks (Neocomian-Albian) dredged from Blake Escarpment: Science, v. 154, p. 1644–1647.

Heezen, B. C., MacGregor, I. D. and others, 1973, Init. Repts. DSDP, v. 20, p. 97–103.

Heirtzler, J. R., Bolli, H. M., Davies, T. A., Saunders, J. B. and Sclater, J. G., eds., 1977, Indian Ocean Geology and Biostratigraphy: American Geophysical Union, Washington, D.C., 616 p.

Hesse, R., 1973, Diagenesis of a seamount oolite from the west Pacific, Leg 20, DSDP: Init. Repts. DSDP, v. 20, p. 363–368.

Jackson, E. D. and Schlanger, S. O., 1976, Regional syntheses, Line Islands chain, Tuamoto Island chain, and Manihiki Plateau, Central Pacific Ocean: Init. Repts. DSDP, v. 33, p. 915–927.

Jackson, E. D., Koizumi, I. and others, 1980, Init. Repts. DSDP, v. 55.

Jackson, E. D., Koizumi, I., Dalrymple, G. B., Kirkpatrick, J. R., Clague, D. A., and Greene, H. G., 1980, Introduction and summary of results from Leg 55, the Hawaiian-Emperor hot-spot experiments: Init. Repts. DSDP, v. 55, p. 5–31.

Jenkyns, H. C., and Schlanger, S. O., *in press,* Significance of plant remains in redeposited Aptian sediments, Hole 462A, Nauru Basin, to Cretaceous oceanic oxygenation models: Init. Repts. DSDP, v. 61.

Jones, E. J. W., 1973, Determination of sedimentary velocities using expendable sonobuoys at DSDP Leg 20 drilling sites, northwest Pacific: Init. Repts. DSDP, v. 20, p. 625–264.

Kumar, N., 1979, Origin of "paired aseismic ridges": Ceará and Sierra Leone Rises in the equatorial, and the Rio Grande Rise and Walvis Ridge in the south Atlantic: Mar. Geology, v. 30, p. 175–191.

Ladd, H. S., Tracey, J. I., Jr. and Gross, G., 1970, Deep drilling on Midway Atoll: U.S. Geol. Survey Prof. Paper 680-A, p. A1–A21.

Ladd, H. S., Newman, W. A. and Sohl, N. F., 1974, Darwin Guyot, the Pacific's oldest atoll: Proc. Sec. Int. Coral Reef Symp., Vol. 2, Brisbane, p. 513–522.

Larson, R. L. and Hilde, T. W., 1975, A revised time scale of magnetic reversals for the early Cretaceous and Late Jurassic: Jour. Geophys. Research, v. 80, p. 2586–2594.

Larson, R. L. and Pitman, W. C., III, 1972, World-wide correlation of Mesozoic anomalies, and its implications, Geol. Soc. Amer. Bull., v. 83, p. 3645–3662.

———, Moberly, R. and others, 1975, Init. Repts. DSDP, v. 32, p. 220.

———, and Schlanger, S. O., *in press,* Cretaceous volcanism and Jurassic magnetic anomalies in the Nauru Basin, Western Pacific Ocean: Geology.

———, ———, and others, *in press*, Init. Repts. DSDP, v. 61.

Laughton, A. S., 1972, The Southern Labrador Sea—a key to the Mesozoic and Early Tertiary evolution of the north Atlantic: Init. Repts. DSDP, v. 12, p. 1155–1180.

Laughton, A. S., Berggren, W. A. and others, 1972, Init. Repts. DSDP, v. 12, p. 33–160, 395–672.

Lees, A., 1975, Possible influences of salinity and temperature on modern shelf carbonate sedimentation: Mar. Geol., v. 19, p. 159–198.

Luyendyk, B. P., 1977, Deep sea drilling on the Ninetyeast Ridge: synthesis and a tectonic model: *In,* Indian Ocean Geology and Biostratigraphy, Heirtzler, J. R., Bolli, H. M., Davies, T. A., Saunders, J. B. and Sclater, J. G., eds., American Geophysical Union, Washington, D.C., p. 165–188.

Malfait, B. and Dinkelman, M. G., 1972, Circum-Caribbean tectonic and igneous activity and the evolution of the Caribbean plate: Geol. Soc. Amer. Bull., v. 83, p. 251–272.

Matter, A. and Gardner, J. V., 1975, Carbonate diagenesis at Site 308, Kōko Guyot: Init. Repts. DSDP, v. 32, p. 521–535.

Matthews, J. L., Heezen, B. C., Catalano, R., Coogan, A., Thorp, M., Natland, J. and Rawson, M., 1974, Cretaceous drowning of reefs on Mid-Pacific and Japanese Guyots: Science, v. 184, p. 462–464.

Mattson, P. H., 1969, The Caribbean: A detached relic of the Darwin Rise: Am. Geophys. Union Trans., v. 50, p. 317.

Maxwell, A. E. and others, 1970, Init. Repts. DSDP, v. 3, p. 413–441.

McKenzie, J., Bernoulli, D. and Schlanger, S. O., 1980, Shallow-water carbonate sediments from the Emperor Seamounts, their diagenesis and paleogeographic significance: Init. Repts. DSDP, v. 55, p. 415–457.

Menard, H. W., 1964, Marine Geology of the Pacific, McGraw-Hill Book Co., New York, 271 p.

Newell, N. D., 1955, Bahamian platforms: *In,* Poldervaart, A., *ed.,* Crust of the Earth, Geol. Soc. America Spec. Paper, no. 62, 303 p.

Parsons, B. and Sclater, J. C., 1977, An analysis of the variation of ocean floor bathymetry and heat flow with age: Jour. Geophys. Research, v. 82, p. 803–827.

Paulus, F. J., 1972, The geology of Site 98 and the Bahama Platform: Init. Repts. DSDP, v. 11, p. 877–897.

Pimm, A. C., McGowan, B. and Gardner, S., 1974, Early sinking history of the Ninetyeast Ridge, northeastern Indian Ocean: Geol. Soc. Amer. Bull., v. 85, p. 1219–1224.

Pitman, W. C., III, 1978, Relationship between eustacy and stratigraphic sequences of passive margins: Geol. Soc. Amer. Bull., v. 89, p. 1389–1403.

Premoli-Silva, I. and Brusa, C., *in press,* Shallow-water skeletal debris and larger foraminifers from DSDP Site 462, Leg 61, Nauru Basin: Init. Repts. DSDP, v. 61.

Pushcharovskiy, Y. M., 1978, Tectonic movements in the oceans: Geotecotonics, v. 12, no. 1, p. 1–9.

Rea, D. K. and Thiede, J., *in press,* Mesozoic and Cenozoic mass accumulation rates of the major sediment components in the Nauru Basin, western Equatorial Pacific: Init. Repts. DSDP, v. 61.

Rothe, P., 1979, Shallow water carbonates from Site 384: Init. Repts. DSDP, v. 43, p. 421–436.

Saunders, J. B., Edgar, N. T., Donnelly, T. W. and Hay, W. W., 1972, Cruise synthesis: Init. Repts. DSDP, v. 15, p. 1077–1112.

Savin, S. M., Douglas, R. G. and Stehli, F. G., 1975, Tertiary marine paleotemperatures, Geol. Soc. Amer. Bull., v. 86, p. 1499–1510.

Sclater, J. G., Abbott, D. and Theide, J., 1977, Paleobathymetry and sediments of the Indian Ocean: *In,* Indian Ocean Geology and Biostratigraphy, Heirtzler, J. R., Bolli, H. M., Davies, T. A., Saunders, J. B. and Sclater, J. G., eds., American Geophysical Union, Washington, D.C., p. 25–60.

Schlanger, S. O., 1954, The petrology of the Vincentown formation: Jour. Sed. Petrology, v. 24, p. 212–217.

Schlanger, S. O., 1963, Subsurface geology of Eniwetok Atoll: U.S. Geol. Survey Prof. Paper, 260-BB, p. 901–1066.

Schlanger, S. O. and Konishi, K., 1966, Contrasting bryozoan content of Plio-Pleistocene to present day carbonates from Guam, Mariana Islands, and Kikai-Kima, Ryukyu Islands and its regional implications: Proc. XI Pac. Science Congress, Tokyo, Japan, Abstract no. 35.

Schlanger, S. O. and Douglas, R. G., 1974, The pelagic ooze-chalk-limestone transition and its implications for marine stratigraphy: *In,* Hsu, K. J. and Jenkyns, H. C., *eds.,* Pelagic Sediments on Land and Under the Sea, Spec. Pub. No. 1, Int. Assoc. Sedimentologists, p. 117–148.

Schlanger, S. O. and Konishi, K., 1975, The geographic boundary between the Coral-Algal and the Bryozoan-Algal facies: A paleolatitude indicator: Theme I, IX Int. Cong. Sedimentology, Nice, p. 187–190.

SCHLANGER, S. O. AND JENKYNS, H. C., 1976, Cretaceous oceanic anoxic events: causes and consequences: Geologie en Mijnbouw, v. 55, p. 179–184.

SCHLANGER, S. O., JACKSON, E. D. AND OTHERS, 1976, Init. Repts. DSDP, v. 33.

SCHLANGER, S. O, JENKYNS, H. C., AND PREMOLI-SILVA, I., 1981, Volcanism and vertical tectonics in the Pacific Basin related to the global Cretaceous transgression: Earth Planetary Science Letters, v. 52, p. 435–449.

SCHLANGER, S. O. AND PREMOLI-SILVA, I., *in press,* Tectonic, volcanic and paleogeographic implications of redeposited reef faunas of Late Cretaceous and Tertiary age from the Nauru Basin and the Line Islands: Init. Repts. DSDP, v. 61.

SHERIDAN, R. E., 1974, Atlantic continental margin of North America: *In,* Burk, C. A. and Drake, C. L., *eds.,* The Geology of Continental Margins, New York (Springer-Verlag), p. 391–407.

SHERIDAN, R. E., BERMAN, R. M. AND CORMAN, D. B., 1971, Faulted limestone block dredged from Blake Escarpment: Geol. Soc. America Bull., v. 82, p. 199–206.

SHERIDAN, R. E. AND ENOS, P., 1979, Stratigraphic evolution of the Blake Plateau after a decade of scientific drilling: *In,* Talwani, M., Hay, W., Ryan, W. B. F., eds., Deep Drilling Results in the Atlantic Ocean: Continental Margins and Paleoenvironment, M. Ewing Series 3, American Geophysical Union, Washington, D.C., p. 109–122.

SUPKO, P. R., PERCH-NIELSON, K. AND OTHERS, 1977, Init. Repts. DSDP, v. 39, p. 231–329, 373–392.

TEICHERT, C., 1958, Cold and deep water coral banks: Amer. Assoc. Petroleum Geologists Bull., v. 42, p. 1064–1082.

THIEDE, J., 1977, Subsidence of aseismic ridges: evidence from sediments on the Rio Grande Rise (southwest Atlantic Ocean), Am. Assoc. Petroleum Geologists Bull., v. 61, p. 929–940.

THIERSTEIN, H., 1979, Paleoceanographic implications of organic carbon and carbonate distribution in Mesozoic deep sea sediments: *In,* Talwani, M., Hay, W. and Ryan, W. B. F., *eds.,* Deep Drilling Results in the Atlantic Ocean: Continental Margins and Paleoenvironment, M. Ewing Series 3, American Geophysical Union, Washington, D.C., p. 249–274.

TUCHOLKE, B. E., VOGT, P. R. AND OTHERS, 1979, Init. Repts. DSDP, v. 43.

VAIL, P. R., MITCHUM, R. M., JR., AND THOMPSON, S., 1977, Seismic Stratigraphy and global changes in sea level, Part 4; Global cycles of relative sea level changes: Amer. Assoc. Petroleum Geologists Mem. 26, p. 83–95.

VAIL, P. R. AND HARDENBOHL, J., 1979, Sea level changes during the Tertiary: Oceanus, v. 22, p. 71–79.

VON DER BORCH, C. C., *ed.,* 1978, Synthesis of Deep Sea Drilling Results in the Indian Ocean: Marine Geology, v. 26, no. 1–2, p. 1–175.

WINTERER, E. L., 1976, Bathymetry and regional tectonic setting of the Line Islands Chain: Init. Repts. DSDP, v. 33, p. 731–748.

WINTERER, E. L., EWING, J. I., AND OTHERS, 1973, Init. Repts. DSDP, v. 17.

APPENDIX—DSDP SITES AT WHICH SHALLOW-WATER CARBONATES AND RELATED SEDIMENTS HAVE BEEN RECOVERED.

Site	DSDP Leg	Published Report*
North Atlantic Ocean		
4, 5	1	Ewing, Worzel and others, 1969
98, 99, 100, 101	11	Hollister, Ewing and others, 1972
111, 116, 117, 118, 119	12	Laughton, Berggren and others, 1972
135, 139	14	Hays, Pimm and others, 1972
338, 341, 343, 349, 352	38	Talwani, Udintsev and others, 1976
370	41	Lancelot, Seibold and others, 1977
384	43	Tucholke, Vogt and others, 1975
390, 391, 392	44	Benson, Sheridan and others, 1978
397	47A	Ryan, von Rad and others, 1976*
398	47B	Ryan, Sibuet and others, 1976*
401, 402, 403, 404	48	Montadert, Roberts and others, 1976*
415, 416	50	Lancelot, Winterer and others, 1977*
417	51	Donnelly, Francheteau and others, 1977*
Equatorial North Atlantic Ocean		
140, 142, 144	14	Hays, Pimm and others, 1972
154	15	Edgar, Saunders and others, 1973
353	39	Supko, Perch-Nielson and others, 1977
366	41	Lancelot, Seibold and others, 1977
Equatorial South Atlantic Ocean		
23, 25	4	Bader: Gerard and others, 1970
364, 365	40	Bolli, Ryan and others, 1978

APPENDIX—DSDP SITES AT WHICH SHALLOW-WATER CARBONATES AND RELATED SEDIMENTS HAVE BEEN RECOVERED (CONTINUED).

Site	DSDP Leg	Published Report*
Gulf of Mexico		
3	1	Ewing, Worzel and others, 1969
85, 86, 91, 94, 95, 97	10	Worzel, Bryant and others, 1973
South Atlantic Ocean		
21	3	Maxwell, von Herzen and others, 1970
327, 330	36	Barker, Dalziel and others, 1976
355, 356, 357, 359	37	Supko, Perch-Nielson and others, 1977
361, 362, 363	40	Bolli, Ryan and others, 1978
Indian Ocean		
212, 213, 214, 216, 217	22	von der Borch, Sclater and others, 1974
219, 222	23	Whitmarsh, Weser and others, 1974
231, 236, 237	24	Fisher, Bunce and others, 1974
239, 240, 241, 246, 247	25	Simpson, Schlich and others, 1974
253, 254, 255, 256, 258	26	Davies, Luyendyk and others, 1974
259, 260, 261, 262, 263	27	Veevers, Heirtzler and others, 1974
North Pacific Ocean		
295, 296, 297	31	Karig, Ingle and others, 1975
308, 309	32	Larson, Moberly and others, 1975
430, 432, 433	55	Jackson, Koimuzi and others, 1978*
439	57	von Huene, Nasu and others, 1978*
445	58	Klein, Kobayashi and others, 1978*
451	59	Kroenke, Scott and others, 1978*
463	62	Vallier, Thiede and others, 1979*
Equatorial North Pacific Ocean		
165, 171	17	Winterer, Ewing and others, 1973
200, 201, 202	20	Heezen, MacGregor and others, 1973
315	33	Schlanger, Jackson and others, 1976
462	61	Larson, Schlanger and others, 1978*
Equatorial South Pacific Ocean		
76	9	Hays and others, 1972
316, 317, 318	33	Schlanger, Jackson and others, 1976
South Pacific Ocean		
205	21	Burns, Andrews and others, 1973
277, 281, 282	29	Kennett, Houtz and others, 1974
285, 286	30	Andrews, Packham and others, 1975

*Geotimes article series. All other references are to the Initial Reports of the Deep Sea Drilling Project, Washington, D.C. (U.S. Gov't. Printing Office) series.

SEPM SPECIAL PUBLICATION NO. 32, P. 227–249, OCTOBER 1981

ORGANIC GEOCHEMISTRY IN THE DEEP SEA DRILLING PROJECT

KEITH A. KVENVOLDEN
U.S. Geological Survey, Menlo Park, California 90425 U.S.A.

ABSTRACT

Since the beginning of the Deep Sea Drilling Project (DSDP) in 1968 and extending through 1975, organic geochemical studies have been undertaken on about 2300 samples recovered on Legs 1 through 44 from sediments beneath the ocean floors. These studies have provided fundamental information regarding the distribution of carbon in oceanic sediments and have yielded a better understanding of the processes that alter and transform organic matter in the marine environment.

Beginning with Leg 38 organic geochemists have been included as members of the scientific staffs of those legs of particular organic geochemical interest. They have advised operations personnel on matters concerning potential hazards, such as drilling into significant accumulations of oil and gas, and have collected samples for the scientific community. The shipboard organic geochemical laboratory is equipped with sophisticated instrumentation to provide those meassurements helpful in advising the drilling operation.

Since Leg 15, samples for organic geochemical studies have been maintained at -18°C at the DSDP sample repository at the Scripps Institution of Oceanography, La Jolla, California. Samples are distributed to the scientific community, but at least one quarter of all cores is retained for possible future studies.

Organic geochemical studies generally relate to one of these topic areas: (1) petroleum potential/source rock evaluation, (2) thermal history and gradients, (3) diagenesis, (4) gas hydrates, (5) novel geochemical fossils, (6) sources of organic materials, (7) paleo-depositional environments, and (8) geochronology.

Much of the organic geochemical work on hydrocarbons and kerogen has been directed to the potential of sediments as sources of petroleum. The organic matter found thus far during drilling is dominantly of terrestrial origin and thermally immature, but much of the organic matter could be a potential source of petroleum (mainly gas) if buried more deeply.

Three new technologies will help advance organic geochemical knowledge of ocean sediments: (1) pressure core barrel for evaluation of gas hydrates, (2) hydraulic piston core to study in detail sources and early diagenesis of organic matter, and (3) riser for the study of late diagenesis and catagenesis.

INTRODUCTION

Since 1968 our knowledge of the evolution of the earth's crust has been increasing at an unprecedented rate due in large part to the scientific efforts associated with the Deep Sea Drilling Project (DSDP). Among the many disciplines benefiting from the extensive DSDP sampling program throughout the world's oceans is the field of organic geochemistry, concerned with the geological occurrence, distribution and fate of organic materials. Prior to DSDP, organic geochemical information from oceanic sediments came from surface samples and short cores usually obtained near the continents. With the advent of DSDP, samples for organic geochemistry have been taken from cores that have penetrated sediments in the major oceans (Fig. 1) and have spanned geologic time from Recent to Jurassic.

Organic geochemistry has been an important part of DSDP since the beginning of the program. Representative sediment samples from most cores have been analyzed for organic carbon, and these measurements alone have provided pertinent information on its worldwide distribution (McIver, 1975; Hunt, 1972), and on the productivity of ancient oceans and the paleo-carbon cycle (Simoneit, 1978c). In addition to these fundamental measurements, more specialized research has been directed to certain classes of organic compounds and to the processes that control their occurrence and distribution. Much of the organic geochemical effort has been directed to determination of the petroleum potential of ocean sediments, and this work probably has had the most immediate practical significance. DSDP results have provided the basic data for evaluation of the possibilities of petroleum in the outer continental margins and small ocean basins (Dow, 1978a; Hedberg and others, 1979; Welte, 1979).

Forty-four sampling legs took place under the banner of DSDP from 1968 to 1975. Although international participation has always been part of the program, in 1975 international participation and support was formally recognized, and the program became DSDP/International Phase of Ocean Drilling (DSDP/IPOD). Organic geochemical information is scattered throughout the Initial Reports that have been generated as part of DSDP/IPOD. This paper considers the organic geochemistry resulting from deep ocean sampling as described

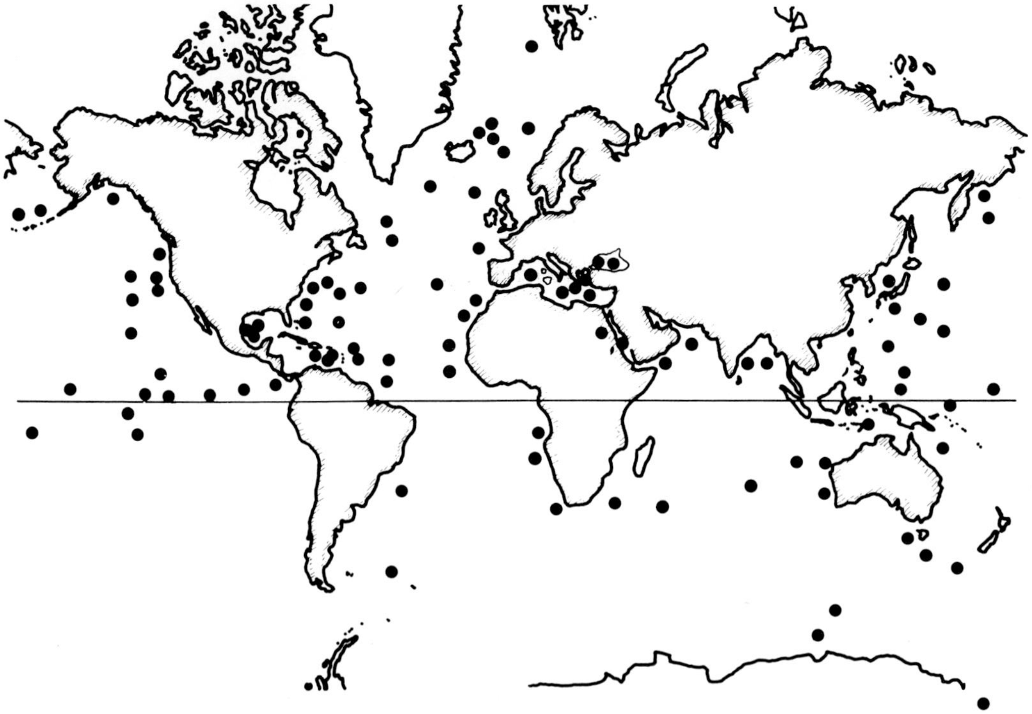

FIG. 1.—Organic geochemical sampling areas, DSDP Legs 1–44.

in the Initial Reports for the first 44 legs of DSDP. These Initial Reports contain 120 articles on topics directly related to organic geochemistry (Appendix I). The articles range in content from visual observations of organic particles to sophisticated chemical analyses of complex organic molecules requiring the most advanced instrumentation available. In addition, 34 articles are devoted specifically to the measurement of organic and carbonate carbon (Appendix II).

This paper reviews three aspects of the DSDP/IPOD organic geochemistry program:

1) the shipboard program, including guidelines and responsibilities of the shipboard organic geochemists and the instrumentation available for organic geochemical studies;

2) procedures for sampling at the DSDP sample repository and for the allocation of samples to the scientific community; and

3) significant organic geochemical results.

SHIPBOARD PROGRAM

Before Leg 38, organic geochemical participation in DSDP was confined to an advisory panel and to investigators in the industrial and academic communities who conducted organic geochemical analyses of sediment samples collected by shipboard staffs. Beginning with Leg 38 shipboard organic geochemists have been regularly included on the scientific staffs of those legs of particular organic geochemical interest. The roster of organic geochemists who have served aboard the *Glomar Challenger* includes representatives of the international community from industrial, academic and government laboratories (Table 1). Three primary duties of the shipboard organic geochemist are: (1) to assist the co-chief scientists, petroleum geologist (if assigned) and cruise operations manager in evaluation of the possibility of encountering significant hydrocarbon accumulations during drilling operations, (2) to collect core and canned sediment samples that will be frozen on shipboard, shipped in a frozen state at the earliest opportunity to the DSDP curatorial facility at Scripps Institution of Oceanography, catalogued by the curatorial staff and distributed to the scientific community, and (3) to collect samples for his own research program and for the research programs of others in so far as these programs and samples have been approved by the co-chief scientists, the scientific party and the chief scientist of DSDP.

A sophisticated organic geochemical laboratory onboard the *Glomar Challenger* provides facilities with which the geochemist can make the measurement required to provide immediate guidance during drilling operations. Six analytical instruments are available (Table 2), and from these instruments not only can quantitative assessment be made of various organic constituents, but also interpretations can be made regarding thermal ma-

TABLE 1.—SHIPBOARD ORGANIC GEOCHEMISTS ON THE *Glomar Challenger*.

Leg	Name	Affiliation
38	D. Morris	Phillips Petroleum Company
40	J. Foresman	Phillips Petroleum Company
41	J. Rankin	N.E. Louisiana University
42B	J. Hunt	Woods Hole Oceanographic Institution
43	J. Kendrick	Shell Oil Company
44	D. Stuermer	University of California, Los Angeles
47A	J. Whelan	Woods Hole Oceanographic Institution
	C. Cornford	Kernforschungsanlage, Germany
47B	D. Johnson	Exxon Production Research Company
48	W. Harrison	Oklahoma Geological Survey
50	E. Galimov	Academy of Sciences, U.S.S.R.
	K. Taguchi	Tohoku University, Japan
56	J. Whelan	Woods Hole Oceanographic Institution
57	S. Sato	Japan Petroleum Development Center, Japan
58	D. Waples	Colorado School of Mines
59	S. Brassell	University of Bristol, England
63	A. Douglas	University of Newcastle-upon-Tyne, England
64	B. Simoneit	University of California, Los Angeles
66	B. Didyk	Empresa Nacional del Petroleo, Chile
67	W. Harrison	Oklahoma Geological Survey

turity, possible migration, sources of organic matter and petroleum potential.

Wherever gassy cores are encountered during drilling, samples are taken for gas analyses. Concentrations of methane and ethane are measured by thermal conductivity on a Carle Gas Chromatograph, although the sensitivity for ethane is low in the presence of high amounts of methane. More accurate concentration values for ethane are obtained on a Hewlett-Packard Gas Chromatograph equipped with a sensitive flame ionization detector. The ratio of methane/ethane is used as a means of differentiating between gas produced biochemically and gas produced thermochemically. Ratios range from greater than 10^5 for purely biochemical gas to less than 5 for some thermally generated gases. When approaching an oil or gas accumulation during the drilling operations, the methane/ethane ratio should decrease, and this ratio, therefore, gives an indication of possible nearby accumulations.

Besides ethane, the Hewlett-Packard gas chromatograph monitors the higher molecular-weight compounds to five carbon atoms including propane, *n*-butane, isobutane, *n*-pentane, isopentane, neopentane, and cyclopentane. These hydrocarbons can be generated during the maturation of organic matter. Analyses of gases from ethane to cyclopentane can provide information on the degree of thermal maturity, the extent of migration and the temperature regime of gas formation. An Exploration Logging Gas Chromatograph also can monitor methane through pentane, but this instrument has lower sensitivity than the Hewlett-Packard gas chromatograph and has been used mostly to monitor ethane.

The most fundamental parameter of organic geochemical investigations of sediments is the organic carbon content; it gives an estimate of the organic matter that is present, and the amount of organic matter can indicate the nature of the depositional environment. A Hewlett-Packard Carbon-Hydrogen-Nitrogen Analyzer is used to estimate the organic carbon and organic nitrogen content of sediments. The accuracy of these determinations is low because of difficulties in calibration and weighing at sea; however, accuracy is sufficient to provide trends and estimates of C/N ratios. Values of organic carbon are used in interpretation of results from the Turner Fluorometer and the Girdel Rock Eval.

The Turner Fluorometer measures the intensity of fluorescence of distillate obtained by pyrolysis of sediment. The results of pyrolysis/fluorescence measurements are indicative of the hydrocarbon-generating potential of sediments and provide a means of differentiating migrated from in-place hydrocarbons.

The Rock Eval is a complicated pyrolysis instrument capable of identifying four characteristics of

TABLE 2.—ANALYTICAL INSTRUMENTATION* IN THE ORGANIC GEOCHEMICAL LABORATORY ON THE *Glomar Challenger*.

Carle 800 Basic Gas Chromatograph
Hewlett-Packard (H-P) Model 5711A Gas Chromatograph
Exploration Logging Model EL-11-94-75Gas Chromatograph
Hewlett-Packard (H-P) Model 185B Carbon-Hydrogen-Nitrogen Analyzer
Turner Model 110 Fluorometer
Girdel Rock Eval

*Any use of trade names and trademarks in this publication is for descriptive purposes only and does not constitute endorsement by the U.S. Geological Survey.

petroleum source rocks:

1) Origin of organic matter—terrestrial, marine or mixed sources

2) Petroleum potential—the total quantity of oil that organic matter could generate given suitable thermal maturation

3) Maturity—the extent to which organic matter has been subjected to the maturation process

4) Oil shows—presence of free hydrocarbons in reservoir rocks

Changes in these parameters can be monitored with increasing depth and thermal maturity. The resulting information leads to a rather complete understanding of the source and conditions of organic matter in the sediments drilled.

Results from gas analyses, carbon determinations, and pyrolysis studies provide fundamental organic geochemical information for the evaluation of potential hazards in the drilling operation that could result if significant hydrocarbon accumulations are ever encountered. The objective, of course, is to avoid such encounters.

Sampling for the scientific community follows guidelines described in the "Shipboard Handbook" that is given to each shipboard scientist before a cruise and is also available through DSDP to other interested scientists. Two kinds of samples are collected: (1) intervals of sediment core, 30 to 150 cm in length, which are frozen immediately in the core liner, and (2) gassy sediments that are extruded into cans which are partially filled with water, sealed, and frozen. These frozen core and canned samples are shipped to the DSDP curatorial facility, located at the Scripps Institution of Oceanography, La Jolla, California, for storage and distribution to the scientific community.

SAMPLING AT DSDP

Prior to Leg 15 in the Caribbean Sea there was no formal or systematic program within DSDP for the collection and preservation of sediment samples for shore-based organic geochemical studies. With Leg 15, provisions had been made for collecting samples of potential interest to organic geochemists and for storing these samples in a frozen condition. Freezing prevents alterations of organic compounds by bacterial action. Since Leg 15, samples have been collected during about half of the legs that followed. A histogram (Fig. 2) of the distribution of frozen samples maintained at the DSDP repository shows the increase in sample collection starting with Leg 38 corresponding to the time when shipboard organic geochemists became part of the scientific staffs of the *Glomar Challenger.* Legs for which no organic geochemical samples were taken are usually legs where the major interest was ocean crust, and little sediment was encountered in drilling. Portions of most of the stored samples have been studied for a variety of organic materials; however, it has been a policy to maintain representative portions of all samples to provide material for future investigators requiring samples preserved in this unique way.

Samples for organic geochemical studies are frozen on shipboard and remain frozen, wherever

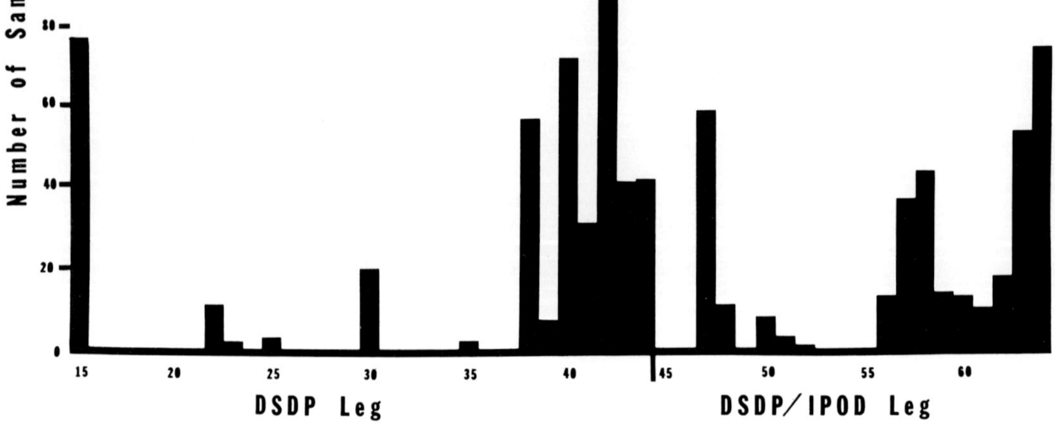

FIG. 2.—Histogram of inventory of organic geochemical samples maintained frozen at the DSDP repository.

possible, during transit from the ship to the core repository at DSDP where the cores are stored in a special locker at −18°C. Upon arrival at the repository the core and canned samples are inventoried. Later, while the material is still frozen, the core liners are split, and the cores are quartered longitudinally using a specially cleaned band saw. The quartered sections are replaced in the liner and returned to freezer storage.

Sampling of the frozen cores is conducted by the curatorial staff of DSDP with assistance from the JOIDES Advisory Panel on Organic Geochemistry. Ideally, a DSDP representative from the leg being sampled is present to describe the cores lithologically. Allocation of samples depends on requests from the scientific community. Anyone interested in the organic geochemistry of these frozen samples should file a form with the Panel at least two months before the leg of interest begins. The form is essentially a letter of intent stating the kind of work proposed and an estimate of the number and size of samples required to carry out the proposed study. These forms are duplicated and sent to DSDP where the information is filed and also made available to the scientific staff on the leg of interest. Leg co-chief scientists are thus alerted to the fact that shore-based studies will be done on samples from their leg, and provisions are made for space in the Initial Reports for the results coming from this work. After the sample inventory is completed at DSDP, each of the individuals who earlier proposed to do organic geochemical studies is provided with the inventory from which he can select specific samples for his proposed work. The resulting sample requests serve as a guide to the DSDP curatorial staff and Panel representative for sample allocation.

During the actual sampling session cores are opened and individual parts of quartered sections are broken with a chisel. Plastic gloves are worn by all who touch the core. The broken quarter sections are packaged and double labeled in Kapak plastic bags that are heat-sealed. Samples are placed, along with dry ice, in styrofoam containers for air shipment to the requestors. Canned samples are shipped in the same manner. The core material remaining after sampling (at least one quarter of the core is always preserved) is replaced in the core liner which is then heat-sealed in plastic shrink tubing. The repackaged core is returned to freezer storage.

Investigators requesting and accepting samples for analyses are obligated to contribute the results of their organic geochemical studies to the Initial Reports of the Deep Sea Drilling Project. After these reports for a specific leg are published, other investigators wishing portions of the frozen samples from that leg can request material through DSDP. On some occasions frozen samples may be released to other investigators prior to publication of the Initial Report. Such an early release can be determined by the leg scientists after consultation with the DSDP Chief Scientist. Examples of the kinds of information resulting from the organic geochemical studies are shown in Table 3. The canned samples have provided material for determinations of the molecular composition of gases and gasoline-range hydrocarbons and for the measurement of the carbon isotopic composition of methane.

ORGANIC GEOCHEMICAL RESULTS

During the first 44 legs of DSDP, organic geochemical techniques were applied to approximately 2300 samples recovered from sites on 35 legs (Appendix III). These sites were located in all of the world's oceans, except the Arctic, and in many of the restricted seas such as the Bering, Caribbean, Mediterranean, Black and Red seas (Fig. 1). Some objectives of the organic geochemical studies have been (1) to measure the abundance and composition of major fractions of organic matter, (2) to characterize significant organic molecules, and (3) to follow the diagenesis of organic matter in sediments. To the first 44 Initial Reports, 126 authors have contributed 120 articles containing information on organic geochemistry. Some of these articles describe for the first time occurrences of several different kinds of organic compounds such as those listed in Table 4 (Appendix IV).

Much of the work on hydrocarbons and kerogen, especially since Leg 38, has been directed to determining the potential of sediments as source rocks of petroleum. In addition, 33 of the first 44 Initial Reports contain articles in which routine organic-carbon analyses are reported for about 10,900 samples. Where organic-carbon determinations have not been included in the Initial Reports, these results are available at DSDP.

Organic geochemical studies of DSDP cores have contributed to at least eight major topic areas as listed in Table 3. These areas are not mutually exclusive, and the results of many studies relate to more than one of the items listed. For example, novel geochemical fossils are of interest in themselves, but these same molecules may be important indicators of paleoenvironments and sources of organic matter. Likewise, petroleum potential depends on thermal maturation which is part of the

TABLE 3.—TOPIC AREAS OF DSDP ORGANIC GEOCHEMISTRY STUDIES.

Petroleum potential/source rock evaluation
Thermal history and gradients
Diagenesis
Gas hydrates
Novel geochemical fossils
Sources of organic materials
Paleo-depositional environments
Geochemical geochronology

processes of diagenesis. Thus the organization of topic areas is done mostly for convenience in showing examples where organic geochemistry has contributed to the DSDP. Other considerations of organic geochemical contributions to DSDP have been given previously (Baker and Eglinton, 1978).

Petroleum Potential and Source Rock Evaluation

The first detailed organic geochemical analyses in DSDP were applied to samples of liquid hydrocarbons recovered from the second hole of the first leg of the project at Challenger Knoll in the Gulf of Mexico. A consensus was reached that these hydrocarbons constituted a bonafide occurrence of crude oil (Davis and Bray, 1969). The oil, judged to be immature, of marine origin, and of Tertiary age, was found in Jurassic/Cretaceous sediments associated with a salt dome. After this first finding of petroleum, every effort has been made to avoid such discoveries. Thus, only on Leg 1 has crude oil been found, although liquid, petroleum-like hydrocarbons have been encountered in samples from Leg 6 (Shatsky Rise, western Pacific Ocean) and Leg 13 (Balearic Abyssal Plain, western Mediterranean Sea). McIver (1975) evaluated these occurrences and concluded that the liquids are not contaminants but represent immature migrated petroleum. The source of these hydrocarbon mixtures has not been identified, however. On Leg 38 in the Norwegian Sea, a petroleum-like mixture was found during shipboard studies at one site (Morris, 1976), but shorebased work failed to indicate anomalous hydrocarbons at the same site (Erdman and Schorno, 1976; Hood and others, 1976), although Simoneit (1976) found droplets of petroleum-like hydrocarbons at another site in the same area. These mixtures probably migrated from more mature sediments elsewhere in the sedimentary section.

Since Volume 38 most papers on organic geochemistry in the Initial Reports have considered aspects of the petroleum potential of sediments as source rocks of petroleum. For example, Foresman (1978) investigated hydrocarbon gases and extractable organic material. He concluded that some sediments from Leg 40 in the south Atlantic are potential source rocks and that petroleum genesis may be taking place in the thicker parts of the rise and slope off Africa. Kendrick and others (1977, 1978a, b, c, 1979) evaluated the petroleum-generating potential of sediments from Legs 40 to 44 in the South and North Atlantic Oceans by pyrolysis techniques and vitrinite reflectance measurements. Potential source rocks were identified in all of these legs, except possibly Leg 44, but the organic matter is still too immature to have generated petroleum. Deeper burial is required to initiate petroleum generation.

Preliminary information on potential source rocks recovered in Legs 47A, 47B, and 48 is given in Pearson and Dow (1979). Sediments from these legs on the southern European and northern Africa continental margins are thermally immature but could produce petroleum if more deeply buried. Welte (1979) described the hydrocarbon potential for sediments from Legs 47A, 47B, 50 and 56/57. Potential but immature source rocks are present in samples from Legs 47A and 50, south of the Canary Islands and offshore Morocco. Leg 47B samples from offshore Spain contain recycled organic matter that is not favored as a petroleum source. No hydrocarbon

TABLE 4.—ORGANIC COMPOUND TYPES IN SEDIMENTS SAMPLED BY DSDP/IPOD.

Acids (organic)	*Hydrocarbons*	*Lipids*
Dicarboxylic acids	Alkanes	
Hydroxy acids	Alkenes	*Pigments*
Isoprenoid acids	Aromatic	Carotenoids
Iso- and Anteiso-fatty acids	Isoprenoid	Tetrapyrroles
Naphthenic acids	Steranes	Chlorins
Saturated fatty acids	Di- sesqui- and triterpanes	Porphyrins
Terpenoid acids		
Unsaturated fatty acids		
	Humic compounds	*Sugars*
Alcohols	*Isotopes*	*Sulfur (organic)*
Fatty alcohols	Carbon	
Sterols	Sulfur	
Amino Compounds	*Kerogen*	
Amino acids		
Amino sugars	*Ketones*	
Amines	Isoprenoid	
	Methyl	
Asphaltenes, Asphaltics (NSO)	Pentacyclic	
	Sterones	

source rocks were found in samples from Legs 56/57 in the Japan Trench.

Thermal History and Gradients

Most papers in the Initial Reports that deal with petroleum potential consider thermal history because this parameter is a primary requisite for the generation of petroleum. Vitrinite reflectance serves as a common guide to thermal history, and this measurement has been used extensively in DSDP studies such as those by Hood and others (1976), Kendrick and others (1977, 1978a, b, c, 1979) and Dow (1978b). Elemental composition of kerogen also can be interpreted in terms of thermal history as demonstrated by Comer and Littlejohn (1976) for sediments of the Falkland Plateau and by Deroo and others (1977) in the Cape Verde area of the eastern North Atlantic. The distributions of *n*-alkanes in sediments can be used to assess thermal histories as was done by Erdman and Schorno (1978) and Dow (1978b) for sediments of the western Atlantic Ocean. Baker and others (1978a) used porphyrin distributions as a means of predicting thermal history in the same area.

Organic geochemical measurements have also been used to measure thermal gradients in sediments from Leg 41 off West Africa. At one site intrusion of diabase sills caused dramatic alteration in the organic matter of adjacent sediments (Simoneit and others, 1978). The rapid change in thermal gradient in the adjacent sediments is reflected in contents of organic carbon and extractable organic material, in the distribution of hydrocarbons and in the reflectance of vitrinite (Dow, 1977). Baker and others (1977) constructed a thermal gradient for the sediments adjacent to the sills based on the distribution, composition, and concentrations of porphyrins.

Diagenesis

One of the main objectives of many organic geochemical studies is to follow the diagenesis of organic matter in sediments by examining the progressive changes that take place with depth and time. Diagenesis is heavily dependent on thermal history because it is temperature that probably has the greatest effect on the chemical reactions of the organic matter. Thus, the studies mentioned previously on thermal histories are also studies of diagenesis. The diagenesis of chlorophyll has been well-studied in DSDP samples, and the changes that take place with time during early, middle and late stages of the diagenetic process have been noted (Baker and others, 1976, 1977, 1978a, b, c; Palmer and others, 1979). Changes in composition and distribution of hydrocarbon gases in sediments offshore South Africa have been interpreted to result from early diagenesis (Hunt, 1978). Also, organic diagenesis in sediments from the Black Sea is evident from studies of amino acids and sugars (Mopper and others, 1978) and humic substances (Huc and others, 1978).

Gas Hydrates

The presence of large quantities of hydrocarbon gas in some sediments recovered by DSDP drilling coupled with the geophysical observations of anomalous acoustic reflection horizons that parallel the seafloor suggest that gas hydrates exist in some sediments of outer continental margins and ocean basins. During core recovery on some DSDP legs, sediment extrusion from the core liners and formation of gas pockets within the liners indicated the expansion of high concentrations of gas (McIver, 1974). This gas is mainly methane, and the presence of gas hydrates is inferred where the quantity and rate of methane evolution is large. On the basis of geophysical and geochemical evidence from Leg 11, gas hydrates are strongly indicated to be on the Blake-Bahama Rise (Lancelot and Ewing, 1972). On Leg 19 there is some evidence for gas hydrates in sediment of the Umnak Plateau of the Bering Sea, but Scholl and Creager (1973) attribute the anomalous bottom simulating acoustic reflector to a lithologic transition rather than to gas hydrate. Preliminary unpublished information from Legs 66 and 67 report that gas-releasing, frozen sediment was recovered during drilling. The sediments released from 7 to 50 times their volume of gas (mainly methane). Up to now the evidence for gas hydrates in marine sediments is circumstantial. Conclusive proof requires the collection and analysis of cores containing hydrates recovered at their original temperature and pressure.

Novel Geochemical Fossils

Geochemical fossils are organic molecules that originally were synthesized by living organisms and have been preserved in sediments unaltered or only slightly altered in structure and composition. Where there has been little or no alteration, the geochemical fossil can, in some instances, be related to its biochemical precursor. As diagenesis proceeds, the information content of geochemical fossils generally decreases, and product-precursor relationships become more obscure. Many of the organic molecules found in DSDP sediments are geochemical fossils and some of them are novel. For example, isoprenoid ketones were identified in sediments from several DSDP legs (Simoneit, 1973). These compounds represent intermediates in the diagenetic transformation of the biochemically important molecule phytol. Dehydroabietic acid was found in sediments from Leg 40 (Simoneit, 1978a), Leg 42B (Simoneit, 1978b) and Leg 43 (Simoneit, 1979). This compound is little altered from its higher plant diterpenoid precursor. Other novel geochemical fossils found in sediments of Leg 42B and 43 are sterenes which are products of sterols and the hopanes, the stereochemistry of which suggests they are extremely immature (Simoneit, 1978b, 1979). Sediments from six DSDP legs (38, 42B, 43, 44, 47A, 47B) contain copper porphyrins (Palmer and Baker, 1978). These compounds had not been confirmed before in marine sediments. They represent an unusual metal complex of intermediate products of chlorophyll diagenesis.

Sources of Organic Material

A major problem in organic geochemistry is the differentiation between various sources of organic matter, particularly between marine and terrestrial sources. Such differentiation is important in studies of petroleum potential because marine organic matter is more prone to generate oil, and terrestrial organic matter is more likely to generate gas. Many geochemical molecules contain information from which source can be interpreted. For example, shorter chain-length *n*-alkanes, dominated by n-C_{17}, are characteristic of marine sources, while longer chain lengths, dominated by odd-carbon-numbered *n*-alkanes, are characteristic of terrestrial higher plants. The distribution of alkanes has been commonly used to determine source of organic matter in DSDP sediments (Simoneit, 1976, 1977, 1978a, b, 1979). Likewise the chain length of *n*-fatty acids is also characteristic of source with shorter molecules coming from marine sources while longer molecules from terrestrial inputs (Simoneit, 1977, 1978a, b, 1979). Diterpenoids and perhaps perylene are characteristic of terrestrially derived organic matter (Simoneit, 1977, 1978b) in sediments of DSDP Legs 41 and 42B. On sediments from Leg 42B, Huc and others (1978) used the composition of humic acids to differentiate between marine and terrestrial sources. Stuermer and Simoneit (1978) used the carbon isotopic composition of humic substances to do the same thing on sediments from Leg 44. In general, most of the organic matter found in the primarily hemipelagic DSDP samples analyzed thus far is dominated by material from terrestrial sources.

Paleo-depositional Environments

Didyk and others (1978) described organic geochemical indicators of paleoenvironmental conditions of sedimentation. Much of their information was derived from considerations of sediments obtained through DSDP. Two indicators are used: ratios of pristane/phytane and the chlorin or porphyrin content. Anoxic environments are characterized by low pristane/phytane ratios and high contents of chlorins or porphyrins. Oxic environments have the opposite characteristics. Application of these criteria to sediments of Leg 40 shows that the Cretaceous sediments were deposited in an anoxic environment, while the later Pleistocene sediments were deposited under oxic environmental conditions (Simoneit, 1978a). Sediment samples from Leg 41 (Simoneit, 1977) and Leg 42B (Simoneit, 1978b) apparently were deposited under anoxic conditions.

Transport mechanisms for bringing organic matter into the depositional environment can be inferred from organic geochemical considerations coupled with sedimentological observations. Simoneit (1976) suggested that ice-rafting played a major role in the deposition of organic matter found in Leg 38 sediments from the Norwegian Sea, while eolian movement of lipids accounts for these compounds in sediments of Leg 41 (Simoneit, 1977) from offshore northern Africa.

Geochronology

Changes in the ratios of optical isomers of amino acids, caused by racemization during diagenesis, have been used to estimate the ages of geologic samples containing these amino acids. A related reaction, the epimerization of isoleucine to alloisoleucine, was first considered relative to DSDP sediment samples from Legs 5, 6, 7, and 9 by Wehmiller and Hare (1972) and later to sediment from Leg 15 by Hare (1973) and Bada and Man (1973). The latter authors were able to show systematic changes in the degree of epimerization of isoleucine with time for several different fractions of sediment extracts. Their success led to the application of this method of age estimation to sediment samples from Leg 37 (Bada and others, 1977) and from Leg 42A (Bada and others, 1978). On Leg 37 the extent of epimerization exceeded predictions: excess extent of epimerization was thought to be caused by the temperature effects of the geothermal gradient. The heat flow, calculated from the epimerization of isoleucine, and the measured values agreed closely. Thus, amino acids appear to be useful in estimating the ages of marine sediments to about four million years before present, and to estimating heat flow where the geothermal gradient has significant effects on chemical reactions.

SUMMARY

In the Initial Reports from the first 44 legs of the Deep Sea Drilling Project there are 120 articles dealing with various aspects of the organic geochemistry of ocean sediments. These articles resulted from analyses of approximately 2300 sediment samples. The samples have come from the major oceans, except the Arctic, and have spanned geologic time from Recent to Jurassic.

Organic geochemistry has been a part of DSDP since the beginning of the program when on the first leg an immature Tertiary crude oil was found in Jurassic/Cretaceous sediments associated with a salt dome. After this finding of petroleum every effort has been made to avoid such discoveries, because there are no provisions in the current drilling technology to stop the flow of gas and oil should significant deposits of these substances be released from the sediment.

Beginning with Leg 38 organic geochemists have been included on the scientific staffs of those legs of organic geochemical interest. The duties of the shipboard organic geochemists are (1) to aid in evaluation of the possibility of encountering significant hydrocarbon accumulations during drilling operations, (2) to collect core and canned samples for the scientific community, and (3) to collect samples for their own research. A sophisticated organic geochemical laboratory on board the *Glomar Challenger* provides facilities with which the geochemist can make measurements required to provide immediate guidance during the drilling operation. Three gas chromatographs, a fluorometer, an elemental analyzer, and an automated pyrolysis unit are available.

Since Leg 15 frozen sediment samples collected during the drilling operation are stored at the DSDP repository. Portions of these frozen samples have been allocated to the scientific community for organic geochemical studies. At least one quarter of the core is reserved for future work.

Organic geochemical results generally relate to one of these topic areas: (1) Petroleum potential/source rock evaluation, (2) thermal history and gradients, (3) diagenesis, (4) gas hydrates, (5) novel geochemical fossils, (6) sources of organic materials, (7) paleo-depositional environments, and (8) geochemical geochronology. Work on hydrocarbons and kerogen, especially since Leg 38, has been directed mainly to determining the potential of sediments as source rocks of petroleum. Much of the organic matter that has been found by DSDP drilling is dominantly of terrestrial origin, immature in terms of its thermal evolution, but could be a potential source of petroleum (mainly gas) if buried to greater depths than at present. This observation is encouraging from the energy standpoint because comparable sediments of great thickness and at greater depth are present on the slopes and particularly the rises of continental margins.

The future of organic geochemistry in DSDP is particularly bright. Three developments will provide new material, not available at present, for increasing understanding of organic geochemical processes: (1) pressure core barrel, (2) hydraulic piston core, and (3) riser capabilities of the *Glomar Explorer*. A working pressure core barrel is needed to verify the presence of gas hydrates in sediments. Only with this information will it be possible to map the extent of gases that are trapped in and beneath these features. Accurate knowledge of gas hydrates is essential for estimation of potential gas resources and reserves. The recently developed hydraulic piston core barrel will provide continuous sediment records from which the sources and early diagenetic history of organic molecules can be followed in great detail without the gaps and sediment disturbance that commonly occur with the present coring technology. The riser will allow deeper drilling into sediments which may be the source rocks of petroleum. Thus far drilling has reached sediment in the stage of late diagenesis. Deeper drilling should penetrate sediment undergoing catagenesis and thus a complete record of transformation of organic matter from diagenesis to metagenesis may be obtained. The organic geochemical consequences of these high temperature stages of organic-matter evolution can be studied in detail. With the application of these three technologies organic geochemistry should be able to advance rapidly in terms of new knowledge about organic matter, its transformation and its various roles in the global carbon cycle.

REFERENCES CITED

NOTE: All references to Deep Sea Drilling Project Initial Reports have been abbreviated; for complete citations see *Appendix I* at the end of this volume, or the individual Initial Reports as published.

BADA, J. L., AND MAN, E. H., 1973, Racemization of isoleucine in cores from Leg 15, Site 148: Init. Repts. DSDP, v. 20, p. 947–951.

———, ———, AND WALKER, A. C., 1977, Racemization of isoleucine in Leg 37, Sites 332 and 333 cores: Init. Repts. DSDP, v. 37, p. 633–636.

———, ———, Katz, B. J., and HSÜ, K. J., 1978, Racemization of isoleucine in foraminifer tests from Leg 42A, Sites 372 and 376: Init. Repts. DSDP, v. 42, p. 489–491.

BAKER, E. W., AND EGLINTON, G., 1978, Organic geochemistry related developments in the Deep Sea Drilling Project: 10th Annual Offshore Technology Conf. OTC 3050, p. 65–72.

———, PALMER, S. E., AND PARRISH, K. L., 1976, Tetrapyrrole pigments in DSDP Leg 38 sediments: Init. Repts. DSDP, v. 38, p. 785–789.

———, ———, AND HUANG, W. Y., 1977, Intermediate and late diagenetic tetrapyrrole pigments, Leg 41: Cape Verde Rise and Basin: Init. Repts. DSDP, v. 41, p. 825–837.

———, ———, AND ———, 1978a, Chlorin and porphyrin geochemistry of DSDP Leg 40 sediments: Init. Repts. DSDP, v. 40 Supplement, p. 639–647.

———, ———, AND ———, 1978b, Early and intermediate chlorophyll diagenesis of Black Sea sediments: Sites 379, 380, and 381: Init. Repts. DSDP, v. 42, p. 707–715.

———, ———, AND ———, 1978c, Miocene and Cretaceous tetrapyrrole pigments from Leg 44, Site 391: Init. Repts. DSDP, v. 44, p. 639–643.

COMER, J. B., AND LITTLEJOHN, R., 1976, Content, composition and thermal history of organic matter in Mesozoic sediments, Falkland Plateau: Init. Repts. DSDP, v. 36, p. 941–944.

DAVIS, J. B., AND BRAY, E. E., 1969, Analyses of oil and caprock from Challenger (Sigsbee) Knoll: Init. Repts. DSDP, v. 1, p. 415–500.

DIDYK, B. M., SIMONEIT, B. R. T., BRASSELL, S. C., AND EGLINTON, G., 1978, Organic geochemical indicators of paleoenvironmental conditions of sedimentation: Nature, v. 272, p. 216–222.

DEROO, G., HERBIN, J. P., ROUCACHE, J., TISSOT, B., ALBRECHT, P., AND SCHAEFFLE, J., 1977, Organic geochemistry of some Cretaceous black shales from Sites 367 and 368, Leg 41, eastern North Atlantic: Init. Repts. DSDP, v. 41, p. 865–873.

DOW, W. G., 1977, Contact metamorphism of kerogen in sediments from Leg 41: Cape Verde Rise and Basin: Init. Repts. DSDP, v. 41, p. 821–824.

———, 1978a, Petroleum source beds on continental slopes and rises: Amer. Assoc. Petroleum Geol. Bull., v. 62, p. 1584–1606.

———, 1978b, Geochemical analyses of samples from holes 391A and 391C, Leg 44: Blake-Bahama Basin: Init. Repts. DSDP, v. 44, p. 625–633.

ERDMAN, J. G., AND SCHORNO, K. S., 1976, Geochemistry of carbon, DSDP Leg 38: Init. Repts. DSDP, v. 38, p. 791–799.

———, AND ———, 1978, Geochemistry of carbon: Deep Sea Drilling Project Leg 44: Init. Repts. DSDP, v. 44, p. 605–615.

FORESMAN, J. B., 1978, Organic geochemistry DSDP Leg 40, continental rise of southwest Africa: Init. Repts. DSDP, v. 40, p. 557–567.

HARE, P. E., 1973, Amino acids, amino sugars, and ammonia in sediments from the Cariaco Trench: Init. Repts. DSDP, v. 20, p. 941–942.

HEDBERG, H. D., MOODY, J. D., AND HEDBERG, R. M., 1979, Petroleum prospects in the deep offshore: Amer. Assoc. Petroleum Geol. Bull., v. 63, p. 286–300.

HOOD, A., CASTÃNO, J. R., AND KENDRICK, J. W., 1976, Petroleum-generating potential and thermal history of DSDP Leg 38 sediments: Init. Repts. DSDP, v. 38, p. 801–803.

HUC, A. Y., DURAND, B., AND MONIN, J. C., 1978, Humic compounds and kerogens in cores from Black Sea sediments, Leg 42B—Holes 379A, B and 380A: Init. Repts. DSDP, v. 42, pt. 2, p. 737–748.

HUNT, J. M., 1972, Distribution of carbon in crust of Earth: Amer. Assoc. Petroleum Geol. Bull., v. 56, p. 2273–2277.

———, 1978, Light hydrocarbons in Holes 361 and 364, Leg 40: Init. Repts. DSDP, v. 40 Supplement, p. 649–650.

KENDRICK, J. W., HOOD, A., AND CASTAÑO, J. R., 1977, Petroleum-generating potential of sediments from Leg 41, Deep Sea Drilling Project: Init. Repts. DSDP, v. 41, p. 817–819.

———, ———, AND ———, 1978a, Petroleum-generating potential of sediments from Leg 40, Deep Sea Drilling Project: Init. Repts. DSDP, v. 40 Supplement, p. 671–676.

———, ———, AND ———, 1978b, Petroleum-generating potential of sediments from the eastern Mediterranean and Black Seas: Init. Repts. DSDP, v. 42, pt. 2, p. 729–735.

———, ———, AND ———, 1978c, Petroleum-generating potential of sediments from Leg 44, Deep Sea Drilling Project: Init. Repts. DSDP, v. 44, p. 599–603.

———, ———, AND ———, 1979, Petroleum-generating potential of Cretaceous sediments from Leg 43, Deep Sea Drilling Project: Init. Repts. DSDP, v. 43, p. 663–668.

LANCELOT, Y., AND EWING, J. I., 1972, Correlation of natural gas zonation and carbonate diagenesis in Tertiary sediment from the north-west Atlantic: Init. Repts. DSDP, v. 11, p. 791–799.

MCIVER, R. D., 1974, Hydrocarbon gas (methane) in canned Deep Sea Drilling Project core samples, *in* Kaplan, I. R., ed., Natural Gases in Marine Sediments: New York, Plenum Publ. Corp., p. 63–69.

———, 1975, Hydrocarbon occurrences from JOIDES Deep Sea Drilling Project: 9th World Petroleum Congress, v. 2, p. 269–280.

MOPPER, K., MICHAELIS, W., GARRASI, C., AND DEGENS, E. T., 1978, Sugars, amino acids and hydrocarbons in Black Sea sediment from DSDP Leg 42B cores: Init. Repts. DSDP, v. 42, p. 697–705.

MORRIS, D. A., 1976, Organic diagenesis of Miocene sediments from Site 341, Vøring Plateau, Norway: Init. Repts. DSDP, v. 38, p. 809–814.

PALMER, S. E., AND BAKER, E. W., 1978, Copper porphyrins in deep-sea sediments: a possible indicator of oxidized terrestrial organic matter: Science, v. 201, p. 49–51.

———, HUANG, W. Y., AND BAKER, E. W., 1979, Tetrapyrrole pigments from Bermuda Rise: DSDP Leg 43: Init. Repts. DSDP, v. 43, p. 657–661.

PEARSON, D. B., AND DOW, W. G., 1979, Petroleum source rocks evaluation of the southern European and northern African Atlantic continental margins—Results of DSDP Legs 47A and B and 48: 11th Annual Offshore Technology Conf., OTC 3407, p. 449–455.

SCHOLL, D. W., AND CREAGER, J. S., 1973, Geologic syntheses of Leg 19 (DSDP) results, far north Pacific and Aleutian Ridge and Bering Sea: Init. Repts. DSDP, v. 19, p. 897–913.

SIMONEIT, B. R. T., 1973, Identification of isoprenoidal ketones in Deep Sea Drilling Project core samples and their geochemical significance: Init. Repts. DSDP, v. 21, p. 909–923.

———, 1976, Sources of the solvent-soluble organic matter in the glacial sequence of DSDP samples from the Nor-

wegian-Greenland Sea, Leg 38: Init. Repts. DSDP, v. 38, p. 805–806.
———, 1977, Leg 41 sediment lipids—search for eolian organic matter in recent samples and examination of a Black shale: Init. Repts. DSDP, v. 41, p. 855–863.
———, 1978a, Lipid analysis of sediments from Site 364 in the Angola Basin, DSDP Leg 40: Init. Repts. DSDP, v. 40 Supplement, p. 659–662.
———, 1978b, Organic geochemistry of terrigenous muds and various shales from the Black Sea, DSDP Leg 42B: Init. Repts. DSDP, v. 42, p. 749–753.
———, 1978c, The organic chemistry of marine sediments, *in* Riley, J. P. and Chester, R., eds., Chemical Oceanography, v. 2, 2nd ed.: New York, Academic Press, p. 233–311.
———, 1979, Organic geochemistry of the shales from the northwestern proto-Atlantic, DSDP Leg 43: Init. Repts. DSDP, v. 43, p. 643–649.
———, Brenner, S., Peters, K. E., and Kaplan, I. R., 1978, Thermal alteration of Cretaceous black shale by basaltic intrusions in the eastern Atlantic: Nature, v. 273, p. 501–504.
Stuermer, D. H., and Simoneit, B. R. T., 1978, Varying sources for the lipid and humic substances at Site 391, Blake-Bahama Basin, DSDP Leg 44: Init. Repts. DSDP, v. 44, p. 587–591.
Wehmiller, J. F., and Hare, P. E., 1972, Amino acid content of some samples from the Deep Sea Drilling Project: Init. Repts. DSDP, v. 9, p. 903–905.
Welte, D. H., 1979, Hydrocarbon source rocks in deep sea sediments: 11th Annual Offshore Technology Conf., OTC 3408, p. 457–464.

Appendix I.—Organic Geochemistry, DSDP Initial Reports

Vol.	Chapter/ Appendix	Pages	Author	Title	Year
I	22	415–500	J.B. Davis & E.E. Bray, (compilers)	Analyses of oil and cap rock from Challenger (Sigsbee) Knoll.	1969
I	22	426–427	E.D. Evans & W.L. Orr	Extraction of Core 5-1.	1969
I	22	456–460	J.G. Erdman, R.L. Borst, & R.S. Scalan	Studies performed on samples from Challenger Knoll.	1969
I	22	461–467	J.G. Erdman, R.L. Borst, & R.S. Scalan	Composition of gas sample 1 by components.	1969
I	22	468–477	P.R. Mommessin, A. Hood, W.E. Ellington, & A.F. Mannel	Analyses of organic matter in core 5, Leg 1, Site 2.	1969
I	22	478–489	B. Koons & P.H. Monaghan	Data and discussion of analyses by Esso Production Research Company, including $^{12}C/^{13}C$ ratios.	1969
I	22	489	J.B. Davis	Petroleum acids.	1969
I	22	492–497	E.D. Evans, E.E. Bray, J.G. Bendoraitis, & H.R. Middleton	Hydrocarbon group type analysis.	1969
I	22	498–499	E.W. Baker	Porphyrins.	1969
I	22	500	S.R. Silverman	Reference to $^{13}C/^{12}C$ ratio of Challenger Knoll oil.	1969
IV	21	431–438	E.W. Baker	Tetrapyrrole pigments.	1970
VI	IV	1327–1329	R.D. McIver	JOIDES cores-evidence of migration of hydrocarbons in Pleistocene sediments of the Shatsky Plateau, Western Pacific Ocean.	1971
VII Pt. 2	14	889–912	B.R. Simoneit & A.L. Burlingame	Some preliminary results on the higher molecular weight hydrocarbons and fatty acids in the Deep Sea Drilling Project cores, Legs 5–7.	1971
VIII	22	871–872	R.D. McIver	Organic geochemical analyses of frozen samples from DSDP Leg 8 cores.	1971
VIII	23	873–900	B.R. Simoneit & A.L. Burlingame	Further preliminary results on the higher molecular weight hydrocarbons and fatty acids in the DSDP cores, Legs 5–8.	1971

Appendix I.—Organic Geochemistry, DSDP Initial Reports (continued)

Vol.	Chapter/ Appendix	Pages	Author	Title	Year
IX	18	857–858	R.D. McIver	Bitumens and organic carbon in samples from DSDP Leg 9 cores.	1972
IX	19	859–901	B.R. Simoneit & A.L. Burlingame	Further preliminary results on the higher molecular weight hydrocarbons and fatty acids in the DSDP cores, Leg 9.	1972
IX	20	903–905	J.F. Wehmiller & P.E. Hare	Amino acid content of some samples from the Deep Sea Drilling Program.	1972
X	22	625–636	B.R. Simoneit, E.S. Scott, & A.L. Burlingame	Preliminary organic analyses of the Deep Sea Drilling Project cores, Leg 10.	1973
XI	24	667–722	F.J. Paulus	Leg 11 measurements of physical properties in sediments of the western North Atlantic and their relationship to sediment consolidation.	1972
XI	36	1013–1045	B.R. Simoneit, E.S. Scott, W.G. Howells, & A.L. Burlingame	Preliminary organic analyses of the Deep Sea Drilling Project cores, Leg 11.	1972
XII	18	1115–1121	M.A. Rogers, J.E. van Hinte, & J.G. Sugden	Organic carbon ^{13}C values from Cretaceous, Tertiary, and Quaternary marine sequences in the North Atlantic.	1972
XIII	32	813–816	R.D. McIver	Geochemical significance of gas and gasoline-range hydrocarbons and organic matter in Miocene samples from Site 134—Balearic abyssal plain.	1973
XVI	17	575–600	B.R. Simoneit, E.S. Scott, & A.L. Burlingame	Preliminary organic analyses of DSDP cores, Leg 14, Atlantic Ocean.	1973
XVII	19	561–590	B.R. Simoneit & A.L. Burlingame	Preliminary organic analyses of DSDP cores, Legs 12 and 13.	1973
XVIII	III	1013–1014	R.D. McIver	Hydrocarbon gases from canned core samples-Sites 174A, 176 and 180.	1973
XIX	34	875–877	R.D. McIver	Hydrocarbons in canned mud from Sites 185, 186, 189, 191-Leg 19.	1973
XIX	35	879–884	G.E. Claypool, B.J. Presley, & I.R. Kaplan	Gas analyses in sediment samples from Legs 10, 11, 13, 14, 15, 18 and 19.	1973
XX	18	905	J.M. Hunt	Organic geochemical studies: Introduction and Summary.	1973
XX	19	907–933	B.R. Simoneit, W.G. Howells, & A.L. Burlingame	Preliminary organic geochemical analyses of the Cariaco Trench Site 147 Deep Sea Drilling Project, Leg 15.	1973
XX	20	935–936	R.D. McIver	Cyclical geochemical properties of organic matter in Cariaco basin cores—Leg 15, Site 147.	1973
XX	21	937–939	T.C. Hoering	Characterization of the organic matter in a Site 147 core from the Cariaco Trench.	1973
XX	22	941–942	P.E. Hare	Amino acids, amino sugars, and ammonia in sediments from the Cariaco Trench.	1973
XX	23	943–946	E.W. Baker & G.D. Smith	Chlorophyll derivatives in sediments, Site 147.	1973
XX	24	947–951	J.L. Bada & E.H. Man	Racemization of isoleucine in cores from Leg 15 Site 145.	1973

APPENDIX I.—ORGANIC GEOCHEMISTRY, DSDP INITIAL REPORTS (CONTINUED)

Vol.	Chapter/ Appendix	Pages	Author	Title	Year
XXI	21	721	R.D. McIver	Low residual gas contents of four Leg 21 canned-sediment samples.	1973
XXI	I	909–923	B. Simoneit	Identification of isoprenoidal ketones in Deep Sea Drilling Project core samples and their geochemical significance.	1973
XXII	33	671	R.D. McIver	Analysis of ten Leg 22 cores for organic carbon and gasoline-range hydrocarbons.	1974
XXII	34	673–675	J.M. Hunt	Hydrocarbon and kerogen studies.	1974
XXII	35	677–679	G.D. Smith & E.W. Baker	Chlorophyll derivatives in DSDP Leg 22 sediments.	1974
XXII	36	681–692	B.R. Simoneit & A.L. Burlingame	Preliminary organic geochemical analyses of the Site 217 core samples in the Bengal Basin, DSDP Leg 22.	1974
XXIII	37	971–973	R.D. McIver	Residual gas contents of organic-rich canned sediment samples from Leg 23.	1974
XXIV	16	791–797	P.J. Cernock	Geochemical analyses of potential petroleum source beds.	1974
XXIV	III	1157–1158	R.D. McIver	Gaseous and heavy hydrocarbons in canned core samples from Leg 24, DSDP.	1974
XXIV	IV	1159–1163	B.R. Simoneit	Complex triterpanoidal acids and hydrocarbons in DSDP core samples and their geochemical significance.	1974
XXIV	V	1165–1167	J.M. Hunt	Hydrocarbon and kerogen studies on Red Sea and Gulf of Aden cores.	1974
XXIV	VI	1169–1176	J.G. Erdman, K.S. Schorno, & R.S. Scalan	Geochemistry of carbon: DSDP Legs 22, 24, 26, 27, and 28.	1974
XXVI	27	613–617	J.A. Calder, G.J. Horvath, D.J. Schultz, & J.W. Newman	Geochemistry of the stable carbon isotopes in some Indian Ocean sediments.	1974
XXVII	17	453–454	R.D. McIver	Methane in canned core samples from Site 262, Timor Trough.	1974
XXVIII	28	815–817	R.D. McIver	Hydrocarbon gases in canned core samples from Leg 28, Sites 271, 272, and 273, Ross Sea.	1975
XXXI	30	621–627	R.L. Ames & R. Littlejohn	Diagenesis of organic matter and estimated temperature history from carbonization measurements, Shikoku Basin.	1975
XXXI	31	629–632	E.W. Baker & G.D. Smith	Chlorophyll derivatives in DSDP Leg 31 sediments.	1975
XXXI	32	633–638	J.G. Erdman, K.S. Schorno, & R.S. Scalan	Geochemistry of carbon: DSDP Leg 31.	1975
XXXI	I	899–900	R.D. McIver	Organic-matter lean sediments of Site 278, Leg 29, DSDP.	1975
XXXI	II	901–903	J.M. Hunt	Hydrocarbon studies.	1975
XXXI	III	905–909	E.W. Baker & G.D. Smith	Chlorophyll derivatives in DSDP Leg 14, 20, 26, 27, and 29 sediments.	1975
XXXI	IV	911–916	J.G. Erdman, K.S. Schorno, & R.S. Scalan	Geochemistry of carbon and sulfur: DSDP Leg 29.	1975

APPENDIX I.—ORGANIC GEOCHEMISTRY, DSDP INITIAL REPORTS (CONTINUED)

Vol.	Chapter/ Appendix	Pages	Author	Title	Year
XXXVI	26	941–944	J.B. Comer & R. Littlejohn	Content, composition and thermal history of organic matter in Mesozoic sediments, Falkland Plateau.	1977
XXXVII	48	633–636	J.L. Bada, E.H. Man, & A.C. Walker	Racemization of isoleucine in Leg 37, Sites 332 and 333 cores.	1977
XXXVIII	18	783–784	K.A. Kvenvolden	Organic geochemistry, Leg 38: Introduction to studies.	1976
XXXVIII	19	785–789	E.W. Baker, S.E. Palmer, & K.L. Parrish	Tetrapyrrole pigments in DSDP Leg 38 sediments.	1976
XXXVIII	20	791–799	J.G. Erdman & K.S. Schorno	Geochemistry of carbon, DSDP Leg 38.	1976
XXXVIII	21	801–803	A. Hood, J.R. Castaño, & J.W. Kendrick	Petroleum-generating potential and thermal history of DSDP Leg 38 sediments.	1976
XXXVIII	22	805–806	B.R. Simoneit	Sources of the solvent-soluble organic matter in the glacial sequence of DSDP samples from the Norwegian-Greenland Sea, Leg 38.	1976
XXXVIII	23	807–808	J.M. Hunt	C_4-C_7 alkane yield.	1976
XXXVIII	24	809–814	D.A. Morris	Organic diagenesis of Miocene sediments from Site 341, Vøring Plateau, Norway.	1976
XXXVIII	25	815–821	L.J. Bogolyvbova & P.O. Timofeev	Plant organic matter in sediments from Hole 336, DSDP Leg 38.	1976
XXXIX	16	497–500	B.R.T. Simoneit	Search for terrigenous lipids in carbonate-rich samples from Site 39-354.	1977
XL	11	557–567	J.B. Foresman	Organic geochemistry: DSDP Leg 40, Continental Rise of Southwest Africa.	1978
XL	Supp.1	627–637	J.J. Boon, F.W. van der Meer, P.J.W. Schuyl, J.W. de Leeuw, P.A. Schenk, & A.L. Burlingame	Organic geochemical analyses of core samples from Site 362, Walvis Ridge, DSDP Leg 40.	1978
XL	Supp.2	639–647	E.W. Baker, S.E. Palmer, & W.Y. Huang	Chlorin and porphyrin geochemistry of DSDP Leg 40 sediments.	1978
XL	Supp.3	649–650	J.M. Hunt	Light hydrocarbons in Holes 361 and 364, Leg 40.	1978
XL	Supp.4	651–658	J.G. Erdman & K.S. Schorno	Geochemistry of carbon: Deep Sea Drilling Project, Leg 40.	1978
XL	Supp.5	659–662	B.R.T. Simoneit	Lipid analyses of sediments from Site 364 in the Angola Basin, DSDP Leg 40.	1978
XL	Supp.6	663–669	J.F. Raynaud & P. Robert	Microscopical survey of organic matter from DSDP Sites 361, 362, and 364.	1978
XL	Supp.7	671–676	J.W. Kendrick, A. Hood, & J.R. Castaño	Petroleum-generating potential of sediments from Leg 40, Deep Sea Drilling Project	1978
XLI	20	815	K.A. Kvenvolden	Organic geochemistry Leg 41, Introduction and summary.	1977
XLI	21	817–819	J.W. Kendrick, A. Hood, & J.R. Castaño	Petroleum-generating potential of sediments from Leg 41, Deep Sea Drilling Project.	1977
XLI	22	821–824	W.G. Dow	Contact metamorphism of kerogen in sediments from Leg 41: Cape Verde Rise and Basin.	1977

APPENDIX I.—ORGANIC GEOCHEMISTRY, DSDP INITIAL REPORTS (CONTINUED)

Vol.	Chapter/ Appendix	Pages	Author	Title	Year
XLI	23	825–837	E.W. Baker, S.E. Palmer, & W.Y. Huang	Intermediate and late diagenetic tetrapyrrole pigments, Leg 41: Cape Verde Rise and Basin.	1977
XLI	24	839–847	E.W. Baker, W.Y. Huang, J.G. Rankin, J.R. Castaño, J.R. Guinn, & A.N. Fuex	Electron paramagnetic resonance study of thermal alteration of kerogen in deep-sea sediments by basaltic sill intrusion.	1977
XLI	25	849–853	J.G. Erdman & K.S. Schorno	Geochemistry of carbon: Deep Sea Drilling Project, Leg 41.	1977
XLI	26	855–858	B.R.T. Simoneit	Leg 41 sediment lipids—search for eolian organic matter in recent samples and examination of a black shale.	1977
XLI	27	859	J.M. Hunt & J.K. Whelan	Light hydrocarbons at Sites 367–370, Leg 41.	1977
XLI	28	861–863	P.R. Doose, M.W. Sandstrom, R.Z. Jodele, & I.R. Kaplan	Interstitial gas analysis of sediment samples from Site 368 and Hole 369A.	1977
XLI	29	865–873	G. Deroo, J.P. Herbin, J. Roucaché, B. Tissot, P. Albrecht, & J. Schaeffle	Organic geochemistry of some Cretaceous black shales from Sites 367 and 368, Leg 41, Eastern North Atlantic.	1977
XLII Pt. 1	13.3	465–472	G. Deroo, J.P. Herbin, & J. Roucaché	Organic geochemistry of some Neogene cores from Sites 374, 375, 377 and 378: Leg 42A, eastern Mediteranean Sea.	1978
XLII Pt. 1	15	483–488	P. Hahn-Weinheimer, F. Fabricius, J. Müller, & W. Sigl	Stable isotopes of oxygen and carbon in carbonates and organic material from Pleistocene to Upper Miocene sediments at Site 374 (DSDP Leg 42A).	1978
XLII Pt. 1	16	489–491	J.L. Bada, E.H. Man, B.J. Katz, & K.J. Hsü	Racemization of isoleucine in foraminifer tests from Leg 42A, Sites 372 and 376.	1978
XLII Pt. 2	25	661–665	J.M. Hunt & J.K. Whelan	Dissolved gases in Black Sea sediments.	1978
XLII Pt. 2	26	667–672	E. Faber, M. Schmitt, & W. Stahl	Carbon isotope analyses of head space methane from samples of Leg 42B, Sites 379, 380 and 381.	1978
XLII Pt. 2	27	673–677	J.K. Whelan & J.M. Hunt	C_1-C_7 hydrocarbons in Holes 378A, 380/380A, and 381.	1978
XLII Pt. 2	28	679–681	R.D. McIver	Residual hydrocarbon gases in canned core material from Holes 379A and 380A, Leg 42B.	1978
XLII Pt. 2	29	683–696	A.A. Geodekyan, G.F. Ulmishek, T.G. Tcherova, V.I. Avilov, A.P. Bokovoy, Z.I. Verkhovskaya, & M.S. Fedorova	Bituminological studies of the samples from Site 379 and laboratory simulation of dispersed organic matter transformation.	1978
XLII Pt. 2	30	697–705	K. Mopper, W. Michaelis, C. Garrasi, & E.T. Degens	Sugars, amino acids and hydrocarbons in Black Sea sediment from DSDP Leg 42B cores.	1978
XLII Pt. 2	31	707–715	E.W. Baker, S.E. Palmer, & W.Y. Huang	Early and intermediate diagenesis of Black Sea sediments: Sites 379, 380, 381.	1978
XLII Pt. 2	32	717–721	J.G. Erdman & K.S. Schorno	Geochemistry of carbon: Deep Sea Drilling Project, Legs 42A and 42B.	1978

Appendix I.—Organic Geochemistry, DSDP Initial Reports (continued)

Vol.	Chapter/ Appendix	Pages	Author	Title	Year
XLII Pt. 2	33	723–728	V.T. Vuchev, C.P. Ivanov, M. St. Kabakchieva, L.L. Petrov, R. Zh. Stojanova, D.D. Stephanov, D.N. Djakova, & L.K. Petrova	Preliminary organic geochemical studies of DSDP cores, Leg 42B, Black Sea.	1978
XLII Pt. 2	34	729–735	J.W. Kendrick, A. Hood, & J.R. Castaño	Petroleum-generating potential of sediments from the eastern Mediterranean and Black Seas.	1978
XLII Pt. 2	35	737–748	A.Y. Huc, B. Durand, & J.C. Monin	Humic compounds and kerogens in cores from Black Sea sediments, Leg 42B-Holes 379A, B, and 380A.	1978
XLII Pt. 2	36	749–753	B.R.T. Simoneit	Organic geochemistry of terrigenous muds and various shales from the Black Sea, DSDP Leg 42B.	1978
XLII Pt. 2	37	755–763	D.F. Anders, G.E. Claypool, C.M. Lubeck, & J.M. Patterson	Preliminary results, organic geochemical investigation of Black Sea sediments: Deep Sea Drilling Project, Leg 42B.	1978
XLII Pt. 2	38	765–770	J.D. King & D.C. White	Muramic acid as a measure of microbial biomass in Black Sea sediments.	1978
XLIII	24	633–642	J.W. Kendrick	Geochemical studies of black clays from Leg 43, Deep Sea Drilling Project.	1979
XLIII	25	643–649	B.R.T. Simoneit	Organic geochemistry of the shales from the northwestern proto-Atlantic, DSDP Leg 43.	1979
XLIII	26	651–655	J.G. Erdman & K.S. Schorno	Geochemistry of carbon: Deep Sea Drilling Project Leg 43.	1979
XLIII	27	657–661	S.E. Palmer, W.Y. Huang, & E.W. Baker	Tetrapyrrole pigments from Bermuda Rise: DSDP Leg 43.	1979
XLIII	28	663–668	J.W. Kendrick, A. Hood, & J.R. Castaño	Petroleum-generating potential of Cretaceous sediments from Leg 43, Deep Sea Drilling Project.	1979
XLIV	19	585	K.A. Kvenvolden	Introduction to organic geochemistry studies, DSDP Leg 44.	1978
XLIV	20	587–591	D.H. Stuermer & B.R.T. Simoneit	Varying sources for the lipids and humic substances at Site 391, Blake-Bahama Basin, DSDP Leg 44.	1978
XLIV	21	593–598	G. Deroo, J.P. Herbin, J.R. Roucaché, B. Tissot, P. Albrecht, & M. Dastillung	Organic geochemistry of some Cretaceous claystones from Site 391, Leg 44, western North Atlantic.	1978
XLIV	22	599–603	J.W. Kendrick, A. Hood, & J. R. Castaño	Petroleum-generating potential of sediments from Leg 44, Deep Sea Drilling Project.	1978
XLIV	23	605–615	J.G. Erdman & K.S. Schorno	Geochemistry of carbon: Deep Sea Drilling Project Leg 44	1978
XLIV	24	617–623	J.N. Cardoso, A.M.K. Wardroper, C.D. Watts, P.J. Barnes, J.R. Maxwell, G. Eglinton, D.G. Mound, & G.C. Speers	Preliminary organic geochemical analyses: Site 391, Leg 44 of Deep Sea Drilling Project.	1978
XLIV	25	625–633	W.G. Dow	Geochemical analyses of samples from Hole 391A and 391C, Leg 44: Blake-Bahama Basin.	1978

APPENDIX I.—ORGANIC GEOCHEMISTRY, DSDP INITIAL REPORTS (CONTINUED)

Vol.	Chapter/ Appendix	Pages	Author	Title	Year
XLIV	27	635–637	G.E. Claypool & J.P. Baysinger	Thermal analysis/pyrolysis of Cretaceous sapropels: DSDP Leg 44, Hole 391C, Blake-Bahama Basin	1978
XLIV	28	639–643	E.W. Baker, S.E. Palmer, & W.Y. Huang	Miocene and Cretaceous tetrapyrrole pigments from Leg 44, Site 391.	1978
XLIV	29	645–649	R.D. McIver & M.A. Rogers	Insoluble organic matter and bitumens in Leg 44 samples.	1978
XLIV	30	651–652	J.M. Hunt & J.K. Whelan	Light hydrocarbons in sediments of DSDP Leg 44 holes.	1978
XLIV	31	653–654	F.M. Swain & J.M. Bratt	Carbohydrate residue in Leg 44 core samples.	1978

APPENDIX II.—DEEP SEA DRILLING PROJECT ORGANIC CARBON DETERMINATIONS

Leg No.	Vol. No.	Chapter/ Appendix	Pages	Authors	No. of samples analyzed
1	I	II	340–347	DSDP	127
2	II	9	319–321	DSDP	37
3	III	15	495–507	A.C. Pimm	404
4	IV	13	307–314	A.C. Pimm	225
5	V	15	431–440	T.L. Vallier	286
6	VI	23	739–752	A.C. Pimm	422
7	VII Pt. 2	10	845–862	E.L. Gealy	513
8	VIII	III	1017–1035	DSDP	688
9	IX	13	797–816	R.E. Boyce & G.W. Bode	645
10	X	24	641–642	R.E. Boyce	93
11	XI	38	1059–1071	R.E. Boyce	466
15	XV	11	1129–1137	G.W. Bode	934
16	XVI	14	521–528	G.W Bode & D.S. Cronan	732
17	XVII	II	927–930	G.W. Bode	376
18	XVIII	VI	1069–1076	G.W. Bode	788
19	XIX	19	663–665	G.W. Bode	217
20	XX	I	741–742	G.W. Bode	80
23	XXIII	II	1131–1135	G.W. Bode	320
24	XXIV	II	1155–1156	G.W. Bode	78
25	XXV	III	841*	W.A. Girdley	196
27	XXVII	21	499–502	G.W. Bode	329
30	XXX	25	687–688	D.H. Cameron	124
32	XXXII	19	561–562	G.W. Bode	123
33	XXXIII	II	959–963	D.H. Cameron	471
34	XXXIV	52	601–602	D.H. Cameron	135
35	XXXV	II	755–756	D.H. Cameron	38
36	XXXVI	III	1047–1050	D.H. Cameron	144
37	XXXVII	49	637–639	G.W. Bode	93
38	XXXVIII Supp.	21	433–436	K. Thompson	309
39	XXXIX	17	501–504	B. Scott	182
42A	XLII Pt. 1	II	1151–1156	R.B. Kidd	321
42A	XLII Pt. 1	IX	1221–1224	W. Siyl	392
43	XLIII	IV	1043–1047	D.H. Cameron	233
44	XLIV	I	983–986	R. Myers	360

*Table of data is missing in this Initial Report.

Appendix III. — Legs and Sites with Organic Geochemical Information

Leg	Sites	No. of Samples	Vol. No.	Pages
1	2	18	I	415–500
4	26,27,30	3	IV	431–438
	30	1	VI	1327–1329
5	34,35	2	VI	1327–1329
	32,33,34,35,36,37,38,40,42	16	VII	889–912
	36,42	2	IX	903–905
6	47,49,50	3	VI	1327–1329
	47,49,50,51,52,53,55,56,58	13	VII	889–912
	47	1	IX	903–905
7	62,64,65,66	8	VII	889–912
	62,64	2	IX	903–905
8	73,74,75	4	VIII	871–872
	72,73,74,75	6	VIII	873–900
9	78,80,82,83,84	6	IX	857–858
	77,78,79,80,82,83,84	10	IX	859–901
	82	1	IX	903–905
10	90,92	2	X	625–636
	88,90,91	14	XIX	879–884
	90,92	2	XXI	909–923
	92	1	XXIV	1159–1163
11	102,103,104,106B	19	XI	667–722
	98,105	2	XI	1013–1045
	102,104,106	8	XIX	879–884
	105	1	XXI	909–923
	105	1	XXIV	1159–1163
12	111,112,116,118,119	153	XII	1115–1121
	112,114	2	XVII	561–590
	112,114	2	XXI	909–923
13	134	1	XIII	813–816
	128,130,134	4	XVII	561–590
	128	3	XIX	879–884
	128,130	3	XXI	909–923
	125	1	XLII Pt. 2	729–735
14	138,144	6	XVI	575–600
	144	2	XIX	879–884
	138,144	5	XXI	909–923
	144	2	XXXI	905–909
15	147,154	12	XIX	879–884
	147	3	XX	907–933
	147	1	XX	937–939
	147	3	XX	941–942
	147	7	XX	943–946
	148	12	XX	947–951
	147	8	XX	935–936
	147	4	XXI	909–923
18	174,176,180	29	XVIII	1013–1014
	174,176	9	XIX	879–884
	174,176,180	29	XXVIII	815–817
19	185,186,189,191	19	XIX	875–877
	185,186,189,191	4	XIX	879–884
	185,186,189,191	19	XXVIII	815–817
20	198	1	XXXI	905–909
21	204,210	4	XXI	721
	204,210	4	XXVIII	815–817

APPENDIX III.—LEGS AND SITES WITH ORGANIC GEOCHEMICAL INFORMATION (CONTINUED)

Leg	Sites	No. of Samples	Vol. No.	Pages
22	217,218	10	XXII	671
	217,218	4	XXII	673–675
	217,218	10	XXII	677–679
	217	4	XXII	681–692
	217,218	8	XXIV	1169–1176
23	222,229	6	XXIII	971–973
	229	1	XXIV	1165–1167
	222,229	6	XXVIII	815–817
24	231	3	XXIV	791–797
	231,232	9	XXIV	1157–1158
	232,233	5	XXIV	1165–1167
	233	4	XXIV	1169–1176
	231,232	9	XXVIII	815–817
26	250,254,257	3	XXIV	1169–1176
	252,253,254,256,257,258	72	XXVI	613–617
	250	1	XXXI	905–909
27	259,263	2	XXIV	1169–1176
	262	11	XXVII	453–454
	262	11	XXVIII	815–817
	259,263	2	XXXI	901–903
	259,263	2	XXXI	905–909
28	267	1	XXIV	1169–1176
	271,272,273	21	XXVIII	815–817
	265	1	XXXI	901–903
29	278	12	XXXI	899–900
	280,281,282	15	XXXI	901–903
	280,281	5	XXXI	905–909
	280,281,282	12	XXXI	911–916
31	297,298	21	XXXI	621–627
	299,302	11	XXXI	629–632
	299,302	11	XXXI	633–638
36	327A,330	10	XXXVI	941–944
37	332A,333	6	XXXVII	633–636
38	336,338,341,343,344,345,346	36	XXXVIII	785–789
	338,339,341,344,346,348,349	42	XXXVIII	791–799
	338,341,342,343,344,346,348,350	39	XXXVIII	801–803
	339,345	4	XXXVIII	805–806
	336,338,345,348	13	XXXVIII	807–808
	341	34	XXXVIII	809–814
	336	27	XXXVIII	815–821
39	354	2	XXXIX	497–500
	356	10	XL	557–569
	356	1	XL Supp.	639–647
40	360,361,362,362A,363,364,365	136	XL	557–567
	362	4	XL Supp.	627–637
	361,362,363,364	19	XL Supp.	639–647
	361,364	10	XL Supp.	649–650
	361,362,362A	40	XL Supp.	651–658
	364	3	XL Supp.	659–662
	361,362,364,	8	XL Supp.	663–669
	361,362,363,364,365	29	XL Supp.	671–676
	364	1	XLIII	643–649
41	367,368,369,370	6	XLI	817–819
	367,368	6	XLI	821–824
	367,368,369,369A,370	28	XLI	825–837

APPENDIX III.—LEGS AND SITES WITH ORGANIC GEOCHEMICAL INFORMATION (CONTINUED)

Leg	Sites	No. of Samples	Vol. No.	Pages
	367,368	30	XLI	839–847
	368	11	XLI	849–853
	367,369	3	XLI	855–858
	367,368,369,370	6	XLI	859
	368,369A	22	XLI	861–863
	367,368	7	XLI	865–873
42A	374,375,377,378	32	XLII Pt. 1	465–472
	374	23	XLII Pt. 1	483–488
	372,376	3	XLII Pt. 1	489–491
	376	4	XLII Pt. 2	717–721
	376,378	4	XLII Pt. 2	729–735
			XLII Pt. 2	
42B	379A,380/380A,381	~300	XLII Pt. 2	661–665
	379A,379B,380A,380B,381	44	XLII Pt. 2	667–672
	379,380/380A,381	38	XLII Pt. 2	673–677
	379A,380A	9	XLII Pt. 2	679–681
	379,379A	23	XLII Pt. 2	683–696
	379A,380A,381	25	XLII Pt. 2	697–705
	379A,379B,380,380A,381	25	XLII Pt. 2	707–715
	380	10	XLII Pt. 2	717–721
	379A,381	6	XLII Pt. 2	723–728
	379,380,381	21	XLII Pt. 2	729–735
	379A,379B,380A	63	XLII Pt. 2	737–748
	379A,380A,381	11	XLII Pt. 2	749–753
	379A,379B,380,380A,381	67	XLII Pt. 2	755–763
	379A,379B,380	10	XLII Pt. 2	765–770
43	386,387	24	XLIII	633–642
	386,387	4	XLIII	643–649
	386	7	XLIII	651–655
	382,386,387	16	XLIII	657–661
	386,387	13	XLIII	663–668
44	391A,391B,391C	8	XLIV	587–591
	391A,391C	8	XLIV	593–598
	391,392	15	XLIV	599–603
	388A,390,390A,391,391A,391C	37	XLIV	605–615
	391A,391C	18	XLIV	617–623
	391A,391C	15	XLIV	625–633
	391C	4	XLIV	635–637
	391A,391B,391C	12	XLIV	639–643
	388A,391A,391C	18	XLIV	645–649
	388A,391C	7	XLIV	651–652
	390,391A,391B,391C,392	15	XLIV	653–654

APPENDIX IV.—COMPOUND TYPES

	Volume Number and Pages
Acids (Organic)	
Dicarboxylic acids	XI 1013–1045. XLIV 617–623.
Hydroxy acids	XL (Supp.) 627–637. XLIV 617–623.
Isoprenoid acids (Also includes iso- and anteiso-)	XL (Supp.) 627–637,659–662. XLI 855–858. XLIII 643–649. XLIV 617–623.
Naphthenic acids	I 415–500, 489.

APPENDIX IV.—COMPOUND TYPES (CONTINUED)

	Volume Number and Pages
	723–728, 749–753, 755–763. XLIII 643–649, 651–655. XLIV 587–591, 593–598, 605–615, 617–623, 625–633, 645–649.
Alkenes (unsaturated)	XX 907–933. XLI 855–858, 865–873. XLII (Pt. 2) 749–753, 755–763. XLIII 643–649. XLIV 617–623.
Aromatic Hydrocarbons	I 478–489, 492–497. IV 1327–1329. XIII 813–816. XX 935–936. XXII 673–675. XXIII 971–973. XXIV 791–797, 1157–1158, 1165–1167. XXXI 901–903. XXXVIII 809–814. XL (Supp.) 627–637, 659–662. XLI 821–824, 849–853, 855–858, 865–873. XLII (pt. 1) 465–472. XLII (Pt. 2) 683–696, 749–753. XLIII 643–649, 651–655. XLIV 587–591, 593–598, 605–615, 625–633, 645–649.
Isoprenoid hydrocarbons (pristane, phytane, etc.)	I 468–477. XVI 575–600. XVII 561–590. XX 907–933. XXII 673–675. XXIV 791–797, 1165–1167. XXXI 901–903. XXXVIII 805–806. XL (Supp.) 627–637, 659–662. XLI 855–858, 865–873. XLII (Pt. 2) 683–696, 723–728, 749–753, 755–763. XLIII 643–649. XLIV 587–591, 645–649.
Steranes	VIII 873–900. IX 859–901. XVI 575–600. XVII 561–590. XX 907–933. XXXVIII 805–806. XLI 855–858, 865–873. XLII (Pt. 2) 749–753, 755–763. XLIII 643–649.
Triterpanes	XVI 575–600. XVII 561–590. XXIV 1159–1163. XXXVIII 805–806. XL (Supp. 659–662. XLI 855–858, 865–873. XLII (Pt. 2) 749–753, 755–763. XLIII 643–649. XLIV 617–623.
Humic compounds	XLI 865–873. XLII (Pt. 1) 465–472. XLII (Pt. 2) 723–728, 737–748. XLIII 643–649. XLIV 587–591, 593–598.
Isotopes	
Carbon	I 415–500, 461–467, 478–489, 500. XII 1115–1121. XIX 879–884. XX 935–936. XXIV 1169–1176. XXVI 613–617. XXXI 633–638, 911–916. XXXVIII 791–799, 809–814. XL 557–567. XL (Supp.) 651–658, 659–662. XLI 839–847, 849–853. 855–858, 861–863. XLII (Pt. 1) 483–488. XLII (Pt. 2) 661–665, 667–672, 673–677, 679–681, 717–721, 749–753. XLIII 643–649, 651–655. XLIV 587–591, 605–615, 645–649.
Sulfur	I 415–500, 456–460.
Kerogen (includes optical studies)	XXII 673–675. XXIII 971–973. XXIV 791–797, 1157–1158, 1169–1176. XXXI 621–627, 633–638, 911–916. XXXVI 941–944. XXXVIII 791–799, 801–803. XL 557–567. XL (Supp.) 651–658, 663–669, 671–676. XLI 817–819, 821–824, 839–847, 849–853, 865–873. XLII (Pt. 1) 465–472, 483–488. XLII (Pt. 2) 679–681, 717–721, 729–735, 737–748, 755–763. XLIII 633–642, 651–655, 663–668. XLIV 593–598, 599–603, 605–615, 617–623, 625–633, 635–637, 645–649.
Ketones	
Isoprenoid ketones	XVII 561–590. XXI 909–923. XLII (Pt. 2) 749–753. XLIII 643–649.
Methyl ketones	XLII (Pt. 2) 749–753. XLIII 643–649.
Pentacyclic ketones	XXIV 1159–1163. XLII (Pt. 2) 749–753.
Sterones	IX 859–901. XVI 575–600.
Lipids	XXIV 1169–1176. XXXI 633–638, 911–916. XXXVIII 791–799, 809–814. XXXIX 497–500. XL 557–567. XL (Supp.) 651–658, 659–662. XLI 849–853, 855–858. XLII (Pt. 2) 697–705, 717–721, 749–753. XLIII 643–649, 651–655. XLIV 587–591, 605–615, 617–623.
Pigments	
Carotenoids	XLIV 617–623.
Tetrapyrroles	
Chlorins	IX 431–438. XX 943–946. XXII 677–679. XXXI 629–632, 905–909. XXXVIII 785–789. XL (Supp.) 639–647. XLI 825–837. XLII (Pt. 2) 707–715. XLIII 657–661. XLIV 639–643.

APPENDIX IV.—COMPOUND TYPES (CONTINUED)

	Volume Number and Pages
Porphyrins	I 415–500, 498–499. IV 431–438. XXII 677–679. XXXI 629–632, 905–909. XXXVIII 785–789. XL (Supp.) 639–647, 659–662. XLI 825–837. XLII (Pt. 2) 707–715. XLIII 657–661. XLIV 639–643.
Sugars	XLII (Pt. 2) 697–705. XLIV 653–654.
Sulfur (Organic)	I 415–500, 456–460. XLIII 633–642.
Summaries	XX 905. XXXVIII 783–784. XLI 815. XLIV 585.

PART 3
BIOSTRATIGRAPHY AND PALEOCEANOGRAPHY

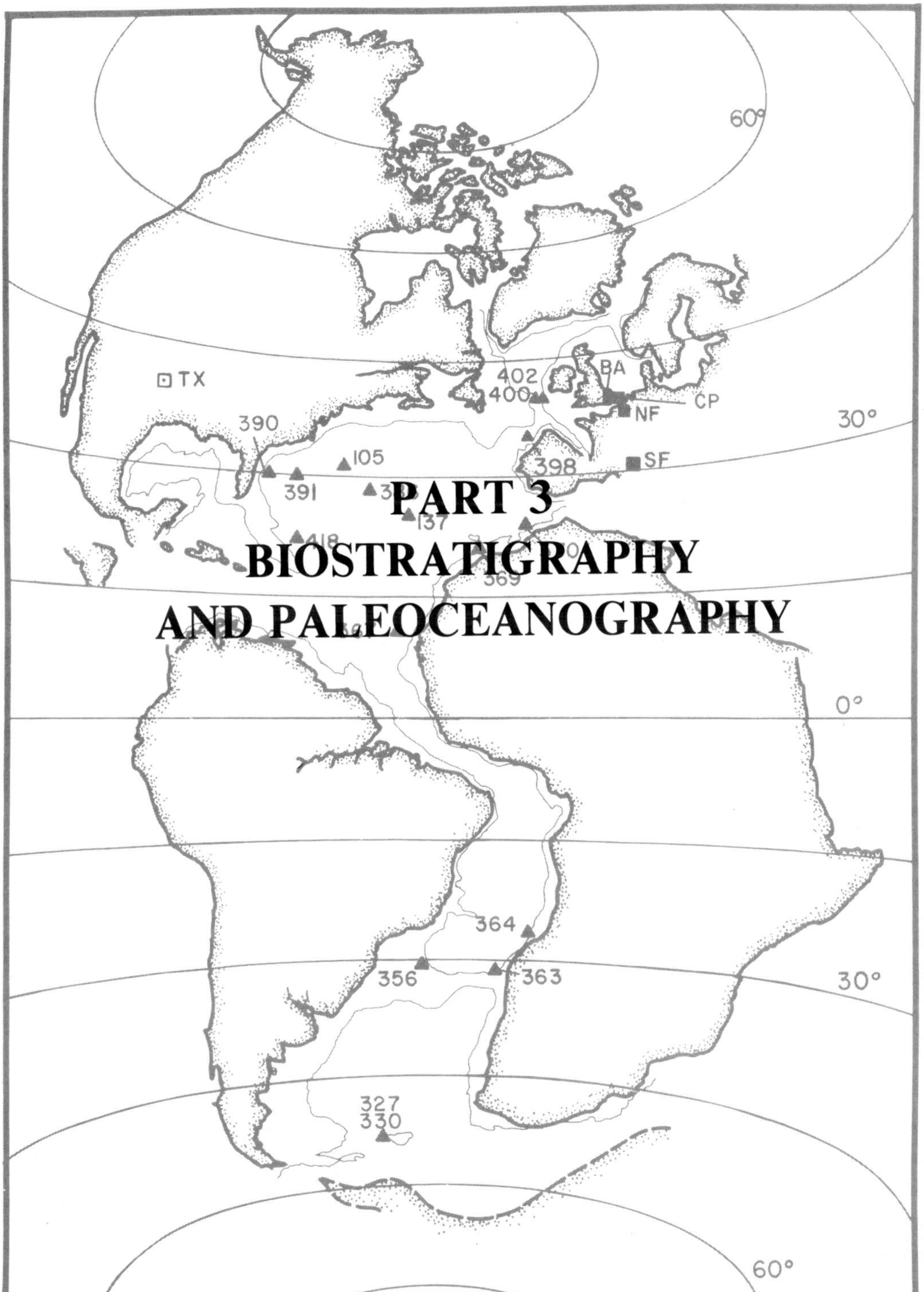

LATE MIDDLE CRETACEOUS

SEPM Special Publication No. 32, p. 253–315, October 1981

DSDP BIOSTRATIGRAPHY IN RETROSPECT AND PROSPECT

W. R. RIEDEL
Scripps Institution of Oceanography, La Jolla, California 92093 U.S.A.

ABSTRACT

As of mid-1978, some 480 papers have appeared on the biostratigraphy of DSDP cores, and the collections of the Project now constitute the basis for a major part of the totality of biostratigraphic studies of pelagic sequences. At least 1,249 new microfossil taxa have been described from DSDP cores. Studies based on the cores from several cruise-legs (as contrasted with single legs) have not increased as rapidly as expected. This remains a field for future exploitation, during which there will be an opportunity to develop further the quantitative aspects, which are inadequate in the biostratigraphic investigations hitherto published.

INTRODUCTION

Biostratigraphy cannot be performed or discussed in isolation from considerations of paleoenvironment, and that context is provided by several of the other chapters in this book, in particular those by Srinivasan and Kennett, Berger and others, and Thiede and others. An especially clear example of the interdependence of biostratigraphy and paleoceanography is provided for the Atlantic Ocean through the Cenozoic by Berggren (1978). So as not to duplicate those other papers, this chapter deals with biostratigraphy in its narrower sense.

Below are described some general characteristics and trends shown by biostratigraphic publications on DSDP materials, with comments on developments that can be expected (or hoped for) in the years immediately ahead. Also presented is the coarsely annotated bibliography on which this survey is based, because such a compilation is not readily accessible elsewhere.

Before commencing the analysis that constitutes the body of this review, it may be helpful to recall some of the characteristics of the DSDP materials that affect their utilization for biostratigraphic studies. The cores represent a rather thorough, though low-resolution, sampling of the Cenozoic oceanic sediment sequences in low and middle latitudes, with incursions into high latitudes in the Southern Ocean, the Bellingshausen Sea and the North Atlantic. The term "low-resolution" is used here because many holes have been cored only intermittently, and the continuity of the sedimentary record is reduced also by natural hiatuses and coring-induced disturbances. Because the rates of sediment accumulation are commonly as low as 5–25 m/million years, a little mechanical disturbance or natural reworking can seriously obscure stratigraphic detail. On the other hand, it is not uncommon to find pelagic sediments of uniform lithology that have accumulated without interruption or admixture for periods as long as 10 million years, which thus provide us with a relatively uncomplicated record of environmental conditions and processes for much longer periods than are usually available in sequences on land. A major complication affecting their interpretation results from movement of the sea floor relative to the overlying water masses, which control the distribution of the plankton. However, there are well-established procedures for evaluating at least the geographic and preservational effects of this sea-floor spreading.

The other notable feature of the DSDP cores is that they are routinely described in a more or less uniform way by the shipboard scientific parties, and the descriptions are published in a continuing series of Initial Reports of the Deep Sea Drilling Project. As will be discussed below, these initial descriptions are not as standardized nor as thorough as one might wish, but biostratigraphers benefit greatly by the availability of descriptions and interpretations of the content of several groups of microfossils in each hole (commonly foraminifera, calcareous nannofossils, and radiolarians, and occasionally diatoms, silicoflagellates and other groups). The stratigraphies of the several fossil groups are thus automatically related to each other, and to lithologic descriptions, at each drilling site.

There is a further concentration of biostratigraphic work on drill-sites that yielded exceptionally long or well preserved sequences of microfossils, and this results in the possibility of interrelating more detailed information on the various fossil groups at those sites. The degree of usefulness of each site for biostratigraphic studies can be estimated roughly from the bibliography below. Indian Ocean sites in general have large numbers of references, because of their inclusion in the chapters of the compilation edited by Heirtzler and others (1977)—twenty or more investigations are listed for Sites 214, 238, 259–261 and 263–266. Among the localities in the other oceans, thirteen or more investigations are listed for Sites 77, 105, 111, 127, 132, 173, 310, 361, 364, 367 and 370. These numbers are of course influenced by cores from early cruise-legs having been available to investigators for a longer time than those from more recent legs. Also, the ranking is likely to change substantially in the next few years because the less disturbed and more complete sequences collected by the recently developed hydraulic piston corer (HPC) will be more attractive to biostratigraphers than the conventional drilled cores.

PUBLICATIONS OF THE PAST TEN YEARS

At this time, there are approximately 480 papers on the biostratigraphy of DSDP materials, of which three quarters are chapters in the Initial Reports, and one quarter are distributed in other books and journals. This high proportion in the Initial Reports is likely to decline somewhat as the existing core collections become more extensive and more widely known, but the biostratigraphic chapters in the Initial Reports will certainly continue to be an important source of information on DSDP microfossils.

At the beginning of the Project, it was planned that the Initial Reports would be a simpler description of the cores than they eventually turned out to be. It proved impractical, and probably undesirable, to restrain competent shipboard biostratigraphers from carrying their studies to any depth they wished, within the constraints of the size of the published volumes. And since the Initial Reports are now, and will continue to be, a major source of DSDP biostratigraphic information, it seems advisable to

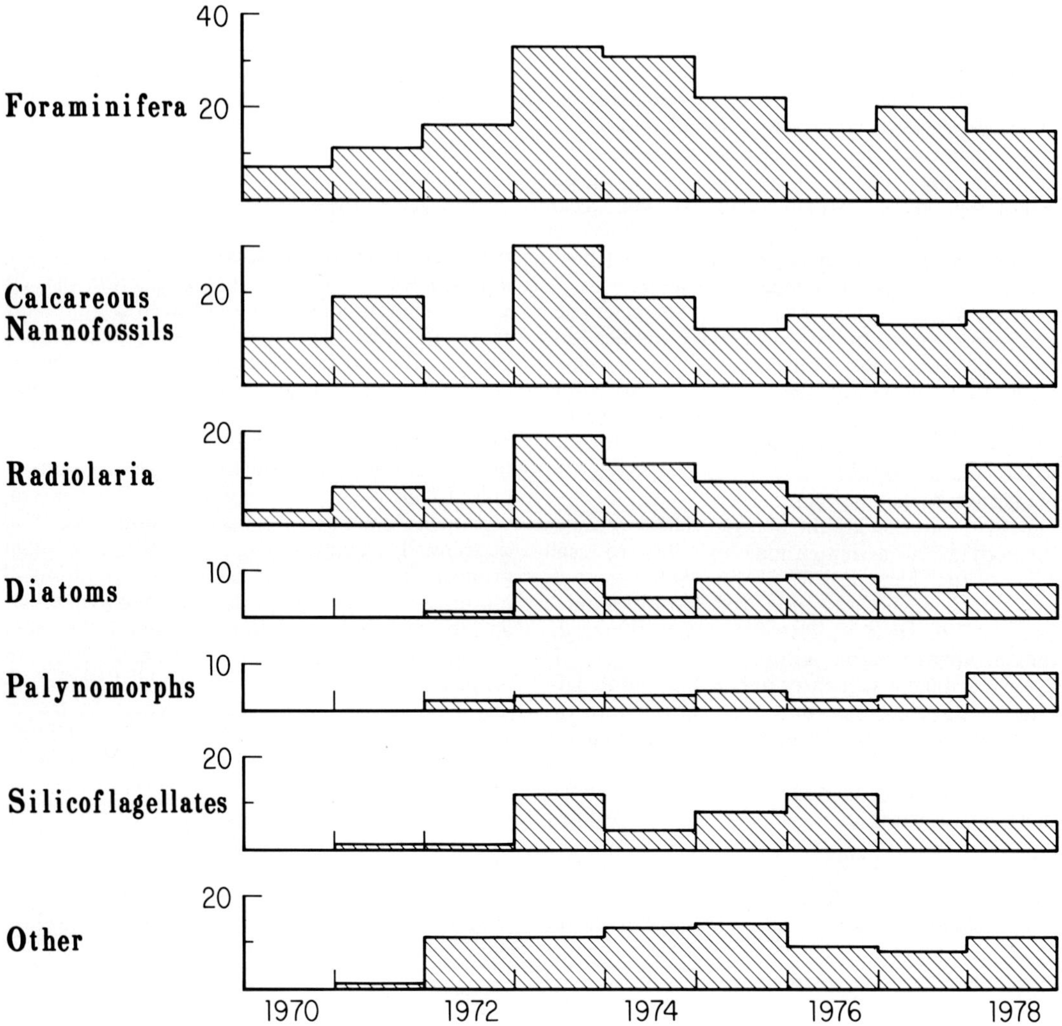

FIG. 1.—Numbers of papers published on the stratigraphy of foraminifera, calcareous nannofossils, radiolarians, diatoms, silicoflagellates, palynomorphs, and other groups of microfossils during the years 1970–78.

analyze their content from time to time in order to determine whether their usefulness or quality could be improved.

The histograms (Fig. 1) illustrating the numbers of papers on the stratigraphy of various microfossil groups for the years 1970–1978 show that initially only the fossil groups most commonly used in pelagic biostratigraphy were investigated (planktonic foraminifera, calcareous nannofossils, and radiolarians). After two or three years, papers began to appear on other groups, including diatoms, silicoflagellates and palynomorphs. Foraminiferal papers are the most numerous, and show a curve that can be interpreted in simple terms. An initial period of increasing awareness of the stratigraphic importance of the DSDP material led to a peak in output in 1973–74. Then came a decline, which may have been a result of the novelty wearing off, or of a temporary exhaustion of the limited supply of planktonic foraminiferal biostratigraphers working on open-ocean materials. For future years one might predict another upward trend, as stratigraphers begin to exploit the accumulated core collections. The studies on calcareous nannofossils and radiolarians show tendencies similar in some respects to the foraminiferal curve.

It might be expected that an upward trend in the future will be accompanied by an increase in the proportion of papers integrating the biostratigraphy of drill sites over large geographic areas (termed "horizontal" studies in Figure 2 and "regions related" in the annotated bibliography) as compared to papers presenting results from a single cruise-leg (for brevity, termed "vertical" studies). However, there is not yet any indication of this in Figure 2, which shows no significant change in the proportion of "at least partly horizontal" studies to "vertical only" studies over the past eight years. The delay in appearance of this expected tendency may be due to the fact that publication dates of biostratigraphic studies lag several years behind the time of initiation of those projects. This delay is substantially longer than that involved in publications on some aspects of lithostratigraphy and magnetostratigraphy, for which data-gathering and synthesis are not so time-consuming.

In preparation for the stratigraphic interpretation of the drilled sequences, a great deal of taxonomic work has been necessary, on microfossil groups that had not previously been available in quantity from Tertiary and Late Mesozoic deep-sea sediments. The extent of this taxonomic work is summarized in Figure 3, which shows the numbers of new taxa described from the DSDP materials. The histograms do not show any particularly noteworthy trends. Most of the spikes result from individual, intensively taxonomic papers. In addition to these new descriptions, there have of course been extensive revisions of the systematics of some groups, as indicated in the annotations to the bibliography.

In an attempt to place DSDP biostratigraphic contributions into a wider context, Figure 4 shows the percentage of papers on marine-microfossil stra-

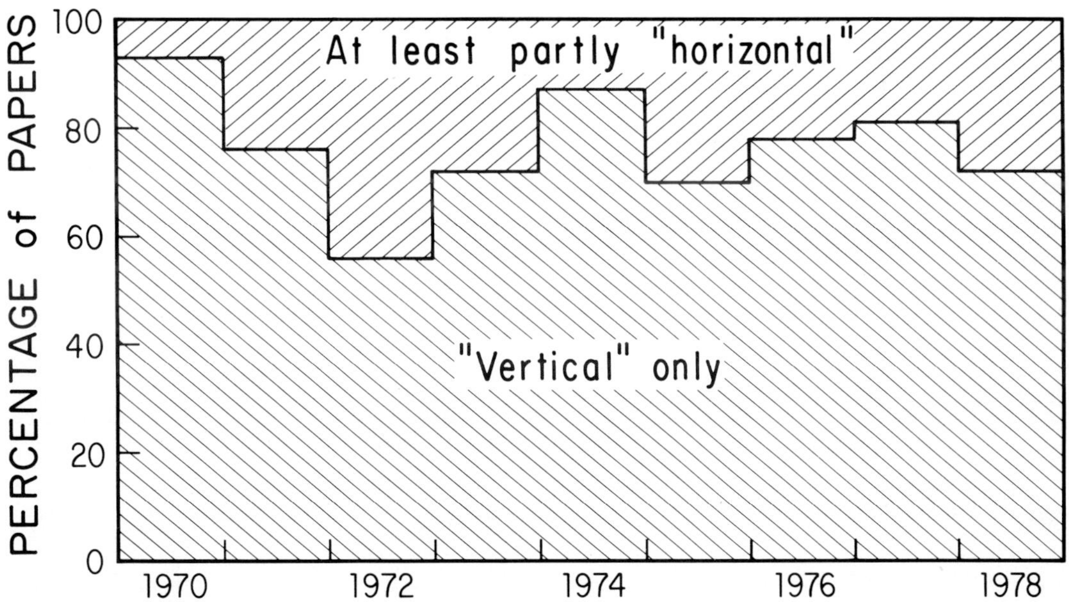

FIG. 2.—Percentages of papers integrating biostratigraphic results over considerable geographic areas (at least partly "horizontal") and those concerned with materials from a single cruise-leg ("vertical" only), for the years 1970–78.

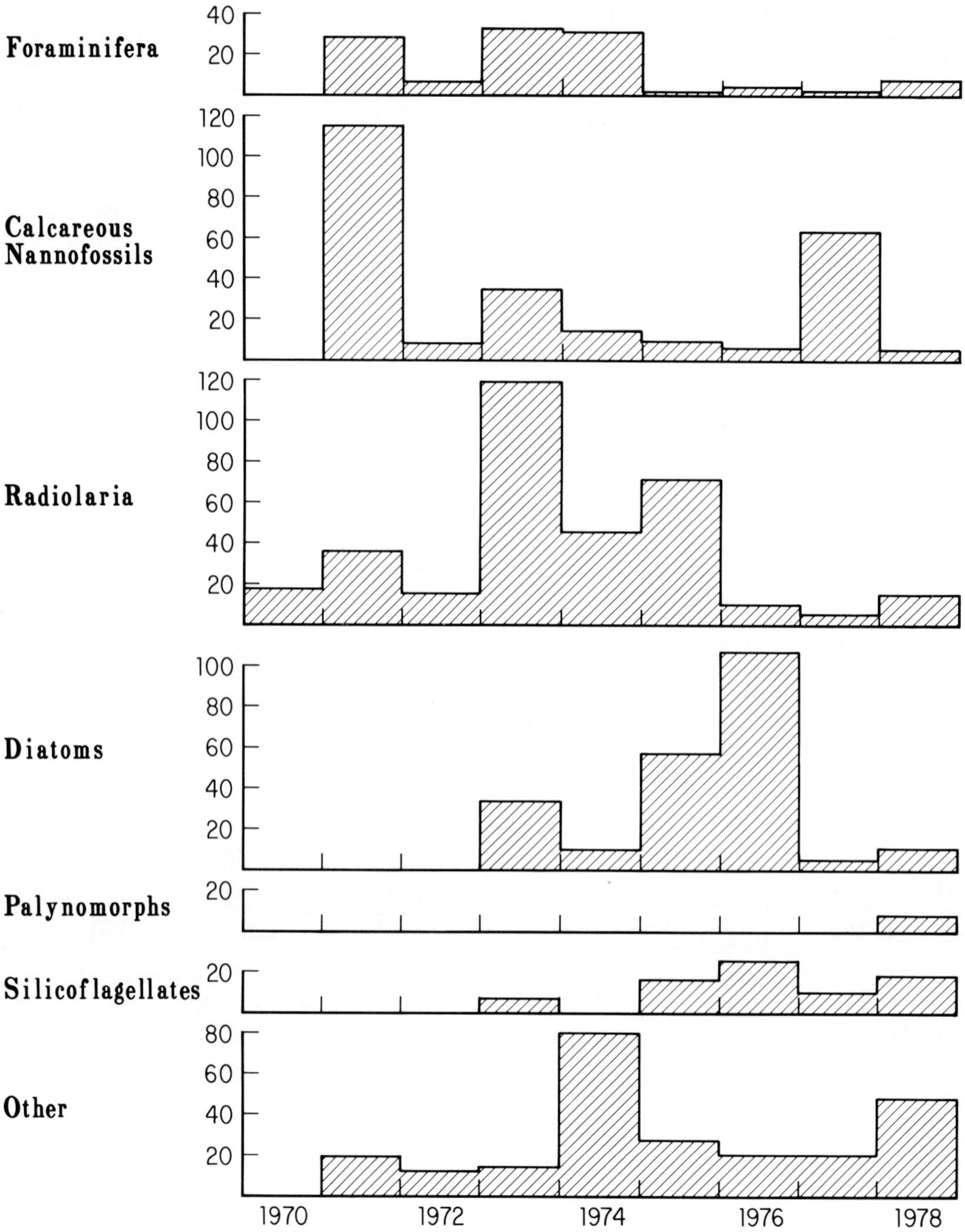

FIG. 3.—Numbers of new taxa of foraminifera, calcareous nannofossils, radiolarians, diatoms, palynomorphs, silicoflagellates and other microfossil groups described on the basis of DSDP materials during the years 1970–78.

tigraphy in the journal *Micropaleontology* which are based wholly, partly or not at all on DSDP materials, for the period 1970 to mid-1978. That journal is considered appropriately representative, because it attracts articles on a wide range of stratigraphic subjects from an international authorship. The steadily increasing proportion of marine biostratigraphic studies based on DSDP cores is evi-

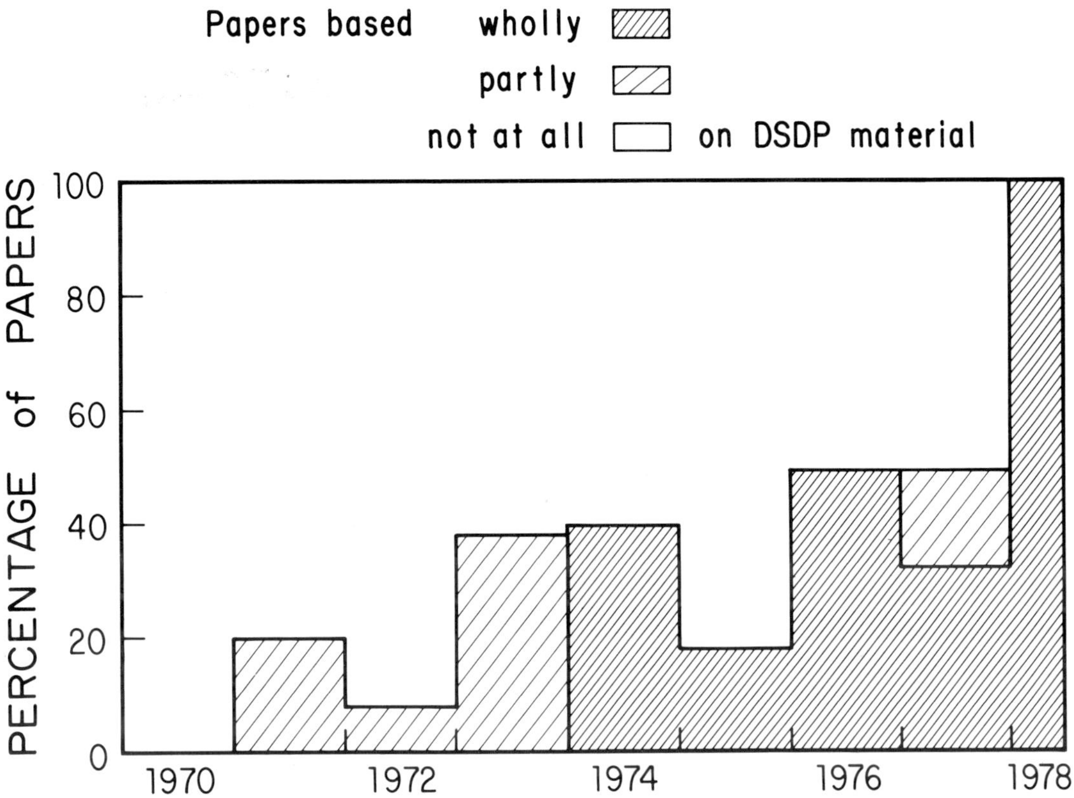

FIG. 4.—Diagram showing the increasing role played by DSDP materials in marine biostratigraphic studies, as represented by the journal *Micropaleontology*.

dent, though the saturation value reached during the first half of 1978 will surely prove to be a temporary spike.

REFERENCE COLLECTIONS OF MICROFOSSILS

Samples from the DSDP cores are readily available to biostratigraphers for the purpose of carrying out investigations leading to publication, but only the first steps have been taken to satisfy the needs of workers who wish to examine specimens for purposes of comparison but not of publishable research. It is planned to establish a small number of identical reference collections of DSDP microfossils for this purpose, one in each continental region having a demonstrated need for it. These reference collections will comprise prepared assemblages of foraminifera, calcareous nannofossils, radiolarians and diatoms from a representative selection of the drilled cores. A beginning has been made on the preparation of one of these collections for Western Europe (at the Naturhistorisches Museum in Basel, Switzerland), and one for North America (at Scripps Institution of Oceanography in La Jolla, California). However, the process will be slow because of limitations of financial support, and it is not expected that the two initial collections will reach a useful size until 1981 or 1982—and even then, only for the two calcareous groups from some parts of the World Ocean.

Part of the importance of these reference collections derives from the fact that DSDP cores are the type localities for a large number of microfossil taxa—1,249 for the nine-year period covered in Figure 3. Topotypic assemblages of many of these taxa will be preserved in the microfossil reference collections, and the remaining topotypic material will be preserved in the core repositories.

PROSPECT

"Horizontal" Studies

As is pointed out above, in the discussion of Figure 2, there has been a surprisingly long delay in the appearance of substantial numbers of studies covering numerous or widely distributed drilling sites, as contrasted to biostratigraphic studies involving single sites or cruise-legs (here designated briefly as "vertical" studies).

"Horizontal" studies for paleoenvironmental objectives will undoubtedly be undertaken spontaneously, but there is a second type of such study that is less attractive and therefore requires active encouragement. I refer to the need for placing all of the DSDP cores into a uniform framework of relative time. A beginning is made toward satisfying this need, by the biostratigraphic contributions to each Initial Report volume. However, the concepts of microfossil species and zones applied by the biostratigraphers on the various cruise-legs differ to such an extent that the placement of cores into a time-frame in the Initial Reports is only very approximate. An additional factor is that our understanding of several pelagic microfossil groups has increased to such an extent during the first ten years of the DSDP that zonal assignments made during the early days of the Project are difficult to correlate with those being made currently.

As a result, any investigator using materials from both early and late cruise-legs of DSDP, for a piece of research requiring a time-frame, must either arrange for a biostratigraphic reassessment of his samples, or work within a stratigraphic frame less refined than is currently possible. The former alternative eventually results in wasteful duplication of effort, and the latter is scientifically unsatisfactory. The cost-effective solution is to conduct a uniform biostratigraphic reassessment of the DSDP cores to present-day standards, and to ensure that the results are promptly available to all researchers conducting investigations in which geologic age is a factor.

Improving the Fundamental Data

The fundamental data forming the basis for biostratigraphic interpretations are records of the occurrence of individual fossil taxa at specified localities. The quality of the interpretations thus depends on the consistency with which taxa can be recognized, and the confidence that can be placed in a record of occurrence or non-occurrence.

It is well known that the morphologic range of specimens included under a particular name varies from one paleontologist to another, and that this is a formidable obstacle to integrating the results of different authors. The problem can be alleviated somewhat by providing opportunities for biostratigraphers to reach a certain degree of agreement by discussing the scope of each stratigraphically important taxon, and by comparing materials, but it cannot be eliminated entirely because there are legitimately different viewpoints as to how much morphological variation should be admitted into each fossil species. Under these circumstances it is of the greatest importance that every author of basic biostratigraphic data specifies clearly the morphologic limits applied in the identification of each taxon. A more radical solution, that may become widely practical as the storage and retrieval of microfossil images and associated alphanumeric information by computer-based devices become less expensive, would be to use objectively defined morphotypes instead of subjectively defined species as

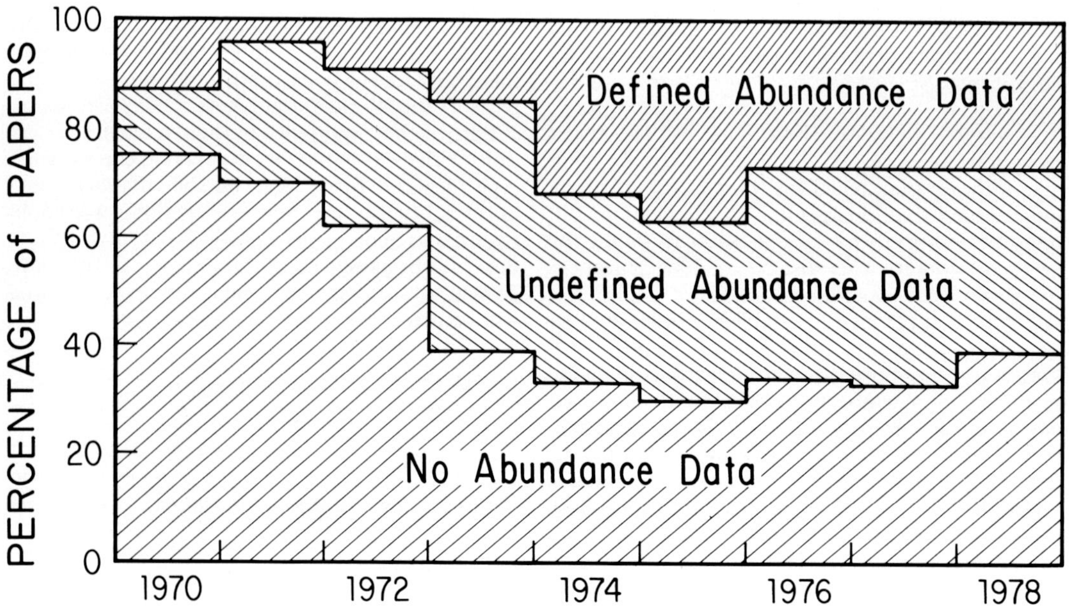

FIG. 5.—Summary diagram of the proportions of papers on DSDP biostratigraphy containing defined or undefined abundance data (for explanation see text), or simply records of presence and absence.

the fundamental unit of paleontologic data (Shaw, 1969; Hughes, 1970; Riedel, 1978; Budai and others, 1980).

The other aspect of the biostratigraphic data base susceptible to improvement is the means of recording whether taxa are present or absent. Such records have little meaning unless they are accompanied by an indication of the size of the sample examined, which implies the recording of some type of relative abundance data. Annotations to the bibliography indicate whether the papers on DSDP biostratigraphy include defined, undefined or no abundance data. Under the category "defined" abundance data are included cases in which abundances are expressed either as relative percentages of the total assemblages, or as absolute numbers of each taxon found. The term "undefined" abundance is used for papers in which categories such as "common", "few" and "rare" are used, without any indication of their numerical connotations. Figure 5 shows the distribution of papers falling into the three different categories, for the years 1970–78. For the first five years of that interval, the number of papers containing abundance data (and especially defined abundance data) increased, possibly as a result of recommendations by the JOIDES Panel on Paleontology and Biostratigraphy. From 1976 onward, however, there is a decrease in the number of papers containing such data, possibly because of a relaxation of vigilance by the JOIDES advisory panels. It will be difficult in the future to evaluate and use the biostratigraphic interpretations unless this recent tendency is reversed.

Estimates of Confidence in Correlations

Biostratigraphy will continue to be regarded as an art rather than a science, until it is possible to attach confidence limits to suggested correlations. The ultimately desired expression would be in such a form as—"With a probability of *p*%, sample A from section X can be said to be equivalent in age to some level between *m* and *n* meters below the top of section Y." Depending on whether *p* is set high or low, the difference between *m* and *n* will be larger or smaller, respectively.

During recent years, substantial progress has been made in this direction by the development of probabilistic procedures for biostratigraphy. Hay (1972) and Hay and Steinmetz (1973) demonstrated a procedure for determining the most probable order of events (earliest and latest occurrences of taxa) within a region small enough to warrant the assumption that the order did not really differ from one locality to another. Worsley and others (1973) applied the same methods to an area large enough to include several sub-areas each with its distinctive order of biostratigraphic events. The procedure used by these authors does not involve any weighting to reflect the different levels of reliability of particular events at particular sites.

In an attempt to overcome some of these inadequacies, my colleague M. J. Westberg and I have elaborated on a concept initiated by a JOIDES Working Group, to develop an index of reliability of events that takes account of abundance of the taxon concerned, the extent of our understanding of the evolution of the taxon, the ease of distinguishing it from co-occurring forms, the preservation of the assemblage relative to the susceptibility of the taxon concerned to solution, low abundances due to dilution, intermittence or constancy of occurrence of the taxon in the sequence, the relation of the locality to the periphery of the area of distribution of the taxon, and the effects of reworking. Some of these factors are numerical, and all are placed arbitrarily on a numerical scale of values 0.1, 0.25, 0.5, 0.75 and 1 (Table 1).

The following paragraphs explain the factors.

1) Abundance of taxon, in three ranges as indicated in Table 1. This refers to the absolute number of specimens of the taxon found in the samples near the limits of the stratigraphic range, not relative abundance. The purpose is to reflect the greater likelihood of having found the true stratigraphic limit if more than five specimens have been observed in each sample.

2) Average relative abundances of ancestor and descendant above and below evolutionary transitions. For example, if the descendant is six times as abundant as the ancestor above the event, and the ancestor is 1.5 times as abundant as the descendant below the event, the average is slightly less than 4, and would score 0.75.

3) Type of event, reflecting whether or not the ancestor or descendant of the taxon concerned is known, and the type of evolutionary interpretation. "Cryptogenetic origin" refers to the earliest occurrence of a taxon for which there are several or no candidate ancestors. An extinction is regarded as "cryptogenetic" when the taxon left no known evolutionary descendant, and "evolutionary" when it is confidently interpreted as terminating its lineage. As with evolutionary extinctions, a score of 0.75 is assigned to morphotypic limits which bracket known evolutionary transitions, as well as to earliest morphotypic occurrences of an evolutionary offshoot from a persistent lineage, because in both of these cases the limit would be extended by searching through much larger samples. By "evolutionary transition" is meant the level at which the numerical dominance changes from ancestor to descendant.

4) Ease of distinguishing the taxon from co-occurring fossils. The reliability of species distinctions is well known to vary widely, and the scoring of this factor therefore ranges from the minimum value to the maximum. This is a less objective factor than some others listed.

5) Assemblage preservation relative to the pre-

servation threshold of the taxon concerned. This is clearly a more pertinent factor than an evaluation of the overall preservation of the assemblage. A median score is given when one is uncertain whether an absence or low abundance is due to solution, or when solution has removed features necessary for identification from some specimens in the sample.

This category is also used to downgrade the score of an upper morphotypic limit above which (or a lower morphotypic limit below which) absence of the taxon is suspected to be due to dilution or dissolution.

6) Constancy of occurrence in the sequence, within the stratigraphic range of the taxon. This is not applicable to evolutionary transitions because these are located by the changing proportions of ancestor and descendant, not by the limit of range of a morphotype. The need to consider intermittence of occurrence is clearly demonstrated by the highly variable abundances of some rather common species throughout their ranges.

7) Relation of locality to area of distribution of the taxon. It seems obvious that the earliest and latest records of a taxon near the periphery of its area of distribution will be less reliable stratigraphically than those well within the boundaries. Also, for localities whose relation to biogeographic distributions have changed through time (because of either changes in oceanic conditions or sea-floor spreading), this factor may be evaluated differently for the upper and lower limits of the same taxon.

8) Effects of reworking. The index of reliability of an event is downgraded slightly if the investigator is confident of having recognized an extension of stratigraphic range as due to reworking, and is downgraded more if he is less confident. This is on the assumption that the reworking is taken into consideration in interpreting the upper or lower limit, and the reworked occurrences are excluded from the range. This is clearly an area of subjective judgement, in which account is taken of sedimentological evidences of disturbance, the abruptness of appearances and disappearances of the taxon in other sequences, evidence of reworking of other taxa into the assemblage, etc.

The "value" of an event in a sequence is expressed as a fraction of unity, obtained by multiplying its scores for each of the eight factors, assigned

TABLE 1.—SCORING OF THE VALUE OF A PARTICULAR BIOSTRATIGRAPHIC EVENT AT A PARTICULAR LOCALITY

Factor \ Multiplier	0.10	0.25	0.50	0.75	1.00
a) Abundance of taxon (numbers of specimens observed in each sample)		1	2–5		>5
b) Average of relative abundances of ancestor and descendant above and below the event (applies only to evolutionary transitions)			<2x	2–5x	>5x
c) Type of event			cryptogenetic origin or extinction	evolutionary extinction, or morphotypic limit, or evolutionary offshoot	evolutionary transition
d) Ease of distinguishing the taxon from co-occurring fossils	difficult		moderately easy		very easy
e) Is the assemblage well preserved, relative to the preservation threshold of the taxon concerned, or are samples immediately above or below the event poorly preserved or highly diluted?	no		uncertain, or some difficulty due to preservation		yes
f) Constancy of occurrence in the sequence, above or below the event (not applicable to evolutionary transitions)				interrupted	constant
g) Relation of locality to periphery of area of distribution of taxon		near periphery	unknown		well within
h) Effects of reworking	the limit as here interpreted may be unreliable due to reworking			occurrences outside this range are certain to be due to reworking	no reworking

according to a scheme such as outlined in Table 1. Using the suggested values, it could range from 0.00001 to 1.00, and would be a much more sensitive index than McCammon's "relative biostratigraphic value" (as applied, for example, by Brower and others 1978, and Millendorf and others 1978). Its application would, however, require the active participation of the biostratigraphers responsible for the original data, to a greater extent than has been necessary for statistical procedures applied heretofore.

Factors *f* and *g* in Table 1 involve paleoenvironmental considerations, and this serves to emphasize the point made in the first paragraph of this chapter, regarding the interdependence of biostratigraphy and paleoceanography. Taxonomy (factors *c* and *d* in Table 1) is the third essential member of this triad.

ACKNOWLEDGMENTS

Part of the work leading to this paper was supported by NSF Grant OCE77-18737. Discussions with N. F. Hughes, J. B. Saunders and David Bukry contributed to the formulation of the reliability index summarized in Table 1. Annika Sanfilippo assisted with the compilation of bibliographic data, and the typing and proof-reading were capably done by Mary A. Hanger.

REFERENCES CITED

NOTE: All references to Deep Sea Drilling Project Initial Reports have been abbreviated; for complete citations see *Appendix I* at the end of this volume, or the individual Initial Reports as published.

BERGGREN, W. A., 1978, Recent advances in Cenozoic planktonic foraminiferal biostratigraphy, biochronology, and biogeography: Atlantic Ocean: Micropaleontology, v. 24, p. 337–370.

BROWER, J. C., MILLENDORF, S. A., AND DYMAN, T. S., 1978, Methods for the quantification of assemblage zones based on multivariate analysis of weighted and unweighted data: Computers and Geosciences, v. 4, p. 221–227.

BUDAI, A., RIEDEL, W. R., AND WESTBERG, M. J., 1980, A general-purpose paleontologic information device: Jour. Paleontology, v. 54, p. 259–262.

HAY, W. W., 1972, Probabilistic stratigraphy, Eclogae Geol. Helvetiae, v. 65/2, p. 255–266.

———, AND STEINMETZ, J. C., 1973, Probabilistic analysis of distribution of Late Paleocene-Early Eocene calcareous nannofossils, Proceedings of Symposium on Calcareous Nannofossils: Gulf Coast Section Soc. Econ. Paleontologists Mineralogists, p. 58–70.

HEIRTZLER, J. R., BOLLI, H. M., DAVIES, T. A., SAUNDERS, J. B., AND SCLATER, J. G., eds., 1977, Indian Ocean Geology and Biostratigraphy: Washington, D.C., Amer. Geophys. Union, 616 p.

HUGHES, N. F., 1970, Remedy for the general data handling failure of palaeontology, *in* Cutbill, J. L., ed., Data Processing in Biology and Geology: Systematics Association Spec. Vol. 3, p. 321–330.

MILLENDORF, S. A., BROWER, J. C., AND DYMAN, T. S., 1978, A comparison of methods for the quantification of assemblage zones: Computers and Geosciences, v. 4, p. 229–242.

RIEDEL, W. R., 1978, Systems of morphologic descriptors in paleontology: Jour. Paleontology, v. 52, p. 1–7.

SHAW, A. B., 1969, Adam and Eve, paleontology, and the non-objective arts: Jour. Paleontology, v. 43, p. 1085–1098.

WORSLEY, T. R., BLECHSCHMIDT, G., RALSTON, S., AND SNOW, B., 1973, Probability-based analysis of the area-time distribution of Oligocene calcareous nannofossils, Proceedings of Symposium on Calcareous Nannofossils: Gulf Coast Section Soc. Econ. Paleontologists Mineralogists, p. 71–79.

ANNOTATED BIBLIOGRAPHY

Column 1: the *fossil groups* treated by each paper are indicated as *F* (foraminifera), *N* (calcareous nannofossils), *R* (radiolarians), *D* (diatoms), *S* (silicoflagellates) and *O* (other groups). Column 2: ages are indicated as *J* (Jurassic), *C* (Cretaceous), *PA* (Paleocene), *E* (Eocene), *O* (Oligocene), *M* (Miocene), *PL* (Pliocene), and *Q* (Quaternary). Column 3: the numbers represent coring Sites investigated. Column 4: *R* indicates taxonomic revisions and *N* indicates new taxa, followed by the number described. Column 5: *D* indicates "defined" abundance data, *U* "undefined", and *N* no abundance data (for further explanation see text); *P* indicates that information on the state of preservation of the microfossils is given. Column 6: *G* indicates that results from several microfossil groups are interrelated, *R* indicates that regions are interrelated (referred to as a "horizontal" study in the text), *V* indicates a "vertical" study (based on cores from a single cruise-leg only), and *F* indicates that the study resulted in increased stratigraphic resolution.

Reference	Fossil groups.	Ages.	Sites.	Revisions. New taxa.	Abundance. \ Preservation.	Groups or regions related. Vertical study. Finer stratigraphy.
Abbott, W. H., 1978, Correlation and zonation of Miocene strata along the Atlantic margin of North America using diatoms and silicoflagellates: Marine Micropaleontology, v. 3, p. 15–34.	D S	M	391		N	G V F R
Akers, W. H., 1972, Larger Foraminifera from Hole 98: Init. Repts. DSDP, v. 11, p. 545–546.	F	O	98			R
———, 1974, Foraminferal range charts for Arabian Sea and Red Sea Sites, Leg 23: Init. Repts. DSDP, v. 26, p. 1013–1050.	F	Q PL M,O E PA	219 to 230		D	V
———, 1974, Results of shore laboratory studies on Tertiary and Quaternary Foraminifera from Leg 26: Init. Repts. DSDP, v. 26, p. 973–982.	F	Q PL M O E	250 to 258		U	V
Alberici, A., Barbieri, F., Iaccarino, S., and Rossi, U., 1973, Considerazioni biostratigrafiche e paleoecologiche sul Neogene del fianco occidentale della Dorsale Medio-Atlantica Meridionale (Foraminiferi dei "Sites" 15 e 16 del "Leg" III DSDP), L'Ateneo Parmense: Acta Naturalia, v. 9, fasc. 2, p. 137–151.	F	Q PL M	15 16		D P	V
Backman, J., 1978, Late Miocene-Early Pliocene nannofossil biochronology and biogeography in the Vera Basin, SE Spain: Stockholm Contributions in Geology, v. 32, no. 2, p. 93–114, 2 pls.	N	PL M	118 123- 125 132 134 135 372 375 376	R	D	R V F
Barbieri, F., Iaccarino, S., and Rossi, U., 1976, Paleoecology of the Pliocene in Sites 62 and 63 DSDP (Leg 7—equatorial western Pacific): Universitá degli Studi di Parma, Istituto di Geologia (Presented to the 1st International Congress on Pacific Neogene Stratigraphy, Tokyo), 25 + ix p.	F N	PL	62 63		D P	G V F R
Barron, J. A., 1976, Revised Miocene and Pliocene diatom biostratigraphy of Upper Newport Bay, Newport Beach, California: Marine Micropaleontology, v. 1, p. 27–63.	D S	PL M	173		U	G V F

Bartenstein, H., 1974, Upper Jurassic–Lower Cretaceous primitive arenaceous foraminifera from DSDP Sites 259 and 261, eastern Indian Ocean: Init. Repts. DSDP, v. 27, p. 683–695.	F	C J	259 261		U	V
Beckmann, J. P., 1971, The Foraminifera of Sites 68 to 75: Init. Repts. DSDP, v. 8, p. 713–725. [Additional information in Site chapters.]	F	Q PL M O E	68 to 75		N P	V
———, 1972, The Foraminifera and some associated microfossils of Sites 135 to 144: Init. Repts. DSDP, v. 14, p. 389–420. [Additional information in Site chapters.]	F	Q PL M O E PA C	135 to 144		U P	V
———, 1976, Shallow-water foraminifers and associated microfossils from Sites 315, 316, and 318, DSDP Leg 33: Init. Repts. DSDP, v. 33, p. 467–489.	F	PL M O E C	315 316 318		N	V
———, 1978, Late Cretaceous smaller benthic foraminifers from Sites 363 and 364, DSDP Leg 40, southeast Atlantic Ocean: Init. Repts. DSDP, v. 40, p. 759–781.	F	C	363 364		U	V
Benson, R. H., 1972, Preliminary report on the ostracods of Holes 117 and 117A: Init. Repts. DSDP, v. 12, p. 427–432.	O	O PA	117		D	R
———, 1973, Psychrospheric and continental Ostracoda from ancient sediments in the floor of the Mediterranean: Init. Repts. DSDP, v. 13, p. 1002–1008.	O	Q PL M	124 to 134		D	V
———, 1974, Preliminary report on the ostracodes of Leg 24: Init. Repts. DSDP, v. 24, p. 1037–1043.	O	Q PL M O E	237 238		D	V
———, 1977, The Cenozoic ostracode faunas of the Sao Paulo Plateau and the Rio Grande Rise (DSDP Leg 39, Sites 356 and 357): Init. Repts. DSDP, v. 39, p. 869–883.	O	Q PL M O E PA	356 357	N12	D	V
Benson, R. N., 1972, Radiolaria, Leg 12, Deep Sea Drilling Project: Init. Repts. DSDP, v. 12, p. 1085–1113. [Additional information in Site chapters.]	R	Q PL M O,E	111 to 119		U P	V

Reference	Fossil groups.	Ages.	Sites.	Revisions. New taxa.	Abundance.	Preservation.	Groups or regions related. Vertical study. Finer stratigraphy.
Berggren, W. A., 1971, Paleogene planktonic foraminiferal faunas on Legs I–IV (Atlantic Ocean), JOIDES Deep Sea Drilling Program—a synthesis: Proceedings of the II Planktonic Conference, Roma 1970, p. 57–77.	F	PA C O E PA	6 to 31				
———, 1972, Cenozoic biostratigraphy of the North Atlantic: Init. Repts. DSDP, v. 12, p. 965–1001. [Additional information in Site chapters.]	F	Q PL M,O E PA C J	111 to 119	N1	N		R V F
———, 1974, Late Paleocene–Early Eocene benthonic foraminiferal biostratigraphy and paleoecology of Rockall Bank: Micropaleontology, v. 20, no. 4, p. 426–448.	F	E PA	117		N		V
———, 1977, Late Neogene planktonic foraminiferal biostratigraphy of DSDP Site 357 (Rio Grande Rise): Init. Repts. DSDP, v. 39, p. 591–614.	F	Q PL M	357	R	N		V
———, and Amdurer, M., 1973, Late Paleogene (Oligocene) and Neogene planktonic foraminiferal biostratigraphy of the Atlantic Ocean (lat. 30°N to lat. 30°S): Riv. Ital. Paleont., v. 79, no. 3, p. 337–392.	F	Q PL M O	2,3 11,12 14,17 18,29 30,31 149,151 154	R	N	P	R V F
———, and Aubert, J., 1976, Late Paleogene (Late Eocene and Oligocene) benthonic foraminiferal biostratigraphy and paleobathymetry of Rockall Bank and Hatton-Rockall Basin: Micropaleontology, v. 22, no. 3, p. 307–326.	F	O E	116 117	R	U		V
———, and ———, 1976, Eocene benthonic foraminiferal biostratigraphy and paleobathymetry of Orphan Knoll (Labrador Sea): Micropaleontology, v. 22, no. 3, p. 327–346.	F	E	111	N1	U		V
———, Lohmann, G. P., and Poore, R. Z., 1974, Shore laboratory report on Cenozoic planktonic Foraminifera: Leg 22: Init. Repts. DSDP, v. 22, p. 635–655.	F	Q PL	212 to		D	P	V

		M O E PA	217			
———, Pessagno, E. A., Jr., and Bukry, D., 1969, Biostratigraphy: Init. Repts. DSDP, v. 1, p. 594–623. [Additional information on Foraminifera in each Site chapter.]	F N R	P M O E C J	1 to 7		N	V
———, and Poore, R. Z., 1974, Late Miocene–Early Pliocene planktonic foraminiferal biochronology: *Globortalia tumida* and *Sphaeroidinella dehiscens* lineages: Riv. Ital. Paleont., v. 80, no. 4, p. 689–698.	F	PL M	214			V F
Bertolani Marchetti, D., and Cita, M. B., 1975, Palynological investigations on late Messinian sediments recorded at DSDP Site 132 (Tyrrhenian Basin) and their bearing on the deep basin desiccation model: Riv. Ital. Paleont., v. 81, no. 3, p. 281–308.	O	M	132		D	V F
Bjørklund, K. R., 1976, Radiolaria from the Norwegian Sea, Leg 38 of the Deep Sea Drilling Project: Init. Repts. DSDP, v. 38, p. 1101–1168.	R	Q PL M O E	336 to 352	N8	U P	V F
Blechschmidt, G., 1976, Biostratigraphy of calcareous nannofossils: Leg 34, Deep Sea Drilling Project: Init. Repts. DSDP, v. 34, p. 705–712.	N	Q PL M O E	319 to 321		U P	V
Blow, W. H., 1969, Deep Sea Drilling Project, Leg 1, Foraminifera from selected samples: Init. Repts. DSDP, v. 1, p. 392–397.	F	Q PL E C	1 to 7		N	V
———, 1970, Deep Sea Drilling Project, Leg 2, Foraminifera from selected samples: Init. Repts. DSDP, v. 2, p. 357–365.	F	Q PL M O E C	9 to 12		N	V
———, 1970, Deep Sea Drilling Project, Leg 3, Foraminifera from selected samples: Init. Repts. DSDP, v. 3, p. 629–661.	F	Q PL M O,E PA C	13 to 22		N	V

Reference	Fossil groups.	Ages.	Sites.	Revisions. New taxa.	Abundance. / Preservation.	Groups or regions related. Vertical study. Finer stratigraphy.
———, 1970, Deep Sea Drilling Project, Leg 4, Foraminifera from selected samples: Init. Repts. DSDP, v. 4, p. 383–400.	F	Q PL M O	23 to 31		N	V
———, 1971, Deep Sea Drilling Project, Leg 6, Foraminifera from selected samples: Init. Repts. DSDP, v. 6, p. 1013–1026.	F	Q PL M O,E PA C	44 to 58		N	V
Boersma, A., 1977, Cenozoic planktonic Foraminifera—DSDP Leg 39 (South Atlantic): Init. Repts. DSDP, v. 39, p. 567–590. [Additional information in Site chapters.]	F	Q PL M O,E PA	353 to 359		N P	V
———, 1977, Eocene to early Miocene benthic Foraminifera, DSDP Leg 39, South Atlantic: Init. Repts. DSDP, v. 39, p. 643–656.	F	M O E	356 to 359		N	V
Bolli, H. M., 1970, The Foraminifera of Sites 23–31, Leg 4: Init. Repts. DSDP, v. 4, p. 577–643.	F	Q PL M,O E,PA C	23 to 31	R	D P	R V F
———, 1974, Jurassic and Cretaceous Calcisphaerulidae from DSDP Leg 27, eastern Indian Ocean: Init. Repts. DSDP, v. 27, p. 843–907.	O	C J	259 260 261 263	N20	N	V F
———, 1974, Calcareous organisms incertae sedis from the Lower Cretaceous of DSDP Leg 27: Init. Repts. DSDP, v. 27, p. 909–913.	O	C	260 261 263		U	V
———, 1974, Synthesis of the Leg 27 biostratigraphy and paleontology: Init. Repts. DSDP, v. 27,	F	Q	259			G V

p. 993–999. [Additional information in Site chapters.]	N R etc.	PL M E PA C	to 263			
———, 1977, Paleontological-biostratigraphical investigations, Indian Ocean Sites 211–269 and 280–282, DSDP Legs 22–29, *in* Heirtzler, and others, eds., Indian Ocean Geology and Biostratigraphy: Washington, D.C., American Geophysical Union, p. 325–338.	F N R D S O		211 to 269 280 to 282			
———, 1978, Cretaceous and Paleogene Calcisphaerulidae from DSDP Leg 40, southeastern Atlantic: Init. Repts. DSDP, v. 40, p. 819–837.	O	PA C	361 363 364		N	V
———, 1978, Upper Jurassic Calcisphaerulidae from DSDP Leg 44, Hole 391C, Blake-Bahama Basin, western North Atlantic: Init. Repts. DSDP, v. 44, p. 911–919.	O	J	391	N2		V
———, and Krasheninnikov, V. A., 1977, Problems in Paleogene and Neogene correlations based on planktonic foraminifera: Micropaleontology, v. 23, no. 4, p. 436–452.	F	Q PL M O,E PA				R V
———, and Premoli-Silva, I., 1973, Oligocene and Recent planktonic Foraminifera and stratigraphy of the Leg 15 sites in the Caribbean Sea: Init. Repts. DSDP, v. 15, p. 475–497.	F	Q PL M O	146 to 154	R N1	N P	V F
———, and Saunders, J. B., 1977, Introduction to stratigraphy and paleontology, *in* Heirtzler, and others, eds., Indian Ocean Geology and Biostratigraphy: Washington, D.C., American Geophysical Union, p. 311–324.	F N S etc.	Q PL M O,E PA C J	211 to 269 280 to 282			V
Boltovskoy, E., 1974, Neogene planktonic Foraminifera of the Indian Ocean (DSDP Leg 26): Init. Repts. DSDP, v. 26, p. 675–741. [Additional information in Site chapters.]	F	Q PL M O	250 to 258	R N1	U P	V F
———, 1977, Neogene deep water benthonic Foraminifera of the Indian Ocean, *in* Heirtzler, and others, eds., Indian Ocean Geology and Biostratigraphy: Washington, D.C., American Geophysical Union, p. 599–616.	F	Q PL M	211 to 269			
———, 1978, Late Cenozoic benthonic Foraminifera of the Ninetyeast Ridge (Indian Ocean): Marine Geology, v. 26, p. 139–175.	F	Q PL M	214 216 217		U	R V F

Reference	Fossil groups.	Ages.	Sites.	Revisions. New taxa.	Abundance. / Preservation.	Groups or regions related. Vertical study. Finer stratigraphy.
		O	253 254			
———, 1978, On the age of the Neogene deposits at Site 329 (DSDP Leg 36): Revista Española de Micropaleontologìa, v. 10, no. 1, p. 145–163.	F	PL	329	R	N	V
Boudreaux, J. E., 1974, Calcareous nannoplankton ranges, Deep Sea Drilling Project Leg 23: Init. Repts. DSDP, v. 23, p. 1073–1090. [Additional information in Site chapters.]	N	Q PL M O E PA	219 to 230		D P	V
Brönnimann, P., Martini, E., Resig, J., Riedel, W. R., Sanfilippo, A., and Worsley, T., 1971, Biostratigraphic synthesis: Late Oligocene and Neogene of the western tropical Pacific: Init. Repts. DSDP, v. 7, p. 1723–1745.	F R N	Q PL M O E	62 to 66			G V F
———, and Resig, J., 1971, A Neogene globigerinacean biochronologic time-scale of the south-western Pacific: Init. Repts. DSDP, v. 7, p. 1235–1469.	F	Q PL M O E	62 to 65	R N28	N P	V
Bukry, D., 1970, Coccolith age determinations, Leg 2, Deep Sea Drilling Project: Init. Repts. DSDP, v. 2, p. 349–355.	N	Q PL M O E C	8 to 12		N P	V
———, 1970, Coccolith age determinations, Leg 4, Deep Sea Drilling Project: Init. Repts. DSDP, v. 4, p. 375–381.	N	Q PL M O E	23 to 31		N	V

———, 1971, Cenozoic calcareous nannofossils from the Pacific Ocean: Transactions of the San Diego Society of Natural History, v. 16, no. 14, p. 303–327.	N	Q PL M,O E,PA	3,36 44,47 54,55 57,63 70,72 74,77 80,83	N18	P	V
———, 1971, Coccolith stratigraphy, Leg 6, Deep Sea Drilling Project: Init. Repts. DSDP, v. 6, p. 965–1004. [Additional information in Site chapters.]	N	Q PL M,O E PA C	44 to 60		N	R V F
———, 1971, Coccolith stratigraphy, Leg 7, Deep Sea Drilling Project: Init. Repts. DSDP, v. 7, p. 1513–1528.	N	Q PL M O E	62 to 65		N	V F
———, 1971, Coccolith stratigraphy, Leg 8, Deep Sea Drilling Project: Init. Repts. DSDP, v. 8, p. 791–807.	N	Q PL M O E	69 to 75		N	V
———, 1971, *Discoaster* evolutionary trends: Micropaleontology, v. 17, no. 1, p. 43–52.	N	PL M,O E PA		N9		V
———, 1972, Coccolith stratigraphy, Leg 9, Deep Sea Drilling Project: Init. Repts. DSDP, v. 9, p. 817–832. [Additional information on calcareous nannofossils in Site chapters.]	N	Q PL M O E	76 to 84		N	V
———, 1973, Coccolith stratigraphy, Leg 10, Deep Sea Drilling Project: Init. Repts. DSDP, v. 10, p. 385–406.	N	Q PL M,O E PA C	85 to 97		N	V
———, 1972, Coccolith stratigraphy, Leg 11, Deep Sea Drilling Project: Init. Repts. DSDP, v. 11, p. 475–482.	N	Q PL M,O E PA	98 to 108		N P	V

Reference	Fossil groups.	Ages.	Sites.	Revisions. New taxa.	Abundance. / Preservation.	Groups or regions related. Vertical study. Finer stratigraphy.
———, 1972, Further comments on coccolith stratigraphy, Leg 12, Deep Sea Drilling Project: Init. Repts. DSDP, v. 12, p. 1071–1083.	N	C,J Q PL M,O E PA C	111 to 119		N P	V
———, 1972, Coccolith stratigraphy, Leg 14, Deep Sea Drilling Project: Init. Repts. DSDP, v. 14, p. 487–494.	N	Q PL M,O E PA C	135 to 144		N P	V
———, 1973, Coccolith stratigraphy, Leg 13, Deep Sea Drilling Project: Init. Repts. DSDP, v. 13, p. 817–822.	N	Q PL M	120 to 134		N	V F
———, 1973, Low-latitude coccolith biostratigraphic zonation: Init. Repts. DSDP, v. 15, p. 685–703.	N	Q PL M,O E C	146 to 154	R	N	V F
———, 1973, Coccolith stratigraphy, eastern equatorial Pacific, Leg 16, Deep Sea Drilling Project: Init. Repts. DSDP, v. 16, p. 653–711.	N	Q PL M,O E C	155 to 163	R N10	N P	V F
———, 1973, Phytoplankton stratigraphy, central Pacific Ocean, Deep Sea Drilling Project, Leg 17: Init. Repts. DSDP, v. 17, p. 871–889.	N D S	Q PL M,O E PA C	164 to 171		N P	G V

———, 1973, Coccolith and silicoflagellate stratigraphy, Deep Sea Drilling Project, Leg 18, eastern North Pacific: Init. Repts. DSDP, v. 18, p. 817–831.	N S	Q PL M	172 to 182		U	G V R
———, 1973, Coccoliths and silicoflagellates from Deep Sea Drilling Project Leg 19, North Pacific Ocean and Bering Sea: Init. Repts. DSDP, v. 19, p. 857–867.	N S	Q PL M O E C	183 to 193		N P	V
———, 1973, Phytoplankton stratigraphy, Deep Sea Drilling Project, Leg 20, western Pacific Ocean: Init. Repts. DSDP, v. 20, p. 307–317.	N S	Q PL M O E	194 to 202	R N5 (n)	N	V
———, 1973, Coccolith and silicoflagellate stratigraphy, Tasman Sea and southwestern Pacific Ocean, Deep Sea Drilling Project, Leg 21: Init. Repts. DSDP, v. 21, p. 885–893.	N S	Q PL M,O E PA C	203 to 210		N	G V
———, 1974, Coccolith and silicoflagellate stratigraphy, eastern Indian Ocean, Deep Sea Drilling Project, Leg 22: Init. Repts. DSDP, v. 22, p. 601–607.	N S	Q PL M,O E PA C	211 to 218		N	V
———, 1974, Coccolith stratigraphy, Arabian and Red Seas, Deep Sea Drilling Project Leg 23: Init. Repts. DSDP, v. 23, p. 1091–1093.	N	Q PL M O E PA	219 to 230		N	V
———, 1974, Stratigraphic value of silicoflagellates in nontropical regions: Geol. Soc. America Bull., v. 85, p. 1905–1906.	S	Q PL M,O E,C PA				V F
———, 1974, Coccolith zonation of cores from the western Indian Ocean and the Gulf of Aden, Deep Sea Drilling Project, Leg 24: Init. Repts. DSDP, v. 24, p. 995–996.	N	Q PL M O E PA	231 to 238		N	V

Reference	Fossil groups.	Ages.	Sites.	Revisions. New taxa.	Abundance. / Preservation.	Groups or regions related. Vertical study. Finer stratigraphy.
———, 1974, Phytoplankton stratigraphy, offshore East Africa, Deep Sea Drilling Project, Leg 25: Init. Repts. DSDP, v. 25, p. 635–646.	N S	Q PL M,O E PA C	239 to 249		N P	G V
———, 1974, Cretaceous and Paleogene coccolith stratigraphy, Deep Sea Drilling Project, Leg 26: Init. Repts. DSDP, v. 26, p. 669–673.	N	Q PL M,O E C	250 to 258		N P	V
———, 1974, Coccolith stratigraphy, offshore Western Australia, Deep Sea Drilling Project, Leg 27: Init. Repts. DSDP, v. 27, p. 623–630.	N	Q O,E PA C J	259 to 263		N	V
———, 1975, Coccolith and silicoflagellate stratigraphy near Antarctica, Deep Sea Drilling Project, Leg 28: Init. Repts. DSDP, v. 28, p. 709–723.	N S	Q PL M O E C	264 to 274		D	V
———, 1975, Silicoflagellate and coccolith stratigraphy, Deep Sea Drilling Project, Leg 29: Init. Repts. DSDP, v. 29, p. 845–872.	N S	Q PL M,O E PA C	275 to 284	R N3 (s)	D	R V F
———, 1975, Phytoplankton stratigraphy, Southwest Pacific, Deep Sea Drilling Project, Leg 30: Init. Repts. DSDP, v. 30, p. 539–547.	N D S	Q PL M,O E	285 to 289		N D	V

		PA C				
———, 1975, Coccolith and silicoflagellate stratigraphy, northwestern Pacific Ocean, Deep Sea Drilling Project, Leg 32: Init. Repts. DSDP, v. 32, p. 677–701. [Additional information in Site chapters.]	N S	Q PL M,O E PA C	303 to 313	R N9 (5s & 4n)	D	V F
———, 1975, New Miocene to Holocene stages in the ocean basins based on calcareous nannoplankton zones, *in* Saito, T., and Burckle, L. H., eds., Late Neogene Epoch Boundaries: New York, Micropaleontology Press, Spec. Pub. 1, p. 162–166.	N	Q PL M O	55,56 71,83 149			V F
———, 1976, Coccolith stratigraphy of Manihiki Plateau, Central Pacific, Deep Sea Drilling Project, Site 317: Init. Repts. DSDP, v. 33, p. 493–501.	N	Q PL M,O E C	317	N1	U	V
———, 1976, Silicoflagellate and coccolith stratigraphy, southeastern Pacific Ocean, Deep Sea Drilling Project, Leg 34: Init. Repts. DSDP, v. 34, p. 715–735.	N S	Q PL M O E	319 to 321	R N2 (s)	D	G V
———, 1976, Comments on some coccoliths and silicoflagellates from Deep Sea Drilling Project Leg 35: Init. Repts. DSDP, v. 35, p. 693–699.	N S	PL M PA	322 to 325		U	V
———, 1976, Cenozoic silicoflagellate and coccolith stratigraphy, South Atlantic Ocean, Deep Sea Drilling Project, Leg 36: Init. Repts. DSDP, v. 35, p. 885–917.	N S	Q PL M O E PA	327 to 331	N13 (s)	D	V
———, 1976, Silicoflagellate and coccolith stratigraphy, Norwegian-Greenland Sea, Deep Sea Drilling Project Leg 38: Init. Repts. DSDP, v. 38, p. 843–855.	N S	M O E	336 to 352	R N6 (s)	D	V
———, 1977, Coccolith and silicoflagellate stratigraphy, central North Atlantic Ocean, Deep Sea Drilling Project Leg 37: Init. Repts. DSDP, v. 37, p. 917–927.	N	Q PL M	332 to 335	N2	D	V F
———, 1977, Coccolith and silicoflagellate stratigraphy, South Atlantic Ocean, Deep Sea Drilling Project Leg 39: Init. Repts. DSDP, v. 39, p. 825–839.	N S	Q PL M,O E PA	353 to 359	R N2 (s)	D	V F

Reference	Fossil groups.	Ages.	Sites.	Revisions. New taxa.	Abundance.	Preservation.	Groups or regions related. Vertical study. Finer stratigraphy.
———, 1978, Biostratigraphy of Cenzoic marine sediment by calcareous microfossils: Micropaleontology, v. 24, no. 1, p. 44–60.	N	PL M,O E,PA		R			V F
———, 1978, Cenozoic silicoflagellate and coccolith stratigraphy, southeastern Atlantic Ocean Deep Sea Drilling Project Leg 40: Init. Repts. DSDP, v. 40, p. 635–649.	N S	Q PL M,O E PA	360 to 365	N2 (s)	D		V
———, 1978, Cenozoic coccolith and silicoflagellate stratigraphy, offshore northwest Africa Deep Sea Drilling Project Leg 41: Init. Repts. DSDP, v. 41, p. 689–707. [Additional informa tion in Site chapters.]	N S	Q PL M,O E PA	366 to 370	R N2 (s)	D	P	V
———, 1978, Cenozoic silicoflagellate and coccolith stratigraphy, northwestern Atlantic Ocean Deep Sea Drilling Project Leg 43: Init. Repts. DSDP, v. 44, p. 775–805.	N S	Q O,E PA	382 to 387	R N 5(s) +1(d)	D		V
———, 1978, Cenozoic coccolith, silicoflagellate, and diatom stratigraphy, Deep Sea Drilling Project Leg 44: Init. Repts. DSDP, v. 44, p. 807–863.	N S D	Q M O E PA	388 to 394	N9 (s)	N D	P	V F
———, 1979, Neogene coccolith stratigraphy, Mid-Atlantic Ridge, Deep Sea Drilling Project Leg 45: Init. Repts. DSDP, v. 45, p. 307–317.	N	Q PL M	395 396	N2	N		V
———, and Bramlette, M. N., 1969, Coccolith age determinations, Leg 1, Deep Sea Drilling Project: Init. Repts. DSDP, v. 1, p. 369–387.	N	Q P,M O,E C,J	1 to 7		N		V
———, and ———, 1970, Coccolith age determinations, Leg 3, Deep Sea Drilling Project: Init.	N	Q	13		N		V F

Repts. DSDP, v. 3, p. 589–611.		PL M,O E PA C	to 22			
———, and ———, 1970, Coccolith age determinations, Leg 5, Deep Sea Drilling Project: Init. Repts. DSDP, v. 5, p. 487–494.	N	Q PL M O E	32 to 42		N	V
———, Douglas, R. G., Kling, S. A., and Krasheninnikov, V., 1971, Planktonic microfossil biostratigraphy of the northwestern Pacific Ocean: Init. Repts. DSDP, v. 6, p. 1253–1300.	F R N	Q PL M O E PA C	44 to 60	R	P	G V R
———, Dinkelman, M. G., and Kaneps, A., 1973, Biostratigraphy of the equatorial East Pacific Rise: Init. Repts. DSDP, v. 16, p. 915–935.	F N R	Q PL M O E	155 to 163			G V
———, and Foster, J. H., 1973, Silicoflagellate and diatom stratigraphy, Leg 16, Deep Sea Drilling Project: Init. Repts. DSDP, v. 16, p. 815–871.	S D	Q PL M	155 to 163	R N2 (s)	U	V F
———, and ———, 1974, Silicoflagellate zonation of Upper Cretaceous to Lower Miocene deep-sea sediment: Jour. Research U.S. Geol. Survey, vol. 2, no. 3, p. 303–310.	S	M O E PA C	6,13 29,66 140,206 208,214 216,274 275,280			V F
———, and Percival, S. F., Jr., 1971, New Tertiary calcareous nannofossils: Tulane Studies in Geology and Paleontology, v. 8, no. 3, p. 123–146.	N	P M O,E PA	14,15 44,47 55–57	R N27		V
Burckle, L. H., 1972, Summary of diatoms in Hole 112: Init. Repts. DSDP, v. 12, p. 176–177.	D	O E	112		U	V
———, 1975, Diatom evidence bearing on the correlation of the Relizian Stage, California: Pacific Geology, v. 9, p. 33–34.	D	M	77		N	R V
———, and Opdyke, N. D., 1977, Late Neogene diatom correlations in the circum-Pacific: Proceedings of the First International Congress on Pacific Neogene Stratigraphy, Tokyo 1976, p. 255–284.	D	Q PL M	157 158 173		N	G V F R

Reference	Fossil groups.	Ages.	Sites.	Revisions. New taxa.	Abundance.	Preservation.	Groups or regions related.	Vertical study.	Finer stratigraphy.
Burns, D. A., 1975, Nannofossil biostratigraphy for Antarctic sediments, Leg 28, Deep Sea Drilling Project: Init. Repts. DSDP, v. 28, p. 589–598. [Additional information in Site chapters.]	N	Q PL M O	264- 274		U			V	
Busen, K. E., and Wise, S. W., Jr., 1977, Silicoflagellate stratigraphy, Deep Sea Drilling Project, Leg 36: Init. Repts. DSDP, v. 36, p. 697–743.	S	Q PL M,O E,PA	326 to 331	R N5	D		R	V	F
Cameron, A., 1978, Neogene benthic foraminifers from DSDP Sites 360 and 362, south eastern Atlantic: Init. Repts. DSDP, v. 40, p. 811–817.	F	Q PL M	360 362		N			V	
Caron, M, 1972, Planktonic Foraminifera from the Upper Cretaceous of Site 98, Leg 11, DSDP: Init. Repts. DSDP, v. 11, p. 551–559.	F	C	98	R N1	N			V	
———, 1975, Late Cretaceous planktonic Foraminifera from the northwestern Pacific: Leg 32 of the Deep Sea Drilling Project: Init. Repts. DSDP, v. 32, p. 719–724.	F	C	305 310 313		N	P		V	
———, 1978, Cretaceous planktonic foraminifers from DSDP Leg 40, southeastern Atlantic Ocean: Init. Repts. DSDP, v. 40, p. 651–678.	F	C	361 363 364	N1	N	P		V	
Čepek, P., 1978, Mesozoic calcareous nannoplankton of the eastern North Atlantic, Leg 41: Init. Repts. DSDP, v. 41, p. 667–687. [Additional information in Site chapters.]	N	C J	366 to 370		U	P		V	
Čepek, P., Johnson, D., Krasheninnikov, V., and Pflaumann, U., 1978, Synthesis of the Leg 41 biostratigraphy and paleontology: Init. Repts. DSDP, v. 41, p. 1181–1198.	F R N	Q PL M,O E,PA C,J	366 to 370				G	V	
Cheetham, A. H., 1975, Preliminary report on Early Eocene cheilostome bryozoans from Site 308 — Leg 32, Deep Sea Drilling Project: Init. Repts. DSDP, v. 32, p. 835–851.	O	E	308		D			V	
———, and Håkansson, E., 1972, Preliminary report on Bryozoa (Site 117): Init. Repts. DSDP, v. 12, p. 432–440.	O	O PA	117		D		R		

Chen, P.-H., 1974, Some new Tertiary Radiolaria from Antarctic deep-sea sediments: Micropaleontology, v. 20, no. 4, p. 480–492.	R	Q PL M	264 to 274	N16	U	V F
———, 1975, Antarctic Radiolaria: Init. Repts. DSDP, v. 28, p. 437–513. [Additional information in Site chapters.]	R	Q PL M O E	264 to 274	R N20	U P	V F
Chevalier, J.-P., 1976, Étude des coraux fossiles récoltés dans le forage DSDP 76 en mer profonde dans les Tuamotu du Nord: Bull. Soc. Géol. France, ser. 7, v. 18, no. 5, p. 1307–1313.	O	Q PL	76	N1		V
Ciesielski, P. F., 1975, Biostratigraphy and paleoecology of Neogene and Oligocene silicoflagellates from cores recovered during Antarctic Leg 28, Deep Sea Drilling Project: Init. Repts. DSDP, v. 28, p. 625–691.	S	Q PL M O	264 to 274	R N4	D	R V F
Cita, M. B., 1972, Biostratigraphy, chronostratigraphy and paleoenvironment of the Pliocene of Cape Verde (North Atlantic): Revue de Micropaléontologie, v. 14, no. 5, p. 17–42.	F	PL N	12	R	N P	G V F
———, 1973, Inventory of biostratigraphical findings and problems: Init. Repts. DSDP, v. 13, p. 1045-1073. [Additional information in Site chapters.]	F N	Q PL M	120 to 134		P	G V F
———, 1973, Pliocene biostratigraphy and chronostratigraphy: Init. Repts. DSDP, v. 13, p. 1343-1379.	F	PL	120 to 134	N2		G V F R
———, 1975, The Miocene/Pliocene boundary: History and definition, *in* Saito, T., and Burckle, L. H., eds., Late Neogene Epoch Boundaries: New York, Micropaleontology Press, Spec. Pub. 1, p. 1–30.	F	PL M	132			R V F
———, 1976, Biodynamic effects of the Messinian salinity crisis on the evolution of planktonic Foraminifera in the Mediterranean: Palaeogeography, Palaeoclimatology, Palaeoecology, v. 20, p. 23–42.	F	PL M	125 132	R		V F
———, Chierici, M. A., Ciampo, G., Moncharmont Zei, M., d'Onofrio, S., Ryan, W. B. F., and Scorziello, R., 1973, The Quaternary record in the Tyrrhenian and Ionian Basins of the Mediterranean Sea: Init. Repts. DSDP, v. 13, p. 1263–1339.	F	Q	125 132		U	R V F
———, Ciampo, G., Ferone, E., Moncharmont Zei, M., Scorziello, R., Taddei Ruggiero, E., 1974, Il Quaternario del Tirreno abissale, Interpretazione stratigrafica e paleoclimatica del pozzo DSDP 132: Revista Española Micropaleontología, v. 6, no. 2, p. 257–326.	F	Q	132	R	D	V F
———, and Gartner, S., Jr., 1971, Deep sea Upper Cretaceous from the western North Atlantic: Proceedings of the II Planktonic Conference, Roma 1970, p. 287–319.	F N	C	10		N	G V F
———, Nigrini, C., and Gartner, S., 1970, Biostratigraphy: Init. Repts. DSDP, v. 2, p. 391–411. [Also information in Site chapters.]	F N R	Q PL M,O E C	8 to 12		N P U	V

Reference	Fossil groups.	Ages.	Sites.	Revisions. New taxa.	Abundance. / Preservation.	Groups or regions related. Vertical study. Finer stratigraphy.
Davey, R. J., 1978, Marine Cretaceous palynology of Site 361, DSDP Leg 40, off southwestern Africa: Init. Repts. DSDP, v. 40, p. 883–913.	O	C	361	N4	N	R V
Dell, R. K., and Fleming, C. A., 1975, Oligocene-Miocene bivalve Mollusca and other macrofossils from Sites 270 and 272 (Ross Sea), DSDP, Leg 28: Init. Repts. DSDP, v. 28, p. 693–703.	O	M O	270 272		D	R
Dengler, A. T., Doyle, P. S., and Riedel, W. R., 1975, Ichthyoliths in some samples from the Philippine Sea, Deep Sea Drilling Project, Leg 31: Init. Repts. DSDP, v. 31, p. 821–833.	O	M O E	293 294 295 297		U	V
Dinkleman, M. G., 1973, Radiolarian stratigraphy: Leg 16, Deep Sea Drilling Project: Init. Repts. DSDP, v. 16, p. 747–813.	R	Q PL M O,E C	155 to 163		U P	V
———, 1973, Correlation between microfossil groups, Panama Basin, eastern equatorial Pacific: Init. Repts. DSDP, v. 16, p. 897–898.	F N R S D	Q PL M	157 158			G
Douglas, R. G., 1971, Cretaceous Foraminifera from the northwestern Pacific Ocean: Leg 6, Deep Sea Drilling Project: Init. Repts. DSDP, v. 6, p. 1027–1053. [Additional information in Site chapters.]	F	C	45 47 to 51		N P	R V
———, 1973, Benthonic foraminiferal biostratigraphy in the central North Pacific, Leg 17, Deep Sea Drilling Project: Init. Repts. DSDP, v. 17, p. 607–671.	F	Q PL M,O E PA C	167 171		D	V
———, 1973, Planktonic foraminiferal biostratigraphy in the central North Pacific Ocean: Init. Repts. DSDP, v. 17, p. 673–694.	F	Q PL	164 to		U P	V

		M,O E PA C	171			
———, and Moullade, M., 1972, Age of the basal sediments on the Shatsky Rise, western North Pacific Ocean: Geol. Soc. America Bull., v. 83, p. 1163–1168.	F	C	49 50		N	V F
Doyle, P. S., Dunsworth, M. J., and Riedel, W. R., 1977, Reworking of ichthyoliths in eastern tropical Pacific sediments: Deep-Sea Research, v. 24, p. 181–198.	O	Q PL M O,E	14,15 19,29 39,41 65,75 111,119 304,310 313		D	V
———, ———, and ———, 1978, Ichthyoliths from some southeast Atlantic sediments, DSDP Leg 40: Init. Repts. DSDP, supplement to v. 38–41, p. 743–759.	O	O PA C	361 365	R N18	D	V
———, Kennedy, G. G., and Riedel, W., 1974, Stratignathy: Init. Repts. DSDP, v. 26, p. 825–905.	O	Q PL M O E	250 251 256	N60	D	R V F
Ducasse, O., and Grékoff, N., 1976, Quelques Ostracodes de l'Éocène inférieur du sudouest de l'océan Indien: Site 246, Croisière 25, "Glomar Challenger", Deep Sea Drilling Project: Revue de Micropaléontologie, v. 19, no. 3, p. 134–152.	O	E	246		D	V
Dumitrica, P., 1973, Cretaceous and Quaternary Radiolaria in deep sea sediments from the northwest Atlantic Ocean and Mediterranean Sea: Init. Repts. DSDP, v. 13, p. 829–901. [Additional information in Site chapters.]	R	Q C	120 121 127 128 129	R N9	N P	V
———, 1973, Miocene and Quaternary silicoflagellates in sediments from the Mediterranean Sea: Init. Repts. DSDP, v. 13, p. 902–933.	S	Q M	124 127 128	R N2	U	V F
———, 1973, Miocene and Quaternary ebridians from the Mediterranean Sea, Deep Sea Drilling Project Leg 13: Init. Repts. DSDP, v. 13, p. 934–939.	O	Q M	124 127 128	N3	U	V
———, 1973, Phytolitharia: Init. Repts. DSDP, v. 13, p. 940–943.	O	Q M	124 130 131		U	V
———, 1973, Phaeodarian Radiolaria in southwest Pacific sediments cored during Leg 21 of the Deep Sea Drilling Project: Init. Repts. DSDP, v. 21, p. 751–785.	R	Q PL M	203 to 206	N4	U	V
———, 1973, Paleocene Radiolaria, DSDP Leg 21: Init. Repts. DSDP, v. 21, p. 787–817.	R	PA	208		N	V

Reference	Fossil groups.	Ages.	Sites.	Revisions. New taxa.	Abundance.	Preservation.	Groups or regions related. Vertical study. Finer stratigraphy.
———, 1973, Cenozoic endoskeletal dinoflagellates in southwestern Pacific sediments cored during Leg 21 of the DSDP: Init. Repts. DSDP, v. 21, p. 819–835.	O	Q PL M O E PA	206 to 208	R N11	U		R V F
———, 1973, Paleocene, Late Oligocene and post-Oligocene silicoflagellates in southwestern Pacific sediments cored on DSDP Leg 21: Init. Repts. DSDP, v. 21, p. 837–883.	S	Q PL M O PA	66 203 to 208	R N3	U D		R V F G
Dunsworth, M. J., Doyle, P. S., and Riedel, W. R., 1975, Ichthyoliths from some NW Pacific sediments, DSDP Leg 32: Init. Repts. DSDP, v. 32, p. 853–863.	O	Q PL M O E	307 311	R N9	D		R V
Dzinoridze, R. N., Jousé, A. P., Koroleva-Golikova, G. S., Kozlova, G. E., Nagaeva, G. S., Petrushevskaya, M. G., and Strelnikova, N. I., 1978, Diatom and radiolarian Cenozoic stratigraphy, Norwegian Basin; DSDP Leg 38: Init. Repts. DSDP, supplement to v. 38–41, p. 289–427.	D R	Q PL M O E	336 to 352		U		G V
Echols, R. J., 1973, Foraminifera, Leg 19, Deep Sea Drilling Project: Init. Repts. DSDP, v. 19, p. 721–735.	F	Q PL M O E,C	184 to 192		U	P	V
Edgerton, C. C., Doyle, P. S., and Riedel, W. R., 1977, Ichthyolith age determinations of otherwise unfossiliferous Deep Sea Drilling Project cores: Micropaleontology, v. 23, no. 2, p. 194–205.	O	Q PL M O,E C	27,29 38,39 50,51 52,66	R	D		V
Edwards, A. R., 1973, Calcareous nannofossils from the southwest Pacific, Deep Sea Drilling	N	Q	203		U	P	V F

Project, Leg 21: Init. Repts. DSDP, v. 21, p. 641–691. [Additional information in Site chapters.]		PL M O E PA C	to 210			
———, and Perch-Nielsen, K., 1975, Calcareous nannofossils from the southern Southwest Pacific, Deep Sea Drilling Project, Leg 29: Init. Repts. DSDP, v. 29, p. 469–539. [Additional information in Site chapters.]	N	Q PL M O E PA	275 to 284	N3	U P	R V
Ellis, C. H., 1975, Calcareous nannofossil biostratigraphy—Leg 31, DSDP: Init. Repts. DSDP, v. 31, p. 655–676.	N	Q PL M O E	290 to 302		U P	V
———, Lohman, W. H., and Wray, J. L., 1972, Upper Cenozoic calcareous nannofossils from the Gulf of Mexico (Deep Sea Drilling Project, Leg 1, Site 3): Quarterly of the Colorado School of Mines, v. 67, no. 3, xi + 103 p.	N	Q PL M	3	R N1	D	V F
Evitt, W. R., 1973, Dinoflagellates from Leg 19, Sites 183 and 192, Deep Sea Drilling Project: Init. Repts. DSDP, v. 19, p. 737–738.	O	E	183 192			V
Fenner, J., 1978, Cenozoic diatom biostratigraphy of the equatorial and southern Atlantic Ocean: Init. Repts. DSDP, supplement to v. 38–41, p. 491–623.	D	Q PL M O E	354 356 358	R N10	D P	V
Ferrer, J., 1975, Neritic Early Eocene smaller Foraminifera from the Kōko Seamount (Emperor Seamounts, central northern Pacific): Init. Repts. DSDP, v. 32, p. 803–823.	F O	E	308		U	V
Fleisher, R. L., 1974, Cenozoic planktonic Foraminifera and biostratigraphy, Arabian Sea, Deep Sea Drilling Project, Leg 23A: Init. Repts. DSDP, v. 23, p. 1001–1072. [Additional information in Site chapters.]	F	Q PL M O E	219 to 224	R N11	U P	V
———, 1974, Preliminary report on Late Neogene Red Sea Foraminifera, Deep Sea Drilling Project, Leg 23B: Init. Repts. DSDP, v. 26, p. 985–1011.	F	Q PL M	225 to 229	R	N	V
———, 1975, Oligocene planktonic foraminiferal biostratigraphy, central North Pacific Ocean, DSDP Leg 32: Init. Repts. DSDP, v. 32, p. 753–763.	F	O	305 310 313		N P	V
———, 1977, Oligocene planktonic foraminiferal assemblages from Deep Sea Drilling Project sites in the Indian Ocean, *in* Heirtzler, J. R., and others, eds., Indian Ocean Geology and Bio-	F	O	211 to		P	

Reference	Fossil groups.	Ages.	Sites.	Revisions. New taxa.	Abundance. / Preservation.	Groups or regions related. Vertical study. Finer stratigraphy.
stratigraphy: Washington, D.C., American Geophysical Union, p. 459–467.			269 280 to 282			
Florer-Heusser, L. E., 1975, Late-Cenozoic marine palynology of northeast Pacific Ocean cores, *in* Suggate, R. P., and Cresswell, M. M., eds., Quaternary Studies: Wellington, Royal Society of New Zealand, p. 133–138.	O	Q	175 176		D	V F
Foreman, H. P., 1971, Cretaceous Radiolaria, Leg 7, DSDP: Init. Repts. DSDP, v. 7, p. 1673–1693.	R	C	61 66	N6	N	V
———, 1973, Radiolaria of Leg 10 with systematics and ranges for the families Amphipyndacidae, Artostrobiidae, and Theoperidae: Init. Repts. DSDP, v. 10, p. 407–474.	R	M O E PA C	86 94 95 96 97	R N33	U P	G V F
———, 1973, Radiolaria from DSDP Leg 20: Init. Repts. DSDP, v. 20, p. 249–305.	R	C	194 to 199	R N31	U	V F
———, 1975, Radiolaria from the North Pacific, Deep Sea Drilling Project, Leg 32: Init. Repts. DSDP, v. 32, p. 579–676. [Additional information in Site chapters.]	R	Q PL M E C	303 to 313	R N15	U P	G V F
———, 1978, Cretaceous Radiolaria in the eastern South Atlantic, Deep Sea Drilling Project, Leg 40: Init. Repts. DSDP, v. 40, p. 839–843.	R	C	361 to 365	N3	N P	V
———, 1978, Mesozoic Radiolaria in the Atlantic Ocean off the northwest coast of Africa, Deep Sea Drilling Project, Leg 41: Init. Repts. DSDP, v. 41, p. 739–761.	R	C J	367 368 369	R N8	U P	V F
———, Hekel, H., Hoskins, R. H., and Krasheninnikov, V. A., 1973, Biostratigraphic synthesis, Leg 20, DSDP: Init. Repts. DSDP, v. 20, p. 645–651.	F N R	Q PL M	194 to 202			G V

		E PA C				
Fütterer, D., 1978, Distribution of calcareous dinoflagellates in Cenozoic sediments of Site 366, eastern North Atlantic: Init. Repts. DSDP, v. 41, p. 709–737.	N	Q PL M O E PA	366	R N2	U	V
Gartner, S., Jr., 1969, Correlation of Neogene planktonic foraminifer and calcareous nannofossil zones: Transactions Gulf Coast Association of Geological Societies, v. 19, p. 585–599.	N F	Q PL M	10 11 12	R N3 (n)	N	G V F
———, 1970, Coccolith age determinations, Leg 3, Deep Sea Drilling Project: Init. Repts. DSDP, v. 3, p. 613–627.	N	Q PL M O,E PA C	13 to 22		N	V
———, 1970, Coccolith age determinations, Leg 5, Deep Sea Drilling Project: Init. Repts. DSDP, v. 5, p. 495–500.	N	Q PL M	32 to 36		N	V
———, 1971, Calcareous nannofossil age determinations, Leg 7, Deep Sea Drilling Project: Init. Repts. DSDP, v. 7, p. 1509–1511.	N	Q PL M O E	62 to 65		N	V
———, 1972, Coccolith age determinations, Leg 9 Deep Sea Drilling Project: Init. Repts. DSDP, v. 9, p. 833–839. [Additional information on calcareous nannofossils in the Site chapters.]	N	Q PL M O E	76 to 83		N	V
———, 1973, Calcareous nannofossil age determinations, Deep Sea Drilling Project, Leg 13: Init. Repts. DSDP, v. 13, p. 822–827.	N	Q PL M C J	120 to 134		N	V
———, 1973, Nannofossil age determinations, DSDP, Leg 16: Init. Repts. DSDP, v. 16, p. 883–889.	N	Q PL M O E,C	155 to 163		N	V
———, 1974, Nannofossil biostratigraphy, Leg 22, Deep Sea Drilling Project: Init. Repts. DSDP,	N	Q	211		N P	V

Reference	Fossil groups.	Ages.	Sites.	Revisions. New taxa.	Abundance. / Preservation.	Groups or regions related. Vertical study. Finer stratigraphy.
v. 22, p. 577–599. [Additional information in Site chapters.]		PL M O E PA C	to 218			
———, and Bukry, D., 1974, *Ceratolithus acutus* Gartner and Bukry, n. sp. and *Ceratolithus amplificus* Bukry and Percival—nomenclatural clarification: Tulane Studies in Geology and Paleontology, v. 11, no. 2, p. 115–118.	N	PL M	22	R N 1		V F
———, and ———, 1975, Morphology and phylogeny of the coccolithophycean family Ceratolithaceae: Jour. Research U.S. Geol. Survey, v. 3, p. 451–465.	N	Q PL M	214 242	R N2	N	V F
———, Johnson, D. A., and McGowran, B., 1974, Paleontology synthesis of Deep Sea Drilling results from Leg 22 in the northeastern Indian Ocean: Init. Repts. DSDP, v. 22, p. 805–813.	F N R	Q PL M O E	211 to 218			G V R
Glass, B. P., and Zwart, M. J., 1977, North American microtektites, Radiolarian Extinction and the age of the Eocene-Oligocene Boundary, *in* Swain, F. M., ed., Stratigraphic Micropaleontology of Atlantic Basin and Borderlands: Amsterdam, Elsevier, p. 553–568.	R	O E	94 149		D	V
Goll, R. M., 1972, Leg 9 synthesis, Radiolaria: Init. Repts. DSDP, v. 9, p. 947–1058, and pl. 88 (in pocket). [Additional information in Site chapters.]	R	Q PL M O E	77 to 84		N	G V
———, 1978, Five trissocyclid Radiolaria from Site 338: Init. Repts. DSDP, supplement to v. 38–41, p. 177–191.	R	M O	338	R N4	N	V
Gombos, A. M., Jr., 1975, Fossil diatoms from Leg 7, Deep Sea Drilling Project: Micropaleontology, v. 21, no. 3, p. 306–333.	D	Q PL M	66		D	V
———, 1976, Diatom investigations in the South Atlantic: Antarctic Journal of the United States,	D	O	327			V F

v. 11, no. 3, p. 175–176.		E PA	328			
———, 1977, Paleogene and Neogene diatoms from the Falkland Plateau and Malvinas Outer Basin: Leg 36, Deep Sea Drilling Project: Init. Repts. DSDP, v. 36, p. 575–687.	D	Q PL M O E PA	326 to 331	N5	U P	V F
Gradstein, F. M., 1978, Foraminifera from DSDP Site 370, Leg 41, eastern North Atlantic Ocean: Init. Repts. DSDP, supplement to v. 38–41, p. 779–782.	F	Q M O E C	370		N	V
———, 1978, Biostratigraphy of Lower Cretaceous Blake Nose and Blake-Bahama Basin foraminifers, DSDP Leg 44, western North Atlantic Ocean: Init. Repts. DSDP, v. 44, p. 663–701.	F	C	390 391 392	R	U P	R V
———, Bukry, D., Habib, D., Renz, R., Roth, P. H., Schmidt, R. R., Weaver, F. M., and Wind, F. H., 1978, Biostratigraphic summary of DSDP Leg 44: western North Atlantic Ocean: Init. Repts. DSDP, v. 44, p. 657–662. [Additional information in Site chapters.]	F N S R etc.	Q M E PA C J	388 to 394			G V
Habib, D., 1972, Dinoflagellates and other palynomorhs in selected samples from Leg 14, Deep Sea Drilling Project: Init. Repts. DSDP, v. 14, p. 649–653.	O	C	135 to 144		N	V
———, 1972, Dinoflagellate stratigraphy, Leg 11, Deep Sea Drilling Project: Init. Repts. DSDP, v. 11, p. 367–425.	O	Q PL M,O E,C J	98 to 108	R N11	N	V
———, 1974, Morphogenetic relationship between *Scriniodinium campanula* Gocht and *Scriniodinium dictyotum* Cookson and Eisenack: Geoscience and Man, v. 9, p. 45–51.	O	C J	99 100 101 105	R		V
———, 1975, Neocomian dinoflagellate zonation in the western North Atlantic: Micropaleontology, v. 21, no. 4, p. 373–392.	O	C	99 100 101 105	R N3	U	R V F
———, 1978, Palynostratigraphy of the Lower Cretacous section at Deep Sea Drilling Project Site 391, Blake-Bahama Basin, and its correlation in the North Atlantic: Init. Repts. DSDP, v. 44, p. 887–897.	O	C	391		N	R V F
———, and Warren, J. S., 1973, Dinoflagellates near the Cretaceous Jurassic boundary: Nature,	O	C	105		N	R V

Reference	Fossil groups.	Ages.	Sites.	Revisions. New taxa.	Abundance.	Preservation.	Groups or regions related.	Vertical study.	Finer stratigraphy.
v. 241, no. 5386, p. 217–218.		J							
Hajós, M., 1973, The Mediterranean diatoms: Init. Repts. DSDP, v. 13, p. 944–969.	D	Q M	124 127 128 130		N			V	
———, 1975, Late Cretaceous Archaeomonadaceae, Diatomaceae, and Silicoflagellateae from the South Pacific Ocean, Deep Sea Drilling Project, Leg 29, Site 275: Init. Repts. DSDP, v. 29, p. 913–1009.	D S etc.	C	275	N61 (54d; 2s; 5o)	U			V	F
———, 1976, Upper Eocene and Lower Oligocene Diatomaceae, Archaeomonadaceae, and Silicoflagellateae in southwestern Pacific sediments, DSDP Leg 29: Init. Repts. DSDP, v. 35, p. 817–883.	S D O	O E	280 281 283	R N19 (17d; 2o)	U			V	
Haq, B. U., 1973, Evolutionary trends in the Cenozoic coccolithophore genus *Helicopontosphaera:* Micropaleontology, v. 19, no. 1, p. 32–52.	N	Q PL M O,E		R				V	
———, 1976, Coccoliths in cores from the Bellingshausen Abyssal Plain and Antarctic Continental Rise (DSDP Leg 35): Init. Repts. DSDP, v. 35, p. 557–567. [Additional information in Site chapters].	N	M O C	322 to 325	N1	N	P	R	V	
Haq, U. Z. B., and Lipps, J. H., 1971, Calcareous nannoplankton: Init. Repts. DSDP, v. 8, p. 777–789. [Additional information in Site chapters.]	N	Q PL M O	68 to 75		N			V	F
Haq, B. U., and Riley, A., 1976, Antarctic silicoflagellates from the Southeast Pacific, Deep Sea Drilling Project, Leg 35: Init. Repts. DSDP, v. 35, p. 673–691.	S	PL M	322 to 325		D			V	
Harris, W. K., 1974, Palynology of Paleocene sediments at Site 214, Ninetyeast Ridge: Init. Repts. DSDP, v. 22, p. 503–519.	O	PA	214		D		R	V	
———, 1977, Palynology of cores from deep sea drilling sites 327, 328 and 330, South Atlantic Ocean: Init. Repts. DSDP, v. 36, p. 761–815.	O	PA C	327 328		N		R	V	

		J	330			
Haskell, T. R., and Wilson, G. J., 1975, Palynology of Sites 280–284, DSDP Leg 29, off south-eastern Australia and western New Zealand: Init. Repts. DSDP, v. 29, p. 723–741.	O	M O E PA	280 to 284		N	R V
Hay, W. W., 1969, Preliminary dating by fossil calcareous nannoplankton, Deep Sea Drilling Project, Leg 1: Init. Repts. DSDP, v. 1, p. 388–391.	N	Q PL M E C J	1 to 7		N	V
———, 1970, Calcareous nannofossils from cores recovered on Leg 4: Init. Repts. DSDP, v. 4, p. 455–501.	N	Q PL M O E	23 to 31	R	D P	V
———, 1971, Preliminary dating by fossil calcareous nannoplankton, Deep Sea Drilling Project: Leg 6: Init. Repts. DSDP, v. 6, p. 1005–1012.	N	Q PL M,O E PA C	44 to 60		N	V
———, 1971, Preliminary dating by fossil calcareous nannoplankton, Deep Sea Drilling Project, Leg 8: Init. Repts. DSDP, v. 8, p. 809–818.	N	Q PL M O E	69 to 75		N	V
———, 1973, Preliminary dating by calcareous nannoplankton, Deep Sea Drilling Project: Leg 10: Init. Repts. DSDP, v. 10, p. 375–383.	N	Q PL M,O E PA C	85 to 97		N	V
———, and Beaudry, F. M., 1973, Calcareous nannofossils—Leg 15, Deep Sea Drilling Project: Init. Repts. DSDP, v. 15, p. 625–683.	N	Q PL M,O E PA C	146 to 154	R N4	U P	V
Hedlund, R. W., and Beju, D., 1977, Stratigraphic palynology of selected Mesozoic samples, DSDP Hole 327A and Site 330: Init. Repts. DSDP, v. 36, p. 817–827.	O	C J	327 330		U	V
Heiman, M. E., Frerichs, W. E., and Vincent, E., 1974, Paleogene planktonic Foraminifera from	F	O	236		U P	V

Reference	Fossil groups.	Ages.	Sites.	Revisions. New taxa.	Abundance.	Preservation.	Groups or regions related.	Vertical study.	Finer stratigraphy.
the western tropical Indian Ocean, Deep Sea Drilling Project, Leg 24: Init. Repts. DSDP, v. 24, p. 851–857. [Additional information in Site chapters.]		E PA	237 238						
Hekel, H., 1973, Nannofossil biostratigraphy, Leg 20, Deep Sea Drilling Project: Init. Repts. DSDP, v. 20, p. 221–247.	N	Q PL M E PA C	194 to 202		U	P		V	
Helms, P. B., and Riedel, W. R., 1971, Skeletal debris of fishes: Init. Repts. DSDP, v. 7, p. 1709–1720.	O	Q PL M O E	7 24 66	N19	U			V	F
Herb, R., 1974, Cretaceous planktonic Foraminifera from the eastern Indian Ocean: Init. Repts. DSDP, v. 26, p. 745–769.	F	C	255 256 257 258	R	U	P	V		
———, and Scheibnerova, V., 1977, Synopsis of Cretaceous planktonic Foraminifera from the Indian Ocean, *in* Heirtzler, and others, eds., Indian Ocean Geology and Biostratigraphy: Washington, D.C., American Geophysical Union, p. 399–415.	F	C	211 to 269 280 to 282			P			
Herman, Y., 1973, Preliminary pteropod results from the Mediterranean Sea: Init. Repts. DSDP, v. 13, p. 993–1001.	O	Q	125 127		U			V	
Hess, H., 1972, Planktonic crinoids of Late Jurassic age from Leg 11, Deep Sea Drilling Project: Init. Repts. DSDP, v. 11, p. 631–643.	O	J	99 100 105				R	V	
Hofker, J., Sr., 1972, The *Sphaeroidinella* gens from Miocene till Recent: Revista Española de Micropaleontología, v. 4, no. 2, p. 119–140.	F	Q PL M	2 16	R				V	

———, 1978, Analysis of a large succession of samples through the Upper Maestrichtian and Lower Tertiary of drill hole 47.2, Shatsky Rise, Pacific, Deep Sea Drilling Project: Jour. Foraminiferal Research, v. 8, no. 1, p. 46–75.	F	PA C	47	R N4	N	V F
Holdsworth, B. K., 1975, Cenozoic Radiolaria biostratigraphy: Leg 30: tropical and equatorial Pacific: Init. Repts. DSDP, v. 30, p. 499–537. [Additional information in Site chapters.]	R	Q PL M O E	285 to 289		U	G V
Hooper, K., and Jones, P., 1977, Preliminary report on benthonic Foraminifera from the Mid-Atlantic Ridge: Leg 37 DSDP: Init. Repts. DSDP, v. 37, p. 963–965.	F	Q PL M	332 to 335		N	V
Hottinger, L., 1972, Campanian larger Foraminifera from Site 98, Leg 11 of the Deep Sea Drilling Project (Northwest Providence Channel, Bahama Islands): Init. Repts. DSDP, v. 11, p. 595–605.	F	C	98	R	N	R
———, 1975, Late Oligocene larger Foraminifera from Kōko Guyot, Site 309: Init. Repts. DSDP, v. 32, p. 825–826.	F	O	309			V
Howe, R. C., 1977, Calcareous nannofossils—Leg 37, Deep Sea Drilling Project: Init. Repts. DSDP, v. 37, p. 909–916. [Additional information in Site chapters.]	N	Q PL M	332 to 335		U	V
———, and Ellis, C. H., 1977, Calcareous nannofossils from Mid-Atlantic basement rocks of DSDP Leg 37: Canadian Jour. of Earth Sciences, v. 14, no. 4, pt. 2, p. 707–715.	N	PL M	332 334		U	V F
Ingle, J. C., Jr., 1973, Neogene Foraminifera from the northeastern Pacific Ocean, Leg 18, Deep Sea Drilling Project: Init. Repts. DSDP, v. 18, p. 517–567.	F	Q PL M O	172 to 182		D P	V
———, 1973, Summary comments on Neogene biostratigraphy, physical stratigraphy, and paleo-oceanography in the marginal northeastern Pacific Ocean: Init. Repts. DSDP, v. 18, p. 949–960.	F N D S R	Q PL M O	172 to 182			G V R
———, 1975, Pleistocene and Pliocene Foraminifera from the Sea of Japan, Leg 31, Deep Sea Drilling Project: Init. Repts. DSDP, v. 31, p. 693–701. [Additional information in Site chapters.]	F	Q PL	299 to 302		D P	V
Ioannides, N. S., and Colin, J. P., 1977, Palynology of Sites 358, 356, 355, DSDP, Leg 39, south-western Atlantic Ocean: Init. Repts. DSDP, v. 39, p. 885–897.	O	O E C	355 356 358		N	V
Jeletzky, J. A., 1972, Mesozoic macrofossils: Site 111: Init. Repts. DSDP, v. 12, p. 67.	O	C	111		R	R
Jenkins, D. G., 1975, Cenozoic planktonic foraminiferal biostratigraphy of the southwestern Pacific and Tasman Sea—DSDP Leg 29: Init. Repts. DSDP, v. 29, p. 449–467. [Additional information in Site chapters.]	F	Q PL M	275 to 284		N	R V F

Reference	Fossil groups.	Ages.	Sites.	Revisions. New taxa.	Abundance. / Preservation.	Groups or regions related. Vertical study. Finer stratigraphy.
		O E PA				
———, 1978, Neogene planktonic foraminifers from DSDP Leg 40, Sites 360 and 362 in the southeastern Atlantic: Init. Repts. DSDP, v. 40, p. 723–739.	F	Q PL M	360 362		N	R V
———, and Orr, W. N., 1971, Cenozoic planktonic foraminiferal zonation and the problem of test solutions: Revista Española de Micropaleontología, v. 3, no. 3, p. 301–304.	F	Q PL M O E	76 to 84			V
———, and ———, 1972, Planktonic foraminiferal biostratigraphy of the eastern equatorial Pacific—DSDP Leg 9: Init. Repts. DSDP, v. 9, p. 1059–1193, plus 63 p. of tabulations omitted from the volume and distributed later. [Additional information in Site chapters.]	F	Q PL M O E	76 to 84		U P	V F
Johnson, D. A., 1974, Radiolaria from the eastern Indian Ocean, DSDP Leg 22: Init. Repts. DSDP, v. 22, p. 521–575.	R	Q PL M O E	211 to 218	R	U P	V
———, 1976, Cenozoic radiolarians from the central Pacific, DSDP Leg 33: Init. Repts. DSDP, v. 33, p. 425–437. [Additional information in Site chapters.]	R	Q PL M O E	314 to 318		U P	V
———, 1978, Cenozoic Radiolaria from the eastern tropical Atlantic, DSDP Leg 41: Init. Repts. DSDP, v. 41, p. 763–789. [Additional information in Site chapters.]	R	Q PL M O E	366 to 370		U P	V
Jones, D. L., and Plafker, G., 1977, Mezozoic megafossils from DSDP Hole 327A and Site 330 on	O	C	327	N2		R V

the eastern Falkland Plateau: Init. Repts. DSDP, v. 36, p. 845–855.		J	330			
Jousé, A. P., and Kazarina, G. H., 1974, Pleistocene diatoms from Site 262, Leg 27, DSDP: Init. Repts. DSDP, v. 27, p. 925–945.	D	Q	262		D	V
Jung, P., 1973, Pleistocene pteropods—Leg 15, Site 147, Deep Sea Drilling Project: Init. Repts. DSDP, v. 15, p. 753–767.	O	Q	147		D	V
———, 1975, Quaternary larval gastropods from Leg 15, Site 147, Deep Sea Drilling Project, Preliminary Report: The Veliger, v. 18, no. 2, p. 109–126, 120 figs.	O	Q	147		D	V F
Kaneps, A. G., 1973, Cenozoic planktonic Foraminifera from the eastern equatorial Pacific Ocean: Init. Repts. DSDP, v. 16, p. 713–745.	F	Q PL M O E	155 to 162		N P	V
———, 1975, Cenozoic planktonic Foraminifera from Antarctic deep-sea sediments, Leg 28, DSDP: Init. Repts. DSDP, v. 28, p. 573–583. [Additional information in Site chapters.]	F	Q PL M O E	264 to 274		N	V
———, 1976, Cenozoic planktonic foraminifers, equatorial Pacific Ocean, Leg 33, DSDP: Init. Repts. DSDP, v. 33, p. 361–367. [Additional information in Site chapters.]	F	Q PL M O E PA	314 to 318		N P	V
Kaufmann, E. G., 1976, Deep-sea Cretaceous macrofossils: Hole 317A, Manihiki Plateau: Init. Repts. DSDP, v. 33, p. 503–535.	O	C	317	N12	N	V
Keany, J., and Kennett, J. P., 1975, Pliocene-Pleistocene radiolarian biostratigraphy and paleoclimatology at DSDP Site 278 on the Antarctic Convergence: Init. Repts. DSDP, v. 29, p. 757–767.	R	Q PL	278		N	V
Keigwin, L. D., Jr., 1976, Late Cenozoic planktonic foraminiferal biostratigraphy and paleoceanography of the Panama Basin: Micropaleontology, v. 22, no. 4, p. 419–442.	F	Q PL M	84 157 158		D P	R V F
———, 1978, Pliocene closing of the Isthmus of Panama, based on biostratigraphic evidence from nearby Pacific Ocean and Caribbean Sea cores: Geology, v. 6, p. 630–634.	F	Q PL	148 154 157		D	R V
Keller, G., 1978, Morphologic variation of *Neogloboquadrina pachyderma* (Ehrenberg) in sediments of the marginal and central northeast Pacific Ocean and paleoclimatic interpretation: Jour. Foraminiferal Research, v. 8, no. 3, p. 208–224.	F	Q PL	173 310		D	V
———, 1978, Late Neogene biostratigraphy and paleoceanography of DSDP Site 310 central North Pacific and correlation with the Southwest Pacific: Marine Micropaleontology, v. 3, p. 97–119.	F	Q PL	310		D	R V F

Reference	Fossil groups.	Ages.	Sites.	Revisions. New taxa.	Abundance.	Preservation.	Groups or regions related. Vertical study. Finer stratigraphy.
Kemp, E. M., 1974, Preliminary palynology of samples from Site 254, Ninetyeast Ridge: Init. Repts. DSDP, v. 26, p. 815–823.	O	M	254		D		
———, 1975, Palynology of Leg 28 drill sites, Deep Sea Drilling Project: Init. Repts. DSDP, v. 28, p. 599–623. [Additional information in Site chapters.]	O	Q PL M O E	264 to 274	R	N		V
Kennett, J. P., 1973, Middle and Late Cenozoic planktonic foraminiferal biostratigraphy of the southwest Pacific—DSDP Leg 21: Init. Repts. DSDP, v. 21, p. 575–639. [Additional information in Site chapters.]	F	Q PL M	203 to 210		N	P	R V F
———, 1975, Neogene planktonic foraminiferal stratigraphy in deep-sea drilling sites, southeast Indian Ocean: Init. Repts. DSDP, v. 29, p. 705–708.	F	Q PL	264 265 266		D		V
———, and Vella, P., 1975, Late Cenozoic planktonic Foraminifera and paleoceanography at DSDP Site 284 in the cool subtropical South Pacific: Init. Repts. DSDP, v. 29, p. 769–799.	F	Q PL M	284		D		R V
Kling, S. A., 1971, Radiolaria: Leg 6 of the Deep Sea Drilling Project: Init. Repts. DSDP, v. 6, p. 1069–1117. [Additional information in Site chapters.]	R	Q PL M C	45 to 60	N1	U		R V F
———, 1973, Radiolaria from the eastern North Pacific, Leg 18: Init. Repts. DSDP, v. 18, p. 617–671.	R	Q PL M	172 to 182	R N5	U	P	V F
Koizumi, I., 1973, The Late Cenozoic diatoms of Sites 183–193, Leg 19, Deep Sea Drilling Project: Init. Repts. DSDP, v. 19, p. 805–855.	D	Q PL M	183 to 193	R	D		R V F
———, 1975, Neogene diatoms from the northwestern Pacific Ocean, Deep Sea Drilling Project: Init. Repts. DSDP, v. 32, p. 865–889.	D	Q PL M	47 194 303 310		D		V
———, 1975, Neogene diatoms from the western margin of the Pacific Ocean, Leg 31, Deep Sea	D	Q	290			P	R V F

Drilling Project: Init. Repts. DSDP, v. 31, p. 779–819.		PL M	to 302			
———, 1975, Late Cenozoic diatom biostratigraphy in the circum-North Pacific region: Journal of the Geological Society of Japan, v. 81, no. 10, p. 611–627.	D	Q PL M	47,173 183,192 299 301 303 310			R V F
Koreneva, E. V., Zalinskaya, E. D., Bratseva, G. M., and Kartashova, G. G., 1976, Palynology studies of Sites 336, 338, 345, 346 and 348, DSDP Leg 38: Init. Repts. DSDP, v. 38, p. 1169–1193.	O	Q PL M O E	336 to 348		N	V
Kotova, I. Z., 1978, Spores and pollen from Cretaceous deposits of the eastern North Atlantic Ocean, Deep Sea Drilling Project, Leg 41, Sites 367 and 370: Init. Repts. DSDP, supplement to v. 38–41, p. 841–881.	O	C	367 370		D	R V
Krasheninnikov, V. A., 1971, Cenozoic Foraminifera: Init. Repts. DSDP, v. 6, p. 1055–1068. [Additional information in Site chapters.]	F	Q PL M O E PA	44 to 60		N	R V
———, 1973, Cretaceous benthonic Foraminifera, Leg 20, Deep Sea Drilling Project: Init. Repts. DSDP, v. 20, p. 205–219.	F	C	195 196 198	R N20	N	V
———, 1974, Upper Cretaceous benthonic agglutinated Foraminifera, Leg 27 of the Deep Sea Drilling Project: Init. Repts. DSDP, v. 27, p. 631–661.	F	C	260 261	R N17	U	V
———, 1974, Cretaceous and Paleogene planktonic Foraminifera, Leg 27 of the Deep Sea Drilling Project: Init. Repts. DSDP, v. 27, p. 663–761. [Additional information in Site chapters.]	F	E PA C	259 260 261 263	R	N	V
———, 1977, Znachenie okeanicheskikh otlozhenie dlya razrabotki stratigraficheskoe shkaly mezozoya i kainozoya (Indieskie Okean): Voprosy Micropaleontologii, v. 19, p. 124–227.	F	Q PL M,O,E PA C	Many			G V F R
———, 1979, Stratigraphy, by means of planktonic foraminifers, of Neogene and Quaternary sediments near the crest of the Mid-Atlantic Ridge, DSDP Sites 395 and 396: Init. Repts. DSDP, v. 45, p. 319–322.	F	Q P M	395 396		N	V
———, 1979, Stratigraphy, and planktonic foraminifers of Neogene and Quaternary sediments of Site 396, Leg 46 of DSDP: Init. Repts. DSDP, v. 46, p. 409–413.	F	Q PL M	396		U	V

Reference	Fossil groups.	Ages.	Sites.	Revisions. New taxa.	Abundance. / Preservation.	Groups or regions related. Vertical study. Finer stratigraphy.
———, and Hoskins, R. H., 1973, Late Cretaceous, Paleogene and Neogene planktonic Foraminifera: Init. Repts. DSDP, v. 20, p. 105–203.	F	Q PL M E PA C	194 to 202	R	U	V
———, and Pflaumann, U., 1978, Cretaceous agglutinated Foraminifera of the Atlantic Ocean off West Africa (Leg 41, Deep Sea Drilling Project): Init. Repts. DSDP, v. 41, p. 565–580.	F	C	367 368 370		N	V
———, and ———, 1978, Zonal stratigraphy and planktonic foraminifers of Paleogene deposits of the Atlantic Ocean to the west off Africa (Deep Sea Drilling Project, Leg 41): Init. Repts. DSDP, v. 41, p. 581–611. [Additional information in Site chapters.]	F	O E PA	366 to 370		N	V
———, and ———, 1978, Zonal stratigraphy of Neogene deposits of the eastern part of the Atlantic Ocean by means of planktonic foraminifers, Leg 41, Deep Sea Drilling Project: Init. Repts. DSDP, v. 41, p. 613–657. [Additional information in Site chapters.]	F	Q PL M	366 to 370		U	V
Kuznetsova, K. I., 1974, Distribution of benthonic Foraminifera in Upper Jurassic and Lower Cretaceous deposits at Site 261, DSDP Leg 27, in the eastern Indian Ocean: Init. Repts. DSDP, v. 27, p. 673–681.	F	C J	261		U	V
———, and Seibold, I., 1978, Foraminifers from the Upper Jurassic and Lower Cretaceous of the eastern Atlantic (DSDP Leg 41, Sites 367 and 370): Init. Repts. DSDP, v. 41, p. 515–537.	F	C J	367 370		U P	V
Labracherie, M., 1975, Description des Bryozoaires Cheilostomes d'âge Éocène inférieur du site 246 (croisière 25, Deep Sea Drilling Project): Bulletin de l'Institute de Géologie du Bassin d' Aquitaine, 1975, no. 18, p. 149–202, 16 pls.	O	E	246			V
———, and Sigal, J., 1975, Les Bryozoaires Cheilostomes des formations Eocène inférieur du Site 246 (croisière 25, Deep Sea Drilling Project): Docum. Lab. Geol. Fac. Sci. Lyon, H.S.3 (fasc. 2), p. 449–466.	O	E	246		D	V
Lehmann, R., 1972, Microfossils in thin sections from the Mesozoic deposits of Leg 11, Deep Sea Drilling Project: Init. Repts. DSDP, v. 11, p. 659–663.	O	C J	99 100 101 105		N	R
Ling, H. Y., 1972, Upper Cretaceous and Cenozoic silicoflagellates and ebridians: Bulletins of	S	Q	6	R	U	R V F

American Paleontology, v. 62, no. 273, p. 135–229.	O	PL M,O E,C	9 13			
———, 1973, Radiolaria: Leg 19 of the Deep Sea Drilling Project: Init. Repts. DSDP, v. 19, p. 777–797.	R	Q PL M	183 to 193		U	V
———, 1973, Silicoflagellates and ebridians from Leg 19: Init. Repts. DSDP, v. 19, p. 751–775.	S O	Q PL M	183 to 193		U	V F
———, 1975, Radiolaria: Leg 31 of the Deep Sea Drilling Project: Init. Repts. DSDP, v. 31, p. 703–761.	R	Q PL M O E	290 to 302		U P	V
———, 1975, Silicoflagellates and ebridians from Leg 31: Init. Repts. DSDP, v. 31, p. 763–777.	S O	Q PL M	290 to 302		U	R V
———, 1976, Distribution and biostratigraphic significance of *Dictyocha subarctios* (silicoflagellate) in the North Pacific: Trans. Proc. Palaeont. Soc. Japan, N. S., no. 101, p. 264–270.	S	Q PL	173 178		N	G V F
———, 1977, Late Cenozoic silicoflagellates and ebridians from the eastern North Pacific region: Proceedings of the First International Congress on Pacific Neogene Stratigraphy, Tokyo 1976, p. 205–322.	S O	Q PL M	33,34 71,77 78,83 54,158 173 179	R N4 (3s; 1o)	U	G V F
Longoria, J. F., 1975, Shore laboratory report on Mesozoic Foraminiferida, Leg 32: Init. Repts. DSDP, v. 32, p. 963–971.	F	C	305 310 313		N	V
Luterbacher, H. P., 1972, Foraminifera from the Lower Cretaceous and Upper Jurassic of the Northwestern Atlantic: Preliminary Note: Init. Repts. DSDP, v. 11, p. 561–593.	F	C J	99 100 101 105		U	R V
———, 1972, Paleocene and Eocene planktonic Foraminifera, Leg 11, DSDP: Init. Repts. DSDP, v. 11, p. 547–550.	F	E PA	98 108		N P	V
———, 1975, Early Cretaceous Foraminifera from the northwestern Pacific: Leg 32 of the Deep Sea Drilling Project: Init. Repts. DSDP, v. 32, p. 703–718. [Additional information in Site chapters.]	F	C	303 to 313		U P	V
———, 1975, Paleocene and early Eocene planktonic Foraminifera, Leg 32, Deep Sea Drilling Project: Init. Repts. DSDP, v. 32, p. 725–733. [Additional information in Site chapters.]	F	E PA	305 308 313		U P	V
———, Caron, M., Fleisher, R. L., Foreman, H. P., Toumarkine, M., and Vincent, E., 1975,	F	Q	303			G V

Reference	Fossil groups.	Ages.	Sites.	Revisions. New taxa.	Abundance. / Preservation.	Groups or regions related. Vertical study. Finer stratigraphy.
Biostratigraphic summary of Leg 32: Init. Repts. DSDP, v. 32, p. 941–944.	N R	PL M O E C	to 303			
Lutze, G. F., 1978, Neogene benthonic Foraminifera from Site 369, Leg 41, Deep Sea Drilling Project: Init. Repts. DSDP, v. 41, p. 659–666.	F	Q PL M	369		D	V
Manum, S. B., 1976, Dinocysts in Tertiary Norwegian-Greenland Sea sediments (Deep Sea Drilling Project Leg 38), with observations on palynomorphs and palynodebris in relation to environment: Init. Repts. DSDP, v. 38, p. 897–919.	O	M O E	336 to 352		N	R V F
Marks, P., 1972, Remarks on *Globotruncana calcarata* Cushman: Proceedings Koninkl. Nederl. Akademie van Wetenschappen Amsterdam, Series B, v. 75, no. 5, p. 408–414.	F	C	10	R		
Martin, S. E., 1972, Reexamination of the Upper Cretaceous planktonic foraminiferal genera *Planoglobulina* Cushman and *Ventilabrella* Cushman: Jour. Foraminiferal Research, v. 2, no. 2, p. 73–92.	F	C	4 21 111	R N4		
Martini, E., 1971, Neogene silicoflagellates from the equatorial Pacific: Init. Repts. DSDP, v. 7, p. 1695–1708.	S	Q PL M	65 66		U	G V F R
———, 1971, Standard Tertiary and Quaternary calcareous nannoplankton zonation: Proceedings of the II Planktonic Conference, Roma 1970, p. 739–785.	N	Q PL M E,O PA	44 62 63 64 65			R V F
———, 1974, Quaternary fish otoliths from Sites 242 and 246 – Leg 25, Deep Sea Drilling Project: Init. Repts. DSDP, v. 25, p. 647–655.	O	Q	242 246		D	
———, 1976, Cretaceous to Recent calcareous nannoplankton from the central Pacific Ocean (DSDP Leg 33): Init. Repts. DSDP, v. 33, p. 383–423. [Additional information in Site chapters.]	N	Q PL M O E	314 to 318	N3	N P	V

		PA C				
———, 1976, Neogene and Quaternary silicoflagellates from the central Pacific Ocean (DSDP Leg 33): Init. Repts. DSDP, v. 33, p. 439–449.	S	Q PL M	315 316	R	U	R V G
———, and Müller, C., 1976, Eocene to Pleistocene silicoflagellates from the Norwegian-Greenland Sea (DSDP Leg 38): Init. Repts. DSDP, v. 38, p. 857–895.	S	Q PL M O E	336 to 352	R N4	U	G V F
———, and Worsley, T., 1971, Tertiary calcareous nannoplankton from the western equatorial Pacific: Init. Repts. DSDP, v. 7, p. 1471–1507. [Additional information in Site chapters.]	N	Q PL M O E	62 to 66	N2	D P	V F
Matsumoto, T., 1978, Notes on *Inoceramus,* Mesozoic bivalves from the southeastern Atlantic, DSDP Sites 361 and 364, Leg 40: Init. Repts. DSDP, supplement to v. 38–41, p. 703–707.	O	C	361 364			V
Maync, W., 1973, Lower Cretaceous foraminiferal fauna from Gorringe Bank, eastern North Atlantic: Init. Repts. DSDP, v. 13, p. 1075–1111.	F	C	120	R N4	N	R V
———, 1973, Lower Cretaceous limestones from the Hellenic Trough, Ionian Basin, (Site 127): Init. Repts. DSDP, v. 13, p. 1112–1135.	F O	C	127			R
McCollum, D. W., 1974, Neogene diatom biostratigraphy of the southern ocean: Antarctic Journal of the United States, v. 9, no. 5, p. 273–274.	D	Q PL M	264 to 274		N	V F
———, 1975, Diatom stratigraphy of the Southern Ocean: Init. Repts. DSDP, v. 28, p. 515–571. [Additional information in Site chapters.]	D	Q PL M O	264 to 274	N3	D	R V F
McGowran, B., 1974, Foraminifera: Init. Repts. DSDP, v. 22, p. 609–627. [Additional information in Site chapters.]	F	Q PL M,O E PA C	211 to 218		N P	V
———, 1977, Maastrichtian to Eocene foraminiferal assemblages in the northern and eastern Indian Ocean region: Correlations and historical patterns: *in* Heirtzler, and others, eds., Indian Ocean Geology and Biostratigraphy: Washington, D.C., American Geophysical Union, p. 417–458.	F	E PA C	211 to 269 280 to 282		Some P	
McLachlan, I. R., and Pieterse, E., 1978, Preliminary palynological results: Site 361, Leg 40, Deep Sea Drilling Project: Init. Repts. DSDP, v. 40, p. 857–881.	O	C	361		D	V

Reference	Fossil groups.	Ages.	Sites.	Revisions. New taxa.	Abundance.	Preservation.	Groups or regions related.	Vertical study.	Finer stratigraphy.
McNulty, C. L., 1976, Cretaceous foraminiferal stratigraphy, DSDP Leg 33, Holes 315A, 316, 317A: Init. Repts. DSDP, v. 33, p. 369–381. [Additional information in Site chapters.]	F	C	315 316 317		N	P		V	
Michael, F. Y., 1975, Mesozoic Foraminifera, Leg 30, Hole 288A and Site 289: Init. Repts. DSDP, v. 30, p. 599–601.	F	C	288 289		U			V	
Miles, G. A., 1977, Planktonic Foraminifera from Leg 37 of the Deep Sea Drilling Project: Init. Repts. DSDP, v. 37, p. 929–961. [Additional information in Site chapters.]	F	Q PL M	332 to 335		N	P		V	
———, and Howe, R. C., 1977, Biostratigraphic summary, Leg 37, Deep Sea Drilling Project: Init. Repts. DSDP, v. 37, p. 977–983.	F N D S	Q PL M	332 to 335				G	V	
Moore, T., 1971, Radiolaria: Init. Repts. DSDP, v. 8, p. 727–775. [Additional information in Site chapters.]	R	Q PL M O E	68 to 75	N8	U	P		V	F
———, 1972, Mid-Tertiary evolution of the radiolarian genus *Calocycletta:* Micropaleontology, v. 18, no. 2, p. 144–152.	R	M O	69 to 74	R N3				V	
———, Jr., 1973, Radiolaria from Leg 17 of the Deep Sea Drilling Project: Init. Repts. DSDP, v. 17, p. 797–869.	R	Q PL M O E,C	164 to 171	R	U	P		V	F
Morgan, R., 1978, Albian to Senonian palynology of Site 364, Angola Basin: Init. Repts. DSDP, v. 40, p. 915–951.	O	C	364		N			V	
Müller, C., 1974, Calcareous nannoplankton, Leg 25 (western Indian Ocean): Init. Repts. DSDP, v. 25, p. 579–633. [Additional information in Site chapters.]	N	Q PL M,O E	239 to 249	R N11	N	P		V	

———, 1976, Tertiary and Quaternary calcareous nannoplankton in the Norwegian-Greenland Sea, DSDP, Leg 38: Init. Repts. DSDP, v. 38, p. 823–841.	N	PA C Q PL M O E	336 to 352		U P	R V
———, 1977, Distribution of calcareous nannoplankton in Oligocene to Holocene sediments of the Red Sea and the Indian Ocean reflecting paleoenvironment: *in* Heirtzler, and others, eds., Indian Ocean Geology and Biostratigraphy: Washington, D.C., American Geophysical Union, p. 371–397.	N	Q PL M O	211 to 269 280 to 282			
Musich, L. F., 1973, Pollen occurrence in eastern North Pacific sediments, Deep Sea Drilling Project Leg 18: Init. Repts. DSDP, v. 18, p. 799–815.	O	Q PL M	172 to 182		D	R V
Nigrini, C., 1974, Cenozoic Radiolaria from the Arabian Sea, DSDP Leg 23: Init. Repts. DSDP, v. 26, p. 1051–1121.	R	Q PL M O E	219 to 227	R N2	U P	V
———, 1977, Tropical Cenozoic Artostrobiidae (Radiolaria): Micropaleontology, v. 23, no. 3, p. 241–269.	R	Q,PL, M,O,E	77 94	R N5	N	V F
———, Fleisher, R. L., and Boudreaux, J. E., 1974, Leg 23 biostratigraphic synthesis: Lower Eocene to Pleistocene of the Arabian Sea: Init. Repts. DSDP, v. 23, p. 1095–1113.	R N R	Q PL M O E	219 220			G V F
Oertli, H. J., 1972, Jurassic ostracodes of DSDP Leg 11 (Sites 100 and 105)—Preliminary account: Init. Repts. DSDP, v. 11, p. 645–657.	O	J	100 105			R V
———, 1974, Lower Cretaceous and Jurassic ostracods from DSDP Leg 27—a preliminary account: Init. Repts. DSDP, v. 27, p. 947–965.	O	C J	259 260 263		D	V
Olsson, R. K., 1971, Pliocene-Pleistocene planktonic foraminiferal biostratigraphy of the northeastern Pacific: Proceedings of the II Planktonic Conference, Roma 1970, p. 921–926.	F	Q PL	36		N	V F
———, 1973, Growth studies on *Globorotalia exilis* Blow and *Globorotalia pertenuis* Beard in the Hole 154A section, Leg 15, Deep Sea Drilling Project: Init. Repts. DSDP, v. 15, p. 617–624.	F	PL	154			V
———, 1974, Pleistocene paleoceanography and *Globigerina pachyderma* (Ehrenberg) in Site 36, DSDP, northeastern Pacific: Jour. Foraminiferal Research, v. 4, no. 2, p. 47–60.	F	Q	36		D P	V
———, 1974, Shore laboratory report on the Foraminifera from Leg 21 sites, Deep Sea Drilling Project: Init. Repts. DSDP, v. 22, p. 865–881.	F	Q PL	203 to		D	V

Reference	Fossil groups.	Ages.	Sites.	Revisions. New taxa.	Abundance. / Preservation.	Groups or regions related. Vertical study. Finer stratigraphy.
		M,O E PA	210			
———, Shore laboratory report on the Foraminifera from Leg 27 sites, Deep Sea Drilling Project: Init. Repts. DSDP, v. 27, p. 967–975.	F	Q PL	259 to 263		D P	V
———, and Goll, R., 1970, Biostratigraphy: Init. Repts. DSDP, v. 5, p. 557–567. [Additional information in Site chapters.]	F R N	Q PL M,O E PA	32 to 43		N	G V F
Parker, F. L., 1973, Late Cenozoic biostratigraphy (planktonic Foraminifera) of tropical Atlantic deep-sea sections: Revista Española de Micropaleontología, v. 5, no. 2, p. 253–289.	F	Q PL M	2,3 12 25 29,30 31	R	U	R V F
———, 1974, Upper Neogene biostratigraphy (planktonic Foraminifera) of DSDP Sites 139 and 141: Jour. Foraminiferal Research, v. 4, no. 1, p. 9–15, 1 pl.	F	PL M	139 141		N	R V F
Partridge, A. D., 1978, Palynology of the Late Tertiary sequence at Site 365, Leg 40, Deep Sea Drilling Project: Init. Repts. DSDP, v. 40, p. 953–961.	O	PL M	365		N	V
Perch-Nielsen, K., 1971, Einige neue Coccolithen aus dem Paleozän der Bucht von Biskaya: Bulletin of the Geographical Society of Denmark, v. 20, p. 347–361, 14 pls.	N	PA	119	R N14	N	V F
———, 1971, Neue Coccolithen aus dem Paleozän von Dänemark, der Bucht von Biskaya und dem Eozän der Labrador See: Bulletin of the Geological Society of Denmark, v. 21, p. 51–66, 7 pls.	N	E PA	111 119	N3	N	V F
———, 1972, Remarks on Late Cretaceous to Pleistocene coccoliths from the North Atlantic: Init. Repts. DSDP, v. 12, p. 1003–1069.	N	Q PL M,O E PA C	111 to 119		N	R V

———, 1975, Late Cretaceous to Pleistocene archaeomonads, ebridians, endoskeletal dinoflagellates, and other siliceous microfossils from the subantarctic Southwest Pacific, DSDP, Leg 29: Init. Repts. DSDP, v. 29, p. 873–907.	O	Q PL M O E C	275 to 284	N10	U	V
———, 1975, Late Cretaceous to Pleistocene silicoflagellates from the southern Southwest Pacific: DSDP, Leg 29: Init. Repts. DSDP, v. 29, p. 677–721. [Additional information in Site chapters.]	S	Q PL M O E C	275 to 284	R N2	D	R V F
———, 1977, Albian to Pleistocene calcareous nannofossils from the western South Atlantic, DSDP Leg 39: Init. Repts. DSDP, v. 39, p. 699–823. [Additional information in Site chapters.]	N	Q PL M,O E PA C	353 to 359	N10	U P	V F
———, 1977, Synthesis of the Leg 39 biostratigraphy (Central and South Atlantic): Init. Repts. DSDP, v. 39, p. 1081–1098.	F N R	Q PL M,O E PA C	353 to 359			G V
———, 1977, Tertiary silicoflagellates and other siliceous microfossils from the western South Atlantic, Deep Sea Drilling Project, Leg 39: Init. Repts. DSDP, v. 39, p. 863–867. [Additional information in Site chapters.]	O S	Q PL M O E	353 to 359		U	V
———, 1978, Eocene to Pliocene archaeomonads, ebridians and endoskeletal dinoflagellates from the Norwegian Sea, DSDP Leg 38: Init. Repts. DSDP, supplement to v. 38–41, p. 147–175.	O	PL M O E	338 to 348	N10	N	V
———, and Franz, H. E., 1977, *Lapideacassis* and *Scampanella,* calcareous nannofossils from the Paleocene at Sites 354 and 356, DSDP Leg 39, southern Atlantic: Init. Repts. DSDP, v. 39, p. 849–862.	N	PA	354 356	R N7	N	V
Percival, S. F., and Saito, T., 1970, [Information on calcareous nannofossils and foraminifera, in each Site chapter.]: Init. Repts. DSDP, v. 3.	F N	Q PL M O E PA	13 to 22		N	V

Reference	Fossil groups.	Ages.	Sites.	Revisions. New taxa.	Abundance. / Preservation.	Groups or regions related. Vertical study. Finer stratigraphy.
Pessagno, E. A., Jr., 1971, Jurassic and Cretaceous Hagiastridae from the Blake-Bahama Basin (Site 5A, JOIDES Leg 1) and the Great Valley sequence, California Coast Ranges: Bulletins of American Paleontology, v. 60, no. 264, p. 1–83.	R	J	5	N4	N	V F
———, 1972, Cretaceous Radiolaria. Part II. *Pseudoaulophacidae* Riedel from the Cretaceous of California and the Blake-Bahama Basin (JOIDES Leg 1): Bulletins of American Paleontology, v. 61, no. 270, p. 281–328.	R	C	4 5	R N1	U	R V F
———, 1975, Upper Cretaceous Radiolaria from DSDP Site 275: Init. Repts. DSDP, v. 29, p. 1011–1029.	R	C	275	N8	D	R V
———, and Longoria T., J. F., 1972, Shore laboratory report on Mesozoic planktonic Foraminiferida: Init. Repts. DSDP, v. 12, p. 119–120.	F	C	111		N	R
———, and ———, 1973, Mesozoic Foraminifera, Leg 15, Deep Sea Drilling Project: Init. Repts. DSDP, v. 15, p. 549–552.	F R	C	146 to 153		N	V
———, and ———, 1973, Shore laboratory report on Mesozoic planktonic Foraminifera—DSDP Leg 16: Init. Repts. DSDP, v. 16, p. 893.	F	C	163		N	
———, and ———, 1973, Shore laboratory report on Mesozoic Foraminiferida, Leg 17: Init. Repts. DSDP, v. 17, p. 891–894.	F	C	165 167 171		N	V
———, and Michael, F. Y., 1974, Mesozoic Foraminifera, Leg 22, Site 217: Init. Repts. DSDP, v. 22, p. 629–634.	F	C	217		U	V
———, and ———, 1974, Results of shore laboratory studies on Mesozoic planktonic Foraminifera from Leg 26, Sites 255, 256, 257, and 258: Init. Repts. DSDP, v. 26, p. 969–972.	F R	C	255 256 257 258		D	V
Petrushevskaya, M. G., 1975, Cenozoic radiolarians of the Antarctic, Leg 29, DSDP: Init. Repts. DSDP, v. 29, p. 541–675.	R	Q PL M O E	275 to 284	R N28	U	V F
———, and Kozlova, G. E., 1972, Radiolaria: Leg 14, Deep Sea Drilling Project: Init. Repts.	R	PL	135	R	U P	V

DSDP, v. 14, p. 495–648.		M O E PA C	to 144	N14		
Plaumann, U., and Krasheninnikov, V. A., 1978, Cretaceous calcisphaerulids from DSDP Leg 41, eastern North Atlantic: Init. Repts. DSDP, supplement to v. 38–41, p. 817–839.	O	C	369 370	N10	N	V
———, and ———, 1978, Early Cretaceous planktonic foraminifers from eastern North Atlantic, DSDP, Leg 41: Init. Repts. DSDP, v. 41, p. 539–564. [Additional information in Site chapters.]	F	C	367 to 370		U	R V
———, and ———, 1978, Quaternary stratigraphy and planktonic foraminifers of the eastern Atlantic, Deep Sea Drilling Project, Leg 41: Init. Repts. DSDP, Supplement to v. 38–41, p. 883–911.	F	Q	366 to 370	R	U	V
Pisias, N. G., and Moore, T. C, Jr., 1978, Cenozoic radiolarians from Deep Sea Drilling Project Leg 40: Init. Repts. DSDP, v. 40, p. 845–856.	R	PL M	360 to 365		U P	V
Poag, C. W., 1972, Neogene planktonic foraminiferal biostratigraphy of the western North Atlantic: DSDP Leg 11: Init. Repts. DSDP, v. 11, p. 483–543.	F	Q PL M O	98 102 103 104 106	R	U	R V
Pocock, S. A. J., 1972, Jurassic palynomorphs from 12-111-7-1, 4-5cms and 66–67 cms: Init. Repts. DSDP, v. 12, p. 66–67.	O	J	111		N	R
Poore, R. Z., and Berggren, W. A., 1974, Pliocene biostratigraphy of the Labrador Sea: calcareous plankton: Jour. Foraminiferal Research, v. 4, no. 3, p. 91–108.	F N	Q PL	111 112 113	R	U	G V F R
———, and ———, 1975, Late Cenozoic planktonic foraminiferal biostratigraphy and paleoclimatology of Hatton-Rockall Basin: DSDP Site 116: Jour. Foraminiferal Research, v. 5, no. 4, p. 270–293.	F	Q PL M	116			R V F
Premoli Silva, I., and Boersma, A., 1977, Cretaceous planktonic foraminifers—DSDP Leg 39 (South Atlantic): Init. Repts. DSDP, v. 39, p. 615–641. [Additional information in Site chapters.]	F	C	353 to 359		U P	V
———, and Bolli, H. M., 1973, Late Cretaceous to Eocene planktonic Foraminifera and stratigraphy of Leg 15 sites in the Caribbean Sea: Init. Repts. DSDP, v. 15, p. 499–547.	F	E PA C	146 to 154	N1	N P	R V
Proto Decima, F., 1974, Leg 27 calcareous nannoplankton: Init. Repts. DSDP, v. 27, p. 589–621. [Additional information in Site chapters.]	N	Q PL E PA C,J	259 to 263		N P	V
———, 1977, Paleocene to Eocene calcareous nannoplankton of the Indian Ocean, *in* Heirtzler,	N	E	211			

Reference	Fossil groups.	Ages.	Sites.	Revisions. New taxa.	Abundance. / Preservation.	Groups or regions related. Vertical study. Finer stratigraphy.
and others, eds., Indian Ocean Geology and Biostratigraphy: Washington, D.C., American Geophysical Union, p. 353–369.		PA	to 269 280 to 282			
———, and Bolli, H. M., 1978, Southeast Atlantic DSDP Leg 40 Paleogene benthic foraminifers: Init. Repts. DSDP, v. 40, p. 783–809.	F	O E PA	360 to 364	N2	N	V
———, Medizza, F., and Todesco, L., 1978, Southeastern Atlantic, Leg 40, calcareous nannofossils: Init. Repts. DSDP, v. 40, p. 571–634. [Additional information in Site chapters.]	N	Q PL M O E PA C	360 to 365		N P	V
Quilty, P. G., 1976, Planktonic Foraminifera, DSDP Leg 34—Nazca Plate: Init. Repts. DSDP, v. 34, p. 629–703.	F	Q PL M O E	319 to 321	R N3	D	V
Ramsey, C. A., Doyle, P. S., and Riedel, W. R., 1976, Ichthyoliths in Late Mesozoic pelagic sediments, mainly from Italy: Micropaleontology, v. 22, no. 2, p. 129–142.	O	C	196	R N1	D	R V F
Renz, G. W., 1974, Radiolaria from Leg 27 of the Deep Sea Drilling Project: Init. Repts. DSDP, v. 27, p. 769–841.	R	Q C	259 to 263	R N15	U P	V
Renz, O., 1972, Aptychi (Ammonoidea) from the Upper Jurassic and Lower Cretaceous of the western North Atlantic (Site 105, Leg 11, DSDP): Init. Repts. DSDP, v. 11, p. 607–629.	O	C J	105	N1	P	R V
———, 1973, Two lamellaptychi (Ammonoidea) from the Magellan Rise in the Central Pacific: Init. Repts. DSDP, v. 17, p. 895–901.	O	J?	167		D	
———, 1974, A dwarf fauna of Upper Jurassic age in the Western Atlantic. Deep Sea Drilling Project, Leg 11—Site 99A: Verhand. Naturf. Ges. Basel, v. 84, no. 1, p. 510–520. [Ammo-	O	J	99			V

nites, gastropod, and incertae sedis.]						
———, 1978, Aptychi (Ammonoidea) from the Early Cretaceous of the Blake-Bahama Basin, Leg 44, Hole 391C, DSDP: Init. Repts. DSDP, v. 44, p. 899–909.	O	C	391	N8		V
———, 1978, Aptychi (Ammonoidea) from the Late Jurassic and Early Cretaceous of the eastern Atlantic, DSDP Site 367: Init. Repts. DSDP, v. 41, p. 499–513.	O	C J	367			R V
Resig, J. M., 1976, Benthic foraminiferal stratigraphy, eastern margin, Nazca Plate: Init. Repts. DSDP, v. 34, p. 743–759.	F	Q M O E	320 321		D	V
Reynolds, R. A., 1978, Cosmopolitan biozonation for Late Cenozoic radiolarians and paleoceanography from Deep Sea Drilling Project Core 77B of Leg 9: Transactions Gulf Coast Association of Geological Societies, v. 28, p. 423–431.	R	PL M O E	77	R	D	V F
Riedel, W. R., 1971, Radiolarians from Atlantic deep-sea drilling: Proceedings of the II Planktonic Conference, Roma 1970, p. 1957–1068.	R	M O E	6,7,8 9,10 13,17 22,27 30–31		U P	G V F
———, and Hays, J. D., 1969, Cenozoic Radiolaria from Leg 1: Init. Repts. DSDP, v. 1, p. 400–402.	R	E	3 6 7		U	V
———, and Sanfilippo, A., 1970, Radiolaria, Leg 4, Deep Sea Drilling Project: Init. Repts. DSDP, v. 4, p. 503–575. [Figure 4 omitted from volume, but printed and distributed later.]	R	M O E C	24 27 to 31	R N17	U	G V F
———, and ———, 1971, Cenozoic Radiolaria from the Western tropical Pacific, Leg 7: Init. Repts. DSDP, v. 7, p. 1529–1672.	R	Q PL M O E	61 to 67	R N17	U P	R V F
———, and ———, 1973, Cenozoic Radiolaria from the Caribbean, Deep Sea Drilling Project, Leg 15: Init. Repts. DSDP, v. 15, p. 705–751.	R	M O E PA	146 to 153		U P	V
———, and ———, 1974, Radiolaria from the southern Indian Ocean, DSDP Leg 26: Init. Repts. DSDP, v. 26, p. 771–813.	R	PL M E C	250 to 258	R N8	U P	V F
———, and ———, 1978, Stratigraphy and evolution of tropical Cenozoic radiolarians: Micropaleontology, v. 24, no. 1, p. 61–96.	R	Q PL M	40 162	R	U	V F

Reference	Fossil groups.	Ages.	Sites.	Revisions. New taxa.	Abundance. / Preservation.	Groups or regions related. Vertical study. Finer stratigraphy.
———, and ———, and Cita, M. B., 1974, Radiolarians from the stratotype Zanclean (Lower Pliocene, Sicily): Riv. Ital. Paleont., v. 80, no. 4, p. 699–734.	R	O,E Q PL M	77 238	R	U	R V F
Rögl, F., 1974, The evolution of the *Globorotalia truncatulinoides* and *Globorotalia crassaformis* group in the Pliocene and Pleistocene of the Timor Trough, DSDP Leg 27, Site 262: Init. Repts. DSDP, v. 27, p. 743–767.	F	Q PL	262	R	N	V
———, 1976, Danian Calcisphaerulidae of DSDP Leg 35, Site 323, Southeast Pacific Ocean: Init. Repts. DSDP, v. 35, p. 701–711.	O	PA	323	N4		
———, 1976, Late Cretaceous to Plesitocene Foraminifera from the Southeast Pacific Basin, DSDP Leg 35: Init. Repts. DSDP, v. 35, p. 539–555. [Additional information in Site chapters.]	F	Q PL M PA C	322 to 325		N	V
———, and Bolli, H. M., 1973, Holocene to Pleistocene planktonic Foraminifera of Leg 15, Site 147 (Cariaco Basin [Trench], Caribbean Sea): Init. Repts. DSDP, v. 15, p. 553–615.	F	Q	147	R N4	U	V
Rossignol-Strick, M., 1973, Pollen analysis of some sapropel layers from the deep-sea floor of the eastern Mediterranean: Init. Repts. DSDP, v. 13, p. 971–991.	O	Q	125 128 130 131		D	R V
———, 1974, Tectonique de plaques et sédimentation palynologique marine sur la Ride méditerranéenne: Sci. Géol. (Strasbourg), Bull. 27, no. 1–2, p. 135–145.	O	Q	130		D	V
Roth, P. H., 1973, Calcareous nannofossils—Leg 17, Deep Sea Drilling Project: Init. Repts. DSDP, v. 17, p. 695–795.	N	Q PL M,O E PA C	164 to 171	R N6	U P	V
———, 1974, Calcareous nannofossils from the northwestern Indian Ocean, Leg 24, Deep Sea Drilling Project: Init. Repts. DSDP, v. 24, p. 969–994.	N	Q PL M,O	231 to 238		U P	V

		E PA				
———, 1978, Cretaceous nannoplankton biostratigraphy and oceanography of the northwestern Atlantic Ocean: Init. Repts. DSDP, v. 44, p 731–759.	N	C	390 391	R N3	N P	R V F
———, and Thierstein, H., 1972, Calcareous nannoplankton: Leg 14 of the Deep Sea Drilling Project: Init. Repts. DSDP, v. 14, p. 421–485. [Additional information in Site chapters.]	N	Q PL M,O E PA C	135 to 144	R N7	U P	G V
Sachs, K. N., Jr., 1969, Report on larger Foraminifera from Sites 4 and 5: Init. Repts. DSDP, v. 1, p. 398–399.	F	M O E C	4 5		N	V
Sachs, H. M., 1976, Nazca Plate radiolarian distributions: evidence from DSDP Sites 319, 320 and 321: Init. Repts. DSDP, v. 34, p. 713–714. [Additional information in Site chapters.]	R	Q M O E	319 to 321		N P	V
Samtleben, C., 1978, Pliocene-Pleistocene coccolith assemblages from the Sierra Leone Rise—Site 366, Leg 41: Init. Repts. DSDP, supplement to v. 38–41, p. 913–931.	N	Q PL	366		D P	V
Sancetta, C., 1978, Neogene Pacific microfossils and paleoceanography: Marine Micropaleontology, v. 3, p. 347–376.	F R N,D	Q M	Many			G R
Sanfilippo, A., Burckle, L. H., Martini, E., and Riedel, W. R., 1973, Radiolarians, diatoms, silicoflagellates and calcareous nannofossils in the Mediterranean Neogene: Micropaleontology, v. 19, p. 209–234.	R	P M	64 66	R N2	U	G V R
———, and Riedel, W. R., 1973, Cenozoic Radiolaria (exclusive of theoperids, artostrobiids and amphipyndacids) from the Gulf of Mexico, Deep Sea Drilling Project Leg 10: Init. Repts. DSDP, v. 10, p. 475–611.	R	M O E PA	86 94 95 96	R N35	U P	V F
———, and ———, 1974, Radiolaria from the west-central Indian Ocean and Gulf of Aden, DSDP Leg 24: Init. Repts. DSDP, v. 24, p. 997–1035.	R	Q PL M,O E PA	231 to 238 73 77	N4	U P	R V
———, and ———, 1974, Some radiolarian samples from the western Indian Ocean, DSDP Leg 25: Init. Repts. DSDP, v. 25, p. 657–661.	R	Q M E C	239 to 249		U P	V
———, and ———, 1975, Late Tertiary radiolarians from Crete: VI Congress Regional Committee on Mediterranean Neogene Stratigraphy Bratislava 1975, p. 61–74.	R	Q PL M	82 84	N1	U	G V F R

Reference	Fossil groups.	Ages.	Sites.	Revisions. New taxa.	Abundance. / Preservation.	Groups or regions related. Vertical study. Finer stratigraphy.
———, and ———, 1976, Radiolarian occurrences in the Caribbean region: Publication de la VII Conférence géologique des Caraïbes du 30 juin au 12 juillet 1974, p. 145–168.	R	M O E C	27,29 30,31 146 149-153		U P	G V
Saunders, J. B., Beaudry, F. M., Hay, W. W., Bolli, H. M., Rögl, F., Riedel, W. R., Sanfilippo, A., Premoli Silva, I., 1973, Paleocene to Recent planktonic microfossil distribution in the marine and land areas of the Caribbean: Init. Repts. DSDP, v. 15, p. 769–771, 3 charts.	F N R	Q PL M O E PA	146 to 154			G V F R
Scheibnerová, V., 1973, Non-tropical Cretaceous foraminifera in Atlantic deep-sea cores and their implications for continental drift and palaeooceanography of the South Atlantic Ocean: Records of the Geological Survey of New South Wales, v. 15, pt. 1, p. 19–46, 8 pls.	F	C	9 13		N	R
———, 1974, Aptian-Albian benthonic Foraminifera from DSDP Leg 27, Sites 259, 260, and 263, eastern Indian Ocean: Init. Repts. DSDP, v. 27, p. 697–741.	F	C	259 260 263	R N2	N	R V
———, 1977, Synthesis of the Cretaceous benthonic Foraminifera recovered by the Deep Sea Drilling Project in the Indian Ocean, *in* Heirtzler, J.R., and others, eds., Indian Ocean Geology and Biostratigraphy: Washington, D.C., American Geophysical Union, p. 585–597.	F	C	211 to 269 280 to 282			
———, 1978, Aptian and Albian benthic foraminifers of Leg 40, Sites 363 and 364, southern Atlantic: Init. Repts. DSDP, v. 40, p. 741–757.	F	C	363 364	R	N	R V
Schmidt, R. R., 1978, Calcareous nannoplankton from the western North Atlantic, DSDP Leg 44: Init. Repts. DSDP, v. 44, p. 703–729. [Additional information in Site chapters.]	N	Q M E PA C J	388 to 394	R	U P	V

Schrader, H.-J., 1973, Cenozoic diatoms from the northeast Pacific, Leg 18: Init. Repts. DSDP, v. 18, p. 673–797.	D	Q PL M	172 to 182	R N32	U P	R V F
———, 1973, Stratigraphic distribution of marine species of the diatom *Denticula* in Neogene North Pacific sediments: Micropaleontology, v. 19, no. 4, p. 417–430.	D	Q PL M	173 178	R N1	N	V F
———, 1974, Cenozoic marine planktonic diatom stratigraphy of the tropical Indian Ocean: Init. Repts. DSDP, v. 24, p. 887–967.	D	Q PL M	213 215 236 238	R N10	U P	R V F
———, 1976, Cenozoic planktonic diatom biostratigraphy of the Southern Pacific Ocean: Init. Repts. DSDP, v. 35, p. 605–671.	D	Q PL M O	322 to 325	N24	U	V F
———, 1976, Marine diatom biostratigraphy of selected samples from the equatorial Pacific, DSDP Leg 33: Init. Repts. DSDP, v. 33, p. 491–492.	D	Q M	315 316		U	V
———, 1977, Diatom biostratigraphy, Deep Sea Drilling Project, Leg 37: Init. Repts. DSDP, v. 37, p. 967–975.	D	Q PL M	332 to 335		U P	V
———, 1978, Diatoms in DSDP Leg 41 sites: Init. Repts. DSDP, v. 41, p. 791–812.	D	Q PL M,O E,C	366 to 369		U P	V
———, 1978, Opal phytoplankton remains at DSDP Leg 40 sites: Init. Repts. DSDP, supplement to v. 38–41, p. 735–741.	D	Q PL M	362 365		U P	V
———, Bjørklund, K., Manum, S., Martini, E., and van Hinte, J., 1976, Cenozoic biostratigraphy, physical stratigraphy and paleooceanography in the Norwegian-Greenland Sea, DSDP Leg 38 paleontological synthesis: Init. Repts. DSDP, v. 38, p. 1197–1211.	F N R S D O	Q PL M O E	336 to 352			G V
———, and Fenner, J., 1976, Norwegian Sea Cenozoic diatom biostratigraphy and taxonomy, Part I: Norwegian Sea Cenozoic diatom biostratigraphy: Init. Repts. DSDP, v. 38, p. 921–962.	D	Q PL M O E	336 to 352		D	R V F
———, and ———, 1976, Norwegian Sea Cenozoic diatom biostratigraphy and taxonomy, Part II: Diatoms at Leg 38, taxonomic references: Init. Repts. DSDP, v. 38, p. 963–1099.	D	Q PL M O E	336 to 352	R N66		

Reference	Fossil groups.	Ages.	Sites.	Revisions. New taxa.	Abundance.	Preservation.	Groups or regions related. Vertical study. Finer stratigraphy.
Shafik, S., 1975, Nannofossil biostratigraphy of the Southwest Pacific, Deep Sea Drilling Project, Leg 30: Init. Repts. DSDP, v. 30, p. 549–598. [Additional information in Site chapters.]	N	Q PL M,O E PA C	285 to 289		N		V F
Sliter, W. V., 1977, Cretaceous benthic foraminifers from the western South Atlantic, Leg 39, Deep Sea Drilling Project: Init. Repts. DSDP, v. 39, p. 657–697.	F	C	355 to 358		D		V
———, 1977, Cretaceous foraminifers from the southwestern Atlantic Ocean, Leg 36, Deep Sea Drilling Project: Init. Repts. DSDP, v. 36, p. 519–573.	F	C	327 330	N1	U	P	R V
Smali, M., 1973, Foraminiferi campaniano-maastrichtiani della Dorsale Medio-Atlantica: Riv., Ital. Paleont., v. 79, no. 1, p. 35–101.	F	C	10	R	U	P	V F
Smith, L. A., and Beard, J. H., 1973, The Late Neogene of the Gulf of Mexico: Init. Repts. DSDP, v. 10, p. 643–677.	F N	Q PL M	85 to 97		N		G V F R
———, and McNeely, B., 1973, Summary of Leg 10, Biostratigraphy: Init. Repts. DSDP, v. 10, p. 731–736.	F R N	Q PL M,O E PA C	85 to 97				G V
Speden, I. G., 1974, Cretaceous Bivalvia from cores, Leg 27: Init. Repts. DSDP, v. 27, p. 977–981.	O	C	259 260 263		D		
Srinivasan, M. S., and Kennett, J. P., 1976, Evolution and phenotypic variation in the Late Cenozoic *Neogloboquadrina dutertrei* plexus, *in* Takayanagi, Y., and Saito, T., eds., Progress in Micropaleontology Spec. Pub.: New York, Micropaleontology Press, American Museum of Natural History, p. 329–355.	F	Q PL M	206 207 208 209	R	N		R F
Stadum, C. J., and Burckle, L. H., 1973, A silicoflagellate ooze from the east Falkland Plateau: Micropaleontology, v. 19, no. 1, p. 104–109.	S	PL	65		U		

Stevens, G. R., 1974, Leg 27 Cephalopoda: Init. Repts. DSDP, v. 27, p. 983–989.	O	C	263		D	
Stradner, H., 1973, Catalogue of calcareous nannoplankton from sediments of Neogene age in the eastern North Atlantic and Mediterranean Sea: Init. Repts. DSDP, v. 13, p. 1137–1199. [Additional information in Site chapters.]	N	Q PL M	120 to 134	N1	N	V
Sverdlove, M. S., and Habib, D., 1974, Stratigraphy and suggested phylogeny of *Deflandrea vestita* (Brideaux) comb. nov. and *Deflandrea echinoidea* Cookson and Eisenack: Geoscience and Man, v. 9, 1974, p. 53–62. [Dinoflagellates.]	O	C	105	R	D	V F
Swain, F. M., 1978, Notes on Cretaceous Ostracoda from DSDP Leg 44, Sites 390 and 392: Init. Repts. DSDP, v. 44, p. 921–937.	O	C	390 392		N	V
Takayanagi, Y., and Oda, M., 1976, Shore laboratory report on Cenozoic planktonic Foraminifera: Leg 33: Init. Repts. DSDP, v. 33, p. 451–465.	F	Q PL M O E	314 to 318		D	V
Thierstein, H. R., 1971, Tentative Lower Cretaceous calcareous nannoplankton zonation: Eclogae Geologicae Helvetiae, v. 64, no. 3, p. 459–488, 8 pls.	N	C	4 5	R N7	U	R V F
———, 1973, Lower Cretaceous calcareous nannoplankton stratigraphy: Abhandlungen der Geologischen Bundesanstalt Wein, v. 29, 52 pp., 6 pls.	N	C	4,5 135 136 137 144	R N5	U	R V F
———, 1974, Calcareous nannoplankton—Leg 26, Deep Sea Drilling Project: Init. Repts. DSDP, v. 26, p. 619–667. [Additional information in Site chapters.]	N	Q PL M O E C	250 to 259	R N2	U P	G V
———, 1976, Mesozoic calcareous nannoplankton biostratigraphy of marine sediments: Marine Micropaleontology, v. 1, p. 325–362.	N	C J	Many		N P	G V F R
———, 1977, Mesozoic calcareous nannofossils from the Indian Ocean, DSDP Legs 22 to 27, *in* Heirtzler, J. R., and others, eds., Indian Ocean Geology and Biostratigraphy: Washington, D.C., American Geophysical Union, p. 339–351.	N	C J	211 to 263		P	
Tjalsma, R. C., 1977, Cenozoic Foraminifera from the South Atlantic, DSDP Leg 36: Init. Repts. DSDP, v. 36, p. 493–517.	F	Q M O E PA	326 to 329		U P	V
Totten, D. K., Jr., 1976, *Coccolithus pelagicus:* a clarification: Neogene Symposium, Soc. Econ. Paleontologists Mineralogists, Pacific Section mtg., San Francisco 1976, p. 155–160.	N	Q PL M O E	Many	R	R	R F

Reference	Fossil groups.	Ages.	Sites.	Revisions. New taxa.	Abundance. / Preservation.	Groups or regions related. Vertical study. Finer stratigraphy.
Toumarkine, M., 1975, Middle and Late Eocene planktonic Foraminifera from the Northwestern Pacific, Leg 32 of the Deep Sea Drilling Project: Init. Repts. DSDP, v. 32, p. 735–751.	F	E	305 310 313		N P	V
———, 1978, Planktonic foraminiferal biostratigraphy of the Paleogene of Sites 360 to 364 and the Neogene of Sites 362A, 363, and 364, Leg 40: Init. Repts. DSDP, v. 40, p. 679–721.	F	Q PL M O E PA	360 to 364		N P	V
Ujiie, H., 1975, Planktonic foraminiferal biostratigraphy in the western Philippine Sea, Leg 31 of DSDP: Init. Repts. DSDP, v. 31, p. 677–691. [Additional information in Site chapters.]	F	Q PL M O E	290 to 302		N	V
van Andel, T. H., and Bukry, D., 1973, Basement ages and basement depths in the eastern equatorial Pacific from Deep Sea Drilling Project Legs 5, 8, 9, and 16: Geol. Soc. America Bull., v. 84, p. 2361–2370.	N	M O E	74,75 77–83 159–163			F
van Hinte, J. E., 1972, Lists of selected Mesozoic planktonic and benthonic Foraminifera and Ostracoda, and conclusions on age and environment: Init. Repts. DSDP, v. 12, p. 115–118.	F O	C	111		N	V
———, Adams, J. A. S., and Perry, D., 1975, K/Ar age of Lower-Upper Cretaceous boundary at Orphan Knoll (Labrador Sea): Canadian Jour. of Earth Sciences, v. 12, no. 8, p. 1484–1491.	F	C	111		N	V F
———, and Wong, T. E., 1975, *Nummulites rockallensis* n. sp. from the Upper Paleocene of Rockall Plateau (North Atlantic): Jour. Foraminiferal Research, v. 5, no. 2, p. 90–101.	F	PA	117	N1		R V F
Vincent, E., 1974, Cenozoic planktonic biostratigraphy and paleoceanography of the tropical western Indian Ocean: Init. Repts. DSDP, v. 24, p. 1111–1150.	F N R D	Q PL M O E PA	231 to 238			G V

———, 1975, Neogene planktonic Foraminifera from the central North Pacific, Leg 32, Deep Sea Drilling Project: Init. Repts. DSDP, v. 32, p. 765–801.	F	Q PL M	305 310 313		U P	R V G
———, 1977, Indian Ocean Neogene planktonic foraminiferal biostratigraphy and its paleoceanographic implications, *in* Heirtzler, J. R., and others, eds., Indian Ocean Geology and Biostratigraphy: Washington, D.C., American Geophysical Union, p. 469–584.	F N R	Q PL M	211 to 269 280 to 282			
———, Frerichs, W. E., and Heiman, M. E., 1974, Neogene planktonic Foraminifera from the Gulf of Aden and the western tropical Indian Ocean, Deep Sea Drilling Project, Leg 24: Init. Repts. DSDP, v. 24, p. 827–849. [Additional information in Site chapters.]	F	Q PL M O	231 to 238		D P	V
Wall, D. R., and Quilty, P. G., 1976, Ostracodes—Leg 34: Init. Repts. DSDP, v. 34, p. 737–739.	O	Q PL M O	319 to 321		U	V
Wass, R. E., and Yoo, J. J., 1975, Bryozoa from Site 282 west of Tasmania: Init. Repts. DSDP, v. 29, p. 809–831.	O	Q M	282		D	V
Weaver, F. M., 1976, Antarctic Radiolaria from the Southeast Pacific Basin, Deep Sea Drilling Project, Leg 35: Init. Repts. DSDP, v. 35, p. 569–603. [Additional information in Site chapters.]	R	Q PL M	322 to 325	R N2	U P	V
———, and Dinkelman, M. G., 1978, Cenozoic radiolarians from the Blake Plateau and the Blake-Bahama Basin, DSDP Leg 44: Init. Repts. DSDP, v. 44, p. 865–885. [Additional information in Site chapters.]	R	Q M E	388 to 394		U P	V
———, Rögl, F., Haq, B. U., and Schrader, H.-J., 1976, Paleontological summary of deep sea drilling results from Leg 35, Southeast Pacific Basin: Init. Repts. DSDP, v. 35, p. 531–537.	F N S D R	Q PL M PA C	322 to 325			G
Webb, P. N., 1973, Upper Cretaceous-Paleocene Foraminifera from Site 208 (Lord Howe Rise, Tasman Sea), DSDP, Leg 21: Init. Repts. DSDP, v. 21, p. 541–573.	F	PA C	208		D	V F
———, 1975, Paleocene Foraminifera from DSDP Site 283, South Tasman Basin: Init. Repts. DSDP, v. 29, p. 833–843.	F	PA	283		N	V
Westberg, M. J., and Riedel, W. R., 1978, Accuracy of radiolarian correlations in the Pacific Miocene: Micropaleontology, v. 24, no. 1, p. 1–23.	R	M	34,62 71,77 173 206 289	R	D P	R V F
Wezel, F. C., 1974, Primo riesame delle carote raccolte nel Leg 13 del DSDP (Mediterraneo): Giornale di Geologia, ser. 2, v. 39, fasc. 2, p. 447–468.	N	Q PL M	121 to 134		N	G V F

Reference	Fossil groups.	Ages.	Sites.	Revisions. New taxa.	Abundance. / Preservation.	Groups or regions related. Vertical study. Finer stratigraphy.
Wiedmann, J., and Neugebauer, J., 1978, Lower Cretaceous ammonites from the South Atlantic Leg 40 (DSDP), their stratigraphic value and sedimentologic properties: Init. Repts. DSDP, supplement to v. 38–41, p. 709–734.	O	C	363 364	R		V
Wilcoxon, J. A., 1972, Calcareous nannoplankton ranges, Deep Sea Drilling Project: Init. Repts. DSDP, v. 11, p. 459–473.	N	Q PL M O,E PA C	98 to 108		N	V
———, 1972, Upper Jurassic–Lower Cretaceous calcareous nannoplankton from the western North Atlantic Basin: Init. Repts. DSDP, v. 11, p. 427–457.	N	C J	99 100 101 105	R N8	N	R V F
Williams, G. L., 1978, Palynological biostratigraphy, Deep Sea Drilling Project Sites 367 and 370: Init. Repts. DSDP, supplement to v. 38–41, p. 783–815.	O	PL M O E C J	367 370	N4	U	V
Wilson, G. J., 1975, Palynology of deep-sea cores from DSDP Site 275, southeast Campbell Plateau: Init. Repts. DSDP, v. 29, p. 1031–1035.	O	C	275		N	V
Wind, F. H., 1975, *Tetralithus copulatus* Deflandre (Coccolithophyceae) from the Indian Ocean: a possible paleoecological indicator: Antarctic Journal of the United States, v. 10, no. 5, p. 265–268.	N	C	217 249	R	N	V F
———, 1978, Western North Atlantic Upper Jurassic calcareous nannofossil biostratigraphy: Init. Repts. DSDP, v. 44, p. 761–773.	N	J	391	R	N P	R V F
———, Dinkelman, M. G., and Wise, S. W., Jr., 1977, Jurassic scolecodont-like microfossils from the Falkland Plateau (Deep Sea Drilling Project, Site 330): Init. Repts. DSDP, v. 36, p. 829–843.	O	J	330	N5		V
———, and Wise, S. W., Jr., 1976, Jurassic to Holocene calcareous nannofossils from the Falkland (Malvinas) Plateau: Antarctic Journal of the United States, v. 11, no. 3, p. 169–171.	N	Q E	327		N	R V F

		PA C J				
Wise, S. W., Jr., 1973, Calcareous nannofossils from cores recovered during Leg 18, Deep Sea Drilling Project: Biostratigraphy and observations of diagenesis: Init. Repts. DSDP, v. 18, p. 569–615.	N	Q PL M O	172 to 182	R N3	U P	V
———, and Wind. F. H., 1977, Mesozoic and Cenozoic calcareous nannofossils recovered by DSDP Leg 36 drilling on the Falkland Plateau, southwest Atlantic sector of the Southern Ocean: Init. Repts. DSDP, v. 36, p. 269–491.	N	Q M O E PA C J	326 327 329 330	R N44	U P	V F
Wiseman, J. F., 1976, Palynological investigation of samples from Holes 319, 320, 320A and Site 321 of DSDP Leg 34: Init. Repts. DSDP, v. 34, p. 741–742.	O	Q	319 to 321		N	V
———, and Williams, A. J., 1974, Palynological investigation of samples from Sites 259, 261, and 263, Leg 27, Deep Sea Drilling Project: Init. Repts. DSDP, v. 27, p. 915–924.	O	C	259 261 263		N	V
Wolfe, J. A., 1973, Reconnaissance spore and pollen examination, Early Tertiary turbidite beds, Aleutian Abyssal Plain, Site 183: Init. Repts. DSDP, v. 19, p. 739.	O	O E	183			R V
Worsley, T. R., 1971, Calcareous nannofossil zonation of Upper Jurassic and Lower Cretaceous sediments from the western Atlantic: Proceedings of the II Planktonic Conference, Roma 1970, p. 1301–1321.	N	C J	4 5	N36	N	V F
———, 1973, Calcareous nannofossils: Leg 19 of the Deep Sea Drilling Project: Init. Repts. DSDP, v. 19, p. 741–750.	N	Q PL M O E C	183 to 193		U	V
Zaklinskaya, E. D., 1978, Palynology of Paleogene clay from DSDP Site 368, Cape Verde Rise: Init. Repts. DSDP, supplement to v. 38–41, p. 933–937.	O	E	368		N	V
Zobel, B., 1974, Quaternary and Neogene Foraminifera: Biostratigraphy: Init. Repts. DSDP, v. 25, p. 573–578. [Additional information in Site chapters.]	F	Q PL M	239 to 249		N P	V
Zocchi, M., 1974, Variabilità di "*Globorotalia margaritae*" Bolli & Bermudez nel Pliocene inferiore tirrenico: Riv. Ital. Paleont., v. 80, no. 1, p. 101–145.	F	PL	132			V F

SEPM Special Publication No. 32, p. 317–334, October 1981

IN SEARCH OF BIOSTRATIGRAPHIC RESOLUTION

T. C. MOORE, JR. AND KAREN ROMINE
Graduate School of Oceanography, University of Rhode Island, Kingston, RI 02881 U.S.A.

ABSTRACT

The Deep Sea Drilling Project has contributed greatly to the improvement and expansion of marine biostratigraphy. However, if stratigraphic resolution is described in terms of the number of zonal boundaries per million years, it appears that the resolution of zonal stratigraphies varies with time and is related to the degree of variability of the ocean environment. Several problems which contribute to the present limitations of stratigraphic resolution have to be dealt with: a) taxonomic inadequacies and disagreements in taxonomic applications; b) incompleteness of the preserved record; c) incompleteness and disturbed nature of DSDP recovery; d) correct interpretation and correlation of stratigraphically useful signals.

One approach to improving stratigraphic resolution is the development of broad, regional holistic stratigraphies which combine lithostratigraphy, isotopic stratigraphy, paleomagnetic stratigraphy and species assemblage stratigraphy with the use of first and last appearances for species from several fossil groups. Detailed stratigraphic and time-slice studies of the Quaternary serve as guides to both the possibilities and pitfalls of expanding and improving the resolution of Cenozoic stratigraphy. Using composite stratigraphies, the average limit of Quaternary stratigraphic resolution is near 20,000 years as opposed to the present limit of 100,000 years in the Cenozoic.

An improved coring technique recently developed by DSDP, the hydraulic piston corer, should diminish many of the problems with section recovery and disturbance. It offers an opportunity to develop a holistic approach to Cenozoic stratigraphy and ultimately accomplish the goals of high stratigraphic resolution and global chronostratigraphy.

INTRODUCTION

When the Deep Sea Drilling Project first started in 1968 the biostratigraphy of planktonic organisms in marine sections was a rapidly growing field of scientific investigation. Work on the foraminifera was led by Bolli, Blow, Banner, Parker and others. Work on the calcareous nannofossils was led by Bramlette, Hay and many of his students. Riedel was at the head of the field in research on the stratigraphy of Radiolaria in the Cenozoic. When the *Glomar Challenger* put to sea in 1968 it represented the great hope for improving the stratigraphies which had been developed by these investigators and adding to them the stratigraphies of other fossil groups. In this paper we will investigate the troubles that have been encountered with this expansion and improvement and will evaluate the possibilities for future improvements.

The wealth of valuable stratigraphic sections provided by the project gave rise to a rapid growth in stratigraphic development (Fig. 1). In 1968 only the three main microfossil groups (foraminifera, radiolaria, and calcareous nannofossils) were reported. By 1970 thirteen different fossil groups had been reported, and as of 1978, fifteen fossil groups have been presented and discussed in special chapters of the Deep Sea Drilling Project Initial Reports.

In order to see how these stratigraphies improved over the past ten years, the resolution of the three main microfossil groups used in establishing the stratigraphies in the Initial Reports are presented in Figures 2, 3 and 4. The number of zonal boundaries occurring per million years of geologic time are used as a measure of the stratigraphic resolution and are averaged over the major divisions of geologic time. The primary difference between the foraminiferal stratigraphy as it existed in 1968 (Fig. 2A) and as it existed in 1975 (Fig. 2B) lies mainly in the improvement of the resolution in the Late Mesozoic. There are differences in the Neogene part of the stratigraphy, however, these are due largely to minor changes in the ages of the boundaries of the major geologic divisions. It is also interesting to note the oscillatory nature of the density of these boundaries with time. This variation in the degree of resolution achievable as a function of time is common to all the major biostratigraphies.

For the calcareous nannofossils, comparison of the zonation that existed in 1968 (Fig. 3A) with those that existed in the late 1970's (Fig. 3B) shows the greatest improvement in the Neogene and in the Cretaceous. The high resolution available in the Paleogene in 1968 apparently decreased in the '70's. This was due primarily to the poor preservation and poor recovery of these intervals by the Deep Sea Drilling Project cores and thus the lack of a wide applicability for some of these early calcareous nannofossil zonations.

The Radiolaria are more indicative of several other groups which have had a rapid development during the course of the Deep Sea Drilling Project (e.g., the silico-flagellates and the diatoms). In 1968 no formal zonation had been proposed for the Radiolaria, although major divisions of the Ceno-

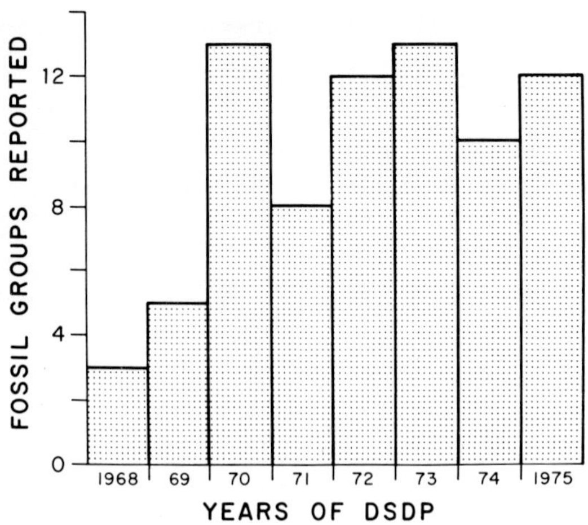

FIG. 1.—Number of fossil groups reported in the Initial Reports of the Deep Sea Drilling Project plotted versus the year of operation.

zoic could be achieved (Riedel and Funnel, 1964). By the mid 1970's a very adequate stratigraphy existed for the Radiolaria down through most of the Cretaceous (Fig. 4B).

RESOLUTION OF THE ZONAL SCHEMES

It might be thought that the combination of these three zonations, each with a resolution of about 2 m.y., would lead to a composite stratigraphy with a resolution three times as good. Unfortunately, this is not so. Although the combination of the stratigraphies of more than one microfossil group does increase the reliability and accuracy of correlation, the overall increase in resolution is only a factor of 2, not a factor of 3. This is due mainly to the fact that many zone boundaries occur at approximately the same point in time. Figure 5 shows a plot of the composite radiolarian, foraminiferal and calcareous nannofossil biostratigraphies. This composite omits all stratigraphic boundaries which are approximately coeval with other zonal boundaries. This compilation shows three main features. First, the overall improvement in the stratigraphic resolution is not as great as one might hope. Second, the oscillatory nature of degree of resolution in biostratigraphy persists even in the combined stratigraphies. For example, the low resolution in the foraminiferal stratigraphies during the Oligocene is not compensated for by high resolution in the Radiolaria or the calcareous nannofossils. Finally, the older sections generally have a lower degree of biostratigraphic resolution. These three problems with biostratigraphic resolution are treated separately in the following paragraphs.

The reasons for older sections showing less biostratigraphic resolution are easy to appreciate. Diagenetic alteration of the biogenic debris obliterates many of the useful microfossils. In addition, the recovery of old material on which these stratigraphies may be established is severely reduced compared to the younger material. The total section in kilometers recovered by DSDP (Fig. 6A) drops quickly from the Quaternary through the Mesozoic, and if this total section is divided by the duration of each geologic interval (Fig. 6B) the total section recovered, in kilometers per million years, drops even more rapidly. Finally, older sections generally contain more hiatuses (Fig. 7). Hiatus abundance increases dramatically with age through the Cenozoic. The presence of hiatuses in pelagic sections is interpreted as being the result of dissolution of biogenic debris and the erosion of marine sediments by bottom currents (Moore and others, 1978). If we use the inverse of the hiatus abundance curve to represent a section survival curve, a model has been developed (Moore and Heath, 1977) which describes this measure of section survival in terms of age of sediment (depth of burial) and distribution of hiatus duration (Fig. 8). Comparison of this model with the observational data shows two features: a) section survival starts to flatten out around 65 to 70 m.y. before present with about 30 percent of original total sections surviving, therefore the effect of section survival on stratigraphic resolution in the Mesozoic should be rather uniform; and b) there are also oscillations in the proportion of section survival around the average survival line projected by the model.

This last point brings us to the second consideration in biostratigraphic resolution: the oscillatory nature in the degree of resolution in biostratigraphic zonations. This feature is usually looked at in terms of evolutionary pulses, or changes in the evolutionary rate of planktonic organisms (e.g., Berggren, 1969a). Three possible explanations for this variation in the degree of resolution are considered here: a) changes in section survival, b) eustatic changes in sea level and c) changes within the oceans. All three explanations involve changes in the history of the oceanic environment, and a comparison of these histories with the history of variation in the biostratigraphic resolution may reveal common causes. First, oscillations in the section survival might lead to changes in zonal resolution either directly or indirectly. The direct effect can be taken as the destruction of the sections or the dissolution of the microfossil assemblages commonly used. An indirect effect might be possible if the variation in section survival is considered as an indicator of changes in the chemical nature of bottom waters and in the patterns of oceanic circulation which were tied to other oceanographic changes affecting the planktonic community. A

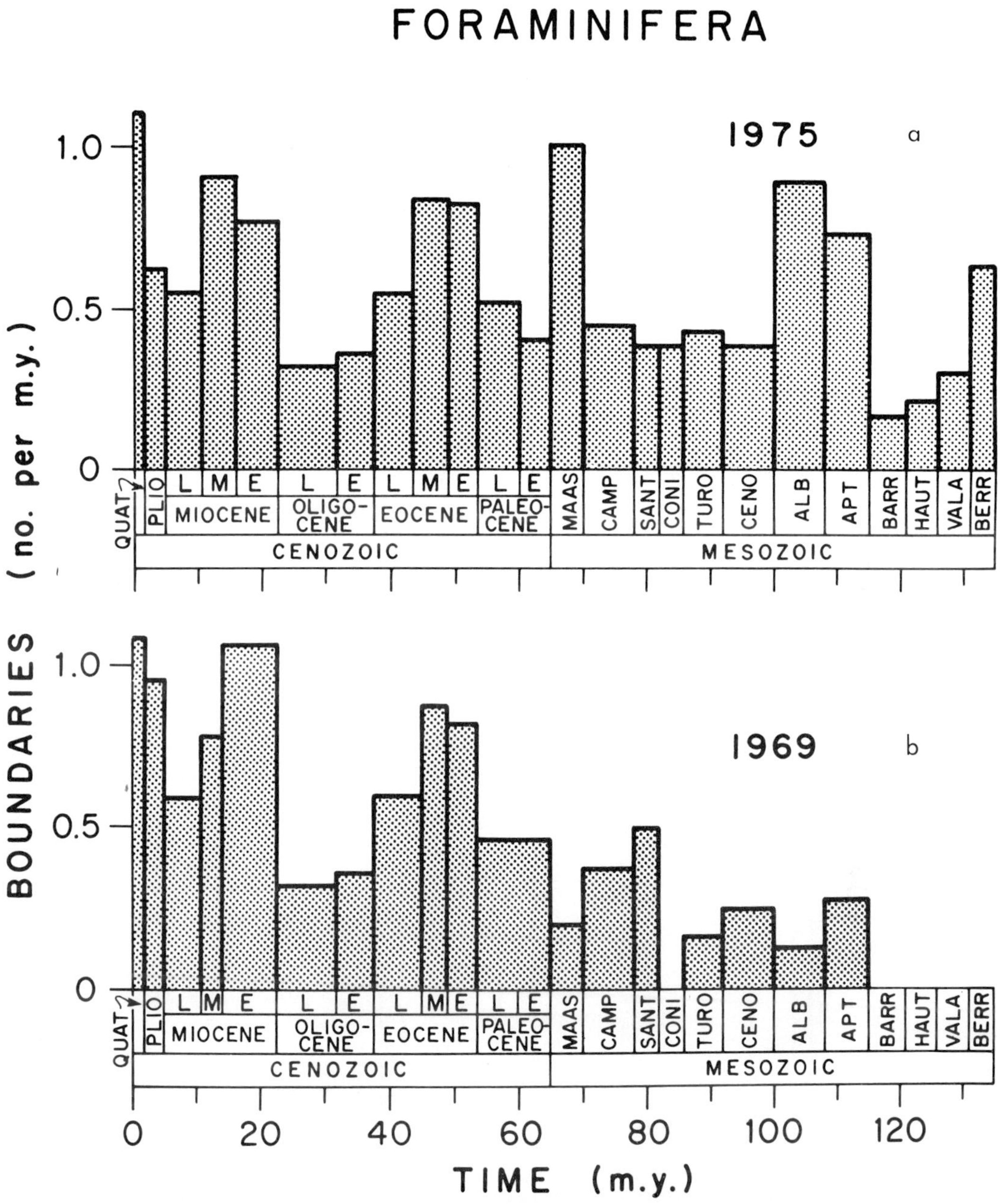

FIG. 2.—The number of biostratigraphic boundaries per million years, averaged over individual epochs and based on the foraminiferal stratigraphy as it existed in 1969(b) and in 1975(a). Biostratigraphic schemes considered in compiling these data include those shown in van Hinte (1976), Orr and Jenkins (1977), Berggren (1969b, 1972, 1977) and Blow, (1969). Time scales after Berggren (1969b, 1972).

comparison of the detrended section survival curve (Fig. 9B) with the biostratigraphic resolution curve of the Cenozoic (Fig. 9A) shows very few similarities. Although there is some slight correspondence between the two curves in the Late Paleocene, the overall correspondence is poor. Maxima in the Pliocene and mid-Miocene of the biostratigraphic resolution curve do not match the maximum at the

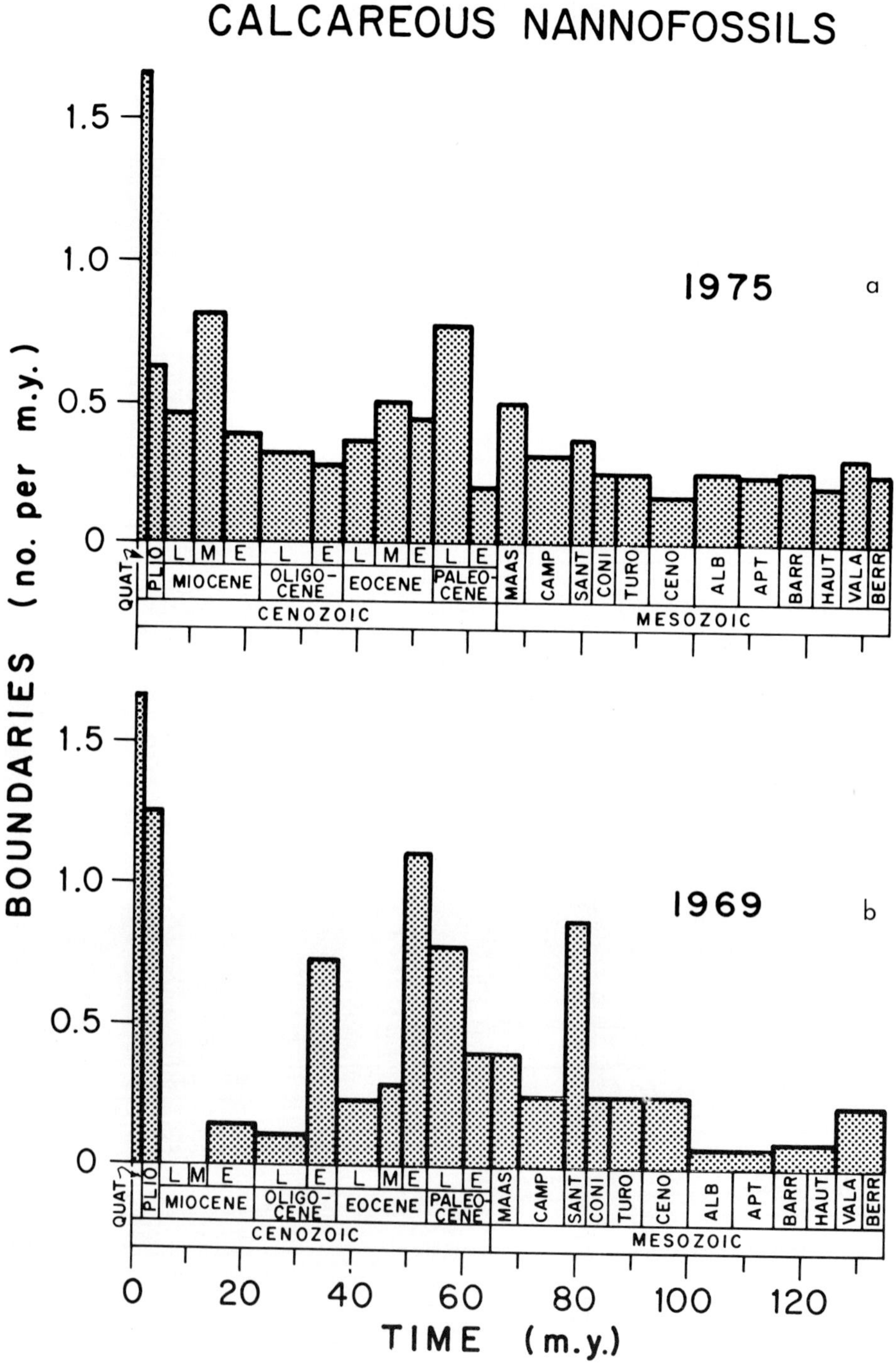

FIG. 3.—The number of biostratigraphic boundaries per million years, averaged over individual epochs and based on the calcareous nannofossil stratigraphy as it existed in 1969(b) and in 1975(a). Biostratigraphic schemes considered in compiling these data include those shown in Hay and others (1967), Cepek and Hay (1969), Berggren (1972), van Hinte (1976), Gartner, (1977) and Bukry and others (1977). Time scales after Berggren (1969b, 1972).

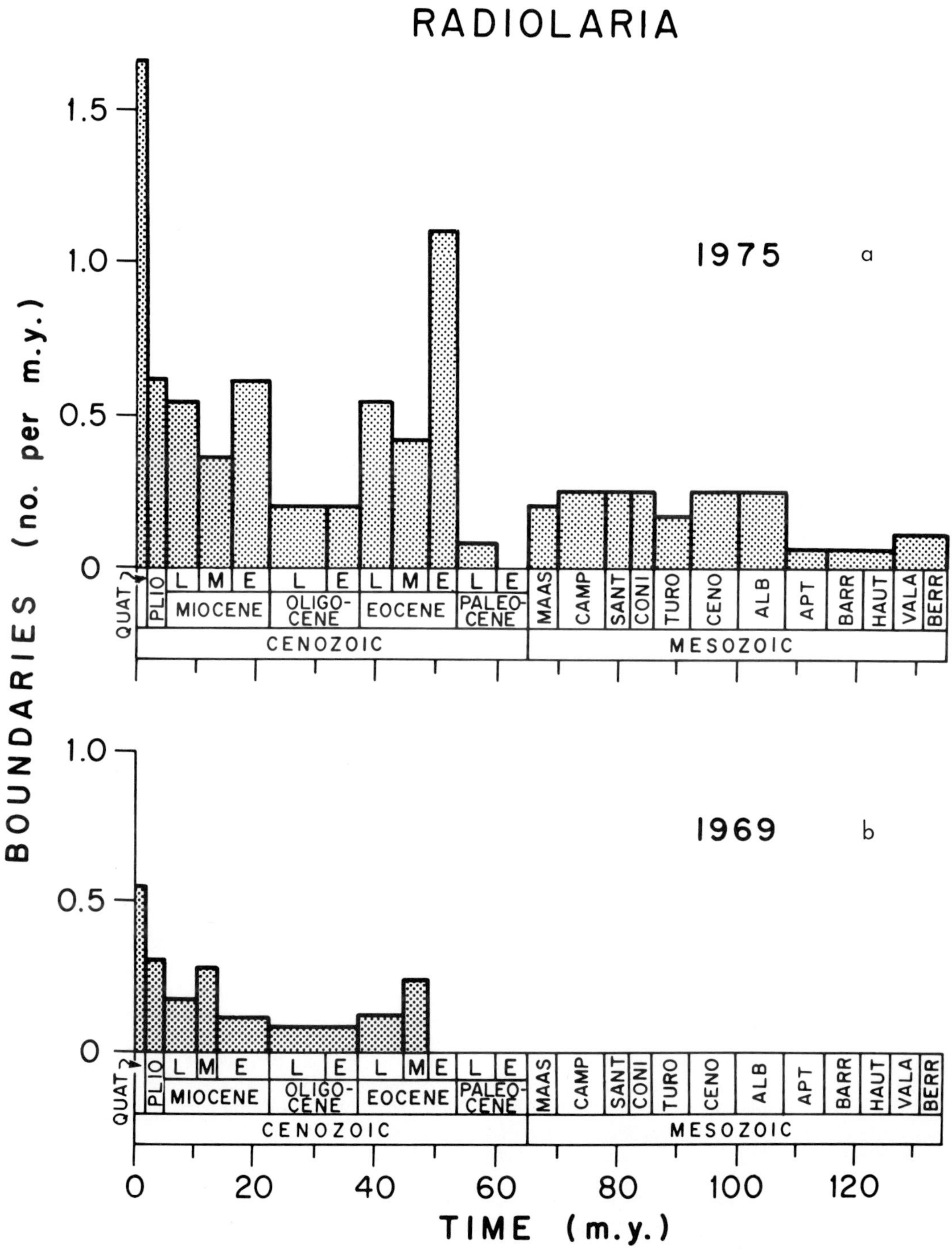

FIG. 4.—The number of biostratigraphic boundaries per million years, averaged over individual epochs and based on the radiolarian stratigraphy as it existed in 1969(b) and in 1975(a). Biostratigraphic schemes considered in compiling these data include those shown in Ridel and Sanfilippo (1970, 1971, 1974) Berggren, (1972), and van Hinte (1976). Time scale after Berggren (1969b, 1972).

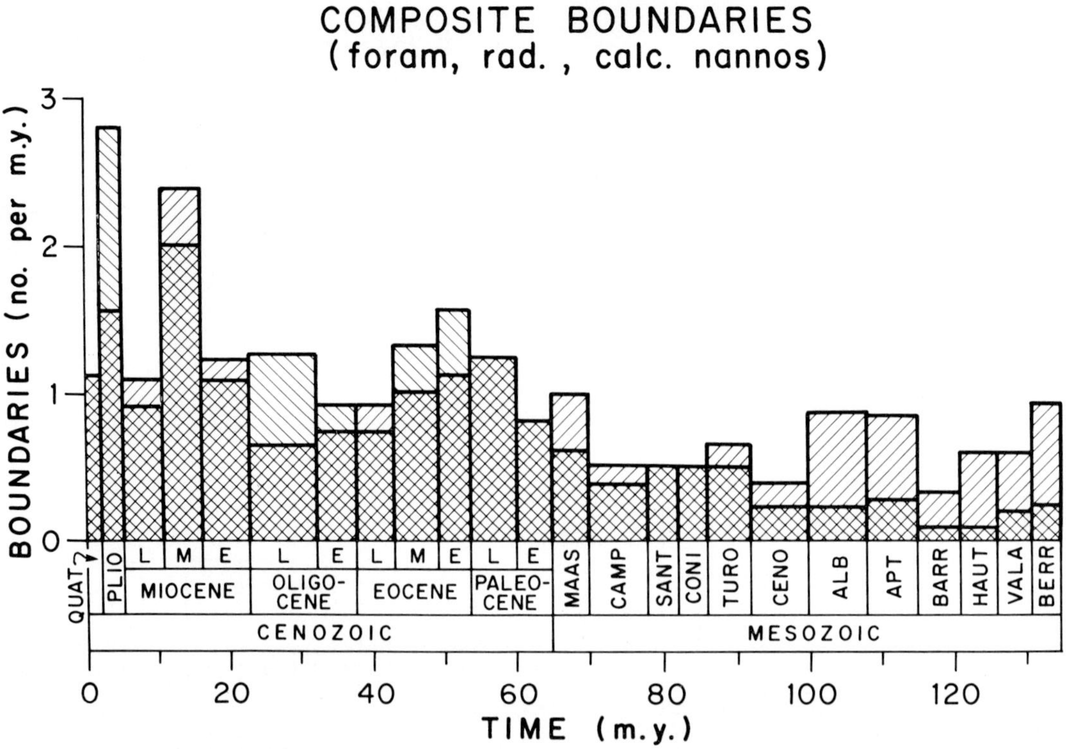

FIG. 5.—Number of biostratigraphic boundaries per million years (combined foraminiferal, calcareous nannofossil and radiolarian stratigraphies) averaged over individual epochs. Where boundaries from two or more microfossil groups appear to be approximately coeval (*i.e.*, are within 0.5 m.y. of each other) only one of the boundaries is used in determining the average number of boundaries per million years. Average resolution as it existed in 1969 is shown by left-diagonal shading; 1975 is shown in right-diagonal shading. Time scales after Berggren (1969b, 1972).

Miocene-Oligocene boundary in the section survival curve.

Changes in relative sea-level might cause changes in the shallow-water habitats and niches available to the near-surface flora and fauna which might in turn change diversity and biostratigraphic resolution. Such relative sea-level changes would also affect deep- to shallow-water fractionation of the biogenic debris (Berger, 1970). A comparison of the relative sea-level curve produced by Vail, and others (1977) also shows little correspondence to the biostratigraphic resolution curve. The only visible similarity between the two curves (Fig. 9A and 9C) is the presence of minima in the Oligocene and Late Miocene for both sea-level and stratigraphic resolution. This is the opposite result which would be expected if the effects of lowered sea level improve the preservation of carbonate in the deep sea, but it is the expected relationship if lowered sea level decreased biotic diversity.

As a broad representation of general changes taking place within the Cenozoic ocean, we have used the $\varrho 0^{18}$ measurements made on benthic foraminifera (Shackleton and Kennett, 1975). These measurements (Fig. 9D) are taken as indicators of deep-water temperatures; and if it is assumed that the temperature of the low latitude surface waters are rather constant, the variation in deep-water temperature can be used as an indicator of changes in low- to high-latitude temperature gradients, as well as vertical (deep-to-shallow) gradients in the oceans. The general trend of this curve shows an overall cooling of the deeper waters through the Cenozoic; and correspondence with the biostratigraphic resolution curve (Fig. 9A) is poor. There is considerable variation, however, in the individual isotopic measurements around the overall trend of this curve. If the degree of this variation is taken as a measure of variability, then the range of variability per unit time provides an apt indicator for comparison with the degree of resolution provided by biostratigraphic zonations. When this comparison is made (Fig. 10), then the match between these two curves is good. If the extremely high variability of the Pleistocene (which is due primarily to variations in icecap volume) is ignored,

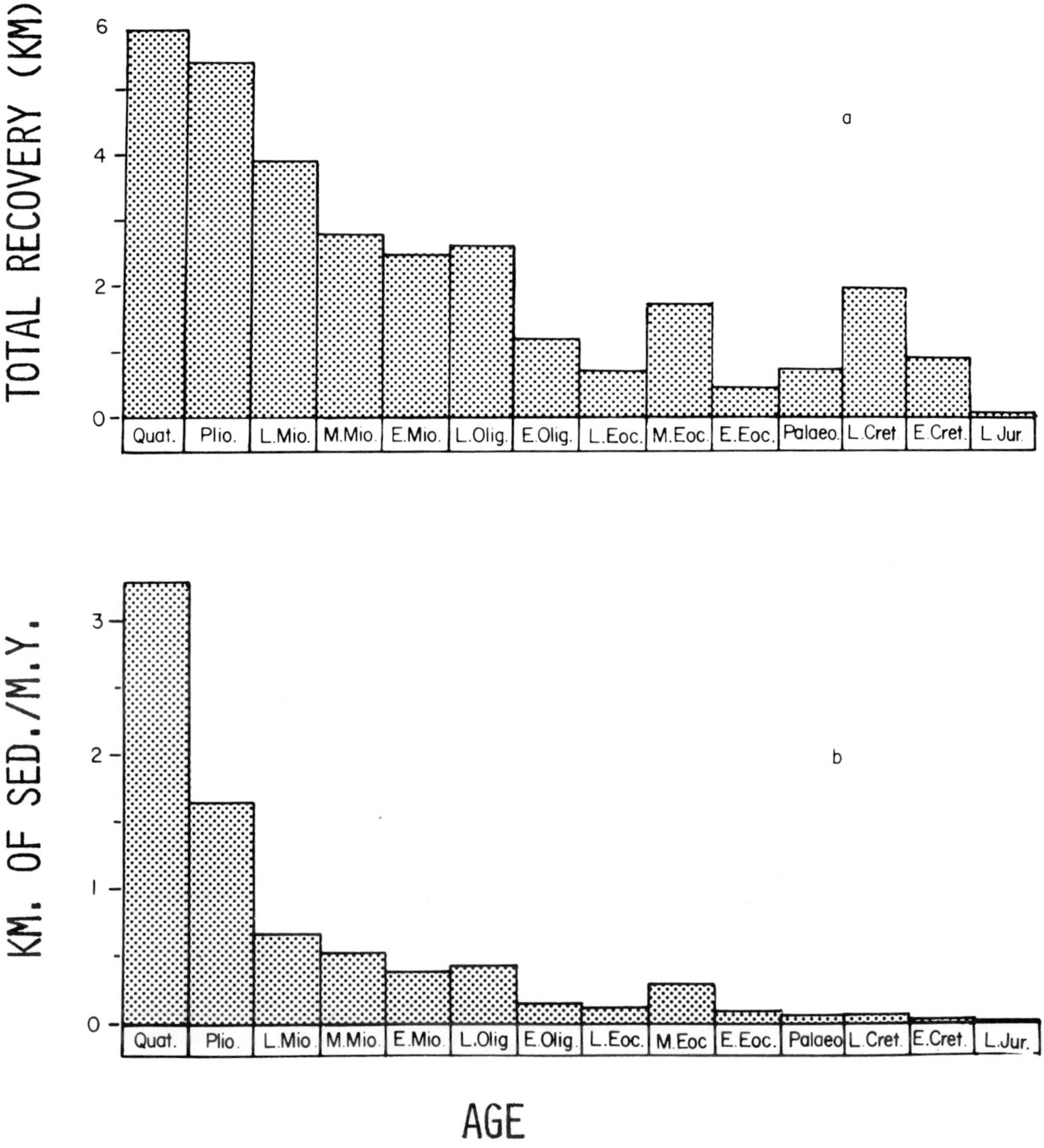

FIG. 6.—A, Total section, in kilometers of sediment, recovered by the Deep Sea Drilling Project (Legs 1–33) shown as a function of geologic age. B, Average thickness of sediment per million years of geologic time recovery by DSDP (Legs 1-33) grouped by geologic intervals; after Moore and Heath (1978).

then the match is excellent for most of the Cenozoic. The somewhat poorer match in the Early Paleogene may be accounted for by either the sparseness of oxygen isotope data or the poor recovery of this interval.

From this study we conclude that the number of zonal boundaries per million years does vary with time and that this variation is probably related to the degree of variability in the oceanic environment. If this conclusion is valid, the intervals of low resolution cannot be greatly improved. In fact there is a practical upper limit to the number of microfossil groups present in most sections; and thus a significant increase in biostratigraphic resolution cannot be achieved as long as we are limited to the standard biostratigraphic zonal techniques.

IMPROVING STRATIGRAPHIC RESOLUTION

The shortcomings in resolution using standard biostratigraphic techniques have been realized for

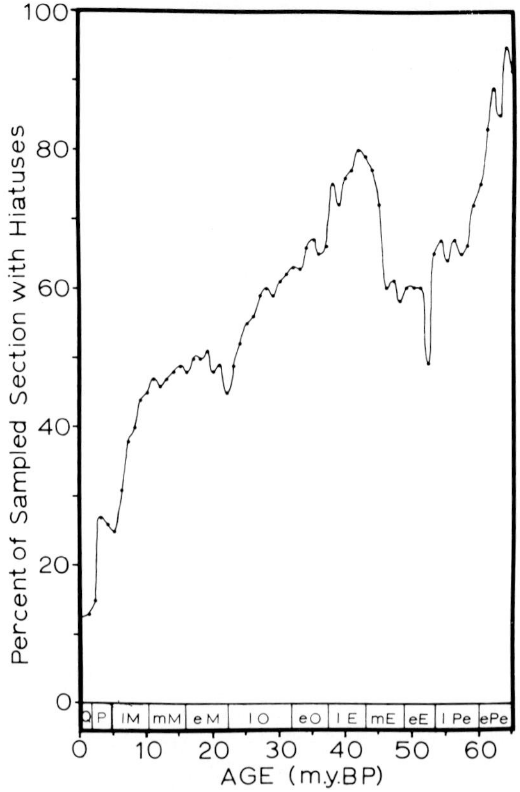

FIG. 7.—Percentage of sampled deep-sea sections (DSDP Legs 1–33) represented by hiatuses (after Moore and Heath, 1977).

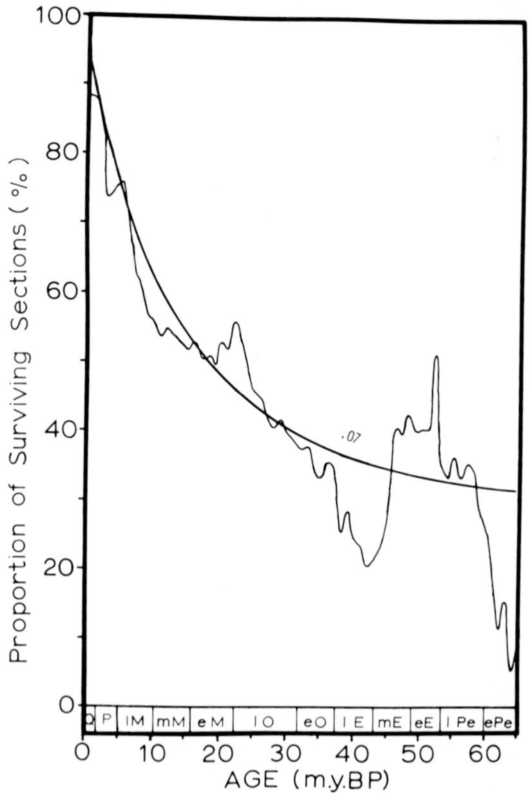

FIG. 8.—Proportion of surviving sections (inverse of Fig. 7) and model estimate of section survival (smooth curve) based on observed duration of hiatuses and age of the sedimentary section (after Moore and Heath, 1977).

several years now. Shaw (1964) and later Southam and others (1975) have proposed approaches which extend this resolution by making use of every reliable species range. They have pointed out the desirability of using all information on species' ranges rather than just information on those species which define individual biostratigraphic zones. If our goal is the best stratigraphic resolution possible, it is logical to extend even further the ideas proposed and used by Shaw to include not only biostratigraphic ranges, but to combine all stratigraphies in a holistic approach to stratigraphic problems. Other stratigraphies, such as those based on fluctuations in lithologies (Hays and others, 1969; Kaneps, 1973) and in the abundance of individual assemblages (e.g., Ruddiman, 1971), can also be useful in broad regional correlations. Stratigraphies based on the stable isotopes of oxygen (Shackleton and Opdyke, 1973; 1976) and carbon (Keigwin, 1979; Loutit and Kennett, 1979) have been useful in both regional and global correlations. Finally, paleomagnetic stratigraphy has proved to be one of the most valuable tools in establishing a global chronostratigraphy.

The advantages of the holistic approach go beyond increased resolution, for each of these stratigraphies have their own strong and weak points. For example, the evolutionary progression of biostratigraphy allows an approximate ordering of widely-separated sections which cannot be easily matched by other stratigraphies. However, the likelihood of a first or last appearance being diachronous always thwarts our ability to achieve a global chronostratigraphy using only biostratigraphic techniques.

The most convincing inter-ocean and global correlations rely on magnetic reversals in the sediments to tie together other types of stratigraphy (e.g., Ryan and others, 1974). These reversals are globally synchronous and represent the single most important tool in establishing the degree of diachronism in these other stratigraphies. But paleomagnetic stratigraphy has its own shortcomings. First, it is nearly impossible to tell one sequence of reversals from another without the use of biostratigraphy. Second, reversals are not uniformly spaced in time; thus, there may not be a reversal within the particular interval you wish to correlate. The most

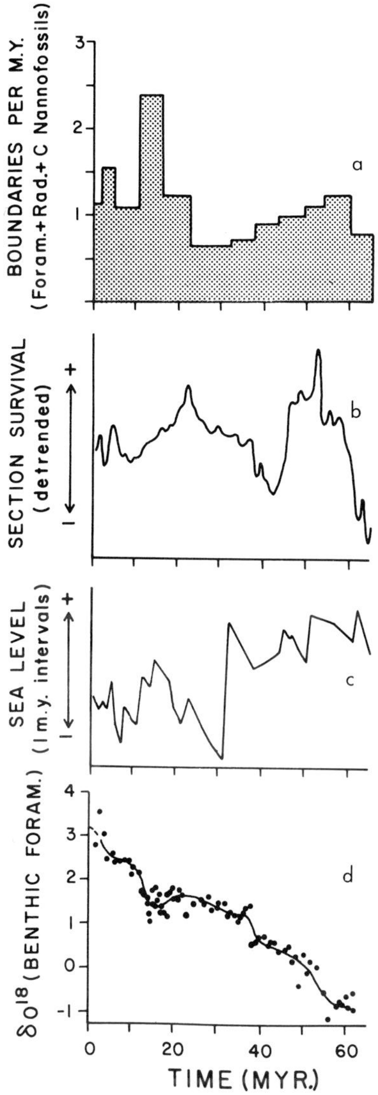

FIG. 9.—A, Cenozoic biostratigraphic resolution as described by the number of boundaries per million years of the combined foraminiferal, calcareous nannofossil and radiolarian stratigraphies (from Fig. 4). B, Section survival curve expressed as a deviation from the model of Moore and Heath (1977) (from Fig. 8). C, Curve of estimated sea level fluctuations (after Vail and others, 1978). D, Measurements of $\delta^{18}O$ made on benthic foraminifera from DSDP Sites 277, 279, and 282 (after Shackleton and Kennett, 1975).

recent Quaternary reversal occurred 730,000 years ago; and in the Cretaceous, there was an interval representing about 20 million years of time where no reversals have been detected.

In the Quaternary and Late Neogene, stratigraphies based on stable isotopes measured in calcar-

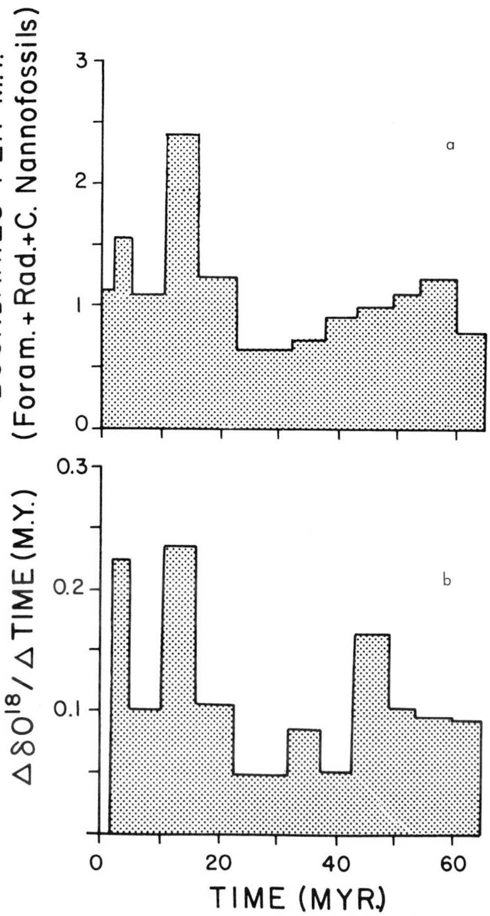

FIG. 10.—A, Cenozoic biostratigraphic resolution (as shown in Fig. 9a). B, Variation of $\delta^{18}O$ measurements expressed as range averaged over epochs of the Cenozoic. Data from Figure 9d (Shackleton and Kennett, 1975). The large Pleistocene variations associated with fluctuations of the northern hemisphere ice sheets are ignored in this plot.

eous organisms have been very effective in establishing global correlations. As previously discussed, Quaternary variations in oxygen isotopes are thought to represent changes in the volume of the continental ice-caps. Thus, within the mixing time of the oceans ($\simeq$1,000 yr.), the isotopic signal measured in calcareous microfossils (particularly deep-sea benthics) should be synchronous. These fluctuations extend back into Upper Pliocene sections (Shackleton and Opdyke, 1973, 1976; Shackleton, 1967); however, the amplitude and spacing of these fluctuations appear to diminish into the Lower Pliocene and Late Miocene even though a Southern Hemisphere ice-cap existed as early as the Middle Miocene (Shackleton and Kennett, 1975). Herein lies one of the problems in applying what we have learned about oxygen isotope stratigraphy in the Quaternary to the problems of Cenozoic stratigraphy. Very few Neogene sections have been looked at in great detail, and it is still unclear a) whether or not rather low-amplitude fluctuations in the oxygen isotopic signal of the Late Miocene can be used for global or even regional correlations and, b) at what point in time fluctuations in oxygen isotope ratios change over from being dominated by temperature effects to being dominated by ice-volume effects. As long as ice-volume dominates, the signal may be useful in global correlations. When temperature dominates, then oxygen isotope measurements on deep-sea benthic microfossils may still be very useful for regional correlations, but isotopic stratigraphies based on planktonic organisms may be difficult to evaluate.

Any change in the deep waters of the oceans (which make up about 97% of the modern ocean) that leaves a record in the microfossils or biogenic sediments of the deep sea, provides a potential tool for broad regional correlations. This is because the deep water-masses are rather uniform in character and cover very large areas of the sea floor; while surface water-masses, each with its particular flora and fauna, are more restricted in their geographical extent. To the degree that the changes in the chemistry of the deep waters control the dissolution and concentration of biogenic debris in pelagic sediments, fluctuations in the lithology of pelagic sections can also be used for stratigraphic correlation. Such fluctuations in carbonate concentration have been used very effectively in the Pliocene and Quaternary sections of the Pacific (Ahrrenius, 1952; Hays and others, 1969; Kaneps, 1973), and initial results of the CENOP project indicate that carbonate concentrations may provide a very useful regional stratigraphy of the Late Miocene (Fig. 11).

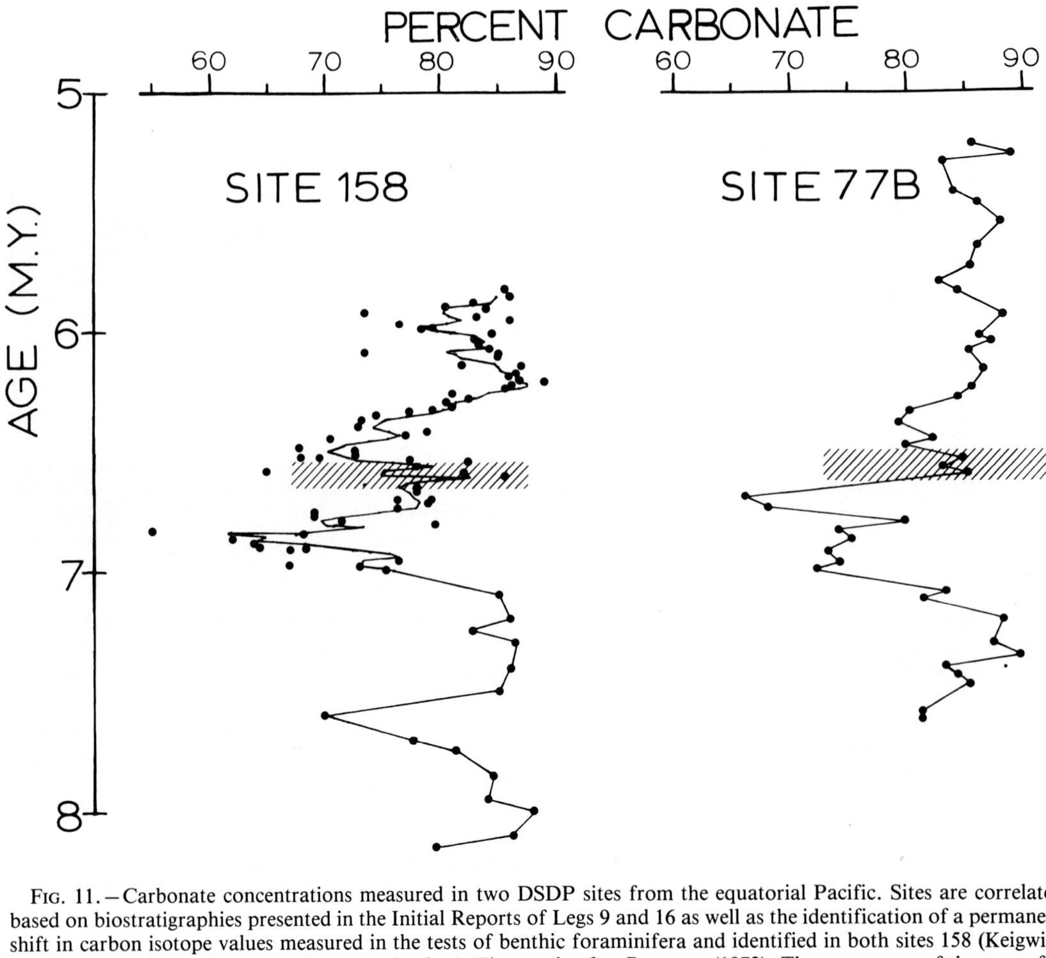

FIG. 11.—Carbonate concentrations measured in two DSDP sites from the equatorial Pacific. Sites are correlated based on biostratigraphies presented in the Initial Reports of Legs 9 and 16 as well as the identification of a permanent shift in carbon isotope values measured in the tests of benthic foraminifera and identified in both sites 158 (Keigwin, 1979) and 77B (S. Savin, personal communication). Time scale after Berggren (1972). The upper part of the curve for site 158 represents a three-point moving average of the data (indicated by dots). This averaging was performed in order to reduce the detail of part of the curve and make it more comparable to that available for Site 77B.

Variations in the relative abundance of indicator species (e.g., Ericson and Wollin, 1968; Hays and others, 1976) have also been used to great advantage in regional correlations of Quaternary sections. This technique can be very limited, however, in both the area and time interval in which it is useful. At any given time an individual species may belong to more than one planktonic community, and with changing environmental conditions the relative abundance of this species may respond differently in each of these communities (Wollin and others, 1971). In addition, a species may change its response to environmental conditions through evolution (Kennett, 1978).

The use of variations in the relative abundance of groups of species, or assemblages (*e.g.,* Ruddiman, 1971), is much less sensitive to the problems faced with the use of individual species. This technique has provided another tool for both the detailed correlation of Quaternary sections and the paleoceanographic interpretation of these fluctuations in terms of shifting water mass boundaries and changing current intensities.

Each of the stratigraphies mentioned above have advantages over biostratigraphy in either greater resolution or less problems with diachronism, but as with paleomagnetic stratigraphy all depend on biostratigraphy for the ordering of sequences. Fortunately, in many pelagic sections, it is possible to use more than one or two of these stratigraphies at a time.

Initial results in the CENOP project indicate that the biostratigraphy of any one group will yield zonal boundaries on the order of one per million years for the Neogene; other biostratigraphic events or datums provide a resolution on the order of one to two events per million years; fluctuations in individual assemblages provide resolution of approximately one million years; magnetic reversals occur on the average of three times per million years; and large isotopic fluctuations occur on the order of three to four times per million years. The sum of all these methods provides about nine to ten events per million years or a resolution of approximately one-hundred-thousand years. This summation does not consider the use of other stratigraphies such as lithostratigraphy, or the use of more than one microfossil group.

THE QUATERNARY AS A GUIDE TO RESOLUTION

If 100,000 years represent some sort of lower limit on achievable resolution using multiple stratigraphic techniques, what is the upper limit? The Quaternary has been looked upon as a time of extremely high variability and has received much attention in marine stratigraphic studies. As pointed out earlier, the high frequency variation characteristic of the Quaternary (*i.e.,* periods $\leq 10^5$ yr) is most commonly associated with relatively large amplitude changes in the oxygen isotope ratio ($^{18}O/^{16}O$) of sea water (Shackleton, 1967; Shackleton and Opdyke, 1973). These isotopic changes reflect the waxing and waning of continental ice-sheets and have long been used as an indicator of climatic change in the Quaternary (Emiliani, 1955). The degree of change in the oceans during the ice-ages, however, was comparatively small. The bottom waters (at 2°C) are very close to freezing during inter-glacials; thus, it is unlikely that they get much colder during glacials. The average global change in sea surface temperature between glacial and interglacial intervals was only about 2°C (CLIMAP, 1976). When compared to the long-term Cenozoic change in bottom-water temperature of about 10°C (Savin and others, 1975), the Quaternary oscillations seem rather puny.

Still, within the comparatively narrow time interval of a million years or so the Quaternary may provide a very good guide to what may be achieved in terms of stratigraphic resolution. Figure 12 shows a composite of biostratigraphic zonations, biodatums, magnetic stratigraphy, isotopic stratigraphy and biotic variation stratigraphy for the Quaternary. The cumulative resolution provided by these stratigraphic techniques is on the order of sixty events per million years, or a resolution of approximately twenty-thousand years.

The rather high degree of stratigraphic resolution in the Quaternary derives not only from the short-period oscillations in climate, but also from the fact that marine Quaternary sections are so readily available for study. The availability of sample material has allowed researchers to investigate not only the time-stratigraphic variation in the Quaternary, but also spatial variability. The reconstruction of biotic distribution patterns in modern and ice-age oceans (e.g., Kipp, 1976; McIntyre and others, 1976; Hays and others, 1976; Moore, 1978), and the conversion of these biotic assemblages into estimates of past sea-surface temperatures (CLIMAP, 1976), have allowed these investigators to assess which particular regions of the world's oceans are most sensitive to climatic change. In terms of sea-surface temperature differences between glacial and interglacial conditions, the subpolar areas experienced the largest change (Fig. 13). Regions of divergence and upwelling, such as the eastern boundary currents and equatorial areas, also experienced marked cooling during glacial intervals; but the central gyre regions had very little change in temperature, and in some areas glacial temperatures may have been slightly warmer than those of the present interglacial. When divided roughly along latitudinal zones of applicability, the resolution of Quaternary biostratigraphies for siliceous organisms (Fig. 14) (Burckle and Opdyke, 1977; Koizumi, 1977; Nigrini, 1971; Robertson,

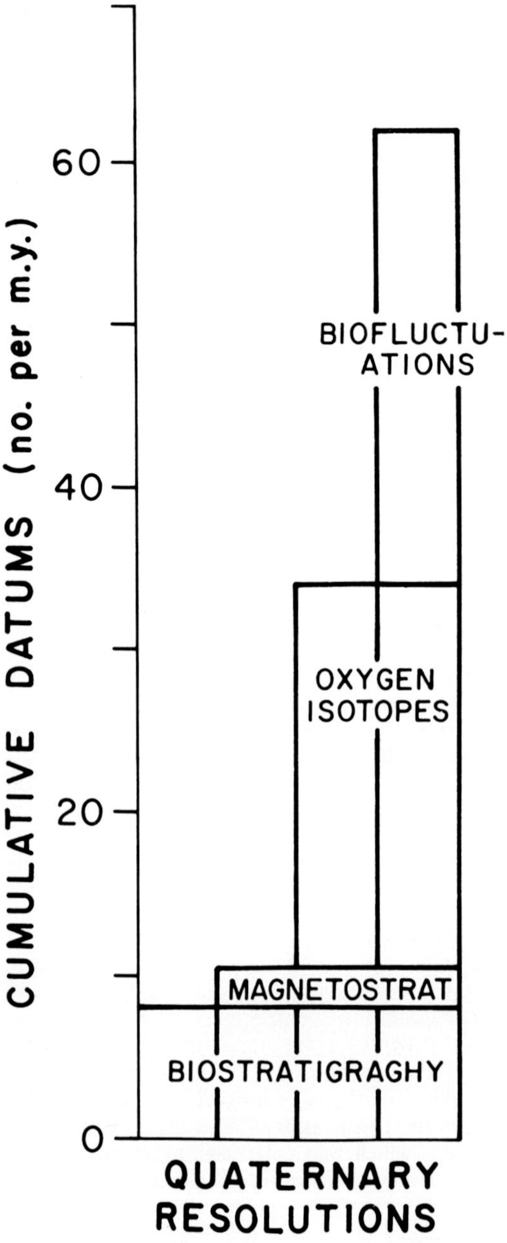

FIG. 12.—Quaternary stratigraphic resolution meas ured as cumulative number of datums per million years using biostratigraphy, magnetic stratigraphy, oxygen isotopic stratigraphy, and biotic variation (biofluctuation) stratigraphy.

1975), roughly mirror this degree of spatial variability in sea surface temperatures (Fig. 13).

The modern oceanic environment was arrived at through a long series of changes in its lateral boundaries, which led to changes in surface circulation, vertical structure, and deep circulation. The resulting modern ocean is marked by strong temperature gradients from high to low latitudes and from surface to deep waters. Furthermore the basins are isolated by longitudinal land bodies at low latitudes and interconnected only in the polar regions. The gradual subdivision of the oceans into isolated environments has led to an increase in endemism of the biota (e.g., Kennett, 1977, 1978) and to increased difficulty in correlating from basin to basin and from low to high latitude.

Here again Quaternary studies may provide insights into both the pitfalls in, and solutions to, the task of correlating from one region of the world ocean to another. For example, results of CLIMAP studies (Fig. 13) indicate that the appearance of a cooler-water assemblage in one part of an ocean basin may coincide with the appearance of a warmer-water assemblage in another part of the basin. This happened at 18,000 yr BP. when temperatures off northwestern Australia are estimated to be warmer than present while off southern New Zealand they were cooler. This synchronous, but incongruent, change in assemblages was brought out by the use of oxygen isotope stratigraphy. Use of multiple stratigraphies in the Cenozoic would also help guard against the misinterpretation of assemblage changes in regions which are oceanographically complex. If the Quaternary can be used as a guide in identifying such complex areas, it appears that the western boundary-current regions are the most difficult to deal with stratigraphically. This is particularly true of the southwestern Pacific and western Indian Oceans where surface flow is seasonally very variable or not well-defined. These studies also indicate that sections recovered from beneath the eastern boundary currents would probably be the most useful in tying together the stratigraphies of low and high latitudes. Here currents are well-defined and appear to have responded to climatic change in a similar way at all latitudes.

The central water-mass areas showed little temperature change during the Quaternary climatic extremes and may have served as refuges for some of the warm-water species. However, these areas tended to be isolated from one another during glacial times by intensified zonal and boundary currents and increased divergence and upwelling of cooler waters. This intermittent isolation could foster diachronous first and last appearances and evolutionary divergence between those regions of the oceans which appear to be the most stable.

USING DSDP MATERIAL

Before undertaking a study of cores collected by the Deep Sea Drilling Project, it is important to know the problems which exist with the sections, their recovery, and the data which have been given

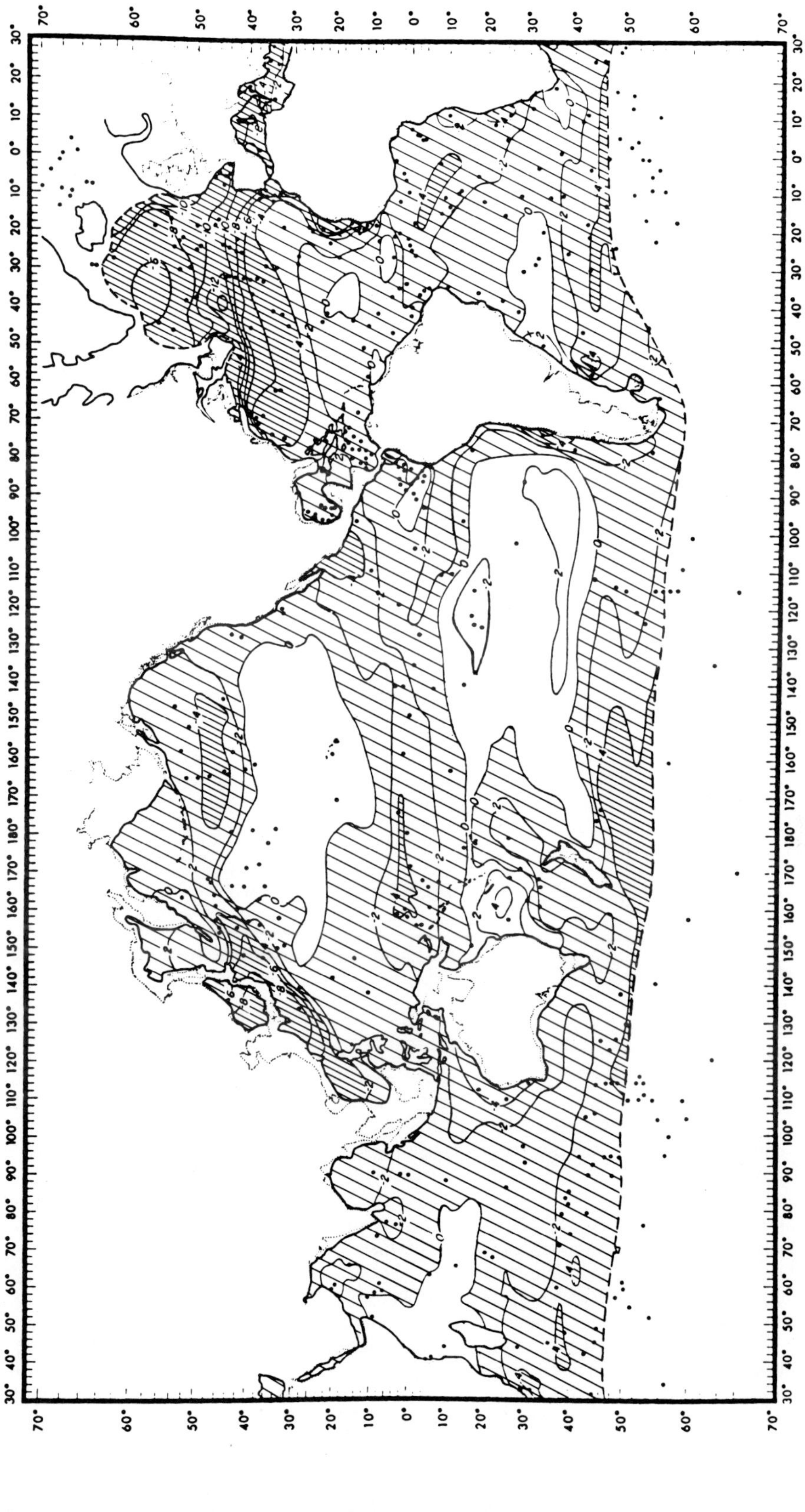

FIG. 13.—Difference between 18,000 yr BP and modern August sea-surface temperatures (after CLIMAP, 1980, *in press*). Dots indicates 18,000 yr BP (ice-age) sample locations. Dense hatchuring indicates oceanic regions cooler than 4°C during the ice-age; light hatchuring indicates regions from 0°C to 4°C cooler. Unshaded regions indicate those ocean areas which are estimated to have been warmer 18,000 yr BP than they are in modern times.

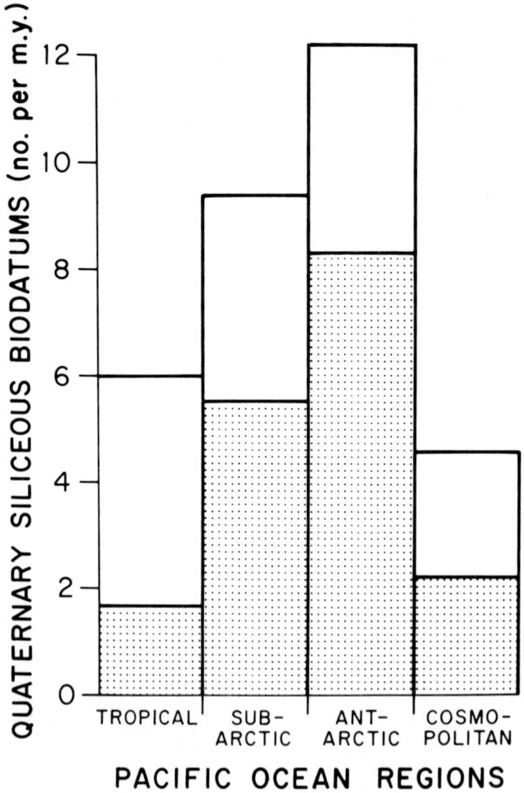

FIG. 14.—Resolution of Quaternary biostratigraphies for siliceous organisms in four Pacific Ocean zones. Shaded bars represent diatom stratigraphies; unshaded bars represent radiolarian stratigraphies. The Antarctic zone includes both Antarctic and Subantarctic assemblages. (Compiled from Burckle and Opdyke, 1977; Hays and others, 1976; Nigrini, 1971).

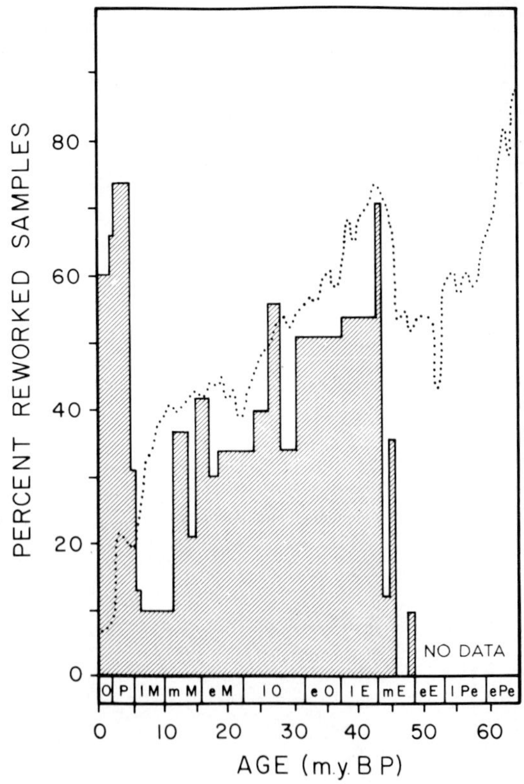

FIG. 15.—Percent of samples from the equatorial Pacific drill sites containing reworked older Radiolaria averaged over each radiolarian zone (shaded region). Dotted line indicates proportion of sections containing hiatuses (from Fig. 7). Radiolaria are absent from most Lower Eocene and Paleocene sections in the Pacific. (After Moore and others, 1978.)

in the Initial Reports. There are three basic problems encountered in dealing with the pelagic sediments recovered and reported on by DSDP. These problems lie with the sections themselves, the various taxonomies and the stratigraphers who use them, and with the nature of the recovery that the Deep Sea Drilling Project has achieved. The trouble with the recovered sections lies mainly with the fact that they are not complete. Not only do they contain abundant hiatuses (Fig. 7) but lateral reworking of older material into younger material varies directly with the abundance of hiatuses. This parallelism of hiatus abundance and reworked material indicates that the hiatuses are associated with changes in the intensity of bottom flow (Fig. 15). In some areas and for some time intervals it may be difficult to obtain a complete, unreworked section; therefore, the sections recovered from such intervals may have a low biostratigraphic resolution.

The problems with the taxonomies and the stratigraphers who apply them lies in the poorly-defined species, poor preservation during some parts of the record and personal differences in the application of the taxonomies—all of which lead to some confusion in the Initial Reports. Furthermore, different investigators have applied different degrees of attention to those species which are not required in defining the classical zonal schemes. Thus, it is difficult to refine the stratigraphy reported in the Deep Sea Drilling Initial Reports if all of the species are not consistently reported. It would seem to be an easy task to simply require that all microfossil species be reported, but it is not. The Radiolaria, for example, have almost two-hundred species living at any one time during the Cenozoic. Many of these species are undescribed and unnamed. At present it is difficult to know which are the most important in achieving a high degree of resolution in the radiolarian stratigraphy.

As an example of the problem encountered, we have looked at five different Neogene sections in the equatorial Pacific (Sites 71, 73, 77, 158 and 289). These sections are studied by four different investigators, and from their reports we have selected fifty-five radiolarian first and last appearances which were listed in two or more of the reports. From these fifty-five events we have selected eleven of the most commonly used datums (*i.e.,* commonly used as zonal markers) and have fixed them in time according to the time scale of Berggren (1972). Using these fixed age points, we have interpolated the age of the remaining forty-four radiolarian events in each of the five sections. The spread of the interpolated ages is used as a measure of the reliability of the individual datums (Fig. 16). The age of sixteen percent of these events agree within 500,000 years. Sixty percent of the events agree within two million years or less. Forty percent disagree by greater than two-million years. This selection of sites is from a relatively small, homogenous area; thus the differences noted probably result from differences in the application of the radiolarian taxonomy, as well as differences in the reworking and preservation of the assemblages.

In the past there have been several problems with the recovery of pelagic sections by the *Glomar Challenger.* In the early years the complaints revolved around too much drilling and not enough coring to suit most stratigraphers and paleoceanographers (Moore, 1972). The cores recovered were often badly disturbed or incomplete. In soft sediments layering is evidenced more often by a marbled texture of the recovered cores than it is by distinct layers. In intervals which have a mixture of lithified and nonlithified sediments the lithified portions are usually all that is recovered. Because of the disturbance of the soft sediments magnetic stratigraphy cannot be usefully performed on most of the Late Cenozoic sections. Occasionally the drilling process disturbs sediments in the upper part of the holes and causes younger material to be mixed down into the older sediments. This may lead to considerable difficulty in establishing the lower ranges of some of the species. Finally it has been noticed that even when coring continuously and recovering full barrels of sediment every time the core is taken, much of the section is missing between the cores. This missing portion of the section is evidenced by the abundant occurrence of biostratigraphic boundaries between individual cores in a continuously cored section (Fig. 17).

Recently the Deep Sea Drilling Project engineers have made a great improvement in coring technique that may overcome many of the problems with recovery that are noted above. This development is a hydraulic piston core that is driven ahead of the bit and recovers undisturbed sections which can be used for detailed biostratigraphy and paleomagnetic stratigraphy. Thus, with a little additional effort, there is hope of recovering a very good suite of Cenozoic sections which will be extremely valuable not only for stratigraphy and paleoceanography, but also for the study of geochemical processes

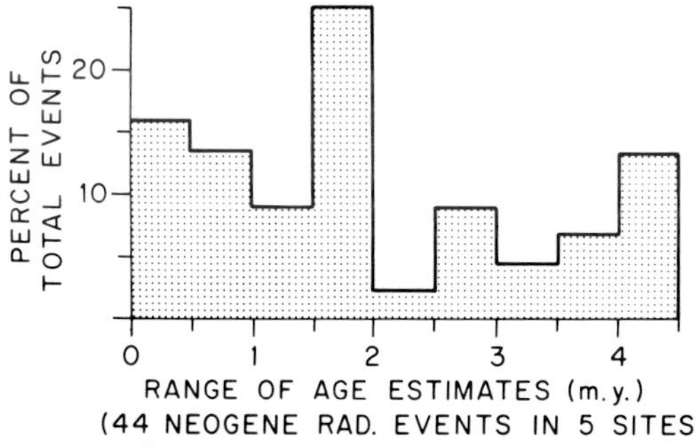

FIG. 16.—Range of age estimates of 44 Neogene radiolarian events in five DSDP sites expressed as a percentage of the total number of events. See text for explanation of the method by which the relative time scale and age of each event was established in each site.

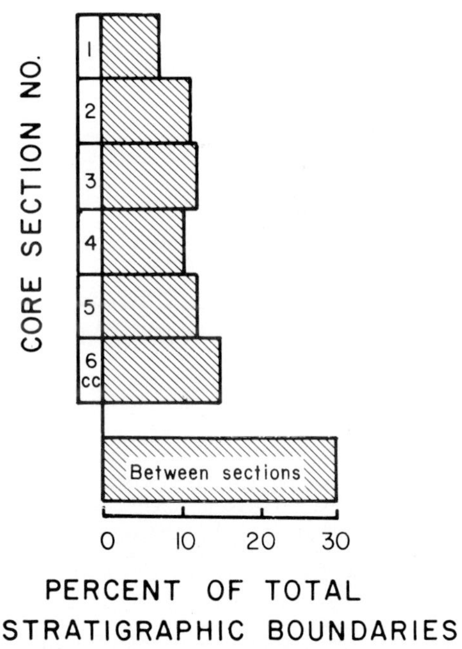

FIG. 17.—Percent of total biostratigraphic boundaries occurring in core sections and between cores in those DSDP sequences which were continuously cored and 100 percent recovery was achieved in all contiguous barrels. (After Moore and Heath, 1978.)

in pelagic sections and for the nature of the physical properties of these sections.

CONCLUSION

The Deep Sea Drilling Project has contributed greatly to the improvement and expansion of marine biostratigraphies. The recovery of many kilometers of pelagic section has not only allowed us to add several new microfossil groups to our list of biostratigraphic tools; it has also provided us with the opportunity to integrate several stratigraphic techniques and to apply them to paleoceanographic studies. There are many problems that must be dealt with in trying to improve the resolution and applicability of existing stratigraphies. These include: a) taxonomic inadequacies and disagreement, b) incompleteness of the preserved record, c) incompleteness and disturbed nature of DSDP recovery, and d) correct interpretation and correlation of stratigraphically useful signals (*i.e.,* lithologic, isotopic, and biotic variations).

With the advent of the hydraulic piston corer we now have an opportunity to develop a holistic stratigraphy which combines lithostratigraphy, isotopic stratigraphy, paleomagnetic stratigraphy, and assemblage stratigraphy with the use of first and last appearances for several fossil groups. This holistic approach to stratigraphy has two primary goals: a) high stratigraphic resolution, and b) a global chronostratigraphy.

Considerable progress is being made in certain regions of the ocean toward increasing the stratigraphic resolution that is achievable. If the Quaternary can be used as a guide for the Cenozoic, stratigraphic resolution based on the occurrence of individual events should reach 20,000 years or less. Our ability to tie these high-resolution, regional stratigraphies into a global framework of chronostratigraphy rests largely on our ability to tie them to a paleomagnetic time-scale. This is why the development and use of the hydraulic piston core by DSDP is so crucial to the advancement of stratigraphic research and the achievement of our second goal. Combining paleomagnetic stratigraphy and biostratigraphy has already proven its worth in the study of standard piston cores (Theyer and others, 1978; Burckle and Opdyke, 1977). With the recovery by DSDP of more complete sections, having higher accumulation rates and coming from several ocean basins, the ultimate in resolution will be better realized.

REFERENCES CITED

NOTE: All references to Deep Sea Drilling Project Initial Reports have been abbreviated; for complete citations see *Appendix I* at the end of this volume, or the individual Initial Reports as published.

ARRHENIUS, G., 1972, Sediment cores from the east Pacific: Reports of the Swedish Deep Sea Expeditions 1947–1948: Goteborg, v. 5, p. 6–227.

BERGER, W. H., 1970, Biogenous deep-sea sediments: fractionation by deep-sea circulation: Geol. Soc. America Bull., v. 81, p. 1385–1402.

BERGGREN, W. A., 1969a, Rates of evolution in some Cenozoic planktonic foraminifera: Micropaleontology, v. 15, p. 351–365.

———, 1969b, Cenozoic chronostratigraphy, planktonic foraminiferal zonation and the radiometric time scale: Nature, v. 224, p. 1072–1075.

———, 1972, A Cenozoic time-scale—some geologic implications for regional geology and paleobiogeography: Lethaia, v. 5, p. 195–215.

———, 1977, Atlas of Paleogene foraminifera, *in* Ramsay, A. T., ed., Oceanic micropaleontology, v. 1: New York, Academic Press, p. 205–300.

BLOW, W. H., 1969, Late middle Eocene to recent planktonic foraminiferal biostratigraphy, *in* Bronniman, P., and Renz, H. H., eds., Proceedings of the first international conference on planktonic microfossils, v. 1: Leiden, Brill.

BUKRY, D., BRABB, E. E., AND VEDDER, J. G., 1977, Correlation of Tertiary nannoplankton assemblages from the coast and peninsular ranges of California: Geol. Bull., Spec. Publ. 7, p. 1410–1483.

BURCKLE, L. H., AND OPDYKE, N. D., 1977, Late Neogene diatom correlations in the circum-Pacific: Proc. First International Congress on Pacific Neogene stratigraphy, Tokyo, p. 225–284.

CEPEK, P., AND HAY, W. W., 1969, Calcareous nannoplankton and biostratigraphic subdivision of the upper Cretaceous: Trans. Gulf Coast Assoc. Geol. Soc., v. 19, p. 323–336.

CLIMAP, 1976, The surface of the ice-age earth: Science, v. 191, p. 1131.

CLIMAP, *in press,* Seasonal reconstruction of the Ice-Age Earth: Geol. Soc. America Spec. Paper.

EMILIANI, C., 1955, Pleistocene temperatures: Jour. Geology, v. 63, p. 538–578.

ERICSON, D. B., AND WOLLIN, G., 1968, Pleistocene climates and chronology in deep-sea sediments: Science, v. 162, p. 1227–1234.

GARTNER, S., 1977, Calcareous nannofossil biostratigraphy and revised zonation of the Pleistocene: Marine Micropaleontology, v. 2, p. 1–25.

HAYS, J. D., SAITO, T., OPDYKE, N. D., AND BURCKLE, L. H., 1969, Pliocene-Pleistocene sediments of the equatorial Pacific: their paleomagnetic, biostratigraphic, and climatic record: Geol. Soc. America Bull., v. 80, p. 1481–1513.

———, LOZANO, J. A., SHACKLETON, N. J., AND IRVING, G., 1976, Reconstruction of the Atlantic and western Indian

Ocean sectors of the 18,000 B.P. Antarctic Ocean, *in* Cline, R. M., and Hays, J. D., eds., Investigations of Late Quaternary paleoceanography and paleoclimatology: Geol. Soc. America Mem. 145, p. 303-365.

HAY, W. W., MOHLER, H. P., ROTH, P. H., SCHMIDT, R. R., AND BOUDREAUX, J. E., 1967, Calcareous nannoplankton zonation of the Cenozoic of the Gulf Coast and Caribbean-Antillean area and transoceanic correlation: Trans. Gulf Coast Assoc. Geol. Soc., v. 17, p. 428-480.

KANEPS, A. G., 1973, Carbonate chronology for Pliocene deep-sea sediments: Init. Repts. DSDP, v. 16, p. 873-882.

KEIGWIN, L. D., *in press,* Late Cenozoic stable iosotope stratigraphy and paleoceanography of DSDP sites from the east equatorial and north central Pacific Ocean: Earth Planet. Sci. Letters.

KENNETT, J. P., 1977, Cenozoic evolution of Antarctic glaciation, the Circum-Antarctic Ocean and their impact on global paleoceanography: Jour. Geophys. Res., v. 82, p. 3843-3860.

———, 1978, The development of planktonic biogeography in the Southern Ocean during the Cenozoic: Marine Micropaleontology, v. 3, p. 301-345.

KIPP, N. G., 1976, New transfer function for estimating past sea-surface conditions from seabed distribution of planktonic foraminiferal assemblages in the North Atlantic, *in* Cline, R. M., and Hays, J. D., eds., Investigations of Late Quaternary paleoceanography and paleoclimatology: Geol. Soc. America Mem. 145, p. 3-41.

KOIZUMI, I., 1977, Diatom biostratigraphy in the North Pacific region: Proc. First International Congress on Pacific Neogene Stratigraphy, Tokyo, 1976, p. 235-253.

LOUTIT, T. S., AND KENNETT, J. P., 1979, Application of carbon isotope stratigraphy of late Miocene shallow marine sediments, New Zealand: Science, v. 204, p. 1196-1199.

MCINTYRE, A., KIPP, N. G., BE, A. W., CROWLEY, T., KELLOGG, T., GARDNER, J. V., PRELL, W., AND RUDDIMAN, W. F., 1976, Glacial north Atlantic 18,000 years ago: a CLIMAP reconstruction, *in* Cline, R. M., and Hays, J. D., eds., Investigation of Late Quaternary paleoceanography and paleoclimatology: Geol. Soc. America Mem. 145, p. 43-76.

MOORE, T. C., JR., 1972, DSDP: Successes, failures, proposals: Geotimes, v. 17, p. 27-31.

———, 1978, The distribution of radiolarian assemblages in the modern and ice-age Pacific: Marine Micropaleontology, v. 3, p. 229-266.

———, AND HEATH, G. R., 1977, Survival of deep-sea sedimentary sections: Earth Planet. Sci. Letters, v. 37, p. 71-80.

———, AND ———, 1978, Sea-floor sampling techniques, *in* Riley, J. P., and Chester, R., eds., Chem. oceanography, 2nd Ed., v. 7: New York, Academic Press, p. 75-126.

———, VAN ANDEL, T. H., SANCETTA, C., AND PISIAS, N., 1978, Cenozoic hiatuses in pelagic sediments: Micropaleontology, v. 24, p. 113-138.

NIGRINI, C. A., 1971, Radiolarian zones in the Quaternary of the equatorial Pacific Ocean, *in* Funnel, B. M., and Riedel, W. R., eds., The micropaleontology of oceans: Cambridge, Cambridge Univ. Press, p. 443-462.

ORR, W. N., AND JENKINS, D. G., 1977, Cenozoic planktonic foraminifera zonation and selective test solution, *in* Ramsey, A. T., ed., Oceanic micropaleontology, v. 1: New York, Academic Press, p. 163-204.

RIEDEL, W. R., AND FUNNEL, B. M., 1964, Tertiary sediment cores and microfossils from the Pacific Ocean floor: Geol. Soc. London Quart. Jour., v. 120, p. 305-368.

———, AND SANFILIPPO, A., 1970, Radiolaria, Leg 4, Deep-Sea Drilling Project: Init. Repts. DSDP, v. 4, p. 503-575.

———, AND ———, 1978, Stratigraphy and evolution of tropical Cenozoic radiolarians: Micropaleontology, v. 24, p. 61-96.

ROBERTSON, J. H., 1975, Glacial to interglacial oceanographic changes in the northwest Pacific, including a continuous record of the last 400,000 years: Ph.D. Thesis, Columbia University, New York.

RUDDIMAN, W. F., 1971, Pleistocene sedimentation in the equatorial Atlantic: stratigraphy and faunal paleoclimatology: Geol. Soc. America Bull., v. 82, p. 283-302.

RYAN, W. B., CITA, M. B., RAWSON, M. D., BURKLE, L. H. AND SAITO, T., 1974, A paleomagnetic assignment of stage boundaries and the development of isochronous datum planes between the Mediterranean, the Pacific and Indian Oceans in order to investigate the response of the world ocean to the Mediterranean salinity consistency: Riv. Ital. Paleont., v. 80, p. 631-688.

SAVIN, S. M., DOUGLAS, R. G., AND STEHLI, F. G., 1975, Tertiary marine paleotemperatures: Geol. Soc. America Bull., v. 86, p. 1499-1510.

SHACKLETON, N. J., 1967, Oxygen isotope analysis and Pleistocene temperatures reassessed: Nature, v. 215, p. 15-17.

———, AND KENNETT, J. P., 1975, Paleotemperature history of the Cenozoic and the initiation of Antarctic glaciation: oxygen and carbon isotope analyses in DSDP Sites 277, 279 and 281: Init. Repts. DSDP, v. 29, p. 743-756.

———, AND OPDYKE, N. D., 1973, Oxygen isotope and paleomagnetic stratigraphy of equatorial Pacific core V28-238: oxygen isotope temperatures and ice volumes on a 10^5year and 10^6 year scale: Quaternary Res., v. 3, p. 39-55.

———, AND ———, 1976, Oxygen isotope and paleomagnetic stratigraphy of Pacific core V28-239, late Pliocene to latest Pleistocene, *in* Cline, R. M., and Hays, J. D., eds., Investigations of Late Quaternary paleoceanography and paleoclimatology: Geol. Soc. America Mem. 145, p. 449-464.

SHAW, A. B., 1964, Time in stratigraphy: New York, McGraw-Hill, Inc., 365 p.

SOUTHAM, J. R., HAY, W. W., AND WORSLEY, T. R., 1975, Quantitative formulation of reliability in stratigraphic correlation: Science, v. 188, p. 357-359.

THEYER, F., MATO, C. Y., AND HAMMOND, S. R., 1978, Paleomagnetic and geochronologic calibration of latest Oligocene to Pliocene radiolarian events, equatorial Pacific: Marine Micropaleontology, v. 3, p. 377-395.

VAIL, P. R., MITCHUM, R. M., JR., TODD, R. G., WIDMIER, J. M., THOMSPON, S., III, SANGREE, J. B., BUBB, J. N., AND HATLELID, W. G., 1977, Seismic stratigraphy and global changes in sea level, *in* Payton, C. E., ed., Seismic

stratigraphy—applications to hydrocarbon exploration: Tulsa Okla., Amer. Assoc. Petroleum Geologists, p. 49–212.
van Hinte, J. E., 1976, A Cretaceous time-scale: Amer. Assoc. Petroleum Geologists Bull., v. 60, p. 498–516.
Wollin, G., Ericson, D. B., and Ewing, M., 1971, Late Pleistocene climates recorded in Atlantic and Pacific deep-sea sediments, *in* Turekian, K. K., ed., Late Cenozoic glacial ages: New Haven, Conn., Yale University Press, p. 199–214.

SEPM Special Publication No. 32, p. 335–353, October 1981

CENOZOIC COCCOLITHS FROM THE DEEP SEA DRILLING PROJECT

DAVID BUKRY
U.S. Geological Survey, Scripps Institution of Oceanography, La Jolla, California 92093 U.S.A.

ABSTRACT

Coccoliths, as the dominant constituent of many Deep Sea Drilling Project cores, have provided the means of rapid and detailed biostratigraphic zonation to help guide ocean-sediment coring operations aboard D.V. *Glomar Challenger.* The Cenozoic has been divided into 50 to 60 zones and subzones which are most effective for middle- and low-latitude sites. Because key stratigraphic coccoliths have proved more resistant to diagenetic change than planktic foraminifers, they contributed to dating of sedimentary beds within and just above basalt, directly demonstrating the correctness of the sea-floor spreading hypothesis. An unexpected discovery of *Braarudosphaera* beds in the South Atlantic produced valuable speculation on the response of coccolithophorids to paleoceanographic changes.

Samples from the diversity of oceanic terrains explored by early cruises of *Glomar Challenger* helped in the recognition of the effects of differential preservation—etching and overgrowth from species to species—and in the determination of biogeographic ranges which would affect the taxonomic, paleoecologic and biostratigraphic assignments of coccoliths. DSDP sites have provided the matrix needed to outline depth sequences based on dissolution rank, and paleotemperature sequences based on abundance along latitudinal traverses. New Hydraulic Piston Corer capabilities will make possible application of magnetic reversal stratigraphy to undisturbed cores where coccolith evolutionary transitions and datum-interval stratigraphy can be established.

INTRODUCTION

Coccoliths are the microscopic calcite skeletal elements (Fig. 1) produced by planktic unicellular marine algae and are most numerous and diverse in sediment deposited under tropical and subtropical conditions.

Before Leg 1 of the Deep Sea Drilling Project (August–September 1968), there was some doubt whether coccoliths would be consistently present in the sediment to be cored. Of the 56 drilled cores collected from Leg 1, 44 contain age-diagnostic coccoliths, and in 17 of these recoveries coccoliths are the only source of biostratigraphic information (Bukry and Bramlette, 1969b). Because of their great abundance in low- and mid-latitude sediment and their superior dissolution-resistance relative to planktic foraminifers, coccoliths became a primary tool for DSDP correlation. But their greatest contribution was the rapid age-determination (within five minutes of core recovery) which permitted coccoliths to guide the coring operations. This was critical to early legs which generally spot-cored selected intervals down through the drill site, instead of the continuous coring operations of later legs. The small size of coccoliths permitted favorable biostratigraphic results for tiny dabs of sediment from core catchers and from tiny vugs in chert cobbles which often contributed the only recovery from a 9-meter coring run. In terms of biostratigraphy, paleoecology, and recognizing reworking, coccoliths have been the most useful planktic fossil group for low- and mid-latitude areas. Only in high-latitude areas, where coccolith diversity becomes poorer sequentially through the Cenozoic, do siliceous microfossils provide better overall geologic information.

When the Deep Sea Drilling Project began in 1968, there were no comprehensive studies of Cenozoic oceanic zonation, preservation, or paleoecology. Several excellent biostratigraphic studies of land-sections and shallow marine cores (Stradner, 1959; Bramlette and Sullivan, 1961; Hay and others, 1967; Bramlette and Wilcoxon, 1967) had, nevertheless, provided a framework for assembling the discoveries from the extensive deep-sea sections to be cored. Early study of DSDP coccoliths had to focus on two coordinate aspects: (1) how carbonate dissolution and overgrowth affected taxonomic identification and diversity of assemblages, and (2) which sets of taxa provided the best consistent biostratigraphic sequences in the most ocean areas for zonation. Following analyses of these problems through Leg 17, the drilling operations began exploration of high-latitude cool-water areas, where ecologic limitations on coccolith ranges led coccolith workers into studies and commentaries on paleogeography and the limitations of lower-latitude relations. Parenthetically, silicoflagellate studies were adopted by several coccolith workers to augment zonation at high latitude (Bukry, 1973f; Wise, 1973; Perch-Nielsen, 1975; Haq and Riley, 1976; Martini and Müller, 1976).

BIOSTRATIGRAPHIC ZONATION

Biostratigraphic study of coccoliths was originated by Bramlette and Riedel (1954) and was only beginning to achieve recognition in the 1960's through the pioneering efforts of M. N. Bramlette,

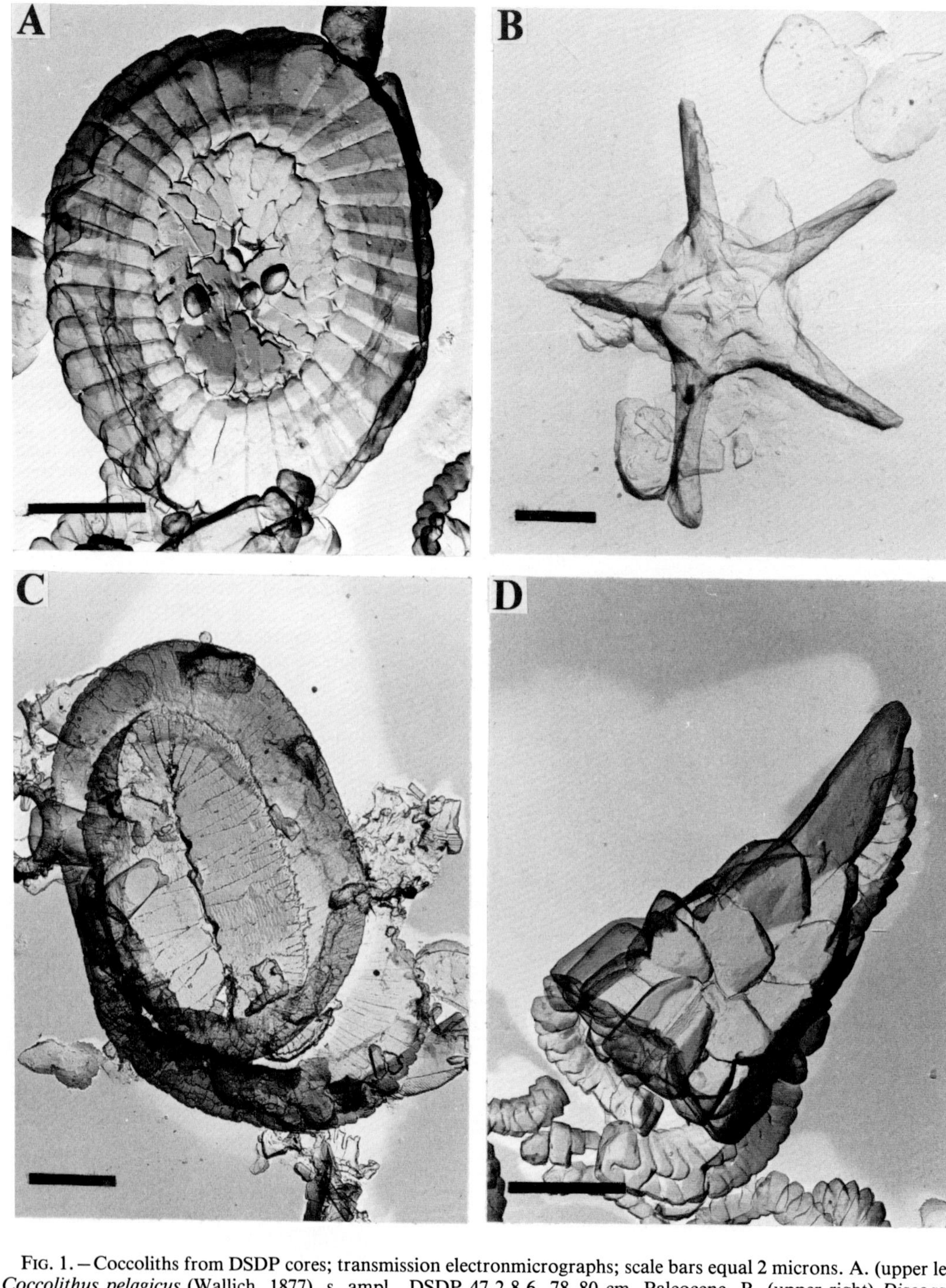

FIG. 1.—Coccoliths from DSDP cores; transmission electronmicrographs; scale bars equal 2 microns. A. (upper left) *Coccolithus pelagicus* (Wallich, 1877), s. ampl., DSDP 47.2-8-6, 78–80 cm, Paleocene. B. (upper right) *Discoaster berggrenii* Bukry, 1971, DSDP 3-9-3, 75 cm, Miocene. C. (lower left) *Helicosphaera carteri* (Wallich, 1877), DSDP 47.0-1-4, 77–78 cm, Pleistocene. D. (lower right) *Sphenolithus anarrhopus* Bukry and Bramlette, 1969, DSDP 47.2-8-6, 78–80 cm, Paleocene.

W. W. Hay, E. Martini, and H. Stradner, when DSDP began.

The Cenozoic zonation and biostratigraphy of coccoliths used for the early legs of DSDP were developed from studies of discontinuous land-based and nearshore marine sections (Martini and Bramlette, 1963; Hay, 1964; Bramlette and Wilcox-on, 1967; Hay and others, 1967; Bramlette and Bukry, 1968; Gartner, 1969). With the recovery of more continuous and complete sequences from deep-ocean localities the means and the need for a Cenozoic oceanic zonation led to several early zonations (Bukry and Bramlette, 1970a; Milow, 1970; Bukry, 1971c; Martini and Worsley, 1971). As more experience with oceanic assemblages was gained, relative ranges of species became better established, new key species were identified from intervals not previously available for detailed study onshore, and the limitations of species-selective diagenetic affects were recognized. These developments produced a series of more complete and detailed zonations more atuned with new oceanic relations (Bukry, 1971b, 1973c, 1975b; Perch-Nielsen, 1972, 1977; Roth, 1973; Martini, 1971, 1976; Gartner, 1977; Okada and Bukry, 1980). Because of selective diagenesis and provinciality, some species, formerly used as zonal guide species in nearshore and ashore, were generally lacking in deep-sea sediment. For example, secondary marker species events had to be selected for the Miocene *Helicosphaera ampliaperta* Zone (Bukry, 1971a, 1973c, unpublished data from Site 292, 1980) or the original *Helicosphaera ampliaperta* Zone would be difficult to identify in ocean sediment (Hay, 1971; Martini, 1976) because of the scarcity of *H. ampliaperta*.

While investigating and discovering the various affects of selective diagenesis and provinciality on biostratigraphy it was expedient to use multiple concurrent range zone relations (Bukry, 1971b). With better understanding of species distributions, and the advent of paleomagnetic data for DSDP sediments it will become increasingly practical to identify the ages associated with the lowest occurrence surface (LOS) and highest occurrence surface (HOS) of species (Hay and Beaudry, 1973) in given regions.

The Cenozoic zonations compiled by Martini (1971) and Bukry (1973c) have been most frequently applied in DSDP coccolith stratigraphy. Modifications by the authors (Bukry, 1975b; Martini, 1976) and others have been offered to improve the usefulness of these systems. The use of different guide species between the 46-zone system of Martini (1971) and the 59-zone and subzone system of Okada and Bukry (1980) produces offset correlations between zones (Fig. 2).

During the early legs of DSDP, new oceanic species relations were discovered and incorporated into several developing systems of correlation. Leg 1 showed that a significant acme of *Discoaster triradiatus* would help to identify the late Pliocene and that coccolith assemblages could be dated rapidly enough to permit chief scientists to core or wash successive sediment intervals without any loss of operating time. Both shorelab and shipboard coccolith reports relied on subseries instead of zones for Legs 1 and 2. For Leg 3, an amalgamation of pre-DSDP zones and new DSDP assemblage relations produced the first of many zonations to be published in DSDP volumes (Bukry and Bramlette, 1970a). Preprints of this zonation and Gartner's (1969) zonation were used in preparing the first synthesis of a complete Cenozoic zonation for DSDP (Martini and Worsley, 1971). An especially innovative zonation by Milow (in McManus and others, 1970) for Leg 5 off California, incorporated a mixture of coccoliths, diatoms, and silicoflagellates to identify biostratigraphic units, and proposed a system of subzones to recognize potential differences in zonal assemblages. Although not explicitly defined, some of the subzonal units were based on coccolith species which had not been used before for zonation. Both *Chiasmolithus gigas* and *Discoaster variabilis* became part of subsequent zonation (Bukry, 1971c, 1973f).

Each DSDP cruise generated new information about coccolith ranges from different parts of the stratigraphic column. The first occurrence of *Discoaster exilis* was proposed as the base of the *Discoaster exilis* Zone in the Atlantic (Hay, 1970); but secondary overgrowth confused identification of this species with other overgrown long-rayed discoasters in the Pacific, so an alternative *Sphenolithus heteromorphus* extinction has been used. Alternative criteria involving short-rayed versus long-rayed discoaster abundances were applied (Bukry, 1971b) for the base of the *S. heteromorphus* Zone because of a lack of nearshore marker *Helicosphaera ampliaperta.* Philosophical differences between datum (isochronous guide species) zonations and concurrent-assemblage zonations, such as the reports of Pliocene "datum" species *D. asymmetricus* (4 m.y., Meulenkamp and others, 1979) in the Miocene (Bukry, 1971b) or first appearing above the *A. tricorniculatus* extinction (Gartner, 1972; Roth and Thierstein, 1972) were presented in Leg 7.

Early zonations were continually upgraded. For example the Miocene *Discoaster druggii* Zone base, identified by the first *Discoaster druggii*, was put into doubt by discontinuous occurrences from Legs 8, 30, and 33 (Bukry, 1976), where two abundance peaks were separated by a nonoccurrence interval.

AGE	ZONE		SUBZONE		Martini (1971) Zone
Quaternary	CN15	*Emiliania huxleyi*			NN21
	CN14	*Gephyrocapsa oceanica*	CN14b	*Ceratolithus cristatus*	NN20
			CN14a	*Emiliania ovata*	NN19
	CN13	*Crenalithus doronicoides*	CN13b	*Gephyrocapsa caribbeanica*	
			CN13a	*Emiliania annula*	
Pliocene	CN12	*Discoaster brouweri*	CN12d	*Calcidiscus macintyrei*	NN18
			CN12c	*Discoaster pentaradiatus*	NN17
			CN12b	*Discoaster surculus*	NN16
			CN12a	*Discoaster tamalis*	
	CN11	*Reticulofenestra pseudoumbilica*	CN11b	*Discoaster asymmetricus*	NN15
			CN11a	*Sphenolithus neoabies*	
	CN10	*Amaurolithus tricorniculatus*	CN10d	*Amaurolithus delicatus*	NN14
			CN10c	*Ceratolithus rugosus*	NN13
			CN10b	*Ceratolithus acutus*	NN12
Miocene			CN10a	*Triquetrorhabdulus rugosus*	
	CN9	*Discoaster quinqueramus*	CN9b	*Amaurolithus primus*	NN11
			CN9a	*Discoaster berggrenii*	
	CN8	*Discoaster neohamatus*	CN8b	*Discoaster neorectus*	NN10
			CN8a	*Discoaster bellus*	
	CN7	*Discoaster hamatus*	CN7b	*Catinaster calyculus*	NN9
			CN7a	*Helicosphaera carteri*	
	CN6	*Catinaster coalitus*			NN8
	CN5	*Discoaster exilis*	CN5b	*Discoaster kugleri*	NN7
			CN5a	*Coccolithus miopelagicus*	NN6
	CN4	*Sphenolithus heteromorphus*			NN2/NN5
	CN3	*Helicosphaera ampliaperta*			
	CN2	*Sphenolithus belemnos*			
	CN1	*Triquetrorhabdulus carinatus*	CN1c	*Discoaster druggii*	
			CN1b	*Discoaster deflandrei*	NP25/NN1
Oligocene			CN1a	*Cyclicargolithus abisectus*	
	CP19	*Sphenolithus ciperoensis*	CP19b	*Dictyococcites bisectus*	
			CP19a	*Cyclicargolithus floridanus*	NP24
	CP18	*Sphenolithus distentus*			NP23
	CP17	*Sphenolithus predistentus*			
	CP16	*Helicosphaera reticulata*	CP16c	*Reticulofenestra hillae*	NP22
			CP16b	*Coccolithus formosus*	NP21
			CP16a	*Coccolithus subdistichus*	
Eocene	CP15	*Discoaster barbadiensis*	CP15b	*Isthmolithus recurvus*	NP19/NP20
			CP15a	*Chiasmolithus oamaruensis*	NP18
	CP14	*Reticulofenestra umbilica*	CP14b	*Discoaster saipanensis*	NP17
			CP14a	*Discoaster bifax*	NP15/NP16
	CP13	*Nannotetrina quadrata*	CP13c	*Coccolithus staurion*	
			CP13b	*Chiasmolithus gigas*	
			CP13a	*Discoaster strictus*	
	CP12	*Discoaster sublodoensis*	CP12b	*Rhabdosphaera inflata*	NP14
			CP12a	*Discoasteroides kuepperi*	
	CP11	*Discoaster lodoensis*			NP12/NP13
	CP10	*Tribrachiatus orthostylus*			
	CP9	*Discoaster diastypus*	CP9b	*Discoaster binodosus*	NP11
			CP9a	*Tribrachiatus contortus*	NP10
Paleocene	CP8	*Discoaster multiradiatus*	CP8b	*Campylosphaera eodela*	NP9
			CP8a	*Chiasmolithus bidens*	
	CP7	*Discoaster nobilis*			NP7/NP8
	CP6	*Discoaster mohleri*			
	CP5	*Heliolithus kleinpellii*			NP6
	CP4	*Fasciculithus tympaniformis*			NP5
	CP3	*Ellipsolithus macellus*			NP4
	CP2	*Chiasmolithus danicus*			NP3
	CP1	*Zygodiscus sigmoides*	CP1b	*Cruciplacolithus tenuis*	NP2
			CP1a	*Cruciplacolithus primus*	NP1

Improvement of the identification criteria for the Miocene *Amaurolithus tricorniculatus* Zone and underlying *Discoaster quinqueramus* Zone was needed for Leg 9, because of apparent overlap and nonoverlap of the guide species through the same section. Definition of the *Amaurolithus primus* Subzone, on the basis of the first *Amaurolithus primus,* within the *D. quinqueramus* Zone resolved the problem (Bukry, 1971c).

A useful improvement for chronostratigraphic application of coccolith stratigraphy was the introduction of estimated ages and durations for coccolith zones of Leg 12 (Perch-Nielsen, 1972). The zonation described is a combination of prior zonations with the new addition of the *Discoaster nobilis* Zone for the Paleocene. The lack of many Neogene zonal guide species at high-latitude was noteworthy.

Leg 13 in the Mediterranean Sea provided an opportunity to test the relative ranges of oceanic zonal guide species in a restricted setting. The resulting range charts (Bukry, 1973b) are similar except for more common Quaternary *Braarudosphaera,* a shorter upper range for warm-water species *Ceratolithus rugosus* and a longer range for cosmopolitan *Calcidiscus macintyrei;* also, *Discoaster* is relatively sparser than at oceanic sites. A general absence or scarcity of *Helicosphaera sellii* above the *Gephyrocapsa caribbeanica* Subzone in the Mediterranean (Bukry, 1973b), was later extended to DSDP oceanic cores and used to identify an *H. sellii* Zone for a revised Quaternary zonal system (Gartner, 1977). [Aside from Katharina Perch-Nielsen's (1979) Danian zonation, nearly all published improvements to transoceanic Cenozoic zonation in the 1970 decade were based on DSDP cores.]

Borrowing and modification of elements (species and zonal criteria) between zonal systems was commonplace through the early DSDP. For example, the base of the *H. ampliaperta* Zone was modified from the last *Sphenolithus belemnos* to the first *S. heteromorphus* for Leg 14 (Bukry, 1972b).

A general commitment to continuous coring on Legs 15 and 16 helped to produce a low-latitude biostratigraphic zonation and a sequence of stages and a discoaster range chart, all based principally on DSDP coccolith distributions (Bukry, 1973c, d). Many new stratigraphic species (distinct morphologies with brief stratigraphic ranges) were identified from these legs (Hay and Beaudry, 1973; Bukry, 1973d).

More northerly drilling along the Pacific Coast on Leg 18 produced impoverished assemblages where most tropical guide species were missing. Silicoflagellates and cosmopolitan coccoliths were used to design a new local zonation which contained some duplication of earlier zone names, justified by noting that usage takes precedence over priority in zone names (Wise, 1973). There is no serious difficulty in this approach for high-latitude assemblages, but the confusion between the high-latitude *Triquetrorhabdulus carinatus* Zone definition and numerous low-latitude definitions (Bukry, 1978b) would seem to make a local numbered or lettered system of zone titles more unique. Naming of zones would appear to be a trivial item, but because zone names are a means of general communication it is preferable to reduce confusion where possible. But individual taste in nomenclature still plays an important role in biostratigraphy. For example, the uppermost Pliocene *Calcidiscus macintyrei* Subzone or Zone has been used routinely in DSDP reports since 1971. But a revised zonation (Gartner, 1977) designated a lowermost Quaternary unit with the same name, justified because the extinction of the nominative species should logically be used as a zone boundary. The same criteria could not, however, be applied to the *Gephyrocapsa oceanica* Zone in the same zonation proposal (Gartner, 1977). Therefore, it would seem expedient to clearly identify the year, author, and source of all DSDP zone names to avoid ambiguous or illogical correlations. Of course, using unique codes of letters and numbers (G77-1, G77-2, etc.) for individual zonations would also preserve the authors intent without compromising correlations for other systems.

A more effective alternative solution to the duplication of zone names having different guide species, or stratigraphic significance, is possible. Through DSDP coring, the relative sequence of coccolith stratigraphic ranges has become generally well known. This situation permits the use of a datum interval (DI) to convey stratigraphic information directly in the format of a zone. For example, a core, or any stratigraphic interval, containing the last *Discoaster hamatus* at the base and the first *D. quinqueramus* at the top would presently be designated the *Discoaster neohamatus* Zone (CN8) of Bukry (1971c) and Okada and Bukry (1980) or the *Discoaster calcaris* Zone (NN10) of Martini (1971). Using the concept of a datum-interval name this biostratigraphic unit would be designated

DI: *D. hamatus*†/*D. quinqueramus**.

In well-diversified assemblages within that DI, even smaller intervals could be identified, such as

DI: *D. pentaradiatus**/*D. loeblichii**.

This would make labelling of biostratigraphic units more direct and specific.

FIG. 2.—Comparison between the Okada and Bukry (1980) coded coccolith zonation from DSDP Leg 63 (Bukry, *in press*) and Martini's (1971) standard coccolith zonation. Intervals having noncorresponding boundaries shown by slashes.

In the poorest, etched assemblages, even monospecific situations, a DI could be designated as an incomplete (I) unit—

IDI: *Dictyococcites bisectus*

—which would connote the full range of the taxon (name shown without event symbol).

Use of the DI concept would emphasize the presence of species relative to the present system of zones which effectively uses the absence of species for several intervals. For example a core within the *Discoaster pentaradiatus* Subzone (CN12c) defined between the extinctions of *D. surculus* and *D. pentaradiatus* might be designated

DI: *Rhabdosphaera claviger*/D. pentaradiatus*†

because both species were present throughout the core.

The key to using the DI biostratigraphic system is establishing a sequence of coccolith datum events. The new Hydraulic Piston Corer (HPC) of DSDP will provide the best opportunity to recover undisturbed sediment sequences from various oceanographic areas so the coccolith sequence can be most accurately assembled. When this system of events is eventually completed it could be numbered for optional reference. For the moment, the species sequences, based on DSDP rotary-cored sediment, which are detailed as part of zonal systems such as Martini (1971), Bukry (1973c), and Gartner (1977) can be assembled.

The opportunity to deal with a biostratigraphic unit and datums that simultaneously convey direct information without the ambiguity and misquotation that is possible with traditional zone names is an appealing possibility that has been presented through DSDP studies. Although datum-sequence stratigraphy has been advocated previously by Burckle (1978) for diatoms, the concept of the datum interval, instead, produces more familiar biostratigraphic units (surrogate zones) for reporting. The DI arises from new data and is not a previous construct that must be imposed onto the new data. Beyond the identification of species, less subjective authorship is involved.

Application of low-latitude zonation was not successful for North Pacific high-latitude Legs 18 and 19. For later high-latitude drilling off New Zealand on Leg 21, Edwards (1973) recommended use of regional zonations instead of trying to massage tropical zonations to fit. At high-latitude in the North Atlantic on Leg 38, Müller (1976) also had to design a special local zonation which used cosmopolitan species and unusual taxa such as *Imperiaster*. Coccolith assemblages from adverse environments provide only generalized biostratigraphic units of long duration. Although the list of new stratigraphic coccolith species which are limited to cold-water has grown as a result of DSDP studies—new cold-water species include:

Discoaster intercalaris Bukry, 1971
Coccolithus pliopelagicus Wise, 1973
Cyclicargolithus bukryi Wise, 1973
Discoaster mendomobensis Wise, 1973
Ericsonia tasmaniae Edwards and Perch-Nielsen, 1975
Helicosphaera? subantarctica (Edwards and Perch-Nielsen, 1975)
Hornibrookina australis Edwards and Perch-Nielsen, 1975
Crenalithus productellus Bukry, 1975
Dictyococcites antarcticus Haq, 1976
Heliolithus universus Wind and Wise, 1977—the majority of all species and, particularly, short-ranged species, are from subtropical and tropical areas. Therefore, there is a low priority to coccolith zonation at high-latitude, compared to siliceous diatoms, radiolarians, and silicoflagellates that have diverse assemblages at high latitude. In high-latitude Site 277 of Leg 29, Edwards and Perch-Nielsen (1975) noted that the order between the appearances of *Dictyococcites bisectus* and *Chiasmolithus oamaruensis* in the Eocene was reversed between Denmark and Site 277. High-latitude correlations may be less consistent than low-latitude.

In publications for Pacific Leg 32, new subzones were added to the low-latitude zonation to improve resolution (Bukry, 1975b) and for Leg 33, new consolidations were made to the standard zonation to improve the zonal sequence for the Pacific (Martini, 1976). Both groups of modifications reflect the continued addition of biostratigraphic data generated from new DSDP coring. Perch-Nielsen (1977) made numerous stratigraphic observations for Leg 39 in the South Atlantic, including the abundance of thoracospheres marking the Cretaceous/Tertiary boundary, the absence of *Biantholithus* and NP1 in many sections, and discussions of relative species ranges within zones. Also from South Atlantic cores of Leg 40, Proto Decima and others (1978) combined and cross-correlated coccolith zonations and also reconfirmed that *Braarudosphaera* and thoracospheres characterized the Danian for Leg 40. Bukry (1978a) reported that analysis of the *Sphenolithus distentus* Zone coccolith assemblages in the Oligocene *Braarudosphaera* pentalith chalk beds of the South Atlantic (discovered on Leg 3) were found to be nearly identical in and around the beds. The unusual events creating the beds seem to have mainly affected *Braarudosphaera* production.

The dramatic change between Maestrichtian and Danian coccolith assemblages (Bramlette and Martini, 1964) has been a continuing interest with each new DSDP recovery, from Leg 3, considered discontinuous (Maxwell and others, 1970), and Leg 6, where the foraminifer change preceded the coccolith change (Bukry and others, 1971), through Leg 43, where continuous detailed range relations were described (Thierstein and Okada, 1979). Description of Maestrichtian/Danian biostratigraphic zonation (Perch-Nielsen, 1979) and the analysis of the significance of boundary assemblages (Thierstein, this volume) including *Braarudosphaera,* are important new contributions.

Further drilling in the Mediterranean Sea by Leg 42A showed that many oceanic guide species were sparse or missing in the Neogene. It was necessary to change the definition for the top of Martini's (1971) *Discoaster brouweri* Zone (NN18) from the last *Discoaster brouweri* to the last *Calcidiscus macintyrei* because discoasters were so rare (Müller, 1978). This new correlation moves the Pliocene/Pleistocene boundary to 1.51 m.y. if the *C. macintyrei* disappearance is isochronous between the Mediterranean and Pacific (Gartner, 1977), and should therefore be reevaluated from the standpoint of classical Italian sections (Rio, 1974) and paleomagnetic correlations (Haq and others, 1977a) to determine if the indicated 100,000-year age difference between the Pacific and Mediterranean Pliocene/Pleistocene boundaries is reproducible or an artifact of magnetic time scales. Correlation of coccolith stratigraphy with magnetic reversal stratigraphy holds great promise for establishing the relative time sequence between the most consistent stratigraphic guide species. Initial results on the Neogene off northwest Africa were recently presented for Leg 47A (Mazzei and others, 1979). There is good correlation to Gartner's (1977) paleomagnetic determinations for the Pacific and to relative species ranges in Italian sections.

PRESERVATION

Prior to DSDP it was believed that large planar crystal faces on some discoasters were primary biologic features (Black, 1963, 1972; Hay and Mohler, 1967; and see *Discoaster woodringii* Bramlette and Riedel, 1954). The extensive coring of carbonate-rich mid Cenozoic sediment by early DSDP legs recovered complete gradation sets showing sequences of secondary overgrowth over biologic crystallite forms (Bukry, 1971a; Martini and Worsley, 1971; Roth and Thierstein, 1972; Wise, 1973). A synopsis of this important effect is shown in San Miguel Arribas (1976) and Wise (1977), and DSDP samples were used to simulate the diagenetic overgrowth process in a laboratory high-pressure bomb experiment (Adelseck and others, 1973).

The overgrowth effect has generally been linked to differential solution from other coccoliths and foraminifers which may be thinned and ragged in the same sample. Relative sequences have been assembled that show ortholithids such as *Discoaster* to be most resistant to dissolution; however, the perforation and disaggregation which can affect *Discoaster* (Roth and Thierstein, 1972) has been traced to show that some of the small carbonate particles of indeterminate origin in marine sediment can be solution remnants of *Discoaster* rays (Bukry, 1973e, pl. 1, 2). Differential solution of Paleogene assemblages can concentrate *Discoaster*, a warm-water indicator (Hay, 1970), whereas after the extinction of *Discoaster,* in the Quaternary, intense solution would concentrate cool-water taxa such as *Coccolithus pelagicus* (Berger, 1973). Less differential solution of coccoliths relative to planktic foraminifers is the basis for the great significance of coccoliths to DSDP stratigraphy. Key biostratigraphic guide coccoliths, such as *Discoaster*, simply survive dissolution longer than key foraminifers, or to a greater depth. Other coccoliths such as holococcoliths are preserved only under shallow marine conditions (Gartner and Bukry, 1969). *Zygrhablithus bijugatus* probably is preserved well above the calcite-compensation depth (Bramlette, 1961) at less than 1000-2000 m (Edwards and Perch-Nielsen, 1975; Wise and Wind, 1977; Perch-Nielsen, 1977). General solution sequences and indices have been proposed to help determine the solution status of an assemblage (Bukry, 1971c; Scheidermann, 1973; Roth and Berger, 1975), because this knowledge is necessary in order to draw accurate conclusions about the biostratigraphy and paleoecology of an assemblage (Roth, 1974). For example, the exclusion of *Discoaster* from North Pacific/Bering Sea Leg 19 was attributed to ecologic tolerance, because of the high-solution resistance of the genus. Preservation state has also been used in DSDP studies to suggest phylogenetic relation. Hekel (1973) noted the character of little overgrowth on *Hayaster perplexus* in tropical Leg 20 sites was more similar to placoliths than to *Discoaster. H. perplexus* had originally been described in the genus *Discoaster*. The most useful aspect of coccolith dissolution is that it provides a means to indicate relative paleodepth in a group of planktic organisms (Bukry, 1971c). The coccolith indicators are not as precise as the planktic foraminiferal depth-solution guide (Berger, 1967; Roth and Berger, 1975; Ruddiman and Heezen, 1967), perhaps because of more complex size and transport relations (Honjo, 1975). Empirically, DSDP recoveries from various depths showed a depth and solution hierarchy (Bukry, 1971c) for the lower Cenozoic which, on the average, produced the following order from deepest to shallowest:

Discoaster
Coccolithus
Calcidiscus
Dictyococcites
Reticulofenestra
Chiasmolithus
Sphenolithus

Discolithina (imperforate)
Helicosphaera
Scyphosphaera
Lophodolithus
Braarudosphaera
Micrantholithus
Discolithina (perforate)
Rhabdosphaera
Syracosphaera
Transversopontis

The Quaternary ranking by Roth and Berger (1975) is similar for surviving genera and equivalents, such as *Scyphosphaera, Pontosphaera* (= *Discolithina*), and *Syracosphaera,* being among the most susceptible to solution at depth.

Study of DSDP assemblages from many different sedimentary facies and diagenetic situations forced an early recognition of the significance of differential preservation to relevant studies of marine microfossils (Bukry and others, 1971, 1973). A convenient notation for overgrowth and solution (Roth and Thierstein, 1972) helped to focus continuing attention on the preservation effect. However, considerable work is still needed to establish preservation sequences for various taxa so identification keys can be made widely available. Preservational changes to the form of coccoliths exceed any changes of form based on intraspecific variation. The correlated array of DSDP cores will make efficient preservation studies possible.

COCCOLITH PALEOECOLOGY

The prime aim of DSDP Legs 1 to 11 was to establish deep-ocean stratigraphic reference sections and test the hypothesis of sea-floor spreading in areas where the new drillship systems could be proven in safety. Therefore, these legs investigated tropical and subtropical regions that provided excellent coccolith assemblages for zonation studies. Although variations in species occurrences were indirectly credited partly to ecologic controls in Legs 1 to 4, the Leg 5 assemblages off California provided the first relatively coolwater comparisons with the tropics. The younger Neogene assemblages lack diversity such that there are fewer guide species available for zonation. In the Upper Miocene, robust, long-ranged *Discoaster* species, *D. intercalaris* and *D. variabilis* predominate, whereas tropical biostratigraphic guides, such as *D. bellus, D. neohamatus, D. pentaradiatus,* and *D. quinqueramus* are missing or scarce. The birefringent ceratoliths, which provide excellent evidence for the Miocene/Pliocene boundary in the tropics, are missing (Bukry and Bramlette, 1970b). The presence of some members of tropical assemblages, such as *Amaurolithus tricorniculatus* s. ampl. and *D. variabilis,* suggested that ecology (probably temperature), and not solution, was responsible for the exclusion of other species. This early encounter with ecological limitations to the application of coccolith biostratigraphy helped to influence subsequent biostratigraphic systems, as in the recognition of subzones (Milow, 1970; Bukry, 1971c) and maintenance of the multiple concurrent-range zone concept (Bukry, 1971b). More experience with watermass or latitudinal temperature transitions was needed.

Preliminary comparisons of DSDP assemblages showed species which favored warm-water, seen in the abundances of *Scyphosphaera globulata* and *Sphenolithus abies,* and those which tolerated broader ranges of conditions such as *Calcidiscus leptoporus* (Bukry, 1971a). Factors which might complicate establishing the paleoecologic significance of species were differential preservation (Bukry and others, 1971) and shifting ocean-current patterns (Bukry, 1971b).

Early in phase two of DSDP, Leg 12 was drilled in the North Atlantic at high latitude. Significant ecological restrictions on biostratigraphic guide species were noted (Perch-Nielsen, 1972). The extinction of *Discoaster* which had long been used to indicate the Pliocene/Pleistocene boundary (Ericson and Wollin, 1962; Riedel and others, 1963) was found to occur earlier in the middle Pliocene. Similarly, the extinctions of *Reticulofenestra pseudoumbilica* and *S. abies* were reported to occur earlier at high latitude (Perch-Nielsen, 1972). Another, older, Eocene-Oligocene biostratigraphic guide, *Sphenolithus pseudoradians* was missing—a high-latitude relation later confirmed for the Southern Ocean on DSDP Leg 28 (Burns, 1975). The Leg 12 assemblages also showed lower diversity through the lack of many discoasters, such as *D. tanii,* thus accentuating the generally low Oligocene discoaster diversity (Haq, 1971; Bukry, 1972a). Domination of assemblages by a few cosmopolitan species, mainly placoliths, was noted, as were the especially great abundances of *Chiasmolithus altus* and *Zygrhablithus bijugatus* in the Oligocene and *Coccolithus pelagicus* and *Helicosphaera sellii* near the Pliocene/Pleistocene boundary (Bukry, 1972a). These observations helped to identify the nontropical significance of these species during those time periods.

Following the association of Quaternary *Coccolithus pelagicus* with coldwater conditions of less than 14°C (McIntyre and Bé, 1967), the abundance of this species was used as a paleotemperature guide for Quaternary and older assemblages at DSDP Site 62 in the Pacific (Martini and Worsley, 1971). The temperature significance of *C. pelagicus* appears to falter in the late Miocene because one of the intervals of greatest *C. pelagicus* abundance at DSDP Site 157 is also associated with warm-water silicoflagellate assemblages (Bukry, 1973d).

High-latitude drilling in the North Pacific on Leg 18 showed that among discoasters only *Discoaster deflandrei* occurred in the Miocene as far north as the Gulf of Alaska at Site 178 (Wise, 1973). But even this predominant cosmopolitan middle Cenozoic species probably arrived at high-latitudes via tectonic plate motions. Younger species of discoasters are missing there. In sites near the Aleutians, drilled on Leg 19, ecologic conditions excluded the solution-resistant genera *Discoaster* and *Sphenolithus* from the younger Cenozoic (Bukry, 1973g). The coolwater assemblages there were noted to be dominated by placoliths, a relation later shown to be bipolar by results from the Southern Ocean (Wise and Wind, 1977). For Leg 21 sites, between Australia and New Zealand, a *Discoaster/Chiasmolithus* ratio was suggested as a paleotemperature guide for the Paleogene (Bukry, 1973h). The ratio relied on the observation that *Discoaster* was, again, scarce at cooler sites, whereas *Chiasmolithus* was more common. The relation has proved only broadly useful because of reversals in detailed studies (Wise and Wind, 1977) and nonconcordance with benthic foram isotope ratios (Perch-Nielsen, 1977); nevertheless, beguiling examples of consistent trends across the Eocene-Oligocene boundary, as at Site 360 off South Africa, do exist (Bukry, 1978a). Finally, Antarctic drilling revealed that coccolith assemblages of the earliest Cenozoic (Danian) were basically cosmopolitan throughout the oceans (Haq, 1976). In Oligocene assemblages, a new local species, *Dictyococcites antarctica* (Haq, 1976), was described. Bipolar species, such as *Coccolithus pliopelagicus,* were noted in younger strata (Bukry, 1975a).

Observations on ecologic distributions at lower latitudes included *Discoaster brouweri* and *D. blackstockae* being more prominent than *D. surculus* in tropical areas and the preliminary application of a *Discoaster/Coccolithus* ratio for Pliocene paleotemperature trends (Bukry, 1976). Off Africa, Paleogene *Zygolithus dubius* was noted to be rare in the tropics (Proto Decima and others, 1978), much like its probable descendant, *Isthmolithus recurvus* (Bukry, 1975b), which confirmed earlier DSDP reports about these species in the tropical Pacific (Martini and Worsley, 1971). Drilling of the Mid-Atlantic Ridge by Leg 37 showed that the Pliocene biostratigraphic guide *Ceratolithus* was scarce poleward of about latitude 35° (Bukry, 1977), thus providing some limitation to the area where tropical zonations could be applied.

Because of the number and far-flung distribution of DSDP sites, it has been possible to rapidly distinguish between preservational bias and actual ecologically controlled occurrence patterns through time and to synthesize the results in terms of floras. Integration of plate tectonic motions and oceanographic evolution through the Cenozoic is providing better understanding of past ecologic relations (Bukry, 1978b). Studies of changing coccolith floras through the Cenozoic (Haq and Lohmann, 1976; Haq and others, 1977b; and Haq, *in press*), following the pioneering studies on Quaternary floral migrations (McIntyre and Jantzen, 1969; McIntyre and others, 1972) will provide valuable feedback information for revealing and understanding changes in Cenozoic ocean circulation and temperature patterns.

UPPER CENOZOIC COCCOLITH PALEOECOLOGY FROM A PACIFIC COAST TRAVERSE

The DSDP Pacific Coast Sites 173 and 467 to 473 provide the opportunity to compare species distributions in assemblages of similar age at various latitudes from about 40°N to 20°N through the California Current area. For the middle Miocene to Quaternary many assemblage differences follow North-South trends, presumably relating to temperature gradations along the Coast (Lynn, 1967; Addicott, 1970). A group of coccolith genera was selected for survey counts of 100 specimens to assess changing ecologic controls on key stratigraphic taxa off the Pacific Coast. The seven genera selected are the source of the majority of the zonal guide species for the Miocene to Quaternary. The ratio of warm-indicator *Discoaster* to cool-indicator *Coccolithus* was calculated from this first survey. A second independent survey of 100 specimens of discoaster counted all of the *Discoaster* species present, because some species like *D. brouweri* and *D. signus* are abundant under warmwater conditions, whereas others like *D. intercalaris* or *D. variabilis* are most numerous under temperate to cool conditions.

Seven biostratigraphic zones or subzones through the Miocene to Quaternary were selected for sample counts to provide maximum representation of the sites across latitude (Fig. 3). Calcareous microfossil assemblages of similar age for each zone or subzone are available for comparison at Sites 173 and 467 to 470. Several gaps occur in the stratigraphy of southern Sites 471 to 473 because of dissolution and erosion.

The Quaternary *Emiliania ovata* Subzone (CN14a) is the youngest unit studied. Among the genera selected for counting, *Coccolithus* is considered the cool-indicating taxon because of its Holocene limitation to waters colder than 14°C in the Atlantic (McIntyre and others, 1970). *Helicosphaera* is generally considered a warmwater taxon (20°C to 28°C), however, its range does extend as low as 8°C (Roth and Berger, 1975). Its high relative abundance at southern Sites 470 and 473 (Fig. 4) supports warmwater conditions, but the additional presence of numerous *Pontosphaera* and

Age	Zone or Subzone	Northern		DSDP Sites				Southern
		173	467	469	470	471	472	473
		Latitude N.						
		39°57.71'	33°50.97'	32°37.00'	28°54.46'	23°28.93'	23°00.35'	20°57.92'
Quaternary	*Emiliania ovata* CN14a	3-2 (70-71 cm)	5-3 (50-51 cm)	4-2 (50-51 cm)	1-CC	-	-	5-CC
Late Pliocene	*Calcidiscus macintyrei* CN12d	12-6 (20-21 cm)	17-3 (50-51 cm)	7-1 (50-51 cm)	4-1 (30 cm)	-	-	11-CC
Late Pliocene	*Discoaster tamalis* CN12a	13-6 (17-18 cm)	25-3 (9-10 cm)	8-3 (100-102 cm)	5-3 (50-51 cm)	4-5 (100-102 cm)	-	-
Late Miocene	*Amaurolithus primus* CN9b	15-4 (20-21 cm)	52-1 (50-51 cm)	13-1 (50 cm)	8-1 (75 cm)	-	-	27-1 (76 cm)
Late Miocene	*Discoaster bellus* CN8a	19-4 (20-21 cm)	73-1 (45 cm)	17-CC	5A-1 (22 cm)	-	-	-
Mid Miocene	*Discoaster kugleri* CN5b	21-3 (20-21 cm)	90-2 (25 cm)	19-1 (20-21 cm)	14-1 (7 cm)	-	-	-
Mid Miocene	*Sphenolithus heteromorphus* CN4	29-2 (20-21 cm)	100-CC	25-CC	18-2 (69-70 cm)	79-1 (34 cm)	12-2 (59-61 cm)	-

FIG. 3.—Array of samples used to compare changes in coccolith assemblages from DSDP Pacific Coast sites for different zonal time slices.

Thoracosphaera and reworked Cenozoic coccoliths could also indicate increased coastal influence as a contributing factor.

A dramatic change in the *Discoaster/Coccolithus* ratio from 173 southward to 473 in the latest Pliocene *Calcidiscus macintyrei* Subzone CN12d (Fig. 5) shows a strong temperature gradient of warming to the south, second only to that indicated for early Late Pliocene CN12a. The large 50 percent abundance of *Helicosphaera* at northern Site 173 is based on a great bloom of the species *H. sellii* at high Middle latitudes. This same nannofacies, with *Coccolithus pelagicus* and *H. sellii* dominant, also occurs in DSDP Leg 12 in the North Atlantic (Bukry, 1972a) near the Pliocene/Pleistocene boundary. The variance in the proportions of *Discoaster brouweri* and *D. triradiatus* could represent a bracketing of a fairly narrow habitat for *D. triradiatus* by this traverse, a random selection of samples from various levels in the Late Pliocene acme of *D. triradiatus*, or, less likely, variations in *D. brouweri* productivity at sites within the traverse.

The reversal in the *Discoaster/Coccolithus* ratio for the *Discoaster tamalis* Subzone CN12a occurs between Sites 467 and nearby Site 469 (Fig. 6). The degree of change in the ratio is large, as are the end-member values for Sites 173 and 471, indicating a strong North/South temperature gradient of warming. The restriction of warmwater species *Discoaster blackstockae, D. pentaradiatus,* and *Hayaster perplexus* to southerly Sites 469, 470, and 471 further emphasizes the temperature trend present in the generic count.

The *Discoaster/Coccolithus* ratio is less discriminating in the Late Miocene *Amaurolithus primus* Subzone CN9b than in younger levels, but suggests that the traverse region was relatively warm. Although discoasters were abundant, the count of

Quaternary *Emiliania ovata* Subzone Assemblages CN14a	Genera						
DSDP Sample	*Amaurolithus*	*Ceratolithus*	*Coccolithus*	*Discoaster*	*Helicosphaera*	*Sphenolithus*	*Triquetrorhabdulus*
173-3-2, 70-71 cm	-	-	99	-	1	-	-
467-5-3, 50-51 cm	-	2	97	-	1	-	-
469-4-2, 50-51 cm	-	3	88	-	9	-	-
470-1-CC	-	1	26	-	72	-	-
473-5-CC	-	-	-	-	100	-	-

FIG. 4.—Percentages of selected coccolith genera in Pacific Coast assemblages of Subzone CN14a. Values for each sample based on a survey of 100 specimens of selected genera. Samples arrayed according to latitudinal location with north at the top.

Pliocene *Calcidiscus macintyrei* Subzone CN12d													
	Selected genera								*Discoaster* species				
DSDP Sample	*Amaurolithus*	*Ceratolithus*	*Coccolithus*	*Discoaster*	*Helicosphaera*	*Sphenolithus*	*Triquetrorhabdulus*	*Discoaster/Coccolithus* Ratio	*Discoaster browweri*	*D. triradiatus*	**D. barbadiensis*	**D. deflandrei*	**D. variabilis*
173-12-6, 20-21 cm	-	-	50	-	50	-	-	0.0					
467-17-3, 50-51 cm	-	-	95	4	1	-	-	0.04	90	4	X	2	4
469-7-1, 50-51 cm	-	X	57	23	20	-	-	0.4	83	17			
470-4-1, 30 cm	-	-	29	52	19	-	-	1.8	63	37			
473-11-CC	-	-	7	84	9	-	-	12.0	81	3			

FIG. 5.—Percentages of selected coccolith genera in Pacific Coast assemblages of Subzone CN12d. Values for each sample based on a survey of 100 specimens of the selected genera. The *Discoaster/Coccolithus* ratios calculated from these values suggest a north-south paleotemperature trend. All of the *Discoaster* species encountered in a separate 100-specimen count are enumerated. * = Reworked taxon. X = Recorded as present after 100-count completed.

species shows that temperate species *D. intercalaris* and *D. variabilis* are dominant in all but the southernmost site (Fig. 7) were *D. quinqueramus* is predominant.

Discoaster, dominated by temperate *D. variabilis,* is abundant throughout the traverse for the early Late Miocene *Discoaster bellus* Subzone CN8a (Fig. 8). The concentration of warmwater species *Catinaster calyculus, Discoaster bellus, D. blackstockae,* and *D. neohamatus* in southern sites and their absence at northernmost Site 173 suggests a temperature gradient. But the *Discoaster/Coccolithus* ratio gives a cooler value for Site 470 than for 467 or 469. Therefore, the generic ratio which was a convenient and consistent measure for the Pliocene is misleading for the Late Miocene of these sites. The change is likely the result of the evolution of different habitat requirements by new biologic species of the genera upsection.

Another factor affecting the ratio could be carbonate solution, which may differentially remove somewhat more *Coccolithus* than *Discoaster.* Comparison of carbonate percentages from cored sediments to respective *Discoaster/Coccolithus* ratios is not a straightforward measure of carbonate solution, because of the terrigenous and volcanogenic dilution at many sites. But with this proviso in mind, a paleotemperature significance for the ratio does seem to override any solution effect for Subzone CN8a. The carbonate percentage and *Discoaster/Coccolithus* ratio for Sites 173, 467, 469, and 470 are, respectively, 12% and 2.4, 61% and 14.5, 77% and 17.7, 41% and 6.3. This relation shows the higher proportion of *Discoaster* with the higher carbonate percentages. The results for other zones are mixed or show the opposite trend. For example Subzone CN12d has the following values for Site 173, 467, 469, 470, and 473: 14% and 0.0, 40% and 0.04, 38% and 0.4, 16% and 1.8, and 8% and 12.0. The last three sets match higher proportions of *Discoaster* with lower carbonate percentages, which could be caused by paleotemperature and/or solution. The values for Site 467 suggest the opposite trend, with temperature overriding any solution effect. Further work on this relation should be done in future studies that use coccoliths to analyze paleoenvironment.

An additional factor that might account for the reversals in Zone CN8, for example, are high-frequency climatic oscillations during the interval

DSDP Sample	*Amaurolithus*	*Ceratolithus*	*Coccolithus*	*Discoaster*	*Helicosphaera*	*Sphenolithus*	*Triquetrorhabdulus*	*Discoaster/Coccolithus* Ratio	*Discoaster asymmetricus*	*D. blackstockae*	*D. brouweri*	*D. challengeri*	*D. intercalaris*	*D. pentaradiatus*	*D.* sp. aff. *D. surculus*	*D. tamalis*	*D. triradiatus*	*D. variabilis*	*D.* spp. (indeterminate)	*Hayaster perplexus*
Pliocene *Discoaster tamalis* Subzone Assemblages CN12a																				
	Selected genera								Selected species											
173-13-6, 17-18 cm	-	-	94	1	5	-	-	0.01							100					
467-25-3, 9-10 cm	-	-	57	43	-	-	-	0.75	2		23		35		9			5	26	
469-8-3, 100-102 cm	-	-	16	83	1	-	-	5.2	5		17		14	29	11	18	5		1	
470-5-3, 50-51 cm	-	7	2	91	-	-	-	46.0	11		32	1	4	12	18	14	1		6	
471-4-5, 100-102 cm	-	-	2	86	12	-	-	43.0	1	1	6			56	12	16			6	2

FIG. 6.—Percentages of selected coccolith genera in Pacific Coast assemblages of Subzone CN12a. Values for each sample based on a survey of 100 specimens of the selected genera. Large differences in the *Discoaster/Coccolithus* ratios occurred in CN12a from north to south. All of the *Discoaster* and tropical *Hayaster* species encountered in a separate 100-specimen count are enumerated (the lone *Discoaster* specimen in the generic count for Site 173 precluded a 100 count).

DSDP Sample	*Amaurolithus*	*Ceratolithus*	*Coccolithus*	*Discoaster*	*Helicosphaera*	*Sphenolithus*	*Triquetrorhabdulus*	*Discoaster/Coccolithus* Ratio	*Discoaster brouweri*	*D.* sp. aff. *D. challengeri*	*D. intercalaris*	*D. pentaradiatus*	*D. quinqueramus* s. ampl.	*D. triradiatus*	*D. variabilis*	*D.* spp. (indeterminate)
Miocene *Amaurolithus primus* Subzone Assemblages CN9b																
	Selected genera								*Discoaster* species							
173-15-4, 20-21 cm	2	-	22	74	-	2	-	3.4	3	3	14		5		76	
467-52-1, 50-51 cm	1?	-	-	99	-	-	-	>99.0			60				40	
469-13-1, 50 cm	1	-	1	98	-	-	-	98.0	9		11		6		53	21
470-8-1, 75 cm	5	-	1	92	-	1	1	92.0	9		4	4	4	3	76	
473-27-1, 76 cm	-	-	4	20	10	66	-	5.0	10			10	50		30	

FIG. 7.—Percentages of selected coccolith genera in Pacific Coast assemblages of Subzone CN9b. Values for each sample based on a survey of 100 specimens of the selected genera. A predominance of temperate species *Discoaster intercalaris* and *D. variabilis* among the discoasters suggests that the temperature-indicating capability of the *Discoaster/Coccolithus* ratio is poorer for the Miocene than for the Pliocene. The *Discoaster* species count for Site 473 is part of the generic count owing to the sparcity of specimens in the terrigenous-rich sediment.

	Miocene *Discoaster bellus* Subzone CN8a																					
	Selected genera								*Catinaster* and *Discoaster* species													
DSDP Sample	*Amaurolithus*	*Ceratolithus*	*Coccolithus*	*Discoaster*	*Helicosphaera*	*Sphenolithus*	*Triquetrorhabdulus*	*Discoaster/Coccolithus* Ratio	*Catinaster calyculus*	*Discoaster bellus*	*D.* sp. cf. *D. blackstockae*	*D. braarudii*	*D. brouweri* s. ampl.	*D. challengeri*	*D. intercalaris*	*D. neohamatus*	*D. pansus*	*D. pseudovariabilis*	*D. triradiatus*	*D. variabilis*	*D.* spp. (indeterminate)
173-19-4, 20-21 cm	-	-	24	57	-	19	-	2.4					2	3	7					79	6
467-73-1, 45 cm	-	-	6	87	-	6	1	14.5		8		3					2			59	28
469-17-CC	-	-	3	53	-	43	1	17.7	1	38			5	6						42	8
470A-5-1, 22 cm	-	-	10	63	-	26	1	6.3		27	1	4	8	1		4	1	1	2	42	9

FIG. 8.—Percentages of selected coccolith genera in Pacific Coast assemblages of Subzone CN8a. Values for each sample based on a survey of 100 specimens of the selected genera. Relative values for the *Discoaster/Coccolithus* ratio do not show a consistent north-south trend, but the high value for Site 469 is supported by the occurrences of warmwater *Catinaster calyculus* and *Discoaster bellus.* All of the *Catinaster* and *Discoaster* species encountered in a separate 100-specimen count are enumerated.

examined. Preliminary comparisons of climatic indices from foraminifers, diatoms, and silicoflagellates indicate a high-frequency oscillation during Subzone CN8a (Richard Z. Poore, personal communication, 1980). Therefore, broad sampling patterns of different parts of the climatic curve at different sites could easily produce apparent reversals. Establishing these relations will require more detailed work.

The northern four sites of the Pacific Coast traverse recovered assemblages of, or correlative to, the relatively brief mid-Miocene *Discoaster kugleri* Subzone CN5b. Although *D. kugleri* is missing at Site 173, the assemblage is correlative on the basis of diatom zonation (John A. Barron, personal communication, 1980). A high *Discoaster/Coccolithus* ratio at Site 173 (Fig. 9) seems to indicate warmwater conditions, whereas the low diversity of the assemblages is more typical of coolwater conditions. The exceptionally high *Discoaster/Coccolithus* ratios at all four sites were checked against tropical Pacific Site 77 (lat 00°28.90′N). This tropical *Discoaster/Coccolithus* ratio for CN5b is distinctly lower than for the Pacific Coast traverse sites, indicating that the *Coccolithus* species *C. miopelagicus* and *C. pelagicus* (s. ampl., small) tolerated warm water and cannot be employed as strictly coolwater indicators for Middle Miocene. In fact, during CN5b deposition, *Coccolithus* might be used as a warmwater indicator. Clearly, the abundance center of the biogeographic range of *Coccolithus* changed considerably from Middle Miocene to Quaternary.

An interval from the upper *Sphenolithus heteromorphus* Zone CN4 of the mid-Miocene was compared across six DSDP sites included in a Pacific Coast traverse (Fig. 10). *Discoaster exilis* is dominant in all of the discoaster assemblages; however, warmwater *D. signus* is limited to the three southern sites, suggesting the existence of a temperature gradient. Dissolution was strongest at northern Site 173, where *Coccolithus* may have been differentially dissolved because the highly resistant placoliths of *Cyclicargolithus* and *Reticulofenestra* are fragmented and etched. As in other Miocene intervals examined, the *Discoaster/Coccolithus* ratio for CN4 does not show a consistent gradient across latitude; high and low ratios are interspersed.

A comparison of Zone CN4 at California Continental Borderland Site 470 and nearby Sites 471 and 472 to equatorial Pacific Site 77 (Hole 77B) shows that the numbers of *C. miopelagicus* and *C. pelagicus* s. ampl. are as high at the equatorial site as at the southernmost traverse sites: 12% to 25% of the selected genera count at Hole 77B (Cores 25–27), 11% to 13% at Sites 471 and 472, and up to 60% at Site 470. The *Discoaster/Coccolithus* ratios for Zone CN4 at Hole 77B (Cores 25–27) are also similar but lower, with values of 1.0 to 2.4. These similarities between the assemblages (owing to the broader range of *Coccolithus* in the lower Middle Miocene) are only apparent because the equatorial assemblages can be discriminated by using their much greater abundances of *Sphenolithus*. For example, compare the following *Sphenolithus* (S), *Discoaster* (D), *Coccolithus* (C) values from the selected genera counts to those of Leg 63: Samples 77B-25-2, 20–22 cm, S = 64%, D = 14%, C =

DSDP Sample	Miocene *Discoaster kugleri* Subzone CN5b: Selected genera: *Amaurolithus*	*Ceratolithus*	*Coccolithus*	*Discoaster*	*Helicosphaera*	*Sphenolithus*	*Triquetrorhabdulus*	*Discoaster/Coccolithus* Ratio	*Discoaster* species: *Discoaster* sp. aff. *D. bollii*	*D. braarudii*	*D.* sp. cf. *D. deflandrei* (tiny)	*D. exilis*	*D. kugleri*	*D. moorei*	*D. variabilis*	*D.* sp. (5-rayed)	*D.* spp. (indeterminate)
173-21-3, 20-21 cm	-	-	-	78	-	22	-	>78.0				96					4
467-90-2, 25 cm	-	-	X	80	-	20	-	>80.0	1			37	17		9		36
469-19-1, 20-21 cm	-	-	1	89	-	10	-	89.0		2	3	34	29	2	12		18
470-14-1, 7 cm	-	-	1	93	-	6	-	93.0		3		39	2	1	18	1	36
77B-20-4, 63-64 cm	-	-	30	41	-	26	3	1.4		1		27	27		11		34

FIG. 9.—Percentages of selected coccolith genera in Pacific Coast and tropical Pacific (Site 77) assemblages of Subzone CN5b. Values for each sample based on a survey of 100 specimens of the selected genera. All of the *Discoaster* species encountered in a separate 100-specimen count are enumerated. *Discoaster kugleri* is exceptionally common at Sites 77, 467, and 469. All four Pacific Coast assemblages are dominated by *Reticulofenestra* placoliths and fragments; *Coccolithus* is extremely sparse. Tropical Site 77, however, contains a significant number of *Coccolithus miopelagicus,* and *C. pelagicus* (s. ampl., small).

DSDP Sample	Miocene *Sphenolithus heteromorphus* Zone CN4: Selected genera: *Amaurolithus*	*Ceratolithus*	*Coccolithus*	*Discoaster*	*Helicosphaera*	*Sphenolithus*	*Triquetrorhabdulus*	*Discoaster/Coccolithus* Ratio	*Discoaster* species: *Discoaster aulakos*	*D. braarudii*	*D. challengeri*	*D. deflandrei*	*D. exilis*	*D.* sp. aff. *D. sanmiguelensis*	*D. signus*	*D. variabilis* s. ampl.	*D.* spp. (indeterminate)
173-29-2, 20-21 cm	-	-	17	82	-	1	-	4.8		1	1	6	48	9		35	
467-100-CC	-	-	65	21	1	13	-	0.32				5	62			10	24
469-25-CC	-	-	22	72	1	5	-	3.3	4			1	57			19	35
470-18-2, 69-70 cm	-	-	60	27	2	11	-	0.45					56	3	8	20	13
471-79-1, 34 cm	-	-	11	37	17	35	-	3.4	3			3	32		24	8	32
472-12-2, 59-61 cm	-	-	13	69	-	18	-	5.3	3			12	35		18	8	24

FIG. 10.—Percentages of selected coccolith genera in Pacific Coast assemblages of Zone CN4. Values for each sample based on a survey of 100 specimens of the selected genera. *Discoaster/Coccolithus* ratios show a narrow range of variation, but also show apparent reversals. The tropical *Discoaster, D. signus* is limited to southern sites off Baja California. All of the *Discoaster* species encountered in a separate 100-specimen count are enumerated.

19%; 77B-25-6, 94–98 cm, S = 50%, D = 24%, C = 25%; 77B-27-4, 67–68cm, S = 59%, D = 29%, C = 12%.

Evidence of a change, or evolution, in the optimum temperature preference of *Coccolithus* species can be seen in the total absence of *Coccolithus* in the Upper Pliocene Subzone CN12a at equatorial Site 77 (Sample 77B-4-6, 70–71 cm). On the other hand, *Coccolithus* is present at all Pacific Coast traverse sites with increasing abundance northward. The 14°C limitation of modern *Coccolithus* certainly did not apply to the substantial Middle Miocene (Zones CN4 and CN5) populations of *Coccolithus* found at equatorial Site 77. The difference in representation of the genus at equatorial sites between the Miocene and Pliocene is a matter of changing temperature tolerance of biologic species evolving in the genus.

This brief survey of 35 samples from eight upper Cenozoic sites involved counts of only 7,000 specimens from seven selected biostratigraphic units, but shows north-south gradients for all intervals. The distribution of species of *Discoaster* is most helpful in identifying warm- and coolwater regimes at different times. The affinity of *Coccolithus* for coolwater (McIntyre and others, 1970) had evolved by the Pliocene, making the *Discoaster/Coccolithus* ratio an effective measure of paleotemperature trends through the 20° latitude of the traverse region for the Pliocene. In older Miocene assemblages the temperature affinities of the species of *Coccolithus* appear to have been broader, with persistent occurence into the tropics. Variable depths and dissolution histories at the sites, and even potential tectonic translations across latitude (Crouch, 1979) could have altered the trends, but the *Discoaster* species appear to be a consistent guide through the space-time matrix surveyed.

The key area in the traverse is between Sites 467 and 470, near the outer California Continental Borderland. Reversals in abundance trends for indicator taxa for all the zones and subzones above the Late Miocene CN8a are located between those sites. Older intervals are less clear. DSDP sites off central California are needed to assess biogeographic ranges for older intervals. The strongest gradients occur in the intervals examined from the Quaternary and Upper Pliocene.

BRAARUDOSPHAERA CHALKS

A most remarkable discovery of the DSDP is the presence of thin beds of *Braarudosphaera* chalk in the Oligocene *Sphenolithus distentus* Zone (CP18) of the South Atlantic. The *Braarudosphaera* pentaliths are produced by a one-celled organism, probably related to other coccolithophores (Lefort, 1972), which must have experienced tremendous blooms so that it far outnumbered coeval coccoliths in the preserved *Braarudosphaera* chalk beds. There are no aberrations in the preservation or diversity of the of the coeval coccoliths to suggest a cause for the beds (Bukry, 1978a). Oxygen isotope study of the pentaliths, coeval coccoliths, and foraminifers may help to explain the formation of these beds, by detecting whether the *Braarudosphaera* were part of the surface plankton or part of a deeper watermass. Preliminary isotope studies indicated there could be an interpretation problem with formation of pentaliths in isotopic equilibrium with sea water, and with re-equilibrium in bottom water, because the generally cold 3°C temperatures that were determined seemed colder than anticipated upwelling (Lloyd and Hsü, 1972). The situation is also complicated because little is known about living *Braarudosphaera*; however, it does favor lower-salinity waters (Bukry, 1974). Therefore, because of the location and cyclicity of the beds there should be a connection between the beds and Antarctic meltwater or bottomwater pulses in the Oligocene of the South Atlantic (Lloyd and Hsü, 1972; Hsü, personal communication, 1977). The direct or indirect nature of this relationship needs to be established.

Although sediment rich in *Braarudosphaera* is known from the Quaternary of the Black Sea (Percival, 1978), upper Oligocene in the North Atlantic (Bukry, 1978a), the Middle Eocene in the Gulf of Mexico (Bukry, 1973a), the Danian of Spain (Percival, 1972), and the Hauterivian in the North Atlantic (Bukry and Bramlette, 1969a), the most extensive beds of Oligocene age have been sampled in an ENE direction from the Rio Grande Rise (Site 22, lat 30°S) to the Walvis Ridge (Site 362, lat 20°S). This trend has been established from *Braarudosphaera*-rich samples at Sites 14, 17, 19, 20, 22, 362, and 522 and appears to represent a trans Atlantic belt. Late Oligocene coccoliths are not preserved at Sites 358 and 359 (Perch-Nielsen, 1977) which are south of the belt.

This belt is probably the trace of a middle Oligocene oceanographic front related to bottom topography of the Rio Grande Rise and the Walvis Ridge. New DSDP drilling in the South Atlantic (Legs 71 to 74) should provide the evidence needed to understand the Oligocene *Braarudosphaera* events. Earlier and later *Braarudosphaera* events could have different causes if *Braarudosphaera* were encystment stage, reacting to environmental stress. But modern *Braarudosphaera* has bloomed in concert with coeval coccolithophores in the Panama Basin (Smayda, 1966). The importance of *Braarudosphaera* to understanding the Cretaceous/Tertiary boundary (Thierstein and Okada, 1979) assemblages assures more detailed study of the genus as DSDP drills Cretaceous strata off eastern North America during future legs.

CONCLUSIONS

The success of DSDP and the development of coccolith stratigraphy and paleontology into a fundamental element of micropaleontology are mutually related. A unified approach to coccolith studies including taxonomy, preservation, paleoecology, and stratigraphy (evolution sequences) had to be considered relative to plate tectonic, paleoceanographic, and paleomagnetic determinations arising from succeeding DSDP legs. The ability and the need to correlate coccolith results through nearly all the oceans, because of DSDP drilling, has produced a basic understanding of coccolith paleontology in a short time.

Because answers raise new questions and reveal new avenues for thought, future DSDP drilling is a fundamental basis of coccolith research. Establishing and refining the HOS and LOS for many species needed in datum and magnetic stratigraphy correlations rests directly on the continuing recovery of DSDP Hydraulic Piston Core material. New undisturbed cores can provide sediment from warm-cold transition areas (Sites 310 and 362) where paleoclimatic fluctuations can be established. In addition, evolutionary details of the important paleoecologic and biostratigraphic guide genus *Discoaster* are not well understood because of a lack of studies of transitional forms. Although the morphotypic limits and general ranges of species of *Discoaster* have been established through heavy reliance on DSDP cores, further research on evolution between species will require undisturbed, well-preserved ocean sediment which can best be obtained by continuation of HPC operations in the tropical Atlantic and Indo-Pacific.

ACKNOWLEDGMENTS

I thank R. Z. Poore and J. A. Barron, U.S. Geological Survey, for their helpful reviews of this paper. W. W. Hay and the late M. N. Bramlette are due special appreciation for providing many sound stratigraphic and taxonomic methods and ideas that promoted rapid recognition of new coccolith relations from DSDP materials. I also thank D. M. Raup, Field Museum of Natural History, Chicago, and A. G. Fischer, Princeton University, for their early instruction and insight in the dynamics of marine paleontology and stratigraphy which have made coccolith studies so enjoyable.

REFERENCES CITED

NOTE: All references to Deep Sea Drilling Project Initial Reports have been abbreviated; for complete citations see *Appendix I* at the end of this volume, or the individual Initial Reports as published.

Addicott, W. O., 1970, Latitudinal gradients in Tertiary molluscan faunas of the Pacific coast: Palaeogeography, Palaeoclimatology, Palaeoecology, v. 8, p. 287–312.

Adelseck, C. G., Geehan, G. W., and Roth, P. H., 1973, Experimental evidence for the selective dissolution and overgrowth of calcareous nannofossils during diagenesis: Geol. Soc. America Bull., v. 84, p. 2755–2762.

Barron, J. A., *in press,* Late Cenozoic diatom biostratigraphy and paleoceanography of middle-latitude eastern North Pacific, DSDP Leg 63: Init. Repts. DSDP, v. 63.

Berger, W. H., 1967, Foraminiferal ooze: solution at depths: Science, v. 156, p. 383–385.

———, 1973, Deep-sea carbonates: evidence for a coccolith lysocline: Deep-sea Research, v. 20, p. 917–921.

Black, M., 1963, The fine structure of the mineral parts of Coccolithophoridae: Linnean Soc. London Proc., v. 174, p. 41–46.

———, 1972, Crystal development in Discoasteraceae and Braarudosphaeraceae (planktonic algae): Palaeontology, v. 15, p. 476–489.

Bramlette, M. N., 1961, Pelagic sediments, *in* Sears, M., (ed.), Oceanography: Washington, Am. Assoc. Adv. Sci., p. 345–366.

———, and Riedel, W. R., 1954, Stratigraphic value of discoasters and some other microfossils related to recent coccolithophores: Jour. Paleontology, v. 28, p. 385–403.

———, and Sullivan, F. R., 1961, Coccolithophorids and related nannoplankton of the early Tertiary in California: Micropaleontology, v. 7, p. 129–188.

———, and Martini, E., 1964, The great change in calcareous nannoplankton fossils between the Maestrichtian and Danian: Micropaleontology, v. 10, p. 291–322.

———, and Wilcoxon, J. A., 1967, Middle Tertiary calcareous nannoplankton of the Cipero Section, Trinidad, W. I.: Tulane Studies Geology, v. 5, p. 93–131.

———, and Bukry, D., 1968, Illustrations of Tertiary nannofossils: JOIDES Deep Sea Drilling Project Core Description Manual, pt. 7, chap. 2, 26 p.

Bukry, D., 1971a, Coccolith stratigraphy Leg 6, Deep Sea Drilling Project: Init. Repts. DSDP, v. 6, p. 965–1004.

———, 1971b, Coccolith stratigraphy Leg 7, Deep Sea Drilling Project: Init. Repts. DSDP, v. 7, pt. 2, p 1513–1528.

———, 1971c, Cenozoic calcareous nannofossils from the Pacific Ocean: San Diego Soc. Nat. History Trans., v. 16, p. 303–327.

———, 1972a, Further comments on coccolith stratigraphy Leg 12, Deep Sea Drilling Project: Init. Repts. DSDP, v. 12, p. 1071–1083.

———, 1972b, Coccolith stratigraphy—Leg 14, Deep Sea Drilling Project: Init. Repts. DSDP, v. 14, p. 487–494.

———, 1973a, Coccolith stratigraphy, Leg 10, Deep Sea Drilling Project: Init. Repts. DSDP, v. 10, p. 385–406.

———, 1973b, Coccolith stratigraphy Leg 13, Deep Sea Drilling Project: Init. Repts. DSDP, v. 13, p. 817–822.

———, 1973c, Low-latitude coccolith biostratigraphic zonation: Init. Repts. DSDP, v. 15, p. 685–703.
———, 1973d, Coccolith stratigraphy, eastern equatorial Pacific, Leg 16 Deep Sea Drilling Project: Init. Repts. DSDP, v. 16, p. 653–711.
———, 1973e, Phytoplankton stratigraphy, central Pacific Ocean, Deep Sea Drilling Project Leg 17: Init. Repts. DSDP, v. 17, p. 871-889.
———, 1973f, Coccolith and silicoflagellate stratigraphy, Deep Sea Drilling Project Leg 18, eastern North Pacific: Init. Repts. DSDP, v. 18, p. 817–831.
———, 1973g, Coccoliths and silicoflagellates from Deep Sea Drilling Project Leg 19, North Pacific Ocean and Bering Sea: Init. Repts. DSDP, v. 19, p. 857–867.
———, 1973h, Coccolith and silicoflagellate stratigraphy, Tasman Sea and southwestern Pacific Ocean, Deep Sea Drilling Project Leg 21: Init. Repts. DSDP, v. 21, p. 885–893.
———, 1974, Coccoliths as paleosalinity indicators—evidence from Black Sea: Am. Assoc. Petroleum Geologists Mem. 20, p. 353–363.
———, 1975a, Silicoflagellate and coccolith stratigraphy, Deep Sea Drilling Project Leg 29: Init. Repts. DSDP, v. 29, p. 845–872.
———, 1975b, Coccolith and silicoflagellate stratigraphy, northwestern Pacific Ocean, Deep Sea Drilling Project Leg 32: Init. Repts. DSDP, v. 32, p. 677–701.
———, 1976, Coccolith stratigraphy of Manihiki Plateau, central Pacific, Deep Sea Drilling Project Site 317: Init. Repts. DSDP, v. 33, p. 493–501.
———, 1977, Coccolith and silicoflagellate stratigraphy, central North Atlantic Ocean, Deep Sea Drilling Project Leg 37: Init. Repts. DSDP, v. 37, p. 917–927.
———, 1978a, Cenozoic silicoflagellate and coccolith stratigraphy, southeastern Atlantic Ocean, Deep Sea Drilling Project Leg 40: Init. Repts. DSDP, v. 40, p. 635–649.
———, 1978b, Biostratigraphy of Cenozoic marine sediment by calcareous nannofossils: Micropaleontology, v. 24, p. 44–60.
———, *in press,* Pacific Coast coccolith stratigraphy of Deep Sea Drilling Project Leg 63: Init. Repts. DSDP, v. 63.
———, and Bramlette, M. N., 1969a, Coccolith age determinations Leg 1, Deep Sea Drilling Project: Init. Repts. DSDP, v. 1, p. 369–387.
———, and ———, 1969b, Summary of coccolith biostratigraphy: Init. Repts. DSDP, v. 1, p. 621–623.
———, and ———, 1970a, Coccolith age determinations Leg 3, Deep Sea Drilling Project: Init. Repts. DSDP, v. 3, p. 589–611.
———, and ———, 1970b, Coccolith age determinations Leg 5, Deep Sea Drilling Project: Init. Repts. DSDP, v. 5, p. 487–494.
———, and others, 1971, Planktonic microfossil biostratigraphy of the northwestern Pacific Ocean: Init. Repts. DSDP, v. 6, p. 1253–1300.
———, Dinkelman, M. B., and Kaneps, A., 1973, Biostratigraphy of the equatorial East Pacific Rise: Init. Repts. DSDP, v. 15, p. 915–935.
Burckle, L. H., 1978, Early Miocene to Pliocene diatom datum levels for the equatorial Pacific: Indonesia Ministry of Mines and Energy Geological Research and Development Centre Spec. Pub. 1, p. 25–44.
Burns, D. A., 1975, Nannofossil biostratigraphy for Antarctic sediments, Leg 28, Deep Sea Drilling Project: Init. Repts. DSDP, v. 28, p. 589–598.
Crouch, J. K., 1979, Neogene tectonic evolution of the California Continental Borderland and western Transverse Ranges: Geol. Soc. America Bull., pt. 1, v. 90, no. 4, p. 338–345.
Edwards, A. R., 1973, Calcareous nannofossils from the southwest Pacific, Deep Sea Drilling Project Leg 21: Init. Repts. DSDP, v. 21, p. 641–691.
———, and Perch-Nielsen, K., 1975, Calcareous nannofossils from the southern southwest Pacific, Deep Sea Drilling Project, Leg 29: Init. Repts. DSDP, v. 29, p. 469–539.
Ericson, D. B., and Wollin, G., 1962, Micropaleontology: Sci. American, v. 207, no. 1, p. 96–108.
Gartner, S., jr., 1969, Correlation of Neogene planktonic foraminifer and calcareous nannofossil zones: Gulf Coast Assoc. Geol. Socs. Trans., v. 19, p. 585–599.
———, 1972, Coccolith age determinations, Leg 9 Deep Sea Drilling Project: Init. Repts. DSDP, v. 9, p. 833–839.
———, 1977, Calcareous nannofossil biostratigraphy and revised zonation of the Pleistocene: Marine Micropaleontology, v. 2, p. 1–25.
———, and Bukry, D., 1969, Tertiary holococcoliths: Jour. Paleontology, v. 43, p. 1213–1221.
Haq, B. U., 1971, Paleogene calcareous nannoflora. Part IV: Paleogene nannoplankton biostratigraphy and evolutionary rates in Cenozoic calcareous nannoplankton: Stockholm Contr. Geology, v. 25, p. 129–158.
———, 1976, Coccoliths in cores from the Bellingshausen Abyssal Plain and Antarctic Continental Rise (DSDP Leg 35): Init. Repts. DSDP, v. 35, p. 557–567.
———, *in press,* Miocene biogeographic evolution of calcareous nannoplankton and paleoceanography of the Atlantic Ocean: Micropaleontology.
———, and Lohmann, G. P., 1976, Early Cenozoic calcareous nannoplankton biogeography of the Atlantic Ocean: Marine Micropaleontology, v. 1, p. 119–194.
———, and Riley, A., 1976, Antarctic silicoflagellates from the southeast Pacific, Deep Sea Drilling Project Leg 35: Init. Repts. DSDP, v. 35, p. 673–691.
———, Berggren, W. A., and van Couvering, J. A., 1977a, Corrected age of the Pliocene/Pleistocene boundary: Nature, v. 269, p. 483–488.

———, LOHMANN, G. P., AND WISE, S. W., JR., 1977b, Calcareous nannoplankton biogeography and its paleoclimatic implications: Cenozoic of the Falkland Plateau (DSDP Leg 36) and Miocene of the Atlantic Ocean: Init. Repts. DSDP, v. 36, p. 745–759.

HAY, W. W., 1964, Utilisation stratigraphique des Discoastéridés pour la zonation du Páleocène et de l'Eocene inferieur: Bureau Réch. Géol. et Min. Mém., v. 28, p. 885–889.

———, 1970, Calcareous nannofossils from cores recovered on Leg 4: Init. Repts. DSDP, v. 4, p. 455–501.

———, 1971, Preliminary dating by fossil calcareous nannoplankton, Deep Sea Drilling Project, Leg 8: Init. Repts. DSDP, v. 8., p. 809–818.

———, AND MOHLER, H. P., 1967, Calcareous nannoplankton from early Tertiary rocks at Pont Labau, France, and Paleocene-early Eocene correlations: Jour. Paleontology, v. 41, p. 1505–1541.

———, AND OTHERS, 1967, Calcareous nannoplankton zonation of the Cenozoic of the Gulf Coast and Caribbean-Antillean area, and transoceanic correlation: Gulf Coast Assoc. Geol. Socs. Trans., v. 17, p. 428–480.

———, AND BEAUDRY, F. M., 1973, Calcareous nannofossils—Leg 15, Deep Sea Drilling Project: Init. Repts. DSDP, v. 15, p. 625–683.

HEKEL, H., 1973, Nannofossil biostratigraphy, Leg 20, Deep Sea Drilling Project: Init. Repts. DSDP, v. 20, p. 221–247.

HONJO, S., 1975, Dissolution of suspended coccoliths in the deep-sea water column and sedimentation of coccolith ooze: Cushman Found. Foram. Res. Spec. Pub. 13, p. 114–128.

LEFORT, F., 1972, Quelques caractères morphologiques de deux espèces actuelles de *Braarudosphaera* (Chrysophycées, Coccolithoracées): Botaniste, v. 55, p. 81-93.

LLOYD, R. M., AND HSÜ, K. J., 1972, Stable-isotope investigations of sediments from the DSDP III cruise to South Atlantic: Sedimentology, v. 19, p. 45–58.

LYNN, R. J., 1967, Seasonal variation of temperature and salinity at 10 meters in the California Current: California Cooperative Oceanic Fisheries Investigations Rept., v. 11, p. 157–186.

MCINTYRE, A., AND BÉ, A. W. H., 1967, Modern Coccolithophoridae of the Atlantic Ocean—I. Placoliths and cyrtoliths: Deep-sea Research, v. 14, p. 561–597.

———, AND JANTZEN, R. E., 1969, North Atlantic Pleistocene climatic fluctuations deduced from coccolith floras: Geol. Soc. America Abstracts with Programs for 1969, pt. 7, p. 146.

———, BÉ, A. W. H., AND ROCHE, M. B., 1970, Modern Pacific Coccolithophorida: A paleontological thermometer: New York Acad. Sci. Trans., v. 32, p. 720–731.

———, RUDDIMAN, W. F., AND JANTZEN, R., 1972, Southward penetrations of the North Atlantic Polar Front: faunal and floral evidence of large-scale surface water mass movements over the last 225,000 years: Deep-sea Research, v. 19, p. 61–77.

MCMANUS, D. A., AND OTHERS, 1970, Introduction: Init. Repts. DSDP, v. 5, p. 3–14.

MARTINI, E., 1971, Standard Tertiary and Quaternary calcareous nannoplankton zonation: 2nd Planktonic Conf. Proc., p. 739–785.

———, 1976, Cretaceous to recent calcareous nannoplankton from the central Pacific Ocean (DSDP Leg 33): Init. Repts. DSDP, v. 33, p. 383–423.

———, AND BRAMLETTE, M. N., 1963, Calcareous nannoplankton from the experimental Mohole drilling: Jour. Paleontology, v. 37, p. 845–856.

———, AND WORSLEY, T. R., 1971, Tertiary calcareous nannoplankton from the western equatorial Pacific: Init. Repts. DSDP, v. 7, p. 1471–1507.

———, AND MÜLLER, C., 1976, Eocene to Pleistocene silicoflagellates from the Norwegian-Greenland Sea (DSDP Leg 38): Init. Repts. DSDP, v. 38, p. 857–895.

MAXWELL, A. E., AND OTHERS, 1970, Site 20: Init. Repts. DSDP, v. 3, p. 319–366.

MAZZEI, R., AND OTHERS, 1979, Calibration of late Neogene calcareous plankton datum planes with the paleomagnetic record of Site 397 and correlation with Moroccan and Mediterranean sections: Init. Repts. DSDP, v. 47, pt. 1, p. 375–389.

MUELENKAMP, J. E., AND OTHERS, 1979, Late Miocene - Pliocene climatic fluctuations and marine "cyclic" sedimentation patterns: Annales Géol. Pays Helléniques, Tome hors série, 1979, fasc. II. VIIth International Congress on Mediterranean Neogene, Athens, p. 831–842.

MILOW, E. D., 1970, Tentative nannofossil zones and subzones and their radiometric age, northeast Pacific: Init. Repts. DSDP, v. 5, p. 8–10.

MÜLLER, C., 1976, Tertiary and Quaternary calcareous nannoplankton in the Norwegian-Greenland Sea, DSDP, Leg 38: Init. Repts. DSDP, v. 38, p. 823–841.

———, 1978, Neogene calcareous nannofossils from the Mediterranean—Leg 42A of the Deep Sea Drilling Project: Init. Repts. DSDP, v. 42, pt. 1, p. 727–751.

OKADA, H., AND BUKRY, D., 1980, Supplementary modification and introduction of code numbers to the Low-latitude coccolith biostratigraphic zonation (Bukry, 1973; 1975): Marine Micropaleontology, v. 5.

PERCH-NIELSEN, K., 1972, Remarks on Late Cretaceous to Pleistocene coccoliths from the North Atlantic: Init. Repts. DSDP, v. 12, p. 1003–1069.

———, 1975, Late Cretaceous to Pleistocene silicoflagellates from the southern southwest Pacific, DSDP Leg 29: Init. Repts. DSDP, v. 29, p. 677–721.

———, 1977, Albian to Pleistocene calcareous nannofossils from the western South Atlantic, DSDP Leg 39: Init. Repts. DSDP, v. 39, p. 699–823.

———, 1979, Calcareous nannofossil zonation at the Cretaceous/Tertiary boundary in Denmark, *in* Birkelund, T., and

Bromley, R. G., (eds.), Cretaceous - Tertiary Boundary Events, Symposium 18–24 September 1979. I. The Maastrichtian and Danian of Denmark: University of Copenhagen 1979, p. 115–135.

Percival, S. F., jr., 1972, Changes in calcareous nannoplankton in the Cretaceous-Tertiary biotic crisis at Zumaya, Spain: Ph.D. Dissertation, Princeton Univ., p. 1–99.

———, 1978, Indigenous and reworked coccoliths from the Black Sea: Init. Repts. DSDP, v. 42, pt. 2, p. 773–781.

Proto Decima, F., Medizza, F., and Todesco, L., 1978, Southeastern Atlantic Leg 40 calcareous nannofossils: Init. Repts. DSDP, v. 40, p. 571–634.

Riedel, W. R., Bramlette, M. N., and Parker, F. L., 1963, "Pliocene-Pleistocene" boundary in deep-sea sediments: Science, v. 140, p. 1238–1240.

Rio, D., 1974, Remarks on late Pliocene-early Pleistocene calcareous nannofossils stratigraphy in Italy: L'Ateneo Parmense-Acta Naturalia, v. 10, no. 3, p. 409–449.

Roth, P. H., 1973, Calcareous nannofossils—Leg 17, Deep Sea Drilling Project: Init. Repts. DSDP, v. 17, p. 695–795.

———, 1974, Calcareous nannofossils from the northwestern Indian Ocean, Leg 24, Deep Sea Drilling Project: Init. Repts. DSDP, v. 24, p. 969–994.

———, and Thierstein, H., 1972, Calcareous nannoplankton: Leg 14 of the Deep Sea Drilling Project: Init. Repts. DSDP, v. 14, p. 421–485.

———, and Berger, W. H., 1975, Distribution and dissolution of coccoliths in the south and central Pacific: Cushman Found. Foram. Res. Spec. Pub. 13, p. 87–113.

Ruddiman, W. F., and Heezen, B. C., 1967, Differential solution of planktonic foraminifera: Deep-sea Research, v. 14, p. 801–808.

San Miguel Arribas, M., 1976, Importancia de los cocolitofóridos en la formación de las rocas carbonatadas: Instituto de Investigaciones Geológicas Universidad de Barcelona, v. 31, p. 133–146.

Schneidermann, N., 1973, Deposition of coccoliths in the compensation zone of the Atlantic Ocean: Soc. Econ. Paleontologists Mineralogists, Gulf Coast Sec., Calcareous Nannofossils Symposium Proc., p. 140–151.

Smayda, T. J., 1966, A quantitative analysis of the phytoplankton of the Gulf of Panama. III. General ecological conditions and the phytoplankton dynamics at 8°45′N, 79°23′W from November 1954 to May 1957: Inter-Amer. Trop. Tuna Comm. Bull., v. 11, p. 355–612.

Stradner, H., 1959, First report on the discoasters of the Tertiary of Austria and their stratigraphic use: Fifth World Petrol. Cong. Proc., New York, sec. 1, p. 1081–1095.

Thierstein, H. R., this volume, Late Cretaceous calcareous nannoplankton and the change at the Cretaceous-Tertiary boundary.

———, and Okada, H., 1979, The Cretaceous/Tertiary boundary event in the North Atlantic: Init. Repts. DSDP, v. 43, p. 601–616.

Wise, S. W., jr., 1973, Calcareous nannofossils from cores recovered during Leg 18, Deep Sea Drilling Project: Biostratigraphy and observations of diagenesis: Init. Repts. DSDP, v. 18, p. 569–615.

———, 1977, Chalk formation: early diagenesis, *in* Andersen, N. R., and Malahoff, A., (eds.), The fate of fossil fuel CO_2 in the oceans: Plenum Publishing Corporation, p. 717–739.

———, and Wind, F. H., 1977, Mesozoic and Cenozoic calcareous nannofossils recovered by DSDP Leg 36 drilling on the Falkland Plateau, southwest Atlantic sector of the Southern Ocean: Init. Repts. DSDP, v. 36, p. 269–491.

SEPM Special Publication No. 32, p. 355–394, October 1981

LATE CRETACEOUS NANNOPLANKTON AND THE CHANGE AT THE CRETACEOUS-TERTIARY BOUNDARY

HANS R. THIERSTEIN
Scripps Institution of Oceanography, University of California,
San Diego, La Jolla, California 92093 U.S.A.

ABSTRACT

Quantitative taxonomic analyses of Late Campanian to Early Danian calcareous nannofossil assemblages suggest worldwide stable ecological conditions in the photic zone during the last 15 million years of the Mesozoic. Late Cretaceous bioprovincialism was strongly paleolatitudinal. The sudden extinction of the Cretaceous assemblages was followed by low supply of phytoplankton carbonate and by blooms of neritic taxa in the Early Danian, indicating low productivity and ecological instability. The most complete Cretaceous-Tertiary boundary record studied is at DSDP Site 356 (South Atlantic), where thicknesses of zones close to the boundary are twice those at Gubbio, Italy. The mode of calcareous phytoplankton evolution appears unsuitable for tests of theoretical evolutionary models.

INTRODUCTION

The Significance of the Deep Sea Record

The exploration of the deep ocean floors, initially by remote geophysical methods and subsequently by over ten years of deep ocean drilling, is, for the first time, providing the opportunity to study the earth's geological history of the past 100 million years on a truly global scale. These marine geological and geophysical investigations have contributed significantly to the establishment of a new paleogeographic framework (*e.g.,* Berger and Winterer, 1974; Smith and Briden, 1977; Lancelot, 1978) and of a greatly improved chronology (*e.g.,* Berggren and van Couvering, 1974; Thierstein, 1976; Hardenbol and Berggren, 1978). Preliminary stratigraphic results have led to global paleoecological and evolutionary interpretations, which range from essentially "uniformitarian" or "gradualistic" (Fischer and Arthur, 1977) to more "catastrophic" views (Thierstein and Berger, 1978; Berger and others, this volume).

In particular, the mass extinctions observed at the Cretaceous-Tertiary boundary have been recognized as one of the major late Phanerozoic evolutionary discontinuities. Considerable interest has focused recently on the chronostratigraphy, environmental selectivity of extinctions and geochemistry in associated sediments. Most of the proposed models, extraterrestrial (solar flare, impact of asteroid or comet) and terrestrial (magnetic reversal, fresh-water injection from Arctic, carbon dioxide buildup), imply constraints on the timing and environmental selectivity of the evolutionary events and on the associated geochemical phenomena preserved in fossils and sediments. Pertinent questions for tests of the proposed extinction models include: Are the mass extinctions real? Were they selective? Are they synchronous in all environments or did they occur in succession? How rapidly did they happen?

Microfossils are uniquely suited to address some of these questions because their common occurrence in large numbers allows the study of their population dynamics and avoids the problems of representativeness of available sample sizes that beset most macrofossil groups. Microfossil assemblages in some deep-sea sections are particularly valuable because of their exceptional preservation and continuous abundance, which are unmatched in any of the land sections known to cover the Cretaceous-Tertiary boundary.

It is the purpose of this paper to document the global biogeography of the calcareous nannoplankton communities and their evolutionary changes across the Cretaceous/Tertiary boundary, using the best preserved sections and samples available.

Previous Work

The great change in calcareous nannofossils between the Maastrichtian and Danian was first pointed out by Bramlette and Martini (1964). Subsequently, stratigraphic studies across the Cretaceous-Tertiary boundary were done in New Zealand (Edwards, 1966), southwestern France (Hay and Mohler, 1967), Denmark (Perch-Nielsen, 1969), and Alabama (Worsley, 1974). The emphasis in these studies was on taxonomy and the establishment of stratigraphic ranges for the recognized species. A quantitative taxonomic analysis of the boundary in Zumaya (northern Spain) led Percival and Fischer (1977) to propose a transitory event of conditions unfavorable for calcareous nannoplankton growth. Diagenetic alteration and reworking by lateral transport, however, were suspected as contributing factors by these authors. Romein (1977) reported the relative abundances of dominant species across the Cretaceous-Tertiary boundary in the

Barranco del Gredero section (southern Spain), which showed an earliest Tertiary *Thoracosphaera* acme and a subsequent dominance of *Braarudosphaera* extending over several meters of sediment thickness. Elaborating on earlier ideas by Hay (1960) and Worsley (1974), Romein further hypothesized a temperature decrease in the late Maastrichtian, leading to increased latitudinal provincialism and finally to extinction of the Cretaceous taxa with presumably elevated lower temperature tolerances.

Gartner and Keany (1978) described the nannofossil stratigraphy of bore hole 2/4 - A8 in the Ekofisk field (North Sea) and found a unique repetition of Cretaceous and Tertiary assemblages spread over 60 m of sediment thickness, which they explained by repeated invasion of a Danian brackish water assemblage from the nearby Arctic Ocean. They assumed that sedimentation at their site was continuous and that no lateral transport had occurred. The presence, however, in their lowermost Danian assemblages, of two species *Chiasmolithus danicus* and *Coccolithus cavus,* which appear consistently a few meters above the Cretaceous-Tertiary boundary elsewhere, suggest the presence of a hiatus below their lowermost Danian assemblage. In addition, Perch-Nielsen and others (1979) and Watts and others (1980) have shown with additional data that the stratigraphic repetition is due to large-scale submarine debris-flows known to occur in that area. Based on other evidence, such as the sudden extinction of Cretaceous plankton assemblages at DSDP Sites 384 and 356, the subsequent blooms of neritic taxa, and preliminary data showing strong shifts towards lighter values in the stable isotopic compositions of fine-fraction carbonate observed at DSDP Site 356, and the similarity of these signals to other events in ocean history led Thierstein and Berger (1978) to suggest that injection of salinity-fractionated waters from reconnecting, temporarily isolated ocean basins may account for a number of paleoceanographic events preserved in the stratigraphic record. The effects of these events that are to become preserved in the sedimentary record are variable and would depend on whether the isolated basin was located in low or high paleolatitudes and on whether the event was superimposed on a general trend (*e.g.,* amplification of anoxia by brine injection in the Aptian in a general setting of sluggish deep-water circulation) or was sudden and unannounced, such as the Cretaceous-Tertiary boundary event. The basic idea of Thierstein and Berger's (1978) model of temporary storage and release of fractionated waters is developed further in Berger and others (this volume) to include other temporary reservoirs of geophysical energy that may have played a major role in late Phanerozoic paleoceanographic and paleoclimatic history.

Previous stratigraphic studies of the stable isotopic composition of Maastrichtian and Paleocene planktonic and benthic foraminifera by Saito and van Donk (1974) and Savin and others (1975) showed only a minor change across the Cretaceous-Tertiary boundary. The shift towards lighter carbon isotopes at the Cretaceous-Tertiary boundary at DSDP Site 356 reported by Thierstein and Berger (1978) has been found also in the planktonic foraminifera (Boersma and others, 1979; Thierstein, unpublished data) and in whole-rock samples from other sites (Scholle and Arthur, 1980). The oxygen isotopic shift at DSDP Site 356, however, is neither recorded by the planktonic foraminifera (Boersma and others, 1979; Thierstein, unpublished data) nor in the less than 38 μm fractions from other sites (Thierstein, unpublished data). Investigations are underway to determine the cause of the apparently unique oxygen isotopic shift recorded in the original less than 62 μm carbonate fractions from DSDP Site 356.

Little work has been done to date on the paleobiogeography and paleoceanography of the Late Cretaceous oceans. Masters (1977) summarized the reported occurrences of Mesozoic foraminifera and plotted their known presence on geographic maps. These patterns revealed a strong decrease in diversity with increasing latitude, and have been confirmed in a study by Sliter (1977) on the presence and absence of Late Cretaceous planktonic foraminifera in the southern hemisphere. After the extinction, midlatitudinal Paleogene planktonic foraminifera apparently dominated most Early Paleocene assemblages according to Haq and others (1977). Late Cretaceous latitudinal nannofossil provincialism was recognized by Worsley and Martini (1970) based on occurrence of *Micula murus* and *Nephrolithus frequens* and by Bukry (1973) based on the absence of *Watznaueria barnesae* in high latitudes. A quantitative analysis of earliest Danian nannofossil assemblages in the Atlantic by Haq and Lohman (1976) yielded two characteristic communities: a neritic one, dominated by braarudospherids, and an oceanic one, dominated by thoracosphaerids. Late Cretaceous ocean surface current patterns were estimated by Gordon (1973) by analogy with today's current patterns and using the known distribution patterns of molluscs in outcrops on the continents.

Census Data

The analysis is based on counts of at least 300 specimens per sample in both Cretaceous and Tertiary assemblages in smear slides of untreated sediment under the light microscope. The evolutionary analysis has to be interpreted in the context of this sample size with its statistically limited representativeness of a true fossil population. The data are therefore mainly of an ecological nature and have limited bearing on diversity or the timing of true origination or extinction of the individual taxa. The utilized sample size of 300 specimens provides a probability of 95% of encountering a taxon whose true abundance in the population is 1%. Nevertheless, numerous nannofossil species with considerably lower abundances than 1% have been successfully used as stratigraphic markers in Mesozoic and Cenozoic sediments, since a routine scan of a smear slide in the light microscope can easily include tens of thousands of specimens.

Geographic and Stratigraphic Distribution of Samples

The data base for this analysis consists of taxonomic counts of 243 Late Cretaceous and 58 earliest Tertiary nannofossil assemblages in samples recovered mostly by the Deep Sea Drilling Project. A list of the samples studied is given Appendix I.

The assemblages range in age from the Late Campanian (base of *Tetralithus gothicus* Zone, Martini, 1976) to the Early Danian (middle part of NP2 *Cruciplacolithus tenuis* Zone, Martini, 1971). The paleogeographic distribution of the studied assemblages is shown in Figures 1-3, subdivided into three time-slices: Late Campanian to Early Maastrichtian (base of *Tetralithus gothicus* to top of *Tetralithus trifidus*), Early Maastrichtian to Late Maastrichtian (top *Tetralithus trifidus* to base *Micula murus* or *Nephrolithus frequens*), and latest Maastrichtian (base *Micula murus* or *Nephrolithus frequens* to first appearance of Tertiary nannofossil and/or foraminifera assemblage). The first appearance of a Tertiary assemblage is defined as the base of the mixed or transitional interval, with a presence of over one percent of those taxa that persist beyond the earliest Danian assemblages, *i.e.* NP1, NP2, and *Globigerina eugubina* Zones. The continental outlines on all maps are those of 70 million years ago; they have been interpolated from the maps given in Smith and Briden (1977) for 80 m.y. and 60 m.y.BP. The sample localities were plotted on the 80 m.y. and 60 m.y. maps relative to the closest continental outline near passive margins. Sample localities on oceanic plates were plotted following the paleopositions given by Sclater and others (1977) in the eastern Indian Ocean and Lancelot (1978) in the Central Pacific. The paleoposition for each assemblage studied was then interpolated linearly using its vertical distance in the section from the nearest biohorizons and absolute age estimates for the latter as given in Thierstein (1976). These paleopositions were plotted on maps by a computer line printer with a resolution of 2° latitude and 3° longitude. This limited resolution led to a stepwise movement of localities, occasionally resulting in two paleopositions for one locality within an individual time-slice, as observed in Figures 1–3.

The numerical data (including taxonomic raw counts, measured carbonate and coarse-fraction contents, computed age, paleolatitude and paleolongitude, and other information on each sample) will be published in Thierstein and Woodward (*in*

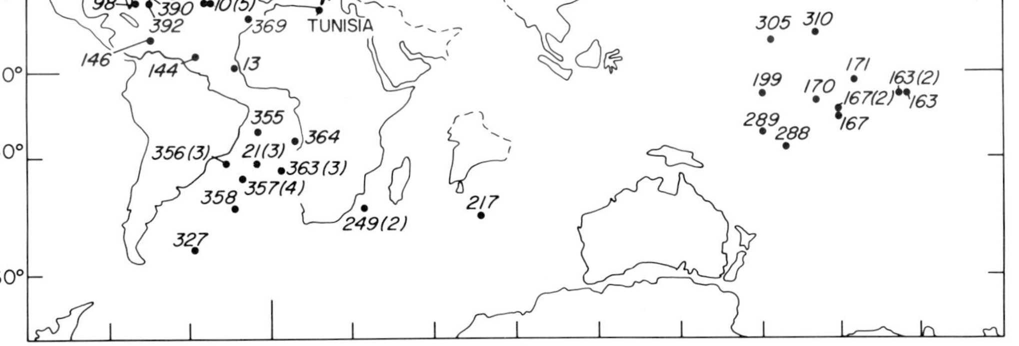

FIG. 1.—Paleogeographic distribution of 54 Late Campanian to Early Maastrichtian calcareous nannofossil assemblages of the *Tetralithus gothicus* and *Tetralithus trifidus* Zones. Number of assemblages at each site given in parenthesis, if over one.

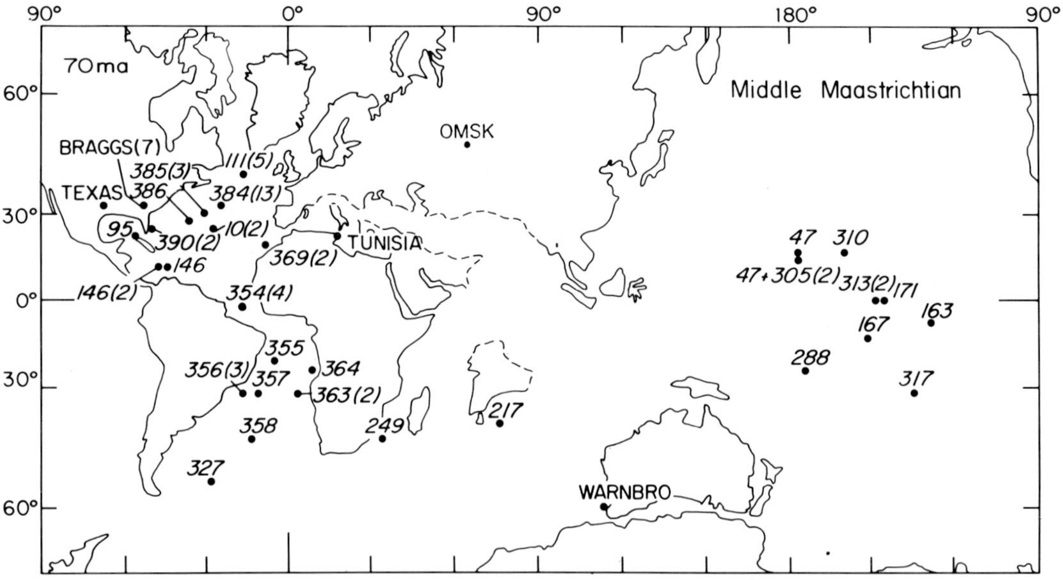

FIG. 2.—Paleogeographic distribution of 71 Middle Maastrichtian calcareous nannofossil assemblages of the *Arkhangelskiella cymbiformis* and *Lithraphidites quadratus* Zones. Number of assemblages at each site given in parenthesis, if over one.

press). Additional information on geographic location, lithology, and stratigraphy of the DSDP material can be found in the Initial Reports of the Deep Sea Drilling Project. References to information on the studied land-based samples are given in Appendix I.

Taxonomic Concept

The taxonomic concepts used in gathering the species counts, which form the basis of the present analysis, were developed during earlier studies (Thierstein, 1974, 1976). They have not been changed after data-gathering was initiated,

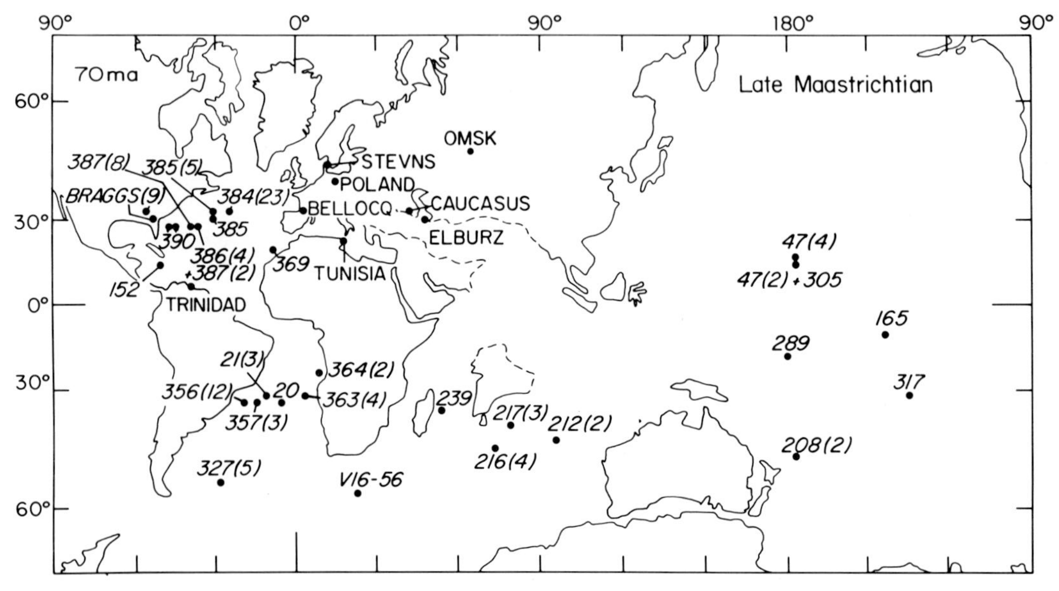

FIG. 3.—Paleogeographic distribution of 118 Late Maastrichtian calcareous nannofossil assemblages of the *Micula murus* and *Nephrolithus frequens* Zones. Number of assemblages at each site given in parenthesis, if over one.

therefore the internal consistency of the census data has been preserved. The occurrence of 73 Late Cretaceous and 16 Early Tertiary species and morphotypes was recorded while counting. After lumping the morphotypes (Thierstein, 1974) and ignoring isolated shields, 54 of the Cretaceous taxa attained a relative abundance of over two percent in at least one assemblage. Thirty-six taxa reached maximum relative abundances of over 4% and are shown in this analysis. Ten of the exclusively Tertiary taxa were consistently recognizable and are used in the stratigraphic plots. Broken nannofossils were counted as one specimen if more than half was preserved. *Thoracosphaera* pieces were added up to more than half a sphere before they were recorded as one specimen. An alphabetical list of the taxa, references to descriptions, illustrations, adopted taxonomic discussions, and additional remarks are given in Appendix II.

Preservation

A considerable range in the intensity of etching and overgrowth features was observed during light-microscopic examinations of the samples, despite the fact that they were chosen selectively and carefully to include the best-preserved assemblages available. The distorting effects of carbonate dissolution on the taxonomic composition of modern calcareous nannofossil assemblages are well known (McIntyre and McIntyre, 1971; Schneidermann, 1973; Roth and Berger, 1975). A similar preservational bias must be expected in older assemblages and may be further complicated by the diagenetic history of the individual samples. Three tests have so far been made to detect preservational trends in the taxonomic variation of the present sample set. These tests consisted of ranking the individual Late Cretaceous and earliest Tertiary nannofossil taxa based on their abundance changes with increasing paleodepth, with increasing burial depth, and of results from an *in-situ* dissolution experiment. The dissolution experiment was judged the most straightforward approach and is reported in detail elsewhere (Thierstein, 1980). The experiment consisted of exposing several Late Cretaceous and earliest Tertiary assemblages of varying taxonomic compositions to corrosive deep waters for about four months at different depths along a deep mooring deployed in the western North Atlantic in 1977. The observed taxonomic changes in each deployed assemblage with increasing depth were used to rank the counted taxa with respect to their susceptibility and resistance to dissolution. The ranking clearly dissociated two taxa from all others, *Cruciplacolithus primus* as the most susceptible and *Micula staurophora* as the most resistant to dissolution (Fig. 4). A dissolution ranking using paleodepth estimates for assemblages from DSDP samples identified *Micula staurophora* as the most resistant, although the relative positions of some other taxa were different. The third test was for susceptibility to diagenetic recrystallization and consisted of a similar ranking using taxonomic changes with increasing burial depth. The results also identified *Micula staurophora* as the most resistant and indicated that diagenetic overgrowth is accompanied by dissolution of more delicately-structured taxa.

EVOLUTION

A meaningful interpretation of the biogeography of fossil organisms for any geological interval requires some knowledge of their evolutionary or stratigraphic variability in order to assess the biogeographic signal-to-noise ratio within the defined time slice, a procedure that is often ignored in biogeographic studies. The relative biochronologic resolution in uppermost Cretaceous and lowermost Tertiary pelagic oozes is judged to be approximately one to several million years. The paleoecological evolution of the calcareous nannofossil assemblages within that interval has been assessed by analyzing numerous assemblages within each of the time slices in sections from the North Atlantic, South Atlantic, Pacific, and from continental North America.

Late Campanian to Early Maastrichtian in the North Atlantic: DSDP Site 10

Site 10 was drilled in the western North Atlantic at 32°52′N, 52°13′W in 4697 m of water and was at an estimated paleodepth of 2.1 to 2.5 km and paleolatitude of about 23°N (Peterson and others, 1970; Thierstein, 1979). The biostratigraphy of planktonic foraminifera and calcareous nannofossils in the over 80 m-thick Late Campanian to Early Maastrichtian sequence has been studied by Cita and Gartner (1971) and Thierstein (1976).

The bulk carbonate content, the percent coarse fraction, and the relative abundance of the nannofossil taxa have been determined in nine samples (Fig. 5). The nannofossil assemblages are well preserved throughout the sequence and show only slight evidence of etching and overgrowth (see also illustrations in Cita and Gartner, 1971). The diverse assemblages are not strongly dominated by any individual taxon, and the abundance fluctuations occur mainly between relatively dissolution-resistant taxa (*e.g. Micula staurophora, Tetralithus gothicus, Tetralithus trifidus*) and dissolution-susceptible taxa (*e.g., Biscutum constans, Prediscosphaera cretacea*). The paleoecological conditions at this location appear to have been quite stable throughout the interval represented.

Late Maastrichtian to Early Danian in the Western Pacific: DSDP Site 47.2

Drilled on the rim of Shatsky Rise (32°27′N, 157°43′E) in 2689 m of water, Site 47.2 yielded an

FIG. 4.—Dissolution ranking of Late Cretaceous and earliest Tertiary calcareous nannofossils (from Thierstein, 1980.)

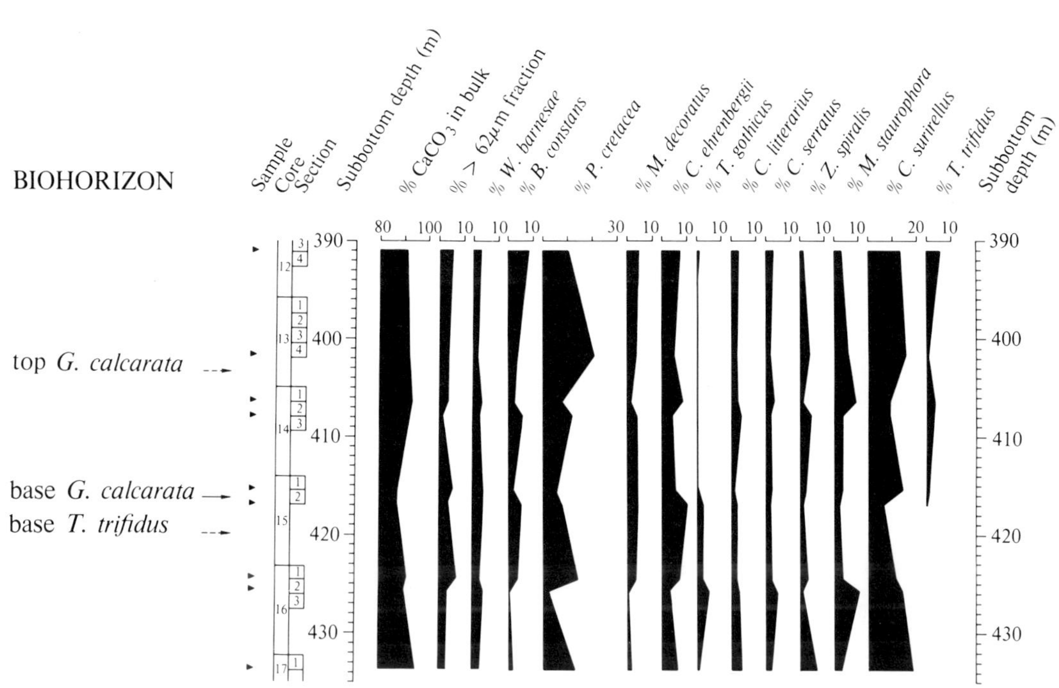

FIG. 5.—Stratigraphy and calcareous nannofossil abundances in Upper Campanian to Lower Maastrichtian sediments at DSDP Site 10 (North Atlantic).

Upper Maastrichtian through Paleocene cherty chalk sequence (Fischer and others, 1971), which was deposited at an estimated paleodepth of about 2400 m and at an estimated paleolatitude of about 15°N.

Eleven nannofossil assemblages analyzed quantitatively from a 30 m thick interval covering the Late Maastrichtian and Early Danian (Fig. 6) were selected from a total of 28 samples examined. Moderate overgrowth and etching of most nannoliths examined, together with considerable drilling disturbance evidenced clearly in the lithology of the archive halves in the Deep Sea Drilling Project repository, made an analysis of more closely spaced samples across the Maastrichtian-Danian boundary unattractive. The highest assemblage composed exclusively of Cretaceous taxa was found in Sample 47.2-11-4, 15 cm (for sample annotation, see Appendix I). Assemblages from the interval 47.2-11-4, 8 cm to 47.2-11-3, 136 cm contain less than 1 percent Early Danian nannoliths, the composition of which is equivalent to *Braarudosphaera*-free Atlantic assemblages older than the middle part of the *Cruciplacolithus tenuis* (NP2) Zone. From sample 47.2-11-4, 33 cm to the top of Core 11, the Tertiary taxa usually amount to more than one percent of the assemblage, but their abundance fluctuates wildly throughout Section 3, and the presence of *Prinsius martinii* indicates an age corresponding to the later part of the *Cruciplacolithus tenuis* (NP2) Zone. Assemblages from all 6 sections recovered in Core 10 contain a mixed nannoflora composed of taxa equivalent to the late *Cruciplacolithus tenuis* to *Fasciculithus tympaniformis* (NP5) Zones, but with *Chiasmolithus danicus* missing. Late Paleocene nannofossil assemblages of normal diversity are found at the base of Core 9 and upward. Drilling disturbance leading to these unusual taxonomic compositions is clearly illustrated for Core 10 in Fischer and others (1971, p. 127). The lithologic as well as the paleontologic evidence strongly oppose Hofker's (1978) claim of an undisturbed succession across the Cretaceous-Tertiary boundary at this site.

Despite the problems introduced by mechanical disturbance, the replacement of the Late Maastrichtian by the Danian nannoflora can be located between samples 47.2-11-1, 111 cm and 47.2-11-4, 21 cm, *i.e.,* between 101.71 m and 105.31 m subbottom depth. A strong increase in the proportions of dissolution-resistant taxa, such as *Micula staurophora* and *Micula murus* is observed in the middle part of Core 11 and reflects a short-term dissolution pulse near or at the Cretaceous-Tertiary boundary.

The correlation between preservational change of nannofossils, foraminifera, and carbonate contents in the fine-fraction strongly suggests a preservational rather than an evolutionary cause for the "gradually extinguishing" Late Cretaceous planktonic foraminifera and the "aplanktonic zone" pro-

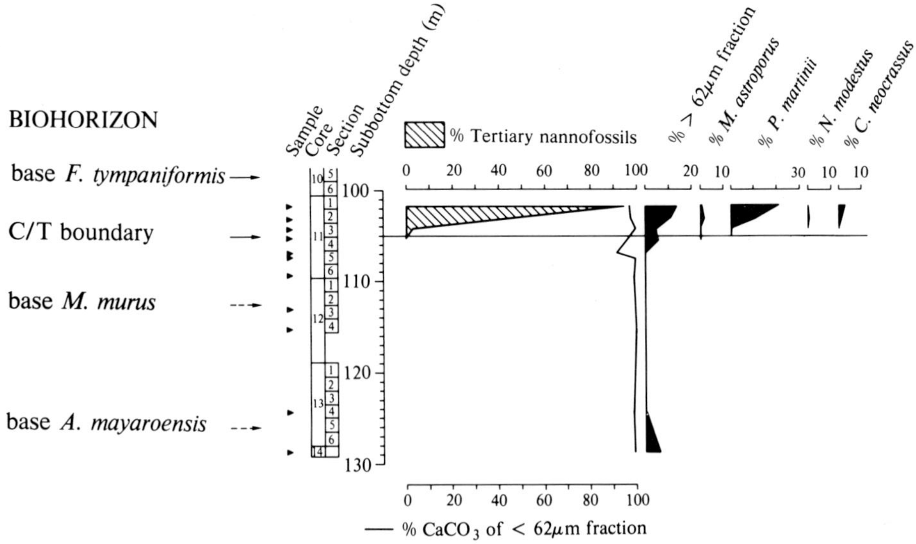

FIG. 6.—Stratigraphy and calcareous nannofossil abundances in Upper Maastrichtian to Danian sediments at DSDP Site 47.2 (western Pacific).

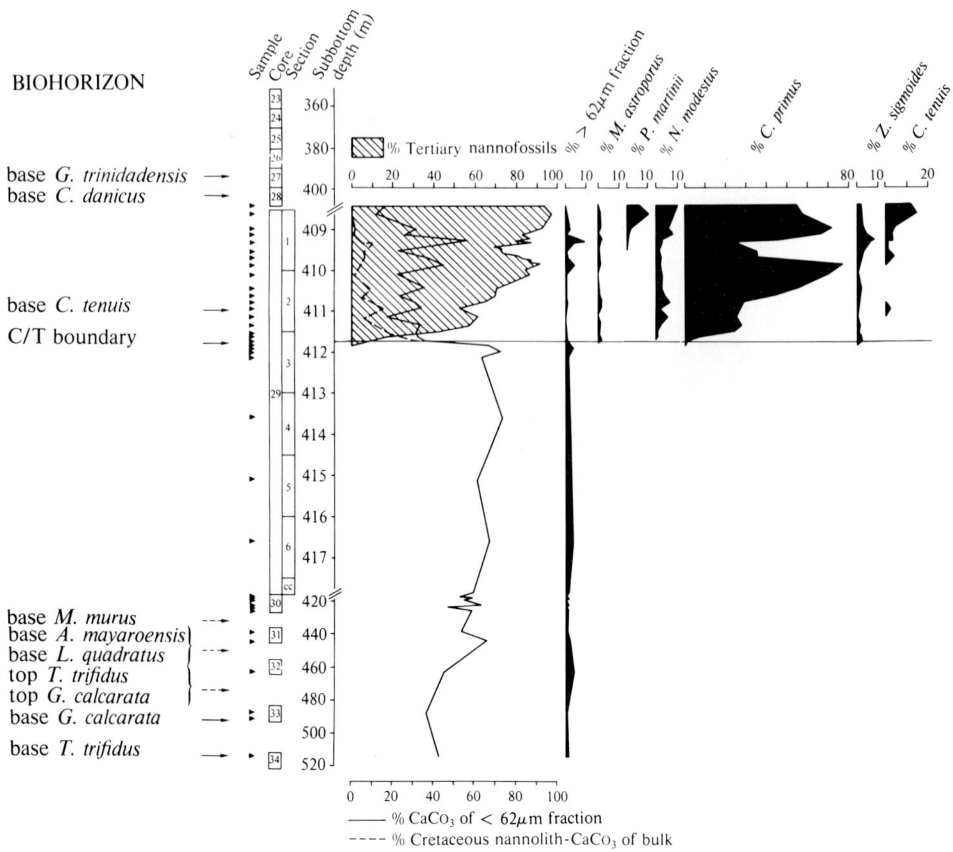

FIG. 7.—Stratigraphy and calcareous nannofossil abundances in Upper Campanian to Danian sediments at DSDP Site 356 (western South Atlantic).

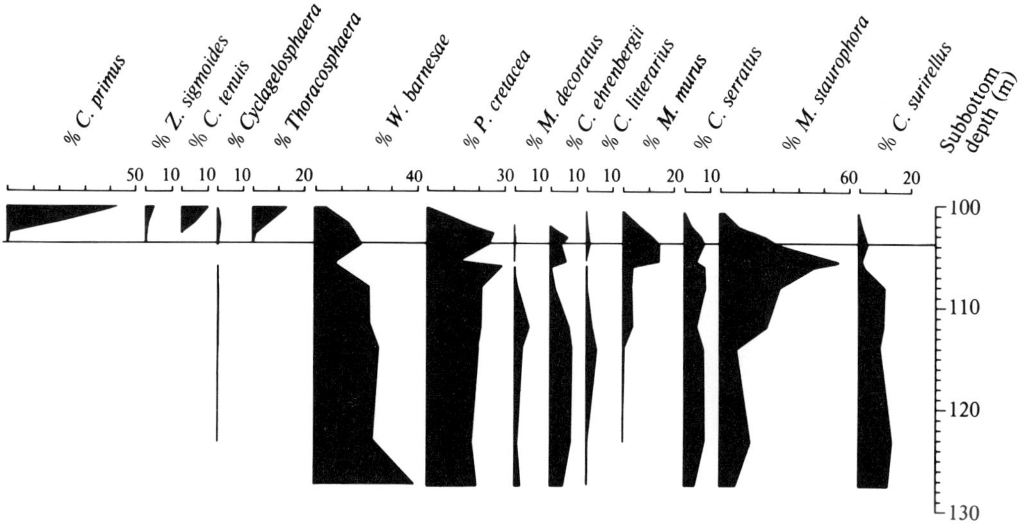
% C. primus
% Z. sigmoides
% C. tenuis
% Cyclagelosphaera
% Thoracosphaera
% W. barnesae
% P. cretacea
% M. decoratus
% C. ehrenbergii
% C. litterarius
% M. murus
% C. serratus
% M. staurophora
% C. surirellus
Subbottom depth (m)
50
10
10
10
20
40
30
10
10
10
20
10
60
20
100
110
120
130

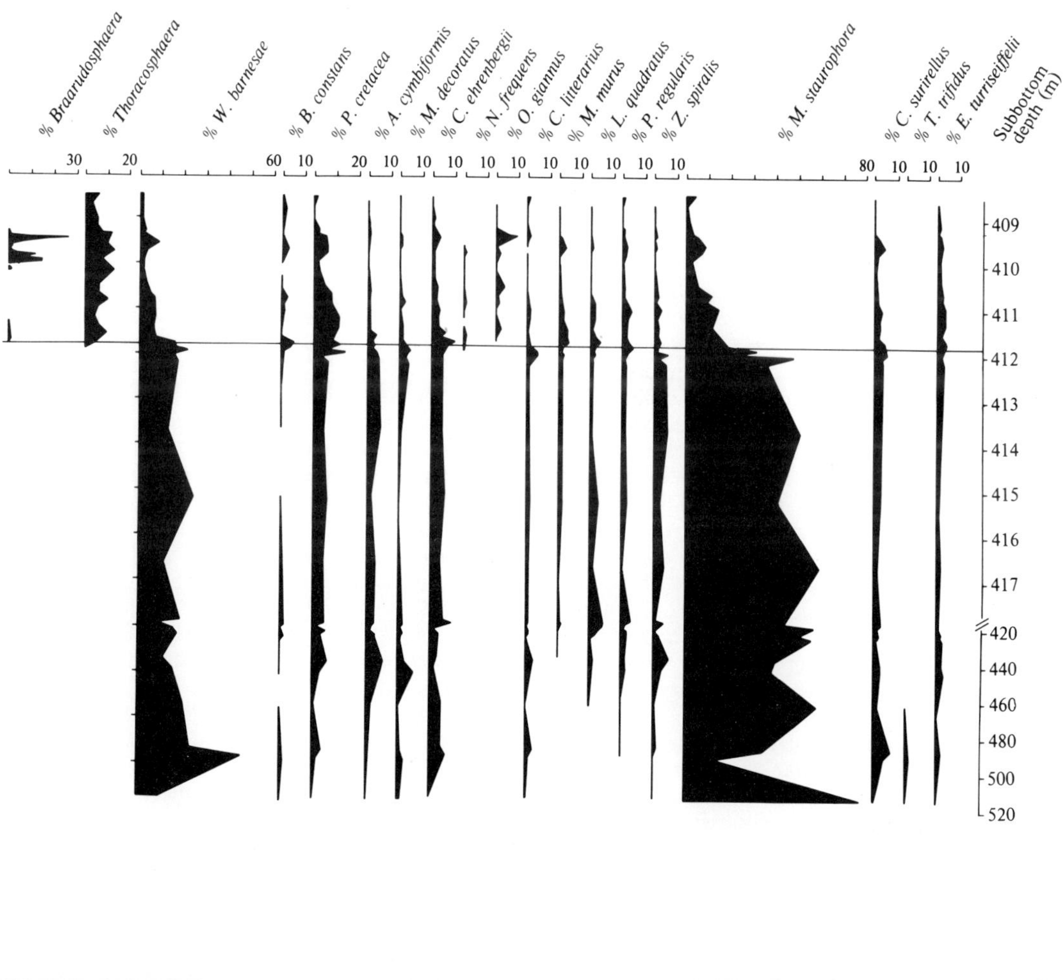
% Braarudosphaera
% Thoracosphaera
% W. barnesae
% B. constans
% P. cretacea
% A. cymbiformis
% M. decoratus
% C. ehrenbergii
% N. frequens
% O. giannus
% C. litterarius
% M. murus
% L. quadratus
% P. regularis
% Z. spiralis
% M. staurophora
% C. surirellus
% T. trifidus
% E. turriseiffelii
Subbottom depth (m)
30
20
60
10
20
10
10
10
10
10
10
10
10
10
10
80
10
10
10
409
410
411
412
413
414
415
416
417
420
440
460
480
500
520

posed by Hofker (1978). The earliest Tertiary nannoplankton assemblages differ from those known in the Atlantic by a complete absence of *Braarudosphaera* and the presence of *Prinsius martinii* in the very earliest Tertiary assemblages. Whether these peculiarities are caused by a hiatus at the base of the Danian, by dissolution before burial, by drilling disturbance, or by a combination of these can only be decided when other continuous sections across the Cretaceous-Tertiary boundary become available in the Pacific Ocean.

Late Campanian to Danian in the South Atlantic: DSDP Site 356

Site 356 was drilled on the Sao Paolo Plateau at 28°17′S, 41°05′W in 3175 m of water (Supko, Perch-Nielsen and others, 1977). A carbonate sequence over 120 m thick of Late Campanian through Paleocene age was recovered, which had been deposited at an estimated paleodepth of about 2200 m to 2600 m and at an estimated paleolatitude of about 31°S to 32°S.

Detailed descriptions of the lithology and biostratigraphy are found in Supko, Perch-Nielsen and others, (1977). The evolutionary transition from *Micula staurophora* to delicate morphotypes of *Micula murus* (recently designated *Micula prinsii* by Perch-Nielsen, 1979a), however, was observed in Cores 29 and 30, *i.e.*, lower than the first occurrence of *Micula murus* reported by Perch-Nielsen (1977). Stable isotope variations in biogenic carbonate particles have been analyzed by Thierstein and Berger (1978), Boersma and others, (1979), and Thierstein and others, (in preparation).

The relative abundances of most Cretaceous species are quite constant from the Upper Campanian to the top of the Maastrichtian (Fig. 7); fluctuations occur in the resistant *Micula staurophora* and are compensated by inverse changes in *Watznaueria barnesae* and numerous other dissolution susceptible taxa. These Late Cretaceous abundance fluctuations are therefore most likely caused by differential diagenesis. The Cretaceous-Tertiary boundary was recovered at 411.83 m subbottom depth (356-29-3, 31 cm) and is characterized by a marked lithologic change from pinkish-gray to bluish-gray nannochalk (Supko, Perch-Nielsen and others, 1977, p. 184) and by a sharp decrease of the fine-fraction carbonate content from a Late Cretaceous average of about 70% to a Danian average of about 30% (Fig. 7). The preservation of nannofossils improves across the Cretaceous-Tertiary boundary. Upper Cretaceous assemblages are slightly to strongly etched and slightly to moderately overgrown, whereas Lower Danian assemblages show mostly excellent preservation, with only occasional signs of slight etching and overgrowth. This preservational improvement is reflected in a decrease in the relative proportion of *Micula staurophora* and an increase in the relative proportions of *Prediscosphaera cretacea, Cribrosphaerella ehrenbergii,* and *Nephrolithus frequens* within the Cretaceous assemblages, which have been reworked into the Danian sediments. The relative abundances of the individual Cretaceous taxa in samples from above the Cretaceous-Tertiary boundary change quite monotonically and there is no evidence for significant fluctuations within the surviving or reworked Cretaceous assemblage, nor is there evidence for sequential extinction events of individual taxa (Fig. 7). These features are also observed with an additional eight Cretaceous taxa with persistent occurrences at Site 356, but whose maximum relative abundance never exceeds 2 percent and which, therefore, have not been graphed in Figure 7. A similar pattern of stability within the Cretaceous assemblage has been observed previously in the lowermost Danian sediments at DSDP Site 384 (Thierstein and Okada, 1979, fig. 4). The sudden fine-fraction carbonate decrease across the Cretaceous-Tertiary boundary coincides with the first appearance of Tertiary taxa. If the fine-fraction carbonate is divided into Cretaceous and Tertiary portions using the respective nannofossil abundances, a rather steep decline of Cretaceous nannofossil carbonate becomes apparent (Fig. 8). The steepness of the decline can be used to put some constraints on the duration of the Cretaceous nannofossil extinction event.

Some dilution of the original steepness of the fine-fraction carbonate decline may have resulted from the lateral influx of reworked Cretaceous sediment particles from nearby outcrops. Such reworking is evident from the occasional occurrence of Early and Middle Cretaceous nannofossils, such as *Podorhabdus albianus* Black and *Ellipsochiastus punctatus* Worsley, and from the fluctuating frequency of Cretaceous species above the Cretaceous-Tertiary boundary. Such a transport mechanism is plausible considering the topographic setting of Site 356 in a valley behind the outer Sao Paola Basement Ridge (Supko, Perch-Nielsen, and others, 1977, p. 144). Another diluting effect must be expected from the presence of significant numbers of very small earliest Tertiary globigerinid planktonic foraminifera in the less than 62 μm fraction. Assemblages of such small foraminifera are also known to occur in sediments belonging to the *Globigerina eugubina* Zone elsewhere (Luterbacher and Premoli Silva, 1964; von Hillebrandt, 1965; Smit, 1977). Therefore, the weight percent of the 38–62 μm and of the less than 38 μm fractions were determined in nine samples closest to the Cretaceous-Tertiary boundary. The average value of the 38–62 μm fraction in five Cretaceous samples was 1.1 $\pm$ 0.4 percent and in the four overlying Danian samples 3.6 $\pm$ 1.0 percent, and the less than 38 μm fractions changed from an average of 95.9 $\pm$ 1.6 percent below to 96.0 $\pm$ 1.1 percent above the boundary. The total foraminifera percentage there-

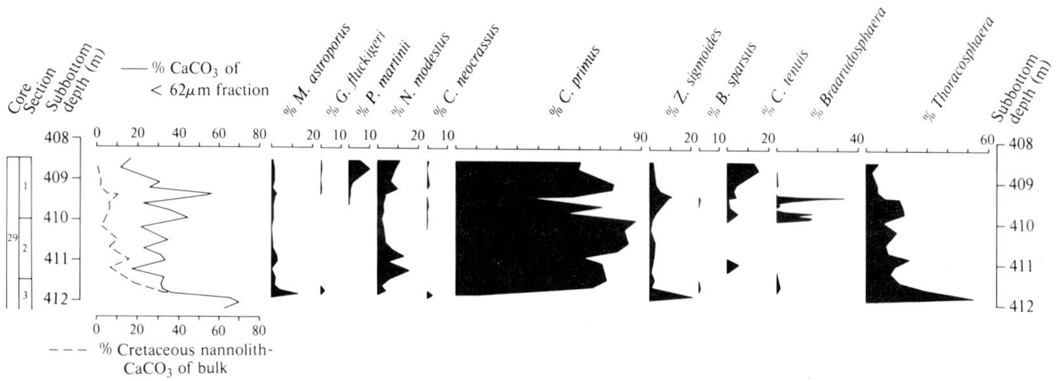

FIG. 8.—Evolutionary patterns of earliest Tertiary calcareous nannofossils at DSDP Site 356.

fore remained constant across the Cretaceous-Tertiary boundary with only a change from Late Cretaceous globotruncanid assemblages to the smaller Danian globigerinid assemblages. Because all samples in the transition consist of over 95% particles of less than 38 μm size, no noticeable error is introduced in the nannolith carbonate estimates.

If we were to assume that the extinction of the Cretaceous plankton assemblages was geologically instantaneous, the decrease of the Cretaceous nannolith carbonate could be used to estimate the thickness of the bioturbated layer. Guinasso and Schink (1975) have shown that the weighted mean depth of a tracer after mixing is at the level of the original tracer impulse. The weighted mean depth of the reworked Cretaceous nannolith carbonate therefore will be the original level of extinction, and the downward displacement of the earliest sharp carbonate decrease will be equal to the fossilized benthic mixed-layer thickness. The weighted mean depth of the Cretaceous nannolith carbonate distribution between 411.20 m and 411.91 m subbottom depth is 411.74 m (*i.e.*, at 356-29-3, 24 cm) or about 6 to 8 cm above the lithologic and paleontologic Cretaceous-Tertiary boundary in Section 3 of Core 29 (Supko, Perch-Nielsen and others, 1977). The similarity of this derived thickness of the bioturbated layer with carbon-14 dated thicknesses of benthic mixed layers in present deep-ocean environments (Nozaki and others, 1977; Berger and Johnson, 1978; Peng and others, 1979) supports an instantaneous phytoplankton extinction hypothesis. A comparable mixed-layer thickness can be derived from two *Braarudosphaera* blooms as discussed below.

The evolutionary patterns within the newly evolving Tertiary assemblage are characterized intially by a dominance of *Thoracosphaera, Zygodiscus sigmoides,* and *Markalius astroporus,* quickly superseded by that of *Cruciplacolithus primus* (Fig. 8). Two to three meters above the Cretaceous-Tertiary boundary, two short-term *Braarudosphaera* blooms can be clearly recognized lithologically at 356-29-1, 80–90 cm and 125–145 cm. The relative proportions of *Braarudosphaera* in this interval have been determined at additional levels using toothpick-sized samples for smear slides. The depth difference between the weighted mean depth and the earliest increase of the two *Braarudosphaera* blooms are 10 cm for the earlier bloom (409.65–409.94 m) and 8 cm for the later bloom (409.30–409.50 cm). The upper, decreasing slopes of the two *Braarudosphaera* abundance peaks yield virtual mixing lengths of around 2 cm, using the exponential mixing model of Berger and Heath (1968). These upper slopes may be oversteepened by incomplete homogenization, winnowing, lateral transport, or a combination of these, as evidenced in the laminated texture of the sediments in these intervals. Presence of the *Braarudosphaera* blooms cannot be due entirely to differential preservation, because the assemblages are still dominated by the very dissolution-susceptible *Cruciplacolithus primus* during the blooms. The *Braarudosphaera* blooms, however, coincide with levels of increased fine-fraction carbonate contents, which may reflect both the larger size of its coccoliths as well as the known diagenetic potential of *Braarudosphaera* to collect secondary calcite by preferential overgrowth.

The differences in the evolutionary stratigraphy of Site 356 from other sections may in principle be related to regional changes in sedimentation rates or preservation, to hiatuses, or to paleoecologic differences and will be evaluated after a discussion of the evolutionary patterns of Site 384 and the Braggs Section.

Middle Maastrichtian to Early Danian in the North Atlantic: DSDP Site 384

This site, located at the northern end of the J-Anomaly Ridge (40°22′N, 51°40′W) was drilled in 3909 m of water (Tucholke, Vogt and others, 1979) and had an estimated paleodepth of about 3 km

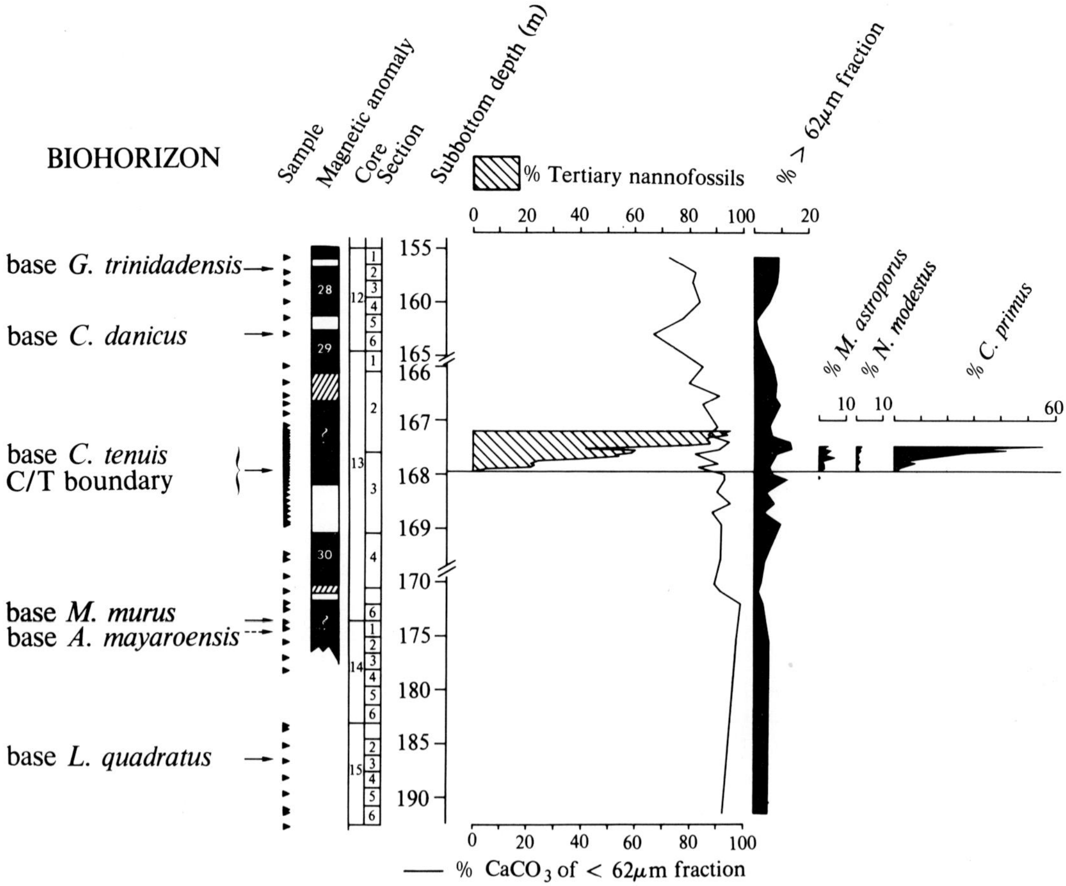

FIG. 9.—Stratigraphy and calcareous nannofossil abundances in Middle Maastrichtian to Danian sediments at DSDP Site 384 (western North Atlantic).

(Thierstein and Okada, 1979; Tucholke, Vogt and others, 1979) and an estimated paleolatitude of about 33°N at the end of the Cretaceous. The Cretaceous-Tertiary boundary was recovered in Section 3 of Core 13 and is illustrated in the frontispiece and on p. 602 in Tucholke, Vogt and others, (1979). The biostratigraphy of planktonic foraminifera and calcareous nannofossils suggests that the Cretaceous-Tertiary transition at Site 384 is as continuous and complete as any other known section (Boersma and others, 1979; McNulty, 1979; Okada and Thierstein, 1979). The paleomagnetic data across the Cretaceous-Tertiary boundary are not of very good quality and, based on a comparison with the paleomagnetic records from Gubbio and the seafloors, have been interpreted by Larson and Opdyke (1979) to include a short hiatus at the Cretaceous-Tertiary boundary. Thierstein and Okada (1979) argued for a slightly different interpretation of the magnetic reversal pattern on biostratigraphic grounds, which left a short, unexplained normally magnetized interval straddling the plankton extinction event within the polarity epoch Gubbio G-. The presence of an additional dominantly normally magnetized interval between magnetic anomalies 29 and 30 has received additional support from several paleomagnetic investigations in sections from continental North America recording the palynofloral changes and extinctions of large vertebrates. The last dinosaur remains were found in a normally magnetized interval in the San Juan Basin of New Mexico (Butler and others, 1977; Lindsay and others, 1978). Although the chronology of the terrestrial sequences of Late Cretaceous and Early Paleocene age in the San Juan Basin seems controversial (Alvarez and Vann, 1979; Fassett, 1979; Lindsay and others, 1979a, b; Lucas and Rigby, 1979), the existence of an additional normal interval between anomalies 29 and 30 would eliminate the inferred diachrony between terrestrial and marine extinctions. Dominantly normally-magnetized sedimentary rock samples across the Cretaceous-Tertiary boundary were found in sections in the Red Deer Valley, Alberta by Lerbekmo

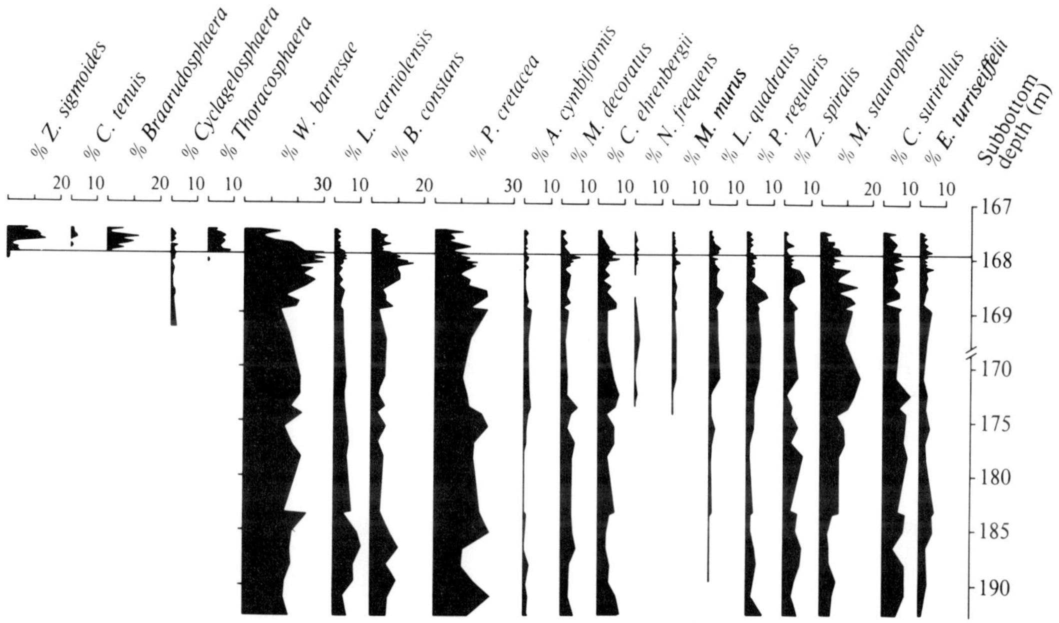

and others, (1979) and in sections at Hell Creek (Montana), Morgan Creek (Saskatchewan), and Braggs (Alabama) by Danis and others, (*in press*).

The evolutionary abundance patterns of the calcareous nannofossil assemblages across the Cretaceous-Tertiary boundary at Site 384 have been analyzed in detail by Thierstein and Okada (1979) and are shown in Figure 9, together with other pertinent information. The comparatively rapid replacement of Cretaceous by Tertiary taxa indicates less persistent lateral reworking than at Site 356, and could be explained by an instantaneous extinction of Cretaceous nannoliths and subsequent benthic mixing, with a necessary mixed layer thickness of about 21 cm. The evolutionary patterns of the Danian assemblages at this site are similar to those in the other sections investigated, with the exception of the *Braarudosphaera* bloom occurring

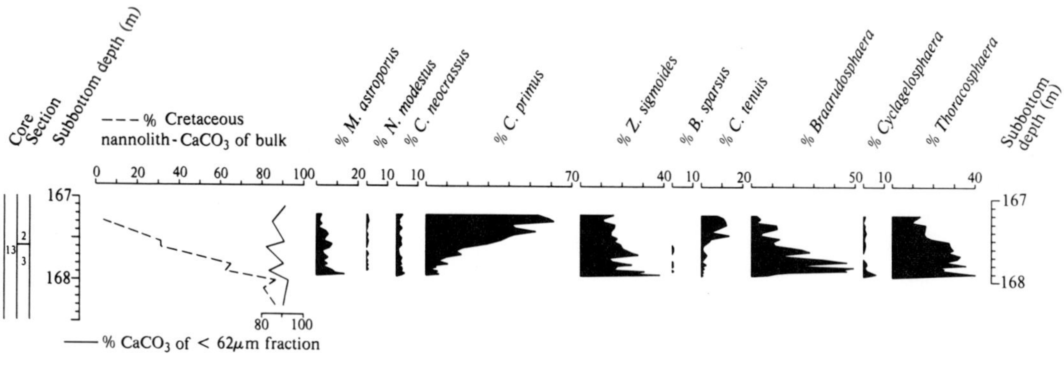

FIG. 10.—Evolutionary patterns of earliest Tertiary calcareous nannofossils at DSDP Site 384.

apparently closer to the Cretaceous-Tertiary boundary (Fig. 10). This apparent shift may be due to decreased sedimentation rate or sediment disturbance at the Cretaceous-Tertiary boundary (Tucholke, Vogt and others, 1979; see also discussion below) or paleo-provincialism of *Braarudosphaera.* The nannofossil assemblages at Site 384 are well-preserved throughout the studied interval, which is reflected in the constant abundances of dissolution-susceptible and dissolution-resistant Cretaceous taxa across the boundary (Thierstein and Okada, 1979, fig. 4) and in the absence of any significant change in carbonate contents.

Middle Maastrichtian to Early Danian in Central Alabama: Braggs Section

A detailed lithologic description of this section along Alabama Highway 7 near Braggs, Lowndes County, Alabama is found in Worsley (1974). The samples investigated in this study have been kindly supplied by Dr. C. W. Copeland (Geological Survey of Alabama) and by Dr. D. A. Russell (National Museum, Ottawa) and are numbered by distance from the top of the section measured by Danis and others (*in press*), which begins 4.43 m above the Cretaceous-Tertiary boundary as defined by Worsley (1974). The lithology of the section consists of "chalky, silty clay, with irregularly indurated layers of sandstone, siltstone, and limestone", with a molluscan macrofauna indicating a shallow neritic depositional paleoenvironment (Worsley, 1974). The inferred latest Mesozoic paleolatitude for Braggs is about 33°N.

The chronology used in the Braggs section is based on calcareous nannofossils. The placement of some of the biohorizons is rather imprecise due to the wide spacing of fossiliferous samples in some intervals (Fig. 11). The younger end of the sampled section is still near the middle part of the *Cruciplacolithus tenuis* Zone (NP2), since neither *Prinsius martinii,* nor *Chiasmolithus danicus* have been encountered.

The evolutionary patterns of the Late Cretaceous taxa are governed mainly by fluctuations in the relative abundance of dissolution-resistant *Micula staurophora.* The gradual increase in the abundance of this species toward the top of the Cretaceous section has been noted by Worsley (1974), who used the increase of this "last Cretaceous survivor" to estimate the thickness of the sediment interval missing at the top of the Maastrichtian. Experimental data (Thierstein, 1980), the records from other sections, and the preservational changes observed during the light-microscopic examination of the Braggs samples, however, suggest that the increase in the abundance of *Micula staurophora* is the result of deteriorating preservation, rather than gradual extinction of other taxa.

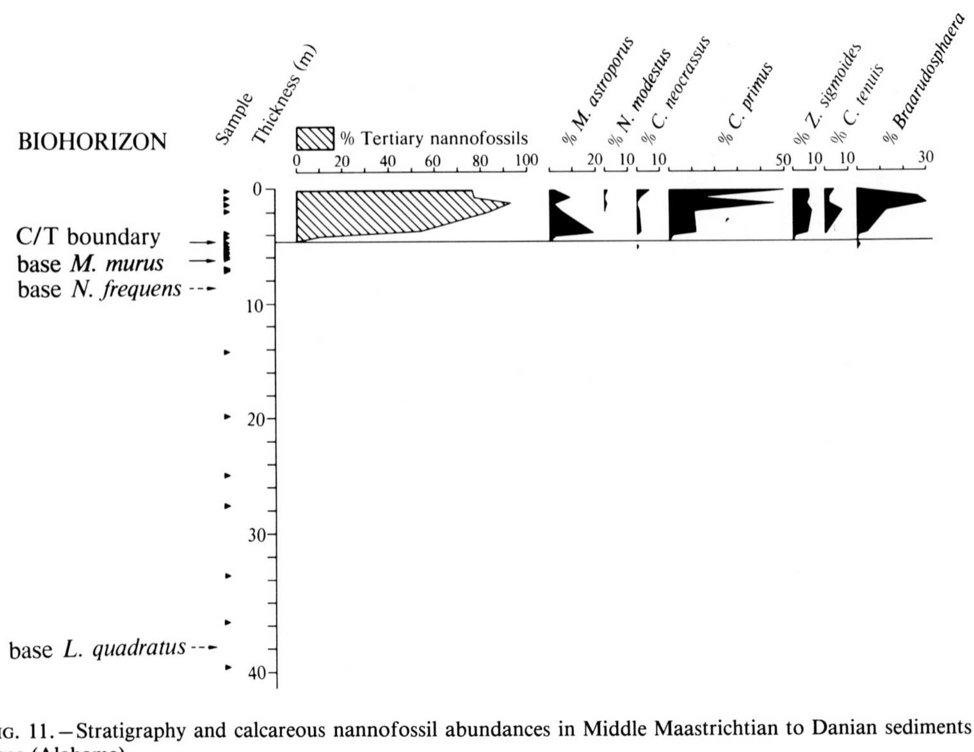

FIG. 11.—Stratigraphy and calcareous nannofossil abundances in Middle Maastrichtian to Danian sediments near Braggs (Alabama).

The evolution of the earliest Tertiary assemblages at Braggs is distinguished by an initial dominance of circular placoliths (*Markalius astroporus* and *Cyclagelosphaera*) and *Thoracosphaera*, superseded by that of *Braarudosphaera* and *Cruciplacolithus primus* (Fig. 12). If the peculiarly high proportion of *Cyclagelosphaera*, which may be related to the neritic paleoenvironment (see also Thierstein, 1973, p. 35), is ignored, the evolutionary patterns at Braggs are intermediate between those found at Site 356 and 384, which differ mainly in their relative position of the *Braarudosphaera* blooms and the rise of *Cruciplacolithus primus* to dominance.

The relatively widely-spaced samples from the Braggs section and the variable lithology, implying variable sedimentation rates, prevent a meaningful estimate of the thickness of the bioturbated layer.

On the Stratigraphic Continuity of the Sections at DSDP Sites 356, 384, and Braggs

The long known and widespread occurrence of unconformities and hardgrounds at the Cretaceous-Tertiary boundary has been reviewed by Cita and Premoli Silva (1974) and by Worsley (1974). Based on comparisons between magnetic reversals preserved in sedimentary records (Alvarez and others, 1977; Butler and others, 1977; Alvarez and Lowrie, 1978; Lerbekmo and others, 1979; Larson and Opdyke, 1979; Danis and others, *in press*) and the sea-floor magnetic reversal patterns (LaBrecque and others, 1977), the terminal Mesozoic plankton extinction event is estimated to have happened within 10,000 years or less (Kent, 1977). Even that number is an approximation based on numerous assumptions about the sedimentation regime at the Gubbio section. The estimated sedimentation rates in the most complete sections range from 1.25 cm per thousand years in Gubbio (Kent, 1977) to 0.6 cm per thousand years at DSDP Site 384 (Tucholke, Vogt and others, 1979, p. 126). The smallest possible sampling intervals and the distorting effects of sediment burrowers clearly limit the minimum absolute duration that can be estimated for the extinction event to a few thousand years at best.

The highest relative resolution obtainable from a stratigraphic analysis of paleoenvironmental parameters in a section can be reckoned by comparisons of sediment thicknesses between bio- and magnetostratigraphic events close to the Cretaceous-Tertiary boundary. A comparison of thicknesses between biostratigraphic events among DSDP Sites 356, 384, and the Braggs section is graphically displayed in Figure 13. The relative sequence of the biostratigraphic events identifiable is consistent in all sections with the exception of the early occurrence of the *Braarudosphaera* bloom relative to the increase in the ratio of Tertiary/Cretaceous nannofossil taxa. The measured or estimated thicknesses

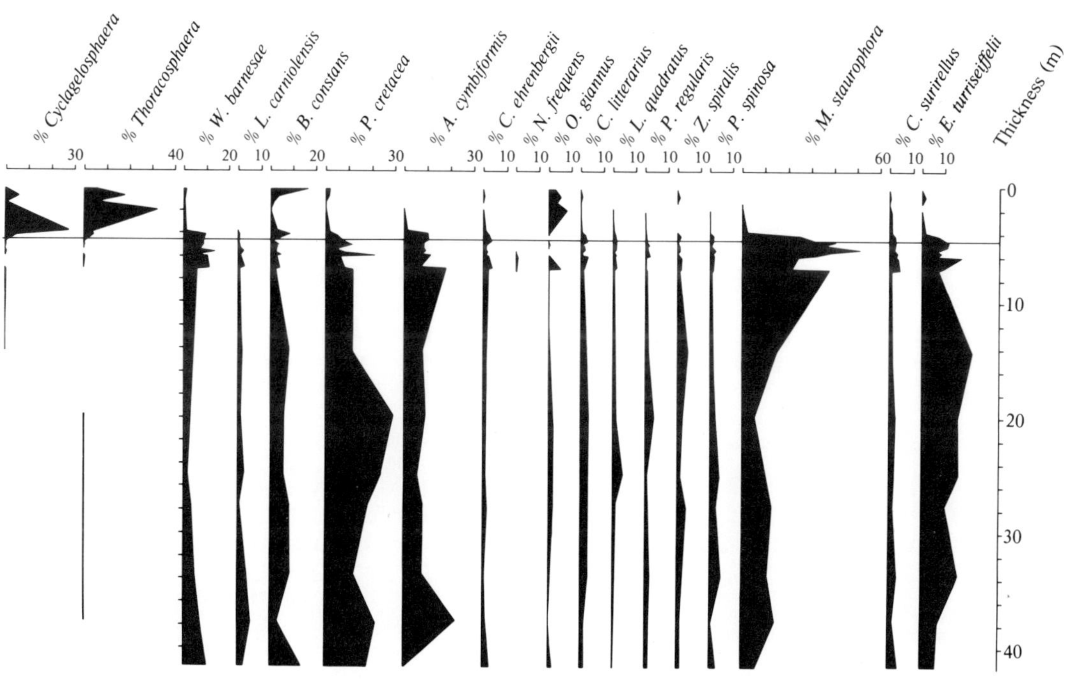

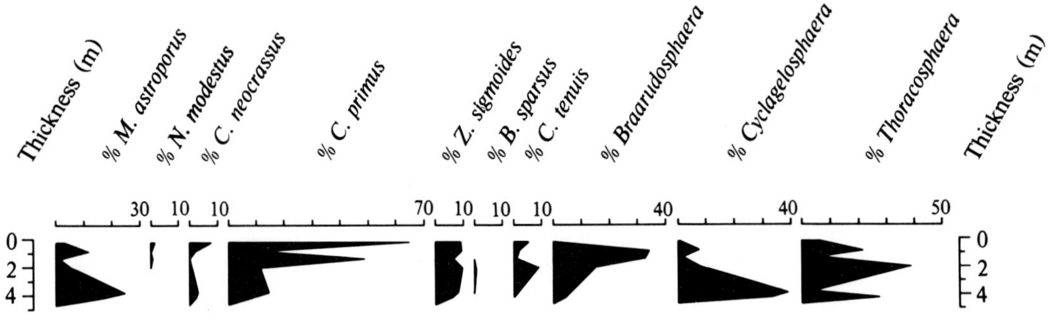

FIG. 12.—Evolutionary patterns of earliest Tertiary calcareous nannofossils in the Braggs section.

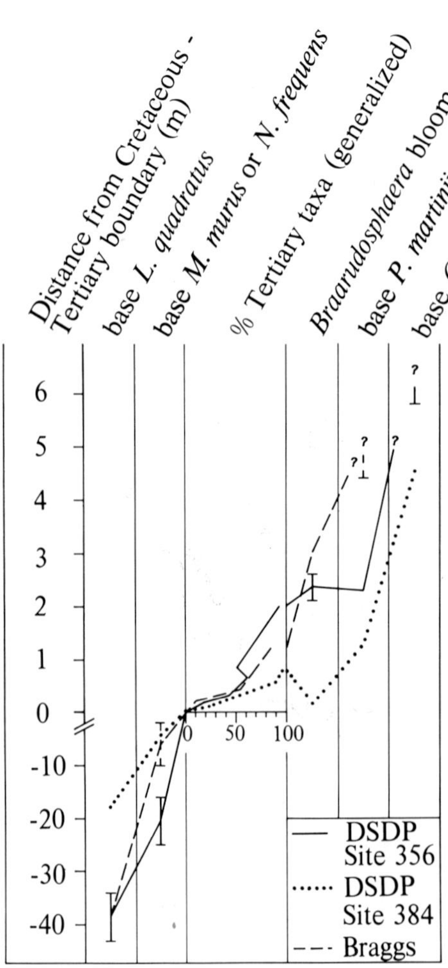

FIG. 13.—Thickness of sediments between biostratigraphic events and the Cretaceous-Tertiary boundary and of the earliest increase of Tertiary taxa immediately above the boundary in sections at DSDP Sites 356 and 384 and at Braggs. Vertical bars indicate uncertainty introduced by wide sample spacing.

between biostratigraphic events show that the Braggs section and DSDP Site 356 accumulated at similar rates and about twice as fast as the sediments at DSDP Site 384. Figure 13 further suggests that the top of the Maastrichtian at Braggs may be missing, as suspected by Worsley (1974) and that the evolution of the earliest Tertiary nannofossil assemblages, particularly blooms of *Braarudosphaera,* may be time-transgressive.

The succession as well as the observed thicknesses of calcareous nannofossil and planktonic foraminifera zones in the studied deep-sea sections are similar to those reported from other sections (Birkelund and Bromley, 1979; Christensen and Birkelund, 1979). A comparison with the Gubbio reference section, whose magnetic record has been shown to closely match the seafloor anomaly patterns (Alvarez and others, 1977), indicates that sediments at Site 384 accumulated at comparable rates and those at Site 356 at double the rate of the sediments at Gubbio. The observed earliest Danian sediment thicknesses, between the Cretaceous-Tertiary boundary and the first appearance of *Chiasmolithus danicus* are 1.7 m at Gubbio (Monechi, 1977), 4.6 m at DSDP Site 384 (Thierstein and Okada, 1979) at 6.8 m at DSDP Site 356 (Perch-Nielsen, 1977). The observed sediment thicknesses between the Cretaceous-Tertiary boundary and the first appearance of the Middle Danian foraminifer *Globorotalia trinidadensis* are 5.4 m at Gubbio (Premoli Silva, 1977), 10.4 m at DSDP Site 384 (Kaneps, personal communication, 1977), and 11 m at DSDP Site 356 (Boersma, 1977; Boersma and others, 1979). The Late Maastrichtian *Abathomphalus mayaroensis* foraminifer Zone has a thickness of 17.6 m at Gubbio and of over 28 m at DSDP Site 356. At this latter site well-preserved, bifurcated morphotypes of *Micula murus* (recently designated *Micula prinsii* by Perch-Nielsen 1979a) are present in Cores 30 and 29, *i.e.,* lower than the first occurrence of *Micula murus* reported by Perch-Nielsen (1977). The latest Maastrichtian sedimentation rates at DSDP Site 384 are less than those at

Gubbio and Site 356, although both, the *Abathomphalus mayaroensis* Zone (7 m thick) and the *Micula murus* Zone (6 m thick, based on *Micula prinsii* morphotypes) are present (Tucholke, Vogt and others, 1979).

All studied sections show the size increase of *Cruciplacolithus primus* into delicately structured, but somewhat larger *Cruciplacolithus tenuis* closer to the Cretaceous-Tertiary boundary than has been reported in other land-based sections (see various contributions in Christensen and Birkelund, 1979). The resulting variability in the reported thicknesses of the *Markalius astroporus* (NP1) Interval-Zone appears to be related to the commonly poorer preservation of nannofossils in land-based sections, obliterating delicate *Cruciplacolithus tenuis* and to possible bioprovincialism of earliest Danian nannofossil assemblages (Perch-Nielsen, 1979c; Thierstein and Okada, 1979). Paleomagnetic data from the studied sections, although of moderate quality, do not suggest the presence of a hiatus at any of the three sections shown in Figure 13 (D. Kent and D. A. Russell, personal communications, 1979; Thierstein and Okada, 1979). The identification, however, of magnetic anomalies in sedimentary sections of short stratigraphic extent is often unreliable and tends to compound errors and disagreements on biostratigraphic zonations, which may arise from differences of taxonomic concepts and utilized techniques by various biostratigraphers (see also Roggenthen, 1977).

The Cretaceous-Tertiary boundary section at DSDP Site 356 appears to be the most complete and expanded sequence yet investigated.

The Population Dynamics of a Mass-Extinction

The fossil record of the calcite-secreting phytoplankton, preserved under optimal conditions in the deep sea, shows a gradual replacement of Late Maastrichtian by earliest Tertiary taxa over sediment thicknesses ranging from less than one to a few meters. The population dynamics of the assemblages deposited prior to the transition at several sites attest to the evolutionary and paleoecological stability through the last few million years of Cretaceous time. This stasis within the Cretaceous assemblage persists into the earliest Danian. The abundances of Cretaceous taxa decrease in the transitional interval in proportion to their abundances during the Late Maastrichtian. No sequential extinctions of Cretaceous taxa can be established in the Danian (Thierstein and Okada, 1979; previous discussion). Such a gradual sympatric replacement of all species in an assemblage is incompatible with the Darwinian principle of natural selection as the dominant mechanisms controlling extinctions. A gradual replacement seems also unlikely in view of the pronounced seasonality of living coccolithophores (Okada and McIntyre, 1979) and in view of the known competitive exclusion of modern taxa under environmental stress conditions (*e.g.,* Bukry, 1974; Okada and Honjo, 1975). An alternative interpretation is that the supply of Cretaceous nannofossils from the surface waters suddenly terminated and that this event became smeared in the sedimentary record by bioturbation. It has been shown that the "transitional" record can be deconvoluted to an instantaneous extinction of the entire Cretaceous assemblage assuming mixing parameters similar to those derived from Holocene pelagic environments. The possibility, that the sudden disappearance of the Cretaceous nannofossil assemblages in the studied deep-sea sections is caused by a hiatus in carbonate sedimentation, has been shown to be extremely unlikely in the preceding paragraph, despite the persisting popularity of such inferences (*e.g.,* Hsü, 1980). The base of the nannofloral transition interval coincides with a sharp decrease of the fine-fraction carbonate contents at DSDP Site 356, the most complete and expanded among the investigated transitions. The drastic decrease of the nannofossil carbonate is not due to dissolution but is readily explained by a sudden demise of the Cretaceous holopelagic phytoplankton assemblage. The observed extinction patterns strongly suggest a catastrophic environmental deterioration and defy the law of constant extinction rates and thus would fall under the category of "unusual" processes of extinctions (Van Valen, 1973).

Another aspect of a mass-extinction event, of course, is the mode of recovery of the affected assemblages. The earliest Danian nannofossil assemblages are dominated by blooms of *Thoracosphaera, Braarudosphaera,* and, in some sections *Cyclagelosphaera.* All three taxa are known from Cretaceous sediments. Why did they survive the extinction event? Can their Cretaceous distribution yield clues to their environmental preferences and tolerances? When and where did the other earliest Tertiary nannofossils evolve?

The three species of *Thoracosphaera* observed in the Early Danian blooms are listed in Appendix II. They have been reported as rare occurrences in Maastrichtian sediments from Tunisia, Bellocq, Arkansas, Alabama (Bramlette and Martini, 1964) and from the Lower Cretaceous of northern Germany (Keupp and Michael, 1979). In this study, rare *Thoracosphaera* have been encountered in Upper Cretaceous sediments at the following localities: DSDP Sites 327A, 354, 356, 384, 390A, Braggs, Omsk, Caucasus (for details see Thierstein and Woodward, *in press*). Some of these occurrences can be explained by contamination or drilling disturbance, but there is no doubt that the species blooming in the Danian must have evolved in the Cretaceous. *Thoracosphaera* are considered calcified dinoflagellate cysts (Fütterer, 1976) and have recently been included with the calcisphere genus *Pithonella* in the dinoflagellate family *Calciodinelloidea* (Keupp and Michael, 1979). Paleoenvironmental preferences of either genus, however, are not well known. The closest living relatives of *Thoracosphaera* are *Peridinium trochoideum,* and *Scrippsiella sweenyae* whose calcified cysts were found in Holocene sediments off Woods Hole, Miami, Bermuda, Kingston (Jamaica), and from the Gulf of Paria (Trinidad), and which, after incubation, germinated to produce motile thecate dinoflagellates (Wall and Dale, 1968). Incubation experiments with calcitic cysts of *Peridinium trochoideum* by Wall and others (1970) showed that encystment can occur at temperatures of 13°–30°C and that ungerminated cysts can remain on the seafloor for up to nine months if temperatures fall below 10°C. In cultures that were kept at a constant temperature of 21.5°C following encystment, cyst germination occurred within a maximum time of four months. Wall and others (1970) also summarized evidence from earlier studies that had demonstrated the high temperature, salinity and nutrient tolerance of *Peridinium trochoideum.*

Most of the *Braarudosphaera* observed in Early Danian assemblages are *Braarudosphaera bigelowii,* occasionally occurring *Braarudosphaera discula* and *Braarudosphaera imbricata* show signs of overgrowth (see illustrations in Okada and Thierstein, 1979). In the present study *Braarudosphaera bigelowii* has been encountered in Late Cretaceous assemblages at maximum abundances of less than 1% from DSDP Sites 21, 95, Lueneburg, Poland, and at higher abundances at DSDP Site 111A (maximum 3.4%), Braggs (maximum 1.2%), Omsk (maximum 1.1%), and Elburz (maximum 2.1%). The earliest *Braarudosphaera bigelowii* are known from the Berriasian (lowermost Cretaceous) in Tethyan areas (Thierstein, 1973), where they occur consistently up to the Lower Aptian. From the Albian to the Holocene *Braarudosphaera* has been reported sporadically in well-preserved assemblages from neritic environments, and blooms are known to have occurred in the earliest Danian, the Oligocene, the Late Miocene and the Holocene (Stradner, 1960; Bukry, 1974; Roth, 1974). *Braarudosphaera* has been found living in the Gulf of Maine and Bay of Fundy (Gran and Braarud, 1935), off Morocco (Lecal-Schlauder, 1950), the central North Atlantic (Gaarder, 1954; Hulburt, 1962), the Gulf of Panama (Smayda, 1966), and Brittany (Lefort, 1972). An abundance increase of *Braarudosphaera* in neritic zones has been demonstrated in surface sediments of the Persian Gulf (Martini, 1967) and the Bay of Sendai (Takayama, 1972). The abundance of *Braarudosphaera bigelowii* in surface and Holocene sediments of the Black Sea (Bukry, 1974) and its occurrence in the phytoplankton during times of lowered salinities (Smayda, 1966) suggest that fresh water influx and surface stratification, rather than water depth, nutrients or temperature are the dominant factors promoting *Braarudosphaera* blooms. The distribution of *Braarudosphaera* in sediments also has a strong preservational overprint, since it is very susceptible to pre-burial dissolution (Thierstein, 1980), a fact that may explain its absence in slightly to moderately dissolved Danian assemblages from the Pacific and Indian Oceans.

Cyclagelosphaera, mostly *Cyclagelosphaera reinhardtii* occur in relatively high abundances at the very base of the Danian in neritic settings such as Braggs (Fig. 12) and Denmark (Perch-Nielsen, 1979b).

In both areas they were also observed very rarely in Late Maastrichtian assemblages. In all other localities rare *Cyclagelosphaera margerelii* were observed in Late Cretaceous assemblages. *Cyclagelosphaera reinhardtii* appears to be a neritic species that evolved during Maastrichtian time.

Several other species, such as *Markalius astroporus, Neochiastozygus modestus, Goniolithus fluckigeri, Crepidolithus neocrassus, Cruciplacolithus primus, Zygodiscus sigmoides,* and *Biantholithus sparsus* are present in the very earliest Danian assemblages. Their relative abundances in the Danian vary greatly in space and time. Where and when did they originate? *Markalius astroporus* have been reported as rare in Maastrichtian sediments from Denmark, Holland, Alabama, southeastern France (Bramlette and Martini, 1964) and have been observed rarely in this study in Late Maastrichtian assemblages from DSDP Sites 47.2, 208, 216, 384, 387, from the Elburz and from Braggs. The occurrences at DSDP Sites 47.2 and 384 are in intervals of known or suspected drilling disturbance and thus may be explained by contamination. Contamination for the other occurrences appears unlikely, since no other Early Tertiary taxa were found in these assemblages *Markalius astroporus* therefore must have evolved during the latest Cretaceous.

The presence of very rare *Neochiastozygus modestus* in only two Upper Cretaceous samples, one from DSDP Site 208 and one from Omsk may be due to contamination. This taxon has not been reported in pre-Danian sediments in the literature.

Goniolithus fluckigeri is present in the lowermost Danian at DSDP Site 356, but has no known occurrence in the Late Cretaceous.

Crepidolithus neocrassus has been reported from Maastrichtian sediments in Denmark (Perch-Nielsen, 1968) and is present in the topmost Maastrichtian sample at Braggs. These forms may be overgrown specimens of *Biscutum notaculum,* reported by Wise and Wind (1977) from DSDP Site 327A.

Cruciplacolithus primus is present in the well-preserved lowermost Danian assemblages at DSDP Sites 47.2, 356, 384, and at Braggs. Its absence in earliest Danian assemblages in Denmark, France, and Spain appears to be due to poorer preservation of the earliest Danian nannofossils in these sections (Perch-Nielsen, 1979b, c; Romein, 1977; and personal observations). *Cruciplacolithus primus* has not been observed in Upper Cretaceous sediments.

Zygodiscus sigmoides evolved in the Late Cretaceous. Late Cretaceous specimens were illustrated by Wise and Wind (1977) and were encountered at DSDP Sites 47.2, 111A, 249, 327A, 384, and Omsk. Only the occurrences at Sites 47.2, 384, and Omsk suggest contamination by drilling disturbance.

Biantholithus sparsus is a truly Tertiary taxon. It has no known Cretaceous occurrence, but is usually present from the very base of the Danian upwards.

Tertiary taxa that evidently evolved after the extinction of the Cretaceous assemblages are *Cruciplacolithus tenuis, Biscutum? dimorphosum, Prinsius martinii, Coccolithus cavus, Toweius* sp., and *Chiasmolithus danicus.*

Summarizing the preceding discussion, it must be concluded that the majority of the calcareous nannofossil taxa present in the earliest Danian are Cretaceous survivors. Most of these survivors are known only from uppermost Cretaceous sediments and do not have a pre-Maastrichtian record. They usually are extremely rare and never amount to more than a few percent of the entire Late Cretaceous assemblage and are therefore at the limit of detectability given the sample sizes used in the present study. These surviving taxa were more consistently encountered in Maastrichtian assemblages from high than from low paleolatitudes. Their survival may have been possible because of greater tolerance to lowered salinities, cool temperatures or low light incidence during winter seasons. It is possible that some of the truly Danian taxa may have evolved in the latest Cretaceous, but, because of their very low abundances, may so far have escaped detection in Late Cretaceous assemblages. Speciation rates of calcareous nannofossils therefore do not seem to have changed appreciably from Late Campanian to Middle Danian times.

The large relative abundance changes of individual taxa occurring in rapid succession within the earliest Tertiary nannofossil assemblages in all sections studied indicate a prolonged interval of ecological instability following the extinction event. The low carbonate contents in basal Danian sediments suggest either a lowered productivity of calcite secreting phytoplankton, or a drastically increased supply of non-carbonate detritus in Early Danian. An increase in non-carbonate supply could be caused by changes of volcanic activity, climate, sea level or tectonism, and would presumably be reflected in a change in the clay-mineralogic composition of the deep-sea sediments. Preliminary x-ray mineralogic data on closely spaced samples in Core 29 at DSDP Site 356 however, show no appreciable change across the Cretaceous-Tertiary boundary in the mineralogic composition of the sediments. The decreased carbonate contents in Danian sediments, therefore, are most likely caused by lowered carbonate supply.

The Terminal Cretaceous Calcareous Phytoplankton Mass-Extinction Event and Evolutionary Theory

The utilized sample sizes, their wide geographic distribution, and the stratigraphic interval covered, during which numerous new taxa appear for the first time in the fossil record, constitute an almost ideal data set for an evaluation of tempo and mode of phytoplankton evolution.

Numerous speculative papers on possible phylogenetic relationships among calcareous nannofossil taxa have been published (*e.g.,* Gartner, 1970; Prins, 1971; Haq, 1973; Perch-Nielsen, 1979a), but only very few have included detailed morphometric analyses (*e.g.,* Verbeek, 1977; Towe, 1979). The dearth of morphometric studies of calcareous nannofossils may, at least in part, be due to an almost complete lack of studies of geographical morphological variability of extant coccolithophores and to uncertainties related to their polymorphism, dimorphism, and possibly dithecatism (*e.g.,* Halldal and Markali, 1954; Parke and Adams, 1960; Winter and others, 1979). No gradual morphological changes leading from Cretaceous to earliest Tertiary taxa could be demonstrated in the present study, although occasional intermediate morphotypes between *Biscutum magnum* Wind and Wise and *Markalius astroporus* were observed.

The reason why gradual phylogeny could not be demonstrated for any calcareous phytoplankton lineage lies in the scarcity of new taxa in the early part of their stratigraphic range. Most of the new taxa characteristic of Early Danian assemblages are undoubtedly present in Upper Maastrichtian sediments, but they must occur in such small numbers that they are only occasionally encountered and no statistically valid sample size for morphometric studies of these taxa are available. Their rare occurrence makes their appearances in the geological record seem punctuational at a stratigraphic level considerably above their original gradual or punctuational speciation. This particular evolutionary pattern has also been observed in the Late Pleistocene appearance of *Emiliania huxleyi* (Lohman) whose relative abundance is extremely low in the first 0.05–0.1 million years of its range (Thierstein

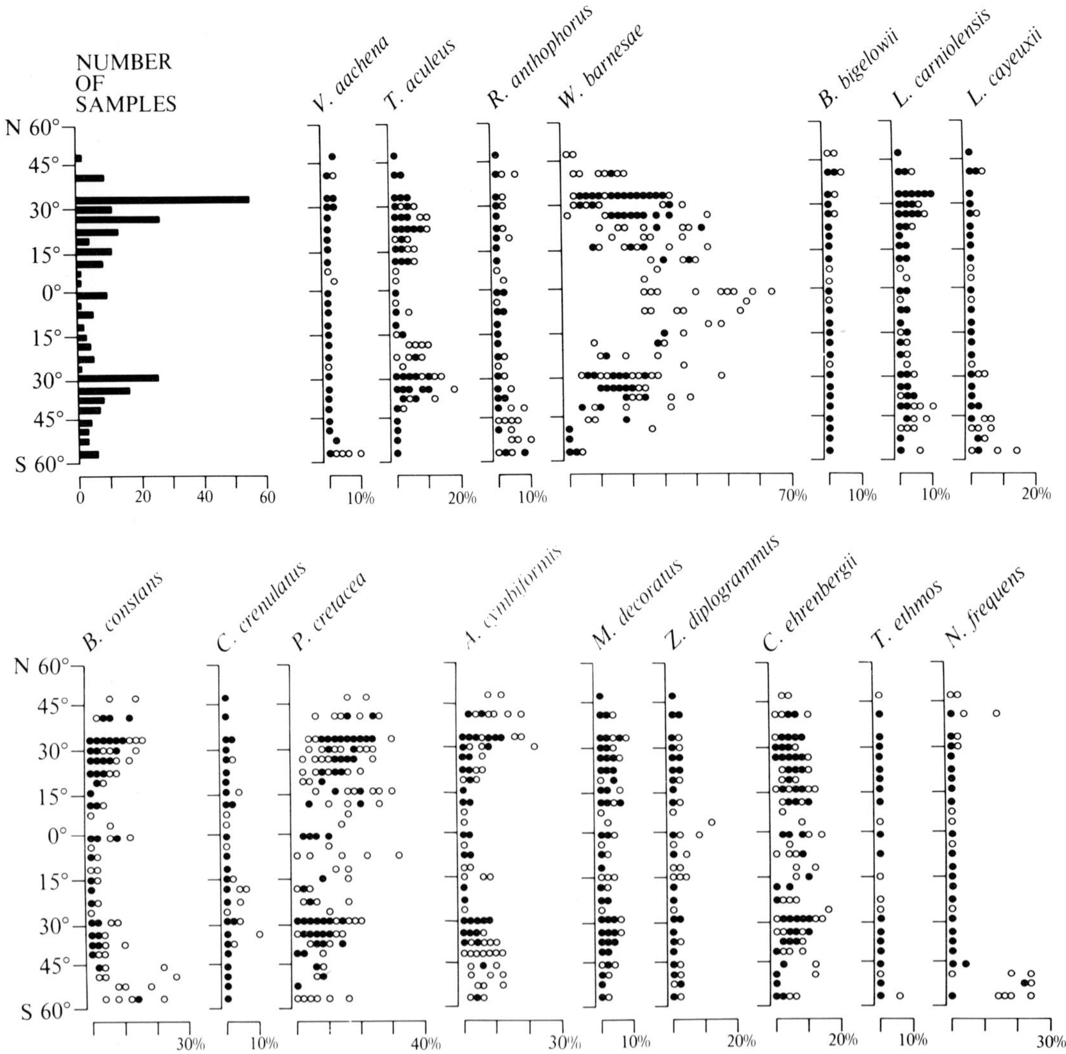

FIG. 14.—Paleolatitudinal distribution of Upper Campanian through uppermost Maastrichtian samples and calcareous nannofossil abundances. Circles indicate one, dots indicate two or more samples.

and others, 1977). Its possible Pleistocene phyletic ancestor *Emiliania protohuxleyi* McIntyre (1970) has only recently been found living in the Gulf of Elat (Winter and others, 1978).

It is not obvious why evolution in other organisms should not proceed in a similar way by speciation (punctuational or gradual) in very low numbers and possibly in marginal ecologic and geographic settings with a prolonged initial existence as an almost unrecognizable minority within an established community. What are then, one wonders, the chances to prove or disprove the existence of true phyletic gradualism as opposed to growth of an existing subpopulation to dominance (see also discussions in Gould and Eldredge, 1977)? The limitation to evolutionary studies by the often quoted "incompleteness of the geological record" does not only apply to stratigraphic continuity but possibly more so to the vanishingly small probability that evolutionary processes in peripheral populations (in an ecological, *e.g.* seasonal, as well as in a geographical sense) are preserved in the fossil record.

BIOGEOGRAPHY

The relative abundance distributions of 54 Late Cretaceous nannofossil taxa were plotted on maps and versus paleolatitude for three time-slices. The first time-slice included 54 assemblages from the first appearance of *Tetralithus gothicus* to the last

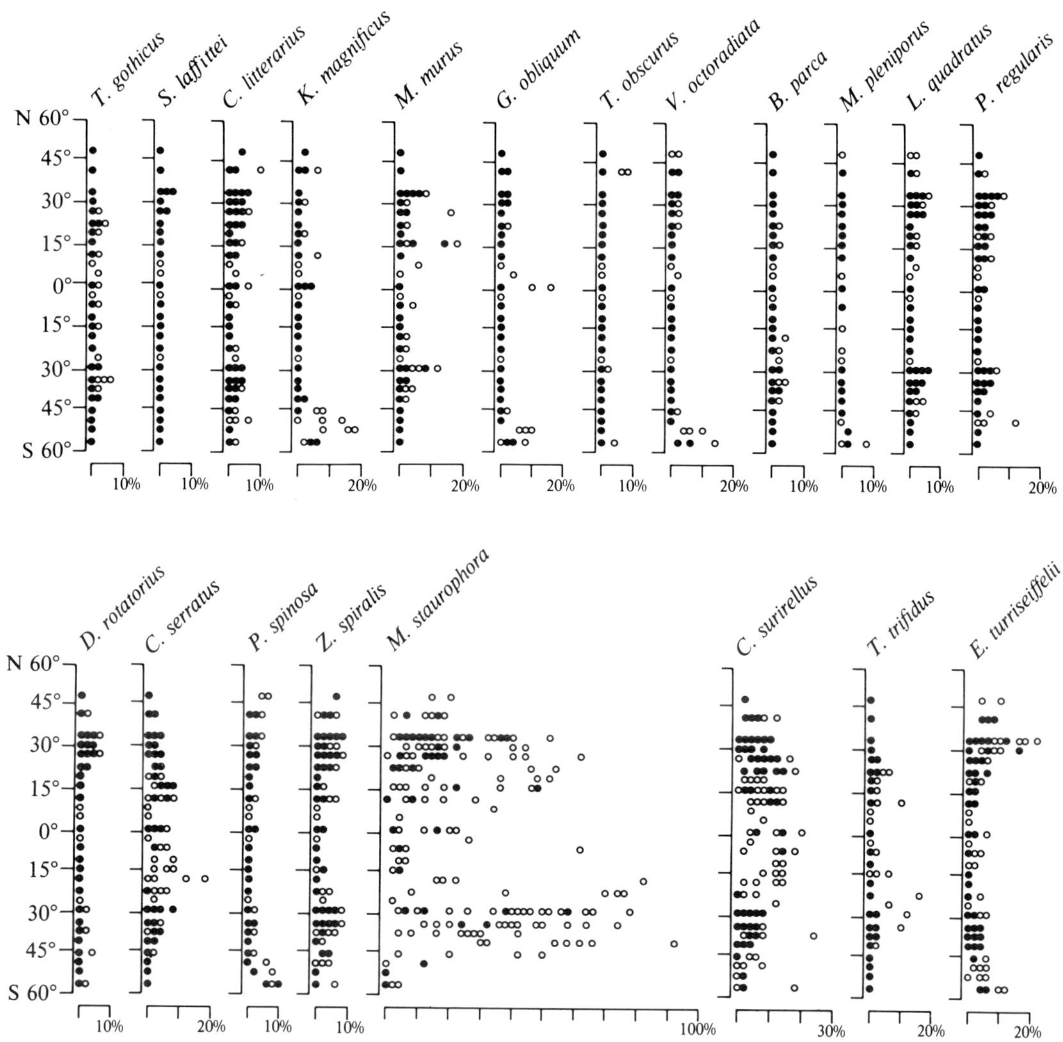

occurrence of *Tetralithus trifidus* (Late Campanian to Early Maastrichtian), the second time-slice included 71 assemblages from the last occurrence of *Tetralithus trifidus* to the first appearance of either *Micula murus* or *Nephrolithus frequens* (Early to Late Maastrichtian), and the third time-slice comprised 118 assemblages with either *Micula murus* or *Nephrolithus frequens* present (latest Maastrichtian). No significant changes in the distribution patterns of any of the taxa were detected between the individual time-slices which represent approximately the last 15 million years of Cretaceous time. The observed stasis of biogeographic distributions through the latest Cretaceous is in agreement with the stable stratigraphic abundance patterns found at the individual sites discussed previously. The biogeography of latest Cretaceous nannofossils is therefore not broken down into different time-slices in this documentation, but is shown for the entire latest Cretaceous interval (Late Campanian through Late Maastrichtian), which includes 243 counted assemblages.

The biogeography of Early Danian assemblages is not documented here because it shows highly irregular patterns. The reason for this lies in the rapid evolutionary abundance changes observed in all studied sections (Figs. 8, 10, 12). These changes occur within the first few meters of Danian sediments up to at least the middle of the *Cruciplacolithus tenuis* (NP2) Zone. The paleogeographic

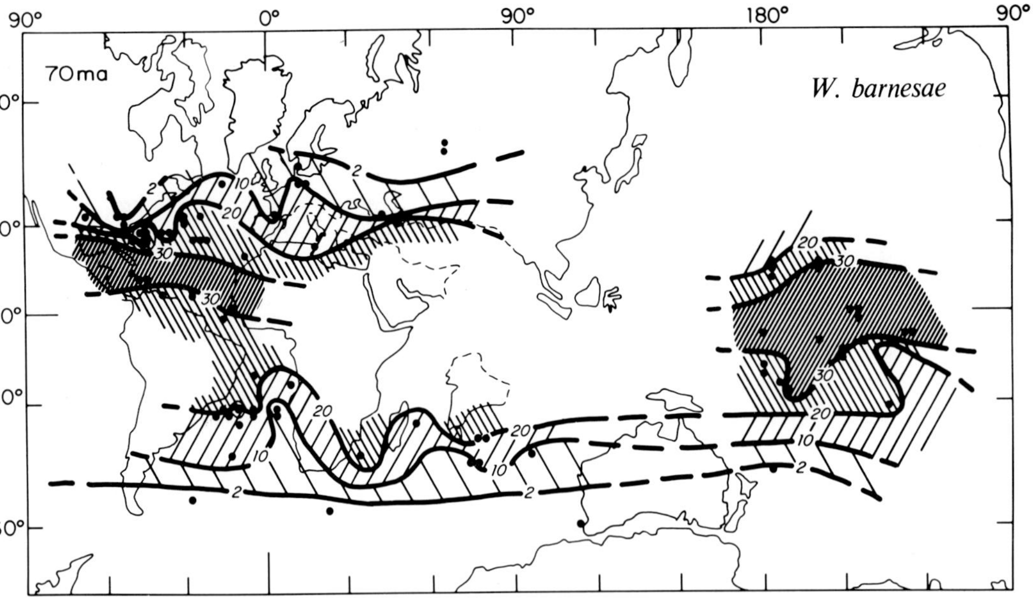

FIG. 15.—Averaged percent abundance of *Watznaueria barnesae* in the 243 Late Campanian through latest Maastrichtian assemblages investigated.

abundance distribution of any earliest Tertiary taxon based on single samples therefore necessarily reflects mostly the stratigraphic provenance of assemblages within the earliest Danian, rather than true geographic patterns. As stated in the previous section, the evolutionary patterns are not exactly the same in all sections; a geographic interpretation of these, however, can only be made after more sections, particularly from high and low paleolatitudes have been studied with a comparatively high stratigraphic resolution.

The geographic distribution patterns of Late Cretaceous calcareous nannofossils are predominantly latitudinal (Fig. 14). Only two species ever reach

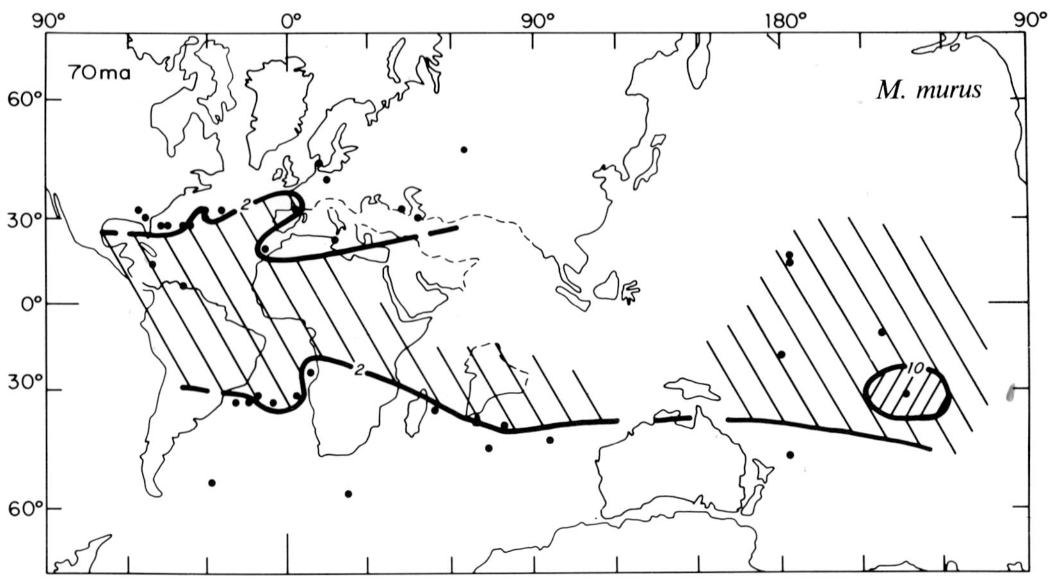

FIG. 16.—Averaged percent abundance of *Micula murus* in the 118 latest Maastrichtian assemblages investigated.

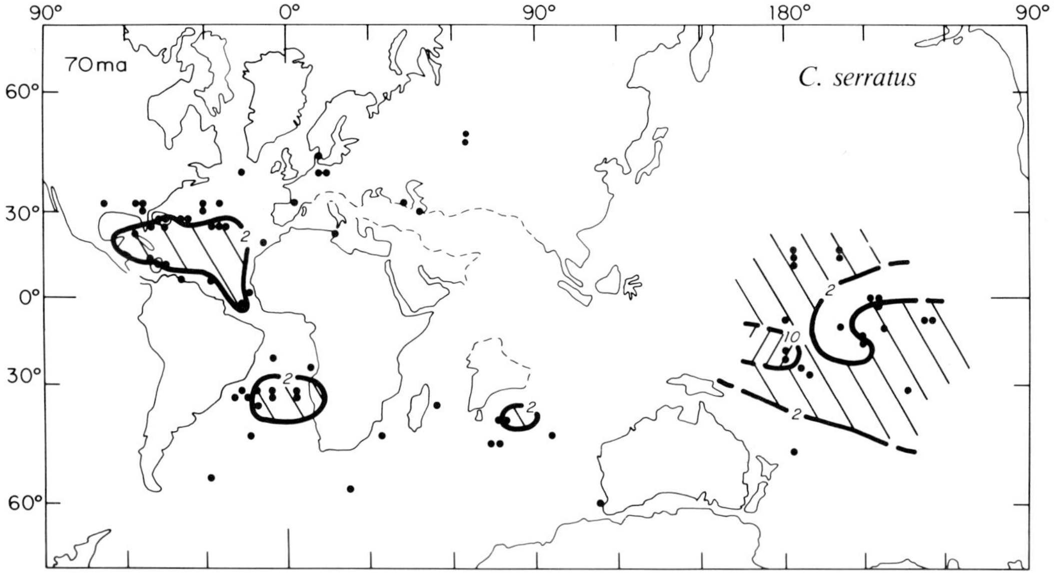

FIG. 17.—Averaged percent abundance of *Cylindralithus serratus* in the 243 Late Campanian through latest Maastrichtian assemblages investigated.

abundances of over 40%: *Watznaueria barnesae* and *Micula staurophora.* Five additional species have maximum abundances of over 20%: *Biscutum constans, Prediscosphaera cretacea, Nephrolithus frequens, Cretarhabdus surirellus,* and *Eiffellithus turriseiffelii.* None of the other 28 taxa with highest abundances over 4% ever dominate any assemblage. From a statistical point of view, Late Cretaceous nannoplankton thanatocoenoses are about twice as diverse as modern ones (Geitzenauer and others, 1977). Coccolith diversity changes in the geological past, however, are not readily interpretable in an evolutionary or paleoecological sense, because large discrepancies occur in abundance and

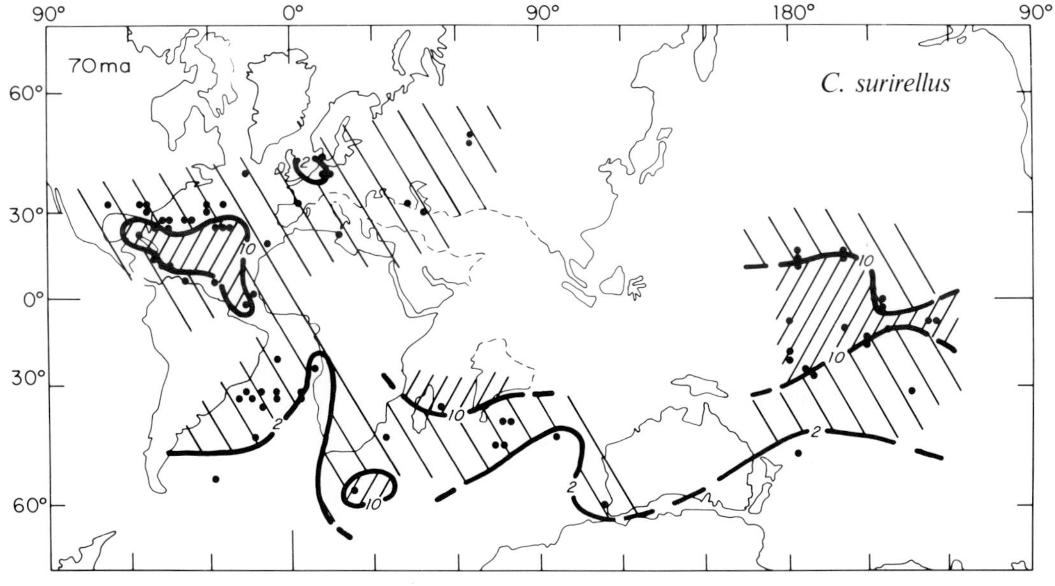

FIG. 18.—Averaged percent abundance of *Cretarhabdus surirellus* in the 243 Late Campanian through latest Maastrichtian assemblages investigated.

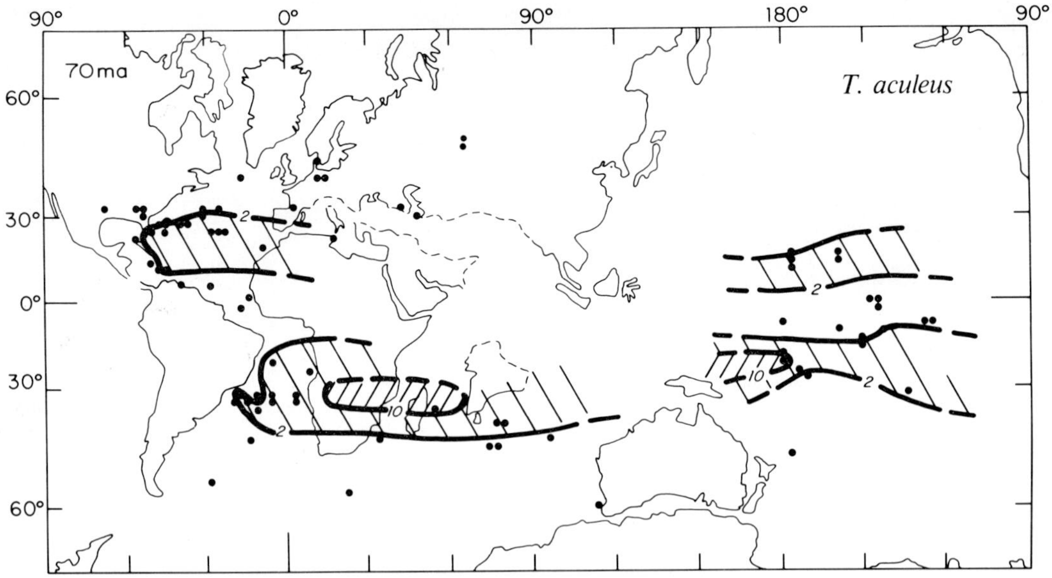

FIG. 19.—Averaged percent abundance of *Tetralithus aculeus* in the 243 Late Campanian through latest Maastrichtian assemblages investigated.

diversity between coccolithophores living in today's surface waters (McIntyre and Bé, 1967; Okada and Honjo, 1973; Okada and McIntrye, 1977) and coccoliths found in surface sediments (McIntyre and Bé, 1967; Roth and Berger, 1975; Geitzenauer and others, 1977). These differences are thought to be mainly due to differential dissolution and possibly to selective grazing by herbivorous plankton. The apparently higher diversities of Cretaceous nannofossil assemblages could therefore reflect changes in either or both of these processes.

A considerable number of Late Cretaceous

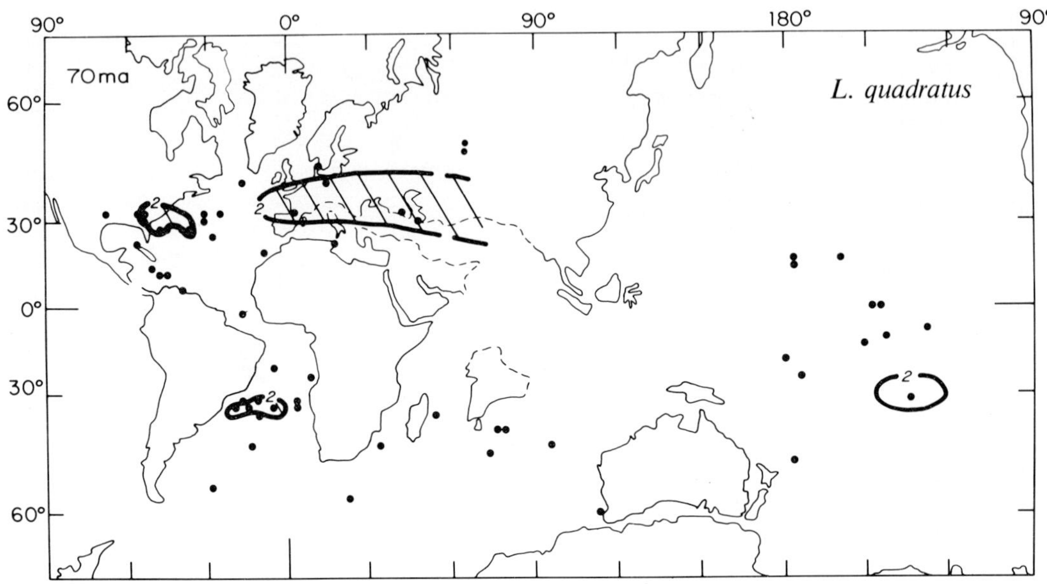

FIG. 20.—Averaged percent abundance of *Lithraphidites quadratus* in the 189 Middle to latest Maastrichtian assemblages investigated.

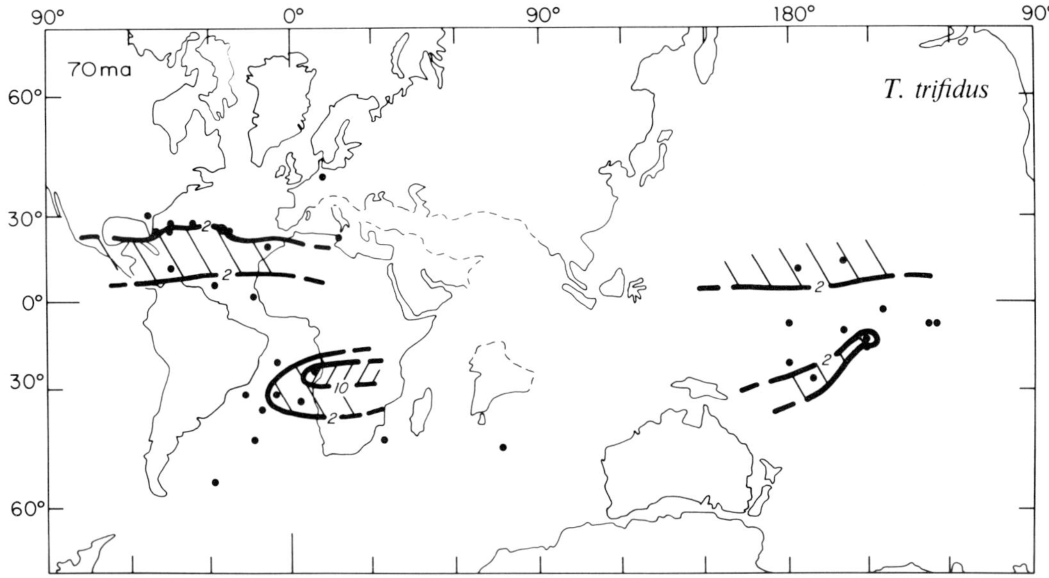

FIG. 21.—Averaged percent abundance of *Tetralithus trifidus* in the 54 Late Campanian to earliest Maastrichtian assemblages investigated.

nannoplankton taxa do not reveal characteristic distribution patterns in paleolatitudinal plots or on paleogeographic maps because of their consistently low relative abundances. Because of the inherent negative correlation of variables expressed as ratios of a total (Koch and Link, 1971, p. 153), the relative abundance increase of dominant taxa necessarily will lead to a decrease in the abundance of all others. If preservational variability is influencing the abundance patterns of dominant taxa, the introduced noise may obscure real original patterns of the minor taxa.

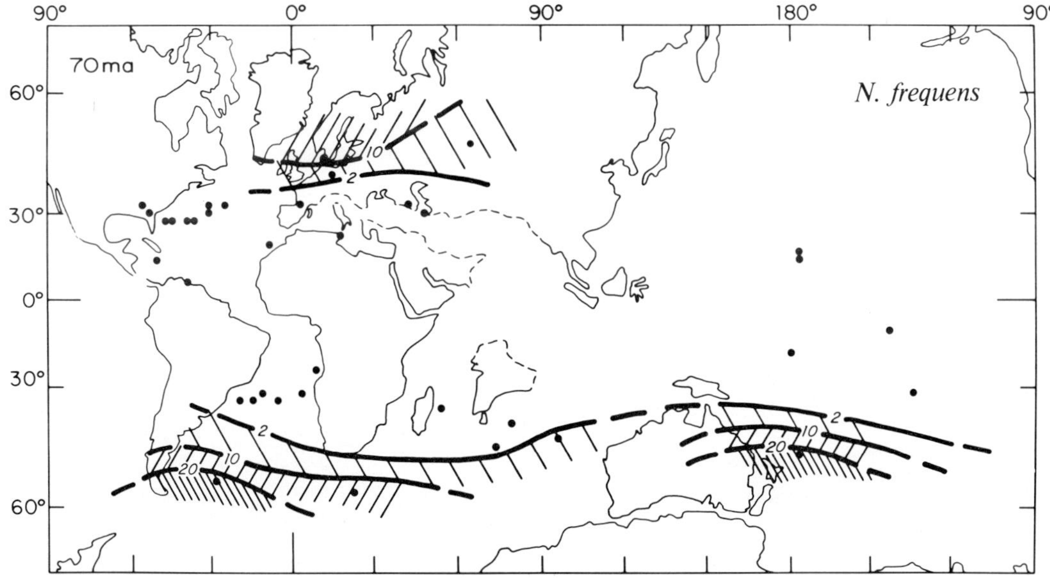

FIG. 22.—Averaged percent abundance of *Nephrolithus frequens* in the 118 latest Maastrichtian assemblages investigated.

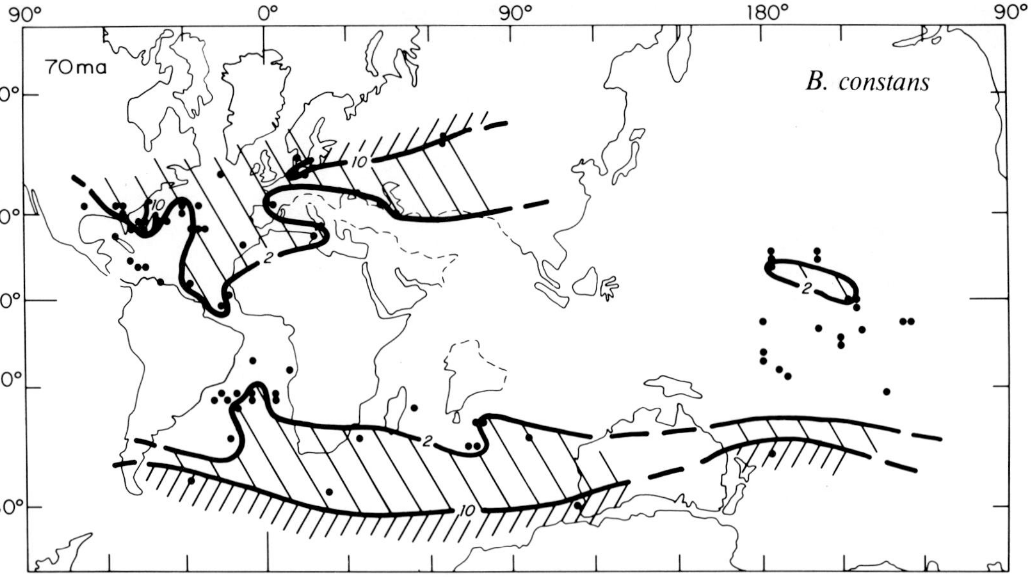

FIG. 23.—Averaged percent abundance of *Biscutum constans* in the 243 Late Campanian through latest Maastrichtian assemblages investigated.

Dominant taxa and those with recognizable distribution patterns are discussed below with reference to their paleolatitudinal trends (Fig. 14) and to their global biogeographic distributions, which are contoured on paleogeographic maps (Figs. 15–28).

Tropical and Subtropical Nannofossils

Watznaueria barnesae, one of the most dominant species, clearly shows a strong inverse correlation with paleolatitude (Fig. 14). The scarcity or com-

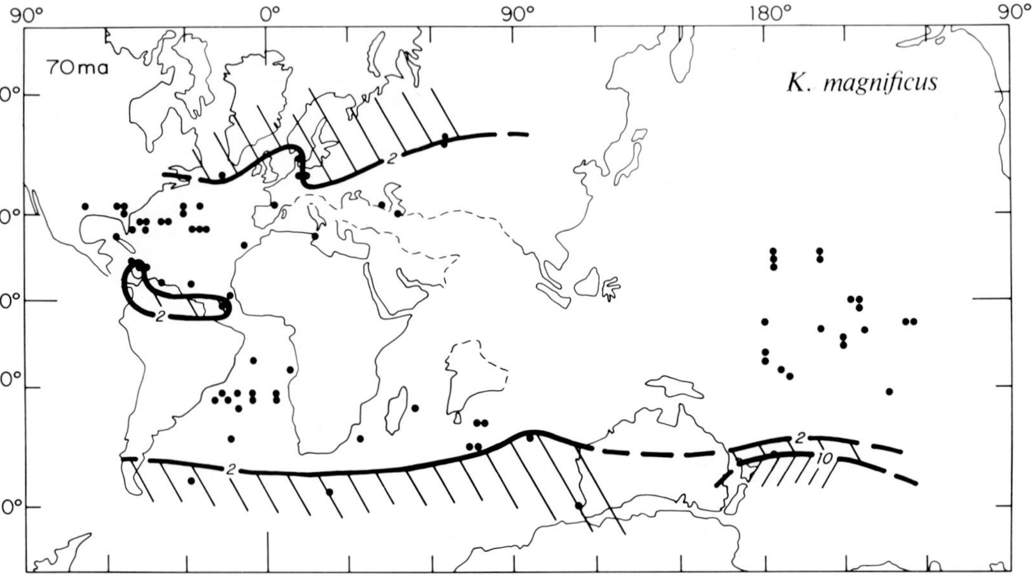

FIG. 24.—Averaged percent abundance of *Kamptnerius magnificus* in the 243 Late Campanian through latest Maastrichtian assemblages investigated.

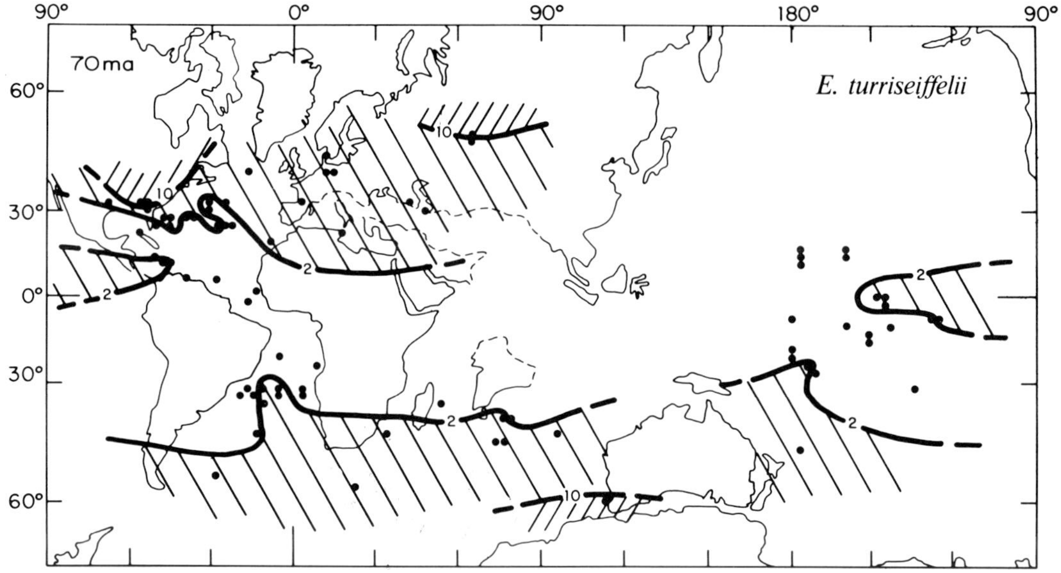

FIG. 25.—Average percent abundance of *Eiffellithus turriseiffelii* in the 243 Late Campanian through latest Maastrichtian assemblages investigated.

plete absence of *Watznaueria barnesae* in some Late Cretaceous assemblages from high latitudes was noted by Bukry (1973) and appears to be a global phenomenon. The irregularities in the abundance contours in the southern mid-latitudes evident on the paleogeographic distribution map (Fig. 15) are compensated mainly by increases in the proportions of *Micula staurophora* and are caused by poorer preservation.

The geographic distribution patterns of *Micula*

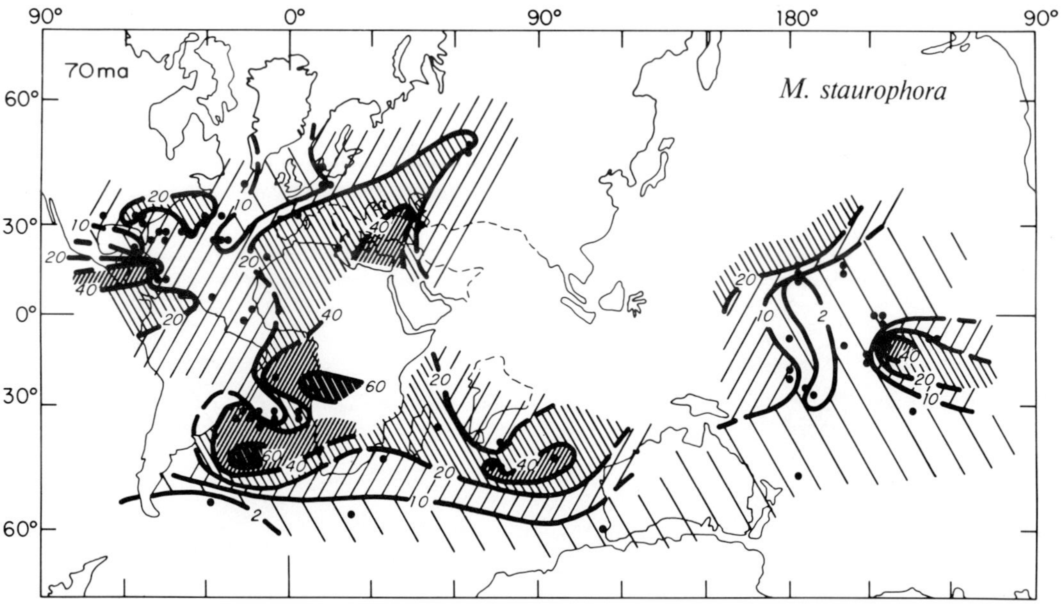

FIG. 26.—Average percent abundance of *Micula staurophora* in the 243 Late Campanian through latest Maastrichtian assemblages investigated.

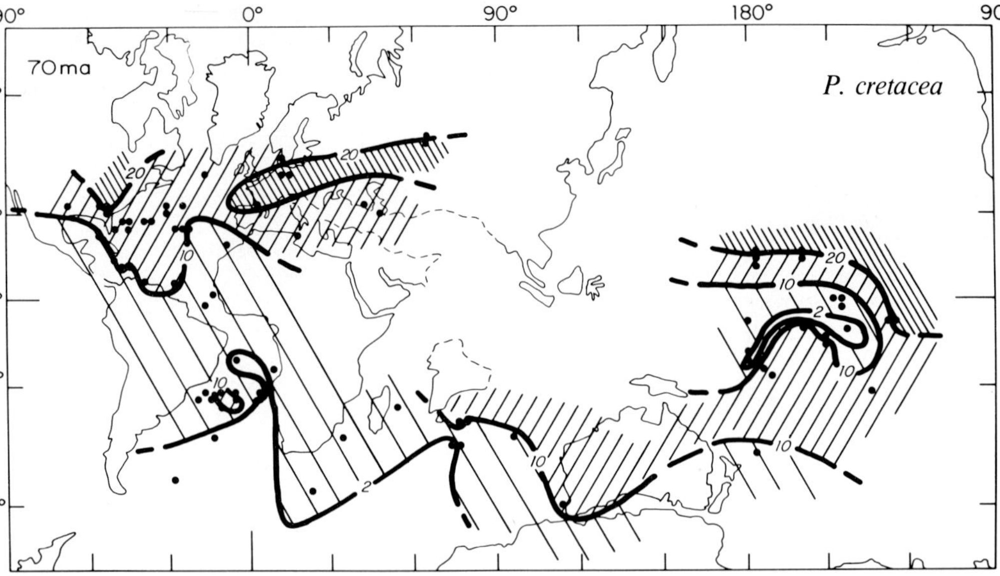

FIG. 27.—Averaged percent abundance of *Prediscosphaera cretacea* in the 243 Late Campanian through latest Maastrichtian assemblages investigated.

murus, Cylindralithus serratus, Cretarhabdus surirellus (Figs. 16–18) follow dominantly tropical and subtropical paleolatitudes. The highest abundances of *Micula murus* in the paleolatitudinal distribution patterns are over 10% in both hemispheres (Fig. 14), but become averaged to lower values when other samples from the latest Maastrichtian *Micula murus* Zone at the same sites are included (Fig. 16). The paleobiogeographic distribution of *Cylindralithus serratus,* although clearly tropical and subtropical (Fig. 14), may include an affinity for open ocean environments as evidenced in Fig. 17. *Cretarhabdus surirellus* abundances are inversely correlated with paleolatitude (Fig. 14), with the excep-

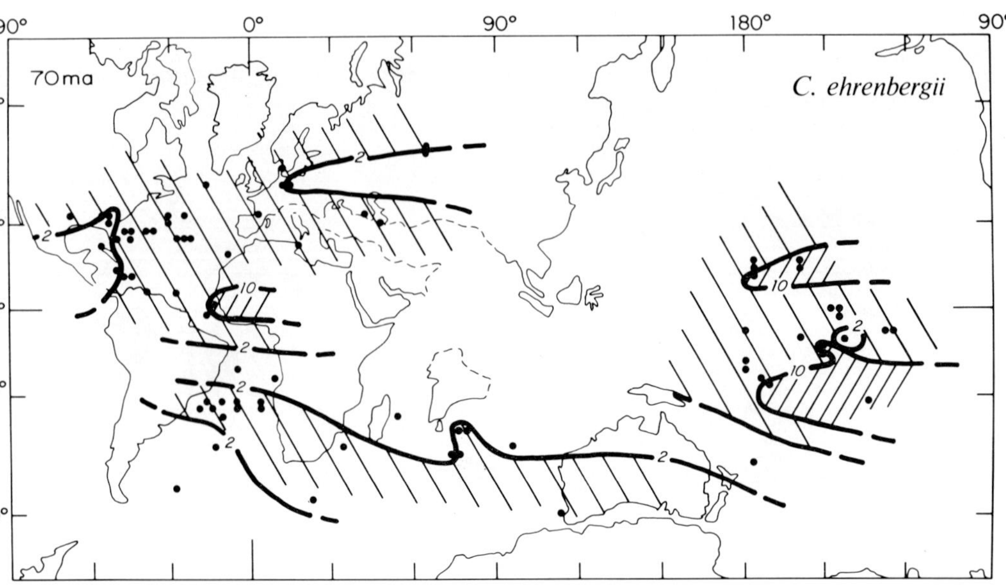

FIG. 28.—Averaged percent abundance of *Cribrosphaerella ehrenbergii* in the 243 Late Campanian through latest Maastrichtian assemblages investigated.

tion of two latest Maastrichtian assemblages from the southwestern Indian Ocean (Fig. 18).

The paleolatitudinal distribution patterns of three stratigraphic marker species, *Tetralithus aculeus, Lithraphidites quadratus,* and *Tetralithus trifidus* show subtropical abundance peaks in both hemispheres (Fig. 14). These patterns follow somewhat the paleolatitudinal sample density, but they are judged sufficiently different from the latter to be considered real (Figs. 19–21). *Lithraphidites quadratus* occurs more consistently in Tethyan and Atlantic than in Pacific assemblages, although it is present in the Pacific in very low numbers also. The apparent bimodal latitudinal distribution of *Micula staurophora* (Fig. 14) is considered to be mainly due to preservational effects and will be discussed below.

Boreal and Austral Nannofossils

Vagalapilla aachena, Reinhardtites anthophorus, Lucianorhabdus cayeuxii, Biscutum constans, Arkhangelskiella cymbiformis, Nephrolithus frequens, Kamptnerius magnificus, Gartnerago obliquum, Vagalapilla octoradiata, Misceomarginatus pleniporus, Prediscosphaera spinosa, and *Eiffellithus turriseiffelii* show more or less distinct abundance peaks in high latitudes (Fig. 14). The most pronounced boreal and austral affinities are found with *Nephrolithus frequens* (Fig. 22), although its presence in very low numbers has also been reported from Egypt (Shafik and Stradner, 1971). Some of the taxa listed above show an additional peculiar abundance increase in tropical paleolatitudes: *Biscutum constans* (Fig. 23), *Kamptnerius magnificus* (Fig. 24), *Gartnerago obliquum,* and *Eiffellithus turriseiffelii* (Fig. 25). It remains to be tested whether these patterns are related to high productivity, cooler waters upwelling along equatorial divergences, or peculiar surface salinity distributions related to evaporation and precipitation patterns.

Nannofossils with Other Distribution Patterns

Among the remaining taxa, none of which shows obvious latitudinal or longitudinal patterns, three dominant species are documented further: *Prediscosphaera cretacea, Cribrosphaerella ehrenbergii,* and *Micula staurophora.*

Micula staurophora is the most dominant species encountered; its highest abundance reaches 92.2% in sample 358-15-2, 142 cm (South Atlantic). Its paleolatitudinal distribution pattern shows two subtropical peaks whose maximum heights do not follow those of the sample distribution pattern (Fig. 14) and are therefore real. The abundance of *Micula staurophora* in subtropical assemblages is not consistently high, as in the case of *Watznaueria barnesae* in the tropics, and latitudinally distributed parameters such as light, temperature, and seasonality do not appear to be the only controlling influences. The geographic distribution of *Micula staurophora* (Fig. 26) also confirms its erratic abundance variation. A strong increase in the relative abundance of *Micula staurophora* with enhanced dissolution has been shown to occur in assemblages deployed during a deep mooring experiment in the North Atlantic (Thierstein, 1980). The relative abundance of *Micula staurophora* therefore is a good indicator of coccolith preservation. The geographically inconsistent dominance of *Micula staurophora* in poorly preserved assemblages may introduce considerable noise in the abundance patterns of minor taxa because of the closed sum effect discussed above. In particular, *Prediscosphaera cretacea* and *Watznaueria barnesae* show distribution patterns which are locally complementary to those of *Micula staurophora* and thus may have their biogeographic contours distorted by preservational effects.

The paleolatitudinal distribution of *Prediscosphaera cretacea* shows a slight asymmetry (Fig. 27), possibly related to the uneven sample distribution. Its biogeography shows no distinct patterns and appears to be influenced by closed sum effects of the dominant taxa, such as *Watznaueria barnesae* and *Micula staurophora.*

Cribrosphaerella ehrenbergii has a quite even paleolatitudinal frequency distribution, with averaged abundances of over 10% in tropical and subtropical latitudes only (Figs. 14 and 28).

SUMMARY

1. The deep-sea record is uniquely suited for detailed analysis of the marine plankton mass-extinction at the Cretaceous-Tertiary boundary because of optimal preservation of faunas and floras.

2. Quantitative taxonomic analyses of Late Campanian to Early Danian nannofossil assemblages are documented to elucidate evolutionary and biogeographic patterns associated with this most severe plankton extinction event known in the geologic record.

3. The observed stratigraphic stasis of Late Cretaceous nannofossil abundances in several sections suggests stable oceanic surface environments through the last several million years of Cretaceous time.

4. The replacement of latest Cretaceous by the "new" earliest Tertiary nannofossil assemblages typically occurs over a sediment thickness of a few tens of centimeters. This "transition" can be deconvoluted to an instantaneous event assuming bioturbation and using mixing parameters similar to those observed in Holocene deep-sea environments.

5. No subsequent extinction events of taxa within the vanishing Cretaceous assemblages can be observed.

6. Comparisons of sediment thicknesses between

stratigraphic events indicate that the Cretaceous-Tertiary boundary interval at DSDP Site 356 is complete and about twice as thick as that near Gubbio, Italy.

7. Sedimentological and paleontological data from DSDP Site 356 suggest that carbonate fixation by phytoplankton decreased dramatically at the Cretaceous-Tertiary boundary.

8. The earliest Danian nannofossil communities are characterized by strongly fluctuating absolute and relative abundances and suggest an interval of ecological instability lasting over hundreds of thousands of years.

9. Modern relatives and descendants of the taxa occurring in blooms in earliest Danian holopelagic nannofossil assemblages (*Thoracosphaera, Braarudosphaera*) thrive in neritic environments.

10. The majority of the typical earliest Danian nannofossil taxa have been recorded in Late Cretaceous sediments in extremely low abundances and mainly at high latitude sites. Their survival into the Tertiary may have been due to their adaptation to higher seasonality of temperature, salinity or light.

11. There is therefore no conclusive evidence for any change in speciation rates between the Late Cretaceous and earliest Tertiary.

12. Because of the extremely low abundances of newly evolving nannofossil species in the early part of their stratigraphic ranges, as observed in the latest Cretaceous and the Late Pleistocene, phylogenetic gradualism cannot be demonstrated or rejected. Whether similar modes of evolution apply to other fossil groups remains to be tested.

13. Latest Cretaceous nannofossil biogeography was strongly paleolatitudinal. Tropical and subtropical floras are dominated by *Watznaueria barnesae, Micula murus, Cylindralithus serratus,* and *Cretarhabdus surirellus.* Boreal and austral assemblages consist mainly of *Vagalapilla aachena, Reinhardtites anthophorus, Lucianorhabdus cayeuxii, Biscutum constans, Arkhangelskiella cymbiformis, Nephrolithus frequens, Kamptnerius magnificus, Gartnerago obliquum, Vagalapilla octoradiata, Misceomarginatus pleniporus, Prediscosphaera spinosa,* and *Eiffellithus turriseiffelii. Micula staurophora* becomes strongly dominant in poorly preserved samples.

ACKNOWLEDGMENTS

Most of the samples used in this study were recovered by the Deep Sea Drilling Project, which is sponsored by the National Science Foundation, and were kindly made available by the Curator. Additional samples were received from Professor H. M. Bolli (Zurich), Dr. E. G. Cartier (The Hague), Dr. C. W. Copeland (University of Alabama), Dr. B. U. Haq (Woods Hole), Dr. H. Hiltermann (Hannover), Dr. P. G. Quilty (Perth, S.W. Australia), Dr. D. A. Russell (Ottawa), Dr. J. Smit (Amsterdam), Dr. M. Toumarkine (Zurich), Dr. V. N. Vekshina (USSR), and the Curator of the core collection at the Lamont-Doherty Geological Observatory. Some of the studied material had been sent to the late Dr. M. N. Bramlette. Dr. Y. Lancelot (Paris) provided the data to compute paleopositions of the Pacific DSDP sites. I am grateful for informative discussions with W. H. Berger, S. Gartner, K. Perch-Nielsen, I. Premoli-Silva, W. V. Sliter, F. H. Wind and T. R. Worsley. Rachel Haymon, M. C. Chavez and D. L. Woodward helped with the computation of paleopositions, sample preparation and sedimentological analyses. Financial support was provided through NSF Grant No. OCE 76-22150.

REFERENCES CITED

NOTE: All references to Deep Sea Drilling Project Initial Reports have been abbreviated; for complete citations see *Appendix I* at the end of this volume, or the individual Initial Reports as published.

Alvarez, W., Arthur, M. A., Fischer, A. G., Lowrie, W., Napoleone, G., Premoli-Silva, I., and Roggenthen, W. M., 1977, Upper Cretaceous-Paleocene magnetic stratigraphy at Gubbio, Italy. V. Type section for the Late Cretaceous-Paleocene geomagnetic reversal time scale: Geol. Soc. America Bull., v. 88, p. 367–389.

Alvarez, W., and Lowrie, W., 1978, Upper Cretaceous paleomagnetic stratigraphy at Moria (Umbrian Apennines, Italy): verification of the Gubbio section: Geophysical Jour. Royal Astronomical Soc., v. 35, p. 1–17.

Alvarez, W., and Vann, D. W., 1979, Comments and replies on "Biostratigraphy and magnetostratigraphy of Paleocene terrestrial deposits, San Juan Basin, New Mexico.": Geology, v. 7, p. 66–71.

Berger, W. H., and Heath, G. R., 1968, Vertical mixing in pelagic sediments: Jour. Marine Res., v. 26, p. 134–143.

Berger, W. H., and Johnson, R. F., 1978, On the thickness and the ^{14}C age of the mixed layer in deep-sea carbonates: Earth Planet. Sci. Letters, v. 41, p. 223–227.

Berger, W. H., and Winterer, E. L., 1974, Plate stratigraphy and the fluctuating carbonate line, *in,* Hsü, K. J. and Jenkyns, H., C., eds., Pelagic sediments on land and under the sea: International Assoc. Sedimentologists Spec. Pub. No. 1, p. 11–48.

Berggren, W. A., and van Couvering, J. A., 1974, Biostratigraphy, geochronology and paleoclimatology of the last 15 million years in marine and continental sequences: Palaeogeography, Palaeoclimatology and Palaeoecology, v. 16, 216 p.

BIRKELUND, T., AND BROMLEY, R. G., 1979, Cretaceous/Tertiary boundary events symposium. I. The Maastrichtian and Danian of Denmark: University of Copenhagen, 210 pp.

BOERSMA, A., 1977, Cenozoic planktonic foraminifera—DSDP Leg 39 (South Atlantic): Init. Repts. DSDP, v. 39, p. 567-587.

BOERSMA, A., SHACKLETON, N., HALL, M., AND GIVEN, Q., 1979, Carbon and oxygen isotope records at DSDP Site 384 (North Atlantic) and some Paleocene paleotemperatures and carbon isotope variations in the Atlantic Ocean: Init. Repts. DSDP, v. 43, p. 695-717.

BOLLI, H., 1957, The genera Praeglobotruncana, Rotalipora, Globotruncana, and Abathomphalus in the Upper Cretaceous of Trinidad, B.W.I., *in,* Loeblich, A. R., and others, Studies in Foraminifera: U.S. National Museum Bull., v. 215, p. 51-60.

BRAMLETTE, M. N. AND MARTINI, E., 1964, The great change in calcareous nannoplankton fossils between Maastrichtian and Danian: Micropaleontology, v. 10, p. 291-322.

BUKRY, D., 1973, Coccolith and silicoflagellate stratigraphy, Tasman Sea and Southwestern Pacific Ocean, Deep Sea Drilling Project Leg 21: Init. Repts. DSDP, v. 21, p. 885-893.

BUKRY, D., 1974, Coccoliths as paleosalinity indicators—evidence from Black Sea: Am. Assoc. Petroleum Geologists, Mem. 20, p. 353-363.

BUTLER, R. F., LINDSAY, E. H., JACOBS, L. L., AND JOHNSON, N. M., 1977, Magnetostratigraphy of the Cretaceous-Tertiary boundary in the San Juan Basin, New Mexico: Nature, v. 267, p. 318-323.

CARTIER, E. G., 1971, Die Geologie des unteren Chalus Tals, Zentral-Alborz/Iran: Ph.D. thesis, Univ. Zürich. Mitteilungen Geologisches Institut ETH und Universität Zürich, N.F. 164, 130 pp.

CHRISTENSEN, K. W., AND BIRKELUND, T., 1979, Cretaceous/Tertiary boundary events symposium. II. Proceedings: University of Copenhagen, 250 pp.

CHRISTENSEN, L., FREGERSLEV, S., SIMONSEN, A., AND THIEDE, J., 1973, Sedimentology and depositional environment of Lower Danian fish clay from Stevns Klint, Denmark: Bull. Geological Soc. Denmark, v. 22, p. 193-212.

CITA, M. B., AND GARTNER, S., 1971, Deep sea Upper Cretaceous from the western North Atlantic, *in,* Farinacci, A. ed., Proceedings II. Planktonic Conference, Roma, 1970: Rome, Edizioni Tecnoscienza, v. 1, p. 287-319.

CITA, M. B., AND PREMOLI SILVA, I., 1974, Il limite Cretaceo-Paleocene e l'evoluzione della Tetide: Rivista Italiana Paleontologia Stratigrafia, Memoria XIV, p. 193-249.

DANIS, G. P. L., FOSTER, J. H., AND RUSSELL, D. A., *in press,* The paleomagnetic record and the Cretaceous-Tertiary boundary in North America: Syllogeus.

EDWARDS, A. R., 1966, Calcareous nannoplankton from the uppermost Cretaceous and lowermost Tertiary of the Mid-Waipara section, South Island, New Zealand: New Zealand Jour. Geol. and Geophys., v. 9, p. 481-490.

FASSETT, J. E., 1979, Comments and replies on "Biostratigraphy and magnetostratigraphy of Paleocene terrestrial deposits, San Juan Basin, New Mexico": Geology, v. 7, p. 66-71.

FISCHER, A. G., AND ARTHUR, M. A., 1977, Secular variations in the pelagic realm, *in,* Cook, H. E. and Enos, P. eds., Deep-water carbonate environments: Soc. Econ. Paleontologists and Mineralogists Spec. Pub. No. 25, p. 19-50.

FISCHER, A. G., AND OTHERS, 1971, Init. Repts. DSDP, v. 6, 1329 p.

FÜTTERER, D., 1976, Kalkige Dinoflagellaten (*"Calciodinelloidae"*) und die systematische Stellung der *Thoracosphaeroidea:* Neues Jahrb. Geol. Paleont. Abh., v. 151, p. 119-141.

GAARDER, K. R., 1954, *Coccolithineae, Silicoflagellatae, Pterospermataceae* and other forms from the "Michael Sars" North Atlantic Deep-Sea Expedition 1910, *in,* Report on the Scientific Results of the "Michael Sars" North Atlantic Deep-Sea Expedition 1910: University of Bergen, Norway, v. 2, p. 1-21.

GARTNER, S., 1970, Phylogenetic lineages in the Lower Tertiary Coccolith Genus *Chiasmolithus*: Proc. North Amer. Paleontological Convention, Sept. 1969, Part G, p. 930-957.

———, AND KEANY, J., 1978, The terminal Cretaceous event: A geologic problem with an oceanographic solution: Geology, v. 6, p. 708-712.

GEITZENAUER, K. R., ROCHE, M. B., AND MCINTYRE, A., 1977, Coccolith biogeography from North Atlantic and Pacific surface sediments, *in,* Ramsay, A. T. S. ed., Oceanic Micropaleontology: London, Academic Press, v. 2, p. 973-1008.

GORDON, W. A., 1973, Marine life and ocean surface currents in the Cretaceous: Jour. Geology, v. 81, p. 269-284.

GOULD, S. J., AND ELDREDGE, N., 1977, Punctuated equilibria: the tempo and mode of evolution reconsidered: Paleobiology, v. 3, p. 115-151.

GRAN, H. H., AND BRAARUD, T., 1935, A quantitative study of the phytoplankton in the Bay of Fundy and the Gulf of Maine (including observations on hydrography, chemistry and turbidity): Jour. Biological Board, Canada, v. 1, p. 279-467.

GUINASSO, N. L., AND SCHINK, D. R., 1975, Quantitative estimates of biological mixing rates in abyssal sediments: Jour. Geophys. Res., v. 80, p. 3022-3043.

HALLDAL, P., AND MARKALI, J., 1954, Observations on coccoliths of *Syracosphaera mediterranea* Lohm., *S. pulchra* Lohm., and *S. molischi* Schill. in the electron microscope: Journal Conseil International pour l'Exploitation de la Mer, v. 19, p. 329-336.

HAQ, B. U., 1971, Paleogene calcareous nannoflora. Part IV: Paleogene nannoplankton biostratigraphy and evolutionary rates in Cenozoic calcareous nannoplankton: Stockholm Contributions in Geology, v. 25, p. 129-158.

———, 1973, Evolutionary trends in the Cenozoic coccolithophore genus *Helicopontosphaera:* Micropaleontology, v. 19, p. 32-52.

———, AND LOHMAN, G. P., 1976, Early Cenozoic calcareous nannoplankton biogeography of the Atlantic Ocean: Marine Micropaleontology, v. 1, p. 119–194.

———, PREMOLI SILVA, I., AND LOHMAN, G. P., 1977, Calcareous plankton paleobiogeographic evidence for major climatic fluctuations in the early Cenozoic Atlantic Ocean: Jour. Geophys. Res., v. 82, p. 3861–3876.

HARDENBOL, J., AND BERGGREN, W. A., 1978, A new Paleogene numerical time scale, *in,* Cohee, G. V. and others eds., Contributions to the geologic time scale: Am. Assoc. Petroleum Geologists, Studies in Geology, v. 6, p. 213–234.

HAY, W. W., 1960, The Cretaceous-Tertiary boundary in the Tampico Embayment, Mexico: 21st International Geological Congress Proceedings, pt. 5, p. 70–77.

———, AND MOHLER, H. P., 1967, Calcareous nannoplankton from early Tertiary rocks at Pont-Labau, France, and Paleocene–Early Eocene correlations: Jour. Paleontology, v. 41, p. 1505–1541.

HOFKER, J., 1978, Analysis of a large succession of samples through the Upper Maastrichtian and the Lower Tertiary of drill hole 47.2, Shatsky Rise, Pacific, Deep Sea Drilling Project: Jour. Foraminiferal Res., v. 8, p. 46–75.

HSÜ, K. J., 1980, Terrestrial catastrophe by cometary impact at the end of Cretaceous: Nature, v. 285, p. 201–203.

HULBERT, E. M., 1962, Phytoplankton in the southwestern Sargasso Sea and North Equatorial Current, February 1961: Limnology Oceanography, v. 7, p. 307–315.

JOHNSTONE, M. H., LOWRIE, D. C., AND QUILTY, P. G., 1973, The geology of southwestern Australia—a review: Jour. Royal Soc. Western Australia, v. 56, p. 5–15.

KEUPP, H., AND MICHAEL, E., 1979, *Calciodinelloidea* aus der Blätterton Fazies des nordwestdeutschen Unter-Barremium: Berichte Naturhistorischen Gesell. Hannover, v. 122, p. 7–69.

KENT, D. V., 1977, An estimate of the duration of the faunal change at the Cretaceous-Tertiary boundary: Geology, v. 5, p. 769–771.

KOCH, G. S., AND LINK, R. F., 1971, Statistical analysis of geological data: New York, John Wiley & Sons, v. 2, 438 pp.

LABRECQUE, J. L., KENT, D. V., AND CANDE, S. C., 1977, Revised magnetic polarity time scale for Late Cretaceous and Cenozoic time: Geology, v. 5, p. 330–335.

LANCELOT, Y., 1978, Relations entre évolution sédimentaire et tectonique de la plaque pacifique depuis le Crétacé inférieur. Société Géologique France, Nouv. Sér. 57, Mémoire 134, p. 1–40.

LARSON, P. A., AND OPDYKE, N. D., 1979, Paleomagnetic results from Early Tertiary/Late Cretaceous sediments of Site 384: Init. Repts. DSDP, v. 43, p. 785–788.

LECAL-SCHLAUDER, J., 1950, Notes préliminaires sur les Coccolithophoridés d'Afrique du Nord: Société de l' Histoire Naturelle d'Afrique du Nord, Bulletin, v. 40, p. 160–167.

LEFORT, F., 1972, Quelques caractères morphologiques de deux espèces actuelles de *Braarudosphaera* (Chrysophycées, Coccolithacées); Botaniste, v. 55, p. 81–93.

LERBEKMO, J. F., EVANS, M. E., AND BAADSGAARD, H., 1979, Magnetostratigraphy, biostratigraphy and geochronology of Cretaceous-Tertiary boundary sediments, Red Deer Valley: Nature, v. 279, p. 26–30.

LINDSAY, E. H., JACOBS, L. L., AND BUTLER, R. F., 1978, Biostratigraphy and magnetostratigraphy of Paleocene terrestrial deposits, San Juan Basin, New Mexico: Geology, v. 6, p. 425–429.

———, BUTLER, R. F., JOHNSON, N. M., AND JACOBS, L. L., 1979a, Comments and replies on "Biostratigraphy and magnetostratigraphy of Paleocene terrestrial deposits, San Juan Basin, New Mexico": Geology, v. 7, p. 66–71.

———, ———, AND ———, 1979b, Comment and reply on "Biostratigraphy and magnetostratigraphy of Paleocene terrestrial deposits, San Juan Basin, New Mexico.": Geology, v. 7, p. 323–327.

LOEBLICH, A. R., AND TAPPAN, H., 1966, Annotated index and bibliography of the calcareous nannoplankton: Phycologia, v. 5, p. 81–216.

———, AND ———, 1968, Annotated index and bibliography of the calcareous nannoplankton II: Jour. Paleontology, v. 42, p. 584–598.

———, AND ———, 1969, Annotated index and bibliography of the calcareous nannoplankton III: Jour. Paleontology, v. 43, p. 568–588.

———, AND ———, 1970a, Annotated index and bibliography of the calcareous nannoplankton IV: Jour. Paleontology, v. 44, p. 558–574.

———, AND ———, 1970b, Annotated index and bibliography of the calcareous nannoplankton V: Phycologia, v. 9, p. 157–174.

———, AND ———, 1971, Annotated index and bibliography of the calcareous nannoplankton VI: Phycologia, v. 10, p. 315–339.

———, AND ———, 1973, Annotated index and bibliography of the calcareous nannoplankton VII: Jour. Paleontology, v. 47, p. 715–759.

LUCAS, S. G., AND RIGBY, J. K., JR., 1979, Comment and reply on "Biostratigraphy and magnetostratigraphy of Paleocene terrestrial deposits, San Juan Basin, New Mexico": Geology, v. 7, p. 323–327.

LUTERBACHER, H. P., AND PREMOLI SILVA, I., 1964, Biostratigrafia del limite Cretaceo-Terziario nell'Appennino Centrale: Riv. Ital. Paleontol., v. 70, p. 67–129.

MARTINI, E., 1961, Nannoplankton aus dem Tertiär und der obersten Kreide von von SW-Frankreich: Senckenberg. Lethaea, v. 42, p. 1–41.

———, 1967, Nannoplankton und Umlagerungserscheinungen im Persischen Golf und im nördlichen Arabischen Meer: Neues Jahrb. Geol. Paleont. Monatshefte, v. 10, p. 597–607.

———, 1971, Standard Tertiary and Quaternary calcareous nannoplankton zonation, *in* Farinacci, A. ed., Proceedings

II. Planktonic Conference, Roma, 1970: Rome Edizioni Tecnoscienza, v. 2, p. 739–785.

———, 1976, Cretaceous to recent calcareous nannoplankton from the central Pacific Ocean (DSDP Leg 33): Init. Repts. DSDP, v. 33, p. 383–423.

MASTERS, B. A., 1977, Mesozoic planktonic foraminifera, *in,* Ramsay, A. T. S. ed, Oceanic Micropaleontology: London, Academic Press, p. 301–731.

MCINTYRE, A., 1970, *Gephyrocapsa protohuxleyi* sp. n. a possible phyletic link and index fossil for the Pleistocene: Deep-sea Research, v. 17, p. 187–190.

———, AND BÉ, A. W. H., 1967, Modern *Coccolithophoridae* of the Atlantic Ocean—I. Placoliths and Cyrtoliths: Deep-sea Research, v. 14, p. 561–592.

———, AND MCINTYRE, R., 1971, Coccolith concentrations and differential solution in oceanic sediments, *in,* Funnell, B. M. and Riedel, W. R. eds., The Micropaleontology of Oceans: Cambridge University Press, p. 253–261.

MCNULTY, C. L., 1979, Smaller Cretaceous foraminifers of Leg 43, Deep Sea Drilling Project: Init. Repts. DSDP, v. 43, p. 487–505.

MONECHI, S., 1977, Upper Cretaceous and Early Tertiary nannoplankton from the Scaglia Umbra Formation (Gubbio, Italy): Riv. Ital. Paleontol., v. 83, p. 759–802.

NOZAKI, Y., COCHRAN, J. K., AND TUREKIAN, K. K., 1977, Radiocarbon and ^{210}Pb distribution in submersible-taken deep-sea cores from project FAMOUS: Earth Planet. Sci. Letters, v. 34, p. 167–173.

OKADA, H., AND HONJO, S., 1973, The distribution of oceanic coccolithophorids in the Pacific: Deep-sea Research, v. 20, p. 355–374.

———, AND ———, 1975, Distribution of coccolithophores in marginal seas along the western Pacific Ocean and in the Red Sea: Marine Biology, v. 31, p. 271–285.

———, AND MCINTYRE, A., 1977, Modern coccolithophores of the Pacific and North Atlantic Oceans: Micropaleontology, v. 23, p. 1–55.

———, AND ———, 1979, Seasonal distribution of modern coccolithophores in the western North Atlantic Ocean: Marine Biology, v. 54, p. 319–328.

———, AND THIERSTEIN, H. R., 1979, Calcareous nannoplankton Leg 43, Deep Sea Drilling Project: Init. Repts. DSDP, v. 43, p. 507–573.

PARKE, M. AND ADAMS, I., 1960, The motile *(Crystallolithus hyalinus* Gaarder and Markali) and non-motile phases in the life history of *Coccolithus pelagicus* (Wallich) Schiller: Jour. Mar. Biol. Assoc. U.K., v. 39, p. 263–274.

PENG, T. H., BROECKER, W. S., AND BERGER, W. H., 1979, Rates of benthic mixing in deep-sea sediment as determined by radioactive tracers: Quaternary Res., v. 11, p. 141–149.

PERCH-NIELSEN, K., 1968, Der Feinbau und die Klassifikation der Coccolithen aus dem Maastrichtien von Dänemark. Det Kongelige Danske Videnskabernes Selskab: Biologiske Skrifter, v. 16, 96 pp.

———, 1969, Die Coccolithen einiger Dänischer Maastrichtien-und Danienlokalitäten: Meddelelser fra Dansk Geologisk Forening (Bulletin of the Geological Society of Denmark), v. 19, p. 51–68.

———, 1977, Albian to Pleistocene calcareous nannofossils from the western South Atlantic, DSDP Leg 39: Init. Repts. DSDP, v. 39, p. 1099–1132.

———, 1979a, Calcareous nannofossils from the Cretaceous between the North Sea and the Mediterranean, *in,* Wiedmann, J. ed., Aspekte der Kreide Europas: Stuttgart, IUGS Series A, No. 6, p. 223–272.

———, 1979b, Calcareous nannofossils in Cretaceous/Tertiary boundary sections in Denmark, *in,* Christensen, W. K. and Birkelund, T., eds., Cretaceous-Tertiary boundary events symposium. II. Proceedings: University of Copenhagen, p. 120–126.

———, 1979c, Calcareous nannofossils at the Cretaceous-Tertiary boundary near Biarritz, France, *in,* Christensen, W. K. and Birkelund, T. eds., Cretaceous-Tertiary boundary events symposium. II. Proceedings: University of Copenhagen, p. 151–155.

———, ULLEBERG, K., AND EVENSEN, J. E., 1979, Comments on "The terminal Cretaceous event: A geologic problem with an oceanographic solution" (Gartner and Keany, 1978), *in,* Christensen, W. K. and Birkelund, T. eds., Cretaceous-Tertiary boundary events symposium. II. Proceedings: University of Copenhagen, p. 106–111.

PERCIVAL, S. F., AND FISCHER, A. G., 1977, Changes in calcareous nannoplankton in the Cretaceous-Tertiary biotic crisis at Zumaya, Spain: Evolutionary Theory, v. 2, p. 1–35.

PETERSON, M. N. A., AND OTHERS, 1970, Init. Repts. DSDP, v. 2, 501 pp.

PLUMMER, H. J., 1931, Some Cretaceous foraminifera in Texas: University of Texas Bulletin 3101, p. 109–203.

POZARYSKA, K., 1967, The Upper Cretaceous and the Lower Paleogene in Central Poland: Instytut Geologiczny Warszawa, Biuletyn 211, p. 41–67.

PREMOLI SILVA, I., 1977, Upper Cretaceous-Paleocene magnetic stratigraphy at Gubbio, Italy. II. Biostratigraphy: Geol. Soc. America Bull., v. 88, p. 367–389.

PRINS, B., 1971, Speculations on relations, evolution and stratigraphic distribution of discoasters, *in,* Farinacci, A., ed., Proceedings II Planktonic Conference, Roma 1970: Rome, Edizioni Tecnoscienza, v. 2, p. 1017–1037.

RISATTI, J. B., 1973, Nannoplankton biostratigraphy of the Upper Bluffport Marl–Lower Prairie Bluff Chalk interval (Upper Cretaceous), in Mississippi, *in,* Smith, L. A., and Hardenbol, J., eds., Proceedings Symposium Calcareous Nannofossils: Houston, Society Economic Paleontologists Mineralogists, p. 8–57.

ROGGENTHEN, W. M., 1977, Magnetic stratigraphy of the Paleocene: A comparison between Spain and Italy: Memorie della Società Geologica Italiana, v. 15, p. 73–82.

ROMEIN, A. J. T., 1977, Calcareous nannofossils from the Cretaceous/Tertiary boundary interval in the Barranco del Gredero (Caravaca, Prov. Murcia, S.E. Spain): I. Proceedings Koninklijke Nederlandse Akademie van Wetenschappen Amsterdam, Ser. B., v. 80, p. 256–278.

ROTH, P. H., 1974, Calcareous nannofossils from the northwestern Indian Ocean, Leg 24, Deep Sea Drilling Project: Init. Repts. DSDP, v. 24, p. 969–994.

———, 1978, Cretaceous nannoplankton biostratigraphy and paleoceanography of the northwestern Atlantic Ocean: Init. Repts. DSDP, v. 44, p. 731–759.

———, AND BERGER, W. H., 1975, Distribution and dissolution of coccoliths in the South and Central Pacific, *in,* Sliter, W. V., Bé, A. W. H., and Berger, W. H. eds., Dissolution of deep-sea carbonates: Cushman Found. Foram. Res. Spec. Pub. No. 13, p. 87–113.

———, AND BOWDLER, J. L., 1979, Evolution of the calcareous nannofossil genus *Micula* in the Late Cretaceous: Micropaleontology, v. 25, p. 272–280.

SAITO, T., AND VAN DONK, J., 1974, Oxygen and carbon isotope measurements of Late Cretaceous and Early Tertiary foraminifera: Micropaleontology, v. 20, p. 152–177.

SALAJ, J., 1974, *in,* Salaj, J., and others, Livret-guide des excursions, VI Colloque Africain de Micropaléontologie, Tunis, 1974: Editions Service géologique de Tunisie, Tunis, p. 51–57.

SAVIN, S. M., DOUGLAS, R. G., AND STEHLI, F. G., 1975, Tertiary marine paleotemperatures: Geol. Soc. America Bull., v. 86, p. 1499–1510.

SCHMID, F., HILTERMANN, H., AND KOCH, W., 1955, Biostratigraphie der Grenzschichten Maastricht/Campan in Lüneburg und in der Bohrung Brunhilde: Geologisches Jahrb., v. 70, p. 339–384.

SCHNEIDERMANN, N., 1973, Deposition of coccoliths in the compensation zone of the Atlantic Ocean, *in,* Smith, L. A., Hardenbol, J. eds., Proceedings Symposium on Calcareous Nannofossils: Houston, Soc. Econ. Paleontologists Mineralogists, p. 140–151.

SCHOLLE, P. A., AND ARTHUR, M. A., 1980, Carbon isotope fluctuations in Cretaceous pelagic limestones: potential stratigraphic and petroleum exploration tool: Am. Assoc. Petroleum Geologists Bull., v. 64, p. 67–87.

SCLATER, J. G., ABBOTT, D., THIEDE, J., 1977, Paleobathymetry and sediments of the Indian Ocean, *in,* Heirtzler, J. R., and others eds., Indian Ocean Geology and Biostratigraphy: Washington, D.C., Amer. Geophys. Union, p. 25–59.

SHAFIK, S., AND STRADNER, H., 1971, Nannofossils from the Eastern Desert, Egypt, with reference to Maastrichtian nannofossils from the USSR.: Wien, Jahrb. Geologische Bundesanstalt Sonderband 17, p. 69–104.

SISSINGH, W., 1977, Biostratigraphy of Cretaceous calcareous nannoplankton: Geologie en Mijnbouw, v. 56, p. 37–65.

SLITER, W. V., 1977, Cretaceous foraminifers from the southwestern Atlantic Ocean, Leg 36, Deep Sea Drilling Project: Init. Repts. DSDP, v. 36, p. 519–573.

SMAYDA, T. J., 1966, A quantitative analysis of the phytoplankton of the Gulf of Panama. III. General ecological conditions, and the phytoplankton dynamics at 8°45′N, 79°13′W from November 1954 to May 1957: Inter-American Tropical Tuna Commission Bulletin, v. 11, p. 353–612.

SMIT, J., 1977, Discovery of a planktonic foraminiferal association between the *Abathomphalus mayaroensis* Zone and the *"Globigerina" eugubina* Zone at the Cretaceous/Tertiary boundary in the Barranco del Gredero (Caravaca, S.E. Spain): A preliminary report: Proceedings Koninklijke Nederlandse Akademie van Wetenschappen Ser. B, v. 80, p. 280–301.

SMITH, A. G., AND BRIDEN, J. C., 1977, Mesozoic and Cenozoic paleocontinental maps: Cambridge Univ. Press, 63 pp.

STADNER, H., 1960, Über Nannoplankton-Invasionen im Sarmat des Wiener Beckens: Erdöl-Zeitschrift, v. 76, p. 430–432.

SUPKO, P. R., PERCH-NIELSEN, K., AND OTHERS, 1977, Init. Repts. DSDP, v. 30, 1139 pp.

TAKAYAMA, T., 1972, A note on the distribution of *Braarudosphaera bigelowi* (Gran and Braarud) Deflandre in the bottom sediments of Sendai Bay, Japan: Transactions Proceedings Palaeontological Society Japan, N.S., No. 87, p. 429–435.

THIERSTEIN, H. R., 1973, Lower Cretaceous calcareous nannoplankton biostratigraphy: Abhandlungen Geologische Bundes-Anstalt, Wien, Bd. 29, 52 pp.

———, 1974, Calcareous nannoplankton—Leg 26, Deep Sea Drilling Project: Init. Repts. DSDP, v. 26, p. 619–667.

———, 1976, Mesozoic calcareous nannoplankton biostratigraphy of marine sediments: Marine Micropaleontology, v. 1, p. 325–362.

———, 1979, Paleoceanographic implications of organic carbon and carbonate distribution in Mesozoic deep-sea sediments, *in,* Talwani, M., Ryan, W. B. F., and Hay, W., eds., 2nd Maurice Ewing Symposium: Deep Drilling in the Atlantic Ocean, Washington, D. C., Amer. Geophys. Union, p. 249–274.

———, 1980, Selective dissolution of late Cretaceous and earliest Tertiary calcareous nannofossils: experimental evidence: Cretaceous Research, v. 2, p. 165–176.

———, AND BERGER, W. H., 1978, Injection events in ocean history: Nature, v. 276, p. 461–466.

———, AND OKADA, H., 1979, The Cretaceous/Tertiary boundary event in the North Atlantic: Init. Repts. DSDP, v. 43, p. 601–616.

———, AND WOODWARD, D. L., in press, Paleontological, stratigraphic, sedimentological, geochemical, and paleogeographical data of Upper Cretaceous and lowermost Tertiary sediment samples: SIO Reference Series, No. 81–82.

———, GEITZENAUER, K. R., MOLFINO, B., SHACKLETON, N. J., 1977, Global synchroneity of late Quaternary coccolith datum levels: validation by oxygen isotopes: Geology, v. 5, p. 400–404.

Towe, K. M., 1979, Variation and systematics in calcareous nannofossils of the genus *Sphenolithus:* Am. Zoologist, v. 19, p. 555–572.

Tucholke, B. E., Vogt, P. R., and others, 1979, Init. Repts. DSDP, v. 43, 1115 pp.

Van Valen, L., 1973, A new evolutionary law: Evolutionary Theory, v. 1, p. 1–30.

Vekshina, V. N., 1956, Novye dannye o svyazi morskogo basseina Zapadnoi Sibiri s yuzhno-evropeiskimi moryami v. kampan(?)-maastrikhtskoe vremya (New data concerning the communication which existed between the west Siberian Sea basin and the seas of southern Europe during Campanian(?)–Maastrichtian time): Doklady Akad. Nauk SSSR, v. 110, p. 1057–1059.

Verbeek, J. W., 1976, Upper Cretaceous calcareous nannoplankton from Ballon and Théligny, in the type area of the Cenomanian Stage (Sarthe, France): Proceedings, Koninklijke Nederlandse Akademie van Wetenschappen Amsterdam, Ser. B, v. 79, p. 69–82.

———, 1977, Calcareous nannoplankton biostratigraphy of middle and upper Cretaceous deposits in Tunisia, southern Spain and France: Utrecht Micropaleontological Bulletins, v. 16, 157 pp.

von Hillebrandt, A., 1965, Foraminiferen-Stratigraphie im Alttertiär von Zumaya (Provinz Guipuzcoa, N.W.-Spanien) und im Vergleich mit anderen Tethys-Gebieten: Abhandlungen Bayerische Akademie Wissenschaften Mathematisch-Naturwissenschaftliche Klasse N.F. 123, p. 1–62.

Wall, D., and Dale, B., 1968, Quaternary calcareous dinoflagellates *(Calcidinellideae)* and their natural affinities: Jour. Paleontology, v. 42, p. 1359–1408.

———, Guillard, R. R. L., Dale, B., Swift, E., and Watabe, N., 1970, Calcitic resting cysts in *Peridinium trochoideum* (Stein) Lemmermann, an autotrophic marine dinoflagellate: Phycologia, v. 9, p. 151–156.

Watts, N. L., Lapré, J. F., Van Schijnel-Goester, F. S., and Ford, A., 1980, Upper Cretaceous and Lower Tertiary chalks of the Albuskjell area, North Sea: deposition in a slope and base-of-slope environment: Geology, v. 8, p. 217–221.

Winter, A., Reiss, Z., and Luz, B., 1978, Living *Gephyrocapsa protohuxleyi* McIntyre in the Gulf of Elat ('Aqaba): Marine Micropaleontology, v. 3, p. 295–298.

———, ———, and ———, 1979, Distribution of living coccolithophore assemblages in the Gulf of Elat ('Aqaba): Marine Micropaleontology, v. 4, p. 197–223.

Wise, S. W., Jr., and Wind, F. H., 1977, Mesozoic and Cenozoic calcareous nannofossils recovered by DSDP Leg 36 drilling on the Falkland Plateau, southwest Atlantic sector of the southern ocean: Init. Repts. DSDP, v. 36, p. 269–491.

Worsley, T., 1974, The Cretaceous-Tertiary boundary event in the ocean, *in,* Hay, W. W., ed., Studies in Paleo-oceanography: Soc. Econ. Paleontologist Mineralogists, Spec. Pub. No. 20, p. 94–175.

———, and Martini, E., 1970, Late Maastrichtian nannoplankton provinces: Nature, v. 225, p. 1242–1243.

Appendix I. List of Assemblages Studied

Deep Sea Drilling Project samples given as leg-site-core-section, cm from top of section; for additional information, see Initial Reports of the Deep Sea Drilling Project. Other samples are listed with sources or references to additional information on outcrops, stratigraphy, etc. Taxonomic census data, geographic paleocoordinates, sedimentologic and stable isotope data will be published in Thierstein and Woodward (*in press*). cc = core catcher.

DSDP Leg	Site	Core	Sect.	cm From Top
Late Campanian — Early Maastrichtian				
2	10	12	3	120
2	10	13	4	120
2	10	14	1	130
2	10	14	2	120
2	10	15	1	125
2	10	15	2	129
2	10	16	1	121
2	10	16	2	94
2	10	17	1	125
3	13A	3	4	102
3	21	5	2	100
3	21	5	5	98
3	21	6	3	127
11	98	13	1	128
14	144A	4	2	90
15	146	19	7	CC
16	163	16	4	52
16	163	19	5	49
16	163	22	1	95
17	167	44	7	CC
17	167	45	6	21
17	167	50	1	122
17	170	5	7	CC
17	171	16	7	CC
20	199	11	2	18
22	217	26	3	67
25	249	18	1	97
25	249	19	1	104
30	288A	11	1	133
30	289	126	1	143
32	305	20	5	99
32	310	15	1	100
36	327A	13	2	134
39	355	18	3	138
39	356	33	3	69
39	356	33	6	43
39	356	34	2	69

Appendix I. (continued)

DSDP Leg	Site	Core	Sect.	cm From Top
39	357	34	3	21
39	357	35	1	142
39	357	36	1	82
39	357	36	3	100
39	358	16	7	CC
40	363	21	2	10
40	363	21	6	14
40	363	23	1	89
40	364	14	1	117
41	369A	38	7	CC
43	387	29	2	57
44	390A	14	5	48
44	392A	1	2	71

Others

Braggs, Alabama, 41.42 m from top of sequence measured by Danis and others, *(in press)*.

Lueneberg-Zeltberg, northern Germany, sample No. 25582 from Dr. H. Hiltermann, *Belemnitella langei* Zone (Schmid and others, 1955).

Lueneburg-Zeltberg, northern Germany, sample, No. 25583, From Dr. H. Hiltermann, *Bostrychoceras polyplocum* Zone (Schmid and others, 1955).

Tunisia, Qued el Abiod 621 (Bramlette and Martini, 1964).

Appendix I.
List of Assemblages Studied (continued)

DSDP Leg	Site	Core	Sect.	cm From Top
Middle Maastrichtian				
2	10	10	1	104
2	10	10	4	14
6	47.2	13	4	100
6	47.2	14	4	100
10	95	13	3	100
12	111A	11	1	139
12	111A	11	2	97
12	111A	11	3	46
12	111A	11	4	7
12	111A	11	5	46
15	146	13	1	130
15	146	13	3	60
15	146	15	7	CC
16	163	15	1	19
17	167	42	1	140
17	171	10	4	30
22	217	20	3	112
25	249	17	1	53
30	288A	9	3	18
32	305	17	5	100
32	305	18	5	96
32	310	12	2	98
32	313	15	2	52
32	313	17	5	68
33	317A	3	3	61

DSDP Leg	Site	Core	Sect.	cm From Top
36	327A	13	1	148
39	354	17	1	70
39	354	17	3	116
39	354	18	2	98
39	354	18	6	136
39	355	17	2	80
39	356	31	2	69
39	356	31	6	33
39	356	32	6	47
39	357	33	1	76
39	358	15	2	142
40	363	20	1	133
40	363	20	2	12
40	364	13	1	142
41	369A	36	5	48
41	369A	37	2	102
43	384	13	7	CC
43	384	14	1	70
43	384	14	2	39
43	384	14	3	40
43	384	14	4	10
43	384	14	7	CC
43	384	15	1	30
43	384	15	2	40
43	384	15	3	40
43	384	15	4	40
43	384	15	5	40
43	384	15	6	22
43	384	15	7	CC
43	385	13	3	26
43	385	13	4	141
43	385	13	7	CC
43	386	35	5	48
44	390A	12	6	134
44	390A	13	6	16

Others

Braggs, Alabama, 7.13 m from top of section measured by Danis and others, *(in press)*.

Braggs, Alabama, 14.21 m from top of section measured by Danis and others, *(in press)*.

Braggs, Alabama, 19.76 m from top of section measured by Danis and others, *(in press)*.

Braggs, Alabama, 24.83 m from top of section measured by Danis and others, *(in press)*.

Braggs, Alabama, 27.49 m from top of section measured by Danis and others, *(in press)*.

Braggs, Alabama, 33.51 m from top of section measured by Danis and others, *(in press)*.

Braggs, Alabama, 37.57 m from top of section measured by Danis and others, *(in press)*.

Omsk 724 m depth, Bore P-1, western Siberia (Vekshina, 1956).

Texas, A23-4, 65 feet above Onion Creek bed, Travis County, Sta. 226-T-9 of Plummer (1931).

Tunisia, Tou 74/252, *Racemiguembelina varians varians* Subzone, El Haria 8, b2, near El Kef (Salaj, 1974).

Warnbro No. 1 Bore, 2194 feet depth, W-Australia, (Johnstone and others, 1973).

APPENDIX I. (CONTINUED)

DSDP Leg	Site	Core	Sect.	cm From Top
Late Maastrichtian				
3	20C	6	5	132
3	21	3	1	105
3	21	3	4	100
3	21	3	6	100
6	47.2	11	4	15
6	47.2	11	5	35
6	47.2	11	5	86
6	47.2	11	6	135
6	47.2	12	3	50
6	47.2	12	4	110
15	152	10	1	142
17	165A	16	7	CC
21	208	34	2	28
21	208	34	3	123
22	212	31	4	75
22	212	35	4	138
22	216	23	2	118
22	216	23	2	126
22	216	23	2	139
22	216	24	3	50
22	217	17	1	20
22	217	17	3	29
22	217	18	2	95
25	239	19	2	149
30	289	123	1	80
32	305	15	1	116
33	317A	2	1	142
36	327A	10	2	82
36	327A	10	3	70
36	327A	11	2	120
36	327A	12	1	125
36	327A	12	4	14
39	356	29	3	41
39	356	29	3	47
39	356	29	3	56
39	356	29	3	70
39	356	29	4	70
39	356	29	5	70
39	356	29	6	66
39	356	30	1	69
39	356	30	2	68
39	356	30	4	70
39	356	30	5	71
39	356	30	6	125
39	357	31	2	22
39	357	32	1	122
39	357	32	3	87
40	363	18	3	80
40	363	18	4	66
40	363	19	2	130
40	363	19	5	51
40	364	11	1	137
40	364	12	1	61
41	369A	35	5	125
43	384	13	3	34
43	384	13	3	35
43	384	13	3	36
43	384	13	3	38

DSDP Leg	Site	Core	Sect.	cm From Top
Late Maastrichtian (continued)				
43	384	13	3	40
43	384	13	3	42
43	384	13	3	44
43	384	13	3	46
43	384	13	3	48
43	384	13	3	49
43	384	13	3	55
43	384	13	3	60
43	384	13	3	70
43	384	13	3	80
43	384	13	3	93
43	384	13	3	100
43	384	13	3	110
43	384	13	3	120
43	384	13	3	130
43	384	13	3	132
43	384	13	4	40
43	384	13	5	40
43	384	13	6	39
43	385	12	1	99
43	385	12	1	100
43	385	12	1	149
43	385	12	2	15
43	385	12	7	CC
43	385	13	2	90
43	386	35	4	13
43	386	35	4	30
43	386	35	4	46
43	386	35	4	115
43	387	27	2	70
43	387	27	2	149
43	387	27	3	74
43	387	27	4	70
43	387	27	5	77
43	387	27	5	80
43	387	27	6	83
43	387	27	7	CC
43	387	28	1	120
43	387	28	7	CC
44	390A	11	6	26
44	390A	12	1	63

Others:

Bellocq 1, southwestern France (Bramlette and Martini, 1964).

Braggs, Alabama, 4.49 m from top of sequence measured by Danis and others, *(in press)*.

Braggs, Alabama, 4.70 m from top of sequence measured by Danis and others, *(in press)*.

Braggs, Alabama, 4.72 m from top of sequence measured by Danis and others, *(in press)*.

Braggs, Alabama, 4.99 m from top of sequence measured by Danis and others, *(in press)*.

Braggs, Alabama, 5.02 m from top of sequence measured by Danis and others, *(in press)*.

Braggs, Alabama, 5.41 m from top of sequence measured by Danis and others, *(in press)*.

APPENDIX I. (CONTINUED)

Late Maastrichtian (others continued)

Braggs, Alabama, 5.63 m from top of sequence measured by Danis and others, *(in press).*
Braggs, Alabama, 5.91 m from top of sequence measured by Danis and others, *(in press).*
Braggs, Alabama, 6.01 m from top of sequence measured by Danis and others, *(in press).*
Braggs, Alabama, 7.02 m from top of sequence measured by Danis and others, *(in press).*
Caucasus, sample 3000, Cherkessk by Kuban River (Haq, 1971).
Elburz, Galanderud Valley, Sample 438 (Cartier, 1971).
Omsk 600 m depth, Bore P-1, western Siberia (Vekshina, 1956).
Poland 121, *Belemnella casimiroviensis* Zone, Kasimierz Quarry, central Poland (Pozaryska, 1967).
Stevns Klint 4, gray chalk, 150 cm below contact with Danian (Christensen and others, 1973).
Trinidad, *Abathomphalus mayaroensis* Zone, Guayaguare Bore 163 at 5591 feet (Bolli, 1957).
Tunisia, Qued el Abiod 641 (Bramlette and Martini, 1964).
V16-56, 230 cm, Agulhas Plateau (Thierstein, 1976).

APPENDIX I.
LIST OF ASSEMBLAGES STUDIED (CONTINUED)

DSDP Leg	Site	Core	Sect.	cm From Top
Early Danian				
6	47.2	11	1	111
6	47.2	11	2	95
6	47.2	11	3	65
10	95	12	4	72
10	95	12	4	144
10	95	13	1	63
10	95	13	1	80
10	95	13	1	111
20	199	11	1	140
21	208	33	1	39
22	216	23	2	113
22	216	23	2	115
22	217	16	6	140
30	288A	8	2	145
35	323	16	1	80
39	356	28	7	CC
39	356	29	1	20
39	356	29	1	55
39	356	29	1	70
39	356	29	1	85
39	356	29	1	88
39	356	29	1	110
39	356	29	1	125
39	356	29	1	144
39	356	29	2	18
39	356	29	2	51
39	356	29	2	70
39	356	29	2	85
39	356	29	2	100
39	356	29	2	120

Early Danian (continued)

DSDP Leg	Site	Core	Sect.	cm From Top
39	356	29	2	141
39	356	29	3	8
39	356	29	3	20
39	356	29	3	30
39	358	14	4	140
40	363	18	2	20
43	384	13	2	138
43	384	13	2	144
43	384	13	2	146
43	384	13	3	0
43	384	13	3	4
43	384	13	3	8
43	384	13	3	12
43	384	13	3	16
43	384	13	3	18
43	384	13	3	24
43	384	13	3	28
43	384	13	3	30
43	384	13	3	32
44	390A	11	4	102
44	390A	11	5	109
44	390A	11	6	26

Others

Braggs, Alabama, 0.18 m from top of sequence measured by Danis and others, *(in press).*
Braggs, Alabama, 0.83 m from top of sequence measured by Danis and others, *(in press).*
Braggs, Alabama, 1.37 m from top of sequence measured by Danis and others, *(in press).*
Braggs, Alabama, 2.03 m from top of sequence measured by Danis and others, *(in press).*
Braggs, Alabama, 3.87 m from top of sequence measured by Danis and others, *(in press).*
Braggs, Alabama, 4.15 m from top of sequence measured by Danis and others, *(in press).*

APPENDIX II

The taxa shown in the stratigraphic and paleogeographic plots are listed here in alphabetical order of their specific epithets, with authors and years of publication. Included are references to additional pertinent discussions and illustrations, as well as lists of other taxa whose abundances were lumped here with a given taxon. Detailed references to the used taxonomic literature can be found in Loeblich and Tappan (1966, 1968, 1969, 1970a, 1970b, 1971, 1973). Later references used can be found in references cited for this paper.

Vagalapilla aachena Bukry, 1969
For light micrographs, see Thierstein, 1974, pl. 3, figs. 1–4.

Tetralithus aculeus (Stradner, 1961) Gartner, 1968
Included in this taxon is *Ceratolithoides arcuatus* Prins and Sissingh, 1977, in Sissingh, 1977, p. 60, pl. 1, fig. 6.

APPENDIX II. (CONTINUED)

Reinhardtites anthophorus (Deflandre, 1959) Perch-Nielsen, 1968
Included in this taxon is *Reinhardtites levis* Prins and Sissingh, 1977 in Sissingh, 1977, p. 61, pl. 1, figs. 1-3.

Markalius astroporus (Stradner, 1963) Hay and Mohler, 1967
Includes *Markalius* sp. 1 in Okada and Thierstein, 1979 p. 522, pl. 3, fig. 2; pl. 12, figs. 6, 7.

Watznaueria barnesae (Black, 1953) Perch-Nielsen, 1968

Lithraphidites carniolensis Deflandre, 1963

Lucianorhabdus cayeuxii Deflandre, 1959
Included in this taxon are all central processes of the genus encountered and previously described as: *Lucianorhabdus compactus* (Verbeek, 1976) Prins and Sissingh, 1977; *Lucianorhabdus quadrifidus,* Forchheimer, 1972; *Lucianorhabdus scotus* (Risatti, 1973) Verbeek, 1977.

Biscutum constans (Gorka, 1957) Black, 1967
This taxon includes *Biscutum? dimorphosum* Perch-Nielsen, 1977 in Tertiary samples.

Cretarhabdus crenulatus Bramlette and Martini, 1954
Includes all Late Cretaceous species of *Cretarhabdus* with 8 radial arms in central area. Early Cretaceous *Cretarhabdus angustiforatus* have a much larger central area.

Prediscosphaera cretacea (Arkhangelsky, 1912) Gartner, 1968

Arkhangelskiella cymbiformis Vekshina, 1959
Includes also late Campanian morphotypes transitional to *Broinsonia enormis* (Schumenko, 1968) Manivit, 1971.

Microrhabdulus decoratus Deflandre, 1959
Includes occasionally occurring *Microhabdulus stradneri* Bramlette and Martini, 1964.

Zygodiscus diplogrammus (Deflandre, 1954) Gartner, 1968

Cribrosphaerella ehrenbergii (Arkhangelsky, 1912) Deflandre, 1952
For synonyms, see Thierstein (1976). Care was taken to exclude poorly preserved *Cribrosphaerella? daniae* Perch-Nielsen, 1973.

Teichorhabdus ethmos Wind and Wise, 1977
In Wise and Wind, 1977, p. 306, pl. 44, fig. 6

Nephrolithus frequens Gorka, 1957
For synonyms, see Thierstein (1976).

Ottavianus giannus Risatti, 1973
Included is *Ottavianus terrazetus,* Risatti, 1973.

Tetralithus gothicus Deflandre, 1959
For synonyms, see Thierstein (1976).

Stephanolithion laffittei Noel, 1957

Chiastozygus litterarius (Gorka, 1957) Manivit, 1971

Kamptnerius magnificus Deflandre, 1959
For synonyms, see Thierstein (1976).

Prinsius martinii (Perch-Nielsen, 1969) Haq, 1971
This species includes here early evolutionary placoliths with highly birefringent central areas leading to the following species: *Prinsius martinii* (Perch-Nielsen, 1969) Haq, 1971, in Haq and Lohman, 1976, p. 122, pl. 14, figs. 1, 2; *Prinsius bisulcus* (Stradner, 1963) Hay and Mohler, 1967, in Okada and Thierstein, 1979, pl. 3, figs. 4, 5; *Cocclithus pelagicus* (Wallich, 1877) Schiller, 1930, in Okada and Thierstein, 1979, pl. 2, fig. 6.

Neochiastozygus modestus Perch-Nielsen, 1971c
This taxon includes small morphotypes of the genus *Neochiastozygus,* such as *Neochiastozygus modestus* in Perch-Nielsen, 1971c, pl. 5, figs. 5-8; *Neochiastozygus* sp. 2 in Perch-Nielsen, 1977, pl. 38, figs. 13-16; *Neochiastozygus concinnus* in Okada and Thierstein, 1979, pl. 17, fig. 9.

Micula murus (Martini, 1961) Bukry, 1973
Micula prinsii Perch-Nielsen, 1979 is synonymous with *Micula murus* as used in this study. The stratigraphic range of *Micula murus* covers the upper half of the foraminiferal *Abathomphalus mayaroensis* Zone. The transition from *Micula staurophora* to *Micula murus,* i.e. the reduction in size of one of the two element cycles (Martini, 1961; Roth and Bowdler, 1979), is observed in Core 30 at DSDP Site 356 (see Fig. 7), about 8 m below the level reported by Perch-Nielsen (1977). The presence of bifurcations at the tips of the arms is only observed on exceptionally well-preserved specimens.

Crepidolithus neocrassus Perch-Nielsen, 1968
Includes *Crepidolithus* sp. in Okada and Thierstein, 1979, pl. 5, fig. 1.

Gartnerago obliquum (Stradner, 1963) Noel, 1970
For synonyms, see Thierstein (1976).

Tetralithus obscurus Stradner, 1963
For synonyms, see Thierstein (1976).

Vagalapilla octoradiata (Gorka, 1957) Bukry, 1969

Broinsonia parca (Stradner, 1963) Bukry, 1969

Misceomarginatus pleniporus Wind and Wise, 1977
In Wise and Wind, 1977, p. 301, pl. 18, figs. 1-8.

Cruciplacolithus primus Perch-Nielsen, 1977
See also Okada and Thierstein, 1979, p. 522, pl. 1, figs. 5, 6; pl. 9, figs. 12-14; pl. 10, figs. 1, 2.

Lithraphidites quadratus Bramlette and Martini, 1964
Includes *Lithraphidites praequadratus* Roth, 1978.

Parhabdolithus regularis (Gorka, 1957) Bukry, 1969

Discorhabdus rotatorius (Bukry, 1969) Thierstein, 1973

Cylindralithus serratus Bramlette and Martini, 1964
This taxon includes *Cylindralithus gallicus* (Stradner, 1963) Bramlette and Martini, 1964.

Zygodiscus sigmoides Bramlette and Sullivan, 1961

Prediscosphaera spinosa (Bramlette and Martini, 1964) Gartner, 1968

Zygodiscus spiralis Bramlette and Martini, 1964

Micula staurophora (Gardet, 1955) Stradner, 1963
For synonyms, see Thierstein (1976).

Cretarhabdus surirellus (Deflandre, 1954) Reinhardt, 1970
This taxon includes *Cretarhabdus* morphotypes with more than eight radial arms and one cycle of perforations in the central area (see Thierstein, 1973).

APPENDIX II. (CONTINUED)

Cruciplacolithus tenuis (Stradner, 1961) Hay and Mohler, 1967

Tetralithus trifidus (Stradner, 1961) Bukry 1973
For synonyms, see Thierstein (1976).

Eiffellithus turriseiffelii (Deflandre, 1954) Reinhardt, 1965
This taxon includes *Eiffellithus eximius* (Stover, 1966) Perch-Nielsen, 1968.

Braarudosphaera
Includes: *Braarudosphaera bigelowii* (Gran and Braarud, 1935) Deflandre, 1947, *Braarudosphaera discula* Bramlette and Riedel, 1954, and *Braarudosphaera imbricata* Manivit, 1966, all illustrated in Okada and Thierstein (1979).

Cyclagelosphaera
This taxon includes the rare occurrences of *Cyclagelosphaera margerelii* Noel, 1965 in the late Maastrichtian, and of dominantly *Cyclagelosphaera reinhardtii* (Perch-Nielsen, 1968) in the Danian. Some intermediate forms (e.g. *Cyclagelosphaera* sp. 1 of Perch-Nielsen, 1979b, plate 3, figure 8) appear to persist into the Danian.

Thoracosphaera
Included in *Thoracosphaera* are: *Thoracosphaera operculata* Bramlette and Martini, 1964; *Thoracosphaera saxea* Stradner, 1961; *Thoracosphaera deflandrei* Kamptner, 1956, all also illustrated in Okada and Thierstein, 1979, pl. 8, figs. 6–8.

SEPM Special Publication No. 32, p. 395–432, October 1981

A REVIEW OF NEOGENE PLANKTONIC FORAMINIFERAL BIOSTRATIGRAPHY: APPLICATIONS IN THE EQUATORIAL AND SOUTH PACIFIC

M. S. SRINIVASAN AND J. P. KENNETT
Graduate School of Oceanography, University of Rhode Island, Kingston, R.I. 02881 U.S.A.

ABSTRACT

The Deep Sea Drilling Project has been fundamental in providing carbonate sedimentary sequences throughout the oceans for the study of planktonic foraminiferal biostratigraphy and evolution. We have reviewed the principal developments in the establishment of Neogene planktonic foraminiferal biostratigraphy and later applications, extensions and modifications due to deep-sea drilling. A review of the principal Neogene evolutionary lineages of planktonic foraminifera shows that only a few have formed the basis for biostratigraphic subdivision. A new Early Miocene lineage (*G. kugleri–G. peripheroronda*) is reported from mid-latitude southwest Pacific DSDP Site 208. Major evolutionary lineages within the Neogene globorotaliids are recognized as separate taxonomic sub-divisions following Bandy (1972). The subgenus *Globoconella* Bandy, which represents the major globorotaliid lineage in temperate areas, is redefined, and we restrict the subgenus *Turborotalia* to Paleogene lineages. The lineage concept of taxonomy is recommended to be employed as phylogenies become well known.

The southwest Pacific represents an optimal area for the study of planktonic foraminiferal biostratigraphic correlations between equatorial and temperate areas. A relatively detailed analysis of five Neogene southwest Pacific DSDP sites ranging from cool-subtropical (Sites 284, 207A) to warm-subtropical (Sites 206, 208) to tropical (Site 289), provides a strong basis for correlation between these distinct water masses. Studies of equatorial site 289 necessitate some emendments of the widely used tropical zonation of Blow (1969). Since the last appearance of *G. fohsi robusta* and of the entire *Fohsella* lineage provides an easily identifiable, useful Middle Miocene datum, it is recommended for defining the Zone N12/N13 boundary in preference to the first evolutionary appearance of *S. subdehiscens* (Blow, 1969). Zone N17 is subdivided into new zones, Zone N17A and N17B, based on the first appearance of *Pulleniatina,* a solution-resistant form occurring in deep-sea sequences from tropical to subtropical areas.

A new zonation is recognized in the warm-subtropics but not for the temperate area. The warm-subtropical zones at Site 208 represent a combination of tropical and temperate zonal schemes. Because Site 208 contains a mixture of tropical and temperate elements, a greater number of faunal events is recorded compared with both tropical and temperate regions. However, faunal events at warm-subtropical Site 208 do not necessarily represent datums, since the appearance or disappearance can be diachronous with respect to either tropical or temperate regions. A zonation based on tropical species is applicable in the Early Neogene until the early Middle Miocene, after which temperate species are more valuable for biostratigraphic subdivision. This reflects global cooling related to the major development of Antarctic ice beginning in the Middle Miocene, concommitant with latitudinal shift of cool waters towards the equator.

The duration of Neogene zones varies from 0.2 m.y. in the case of Zone N18 to as much as 2.7 m.y. for Zone N17 (as undivided). Only two zones have an estimated duration of less than 0.5 m.y. (Zones N13 and N18), and only two zones are longer than 2 m.y. (Zones N16 and N17). The Late Miocene exhibits zones of longest duration, the Middle Miocene zones of shortest duration and the Early Miocene, Pliocene-Pleistocene of intermediate durations. These changes in zonal duration seem to be directly related to global paleoceanographic evolution and the resultant faunal diversities in planktonic foraminifera through the Neogene.

INTRODUCTION

Planktonic foraminifera represent one of the three major microfossil groups (with calcareous nannofossils and radiolaria) which have been employed for biostratigraphic subdivision and correlation of Late Phanerozoic sections drilled by the Deep Sea Drilling Project (DSDP). This in turn has led to some refinement of existing foraminiferal zonations, the development of new schemes, the continuing advancement of knowledge on the nature of evolutionary sequences and the dating of biostratigraphic events. Because of the vast amount of new data on Cenozoic planktonic foraminifera resulting from the Deep Sea Drilling Project, it is appropriate at this time to briefly review the status of planktonic foraminiferal biostratigraphy. This review is restricted to the Neogene, because it relates to new data which we also now present for the equatorial and South Pacific.

Glaessner (1937) was one of the first to emphasize the importance of planktonic foraminifera in Cenozoic biostratigraphy. It was Bolli (1957a, b), however, who first successfully used planktonic foraminifera for the biostratigraphic subdivision of the Cenozoic. Since his early work was confined to Trinidad marine sequences, the zonation was largely applicable to the tropical areas. Nevertheless, it provided the required fundamental basic framework for the further subdivision of the Cenozoic by other workers throughout the world. For instance, further biostratigraphic schemes were established soon after by Banner and Blow (1965) and Blow

(1969) for the tropics and Jenkins (1966a, 1967) for the temperate area of New Zealand. The developments of all of these classical biostratigraphic schemes resulted from studies of only land-based marine sequences. Parker (1967) was the first to use Tertiary marine sequences from the deep ocean basin for planktonic foraminiferal zonation. This was made possible because of the availability at Scripps Institution of Oceanography of a set of pre-Quaternary carbonate piston cores from the tropical Pacific Ocean. Parker (1967) used the zonal scheme of Banner and Blow (1965) which was readily applicable to her studies.

Oceanic Biostratigraphy

The primary basis for oceanic biostratigraphy has been to determine the stratigraphic ranges of index taxa. The zonal boundaries were defined by initial appearances and last appearances of selected forms and successive evolutionary appearances of selected groups observed by the study of sequential samples. By the late 1960's the planktonic foraminifera provided a basic framework for subdivision of the marine Cenozoic sequence both in the tropical and temperate areas. The planktonic foraminifera represented the only marine microfossil group of Cenozoic age for which much was known. Biostratigraphic subdivisions were only beginning to be established using the calcareous nannofossils (Bramlette and Martini, 1964; Bramlette and Wilcoxon, 1967) and the radiolarians (Riedel, 1957a, b). The Deep Sea Drilling Project began in 1968, and fortunately planktonic foraminiferal biostratigraphies were sufficiently well-established to provide the necessary paleontological control for biostratigraphic subdivision and correlation of the drilled sequences. A chronological framework was immediately required for dating these sequences and Berggren (1969, 1971) began to integrate the planktonic foraminiferal zones with the radiometric time scale. This work continued as new data were generated (Berggren, 1972a; Berggren and Van Couvering, 1974, 1978).

The Deep Sea Drilling Project required workable biostratigraphic schemes for the different microfossil groups preserved in deep-sea sediments and thus provided an unprecedented opportunity for examination of these microfossil groups from the tropics to poles and from Late Mesozoic to the present day. The development of new workable zonations resulted for the calcareous nannofossils, the radiolarians and the diatoms. Planktonic foraminiferal biostratigraphies were sufficiently well-established before the project began, to retain much of their original identity. Planktonic foraminiferal zones distinguished in tropical areas have largely been those originally recognized by Bolli (1957b, 1966) and Blow (1969). Temperate-zonal schemes retain much of their original character as established by Jenkins (1967, 1971). Nevertheless, the DSDP has been critical in providing a network of Cenozoic sequences from areas which otherwise would not have been represented because of an absence of land-based marine sequences. This is particularly true for the Southern Hemisphere, although certainly not confined to that region. Furthermore, the DSDP has provided a large number of carbonate sequences which have enabled some modifications of the previous tropical and temperate zones as well as the establishment of the zonal schemes in transitional and in polar and subpolar water masses. The first high-latitude drilling, carried out in the North Atlantic (DSDP Leg 12) soon showed that existing temperate and tropical zonal schemes could not be easily applied to that region (Berggren, 1972b). Many of the age-diagnostic species, including entire lineages, were absent, and the resolving power of stratigraphy was sharply reduced compared with the low-latitudes. Thus, it became necessary to devise complementary zonal schemes applicable to certain high and middle latitude regions. This in turn required an increase in knowledge of the faunas of these areas.

A further major accomplishment of the DSDP has been towards the intercalibration of zonal schemes and datum levels of the important oceanic microfossil groups: namely planktonic foraminifera, calcareous nannofossils, radiolaria and diatoms. Such an intercalibration came as a natural consequence of the staffing of the *Glomar Challenger,* because several micropaleontological specialists worked together as a team. This has led to well-developed intercalibration within the groups employing calcium carbonate for test construction (i.e., the planktonic foraminifera and calcareous nannofossils) and those that use opaline silica (i.e., radiolaria and diatoms). Intercalibration is still not well-developed between the carbonate and siliceous groups because calcium carbonate and opaline siliceous sediments are largely mutually exclusive in the ocean basins. However, few DSDP sites (such as Sites 71, 151, 153; Cores V24–58; V24–59) do contain both relatively well-preserved carbonate and siliceous microfossil groups, and hence have become important for stratigraphic intercalibration of these diverse microfossil groups.

Optimal Areas for Foraminiferal Zonation

The development of planktonic foraminiferal zonations in the oceans has not been without considerable difficulty. Certain sedimentary and geochemical processes have occurred, which have affected our ability to use planktonic foraminifera for stratigraphy and correlation. Those areas which have been most valuable for examination of foraminiferal sequences possess the following characteristics:

1) Minimal dissolution.
2) Minimal disruption due to unconformities.
3) Adequate sedimentation rates for enhancing stratigraphic resolution.
4) Latitudinal traverses of cores can be obtained. These provide a basis for the study of biostratigraphy and correlation with changing latitude.

Areas of Minimal Dissolution.—Carbonate dissolution is controlled by sediment paleodepth and geographic location in relation to bottom-water flow. Carbonate sequences shallower than the lysocline are preferable for optimal preservation. Sequences from depths shallower than 3,000 m have generally been of high quality, and those drilled at depths shallower than 2,000 m have been superior, such as those on aseismic ridges (*e.g.,* Ontong-Java Plateau), certain continental margins and oceanic rises (*e.g.,* Lord Howe Rise).

Carbonate dissolution also increases in Antarctic bottom waters and with increasing age of bottom waters. Because Antarctic bottom waters are produced in the Antarctic, the depths of these water-masses become shallower with increasing latitude, and thus carbonate sequences become more limited to shallower depths. As a result, suitable carbonate sequences of Neogene age are almost non-existent in the Antarctic region. A few suitable, shallow-water, elevated areas such as the Kerguelen-Gaussberg Ridge are known to be mantled by carbonate sediments, but these have not yet been drilled.

The youngest bottom-waters are largely formed in the Atlantic sector of the Polar areas. Bottom-waters flow from the Atlantic into the Indian Ocean, to the South Pacific and hence to the North Pacific, while increasing the CO_2 content. Carbonate dissolution increases accordingly. Because of this circulation pattern, the Atlantic Ocean is the most carbonate-rich ocean and exhibits the deepest carbonate compensation depth (CCD), followed in turn by the Indian Ocean, South Pacific and North Pacific.

Adequate Sedimentation Rate and Core Quality.—Carbonate sequences exhibiting sedimentation rates higher than 2 cm/1,000 yr. are common in the ocean basins and provide adequate biostratigraphic resolution. The stratigraphic resolution possible in the study of DSDP carbonate sections has, however, rarely been controlled by inadequate high sedimentation rates. Instead, the primary control on biostratigraphic resolution has resulted from limitations in drilling technology. The ultimate resolution at which stratigraphic studies can be conducted is largely controlled by the continuity of the coring, the percentage of core recovery, and core quality. Unfortunately, continuous coring was not generally implemented during the early years of the Deep Sea Drilling Project. Fortunately, this changed during later stages of the Deep Sea Drilling Project, and during the International Decade of Ocean Drilling (IPOD). More recently, hydraulic piston coring has provided sequences with better than 80% or even 90% core recovery, thus generating much potential for high resolution stratigraphy for the Neogene and even for the Paleogene in certain areas.

Latitudinal Traverses.—Biostratigraphic studies are enhanced when traverses of cores can be obtained over wide latitudes. Because the distribution of carbonate sediments is largely controlled by water depth, the potential for obtaining latitudinal traverses of carbonate sequences is controlled by sea-floor topography. Optimal areas for such work are therefore on ridges that extend for large distances in the north-south direction. The North Pacific contains few such structures. A few seamounts and rises, such as the Hess Rise, have been cored but even these sequences exhibit relatively strong effects of calcium carbonate dissolution. The South Atlantic contains no shallow north-south trending ridges, only east-west trending ridges such as the Rio Grande Rise and the Walvis Ridge. The Indian Ocean does contain a sequence of north-south trending ridges including the Kerguelen-Gaussberg Plateau, Broken Ridge and Ninety-East Ridge, but drilled cores have not been optimally sited to obtain continuous, high-quality carbonate sequences. The optimal region for such latitudinal traverses is in the southwest Pacific. South-to-north trending shallow topographic features include the Macquarie Ridge, South Tasman Rise, Lord Howe Rise and the Ontong-Java Plateau. These shallow pedestals have already provided a sequence of high quality, relatively continuously cored, carbonate Neogene sections, exhibiting minimum calcium carbonate dissolution.

Purpose of This Contribution.—This contribution has several purposes. We briefly review the main developments in the establishment of Neogene planktonic foraminiferal biostratigraphies at different latitudes. This is followed by discussion of the impact of the Deep Sea Drilling Project on our understanding of Neogene planktonic foraminiferal zonal schemes and datum levels. A brief summary is made of the primary evolutionary lineages which have played a major role in the establishment of Neogene planktonic foraminiferal biostratigraphy. A relatively detailed analysis is provided of planktonic foraminiferal biostratigraphy in five Neogene sites in the southwest Pacific ranging from temperate latitudes to the Equator (Sites 284, 207A, 206, 208 and 289). Finally, the new data allow several modifications of the existing zonal schemes.

REVIEW OF THE DEVELOPMENT OF NEOGENE PLANKTONIC FORAMINIFERAL BIOSTRATIGRAPHY

The Deep Sea Drilling Project has led to an extraordinary increase in the study of Cenozoic planktonic foraminifera. Hundreds of drilled sequences throughout the ocean basins have been studied and stratigraphically subdivided by a large number of

EPOCH		PLANKTONIC FORAMINIFERAL ZONES	FIRST APPEARANCES (F.A.) and LAST APPEARANCES (L.A.) of SPECIES and SUBSPECIES used to DELIMIT ZONES	
PLEISTOCENE		*Globorotaila truncatulinoides*: *G. bermudezi* / *G. calida* / *G. hessi* / *G. viola*	*G. fimbriata*	F.A.
			G. tumida flexuosa	L.A.
			G. calida calida	F.A.
			G. hessi	F.A.
			G. truncatulinoides	F.A.
PLIOCENE	LATE	*Globorotalia truncatulinoides cf. tosaensis*	*G. miocenica*	L.A.
	MIDDLE	*Globorotalia miocenica*: *G. exilis* / *G. fistulosus*	*G. trilobus fistulosus*	L.A.
			G. margaritae evoluta	L.A.
	EARLY	*Globorotalia margaritae*: *G. evoluta* / *G. marg.*	*G. margaritae evoluta*	F.A.
			G. margaritae margaritae	F.A.
MIOCENE	LATE	*Neogloboquadrina dutertrei*	*N. dutertrei*	F.A.
		Globorotalia acostaensis	*G. acostaensis*	F.A.
	MIDDLE	*Globorotalia menardii*	*G. mayeri*	L.A.
		Globorotalia mayeri	*G. ruber*	L.A.
		Globigerinoides ruber	*G. fohsi robusta*	L.A.
		Globorotalia fohsi robusta	*G. fohsi robusta*	F.A.
		Globorotalia fohsi lobata	*G. fohsi lobata*	F.A.
		Globorotalia fohsi fohsi	*G. fohsi fohsi*	F.A.
		Globorotalia fohsi peripheroronda	*G. insueta*	L.A.
	EARLY	*Praeorbulina glomerosa*	*P. glomerosa*	F.A.
		Globigerinatella insueta	*C. dissimilis*	L.A.
		Catapsydrax stainforthi	*G. insueta*	F.A.
		Catapsydrax dissimilis	*G. kugleri*	L.A.
		Globigerinoides primordius	*G. primordius*	F.A.
LATE OLIGOCENE		*Globorotalia kugleri*		

workers. Despite this, there has been much constraint in the development of new biostratigraphic schemes. Almost all of the biostratigraphers have employed existing zonal schemes or made only relatively minor modifications of the schemes. The most widely employed have been those developed by Bolli (1957b, 1966) and by Banner and Blow (1965) and Blow (1969) for the tropical regions and by Jenkins (1967, 1971) for temperate regions. These zonal schemes were mainly established on faunas from uplifted sediments originally laid down on continental shelves and upper continental slopes.

Tropical Zonation

Bolli (1957a, b) first published a planktonic foraminiferal zonal scheme for the Trinidad Paleocene to Miocene, and has subsequently subdivided the Pliocene and Pleistocene into zones and subzones (Bolli and Bermudez, 1965; Bolli, 1966). Initially Bolli (1966) divided the Miocene into 15 zones and the Pliocene into two zones. Later, with additional data from the DSDP Legs 4 and 15, Bolli (1970) and Bolli and Premoli Silva (1973) presented a revised zonal scheme with 14 zones in the Miocene, subdivided the Pliocene into three zones, the Pleistocene *G. truncatulinoides* Zone into four subzones, and a Holocene subzone (Fig. 1). The scheme of Bolli has been emended by a few later workers. For instance, Lamb and Beard (1972) suggested modifications to enhance correlations between the Caribbean sequences and the European stages. This modified scheme was later adopted by Smith and Beard (1973). Likewise Cita (1973) proposed new Pliocene zones in the Mediterranean basin using evolution of the *Globorotalia margaritae* plexus. Overall, most workers have found Bolli's zones broadly applicable, although few have directly employed them despite their priority (Stainforth and others, 1975; Srinivasan, 1977). The most commonly used biostratigraphic scheme is that of Banner and Blow (1965) and Blow (1969). In many ways their scheme is similar to that of Bolli's and has not led to any greater stratigraphic refinements. Both zonal schemes, for instance, contain approximately 45 zones for the entire Cenozoic (Bolli, 1966, 42 zones; Blow, 1969, 43 zones). The widespread application of Blow's (1969) zones has largely resulted because of the numerical designation for each zone. The Paleogene thus contains 22 "P" zones from P1 in the Early Paleocene to P22 in the latest Oligocene. Likewise the Neogene is subdivided into 20 "N" zones from N4 in the earliest Miocene to N23 in the Late Quaternary (Fig. 2). It is unfortunate that Banner and Blow (1965) and Blow (1969) did not designate their earliest Neogene zone as N1. The widespread acceptance of this scheme has resulted from its convenience because taxonomic names are not used for zonal identifications and numerical identification provides a general idea about stratigraphic sequence. This scheme has been accepted in spite of the fact that it is contrary to the existing rules of the stratigraphic code (Bolli, 1966; Jenkins and Orr, 1971, 1972). Studies of deep drilled sequences that have employed Blow's (1969) zonal scheme include Parker (1967), Berggren (1969, 1971), Bronnimann and Resig (1971), Kaneps (1973), Fleisher (1974), Vincent (1975, 1977), McGowran (1974), and Saito (in Andrews and others, 1975.

A comparison of Bolli's (1957b, 1966, 1970; Bolli and Premoli Silva, 1973) with Blow's (1969) Neogene zonation shows that only eight criteria for the definition of the zonal boundaries are common to both schemes (Fig. 3). The fundamental difference between the schemes is that Blow (1969) has placed much greater emphasis on placing boundaries at critical levels within evolutionary lineages such as the *Globorotalia* group, *Orbulina* and *Sphaeroidinella.* Because these zones are established on evolutionary first appearances in many cases, boundaries are often distinguished by rather subtle morphologic changes. In retrospect this has been valuable because it required later workers to develop rather intensive knowledge of the morphological and evolutionary changes exhibited in Cenozoic foraminifera.

In general, Bolli's (1957b, 1966, 1970) zonal boundaries are more easily identifiable than Blow's (1969) because they are not based so much on changes within evolutionary lineages. For instance, in Blow's (1969) zonation the N18/N19 boundary is marked by the first evolutionary appearance of *Sphaeroidinella dehiscens* from *Sphaeroidinellopsis paenedehiscens.* The detection of this evolutionary event is often difficult, especially when specimens are rare, because *Sphaeroidinella* initially evolves merely by the addition of a minute secondary aperture which is often difficult to recognize. Thus the level at which the *Sphaeroidinella* datum is placed may be the result of a greater number of specimens examined rather than its valid evolutionary first appearance. Similar difficulties are also encountered in dealing with the evolution of *Globorotalia merotumida* to *Globorotalia plesiotumida,* a change which marks the N16/N17 boundary.

It needs to be emphasized that a high proportion of the species and lineages employed for both Bolli's (1957b, 1966, 1970) and Blow's (1969) zonal schemes are restricted to tropical water-masses. These zonal schemes are not globally applicable despite such earlier claims (Blow, 1969). Attempts

FIG. 1.—Tropical Neogene planktonic foraminiferal zonation of Bolli (1957b, 1966, 1970) and Bolli and Premoli Silva (1973), and species used for zone definitions.

EPOCH		PLANKTONIC FORAMINIFERAL ZONES		FIRST APPEARANCES (F.A.) and LAST APPEARANCES (L.A.) of SPECIES and SUBSPECIES used to DELIMIT ZONES	
PLEISTOCENE		N23	*G. calida calida - S. dehiscens excavata* A-Z	*G. calida calida*	F.A.
PLEISTOCENE		N22	*G. truncatulinoides* P-R-Z	*G. truncatulinoides*	F.A.
PLIOCENE		N21	*G. tosaensis tenuitheca* cons-R-Z	*G. tosaensis tenuitheca*	F.A.
PLIOCENE		N20	*G. multicamerata - P. obliquiloculata* P-R-Z	*G. pseudopima*	F.A.
PLIOCENE		N19	*S. dehiscens - G. altispira* P-R-Z	*S. dehiscens*	F.A.
MIOCENE	LATE	N18	*G. tumida - S. subdehiscens paenedehiscens* P-R-Z	*G. tumida tumida*	F.A.
MIOCENE	LATE	N17	*G. tumida plesiotumida* cons-R-Z	*G. tumida plesiotumida*	F.A.
MIOCENE	LATE	N16	*G. acostaensis - G. merotumida* P-R-Z	*G. acostaensis*	F.A.
MIOCENE	MIDDLE	N15	*G. continuosa* R-Z	*G. siakensis*	L.A.
MIOCENE	MIDDLE	N14	*G. nepenthes - G. siakensis* cons-R-Z	*G. nepenthes*	F.A.
MIOCENE	MIDDLE	N13	*S. subdehiscens - G. druryi* P-R-Z	*S. subdehiscens*	F.A.
MIOCENE	MIDDLE	N12	*G. fohsi* P-R-Z	*G. fohsi (s.l.)*	F.A.
MIOCENE	MIDDLE	N11	*G. praefohsi* cons-R-Z	*G. praefohsi*	F.A.
MIOCENE	MIDDLE	N10	*G. peripheroacuta* cons-R-Z	*G. peripheroacuta*	F.A.
MIOCENE	MIDDLE	N9	*O. suturalis - G. peripheroronda* P-R-Z	*O. suturalis*	F.A.
MIOCENE	EARLY	N8	*G. sicanus - G. insueta* P-R-Z	*G. bisphericus*	F.A.
MIOCENE	EARLY	N7	*G. insueta - G. quadrilobatus trilobus* P-R-Z	*C. dissimilis (s.l.)*	L.A.
MIOCENE	EARLY	N6	*C. insueta - G. dissimilis* cons-R-Z	*G. insueta*	F.A.
MIOCENE	EARLY	N5	*G. dehiscens praedehiscens - G. dehiscens* P-R-Z	*G. kugleri*	L.A.
MIOCENE	EARLY	N4	*G. quadrilobatus primordius - G. kugleri* cons-R-Z	*G. primordius*	F.A.
OLIGOCENE		N3/P22			

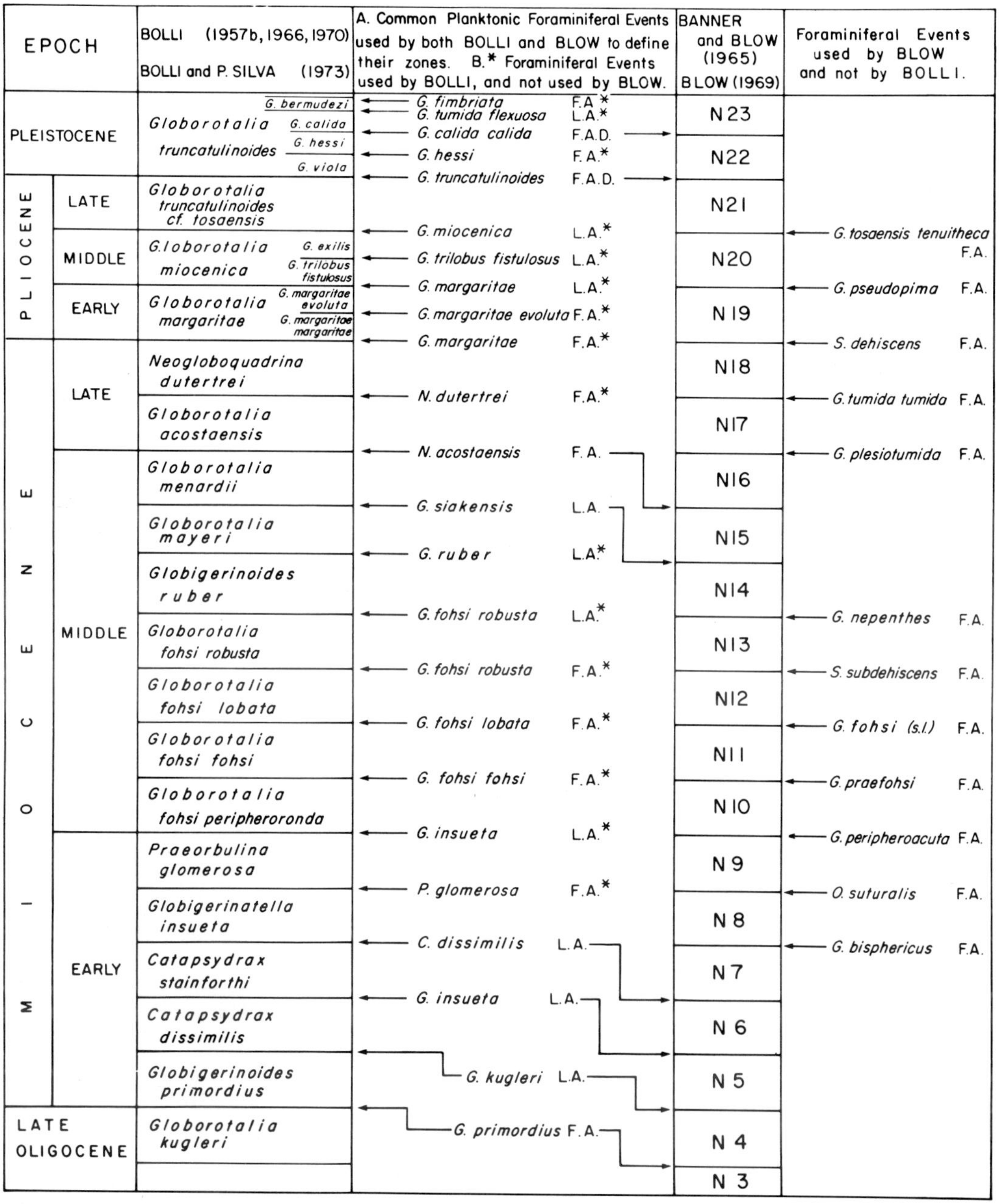

FIG. 3.—Comparison of zonal definitions between Bolli's (left) and Blow's (right) zonations. F.A. and L.A. denote first appearances and last appearances, respectively.

to use the numerical zonal scheme of Blow outside of the tropics have been incorrect and confusing. The practicality of a number of Blow's zones are further diminished by their restriction even within the warm-water sphere of the oceans. For instance, evolutionary members of the *Globoratalia fohsi*

←

FIG. 2.—Tropical Neogene planktonic foraminiferal zonation of Banner and Blow (1965) and Blow (1969) and species used for zone definitions. (A-Z = Assemblage zone; P-R-Z = partial range zone; cons.-R-Z = consecutive range zone; R-Z = range zone).

lineage and *Globorotalia merotumida–plesiotumida* are only well represented in the equatorial region (within about 15° north and south of the Equator). Even in the adjacent warm-subtropics the first appearance of these forms, if present, is often non-evolutionary. Interocean differences also exist in species distribution used in the zonal schemes. For instance, the evolution from *G. plesiotumida* to *G. tumida* has been traced in detail in the Indo-Pacific region, but apparently not in the Atlantic (Bolli, 1970; Parker, 1973, 1974; Bolli and Krasheninnikov, 1977). Its evolution has been identified in the Labrador Sea which was much warmer during the Late Miocene (Laughton, Berggren and others, 1972; Poore and Berggren, 1975). Few modifications have been made of Blow's (1969) zonation; although some workers such as Bronnimann and Resig (1971) have modified the ranges of some species, the zones themselves have remained intact. The zonal scheme is partly inapplicable in certain areas such as the eastern North Atlantic (Cape Bodjador) according to Salvatorini and Cita (1979) who proposed two new zones for the Late Miocene–Early Pliocene and emended the definition of Zone N16.

The part of Blow's (1969) zonation that has been most subject to change has been the Pliocene. Parker (1973) distinguished six zones from the Late Miocene to the Quaternary in tropical Atlantic deep-sea cores (DSDP Legs 1, 2, 4). She did not consider these to be applicable to the tropical Pacific. Berggren (1973) also developed a new Pliocene zonation for the North Atlantic based on extinction levels of *G. nepenthes, G. margaritae, S. subdehiscens, G. altispira, G. exilis, G. miocenica* and *G. obliquus extremus.* Blow's zonation is also not applicable in equatorial areas dominated by cold eastern boundary currents and upwelling such as the Panama Basin. For this reason, Kaneps (1973) developed a new zonal scheme for the Plio-Pleistocene of the Panama Basin.

Zonation Based on Solution-Resistant Forms

Planktonic foraminifera vary considerably with respect to their resistance to carbonate dissolution. Modern forms have been ranked accordingly by Berger (1967, 1970), but much less is known about the relative susceptibility to dissolution of older forms. Yet carbonate dissolution plays a major role in effecting the quality, continuity and aerial distribution of planktonic foraminiferal sequences in the world oceans. Both Bolli (1957b, 1966, 1970) and Blow (1969) did not take dissolution into consideration in their selection of zonal marker species. As a result many of the species used in the zonal designations are solution-susceptible forms. Thus, carbonate sequences deposited below the lysocline often lack the critical marker species, and zonal designations are less accurate. Jenkins and Orr (1971, 1972) recognized this problem and established a zonal scheme for the tropical regions based on solution-resistant species (Fig. 4), many of which are different from those of Bolli (1957b, 1966, 1970) and Blow (1969). In general, thickened forms are more resistant to dissolution; these include most *Globorotalia, Catapsydrax, Neogloboquadrina, Pulleniatina,* and *Sphaeroidinellopsis–Sphaeroidinella.* Solution-susceptible forms include most *Globigerina, Globigerinoides, Orbulina* and *Globigerinatella.* Evolutionary first appearances are most difficult to place stratigraphically in sequences that exhibit much dissolution, because thickened, solution-resistant morphotypes are concentrated.

The relationship between tropical zonal schemes of Bolli (1957b, 1966, 1970), Blow (1969) and Jenkins and Orr (1972) are shown in Figure 5. Additional tropical planktonic foraminiferal sequences have been described using DSDP materials by the following workers: Parker (1967) and Bronnimann and Resig (1971) for the southwestern Pacific Ocean; Hays and others (1969), Jenkins and Orr (1972), Kaneps (1973), Saito (1977), and Keller (*in press* a, b, c) for the Equatorial Pacific; Kennett (1973) for the warm-subtropical South Pacific; Parker (1973), Berggren and Amdurer (1973), Boersma (1977) and Berggren (1977a, b) for the Atlantic; and McGowran (1974), Fleisher (1974), and Vincent (1975, 1977) for the northern Indian Ocean.

Temperate Zonation

A largely different suite of planktonic foraminifera mark the temperate areas of the world. This fact has required entirely different biostratigraphic schemes than those for the tropics. The most important scheme is that of Jenkins (1966a, 1967) based on New Zealand marine sequences. Jenkins (1966a) distinguished 11 zones for the Paleogene and eight zones for the Neogene (Jenkins, 1967). In addition he distinguished nine subzones in the Pliocene and Quaternary based on switches in coiling directions of *Neogloboquadrina pachyderma* (Fig. 6). A comprehensive account of the Cenozoic zones in New Zealand and related planktonic foraminiferal taxonomy is summarized by Jenkins (1971).

Since the work of Jenkins (1971) a number of investigators have examined DSDP sequences in various temperate areas of world oceans. This has led to the extension and modification of Jenkins (1971) zones, and in some areas, the development of new zonal schemes, for example Berggren (1972b, 1977a, b), Berggren and Amdurer (1973), Poore and Berggren (1974, 1975), Kennett (1973, 1978), Kennett and Vella (1975), Jenkins (1975, 1977, 1978), and Vincent (1975).

The quality of Late Neogene DSDP cores obtained from the southwest Pacific enabled Kennett (1973) to establish a somewhat more detailed zona-

EPOCH		Planktonic Foraminiferal Zones	Definition of Zonal Boundaries (F.A.-First Appearance, L.A.-Last Appearance)	
PLEISTOCENE		*Pulleniatina obliquiloculata*	*G. fistulosus*	L.A.
PLIOCENE	Upper	*Globigerinoides fistulosus*	*G. fistulosus*	F.A.
	Lower	*Sphaeroidinella dehiscens*	*S. dehiscens*	F.A.
MIOCENE	Upper	*Globorotalia tumida*	*G. tumida*	F.A.
		Globorotalia plesiotumida	*G. plesiotumida*	F.A.
	Middle	*Globoquadrina altispira*	*G. fohsi lobata*	L.A.
		Globorotalia fohsi lobata	*G. fohsi lobata*	F.A.
		Globorotalia fohsi fohsi – Globorotalia peripheroacuta	*G. peripheroacuta*	F.A.
		Globorotalia peripheroronda	*P. glomerosa curva*	L.A.
		Praeorbulina glomerosa curva subzone	*P. glomerosa curva*	F.A.
	Lower	*Globigerinoides bispericus subzone*	*G. bisphericus*	F.A.
		Globoquadrina venezuelana	*C. dissimilis*	L.A.
		Catapsydrax dissimilis	*G. kugleri*	L.A.
		Globorotalia kugleri	*G. kugleri*	F.A.

FIG. 4.—Tropical Neogene planktonic foraminiferal zonation of Jenkins and Orr (1971, 1972) using solution-resistant species shown at right. F. A. and L.A. denote first and last appearances, respectively.

tion than that of Jenkins (1971). Ten zones were established from the late Middle Miocene to the Late Quaternary based almost entirely on species of *Globorotalia* (Fig. 7). Even Jenkins' (1971) zones are largely based on *Globorotalia,* reflecting the importance of these evolutionary lineages in temperate areas.

The zonal schemes of Jenkins (1967, 1971) and Kennett (1973) cannot be uniformly applied to areas outside the South Pacific, which are marked by differences in biogeographic provinciality and paleoceanography. For the North Atlantic area Poore and Berggren (1975) developed a modified version of Kennett's (1973) Neogene zonal scheme in which they employed eight zones; later Poore (1979) developed a somewhat different 11-fold subdivision of the Neogene for the North Atlantic region.

The Subantarctic Neogene zones are largely those of the temperate regions of Jenkins (1975), although the Late Pliocene and Quaternary zones are based on entirely different criteria (Keany and Kennett, 1972). Diversity of planktonic foraminifera has been extremely low in the Antarctic region since the Early Oligocene. A sequence of zones has been established but they are extremely broad and virtually of no value in stratigraphic subdivision and

EPOCH		BOLLI (1957b, 1966, 1970) BOLLI and P. SILVA (1973)		BLOW (1969)	JENKINS and ORR (1971, 1972)
PLEISTOCENE		*Globorotalia truncatulinoides*	*G. bermudezi*	N23	*Pulleniatina obliquiloculata*
			G. calida		
			G. hessi	N22	
			G. viola		
PLIOCENE	LATE	*Globorotalia truncatulinoides cf. tosaensis*		N21	*Globigerinoides fistulosus*
	MIDDLE	*Globorotalia miocenica*	*G. exilis*	N20	*Sphaeroidinella dehiscens*
			G. trilobus fistulosus		
	EARLY	*Globorotalia margaritae*	*G. margaritae evoluta*	N19	
			G. margaritae margaritae		
MIOCENE	LATE	*Neogloboquadrina dutertrei*		N18	*Globorotalia tumida*
				N17	*Globorotalia plesiotumida*
		Globorotalia acostaensis		N16	*Globoquadrina altispira*
	MIDDLE	*Globorotalia menardii*		N15	
		Globorotalia mayeri		N14	
		Globigerinoides ruber		N13	
		Globorotalia fohsi robusta			*Globorotalia fohsi lobata*
		Globorotalia fohsi lobata		N12	
		Globorotalia fohsi fohsi		N11	*Globorotalia fohsi fohsi*
				N10	*Globorotalia peripheroacuta*
		Globorotalia fohsi peripheroronda		N9	*Globorotalia peripheroronda*
	EARLY	*Praeorbulina glomerosa*			*Praeorbulina glomerosa curva* subzone
		Globigerinatella insueta		N8	*Globigerinoides bisphericus* subzone
				N7	*Globoquadrina venezuelana*
		Catapsydrax stainforthi		N6	*Catapsydrax dissimilis*
		Catapsydrax dissimilis		N5	
		Globigerinoides primordius		N4	*Globorotalia kugleri*
LATE OLIGOCENE		*Globorotalia kugleri*		N3	
					Globigerina angulisuturalis

correlation (Kaneps, 1975; Jenkins, 1975). Biostratigraphic correlation in the Antarctic region is largely based on radiolarians and diatoms.

Middle-Latitude Zonation

The first Neogene planktonic foraminiferal zonal scheme for the areas transitional between the temperate and tropical regions was developed by Kennett (1973) based on DSDP cores in the southwest Pacific (Fig. 7). Thirteen Neogene zones were established. The warm-subtropical zones represent a combination of tropical and temperate zonal schemes. A zonation based on tropical species is applicable in the Early Neogene until the early Middle Miocene, after which temperate species are more valuable for biostratigraphic subdivision. This reflects global cooling related to the major development of Antarctic ice formation beginning in the Middle Miocene, concommitant with latitudinal shift of cool waters towards the Equator. Tropical elements continue to be persistent through the Late Neogene, but are less common. The combination of temperate and tropical elements in these intermediate latitudinal sites is assisting in our correlation between tropical and temperate sequences. Very few species seem to be restricted to the transitional areas, indicating that almost all the evolution is occurring in either the true temperate or tropical areas. The only other zonal scheme established for the transitional latitudes is that of Berggren (1977b) for the South Atlantic, based on the changing sequence of *Globorotalia.* Three zones were established for the Late Miocene and five for the Early Pliocene.

MAJOR NEOGENE LINEAGES

Planktonic foraminifera have evolved rapidly during the Cretaceous and Cenozoic. During the Cenozoic there have been two major radiations, one occurring during the Eocene, one during the Miocene. These two radiations are separated by a period of at least 15 m.y., during the Oligocene, when the number of evolutionary lineages was severely reduced (Cifelli, 1969). Similar morphotypes evolved from distinctly different lineages during the Paleogene and the Neogene radiations. The particular paleoceanographic conditions that led to this repetitive evolution is not clearly understood, although both of these radiations coincide with globally warm intervals (Shackleton and Kennett, 1975; Savin, 1977).

We are concerned here only in summarizing the major evolutionary lineages of planktonic foraminifera that occurred during the Neogene. These now provide the fundamental basis for the biostratigraphic subdivision of the Neogene sequences (Wade, 1964). The evolution in Neogene planktonic foraminifera is now quite well-understood and has resulted from the efforts of a large number of investigators. However, almost all our knowledge about the evolution in the tropical regions stems from the work of Bolli (1950, 1957a, 1967), Blow (1956), Blow and Banner (1966), and Banner and Blow (1967) who unravelled the evolution in such important lineages as the *Globorotalia fohsi* group, *Orbulina, Pulleniatina, Sphaeroidinella* and the *Globoquadrina dehiscens* group. Important additional understanding resulted from the work of Bandy and others (1967), Bandy (1972), Parker (1967), Berggren and others (1967), and Lamb and Beard (1972).

Our knowledge of evolution in the temperate planktonic foraminifera has stemmed largely from the investigations of Jenkins (1965, 1966b, 1971). He unravelled phylogenies of the *Globigerina woodi* group, the ancestry of *Globigerinoides trilobus* via *G. woodi,* the *Globorotalia mayeri* lineage, *Globorotalia praescitula–G. miozea* group, and the ancestry of *Neogloboquadrina pachyderma* via *Globorotalia continuosa.* Walters (1965) further provided basic understanding of the evolution of the *Globorotalia zealandica incognita–G. zealandica pseudomiozea* group. The important Late Cenozoic lineage of *Globorotalia conoidea* through *G. puncticulata* to *G. inflata* was largely unravelled by Kennett (1966, 1973) and Kennett and Vella (1975). Further research on this group was conducted by Berggren (1977b) for the South Atlantic region.

Much of the classical work that has provided our basic knowledge of foraminiferal evolution was carried out on marine sequences on land. Nevertheless, these sequences have not been optimal for examining the actual gradations that occur from one form to the next. Ancestor-descendent relationships were often inferred from the broader morphological changes exhibited in sequences of known stratigraphic order. The major contribution that the Deep Sea Drilling Project has made has been in providing relatively continuous sequences containing abundant foraminifera enabling direct observation of evolutionary gradations from one form to the next. In most cases this has not changed our concept of evolutionary succession (Salvatorini and Cita, 1979), but in a number of cases new information has been provided about lineages. For example Cita (1973) described details about the evolution of *Globorotalia margaritae* group in the Mediterranean Pliocene sequence. Likewise, Kennett (1973) and Kennett and Vella (1975) described the evolution of the *Globorotalia puncticulata–G. inflata* group and Srinivasan and Kennett (1976) the Late

FIG. 5.—Correlation between tropical Neogene planktonic foraminiferal zonal schemes of Bolli (1957b, 1966, 1970), Blow (1969) and Jenkins and Orr (1971, 1972).

EPOCH	Planktonic Foraminiferal Zones and Subzones D-Dextral; S-Sinistral		First Appearances (F.A.) and Last Appearances (L.A.) of species and subspecies used to delimit zones
RECENT	*Globorotalia (T.) inflata*	D	Nine subzones based on the sinistral (S) and dextral (D) coiling changes of *Neogloboquadrina pachyderma*
PLEISTOCENE		D_5	
		S_4	
		D_4	
		S_3	
		D_3	
		S_2	
PLIOCENE		D_2	
		S_1	
		D_1	
			G. inflata F.A.
MIOCENE	*Globorotalia (G.) miozea sphericomiozea*		
			G. miozea sphericomiozea F.A.
	Globorotalia (G.) miotumida miotumida		
			G. mayeri mayeri L.A.
	Globorotalia (T.) mayeri mayeri		
			G. mayeri mayeri F.A.
	Orbulina suturalis		
			O. suturalis F.A.
	Praeorbulina glomerosa curva		
			P. glomerosa curva F.A.
	Globigerinoides trilobus trilobus		
			G. trilobus F.A.
	Globigerina (G.) woodi connecta		
			G. woodi connecta F.A.
	Globigerina (G.) woodi woodi		
			G. woodi woodi F.A.
	Globoquadrina dehiscens		
			G. dehiscens F.A.
OLIGOCENE	*Globigerina (G.) euapertura*		

FIG. 6.—Temperate Neogene planktonic foraminiferal zonation (zones and subzones) of Jenkins (1967, 1971).

Age	DSDP Leg 21, (Kennett, 1973) Warm Subtropical Planktonic Foraminiferal Zones	DSDP Leg 21, (Kennett, 1973) Cool Subtropical Planktonic Foraminiferal Zones	New Zealand Zones (Jenkins, 1971)	New Zealand Stages
Pleistocene	*Globorotalia truncatulinoides*	*Globorotalia truncatulinoides*		
	Globorotalia truncatulinoides Globorotalia tosaensis overlap	*Globorotalia truncatulinoides Globorotalia tosaensis* overlap		
Late Pliocene	*Globorotalia tosaensis*	*Globorotalia tosaensis*	*Globorotalia inflata*	*Opoitian*
	Globorotalia inflata	*Globorotalia inflata*		
	Globorotalia crassaformis	*Globorotalia crassaformis*		
Early Pliocene	*Globorotalia puncticulata*	*Globorotalia puncticulata*		
	Globorotalia margaritae			
Late Miocene	*Globorotalia conomiozea*	*Globorotalia conomiozea*	*Globorotalia miozea sphericomiozea*	*Kapitean*
	Globigerina nepenthes	*Globigerina nepenthes*	*Globorotalia miotumida miotumida*	*Tonga-porutuan*
	Globorotalia continuosa	*Globorotalia continuosa*		
Middle Miocene	*Globorotalia mayeri*	*Globorotalia mayeri*	*Globorotalia mayeri mayeri*	*Waiauan – Lillburnian*
	Orbulina suturalis		*Orbulina suturalis*	
Early Miocene	*Globigerinoides trilobus*			

FIG. 7.—Correlation of Neogene warm-subtropical and cool-subtropical planktonic foraminiferal zonation of Kennett (1973) with temperate zonal scheme established by Jenkins (1967; 1971) for New Zealand and New Zealand stages (after Kennett, 1973).

Miocene through Quaternary *Neogloboquadrina* plexus in the southwest Pacific.

It is clear from the large number of investigations on planktonic foraminifera at all latitudes that almost all of the evolution is occurring in two rather well-defined biogeographic provinces; the tropical-equatorial water mass and the temperate (cool-subtropical) water mass. Although the intervening warm-subtropical sequences contain many of the tropical and temperate elements, there is little evidence to indicate that much evolution is taking place in this region. The central seats of planktonic foraminiferal evolution are well established in the tropical and temperate areas. Exceptions exist of course. For instance, we have observed the evolution of *Globorotalia peripheroronda* from *G. kugleri* in warm-subtropical Site 208 (southwest Pacific), but not in adjacent tropical sites.

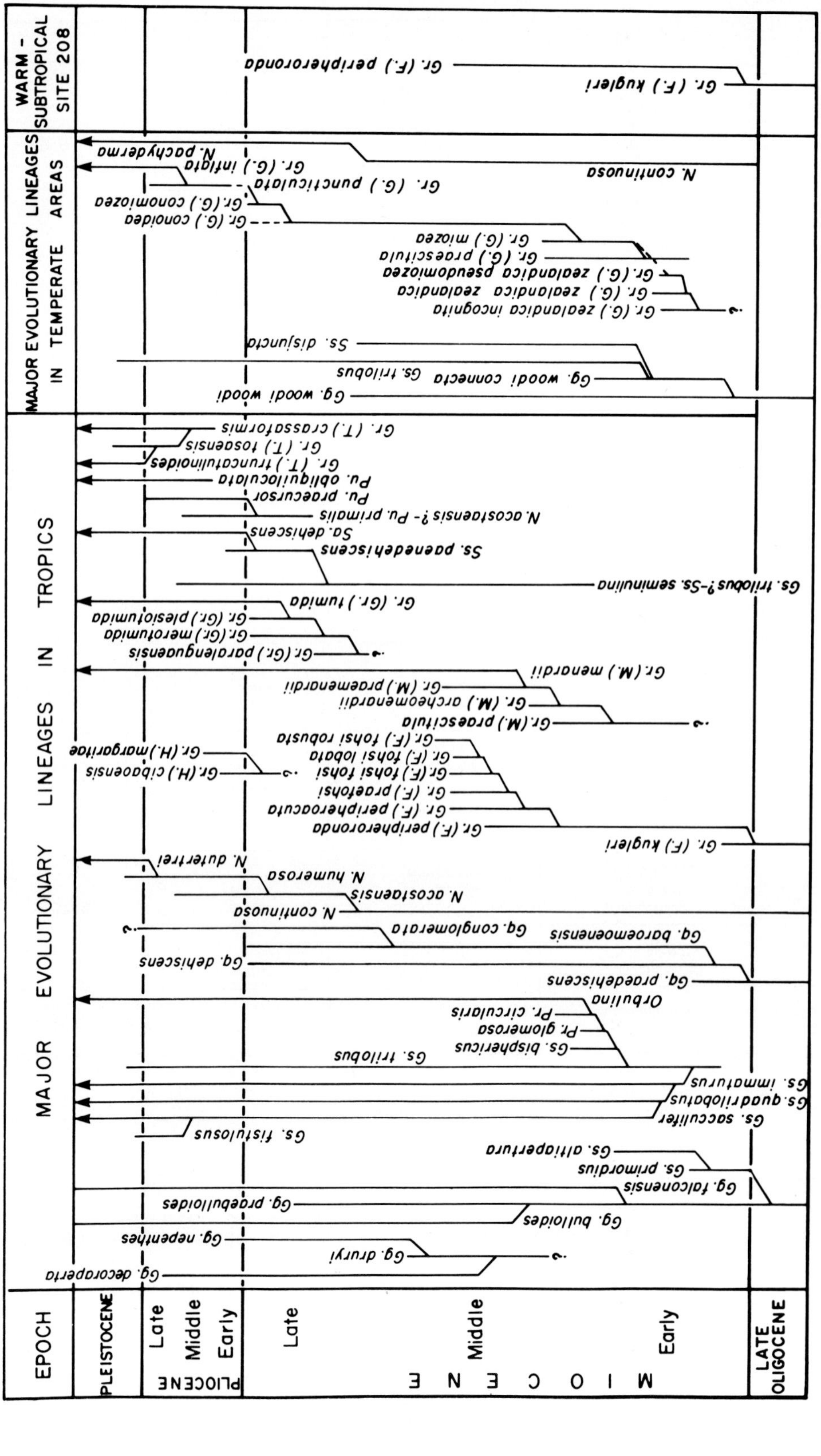

FIG. 8.—Major Neogene planktonic foraminiferal lineages for tropical to temperate areas. (Gg. = *Globigerina*; Gs. = *Globigerinoides*; Gr. = *Globorotalia*; Gq. = *Globoquadrina*; N. = *Neogloboquadrina*; Pr. = *Praeorbulina*; Pu. = *Pulleniatina*; Ss. = *Sphaeroidinellopsis*; Sa. = *Sphaeroidinella*; F. = *Fohsella*; H. = *Hirsutella*; M. = *Menardella*; T. = *Truncorotalia*; and G. = *Globoconella*.)

No examples of planktonic foraminiferal evolution have been described from Antarctic and southern Subantarctic water masses (Kennett, 1978). All species evolved in the temperate water masses to the north and later migrated into the high-latitude areas (Kennett, 1970; Keany and Kennett, 1972). Clearly the Antarctic environment does not sponsor planktonic foraminiferal evolution.

Tropical Lineages

Eleven major phylogenies mark the tropical Neogene planktonic foraminiferal sequence. These are listed below for convenience and are also shown in Figure 8. The age ranges of the individual species that make up each of the lineages is shown in Figure 8.

1. *Globigerina* (Late Oligocene to Middle Miocene):
   a) *G. praebulloides* → *G. bulloides* → *G. falconensis.*
   b) *G. druryi* < *G. decoraperta* / *G. nepenthes*
2. *Globigerinoides* (Early Miocene–Late Pliocene):
   The possible ancestor of *Globigerinoides primordius* appears to be *Globigerina praebulloides* (Blow, 1969). *Globigerinoides trilobus trilobus* evolved from *Globigerina woodi connecta* in cool-subtropical latitudes, then migrated to the tropical regions (Jenkins, 1965).
   a) *Globigerina praebulloides* → *Globigerinoides primordius* → *G. altiapertura.*
   b) *G. woodi connecta* → *G. trilobus trilobus* → *G. immaturus* → *G. quadrilobatus* → *G. sacculifer* → *G. fistulosus.*
3. *Praeorbulina–Orbulina* (late Early Miocene to early Middle Miocene):
   G. bisphericus < *Praeorbulina glomerosa* → *P. circularis* → *O. suturalis* / *P. transitoria* → *O. bilobata*
4. *Globoquadrina* (Late Oligocene to Middle Miocene):
   G. praedehiscens → *G. dehiscens* → *G. baroemoenensis* → *G. conglomerata.*
5. *Neogloboquadrina* (Late Miocene to Quaternary):
   Srinivasan and Kennett (1976) recognized two evolutionary lineages within *Neogloboquadrina,* both having been originally derived from *Globorotalia continuosa.* In tropical areas *G. continuosa* gave rise to *N. acostaensis* which developed into *N. humerosa* and in turn to *N. dutertrei* Group A (microcrystalline ultrastructure). In subtropical areas, *N. pachyderma* developed into *N. dutertrei subcretacea* which developed into *N. dutertrei dutertrei* Group B (crystalline ultrastructure).
6. *Sphaeroidinella* (Early Miocene to Late Pliocene):
   S. seminulina → *S. paenedehiscens* → *S. dehiscens immatura* → *S. dehiscens excavata.*
7. *Pulleniatina* (Late Miocene to Middle Pliocene):
   P. primalis → *P. praecursor* → *P. obliquiloculata.*
8. *Globorotalia fohsi* group (Early to Middle Miocene):
   G. peripheroronda → *G. peripheroacuta* → *G. praefohsi* → *G. fohsi fohsi* → *G. fohsi lobata* → *G. fohsi robusta.*
9. *Globorotalia menardii* group (Early to Middle Miocene):
   G. praescitula → *G. archeomenardii* → *G. praemenardii* → *G. menardii.*
10. *Globoratalia tumida* group (late Middle Miocene to Pliocene):
    G. paralenguaensis → *G. merotumida* → *G. plesiotumida* → *G. tumida.*
11. *Globorotalia truncatulinoides* group (Middle Pliocene to Quaternary).
    G. crassaformis → *G. tosaensis* → *G. truncatulinoides.*

Temperate Lineages

Five major phylogenies mark the temperate Neogene planktonic foraminiferal sequence. Species ranges are shown in Figure 8.

1. *Globigerina* (Early Miocene):
   G. woodi woodi → *G. woodi connecta* → *S. disjuncta.*
2. *Globigerinoides* (Early Miocene):
   G. woodi connecta → *G. trilobus trilobus.*
3. *Globorotalia zealandica* group (Early Miocene):
   G. zealandica incognita → *G. zealandica zealandica* → *G. zealandica pseudomiozea*
4. *Globorotalia miozea–inflata* group (Early Miocene to Late Pliocene):
   G. praescitula → *G. miozea* → *G. conoidea* → *G. conomiozea* → *G. puncticulata* → *G. inflata.*
5. *Neogloboquadrina* (Late Miocene):
   N. continuosa → *N. pachyderma*

Middle Latitude Lineage

One lineage is here reported from middle latitudes.

1. *Globorotalia kugleri* to *G. peripheroronda* in the Early Miocene.

We have observed morphological gradation in the lineage between *G. kugleri* and *G. peripheroronda* in the early part of the Early Miocene in DSDP Site 208. This is considered to represent an evolutionary gradation, not a surprising observation since Fleisher (1974) and Stainforth and others (1975) both suggested that from morphological and stratigraphic considerations *G. kugleri* should be the ancestor of *G. peripheroronda.* Ours is the first reported gradation between these two forms. The upward evolutionary transition from *G. kugleri* to *G. peripheroronda* involves a reduction from 6–7 chambers in the final-whorl to 5–6 chambers, a gradual loss in the umbilical pit (belly-button) in *G. kugleri,* and the development of a flange-like lip in *G. peripheroronda.* The upward gradation occurs within a 6 m interval in Core 24 (24 cc to 24-2-50 cm) in Site 208. This occurs within the *G. primordius* Zone (Zone N4) of Blow (1969).

In the Caribbean region, no evolutionary gradation has been observed between *G. kugleri* and *G. peripheronda.* Bolli (1957b) showed a slight overlap between their ranges but no gradation. Bolli and Premoli Silva (1973) showed a short gap between the ranges of these two forms in DSDP Site 149 in the Caribbean and no apparent gradation. It is likely therefore that the evolution between *G. kugleri* and *G. peripheronda* did not take place in the Caribbean region even though, during the Early Miocene, the equatorial Pacific and Atlantic Oceans were in communication.

We have observed morphological gradation in the lineage between *G. acrostoma* and *G. mayeri* in the Early Miocene of Site 208. At the base of Zone N4 (Core 25; 430 m) in Site 208 a gradation exists between morphotypes referable to *G. acrostoma* and morphotypes referable to *G. mayeri.* Over the next 30 m upward through the sequence, the *G. acrostoma* form diminishes in importance in relation to the *G. mayeri* form. This involves a reduction in the forms exhibiting 4 to 4.5 chambers in the final whorl (*G. acrostoma* morphotype) to a complete dominance of forms with 5 chambers (*G. mayeri* morphotype). A reduction also occurs in apertural size from a wide open, arched aperture to a narrowly-arched, comma-shaped aperture in *G. mayeri.* The upward gradation, therefore, represents a change from samples containing intergrading morphotypes of *G. acrostoma* and *G. mayeri* to samples containing only the *G. mayeri* morphotype. Poore (1979) commented on the close morphological similarity between *G. acrostoma* and *G. mayeri* and treated *G. acrostoma* as a subspecies of *G. mayeri.* The fact that this gradation was reported by Poore (1979) in North Atlantic DSDP sites of Middle Miocene age represents a biostratigraphic problem. Perhaps the two forms represent ecophenotypes or subspecies that inhabited different water masses in the Early Miocene, and gradations involving two forms may represent surface-water paleoceanographic oscillations rather than evolution. Jenkins (1966a) suggested the ancestry of *G. mayeri* to be *G. peripheroronda* in New Zealand Middle Miocene sequences. Since *G. mayeri* occurs much earlier in warm-subtropical and tropical areas, it clearly has different ancestry than that suggested by Jenkins (1966a).

The Lineage Concept in Globorotaliid Taxonomy

The Deep Sea Drilling Project has provided sequences needed for detailed evolutionary studies of the planktonic foraminifera. An excellent understanding now exists about the phylogeny of nearly all of the major groups of *Globorotalia* through the Neogene. It is clear that several groups of forms referred to as *Globorotalia* have distinct ancestry. From a phylogenetic approach, the designation of these different forms to *Globorotalia* is therefore somewhat artificial. Because of this Bandy (1972) and later Fleisher (1974) proposed a suite of new subgenera to represent the different major globorotaloid lineages through the Neogene. These taxa are based solely on phylogenetic considerations. Any particular lineage may contain a wide range of morphological forms, including keeled or non-keeled forms, forms with sharp or rounded peripheries, inflated or compressed forms and forms with relatively large numbers of chambers in the final whorl to few chambers. Therefore, the taxa cannot be easily defined using specific morphological characteristics as in traditional taxonomic classification. Nevertheless, the gradations exhibited within the individual lineages represent a morphological continuum from ancestor to descendent. The formal taxonomic designation of these lineages as subgenera is important for defining their evolutionary relationships. The subdivisions are of no value for enhancing biostratigraphic resolution, but this is not their purpose. The purpose of the subgenera is to identify phylogenetic differences between different groups of species. This classification is, therefore, natural and in accordance with the objectives of taxonomic classification. Bandy (1972) proposed four phylogenetically based Neogene subgenera as follows: *Fohsella, Menardella, Globorotalia* and *Hirsutella.* Fleisher (1974) added another taxon *Tenuitella.* Later Bandy (1975) erected a new subgenus *Globoconella* for forms related to the *Globorotalia miozea* lineage. The taxonomic content of these phylogenetic subdivisions are summarized as follows; in addition we have redefined the subgenus *Globoconella* based on phylogenetic considerations:

Genus *Globorotalia* Cushman, 1927

1. Subgenus *Fohsella* Bandy 1972
   Type species *Globorotalia praefoshi* Blow
   This lineage includes *Globorotalia (Fohsella) kugleri* → *G. (F.) peripheroronda* → *G. (F.) peripheroacuta* → *G. (F.) praefohsi* → *G. (F.) fohsi fohsi* → *G. (F.) fohsi lobata* → *G. (F.) fohsi robusta.*

2. Subgenus *Menardella* Bandy, 1972
   Type species *Globorotalia menardii* (d'Orbigny)
   This lineage includes *Globorotalia* (Menardella) *praescitula* → *G. (M.) archeomenardii* → *G. (M.) praemenardii* → *G. (M.) menardii.*

3. Subgenus *Globorotalia* Bandy, 1972
   Type species *Globorotalia tumida* (Brady)
   This lineage includes *Globorotalia (Globorotalia) lenguaensis* → *G. (G.) paralenguaensis* → *G. (G.)*

merotumida → *G. (G.) plesiotumida* → *G. (G.) tumida*.

4. Subgenus *Hirsutella* Bandy, 1972
Type species *Globorotalia hirsuta* (d'Orbigny)
This lineage includes *Globorotalia (Hirsutella) scitula* → *G. (H.) cibaoensis* → *G. (H.) margaritae* → *G. (H.) praehirsuta* → *G. (H.) hirsuta* series.

5. Subgenus *Tenuitella* Fleisher, 1974
Type species *Globorotalia gemma* Jenkins
This lineage includes *Globorotalia (Tenuitella) gemma, G. (T.) cifelli, G. (T.) browni, G. (T.) munda, G. (T.) clemenciae, G. (T.) minutissima, G. (T.) anfracta,* and *G. (T.) parkerae* series.

6. Subgenus *Truncorotalia* Cushman and Bermudez, 1949
Type species *Rotalina truncatulinoides* d'Orbigny
This lineage includes *Globorotalia (Truncorotalia) crassaformis* → *G. (T.) tosaensis* → *G. (T.) truncatulinoides*.

7. Subgenus *Globoconella* Bandy, 1975 (Redefined)
Type species *Globorotalia conomiozea* Kennett
This lineage includes the phylogeny of species from *Globorotalia (Globoconella) zealandica incognita* → *G. (G.) zealandica zealandica* → *G. (G.) pseudomiozea* → *G. (G.) miozea* → *G. (G.) conoidea* → *G. (G.) conomiozea* → *G. (G.) sphericomiozea* → *G. (G.) puncticulata* → *G. (G.) inflata* series (Fig. 8). *G. (G.) zealandica incognita* the earliest member of this lineage evolves in the Early Miocene from Oligocene globorotalid stock. Jenkins (1971) has suggested its ancestry to be from *G. continuosa.*

The *Globoconella* series is the major lineage in temperate areas and the most useful for biostratigraphic subdivision of the Neogene temperate sequences. The continuum of forms in this lineage include both keeled and non-keeled forms, inflated and compressed forms. Most of the elements are thick-walled and thus dissolution-resistant. In some cases thin-walled forms have been designated as different species from thick-walled forms. For instance, we consider *Globorotalia (Globoconella) miotumida* to be thin-walled forms of *G. (G.) conoidea*.

The above subgenera of *Globorotalia* encompass most of the Neogene globorotalids.

Bandy (1972) assigned *Globorotalia continuosa* Blow to the subgenus *Turborotalia* but, curiously at the same time, assigned *Turborotalia* to be restricted to the Eocene evolutionary lineage from *Globorotalia centralis* to *G. cerroazulensis*. We restrict *Turborotalia* to Paleogene lineages. *G. continuosa* is an important Neogene form which gave rise to the *Neogloboquadrina* plexus in the Late Miocene. From a phylogenetic point of view Berggren (1977a, b) considered *G. continuosa* to be *Neogloboquadrina*. We concur with this designation.

Of the Neogene *Globorotalia* phylogenetic subgenera, only one—*Fohsella*—has become extinct; the last species within this lineage was *Globorotalia (Fohsella) fohsi robusta* which disappeared in the Middle Miocene at 12.4 Ma. All of the other subgenera are represented by modern forms. Like Bandy (1972, 1975) and Fleisher (1974) we consider the use of such taxonomic subdivision to be extremely valuable because they reflect the true phylogenies of the planktonic foraminifera. The primary objective of taxonomic subdivision is to reflect phylogenic relations, and these subdivisions meet with this objective. Ultimately, the specific and subspecific subdivisions within these lineages will be accurately defined by statistical characteristics of the populations. A new numerical system will eventually evolve from our present taxonomically based system, but this will require enormous amounts of biometric work. Similar separate lineages occur in other major planktonic foraminiferal groups, which also require phylogenetic recognition. The genus *Globigerina* contains a number of quite separately evolving groups through the Neogene. The lumping of all these forms within one genus is artificial. Several important groups can already be phylogenetically subdivided using ultrastructural characteristics.

RELATIONSHIP BETWEEN ZONAL BOUNDARIES AND THE EVOLUTIONARY LINEAGES

The zonal schemes used in both temperate and tropical areas are largely based on the evolutionary lineages within the planktonic foraminifera. Changes in a relatively small number of lineages define most of the zones in both temperate and tropical areas.

In low-latitude areas, the Early Miocene zones are based largely on the evolutionary sequence from *Globigerinoides* to *Orbulina (G. trilobus* → *G. bisphericus* → *Praeorbulina* → *O. suturalis)*. The Middle Miocene is largely based on evolution within *Fohsella*. Late Miocene zones are largely based on evolution within the subgenus *Globorotalia* {*G. (G.) merotumida* → *G. (G.) plesiotumida* → *G. (G.) tumida*} and *Neogloboquadrina (N. acostaensis* → *N. humerosa)*. The Pliocene zones are based on evolution within *Hirsutella* especially *G. (H.) margaritae*, and the Quaternary on evolution within *Truncorotalia* (i.e. the evolution from *G. (T.) crassaformis* → *G. (T.) tosaensis* → *G. (T.) truncatulinoides)*.

In temperate areas the Early to Middle Miocene zones are largely based on evolution within *Globigerina woodi* → *Globigerinoides trilobus* → *Orbulina suturalis* → and *Globorotalia mayeri*. Late Miocene zones are based on evolution within the *Neogloboquadrina (N. continuosa* → *N. pachyderma)* and *Globoconella* lineages. Subdivision using the *Globoconella* lineage continues through the Pliocene. As in tropical areas, the Quaternary is defined by evolution within *Truncorotalia* (i.e. *G. (T.) tosaensis* → *truncatulinoides)*.

The major evolutionary lineages in the tropical and temperate areas are sufficiently different to require maintenance of separate zonal schemes. Attempts to apply tropical N-Zones in temperate regions, although indirectly correlated, have met only with limited success, and are unsatisfactory. Certain evolutionary lineages do occur in both temperate and tropical areas allowing direct correlation. The useful lineages common to both regions include *G. trilobus* to *O. suturalis* and *G. (T.) tosaensis* to *G. (T.) truncatulinoides.*

The tropical lineages *Fohsella, Menardella, Globorotalia (G.) merotumida* to *G. (G.) tumida* and *Hirsutella* are largely absent in temperate regions and the *Globoconella* lineage is largely absent in the tropics. Srinivasan and Kennett (1976) discovered that certain lineages such as *Neogloboquadrina* that occur as different forms in tropical and temperate areas are linked by intermediate forms in transitional waters. Both *Neogloboquadrina acostaensis* and *Neogloboquadrina pachyderma* evolved from *Neogloboquadrina continuosa* in tropical and temperate areas respectively. In tropical areas, *N. acostaensis* gave rise to *N. dutertrei humerosa* which in turn developed into *N. dutertrei dutertrei* Group A (microcrystalline wall with pore pits and distinct apertural plate). In temperate areas, *N. pachyderma* evolved into *N. dutertrei subcretacea* which developed into *N. dutertrei dutertrei* Group B. (rosette pattern surface with no apertural plates). Throughout the development from *N. pachyderma* to *N. dutertrei dutertrei* Group B, the surface ultrastructure has the characteristic rosette pattern. On the other hand, throughout the development from *N. acostaensis* to *N. dutertrei dutertrei* Group A the surface ultrastructure has the characteristic microcrystalline surface with circular pores. Integration through forms with intermediate surface ultrastructure suggest that the two groups are phenotypic variants of one taxon. Therefore evolution in this plexus proceeds in similar fashion in both tropical and temperate regions but at any particular time during this evolution, temperate and tropical elements are represented by slightly different morphotypes. Such evolution over broad areas involving either different geographic subspecies or ecophenotypes may occur in other groups of planktonic foraminifera. For instance, during the Middle Miocene, there may be morphotypic gradation between the *G. (F.) peripheroronda-G. (F.) peripheroacuta* group in temperate areas and the *G. (F.) fohsi fohsi* group in the tropics. Specifically, in the tropics *G. (F.) peripheroronda* evolves into *G. (F.) fohsi fohsi* with the addition of a keel, while in the temperate areas, *G. (F.) fohsi fohsi* is absent. Instead *G. (F.) peripheroronda* seems to grade upwards into a non-keeled form similar to *Globorotalia mayeri.* This lineage is poorly defined and needs careful investigation.

PLANKTONIC FORAMINIFERAL DATUMS

The datum concept, based on abrupt evolutionary appearances or last appearances of planktonic foraminifera, have been widely applied in correlating Neogene deep sea sequences. Datums represent correlation planes joining levels in sediment sequences which on paleontological or other grounds appear to be isochronous (Hornibrook, 1966). The concept was well-established before the beginning of the Deep Sea Drilling Project (Allan, 1956; Hornibrook, 1966; Jenkins, 1966c). In a practical sense, datums represent conspicuous paleontological events which are considered to be correlatable and isochronous over reasonable distances. However, the distances vary considerably over which individual datum levels can be correlated, ranging from within individual water-masses to those applied in correlation between tropical and temperate areas. Datums, therefore, differ from zonal boundaries, although many are based on the same paleontological event.

The ultimate objective of stratigraphy is to place rocks and geological events within a radiometric time-framework. Quantitative stratigraphy enables rates of change to be established and in turn provides a better frame of reference for understanding geological processes. The planktonic foraminiferal datums and zones have been integrated into a chronological framework largely by Berggren (1969, 1971, 1972a), and Berggren and Van Couvering (1974, 1978). This has been largely accomplished by intercalibration of the available radiometric dates with the marine fossil sequence and by the direct application of paleomagnetic stratigraphy. Such dating of paleontological events has become known as biochronology (Berggren and Van Couvering, 1974, 1978). In addition to Berggren's work, other contributions which deal with the dating of planktonic foraminiferal datum levels and zones include Glass and others (1967), Hays and others (1969), Ikebe and others (1972), Saito (1972, 1977) and Ryan and others (1974). Berggren and Van Couvering (1974) provided a sequence of 13 radiometrically dated Neogene planktonic foraminiferal datums, whereas Saito (1977) increased the number of radiometrically dated datums to 37 in the tropical Pacific Neogene. Recently, Keller (1978 and *in press* a, b, c) identified a sequence of 45 datum levels for the Neogene equatorial and northern subtropical Pacific (DSDP Sites 77B, 71, 55, 319, 296, 173 and 310).

For the temperate region, Hornibrook and Edwards (1971) established 34 planktonic foraminiferal datum levels for the Neogene of New Zealand. These still require integration into the Neogene time-scale. Three orders of reliability were established for the set of datums based on ease of identification, relative abundance, geographic distribution and understanding of phylogenetic relations. Hornibrook and Edwards (1971) were also among the first workers to integrate datum levels based on two distinct paleontological groups (planktonic foraminifera and calcareous nannofossils). Since then, the Deep Sea Drilling Project has provided an important basis for the integration of different microfossil groups.

Although most datums are based on faunal evolution or extinctions, another type of datum level employed by some workers is coiling direction change in planktonic foraminifera (Bolli, 1950; Ericson, 1953; Ericson and Wollin, 1956; Saito, 1976). Examples of the use of planktonic foraminiferal coiling direction as datum levels include *Globorotalia menardii, Pulleniatina* and *Neogloboquadrina pachyderma* (Saito, 1976; Thompson and Sciarillo, 1978). Coiling direction changes in planktonic foraminifera have also been employed extensively for general correlation and for subzonal division (Bandy, 1960; Ingle, 1967; Kennett, 1968; Jenkins, 1971; Kennett and Vella, 1975).

DURATION OF THE NEOGENE ZONES

The duration of the Neogene has been estimated to be 23.5 Ma (Van Couvering and Berggren, 1977) and 22.5 Ma by Saito (1977). A total of 19 planktonic foraminiferal zones have been established for the tropical regions (Zone N4 to N22). On average this provides a stratigraphic resolution of just over 1 m.y. per zone. Experience shows that this resolution seems to have reached its practical limit. This practical limit is largely constrained by the evolution of important new species within distinctive and useful lineages. Further subdivision of the existing zones is of course possible when additional criteria are employed, but further subdivision into zones of shorter time-intervals does not guarantee a practical scheme for biostratigraphic subdivision; that is, such zones may not be widely applicable. One clear exception is the needed subdivision of Zone N17 within the Late Miocene. This zone is considered to have a duration of 2.7 m.y. (Saito, 1977), and is thus the longest zone in the Neogene. Zone N17 can easily be subdivided into two zones, since *Pulleniatina* first appears near its middle part (Banner and Blow, 1967; Saito in Andrews, Packham and others, 1975; Srinivasan, 1977).

We also have observed this event in DSDP Site 289 and 208. The evolution of *Pulleniatina* is from *Neogloboquadrina acostaensis* (Banner and Blow, 1967) and dated at 6.2 m.y. (Saito, 1977). The first appearance of *Pulleniatina* is well-recorded in deep-sea sequences in tropical and warm-subtropical areas. Because this form is solution-resistant its first appearance is also well represented in deeper oceanic sequences below the lysocline. The required subdivision of N17 into two new zones raises a major problem of using a numerically based zonal scheme such as that of Banner and Blow (1965) and Blow (1969). Continued use of the established numerical zonal framework requires that Zone N17 be subdivided into two new zones designated as Zone N17A and Zone N17B. The definition of these zones are as follows:

Zone N17A: *Globorotalia (G.) plesiotumida* consecutive range zone of Blow (1969), here emended.

DEFINITION:

Top: Initial evolutionary appearance of *Pulleniatina primalis.*

Base: Initial evolutionary appearance of *Globorotalia (G.) plesiotumida* from its immediate ancestor *G. (G.) merotumida.*

Age: Late Miocene

Reference Section: Site 289: 0°29.92′S, 158°30.69′E.

Occurrence of Zone: Continuously cored at Site 289 (25-4-41 cm to 21-2-53 cm).

Zone N17B: *Pulleniatina primalis–Globorotalia (G.) plesiotumida* partial-range zone, here proposed.

DEFINITION:

Top: Initial evolutionary appearance of *Globorotalia (G.) tumida* from its immediate ancestor *G. (G.) plesiotumida*

Base: First evolutionary appearance of *Pulleniatina primalis*

The zone, therefore, is characterized by that part of the concurrence of the ranges of *Pulleniatina primalis* and *G. (G.) plesiotumida* which occurs before *G. (G.) tumida*

Age: Late Miocene

Reference Section: Site 289: 0°29.92′S, 158°30.69′E

Occurrence of Zone: Continuously cored at Site 289 (21-2-53 cm to 18-4-104 cm).

The duration of each of the tropical Neogene zones is indicated in Figure 9. The duration of the zones is taken from Saito (1977). The duration of the zones varies considerably from 0.2 m.y. in the case of Zone N18 to as much as 2.7 m.y. for Zone N17 (as undivided). Only two zones have an estimated duration of less than 0.5 m.y. (Zones N13 and N18). Only two zones are longer than 2 m.y. (Zones N16 and N17). Therefore, the duration of 15 of the 19 zones is between 0.5 and 2.0 m.y.

Zones are not of similar duration throughout the Neogene, but exhibit groupings related to the amount of evolution taking place within the planktonic foraminifera. The Early Miocene has been divided into five planktonic foraminiferal zones (Zone N4 to N8) spanning 6.5 m.y. giving an average zone duration of 1.3 m.y. The Middle Miocene consists of seven zones (Zone N9 to N15) spanning 6 m.y. with an average of 0.85 m.y. per zone. The Late Miocene is divided into three zones (N16

CHRONOLOGY AFTER BERGGREN and VAN COUVERING, 1974; VAN COUVERING and BERGGREN, 1977			N-ZONES	DURATION of ZONES in M.Y.	CHRONOLOGY after SAITO, 1977		
			23				
			22	1.95			
G. truncatulinoides	F.A.	1.85 →			← 1.95	*G. truncatulinoides*	F.A.
			21	1.15			
G. altispira/S. seminulina	L.A.	3.0 →			← 3.1	*G. tosaensis*	F.A.
			20	0.6			
G. nepenthes	L.A.	3.7 →					
			19	1.1			
S. dehiscens	F.A.	5.0 →			← 4.8	*S. dehiscens*	F.A.
			18	0.2			
					← 5.0	*G. tumida*	F.A.
			17B	1.2			
					← 6.2	*P. primalis*	
			17A	1.5			
					← 7.7	*G. plesiotumida*	F.A.
			16	2.3			
N. acostaensis	F.A.	10.5 →			← 10.0	*N. acostaensis*	F.A.
			15	1.2			
					← 11.2	*G. siakensis*	L.A.
			14	0.8			
G. nepenthes	F.A.	12.0 →					
			13	0.4			
					← 12.4	*G. lobata/robusta*	L.A.
			12	1.5			
G. fohsi	F.A.	12.8 →			← 13.9	*G. fohsi*	F.A.
			11	0.8			
					← 14.7	*G. praefohsi*	F.A.
			10	0.6			
					← 15.3	*G. peripheroacuta*	F.A.
			9	0.7			
Orbulina	F.A.	16.0 →			← 16.0	*Orbulina spp.*	F.A.
			8	1.2			
					← 17.2	*G. bisphericus*	F.A.
			7	0.8			
C. dissimilis	L.A.	17.5 →			← 18.0	*C. dissimilis*	L.A.
			6	0.6			
G. insueta	F.A.	19.0 →			← 18.6	*G. insueta*	F.A.
			5	1.9			
					← 20.5	*G. kugleri*	L.A.
			4	2.0			
Globigerinoides	F.A.	25.0 →			← 22.5	*Globigerinoides spp.*	F.A.
			3				

FIG. 9.—Chronology of Neogene zones of Blow (1969) and datum levels from Berggren and Van Couvering (1974), Van Couvering and Berggren (1977) and Saito (1977). F.A. and L.A. denote first and last appearances, respectively.

through N18) spanning 5.2 m.y. with an average zonal duration of 1.73 m.y. The Pliocene-Pleistocene has been divided into four zones (N19–N22) spanning an interval of 4.8 m.y. with an average duration of 1.2 m.y. Thus, the Late Miocene has zones of longest duration, the Middle Miocene has zones of shortest duration, and the Early Miocene and Pliocene-Pleistocene are of intermediate durations. These changes in zonal duration seem to be directly related to changes in levels of diversity and in rates of evolution of the planktonic foraminifera through the Neogene. The Middle Miocene diversities and evolutionary rates were the highest exhibited in the Neogene. High diversities are related to the Middle Neogene climatic optimum (Shackleton and Kennett, 1975; Savin, 1977). In contrast, the Late Miocene was the time of major cooling, and diversities decreased from a Neogene maximum in the Middle Miocene. Although paleotemperatures were increasing through the Early Miocene, diversities were not as high as in the Middle Miocene assemblages because the Neogene inherited the dramatically low faunal diversities that marked the oligotaxic oceans of the Oligocene (Fisher and Arthur, 1977). It is not clear why durations for zones in the Plio-Pleistocene increased slightly from those of the Late Miocene, although the return of slightly warmer conditions during the Early and Middle Pliocene may have led to an increase in planktonic foraminiferal diversities in the oceans.

The duration of planktonic foraminiferal zones in the temperate areas is still poorly known because of the lack of an adequate chronological framework. Slightly fewer zonal subdivisions have been made for the temperate Neogene sequence. Specifically, at least 14 zones have been solidly recognized and employed in temperate biostratigraphic subdivision, with perhaps two more zones recognized but not widely applied. Thus the number of Neogene zones recognized in temperate areas is only slightly less than in tropical areas, and their duration only slightly longer.

SOUTHWEST PACIFIC EQUATORIAL TO TEMPERATE NEOGENE ZONE

The Southwest Pacific represents one of the few optimal areas of the world for the study of a traverse of planktonic foraminiferal biostratigraphic sequences ranging from the Equator to the temperate regions. We have examined a traverse of DSDP cores ranging from the Equator to temperate latitudes at nearly 40°S. Sites examined include 289, 208, 206, 207A, and 284 (Figure 10). Planktonic foraminiferal zonal schemes range from the cool subtropical water mass to the south (Sites 284, 207A) to the warm-subtropical (Sites 206 and 208)

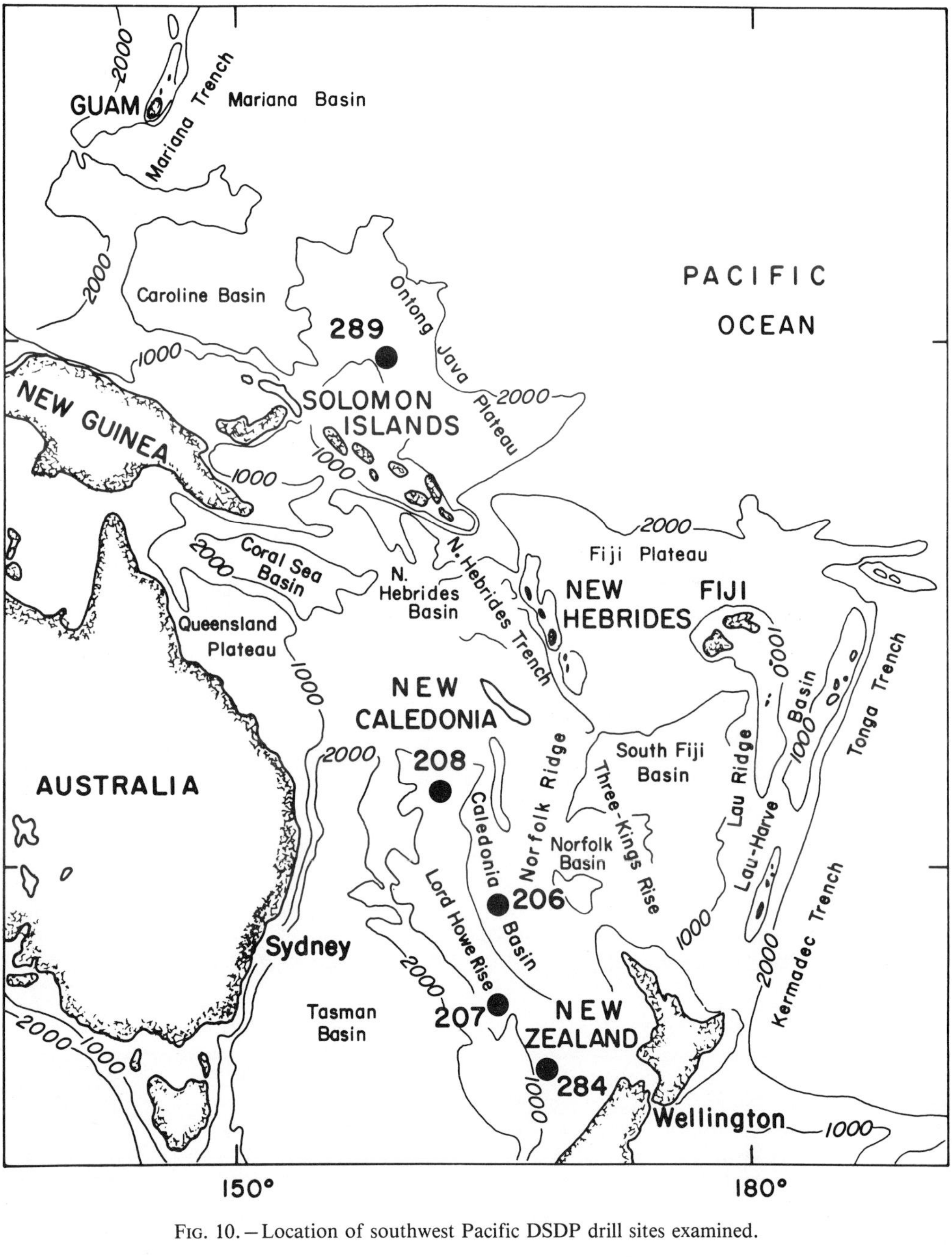

FIG. 10.—Location of southwest Pacific DSDP drill sites examined.

and the tropical water mass (Site 289). Table 1 shows location, water depth and thickness of Neogene sections. Details on the lithologies, physical properties, core-photographs and other parameters are provided in the Initial Reports of the Deep Sea Drilling Project.

Equatorial Site 289

Site 289 is located close to the Equator, to the northeast of New Guinea on the Ontong Java Plateau (Fig. 10). This site has a superb, continuous sequence of abundant planktonic foraminifera through the entire Neogene. The site consists of 660 m of Neogene foraminiferal-rich nanno-ooze (289-1-1 to 289-70-2). Deposition in shallow depths (2,224 m) has assured excellent preservation of faunas throughout. We have noticed a brief interval of dissolution over an approximate thickness of 18 m close to the Early–Middle Miocene boundary. This led to the elimination of a few dissolution-susceptible forms such as *Orbulina.* A preliminary report on the planktonic foraminiferal biostratigraphy was published by Saito (in Andrews, Packham and others, 1975). All of the equatorial zones of Blow (1969) are recognized in this site, and hence we have been able to make a careful examination of this zonal scheme (Fig. 11). Relationships between Blow's zonal scheme and that of Bolli's un-numbered zonal scheme are shown in Figure 3. We have found that Blow's (1969) zonation is practical in all but two of its features. Firstly, Zone N17 requires subdivision as previously discussed. Secondly, particular difficulties are encountered in the definition of boundaries between Zone N12 and N13. Blow (1969) defined this boundary on the first evolutionary appearance of *Sphaeroidinellopsis subdehiscens* from *Sphaeroidinellopsis seminulina.* Aside from taxonomic problems, there is no conspicuous evolution in the *Sphaeroidinellopsis* lineage at or near this boundary. Evolution of *S. seminulina* occurs later in the Neogene (Fig. 8). Therefore, the boundary between the Zones N12/N13 requires redefinition. We recommend that the extinction of *Fohsella* be used because this is an easily recognizable datum. Blow (1969) found that the extinction of *G. foshi* (s.l.) occurred in the lower part of Zone N13. Bolli (1967) previously recognized the problem of distinguishing evolution of *Sphaeroidinellopsis* in this time interval and recommended the use of the extinction of *G. fohsi* lineage as a more practical zonal boundary. We, therefore, formally recommend emending the definition of the Zone N12/N13 boundary as follows:

Zone N12: *Globorotalia fohsi* partial range zone of Blow (1969), here emended. Interval from the first evolutionary appearance of *G. (F.) fohsi* (s.l.) from its immediate ancestor *G. (F.) praefohsi* to the last appearance of the zonal marker. Since *G. (F.) fohsi* (s.l.) is the last member of its lineage it also represents the extinction of *Fohsella.*

Zone N13: *Sphaeroidinellopsis subdehiscens subdehiscens -Globigerina druryi* partial range zone of Blow (1969), here emended. Interval from the last appearance of *G. (F.) fohsi* (s.l.) to the first appearance of *Globigerina nepenthes.*

Warm Subtropical Site 208

Site 208 is located on the northern part of the Lord Howe Rise at 26°06.61′S in the Southwest Pacific (Fig. 10). This site is one of the most complete available for the Neogene in warm-subtropical waters intermediate between tropical and temperate areas. The Neogene consists of 435 m (208-1-1 to 208-25-3-50 cm) of foraminiferal nanno-ooze containing rich planktonic foraminiferal assemblages throughout with an excellent preservation. Although the site was discontinuously cored at every 18 m below the Late Miocene, an almost complete biostratigraphic zonation seems to be present. A relatively minor disconformity separates the late Early Miocene from the early Middle Miocene (Fig. 13).

The Neogene sequence at Site 208 contains a mixture of tropical and temperate planktonic foraminiferal elements. These elements do not occur in the same proportions, however, throughout the sequence. In the Early and early Middle Miocene, tropical elements dominate. In the sequence younger than Middle Miocene, temperate elements dominate although many tropical elements persist. Thus, biostratigraphic subdivision of the Early Miocene requires use of the tropical zones while the Late Middle and younger sequence requires the use of temperate zones. The early Middle Miocene was subdivided using the evolution of *Orbulina* which transcends both tropical and temperate areas. The only true temperate elements present in the Early and early Middle Miocene are the early forms of *Globoconella* (including *Globorotalia zealandica*

TABLE 1.—LOCATION OF DSDP SITES STUDIED IN THE EQUATORIAL AND SOUTHWEST PACIFIC.

Site	Latitude	Longitude	Water Depth (m)	Thickness of Neogene
289	00°29.92′S	158°30.69′E	2224	660 m
208	26°06.61′S	161°13.27′E	1545	435 m
206	32°00.75′S	165°27.15′E	3196	300 m
207A	36°57.75′S	165°26.06′E	1389	120 m
284	40°30.48′S	167°40.81′E	1068	208 m

incognita–G. miozea). Only the early forms of *Fohsella* (*G. peripheroacuta–G. fohsi* s.l.) are present and thus are only useful for subdivision of the early Middle Miocene. The later forms of *Fohsella* are missing at this latitude and thus biostratigraphic subdivision is based on temperate zones. Some species used in tropical zonation are present in the late Middle and Late Miocene such as *Globigerina nepenthes, Globorotalia (M.) menardii, Neogloboquadrina acostaensis* and *Pulleniatina primalis.* The zones based on these species can be recognized but are not employed because temperate forms are more dominant. Major tropical elements that are largely absent in Site 208 include forms of *Globorotalia* (including *G. (G.) plesiotumida, G. (G.) tumida*) and *Sphaeroidinella.* Thus, the tropical zones based on these species are not recognized. The change-over of applicability of early Neogene tropical zones to late Neogene temperate zones result from global paleoclimatic change. The middle Middle Miocene represents the time of formation of the Antarctic ice cap accompanied by surface-water oceanic cooling, and thus the migration of temperate assemblages towards the tropical regions. Because Site 208 does represent a mixture of tropical and temperate elements, a greater number of faunal events is recorded compared with both tropical and temperate regions. However, faunal events at Site 208 do not necessarily represent datums, because the appearance or disappearance can be diachronous with either tropical or temperate regions. For example, the appearance of *Globigerinatella insueta* is much later than in tropical sequences, and thus this species is restricted to a narrow stratigraphic interval. Likewise, the appearance of *Globorotalia (G.) conomiozea* in Site 208 is a non-evolutionary event and is slightly later than its evolutionary appearance in temperate areas (Kennett, 1973). The stratigraphic distribution of selected planktonic foraminifera in Site 208 and the zonation we have adopted is shown in Figure 12.

The zonation which we have employed for Site 208 is that of Kennett (1973) for the late Middle Miocene to Present Day. Because no zonation has been previously proposed for the early Neogene at Site 208, six new zones are defined. Furthermore, the *Globorotalia mayeri* Zone of Kennett (1973) is emended.

ZONAL DEFINITIONS

Early Miocene Zones

Globigerinoides primordius Zone

DEFINITION:

Top: Extinction of *Globorotalia kugleri* Bolli

Base: Initial appearance of *Globigerinoides primordius* Blow and Banner

Age: Early Miocene

TAXA:

Extinction within Zone:
Globorotalia opima nana

Initial Appearances within Zone:
Globigerina woodi
Globorotalia acrostoma
Globorotalia peripheroronda
Globigerinoides trilobus
G. altiapertura
G. subquadratus
Globoquadrina dehiscens

Important Species Ranging throughout the Zone:
Neogloboquadrina continuosa
Catapsydrax dissimilis
C. unicava
Globoquadrina binaiensis
Globorotalia siakensis

Reference Section: Site 208: 26°06.61′S, 161°13.27′E

Occurrence of Zone: Alternate coring for 9 meters every 18 meters (25-3-99 to 23-3-100 cm).

Correlation: Equivalent to Zone N4 of Blow (1969) and *G. primordius* Zone of Bolli and Premoli Silva (1973). Includes upper part of *G. kugleri* Zone of Bolli (1957b) and Jenkins and Orr (1972), above the appearance of *Globigerinoides trilobus.*

Catapsydrax dissimilis Zone

DEFINITION:

Top: Extinction of *Catapsydrax dissimilis* (Cushman and Bermudez)

Base: Extinction of *Globorotalia kugleri* Bolli

Age: Early Miocene

TAXA:

Extinctions within Zone:
Globorotalia acrostoma
Globigerinoides primordius
Globoquadrina binaiensis

Initial Appearances within Zone:
Globorotalia zealandica incognita
G. zealandica zealandica
G. zealandica pseudomiozea
G. miozea
G. praescitula
G. archeomenardii
Globoquadrina altispira
Globigerinatella insueta

Important Species Ranging throughout the Zone:
Globigerina woodi
Globigerinoides trilobus
G. altiapertura
G. subquadratus
G. peripheroronda
G. mayeri
Globoquadrina dehiscens
Neogloboquadrina continuosa

Reference Section: Site 208: 26°06.61′S, 161°13.27′E

Occurrence of Zone: Alternate coring for 9 meters every 18 meters (23-3-100 to 22-1-107 cm).

Correlation: Equivalent to the combined *C. dissimilis* and *C. stainforthi* Zones of Bolli (1966); Equivalent to the Zones N5 and N6 of Blow (1969) and *C. dissimilis* Zone of Jenkins and Orr (1972).

Globorotalia miozea Zone

DEFINITION:

Top: Initial appearance of *Praeorbulina glomerosa curva* (Blow)

Base: Extinction of *Catapsydrax dissimilis* (Cushman and Bermudez)

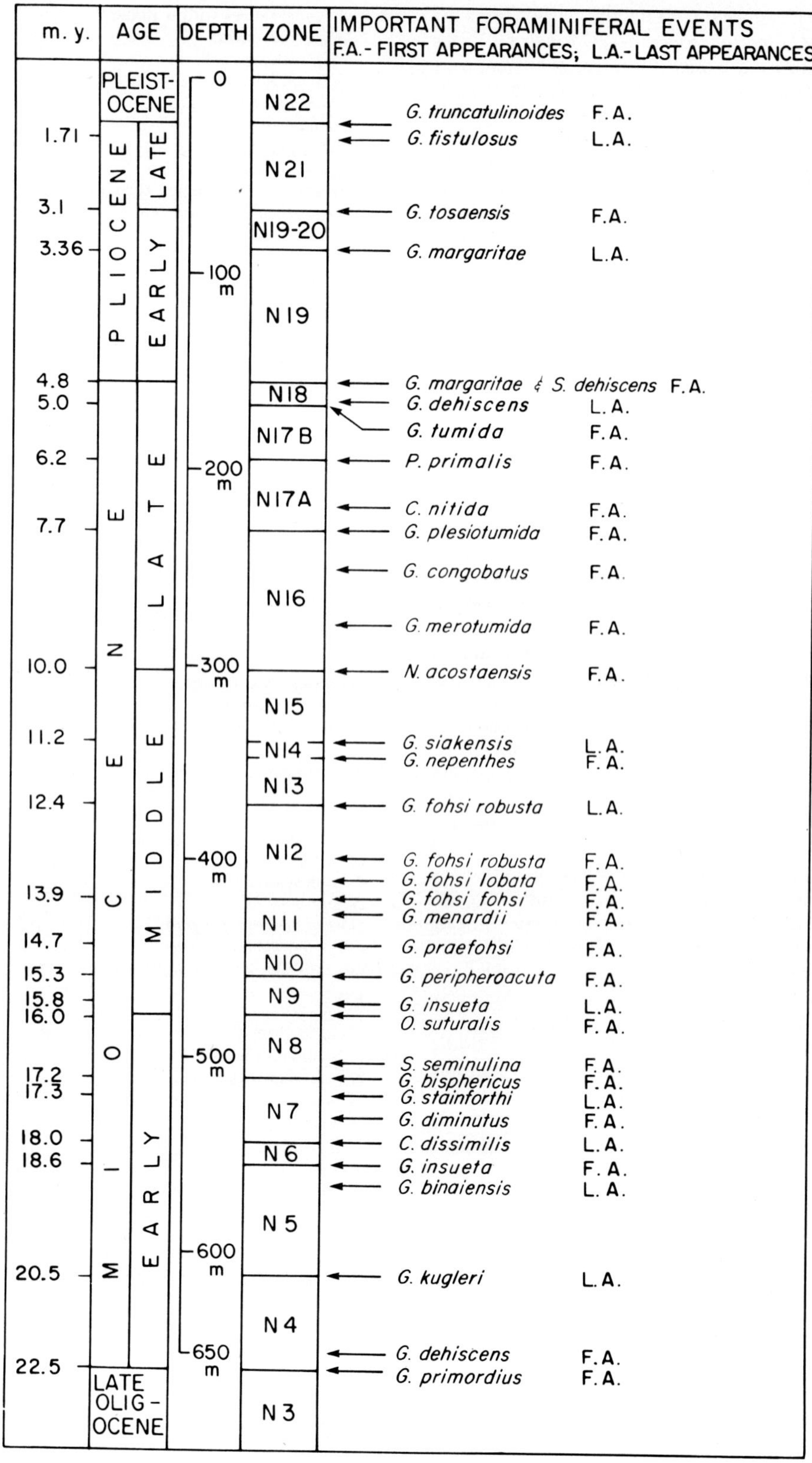
m. y.
AGE
DEPTH
ZONE
IMPORTANT FORAMINIFERAL EVENTS
F.A.- FIRST APPEARANCES; L.A.- LAST APPEARANCES
PLEISTOCENE
PLIOCENE
LATE
EARLY
MIOCENE
LATE
MIDDLE
EARLY
LATE OLIGOCENE
1.71
3.1
3.36
4.8
5.0
6.2
7.7
10.0
11.2
12.4
13.9
14.7
15.3
15.8
16.0
17.2
17.3
18.0
18.6
20.5
22.5
0
100 m
200 m
300 m
400 m
500 m
600 m
650 m
N22
N21
N19-20
N19
N18
N17B
N17A
N16
N15
N14
N13
N12
N11
N10
N9
N8
N7
N6
N5
N4
N3
G. truncatulinoides F.A.
G. fistulosus L.A.
G. tosaensis F.A.
G. margaritae L.A.
G. margaritae & S. dehiscens F.A.
G. dehiscens L.A.
G. tumida F.A.
P. primalis F.A.
C. nitida F.A.
G. plesiotumida F.A.
G. congobatus F.A.
G. merotumida F.A.
N. acostaensis F.A.
G. siakensis L.A.
G. nepenthes F.A.
G. fohsi robusta L.A.
G. fohsi robusta F.A.
G. fohsi lobata F.A.
G. fohsi fohsi F.A.
G. menardii F.A.
G. praefohsi F.A.
G. peripheroacuta F.A.
G. insueta L.A.
O. suturalis F.A.
S. seminulina F.A.
G. bisphericus F.A.
G. stainforthi L.A.
G. diminutus F.A.
C. dissimilis L.A.
G. insueta F.A.
G. binaiensis L.A.
G. kugleri L.A.
G. dehiscens F.A.
G. primordius F.A.

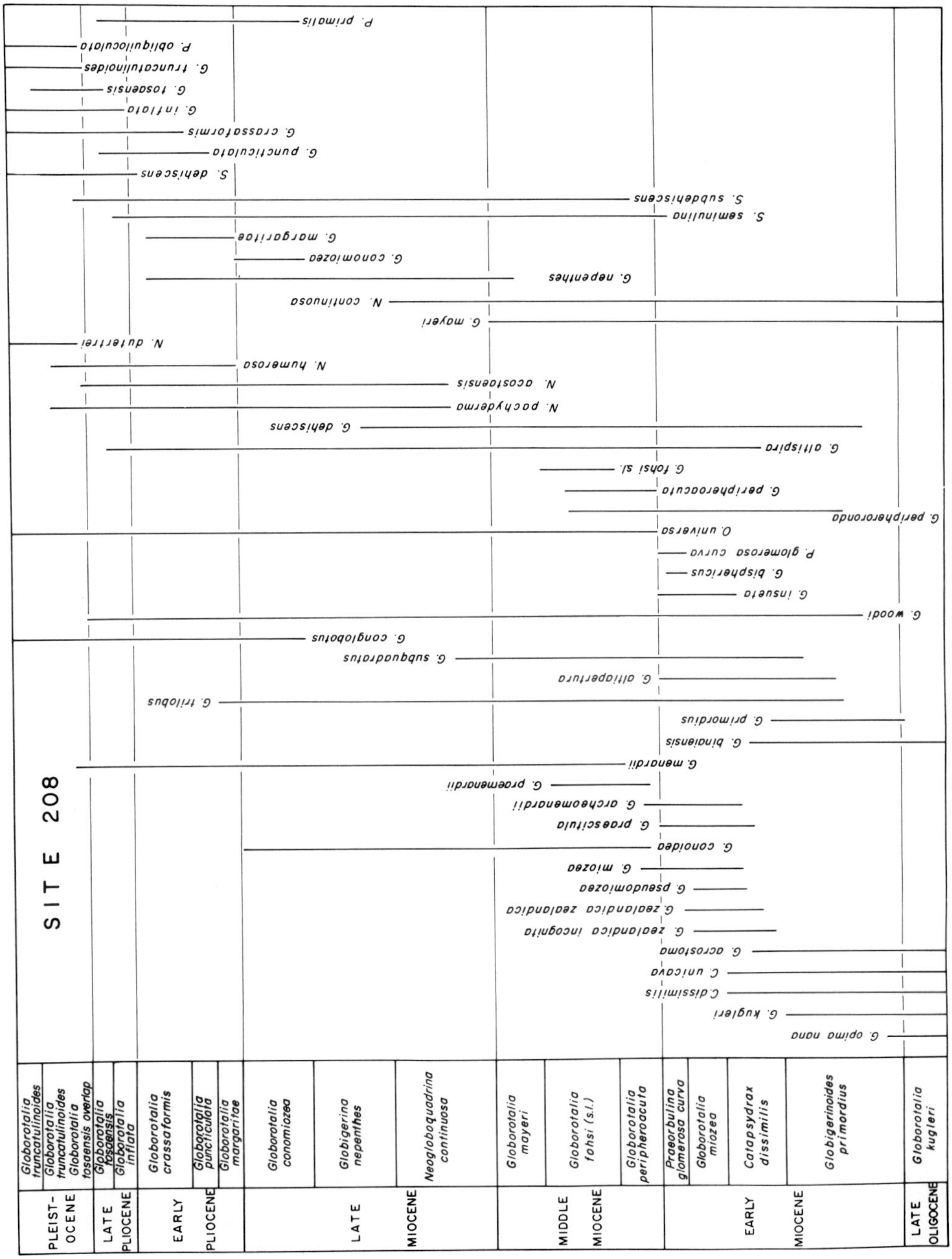

FIG. 12.—Stratigraphic range of selected planktonic foraminifera in the warm-subtropical DSDP Site 208 and zonation.

←

FIG. 11.—Neogene planktonic foraminiferal zones and important foraminiferal events at DSDP Site 289. Estimated ages after Saito (1977). F.A. and L.A. denote first and last appearances, respectively.

Remark: *G. miozea* serves as the name bearer for this zone due to its abundance in the assemblage, perhaps as a result of resistance to solution.

Age: Early Miocene

TAXA:

Extinctions within Zone:
Globorotalia zealandica incognita
G. pseudomiozea

Initial Appearance within Zone:
Globigerinoides bisphericus

Important Species Ranging throughout the Zone:
Globigerina woodi
Globorotalia peripheroronda
G. mayeri
G. miozea
G. praescitula
G. archeomenardii
Globigerinoides trilobus
G. altiapertura
G. subquadratus
Globoquadrina altispira
G. dehiscens
Neogloboquadrina continuosa

Reference Section: Site 208: 26°06.61′S, 161°13.27′E

Occurrence of Zone: Alternate coring for 9 meters every 18 meters (22-1-107 to 21-6-105 cm).

Correlation: Equivalent to *G. insueta* Zone of Bolli (1966). Includes Zone N7 and Zone N8 of Blow (1969) prior to the first appearance of *Praeorbulina*. Broadly equivalent to *Globoquadrina venezuelana* Zone of Jenkins and Orr (1972).

Praeorbulina glomerosa curva Zone

DEFINITION:

Top: Not defined, but the zone includes all faunas prior to the initial appearance of *Orbulina suturalis*.

Base: Initial appearance of *Praeorbulina glomerosa curva* (Blow)

Age: Early Miocene

TAXA:

Extinctions within Zone:
Globorotalia zealandica zealandica
G. praescitula
Globigerinoides bisphericus
G. altiapertura

Initial Appearance within Zone:
Sphaeroidinellopsis seminulina

Important Species Ranging throughout the Zone:
Globorotalia peripheroronda
G. mayeri
G. archeomenardii
Globigerinoides trilobus
G. subquadratus
Globoquadrina dehiscens
G. altispira
Neogloboquadrina continuosa

Reference Section: Site 208: 26°06.61′S, 161°13.27′E

Occurrence of Zone: Alternate coring for 9 meters every 18 meters (21-6-105 to 21-3-55 cm). The top of the zone is missing due to a disconformity.

Correlation: Equivalent to *Praeorbulina glomerosa* Zone of Bolli (1966) prior to the first appearance of *Orbulina*. Equivalent to Zone N8 (Blow, 1969) above the initial appearance of *Praeorbulina*. The zone also corresponds to *P. glomerosa curva* Zone of Jenkins (1967, 1971).

Middle Miocene Zones

Globorotalia peripheroacuta Zone

DEFINITION:

Top: Initial appearance of *G. fohsi* (s.l.). Cushman and Ellisor

Base: Not defined, but the zone includes all faunas above the initial appearance of *G. peripheroacuta* Blow and Banner.

Age: Middle Miocene

TAXA:

Extinctions within Zone:
Globorotalia miozea
G. archeomenardii

Initial Appearances within Zone:
Globorotalia conoidea
G. praemenardii
G. menardii

Important Species Ranging throughout the Zone:
Globorotalia conoidea
G. praemenardii
G. peripheroronda
G. peripheroacuta
G. mayeri
Globigerinoides trilobus
Orbulina universa
Globoquadrina dehiscens

Reference Section: Site 208: 26°06.61′S, 161°13.27′E

Occurrence of Zone: Alternate coring for 9 meters every 18 meters (21-3-55 to 20-2-50 cm). At Site 208 the base of the zone is a disconformity and the base of the zone is thus missing. The disconformity represents an interval extending from upper part of *P. glomerosa curva* Zone to lower part of *G. peripheroacuta* Zone.

Correlation: Equivalent to the *G. fohsi fohsi* Zone of Bolli (1966) and to Zone N10 of Blow (1969). Broadly equivalent to the *G. peripheroacuta* Zone of Jenkins and Orr (1972).

Globorotalia fohsi (s.l.) Zone

DEFINITION:

Top: Extinction of *Globorotalia fohsi* (s.l.)

Base: Initial appearance of *Globorotalia fohsi* (s.l.)

Age: Middle Miocene

TAXA:

Extinctions within Zone:
Globorotalia peripheroronda
G. peripheroacuta
G. praemenardii

Important Species Ranging throughout the Zone:
Globorotalia conoidea
G. menardii
G. mayeri
Neogloboquadrina continuosa
Globigerinoides trilobus
G. subquadratus
Globoquadrina altispira
G. dehiscens
Sphaeroidinellopsis seminulina

Reference Section: Site 208: 26°06.61′S, 161°13.27′E

Occurrence of Zone: Alternate coring for 9 meters every 18 meters (20-2-50 to 19-2-50 cm).

Remarks: *Globorotalia fohsi lobata* Bolli and *Globorotalia fohsi robusta* Bolli, which are well represented at DSDP Site 289, are not present at Site 208.

Correlation: Broadly equivalent to the *G. foshi lobata* and *G. fohsi robusta* Zones of Bolli (1966) and to Zone N11 to lower part of Zone N13 of Blow (1969). Includes the *G. fohsi lobata* Zone of Jenkins and Orr (1972).

Globorotalia mayeri Zone

DEFINITION:

Top: Extinction of *Globorotalia mayeri* Cushman and Ellisor

Base: Extinction of *Globorotalia fohsi* (s.l.)

Age: Middle Miocene

TAXA:

Extinction within Zone:
Globorotalia siakensis

Initial Appearances within Zone:
Globigerina nepenthes

Important Species Ranging throughout the Zone:
Globigerina woodi
G. decoraperta
Globorotalia mayeri
Orbulina universa
Globigerinoides trilobus
Globoquadrina altispira
G. dehiscens

Reference Section: Site 208: 26°06.61′S, 161°13.27′E

Occurrence of Zone: Alternate coring for 9 meters every 18 meters (19-2-50 to 18-3-50 cm).

Correlation: Equivalent to Zone N13 above the last appearance of *Globorotalia fohsi* s.l. to the top of Zone

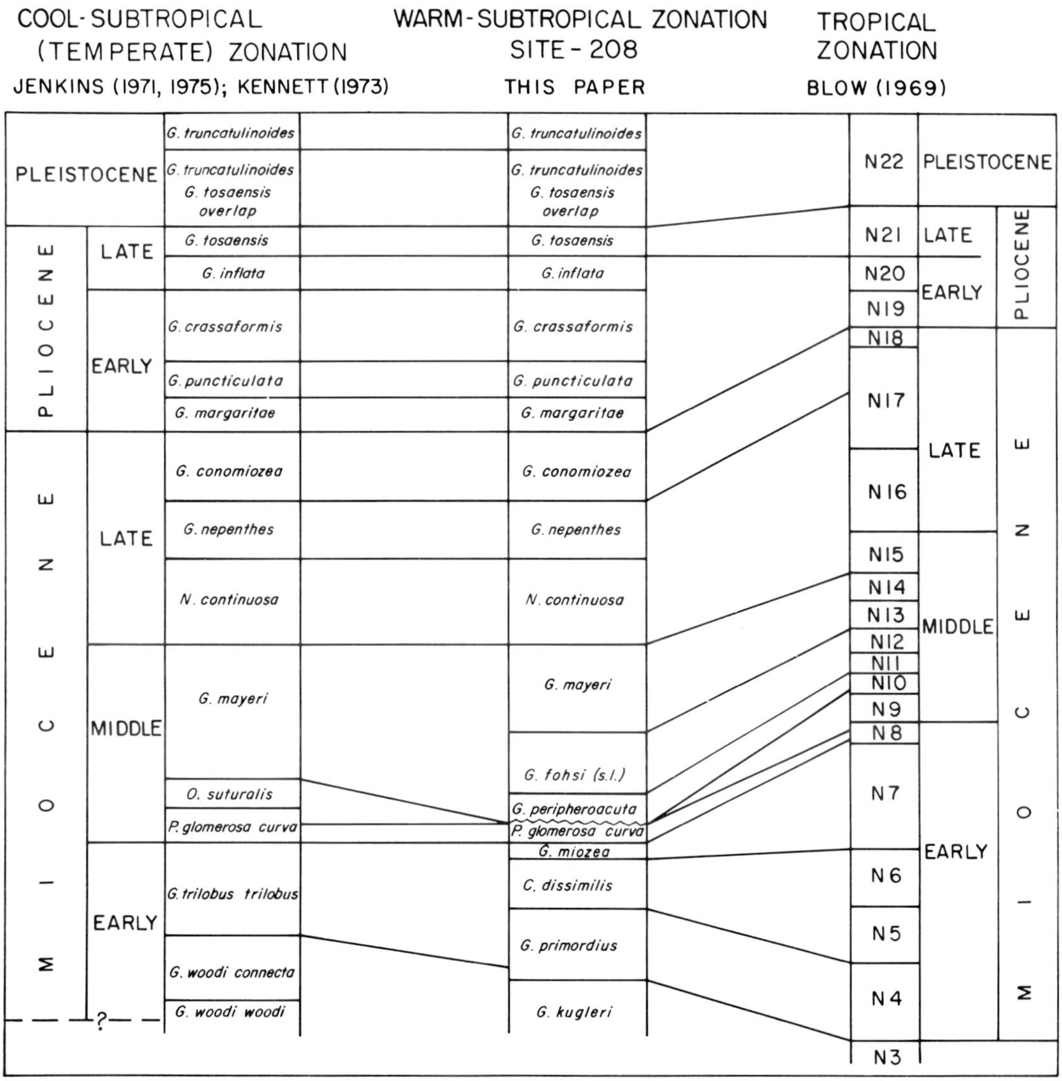

FIG. 13.—Planktonic foraminiferal zonation of warm-subtropical DSDP Site 208 correlated with temperate zonal scheme of Jenkins (1971, 1975) and Kennett (1973) and tropical zonal scheme of Blow (1969).

N14 of Blow (1969). Equivalent to the *G. ruber* and *G. mayeri* Zones of Bolli (1966). Broadly equivalent to the upper part of the *G. mayeri* Zone of Jenkins (1971).

Late Miocene through Quaternary Zones

The late Neogene zones established by Kennett (1973) are used in this study without modification (Fig. 13).

Transitional Site 206.—DSDP Site 206 is located in the Caledonia Basin to the east of Lord Howe Rise in the southwest Pacific (Fig. 10). The Neogene is represented by 300 m of foraminiferal-nanno ooze containing abundant and moderately well-preserved planktonic foraminifera throughout. Preservation is of slightly lower quality compared with the other sites of the region, because of the greater depth of deposition (3,196 m). Site 206 is strategically located between temperate and warm-subtropical water masses. Although tropical elements are persistant through the sequence, the assemblages are clearly dominated by temperate elements. The Neogene sequence at Site 206 is continuous except for an important unconformity centered over the Miocene/Pliocene boundary. The presence of this unconformity has complicated correlations with the latest Miocene and earliest Pliocene of other sites in the traverse. A minor disconformity also occurs at the boundary between the Lower and Middle Miocene. High sedimentation rates (6 cm/1,000 yr) allow relatively high-resolution stratigraphy to be carried out in the remaining parts of the sequence. The zonation of Kennett (1973) is employed (Fig. 14), but our more detailed work may require future modifications of it.

Temperate Site 207A.—DSDP Site 207A is located in the southern end of Lord Howe Rise in the

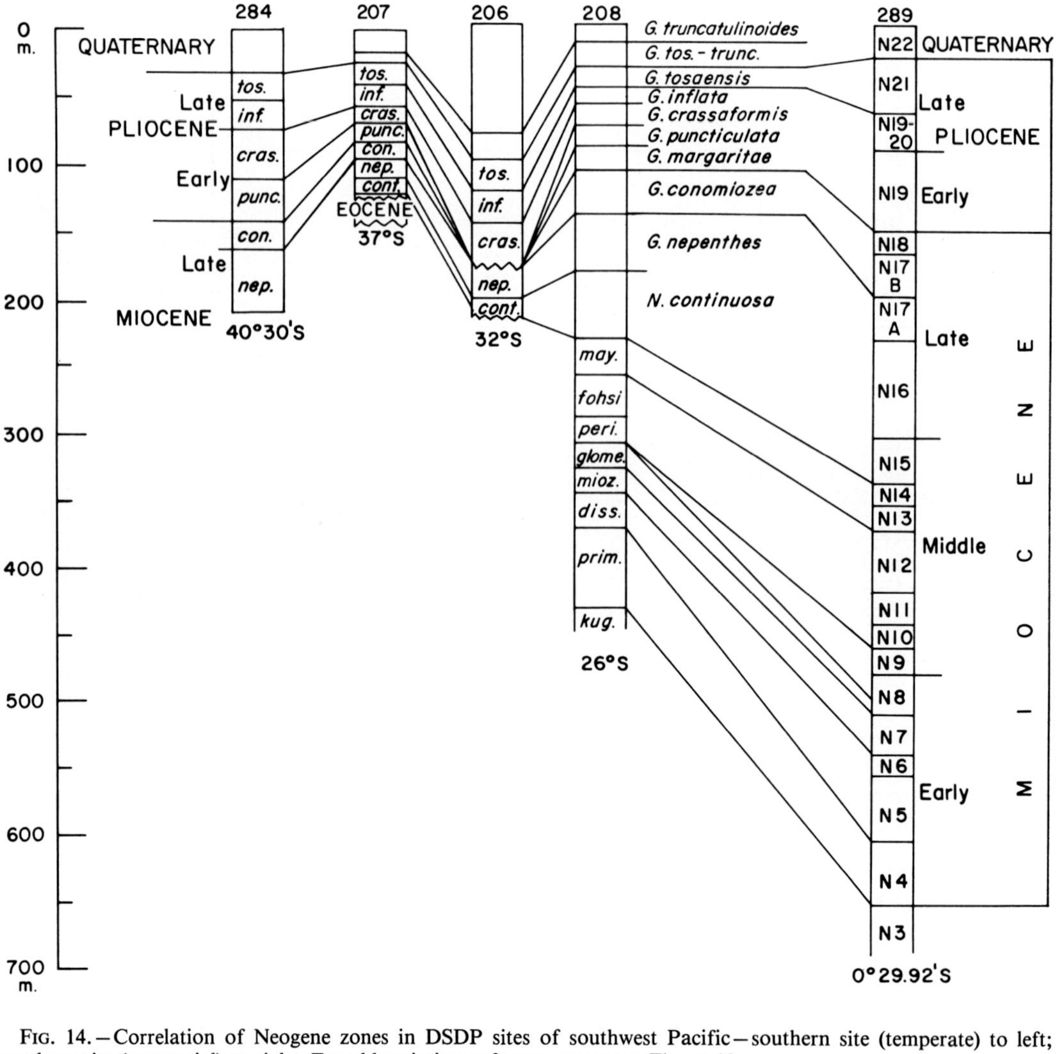

FIG. 14.—Correlation of Neogene zones in DSDP sites of southwest Pacific—southern site (temperate) to left; northern site (equatorial) to right. For abbreviations of zone names see Figure 13.

southwest Pacific (Fig. 10). Its relatively shallow depth (1,389 m) has assured excellent preservation throughout. The planktonic foraminiferal assemblages are completely dominated by temperate forms. Tropical and even sub-tropical elements are rare. The zonation employed for this site by Kennett (1973) is shown in Figure 14.

Temperate Site 284.—DSDP Site 284 is located near the edge of the Challenger Plateau, west of Wellington, New Zealand, in shallow water (1,068 m). Site 284 is only 3° latitude south of Site 207A, and hence temperate planktonic foraminiferal assemblages dominate. However, even though only 3° of latitude separates Site 284 from Site 207A, the water-masses were different enough to create slight differences in the biostratigraphic sequence (Kennett and Vella, 1975). The same set of informal zones have been employed in Site 284 as in 207A (Fig. 14).

PACIFIC INTERSITE DATUM CORRELATIONS

Zonal correlations between the southwest Pacific DSDP sites are shown in Figures 13 and 14. In this section we are concenred with intersite correlations of planktonic foraminiferal datums in DSDP sites throughout the Pacific basin, particularly as determined by Keller (*in press* a, b). These are discussed in sequential order beginning from the earliest Neogene. In the following section the abbreviations F.A. and L.A. stand for first appearance and last appearance, respectively.

Early Miocene

Globigerinoides primordius F.A.

The base of the Miocene is normally taken at the first appearance of *G. primordius,* the oldest known form of *Globigerinoides*. First appearance of *G. primordius* is well marked in Site 289 (Core 70) and in Site 208 (Core 25), providing an excellent correlation datum. The *G. primordius* datum has been dated at 25.0 Ma by Van Couvering and Berggren (1977) and at 22.5 Ma by Saito (1977). This datum is of restricted value because *G. primordius* is solution-susceptible, and thus absent in a large number of Pacific DSDP sequences. For instance, Keller (*in press* a, b) did not recognize this datum in DSDP sites 71 and 77B in the equatorial north Pacific she examined. However, these sites exhibit much greater dissolution than Sites 289 and 208. The Oligocene/Miocene boundary is difficult to place when *Globigerinoides* is rare or absent, in which case the boundary is often based on nannofossil data. In low-latitude Indian Ocean sites where the *Globigerinoides* datum cannot be recognized, the Oligocene/Miocene boundary has been arbitrarily placed at the initial appearance of *Globorotalia kugleri* (Berggren and others, 1974; Fleisher, 1977; Vincent, 1977).

Globorotalia kugleri L.A.

The last appearance of *G. kugleri* at both Sites 289 and 208 marks the top of Zone N4. Saito (1977) estimated this datum to be 20.5 Ma, but Keller (*in press* a) provides an age of 21.7 based on indirect correlations between Site 77B using radiolarian datums with similar datums in paleomagnetically dated piston cores (Theyer and Hammond, 1974).

Globoquadrina dehiscens F.A.

The first appearance datum of *G. dehiscens* at Site 289 and in Site 208 falls within the upper stratigraphic range of *G. kugleri,* above the initial appearance of *G. primordius.* Keller (*in press* a) also recorded a similar first appearance of *G. dehiscens* at Site 77B, coincident with the base of the radiolarian Zone *C. virginis* (Goll, 1972), which has been dated at 22.2 Ma within the earliest magnetic Epoch 20 (Theyer and others, 1978).

At DSDP Site 296, *G. dehiscens* first appears within the upper part of the nannofossil Subzone *C. abisectus* and radiolarian Zone *L. elongata* (Keller, *in press* a). The top of these zones have estimated ages of 23.0 M and 22.2 Ma respectively (Bukry, 1975; Theyer and others, 1978). Thus the first appearance of *G. dehiscens* in all these four Pacific sites (Sites 289, 208, 77B and 296) is at the same stratigraphic level, providing an important datum. Because this species is solution resistant and its first datum is close to the *Globigerinoides* datum, it may prove to be a more valuable criterion for recognition of the Miocene/Oligocene boundary.

Globorotalia zealandica incognita F.A.

Walters (1965) recorded the first evolutionary appearance of *G. zealandica incognita* about 60 meters below the first appearance of *G. trilobus* in the Waiomoko River section, eastern New Zealand. The first appearance of *G. zealandica incognita* at Site 208 occurs at 372 m (3 m above the last appearance datum of *G. kugleri*) within the lower part of *Catapsydrax dissimilis* Zone, which correlates with the lower part of Zone N5 (Figs. 12, 13). However, Keller (*in press* a, b) recorded the first appearance of *G. zealandica incognita* in the upper part of Zone N4 at Sites 71, 77B and 292. This species is not recorded in equatorial Site 289. Thus, the first occurrence of this temperate species at Site 208 reflects later migration, following its evolution at higher latitudes.

Globorotalia zealandica pseudomiozea–G. miozea F.A.

Walters (1965) described the evolutionary lineage of *G. zealandica incognita* to *G. pseudomiozea* to *G. miozea* from the Early Miocene Waiomoko River section of New Zealand. Later, Keller (*in press* a) documented this temperate bioseries at DSDP Site 296. *G. zealandica pseudomiozea* and *G. miozea* first appear in the middle and upper part respectively of *Catapsydrax dissimilis* Zone at Site 208, which correlates within the lower part of Zone

N6 (Fig. 12). This level is broadly equivalent to the horizon of its first appearance in Site 296 (Keller, *in press* a). The entire *G. zealandica incognita* and *G. miozea* bioseries is not represented at Site 289.

Globoquadrina altispira F.A.

At Site 208 *G. altispira* first appears at the same stratigraphic level as *G. pseudomiozea* or slightly below the first appearance of *G. praescitula,* as has been documented at Site 296 by Keller (*in press* a). At Site 296, *G. altispira's* first appearance is not a true first appearance because much of Zone N4 is missing. However, Keller (*in press* b) shows that in Sites 292, 55, 71 and 77B *G. altispira* appears in Zone N4 above the *G. dehiscens* F.A. datum, as has been documented at Site 289. The appearance of *G. altispira* at Site 208 appears to be due to later migration, thus this datum is less useful.

Globoquadrina binaiensis L.A.

G. binaiensis is a useful zonal marker for the Late Oligocene and Early Miocene of the tropical Indo-Pacific region. Its last appearance within Zone N5 at Site 289 (Core 60) and in the lower part of *C. dissimilis* Zone at Site 208 (Core 22) represents a useful datum for correlation between these two Pacific sites.

Globigerinatella insueta F.A.

The first appearance datum of *G. insueta* in Site 289 occurs in Core 59, defining the base of Zone N6. Because this taxon is very solution-susceptible, its stratigraphic distribution is sporadic, especially in the lower part of its range at Site 208. Thus the first appearance of *G. insueta* may not represent the true first appearance datum at Site 208. Berggren and Van Couvering (1974) estimated an age of 19.0 Ma for the *G. insueta* F.A. datum, whereas Saito (1977) provided an age at 18.6 Ma.

The last appearance datum of *G. insueta* at Site 289 occurs in Core 51 (upper part of Zone N9). This datum is dated at 15.8 Ma by Saito (1977). The extinction level of *G. insueta* is not recorded at Site 208 because of a disconformity. Keller (*in press* a) records this species only in Site 319 where it is consistently present between Zone N8 to the middle of Zone N10.

Catapsydrax dissimilis and *Catapsydrax unicavus* L.A.

C. dissimilis is easily recognizable and is one of the most important early Neogene solution-resistant species. This species occurs abundantly at both Sites 289 and 208. It is an important Oligocene form but ranges upwards into the Early Miocene. Its last appearance in the Early Miocene together with *C. unicavus* marks a well-defined datum level at both Sites 289 and 208 (Fig. 13), delineating the Zone N6/N7 boundary (Blow, 1969). The extinction level of *C. dissimilis* has been paleomagnetically dated by Berggren and Van Couvering (1974) at 17.5 Ma, by Ryan and others (1974) at 18.5 Ma and by Saito (1977) at 18.0 Ma. However, Keller (*in press* a) noted the *C. dissimilis* extinction datum nearly coincides with the base of the radiolarian datum *C. leptetrum* (Goll, 1972), dated at 21.6 Ma in the middle of magnetic Epoch 20 (Theyer and others, 1978).

Catapsydrax stainforthi L.A.

The last appearance datum of *C. stainforthi* at Site 289 occurs in Core 55 in the upper part of Zone N7. The species is rare in Site 208. This datum has been dated at 17.3 Ma by Saito (1977). The last appearance of *C. stainforthi* recorded by Keller (*in press* a) in Sites 77B and 296 may be earlier than its genuine extinction in Site 289, because of dissolution at this level in these two sites.

Globigerinoides bisphericus F.A.

At Site 289 the *G. bisphericus* (= *G. sicanus*) first appearance datum occurs above the extinction level of *C. stainforthi* and marks the base of foraminiferal Zone N8. In Site 208, this species first appears slightly later and approximately at the same stratigraphic level as *Praeorbulina glomerosa curva.* Thus the delayed first appearance of *G. bisphericus* in Site 208 may be due to a later migration of warmer waters to the warm subtropics. There are a number of studies documenting similar diachronous first appearances of this species in DSDP sites (Bronnimann and Resig, 1971; Fleisher, 1974).

This datum has received a number of dates as follows: Ikebe and others (1972) – 16.5–17.0 Ma; Ryan and others (1974) – 16.8 Ma; Keller (*in press* a) – 16.1; and Saito (1977) – 17.2 Ma. These dates suggest a total range of about one million years.

Globorotalia peripheroronda and
Globigerinoides subquadratus F.A.

G. peripheroronda and *G. subquadratus* first appear in Zone N4 in Sites 289 and 208. However, Keller (*in press* a) noted the first appearance of these two species above the *G. bisphericus* first appearance datum in Site 296. The diachronous first appearance in Site 296 is due to selective dissolution below this level. Thus the first appearance datum of *G. peripheroronda* and *G. subquadratus,* although useful in correlating Site 289 and 208, is of limited value for biostratigraphic correlation.

Sphaeroidinellopsis seminulina F.A.

The first appearance of *S. seminulina* occurs simultaneously in Site 289 (lower Zone N8) and in Site 208 (*P. glomerosa curva* Zone) providing a useful datum level for intersite correlation. Jenkins and Orr (1972) recorded the first appearance of *S. seminulina* in the *P. glomerosa* Zone at Site 77B, in agreement with the stratigraphic range in the western Pacific.

Middle Miocene

All the zonal diagnostic species established by Blow (1969) for the Middle Miocene were encountered at the equatorial Pacific Site 289. At Site 208 the *Globorotalia fohsi* lineage is represented only by *G. peripheroacuta* and *G. fohsi* (s.l.) because of cooler conditions.

Orbulina F.A.

The first evolutionary appearance of *Orbulina,* the "*Orbulina* datum" delineates the Early/Middle Miocene boundary (Zone N8/N9 boundary of Blow, 1969). *Orbulina* is one of the common solution-susceptible forms, and therefore in most DSDP sites its first evolutionary appearance is difficult to determine (Vincent, 1977; Keller, *in press* a). In Site 289 *Orbulina* is rare but still can be used to delineate the Early/Middle Miocene boundary. In Site 208, due to a disconformity, the first evolutionary appearance is not recorded and the species first appears accompanied by *G. peripheroacuta* (Fig. 13). Both Berggren and Van Couvering (1974) and Saito (1977) estimated an age at 16.0 Ma for the first evolutionary appearance of *Orbulina.* Later Van Couvering and Berggren (1978) have revised it to 15 Ma. Keller (*in press* a) also dated this datum at 15.0 Ma (Bukry, 1975).

Globorotalia fohsi lineage

The *Fohsella* lineage forms the basis for the standard tropical planktonic foraminiferal zonation in the Middle Miocene. The succession of the component species at Site 289 is similar to those established by Bolli (1957b, 1967) and Banner and Blow (1965). At Site 208 only *G. peripheroacuta* and forms referable to *G. fohsi* (s.l.) are present because of the cooler conditions.

Globorotalia peripheroacuta F.A.

The first evolutionary appearance of *G. peripheroacuta* in Core 49 at Site 289 marks the base of Zone N10. At Site 208, the base of *G. peripheroacuta* Zone is missing due to a disconformity (Fig. 13). Saito (1977) estimated the first appearance of *G. peripheroacuta* to be 15.3 Ma, while Keller (*in press* a) estimated its age at 14.6 Ma based on correlation with DSDP Site 77B.

Globorotalia praefohsi F.A.

The first appearance of *G. praefohsi,* which defines the Zone N10/N11 boundary appears (298 m) in Core 47 at Site 289 (446 m) and Core 20 cc; in Site 208. Saito (1977) estimated an age of 14.7 Ma for the first appearance datum of *G. praefohsi.* Keller (*in press* a) recorded the first appearance of *G. praefohsi* in Site 77B virtually coincident with the base of the *D. exilis* nannoplankton Zone dated at 14.0 Ma (Bukry, 1972, 1975) and the first appearance of the radiolarian *L. neotera* dated at 14.1 Ma (Theyer and others, 1978).

Globorotalia fohsi fohsi F.A.

The first evolutionary appearance of *G. fohsi fohsi* from *G. praefohsi* at Site 289 occurs at 421 m delineating the N11/N12 boundary (Fig. 11). At Site 208 on the other hand, *G. fohsi* s.l. develops from *G. peripheroacuta.* At this site, forms referable to *G. fohsi lobata* and *G. fohsi robusta* are absent and the entire assemblage referable to *G. fohsi* s.l. Zone broadly correlates with Zones N11 and N12 at Site 289.

Saito (1977) provided an estimated age of 13.9 Ma for the first appearance datum of *G. fohsi fohsi.* Keller (*in press* a) estimated an age of 13.6 Ma for the *G. fohsi fohsi* first appearance at Site 77B, which is close to Saito's (1977) estimate, but considerably younger than an age of 15.5 Ma estimated by Ryan and others (1974).

Globorotalia fohsi lobata F.A.

G. fohsi lobata is a distinct and short-ranging subspecies of *Fohsella,* and its first appearance level is a useful datum. In Site 289, *G. fohsi lobata* first appears shortly after *G. fohsi fohsi.* A similar stratigraphic range for *G. fohsi lobata* has been documented by Keller (*in press* a) at Site 77B and 319. *G. fohsi lobata* is absent in the cooler Site 208. Keller (*in press* a) estimated an age of 13.1 Ma for the first appearance of *G. fohsi lobata* in Site 77B.

Globorotalia menardii F.A.

The first evolutionary appearance of *G. menardii* from *G. praemenardii* occurs in Site 289 at 425 m, in the upper part of Zone N11. In Site 208, *G. menardii* first appears in Core 20, approximately at the same level as the first appearance of *G. fohsi* s.l., providing a useful Middle Miocene datum.

Globorotalia fohsi robusta L.A.

The top of the *G. fohsi robusta* Zone as determined by the last appearance of *G. fohsi robusta* and of *Fohsella* provides an easily identifiable, useful Middle Miocene datum. As discussed previously, this datum is recommended for defining the Zone N12/N13 boundary in preference to the first appearance of *S. subdehiscens* (Blow, 1969).

Berggren and Van Couvering (1974) estimated the extinction level of *Fohsella* at 12.8 Ma. Saito (1977) estimated an age of 12.4 Ma for this datum based on Site 289 data.

Globigerina nepenthes F.A.

The first appearance of *G. nepenthes* defines the base of Zone N14 (Blow, 1969). This datum is well defined in Site 289, where *G. nepenthes* makes its initial appearance at 348 m. In Site 208, this datum occurs in 18 cc, within the upper part of the *G. mayeri* Zone. Recently, Berggren and Van Couvering (1978) discussed the chronology of *G. nepenthes* datum in detail and estimated an age of 12.7 Ma for this event.

Globorotalia mayeri and *G. siakensis* L.A.

The last appearances of *G. mayeri* and *G. siakensis* mark the top of Zone N14 (Blow, 1969) and form a useful datum in many Pacific DSDP sites (Saito, 1977; Keller, *in press* b, c). This datum is easily recognized at both sites 289 and 208. Saito (1977) estimated an age of 11.2 Ma for the *G. mayeri/G. siakensis* extinction datum.

Late Miocene

Zone N15 and most of the younger N zones of Blow (1969) recognized in Site 289 cannot be distinguished in Site 208 because of its predominantly temperate assemblages (Fig. 13). Hence, a precise correlation between these two sites is not possible for the Late Miocene and Pliocene. The Late Miocene and younger faunal assemblages of Site 208, instead, show closer affinities to the temperate assemblages in Site 206, 207A and 284. Thus intersite correlations are more easily invoked between Site 208 and sites to the south (Fig. 14). Since this discussion is concerned with discussing only those datums that are applicable for correlation with the warm subtropical (Site 208) sequence, this means that the most useful datums are not tropical, as in the Early Neogene, but related to temperate taxa. The tropical Late Miocene and Pliocene datums are all well represented in Site 289 but not at the other sites to the south. These include the following datums, *N. acostaensis* F.A.; *G. plesiotumida* F.A.; *P. primalis* F.A.; *G. tumida* F.A.; *S. dehiscens* F.A.; *G. margaritae* F.A. and L.A.; *G. nepenthes* L.A.; *G. fistulosus* F.A. and L.A.; *S. seminulina* L.A.; and *G. obliquus* L.A. The ages of these datums are discussed by Saito (1977).

Globorotalia conomiozea F.A.

Just as the *Fohsella* lineage provides useful datums in the Middle Miocene and the *Globorotalia* lineage (*G. (G.) merotumida, G. (G.) plesiotumida* and *G. (G.) tumida*) in the Late Miocene of tropical Site 289, species of *Globoconella* are of value in the Neogene of the cool subtropical to temperate regions (Kennett, 1966, 1967, 1970, 1973; Kennet and Vella, 1975; Collen and Vella, 1973; Jenkins, 1967, 1971).

In contrast to tropical regions, only a few radiometric and/or paleomagnetic age determinations are available for the subtropical to temperate datums of the South Pacific.

The first evolutionary appearance of *G. conomiozea* has been paleomagnetically dated at 6.3 Ma by Ryan and others (1974) and Loutit and Kennett (1979). Within the middle of paleomagnetic Epoch 6 this datum is thus almost coeval with the *Pulleniatina primalis* datum in the tropics at 6.2 Ma (Saito, 1977).

D'Onofrio and others (1975) and Van Couvering and others (1976) have considered the first appearance datum of *G. conomiozea* to be almost coincident with the base of the Messinian Stage and estimated an age of 6.5 Ma for the first appearance datum of this taxon, correlating it with the middle of Zone N17. In the South Pacific, the first evolutionary appearance of *G. conomiozea* is at the base of the Kapitean Stage (Kennett, 1967; Loutit and Kennett, 1979). The first appearance of *G. conomiozea* in Site 208 is non-evolutionary, and is slightly after the first appearance of *P. primalis*. Thus, the *G. conomiozea* Zone of Site 208 broadly corresponds to Zone 17B in Site 289. A similar range was documented for *G. conomiozea* by Keller (1978) in the central North Pacific Site 310. The first appearance of *G. conomiozea* in Sites 207A and 284 are clear evolutionary gradations from *G. conoidea* (Kennett, 1973; Kennett and Vella, 1975).

Pliocene

Globorotalia margaritae F.A.

Blow (1969) considered the evolutionary first appearance of *Sphaeroidinella dehiscens* at N18/N19 boundary to mark the Miocene/Pliocene boundary. Bolli (1970), based on DSDP cores from Leg 4 in the central Atlantic and Caribbean, considered the *Globorotalia margaritae* Zone to be entirely of Pliocene age. Bolli considered the first appearance of *G. margaritae* to mark the Miocene/Pliocene boundary. Furthermore, Cita (1973) recorded the first appearance of *G. margaritae* to be slightly above the Miocene/Pliocene boundary in Mediterranean deep drilled sites, and to range up to the end of the Early Pliocene. Berggren (1977a) placed the approximate position of the Miocene/Pliocene boundary in DSDP Site 357 (Rio Grande Rise) at the initial appearance of *G. margaritae* and the extinction of *Globoquadrina dehiscens*. Krasheninnikov and Pflaumann (1978) regarded the first evolutionary appearance of *G. margaritae margaritae* from *G. margaritae primitiva* to mark the base of the Pliocene.

At Site 289 *Globorotalia tumida* first appears at 68 m (within Core 18) while *G. margaritae* (s.s.) and *S. dehiscens* both first appear 10 m higher at 58 m (within Core 17). The last appearance of *G. dehiscens* is in Core 18 at 65 m. Thus there is a stratigraphic overlap in the upper range of *G. dehiscens* and the lower range of *G. tumida*. Berggren (1973) regarded the initial appearance of *G. tumida* and the extinction of *G. dehiscens* (at 5.0 Ma) as the basis for recognition of the Miocene/Pliocene boundary. We consider the first appearance of *G. margaritae* (s.s.) to represent the Miocene/Pliocene boundary rather than *G. tumida* which overlaps the terminal part of the range of *G. dehiscens* at Site 289.

At Site 208, *G. tumida* is rare, and the first appearance of *S. dehiscens* is non-evolutionary, occurring near the base of Late Pliocene, thereby diminishing the value of these two species to be used as datums in the warm-subtropics. Nevertheless, the first appearance of *G. margaritae* marks the base of the Early Pliocene in Site 208 as well as in other South Pacific sites (Kennett, 1973), where

S. dehiscens and *G. tumida* are not present. Thus, the *G. margaritae* first appearance datum seems to be a much more reliable indicator for the base of the Early Pliocene, and serves as a common datum for correlating Site 289 and 208 (Fig. 14).

In recent years, there has been a number of papers dealing with the Miocene/Pliocene boundary with respect to the paleomagnetic stratigraphy (Ryan and others, 1974; Kennett and Watkins, 1974; Berggren and Van Couvering, 1974; Cita, 1975, 1976; Van Couvering and others, 1976; Burckle and Opdyke, 1977). It is now generally agreed that this boundary approximately lies at 5.0 to 5.1 Ma between the top of magnetic Epoch 5 and the lower part of the 'C' Event of the Gilbert. Berggren and Van Couvering (1974) estimated an age of 5.0 Ma for *S. dehiscens* first appearance datum, while Saito (1977) estimated an age of 5.0 Ma for the first appearance datum of *G. tumida* and 4.8 Ma for the *S. dehiscens* first appearance datum.

Globorotalia puncticulata F.A.

In New Zealand *G. puncticulata* first appears at the base of the Opoitian Stage, which has been paleomagnetically dated at 5.3 Ma (Loutit and Kennett, 1979). The first evolutionary appearance of *G. puncticulata* from its immediate ancestor *G. conomiozea* is clearly represented in Site 284 and 207A by very abundant specimens. In Site 284 *G. puncticulata* first appears at 142 m while in Site 207A it appears at 76 m (Core 4, Section 3). The Miocene/Pliocene boundary in these sites has been placed at the first appearance of *G. puncticulata* (Kennett, 1973). *G. margaritae* is absent in Site 284 because of very cool conditions. Nevertheless, it first appears in Site 207A approximately at the same level as *G. puncticulata.*

In Site 206 *G. puncticulata, G. margaritae* and *G. crassaformis* appear together because of a distinct unconformity representing a gap from latest Miocene to Early Pliocene (Fig. 14). This site, thus, does not assist in intersite correlation of the Miocene/Pliocene boundary between these different water masses.

In Site 208 the first appearance of *G. puncticulata* is later than the last appearance of *G. conomiozea* and the first appearance of *G. margaritae,* reflecting later migration to warmer waters after its initial evolutionary appearance in cooler waters to the south. Therefore, the Miocene/Pliocene boundary is placed at the first appearance of *G. margaritae* rather than *G. puncticulata.* In Site 289 *G. puncticulata* is absent.

Globorotalia crassaformis F.A.

The first appearance of *G. crassaformis* in South Pacific Sites 284, 207A, 206 and 208 is recorded above the first evolutionary appearance of *G. puncticulata,* providing a useful datum within the Pliocene in temperate areas. However, in a number of instances, these two events have been recorded almost at the same stratigraphic level (*e.g.,* at Site 111 in the North Atlantic, Poore and Berggren, 1974; at Site 310 in the North Pacific, Vincent, 1975 and Keller, 1978). *G. crassaformis* is one of the most widely distributed Pliocene planktonic species which occurs both in tropical and temperate regions. In tropical areas, however, it first appears in the Middle Pliocene, later than the temperate areas (Fleisher, 1974; Srinivasan, 1977; Vincent, 1977; Srinivasan and others, *in press*).

The first appearance of *G. crassaformis* predates the Gilbert A Event in the Indian Ocean Core V20-163 (Saito and others, 1975), indicating this datum to be at least 4.0 Ma. Berggren (1977b) suggested that the evolutionary first appearance of *G. crassaformis* from *G. puncticulata* occurred at 4.0 Ma.

Globorotalia inflata F.A.

The first evolutionary appearance of *G. inflata* at Site 208, 206, 207A and 284 provides a useful datum for correlation of DSDP sections in the South Pacific (Kennett, 1973; Kennett and Vella, 1975). Cita (1973) recorded the first evolutionary appearance of *G. inflata* in association with *G. tosaensis* in Mediterranean DSDP sites. In the South Pacific DSDP sites, however, the first appearance of *G. inflata* precedes the first occurrence of *G. tosaensis* as it does in most temperate sections (Vincent, 1977). Keller (1978) regarded the first appearance of *G. inflata* at North Pacific DSDP Site 310 to be coincident with the first appearance of *G. tosaensis* in the South Pacific. The base of the *G. puncticulata* Zone of Kennett (1973) was considered by Keller (1978) to be equivalent to the first appearances of *G. crassaformis* and *G. puncticulata* and corresponding to the base of Zone N19. The *G. crassaformis* Zone and the *G. tosaensis* Zone distinguished in the South Pacific (Kennett, 1973) were not distinguished by Keller (1978) in the North Pacific Site 310. Keller (1978) also pointed out the striking similarity in stratigraphic range and frequency distribution of *G. crassaformis* and the later forms of *Globoconella* in temperate regions of the North and South Pacific. Similar ranges have also been recorded in the North Atlantic (Poore and Berggren, 1974, 1975) and South Atlantic (Berggren, 1977a, b; Jenkins, 1978).

The first evolutionary appearance of *G. inflata* is at or just below the NN16/NN15 nannofossil zonal boundary. This datum at 3.0 Ma is slightly younger than the evolutionary appearance of *G. tosaensis* (Vincent, 1977).

Quaternary

Globorotalia truncatulinoides F.A.

The base of the Quaternary in all of the South Pacific sites examined (Site 289, 208, 207A, 206 and 284) is taken to be at the first appearance of *Globo-*

rotalia truncatulinoides (Fig. 14). In these sites, the transition from *G. tosaensis* to *G. truncatulinoides* is not always simple. For instance, at Sites 284 and 207A (temperate) *G. truncatulinoides* appears suddenly and is continuously present throughout the Pleistocene. Whereas a fairly simple gradation is present between *G. tosaensis* and *G. truncatulinoides* in Site 206 (Kennett, 1973), in Site 208 the transition from *G. tosaensis* to *G. truncatulinoides* is not a simple gradational evolutionary change (Kennett, 1973). Instead the relative abundance of the two forms fluctuates significantly within the interval of overlap, as in a South Pacific piston core described by Kennett and Geitzenauer (1969).

The first appearance of *G. truncatulinoides* has been dated 1.85 Ma (Olduvai Event) by Berggren and Van Couvering (1974) and 1.95 Ma by Saito (1977) based on magnetic stratigraphy.

ACKNOWLEDGMENTS

This research was supported by U.S. National Science Foundation grants OCE76-81489 (CENOP) and DPP78-08512 (Division of Polar Programs).

We thank N. Penrose for technical assistance, M. Leonard for drafting the figures, D. Scales for photography and N. Meader for typing the manuscript. The Deep Sea Drilling Project supplied the necessary materials for this research.

We wish to thank Drs. H. M. Bolli, W. A. Berggren and Gerta Keller who read the manuscript and offered many helpful comments and criticisms.

REFERENCES CITED

NOTE: All references to Deep Sea Drilling Project Initial Reports have been abbreviated; for complete citations see *Appendix I* at the end of this volume, or the individual Initial Reports as published.

Allan, R. S., 1956, Report of the Standing Committee on datum planes in the geological history of the Pacific region: Proc. 8th Pacific Sci. Cong. 2: p. 325–423.

Andrews, J. E., Packham, G. and others, 1975: Init. Repts. DSDP, v. 30, p. 231–398.

Bandy, O. L., 1960, The geologic significance of coiling ratios in the foraminifer *Globigerina pachyderma* (Ehrenberg): Jour. Paleontology, v. 34, no. 4, p. 671–681.

———, 1972, Origin and development of *Globorotalia (Turborotalia) pachyderma* (Ehrenberg): Micropaleontology, v. 18, p. 294–318.

———, 1975, Messinian evaporite deposition and the Miocene/Pliocene boundary, Pasquasia-Capodarso sections, Silicy, *in* Saito, T., and Burckle, L. H., eds., Late Neogene epoch boundaries: Micropaleontology Spec. Pub. No. 1, (Micropaleontology Press, N. Y.), p. 49–63.

———, Frerichs, W. E., and Vincent, E., 1967, Origin, development and geologic significance of *Neogloboquadrina* Bandy, Frerichs and Vincent, gen. nov.: Cushman Found. Foram. Res. Contr., v. 18, p. 152–157.

Banner, F. T., and Blow, W. H., 1965, Progress in the planktonic foraminiferal biostratigraphy of the Neogene: Nature, v. 208, p. 1164–1166.

———, and ———, 1967, The origin, evolution and taxonomy of the foraminiferal genus *Pulleniatina* Cushman, 1927: Micropaleontology, v. 13, p. 133–162.

Berger, W. H., 1967, Foraminiferal ooze: Solution at depths: Science, v. 156, p. 383–385.

———, 1970, Planktonic foraminifera: selective solution and the lysocline: Marine Geology, v. 8, p. 111–138.

Berggren, W. A., 1969, Cenozoic chronostratigraphy, planktonic foraminiferal zonation and the radiometric time scale: Nature, v. 224, p. 1072–1075.

———, 1971, Tertiary boundaries and correlations, *in* Funnell, B. M., and Riedel, W. R., ed., Micropaleontology of the Oceans: Cambridge Univ. Press., p. 693–809.

———, 1972a, A Cenozoic time-scale—some implications for regional geology and paleobiology: Lethaia, v. 5, p. 195–215.

———, 1972b, Cenozoic biostratigraphy and paleobiogeography of the North Atlantic: Init. Repts. DSDP, v. 12, p. 965–1001.

———, 1973, The Pliocene time scale; calibration of planktonic foraminiferal and calcareous nannoplankton zones: Nature, v. 243, p. 391–397.

———, 1977a, Late Neogene planktonic foraminiferal biostratigraphy of DSDP Site 357 (Rio Grande Rise): Init. Repts. DSDP, v. 39, p. 591–614.

———, 1977b, Late Neogene planktonic foraminiferal biostratigraphy of the Rio Grande Rise (South Atlantic): Marine Micropaleontology, v. 2, p. 251–265.

———, and Amdurer, M., 1973, Late Paleogene (Oligocene) and Neogene planktonic foraminiferal biostratigraphy of the Atlantic Ocean (Lat. 30°N to 30°S): Riv. Ital. Paleontol., v. 79, 337 p.

———, Lohmann, G. P., and Poore, R. Z., 1974, Shore laboratory report on Cenozoic planktonic foraminifera: Leg 22: Init. Repts. DSDP, v. 22, p. 635–655.

———, Phillips, J. D., Bertels, A., and Wall, D., 1967, Late Pliocene-Pleistocene stratigraphy in deep-sea cores from the South Central North Atlantic: Nature, v. 216, p. 253–254.

———, and Van Couvering, J. A., 1974, The Late Neogene: biostratigraphy, geochronology and paleoclimatology of the last 15 million years in marine and continental sequences: Paleogeography, Paleoclimatology, Paleoecology, v. 16, p. 1–126.

———, and ———, 1978, Biochronology, *in* Cohee, G. V., Glaessner, M. F., and Hedberg, H. D., The geologic time scale, studies in geology: Am. Assoc. Petroleum Geologists Bull., n. 6, p. 39–55.

Blow, W. H., 1956, Origin and evolution of the foraminiferal genus *Orbulina* d'Orbigny: Micropaleontology, v. 2, p. 57–70.
———, 1969, Late middle Eocene to Recent planktonic foraminiferal biostratigraphy, *in* Bronnimann, P., and Renz, H. H., eds., First international conference on planktonic microfossils, Geneva, 1967: p. 199–421.
———, and Banner, F. T., 1966, The morphology, taxonomy and biostratigraphy of *Globorotalia barisanensis* LeRoy, *Globorotalia fohsi* Cushman and Ellisor, and related taxa: Micropaleontology, v. 12, p. 286–302.
Boersma, A., 1977, Cenozoic planktonic foraminifera—DSDP Leg 39 (South Atlantic): Init. Repts. DSDP, v. 39, p. 567–590.
Bolli, H. M., 1950, The direction of coiling in the evolution of some Globorotalidae: Contr. Cushman Found. Foram. Res., v. 1, no. 3–4, p. 82–89.
———, 1957a, The genera *Globigerina* and *Globorotalia* in the Paleocene-lower Eocene Lizard Springs formation of Trinidad, B. W. I.: U. S. Nat. Mus. Bull. 215, p. 61–81.
———, 1957b, Planktonic foraminifera from the Oligocene-Miocene Cipero and Lengua Formations of Trinidad., B. W. I.: U. S. Nat. Mus. Bull. 215, p. 97–213.
———, 1966, Zonation of Cretaceous to Pliocene marine sediments based on planktonic foraminifera: Assoc. Venezolana Geol. Min., Bol., v. 9, no. 1, p. 2–32.
———, 1967, The subspecies of *Globorotalia fohsi* Cushman and Ellisor and the zones based on them: Micropaleontology, v. 13, p. 502–512.
———, 1970, The foraminifera of Sites 23–31, Leg 4: Init. Repts. DSDP, v. 4, p. 577–643.
———, and Bermudez, P. J., 1965, Zonation based on planktonic foraminifera of Middle Miocene to Pliocene warm water sediments: Assoc. Venezolana Geol. Min., Bol., v. 8, no. 4, p. 121–149.
———, and Krasheninnikov, A., 1977, Problems in Paleogene and Neogene Correlations based on planktonic foraminifera: Micropaleontology, v. 23, p. 436–452.
———, and Premoli Silva, I., 1973, Oligocene to Recent planktonic foraminifera and stratigraphy of the Leg 15 sites in the Caribbean Sea: Init. Repts. DSDP, v. 15, p. 475–497.
Bramlette, M. N., and Martini, E., 1964, The great change in calcareous nanno-plankton fossils between the Maastrichtian and Danian: Micropaleontology, v. 10, p. 291.
———, and Wilcoxon, J. A., 1967, Middle Tertiary calcareous nanno-plankton of the Cipero section, Trinidad, W. I.: Tulane Studies Geology, v. 5, p. 93.
Bronnimann, P., and Resig, J., 1971, A Neogene globigerinacean biochronologic time scale of the south-western Pacific: Init. Repts. DSDP, v. 7, p. 1235–1469.
Bukry, D., 1972, Coccolith stratigraphy, Leg 9 Deep Sea Drilling Project: Init. Repts. DSDP, v. 9, p. 817–832.
———, 1975, Coccolith and silicoflagellate stratigraphy, North Western Pacific Ocean, Deep Sea Drilling Project Leg 32: Init. Repts. DSDP, v. 32, p. 677–692.
Burckle, L. H., and Opdyke, N. D., 1977, Late Neogene diatom correlations in the Circum-Pacific, *in* Proceedings of the First International Congress on Pacific Neogene Stratigraphy, Tokyo, 1976, p. 255–284.
Cifelli, R. J., 1969, Radiation of Cenozoic planktonic foraminifera: Systematic Zoology, v. 18, p. 154–168.
Cita, M. B., 1973, Pliocene biostratigraphy and chronostratigraphy: Init. Repts. DSDP, v. 4, p. 1343–1378.
———, 1975, The Miocene/Pliocene boundary: History and definition, *in* Saito, T., and Burckle, L. H., eds., Late Neogene Epoch Boundaries, Spec. Publ. No. 1: Micropaleontology Press, N. Y., p. 1–30.
———, 1976, Planktonic foraminiferal biostratigraphy of the Mediterranean Neogene, *in* Takayanagi, Y., and Saito, T., eds., Progress in Micropaleontology, Selected Papers in Honor of Prof. K. Asano: Micropaleontology Press, N. Y., p. 47–68.
Collen, J. D., and Vella, P., 1973, Pliocene planktonic foraminifera, southern North Island, New Zealand: Jour. Foraminiferal Res., v. 3, p. 13–29.
D'Onofrio, S., Gianelli, L., Iaccarino, S., Morlotti, E., Romeo, M., Salvatorini, G., Sampo, M., and Sprovieri, R., 1975, Planktonic foraminifera of the Upper Miocene from some Italian sections and the problem of the lower boundary of the Messinian: Boll. Soc. Paleontol. Ital., v. 14, n. 2, p. 177–196.
Ericson, D. B., 1953, Sediments of the Atlantic Ocean: Columbia Univ., Lamont Geol. Observ. Tech. Rept., no. 1, p. 2–34.
———, and Wollin, G., 1956, Correlation of six cores from the Equatorial Atlantic and the Caribbean: Deep-sea Research, v. 3, n. 2, p. 104–125.
Fischer, A. G., and Arthur, M. A., 1977, Secular variations in the Pelagic Realm: Soc. Econ. Paleontologists Mineralogists Spec. Pub. No. 25, p. 19–50.
Fleisher, R. L., 1974, Cenozoic planktonic foraminifera and biostratigraphy, Arabian Sea Deep Sea Drilling Project, Leg 23A: Init. Repts. DSDP, v. 23, p. 1001–1072.
———, 1977, Oligocene planktonic foraminiferal assemblages from Deep-Sea Drilling Project sites in the Indian Ocean, *in* Heirtzler, J. R., *et al.*, eds., Indian Ocean geology and biostratigraphy: Amer. Geophys. Union, p. 459–467.
Glass, B., Ericson, D. B., Heezen, B. C., Opdyke, N. D., and Glass, J. A., 1967, Geomagnetic reversals and Pleistocene chronology: Nature, v. 216, p. 437–442.
Glaessner, M. F., 1937, Plankton foraminiferen aus der Kreide und dem Eozaen und ihre stratigraphische Bedeutung: Moscow Univ. Lab. Paleont., Studies Micropaleont., v. 1, p. 27–46.
Goll, R. M., 1972, Section on radiolaria for synthesis chapter, Leg. 9: Init. Repts. DSDP, v. 9, p. 947–1058.
Hays, J. D., Saito, T., Opdyke, N. D., and Burckle, L. H., 1969, Pliocene-Pleistocene sediments of the equatorial Pacific: Their paleomagnetic, biostratigraphic, and climatic record: Geol. Soc. America Bull., v. 80, p. 1481–1513.

Hornibrook, N. de B., 1966, New Zealand Tertiary microfossil zonation, correlation and climate, *in* Hatai, K., ed., Tertiary correlations and climatic changes in the Pacific, Symposium No. 25: 11th Pacific Sci. Congress—Tokyo, p. 29–39.

———, and Edwards, A. R., 1971, Integrated planktonic foraminiferal and calcareous nannoplankton datum levels in the New Zealand Cenozoic: Proc. 2, Int. Conf. Planktonic Microfossils, 2nd (Roma, 1970), p. 649–657.

Ikebe, N., Takayanagi, Y., Chiji, M., and Chinzei, K., 1972, Neogene biostratigraphy and radiometric time scale of Japan—an attempt at intercontinental correlation: Pacific Geology, v. 4, p. 39–78.

Ingle, J. C., Jr., 1967, Foraminiferal biofacies variation and the Miocene-Pliocene boundary in Southern California: Bull. Amer. Paleontology, v. 52, n. 236, p. 217–394.

Jenkins, D. G., 1965, The origin of the species *Globigerinoides trilobus* (Reuss) in New Zealand: Contr. Cushman Found. Foram. Res., v. 16, n. 3, p. 116–120.

———, 1966a, Planktonic foraminiferal zones and new taxa from the Danian to Lower Miocene of New Zealand: New Zealand Jour. Geol. Geophys., v. 8, p. 1088–1126.

———, 1966b, Two lineages from the Neogene planktonic foraminifera of the Australasian region: Proc. Int. un. Geol. Sci. Comm. Mediterranean Neogene Stratgr. 3rd sess., p. 23–29.

———, 1966c, Planktonic foraminiferal datum planes in the Pacific and Trinidad Tertiary: New Zealand Jour. Geol. Geophys, v. 9, p. 424–427.

———, 1967, Planktonic foraminiferal zones and new taxa from the lower Miocene to the Pleistocene of New Zealand: New Zealand Jour. Geol. Geophys., v. 10, p. 1064–1078.

———, 1971, New Zealand Cenozoic planktonic foraminifera: New Zealand Geol. Surv. Paleontol. Bull., No. 42, 278 p.

———, 1975, Cenozoic planktonic foraminiferal biostratigraphy of the Southwestern Pacific and Tasman Sea—DSDP Leg 29: Init. Repts. DSDP, v. 29, p. 449–467.

———, 1977, Lower Miocene planktonic foraminifera from a borehole in the English Channel: Micropaleontology, v. 23, p. 297–318.

———, 1978, Neogene planktonic foraminifers from DSDP Leg 40 sites 360 and 362 in the Southeastern Atlantic: Init. Repts. DSDP, v. 40, p. 723–739.

———, and Orr, W. N., 1971, Cenozoic planktonic foraminiferal zonation and the problem of test solutions: Revista Espanola de Micropaleontologia, v. 3, p. 301–304.

———, and ———, 1972, Planktonic foraminiferal biostratigraphy of the eastern equatorial Pacific, Leg 9: Init. Repts. DSDP, v. 9, p. 1059–1196.

Kaneps, A. G., 1973, Cenozoic planktonic foraminifera from the eastern equatorial Pacific Ocean: Init. Repts. DSDP, v. 16, p. 713–745.

———, 1975, Cenozoic planktonic foraminifera from Antarctic Deep Sea Sediments, Leg. 28, DSDP: Init. Repts. DSDP, v. 28, p. 573–583.

Keany, J., and Kennett, J. P., 1972, Pliocene-early Pleistocene paleoclimatic history recorded in Antarctic-Subantarctic deep-sea cores: Deep-sea Research, v. 19, p. 529–548.

Keller, G., 1978, Late Neogene biostratigraphy and paleoceanography of DSDP Site 310 Central North Pacific and correlation with the Southwest Pacific: Marine Micropaleontology, v. 3, p. 97–119.

———, *in press,* a, Early to Middle Miocene planktonic foraminiferal datum levels of the Equatorial and Subtropical Pacific: Micropaleontology.

———, *in press,* b, Planktonic foraminiferal faunas of the Equatorial Pacific suggest Early Miocene origin of present oceanic circulation: Marine Micropaleontology.

———, *in press,* c, Middle to Late Miocene datum levels and paleoceanography of the North and Southeastern Pacific Ocean: Marine Micropaleontology.

Kennett, J. P., 1966, The *Globorotalia crassaformis* bioseries in north Westland and Marlborough, New Zealand: Micropaleontology, v. 12, p. 235–245.

———, 1967, Recognition and correlation of the Kapitean Stage (Upper Miocene, New Zealand): New Zealand Jour. Geol. Geophys., v. 10, p. 1051–1063.

———, 1968, Paleoceanographic aspects of the foraminiferal zonation in the Upper Miocene-Lower Pliocene of New Zealand: Committee Mediterranean Neogene Stratigraphy Proc. IV Session. Giornale di Geologia Annali Del Museo Geologia de Bologna 35, p. 143–156.

———, 1970, Pleistocene paleoclimates and foraminiferal biostratigraphy in Subarctic deep-sea cores: Deep-sea Research, v. 17, p. 125–140.

———, 1973, Middle and Late Cenozoic planktonic foraminiferal biostratigraphy of the southwest Pacific—DSDP Leg 21: Init. Repts. DSDP, v. 21, p. 575–640.

———, 1978, The development of planktonic biogeography in the Southern Ocean during the Cenozoic: Marine Micropaleontology, v. 3, p. 301–345.

———, and Geitzenauer, K. R., 1969, The Pliocene-Pleistocene boundary in a South Pacific deep-sea core: Nature, v. 224, p. 899.

———, and Vella, P., 1975, Late Cenozoic planktonic foraminifera and paleoceanography at DSDP Site 284 in the cool subtropical South Pacific: Init. Repts. DSDP, v. 29, p. 769–799.

———, and Watkins, N. D., 1974, Late Miocene–Early Pliocene paleomagnetic stratigraphy, paleoclimatology and biostratigraphy in New Zealand: Geol. Soc. America Bull., v. 85, p. 1385–1398.

Krasheninnikov, V. A., and Pflaumann, U., 1978, Zonal stratigraphy of Neogene deposits of the eastern part of the

Atlantic Ocean by means of planktonic foraminifers, Leg 41 DSDP: Init. Repts. DSDP, v. 41, p. 613–657.
LAMB, J. L., AND BEARD, J. H., 1972, Late Neogene planktonic foraminifers in the Caribbean, Gulf of Mexico, and Italian stratotypes: Kansas Univ. Paleont. Contr., Art. 57 (Protozoa 8), 67 p., 36 pl.
LAUGHTON, A. S., BERGGREN, W. A. AND OTHERS, 1972, Site 116 and 117: Init. Repts. DSDP, v. 12, p. 395–672.
LOUTIT, T. S., AND KENNETT, J. P., 1979, Application of carbon isotope stratigraphy of Late Miocene shallow marine sediments, New Zealand: Science, v. 204, p. 1196–1199.
MCGOWRAN, B., 1974, Foraminifera: Init. Repts. DSDP, v. 22, p. 609–627.
PARKER, F. L., 1967, Late Tertiary biostratigraphy (planktonic foraminifera) of tropical Indo-Pacific deep sea cores: Bull. Amer. Paleontology, v. 52, n. 235, p. 115–208.
———, 1973, Late Cenozoic biostratigraphy (planktonic foraminifera) of tropical Atlantic deep-sea sections: Revista Espanola de Micropaleontologia, v. 5, p. 253–289.
———, 1974, Upper Neogene biostratigraphy (planktonic foraminifera) of DSDP Sites 139 and 141: Jour. Paleontology, v. 4, p. 9–15.
POORE, R. Z., 1979, Oligocene through Quaternary planktonic foraminiferal biostratigraphy of the North Atlantic: DSDP Leg 49: Init. Repts. DSDP, v. 49, p. 447–517.
———, AND BERRGREN, W. A., 1974, Pliocene biostratigraphy of the Labrador Sea: calcareous plankton: Jour. Foraminiferal Res., v. 4, p. 91–108.
———, AND BERGGREN, W. A., 1975, Late Cenozoic planktonic foraminiferal biostratigraphy and paleoclimatology of Hatton-Rockall Basin: DSDP Site 116: Jour. Foraminiferal Res., v. 5, p. 270–293.
RIEDEL, W. R., 1957a, Radiolaria: a preliminary stratigraphy: Rept. Swed. Deep Sea Exped., v. 6, no. 3, 59 p.
———, 1957b, Eocene Radiolaria: U. S. Geological Survey Prof. Paper 280 G., 257 p.
RYAN, W. B. F., CITA, M. B., DREYFUS, R. M., BURCKLE, L. H., AND SAITO, T., 1974, A paleomagnetic assignment of Neogene stage boundaries and the development of isochronous datum planes between the Mediterranean, the Pacific and Indian Oceans in order to investigate the response of the world ocean to the Mediterranean "salinity crisis.": Riv. Ital. Paleontol., v. 80, p. 631–688.
SAITO, T., 1972, Late Neogene epoch boundaries in deep sea sediments: Internatl. Geol. Cong. 24th, Canada, Abstr., p. 540.
———, 1976, Geologic significance of coiling direction in the planktonic foraminifera *Pulleniatina:* Geology, v. 4, p. 305–309.
———, 1977, Late Cenozoic planktonic foraminiferal datum levels: the present state of knowledge toward accomplishing pan-Pacific correlation, *in* Proceed. of the First International Congress of Pacific Neogene Stratigraphy, Tokyo, 1976, p. 61–80.
SAITO, T., BURCKLE, L. H., AND HAYS, J. D., 1975, Late Miocene to Pleistocene biostratigraphy of equatorial Pacific sediments, *in* Saito, T., and Burckle, L. H., eds., Late Neogene epoch boundaries, Micropaleontology Spec. Pub. No. 1: New York, Micropaleontology Press, p. 226–244.
SALVATORINI, G., AND CITA, M. B., 1979, Miocene foraminiferal stratigraphy, DSDP Site 397 (Cape Bojador, North Atlantic): Init. Repts. DSDP, v. 47, p. 317–373.
SAVIN, S. M., 1977, The history of the earth's surface temperature during the past 100 million years: Annual Review Earth Planet. Sci., v. 5, p. 319–344.
SHACKLETON, N. J., AND KENNETT, J. P., 1975, Late Cenozoic oxygen and carbon isotopic changes at DSDP Site 284: Implications for glacial history of the Northern Hemisphere and Antarctica: Init. Repts. DSDP, v. 29, p. 801–807.
SMITH, L. A., AND BEARD, J. H., 1973, The Late Neogene of Gulf of Mexico: Init. Repts. DSDP, v. 10, p. 643–677.
SRINIVASAN, M. S., 1977, Standard planktonic foraminiferal zones of the Andaman-Nicobar Late Cenozoic: Recent Res. Geol., v. 3, p. 23–29, Hindustan Publ. Corp, Delhi.
———, AND KENNETT, J. P., 1976, Evolution and phenotypic variation in the Late Cenozoic *Neogloboquadrina dutertrei* plexus, *in* Takayanagi, Y., and Saito, T., eds., Progress in Micropaleontology: Micropaleontology Press Spec. Pub. 2, p. 329–355.
———, ———, AND RODDA, P., *in press,* Late Neogene planktonic foraminiferal biostratigraphy, Suva, Fiji: Journal Paleontology.
STAINFORTH, R. M., LAMB, J. L., LUTERBACHER, H., BEARD, J. H., AND JEFFORDS, R. M., 1975, Cenozoic planktonic foraminiferal zonation and characteristics of index forms: University of Kansas Paleontological Contributions, Article 62, 425 p.
THEYER, F., AND HAMMOND, S. R., 1974, Paleomagnetic polarity sequence and radiolarian zones, Brunhes to polarity Epoch 20: Earth Planet. Sci. Letters, v. 22, p. 307–319.
———, MATO, C. Y., AND HAMMOND, S. R., 1978, Paleomagnetic and geochronologic calibration of latest Oligocene to Pliocene radiolarian events, Equatorial Pacific: Marine Micropaleontology, v. 3, p. 377–395.
THOMPSON, P. R., AND SCIARRILLO, J. R., 1978, Planktonic foraminiferal biostratigraphy in the Equatorial Pacific: Nature, v. 275, n. 5683, p. 29–33.
VAN COUVERING, J. A., AND BERGGREN, W. A., 1977, Biostratigraphical basis of the Neogene time scale, *in* Kauffman and Hazel, eds., Concepts and methods of biostratigraphy: p. 283–306.
———, ———, DRAKE, R. E., AGUIRRE, E., AND CURTIS, G. H., 1976, The terminal Miocene event: Marine Micropaleontology, v. 1, p. 263–286.
VINCENT, E., 1975, Neogene planktonic foraminifera from the central North Pacific Leg 32, Deep Sea Drilling Project: Init. Repts. DSDP, v. 32, p. 165–801.
———, 1977, Indian Ocean Neogene planktonic foraminiferal biostratigraphy and its paleoceanographic implications,

in Heirtzler, J. R., *et al.*, eds., Indian Ocean geology and biostratigraphy: Amer. Geophys. Union, p. 469–484.
WADE, M. E., 1964, Application of the lineage concept to biostratigraphic zoning based on planktonic foraminifera: Micropaleontology, v. 10, p. 273–290.
WALTERS, R., 1965, The *Globorotalia zealandica* and *G. miozea* lineages: New Zealand Jour. Geol. Geophys., v. 8, p. 109–127.

SEPM SPECIAL PUBLICATION NO. 32, P. 433–444, OCTOBER 1981

SYNTHESIS OF SILICOFLAGELLATE STRATIGRAPHY FOR MAESTRICHTIAN TO QUATERNARY MARINE SEDIMENT

DAVID BUKRY
United States Geological Survey, Scripps Institution of Oceanography,
La Jolla, California 92093 U.S.A

ABSTRACT

Cosmopolitan and low-latitude silicoflagellate distributions determined from Deep Sea Drilling Project cores are synthesized into a sequence of 23 biostratigraphic zones and subzones for tropical and subtropical ocean areas. The stratigraphic relation of major morphologic changes in silicoflagellates from all areas is summarized. Two new Neogene subzones, one new Paleogene subzone, one new species, and one new variety are defined: *Dictyocha perlaevis ornata* Subzone, *Dictyocha neonautica* Subzone, *Distephanus speculum haliomma* Subzone, *Dictyocha neonautica* Bukry, n. sp., and *D. neonautica* var. *cocosensis* Bukry, n. var.

INTRODUCTION

Silicoflagellates are unicellular, flagellated, marine algae producing microscopic, tubular skeletons of opal. They are most abundant in modern ocean upwelling areas and are known in fossil deposits that are as old as Late Cretaceous. Typically, silicoflagellates are associated with more abundant diatoms. The greatest value of silicoflagellates to Deep Sea Drilling Project (DSDP) research has been in dating sediment from high-latitude areas and in providing a guide to paleoenvironmental conditions.

Biostratigraphic study of silicoflagellates from the best reference sections of DSDP has resulted in numerous individual interpretations for regional zonal sequences (for example, see Ciesielski (1975) fig. 2 for the Neogene, and Bukry and Foster (1974), and Martini (1972, 1974) for the Paleogene). Water-mass changes over different sites have also contributed to local signatures in silicoflagellate assemblages.

PROCESSING AND RECORDING

Another factor bearing on stratigraphic variation is the unstandardized technique of data acquisition and portrayal between workers. For example, cleaning and fractionation procedures differed considerably for Leg 28, where three size-fraction slides were employed for some samples (Ciesielski, 1975), for Leg 29 and 36, where many unprocessed smear slide preparations were used (Bukry, 1975b, 1976b), and for Leg 49 (Bukry, 1979b) where only hydrogen peroxide and hydrochloric acid residues were used. Several styles are evident in establishing and reporting the species arrays in silicoflagellate assemblages. This variation is illustrated by Leg 31 (Ling, 1975) and Leg 38 (Martini and Müller, 1976) where relative abundances are recorded by code letters (ACFR or X and O) which are not convertible into a real count; neither is a total count provided for weighing the merits of results between samples and sites. Other philosophies of notation include Ciesielski (1975) and Martini (1971), who provided direct counts and totals to allow readers to make their own later interpretations of data, although there is wide variation in the size of the counts. Most DSDP silicoflagellate data are presented in terms of total count and percentage figures for individual species in an assemblage. With counts of 200 to 300 per sample, the use of percentages provides a convenient way for the author and reader to establish trends and ratios on a consistent basis within and between sites. Owing to fluctuations in the abundance of silicoflagellates and to the short time available due to DSDP publication deadlines, authors often must compromise using a goal of 100 specimens per sample (Ciesielski, 1975) instead of attempting counts of 200 to 300 which would provide a more satisfactorily reproducible census.

MORPHOLOGY AND STRATIGRAPHY

With the preceding representational constraints in mind, the studies of DSDP silicoflagellates, nevertheless, provide an excellent basis for synthesizing the stratigraphy and latitudinal variety of silicoflagellates. The distribution of silicoflagellates through DSDP Legs 1 to 43 has been tabulated (Bukry, 1978c) and shows an abundance of Upper Miocene and Upper Pliocene to Pleistocene occurrences. Paleogene intervals are known from few sites, making zonal conclusions simpler, but less accurate, because of more limited comparisons.

Some of the more stratigraphically useful morphologic events of silicoflagellates are summarized in Figure 1. Acknowledging that only two Cretaceous and three Paleocene DSDP sites have been studied, the earliest major silicoflagellate event is

AGE	COCCOLITHS ZONE	COCCOLITHS SUBZONE	SILICOFLAGELLATES ZONE	SILICOFLAGELLATES SUBZONE
Quaternary	*Emiliania huxleyi*		*Dictyocha aculeata aculeata*	
	Gephyrocapsa oceanica	*Ceratolithus cristatus*		
		Emiliania ovata	*Mesocena quadrangula*	
	Crenalithus doronicoides	*Gephyrocapsa caribbeanica*	*Dictyocha stapedia stapedia*	*Dictyocha perlaevis delicata*
		Emiliania annula		
Pliocene	*Discoaster brouweri*	*Cyclococcolithina macintyrei*		
		Discoaster pentaradiatus		*Dictyocha perlaevis ornata*
		Discoaster surculus		
		Discoaster tamalis		
	Reticulofenestra pseudoumbilica	*Discoaster asymmetricus*	*Dictyocha fibula*	-
		Sphenolithus neoabies		
	Amaurolithus tricorniculatus	*Ceratolithus rugosus*		
		Ceratolithus acutus		
Miocene		*Triquetrorhabdulus rugosus*		
	Discoaster quinqueramus	*Amaurolithus primus*		*Dictyocha neonautica*
		Discoaster berggrenii		
	Discoaster neohamatus	*Discoaster neorectus*	*Dictyocha brevispina*	
		Discoaster bellus		
	Discoaster hamatus	*Catinaster calyculus*		
		Helicosphaera carteri		
	Catinaster coalitus			
	Discoaster exilis	*Discoaster kugleri*	*Corbisema triacantha*	*Distephanus stauracanthus*
		Coccolithus miopelagicus		-
	Sphenolithus heteromorphus			
	Helicosphaera ampliaperta		*Naviculopsis ponticula*	
	Sphenolithus belemnos		*Naviculopsis quadrata*	
	Triquetrorhabdulus carinatus	*Discoaster druggii*	*Naviculopsis lata*	
		Discoaster deflandrei		
Oligocene		*Cyclicargolithus abisectus*	*Naviculopsis biapiculata*	*Distephanus speculum haliomma*
	Sphenolithus ciperoensis	*Dictyococcites bisectus*		
		Cyclicargolithus floridanus		-
	Sphenolithus distentus			
	Sphenolithus predistentus			
	Helicosphaera reticulata	*Reticulofenestra hillae*	*Corbisema apiculata*	
		Coccolithus formosus		
		Coccolithus subdistichus		
Eocene	*Discoaster barbadiensis*	*Isthmolithus recurvus*		
		Chiasmolithus oamaruensis		
	Reticulofenestra umbilica	*Discoaster saipanensis*	*Dictyocha hexacantha*	
		Discoaster bifax		
	Nannotetrina quadrata	*Coccolithus staurion*	*Naviculopsis foliacea*	*Dictyocha spinosa*
		Chiasmolithus gigas		
		Discoaster strictus		
	Discoaster sublodoensis	*Rhabdosphaera inflata*		
		Discoasteroides kuepperi		
	Discoaster lodoensis			
	Tribrachiatus orthostylus			*Naviculopsis robusta*
	Discoaster diastypus	*Discoaster binodosus*	*Naviculopsis constricta*	
		Tribrachiatus contortus		
Paleocene	*Discoaster multiradiatus*	*Campylosphaera eodela*		
		Chiasmolithus bidens		
	Discoaster nobilis			
	Discoaster mohleri			
	Heliolithus kleinpellii		*Corbisema hastata*	
	Fasciculithus tympaniformis			
	Cruciplacolithus tenuis			
Cretaceous	*Micula mura*		*Lyramula furcula*	
	Lithraphidites quadratus			
	Tetralithus trifidus			

FIG. 1.—Correlation of tropical/subtropical silicoflagellate and coccolith zones from Deep Sea Drilling Project sites.

the disappearance of the dominant Cretaceous genera at the base of the Paleocene. The abrupt disappearance of *Vallacerta* and *Lyramula* may be related to physical and biological effects on plankton associated with dramatic cosmic (West, 1979; Alvarez and others, 1980) or oceanographic events (Thierstein and Berger, 1978; Gartner and Keany, 1978). But the long-ranging genera *Corbisema* and *Dictyocha* cross the boundary.

Paleocene silicoflagellates showed a great diversification in *Corbisema* and the successful appearance of two genera, *Naviculopsis* and *Mesocena.* Major structural changes in these genera, such as acquisition and changes in spine proportions and positions, are useful stratigraphic guides through the Cenozoic.

Important diversification of silicoflagellate morphologies continued through the Eocene. The new genus *Distephanus,* first recorded from the Lower Eocene, has an apical ring instead of a bar, as in the related genus *Dictyocha.* In the Middle Eocene, sparse *Distephanus* with divided rings appear, but the first consistent populations of *Distephanus* with multiply-divided rings occur much higher in the Upper Oligocene. Spined and septate *Mesocena* also appear in the Eocene.

Although speciation continued in the Oligocene (Fig. 2, Table 1), no major new structural groups appeared. Extinctions provide many of the stratigraphic guides above the Eocene.

In the Upper Miocene, at low latitudes, Martini (1971) pointed out a reversal in the bar orientation of *Dictyocha* that could be used stratigraphically. Analysis of DSDP occurrence charts published from 1969–1979 shows that this is a fundamental change, because the asperoid orientation (Fig. 1) was dominant at low latitude at essentially all levels below the Upper Miocene. High-latitude sites, however, show fluctuation in dominance between asperoid and fibuloid forms. This suggests that within

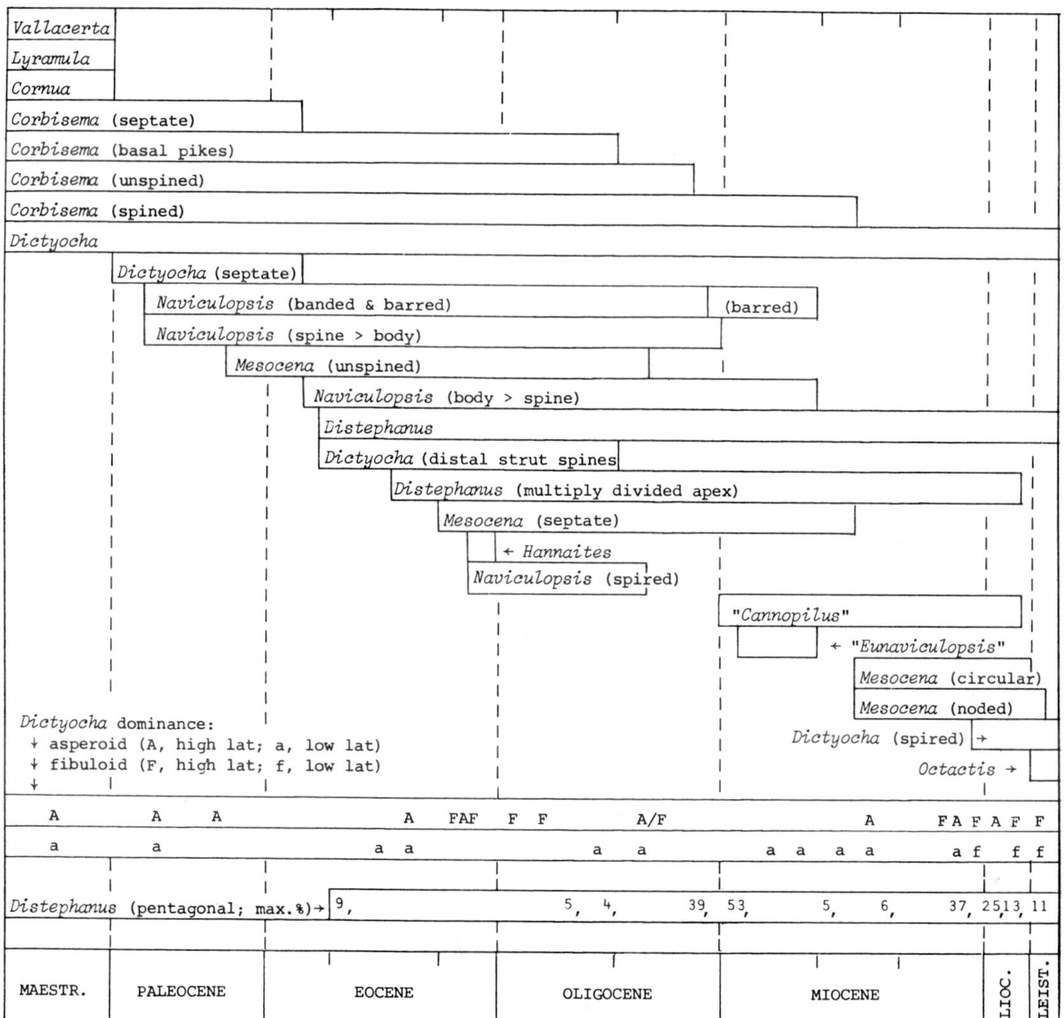

FIG. 2.—Approximate stratigraphic relations of silicoflagellate generic and skeletal morphologic events, from Maestrichtian to Quaternary, based on Deep Sea Drilling Project Legs 1 to 63. Asperoid (–) and fibuloid (ı) classes of *Dictyocha* are based on the orientation of the bar to the major axis of the ring, respectively, perpendicular or parallel. Taxonomy of "*Cannopilus*" and "*Eunaviculopsis*" is unsettled (Lipps, 1970; Bukry, 1976b, *in press* a; Ling, 1977).

Dictyocha the asperoid mode reflected stable conditions, whereas fibuloids are more of an adaptation to stress conditions.

Pentagonal specimens of *Distephanus* range from Eocene to Pleistocene, with major abundance peaks (Fig. 1) occurring near the Oligocene/Miocene and Miocene/Pliocene boundaries at high-latitude sites (Bukry, 1975b; Ling, 1973). Most significant abundances (greater than 2%) in the Oligocene and Lower Miocene also occur at high latitude (though the limited number of low-latitude sites could exaggerate this relation). Large-abundance peaks at low-latitude sites occur in the Pliocene and Lower Pleistocene, but pentagonal *Distephanus* is virtually absent at high-latitude sites of Legs 19, 35, and 49 in the Upper Pliocene and Lower Pleistocene. The last common occurrence of pentagonal *Distephanus* from Legs 19, 31, 35 and 49 is consistently in the Lower Pliocene according to diatom correlations (John A. Barron, oral commun., 1979). Therefore, it is proposed that pentagonal *Distephanus* had their optimal cool conditions only at high latitudes during the Miocene, but moved into midlatitudes during the Late Pliocene and Early Pleistocene when previously restrictive factors eased.

PALEOECOLOGY

Paleogeographic distribution of genera through the DSDP array of sites suggests paleoecologic significance. The great dominance of *Distephanus* over *Dictyocha* in the Upper Miocene through Pleistocene at high latitude supports the cold-water indicator status of *Distephanus* (Lipps, 1970; Ciesielski, 1975); *D. speculum* is especially useful (Bukry, 1978a). Similarly, the high proportion of *Corbisema* at equable Paleocene sites, at low-latitude Oligocene DSDP Site 369, and the extinction of *Corbisema* at the end of a thermal high in the Middle Miocene, all indicate warm-water affinities. *Mesocena* and *Naviculopsis,* the other long-ranged genera, seem more cosmopolitan and diverse in their response to cool or warm conditions. For example, *Mesocena apiculata* and *M. oamaruensis* occur in the Oligocene of high-latitude Leg 29, as do *Naviculopsis biapiculata* and *N. trispinosa,* but at DSDP 369 only the former species of each pair is present. Total percentages of the two genera are only slightly reduced at low-latitude; however, of the two, *Naviculopsis* seems slightly more adaptable.

The importance of DSDP results in understanding the paleoecologic significance of silicoflagellates is exemplified by the Upper Pliocene species *Dicty-*

TABLE 1.—Silicoflagellate diversities of Lipps (1970) and Bukry (1974b) compared to more completely documented DSDP assemblages. The stratigraphic distribution of new taxa generally matches the established diversity, with increases from Cretaceous to Eocene probably reflecting the increase in variety and number of sites available. The decline from Miocene to Quaternary, however, is counter to a corresponding increase in numbers and variety of sites. The number of newly-described taxa per unit time is highest in the Pliocene and Pleistocene, supporting Lipps' (1970) observation on a lack of data from those intervals.

	Total subgeneric diversity (Lipps, 1970)	Species diversity (Bukry, 1974b; based on Ling, 1972)	Average (and maximum) species diversity at DSDP sites. Includes cf. and aff. From Bukry DSDP reports		New species and subspecies described from DSDP cores 1969-1979
			High latitude	Low latitude	
QUATERNARY	5	-	8 (10)	16 (32)	8
PLIOCENE	1	-	16 (22)	18 (25)	13
MIOCENE	98	15	27 (40)	23 (39)	16
OLIGOCENE	11	-	20 (34)	26 (39)	7
EOCENE	33	14	17 (24)	28 (30)	18
PALEOCENE	9	-	10 (19)	22 (22)	11
MAESTRICHTIAN	∿12	5	7 (8)	- (-)	1

ocha perlaevis ornata which was described from Mid-Atlantic Ridge cores where it provided a good local abundance horizon to correlate Sites 333 and 335 (Bukry, 1977a). DSDP provided the correlative data needed to look for this species at other sites. Further DSDP Atlantic coring showed the subspecies ranged north to Site 410 (lat 45°N) and south along the African coast to Site 362 (lat 20°S) within the Upper Pliocene. It could also be determined that two phenotypic groups existed, the original mid-Atlantic group and the western-Africa group (Bukry, 1979b). Stradner and Bachmann (1978) described Upper Pliocene *Dictyocha aegea* from the Aegean Sea; it may represent a phenotype of *D. perlaevis ornata* from a stressed environment which also suggests the reason for the phenotypic change between the Mid-Atlantic group and the coastal western-Africa group.

The paleoecologic-indicator potential of silicoflagellates at the species level is high. DSDP sites directed at studying the silica history of the ocean basins would greatly enhance this aspect of silicoflagellate studies by providing material from a wide geographic area.

SILICOFLAGELLATE STRATIGRAPHIC ZONATIONS

Excellent summaries of the knowledge of silicoflagellate stratigraphy and taxonomy prior to DSDP studies are available in Loeblich and others (1968), Mandra (1968), and Lipps (1970). Between 1971 and 1974, biozonations based on a combination of existing land studies and newly studied DSDP cores were published. In the following years, a series of DSDP studies from newly-cored ocean areas has helped to refine taxonomic and occurrence interpretations. A number of zonations have been proposed, usually on the basis of species occurrences from only a local region. The zonation presented here is a composite tropical/subtropical zonation that is based on low-latitude or cosmopolitan marker species. Middle latitude zones such as the North Pacific *Distephanus pseudofibula* Zone (Bukry, 1973b) or the finely divided Paleocene/Eocene high-latitude zonations of northern Europe (Martini, 1974; Perch-Nielsen, 1976) are not treated. Summary charts comparing other zonations are available in several DSDP publications (Ciesielski, 1975; Perch-Nielsen, 1975; Bukry, 1977b).

ZONATION FOR TROPICAL/SUBTROPICAL REGIONS

Boundary species events in the zone descriptions that follow include: * = appearance, † = disappearance, A* = acme beginning, A† = acme end.

Lyramula furcula Zone

Boundary species: Top—*Corbisema hastata**; bottom—*Lyramula furcula**.

Assemblage: *Corbisema archangelskiana, C. geometrica, C. lateradiata, Cornua aculeifera, C. trifurcata, Dictyocha quadralta, Lyramula deflandrei, L. furcula, L. minor, L. simplex, Vallacerta hannai, V. hortonii, V. siderea, V. tumidula.*

Comment: *Lyramula furcula* (76 to 94%) dominates the zonal assemblages at DSDP Sites 216 and 275, off Sumatra and New Zealand. But, Arctic assemblages commonly have *Vallacerta* dominant (Ling and others, 1973; Gartner and Keany, 1979; Bukry, unpublished data). The zone is known from few localities and is potentially divisible into subzones based on a first occurrence sequence of genera and dominant cosmopolitan species.

Top identified by first *C. hastata,* but the disappearance of genera *Lyramula* and *Vallacerta* is nearly coincident with or slightly above the Cretaceous-Tertiary boundary, also. Correlation with other microfossils suggests the zone is Maestrichtian at DSDP Site 216 and in California, and Campanian or Maestrichtian at DSDP Site 275.

Original definition: Bukry and Foster, 1974.

Other discussion: Glezer, 1966; Cornell, 1972; Ling, 1972; Bukry, 1974a, 1975b; Perch-Nielsen, 1975.

Corbisema hastata Zone

Boundary species: Top—*Naviculopsis constricta**; bottom—*Corbisema hastata**.

Assemblage: *Corbisema archangelskiana, C. disymmetrica communis, C. disymmetrica disymmetrica, C. geometrica, C. glezerae, C. hastata globulata, C. hastata hastata, C. inermis inermis, C. triacantha* s. ampl., *Dictyocha precarentis, D.* spp., and precursors of *Naviculopsis constricta* and *N. danica.*

Comment: The lower assemblages of the zone at DSDP Sites 208 and 214 and the upper at DSDP Site 327 are dominated by *Corbisema,* suggesting equable warm-water conditions. *Corbisema* also dominates the Early Paleocene from the U.S.S.R. (Glezer, 1966). The top of the zone is restricted to the Late Paleocene based on coccolith correlation and the first appearance of *Naviculopsis constricta* at Site 327 (Bukry, 1976b).

Original definition: Bukry and Foster, 1974.

Other discussion: Glezer, 1966.

Naviculopsis constricta Zone

Boundary species: Top—Naviculopsis foliacea;* bottom—*Naviculopsis constricta.**

Assemblage: *Corbisema apiculata, C. archangelskiana, C. disymmetrica angulata, C. disymmetrica communis, C. disymmetrica disymmetrica, C. falklandensis, C. glezerae, C. hastata cunicula, C. hastata globulata, C. hastata hastata, C. inermis inermis, C. inermis minor, C. neoparallela, C. triacantha* s. ampl., *Dictyocha elongata, D. fibula* s. ampl., *D. precarentis, Naviculopsis* sp. cf. *N. aspera, N. constricta, N. danica.*

Comment: A transition from the underlying *Corbisema hastata* Zone to the *Naviculopsis constricta* Zone occurs at DSDP Site 327 on the basis of the sporadic, low frequency occurrence of the earliest *Naviculopsis constricta.* The lower boundary was recorrelated within the Upper Paleocene based on coccoliths at DSDP Site 327. The upper boundary has not been observed. *Naviculopsis danica* is common in North Atlantic assemblages (Perch-Nielsen, 1976; Bukry, 1978c) of Paleocene age. The zone is mainly Paleocene, because the overlying *Naviculopsis foliacea* Zone has been recorded as low as the Lower Eocene *Tribrachiatus orthostylus* Zone of coccoliths at DSDP Site 390.

Zone definition: Bukry and Foster, 1974; modified in Bukry, 1977b.

Other discussion: Perch-Nielsen, 1976; Busen and Wise, 1977; Bukry, 1978c.

Naviculopsis foliacea Zone

Boundary species: Top—*Dictyocha hexacantha**; bottom—*Naviculopsis foliacea**.

Assemblage: *Corbisema apiculata, C. bimucronata, C. disymmetrica communis, C. hastata globulata, C. hastata hastata, C. hastata minor, C. inermis inermis, C.* sp. aff. *C. schulzii, C. toxeuma, C. triacantha* s. ampl., *C. triacantha convexa, Dictyocha deflandrei lobata, D. rotundata secta, D. spinosa, Distephanus antiquus, D. crux* s. ampl., *D. quinquangellus* s. ampl., *Mesocena apiculata inflata, M. occidentalis, M. ovata, M. venusta, Naviculopsis constricta, N. eobiapiculata, N. foliacea foliacea, N. foliacea tumida, N. robusta.*

Comments: This zone is easily recognized by the occurrence of *Naviculopsis foliacea foliacea,* a cosmopolitan Eocene taxon, in the absence of *Dictyocha hexacantha.* The appearance of *Dictyocha spinosa* within the zone is the basis for division into an upper *Dictyocha spinosa* Subzone (Bukry, 1977b) and lower *Naviculopsis robusta* Subzone (Bukry, 1978d). In the North Atlantic the *N. robusta* Subzone occurs with the Lower Eocene *Tribrachiatus orthostylus* Zone of coccoliths, and the *D. spinosa* Subzone with the Lower Eocene *Discoaster lodoensis* Zone of coccoliths at DSDP Site 390. The *D. spinosa* Subzone extends as high as the Middle Eocene *Coccolithus staurion* Subzone at DSDP Site 356 in the South Atlantic, thus, it bridges the Lower to Middle Eocene boundary.

Original definition: Martini, 1974; emended Bukry, 1977b.

Other discussion: Ling, 1972; Martini and Müller, 1976; Bukry, 1978c, 1978d.

Dictyocha hexacantha Zone

Boundary species: Top—*Dictyocha hexacantha*†; bottom—*D. hexacantha**.

Assemblage: *Corbisema apiculata, C. archangelskiana, C. bimucronata, C. flexuosa, C. hastata hastata, C. hastata minor, C. inermis, C. lamellifera, C. ovalis, C. triacantha* s. ampl., *Dictyocha deflandrei, D. fibula* s. ampl., *D. hexacantha, D. spinosa, D.* sp. A (Bukry, 1978c), *Distephanus crux* s. ampl., *D. norvegiensis, D. quinquangellus* s. ampl., *Macrora barbadiensis, M. najae, Mesocena apiculata, M. oamaruensis, M. occidentalis, Naviculopsis constricta, N. eobiapiculata, N. foliacea,* [*N. nordica, N. punctilia, N. vemae,* only at northern high latitude].

Comment: *Dictyocha hexacantha* is a rather cosmopolitan Late Eocene silicoflagellate zonal guide species for the *D. hexacantha* Zone ranging from the Tasman Sea (DSDP Site 283) to the Norwegian-Greenland Sea (DSDP Site 340). But, it is most useful at low and mid latitudes because the Late Eocene is the beginning of differentiation into distinctly provincial species arrays. Succeeding zonations through the Oligocene show no cosmopolitan, short-ranged taxa. Sites from the Antarctic (DSDP Sites 274 and 280; Bukry, 1975a) show silicoflagellates dissimilar to coeval Site DSDP 369 off northwest Africa (Bukry, 1978b).

The differentiation in the *D. hexacantha* Zone is most notable in the upper part of the zone at high latitudes. The Norwegian-Greenland Sea assemblages (Bukry, 1976c; Martini and Müller, 1976; Perch-Nielsen, 1976) contain several species, such as *Naviculopsis punctilia* and *N. nordica,* which are unknown from warmer regions. Through much of the Atlantic Basin the lower part of the zone is characterized by the common occurrence of *Macrora barbadensis,* usually within the Upper Middle Eocene *Reticulofenestra umbilica* Zone of coccoliths (Bukry, 1978c), however it was not recorded at high latitude sites (Martini and Müller, 1976). Therefore, silicoflagellates can provide direct evidence of the Eocene-Oligocene cooling (Kennett and Shackleton, 1976) identified by oxygen-isotope techniques.

Original definition: Bukry and Foster, 1974; modified by Bukry, 1977b.

Other discussion: Bukry, 1975b, 1978c; Perch-Nielsen, 1976.

Corbisema apiculata Zone and
Naviculopsis biapiculata Zone

Latest Eocene and Oligocene zonation schemes reflect strong provincialism of species (Bukry, 1978b), with high-latitude marker species absent at mid-latitude sites. Alternate schemes for different areas have been proposed by different authors: Southern Ocean (Bukry, 1974c; Ciesielski, 1975), Tasman Sea (Bukry, 1975b; Perch-Nielsen, 1975), South Atlantic (Bukry, 1976b, 1977b; Busen and Wise, 1977), North Atlantic (Bukry 1976c, 1978b, 1979b; Martini, 1979; Martini and Müller, 1976). The zonal sequences presented here are for the warm-water assemblages in the Atlantic. The zonation for cool-water assemblages (Bukry, 1974c) has been recognized in California (J. Dumoulin, written commun., 1979), and the Bering Sea (J. Barron, written commun., 1979), in addition to the South Atlantic.

Corbisema apiculata Zone

Boundary species: Top—*Corbisema apiculata*† or *C. hastata hastata* s. ampl.†; bottom—*Dictyocha hexacantha*†.

Assemblage: *Corbisema apiculata* (some isosceles), *C. archangelskiana, C. bimucronata, C. hastata globulata, C. hastata hastata* s. ampl., *C. triacantha mediana, C. triacantha triacantha, Dictyocha aspera martinii* s. ampl., *Dictyocha deflandrei* [rare], *Distephanus crux* s. ampl., *M. apiculata* [rare], *Naviculopsis biapiculata, N. constricta, N. foliacea.*

Comment: The assemblage is a composite from samples at DSDP Site 369 off northwest Africa and represents the emended *Corbisema apiculata* Zone (Bukry, 1978b) based on the boundary species selected by Perch-Nielsen (1975) for the original zone in the Tasman Sea. This usage was necessitated by the lack of the cool-water guide species: *Naviculopsis trispinosa, Dictyocha deflandrei, D. fischeri,* and the local extent of Tasman Sea acmes. The cool-water correlative of the *C. apiculata* Zone is the *Dictyocha deflandrei* Zone (Bukry and Foster, 1974) which has been subdivided into three subzones (Bukry, 1975b, 1977b). The *D. deflandrei* Zone or its Subzones have been identified in the Southern Ocean, Tasman Sea, Falkland Plateau (Busen and Wise, 1977), and California (J. Dumoulin, written commun., 1979).

Original definition: Perch-Nielsen, 1975; emended, Bukry, 1978b.

Naviculopsis biapiculata Zone

Boundary species: Top—*Naviculopsis lata**; bottom—*Corbisema apiculata*† or *C. hastata*† [warm]; *Naviculopsis biapiculata* A* [cool].

Assemblage: *Corbisema archangelskiana, C. bimucronata, C. triacantha mediana, C. triacantha triacantha, Distephanus crux crux* s. ampl., *D. crux loeblichii, D. raupii, D. speculum haliomma, D. speculum hemisphaericus, D. speculum speculum, D. speculum triommata, Mesocena apiculata apiculata, M. apiculata curvata, M. apiculata glabra, N. biapiculata, N. constricta, N.* spp., (*Rocella gelida* and *R. schraderi* of Bukry, 1978c).

Comment: Several authors have applied some type of *Naviculopsis biapiculata* Zone to the Oligocene and Eocene. The zone identified here is detailed in Bukry (1978b) and incorporates the latest occurrence of *Naviculopsis constricta,* the earliest *Distephanus speculum hemisphaericus;* it also contains the *Rocella gelida* acme zone of Bukry and Foster (1974) which correlates with the *Sphenolithus ciperoensis* Zone and *Triquetrorhabdulus carinatus* Zone of coccoliths. The *N. biapiculata* Zone is, herein, divided into a lower unnamed subzone and an upper *Distephanus speculum haliomma* Subzone on the basis of the appearance, within the zone, of *Distephanus speculum haliomma* and *D. speculum hemisphaericus,* both with divided apical rings. This relatively cosmopolitan event can be recognized in the Bering Sea, off northwestern Africa, and the Tasman Sea.

The *Distephanus raupii* Zone of cool-water areas (Bukry, 1975b), lying above the *N. biapiculata* Zone, probably correlates to part of the *N. lata* Zone or upper *N. biapiculata* Zone in warmer areas. The composite assemblage shown here is based mainly on DSDP Sites 369 and 385; compare to DSDP Sites 267, 278, 328, 338, and 407 for high-latitude variations.

Original definition: Bukry, 1974c, modified by Bukry, 1978b.

Other discussion: Martini, 1974; Ciesielski, 1975; Bukry, 1975a, 1975b, 1976b, 1978c; Martini and Müller, 1976; Busen and Wise, 1977.

Naviculopsis lata Zone

Boundary species: Top—*Naviculopsis quadrata**; bottom—*Naviculopsis lata**.

Assemblage: *Corbisema flexuosa, C. triacantha, Dictyocha brevispina, Distephanus crux* s. ampl., *D. hannai, D.* sp. cf. *D. longispinus, D. speculum haliomma, D. speculum hemisphaericus, D. speculum speculum, D. speculum triommata, D. staurodon, Mesocena apiculata apiculata, M. apiculata curvata, Naviculopsis biapiculata, N. lata, Macrora stella.*

Comment: The *Naviculopsis lata* Zone and succeeding lower Miocene zones are identified by a sequence of *Naviculopsis* species events, the relative order and intervals of which are not yet fully established (Bukry, 1978b, *in press* a; Martini, 1979). The Lower Miocene zones typically can be identified at both mid- and high-latitude DSDP sites from the Atlantic where silicoflagellates have been most thoroughly studied.

Potential difficulty with the *Naviculopsis* lineage is suggested by some DSDP results. A Late Oligocene or Early Miocene age was originally assigned to the *N. lata* Zone (Martini, 1972; Sanfilippo and others, 1973). DSDP Leg 38 results prompted Martini and Müller (1976) to draw the base of the zone as low as the Upper Oligocene *Sphenolithus distentus* Zone of coccoliths (Martini, 1971). But Leg 41 coccolith correlation shows the *N. lata* Zone ranging from the Lower Miocene *Discoaster deflandrei* Subzone to the *Sphenolithus belemnos* Zone. However, some isolated specimens of *N. lata* were found in a sample from the Upper Oligocene (*Dictyococcites bisectus* Subzone) of DSDP Site 369A. Subsequent studies of Leg 49 (Bukry, 1979b; Martini, 1979) showed an Early Miocene correlation, again with the *Triquetrorhabdulus carinatus* Zone. There has been disagreement about the upper extent of the zone because of the use of *Naviculopsis quadrata* in the present zonal scheme but not in Martini's. Such distinctions will be resolved as more sections are studied and the stratigraphic reliability of the various species of *Naviculopsis* can be established.

Original definition: Martini, 1972; emended, Bukry, 1978b.

Other discussion: Martini and Müller, 1976; Ling, 1977; Bukry, 1979b.

Naviculopsis quadrata Zone

Boundary species: Top—*Naviculopsis ponticula**, *N. quadrata*†; bottom—*Naviculopsis quadrata**.

Assemblage: *Corbisema flexuosa, C. triacantha, Dictyocha brevispina, D. fibula, Distephanus crux* s. ampl., *D. hannai, D. speculum haliomma, D. speculum hemisphaericus, D. speculum speculum, D. speculum triommata, Macrora stella, Mesocena apiculata apiculata, M. apiculata curvata, Naviculopsis lata, N. quadrata.*

Comment: The upper Lower Miocene *Naviculopsis quadrata* Zone occurs in the range of the upper *Triquetrorhabdulus carinatus* Zone or *Sphenolithus belemnos* Zone (Bukry, 1976c, 1978b, 1979b). The *N. quadrata* Zone corresponds to parts of Martini's *N. lata* Zone and *N. navicula* Zone based on DSDP Leg 49. Sample 370-4-2 could questionably be assigned to the *N. quadrata* Zone which would associate it with the ?*S. belemnos* Zone coccolith assemblage.

Original definition: Bukry and Foster, 1974.

Naviculopsis ponticula Zone

Boundary species: Top—*Naviculopsis*†; bottom—*N. ponticula**, *N. quadrata*†.

Assemblage: *Cannopilus depressus, C. schulzii, Corbisema triacantha, Dictyocha brevispina ausonia, D. brevispina brevispina, D. fibula, D. pulchella, Distephanus crux* s. ampl., *D. hannai, D.* sp. cf. *D. longispinus, D. speculum diommata, D. speculum haliomma, D. speculum speculum, D. speculum triommata, D. staurodon, Mesocena apiculata curvata, M. elliptica, Naviculopsis lata, N. navicula, N. ponticula, N. quadrata.*

Comment: This new zone was suggested for DSDP Leg 50 assemblages from offshore northwest Africa (Bukry, *in press* a). It is predicated on the taxonomic distinction of *Naviculopsis ponticula* s. str. from *N. navicula* s. str. (see Locker, 1974; Bukry, 1978d). The *Naviculopsis ponticula* Zone is defined, herein, as the interval from the first *N. ponticula* or last *N. quadrata* to the disappearance of *Naviculopsis* near the Lower to Middle Miocene boundary. The most cosmopolitan taxa at four DSDP reference sites (369A, 370, 407, and 415) are *D. crux, D. speculum speculum, D. speculum triommata, Mesocena apiculata curvata,* and *N. ponticula.* This short zone occurs with the *Helicosphaera ampliaperta* Zone of coccoliths at Sites 369A, 407, and 415. By inference it belongs to the same zone at Site 370. All four cited occurrences are in the Atlantic Ocean: Sites 369A-11-3, 370-3-CC, 407-19-2, and 415-5-6. This zone is of much shorter duration than the *Eunaviculopsis ponticula* Zone proposed by Ling (1977) as a gap zone between the last *Rocella gelida* and the first *Distephanus stauracanthus* at DSDP Site 71.

Corbisema triacantha Zone

Boundary species: Top—*Corbisema triacantha*†; bottom—*Naviculopsis*†.

Assemblage: *Cannopilus depressus, C. schulzii, Corbisema triacantha, Dictyocha brevispina ausonia, D. brevispina brevispina, D. fibula, D. pulchella, Distephanus crux* s. ampl., *D.* sp. cf. *D. longispinus, D. speculum haliomma, D. speculum hemisphaericus, D. speculum speculum, D. speculum triommata, D. stauracanthus, D. stradneri, Mesocena apiculata apiculata, M. apiculata curvata, M. circulus, M. diodon, M. elliptica.*

Comment: The *Corbisema triacantha* Zone, like the *Naviculopsis* zones, is rather cosmopolitan, being recorded from DSDP Leg 28 in the Southern Ocean to Leg 49 in the North Atlantic Ocean. It represents the last widespread occurrence of genus *Corbisema* which was prominent in most Paleogene assemblages. Silicoflagellate assemblages become more provincial above this zone, leading to the recognition of many zonal systems for individual cruises and areas (see Perch-Nielsen, 1975).

The *Corbisema triacantha* Zone correlates with the upper *Helicosphaera ampliaperta* Zone, *Sphenolithus heteromorphus* Zone, and *Discoaster exilis* Zone of coccoliths in the lower Middle Miocene. This is an interval of considerable taxonomic change in nannofloras as older lineages of the Paleogene are replaced by new stocks that will persist through the Neogene (Bukry, 1973a). A similar effect is seen in silicoflagellates of the *C. triacantha* Zone, as genera *Naviculopsis* and *Corbisema* disappear and the *Distephanus speculum hemisphaericus* and *Mesocena apiculata* groups dwindle at or in this zone. Notable appearances within the zone include the noded mesocenoids (see Bukry, 1978b) and *Distephanus stauracanthus*. Studies of assemblages from the Atlantic area (Ernissee and others, 1977; Bukry, 1979b, *in press* a) and from the Pacific sites of DSDP Leg 63 (Bukry, *in press* c) show that the interval where *D. stauracanthus* occurs is within the upper *C. triacantha* Zone and correlates with the *Coccolithus miopelagicus* Subzone of coccoliths. The *Distephanus stauracanthus* Subzone is, herein, defined as the upper subzone of the *C. triacantha* Zone, identified by the appearance of *D. stauracanthus* and the presence of *M. circulus*. The lower subzone is unnamed.

Original definition: Martini, 1971, 1972.

Other discussion: Bukry and Foster, 1973; Bukry, 1975b, 1978b, 1978d, 1978c, *in press* a, *in press* c; Haq and Riley, 1976; Ling, 1977; Martini and Müller, 1976.

Dictyocha brevispina Zone

Boundary species: Top—asperiod/fibuloid dominance reversal in *Dictyocha;* bottom—*Corbisema triacantha*†.

Assemblage: *Dictyocha aspera, D. brevispina, D. fibula* s. ampl., *D. neonautica* [trace], *D. pulchella, Distephanus crux* s. ampl., *D. pseudocrux, D. pseudofibula, D. speculum minutus, D. speculum speculum, Mesocena circulus, M. diodon nodosa, M.* sp. aff. *M. quadrangula.*

Comment: The unusual upper boundary of the *D. brevispina* Zone was defined by Martini (1971) on the basis of cores from DSDP Leg 7 in the Pacific. On DSDP Leg 16 in the Pacific the asperoid/fibuloid reversal was found to vary from the upper *Discoaster quinqueramus* Zone to the upper *Discoaster neohamatus* Zone of coccoliths (Bukry and Foster, 1973). At DSDP Site 310 of Leg 32 asperoid-dominated assemblages range from Pliocene *Discoaster tamalis* Subzone to the Miocene *D. berggrenii* Subzone of coccoliths (Bukry, 1975d, 1978a) and represent a cooler environment which reduces the anticipated range of *Dictyocha fibula* or *D. perlaevis* domination in *Dictyocha.* On Leg 63 the reversal was found to fluctuate in a single Upper Miocene section, suggesting that the time value of the event may be quite variable. More analyses of warm-water Upper Miocene sequences are needed to explain and improve this situation. The disappearance of *Distephanus longispinus* and the appearance and disappearance of *Distephanus pseudofibula* which aid in temperate stratigraphy (Bukry, 1973b; Barron, 1976) may have warm-water corollaries, but present data on DSDP cores is not sufficiently detailed and updated (Ling, 1977). *Mesocena hexalitha* may also be stratigraphically useful for Middle to Upper Miocene Pacific correlation (see Dumitrică, 1973; Barron, 1976; Barron, verbal commun., 1978).

Original definition: Martini (1971).

Other discussion: Bukry and Foster (1973), Bukry (1975c, 1977a), Martini (1976), Ling (1977).

Dictyocha fibula Zone

Boundary species: Top—*Dictyocha stapedia stapedia**; bottom—asperoid/fibuloid dominance reversal in *Dictyocha.*

Assemblage: *Dictyocha aspera aspera, D. aspera clinata, D. brevispina, D. fibula* s. ampl., *D. neonautica, D. perlaevis, D. pulchella, D. varia, Distephanus crux* s. ampl., *D. polyactis, D. quinquangellus, D. speculum minutus, D. speculum speculum, Mesocena circulus, M. diodon nodosa, M.* sp. aff. *M. quadrangula.* May contain *Distephanus boliviensis, D. frugalis,* and *D. jimlingii,* in cool areas.

Comment: *Dictyocha fibula* s. ampl. and *D. perlaevis* are prominent in this zone in warm areas. *Distephanus* spp. replace them when, or where, cooler conditions prevail. The nature of the lower boundary of the *Dictyocha fibula* Zone is discussed with the *Dictyocha brevispina* Zone. The boundary with the overlying *Dictyocha stapedia stapedia* Zone has not been well established due to a lack of DSDP Lower Pliocene siliceous sections. Some of the most interesting species events in the *Dictyocha fibula* Zone occur in the Upper Miocene near the Miocene-Pliocene boundary. *Dictyocha aspera clinata* occurs in the uppermost Miocene through the North Pacific from DSDP 158 in the east (Bukry and Foster, 1973) to DSDP Site 303 in the west (Bukry, 1975c) and even into the Sea of Japan at DSDP Site 302 (Ling, 1975). Although the association of this species with *Dictyocha neonautica* n. sp. probably defines a brief interval in the Upper Miocene through the Pacific, the asperoid/fibuloid zonal criteria cannot be unequivocally applied because of reversals occurring near this interval at DSDP Sites 158 and 304, or asperoid dominance persisting upward through the interval at DSDP Site 303. The *Dictyocha neonautica* Subzone, defined herein by the range of *D. neonautica* (formerly recorded as *D. navicula,* etc.), is generally associated with the appearance and acme of *D. aspera clinata*, in the lower *D. fibula* Zone, and is abundant at reference localities DSDP Sites 158 and 303. *D. neonautica* is correlated with the *Amaurolithus primus* Subzone of coccoliths. The substantial recurrence of naviculopsoid silicoflagellates following the extinction of *Naviculopsis* in Early Miocene is exceptional. The *D. neonautica* Subzone apparently also occurs just above the level of a significant change in carbon ratios (G. Keller, written commun., 1979; Loutit and Kennett, 1979) and could, with the asperoid/fibuloid ratio, signify a combined geochemical and climatic effect through the Pacific. The trace occurrence of *D. neonautica* in the upper *D. brevispina* Zone (top of *D. berggrenii* Subzone) at DSDP Site 334 is suggestive of the presence of the *D. neonautica* Subzone in the Atlantic, but no definite development of the subzone is known owing to a lack of DSDP siliceous cores from warm-water areas, there. The known stratigraphic range of the *D. neonautica* Subzone in the Pacific is brief, within the *A. primus* Subzone: DSDP 310-8-1, DSDP 158-11-CC, DSDP 77-9B-2 to 77-9B-4. However, Koizumi's (1975) correlation of DSDP 303-3 to the diatom *Denticula kamtschatica* Zone could extend the zone upwards into the Lower Pliocene. The upper boundary of the *D. fibula* Zone was lowered from the original first Quaternary *Mesocena quadrangula* (Martini, 1971) to the first Pliocene to Quaternary *D. stapedia stapedia* (Bukry, 1977a).

Original definition: Martini (1971); modified by Bukry (1977a).

Other discussion: Bukry and Foster (1973); Ling (1977).

Dictyocha stapedia stapedia Zone

Boundary species: Top—*Mesocena quadrangula* A*; bottom—*Dictyocha stapedia stapedia**.

Assemblage: *Dictyocha aspera aspera, D. brevispina, D. calida calida, D. calida ampliata, D. fibula* s. ampl., *D. perlaevis delicata, D. perlaevis flexatella, D. perlaevis ornata, D. perlaevis perlaevis, D. stapedia aspinosa, D. stapedia stapedia, D. vexativa, Distephanus boliviensis, D. boliviensis* (cannopilean), *D. quinquangellus, D. speculum speculum, D.* sp. A (Bukry, 1979a), *Mesocena circulus, M. quadrangula, Octactis pulchra.*

Comment: The *Dictyocha stapedia stapedia* Zone was defined from the North Atlantic DSDP sites of Leg 37 because the *Distephanus boliviensis* Zone was too indefinite at mid latitudes and the original *Dictyocha fibula* Zone was too broadly based. In the eastern equatorial Pacific sites of DSDP Legs 16 and 54, the *D. stapedia stapedia* Zone can be subzoned by the presence and acme of *Dictyocha perlaevis delicata* below the *Mesocena quadrangula* Zone. The *D. perlaevis delicata* Subzone has only been identified from the Pacific and is largely Lower Quaternary based on coccolith correlation (Bukry, 1976a, *in press* b). Several subspecies of the *Dictyocha perlaevis plexus* appear and disappear within the *D. stapedia stapedia* Zone, including *D. perlaevis delicata, D. perlaevis flexatella*, and *D. perlaevis ornata. D. perlaevis ornata* has aided in correlation near the Pliocene-Quaternary boundary through the Atlantic Legs 37, 40, 47, and 49. Its distribution is the opposite of *D. perlaevis delicata* being sparse in the Pacific and fairly common in the Atlantic for a short time. The lower interval of the *D. stapedia stapedia* Zone, below the *D. perlaevis delicata* Subzone is, herein, named the *Dictyocha perlaevis ornata* Subzone and defined as the interval from the appearances of *D. stapedia stapedia* at the base to *D. perlaevis delicata* at the top, the interval of most common *D. perlaevis ornata. D. perlaevis ornata* is identified in substantial numbers in Atlantic DSDP Sites 333 and 335, and in lesser numbers in Pacific Site 157 (Core 18), primarily in Upper Pliocene assemblages of the *Discoaster tamalis* Subzone or *D. surculus* Subzone of coccoliths.

Coccolith correlation of the *D. stapedia stapedia* Zone suggests the zone extends from the Upper Pliocene *Discoaster tamalis* Subzone (Sites 332 and 333) or *D. surculus* Subzone (Sites 335 and 397) to the *Emiliania annula* Subzone (Site 157, 333, 397, 410, and 425). It may range slightly higher in the Quaternary at Sites 310 and 362.

Original definition: Bukry (1977a).

Other discussion: Bukry (1978a, 1979a, 1979b, *in press* b).

Mesocena quadrangula Zone

Boundary species: Top—*Mesocena quadrangula* A†; bottom—*M. quadrangula* A*.

Assemblage: *Dictyocha aculeata subaculeata, D. calida ampliata, D. calida calida, D. hessii, D. lingii, D. perlaevis perlaevis, D. stapedia aspinosa, D. stapedia stapedia, D. subarctios* (only at 310), *D. vexativa, Distephanus boliviensis, D. quinquangellus, D. speculum speculum, D.* sp. A (Bukry, 1979a), *Mesocena quadrangula, Octactis pulchra.*

Comment: The Middle Quaternary acme of *Mesocena quadrangula* provides the most widely recognized warm-water silicoflagellate event. It has been correlated, calibrated, described, and codified by various authors (for example, Hays and others, 1969; Martini, 1971; Dumitrică, 1973; Bukry and Foster, 1973; Jousé and Muhina, 1973; Burckle, 1977; Bukry, 1979a, *in press* b). Because *M. quadrangula* was extant through the Pliocene and Early Quaternary, the beginning of the acme may be less abrupt than the end. Burckle (1977) supported Ling's (1970) contention that the extinction of *M. quadrangula* is isochronous. Burckle correlates the extinction to early glacial isotope Stage 22, just above the Jaramillo magnetic event (approximately 0.79 m.y.BP). The disappearance of *Dictyocha lingii* and *D. subarctios*, together, with *M. quadrangula*, noted by Ling (1972) and Dumitrică (1973), has been generally confirmed by subsequent DSDP drilling at Hess Rise (Bukry, 1978a) and in the North Atlantic (Bukry, 1979b).

Original definition: Bukry and Foster, 1973.

Other discussion: Martini (1971, 1976); Ling (1972, 1977); Dumitrică (1973); Bukry (1976a, 1977a, 1979a, 1979b); Saito and Burckle (1977).

Dictyocha aculeata aculeata Zone

Boundary species: Top—None; bottom—*Mesocena quadrangula* A†.

Assemblage: *Dictyocha aculeata aculeata, D. aculeata subaculeata, D. calida ampliata, D. calida calida, D. perlaevis perlaevis, D. stapedia aspinosa, D. stapedia stapedia, D. speculum speculum, D.* sp. A (Bukry, 1979a), *Mesocena quadrangula* (reworked?), *Octactis pulchra.*

Comment: The *Dictyocha aculeata aculeata* Zone is recognized by the common occurrence of *Dictyocha aculeata aculeata* above the Quaternary acme of *Meso-*

cena quadrangula (Bukry, 1979a, 1979b). This is an emendation of the original *Dictyocha epiodon* (= *D. aculeata aculeata* Zone) of Bukry and Foster (1973) to formalize the end of the *M. quadrangula* acme, in addition to common *D. aculeata aculeata* as a zonal guide. Generally, there is lower diversity in the *D. aculeata aculeata* Zone than in preceding zones and specimens tend to be smaller. The assemblages are further distinguished by the sparseness or absence of asperoid specimens of *Dictyocha*. The zone is fairly cosmopolitan, occurring at DSDP Sites 328, 412 in the Atlantic (Bukry, 1976b, 1979b), Sites 281, 310, 315 in the Pacific (Perch-Nielsen, 1975; Martini, 1976; Bukry, 1978a), and at Sites 213 and 241 in the Indian and Southern Oceans (Bukry, 1974a and 1974b) at latitudes from the equator to 40° or 50° lat. The taxonomic and stratigraphic problems of *D. aculeata aculeata* and *D. aculeata subaculeata* are described in connection with DSDP Leg 54 (Bukry, *in press* b).

TAXONOMY OF NEW SILICOFLAGELLATES

Genus DICTYOCHA Ehrenberg, 1837

Dictyocha neonautica Bukry, n. sp.

Dictyocha navicula Ehrenberg, Bukry and Foster, 1973, p. 827, pl. 3, figs. 6–8.

Description.—*Dictyocha neonautica* is a naviculopsoid species with only two short spines aligned at, or near, the major axis of the elongate ring. The ring is generally oblong, but a few specimens range to acicular outlines, both forms showing major/minor axis ratios of approximately 4/1. A single straight to slightly arched bar connects the subparallel sides of the ring along the minor axis. Slight, irregular ornamentation of the distal periphery is present in light microscopy.

A minor variant population, noted only at DSDP Site 158, Core 11, in the Panama Basin, is generally wider, with axis ratios 2/1 to 3/1, and is broadly acicular. Also, the bar is narrower than the ring and the bar is slightly offset to one end of the ring in many specimens. This form is identified as a new variety: *Dictyocha neonautica* var. *cocosensis* n. var.

Remarks.— *Dictyocha neonautica* and *D. neonautica* var. *cocosensis* resemble the Lower Miocene species *Naviculopsis quadrata* and *N. lata,* respectively. But the last species of *Naviculopsis,* a genus with continuity from Lower Paleocene through Lower Miocene, is recorded from about 16.5 Ma. No species are known from the Middle and lower Upper Miocene, an interval spanning about 10 m.y. Therefore, if species of *Naviculopsis* evolved within the genus, as seems to be true in the *N. foliacea* plexus including *N. foliacea tumida, N. nordica hyalina,* and *N. nordica nordica,* and were not repeatedly derived from *Dictyocha,* then there is no direct way to attribute upper Upper Miocene naviculopsoid specimens to *Naviculopsis.* Further, the high percentages of *D. neonautica* and lack of any Lower Miocene reworking argues against mixing. *D. neonautica* is distinguished from *N. quadrata* by a narrower ring and shorter spines, *D. neonautica* var. *cocosensis* from *N. lata* by shorter spines, thinner offset bar and acicular tendencies in ring form. *D. neonautica* is distinguished from *N. ponticula* by lack of widened ends. *D. neonautica* was probably derived from a *D. brevispina* stock with naviculopsoid tendencies (Deflandre, 1941), but was widespread through the Pacific, instead of a local variant population and therefore is taxonomically discriminated.

Occurrence.—*Dictyocha neonautica* and *D. neonautica* var. *cocosensis* occur in upper Upper Miocene and possibly lowermost Pliocene cores at Pacific DSDP Sites 157, 158, 303, 304, 310 at levels correlative with the *Amaurolithus primus* Subzone of coccoliths. See *D. neonautica* Subzone discussion under *Dictyocha fibula* Zone (this paper). Reworked *D. neonautica* and *D. neonautica* var. *cocosensis* are illustrated from DSDP Sites 157 and 425 (Bukry, 1976a; *in press* b).

Dictyocha neonautica Bukry, n. sp.

SIZE: Length 70–90 μm, width 15–25 μm.

HOLOTYPE: Plate 3, figure 6 of Bukry and Foster, 1973.

ISOTYPE: Plate 3, figure 7 of Bukry and Foster, 1973.

TYPE LOCALITY: Pacific Ocean, Carnegie Ridge, DSDP 157-38-CC (343 m).

Dictyocha neonautica var. *cocosensis* Bukry, n. var.

SIZE: Length 50–65 μm, width 25–35 μm.

HOLOTYPE: Plate 3, figure 8 of Bukry and Foster, 1973.

TYPE LOCALITY: Pacific Ocean, Cocos Ridge DSDP 158-11-CC (99 m).

REFERENCES CITED

NOTE: All references to Deep Sea Drilling Project Initial Reports have been abbreviated; for complete citations see *Appendix I* at the end of this volume, or the individual Initial Reports as published.

ALVAREZ, L. W., ALVAREZ, W., ASARO, F., AND MICHEL, H. V., 1980, Extraterrestrial cause for the Cretaceous-Tertiary extinction: Science, v. 208, p. 1095–1108.

BARRON, J.A., 1976, Revised Miocene and Pliocene diatom biostratigraphy of Upper Newport Bay, Newport Beach, California: Marine Micropaleontology, v. 1, p. 27–63.

BUKRY, D., 1973a, Coccolith stratigraphy, eastern equatorial Pacific, Leg 16 Deep Sea Drilling Project: Init. Repts. DSDP, v. 16, p. 653–711.

———, 1973b, Coccolith and silicoflagellate stratigraphy, Deep Sea Drilling Project Leg 18, eastern North Pacific: Init. Repts. DSDP, v. 18, p. 817–831.
———, 1974a, Coccolith and silicoflagellate stratigraphy, eastern Indian Ocean, Deep Sea Drilling Project Leg 22: Init. Repts. DSDP, v. 22, p. 601–607.
———, 1974b, Phytoplankton stratigraphy, offshore East Africa, Deep Sea Drilling Project Leg 25: Init. Repts. DSDP, v. 25, p. 635–646.
———, 1974c, Stratigraphic value of silicoflagellates in nontropical regions: Geol. Soc. America Bull., v. 85, p. 1905–1906.
———, 1975a, Coccolith and silicoflagellate stratigraphy near Antarctica, Deep Sea Drilling Project Leg 28: Init. Repts. DSDP, v. 28, p. 709–723.
———, 1975b, Silicoflagellate and coccolith stratigraphy, Deep Sea Drilling Project Leg 29: Init. Repts. DSDP, v. 29, p. 845–872.
———, 1975c, Phytoplankton stratigraphy, southwest Pacific, Deep Sea Drilling Project Leg 30: Init. Repts. DSDP, v. 30, p. 539–547.
———, 1975d, Coccolith and silicoflagellate stratigraphy, northwestern Pacific Ocean, Deep Sea Drilling Project Leg 32: Init. Repts. DSDP, v. 32, p. 677–701.
———, 1976a, Silicoflagellate and coccolith stratigraphy, southeastern Pacific Ocean, Deep Sea Drilling Project Leg 34: Init. Repts. DSDP, v. 34, p. 715–735.
———, 1976b, Cenozoic silicoflagellate and coccolith stratigraphy, South Atlantic Ocean, Deep Sea Drilling Project Leg 36: Init. Repts. DSDP, v. 35, p. 885–917.
———, 1976c, Silicoflagellate and coccolith stratigraphy, Norwegian-Greenland Sea, Deep Sea Drilling Project Leg 38: Init. Repts. DSDP, v. 38, p. 843–855.
———, 1977a, Coccolith and silicoflagellate stratigraphy, central North Atlantic Ocean, Deep Sea Drilling Project Leg 37: Init. Repts. DSDP, v. 37, p. 917–927.
———, 1977b, Coccolith and silicoflagellate stratigraphy, South Atlantic Ocean, Deep Sea Drilling Project Leg 39: Init. Repts. DSDP, v. 39, p. 825–839.
———, 1978a, Cenozoic silicoflagellate and coccolith stratigraphy, southeastern Atlantic Ocean, Deep Sea Drilling Project Leg 40: Init. Repts. DSDP, v. 40, p. 635–649.
———, 1978b, Cenozoic coccolith and silicoflagellate stratigraphy, offshore northwest Africa, Deep Sea Drilling Project Leg 41: Init. Repts. DSDP, v. 41, p. 689–707.
———, 1978c, Cenozoic silicoflagellate and coccolith stratigraphy, northwestern Atlantic Ocean, Deep Sea Drilling Project Leg 43: Init. Repts. DSDP, v. 44, p. 775–805.
———, 1978d, Cenozoic coccolith, silicoflagellate, and diatom stratigraphy, Deep Sea Drilling Project Leg 44: Init. Repts. DSDP, v. 44, p. 807–863.
———, 1979a, Comments on opal phytoliths and stratigraphy of Neogene silicoflagellates and coccoliths at Deep Sea Drilling Project Site 397 off northwest Africa: Init. Repts. DSDP, v. 49, p. 977–1009.
———, 1979b, Coccolith and silicoflagellate stratigraphy, northern Mid-Atlantic Ridge and Reykjanes Ridge, Deep Sea Drilling Project Leg 49: Init. Repts. DSDP, v. 49, p. 551–581.
———, *in press,* a, Miocene *Corbisema triacantha* Zone phytoplankton from Sites 415 and 416 off northwest Africa, Deep Sea Drilling Project Leg 50: Init. Repts. DSDP, v. 50.
———, *in press,* b, Silicoflagellate biostratigraphy and paleoecology in the eastern equatorial Pacific, DSDP Leg 54: Init. Repts. DSDP, v. 54.
———, *in press,* c, Pacific Coast silicoflagellates, Deep Sea Drilling Project Leg 63: Init. Repts. DSDP, v. 63.
———, AND FOSTER, J. H., 1973, Silicoflagellate and diatom stratigraphy, Leg 16, Deep Sea Drilling Project: Init. Repts. DSDP, v. 16, p. 815–871.
———, and ———, 1974, Silicoflagellate zonation of Upper Cretaceous to lower Miocene deep-sea sediment: U.S. Geological Survey Jour. Res., v. 2, p. 303–310.
BURCKLE, L. H., 1977, Pliocene and Pleistocene diatom datum levels from the equatorial Pacific: Quaternary Res., v. 7, p. 330–340.
BUSEN, K. E., AND WISE, S. W., JR., 1977, Silicoflagellate stratigraphy Deep Sea Drilling Project, Leg 36: Init. Repts. DSDP, v. 36, p. 697–743.
CIESIELSKI, P. F., 1975, Biostratigraphy and paleoecology of Neogene and Oligocene silicoflagellates from cores recovered during Antarctic Leg 28, Deep Sea Drilling Project: Init. Repts. DSDP, v. 28, p. 625–691.
CORNELL, W. C., 1972, Chrysomonad cysts and silicoflagellates from the Marca Shale Member, Moreno Formation (Maestrichtian), Fresno County, California: Ph.D. Dissertation, Univ. California, Los Angeles, 235 p.
DEFLANDRE, G., 1941, Les notions de genre et de grade chez les Silicoflagellidées et la phylogenèse des mutants naviculaires: Acad. Sci. Comptes Rendus, v. 212, p. 100–102.
DUMITRICĂ, P., 1973, Paleocene, late Oligocene and post-Oligocene silicoflagellates in southwestern Pacific sediments cored on DSDP Leg 21: Init. Repts DSDP, v. 21, p. 837–883.
ERNISSEE, J. J., ABBOTT, W. H., AND HUDDLESTUN, P. F., 1977, Microfossil correlation of the Coosawhatchie Clay (Hawthorn Formation, Miocene) of South Carolina, and its equivalent in Georgia: Marine Micropaleontology, v. 2, p. 105–119.
GARTNER, S., AND KEANY, J., 1978, The terminal Cretaceous event: A geologic problem with an oceanographic solution: Geology, v. 6, p. 708–712.

———, AND ———, 1979, Reply on the terminal Cretaceous event: A geologic problem with an oceanographic solution: Geology, v. 7, p. 229.

GLEZER, Z. I., 1966, Silicoflagellatophyceae: *In* Gollerbakh, M. M., (ed.), Cryptogamic plants of the U.S.S.R.: Akad. Nauk SSSR, V. A. Komarova Bot. Inst. (Translated from Russian by Israel Program for Scientific Translations Ltd., Jerusalem, 1970), v. 7, p. 1-363.

HAQ, B. U., AND RILEY, A., 1976, Antarctic silicoflagellates from the southeast Pacific, Deep Sea Drilling Project Leg 35: Init. Repts. DSDP, v. 35, p. 673-691.

HAYS, J. D., SAITO, T., OPDYKE, N. D., AND BURCKLE, L. H., 1969, Pliocene-Pleistocene sediments of the equatorial Pacific: their paleomagnetic, biostratigraphic, and climatic record: Geol. Soc. America Bull., v. 80, p. 1481-1514.

JOUSÉ, A. P., AND MUHINA, V. V., 1973, The *Mesocena elliptica* Ehr. Zone in the Pleistocene sediments of the Pacific Ocean: Oceanology [AGU Translation], v. 13, p. 386-394.

KENNETT, J. P., AND SHACKLETON, N. J., 1976, Oxygen isotopic evidence for the development of the psychrosphere 38 Myr ago: Nature, v. 260, p. 513-515.

KOIZUMI, I., 1975, Neogene diatoms from the northwestern Pacific Ocean, Deep Sea Drilling Project: Init. Repts. DSDP, v. 32, p. 865-889.

LING, H. Y., 1970, Silicoflagellates from central North Pacific core sediments: Bull. Amer. Paleontology, v. 58, p. 85-129.

———, 1972, Upper Cretaceous and Cenozoic silicoflagellates and ebridians: Bull. Amer. Paleontology, v. 62, p. 135-229.

———, 1973, Silicoflagellates and ebridians from Leg 19: Init. Repts. DSDP, v. 19, p. 751-775.

———, 1975, Silicoflagellates and ebridians from Leg 31: Init. Repts. DSDP, v. 31, p. 763-777.

———, 1977, Late Cenozoic silicoflagellates and ebridians from the eastern North Pacific region: 1st Internat. Cong. Pacific Neogene Stratigraphy Proc., Tokyo, p. 205-233.

———, MCPHERSON, L. M., AND CLARK, D. L., 1973, Late Cretaceous (Maestrichtian?) silicoflagellates from the Alpha Cordillera of the Arctic Ocean: Science, v. 180, p. 1360-1361.

LIPPS, J. H., 1970, Ecology and evolution of silicoflagellates: North American Paleontol. Conv. Proc., pt. G, p. 965-993.

LOCKER, S., 1974, Revision der Silicoflagellaten aus der Mikrogeologischen Sammlung von C. G. Ehrenberg: Eclogae Geol. Helvetiae, v. 67, p. 631-646.

LOEBLICH, A. R. 3D, LOEBLICH, L. A., TAPPAN, H., AND LOEBLICH, A. R., JR., 1968, Annotated index of fossil and recent silicoflagellates and ebridians with descriptions and illustrations of validly proposed taxa: Geol. Soc. America Mem. 106, 319 p.

LOUTIT, T. S., AND KENNETT, J. P., 1979, Application of carbon isotope stratigraphy to late Miocene shallow marine sediments, New Zealand: Science, v. 204, no. 4398, p. 1196-1199.

MANDRA, Y. T., 1968, Silicoflagellates from the Cretaceous, Eocene, and Miocene of California, U.S.A.: California Acad. Sci. Proc., v. 36, p. 231-277.

MARTINI, E., 1971, Neogene silicoflagellates from the equatorial Pacific: Init. Repts. DSDP, v. 7, p. 1695-1708.

———, 1972, Silicoflagellate zones in the late Oligocene and early Miocene of Europe: Senckenberg. Lethaea, v. 53, p. 119-122.

———, 1974, Silicoflagellate zones in the Eocene and early Oligocene: Senckenberg. Lethaea, v. 54, p. 527-532.

———, 1976, Neogene and Quaternary silicoflagellates from the central Pacific Ocean (DSDP Leg 33): Init. Repts. DSDP, v. 33, p. 439-449.

———, 1979, Calcareous nannoplankton and silicoflagellate biostratigraphy at Reykjanes Ridge, northeastern North Atlantic (DSDP Leg 49, Sites 407 and 409): Init. Repts. DSDP, v. 49, p. 533-549.

———, AND MÜLLER, C., 1976, Eocene to Pleistocene silicoflagellates from the Norwegian-Greenland Sea (DSDP Leg 38), Init. Repts. DSDP, v. 38, p. 857-895.

PERCH-NIELSEN, K., 1975, Late Cretaceous to Pleistocene silicoflagellates from the southern southwest Pacific, DSDP Leg 29: Init. Repts. DSDP, v. 29, p. 677-721.

———, 1976, New silicoflagellates and a silicoflagellate zonation in north European Palaeocene and Eocene diatomites: Geol. Soc. Denmark Bull., v. 25, p. 27-40.

SAITO, T., AND BURCKLE, L. H., 1977, Occurrence of silicoflagellate *Mesocena elliptica:* Geol. Soc. Japan Jour., v. 83, p. 181-186.

SANFILIPPO, A., BURCKLE, L. H., MARTINI, E., AND RIEDEL, W. R., 1973, Radiolarians, diatoms, silicoflagellates and calcareous nannofossils in the Mediterranean Neogene: Micropaleontology, v. 19, p. 209-234.

STRADNER, H., AND BACHMANN, A., 1978, Late Pliocene and early Pleistocene silicoflagellates and ebridians from DSDP Site 378 in the Aegean Basin, north of Crete: Init. Repts. DSDP, v. 42, pt. 1, p. 805-815.

THIERSTEIN, H. R., AND BERGER, W. H., 1978, Injection events in ocean history: Nature, v. 276, p. 461-466.

WEST, S., 1979, Dinosaur Demise: Extraterrestrial Source?: Science News, v. 115, no. 22, p. 356.

SEPM SPECIAL PUBLICATION NO. 32, P. 445–470, OCTOBER 1981

SOUTHERN HIGH-LATITUDE DIATOM BIOSTRATIGRAPHY

F. M. WEAVER AND A. M. GOMBOS, JR.
Exxon Production Research Company, P.O. Box 2189, Houston, Texas 77001 U.S.A.

ABSTRACT

The development of Cenozoic and Mesozoic southern high-latitude diatom biostratigraphy over the past fifteen years is summarized with emphasis on significant advancements resulting from deep-sea drilling in the Antarctic and Sub-Antarctic regions.

Regional differences in diatom distribution and occurrence in Neogene sediments recovered at DSDP Sites 266 and 278 are compared. Spatial and temporal variance in species as reported in earlier investigations results from several factors: 1) the lack of continuous coring at Site 266, 2) relative incompleteness of the Neogene record recovered at Sites 266 and 278, 3) unrecognized unconformities, 4) regional paleobiogeographic differences, 5) applied taxonomy, and 6) incomplete species range documentation.

We propose a regionally applicable Neogene diatom zonation utilizing easily recognizable species with broad geographic distribution. This biostratigraphic framework consists of five previously defined and six modified zones from DSDP Legs 28 and 29 reports. In addition, we propose five new zones and document sixteen datums having apparent regional significance.

We herein provide the datums used for geochronologic subdivision and the correlation of epoch boundaries in Sites 266 and 278, floral references for all stratigraphically significant species, and a comprehensive bibliography on Southern Ocean diatoms.

INTRODUCTION

Early diatom biostratigraphic studies in the Southern Ocean were based on numerous long piston cores that rarely penetrated pre-Pliocene sediments. Therefore, the initial zonations concentrated only on paleontological events of the last four to five million years. Not until deep-sea drilling became a reality in the Antarctic region, were older, continuously cored Neogene and Paleogene sedimentary sequences available for detailed examination. Deep Sea Drilling Project Legs 28 (The Ross Sea region), 29 (Sub-Antarctic–New Zealand area), 35 (Bellingshausen Basin) and 36 (Falkland Plateau area) all provided important sample material, which has allowed us to significantly expand our knowledge and understanding of Cenozoic diatom distribution throughout southern high latitude regions (Table 1).

The purpose of this report is to compare and summarize existing diatom biostratigraphic data from the Southern Ocean, and to propose a zonation based on easily recognizable datums with maximum regional applicability. The proposed zonation is in part based on the detailed reexamination of several important stratigraphic sequences.

The locality map (Fig. 1) shows the positions of core sites throughout the Sub-Antarctic and Antarctic region used in biostratigraphic studies over the last fifteen years (Table 1). Core material examined in detail during this investigation include DSDP Sites 266 and 328B, and *Eltanin* cores E16-4, E36-33, E50-28, E38-8, E14-8, E34-19 and E38-7. All other data synthesized herein came directly from published reports.

NEOGENE DIATOM BIOSTRATIGRAPHY

Pleistocene–Pliocene[1]

Jousé and others (1963) were the first to experiment in biostratigraphic development and correlation in the Antarctic region. In *Ob* core 256 taken in the Indian Ocean sector of the Southern Ocean (Fig. 1; Table 2), they defined four stratigraphically significant assemblage horizons from the Pliocene and Pleistocene. However, their correlation with these standard ages was poorly documented and somewhat uncertain at the time (see McCollum, 1975, fig. 12).

Donahue (1970a) completed the first comprehensive examination of diatom distribution in the Southern Ocean using long piston core samples of Pliocene and Pleistocene age recovered from the western Bellingshausen Basin area. She established four zones for the last two and one-half million years (Fig. 2), and correlated this framework with paleomagnetic polarity zones and with the geochronometric time scale.

Later, Abbott (1972) developed a similar zonation from piston core samples in the Indian Ocean sector of the Southern Ocean. He defined four zones based upon the extinction of several stratigraphically significant diatom species and correlated these to paleomagnetic polarity zones. The difference between Abbott's zonation and Donahue's is that the former provides, with addition of

[1] Absolute ages for the Pliocene/Pleistocene boundary follow Stainforth and others (1975), *i.e.*, at 2.8 m.y.BP.

Table 1.—Diatom Reports from Southern High Latitudes of the Deep Sea Drilling Project.

Leg	Authors and Volumes in Which Reports Appear
28	McCollum (Vol. 28)
29	Hajos (Cretaceous, Vol. 29; Eocene and Oligocene, Vol. 35); Schrader (Neogene, Late Oligocene, Vol. 35)
35	Schrader (Vol. 35)
36	Gombos (Vol. 36)

the *Hemidiscus karstenii* Zone (Fig. 3), increased biostratigraphic resolution within the Brunhes paleomagnetic polarity zone.

McCollum (1975), in conjunction with his DSDP Leg 28 diatom studies, also examined numerous long piston core samples from the Southern Ocean. These samples ranged in age from the early Gilbert through the Brunhes paleomagnetic polarity zones (see McCollum 1975, his table 2). Using these and other DSDP Leg 28 data, together with information from previous diatom biostratigraphic studies, McCollum (1975) presented a revised Pliocene and Pleistocene zonal scheme. He defined eight zones and correlated these with the paleomagnetic timescale (Fig. 4).

In our experience from studying numerous piston core samples from the Southern Ocean, we find that McCollum's (1975) zonation is regionally ap-

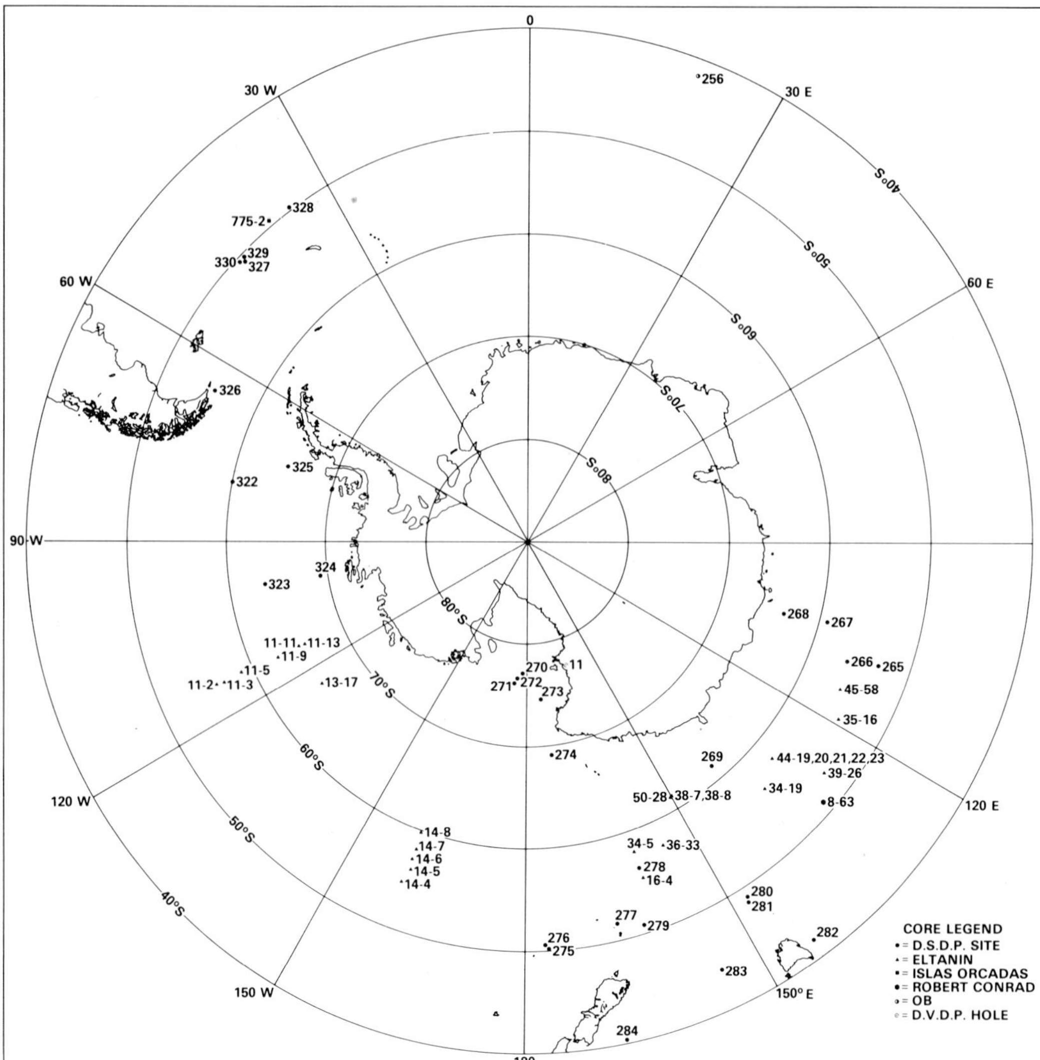

Fig. 1.—Core Locality Map.

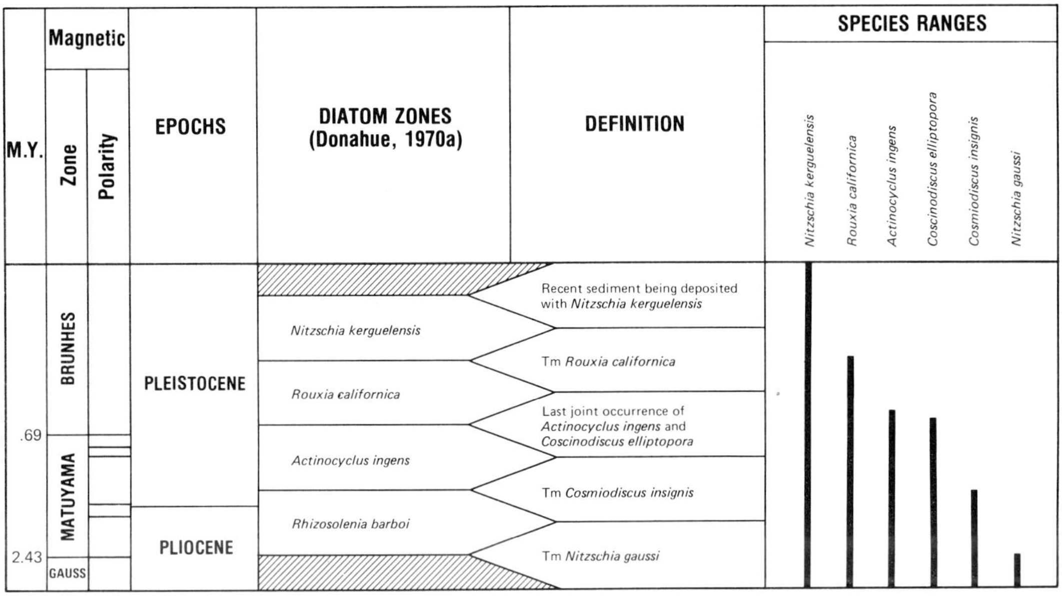

FIG. 2.—Late Pliocene and Pleistocene diatom biostratigraphy of Donahue (1970a).

plicable and useful for the Pleistocene, but somewhat less so for the Pliocene. Therefore, we propose replacement of McCollum's (1975) *Nitzschia praeinterfrigidaria* Zone by the *N. reinholdii* and *N. angulata* Zones (Fig. 7). The reason for this modification is that the *N. praeinterfrigidaria* Zone is difficult to recognize within the Southern Ocean because the last occurrence of both *N. praeinterfrigidaria* and *Denticulopsis hustedtii* is often obscured. Both species are easily and often reworked into younger sediments, and the identification of *N. praeinterfrigidaria* can be confused with similar species occurring within the upper part of its range, because it has not been precisely defined in

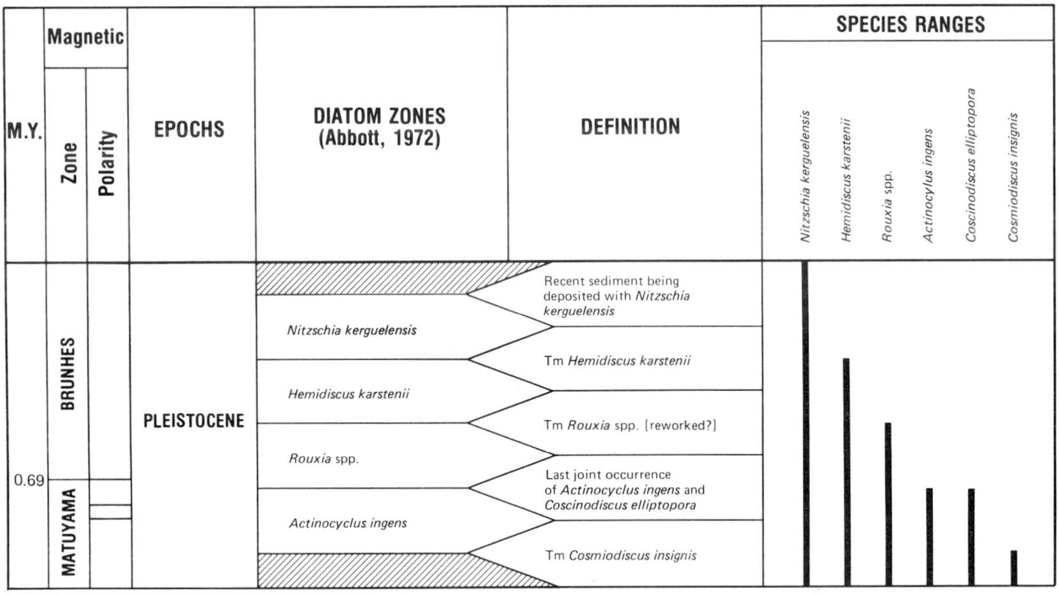

FIG. 3.—Pleistocene diatom biostratigraphy of Abbott (1972).

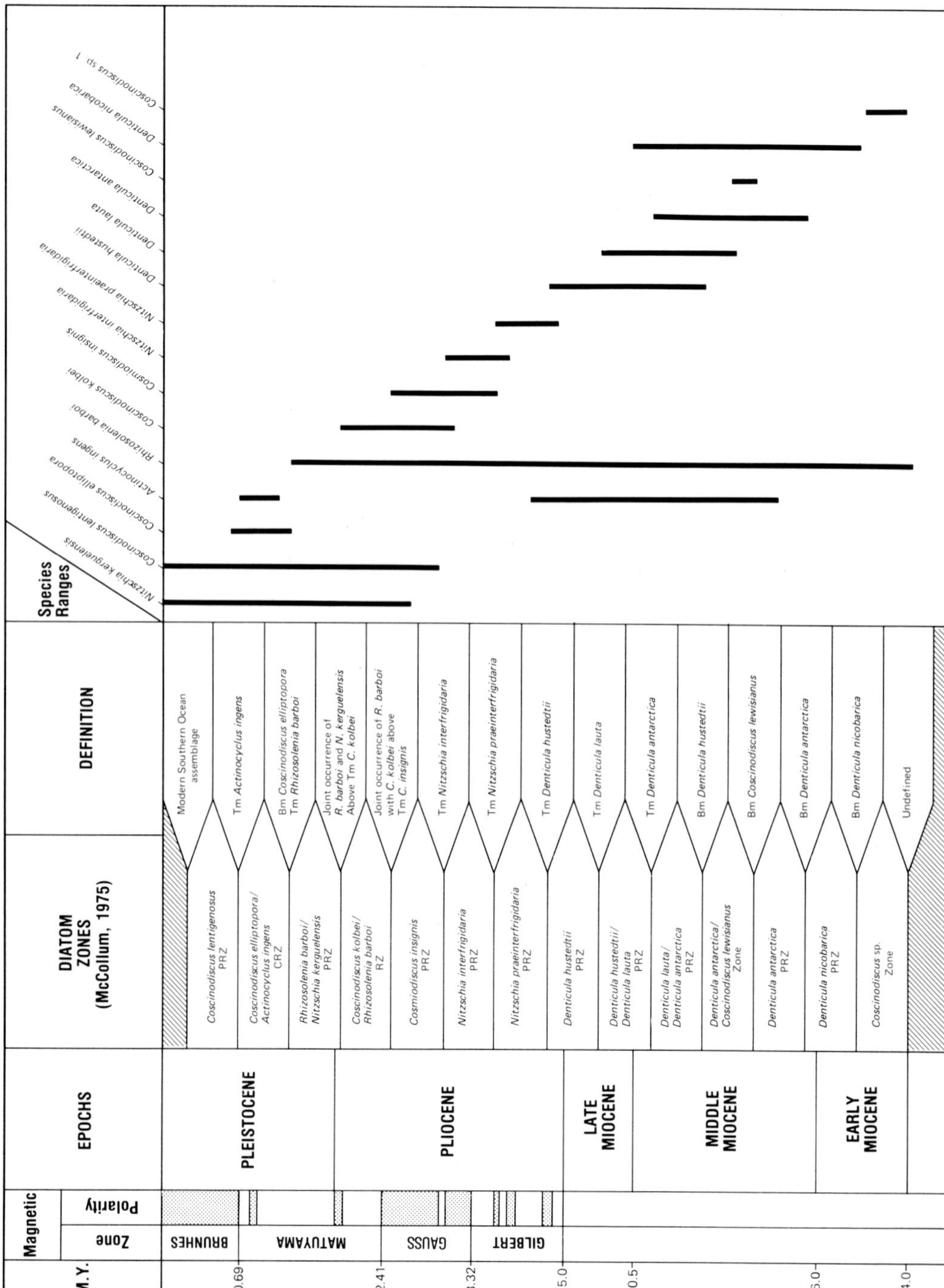

FIG. 4.—Neogene diatom biostratigraphy of McCollum (1975), Deep Sea Drilling Project Leg 28.

regard to other closely related species.

Biostratigraphic datums used to define the *Nitzschia reinholdii* and *N. angulata* Zones (Fig. 7), include the last abundance appearance datum (LAAD) of *Denticulopsis hustedtii,* the first appearance datum (Bm) of *Nitzschia angulata,* and the rapid evolutionary change from *N. praeinterfrigidaria* to *N. interfrigidaria.* These events occur within the middle to upper Gilbert paleomagnetic polarity Zone, and are all easily recognizable regionally within Pliocene Southern Ocean sediments (Weaver, 1976; Ciesielski, 1978).

Two additional regional datums that we include in McCollum's (1975) biostratigraphy are the last occurrences of *Coscinodiscus vulnificus* within the *Coscinodiscus kolbei/Rhizosolenia barboi* Zone and the LAAD of *Hemidiscus karstenii* within the *Coscinodiscus lentigenosus* Zone (see Burckle and others, 1978). A discussion of our emendations of McCollum's (1975) zonation follows in our proposed regional Neogene diatom zonation.

Miocene[1]

McCollum's (1975) DSDP Leg 28 diatom studies allowed him to extend his biostratigraphy through the Miocene for the first time. From the numerous Leg 28 sites (Fig. 1, Table 2), drilled throughout the Ross Sea and contiguous areas, McCollum (1975) defined seven Miocene diatom zones (Fig. 4). These zones are based on the upward sequential changes in the succession of *Denticulopsis* species (Fig. 4). No paleomagnetic stratigraphy was available for these sediments, and correlation of epoch boundaries were made from limited information about co-occurring rare calcareous microfossils.

Schrader (1976) later completed a comprehensive diatom study of DSDP Site 278, collected during Leg 29 in the Sub-Antarctic region south of New Zealand. The continuously cored Neogene section at Site 278 provided the most complete record of the Miocene yet recovered by deep-sea drilling in the Southern Ocean. Schrader (1976) was able to apply McCollum's (1975) Pliocene–Pleistocene zonation at Site 278. However, he had difficulty using McCollum's Miocene zones because his data showed different ranges for several index species. Schrader therefore chose to establish a new zonation, which he used for a biostratigraphic interpretation of the Miocene interval in the Sub-Antarctic region (Fig. 5). At DSDP Site 278, Schrader defined thirteen zones within Late to Early Miocene sediments. These zones are correlated with the time-scale based on a comparison with other zonations developed in several subglobal regions. No direct correlation with paleomagnetic stratigraphy was possible at DSDP Site 278.

In addition to the differences in several index species ranges between Legs 28 and 29, Schrader illustrated that the overall Miocene diatom assemblage from the Sub-Antarctic varies considerably from that reported by McCollum (1975). These differences are especially pronounced within the Middle and Late Miocene intervals.

Because of these discrepancies, we reexamined the diatom distribution in DSDP Leg 28, Site 266, using Schrader's (1976) more detailed taxonomic treatment of Neogene diatom species. Two questions needed to be resolved: (1) Are the differences between Miocene diatom assemblages in the Sub-Antarctic and Antarctic significant? (2) Is a regional zonation possible? Schrader did not believe a standard diatom stratigraphy is possible because of differences in the ranges of diatom species, ecological variability, and paleogeographic distribution.

Deep Sea Drilling Project: Sites 266 and 278

The two most complete Neogene sedimentary sequences collected during deep-sea drilling in the Southern Ocean occur at Leg 28–Site 266 and Leg 29–Site 278 (Fig. 1; Table 2). On Legs 35 and 36, by contrast, Neogene sections that were incompletely recovered failed to provide sediments useful for comprehensive biostratigraphic studies. DSDP Sites 266 and 278 therefore become critical to our understanding of Neogene diatom biostratigraphy within the Southern Ocean.

Because substantial differences exist with regard to the distribution of diatom species in the Miocene sediments examined by McCollum (1975) and Schrader (1976), we reexamined DSDP Site 266 and compared our results with those of Schrader (1976). This investigation was directed toward resolving apparent discrepancies in diatom distribution and occurrence between the Sub-Antarctic and Antarctic regions (Fig. 6).

Figure 6 illustrates that a number of diatom index species thought to be restricted to Miocene sediments in the Sub-Antarctic region (Schrader, 1976), do, in fact, occur in the Antarctic. The stratigraphically important diatom species *Nitzschia denticuloides, N. grossepunctata, N. maleinterpretaria,* and *Thalassiosira* sp. 10, although not documented or described by McCollum (1975), occur in the same stratigraphic positions in both DSDP Sites 266 and 278 (Fig. 6).

Reexamination of assemblages from Site 266 suggests that regional differences in diatom distribution and occurrence during the Miocene are not as significant as they had initially appeared to be in Leg 28 and 29 investigation reports (Schrader, 1976; McCollum, 1975). Real and apparent spatial as well

---

[1] The Miocene/Pliocene boundary as used herein follows Burckle's (1978) placement midway between Gilbert Thvera subzone and the top of the E-5 paleomagnetic polarity zone.

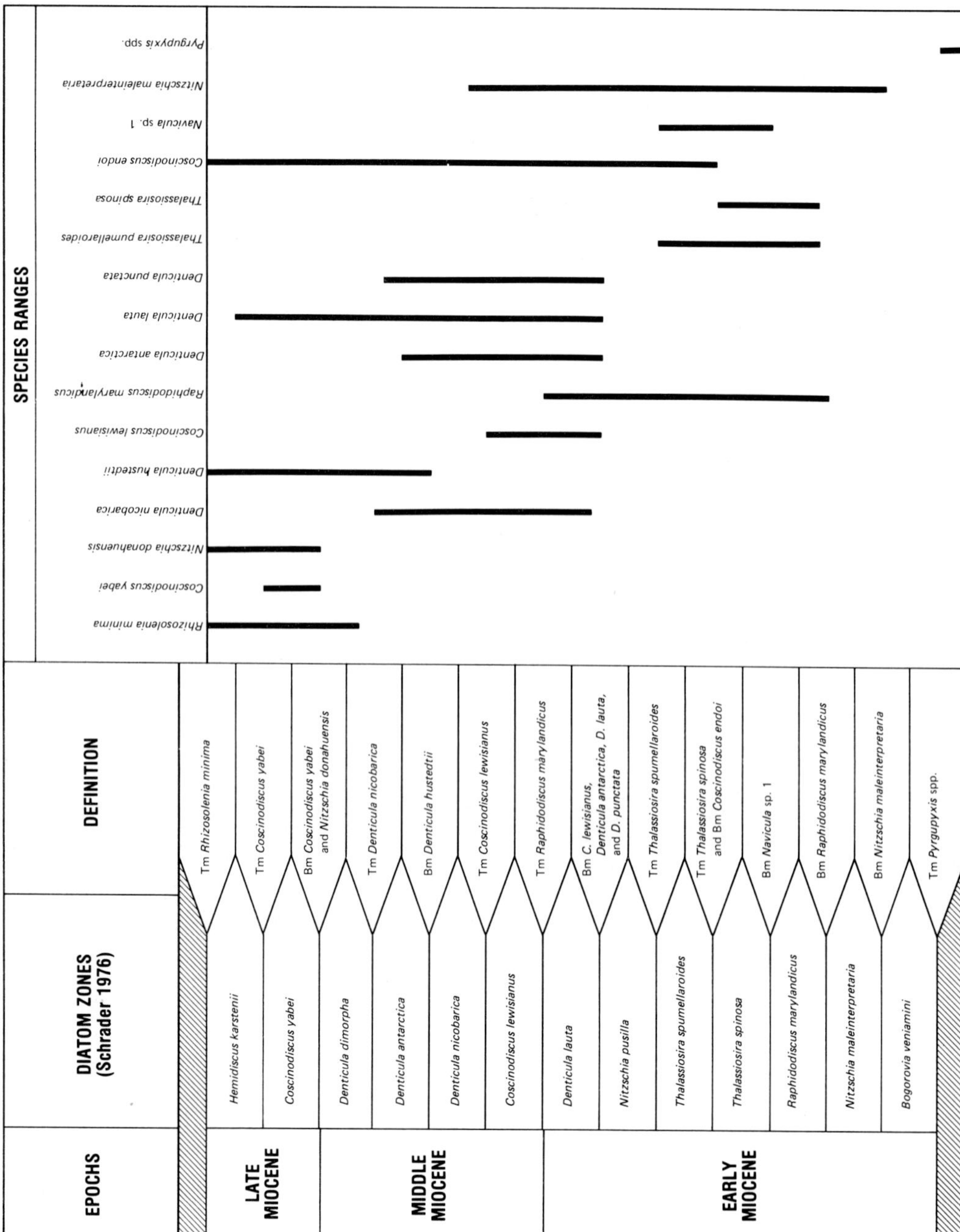

FIG. 5.—Miocene diatom biostratigraphy of Schrader (1976), Deep Sea Drilling Site 278.

TABLE 2.—LOCATION OF ALL DSDP SOUTHERN HIGH-LATITUDE DRILL SITES AND PISTON CORES USED BY VARIOUS WORKERS IN ESTABLISHING DIATOM ZONATIONS.

Core	Latitude	Longitude
Glomar Challenger		
Leg 28		
264	34°58.13′S	112°02.68′E
265	53°32.45′S	109°56.74′E
266	56°24.13′S	110°06.70′E
267	59°15.74′S	104°29.30′E
268	63°56.99′S	105°09.34′E
269	61°40.57′S	140°04.21′E
270	77°26.48′S	178°30.19′W
271	76°43.27′S	175°02.86′W
272	77°07.62′S	176°45.61′W
273	74°32.29′S	174°37.57′E
274	68°59.81′S	173°25.64′E
Leg 29		
275	50°26.34′S	176°18.99′E
276	50°48.11′S	176°48.40′E
277	52°13.43′S	166°11.48′E
278	56°33.42′S	160°04.29′E
279	51°20.14′S	162°38.10′E
280	48°57.44′S	147°14.08′E
281	47°59.84′S	147°45.85′E
282	42°14.76′S	143°29.18′E
283	43°54.60′S	154°16.96′E
284	40°30.48′S	167°40.81′E
Leg 35		
322	60°01.45′S	79°25.49′W
323	63°40.84′S	97°59.69′W
324	69°03.21′S	98°47.20′W
325	65°02.79′S	73°40.40′W
Leg 36		
326	56°35.00′S	65°18.20′W
327	50°52.28′S	46°47.02′W
328	49°48.67′S	36°39.53′W
329	50°39.31′S	46°05.73′W
330	50°55.19′S	46°53.00′W
331	37°53.00′S	38°06.92′W
Eltanin		
11-2	56°03.5′S	115°03.7′W
11-3	56°54.2′S	115°14.6′W
11-5	58°56.6′S	114°43.2′W
11-9	62°49.8′S	115°04.0′W
11-11	64°50.6′S	114°29.5′W
11-13	65°49.0′S	115°01.0′W
13-17	65°41.0′S	124°06.8′W
14-4	54°54.5′S	159°52.2′W
14-5	56°12.0′S	160°32.3′W
14-6	57°00.9′S	160°05.6′W
14-7	58°03.1′S	160°09.0′W
14-8	59°40.0′S	160°17.4′W
16-4	55°36′S	160°12′W
34-5	57°23.3′S	159°59.6′E
34-19	56°40.0′S	135°13.1′E
35-16	56°02.1′S	117°09.8′E
36-33	57°46.3′S	154°55.0′E
38-7	61°48.8′S	149°54.1′E

TABLE 2.—(CONTINUED)

Core	Latitude	Longitude
38-8	61°46.4′S	149°58.7′E
39-26	52°58.5′S	126°08.2′E
44-19	57°59.6′S	130°00.3′E
44-20	58°00.5′S	130°03.8′E
44-21	58°03.6′S	130°06.5′E
44-22	58°05.3′S	130°08.2′E
44-23	58°06.2′S	130°11.0′E
45-58	56°35.0′S	114°06.9′E
50-28	62°54.2′S	150°41.2′E
Robert Conrad		
8-63	51°S	130°E
Islas Orcadas		
775-2	49°27.3′S	39°37.6′W
Ob		
256	42°54′S	20°06′W
Dry Valley Drilling Project		
Hole-11	@77°30′S	@163°30′E

as temporal variance in diatom distribution resulted from a combination of six factors: 1) lack of continuous coring at Site 266, 2) relative incompleteness of the Neogene section recovered at Sites 266 and 278, 3) unrecognized unconformities, 4) regional paleobiogeographic differences, 5) applied taxonomy, and 6) incomplete species-range documentation.

1) Lack of Continuous Coring.—In contrast to Site 278, the coring strategy at Site 266 left numerous gaps in the Neogene stratigraphic sequence (Fig. 6). As a result, differences in the stratigraphic succession of diatom species at both sites are to be expected. This lack of continuous coring apparently accounts for part of the difficulty Schrader (1976) experienced in attempting to utilize McCollum's (1975) Miocene zonation. The coring gap between cores 12 and 13 at Site 266 illustrates the problem (Fig. 6). The last occurrences of *Coscinodiscus lewisianus* and *Nitzschia grossepunctata* are in core 13 followed by the first occurrences of *Denticulopsis hustedtii* and *N. denticuloides* in core 12 (Fig. 6). In contrast, the stratigraphic succession of these diatom species in a continuously cored time equivalent section at Site 278 differs from that of Site 266. The last occurrence of *Coscinodiscus lewisianus* (core 18) is followed successively by the first occurrence of *D. hustedtii* (core 17), the last occurrence of *N. grossepunctata* (core 16), and the first occurrence of *N. denticuloides* (core 16) (Fig. 6).

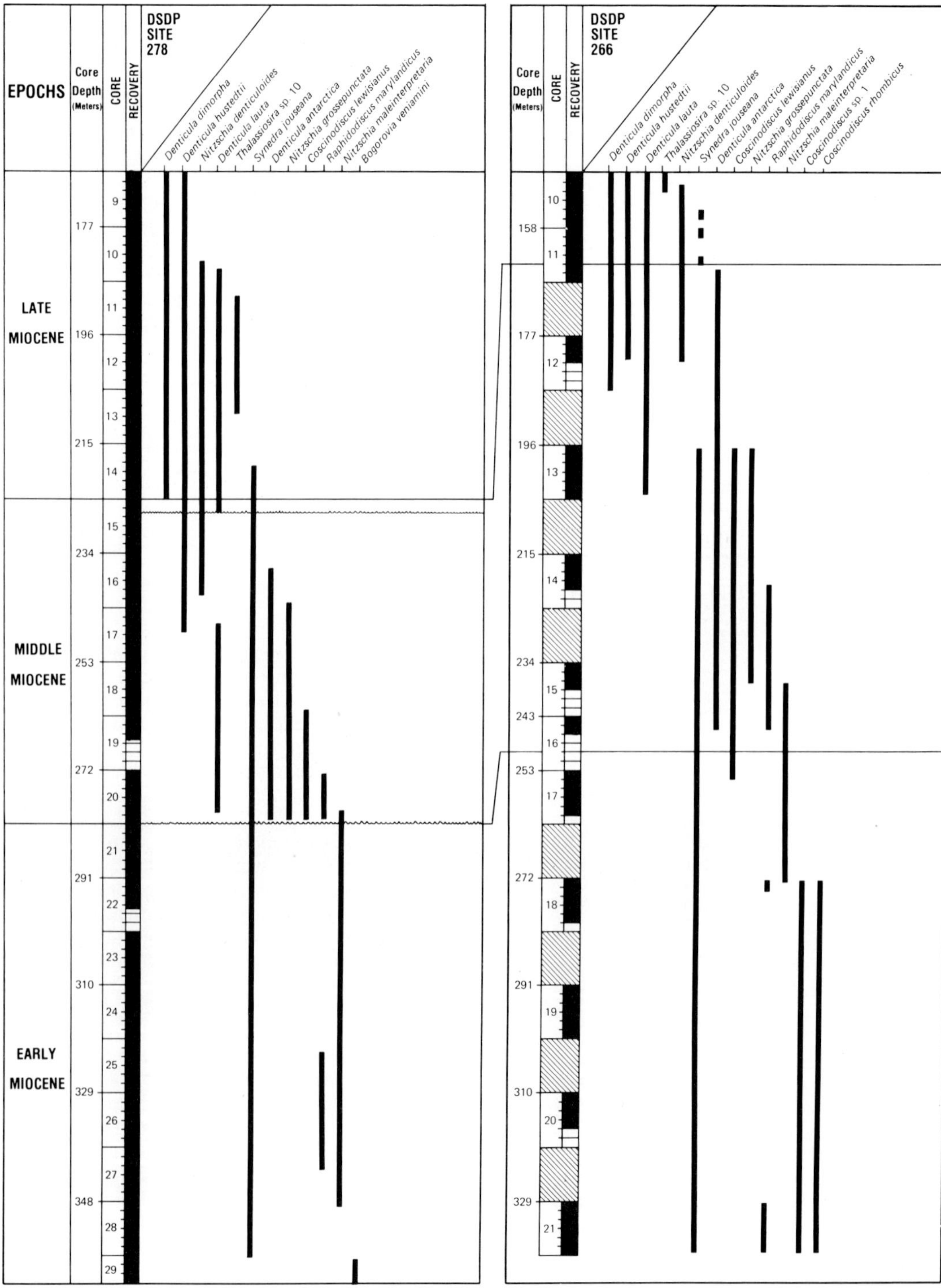

FIG. 6.—Comparison of stratigraphically important diatom species ranges between DSDP Sites 266 and 278.

2) Relative Incompleteness of the Neogene Record.—Studies of other co-occurring microfossils such as radiolarians and silicoflagellates from Sites 266 and 278 provide convincing evidence that a more complete Neogene section was recovered at Site 278 (Weaver, 1976; Perch-Neilsen, 1975; Bukry, 1975; Ciesielski, 1975). Data suggest that cores 9, 10, most of 11, and 29 from Site 278 represent Miocene intervals not collected at Site 266. The possibility therefore exists that many of the Late Miocene diatom species thought to be restricted to the Sub-Antarctic region (Schrader, 1976) may, in fact, co-occur in time equivalent sections in the Antarctic Southern Ocean.

3) Unconformities.—A comparison of Miocene diatom distribution at Sites 266 and 278 suggests that two unconformities exist at Site 278. These occur within core 15, section 2, and between cores 20 and 21. Abrupt faunal and floral changes occur at both horizons (Schrader, 1976; Petrushevskaya, 1975), concurrent with sharp variations in the carbonate content of the sediments. The unconformities at Site 278, which are in part responsible for the difference between Schrader's (1976) and McCollum's (1975) Miocene diatom ranges, clearly disrupt the stratigraphic sequence (Fig. 6).

4) Paleobiogeographic differences.—There is little doubt that many Miocene diatom species reported by Schrader (1976) at Site 278 are restricted to the Sub-Antarctic region. Both McCollum's (1975) data and ours from Site 266 failed to document many of the potentially important species illustrated by Schrader (1976, fig. 14). Species such as *Nitzschia claviceps, N. pseudokerguelensis, N. efferans, Coscinodiscus deformans, C. ruboides, Mediaria splendida, Thalassiosira spinosa, T. spumelleroides, Denticulopsis punctata,* and *Synedra miocenica* were not encountered at Site 266. Schrader (1976) believed that these species, in addition to *Nitzschia denticuloides, N. grossepunctata* and *N. maleinterpretaria* occurred exclusively in the Sub-Antarctic region. Therefore, Miocene paleobiogeographic differences between Sub-Antarctic and Antarctic areas were of such a magnitude that McCollum's (1975) diatom zones were not applicable.

5) Applied Taxonomy.—We conclude from our detailed reexamination of Site 266, that some of the variance in the ranges of Miocene diatom species between McCollum's (1975) and Schrader's (1976) data results from taxonomic difficulties, especially in regard to the application of species concepts.

McCollum's (1975) Leg 28 investigation represents the first comprehensive research on Southern Ocean Miocene diatoms; but, because of time restrictions, he was not able to describe all new species encountered. Moreover, the descriptions of those that were described are very brief and lacking in comparisons with similar or closely related forms. Schrader's (1976) later study at Site 278, however, is much more complete in its taxonomic coverage. These differences in approach have led to some confusion in species concepts, and this problem affected Schrader's (1976) documentation of the stratigraphic occurrences of several species defined by McCollum (1975). For instance, we believe the differences in the range of *Denticulopsis maccollumii (=D. antarctica)* at sites 266 and 278 are a reflection of its imprecise definition (see Schrader, 1976, p. 631).

Another taxonomic problem affects the recorded range of *Denticulopsis nicobarica* at Site 266, a species critical to the definition and application of a portion of McCollum's (1975) Miocene zonation. It is obvious from our reexamination of Site 266 that *Denticulopsis nicobarica* and *Nitzschia maleinterpretaria* of Schrader (1976) had been interpreted differently by McCollum. He illustrated the combined ranges of these two species as that of *D. nicobarica* only. The range of *D. nicobarica* at Site 266 is at variance with the data at Site 278, and this affected Schrader's use of McCollum's (1975) Early Miocene diatom zones.

6) Incomplete Documentation of Ranges.—The ranges of several stratigraphically important diatom species were inadequately documented by McCollum (1975) in Site 266. On the basis of our data, the ranges of the following species are extended: *Coscinodiscus lewisianus* from core 13-1 to core 17-1, *Denticulopsis maccollumii* from core 11-5 to 16-2, *Raphidodiscus marylandicus* from core 14-4 to 16-2, 18-1 to 18-2, and 21-1 to 21-4, and *Coscinodiscus sp 1* McCollum from Core 18-1 to 21-6 (compare our Figures 6 and 8 with McCollum's (1975) figure 3). These adjustments in species ranges at Site 266 substantiate part of Schrader's (1976) observations on local species variability within Miocene sediments of the Sub-Antarctic and Antarctic regions.

The previously discussed problems render the diatom zonations based on material from both Legs 28 and 29 difficult to apply regionally: Schrader's (1976) because he uses numerous species not occurring at Site 266; and McCollum's (1975) because his species ranges lacked complete documentation, and the stratigraphic succession of species is interrupted and confused by coring gaps.

Regional Neogene Diatom Zonation

With a more complete understanding of Miocene diatom distribution and occurrence throughout the Sub-Antarctic and Antarctic region, we propose a new regionally applicable Neogene zonation. It is based on easily recognizable diatom species with widespread distribution throughout the Southern Ocean.

Our biostratigraphic framework contains five

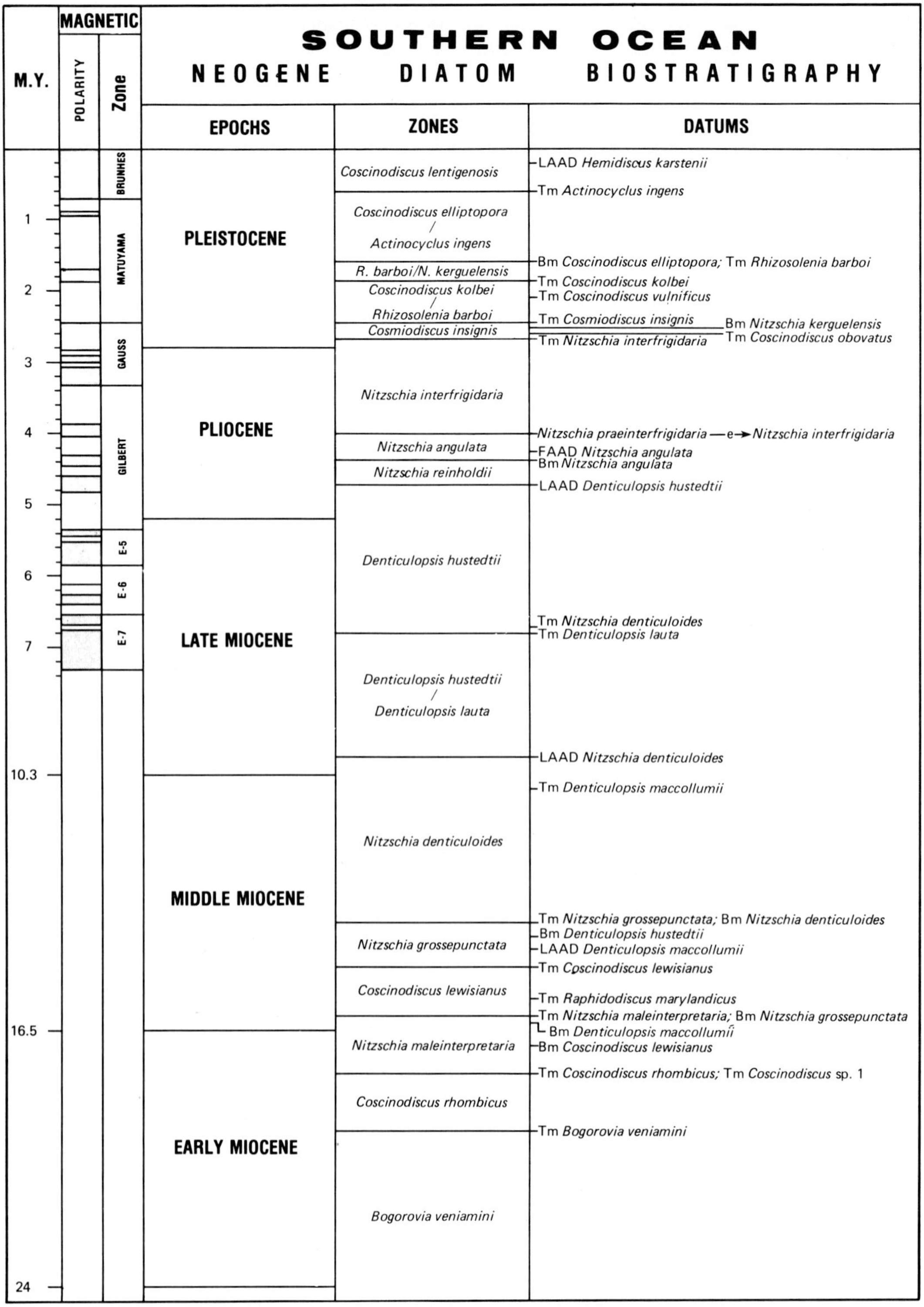

FIG. 7.—Proposed Southern High-Latitude Neogene diatom biostratigraphy.

AGE	D.S.D.P. SITE 266 CORE	SECTION	SAMPLE INTERVAL IN CM
Late Miocene	10	1	90-92
	10	2	40-42
	10	3	NO SAMPLE
	10	4	40-42
	10	5	30-32
	10	5	90-92
	10	6	80-82
	11	1	60-62
	11	2	60-62
	11	3	90-92
	11	4	40-42
Middle Miocene	11	5	30-32
	11	6	70-72
	12	1	30-32
	12	2	120-122
	12	3	110-112
	12	4-6	NO RECOVERY
	13	1	60-62
	13	2	70-72
	13	3	60-62
	13	4	60-62
	13	5	60-62
	13	6	60-62
	14	1	80-82
	14	2	60-62
	14	3	NO SAMPLE
	14	4	60-62
	14	5-6	NO RECOVERY
	15	1	60-62
	15	2	60-62
	15	3	60-62
	15 15	4-6	NO RECOVERY
	16	2	60-62
	16 16	3-6	NO RECOVERY
Early Miocene	17	1	140-142
	17	2	60-62
	17	3	60-62
	17	4	60-62
	17	5	NO SAMPLE
	18	1	63-65
	18	2	60-62
	18	3	60-62
	18	4	60-62
	18	5	60-62
	18	6	NO RECOVERY
	19	1	60-62
	19	2	60-62
	19	3	60-62
	19	4	60-62
	19	5	60-62
	19	6	60-62
	20	1	60-62
	20	2	60-62
	20	3	60-62
	20	4	60-62
	20	5-6	NO RECOVERY
	21	1	61-63
	21	2	62-64
	21	3	60-62
	21	4	60-62

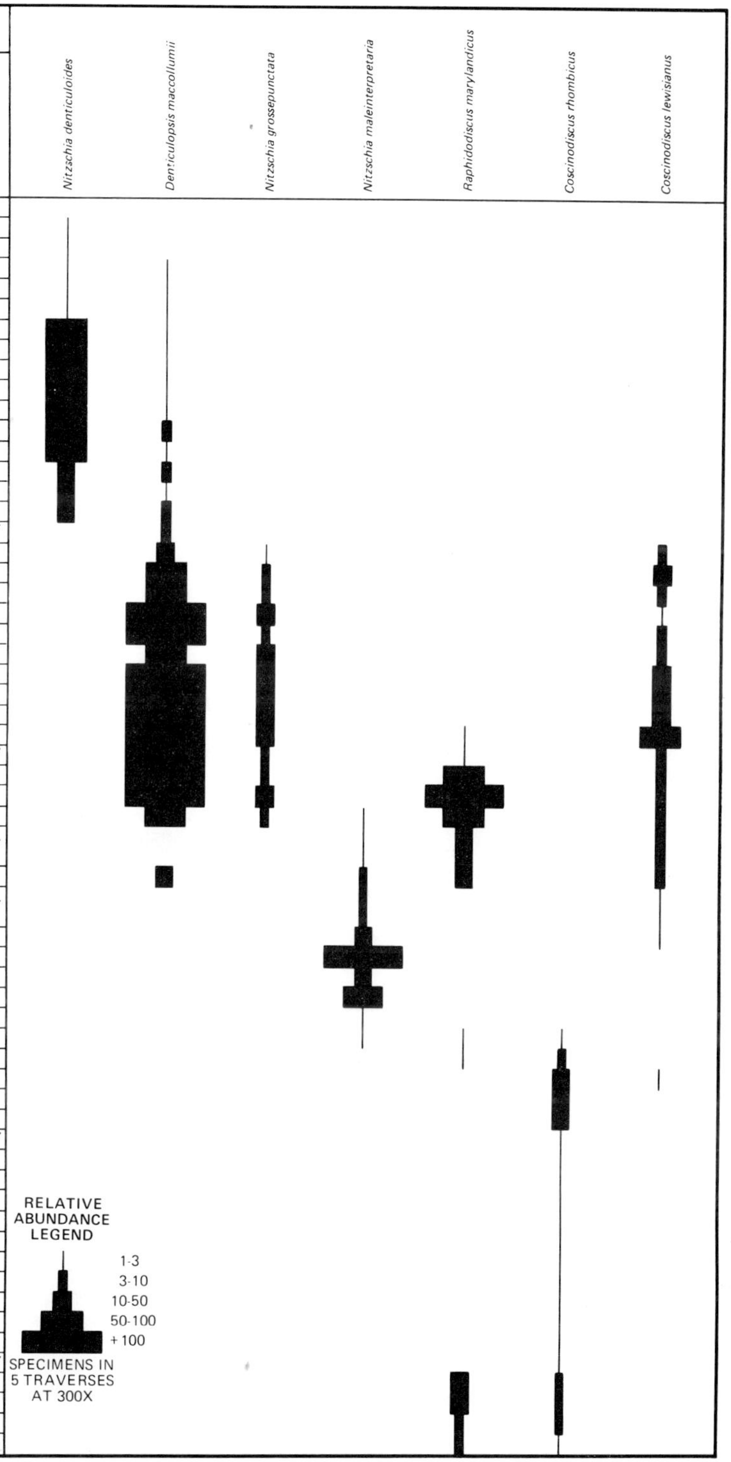

FIG. 8.—Downhole abundance changes in several stratigraphically-important diatom species at DSDP Site 266.

zones as originally proposed by McCollum (1975), modifications of six previously proposed zones (McCollum, 1975; Schrader, 1976), and five new zones. In addition, we document sixteen diatom datums with apparent regional significance (Fig. 7), and illustrate the variability in abundance of several of these datums in Miocene sediments at Site 266 (Fig. 8).

Five types of diatom datums are illustrated. These include 1) Tm—morphotypic last occurrence, 2) Bm—morphotypic first occurrence, 3) FAAD—first abundance appearance datum, 4) LAAD—last abundance appearance datum, and 5) the evolutionary transition of a species "e". We are aware of the difficulty inherent in using abundance changes in biostratigraphy, because these datums can often be time-transgressive. However, the few that we do use are easily recognized regionally. We feel these are more reliable than other closely associated last occurrence datums which are often obscured by reworking.

Epoch Boundaries

Because of the general paucity of stratigraphically important planktonic calcareous microfossils in Neogene sediments of the Southern Ocean region, correlation with standard ages and placement of epoch boundaries are difficult and very subjective. In our zonation, our determination of epoch boundaries and absolute ages at Sites 266 and 278 utilizes data from Vail and Hardenbol (1979). Table 3 summarizes all significant datums used for geochronologic subdivision at Sites 266 and 278. Correlation of planktonic microfossil datums with the paleomagnetic time scale follows that of Ryan and others (1974).

ZONAL DEFINITIONS

Coscinodiscus lentigenosus
Partial Range Zone

Author: McCollum, 1975
Top: Recent
Base: Last occurrence of *Actinocyclus ingens*
Age: Pleistocene
Paleomagnetic Correlation: Basal Brunhes paleomagnetic polarity zone, ~0.6 to 0.5 m.y.BP to recent
Important Datums: Last abundance appearance datum of *Hemidiscus karstenii,* upper part of oxygen isotope stage 7 (≈195,000 m.y.BP) (Burckle and others, 1978)

Coscinodiscus elliptopora/Actinocyclus ingens
Concurrent Range Zone

Author: McCollum, 1975
Top: Coincident with the base of the *Coscinodiscus lentigenosus* Zone
Base: First occurrence of *Coscinodiscus elliptopora*
Age: Pleistocene
Paleomagnetic Correlation: The top of this zone is approximately 0.6 to 0.5 m.y.BP and the base is at approximately 1.6 m.y.BP (middle Matuyama to basal Brunhes paleomagnetic polarity zone).

Rhizosolenia barboi/Nitzschia kerguelensis
Partial Range Zone

Author: McCollum, 1975
Top: Last occurrence of *Rhizosolenia barboi,* coincident with the base of the *C. elliptopora/A. ingens* Zone
Base: Defined by the joint occurrence of *Rhizosolenia barboi* and *Nitzschia kerguelensis* immediately following the last occurrence of *Coscinodiscus kolbei*
Age: Pleistocene
Paleomagnetic Correlation: The top of this zone is approximately 1.6 m.y.BP and the base 1.8 m.y.BP (Just above, to within, the Olduvai subzone of the Matuyama paleomagnetic polarity zone)

Coscinodiscus kolbei/Rhizosolenia barboi
Partial Range Zone

Author: McCollum, 1975
Top: Last occurrence of *Coscinodiscus kolbei*
Base: Last occurrence of *Cosmiodiscus insignis*
Age: Pleistocene
Paleomagnetic Correlation: The top of this zone is approximately 1.8 m.y.BP; and the base 2.45 m.y.BP (basal Matuyama to upper Gauss paleomagnetic polarity zone), ≈Gauss/Matuyama boundary
Important Datums: Last occurrence of *Coscinodiscus vulnificus,* basal Matuyama paleomagnetic polarity Zone below the Olduvai subzone (≈2.0 to 2.2 m.y.BP).

Cosmiodiscus insignis
Partial Range Zone

Author: McCollum, 1975
Top: Last occurrence of *Cosmiodiscus insignis*
Base: Last occurrence of *Nitzschia interfrigidaria*
Age: Pleistocene
Paleomagnetic Correlation: The top of this zone is approximately 2.45 m.y.BP, ≈Gauss/Matuyama boundary; and the base 2.75 m.y.BP (Upper Gauss paleomagnetic polarity zone).
Important Datums: Last occurrence of *Coscinodiscus obovatus* and first occurrence of *Nitzschia kerguelensis* (Fig. 7).

Nitzschia interfrigidaria
Partial Range Zone

Author: McCollum, 1975; modified this report
Top: Last occurrence of *Nitzschia interfrigidaria*
Base: Evolutionary transition from *Nitzschia praeinterfrigidaria* to *N. interfrigidaria*
Age: Pliocene
Paleomagnetic Correlation: The top of this zone is approximately 2.75 m.y.BP, and the base 3.9 m.y.BP (Upper Gauss to the Cochiti subzone of the Gilbert paleomagnetic polarity Zone).
Remarks: McCollum (1975) originally defined the base of this zone as the last occurrence of *Nitzschia praeinterfrigidaria.* We found that this datum was difficult to recognize due to the co-occurrence of several closely related species near the extinction level of *Nitzschia praeinterfrigidaria.* We therefore chose to place the base of the *N. interfrigidaria* Zone at the easily recognizable change in dominance from *N. praeinterfrigidaria* to *N. interfrigidaria.* This level coincides with the Cochiti subzone of the Gilbert paleomagnetic polarity zone.

TABLE 3.—DATUMS USED FOR GEOCHRONOLOGIC SUBDIVISION OF DSDP SITES 266 AND 278.

Epoch	Site	Section	Core	Datum	Refs.	Site	Section	Core	Datum	Refs.	References
EARLY MIOCENE	266	10	5	Bm *Theocalyptra bicornis spongothorax* (~8.8–9.2 m.y.BP)*	1	278	10	5	Tm *Denticulopsis lauta* (~6.8–7.5 m.y.BP)*	1, 13, 2	Weaver (1976) (1) Ciesielski (1975) (2) Bukry (1978) (3)
	266 266	10 11	6 1	Occurrence of *Discoaster calcaris*. Range NN8 to NN11; characteristic of NN10 and NN9, (~8.5 to 12.5 m.y.BP)	5, 8, 9	278	13	3	Bm *Theocalyptra bicornis spongothorax* (~8.8–9.2 m.y.BP)*	1	Bukry (1975) (4) Burns (1975) (5) Kaneps (1975) (6) This report (7)
	266	11	4			278	14	cc			Ryan and others (1974) (8)
MIDDLE MIOCENE	266	11	5	Bm *Didymocyrtis antepenultima* (~10.7 to 11.6 m.y.BP)	1, 8, 15	278	15	1			Martini (1971) (9) Gartner (1977) (10)
	266	12	3	Tm *Cyclicargolithus floridanus* (~13.8 m.y.BP, within NN6)	5, 8, 10	278	15	2	Tm *Denticulopsis nicobarica*	13	Stainforth and others (1975) (11) Jenkins (1974) (12)
	266	14	4	Tm *Sphenolithus heteromorphus* (~14.7 m.y.BP; top of NN5)	4, 5, 8	278	16	1	Tm *Cyclicargolithus floridanus* (~13.8 m.y.BP; within NN6)	3, 8, 10, 14	Schrader (1976) (13) Edwards & Perch-Neilsen (1975) (14)
	266	16	3	First occurrence of *Denticulopsis*	7	278	20	cc	Bm *Discoaster variabilis* (~14.7 m.y.BP)	8, 10, 14	Theyer, Mato and Hammond (1978) (15)
LATE MIOCENE	266	17	1	Bm *Sphenolithus heteromorphus* (~18.2 m.y.BP; top part of NN3)	5, 3, 8, 10	278	21	1			
	266	19	2	Tm *Catapsydrax dissimilis* (~18.5 m.y.BP)	6, 11, 8	278	21	cc	Tm *Catapsydrax dissimilis* (~18.5 m.y.BP)	8, 11, 12	
	266 266	19 19	1 6	Occurrence of *Sphenolithus belemnos*. Range, top half NN2 and NN3; (~18.2 to 20.5 m.y.BP)	3, 5, 8, 10	278	30	cc	Tm *Triquetrorhabdulus carinatus* (~18.5–19.3 m.y.BP)	3, 8, 10, 14	
	266 266	20 21	1 1	Occurrence of *Discoaster druggi* prior to Bm *S. belemnos* (~20.5 to 22 m.y.BP)	3, 5, 8, 9						

Tm = Morphotypic last occurrence—provides the minimum age for a specific datum level.
Bm = Morphotypic first occurrence—provides the maximum age for a specific datum level.
*Calculation based on average sedimentation rates.

Nitzschia angulata (New Zone)
Partial Range Zone

Author: Weaver and Gombos, herein

Top: Coincident with the base of the *Nitzschia interfrigidaria* Zone

Base: First occurrence of *Nitzschia angulata*

Age: Pliocene

Paleomagnetic Correlation: The top of this zone is approximately 3.9 m.y.BP and the base 4.3 m.y.BP (the Cochiti subzone of the Gilbert to approximately the top of the Nunivak subzone of the Gilbert paleomagnetic polarity zone).

Correlation with Other Zonations: The *Nitzschia angulata* Zone correlates in part with the *Nitzschia praeinterfrigidaria* Zone of McCollum (1975).

Occurrence of Zone: This zone is established on the basis of data from DSDP Site 266 and *Eltanin* piston cores E14-8, 50-28, 38-8, and 16-4 (see Table 2 for locations). The *Nitzschia angulata* Zone is applicable in Pliocene Southern Ocean sediments deposited between 53° and 62° S latitude. South of 62° S latitude utilization of McCollum's (1975) *N. praeinterfrigidaria* Zone is preferred due to paleobiogeographic restriction of *N. angulata* from very high latitudes during the Early Pliocene.

Important Datums: First abundance appearance of *Nitzschia angulata* between the Cochiti and Nunivak subzones of the Gilbert paleomagnetic polarity Zone.

Nitzschia reinholdii (New Zone)
Partial Range Zone

Author: Weaver and Gombos, herein

Top: Coincident with the base of the *Nitzschia angulata* Zone

Base: Last abundance appearance datum of *Denticulopsis hustedtii*

Age: Pliocene

Paleomagnetic Correlation: The top of this zone is approximately 4.3 m.y.BP and the base 4.7 m.y.BP (between the top of the Nunivak to the Sidufjall and Thvera subzones of the Gilbert paleomagnetic polarity Zone).

Correlation with Other Zonations: The *Nitzschia reinholdii* Zone correlates in part to the *N. praeinterfrigidaria* Zone and the upper part of the *Denticula hustedtii* Zone of McCollum (1975).

Occurrence of Zone: The *Nitzschia reinholdii* Zone is established on the basis of data from DSDP Site 266, and *Eltanin* piston cores, E16-4, 14-8 and 38-8.

Remarks: The last abundance appearance datum of *D. hustedtii* is used to define the base of the *Nitzschia reinholdii* Zone since it is more easily recognizable regionally than the last occurrence of *Denticulopsis hustedtii,* which is often reworked into younger Southern Ocean sediments. *Nitzschia reinholdii* occurs consistently throughout this zone within Early Pliocene Southern Ocean sediments.

Denticulopsis hustedtii
Partial Range Zone

Author: McCollum (1975); modified herein

Top: Coincident with the base of the *Nitzschia reinholdii* Zone

Base: Last occurrence of *Denticulopsis lauta*

Age: Late Miocene to Pliocene

Paleomagnetic Correlation: The top of this zone is approximately 4.7 m.y.BP and the base ≈6.8–7.0 m.y.BP (lower Gilbert Sidufall and Thvera subzones to approximately the E-7 magnetic polarity Zone). No direct correlation of the basal *D. hustedtii* Zone with the magnetic polarity Zone E-7 is available. Ciesielski (1978) has shown this boundary to be at least as old as basal E-6, and Weaver (1976) extrapolating from average sedimentation rates in DSDP Site 278 estimates this boundary at 6.8 to 7.0 m.y.BP or within E-7.

Correlation with Other Zones: The *Denticulopsis hustedtii* Zone correlates in part to the *Hemidiscus karstenii* Zone of Schrader (1976) and the *Denticula hustedtii* Zone of McCollum (1975).

Important Datums: The last occurrence of *Nitzschia denticuloides* is within the basal *D. hustedtii* Zone (Fig. 7).

Denticulopsis hustedtii/Denticulopsis lauta
Concurrent Range Zone

Author: McCollum, (1975); modified herein

Top: Coincident with the base of the *Denticulopsis hustedtii* Zone

Base: Last abundance appearance datum of *Nitzschia denticuloides*

Age: Late Miocene (6.8–7.0 to 10 m.y.BP based upon extrapolation of average sedimentation rates in DSDP Site 278).

Paleomagnetic Correlation: None available

Correlation with Other Zones: The *Denticulopsis hustedtii/Denticulopsis lauta* Zone correlates to part of the *Hemidiscus karstenii* Zone, the *Coscinodiscus yabei* Zone, and part of the *Denticula dimorpha* Zone of Schrader (1976) and to the *Denticula hustedtii/Denticula lauta* Zone of McCollum (1975).

Remarks: The abrupt demise of *Nitzschia denticuloides* throughout Southern Ocean sediments during the Late Miocene is an easily recognizable event and is observed at DSDP Sites 278 and 266. This datum replaces the last occurrence of *Denticulopsis maccollumii (Denticula antarctica)* used by McCollum (1975) to define the base of his *D. hustedtii/D. lauta* Zone. *Denticulopsis maccollumii* is often reworked in Late Miocene Southern Ocean sediments thereby making it difficult to pick this zonal boundary.

Nitzschia denticuloides (New Zone)
Partial Range Zone

Author: Weaver and Gombos, herein

Top: Coincident with the base of the *Denticulopsis hustedtii/Denticulopsis lauta* Zone

Base: Last occurrence of *Nitzschia grossepunctata*

Age: Middle Miocene to early Late Miocene (approximately 14 to 14.5 m.y.BP to 10.0 m.y.BP). Age estimate is based on extrapolation of sedimentation rates in DSDP Site 278. The first occurrence of *N. denticuloides* in DSDP Site 278 is within core 16, section 5, approximately 5 meters below the last occurrence of the nannofossil *Cyclicargolithus floridanus.* According to Bukry (1978) and Martini (1971), this datum occurs within NN6; the basal part of the *N. denticuloides* Zone therefore is probably older than the top of NN6 at 13.8 m.y.BP.

Important Datums: The last occurrence of *Denticulopsis maccollumii* occurs within the upper part of the *Nitzschia denticuloides* Zone. The first occurrence of *Nitzschia denticuloides* coincides with the base of the *N. denticuloides* Zone.

Paleomagnetic Correlation: None available

Correlation with Other Zones: The *Nitzschia denticuloides* Zone correlates with the part of the *Denticula dimorpha* and *D. antarctica* Zones of Schrader (1976) and with part of the *D. hustedtii/D. lauta* and *D. lauta/D. antarctica* Zones of McCollum (1975).

Nitzschia grossepunctata (New Zone)
Partial Range Zone

Author: Weaver and Gombos herein

Top: Coincident with the base of the *Nitzschia denticuloides* Zone

Base: Last occurrence *Coscinodiscus lewisianus*

Age: Early Middle Miocene; older than 14.0 m.y.BP based upon nannofossil data from Edwards and Perch-Nielsen (1975), and extrapolation of average sedimentation rates in DSDP Site 278.

Paleomagnetic Correlation: The base of the *Nitzschia grossepuncata* Zone correlates with the basal one third of paleomagnetic polarity zone E-16, approximately 14.8 to 15.0 m.y.BP. This determination is based upon our interpretation of data from Eltanin core 36-16 (Watkins and Kennett, 1972). The base of the *N. grossepunctata* Zone occurs between 420 and 450 cm, just above what appears to be the E-15/E-16 paleomagnetic polarity zone boundary.

Correlation with Other Zones: The *Nitzschia grossepunctata* Zone correlates with the lower part of the *Denticula antarctica* Zone, and the *D. nicobarica* Zone of Schrader (1976). The *N. grossepunctata* Zone cannot be correlated to the zonation of McCollum (1975) because of variance in our data regarding the range of *Coscinodiscus lewisianus* in DSDP Site 266 (see text).

Important Datums: The first occurrence of *Denticulopsis hustedtii* and the last abundance appearance datum of *D. maccollumii* occur within the *Nitzschia grossepunctata* Zone.

Coscinodiscus lewisianus
Partial Range Zone

Author: Schrader (1976); modified this report

Top: Coincident with the base of the *Nitzschia grossepunctata* Zone

Base: Last occurrence of *Nitzschia maleinterpretaria*

Age: Early Middle Miocene

Paleomagnetic Correlation: None available

Correlation with Other Zones: The *Coscinodiscus lewisianus* Zone correlates to the *C. lewisianus* Zone and part of the *Denticula lauta* Zone of Schrader (1976). No correlation with the zonation of McCollum is possible because of variance with our data at DSDP Site 266 (see text).

Important Datums: The last occurrence of *Raphidodiscus marylandicus* occurs within the *Coscinodiscus lewisianus* Zone. Also, the first occurrence of *Nitzschia grossepunctata* occurs at the base of the *C. lewisianus* Zone.

Nitzschia maleinterpretaria
Partial Range Zone

Author: Schrader (1976); Redefined this report

Top: Coincident with the base of the *Coscinodiscus lewisianus* Zone

Base: Last occurrence of *Coscinodiscus rhombicus*

Age: Early Miocene to early Middle Miocene. The top of the *N. maleinterpretaria* Zone falls above the Middle/Early Miocene boundary. The base of this zone is younger than 19.3 m.y.BP or the base of NN3 (see Table 3).

Paleomagnetic Correlation: None available

Correlation with Other Zones: The *Nitzschia maleinterpretaria* Zone correlates with part of the *Denticula lauta* Zone and with the *N. pusilla, Thalassiosira spumella*-currence of *B. veniamini* occurs above the range of the *maleinterpretaria* Zones of Schrader (1976). No correlation with the zonation of McCollum (1975) is possible because of his variance with our data at DSDP Site 266 (see text).

Important Datums: The first occurrence of *Denticulopsis maccollumii* and *Coscinodiscus lewisianus* occurs within the *Nitzschia maleinterpretaria* Zone.

Coscinodiscus rhombicus (New Zone)
Partial Range Zone

Author: Weaver and Gombos, herein

Top: Coincident with the base of the *Nitzschia maleinterpretaria* Zone

Base: Last occurrence of *Bogorovia veniamini*

Age: Early Miocene

The last occurrence of *Bogorovia veniamini* in DSDP Site 278 is above the last occurrence of the nannofossil *Triquetrorhabdulus carinatus.* According to Ryan and others (1974) this datum occurs within lower paleomagnetic polarity Zone E-17 at approximately 19.3 m.y.BP. Therefore, the base of the *Coscinodiscus rhombicus* Zone could be as young as 19.3 m.y.BP.

Paleomagnetic Correlation: None available

Correlation with Other Zones: The *Coscinodiscus rhombicus* Zone correlates with the upper part of the *Bogorovia veniamini* Zone of Schrader (1976) and to the *Coscinodiscus sp. 1* Zone of McCollum (1975).

Important Datums: The last occurrence of *Coscinodiscus sp.* 1 McCollum is coincident with the top of the *Coscinodiscus rhombicus* Zone as defined herein.

Bogorovia veniamini
Range Zone

Author: Schrader (1976); modified this report (see text)

Top: Coincident with the base of the *Coscinodiscus rhombicus* Zone.

Base: The first occurrence of *Bogorovia veniamini*

Age: Latest Oligocene to Early Miocene. The last occurrence of *B. veniamini* occurs above the range of the nannofossil *Triquetrorhabulus carinatus* at DSDP Site 278 Edwards and Perch-Neilsen, (1975). Therefore, the top of the *B. veniamini* Zone could be as young as 19.3 m.y.BP (Ryan and others, 1974).

Paleomagnetic Correlation: None available

Correlation with Other Zones: The *Borogovia veniamini* Zone correlates to part of the *Pyrgupyxis* species zone of Schrader (1976). It also correlates to the *Bogorovia veniamini* Zone and part of the *Pyrgupxis prolongata* Zone of Gombos (1976) (see text discussion).

PALEOGENE DIATOM BIOSTRATIGRAPHY

Oligocene

Oligocene diatomaceous sediments were recovered during DSDP Legs 28, 29 and 36. Various attempts to characterize this interval on the basis of diatoms have been made. However, a general lack of detailed knowledge about the stratigraphic distribution of diatoms in the Oligocene, as well as the frequent absence, or sparsity, of associated calcareous microfossils and nannofossils in cored intervals, has hampered attempts to provide reliable stratigraphic datums.

McCollum (1975) defined his *Pyrgupyxis (Pyxilla) prolongata* Zone as the local range of that species in Site 274, Cores 20–34. Calcareous nannofossils are present in that interval and suggest an Oligocene age (Burns, 1975). The presence of the foraminifer *Globigerina angiporoides* in Core 21 likewise suggests an Oligocene age (Kaneps, 1975). On the basis of silicoflagellates, Ciesielski (1975) has assigned the interval from Core 20 to Core 34 to the Early Oligocene *Dictyocha deflandrei* Zone. Within this zone two subzones were recognized: The *Dictyocha frenguelli* subzone from Core 20 through Core 31; and the *Mesocena apiculata* subzone from Core 32 through Core 34. Thus McCollum's *Pyrgupyxis (Pyxilla) prolongata* Zone, as strictly defined, is Early Oligocene. More detailed study of the variety and distribution of diatoms in Site 274 cores is needed in order to determine which species have potential as stratigraphic markers.

In a report on the diatoms of Leg 29, Site 278, Schrader (1976) erected the *Pyrgupyxis (Pyxilla)* spp. Zone for the Late Oligocene section recovered there. He defined the top of this zone as the extinction of *Pyrgupyxis johnsoniana* and all other species of *Pyrgupyxis*; the base was not defined.

Immediately above the *Pyrgupyxis (Pyxilla)* spp. Zone, Schrader defined the *Bogorovia veniamini* Zone, which has its base marked by the extinction of *Pyrgupyxis (Pyxilla)* spp. and its top by the first appearance of *Nitzschia maleinterpretaria* (see Figure 6).

Schrader defined the base of the *Bogorovia veniamini* Zone as being coincident with the Oligocene/Miocene boundary. At Site 278 the base of the *Bogorovia veniamini* Zone is placed between Core 29-3, 50–51cm and 29-4, 50–51cm. It is not clear why Schrader placed the Oligocene/Miocene boundary at the base of the *Bogorovia veniamini* Zone. Presumably this was done on the assumption that *Pyrgupyxis* does not range into the Miocene. The position of the Oligocene/Miocene boundary in Core 29, based on diatoms, is stratigraphically higher than is suggested by other groups of microfossils studied from Site 278.

Foraminifera are unable to provide reliable dates for the interval in question because of the presence in the cores of only a few solution-resistant forms. These suggest a Middle Miocene to Middle Oligocene age for the interval from Core 21 to Core 34. The record for calcareous nannofossils is somewhat better than that for foraminifera. They suggest that the Oligocene/Miocene boundary at Site 278 is between Cores 30 and 31 (which are separated by a 9.5m coring gap), or at the extinction of *Reticulofenestra bisecta* (Edwards and Perch-Nielsen, 1975). Perch-Nielsen (1975) records the acme of the problematical siliceous microfossil *Rocella gemma* in Core 31-2, 110cm at Site 278. The acme of this fossil is known to occur around the Oligocene/Miocene boundary. At Site 278 the *R. gemma* acme occurs just below the boundary as defined by the extinction of *Reticulofenestra bisecta.* It would thus appear that the Oligocene/Miocene boundary, as defined by Schrader on the basis of the extinction of *Pyrgupyxis,* is erroneous, because he illustrates the genus as ranging above the last occurrence of *Reticulofenestra bisecta* at Site 278.

Based on the stratigraphic relationships of the diatoms and calcareous nannofossils observed at Site 278, the Oligocene/Miocene boundary appears to fall within the overlap of the ranges of *Bogorovia veniamini* and *Pyrgupyxis* spp., and just above the acme of *Rocella gemma.*

Gombos (1976b) proposed five zones, summarized in Figure 9, for the Late Eocene to Early Miocene section recovered at Site 328 in the Malvinas Outer Basin. The definitions of the *Bogorovia veniamini* Zone and the *Pyrgupxis prolongata* Zone are herein changed to conform to our new interpretation of Site 278. As originally defined, the top of the *Pyrgupyxis prolongata* Zone was marked by the extinction of that species, and was thought to be coincident with the Oligocene/Miocene boundary as suggested by Schrader (1976). The base of the *Bogorovia veniamini* Zone was defined as being coincident with the top of the *Pyrgupyxis prolongata* Zone and its top was not defined. At Site 328, the ranges of *Bogorovia veniamini* and *Pyrgupyxis prolongata* overlap, a situation similar to that observed at Site 278. The acme of *Rocella gemma* also occurs within the overlap interval at Site 328. No calcareous microfossils or nannofossils are present in the siliceous section at Site 328.

Therefore, in reference to the stratigraphic relationships of diatoms and calcareous nannofossils observed at Site 278, the Oligocene/Miocene boundary at Site 328 is now assumed to lie within the overlap of *Bogorovia veniamini* and *Pyrgupyxis prolongata* and just above the acme of *Rocella gemma*—not at the extinction of *Pyrgupyxis prolongata.*

To avoid confusion of the extinction of *Pyrgupyxis prolongata* with the Oligocene/Miocene boundary, the definitions of the *Pyrgupyxis prolongata* and *Bogorovia veniamini* Zones of Gombos (1976b) have been amended. The *Pyrgupyxis prolongata* Zone is redefined as follows: Base = last occurrence of *Melosira architecturalis* (unchanged from Gombos, 1976b); top = first occurrence of *Bogorovia veniamini.* The *Bogorovia veniamini* Zone is redefined as the range of *Bogorovia veniamini.*

Jousé (written communication) states that *Pyrgupyxis* spp. do not range higher than Lower Oligocene. In addition, preliminary studies of the diatom stratigraphy at DSDP Leg 71, Sites 511 and 513A,

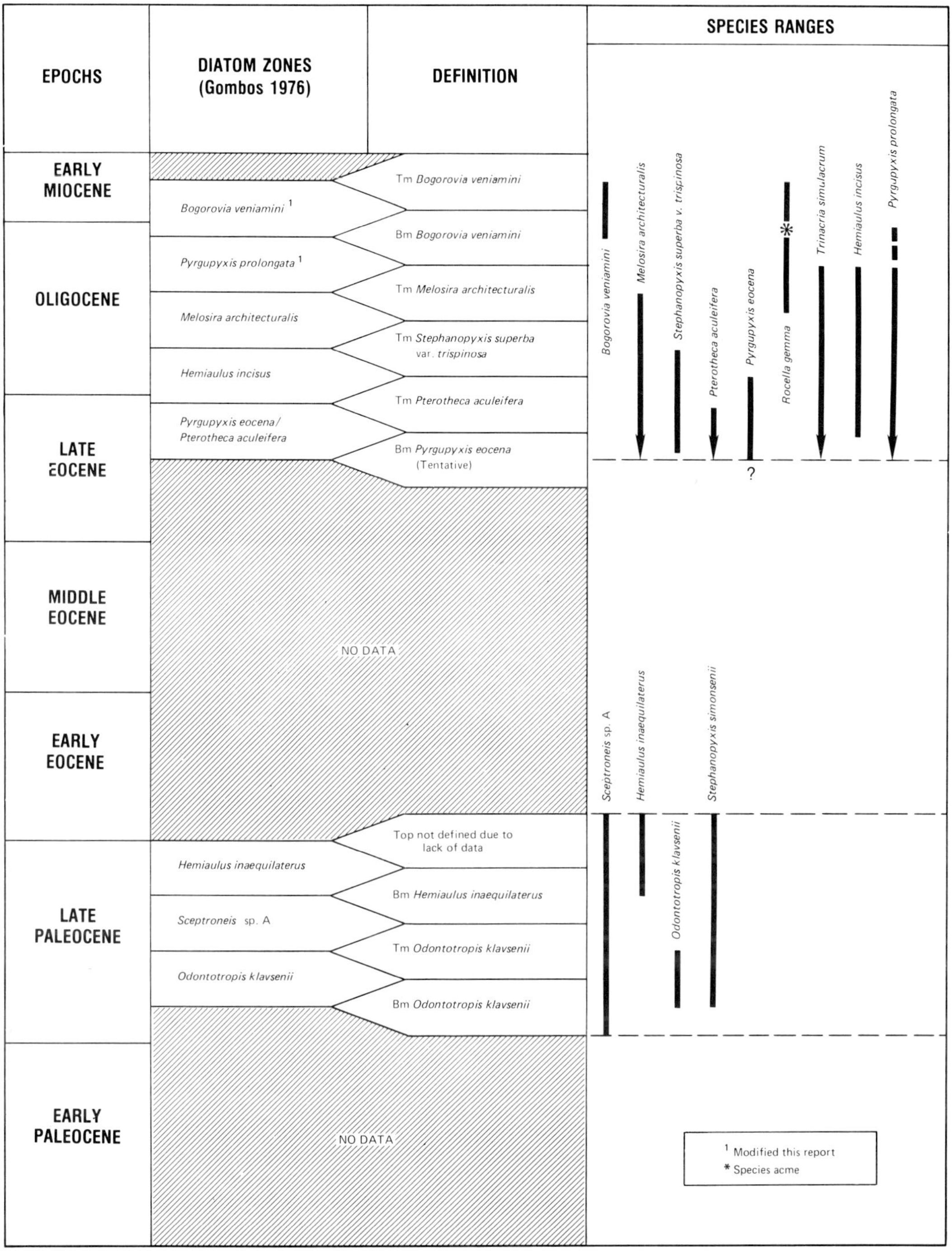

FIG. 9.—Southern High Latitude Paleogene Diatom Biostratigraphy (* = species acme).

which together comprise an apparently complete, Oligocene section rich in siliceous and calcareous nannofossils, also indicates a lower extinction level for *Pyrgupyxis* spp. than suggested by Schrader (1976) and Gombos (1976b). In Hole 513A the extinction of *Pyrgupyxis* spp. occurs in the lower

Upper Oligocene, slightly above the Lower/Upper Oligocene boundary (author's notes). Thus, the extinction of *Pyrgupyxis* spp. at the Oligocene/Miocene boundary as suggested by Schrader (1976) and Gombos (1976b) is erroneous.

When completed, the study of the composite Oligocene section from Holes 511 and 513A will represent the most complete record of Oligocene diatom evolutionary events yet recovered from the deep sea. At that time, a critical re-evaluation of the Oligocene diatom stratigraphy at Sites 278 and 328B will be possible.

Eocene

Very few stratigraphic studies have been done with Eocene diatoms from the southern high latitudes. This is partly due to the low number of workers in the field and partly to the sparsity of long cores recovering siliceous Eocene sections. Hajos (1976) documented the occurrence of diatoms in the Late Eocene and Oligocene at three sites in the South Tasman Sea (Sites 280, 281, 283). No zones were proposed for the Late Eocene at those sites. Gombos (1976b) proposed two diatom zones for the Late Eocene section recovered at Site 328 on the Falkland Plateau. These zones are summarized in Figure 9.

Paleocene

Diatom bearing Paleocene sediments are rare in the Southern Ocean. A 22m thick, Late Paleocene, nannofossil-rich, siliceous ooze was cored at Site 327A on the Falkland Plateau (see Fig. 1). At that site Cores 5 through 7 were dated by means of calcareous nannofossils. Core 5, section 1 to section 4 were assigned to the *Heliolithus universus* Zone and Core 5, section 4 to Core 7, section 2 were assigned to the *Fasciculithus involutus* Zone by Wise and Wind (1976).

The *Heliolithus universus* Zone directly underlies the latest Late Paleocene *Discoaster multiradiatus* Zone (Cores 3 and 4) which is barren of siliceous microfossils at Site 327A. Three local diatom zones were proposed by Gombos (1976b) for the siliceous part of the Late Paleocene section at this Site. These zones are summarized in Figure 9. At present, the lack of more extensive siliceous Paleocene sections hinders further documentation of the stratigraphic distribution of diatoms in that interval in the Southern Ocean.

MESOZOIC DIATOM BIOSTRATIGRAPHY

Cretaceous

The only significant Cretaceous diatomaceous section from the Southern Ocean was recovered at Site 275 located at the southeast margin of the Campbell Plateau (see Fig. 1). There, about 14.5m of Late Cretaceous siliceous ooze, rich in archeomonads, diatoms, radiolarians, silicoflagellates and sponge spicules was recovered. No calcareous microfossils are present in the sediment. The preservation of siliceous microfossils, which constitute 60–90% of the sediment, is generally quite good.

Hajos (1975) provides a list of 158 species which, according to her, is not a complete list. Included in the list are four new genera (three of archeomonads and one of diatoms, *i.e., Incisoria*) and 56 new species of diatoms and archeomonads. Of the 158 species listed by Hajos, 125, or 79%, became extinct in the Cretaceous.

The section at Site 275 from Core 1, section 1 to Core 4, section 2 has been correlated to the *Patulibracchium dickinsoni* Zone (radiolarians) by Pessagno (1975). The *P. dickinsoni* Zone is more or less correlative with the latest Campanian of California. Diatoms occur only in Cores 1 and 2 of Site 275. The section is also included in the *Lyramula furcula* Zone (silicoflagellates) by Perch-Nielsen (1975). This zone is the only Cretaceous silicoflagellate zone yet defined and is considered Late Cretaceous. Hajos defined nine zones and twelve subzones based on first and last occurrences of a number of diatoms and, to a lesser extent, silicoflagellates and archeomonads. At present, these zones are the best available for the Cretaceous and await further refinement and comparison with other Late Cretaceous siliceous sections.

CONCLUSIONS AND RECOMMENDATIONS

The regional Neogene and Paleogene diatom zonations proposed herein not only summarize the current status of diatom stratigraphy in the Sub-Antarctic and Antarctic region, but also form a basis for future research. Further refinement, modification and developments in diatom biostratigraphy require additional study in the following areas: 1) correlation of diatom datums with paleomagnetic and oxygen isotope stratigraphy, 2) calibration of datums to standard calcareous microfossil zones in middle and low latitude regions, 3) regional paleobiogeographic studies and 4) resolution of numerous problems in diatom taxonomy. Efforts along these lines will lead to a more thorough understanding of the biostratigraphy and evolutionary history of diatoms in the Antarctic region.

ACKNOWLEDGEMENTS

Samples from the Deep Sea Drilling Project were obtained through the assistance of the National Science Foundation. *Eltanin* and *Islas Orcadas* piston core samples were provided through the courtesy of the Antarctic Research Facility at Florida State University and the National Science Foundation.

FLORAL REFERENCES

The following is an alphabetical listing of the diatom species used in the new Southern Ocean Neogene biostratigraphy presented in this paper. Also listed are species used in existing Paleogene zonations summarized above. References to the species include the original reference as well as several others which may be more accessible.

Actinocyclus ingens Rattray 1890

Rattray 1890, p. 149, pl. 11, fig. 7; Kanaya 1959, p. 97, pl. 7, figs. 6-9; 1971, p. 554, pl. 40.6, figs. 1-8; Schrader 1973a, p. 701, pl. 18, figs. 1-4, 7; Wornardt 1973, p. 48-49, pl. 1, fig. 2.

Bogorovia veniamini Jousé 1974

Jousé 1974, p. 350-351, pl. 4, figs. 1-3; Schrader 1976, p. 630, pl. 5, figs. 22-23; Gombos 1976b, p. 593, pl. 1, figs. 6-7; pl. 12, figs. 1, 2, 4.

Coscinodiscus elliptopora Donahue 1970a

Donahue 1970a, p. 201, pl. 4, figs. e, i-m; McCollum 1975, p. 526, pl. 16, fig. 10; Gombos 1976b, p. 592, pl. 3, figs. 1-3, 6; pl. 9, fig. 3.

Coscinodiscus kolbei Jousé 1962

Jousé 1962 (illustration only), pl. 2, fig. 4; Donahue 1970a, p. 202, pl. 5, figs. a-c; McCollum 1975, p. 527, pl. 4, figs. 7-9; Gombos 1976b, p. 593, pl. 6, fig. 3.

Coscinodiscus lentigenosus Janisch
in Schmidt *et al.,* 1874-1959

Schmidt *et al.* 1874-1959, pl. 58, fig. 11; Hustedt 1958, p. 116, pl. 4, figs. 22-25; McCollum 1975, p. 527, pl. 5, fig. 1; Gombos 1976b, p. 593, pl. 3, figs. 4, 5.

Coscinodiscus lewisianus Greville 1866

Greville 1861-1866, p. 78, pl. 8, figs. 8-10; Kanaya 1971, p. 555, pl. 40.5, figs. 4-6; Schrader 1973a, p. 703, pl. 8, figs. 1-6, 10, 15; McCollum 1975, p. 527, pl. 5, figs. 2-4; Gombos 1976b, p. 315, pl. 5, figs. 4, 5.

Coscinodiscus obovatus Castracane 1886

Castracane 1886, p. 160, pl. 22, fig. 9; McCollum 1975, p. 534, pl. 5, fig. 6.

Coscinodiscus rhombicus Castracane 1886

Castracane 1886, p. 164, pl. 22, fig. 11; Schrader 1976, p. 972, pl. 21, figs. 1-3, 5.

Coscinodiscus vulnificus Gombos 1976

McCollum 1975 (as *Coscinodiscus* sp. 2), pl. 8, figs. 1, 2; Gombos 1976b, p. 593, pl. 4, figs. 1-3, pl. 42, figs. 1, 2.

Coscinodiscus sp. 1 McCollum 1975

McCollum 1975, p. 527, pl. 8, fig. 3.

Cosmiodiscus insignis Jousé 1961

Jousé 1961, p. 67, pl. 2, fig. 8; Sheshukova-Poretzkaya 1967, p. 175, pl. 25, figs. 2a-b; Koizumi 1973, p. 832, pl. 4, figs. 7-11; McCollum 1975, p. 527, pl. 8, fig. 5; Gombos 1976b, p. 593, pl. 4, figs. 4, 5.

Denticulopsis hustedtii (Simonsen and Kanaya)
Simonsen 1979

Simonsen and Kanaya 1961, p. 501, pl. 1, figs. 19-25; Simonsen 1979, p. 64; Kanaya 1971, p. 555, pl. 40.5, figs. 13, 14; Schrader 1973a, p. 704, pl. 2, figs. 28-34, 36-47; 1973b, p. 418, pl. 1, figs. 12, 13; Gombos 1976b, p. 593, pl. 8, figs. 1-3.

Denticulopsis lauta (Bailey) Simonsen 1979

Bailey 1854, p. 9, figs. 1, 2; Simonsen 1979, p. 64; Simonsen and Kanaya 1961, p. 500-501, pl. 1, figs. 1-8; Schrader 1973a, p. 705, pl. 2, figs. 14-24; 1973b, p. 419, pl. 1, figs. 11, 20, 23, 24; 1974a, p. 545, pl. 1, figs. 16, 17; Gombos 1976b, p. 593, pl. 8, figs. 6, 7, 12.

Denticulopsis maccollumii Simonsen 1979

McCollum 1975, p. 527, pl. 8, figs. 6-10 (as *Denticula antarctica*); Simonsen 1979, p. 65; Schrader 1976, p. 631, pl. 4, figs. 3, 22, 23, 25.
Remarks: see *Nitzschia grossepunctata.*

Hemiaulus inaequilaterus Gombos 1976

Gombos 1976b, p. 594, pl. 20, figs. 5-7.

Hemiaulus incisus Hajos 1975

Hajos 1975, p. 829, pl. 25, figs. 4-9; Gombos 1976b, p. 594, pl. 15, fig. 3; McCollum 1975 (as *Hemiaulus* sp. 1), p. 535, pl. 9, fig. 2.

Hemidiscus karstenii Jousé 1962

Abbott 1972, p. 110-112, pl. 1, figs. D-F; Schrader 1976, p. 71, pl. 14, fig. 2; Gombos 1976b, p. 595, pl. 4, fig. 8.

Melosira architecturalis Brun 1892

Brun 1892 *in* Schmidt *et al.* 1874-1959, pl. 177, figs. 49, 50; Kanaya 1957, (as *Cyclotella hannae*), p. 82-84, pl. 3, figs. 10a, b, 11, 14; Gombos 1976b, p. 595, pl. 26, figs. 5-7.

Nitzschia angulata (O'Meara) Hasle 1972

O'Meara 1877 (as *Diatoma rhombica*), p. 55, pl. 1, fig. 2; Hustedt 1952 (as *Fragilariopsis rhombica*), p. 296, figs. 6, 7; Hasle 1965, p. 24-26, pl. 1, fig. 6; pl. 4, fig. 19; pl. 6, fig. 5; pl. 8, fig. 11; pl. 9, figs. 1-6; pl. 10, figs. 2-6; 1972, p. 115.

Nitzschia denticuloides Schrader 1976

Schrader 1976, p. 633, pl. 3, figs. 7, 8, 10, 12, 18-24; pl. 15, fig. 22; Gombos 1976b, p. 595, pl. 13, figs. 9-11; DeFelice and Miller 1978, p. 115, figs. 10, 11.
Remarks: this species differs from *Denticulopsis maccollumii,* with which it may be confused, by its broader outline and lack of the double rows of pores. Also *Nitzschia denticuloides* has marginal pore fields which are more elliptical than those of *Denticulopsis maccollumii.*

Nitzschia grossepunctata Schrader 1976

Schrader 1976, p. 633, pl. 3, figs. 1-4; McCollum 1975 (as *Rhaphoneis* sp.), p. 535, pl. 11, figs. 9, 10.
Remarks: this species is similar in structure to *Denticulopsis maccollumii* from which it differs by its coarsely punctate valve face. Puncta in the hyaline central area of *Nitzschia grossepunctata* may be few to numerous. All specimens having at least one punctum were considered to be *N. grossepunctata* in this study.

Nitzschia interfrigidaria McCollum 1975

McCollum 1975, p. 535, pl. 9, figs. 7-9.

Nitzschia kerguelensis (O'Meara) Hasle 1972

O'Meara 1877 (as *Terebraria kerguelensis*), p. 56, pl. 1, fig. 4; Hustedt 1952 (as *Fragilariopsis kerguelensis* O'Meara), p. 294, Hasle 1968, p. 205-208, fig. 1, 2, 7-9; 1972, p. 115, pl. 1, fig. 1; Abbott 1974, p. 315, p. 8, figs. G, H; Gombos 1976b, p. 595, pl. 8, figs. 13, 14.

Nitzschia maleinterpretaria Schrader 1976

Schrader 1976, p. 634, pl. 2, figs. 9, 11–19, 21, 24.
Remarks: according to Schrader this species is very close to *Denticulopsis nicobarica* in appearance and may sometimes be confused with it. McCollum (1975) reports *Denticulopsis nicobarica* in cores 17–18 of Site 266. Our reexamination of that Site revealed only *Nitzschia maleinterpretaria* in those Cores.

Nitzschia praeinterfrigidaria McCollum 1975

McCollum 1975, pl. 535, pl. 10, fig. 1; Gombos 1976b, p. 595, pl. 7, figs. 1, 2.

Nitzschia reinholdii Kanaya *sensu* Koizumi

Schrader 1973a, p. 708, pl. 4, figs. 12–16; pl. 5, figs. 1–9; Koizumi 1975, p. 877, pl. 4, figs. 13–17.
Remarks: for a discussion of the nomenclatural problems associated with this species see Harper 1977, p. 90–91.

Odontotropis klavsenii Debes *ex* Hustedt 1930

Hustedt 1930, p. 858, fig. 510a, Gombos 1976b, p. 595, pl. 39, figs. 1–4.

Pterotheca aculeifera Grunow *in* Van Heurck 1882

Grunow *in* Van Heurck 1880–1885, pl. 83, fig. 5; Kanaya 1957, pl. 109, pl. 8, figs. 1, 2; Gombos 1976b, p. 596, pl. 23, figs. 1, 2.

Pyrgupyxis eocena Hendey 1969

Hendey 1969, p. 3, figs. 1–4; Gombos 1976b, p. 596, pl. 22, figs. 6, 7.

Pyrgupyxis prolongata (Brun) Hendey 1969

Brun 1890–1893 (as *Pyxilla prolongata*), p. 176, pl. 24, fig. 7; Hendey 1969, p. 5; McCollum 1975 (as *Pyxilla prolongata*), p. 535, pl. 11, figs. 4–6; Gombos 1976b, p. 596, pl. 21, figs. 1–7; pl. 22, fig. 11.

Raphidodiscus marylandicus Christian 1887

Christian 1887, p. 66–68; Wornardt 1973, p. 47; Andrews 1974, p. 233–235, pl. 1–5; McCollum 1975, p. 535, pl. 11, fig. 8.

Rhizosolenia barboi (Brun) Tempere and Peragallo 1908

Brun 1893–1896 (as *Pyxilla barboi*), p. 87, pl. 5, figs. 16, 17, 23; Tempere and Peragallo 1908, p. 26, No. 47; Donahue 1970b, p. 136, pl. 1, figs. b–c; Jousé 1971 (as *Rhizosolenia curvirostris* variety *inermis*), p. 15, pl. 2, figs. 1–3; Schrader 1973a, p. 709, pl. 24, figs. 4, 7; Koizumi 1973, p. 833, pl. 5, figs. 32, 33; McCollum 1975, p. 535, pl. 11, fig. 13.

Sceptroneis sp. *A* Gombos 1976

Gombos 1976b, p. 597, pl. 13, fig. 6

Stephanopyxis superba (Greville) Grunow variety *trispinosa* Gombos 1976

Gombos 1976b, p. 597, pl. 30, figs. 3–6, 9.

REFERENCES CITED

NOTE: All references to Deep Sea Drilling Project Initial Reports have been abbreviated; for complete citations see *Appendix I* at the end of this volume, or the individual Initial Reports as published.

ABBOTT, W. H., 1972, Vertical and lateral patterns of diatomaceous ooze found between Australia and Antarctica: Unpublished Ph.D. dissertation, University of South Carolina, Columbia, S.C., 130 p.

———, 1974, Temporal and spatial distribution of Pleistocene diatoms from the southeast Indian Ocean: Nova Hedwigia, XXV, p. 291–346.

ANDREWS, G. W., 1974, Systematic position and stratigraphic significance of the marine Miocene diatom *Raphidodiscus marylandicus* Christian: Nova Hedwigia, Beiheft, v. 45, p. 231–243.

BAILEY, J. W., 1854, Notes on new species and localities of microscopical organisms: Smithsonian Contrib. Knowledge, v. 7.

BRUN, J., 1890–1893, Notes sur quelques especes nouvelles: Le Diatomiste, v. 1, p. 173–177.

———, 1893–1896, Diatomees Miocenes: Le Diatomiste, v. 2, p. 229–247.

BUKRY, DAVID, 1975, Coccolith and silicoflagellate stratigraphy near Antarctica, Deep Sea Drilling Project, Leg 28: Init. Repts. DSDP, v. 28, p. 709–723.

———, 1978, Biostratigraphy of Cenozoic marine sediments by calcareous nannofossils: Micropaleontology, v. 24, p. 44–60.

BURKLE, L. H., 1978, Early Miocene to Pliocene diatom datum levels for the equatorial Pacific: Republic of Indonesia, Geol. Research and Development Centre, Spec. Pub. no. 1, p. 25–44.

———, CLARKE, D. B., AND SHACKLETON, N. J., 1978, Isochronous last-abundant appearance datum (LAAD) of the diatom *Hemidiscus karstenii* in the Sub-Antarctic: Geology, v. 6, p. 243–246.

BURNS, D. A., 1975, Nannofossil biostratigraphy for Antarctic sediments, Leg 28, Deep Sea Drilling Project: Init. Repts. DSDP, v. 28, p. 589–598.

CASTRACANE, D. A., 1886, The voyage of H.M.S. *Challenger.* Report on the Diatomaceae: Botany 2 (I-III), 1. Edinburgh (Reprinted 1966).

CHRISTIAN, T., 1887, New diatomaceous deposits: The Microscope, v. 7, p. 65.

CIESIELSKI, P. F., 1975, Biostratigraphy and paleoecology of Neogene and Oligocene silicoflagellates from cores recovered during Antarctic Leg 28, Deep Sea Drilling Project: Init. Repts. DSDP, v. 28, p. 625–692.

———, 1978, The Maurice Ewing Bank of the Malvinas (Falkland) Plateau: Depositional and erosional history and its paleoenvironmental implications: Ph.D. dissertation, Florida State University.

DEFELICE, D. R., AND MILLER, W., 1978, Structure of the Middle Miocene Antarctic diatom *Denticula Antarctica* McCollum: Antarctic J. of U.S., v. 13, no. 4, p. 115.

DONAHUE, J. G., 1970a, Diatoms as Quaternary biostratigraphic and paleoclimatic indicators in high latitudes of the Pacific Ocean: Ph.D. dissertation, Columbia University.

———, 1970b, Pleistocene diatoms as climatic indicators in North Pacific sediments: Geol. Soc. America Mem. 126, p. 126.

EDWARDS, A. T., AND PERCH-NIELSEN, K., 1975, Calcareous nannofossils from the southern southwest Pacific, Deep Sea Drilling Project, Leg 29: Init. Repts. DSDP, v. 29, p. 469–539.

GOMBOS, A. M., JR., 1976b, Paleogene and Neogene diatoms from the Falkland Plateau and Malvinas Outer Basin, Leg 36, Deep Sea Drilling Project: Init. Repts. DSDP, v. 36, p. 575–690.

GREVILLE, R. K., 1861–1866, Description of new and rare diatoms I-XX: Trans. Roy. Microscop. Soc. London. Bibliotheca Phycologica 6; reprinted 1968.

HAJOS, M., 1975, Late Cretaceous Archaemonadaceae, Diatomaceae, and Silicoflagellateae from the South Pacific Ocean, Deep Sea Drilling Project, Leg 29, Site 275: Init. Repts. DSDP, v. 29, p. 913–1011.

———, 1976, Upper Eocene and Lower Oligocene Diatomaceae, Archaeomonadaceae, and Silicoflagellatae in southwestern Pacific sediments, Deep Sea Drilling Project, Leg 29: Init. Repts. DSDP, v. 35, p. 817–883.

HARPER, H. E., 1977, Diatom biostratigraphy of the Miocene/Pliocene boundary in marine strata of the Circum-North Pacific: Ph.D. dissertation, Harvard University.

HASLE, G., 1965, *Nitzschia* and *Fragilariopsis* species studied in the light and electron microscopes. III, The genus *Fragilariopsis:* Skr. Norske Vidensk.-Akad. I. Mat.-Nat. Kl., v. 21, p. 1–49.

———, 1968, Observation on the marine diatom *Fragilariopsis kerguelensis* (O'Meara) Hustedt in the scanning electron microscope: Nytt. Mag. Bot., v. 15, p. 205–208.

———, 1972, *Fragilariopsis* Hustedt as a section of the genus *Nitzschia* Hassal: Nova Hedwigia, Beiheft, v. 30, p. 111–119.

HENDEY, N. I., 1969, *Pyrgupyxis,* a new genus of diatoms from a South Atlantic core: Calif. Acad. Sci. Occas. Paper, no. 72, p. 1–6.

HUSTEDT, F., 1930, Die Kieselalgen Deutschlands, Osterreichs und der Schweiz. Kryptogamen-Flora von Deutsch., Osterr. Schweiz: v. 7, p. 609.

———, 1952, Diatomeen aus der Lebensgemeinschaft des Buckelwals (Negaptera nodosa Bonn.): Archiv Hydrobiol., v. 46, p. 286–298.

———, 1958, Diatomeen aus der Antarktiks und dem Sudatlantik: Deutsche Antarktische Exped., v. 2, p. 103.

JENKINS, D. G., 1974, Cenozoic planktonic foraminiferal biostratigraphy of the southwestern Pacific and Tasman Sea, DSDP Leg 29: Init. Repts. DSDP, v. 29, p. 449–467.

JOUSÉ, A. P., 1961, Miocene and Pliocene marine diatoms from the Far East (Russian): Botanical Material, Spore-Bearing Plants, XIV.

———, 1962, Stratigraphische und palaogeographische Untersuchungen im Nordwestteil des Pazifiks: Verlag Akad. Wiss. USSR, Moscow, v. 5, p. 1–258.

———, 1971, Diatoms in Pleistocene sediments from the Northern Pacific Ocean. *in* Funnell, B. M., and Reidel, W. R., eds., Micropaleontology of the oceans: Cambridge, Cambridge University Press, p. 407.

———, 1974, Diatoms in the Oligocene-Miocene biostratigraphic zones of the tropical areas of the Pacific Ocean: Nova Hedwigia, Beiheft, v. 45, p. 333–357.

———, KOROLEVA, G., AND NAGAIEVA, G., 1963, Stratigraphic and paleographic investigations in the Indian sector of the Southern Ocean: Oceanological Res. Articles, Section X, no. 8, Acad. Sci. USSR, Moscow, p. 137–160.

KANAVA, T., 1957, Eocene diatom assemblages from the Kellogg and "Sidney" shales, Mt. Diablo Area, California: Sci. Rept. Tohoku Univ., 2nd ser., v. 28, p. 1–124.

———, 1959, Miocene diatom assemblages from the Onnagawa formation and their distribution in the correlative formations in northeast Japan: Sci. Rept. Tohoku Univ., 2nd ser. (Geol.), v. 30, p. 1.

———, 1971, Some aspects of pre-Quaternary diatoms in the ocean, *in* Funnell, B. M., and Riedel, W. R., eds., Micropaleontology of the oceans: Cambridge, Cambridge University Press, p. 545–565.

KANEPS, A. G., 1975, Cenozoic planktonic foraminifera from Antarctic deep-sea sediments, Leg 28, DSDP: Init. Repts. DSDP, v. 28, p. 573–583.

KOIZUMI, I., 1973, The late Cenozoic diatoms of Sites 183–193, Leg 19, Deep Sea Drilling Project: Init. Repts. DSDP, v. 19, p. 805–856.

———, 1975, Neogene diatoms from the northwestern Pacific Ocean: Init. Repts. DSDP, v. 32, p. 865–889.

MARTINI, E., 1971, Standard Tertiary and Quaternary calcareous nannoplankton zonation: *in* Farinacci, A., Proceedings of the II Planktonic Conference, Roma 1970, p. 139–785.

MCCOLLUM, D. W., 1975, Diatom stratigraphy of the Southern Ocean: Init. Repts. DSDP, v. 28, p. 515–571.

O'MEARA, E., 1877, On the diatomaceous gatherings made at Kerguelen's Land by H. N. Moseley (M.A.), H.M.S. *Challenger:* J. Linn (Soc. Bot.), v. 15, p. 55–59.

PERCH-NIELSEN, K., 1975, Late Cretaceous to Pleistocene silicoflagellates from the southern southwest Pacific, Deep Sea Drilling Project Leg 29: Init. Repts. DSDP, v. 29, p. 677–720.

PESSAGNO, E. A., 1975, Upper Cretaceous radiolaria from DSDP Site 275: Init. Repts. DSDP, v. 29, p. 1011–1029.

PETRUSHEVSKAYA, M. G., 1975, Cenozoic radiolarians of the Antarctic, Leg 29, DSDP: Init. Repts. DSDP, v. 29, p. 541–675.

RATTRAY, J., 1890, A revision of the Genus *Actinocyclus* Ehrb: J. Quekett Micr. Club, 2nd ser., v. 4, p. 137.

RYAN, W. B. F., AND OTHERS, 1974, A paleomagnetic assignment of Neogene stage boundaries and the development of isochronous datum planes between the Mediterranean, the Pacific, and the Indian Oceans in order to investigate the response of the world oceans to the Mediterranean salinity crisis: Riv. Ital. Paleont., v. 80, no. 4, p. 631–688.

SCHMIDT, A., 1875–1959, Atlas der Diatomaceen-Kinde (continued by Martin Schmidt, Friedrich Fricke, Heinrich

Heiden, Otto Muller, Friedrich Hustedt), pt. 1–120 (1874–1959), Leipzig.
SCHRADER, H. J., 1973a, Cenozoic diatoms from the northeast Pacific, Leg 18: Init. Repts. DSDP, v. 18, p. 673–797.
———, 1973b, Stratigraphic distribution of marine *Denticula* species in Neogene North Pacific sediments: Micropaleontology, v. 19, p. 417–430.
———, 1976, Cenozoic marine planktonic diatom biostratigraphy of the Southern Pacific Ocean: Init. Repts. DSDP, v. 35, p. 605–671.
SHESHUKOVA-PORETZKAYA, V. S., 1967, Neogenovye morskie diatomovye vodorosli Sakhalina i kamchatki (Neogene marine diatoms of Sakhalin and Kamtchatica), Leningrad: Leningrad University (in Russian, with English summary).
SIMONSEN, R., 1979, The diatom system: Ideas on phylogeny: Bacillaria, v. 2, p. 9–72.
———, AND KANAYA, T., 1961, Notes on the marine species of the diatom genus *Denticula* Kutz: Internatl. Rev. Ges. Hydrobiol., v. 46, p. 498–513.
STAINFORTH, R. M., AND OTHERS, 1975, Cenozoic planktonic foraminiferal zonation and characteristics of index forms: Univ. Kansas Paleontol. Contrib., Art. 62, p. 1–425.
TEMPÈRE, J., AND PERAGALLO, H., 1907–1915, Diatomees due monde entier: Ed. 2, 480 p.
THEYER, F., MATO, C. Y., AND HAMMOND, S. R., 1978, Paleomagnetic and geochronologic calibration of latest Oligocene to Pliocene radiolarian events, Equatorial Pacific: Marine Micropaleontology, v. 3, p. 377–395.
VAIL, P. R., AND HARDENBOL, J., 1979, Sea level changes during the Tertiary: Oceanus, v. 22, no. 3, p. 71–79.
VAN HEURCK, H., 1880–1885, Synopsis des diatomees de Belgique: Anvers., p. 235.
WATKINS, N. D., AND KENNETT, J. P., 1972, Regional sedimentary disconformities and Upper Cenozoic changes in bottom water velocities between Australasia and Antarctica: Antarctic Res. Ser., v. 19: Washington, D.C., Am. Geophys. Union, p. 273–293.
WEAVER, F. M., 1976, Late Miocene and Pliocene radiolarian paleobiogeography and biostratigraphy of the Southern Ocean: Ph.D. dissertation, Florida State University.
WISE, S. W., JR., AND WIND, F. H., 1976, Mesozoic and Cenozoic calcareous nannofossils recovered by DSDP, Leg 36: drilling on the Falkland Plateau, Southwest Atlantic sector of the Southern Ocean: Init. Repts. DSDP, v. 36, p. 269–491.
WORNARDT, W. W., JR., 1973, Diatom, silicoflagellate, radiolarian, calcareous, nannofossil, and foraminiferal biostratigraphy of the Middle and Late Miocene and Pliocene of Newport Back Bay, Newport Beach, Calif.: Annual Meeting of Amer. Assoc. Petroleum Geologists, Soc. Econ. Paleontologists Mineralogists: Soc. Exploration Geophysicists Guidebook, SEPM, Trip 1: Miocene sedimentary environments and biofacies, southeastern Los Angeles Basin, p. 39–53.

ADDITIONAL REFERENCES

AGARDH, C. A., 1824, Systems Algarum. Adumbravit C. A. Agardh: Literis Berlingianis (Lundae), XXVII.
ANDREWS, G. W., 1972, Some fallacies of quantitative diatom paleontology: Nova Hedwigia, Beiheft, v. 39, p. 285–294.
ANISSIMOVA, N., 1937, Diatomovye nizhnetretichnykh otlozhenii Srednei Volgi (Diatoms from the Lower Tertiary deposits of the Middle Volga region): VSEGEI (All-Union Geol. Inst.) Library.
BENDA, L., 1972, The diatoms of the Moler formation in Denmark (Lower Eocene): Nova Hedwigia, Beiheft, v. 39, p. 251–266.
BRADY, H. T., 1979, The dating and interpretation of diatom zones in Dry Valley Drilling Project Holes 10 and 11, Taylor Valley, South Victoria Land, Antarctica: Japanese Institute of Polar Studies, Seminar 3, Tokyo.
BRIGHTWELL, T., 1858, Remarks on the genus *"Rhizosolenia"* of Ehrenberg: Quart. J. Micr. Sci., v. 6, p. 93.
BRUN, J., 1891, Diatomees especes nouvelles marine fossiles ou pelagiques: Mem. Soc. Phys. d'Phys. d'Hist. Nat. Geneve, v. 31, p. 1.
———, AND HERIBAUD, J., 1893, Les Diatomees d'Auvergne: Libraire des Sciences Naturelles, Paris, p. 1.
BUKRY, DAVID, 1973, Low latitude coccolith biostratigraphic zonation: Init. Repts. DSDP, v. 15, p. 685–703.
———, 1974, Silicoflagellate and coccolith stratigraphy, Deep Sea Drilling Project, Leg. 29: Init. Repts. DSDP, v. 29, p. 845–871.
BURCKLE, L. H., 1972, Late Cenozoic planktonic diatom zones from the eastern equatorial Pacific: Nova Hedwigia, Beiheft, v. 39, p. 217–250.
BURCKLE, J., AND STANTON, D., 1974, The distribution of displaced Antarctic diatoms in the Argentine Basin (abs.): Third Plankt. Conf., Kiel, p. 14.
BUSEN, K. E., AND WISE, S. W., JR., 1976, Silicoflagellate stratigraphy, Deep Sea Drilling Project, Leg 36: Init. Repts. DSDP, v. 36, p. 697–743.
CHEN, P. H., 1975, Antarctic Radiolaria: Init. Repts. DSDP, v. 28, p. 437–513.
CLEVE, P. T., 1873a, On diatoms from the Arctic Sea: Bihang Svensk. Vetensk. Akad. Handl., v. 13, p. 1.
———, 1873b, Examination of diatoms found on the surface of the Sea of Java: Bihang Kgl. Svensk. Vetensk. Handl., v. 11, p. 1.
———, 1896, Diatoms from Baffin Bay and Davis Strait: Bihang Kgl. Svensk. Vetensk. Handl., Abt. III, v. 4, p. 1–22.
———, 1903, Plankton researches in 1901 and 1902: Kgl. Svensk. Vetensk. Handl., v. 8, p. 36.
CLEVE-EULER, A., 1941, Altertertiare Diatomeen und Silicoflagellaten im Jungeren Schwedens: Paleontogr., v. 92, p. 166.

CLEVE, P. T., AND GRUNOW, A., 1880, Beitrage zur Kenntnis der arktischen Diatomeen: Kgl. Svensk. Vetensk. Akad. Handl., v. 17, p. 3-122.

COMBER, T., 1896, On the occurrence of endocysts in the genus *Thalassiosira:* Trans. Roy. Micr. Soc. London, p. 489-491.

EHRENBERG, C. G., 1830, Beitrage zur Kenntnis der Organisation der Infusorien und ihrer geographischen Verbreitung besonders in Sibirien: Abh. Kgl., Akad. Wiss. Berlin (Preuss. Phys.), p. 1.

———, 1836, Die fossilen Infusorien und die lebendige Dammerde: Abh. Kgl. Akad. Wiss. Berlin.

———, 1838, Die Infusoriontierchen als vollkommeme Organismen. Ein Blick in das tiefere organische Leben der Natur: Leipzig (Leopold Voss), p. 1-548, pls. 1-64.

———, 1839, Uber noch jetzt zahlreich lebende Tierarten der Kreidebildung und den Organismus der Polythalamien: Abh. Kgl. Akad. Wiss. Berlin, p. 81-174.

———, 1841, Uber Verbreitung and Einfluss des mikroskopischen Lebens in Sud-und Nord-Amerika. Der uber die Bekanntmachung geeigneten Verhdlg. der Kgl. Preuss. Akad. Wiss. Berlin, p. 139-144.

———, 1844a, Resultate Untersuchungen der ihm von der Sudpolreise des Kapt. Rofs in den Jahren 1841-1843: Ber. Akad. Wiss. Berlin, p. 182-207.

———, 1844b, Uber zwei neue Lagern von Gebirgmassen aus Infusorien als Meeresabsatz in Nord-Amerika und ein Vergleich derselben mit den organischen Kreidegebilden in Europa und Afrika. Abh. Kgl. Akad. Wiss. Berlin, p. 57-97.

———, 1845, Vorlaufige zweite Mitteilung uber die Beziehungen des kleinsten organischen Lebens zu den vulkanischen Massen der Erde: Ber. Akad. Wiss. Berlin, p. 133-157.

———, 1854, Mikrogeologie des Erden- und Felsen-schaffende Wirken des unsichtbar kleinen selbststandigen Lebens auf der Erde: Leipzig, Leopold Voss.

FENNER, J., SCHRADER, H. J., AND WIENIGK, H., 1976, Diatom phytoplankton studies in the southern Pacific, its composition, correlation to the Antarctic convergence and paleoecological significance: Init. Repts. DSDP, v. 35, p. 757-813.

FORTI, D. A., 1909, Studi per una Monografia del genera *Pyxilla* (diatomee) e die generi affini: Nuova Notarisia, v. 20, p. 5.

———, 1913, Contribuzional Diatomologiche. XIII. Diagnoses Diatomacearum quarumdam fossilium italicarum: Atti del reale instituto Veneto de Scienze, Lettere ed Arti, Anno accademico 1912-1913, Venezia.

FRAKES, L. A., 1975, Oceanographic and biological influences on sediment distributions in the Southern Ocean: Search, v. 6, p. 339-341.

FRENGUELLI, J., 1949, Diatomeas fossiles de los yacimientos chilenos de Tiltil y Mejillones: Darwinia, v. 2, p. 97.

GODDELL, H. G., AND WATKINS, N. D., 1968, The paleomagnetic stratigraphy of the Southern Ocean: 20° west to 160° east longitude: Deep-sea Research, v. 15, p. 89-112.

GOMBOS, A. M., JR., 1974, Diatoms as indicators of Antarctic interglacials (abs.): Symp. Marine Plankt. Sediments, Kiel, p. 27.

———, 1976a, Neogene and Paleogene diatom stratigraphy in the region of the Falkland Plateau: Ph.D. dissertation, Florida State University.

GORDON, A. L., 1967, Structure of Antarctic waters between 20°W and 170°W: Antarctic Map Folio Series, Folio 6, New York: Amer. Geogr. Soc.

———, 1971a, Spreading of Antarctic bottom waters II. *in* Gordon, A. L., ed., Studies in physical oceanography: A tribute to George Wust on his 80th birthday: New York, Gordon and Breach.

———, 1971b, Recent physical oceanographic studies of Antarctic waters: Res. Antarctic, A.A.A.S., p. 609-629.

GRAN, H. H., 1904, Die Diatomeen der Arktischen Meere. I, Die Diatomeen des Planktons in "Fauna Arctica," auf Grund der Ergebnisse der Deutschen Expedition in das Nordliche Eismeer im Jahre 1898, Jena, v. 4, p. 511.

GREVILLE, R. K., 1859, Descriptions of diatomaceae observed in Californian guano: Quart. J. Micr. Sci., v. 7, p. 155-166.

———, 1860, A monograph of the genus *Asterolampra,* including *Asteromphalus* and *Spatangidium:* Trans. Roy. Microscop. Soc. London, v. 8, p. 102-124.

———, 1862, On the Asterolamprae of the Barbados deposit: Trans. Roy. Microscop. Soc. London, v. 10, p. 41-55.

GROVE, E., AND STURT, G., 1866-1887, On a fossil marine diatomaceous deposit from Oamaru, Otago, New Zealand. Parts I-IV: J. Quekett Micr. Club Ser. 11, 111 p.

GRUNOW, A., 1862, Die osterreichischen Diatomaceen nebst Anschluss einger neuen Arten von anderen Lokalitaten und einer kritischen Ubersicht der bisher bekannten Gattungen und Arten. Erste Folge: Epithemiaeae, Meridioneae, Diatomeae, Entopyleae, Surirelleae, Amphiphleureae. Zweite Folge: Familie Nitzschieae: Verh. Kasier-Konigl. Zool.-Bot. Ges. in Wien, v. 12, p. 315-472, 545-585.

———, 1868, Algae in Reise Seiner Majestat Fregatte "Novara" um die Erde in den Jahren 1857, 1858, 1859: Botanischer Teil, v. 1, p. 1-104.

———, 1880, Beitrage zur Kenntnis der Arktischen Diatomeen, *in* Cleve, P. T., and Grunow, A., eds., Kgl. Svensk. Vet. Akad. Handl. (Stockholm), v. 17, p. 3.

———, 1881, *in* Van Heurck, H., Synopsis des diatomees de Belgique: Atlas, Taf. 124.

———, 1884, Die Diatomeen von Franz Josefs-Land: Denkschr. Kgl. Akad. Wiss. Wien, Math, Nat. Wiss. Kl. 48.

HANNA, G. D., 1927a, The lowest known Tertiary diatoms in California: Jour. Paleontology, v. 1, p. 103-127.

———, 1927b, Cretaceous diatoms from California: Calif. Acad. Sci. Occas. Papers, v. 13, p. 1-34.

———, 1930a, A review of the genus *Rouxia:* Jour. Paleontology, v. 4, p. 179-188.

———, 1930b, A new genus of silicoflagellata from the Miocene of Lower California: Jour. Paleontology, v. 4, p. 415–416.

———, AND BRIGGER, A. L., 1964, Some fossil diatoms from Barbados: Calif. Acad. Sci. Occas. Papers, v. 45, p. 1–27.

———, AND GRANT, W. M., 1926, Expedition to the Revillagigedo Islands, Mexico, in 1925. Miocene marine diatoms from Maria Madre Island, Mexico: Calif. Acad. Sci. Proc., v. 5, p. 115–193.

HARGRAVES, P., 1968, Species composition and distribution of net plankton diatoms in the Pacific sector of the Antarctic Ocean: Ph.D. dissertation, Columbia University.

HART, T. J., 1934, On the phytoplankton of the southwest Atlantic and the Bellingshausen Sea, 1929–1931: Discovery Rept., v. 8, p. 1–268.

———, 1942, Phytoplankton periodicity in Antarctic surface waters: Discovery Rept., v. 21, p. 263–348.

HASLE, G. R., 1964, *Nitzschia* and *Fragilariopsis* species studied in the light and electron microscopes. I. Some marine species of the groups Nitzschiella and Lanceolatae: Skr. Norske Vidensk. Akad. I. Mat.-Nat. Kl., v. 16, p. 1–48.

———, 1969, An analysis of the phytoplankton of the Pacific Southern Ocean: Abundance, composition, and distribution during the Brategg Expedition 1947–1948: Hvalradets Skr. Sci. R. Marine Biology Res., v. 52, p. 6–168.

———, AND DE MENDIOLA, B. R. E., 1967, The fine structure of some *Thalassiothrix* and *Thalassionema* species: Phycologia, v. 6, p. 107.

HASSAL, A. H., 1845, A history of the British freshwater algae (including descriptions of Diatomaceae and Desmidiaceae) with upwards of one hundred plates: London, Taylor, Walton and Maberly.

HAYS, J. D., AND BERGGREN, W. A., 1971, Quaternary boundaries and correlations, *in* Funnell, B. M., and Riedel, W. R., eds., Micropaleontology of the oceans: Cambridge, Cambridge University Press, p. 669–691.

HEIBERG, P. A. C., 1863, Conspectus Criticus. Diatomacearum Danicarum. De Danske Diatomeer: Kjoenhaven: Wilhelm Priors Forlag.

HEIDEN, H., AND KOLBE, R. W., 1928, Die marinen Diatomeen der deutschen Sudpolar-Expedition, 1901–1903. *in* Drygalski, E., ed., Deutsche Sudpolar-Expedition, 1903: Berlin und Leipzig, Walter de Gryter & Co., v. 8 (Botank), no. 5, p. 447–715, Taf. 31–43.

HENDEY, N. I., 1937, The plankton diatoms of the southern seas: Discovery Rept., v. 16, p. 151–364.

———, 1964, An introductory account of the smaller algae of British coastal waters. Part V: Bacillariophyceae (Diatoms): London, Her Majesty's Stationery Office, p. 1–317, pl. 1–45.

———, AND WISEMAN, D. H., 1953, The significance and diatom content of a deep-sea floor sample from the neighborhood of the greatest oceanic depth: Deep-sea Research, London, Pergamon Press, v. 1, p. 42–59.

HUSTEDT, F., 1924, Von Sammeln und Praparieren der Kieselalgen sowie Angaben uber Untersuchungs- und Kulturmethoden: Sonderdruck aus Handbuch der Biologischen Arbeitsmethoden, Abt., XI, Teil 4.

———, 1928–1929, Die Kieselalgen Deutschlands, Osterreichs und der Schweiz. *in* L. Rabenhorst Kryptogamen-Flora von Deutschland, Osterreich und der Schweiz, v. 7, p. 1–608.

———, 1931, Bacillariophyta (Diatomaceae). A. Pasher, Die Susswasser-Flora Mitteleuropas, 10 1.2. Auflg. Jena (Fischer).

———, 1959, Die Kieselalgen Deutschlands, Osterreichs und der Schweiz unter Berucksichtigung der ubrigen Lander Europas sowie der angrenzenden Meeresgebiete, Kryptogamen-Flora von Deutsch., Osterr., u.d. Schweiz, Teil 2, v. 7, p. 1.

JORGENSEN, E., 1905, The protist plankton and the diatoms in bottom samples, *in* Nordgaard, O., Hydrographical and biological investigations in Norwegian fjords: Bergen, Bergen Museum Skrift.

JOUSÉ, A. P., 1951, Diatomae aetatis Palaeocaeni Uralii septentrionalis: Botanicheskie Materialy otdela sporovyhk rastenii, v. 7, p. 24.

———, 1968, New species of diatoms in bottom sediments of the Pacific and the Sea of Okhotsk (Russian): Nov. Systemat. Plant non Vascular 1968. Akad. Nauk USSR, v. 3, p. 12.

———, KOZLOVA, O. G. AND MUHINA, V. V., 1971, Distribution of diatoms in the surface layer of sediment from the Pacific Ocean, *in* Riedel, R. W., and Funnell, B. M., eds., Micropaleontology of the oceans: Cambridge, Cambridge University Press, p. 263–269.

———, ———, AND NAGAIEVA, G., 1962, Stratigraphic and paleographic investigations in the Indian sector of the Southern Ocean (in Russian; English summary): Trud. Inst. Okeonol., v. 41, p. 19–42.

KANAYA, T., AND KOIZUMI, I., 1970, The progress in the younger Cenozoic diatom biostratigraphy in the northern Circum-Pacific region: Jour. Marine Geol. Japan, v. 6, p. 47.

KARSTEN, G., 1905, Das Phytoplankton des Antarktischen Meeres. *in* Chun, C., ed., Wissenchaftliche Ergebnisse der Deutschen Tiefsee-Expedition auf dem Dampfer "Valdivia," 1898–1899: Jena, p. 1–136.

———, 1907, Das indische Phytoplankton. *in* Chun, C., ed., Wissenschaftliche Ergebnisse der Deutschen Tiefsee-Expedition auf dem Dampfer "Valdivia," 1898–1899: Jena, p. 223–544.

———, 1928, Abteilung Bacillariophyta (Diatomaceae). *in* Engler, A., and Prantl, K., eds., Die naturlichen Pflanzenfamilien: Auflage 2, 2, 105. Leipzig: Wilhelm Engelmann.

KITTON, F., 1870–1871, Diatomaceous deposits from Jutland: J. Quekett Micr. Club. v. 2, p. 99–102, 168–171.

KOIZUMI, I., 1968, Tertiary diatom flora of Oga Peninsula, Akita Prefecture, northeast Japan: Sci. Rept. Tohoku Univ., 2nd ser. (Geol.), v. 40, no. 3, p. 171–225.

KOLBE, R. W., 1954, Diatoms from equatorial Pacific cores: Rept. Swedish Deep-Sea Exped., v. 6, p. 1–49.

KOZLOVA, O. G., 1964, Diatoms of the Indian and Pacific sectors of the Antarctic: Moskva. Izd. Nauka, p. 1–167.

KUTZING, F. T., 1833, Synopsis Diatomacearum oder Versuch einer systematischen Zusammenstellung der Diatomeen: Linnaea, v. 8, p. 529.

———, 1844, Die kieselschaligen Bacillarien oder Diatomeen: 152 S. Nordhausen.
LISITZIN, A. P., 1960, Bottom sediments of the Eastern Antarctic and southern Indian Ocean: Deep-sea Research, v. 7, p. 89.
LONG, J., FUGE, D., AND SMITH, J., 1946, Diatoms of the Moreno shale: Jour. Paleontology, v. 20, p. 89-118.
MACKINTOSH, N. A., 1946, The Antarctic convergence and the distribution of surface temperatures in Antarctic waters: Discovery Rept., v. 23, p. 177-212.
MANGIN, L., 1915, Phytoplankton antarctique. Exp. Antarc. de la Scotia, 1902-1904: Mem. Acad. Sci. Paris, v. 57, p. 1-134.
MANGUIN, E., 1960, Les diaomees de la Terre Adelie: Ann. Sci. Naturelles, v. 1, p. 223-385.
OPDYKE, N. D., 1972, Paleomagnetism of deep-sea cores: Rev. Geophys. Space Physics, v. 10, p. 213-249.
———, GLASS, B., HAYES, D. J., AND FOSTER, J., 1966, Paleomagnetic study of Antarctic deep-sea cores: Science, v. 154, p. 349-357.
———, BURCKLE, L. H., HAYES, J. D., AND SAITO, T., 1970, Extension of the magnetic stratigraphy to the middle Miocene in deep-sea sediments (abs.): Geol. Soc. Amer. Abstracts with Programs, v. 2, p. 642.
———, ———, AND TODD, A., 1974, The extension of the magnetic time scale in sediments of the central Pacific Ocean: Earth Planet. Sci. Lett., v. 22, p. 300.
PANTOCSEK, J., 1886, Beitrage zur Kenntnis der fossilen Bacillarien Ungarns: Marine Bacillarien, pt. 1.
PERAGALLO, H., 1897-1908, Diatomees marines de France et des districts maritimes voisons: Gerz-zur-Loing.
———, 1921, Diatomees d'eau salee: Deux. Exp. Antarct. Franc, 1908-1910, Sc. Natur. Docum. Sci. Bot., p. 1-98.
PRITCHARD, A., 1852, A history of infusorial animalcules, living and fossil; illustrated by several hundred magnified representations: London, Whittaker and Co.
PROSCHKINA-LAVRENKO, A. I., 1949, Diatomovyi Analiz. Centrales Mediales: Bot. Inst. Akad. Nauk USSR, v. 2, p. 1.
———, 1956, Phytoplankton and silicate species from the tropical Pacific: Skrift. det Norske Videnskapsakad., Oslo. Matemat., Nautrv., v. 2, p. 1.
———, ed., GLESER, S. I., JOUSE, A. P., MAKAROVA, I. V., AND SHESHUKOVA-PORETZKAYA, V. S., 1974, The diatoms of the USSR, fossil and recent: Leningrad (Nauka), v. 1.
RALFES, J., 1861, *in* Pritchard, A., History of infusoria, living and fossil: ed. IV, revised and enlarged; London.
RATTRAY, J., 1889, A revision of the Genus *Coscinodiscus* Ehrb., and some allied genera: Proc. Roy. Soc. Edinburgh, v. 16, p. 449 (Reprint, Neill and Company, Edinburgh).
SAITO, T., 1969, Late Cenozoic stage boundaries in deep-sea sediments (abs.): Geol. Soc. Amer. Abstracts for 1969, p. 289-290.
SCHRADER, H. J., 1969, Die pennaten Diatomeen aus dem Obereozan von Oamaru, Neuseeland: Nova Hedwigia, Beiheft, v. 28, p. 1-124.
———, 1974a, Revised diatom biostratigraphy of the experimental Mohole drilling, Guadalupe Site: Calif. Acad. Sci. Proc., 4th ser., v. 39, p. 517-562.
———, 1974b, Proposal for a standardized method of cleaning diatom-bearing deep-sea and land-exposed marine sediments: Nova Hedwigia, Beiheft, v. 45, p. 403-409.
———, 1974c, Cenozoic marine planktonic diatom biostratigraphy of the tropical Indian Ocean: Init. Repts. DSDP, v. 24, p. 887-967.
SHADBOLT, G., 1854, A short description of some new forms of Diatomaceae from Port Natal: Micr. Soc. London Trans., v. 2, n.s., p. 13-18.
SHIBKOVA, K. G., 1959, *in* Krotov, A. I. and Shibkova, K. G., New species of diatoms from Paleogene deposits in the Urals, Botanicheskie Materialy Otdela Sporovykh rastenii, Botanich: Inst. Akad. Nauk USSR, v. 12, p. 112-129 (in Russian).
SHESHUKOVA-PORETZKAYA, V. S., 1962, Novie i redkie Bacillariophyta iz diatomovoi sviti Sakhalina: Uchenie Zapiski Len. Un. Ser. Biol. Nauk. (Leningrad Univ.), v. 49, p. 203-211.
SIMONSEN, R., 1974, The diatom plankton of the Indian Ocean Expedition of R/V "Meteor" 1964-1965: "Meteor" Forschung. Reihe D., no. 19, pl. 1-66.
SMITH, H. L., 1887, *in* Christian, T., 1887. New diatomaceous deposits: The Microscope, v. 7, p. 65-68.
TEMPERE, J., 1890, Diatomees rares ou nouvelles: Le Diatomiste, v. 1, no. 1.
———, AND PERAGALLO, H., 1910, Diatomees du monde entier: Edit. 2, 30 fasc. Arcachon, Gerz-sur-Loing (S.-et-M.), pt. 1, p. 209.
VAIL, P. R., AND OTHERS, 1977, Seismic stratigraphy and global changes of sea level; Part 3: relative changes of sea level from coastal onlap. *in* Payton, C. E., Seismic stratigraphy—applications to hydrocarbon exploration: Tulsa, Amer. Assoc. Petroleum Geologist Mem. 26, p. 63-97.
VAN HEURCK, H., 1896, A Treatise on the Diatomaceae: London, William Wesley & Son.
———, 1909, Diatomees: Resultats du voyage du S. Y. "Belgica" en 1897-1899: Rapports Sci., Bot., p. 3-126.
VAN LANDINGHAM, S. L., 1967-1971, Catalogue of the fossil and recent genera and species of diatoms and their synonyms. Part I (1967): *Acanthoceras* through *Bacillaria;* Part II (1968): *Bacteriastrum* through *Coscinodiscus;* Part III (1969): *Coscinosphaena* through *Fibula;* Part IV (1971): *Fragilaria* through *Naunema;* Part V (1975): *Navicula;* Part VI (1978): *Neidium* through *Rhoicosigma;* Part VII (1978): *Rhoicosphenia* through *Zygoceros;* Part VIII (1979): Supplement. Lahre: Cramer.
WALLICH, G. C., 1860, On the siliceous organisms found in the digestive cavities of the Salpae, and their relation to the Flint nodules of the Chalk Formation: Micr. Soc. London Trans., v. 8, p. 36.
WEAVER, F. M., AND DINKELMAN, M. G., 1976, Pliocene climatic and glacial history of Antarctica as revealed by

Southeast Indian Ocean deep-sea cores: Discussion and reply. Discussion. Geol. Soc. America Bull., v. 87, p. 1529–1532.

WESTBERG, M. J., 1978, Accuracy of radiolarian correlations in the Pacific, Miocene: Micropaleontology, v. 24, no. 1, p. 1–23.

WITT, O., 1886, Uber den Polierschiefer von Archangelsk-Kurojedowo im Gouv. Simbirsk: Verh. russ-kais. Mineral. Ges., ser. 2, v. 22, p. 137–177.

WORNARDT, W. W., JR., 1967, Miocene and Pliocene marine diatoms from California: Calif. Acad. Sci. Occas. Papers, no. 63, p. 1–108.

ZABELINA, M. M., 1934, Diatoms from the Tertiary deposits of the eastern coast of Kamchatka: Oil Geol. Inst. Trans., v. 48, p. 3–19.

SEPM Special Publication No. 32, p. 471–487, October 1981

DEEP SEA DRILLING IN THE ANTARCTIC: FOCUS ON LATE MIOCENE GLACIATION AND APPLICATIONS OF SMEAR-SLIDE BIOSTRATIGRAPHY

SHERWOOD W. WISE, JR.
Antarctic Marine Geology Research Facility, Department of Geology, Florida State University, Tallahassee, Florida 32306 U.S.A.

ABSTRACT

Exploration of the Southern Ocean by *Glomar Challenger* began in the Pacific sector with a boldly executed set of holes along a cruise track extending into ice-laden waters adjacent to the Antarctic continent. These holes, along with additional sections drilled in subantarctic waters off New Zealand and Australia, established the basis for regional biostratigraphic zonations of the smaller-sized microfossil groups (coccoliths, diatoms, and silicoflagellates). Because these microfossils could be examined readily in smear-slide preparations of raw sediments, a rapid age-dating technique evolved called smear-slide biostratigraphy. In addition, a preliminary history of Antarctic refrigeration leading to severe Late Tertiary continental glaciations was documented.

Exploration of the Atlantic sector, limited primarily by unfortunate weather conditions, provided additional evidence of severe Late Miocene glaciation and its role in shaping the circum-Antarctic current into a powerful geologic force. The successful drilling of critical reference sections on the Falkland Plateau also opened the area to efficient exploration by conventional piston cores sited using techniques of shipboard smear-slide biostratigraphy.

INTRODUCTION

Deep-sea drilling in the Antarctic was an unprecedented feat which took *Glomar Challenger* deep into the reaches of Antarctic ice to obtain the first extensive marine record of the evolution and glacial development of the Southern Ocean. The first two cruises (Legs 28 and 29) into the Pacific sector went relatively smoothly with reasonable weather and penetration by the ship to the borders of the continent, even into the ice-free portions of the expansive Ross Sea. The succeeding legs into the Bellingshausen Sea and the Southwest Atlantic sector (Legs 35 and 36) were met by drilling problems and severe storms which resulted in the recovery of a far more fragmentary Cenozoic record than the previous two legs. Study of the voluminous material derived from Legs 28 and 29, however, established a workable biostratigraphic chronology and a suggested scenario of paleoceanographic events by which the geologic record revealed in the later legs could be compared, measured and interpreted. Perhaps equally important, a firm foundation was established by all the Southern Ocean legs which has made further exploration of the area by conventional piston coring and geophysical techniques especially profitable.

The scientific results of the Pacific legs have been well-summarized in detail by a number of authors and will not be represented here. Notable among these are the comprehensive reviews by Burns (1977) which traced the evolution of lithofacies and water masses through time, by Kemp (1978) which emphasized the evolution of climates and changes in ocean current and atmospheric circulation, and by Kennett (1978) which dealt with the development of the biogeographic and paleoceanographic record. The last of these (Kennett, 1978) was published after all cruise reports were available and included results from the South Atlantic sector as well.

The purpose of this paper is to highlight some of the results of the earlier cruises in the Pacific sector to show how these provided a valuable basis for further exploration in the Atlantic sector by both deep sea drilling and conventional piston core techniques. The story of deep sea drilling in the Southern Ocean illustrates well how the initial establishment of a relatively complete marine geologic record in a remote, unexplored area with little available land outcrop data can stimulate extensive investigation by other methods and allow all results to be integrated in a relatively short period of time. The discussion will concentrate on paleoceanographic events surrounding the particularly severe Late Miocene Southern Hemisphere glaciations and on the development of a technique known as smear-slide biostratigraphy. This technique has become important in subsequent exploration efforts, particularly in exploring submergent outcrops of older (pre-Pliocene) strata.

PRESENT DAY OCEANOGRAPHY

Core sites and land outcrops referred to in this paper are indicated by Figure 1 which also shows the broad expanse of the Southern Ocean. The circulation of that ocean is dominated by the infinite circum-Antarctic current which is driven by the perpetual West Wind Drift (Fig. 2). The primary component of the circum-Antarctic current is the Cir-

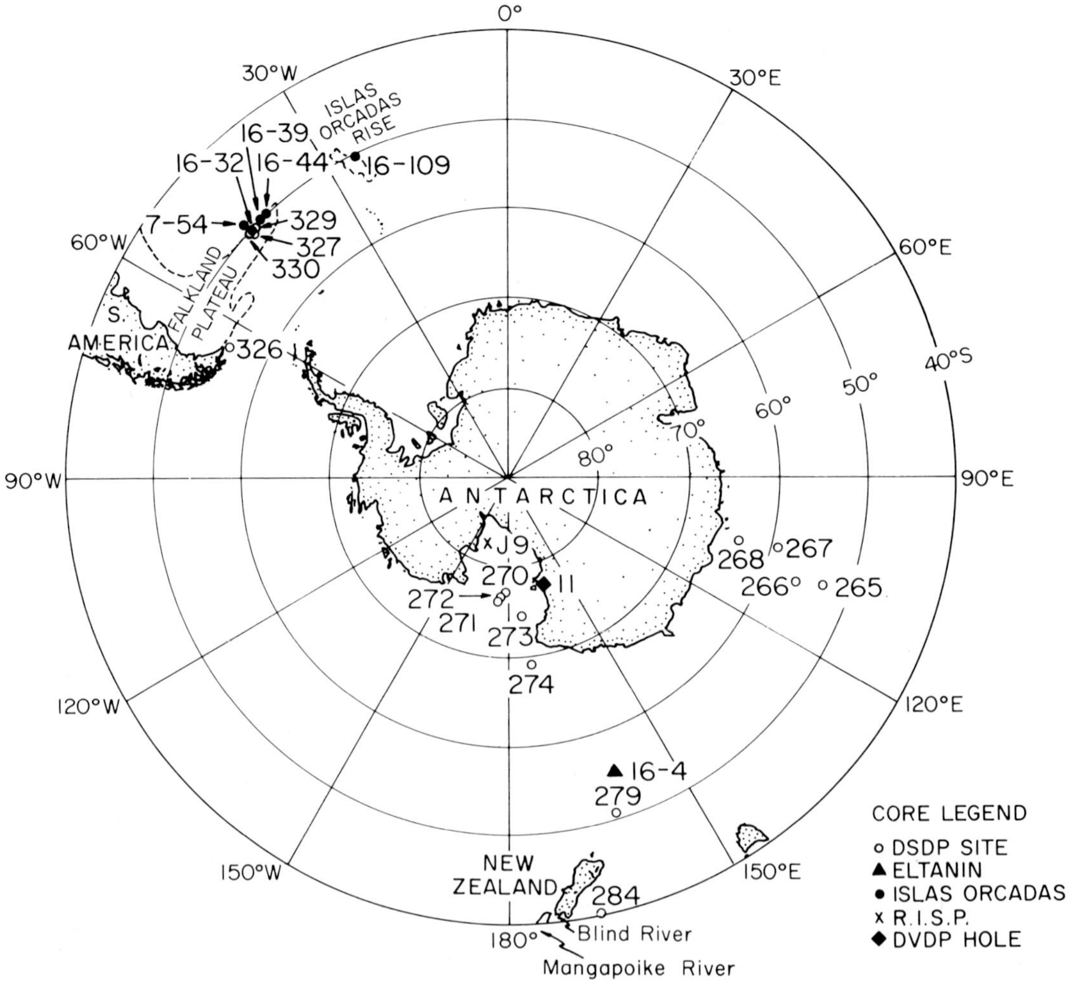

FIG. 1.—Location map.

cumpolar Deep Water which is fed by the world oceans to the north but assumes the dominant clockwise path of rotation around the continent dictated by atmospheric and geostrophic forces. It is important to emphasize the strength and potential erosive power of the circum-Antarctic current which is a bottom-feeling current over most of its path. The role of this current in shaping the depositional regime of bottom sediments has been often overlooked because geologists have traditionally directed much attention to Antarctic Bottom Water as a major erosive force in the area, particularly during glacial episodes. Although Antarctic Bottom Water has been detected far north of the equator in the world's oceans, its primary influence is felt south of the Antarctic Convergence (Polar Front) as shown in Figure 2. Rather than having a strong erosive potential, when strongly developed, Antarctic Bottom Water may actually have a buffering effect on the erosive power of the circum-Antarctic current (Weaver, 1979). That is not to say that Antarctic Bottom Water cannot be an agent of erosion, but it probably functions more in that capacity where it becomes entrained by the Circumpolar Deep Water. Thus, its erosive power is derived more from the circum-Antarctic current than from within itself.

More important for the generation of pelagic sediments in the Antarctic are the surface components of the circum-Antarctic current, particularly: 1) the Antarctic Surface Water which flows north from the continent, and 2) the Subantarctic Surface Water below which the Antarctic Surface Water plunges at the Antarctic Convergence. The Antarctic Convergence is not only an important water-mass boundary, but also a rather sharp dividing line

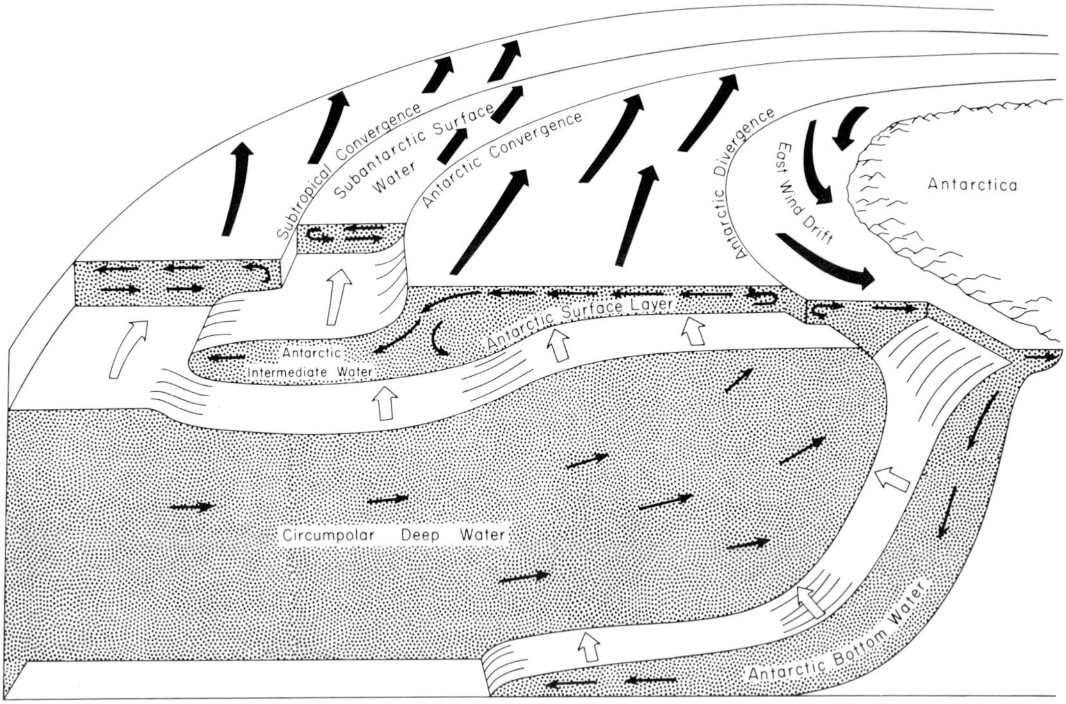

FIG. 2.—Water masses and current flow directions in the Southern Ocean (from Miyajima, 1975).

between predominantly siliceous plankton to the south and predominantly calcareous (or at least a mixture of calcareous and siliceous plankton) to the north. Investigators have long noted a rough correlation between the occurrence of siliceous oozes below the Antarctic Surface Water and carbonate-rich sediments beneath the Subantarctic Surface Water (*e.g.,* Goodell, 1973). This correlation is strongly modified by local topography where the carbonate compensation depth may be a factor and, closer to the continent, by ice cover which inhibits phytoplankton production.

SMEAR SLIDE BIOSTRATIGRAPHY

Except where inhibited by ice cover, phytoplankton productivity is exceptionally high throughout the circum-Antarctic area due to upwelling effects. Diatoms dominate the siliceous phytoplankton populations, whereas coccoliths are dominant in the calcareous belts. These, along with silicoflagellates, are among the smallest planktic microfossils preserved in sediment. They may be readily studied in the light microscope from a smear of raw sediment on a glass slide. Other common fossil groups such as foraminifers (calcareous) and radiolarians (siliceous) are composed of much larger individuals which are better studied after extraction from the sediment by sieving and washing techniques. Because the latter processes are time consuming, Antarctic sediments can be most rapidly age-dated by searching for the smaller microfossils in smear slides. These can be prepared from fresh sediment in about two minutes. All three of the smaller microfossil groups may be present in the same sediment sample as illustrated by the Eocene assemblage in Figure 3.

Before the advent of deep-sea drilling, the marine sedimentary record from the Southern Ocean was so scanty that no comprehensive zonations existed for any of the smaller microfossil groups. Some effort had been made toward establishing a Plio/Pleistocene zonation for diatoms (*e.g.,* Donahue, 1970; Abbott, 1972) and for coccoliths (Geitzenauer, 1972) using piston cores. The lack of suitable sections, however, inhibited further progress. This early work did indicate that the diatoms in the Antarctic, as in most areas, were sufficiently provincial to make a regional diatom zonation necessary for that area. The coccolith datums, however, were largely the same as those used in tropical regions where coccoliths reach their maximum diversity. This indicated that most coccolith index species one would expect to find in high latitudes would be cosmopolitan, coolwater-tolerant forms with extensive geographic ranges.

The basis of a comprehensive diatom zonation was established during Leg 28 by McCollum (1975). He combined some of his unpublished piston core

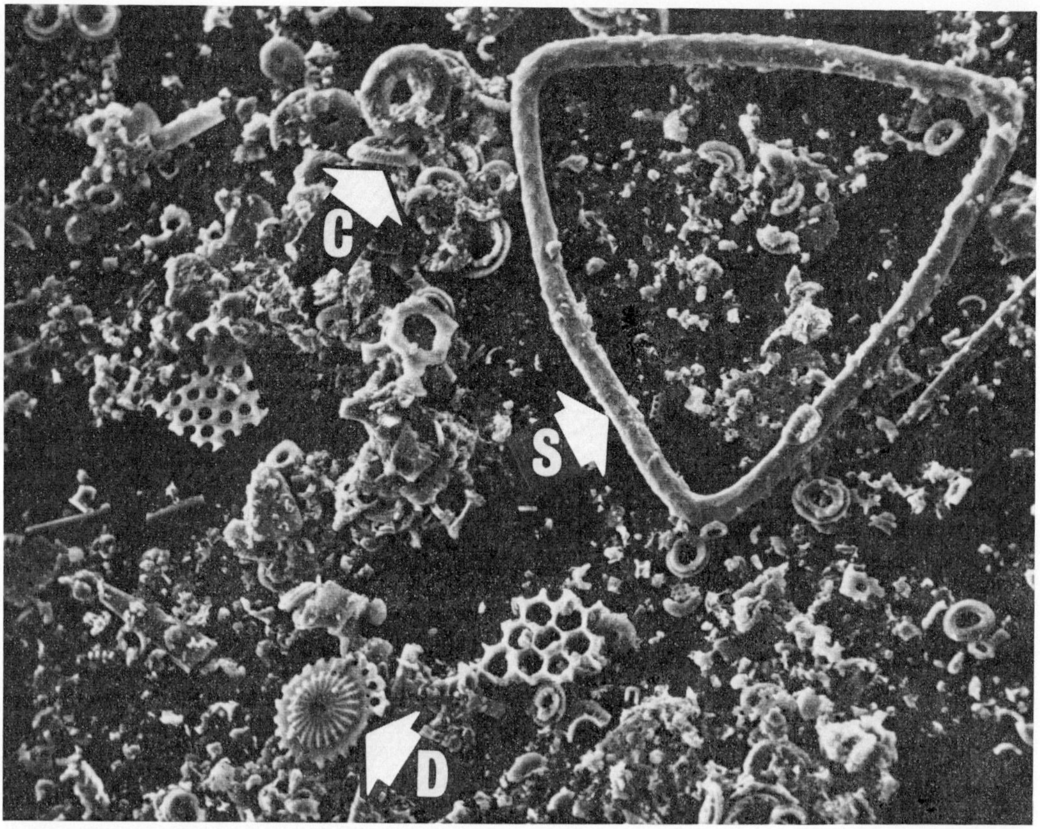

FIG. 3.—Scanning electron micrograph of an Eocene assemblage containing three groups of smaller microfossils used for smear-slide biostratigraphy: coccoliths (C), diatoms (D), and silicoflagellates (S). The assemblage is from *Islas Orcadas* core sample IØ 16-44, 40–41 cm, eastern Falkland Plateau; X 1150.

studies with his drill-hole data to produce a Neogene zonal scheme. His zonation has been widely applied and further improved (see Weaver and Gombos, this volume). Using the same material, Bukry (1975) and Ciesielski (1975) devised a workable silicoflagellate zonation. Coccoliths were not commonly encountered on Leg 28, the southernmost leg, but were used extensively on Leg 29 (Edwards and Perch-Nielsen, 1975) which primarily explored the subtropical regions north of the Antarctic Convergence. The zonation employed by Edwards and Perch-Nielsen (1975) included datums common to the tropical coccolith zonation along with secondary datums defined by Edwards (1971) from New Zealand land sequences.

Most of the important microfossil datums worked out for the Pacific Antarctic and Subantarctic have been recognized in the South Atlantic sector wherever suitable section has been encountered. Drilling to date has been confined nearly exclusively to the area of the eastern Falkland Plateau which underlies the present-day Antarctic Convergence. Attempts to drill further south during DSDP Leg 36 were unsuccessful. Zones and datums recognized in the Falkland Plateau drill- and piston-core sequences are scaled in Figure 4 against the chronology of the tropical coccolith zonation compiled by Bukry (1973). In addition, major paleoceanographic events of the Southern Ocean, as deduced primarily from the study of the DSDP sedimentary sequences, are listed.

EVENTS LEADING TO LATE MIOCENE GLACIATION

As explained in considerable detail by Kennett (1978), the physiography and ocean circulation patterns of the Antarctic have changed radically since the beginning of the Cenozoic. This is primarily due to the rearrangement of the continents as a result of seafloor spreading. The separation of New Zealand and Australia from Antarctica during the Eocene and the opening of Drake Passage during the Late Oligocene were directly responsible for the formation of the circum-Antarctic current. This, along with the concomitant expansion of the Atlantic

Ocean and the disruption of the circumequatorial current, are thought to have altered global ocean circulation patterns sufficiently to establish new modes of heat transfer between low and high latitudes. This developed to the point that Cenozoic climates underwent a deterioration which eventually led to major glaciations, first in the southern and later in the northern hemisphere.

The effect of global cooling on marine sedimentation was profound. As described by Kemp and others (1975) in the DSDP Leg 28 initial report, calcareous phytoplankton, which were responsible for the sedimentation of coccolith ooze close to the continent until the Late Eocene (DSDP Hole 267B), migrated northward as surface waters cooled to inhospitable temperatures during the Oligocene and Miocene. Glacial-marine sedimentation began near the continent, with ice rafting becoming evident in Upper Oligocene sediments of the Ross Sea. Geophysical measurements by Barker and Burrell (*in press*) suggest that the Drake Passage began to open about 27 m.y.BP and reached its present configuration about 5 m.y.BP. This cleared the final barrier to a deep-flowing circum-Antarctic current. Strong fluctuations in oxygen isotope curves for DSDP Site 279 have been interpreted by Shackleton and Kennett (1975a) to be the result of the rapid buildup of a permanent ice cap in East Antarctica between 14 m.y. and 10 m.y.BP. After some further fluctuation, climatic conditions in the Antarctic deteriorated sharply during the Late Miocene, bringing on a severe continental glaciation which may have exceeded all others in intensity. One interpretation by Shackleton and Kennett (1975b) of oxygen isotope data from DSDP Hole 284 suggested an expansion of the ice on the continent to a volume about 50% more than that of present day. Their data supported earlier suggestions by land geologists of major Late Tertiary continental glaciation of Antarctica (Rutford and others, 1972; Mayewski, 1972, 1975).

FURTHER EVIDENCE FROM THE PACIFIC SECTOR

Sedimentologic evidence from the deep sea for climatic deterioration during the Middle and Late Miocene was convincingly demonstrated by Kemp and others (1975). They applied backtracking methods for DSDP Sites 265–267 to show northward shifts in the boundaries between the various sediment lithofacies that ring the continent. Subsequent drilling through the Ross Ice Shelf (Webb and others, 1979; Brady and Martin, 1979) has further added to this history, which is summarized in Figures 5 and 6. In these idealized cross-sections, the DSDP holes have been projected on a straight line running north from the continent in order to better show spatial relationships between the holes. The lithologies have also been simplified to exclude occurrences of pelagic and hemipelagic clay in order to show more clearly the relationships between the accumulations of skeletal oozes and glacial marine sediments.

In the Ross Sea, the Ross Ice Shelf Project (RISP) demonstrated the presence of diatomaceous glacial marine sediments (including small amounts of relatively pure diatom ooze) dated about 14 m.y.BP (late Middle Miocene). The combination of glacial sediments and diverse diatom assemblages, along with a low diversity terrestrial pollen assemblage, suggested deposition under relatively open marine conditions but still in the presence of floating ice. Such ice would have been derived from partially glaciated montane areas peripheral to the Ross embayment and perhaps from the build-up of winter pack ice (Webb and others, 1979; Brady and Martin, 1979).

The above situation changed dramatically during the Late Miocene as an extensive ice cover developed both on the continent and along the marginal marine areas. Hayes and Frakes (1975) believed that ice development in the Ross Sea culminated with a major advance of the Ross Ice Shelf which, by the end of the Miocene, was grounded perhaps out to the margin of the continental shelf, far in advance of its position today (Fig. 5). They based their hypothesis on the presence of a prominent unconformity, which on seismic profiles was seen to truncate the glacial marine sequence within the Ross Sea. In the DSDP drill cores, this unconformity separated Pliocene from Miocene sediments, but the time of its formation has been dated only approximately. In DSDP Hole 272 McCollum dated the youngest of the Miocene sediments below the unconformity as Middle Miocene, although these have been referred to, apparently in error, as "Late Miocene" in some parts of the DSDP Leg 28 initial report. It should be noted, however, that glaciomarine sediments have been dated as Late Miocene in the nearby DVDP Hole 11 of the Dry Valleys (Brady, 1979).

Just beyond the Ross Sea, along the lower continental rise (DSDP Hole 274), Frakes (1975) described a remarkably attenuated Middle Miocene to Lower Pliocene section rich in ferromanganese micronodules and with ice rafted detritus in the upper part. Viewed as a type of lag deposit, these sediments appeared to have been continuously winnowed and scoured by active bottom currents generated during the glacial buildup.

Farther out to sea, the northward shift in facies boundaries during the Late Miocene is reflected by a change from diatom ooze to essentially barren Pliocene glacial marine sediments at DSDP Site 268 (Figs. 5, 6). Most dramatic, perhaps, was an abrupt shift of the diatom ooze/carbonate ooze boundary which, toward the end of the Miocene, moved

AGE	TROPICS — COCCOLITHS Bukry, 1973			ANTARCTIC — COCCOLITHS Wise and Wind, 1977	
HOLOCENE	*Emiliania huxleyi*		*E.h.	*Emiliania huxleyi*	*E.h.
PLEISTOCENE	*Gephyrocapsa oceanica*		*G.o.		
PLEISTOCENE	*Gephyrocapsa doronicoides*	*Gephyrocapsa caribbeanica*	*G.c.		
		Emiliania annula	†D.b.		
LATE PLIOCENE	*Discoaster brouweri*	*Cyclococcolithina macintyrei*	†D.p., D.s.		
		Discoaster pentaradiatus	†D.t.		
		Discoaster tamalis	†R.p., S.n.		
EARLY PLIOCENE	*Reticulofenestra pseudoumbilica*	*Discoaster asymmetricus*	a*D.a.		
		Sphenolithus neoabies	†C.p., C.t.		
	Ceratolithus tricorniculatus	*Ceratolithus acutus*	*C.r., †C.a.		
LATE MIOCENE		*Ceratolithus rugosus*	†T.r.		
		Triquetrorhabdulus rugosus	†D.q.		
	Discoaster quinqueramus	*Ceratolithus primus*	*C.p.		
		Discoaster berggrenii	*D.b.		
	Discoaster neohamatus	*Discoaster neorectus*	*D.n., D.l.		
		Discoaster bellus	†D.h.		
MIDDLE MIOCENE	*Discoaster hamatus*	*Catinaster calyculus*	*C.c.		
		Helicopontosphaera kamptneri	*D.h.		
	Catinaster coalitus		*C.c.		
	Discoaster exilis	*Coccolithus miopelagicus*	*D.k.		
		Discoaster kugleri	†S.h.		
	Sphenolithus heteromorphus		*C.m., a†D.d.		
EARLY MIOCENE	*Helicopontosphaera ampliaperta*		*S.h.		
	Spenolithus belemnos		*S.b.		
	Triquetrorhabdulus carinatus	*Discoaster druggii*	*D.d.		
		Discoaster deflandrei	a†C.a.		a†C.a.
OLIGOCENE		*Cyclicargolithus abisectus*	†S.c., D.b.		†R.b.
	Sphenolithus ciperoensis		*S.c.		
	Sphenolithus distentus		*S.d.		
	Sphenolithus predistentus		†R.h., R.u.		†R.u.
	Helicopontosphaera reticulata	*Reticulofenestra hillae*	†C.f.		
		Cyclococcolithina formosa	a†C.s.		†I.r.
		Coccolithus subdistichus	†D.b., D.s.		†D.s.
LATE EOCENE	*Discoaster barbadiensis*		†C.g.	*Isthmolithus recurvus*	*I.r. †C.g.
MIDDLE EOCENE	*Reticulofenestra umbilica*	*Discoaster bifax*	†D.b., C.s.		
		Discoaster saipanensis	*D.b., R.u.		
	Nannotetrina quadrata	*Coccolithus staurion*	†C.g.		
		Chiasmolithus gigas	*C.g.		
		Discoaster strictus	*N.g., †R.i.		
	Discoaster sublodoensis	*Rhabdosphaera inflata*	*R.i.		
		Discoasteroides kuepperi	*D.s.		
EARLY EOCENE	*Discoaster lodoensis*		*C.c.		†T.o.
	Tribrachiatus orthostylus		*D.l.	*Tribrachiatus orthostylus*	*T.o.
	Discoaster diastypus		*D.d., T.c.		
PALEOCENE	*Discoaster multiradiatus*	*Campylosphaera eodela*	*C.e.		†D.m.
		Chiasmolithus bidens	*D.m.	*Discoaster multiradiatus*	*D.m.
	Discoaster nobilis		*D.n.	*Heliolithus universus*	*H.u.
	Discoaster mohleri		*D.m.		
	Heliolithus kleinpellii		*H.k.		
	Fasciculithus tympaniformis		*F.t.	*Fasciculithus involutus*	*F.i.

* = first occurrence; † = last occurrence; a* = acme beginning; a† = acme ending; ⟶ = evolutionary transition

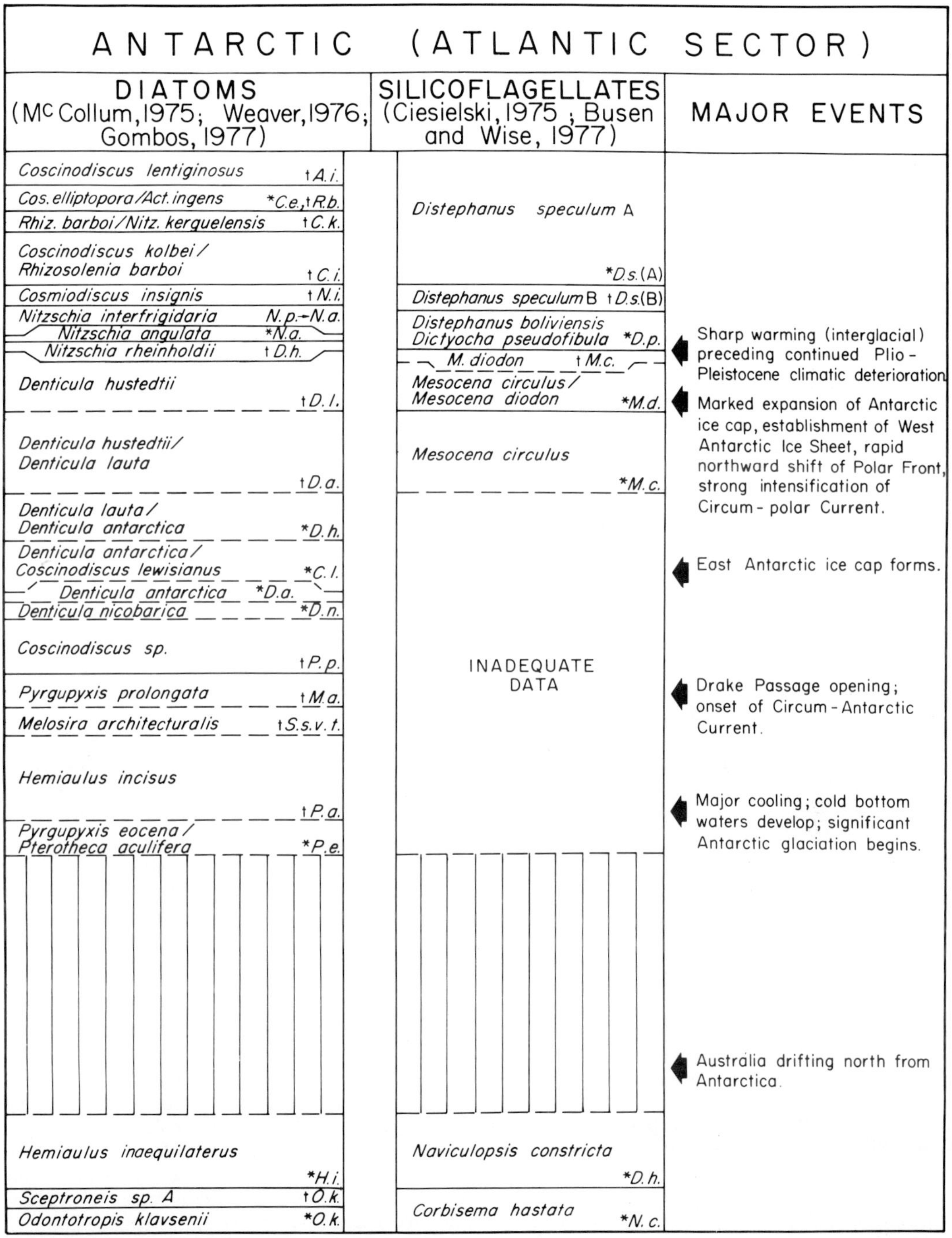

FIG. 4.—Zones and datum levels recognized in cores from the Falkland Plateau area (Atlantic sector of the Southern Ocean) for the three smaller microfossil groups illustrated in Figure 3. The datums are scaled against Bukry's (1973) tropical coccolith zonation. The right-hand column provides a chronology of major regional tectonic, paleoceanographic, or paleoclimatic events.

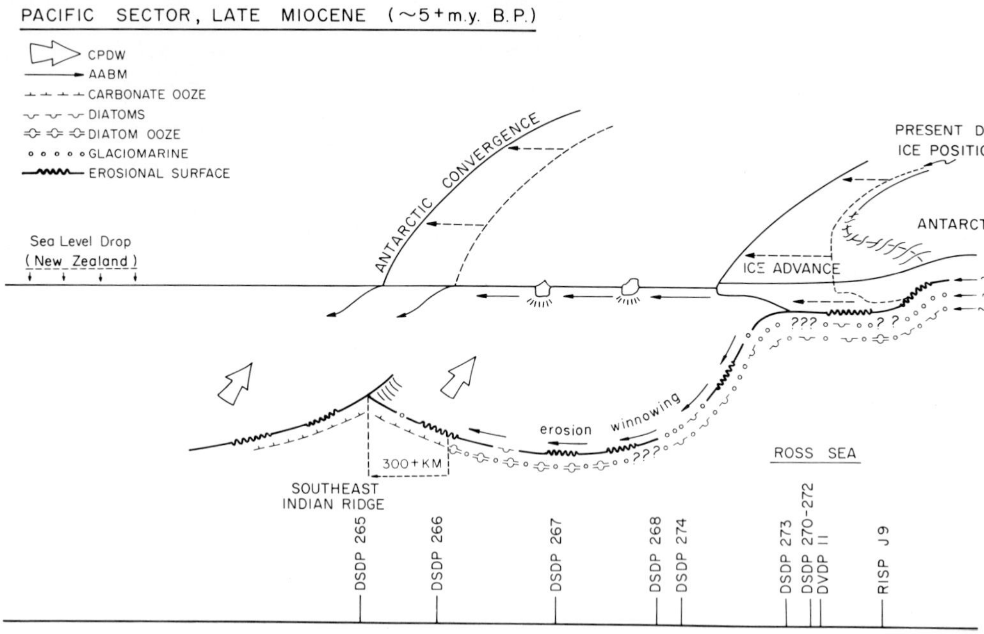

FIG. 5.—Paleoceanographic setting in the Pacific sector of the Southern Ocean during the height of the Late Miocene glacial events (~ 5+ m.y.BP). Drill locations have been projected along a straight line to show in cross-section lithologic facies.

rapidly over a distance of some 300 km to a position near the crest of the Southeast Indian Ridge (Kemp and others, 1975). Kemp and others (1975; see also Hayes and Frakes, 1975) believed this shift occurred coincident with the advance and grounding of the Ross Ice Shelf at the close of the Miocene (about 5 m.y.BP) as a result of rapid cooling and a marked increase in the Antarctic surface water mass. In examining the section most critical to their analysis (DSDP Hole 266), Weaver (1976, figs. 2, 4) documented an apparent erosional disconformity between the Upper Miocene and Lower Pliocene sediments with the hiatus spanning the period of roughly 4 to 7.5 m.y.BP. Even within this expanded time window, the diatom ooze/carbonate ooze boundary had to have moved northward at least three to five times faster than it had moved earlier during the Miocene, and as Kemp and others (1975) have suggested, it may have shifted much more rapidly than that. The ultimate contrast in Upper Miocene and Lower Pliocene sedimentary facies is shown in Figure 6 which depicts the initiation of Pliocene sedimentation that occurred as climatic conditions ameliorated and the ice retreated.

At the conclusion of DSDP Leg 28 the exact chronology of these events associated with Late Tertiary continental glaciation was not entirely clear. This was because new zonation schemes for siliceous microfossils were just being proposed and tested. In addition, the glacial marine sequences recovered near the continent were poorly fossiliferous throughout the critical Upper Miocene-Pliocene interval.

There was another problem. Further to the north, considerable work had been done on equivalent sediments exposed in land sections in New Zealand. Although clear evidence of severe glacial activity had been noted in the microfaunal and lithologic characteristics of these sections (including a regression and disconformity attributed to a glacial eustatic sealevel drop), dating by foraminifers correlated with paleomagnetic data indicated that the most extensive Late Tertiary glaciation had occurred during the Early Pliocene rather than the Late Miocene (Kennett and Watkins, 1975). Thus, a major discrepancy existed in the timing of the critical glacial events.

Van Couvering and others (1976) suggested that the discrepancy was due to a miscorrelation of the foraminiferal data with the paleomagnetic time scale. This was confirmed by Weaver (1976), who undertook an extensive study of the Upper Miocene-Lower Pliocene interval as recorded in piston and drill cores between the Ross Sea area and New Zealand. He noted throughout his study area consistent evidence for a sharp deterioration of climate during the latest Miocene, culminating in an erosional hiatus in most cores. Indeed, the erosional episode near the end of the Miocene was so pervasive that no core recorded an uninterrupted Mio-

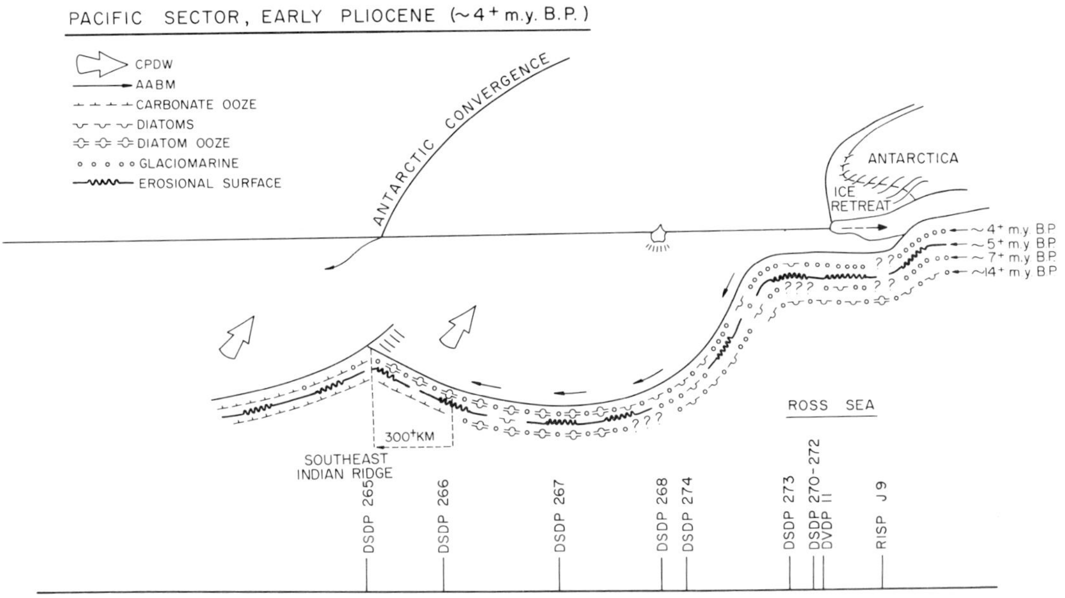

FIG. 6.—Paleoceanographic setting in the Pacific sector of the Southern Ocean as widespread sedimentation resumed during an Early Pliocene amelioration of climate (~ 4+ m.y.BP).

cene-Pliocene transition. He also noted that although the severe glacial conditions of the Late Miocene carried over into the earliest Pliocene, they were soon followed by a significant but short-lived return to relatively warm climatic conditions (see Ciesielski and Weaver, 1974). Weaver (1976) then reinterpreted the New Zealand sequence (Fig. 7) in light of his new evidence and correlated to his time scale other significant events within the Southern Ocean associated with the Late Miocene glaciation and the subsequent Early Pliocene warming event. His interpretation has in essence been concurred with by other workers in the area who have introduced additional substantiating evidence (*e.g.,* Loutit and Kennett, 1979).

EXPLORATION OF THE SOUTH ATLANTIC SECTOR

Against the background of an overwhelmingly successful deep-sea drilling program in the Pacific sector of the Southern Ocean, exploration of the South Atlantic sector began on a keen note of disappointment. In an attempt to obtain sedimentologic evidence to date the opening of Drake Passage, DSDP Leg 36 drilling crews lowered a drill string at a point 60 miles south of the tip of South America in the middle of the Passage and directly in the path of the circum-Antarctic current. The strength of the current was not anticipated, and it drifted the drill string at a sharp angle beneath the ship. Shortly after recovery of the first sediment core, heavy seas generated by the weather worked the ship against the taut 5-inch diameter steel drill pipe, severing it just below the keel of the vessel (Fig. 8). The loss of 3800 m of valuable drill pipe, the only such loss ever sustained by *Glomar Challenger* as a result of natural conditions, strongly impressed on the minds of the scientific party the potential of the circum-Antarctic current as a geologic force.

This impression was further reinforced as drilling began on an elevated portion of the eastern Falkland Plateau (named in the Leg 36 Initial Reports the Maurice Ewing Bank) which today stands in the path of the circum-Antarctic current. Ewing Bank lies some 1500 km east of the apex of Drake Passage where the funnel effect on the circumpolar current is greatest. Nevertheless, Leg 36 drilling results and *Glomar Challenger* seismic profiles taken during the cruise suggested the possibility of large-scale erosion over the entire Ewing Bank. Apparently, broad areas of older (pre-Pliocene) units were exposed beneath a thin (1 to 3 m) cover of glacial marine sediments. This led to the selection of the Ewing Bank as a major target for piston core studies during a five year program of the Argentine Naval Ship *Islas Orcadas* cruises which began the following year.

The piston core technique used to explore outcrops of older strata was simple. The piston core was lowered at a location where seismic profiles correlated to the DSDP drill sites indicated a high probability of striking an outcrop of older sediment. Upon retrieval, core catchers were age-dated immediately using smear-slide biostratigraphy while

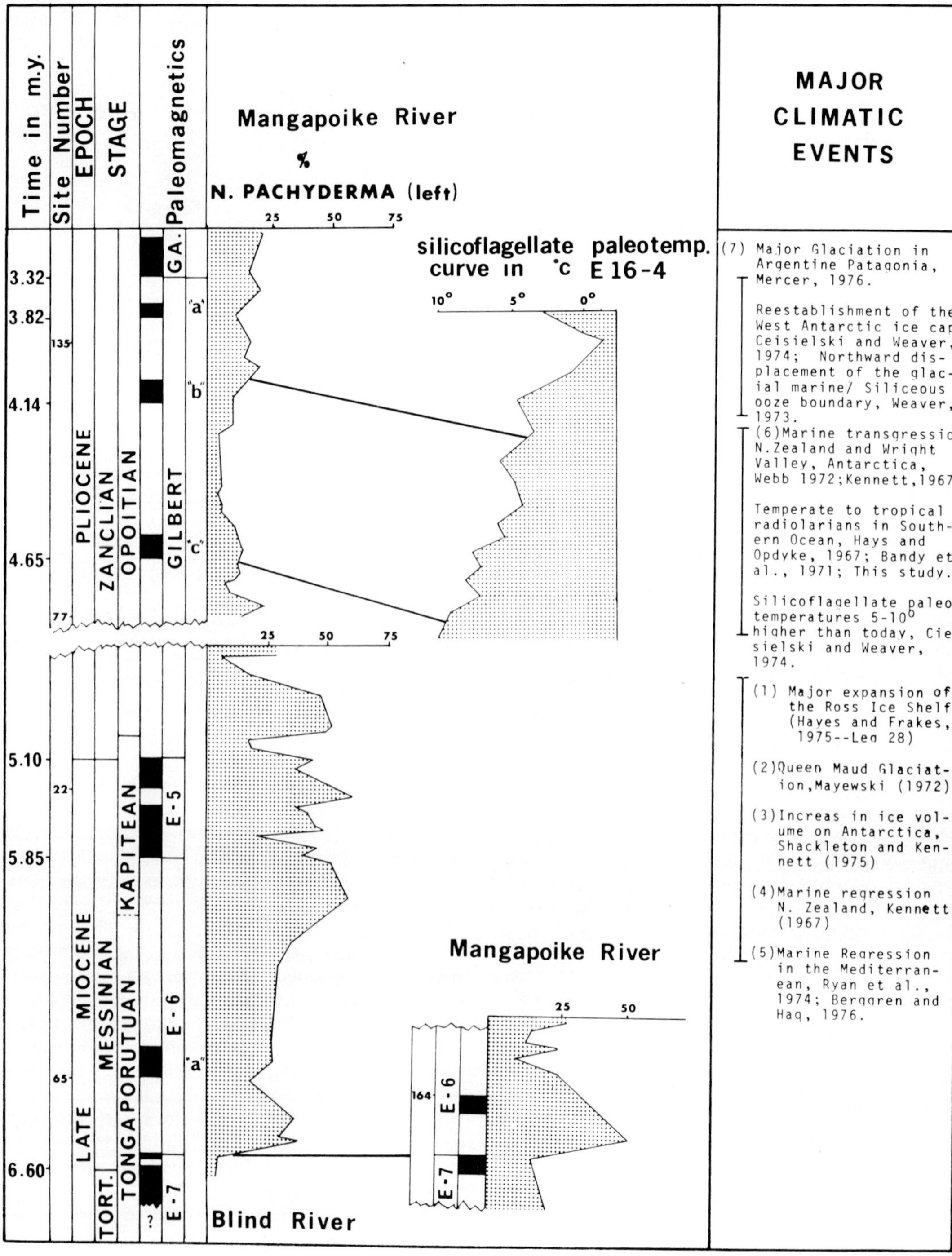

FIG. 7.—Chronology of major climatic events in the Antarctic as interpreted from land sections in New Zealand (from Weaver, 1976).

FIG. 8.—Section of 7/16 inch thick, 5 inch diameter DSDP drill pipe (right) severed under the stress of the circum-Antarctic current during a storm in Drake Passage (DSDP Leg 36, Site 326). Photo courtesy Dennis Graham, Deep Sea Drilling Project.

the core barrel was being changed and the core apparatus re-rigged. The decision as to the next core location was then made based on the information obtained from the freshly retrieved core. In essence, this amounted to a sea-going version of rock hammer geology by which outcrop belts were mapped and data were gathered for the construction of geologic cross-sections such as the one reproduced in Figure 9. The efficiency of this technique was illustrated during a recent *Islas Orcadas* cruise to Ewing Bank where 23 of 34 cores attempted penetrated through the glacial marine cover to recover older sediment (Wise and others, 1978b). By this means the geologic reference sections established by the DSDP drill sites could be used for correlation over extensive geographic areas (Fig. 9; see Ciesielski and Wise, 1977, fig. 4).

The piston cores showed the pre-Pliocene erosional disconformity far more clearly than the drilled cores because of disturbances induced in the uppermost sediments when the *Glomar Challenger* drill string is spudded in. Figures 10A and 10B show this erosional contact above Eocene and Oligocene carbonate oozes, respectively. Note the glacial erratics above the contact in Figure 10B. The core in Figure 10C, taken in the vicinity of DSDP Hole 329 (Fig. 9), shows a more subtle relationship. Here, coarse, laminated glacial sands filling small-scale channels occur within disturbed diatom ooze sequences of the Late Miocene *Denticula lauta*/*Denticula hustedtii* Zone. Below this diatom ooze, piston and drill cores encountered extensive diatomaceous carbonate oozes. Above this disturbed zone is a regional Late Miocene erosional unconformity. In his detailed analysis of *Islas Orcadas* piston cores on Ewing Bank, Ciesielski (1978a) demonstrated that the regional unconformity is overlain in places by rather pure siliceous Pliocene oozes containing warmer-water diatoms. This was predicted by the Ciesielski/Weaver model for an Early Pliocene warming event (Ciesielski and Weaver, 1974; Weaver, 1976).

Located farther to the east and farther from the direct influence of the funnelling effect of the Drake Passage on the circumpolar current, the now familiar Late Miocene-Early Pliocene sedimenta-

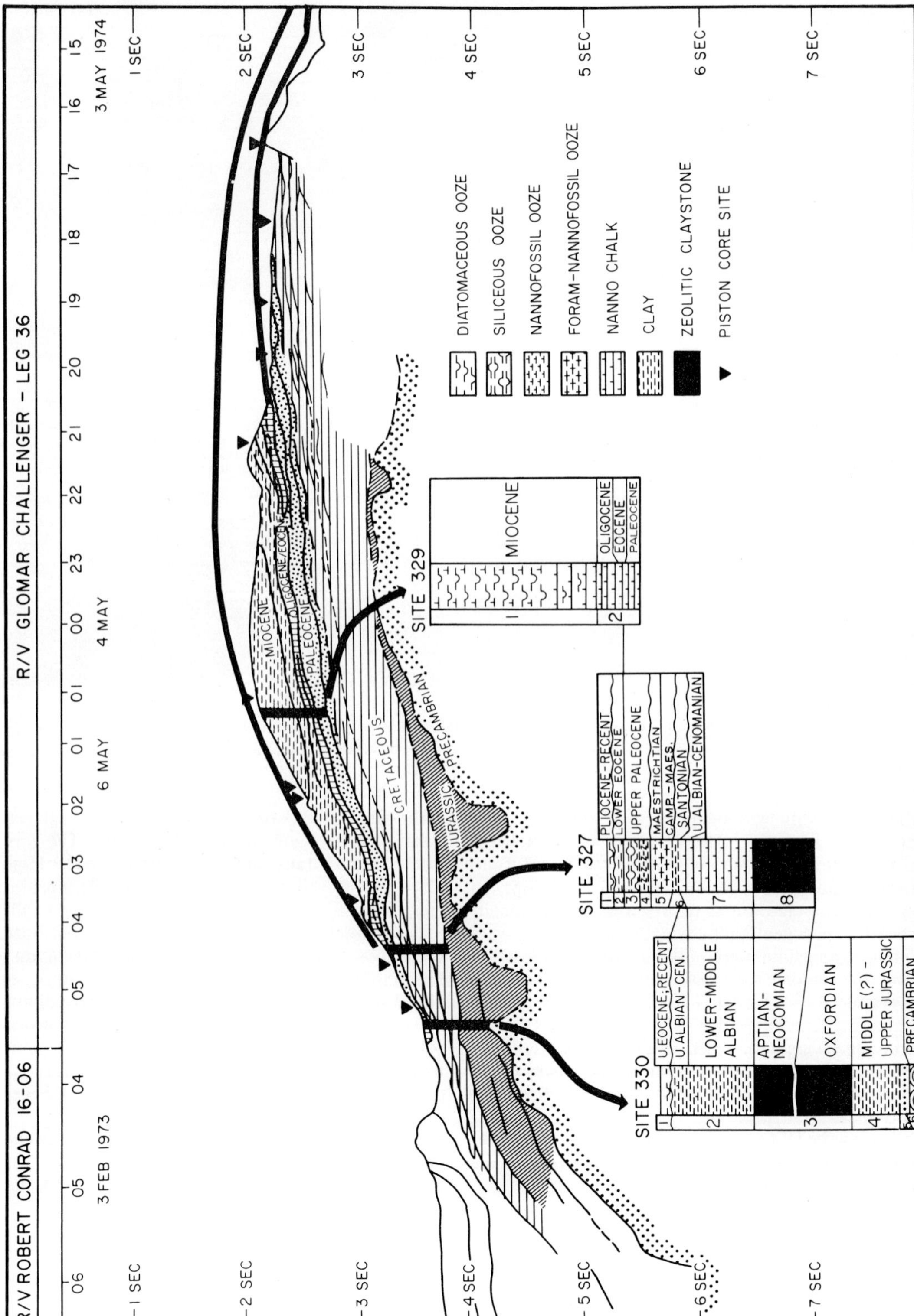
R/V ROBERT CONRAD 16-06
R/V GLOMAR CHALLENGER - LEG 36
3 FEB 1973
6 MAY
4 MAY
3 MAY 1974
1 SEC
2 SEC
3 SEC
4 SEC
5 SEC
6 SEC
7 SEC
MIOCENE
OLIGOCENE/EOCENE
PALEOCENE
CRETACEOUS
JURASSIC
PRECAMBRIAN
SITE 329
SITE 327
SITE 330
OLIGOCENE
EOCENE
PALEOCENE
PLIOCENE-RECENT
LOWER EOCENE
UPPER PALEOCENE
MAESTRICHTIAN
CAMP.-MAES.
SANTONIAN
U. ALBIAN-CENOMANIAN
U. EOCENE; RECENT
U. ALBIAN-CEN.
LOWER-MIDDLE ALBIAN
APTIAN-NEOCOMIAN
OXFORDIAN
MIDDLE (?) - UPPER JURASSIC
PRECAMBRIAN
DIATOMACEOUS OOZE
SILICEOUS OOZE
NANNOFOSSIL OOZE
FORAM-NANNOFOSSIL OOZE
NANNO CHALK
CLAY
ZEOLITIC CLAYSTONE
PISTON CORE SITE

tion pattern was found to be repeated in piston cores taken on the Islas Orcadas Rise (Fig. 1). There, core IØ 16-109 showed a progression upsection from carbonate ooze to siliceous ooze capped by an erosional disconformity which truncated sediments of the latest Miocene portion of the *Denticula hustedtii* Zone (Weaver and others, in preparation). Above this disconformity are post-Miocene glaciomarine sediments.

Thus, the overall patterns of sedimentation first suggested by the DSDP sequences in the Pacific sector appear to hold true in the Atlantic sector as well, although the data there are still fragmentary. The overall Cenozoic deterioration of climate and the northward progression of lithofacies are clearly reflected in the sedimentary patterns of the Maurice Ewing Bank. Here, in general, Lower Tertiary carbonate sediments give way to more diatomaceous Neogene sequences containing coldwater calcareous coccolith assemblages of very restricted diversity. This change in dominance is seen in the comparison of zones and datums recognized in the Ewing Bank sediments presented in Figure 4. Note particularly how the coccolith zones and datums in the Paleogene give way to a plethora of diatom and silicoflagellate zones in the Neogene.

The sudden northward shift in the Antarctic Convergence is seen in the change from carbonate to diatom ooze in nearly all cores which penetrate uppermost Miocene sediments. Where the record is quite complete, this shift may be followed by disturbed and reworked sediments of latest Miocene age containing ice-rafted sands deposited under the influence of an increasingly strong circum-Antarctic current. Finally, after some fluctuations, stratigraphic continuity is lost completely where the Miocene and older units are cut by a regional unconformity eroded by an intensified circumpolar current. Although multiple hiatuses are present in some sections, this current seems to have peaked in intensity just prior to the time of the Miocene/Pliocene boundary (between ~7.4 and 6.2 m.y.BP), and then diminished markely as climatic conditions ameliorated during the Early Pliocene and deposition in the area resumed from about ~4.5–3.9 m.y.BP (Ciesielski, 1978b). The net loss of sediment due to the Late Miocene erosional episode(s) can only be approximated, but mapping and coring along seismic profile lines led Ciesielski and Wise (1977) to consider losses in some locations of several tens of meters of section. A restored geologic section across the Ewing Bank based on their data is suggested in Figure 9.

Once an areally extensive set of piston cores are attained and dated by the smear-slide method, other types of studies are possible. For instance, a plot of lithologies and core depths below sea level against a subsidence curve for the Ewing Bank allowed a preliminary calcium carbonate curve to be constructed (Fig. 10; from Wise and others, 1978a). A comparison of this curve with others from lower latitudes (Fig. 11) shows a major excursion of the CCD in the Late Miocene which is not evident in the other curves. This excursion, of course, resulted from the sharp northward migration of the Antarctic Convergence near the end of the Miocene.

POINT FOR SPECULATION

The reason for the sudden increase in intensity of the circumpolar current near the end of the Miocene is an interesting point for speculation. Ciesielski and Wise (1977) assumed a direct causal relationship between the major regional unconformity and the increased severity of the Late Miocene Antarctic glaciation. The exact connection, however, was not clear. An attractive explanation has recently been advanced by Zinsmeister (1977, 1978) which can be applied in the region of the Falkland Plateau.

Zinsmeister (1978; see also Mercer, 1978) suggested that increasingly severe climatic conditions led to the development of the West Antarctic ice sheet toward the end of Miocene times, thus closing a relatively shallow sea (the "Shackleton Seaway") which existed between East Antarctica and the islands which comprised preglacial West Antarctica. Sea passages through the islands of the Antarctic Peninsula were probably also occluded by ice. In addition, the development of a major ice sheet in West Antarctica would have increased the amount of Antarctic Surface Water formed and thus shifted the Antarctic Convergence northward.

If Zinsmeister (1978) is correct, we can postulate that a primary effect of the formation of the West Antarctic ice sheet would have been a concentration of the flow of the circum-Antarctic current north of the Antarctic Peninsula and through the constricted Drake Passage. This could have suddenly increased the funnel effect of the current against the Ewing Bank and Islas Orcadas Rise, thereby causing wholesale erosion of sediment until Early Pliocene warming briefly opened up some of the island passages and perhaps the "Shackleton Seaway" again.

Following the Early Pliocene, renewed climatic deterioration led to the present cycles of Plio/Pleistocene glaciation. Since Southern Ocean tem-

←

FIG. 9.—Restored geologic section across the Maurice Ewing Bank of the eastern Falkland Plateau indicating the amount of strata that may have been eroded by the circum-Antarctic current during the culmination of the Late Miocene glaciations (after Ciesielski and Wise, 1977).

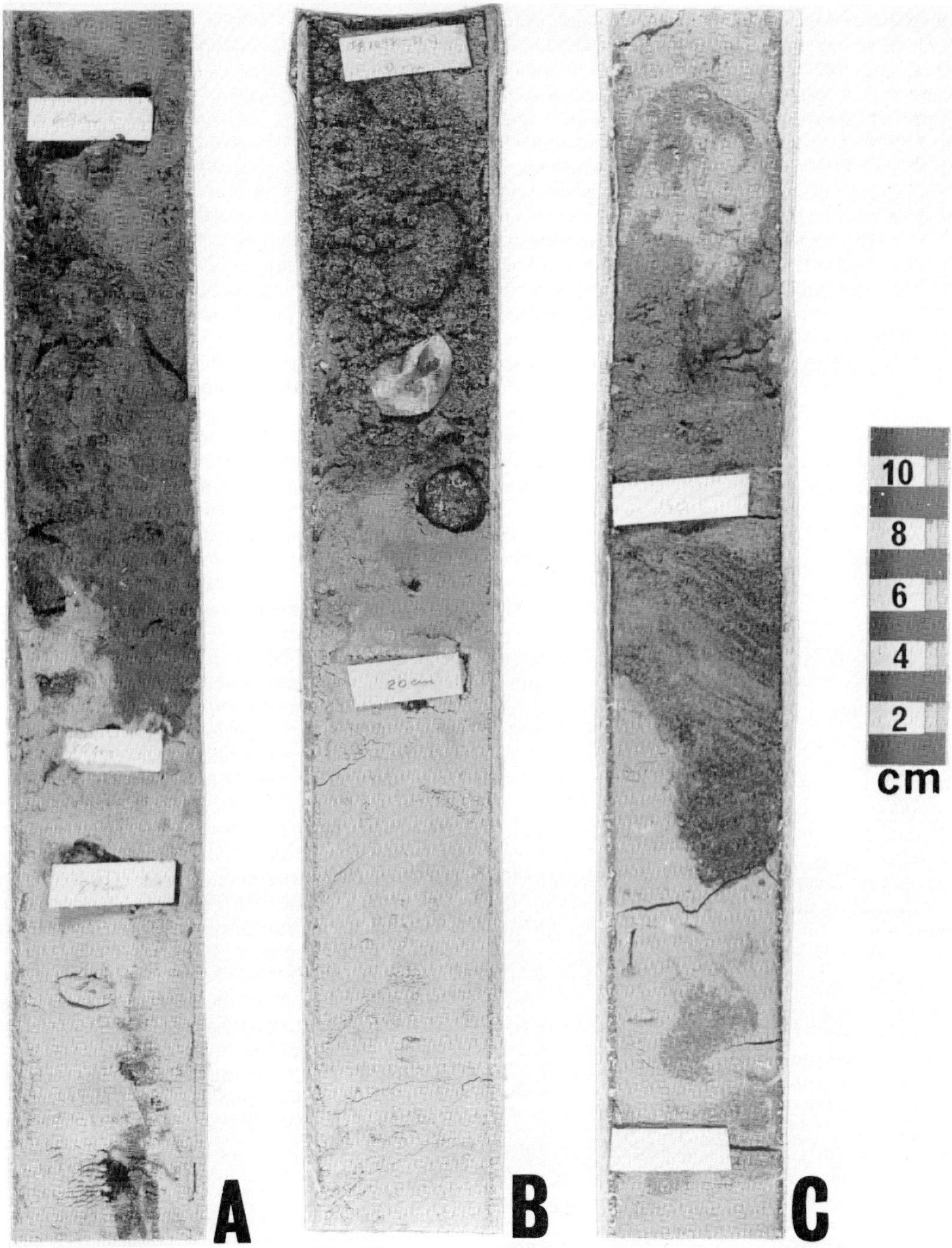

FIG. 10.—Erosional and depositional effects of intensified circum-Antarctic flow along the eastern Falkland Plateau during the Late Miocene glacial episode. *A*, *Islas Orcadas* core 16-32; erosional disconformity between Oligocene nannofossil ooze below (light colored) and Plio/Pleistocene glacial marine clastics above (dark). *B*, *Islas Orcadas* core 16-39; erosional contact between Upper Paleocene nannofossil ooze below (light) and Plio/Pleistocene glacial marine clastics above (dark). Note two glacial erratics near the contact. *C*, *Islas Orcadas* core 7-54 showing an erosional channel cut into diatom ooze. The channel is filled with laminated glacial marine sands deposited under strong current conditions. The diatom ooze is dated within the *Denticula hustedtii*/*Denticula lauta* Zone. Core originally described and dated by Ciesielski (1978a).

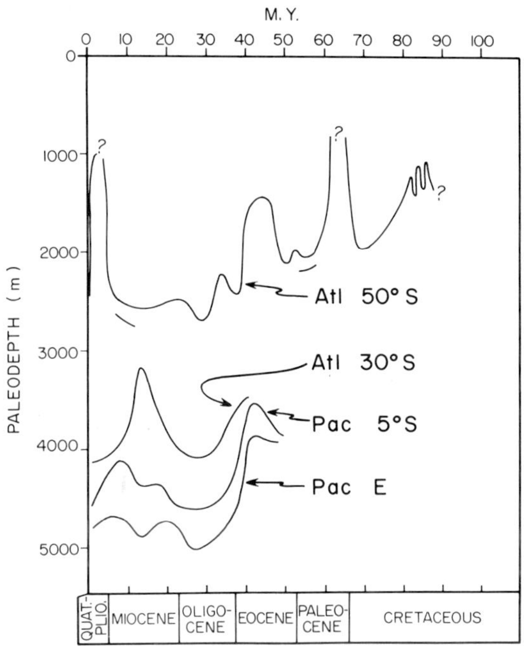

FIG. 11.—Comparison of calcium carbonate compensation curves for the lower latitudes of the Atlantic and Pacific (bottom three curves; from Berger, 1977) with a curve for the Ewing Bank area of the Atlantic sector of the Southern Ocean (top curve; from Wise and others, 1978a). In the top curve, the sharp upward excursion of the CCD near the end of the Miocene resulted from the rapid northward migration of the Antarctic Convergence.

peratures never again rose to levels reached during the Early Pliocene (Ciesielski and Weaver, 1974), the re-establishment of the West Antarctic ice sheet has probably maintained the strong funnel effect on the circum-Antarctic current through the Drake Passage down to the present. Wholesale erosion of the older pelagic sequences on the Ewing Bank and elsewhere in the region, however, seems to have been prevented to a large extent during the past three to four million years. This was accomplished by the build-up of the protective cover of erosion-resistant glacial marine clastics which blanket ("armor") the older sediments throughout most of the area (Ciesielski and Wise, 1977).

CONCLUSION AND EPILOGUE

When the writing of the foregoing discussion was completed, *Glomar Challenger* was in transit to begin what is expected to be her last visit to the area of the Falkland Plateau and its environs. Despite an ambitious drilling program planned to answer some of the unsolved questions which arose during her previous excursion to the area, one could not seriously hope that all major questions would be solved. A drilling program is limited when it still must operate largely in a reconnaissance mode due to the large sample gaps in this remote sector of the world's oceans. A comforting fact of history, however, is that whatever progress is made in establishing new reference sections in the area will open the door for further, meaningful exploration by other means, particularly via seismic profiling and piston coring.

ACKNOWLEDGEMENTS

The invitation for this paper carried with it a request to summarize DSDP activities in the Southern Ocean in which members of the Florida State University Antarctic Marine Geology Research Facility participated. It is a pleasure to acknowledge those who participated aboard ship as members of the FSU facility and on whose work a significant portion of the material for this summary is drawn: Lawrence A. Frakes, Elizabeth M. Kemp, David W. McCollum (DSDP Leg 28); Fred M. Weaver (DSDP Leg 35); Menno G. Dinkelman, Andrew M. Gombos, and the author (DSDP Leg 36). In addition, Paul F. Ciesielski, Karen L. Busen, and Frank H. Wind contributed shore lab reports for the Leg 28 and 36 *Initial Reports.* I thank Dennis S. Cassidy (Curator of the FSU Antarctic Facility), John Hattner, David Watkins, Rosemarie Raymond, and Louise Cox for assistance in preparing the manuscript. Melvin H. Miyajima, Fred M. Weaver, and Dennis Graham kindly provided three of the figures. The study was supported by NSF grant DPP 78-07183.

REFERENCES CITED

NOTE: All references to Deep Sea Drilling Project Initial Reports have been abbreviated; for complete citations see *Appendix I* at the end of this volume, or the individual Initial Reports as published.

ABBOTT, W. H., 1972, Vertical and lateral patterns of diatom ooze found between Australia and Antarctica: Ph.D. Dissertation, University of South Carolina.

BANDY, O. L., CASEY, R. E., AND WRIGHT, R. C., 1971, Late Neogene planktonic zonation, magnetic reversals and radiometric dates, Antarctic to the Tropics, Antarctic Res. Ser., Biol. Antarctic Seas, IV: Washington, Amer. Geophys. Union, v. 15, p. 1–26.

BARKER, P. F., AND BURRELL, JOHN, *in press,* The influence upon Southern Ocean circulation, sedimentation and climate of the opening of Drake Passage, *in* Craddock, Campbell, ed., Antarctic Geoscience: Madison, Wisconsin, Univ. of Wisconsin Press.

Berger, W. H., 1977. Carbon dioxide excursions and the deep-sea record: aspects of the problem, *in* Andersen, N. R. and Malahoff, Alexander, eds., The Fate of Fossil Fuel CO_2 in the Oceans: New York, Plenum Press, p. 505–542.
Berggren, W. A. and Haq, B. U., 1976. The Andalusian Stage (Late Miocene): biostratigraphy, biochronology and paleoecology: Palaeogeography, Palaeoclimatology, Palaeoecology, v. 20, p. 67–129.
Brady, Howard, 1979, The dating and interpretation of diatom zones in Dry Valley Drilling Project Holes 10 and 11 Taylor Valley, South Victoria Land, Antarctica, *in* Nagata Takesi ed., Mem. Nat. Inst. Polar Research Spec. Issue, No. 13: Tokyo, Japan, Yoshida Insatsu, Ltd, p. 150–175.
———, and Martin, Helene, 1979, Ross Sea region in the middle Miocene: a glimpse into the past: Science, v. 203, p. 436–437.
Bukry, David, 1973, Low-latitude coccolith biostratigraphic zonation: Init. Repts. DSDP, v. 15, p. 685–703.
———, 1975, Coccolith and silicoflagellate stratigraphy near Antarctica, Deep Sea Drilling Project Leg 28: Init. Repts. DSDP, v. 28, p. 709–723.
Burns, D. A., 1977, Major features of oceanographic development of the southeast Indian and southwest Pacific oceans interpreted from microfossil evidence: Marine Geology, v. 25, p. 35–59.
Busen, K. E., and Wise, W. W., 1977, Silicoflagellate stratigraphy, Deep Sea Drilling Project, Leg 36: Init. Repts. DSDP, v. 36, p. 697–743.
Ciesielski, P. F., 1975, Biostratigraphy and paleoecology of Neogene and Oligocene silicoflagellates from cores recovered during Antarctic Leg 28, Deep Sea Drilling Project: Init. Repts. DSDP, v. 28, p. 625–691.
———, 1978a, The Maurice Ewing Bank of the Malvinas (Falkland) Plateau: depositional and erosional history and its paleoenvironmental implications: Dissertation, Florida State University, 278 p.
———, 1978b, New evidence on the timing of late Miocene—middle Pliocene glacial fluctuations from the southwest Atlantic sector of the Southern Ocean: Abstracts with Programs, Geol. Soc. America, v. 10, p. 380.
———, and Weaver, F. M., 1974, Early Pliocene temperature changes in the Antarctic Seas: Geology, v. 2, p. 511–516.
———, and Wise, S. W., 1977, Geologic history of the Maurice Ewing Bank of the Falkland Plateau (southwest Atlantic sector of the Southern Ocean) based on piston and drill cores: Marine Geology, v. 25, p. 175–207.
Donahue, J. G., 1970, Diatoms as Quaternary biostratigraphic and paleoclimatic indicators in high latitudes of the Pacific Ocean: Dissertation, Columbia University.
Edwards, A. R., 1971, A calcareous nannoplankton zonation of the New Zealand Paleogene, *in* Farinacci, A. ed., Plankt. Conf. 2nd, Roma, 1970, Proc. v. 1: Tecnoscienza, p. 381–420.
———, and Perch-Nielsen, K., 1975, Calcareous nannofossils from the southern southwest Pacific, Deep Sea Drilling Project, Leg 29: Init. Repts. DSDP, v. 29, p. 469–539.
Frakes, L. A., 1975, Paleoclimatic significance of some sedimentary components at Site 274: Init. Repts. DSDP, v. 28, p. 785–787.
Geitzenauer, K. R., 1972, The Pleistocene calcareous nannoplankton of the Subantarctic Pacific Ocean: Deep-sea Research, v. 19, p. 45–60.
Gombos, A. M., 1977, Paleogene and Neogene diatoms from the Falkland Plateau and Malvinas Outer Basin: Init. Repts. DSDP, v. 36, p. 575–695.
Goodell, H. G., 1973, Marine sediments of the Southern Ocean, *in* Bushnell, V. C. ed., Antarctic Map Folio Series: New York, Amer. Geographical Soc., Folio 17.
Hayes, D. E. and Frakes, L. A., 1975, General synthesis, Deep Sea Drilling Project Leg 28: Init. Repts. DSDP, v. 28, p. 919–942.
Hays, J. D. and Opdyke, N. D., 1967. Antarctic Radiolaria, magnetic reversals, and climatic change: Science, v. 158, p. 1001.
Kemp, E. M., 1978, Tertiary climatic evolution and vegetation history in the southeast Indian Ocean region: Palaeogeography, Palaeoclimatology, Palaeoecology, v. 24, p. 169–208.
———, Frakes, L. A. and Hayes, D. E., 1975, Paleoclimatic significance of diachronous biogenic facies, Leg 28, Deep Sea Drilling Project: Init. Repts. DSDP, v. 28, p. 909–917.
Kennett, J. P., 1967, Recognition and correlation of the Kapitean State (Upper Miocene, New Zealand): New Zealand Jour. Geol. Geophys., v. 10, p. 1051–1063.
———, 1978, The development of planktonic biogeography in the Southern Ocean during the Cenozoic: Marine Micropaleontology, v. 3, p. 301–345.
Kennett, J. P., and Watkins, N. D., 1975, Late Miocene–early Pliocene paleomagnetic stratigraphy in New Zealand. Geol. Soc. America Bull., v. 85, p. 1385.
Loutit, T. S., and Kennett, J. P., 1979, Application of carbon isotope stratigraphy to late Miocene shallow marine sediments, New Zealand: Science, v. 204, p. 1196–1199.
Mayewski, P. A., 1972, Glacial geology near McMurdo Sound and comparison with the central Transantarctic Mountains: Antarctic Jour. U.S., v. 7, p. 103–106.
———, 1975, Glacial geology and late Cenozoic history of the Transantarctic Mountains, Antarctica: Inst. Polar Studies, Ohio State Univ., Rept. 56, 168 p.
McCollum, D. W., 1975, Diatom stratigraphy of the Southern Ocean: Init. Repts. DSDP, v. 28, p. 515–571.
Mercer, J. H., 1976, Glacial history of southernmost South America: Quaternary Res., v. 6, p. 125–166.
———, 1978, Glacial development and temperature trends in the Antarctic and in South America, *in* Van Zinderen, E. M., ed., Antarctic Glacial History and World Paleoenvironments: Bakker, p. 73–94.
Miyajima, M. H., 1975, Subantarctic region, southeast Indian Ocean: absolute chronology of upper Pleistocene

calcareous nannofossil zones and paleoclimatic history determined from silicoflagellate, coccolith, and carbonate analyses: Masters Thesis, Florida State University, 199 p.

RYAN, W. B. F., CITA, M. B., RAWSON, M. D., BURCKLE, L. H. AND SAITO, TSUNEMASA, 1974, A Paleomagnetic assignment of Neogene stage boundaries and the development of isochronous datum planes between the Mediterranean, the Pacific and Indian Oceans in order to investigate the response of the world ocean to the Mediterranean salinity crisis: Riv. Ital. Paleontol., v. 80, p. 631–688.

RUTFORD, R. H., CRADDOCK, C., WHITE, C. M. AND ARMSTRONG, R. L., 1972, Tertiary glaciation in the Jones Mountains, *in* Adie, R. J., ed., Antarctic Geology and Geophysics: Oslo, Universitetsforlaget, p. 239–243.

SHACKLETON, N. J. AND KENNETT, J. P., 1975a. Paleotemperature history of the Cenozoic and the initiation of Antarctic glaciation: oxygen and carbon isotope analysis in DSDP Sites 277, 279, and 281: Init. Repts. DSDP, v. 29, p. 743–756.

———, AND ———, 1975b, Late Cenozoic oxygen and carbon isotope changes at DSDP Site 284: Implications for glacial history of the northern hemisphere and Antarctica: Init. Repts. DSDP, v. 29, p. 801–807.

VAN COUVERNING, J. A., BERGGREN, W. A., DRAKE, R. E., AGUIRRE, E. AND CURTIS, G. H., 1976, The terminal Miocene event: Marine Micropaleontology, v. 1, p. 263–286.

WEAVER, F. M., 1973, Pliocene paleoclimatic and paleoglacial history of East Antarctica recorded in deep-sea piston cores: Florida State Univ., Sedimentol. Res. Lab. Contrib. 36, 142 p.

———, 1976, Late Miocene and Pliocene radiolarian paleobiogeography and biostratigraphy of the Southern Ocean: Dissertation, Florida State University, 175 p.

WEAVER, M. R., 1979, Late Quaternary paleoclimatic and paleoglacial history of the southeast Pacific subantarctic and Antarctic region: analysis of a glacial cycle: Masters Thesis, Florida State University, 141 p.

WEBB, P. N., 1972, Wright Fjord, Pliocene marine invasion of an Antarctic dry valley: Antarctic Jour. U.S., v. 7, p. 225–232.

———, RONAN, T. E., LIPPS, J. H., AND DELACCA, 1979, Miocene glaciomarine sediments from beneath the southern Ross Ice Shelf: Science, v. 203, p. 435–437.

WISE, S. W., AND WIND, F. H., 1977, Mesozoic and Cenozoic calcareous nannofossils recovered by DSDP Leg 36 drilling on the Falkland Plateau, Atlantic sector of the Southern Ocean: Init. Repts. DSDP, v. 36, p. 269–492.

———, JONES, S. C., AND PARODI, R., 1978a, Late Mesozoic-Cenozoic sedimentation on a high latitude oceanic plateau: role of the carbonate compensation depth and bottom current erosion: Abstracts, 10th International Congress on Sedimentology.

———, ———, CIESIELSKI, P. F., GEORGI, D. T., WOODROFFE, D. S., AND JACOBS, S. J., 1978b, Islas Orcadas cruise 16: Antarctic Jour. U.S., v. 13, p. 92–94.

ZINSMEISTER, W. J., 1977, The formation of the West Antarctic Ice Sheet and its effect on Miocene molluscan faunas of southern and central Chile: Abstracts with Programs, Geol. Soc. America, v. 9, p. 1240–1241.

———, 1978, Effect of formation of the West Antarctic Ice Sheet on shallow water marine faunas of Chile: Antarctic Jour. U.S., v. 13, p. 25–26.

SEPM SPECIAL PUBLICATION NO. 32, P. 489–504, OCTOBER 1981

THE DEEP-SEA RECORD: MAJOR STEPS IN CENOZOIC OCEAN EVOLUTION

W. H. BERGER, EDITH VINCENT, AND H. R. THIERSTEIN
Scripps Institution of Oceanography, La Jolla, California 92093 U.S.A.

ABSTRACT

The deep-sea record shows that certain changes in the climate-producing "exogenic" system, provoked through tectonic forces and amplified by internal feedback mechanisms, result in step-like transitions from one climatic state to another. Such steps are important as natural experiments, for the purpose of studying response characteristics of the system, and for establishing high-resolution stratigraphy on a global scale.

An important step-producing mechanism is the buildup of disequilibrium potential within transient reservoirs of hydrosphere, biosphere, or reactive lithosphere, which can collapse under certain conditions. Examples of transient reservoirs are ice caps, the water of isolated basins and their deposits, and readily accessible organic carbon accumulations ("dead" carbon sphere and biosphere).

Event analysis attempts to identify the step-producing mechanism. It considers the following questions: is the event part of a trend? Is it a reversible phenomenon, and does it initiate a more stable climatic regime, or a less stable one? The identification of the nature of positive feedback associated with an event is crucial for such analyses.

INTRODUCTION

A fundamental question that poses itself for the student of geologic history is whether the record represents a series of quasi-catastrophic events, as claimed by Georges Cuvier (1769–1832) or an unbroken chain of gradual change as proposed by Charles Lyell (1797–1875) and Charles Darwin (1809–1882). Darwin especially stressed the incompleteness of the record in defending his thesis of gradualism in the face of evidence for sudden change in the sedimentary strata exposed on land (Darwin, 1859). To test his idea, we must turn to regions where continuous sedimentation is most likely, that is, to the deep ocean floor.

The drilling vessel *Glomar Challenger* completed some 500 holes in eleven years of drilling. Results of paleoceanographic studies suggest that the evolution of circulation and climate is not entirely gradual, but is punctuated by periods of rapid change, which may be called "Climatic Events". Such events may or may not be associated with the classical stratigraphic boundaries established in marine sections exposed on land.

In the following we shall present the available evidence that such events actually occurred, and speculate on the associated geophysical-geochemical processes. Subsequently, we shall explore the idea that the stratigraphic record may be considered the output of an "exogenic system" with inputs from cosmic sources, endogenic processes, and feedback. We shall then introduce the concept of "stability" of this climate-producing system and argue that it is affected by the set-up and close-out of transient reservoirs which act as collapsible storage bins of geophysical energy. These concepts are then used to sketch a procedure for classifying events in the Cenozoic geologic record.

CRETACEOUS/TERTIARY BOUNDARY

The classic break in the stratigraphic record is the event which terminated the Mesozoic and started the Cenozoic. It is familiar to everyone as the time of extinction of the dinosaurs. Many other types of organisms did not survive into the Cenozoic. Among marine forms, well-known examples include nektonic cephalopods (ammonites, belemnoids), and among the shallow water benthos reef-dwelling echinoderms, corals, bivalves, and large benthic foraminifera (Béland and others, 1977).

Thierstein (this volume) presents evidence, based on detailed stratigraphy and quantitative population studies of calcareous nannofossils, which shows that this event was catastrophic and independent of a trend. This evidence supports earlier conclusions (Berggren, 1962; Luterbacher and Premoli-Silva, 1964; Bramlette, 1965) that the plankton extinction occurred within a short time, which makes gradualist explanations unattractive. Thus in this particular case, it would seem that Darwin's gradualism does not apply and Cuvier's catastrophism does.

Gartner and Keany (1978) and Thierstein and Berger (1978) proposed a sudden influx of low-salinity water to the world ocean from a previously isolated Arctic as a possible terrestrial mechanism that could account for the observed change. Two major lines of evidence for that hypothesis have been weakened, however, by the subsequent finding that the stratigraphic record in the North Sea, on which the hypothesis was originally based, is disturbed by slumping (Perch-Nielsen and others, 1979; Watts and others, 1980), and that the drastic oxygen isotopic shift in the carbonate fine-fraction reported by Thierstein and Berger (1978) is not recorded in the planktonic foraminifera (Boersma and Shackleton, 1979; Thierstein, unpubl. data).

Among the proposed notions of a cosmic disturbance (Schindewolf, 1954; Urey, 1973), the meteorite impact hypothesis has recently received strong support by the discovery of increased noble element concentrations found in the boundary clays by Alvarez and others (1980) and Smit and Hertogen (1980). The environmental and climatic effects of such a meteorite impact are far from being established, the claims ranging from a few years of total darkness and concurrent global cooling (Alvarez and others, 1980) to short-term heating and subsequent cyanide poisoning (Hsü, 1980).

Considerable freshening or strong warming of oceanic surface waters, or temporary shielding of sunlight, would all have a detrimental effect on the survival of planktonic organisms, and would result in significant instabilities in the exchange of CO_2 between ocean, biosphere and atmosphere. These instabilities, which are also reflected in the sporadic blooms of surviving planktonic algae (Thierstein, this volume), could lead to large-scale climatic fluctuation, affecting the survival of land organisms and hence transfer of carbon from land to the sea. Regardless of the ultimate cause, the beginning of the Cenozoic is marked by catastrophe, which is reflected in sudden isotopic changes (see also Boersma and others, 1979; Scholle and Arthur, 1980).

CENOZOIC STEPPED COOLING TREND

No such revolutionary event occurred in the Cenozoic. In fact, the events in the Cenozoic appear to be different in character: unlike the Cretaceous termination they are part of historical trends and do not come entirely unexpected. The course of events during the Cenozoic proper is charted by its isotope stratigraphy, a stratigraphy which we owe largely to the work of Douglas and Savin (1975) and Shackleton and Kennett (1975) (Fig. 1).

As is well known, the Cenozoic on the whole is marked by a cooling trend, which is clearly reflected in the trend toward more positive $\delta^{18}O$ values. What is not always appreciated is that this cooling trend is chiefly a high latitude phenomenon. The tropics are not affected very much. Immediately, this discrepancy suggests a clue to the reason for the trend: the cooling happens not because of a change in the sun's output, but because of terrestrial factors. The most likely cause which offers itself is that the land areas in high latitudes became covered with snow and ice, reflecting much of the incoming sunlight back into space. An overall regression provides more land areas (which by itself increases albedo, since land is about three times brighter than the ocean) and continental drift leads to thermal isolation of the Arctic Ocean and the Antarctic Continent, favoring buildup of snow (Hamilton, 1968; Donn and Shaw, 1977).

It will be noted that the trend does not represent a smooth, gradual evolution, but that it has steps, presumably positive feedback events resulting from the attainment of a threshold. It appears that these steps essentially reflect the phase transitions from water to ice, on various parts of the globe.

The single most important modulator of the climatic evolution is sea-level variation. The role of sea level in the temperature evolution on the planet —and hence the effect of Cenozoic regression on the overall cooling trend—is well demonstrated in a comparison of sealevel changes and the stratigraphy of stable isotopes and CCD-level fluctuations (Fig. 2). The oxygen isotope values represent the *anomalies* of the high-latitude trend shown in Fig. 1. These anomalies clearly tend to parallel the sea level variations, indicating that transgressions result in a warming of the ocean and an expansion of the tropical belt (see also Fischer and Arthur, 1977). The low stand of sealevel in the Oligocene and concurrent cooling of the planetary surface is especially noteworthy. This cooling may be largely an effect of the increased planetary albedo due to increased exposure of (light) land at the expense of (dark) ocean. Since a cool atmosphere can hold less moisture than a warm one, conditions of regional drought are favored during regression.

EOCENE-OLIGOCENE TEMPERATURE DROP

Among the steps within the cooling trend, there is one very marked one, which has been studied in detail by Kennett and Shackleton (1976, see Fig. 3). Unfortunately, the condition of the core material on which this curve is based is very poor, so that the time scale cannot be pinned down reliably. In any case, the event seems to have taken less than 500,000 years (Kennett and Shackleton calculate 75,000 to 100,000 years).

Near the time of the sudden change of oxygen isotopes, the deepsea benthic fauna changed drastically (Benson, 1975; Douglas and Woodruff, *in press*), suggesting that the abyssal ocean filled up with cold water. This event established the "psychrosphere" (Kennett and Shackleton, 1976). The Eocene-Oligocene $\delta^{18}O$ change, then, has to be a worldwide event, and this is borne out by data from the North Atlantic both from shallow waters (Buchardt, 1978) and from the deep sea (DSDP Site 398, Vergnaud-Grazzini and others, 1978).

What caused the sudden drop in $\delta^{18}O$ values, and the production of polar bottom waters? In principle, we can invoke two mechanisms to explain

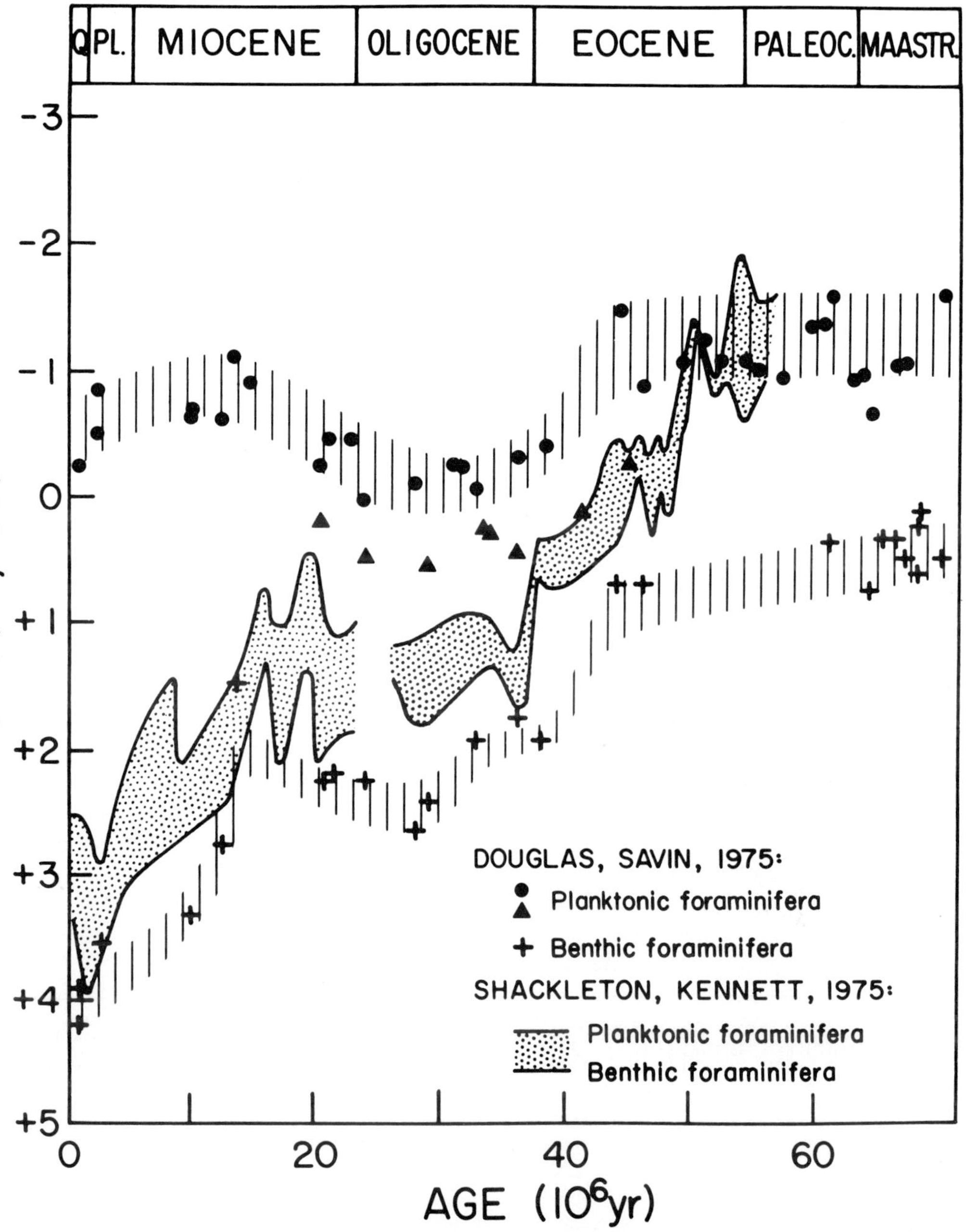

FIG. 1.—Oxygen isotopic composition of planktonic and benthic foraminifera in deep sea sediments. Data of Douglas and Savin (1975), central Pacific, and Shackleton and Kennett (1975), subantarctic Pacific.

changes in $\delta^{18}O$: changes in temperature, and changes in the composition of the water in which the shells in question were precipitated. The following possibilities apply: (1) A strong cooling of Antarctic coastal waters, which results in density increase, sinking, and production of cold bottom waters. (2) An increase in salinity (and therefore ^{18}O content) of the deep water, for example by moving a rainbelt off the Antarctic coast toward the tropics. When this belt moves away from the Antarctic shelf, the shelf waters will become heavier, and will have a better chance of sinking,

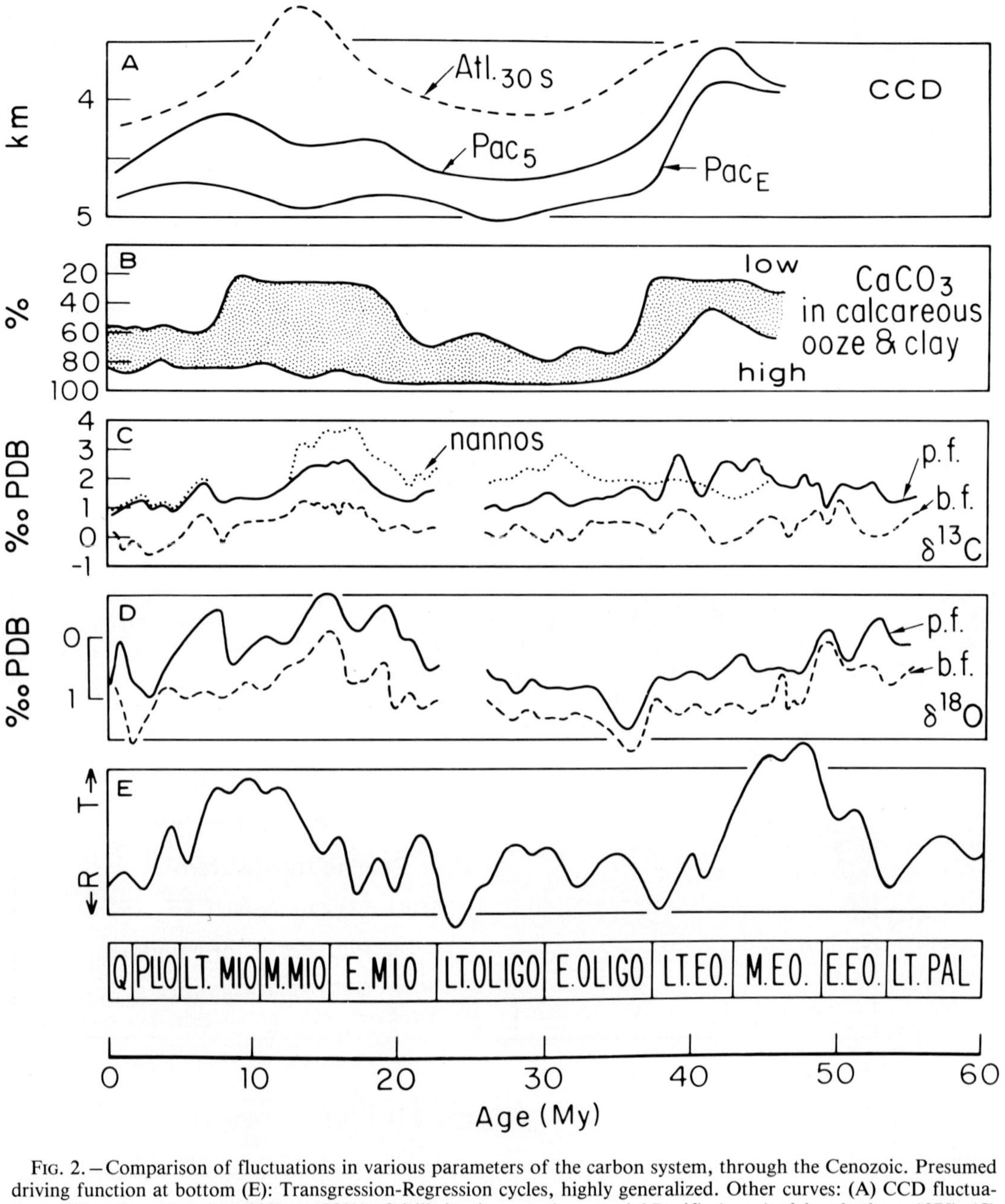

FIG. 2.—Comparison of fluctuations in various parameters of the carbon system, through the Cenozoic. Presumed driving function at bottom (E): Transgression-Regression cycles, highly generalized. Other curves: (A) CCD fluctuations; (B) carbonate scatter (20% to 100%) of drill sites in central equatorial Pacific (van Andel and others, 1975); (C) $\delta^{13}C$ in nannoplankton (Margolis and others, 1975) and in planktonic and benthic foraminifera (Shackleton and Kennett, 1975), DSDP Sites 277, 279, and 281; (D) $\delta^{18}O$ in planktonic and benthic foraminifera, DSDP Sites 277, 278, 281 (Shackleton and Kennett, 1975); plot shows the difference of the signals to the general trend to heavier oxygen isotopes (0.8‰ $\delta^{18}O/10^6$y for both p.f. and b. f. for the last 40 my). Note the correspondence between transgressions, shallow CCD (sequestration of $CaCO_3$ on shelves), abundance of low $CaCO_3$ values (increased dilutants), high $\delta^{13}C$ values (anaerobic C_{org} deposition), low $\delta^{18}O$ values (warming of ocean). From Berger (1977).

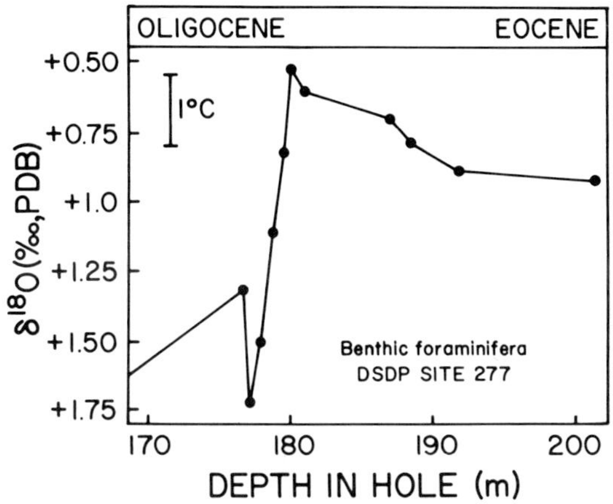

FIG. 3.—Oxygen isotope signal in benthic foraminifera at DSDP Site 277 (Subantarctic Pacific) across the Eocene-Oligocene boundary, after Kennett and Shackleton (1976). Note the indication for warming just before the "crash" to low temperatures, and the possible rebound. The temperature scale is given for comparison; it only applies if the water did not change composition (see text).

even without substantial cooling. (3) Buildup of a large amount of ice on Antarctica, extracting ^{16}O from the ocean, thus increasing the $\delta^{18}O$ values as observed. This latter possibility cannot be dismissed but appears unattractive because of lack of independent evidence for such buildup at that particular time.

The suddenness of the Eocene termination event is striking. Strong positive feedback appears to be involved, presumably from albedo change, and perhaps aided by extraction of CO_2 from the atmosphere by a cooling ocean. A marked drop in fertility appears to be indicated by the sharp lowering of the carbonate compensation depth at the end of the Eocene (van Andel and Moore, 1974; compare Fig. 2). The drop is surprising in that the formation of cold bottom-water is commonly associated with increased mixing and productivity (see Berger and Winterer, 1974, for discussion). The possibility that the opening of a previously isolated Arctic Ocean set off positive feedback mechanisms on a long-term cooling trend was discussed by Thierstein and Berger (1978). The role of the Arctic in climatic evolution merits more attention than it has been given, since the initial speculations of Ewing and Donn (1956), regarding the origin of ice ages.

THE MID-MIOCENE OXYGEN-ISOTOPE STEP

The regressive Oligocene, with its cool and dry climate, was followed by the transgressive Miocene. The transition between these two epochs is marked by increases in the temperature gradient between tropical and polar areas, in cold bottom water production, and in the overall fertility of the ocean. Basically, the tropics are darkened as a result of the Miocene transgression, but the whitening of high latitudes continues. In the mid-Miocene, the $\delta^{18}O$ values of calcareous plankton and of benthic foraminifera increase sharply (Fig. 1). This is generally interpreted as reflecting the buildup of a large ice cap in eastern Antarctica (Savin and others, 1975; Shackleton and Kennett, 1975). This buildup preferentially extracts $H_2{}^{16}0$ from the ocean, thus enriching it with ^{18}O.

The rapidity of the glaciation suggests positive feedback within the system. The warming of the tropics (transgression) and the growth of ice in high latitudes strengthens the temperature gradient, which strengthens wind transport of moisture from the tropics. It is certainly noteworthy that the buildup occurs during maximum transgression and maximum warming of the tropics, which is favorable for supplying moisture to the air.

THE TERMINAL MIOCENE EVENT

The latest Miocene was marked by two important events: (1) a carbon isotope shift marking the onset of the Messinian desiccation event, and (2) the Mediterranean isolation proper, which in turn led to a large salt buildup in the Mediterranean.

The Late Miocene carbon-isotope shift consists of a distinct change of the $^{13}C/^{12}C$ ratios in the shells of deep-sea benthic foraminifera in the Pacific (Keigwin, 1979; Bender and Keigwin, 1979; Keigwin and Shackleton, 1980), in the Indian Ocean (Vincent and others, 1980; see Fig. 4) and in the South Atlantic (Haq and others, 1980). The change is also recorded in planktonic foraminiferal shells from the tropical Indian Ocean (Vincent and others, 1980; see Fig. 4). The shift is toward lighter $\varrho^{13}C$ values, by about 0.8 ‰. It is paleomagnetically dated in Core RC12-66, and occurs within Magnetic Epoch 6 (Keigwin and Shackleton, 1980) at about 6.0 Ma (according to the time scale of LaBrecque and others, 1977, modified by Mankinen and Dalrymple, 1979; see Haq and others, 1980). The event lasted on the order of 200,000 years.

What process or processes can produce such a shift? In principle, there are two ways to shift the isotope ratios of the dissolved carbon in the ocean in the manner observed: (1) Extraction of ^{13}C into adjacent carbon reservoirs, or (2) additional supply of ^{12}C into the ocean, from such reservoirs. The present ocean's inorganic carbon pool has an isotope composition of about $\delta^{13}C = +0.5‰$ PDB (Craig, 1970), that is, near 0‰. The most likely reservoirs available for rapid exchange are the

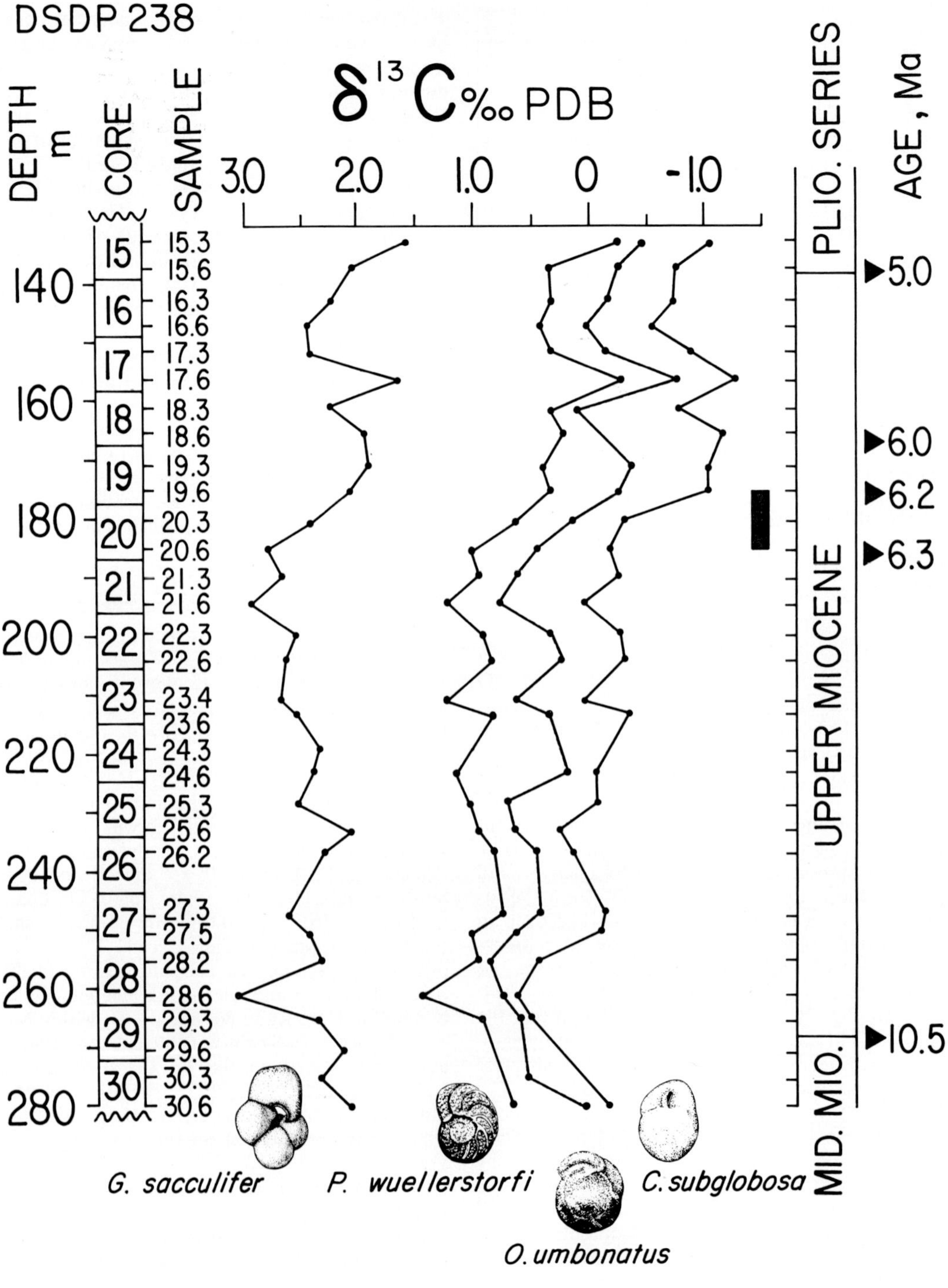

FIG. 4.—The Late Miocene "Carbon Shift" in the tropical Indian Ocean, recorded in benthic and planktonic foraminifera at DSDP Site 238. From left to right $\delta^{13}C$ in *Globigerinoides sacculifer* (planktonic), *Planulina wuellerstorfi, Cassidulina subglobosa, Oridorsalis umbonatus* (benthics). From Vincent and others (1980). Note the carbon isotope shift of approximately 0.8‰ toward lighter values upward in the section between sample 19–6 and 20–6 (vertical black bar). Ages follow the time-scale from Berggren and Van Couvering (1974) and Theyer and others (1978). According to time-scale revisions by La Brecque and others (1977) and Mankinen and Dalrymple (1979) the age of the carbon shift is slightly younger (about 6.0 Ma).

biosphere (mainly terrestrial plant matter, $\delta^{13}C \approx -20$ to -28‰), *"soil" carbon* derived from decay of land plants (including peat and similar deposits, $\delta^{13}C \approx -24$ to -30‰), *marine organic sediments* ($\delta^{13}C \approx -20$ to -26‰, depending on the terrigenous portion), *atmospheric CO_2* ($\delta^{13}C \approx -7$‰), *shallow-water cabonate* ($\delta^{13}C$ mostly from $+4$ to -3‰), and *marine limestones exposed on land* ($\delta^{13}C$ mostly between $+5$ and -5‰) (see Wedepohl (ed.), 1969). The isotopic composition of juvenile carbon may be taken as -6.4‰ (*ibid.*, p. 6–B = 15). The $\delta^{13}C$ of deep-sea carbonate ranges between $+2$ and -1.5‰ (Berger and others, 1978).

From the list of reactive carbon reservoirs and their $\delta^{13}C$ values, it is clear that exchange between carbonate pools and the ocean is of little consequence for this problem, but that we must identify processes which (a) accelerate input from organic carbon pools, or (b) reduce the output of carbon from the ocean into such pools. Input of juvenile carbon, through increased volcanism, and a change of the carbon balance between ocean and atmosphere, in favor of a cooling ocean, also are possible factors, as are internal fractionation processes within the ocean.

Clues to what processes produced the shift may be found in the following: (1) A sharp cooling, indicated in various parts of the world, including DSDP results (see reviews in Adams and others, 1977; Kennett and others, 1979; Keigwin, 1979). This cooling resulted both in a spin-up of ocean circulation and a strengthening of the thermocline, which resulted in a narrowing and intensification of upwelling mixing, as reflected in increased opal sedimentation in fertile regions. (2) A sharp rise in sedimentation rates, evident at many DSDP sites from low- to mid-latitudes (Vincent and others, 1980). Such a general rise can only mean that erosion rates were greatly increased. (3) Increased erosion rates indicate uplift on land, and regression of the sea, which favors a cooling of the climate.

Given these conditions of change, we can immediately deduce ways to accelerate input from organic carbon pools, and reduce output from the ocean to such pools. Erosion associated with regression will first remove organic-rich soil, including peat and related swamp-deposits. The amount of material available is difficult to estimate, but clearly exceeds one atmospheric carbon mass (1 ACM = 7×10^{17}gC) (Björkström, 1979). In addition, drought will decrease the biomass size which is roughly 1 AMC (humid regions: ≈ 400 tons dry organic matter per *ha*, semiarid regions: ≈ 100 tons per *ha;* see Rodin and others, 1975). Hence, at least 1 ACM of ^{13}C-depleted matter ($\delta^{13}C \approx -26$ ‰) will enter the ocean upon regression. The ocean contains about 50 ACM at $\delta^{13}C = 0$. Its composition, therefore, will change by approximately $-26‰/50 = -0.5‰$ for each ACM introduced. The output of the ocean to organic carbon pools could be reduced by increased oxygenation, due to increased bottom-water formation upon cooling. Also, the incorporation of phosphorus within the P-rich deposits of upwelling regions could decrease the overall output of organic carbon, because this output may be associated with a quasi-stoichiometric amount of phosphorus (see Broecker, 1971; Bender and Keigwin, 1979). Furthermore, increased erosion of igneous rocks (and perhaps marine sulfate-rich salts) could deliver the Ca-ion necessary to divert the output of carbon from the organic reservoirs to the carbonate reservoirs.

Input of organic carbon (or blockage of output) should decrease the carbonate saturation of the ocean, hence lead to increased dissolution of deep-sea carbonates. In fact, carbonate values in Core RC 12-66, from the central equatorial Pacific, drop from 90% to 50% (Saito and others, 1975) during the transition. This change represents a decrease in sedimentation rate of carbonate by a factor of nine (90:10 divided by 50:50, assuming constant supply of non-carbonate).

Given that the geochemical effects of regression and cooling can account for the *amplitude* of the shift, why did it happen so fast? And why was it not reversed when a new equilibrium was reached?

Answers to these questions are by no means obvious. It appears to us that we must consider the following possibilities: (1) The shift lies on a general trend, accelerated at the time by effects of increased erosion and aridity. Initially the eroded regions were rich in organic carbon, later less so. (2) The suddenness of the shift may be intimately connected to changes in deep and bottom water production. The newly isolated Mediterranean may have been an important source of heavy deep waters (Thierstein and Berger, 1978). At the same time, there is evidence that the North Atlantic established itself as a deep water source at the time, based on CCD drop and facies change (Berger and von Rad, 1972). Deep circulation changes have been invoked by Bender and Keigwin (1979) and by Vincent and others (1980) as a factor in producing the $\delta^{13}C$ shift: "old" deep waters have a lower $\delta^{13}C$ value than "younger" ones. It is not clear how this information can be transmitted to *planktonic* foraminifera, however.

The Messinian Salinity Event is perhaps one of the most spectacular discoveries of the JOIDES Deep Sea Drilling Project. The record suggests that the Mediterranean dried up at the very end of the Miocene, providing the setting for a gargantuan "Gibraltar Water Fall" 15 times the size of the Victoria Falls of the Zambesi (Hsü and others, 1973). The isolation of the basin started in the early Miocene at its eastern end and culminated in the deposition of enormous amounts of evaporites, in the Messinian. One corollary of this accumulation is a substantial decrease in the salt content of the ocean. Ryan (1973) suggests a drop by 6%; this drop was invoked by Ryan and others (1974) as facilitating ice buildup in the Antarctic. Another implication is that the basin must have acted as a pulsing source of brine to the world ocean, with noticeable effects on ocean stratification, chemistry, and climate (Thierstein and Berger, 1978). Yet another corollary is that a dry Mediterranean (and surrounding drought) results in greatly increased albedo in a sensitive climatic belt—that is, one in which changes from forest to desert and back can proceed rather quickly. Finally, the existence of an isolated Mediterranean that dries up sporadically implies a potential for producing rapid changes in sea level, by about 10 meters, which readily translate into climatic oscillations through albedo feedback. In fact, our oxygen isotope data (Vincent and others, 1980) suggest that the onset of the Messinian Salinity Event marked the change from a climatically quiet interval to one with considerable variability.

It is reasonable to suppose that both changes in geography at the Iberian Portal and a drop in sea level (van Couvering and others, 1976) were responsible for onset of the isolation necessary for salt buildup. Estimates for the timing of this event (Ryan and others, 1974; Adams and others, 1977; Cita, 1979) indicate that it is coeval with the Epoch-6 Carbon Shift. A brief period of strong carbonate dissolution is indicated near the latest Miocene, throughout the global ocean (and apparently even in shallow regions). This dissolution pulse has been related to the Messinian Salinity Event (Ryan and others, 1974; Thierstein and Berger, 1978). Presumably, the wholesale sequestration of carbonate (as well as Ca-sulfate) in the Mediterranean led to undersaturation of ocean waters everywhere else. As mentioned, a large-scale addition of organic carbon to the ocean, from erosion during climatic oscillations, may have to be considered also as a factor influencing carbonate dissolution.

The Messinian Salinity Event was followed by restoration of normal marine conditions at the base of the Pliocene (van Couvering and others, 1976). For the next two million years, things returned to "normal"—but the last great east-west passage, the Panama Straits, kept on closing.

MID-PLIOCENE ICE AGE EVENT

Antarctica, as a continent centered on a pole, was bound to accumulate snow at some point along the overall cooling trend (presumably at the end of the Eocene) and to build thick ice deposits as soon as the thermal gradient and atmospheric moisture transport were favorable (presumably in the Mid-Miocene). In the sub-Arctic regions, the ice arrived much later, in the Mid-Pliocene about three million years ago, and with little warning (Fig. 5). Deep erosion of the northern land masses and rapid transfer of terrigenous material to the northern ocean basins resulted from the glaciations.

What process or processes triggered this step? On land, several possibilities can be envisaged (see e.g., Emiliani and Geiss, 1958; Emiliani, 1978). The rise of mountain ranges during the Pliocene could have

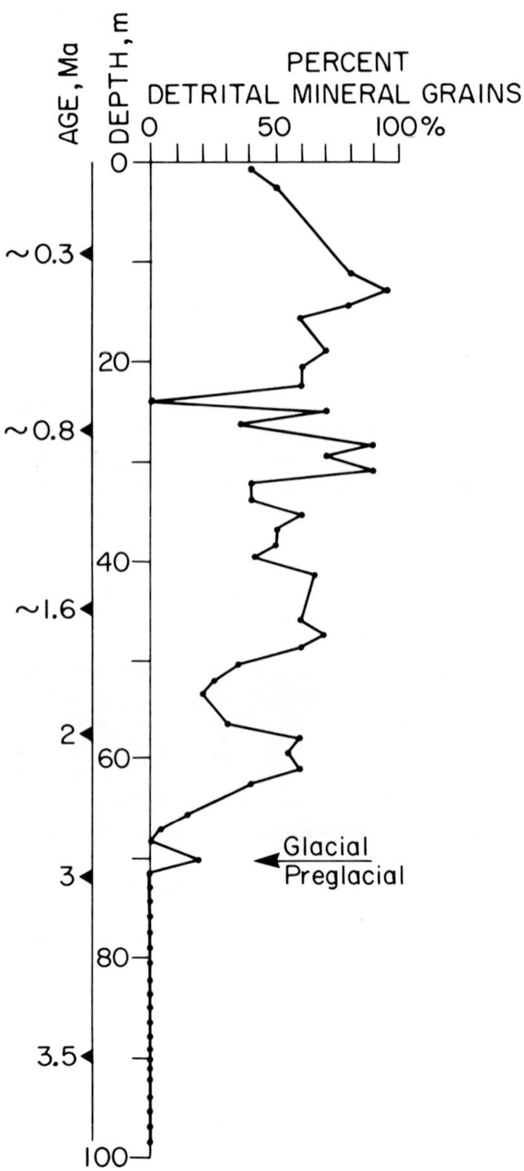

FIG. 5.—Onset of the Northern Hemisphere Ice Age seen in abundance of detrital mineral grains at DSDP Site 116. From Berggren (1972).

reached a critical level. Snow-covered and tree-barren highlands might have increased the albedo to the point where cooling and snow buildup re-enforced each other. In the ocean, the closure of the Panama Straits about three million years ago (Keigwin, 1978) would have obstructed the westward flow of Caribbean surface waters. These waters, then, would have been available for northward transport in the Gulf Stream System. The result would have been a sharp intensification of the Icelandic low pressure region and an increase in atmospheric moisture in high latitudes, due to evaporation from the warm ocean current. In turn, this moisture would have been available for building up ice. A strong positive feedback can be postulated for such a system: increased albedo in high latitudes increases the temperature gradient, which increases wind and ocean currents, and the supply of moisture. It is this type of positive feedback that we believe is a crucial factor in producing the step-like events of both the Mid-Miocene and the Mid-Pliocene.

The onset of the Ice Age was reflected in the deep-sea record in more than one way. A drastic change in the climatic regime, from quiet to highly variable, is seen in both the carbonate record (Saito and others, 1975) and in oxygen isotopes (Shackleton and Opdyke, 1977). The carbonate record (from the east-central equatorial Pacific) is intriguing in that it shows a distinct minimum three million years ago ("GU3"), indicating a very strong dissolution pulse. Why this coincidence?

As mentioned, dissolution can be produced in several ways: (1) introduction of organic carbon, by erosion, (2) introduction of CO_2 through "aging" of the deep water, (3) transfer of carbonate to other sites of deposition. All these factors may be at work: (1) initial glacial erosion products would be rich in carbon; (2) the Atlantic, through isolation, becomes truly anti-estuarine and sends "old" water to the Pacific while (3) collecting carbonate through the same mechanism of "basin-basin fractionation".

Why does the carbonate minimum not persist? The increased continental erosion, which is the hallmark of northern glaciation, greatly accelerates transfer of carbonates from the exposed land to the deep sea: thus, sedimentation rates increase drastically and the CCD is depressed, on a global scale (see van Andel and others, 1977; their Figure 22).

The benthic carbon isotope record in the Pacific DSDP Site 157 (Keigwin, 1979) shows a shift of 0.5‰ toward lower values at the three million year event. Presumably, this is due to the influx of "soil" carbon at the onset of large-scale erosion, as well as a further strengthening of Atlantic bottom water production, analogous to arguments in connection with the Late Miocene Carbon Shift (Vincent and others, 1980). Remarkably, the planktonic values (*Orbulina,* Site 157, Keigwin, 1979) embarked on a trend toward heavier carbon values at that time. This may be interpreted as a strengthening of the thermocline, allowing depletion of the overlying mixed layer in the ^{12}C isotope (which is incorporated preferentially into organic matter and hence leaves this layer as fecal pellets and other particles; see Kroopnick and others, 1977).

THE PLEISTOCENE SAWTOOTH PATTERN

The last million years—and here we leave deep-sea drilling and turn to piston cores—have been characterized by an overabundance of closely-spaced steps, whose significance is as yet obscure. Contemplating the standard isotopic curve of Emiliani (Fig. 6), we note the peculiar "sawtooth" pattern emphasized by Broecker and van Donk (1970). A maximum glaciation—always ending at the same level of intensity—is almost invariably followed by a rapid deglaciation culminating in a peak warm period. There follows an overall increase in ice cover.

The trend toward increasing glaciation on the gentle side of the sawtooth ("gradual buildup") is readily "explained" by positive albedo feedback, that is, by increasing amounts of reradiation from a

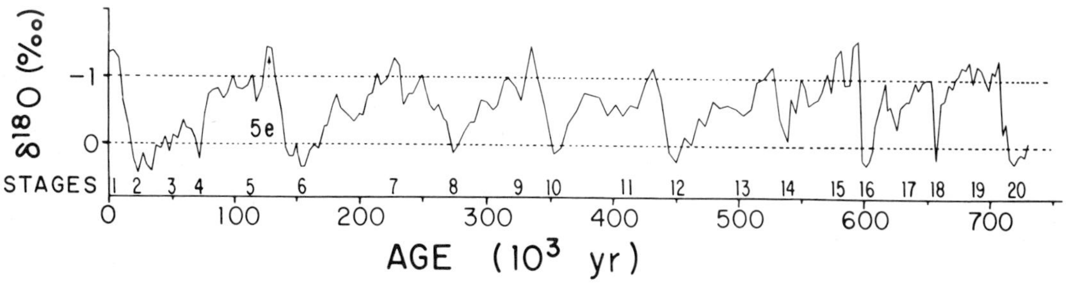

FIG. 6.—Composite $\delta^{18}O$ curve of Emiliani (1978), showing sawtooth pattern and constancy of O-18 maxima and minima. The rapid melting after most ^{18}O maxima suggests a "runaway" effect, once melting starts.

whitening Earth. The rapid deglaciation is more difficult to explain.

How does the deglaciation get started? What feedback processes sustain it? Why is it so rapid? These questions are being addressed by studying box cores from deep-sea sediments, containing the record of the last deglaciation.

Detailed statigraphic analysis of box core material (Berger and others, 1977) suggests that the sought-for positive feedback for deglaciation may be contained in the buildup of a low-salinity layer on top of the ocean, once deglaciation starts, as proposed by Worthington (1968). Once such a layer is established, it would greatly facilitate the transport of heat from the tropics to high latitudes, since tropical surface waters would no longer be cooled from below. Such transport would induce additional melting and thus maintain the low salinity layer—a typical case of positive feedback.

Climatic instabilities, due to the disrupted CO_2 exchange between ocean and atmosphere, would accompany the establishment of a low-density "lid" on the global ocean: the atmosphere would no longer "see" the deep ocean, except at intervals of sporadic mixing.

Our increasing burning of fossil fuels constitutes a global experiment which is now being carried out and which may eventually test some of these ideas (Fig. 7).

THE "EXOGENIC SYSTEM" AS A CYBERNETIC MACHINE

We have enumerated some unusually rapid changes in the ocean-atmosphere system that occurred during the last 70 million years, and we have

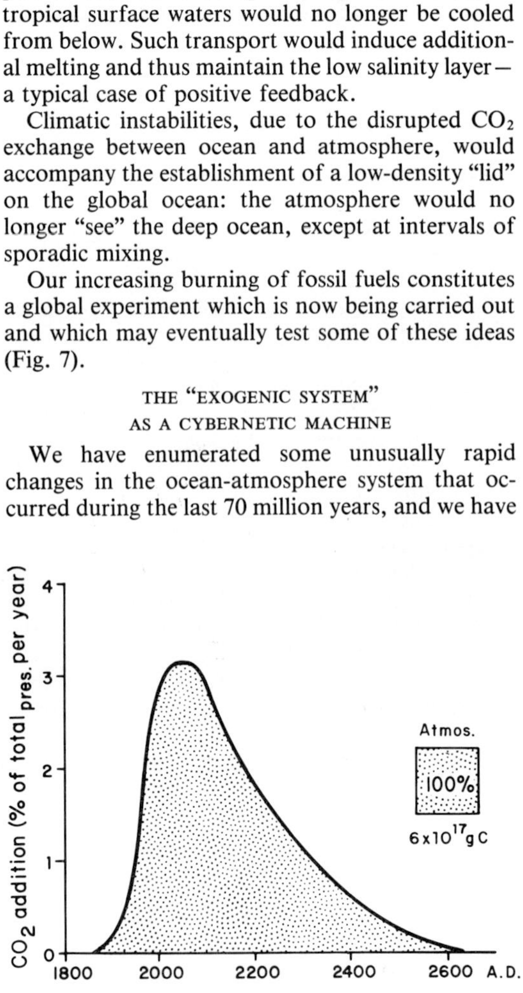

FIG. 7.—Projected input of industrial CO_2 to the atmosphere. From Berger (1977), based on data from Hubbert (1967). The CO_2 increase in the atmosphere will presumably lead to global warming through the "greenhouse effect", and may lead to the development of a "meltwater lid" from melting of polar ice.

attempted to show, with examples, how the record left by such events might be used to gain knowledge about the workings of the "black box" which is the climate-producing ocean-atmosphere-biosphere (reactive)-lithosphere system or briefly: the "exogenic system".

In order to identify the general features that apply to all these events, it is useful to look upon the stratigraphic record as the output from a cybernetically-controlled machine (Scientific American, 1955) with feedback loops and with inbuilt instabilities (Fig. 8).

The exogenic system may be thought of as consisting of three main subsystems: "geologic setting", "climate" and "exogenic production". The "geologic setting" of the sites of the active reservoirs and their interconnections is the "plumbing" so to speak. "Climate" represents the valves and parts; it is the "regulator". Essentially "climate" is the highly variable part of the basic system, which controls the short-term performance of the machine, while the limits are set by the "geologic setting". The third part is the "exogenic production" subsystem, which writes the geologic record.

Feedback, largely as albedo changes and as changes in reservoir sizes, comes from the "exogenic production" subsystem to both "geologic setting" and "climate". Clearly, "climate" is highly responsive to such feedback, while "setting" tends to integrate and dampen the feedback signals. In other words, on the whole the "setting" will not respond to high-frequency stimulation, while "climate" will.

The significance of major events (we think) is that the "geologic setting" suddenly acts as a highly responsive subsystem. This responsiveness can be accomplished by introducing *transient reservoirs* into the geologic setting. When such reservoirs collapse, they greatly augment the disequilibrium present within the system. In fact, they act like a step input to "climate" and "production."

In systems analysis, the character of a (linear) system is uniquely defined by its response to a step or an impulse function (Dransfield 1968). The response to such an input consists of a *transient* and a *steady-state* part (Fig. 9). The steady state part reflects the input, but the transient part reflects properties of the system: the rapidity of response to disturbance, the degree of damping, the ability to sustain disequilibrium.

Of special interest is a system that oscillates after being disturbed. Fundamentally, such oscillation originates from temporary energy storage within parts of the system, such that disequilibrium is maintained by passing energy from one reservoir to another. (Consider the conversions between potential and kinetic energy in standard mechanical systems, or conversions between magnetic and electric fields in circuits.) In the exogenic system,

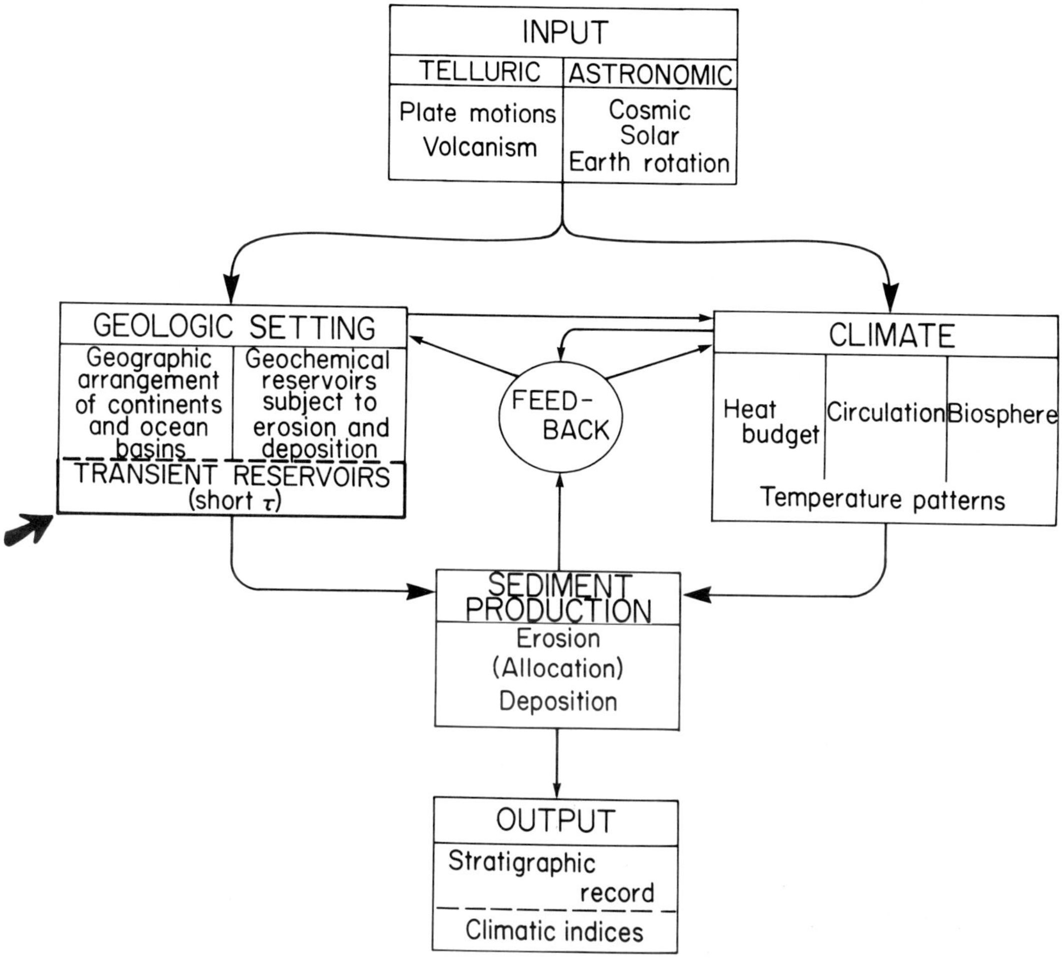

FIG. 8.—The exogenic system as a cybernetically controlled machine, with variable input from external sources and from feedback from climatic variations, and output in terms of sediment patterns, including climatic indices. Instability is introduced by isolated temporary storage bins (arrow) whose contents become available suddenly, mainly through endogenic changes in geography (opening of a strait, rapid regression).

disequilibrium "energy" can be passed back and forth between the hydrosphere and the reactive carbon sphere, for example. In the hydrosphere, disequilibrium potential can be stored by taking water from the ocean to make ice, or by sequestering large fresh water bodies or saline water bodies. In the reactive carbon sphere, potential can be stored by disequilibrium partitioning of CO_2 between ocean and atmosphere, or by buildup of transient carbon reservoirs (biosphere, "dead" carbon). In the exogenic lithosphere, potential can be stored in the temporary deposits of isolated basins (Holser, 1977).

Is there any evidence that oscillations actually occur in the wake of a "step"? There is indeed. The most remarkable example is the last deglaciation event, when the climate fluctuated greatly between warm and cold extremes, from about 15,000 to 9,000 years B.P., with a period of roughly 2,500 years (see Flint, 1971).

Also, there is evidence in at least three of the steps discussed earlier, that fluctuations in the output signal were greater after the event than before (Cretaceous Termination, Epoch-6 Carbon Shift, Mid-Pliocene Ice Age Event). However, we cannot be confident that such fluctuations actually reflect an oscillatory response rather than an increase in the instability of the system. For example, the progressive isolation of the Mediterranean at the end of the Miocene introduced a new transient reservoir which could be used repeatedly to produce step disturbances.

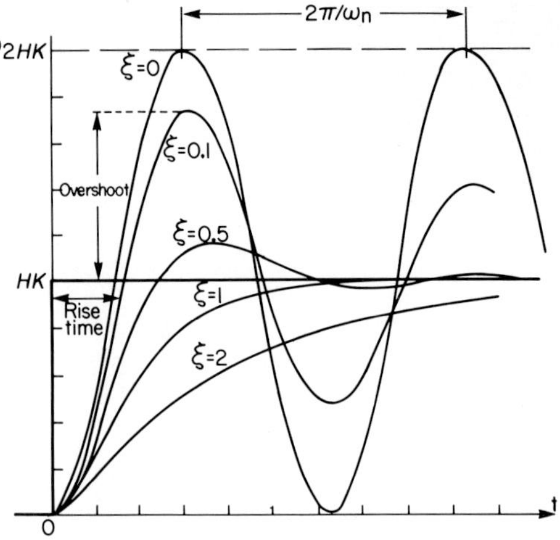

FIG. 9.—Response of a linear second-order system to a step input (HK). The term ξ indicates the amount of damping. It, and the inherent oscillation frequency W_n, defines the system. From Dransfield (1968).

STABILITY AND TRANSIENT RESERVOIRS

A system is said to be stable if it returns to steady-state after a disturbance, within time limits set by the definition. We need not here adopt a strict definition, because we are only interested in *changes* in stability. Clearly, a system that is prone to large and rapid oscillations is less stable than one that is not, and the same is true for a system whose output is characterized by sequences of steps.

On the whole, therefore, the Quaternary is the least stable system of the Cenozoic, and we must ask why this is so. We could simply point to the albedo feedback on a partially glaciated world, and consider this question answered. As the snow- and ice-cover increases, so does the sensitivity of the system to small variations in the forcing functions. However, other factors also enter the problem of transition from stable to less stable regimes, as mentioned, and they need to be addressed explicitly.

In an earlier paper on "events" in the record (Thierstein and Berger, 1978), we have argued that it is possible to view many known stratigraphic events as special cases of a general phenomenon: the rapid merging of previously isolated water reservoirs. This is certainly true for deglaciation, during which addition of water from ice led to a rapid rise of sea level, and, quite possibly, to salinity stratification (à la Worthington, 1968). A similar rapid demise of a fresh-water reservoir may well apply to the termination of the Cretaceous ("Arctic spillover", Gartner and Keany, 1978; Gartner and McGuirk, 1979), as well as for many other terminal events in the Phanerozoic (Berger and Thierstein, 1979).

We now generalize this idea of reservoir buildup and collapse, to include carbon. We suggest that the "exogenic system"—ocean, atmosphere, biosphere, and reactive lithosphere—stores climatic "energy" by separating water or carbon or both from the global ocean pool, into isolated reservoirs which remain potentially reactive. In the case of water, these reservoirs are temporarily isolated basins such as the Arctic, the Mediterranean, the proto-Atlantic basins, or large continental ice masses. For organic carbon, it is especially the "dead" carbon pool ("soil carbon") provided by peat deposits in lowland swamps, and by other loose organic-rich surface sediments, including deposits in oxygen-deficient marginal basins. The biosphere also can build up and release carbon readily (Shackleton, 1977), although amounts are modest compared with the ocean's carbon reservoir. Other large reactive carbon reservoirs are the uppermost few decimeters of deep-sea carbonates (Peng and Broecker, 1978) and, to some degree, shallow-water carbonates.

We suggest that the sudden release of water or carbon from such reservoirs sets off chain reactions with both negative and positive feedback. In the case of water, for example, the release of large amounts of salt water may lead to stable stratification near the ocean bottom, to anaerobic bottom conditions, to nutrient trapping at depth, and to large-scale extinction at the sea floor. Similarly, large-scale release of fresh water may lead to stable stratification in upper waters, to oxygen depletion, nutrient trapping below the wind-stirred layer, and to extinction of surface organisms. In either case, carbon would be able to *enter* the oceans as particles, but not to *leave* it (as gas). Therefore, CO_2 content of the atmosphere should drop. Cooling of the atmosphere should then increase convective mixing of the ocean, releasing CO_2, which raises temperatures again.

In the case of carbon release, the sudden large-scale availability of reactive organic carbon, at the end of a high sea-level stand, is of prime importance. The erosion of shallow marine carbon-rich sediments, the destruction of peatlands, swamps and marshes in the coastal plains, and the removal of "dead" carbon by increased soil erosion in a climate with decreasing precipitation, all could conceivably deliver enough CO_2 to acidify the ocean, and thus lead to increased concentrations of CO_2 in the atmosphere. The attendant warming would initially oppose the cooling expected from increased albedo produced by exposing land (and destroying vegetation). The opposing factors of albedo and atmospheric pCO_2 are not necessarily in phase and thus can foster instability of the system. Hence we might expect rapid fluctuations in climate and high erosion rates during regression.

Large-scale carbon release also would affect the carbonate saturation of the ocean. The nature of major "erosional events" on the deep sea, listed by van Andel and others (1977) must be contemplated under this aspect. Are large hiatuses, produced by carbonate dissolution pulses, provoked by step-release of organic carbon? Interestingly, the "erosion events" occur close to the major steps here listed: van Andel and others (1977) suggest the time spans 2–3 m.y., 4–6 m.y., and Late Eocene-Early Oligocene, among others.

Our "buildup-and-collapse" hypothesis is equally applicable to phosphorous (large amounts of phosphate can be stored in shallow-water basins, on shelves, and in "soil") and to calcium (sulfate *versus* carbonate storage), which are intimately involved in biogenic sedimentation, and hence in geochemical processes relevant to the stratigraphic record. Fischer's (1964) hypothesis of brine buildup-and-release as a mechanism for Permian extinction is another striking example of the viewpoint here advocated. This hypothesis also incorporates the sulfide buildup-and-release in large-scale systems postulated by Holser (1977).

CLASSIFICATION OF EVENTS

Using the concepts of input, stability, transient reservoir, reservoir collapse, and positive feedback, we now have the tools for a systematic classification of events.

The first thing one would like to know about an event is to what degree it was forced on the system, and to what degree it represents instability within the system itself. The temptation is to assign external forcing to each event: a cosmic catastrophe for the Cretaceous Termination, closing (or opening) of straits for the Eocene and Miocene Terminations, and for the Mid-Pliocene onset of northern ice ages.

The closing and opening of straits, although doubtlessly important (Berggren and Hollister, 1977) takes a considerable amount of time. Thus, the assignment of an event to a particular stage in the status of a strait must necessarily be arbitrary. We must still ask why an event occurred at that particular time, not sooner or later.

Instead of matching an (ill-defined) geophysical event with a (well-defined) stratigraphic one, it is operationally easier to ask whether the effects of an event are in the direction of a trend or not.

If yes, we have a *trend-event,* and we must look for (stepped) positive feedback. In essence, all cooling steps of the Cenozoic are of this nature, and the obvious feedback is the phase transition from water to snow in high latitude regions (albedo). The less obvious feedbacks are desertification (albedo), atmospheric pCO_2-decrease by increased weathering and by increased carbonate precipitation under disequilibrium conditions (greenhouse), decreased atmospheric moisture (greenhouse), and greening of the ocean due to increased fertility (albedo). The possible role of volcanism as an external forcing factor is not clear; it may simply be associated with mountain building and regression, which in turn are responsible for the trend.

In the case of a *non-trend event,* which is not on a long range trend or whose effects are opposite to this trend, we must look for external forcing or internal reservoir collapse. The Cretaceous Termination is of this type. In the Cenozoic, since the trend is toward *cooling,* events associated with *warming* are prime candidates for non-trend events, hence all deglaciation events are likely to belong in this category.

Another thing one would like to know about an event is whether it is *reversible.* If yes, it represents a transition between various possible states of the system. It is a *transition or "flip-flop" event.* If no, it represents a transformation of the system itself, that is, an addition or deletion of reservoirs or of the paths of circulation. It is a *transform event.*

In the case of a transform event, it is crucial to know whether stability was increased or decreased. If decreased, it is a *destabilizing* event, because it introduces new oscillation and step potential. If increased, it is a *stabilizing* event. Generally, the closing of an isthmus is likely to be a destabilizing event, the opening of a strait a stabilizing one.

EVOLUTION OF SYSTEM STABILITY

On the whole, the breakup of Pangaea during the Mesozoic produced new opportunities for semi-isolated basins (recognizable from their salt deposits) and had, therefore, a destabilizing influence. It increased the likelihood for steps during the Mesozoic. The closure of the Tethys and the isolated position of the Arctic maintained this potential during the Cenozoic despite the stabilizing influence of the subsequent drifting apart of the continents, which allowed the evolution of a well-mixed world ocean.

The role of the ice versus water albedo feedback in controlling the production of the sedimentary record kept increasing during the Cenozoic, as the continents emerged. The role of the fixed-carbon gaseous-carbon feedback became relatively less important, although it probably remained strong in a subservient role to albedo. The relative importance of the two feedback mechanisms may be gauged from the importance of oxygen isotope fluctuations and of carbon isotope fluctuations in the various parts of the record.

ACKNOWLEDGMENTS

Our research into the nature of events in the

deep-sea record is supported by the National Science Foundation (Oceanography Section), grants OCE78-25587, OCE76-22150, OCE79-19090 (CENOP), and by the Office of Naval Research (USN N00014-75-C-0152).

This paper constitutes a contribution to the International Geological Correlation Program, Project No. 96 (Messinian Correlation, M. B. Cita, project leader).

REFERENCES CITED

NOTE: All references to Deep Sea Drilling Project Initial Reports have been abbreviated; for complete citations see *Appendix I* at the end of this volume, or the individual Initial Reports as published.

ADAMS, C. F., BENSON, R. H., KIDD, R. B., RYAN, W. B. F., AND WRIGHT, R. C., 1977, The Messinian salinity crisis and evidence of Late Miocene eustatic changes in the world ocean: Nature, v. 269, p. 383–386.

ALVAREZ, L. W., ALVAREZ, W., ASARO, F., AND MICHEL, H. V., 1980, Extraterrestrial cause of the Cretaceous-Tertiary extinction: Science, v. 208, p. 1095–1108.

BÉLAND, P., AND OTHERS, 1977, Cretaceous-Tertiary extinctions and possible terrestrial and extraterrestrial causes, Syllogeus No. 12: National Museums of Canada, Ottawa, 162 p.

BENDER, M. L., AND KEIGWIN, L. D. JR., 1979, Speculations about the upper Miocene change in abyssal Pacific dissolved bicarbonate $\varrho^{13}C$: Earth Planet. Sci. Letters, v. 45, p. 383–393.

BENSON, M. L., 1975, The origin of the psychrosphere as recorded in changes of deep-sea ostracode assemblages: Lethaia, v. 8, p. 69–83.

BERGER, W. H., 1977, Carbon dioxide excursions and the deep sea record: aspects of the problem, *in* Andersen, N. R., and Malahoff, A., eds., The Fate of Fossil Fuel CO_2 in the Oceans: New York, Plenum Press, p. 505–542.

———, AND THIERSTEIN, H., 1979, On Phanerozoic mass extinctions: Naturwissenschaften, v. 66, p. 46–47.

———, AND VON RAD, U., 1972, Cretaceous and Cenozoic sediments from the Atlantic Ocean: Init. Repts. DSDP, v. 14, p. 787–954.

———, AND WINTERER, E. L., 1974, Plate stratigraphy and the fluctuating carbonate line, *in* Hsu, K. J., and Jenkyns, H., eds., Pelagic sediments on land and under the sea: Spec. Pub. International Association Sedimentologists, v. 1, p. 11–48.

———, JOHNSON, R. F., AND KILLINGLEY, J. S., 1977, Unmixing of the deep-sea record and the deglacial meltwater spike: Nature, v. 269, p. 661–663.

———, KILLINGLEY, J. S., AND VINCENT, E., 1978, Stable isotopes in deep-sea carbonates: Box Core ERDC-92, west equatorial Pacific: Oceanologica Acta, v. 1, p. 203–216.

BERGGREN, W. A., 1962, Some planktonic foraminifera from the Maestrichtian and the Danian Stages of southern Scandinavia: Stockholm Contributions in Geology, v. 9, p. 1–106.

———, 1972, Late Pliocene-Pleistocene glaciation: Init. Repts. DSDP, v. 12, p. 953–963.

———, AND HOLLISTER, C. D., 1977, Plate tectonics and paleocirculation-commotion in the ocean: Tectonophysics, v. 38, p. 11–48.

———, AND VAN COUVERING, J., 1974, Biostratigraphy, geochronology and paleoclimatology of the last 15 million years in marine and continental sequences: Palaeogeography, Palaeoclimatology, Palaeoecology, v. 16, p. 1–216.

BJÖRKSTRÖM, A., 1979, A model of CO_2 interaction between atmosphere, oceans, and land biota, *in* Bolin, B., and others, eds., Workshop on the carbon cycle, Ratzeburg, Germany, 1977, SCOPE report 13: New York, Wiley & Sons, p. 403–457.

BOERSMA, A. AND SHACKLETON, N., 1979, Some oxygen and carbon isotope variations across the Cretaceous-Tertiary boundary in the Atlantic Ocean, *in* Christensen, K., and Birkelund, T., eds., Cretaceous-Tertiary boundary events symposium II, Proceedings: University of Copenhagen, p. 50–53.

———, ———, HALL, M., AND GIVEN, Q., 1979, Carbon and oxygen isotope records at DSDP Site 384 (North Atlantic) and some Paleocene paleotemperatures and carbon isotope variations in the Atlantic Ocean: Init. Repts. DSDP, v. 43, p. 695–717.

BRAMLETTE, M. N., 1965, Massive extinctions in biota at the end of Mesozoic time: Science, v. 148, p. 1696–1699.

BROECKER, W. S., 1971, A kinetic model for the chemical composition of sea water: Quaternary Res., v. 1, p. 188–207.

———, AND VAN DONK, J., 1970, Insolation changes, ice volumes, and the ^{18}O record in deep-sea cores: Review of Geophysics and Space Physics, v. 8, p. 169–198.

BUCHARDT, B., 1978, Oxygen isotope paleotemperatures from the Tertiary period in the North Sea area: Nature, v. 275, p. 121–123.

CITA, M. B., 1979, Lacustrine and hypersaline deposits in the dessicated Mediterranean and their bearing on paleoenvironment and paleo-ecology, *in* Talwani, M., Hay, W. W., and Ryan, W. B. F., eds., Deep Drilling Results in the Atlantic Ocean: Continental margins and Paleoenvironment: Maurice Ewing Series 3, American Geophysical Union, Washington, D.C., p. 402–419.

CRAIG, H., 1970, Abyssal carbon 13 in the South Pacific: Journal of Geophysical Research, v. 75, p. 691–695.

DARWIN, CHARLES, 1859, The Origin of Species: London, John Murray, p. 490.

DONN, W. L., AND SHAW, D. M., 1977, Model of climate evolution based on continental drift and polar wandering: Geo. Soc. America Bull., v. 88, p. 390–396.

DOUGLAS, R. G., AND SAVIN, S. M., 1975, Oxygen and carbon isotope analyses of Tertiary and Cretaceous microfossils from Shatsky Rise and other sites in the North Pacific Ocean: Init. Repts. DSDP, v. 32, p. 509–520.

———, AND WOODRUFF, F., *in press,* Deep sea benthic foraminifera, *in* Emiliani, C., ed., The Sea, v. 7: New York, Wiley and Sons.

DRANSFIELD, P., 1968, Engineering Systems and Automatic Control: Englewood Cliffs, N. J., Prentice-Hall, Inc.

EMILIANI, C., 1978, The cause of the ice ages: Earth Planet. Sci. Letters, v. 37, p. 349–352.

———, AND GEISS, J., 1958, On glaciations and their causes: Geologische Rundschau, v. 46, p. 576–601.

EWING, M., AND DONN, W. L., 1956, A theory of ice ages: Science, v. 123, p. 1061–1066.

FISCHER, A. G., 1964, Brackish oceans as the cause of the Permo-Triassic marine faunal crisis, *in* Naïrn, A. E. M., ed., Problems in Palaeoclimatology: New York, Interscience, p. 566–579.

———, AND ARTHUR, M. A., 1977, Secular variations in the pelagic realm: Soc. Econ. Paleontologists Mineralogists, Spec. Pub. 25, p. 19–50.

FLINT, R. F., 1971, Glacial and Quaternary Geology: New York, Wiley and Sons, 892 p.

GARTNER, S., AND KEANY, J., 1978, The terminal Cretaceous event: A geologic problem with an oceanographic solution: Geology, v. 6, p. 708–712.

———, AND MCGUIRK, J. P., 1979, Terminal Cretaceous extinction scenario for a catastrophe: Science, v. 206, p. 1272–1276.

HAMILTON, W., 1968, Cenozoic climatic change and its cause: Meteorological Monographs, v. 8, p. 128–133.

HAQ, B. U., WORSLEY, T. R., BURCKLE, L. H., DOUGLAS, R. G., KEIGWIN, L. D., JR., OPDKE, N. D., SAVIN, S. M., SOMMER, II, M. A., VINCENT, E., AND WOODRUFF, F., 1980, The Late Miocene marine carbon-isotopic shift and the synchroneity of some phytoplanktonic biostratigraphic datums: Geology, v. 8, p. 427–431.

HOLSER, W. T., 1977, Catastrophic chemical events in the history of the ocean: Nature, v. 267, p. 403–408.

HSÜ, K. J., 1980, Terrestrial catastrophe caused by cometary impact at the end of Cretaceous: Nature, v. 285, p. 201–203.

———, CITA, M. B., AND RYAN, W. B. F., 1973, The origin of the Mediterranean evaporites: Init. Repts. DSDP, v. 13, p. 1203–1231.

HUBBERT, M. K., 1967, Energy resources, *in* Resources and Man, a study and recommendations by the Committee on Resources and Man of the Division of Earth Sciences, National Academy of Sciences No. 1703, National Research Council: San Francisco, Freeman, p. 157–242.

KEIGWIN, L. D., JR., 1978, Pliocene closing of the Isthmus of Panama, based on biostratigraphic evidence from nearby Pacific Ocean and Caribbean Sea cores: Geology, v. 6, p. 630–634.

———, 1979, Late Cenozoic stable isotope stratigraphy and paleoceanography of DSDP Sites from the east equatorial and north central Pacific Ocean: Earth Planet. Sci. Letters, v. 45, p. 361–382.

———, AND SHACKLETON, N. J., 1980, Uppermost Miocene carbon isotope stratigraphy of a piston core in the equatorial Pacific: Nature, v. 284, p. 613–614.

KENNETT, J. P., AND SHACKLETON, N. J., 1976, Oxygen isotopic evidence for the development of the psychrosphere 38 m.y. ago: Nature, v. 260, p. 513–515.

———, ———, MARGOLIS, S. V., GOODNEY, D. E., DUDLEY, W. C., AND KROOPNICK, P. M., 1979, Late Cenozoic oxygen and carbon isotopic history and volcanic ash stratigraphy: DSDP Site 284, South Pacific: Amer. Jour. Sci., v. 279, p. 52–69.

KROOPNICK, P. M., MARGOLIS, S. V., AND WONG, C. S., 1977, ^{13}C variations in marine carbonate sediments as indicators of the CO_2 balance between the atmosphere and oceans, *in* Andersen, N. R., and Malahoff, A., eds., The Fate of Fossil Fuel CO_2 in the Oceans: New York, Plenum Press, p. 295–321.

LA BRECQUE, J. L., KENT, D. V., AND CANDE, S. C., 1977, Revised magnetic polarity time scale for Late Cretaceous and Cenozoic time: Geology, v. 5, p. 330–335.

LUTERBACHER, H. P., AND PREMOLI-SILVA, I., 1964, Biostratigrafia del limite Cretaceo-Terziario: Rivista Italiana di Paleontologia, v. 70, p. 67–128.

MANKINEN, E. A., AND DALRYMPLE, G. B., 1979, Revised geomagnetic polarity time scale for interval 0–5 m.y.B.P.: Jour. Geophys. Res., v. 84, p. 615–626.

MARGOLIS, S. V., KROOPNICK, P. M., GOODNEY, D. E., DUDLEY, W. C., AND MAHONEY, M. E., 1975, Oxygen and carbon isotopes from calcareous nannofossils as paleoceanographic indicators: Science, v. 189, p. 555–557.

PENG, T.-H., AND BROECKER, W. S., 1978, Effect of sediment mixing on the rate of calcite dissolution by fossil fuel CO_2: Geophys. Res. Letters, vol. 5, (5), p. 349–352.

PERCH-NIELSEN, K., ULLEBERG, K., AND EVENSEN, J. E., 1979, Comments on the terminal Cretaceous event: A geologic problem with an oceanographic solution (Gartner and Keany, 1978), *in* Christensen, W. K., and Birkelund, T. eds., Cretaceous-Tertiary boundary events symposium II, Proceedings: University of Copenhagen, p. 106–111.

RODIN, L. E., BAZILEVICH, N. I., AND ROZOV, N. N., 1975, Productivity of the world's main ecosystems, *in* Productivity of World Ecosystems, Proceedings of a Symposium: Washington, D.C., National Academy of Sciences, p. 13–26.

RYAN, W. B. F., 1973, Geodynamic implications of the Messinian crisis of salinity, *in* Drooger, C. W., ed., Messinian events in the Mediterranean: Koninklijke Nederlandse Akademie van Wetenschappen, North-Holland, Amsterdam, p. 26–38.

———, CITA, M. B., DREYFUS RAWSON, M., BURCKLE, L. H., AND SAITO, T., 1974, A paleomagnetic assignment of Neogene stage boundaries and the development of isochronous datum planes between the Mediterranean, the Pacific and Indian Oceans in order to investigate the response of the World Ocean to the Mediterranean "Salinity Crisis": Rivista Italiana di Paleontologia, v. 80, p. 631–688.

SAITO, T., BURCKLE, L. H., AND HAYS, J. D., 1975, Late Miocene to Pleistocene biostratigraphy of equatorial Pacific sediments, *in* Saito, T., and Burckle, L. H., eds., Late Neogene Epoch Boundaries, Micropaleontology Spec. Pub.

1: New York, Micropaleontology Press, p. 226–244.
SAVIN, S. M., DOUGLAS, R. G., AND STEHLI, F. G., 1975, Tertiary marine paleotemperatures: Geol. Soc. America Bull., v. 86, p. 1499–1510.
SCHINDEWOLF, O. H., 1954, Über die möglichen Ursachen der grossen erdgeschichtlichen Faunenschnitte: Neues Jahrbuch für Geologie und Paleontologie, Monatshefte, v. 10, p. 457–465.
SCHOLLE, P. A., AND ARTHUR, M. A., 1980, Carbon isotope fluctuations in Cretaceous pelagic limestones: potential stratigraphic and petroleum exploration tool: Am. Assoc. Petroleum Geologists Bull., v. 64, p. 67–87.
SCIENTIFIC AMERICAN, 1955, Automatic Control: New York, Simon and Schuster, 148 p.
SHACKLETON, N. J., 1977, Carbon-13 in *Uvigerina:* tropical rainforest history and the equatorial Pacific carbonate dissolution cycles, *in* Andersen, N. R., and Malahoff, A., eds., The Fate of Fossil Fuel CO_2 in the Oceans: New York, Plenum Press, p. 401–427.
———, AND KENNETT, J. P., 1975, Paleotemperature history of the Cenozoic and the initiation of Antarctic glaciation: oxygen and carbon isotope analyses in DSDP Sites 277, 279 and 281: Init. Repts. DSDP, v. 29, p. 743–755.
———, AND OPDYKE, N. D., 1977, Oxygen isotope and paleomagnetic evidence for early northern hemisphere glaciation: Nature, v. 270, p. 216–219.
SMIT, J., AND HERTOGEN, J., 1980, An extraterrestrial event at the Cretaceous-Tertiary boundary: Nature, v. 285, p. 198–200.
THEYER, F., MATO, C. Y., AND HAMMOND, S. R., 1978, Paleomagnetic and geochronologic calibration of latest Oligocene to Pliocene radiolarian events, equatorial Pacific: Marine Micropaleontology, v. 3, p. 377–395.
THIERSTEIN, H. R., AND BERGER, W. H., 1978, Injection events in ocean history: Nature, v. 176, p. 461–466.
UREY, H. C., 1973, Cometary collisions and geological periods: Nature, v. 242, p. 32–33.
VAN ANDEL, T. H., AND MOORE, T. C., 1974, Cenozoic calcium carbonate distribution and calcite compensation depth in the central equatorial Pacific: Geology, v. 2, p. 87–92.
———, HEATH, G. R., AND MOORE, T. C., 1975, Cenozoic History and Paleoceanography of the Central Equatorial Pacific Ocean: Geol. Soc. America, Mem. 143, p. 1-134.
———, THIEDE, J., SCLATER, J. G., AND HAY, W. W., 1977, Depositional history of the South Atlantic ocean during the last 125 million years: Jour. Geology, v. 85, p. 651–698.
VAN COUVERING, J. A., BERGGREN, W. A., DRAKE, R. E., AGUIRRE, E., AND CURTIS, G. H., 1976, The terminal Miocene event: Marine Micropaleontology, v. 1, p. 263–286.
VERGNAUD-GRAZZINI, C., PIERRE, C., AND LETOLLE, R., 1978, Paleoenvironment of the North-East Atlantic during the Cenozoic: oxygen and carbon isotope analyses at Deep Sea Drilling Project Site 398, 400A and 401: Oceanologica Acta, v. 1, p. 381–390.
VINCENT, E., KILLINGLEY, J. S., AND BERGER, W. H., 1980, The magnetic epoch-6 carbon shift: a change in the ocean's $^{13}C/^{12}C$ ratio 6.2 million years ago: Marine Micropaleontology, v. 5, p. 185–203.
WATTS, N. L., LAPRE, J. F., VAN SCHIJNDEL-GOESTER, F. S., AND FORD, A., 1980, Upper Cretaceous and lower Tertiary chalks of the Albuskjell area, North Sea: Deposition in a slope and base-of-slope environment: Geology, v. 8, p. 217–221.
WEDEPOHL, K. H., ed., 1969, Handbook of Geochemistry: New York, Springer-Verlag, II-1, (6).
WORTHINGTON, L. V., 1968, Genesis and evolution of water masses: Meteorological Monographs, v. 8, p. 63–67.

SEPM SPECIAL PUBLICATION NO. 32, P. 505–515, OCTOBER 1981

TEMPORAL AND SPATIAL DISTRIBUTION OF ICE-RAFTED MINERAL GRAINS IN PLIOCENE SEDIMENTS OF THE NORTH ATLANTIC: IMPLICATIONS FOR LATE CENOZOIC CLIMATIC HISTORY

RICHARD Z. POORE
U.S. Geological Survey, 345 Middlefield Rd., Menlo Park, California 94025, U.S.A.

ABSTRACT

An important result of Deep Sea Drilling Project (DSDP) Leg 12 was the discovery of ice-rafted mineral grains in Pliocene sediments of the Labrador Sea and the Hatton-Rockall Basin. The oldest (first) occurrence of ice-rafted mineral grains in these sediments is associated with the evolutionary first appearance of the planktic foraminifer *Globorotalia inflata* (d'Orbigny) and the extinction of the coccolith *Reticulofenestra pseudoumbilica* (Gartner), or the first occurrence of the coccolith *Emiliania ovata* Bukry, or both. This association yields a paleontologic age estimate of 3.0 m.y.BP for the start of low-elevation Northern Hemisphere glaciation extensive enough to produce icebergs in the North Atlantic. Results from subsequent drilling in the North Atlantic (DSDP Legs 37, 48, and 49) confirm that the first ice-rafting is at about 3.0 m.y.BP, and further show that during the Late Pliocene, icebergs penetrated as far south as 45° N. lat in the central North Atlantic (DSDP Hole 410) but not as far south as 37° N. lat (DSDP Holes 333 and 335).

The estimated age of 3.0 m.y.BP for the onset of Northern Hemisphere low-elevation glaciation derived from the North Atlantic is supported by a 3.2 m.y.BP estimate based on paleomagnetically controlled oxygen isotope data from the Equatorial Pacific. North Atlantic planktic foraminiferal assemblages show that warm sea-surface temperatures in the northwestern Atlantic preceded the initiation of Northern Hemisphere glaciation.

INTRODUCTION

The development of continental ice-sheets in the Northern Hemisphere was a significant event in late Cenozoic earth history. Prior to the Deep Sea Drilling Project (DSDP), however, our knowledge of this event was limited principally to the often fragmentary and difficult to interpret record preserved on land. For example, pre-Quaternary climatic fluctuations are recorded in pollen records of northern Europe (van der Hammen and others, 1971), but the age of these records is not well-established. Glacial deposits as old as Late Miocene occur in western Alaska (Denton and Armstrong, 1969) and southeastern Alaska (Bandy and others, 1969), but the late Miocene tillites in the Wrangell Mountains of western Alaska represent montane glaciation rather than the development of continental ice-sheets, and the "Late Miocene" age of glacial deposits in the Yakataga Formation of southeastern Alaska is poorly controlled (see for example Armentrout and others, 1978). Moreover, the glacial events associated with the relatively thin older diamictites of the Yakataga Formation could be local. Tills about 3.0 m.y.BP are known from the Sierra Nevada Mountains of California (Curry, 1966) and from Iceland (McDougall and Wensink, 1966). Although these tills are well-dated, it was difficult to ascertain from the tills alone whether they reflect the buildup of continental ice-sheets or if they represent montane glaciation. Similarly, the sparse and sporadic occurrence of ice-rafted sand grains in arctic sediments as old as 4.5 m.y. (Herman and Hopkins, 1980) represent minor glaciations rather than the development of extensive continental ice-sheets.

The finding of ice-rafted sand grains in upper Pliocene sediments of the Labrador Sea and Hatton-Rockall Basin during DSDP Leg 12 provided direct evidence for the initiation of low-elevation Northern Hemisphere glaciation extensive enough to produce icebergs in the North Atlantic. In the following sections I briefly review the evidence from the Labrador Sea and Hatton-Rockall Basin that suggest a 3.0 m.y.BP age estimate for the initiation of Northern Hemisphere glaciation, and present additional corroborative evidence for this age estimate from DSDP Hole 410. The occurrence of ice-rafted sand grains and faunal data from DSDP North Atlantic cores are used to map the geographic extent of Pliocene ice-rafting and infer the magnitude of Pliocene glaciations. The North Atlantic record is then discussed within a framework of late Cenozoic global climatic evolution.

AGE OF INITIAL ICE-RAFTING

The age for the initiation of ice-rafting in the North Atlantic was first inferred from the association of the earliest occurrence of ice-rafted mineral grains with key microfossil events in Holes 111A and 116A, cored during DSDP Leg 12 (Berggren, 1972).

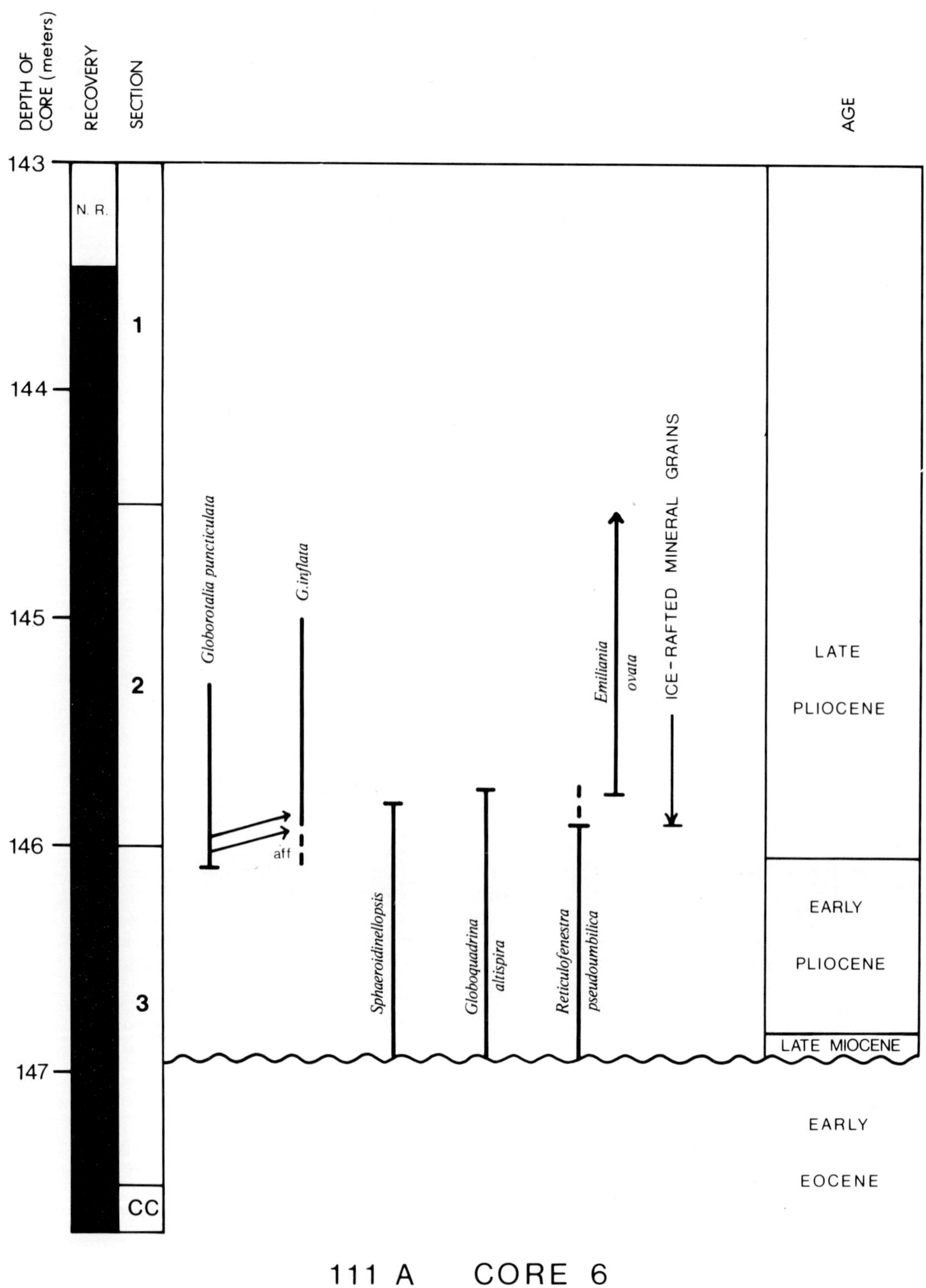

FIG. 1.—Stratigraphic distribution of key microfossils and ice-rafted mineral grains in core 6 of Hole 111A. Figure modified from Poore and Berggren (1974).

Site 111

Site 111 is located in the southern Labrador Sea on a topographic high known as Orphan Knoll. An important aspect of the section recovered from continuously cored Hole 111A is the occurrence of sand-size quartz and other mineral grains in sediments of Late Pliocene age. These grains are too coarse to be windblown and could not have been transported into the area by turbidity currents because Orphan Knoll is a topographic high. In addition, the cores do not show sedimentary structures typical of turbidites. Therefore the sand grains in the Pliocene sediments of Hole 111A are considered ice-rafted detritus. The first occurrence of ice-rafted sand grains is at 140 cm in section 2 of core 111A/6, in a highly condensed Early Pliocene to Late Pliocene sequence containing some tropical and subtropical planktic foraminifers and coccoliths. The stratigraphic distribution of selected planktic foraminifers and coccoliths as well as ice-rafted mineral grains found in detailed studies of 111A/6 (Berggren, 1972; Poore and Berggren, 1974) are summarized in Figure 1. Note that the first ice-rafted material in Hole 111A is closely associated with the last occurrence of *Reticulofenestra pseudoumbilica* (Gartner) and the evolutionary transition of *Globorotalia puncticulata* (Deshayes) to *Globorotalia inflata* (d'Orbigny). Age assignments determined in low-latitude areas for the paleontologic events or datum planes shown on Figure 1 are listed in Table 1. Extrapolation of these age assignments to Hole 111A suggests an age of about 3.0 m.y.BP for the earliest occurrence of ice-rafted sand grains (Fig. 1).

TABLE 1.—Age estimates for paleontologic events (datum planes) used to date the initiation of ice-rafting in the North Atlantic. ↑ = last occurrence; ↓ = first occurrence; pf = planktic foraminifer; c = coccolith. Data are modified from Haq and Berggren (1978) and Gartner (1973). Work done after publication of the Haq and Berggren (1978) study shows that *Emiliania ovata (= Pseudoemiliania lacunosa)* overlaps with *Reticulofenestra pseudoumbilica* (typical variety) (B. U. Haq, pers. commun., 1979). Therefore the age for the first occurrence of *E. ovata* in Table 1 is estimated to be at 3.4 m.y.BP.

Paleontologic event (datum plane)			Age m. y. BP
pf	↑	*Globoquadrina altispira*	3.0
pf	↑	*Sphaeroidinellopsis spp.*	3.0
pf	↓	*Globorotalia inflata*	3.0
c	↑	*Reticulofenestra pseudoumbilica* (typical form)	3.2
c	↓	*Emiliania ovata*	3.4

Site 116

Site 116 is located in the western North Atlantic in the Hatton-Rockall Basin on the Rockall Plateau. Hole 116A was continuously cored from 0 to 99 meters subbottom and recovered Pliocene and Pleistocene sediments. Sand-size quartz grains occurring in cores 1 through 9 are considered to be ice-rafted because of their size and because the Rockall Plateau is a topographic high. In contrast to Hole 111A, Pliocene sediments of Hole 116A do not contain warm-water planktic foraminifers and coccoliths and as a result paleomagnetically calibrated low-latitude datum planes are sparse. The earliest occurrence of ice-rafted detritus in core 9 is, however, associated with the evolutionary transition of *Globorotalia inflata* from *G. puncticulata* (Poore and Berggren, 1975) and is above the first occurrence of *Emiliania ovata* Bukry (Fig. 2). The data from Hole 116A are therefore compatible with an age estimate of 3.0 m.y.BP for the initiation of ice-rafting.

Site 410

Additional sites in the high-latitude North Atlantic were drilled during Legs 48 and 49 of the DSDP and the results obtained from Site 410 of Leg 49 are of particular interest to this study. Site 410 is located on the west side of the Mid-Atlantic ridge at 45°30.1′ N. Hole 410, which was continuously cored into basement, contained ice-rafted sand grains of Late Pliocene age. Aside from an isolated occurrence in core 23, due to down-hole contamination, the earliest ice-rafted grains in Hole 410 are associated with the evolutionary first appearance of *Globorotalia inflata* in core 15 (Poore, 1978; Shor and Poore, 1978). Additional documentation of the stratigraphic distribution of selected planktic foraminifers and coccoliths in cores 18 through 10 of Hole 410 (Table 2) shows that the first evidence of ice-rafting seen in core 15 coincides with or is slightly younger than the extinction of *Reticulofenestra pseudoumbilica* and is definitely younger than the first occurrence of *Emiliania ovata.* The earliest occurrence of ice-rafted grains is older than the last occurrence of *Discoaster tamalis* Kamptner, a form that last occurs late in the Gauss magnetic epoch at about 2.65 m.y.BP (Gartner, 1973). Thus these data from Hole 410 corroborate an age estimate of about 3.0 m.y.BP for the initiation of low-elevation Northern Hemisphere glaciation extensive enough to produce icebergs in the North Atlantic.

Note the anomalous association of the Late Miocene taxon *Discoaster quinqueramus* Gartner in core 18 with the Early Pliocene forms *Ceratolithus rugosus* Bukry and Bramlette and *Globorotalia puncticulata.* This overlap may be due to reworking, but it is interesting that Haq and Berggren (1978) recorded a similar anomalously young upward range for *Discoaster quinqueramus* in a suite of piston cores from the Rio Grande Rise in the South Atlantic.

NORTH ATLANTIC PLIOCENE
PLANKTIC FORAMINIFERAL FAUNAS

Figure 3 shows the composition of planktic foraminiferal assemblages in cores 18 through 10 of Hole 410. The faunas from this interval usually contain significant amounts of right-coiling *Neogloboquadrina pachyderma* (Ehrenberg), *Globigerina bulloides* d'Orbigny (s.1.), *Globigerinita glutinata* (Egger), and *Globorotalia inflata* or its ancestor *G. puncticulata,* and represent subpolar assemblages (Kipp, 1976). An overall cooling trend is indicated for this area of the North Atlantic by the

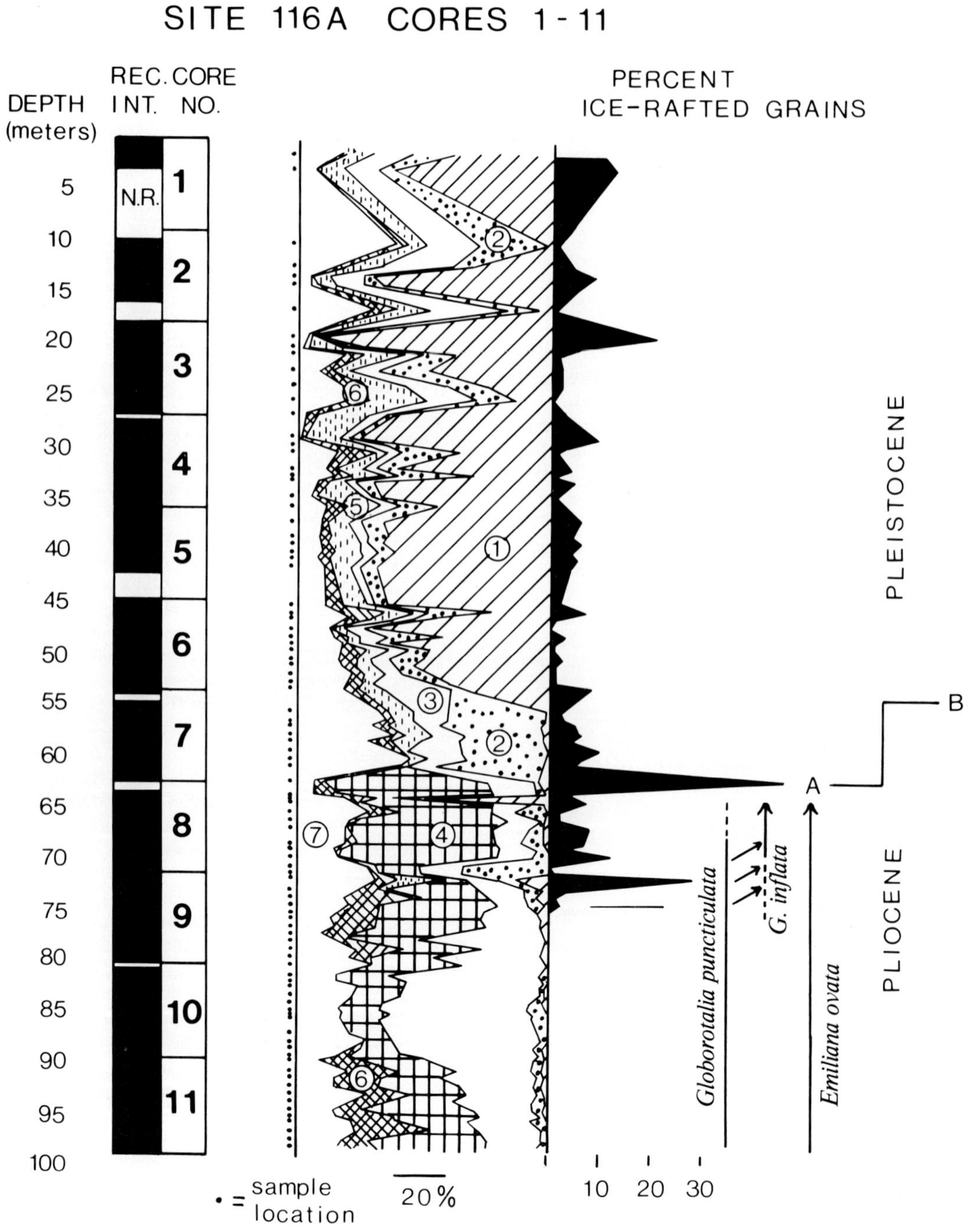

TABLE 2.—Occurrence of selected microfossils, coccolith zone and subzone assignments, and age assignments for samples from cores 18 through 10 of Hole 410. Abundance of ice-rafted mineral grains was estimated from ≥63μm size fraction. O = sparse, Θ = few, ● = common, ? = doubtful identification. Sparse designation for fossil occurrence means only 1 or 2 specimens were seen. Arrows indicate transition between *Globorotalia puncticulata* and *G. inflata.* Sample designation 10/1 39–41 means sample is from core 10, section 1, 39–41 cm depth; CC stands for Core Catcher sample.

Taxa / Samples	Coccoliths																	Planktic Foraminifers									Coccolith Zone or Subzone Bukry, 1975	AGE	
	Amaurolithus primus	*A. delicatus*	*Ceratolithus acutus*	*C. rugosus*	*Discoaster berggrenii*	*D. quinqueramus*	*D. pentaradiatus*	*D. surculus*	*D. asymmetricus*	*D. tamalis*	*D. brouweri*	*Sphenolithus* spp.	*Reticulofenestra* spp.	*R. pseudoumbilica*	*Emiliania annula*	*E. ovata*	*Gephyrocapsa* spp.	*Globorotalia margaritae*	*G. puncticulata*	*G. inflata*	*G. tosaensis*	*G. truncatulinoides*	*Globoquadrina altispira*	*Sphaeroidinellopsis* spp.	*Globigerina nepenthes*	ice-rafted mineral grains			
10/1 39-41															●	●	●			●	●	●				●			
10/3 100-103															●	●	?			●	●	●				Θ	*Crenalithus doronicoides* Zone	Pleistocene	
10 CC															●	●	?			●									
11/1 41-43															●	●	?			●						Θ			
11/3 129-131											●				●	●	?			●	●	●				Θ			
11/6 117-119								?			●				●	●				●						Θ	*Cyclococcolithina macintyrei* Subzone		
11 CC																				●						Θ			
12 CC											●				●	●				●						Θ			
13/1 26-28							●				●									●	●					O			
13 CC																					●					O			
14/2 133-135				●			●		●		●				●	●			?	●						Θ	*Discoaster pentaradiatus* Subzone (? + *D. surculus* Subzone)		
14/4 30-32							●		●		●				●	●			?	●						Θ			
14/5 86-88							●		●		●				●	●										O		Late	
14 CC																			?	●						Θ			
15/1 13-15							●	●	●	●	●		?		●	●			?	●			O			O			
15/2 18-20				●			●	●	●	O	●				●	●			● ↗	●						O			
15/3 55-57							●	●	●	●	●		●	O	●	●			● ↗	●						O	*Discosaster tamalis* Subzone		
15/4 67-69				●	O		●		●	●	●				●	●			●	●						O			Pliocene
15/6 25-27							●	●	●	●	●		●	O					● ↗	●									
15 CC																			●	●			●			O			
16/2 40-42		●	?				●	●	●		●			O		●			●										
16/4 96-98							●	●	●	●	●	●		●	●	●			●	?			●				*Reticulofenestra pseudoumbilica* Zone		
16/5 24-26				●			●	●	●	●	●			●	●	●			●	?									
16 CC																		●	●				●						
17 CC				●			●	●	●	●	●	●		●					●				●	●	●				
18/1 93-95	●	●				●	●	●	●		●			●				●	●				●		●			Early	
18/2 125-127	●					●	●		●		●	●		●				●	●								*Ceratolithus rugosus* Subzone		
18/5 94-96		●		●		●	●	●	●	?	●	●		●				?	●				●	●	●				
18 CC																		●	●				●		●				

general upward increase in abundance of the cool-water taxon *Neogloboquadrina pachyderma* (Ehrenberg). The general cooling trend is punctuated by three Late Pliocene cool pulses shown by *N. pachyderma* abundance peaks in core 16, the top of core 15 and lower 14, and core 13. Ice-rafted sand grains

←

FIG. 2.—Stratigraphic distribution of key microfossils and cumulative percent plot for planktic foraminiferal assemblages in Hole 116A. 1. Left-coiling *Neogloboquadrina pachyderma* (Ehrenberg). 2. Right-coiling *N. pachyderma.* 3. *Globigerina bulloides* d'Orbigny (s.l.). 4. *Neogloboquadrina atlantica* (Berggren). 5. *Turborotalita quinqueloba* (Natland). 6. *Globorotalia inflata* (d'Orbigny), *G. puncticulata* (Deshayes), and *G. sphericomiozea* Walters. 7. Other. Plot at right is an ice-rafting index. Index was calculated by

$$\frac{\text{\# ice-rafted mineral grains}}{\text{\# grains} + \text{\# planktic foraminifers}}$$

from counting splits of approximately 400 specimens. All quantitative data based on ≥149μm size fraction. Figure modified from Poore and Berggren (1975). A indicates position of Pliocene-Pleistocene boundary inferred from last occurrence of *Discoaster* sp. in Hole 116A sediments. B indicates position of Pliocene-Pleistocene boundary inferred from correlation of changes in faunal assemblages in Hole 116A and Hole 410.

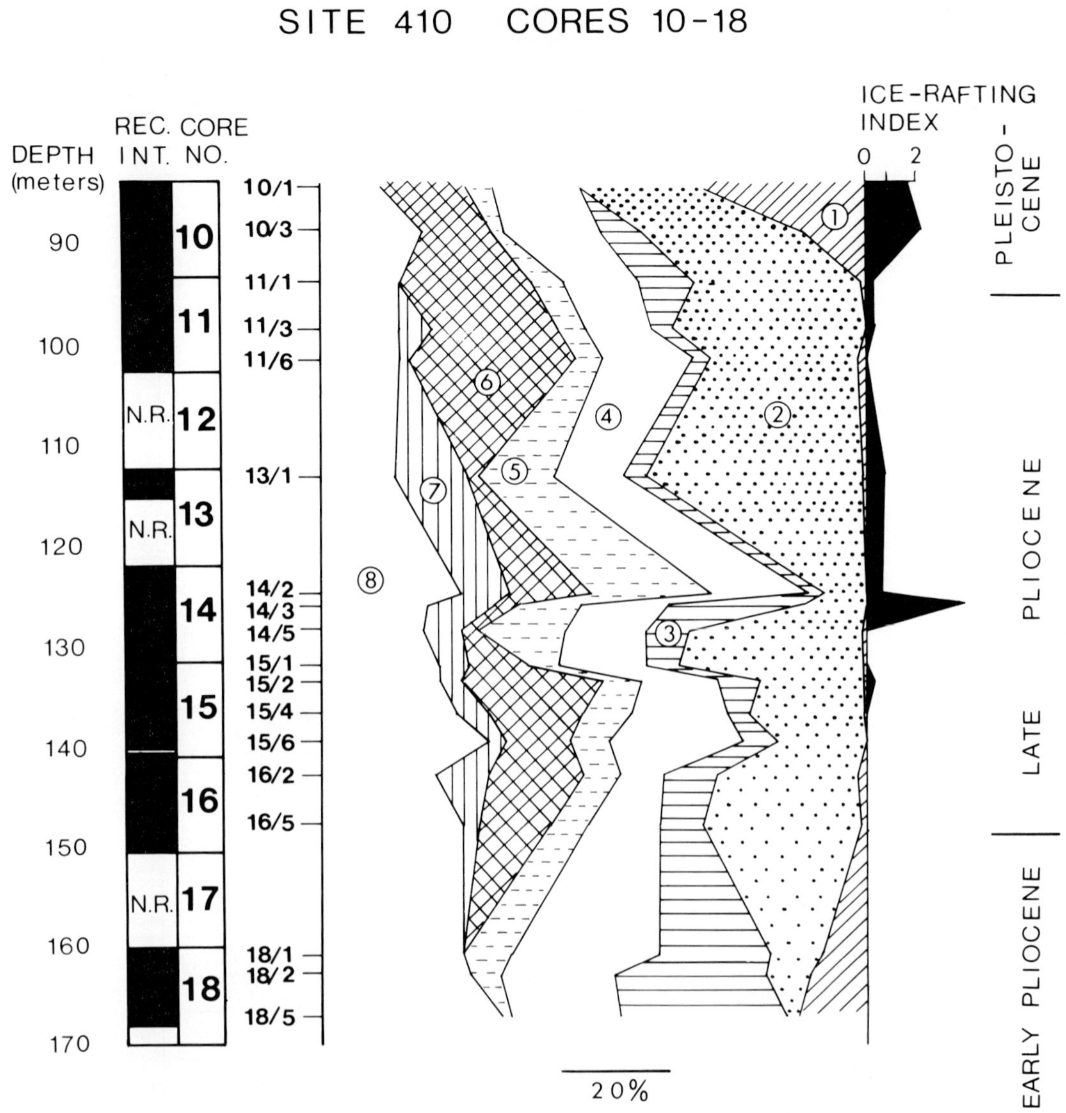

FIG. 3.—Cumulative percent plot for planktic foraminiferal assemblages in cores 18 through 10 of Hole 410. 1. Left-coiling *Neogloboquadrina pachyderma* (Ehrenberg). 2. Right-coiling *N. pachyderma*. 3. *Neogloboquadrina* spp. = *N. atlantica* (Berggren), *N. acostaensis* (Blow), *N. dutertrei* (d'Orbigny). 4. *Globigerina bulloides* d'Orbigny (s.l.) and *G. pseudobesa (Salvatorini)*. 5. *Globigerinita glutinata (Egger)*. 6. *Globorotalia inflata* (d'Orbigny) and *G. puncticulata* (Deshayes). 7. *Globorotalia crassaformis* (Galloway and Wissler). 8. Other. Plot at right is an ice-rafting index. Index was calculated by

$$\frac{\text{\# ice-rafted mineral grains}}{\text{\# planktic foraminifers}}$$

from counting splits of approximately 300 specimens. All quantitative data based on $\geq 149 \mu m$ size fraction.

first appear in the warm interval after the first Late Pliocene cooling in core 16. *Neogloboquadrina atlantica* (Berggren), which is common to abundant in Pliocene sediments of Hole 116A (Fig. 2), reaches its maximum abundance in core 14, and there is a peak in the ice-rafted index in sample 14/3. These data indicate significant southward movement of surface water-masses during the Late Pliocene. An interesting aspect of Figure 3, however, is the lack of evidence for major southward incursions of polar waters during the Late Pliocene. Studies of late Quaternary piston cores show that polar waters, which are characterized by left-coiling *Neogloboquadrina pachyderma,* penetrated well south of 45° N. in this area of the Atlantic during major glacial maxima (McIntyre and others, 1972). Thus the Pliocene foraminiferal assemblages from Hole 410 suggest that Late Pliocene glacial events were of lesser magnitude than major Late Quaternary glacial events. The significant increase of left-coiling *Neogloboquadrina pachyderma* in core 10 indicates that a major southward incursion of polar waters occurred soon after the beginning of the Pleistocene.

Comparison of the Late Pliocene fauna of Hole 410 with the Pliocene fauna of Hole 116A shown on Figure 2 reveals one major difference. *Neogloboquadrina atlantica* is more abundant that right-coiling *N. pachyderma* in Hole 116A whereas the reverse is seen in Hole 410. In other respects the assemblages at the two sites are similar. Because Site 116 is over 10° north of Site 410, *Neogloboquadrina atlantica* is inferred to prefer cooler waters than right-coiling *N. pachyderma.* The absence of *Globorotalia crassaformis* (Galloway and Wissler) and relatively low-diversity nannofossil assemblages from Hole 116A Pliocene sediments supports the interpretation that surface waters in the vicinity of Site 116 were cooler than surface waters near Site 410.

The cores recovered from Hole 116A are highly disturbed, and the details of faunal variation seen in Figure 2 are therefore blurred. Nonetheless, the peak of left-coiling *Neogloboquadrina pachyderma* seen at about 65 m on Figure 2 represents a brief incursion of polar waters into the area of Hatton-Rockall Basin during the Pliocene.

The change at 55 m from assemblages with minor amounts of left-coiling *Neogloboquadrina pachyderma* below to those with significant amounts of left-coiling *N. pachyderma* above is a major feature of the record from Hole 116A. It is tempting to correlate this increase of left-coiling *N. pachyderma* in 116A with the increase of left-coiling *N. pachyderma* seen in core 10 of Hole 410. If this correlation is correct it means that the Pliocene-Pleistocene boundary in Hole 116A, which Poore and Berggren (1975) placed at 62 m on the basis of the last occurrence of discoasters, would be changed to 55 m to match the well-controlled position of the Pliocene-Pleistocene boundary at the top of core 11 in Hole 410. Poore and Berggren (1975) noted that the last occurrence of discoasters in Hole 116A was associated with a major peak in the ice-rafted index, and that the disappearance of discoasters in Hole 116A might not represent the extinction of discoasters associated with the Pliocene-Pleistocene boundary. My preferred interpretation is to correlate the increases of left-coiling *Neogloboquadrina pachyderma* between the two sites and therefore revise the position of the Pliocene-Pleistocene boundary in Hole 116A to 55 m. Regardless of the position of the boundary, the faunal record from Hole 116A shows that climatic fluctuations during the Pliocene were not as intense as Quaternary climatic fluctuations.

Semiquantitative faunal data from Hole 111A (Poore and Berggren, 1974) reveal a significant and relatively uniform component of tropical and subtropical species in Early Pliocene planktic foraminiferal assemblages whereas only subpolar and polar species are known from this area today (Kipp, 1976). Early Pliocene surface waters of the southern Labrador Sea were therefore significantly warmer than they are today and the southern Labrador Sea must have been influenced by the Gulf Stream or proto-Gulf Stream. The abundance of warm-water species begins to decline at about the same level as the first occurrence of ice-rafted mineral grains until a polar assemblage finally takes over in the upper portion of core 6 (Poore and Berggren, 1974, Fig. 7). Polar or subpolar assemblages predominate upsection, although samples selected to be biased towards moderate climatic conditions suggest periodic incursions of transitional faunas into the southern Labrador Sea during the Late Pliocene. The faunas of Hole 111A, which is close to the modern polar front, show much more change associated with the initiation of Northern Hemisphere low-elevation glaciation than those of either Hole 116A or Hole 410.

The warm-water aspect of the Early Pliocene assemblages of Hole 111A is extremely interesting. During the most recent interglacial-to-glacial transition, the two intervals of most rapid Northern Hemisphere continental ice-sheet growth occurred when sea-surface temperatures in the subpolar North Atlantic were relatively warm (Ruddiman and McIntyre, 1979). Ruddiman and McIntyre (1979) reasoned that juxtaposition of a warm subpolar sea surface with the cold, probably snow-covered, Canadian maritime provinces created a strong thermal gradient of optimal configuration for delivering moisture to the Laurentide Ice Sheet. The occurrence of warm surface waters in the western subpolar Atlantic documented by the Early Pliocene faunas of Hole 111A may have "set the stage" for the initial accumulation of continental ice sheets in the Northern Hemisphere. Any event causing a decrease in temperature of North American land masses—such as a decrease in Northern Hemisphere insolation caused by increased explosive volcanism on Iceland, or astronomical changes—could then trigger buildup of an ice sheet.

GEOGRAPHIC EXTENT OF PLIOCENE ICE-RAFTING

The geographic extent and an estimate of the magnitude of Pliocene ice-rafting can be deduced from the distribution of North Atlantic DSDP sites that recovered Pliocene sediments containing ice-rafted mineral grains (see Fig. 4). Ice-rafted mineral grains are sparse in Pliocene samples from Hole 410 (Table 2, Figure 3), and no ice-rafted material is reported from the Upper Pliocene sediments of Holes 333, 335, and 398. Therefore the southern limit of Pliocene ice-rafting must have been close to Site 410.

The Pliocene southern limit of ice-rafting shown on Figure 4 is far north of the Quaternary southern limit of ice-rafting. Moreover, Site 410 is located near the southern limit of relatively common occurrence of ice-rafted sand-sized mineral grains during Late Quaternary warm periods (Conolly and Ewing, 1965; Ruddiman, 1977) and is near the area of greatest deposition of ice-rafted sand during Late Quaternary glacial maxima (Ruddiman, 1977). Since the distribution and total input of ice-rafted sediments to the Atlantic are positively correlated with the size of the Northern Hemisphere ice-sheets (Ruddiman, 1977), the consistently sparse amount of ice-rafted detritus in the Pliocene of Hole 410, like faunal data from Holes 116A and 410, indicates that Pliocene Northern Hemisphere glacial events were not as intense as Late Quaternary glacial events and thus that Pliocene Northern Hemisphere continental ice-sheets were smaller than those of the Late Quaternary.

DISCUSSION

Having reviewed the Pliocene record from the high-latitude North Atlantic, it is now appropriate to see how these results fit with the pattern of late Cenozoic climate history inferred from oxygen isotope studies and ice-rafting records from the North Pacific Ocean.

Following the buildup of a major Antarctic ice-sheet by Middle Miocene time (Shackleton and Kennett, 1975a; Savin and others, 1975), changes in Late Miocene and Early Pliocene average global ice volume appears to have been fairly minor. Global ice volume is best recorded in the oxygen isotope records derived from benthic foraminifers. Because these organisms secrete their tests in a thermally stable environment, changes in their oxygen isotopic composition through time principally represent a record of the isotopic composition of the oceans and thus a record of global ice volume (Shackleton and Kennett, 1975a). A detailed benthic

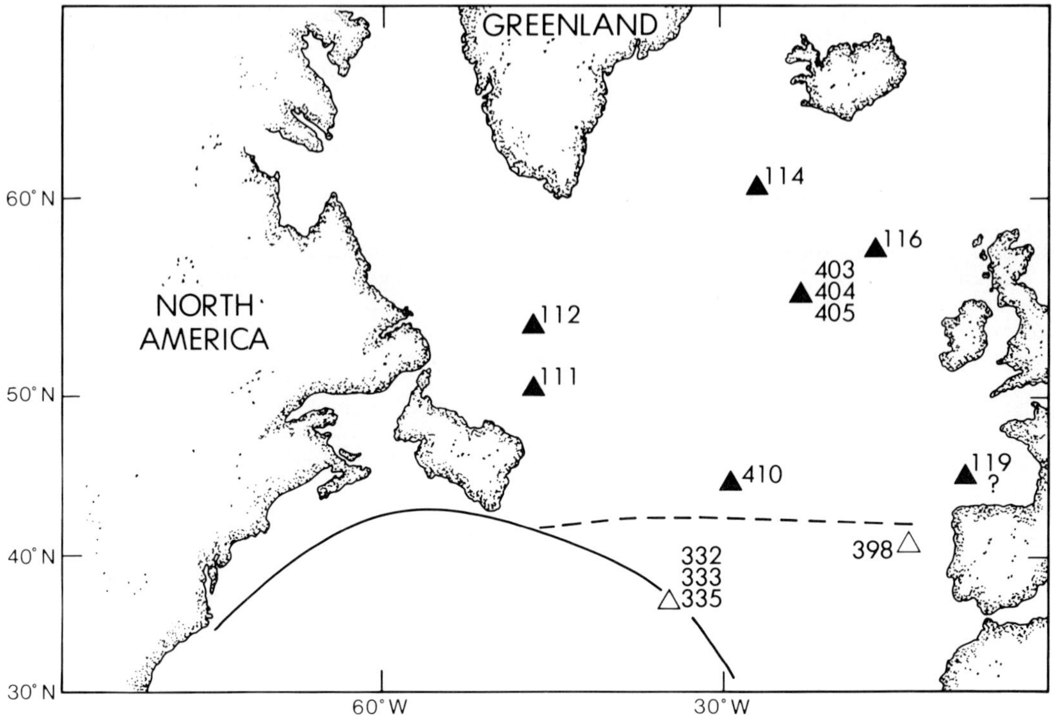

FIG. 4.—Map showing southern limit of North Atlantic Pliocene ice-rafting. Figure modified from Shor and Poore (1978). ▲ = Ice-rafted component in upper Pliocene sediments. △ = No ice-rafted component in upper Pliocene sediments. Solid line shows southern extent of Quaternary ice-rafting (Davis and Laughton, 1972). Dashed line shows inferred southern extent of Late Pliocene ice-rafting.

isotopic record from DSDP Site 284 off of New Zealand that covers about the last 6 million years is given by Shackleton and Kennett (1975b) and Kennett and others (1979). Some variation is seen in the Late Miocene part of the Site 284 record, but this variation is explained by changes in the size of the Antarctic ice sheet (Shackleton and Kennett, 1975b). The striking feature of the Site 284 data is the extremely stable Early to Late Pliocene record that culminates in a marked shift of about 1‰ toward positive values in the Late Pliocene. Shackleton and Kennett (1975b) consider this Late Pliocene shift to represent the buildup of ice sheets in the Northern Hemisphere. An even more detailed benthic isotopic record from paleomagnetically dated equatorial Pacific core V28-179 shows a similar pattern (Shackleton and Opdyke, 1977). Following a period of stable Early Pliocene isotopic values, increased variation that includes excursions towards increasingly positive values begins at 3.2 m.y.BP within the lower part of the Gauss magnetic epoch. Next, isotopic fluctuations suggesting glaciations of about two-thirds the magnitude of Late Quaternary glaciations occur between 2.5 m.y.BP and 1.6 m.y.BP. Younger samples from this core were not analyzed. The positive isotopic excursions seen beginning at 3.2 m.y.BP in core V28-179 represent the initial accumulation of a major Northern Hemisphere ice-sheet (Shackleton and Opdyke, 1977) and correlates well with the initiation of low-elevation glaciation seen in the North Atlantic Basin. Faunal and oxygen-isotope data from the Mediterranean Sea also show evidence for climatic cooling and buildup of Northern Hemisphere ice-sheets beginning at about 3.2 m.y.BP (Keigwin and Thunell, 1979). The increase in intensity of glacial events seen at 2.5 m.y.BP in core V28-179 and in Late Pliocene land sections in New Zealand (Kennett and others, 1971) is not evident in the North Atlantic record. This is not surprising because the record in Hole 116A is smoothed because of drilling disturbance, and core recovery in the Late Pliocene of Hole 410 was poor. Note, however, that both the North Atlantic and equatorial Pacific records indicate that Late Pliocene glacial events were less intense than Late Quaternary glacial maxima. However, the occurrence of tills older than 2.2 m.y.BP in Iowa (Boellstorf, 1978) shows that some late Pliocene glaciations were quite extensive.

Figure 5 is a generalized abundance curve for the occurrence of ice-rafted detritus in sediments of the

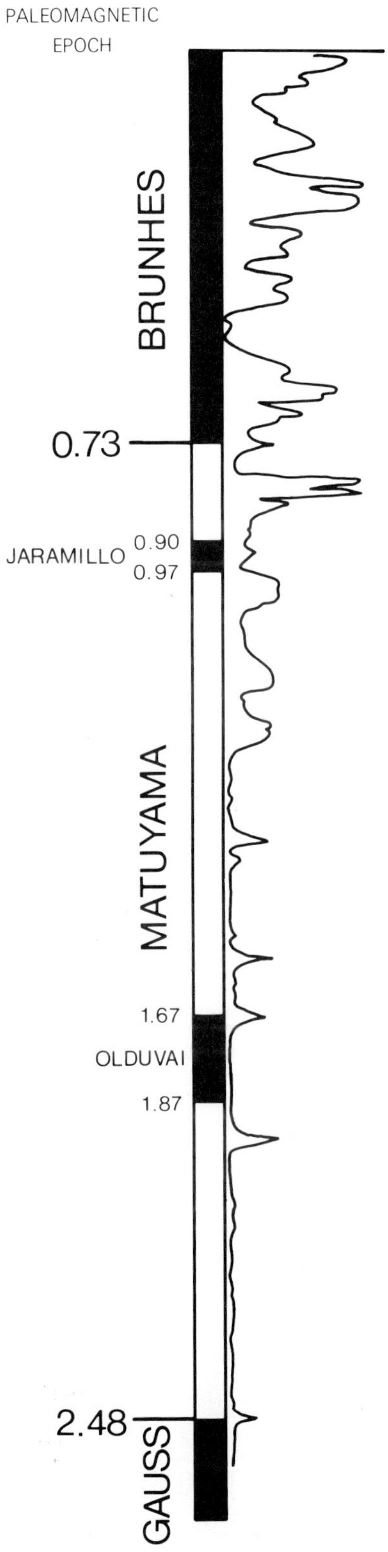

FIG. 5.—Generalized ice-rafting abundance curve for the northeastern Pacific. Redrawn from Kent and others (1971). Age assignments of magnetic events and epochs (in millions of years) have been adjusted to reflect new decay constant for calculating K-Ar dates.

northeastern Pacific Ocean (Kent and others, 1971), and shows that the oldest occurrence of ice-rafted mineral grains there is near the end of the Gauss magnetic epoch, at about 2.5 m.y.BP. Although the record of Kent and others (1971) does not extend far into the Gauss, the age for the oldest ice-rafting shown in Figure 5 is supported by DSDP drilling in the North Pacific. The oldest occurrence of ice-rafted detritus in DSDP Gulf of Alaska sites drilled during Leg 18 is associated with diatoms referable to the upper part of the *Denticula seminae* var. *fossilis*–*D. kamtschatica* Zone (Burckle, 1978), a level Burckle correlates to the upper part of the Gauss. A similar occurrence was found in North Pacific DSDP Site 183. The delay of about 0.5 m.y. in the earliest occurrence of ice-rafted sand grains between the North Pacific and North Atlantic is not surprising. During initial buildup of Northern Hemisphere ice-sheets most ice reaching sea level from western North America or eastern Siberia would be delivered to the Arctic Ocean, where it would be transported eastward toward the Atlantic (Kovacs and Mellor, 1974) or to seas which have limited access to the North Pacific. The earliest occurrence of ice-rafted sand grains in the North Pacific corresponds to the increase in magnitude of glaciations seen at 2.5 m.y.BP in the oxygen isotopic records.

Other important features of Figure 5 include: the sparse occurrence of ice-rafted detritus from the Gauss up into the middle of the Matuyama epoch, the increase of ice-rafted detritus in the Matuyama at about 1.2 m.y.BP, and a further dramatic increase in the amount of ice-rafted detritus at about 0.8 m.y.BP. These northeastern Pacific data show that Late Pliocene and Early Quaternary Northern Hemisphere glaciations were of lesser magnitude than later Quaternary glaciations and that the intensity of Northern Hemisphere glaciations increased significantly at about 0.8 m.y.BP. Long Quaternary planktic foraminifer isotopic records from the Pacific (Shackleton and Opdyke, 1976) and Atlantic (van Donk, 1976) also show that the intensity of Northern Hemisphere glaciations increased to levels comparable with the last glacial maximum at about 0.8 m.y.BP.

SUMMARY AND CONCLUSIONS

The climatic interpretations derived from study of North Atlantic DSDP cores are consistent with and corroborated by isotopic studies and the record of ice-rafting in the North Pacific. Taken together, these data suggest the following general pattern and timing of events in the development of Northern Hemisphere continental glaciations.

1) Following the buildup of a major Antarctic ice sheet by the Middle Miocene, by Late Miocene global ice-volume changes were due to fluctuations in the Antarctic ice sheet. Ice accumulation, if any, in the Northern Hemisphere was minimal and restricted to montane-type glaciation.

2) Average global ice volume during the early to Late Pliocene (5.0–3.2 m.y.BP) was stable. Surface-water temperatures in the northwestern Atlantic were warmer than they are today.

3) At 3.2 m.y.BP, ice sheets began to accumulate in the Northern Hemisphere, and by 3.0 m.y.BP, icebergs were present in the North Atlantic Ocean.

4) Glacial events approximately two-thirds the intensity of Late Quaternary glacial maxima began at 2.5 m.y.BP. Maximum glaciations were of this intensity up to 0.8 m.y.BP.

5) Glaciations of comparable intensity to the last glacial maximum began at 0.8 m.y.BP

It is certain that the sediments on the flanks of the mid-Atlantic Ridge contain a detailed record of Northern Hemisphere climates and glacial cycles back into the Early Pliocene. Hopefully, future DSDP Atlantic drilling utilizing the recently developed hydraulic piston corer will allow us to examine this record.

ACKNOWLEDGMENTS

I thank John Barron, C. A. Repenning, and J. P. Kennett for constructive comments on this manuscript.

REFERENCES CITED

NOTE: All references to Deep Sea Drilling Project Initial Reports have been abbreviated; for complete citations see *Appendix I* at the end of this volume, or the individual Initial Reports as published.

Armentrout, J. M., Echols, R. J., and Nash, K. W., 1978, Late Neogene climatic cycles of the Yakataga Formation, Robinson Mountains, Gulf of Alaska area [abs.]: Stanford University Publications (Geological Sciences), v. 14, p. 3–4.

Bandy, O. L., Butler, E. A., and Wright, R. C., 1969, Alaskan upper Miocene marine glacial deposits and the *Turborotalia pachyderma* datum plane: Science, v. 166, p. 607–609.

Berggren, W. A., 1972, Late Pliocene-Pleistocene glaciation: Init. Repts. DSDP, v. 12, p. 953–963.

Boellstorff, J., 1978, North American Pleistocene Stages reconsidered in light of probable Pliocene-Pleistocene continental glaciation: Science, v. 202, p. 305–307.

Bukry, David, 1975, Coccolith and silicoflagellate stratigraphy, northwestern Pacific Ocean, Deep Sea Drilling Project Leg 32: Init. Repts. DSDP, v. 32, p. 677–701.

Burckle, L. H., 1978, Early Miocene to Pliocene diatom datum levels for the equatorial Pacific: Republic of Indonesia Geological Research Development Centre Special Publication no. 1, p. 24–44.

Conolly, J. R., and Ewing, M., 1965, Pleistocene glacial-marine zones in North Atlantic deep-sea sediments: Nature, v. 208, p. 135–138.

Curry, R. R., 1966, Glaciation about 3,000,000 years ago in the Sierra Nevada: Science, v. 154, p. 770–771.

Davies, T. A., and Laughton, A. S., 1972, Sedimentary processes in the North Atlantic: Init. Repts. DSDP, v. 12, p. 905–934.

Denton, G. H., and Armstrong, R. L., 1969, Miocene-Pliocene glaciations in southern Alaska: Amer. Jour. Sci., v. 267, p. 1121–1142.

Gartner, S., 1973, Absolute chronology of the late Neogene calcareous nannofossil succession in the Equatorial Pacific: Geol. Soc. America Bull., v. 84, p. 2021–2034.

Haq, B. U., and Berggren, W. A., 1978, Late Neogene calcareous plankton biochronology of the Rio Grande Rise (South Atlantic Ocean): Jour. Paleontology, v. 52, p. 1167–1194.

Herman, Y., and Hopkins, D. M., 1980, Arctic oceanic climate in late Cenozoic time: Science, v. 209, p. 557–562.

Keigwin, L. D., Jr., and Thunell, R. C., 1979, Middle Pliocene climatic change in the western Mediterranean from faunal and oxygen isotopic record: Nature, b. 282, p. 294–296.

Kennett, J. P., Shackleton, N. J., Margolis, S. V., Goodney, D. E., Dudley, W. C., and Kroopnick, P. M., 1979, Late Cenozoic oxygen and carbon isotopic history and volcanic ash stratigraphy: DSDP Site 284, South Pacific: Amer. Jour. Sci., v. 279, p. 52–69.

Kennett, J. P., Watkins, N. D., and Vella, P., 1971, Paleomagnetic chronology of Pliocene-early Pleistocene climates and the Plio-Pleistocene boundary in New Zealand: Science, v. 171, p. 276–279.

Kent, D., Opdyke, N. D., and Ewing, M., 1971, Climatic change in the North Pacific using ice-rafted detritus as a climatic indicator: Geol. Soc. America Bull., v. 82, p. 2741–2754.

Kipp, N. G., 1976, New transfer function for estimating past sea-surface conditions from sea-bed distribution of planktonic foraminiferal assemblages in the North Atlantic, *in* Investigation of late Quaternary paleoceanography and paleoclimatology: Geol. Soc. America Mem. 145, p. 3–41.

Kovacs, A., and Mellor, M., 1974, Sea ice morphology and ice as a geologic agent in the southern Beaufort Sea, *in* Reed, J. C., and Sater, J. E., eds., The coast and shelf of the Beaufort Sea: Arctic Institute of North America, Proceedings of a symposium of Beaufort Sea Coast and Shelf Research, p. 113–162.

McDougall, I., and Wensink, H., 1966, Paleomagnetism and geochronology of the Pliocene-Pleistocene lavas in Iceland: Earth Planet. Sci. Letters, v. 1, p. 232–236.

McIntyre, A., Ruddiman, W. F., and Jantzen, R., 1972, Southward penetrations of the North Atlantic Polar Front: faunal and floral evidence of large-scale surface water mass movements over the last 225,000 years: Deep-sea Research, v. 19, p. 61–77.

Poore, R. Z., 1978, Oligocene through Quaternary planktonic foraminiferal biostratigraphy of the North Atlantic: DSDP Leg 49: Init. Repts. DSDP, v. 49, p. 447–518.

Poore, R. Z., and Berggren, W. A., 1974, Pliocene biostratigraphy of the Labrador Sea: Calcareous plankton: Jour. Foraminiferal Res., v. 4, p. 91–108.

———, 1975, Late Cenozoic planktonic foraminiferal biostratigraphy and paleoclimatology of Hatton-Rockall Basin: DSDP Site 116: Jour. Foraminiferal Res., v. 5, p. 270–293.

Ruddiman, W. F., 1977, Late Quaternary deposition of ice-rafted sand in the subpolar North Atlantic (lat 40° to 65° N.): Geol. Soc. America Bull., v. 88, p. 1813–1827.

Ruddiman, W. F., and McIntyre, A., 1979, Warmth of the subpolar North Atlantic Ocean during Northern Hemisphere ice-sheet growth: Science, v. 204, p. 173–175.

Savin, S. M., Douglas, R. G., and Stehli, F. G., 1975, Tertiary marine paleotemperatures: Geol. Soc. America Bull., v. 86, p. 1499–1510.

Shackleton, N. J., and Kennett, J. P., 1975a, Paleotemperature history of the Cenozoic and the initiation of Antarctic glaciation: Oxygen and carbon isotope analyses in DSDP Sites 277, 279, and 281: Init. Repts. DSDP, v. 29, p. 743–755.

———, 1975b, Late Cenozoic oxygen and carbon isotopic changes at DSDP Site 284: Implications for glacial history of the Northern Hemisphere and Antarctica: Init. Repts. DSDP, v. 29, p. 801–807.

Shackleton, N. J., and Opdyke, N. D., 1976, Oxygen-isotope and paleomagnetic stratigraphy of Pacific Core V28-239 Late Pliocene to Latest Pleistocene, *in* Investigation of late Quaternary paleoceanography and paleoclimatology: Geol. Soc. America Mem. 145, p. 449–464.

———, 1977, Oxygen isotope and palaeomagnetic evidence for early Northern Hemisphere glaciation: Nature, v. 270, p. 216–219.

Shor, A. N., and Poore, R. Z., 1978, Bottom currents and ice rafting in the North Atlantic: Interpretation of Neogene depositional environments of Leg 49 cores: Init. Repts. DSDP, v. 49, p. 447–518.

van der Hammen, T., Wijmstra, T. A., and Zagwijn, W. H., 1971, The floral record of the late Cenozoic of Europe, *in* The Late Cenozoic Glacial Ages: New Haven, Conn., Yale University Press, p. 391–424.

van Donk, Jan, 1976, ^{18}O record of the Atlantic Ocean for the entire Pleistocene epoch, *in* Investigation of Late Quaternary Paleoceanography and Paleoclimatology: Geol. Soc. America Mem. 145, p. 147–163.

SEPM Special Publication No. 32, p. 517–546, October 1981

MIDDLE CRETACEOUS CALCAREOUS NANNOPLANKTON BIOGEOGRAPHY AND OCEANOGRAPHY OF THE ATLANTIC OCEAN

PETER H. ROTH and JAY L. BOWDLER*
Department of Geology and Geophysics, University of Utah, Salt Lake City, Utah 84112

ABSTRACT

Quantitative studies of the preservation and biogeographic distribution of calcareous nannofossils in Mid-Cretaceous sediments from the Atlantic are combined with observations on the temporal and spatial distribution of major facies patterns to put constraints on speculations about the paleoceanographic evolution of the Atlantic about 110 to 90 m.y.BP. Organic-rich sediments in deep ocean and continental margin settings are indicative of organic carbon influx from terrestrial and marine sources and considerable carbonate dissolution largely caused by carbon dioxide production during catabolic breakdown of organic matter. Preservation of organic carbon in sediments is the result of rapid burial, especially along continental margins, and/or low concentrations of oxygen in warm and saline bottom waters derived largely from marginal evaporite basins. In the eastern basin of the Atlantic perhaps only the interstitial waters within the sediments were anoxic. There is an overall negative correlation between calcium carbonate and organic carbon content in organic-rich sediments. Coccolith preservation is generally poorest in the most organic carbon-rich shales. Semiquantitative preservation estimates are as sensitive as quantitative estimates using dissolution indices.

Paleogeographic patterns of coccolith distribution show weak latitudinal gradients and more pronounced neritic-oceanic gradients. *Boreal and austral assemblages* first observed in the later half of the Mid-Cretaceous are restricted to latitudes greater than 40°. Rapid spatial and temporal fluctuation in coccolith abundance, including monospecific blooms, are typical of higher latitudes and restricted basins (*e.g.,* Angola Basin). *Neritic assemblages* are found along the continental margin of the eastern Atlantic and over the Walvis-Rio Grande Ridge and are possibly indicative of increased advection of nutrient-rich waters. *Oceanic assemblages* characterize areas far removed from continents such as deep basin and Mid-Atlantic Ridge sites. At the beginning of the Upper Cretaceous, better deepwater connections, the rise in sea-level, and the development of stronger temperature gradients resulted in more vigorous mixing of the deep and surface waters and reduced detrital organic carbon input. Stagnation of deep basins ended, and surface-water temperature-gradients became stronger.

INTRODUCTION

Widespread occurrence of deep-sea sediments rich in organic carbon ("Black Shales") in the Atlantic Ocean are indicative of drastically different oceanographic conditions during the Middle Cretaceous (about 110 to 90 million years before present (m.y.BP)). To this date almost two dozen holes drilled by the Deep Sea Drilling Project-International Phase of Ocean Drilling (DSDP-IPOD) have recovered Middle Cretaceous sediments (Fig. 1). Sufficient core material is now available to attempt to determine the large-scale patterns of sedimentation and the biogeographic distribution of planktic organisms. This will provide a firmer basis for speculations on surface-water currents and fertility patterns, deepwater circulation and chemistry and the climate during Middle Cretaceous time.

The abundance and preservation of nannoliths, the low-Mg calcite plates of calcareous nannoplankton, provide much information on temperature, salinity and nutrient concentration of ocean surface waters where the organisms lived and on the chemistry of the sediment-water interface where their remains accumulate and are either dissolved or buried. In this paper we attempt to contribute to a better understanding of the following paleoceanographic parameters based on distribution patterns of nannolith assemblages in sediments from the Atlantic Ocean and some surrounding epicontinental seas: (1) Surface-water temperature gradients. (2) Surface-water fertility. (3) Calcite dissolution at the sediment-water interface and within the sediments.

The first part of our paper provides a framework for our findings on the preservation and biogeographic distribution of nannolith assemblages. A summary of the paleogeography, bathymetry, climate, circulation, sedimentary facies and biochronology is presented. In the second part we discuss briefly the preservation of nannoliths and treat the biogeography of important species in some detail. Hypotheses on the paleoceanography of the Mid-Cretaceous Atlantic discussed in the first part are then compared with results based on nannolith distribution.

PALEOGEOGRAPHY AND BATHYMETRY

Tectonic reconstructions (Fig. 2) and determination of large-scale bathymetry of the Atlantic Ocean (Smith and others, 1973; Smith and Briden, 1977; Sclater and others, 1977) allow placing sedi-

*Present address: Union Oil Company of California, Houston, Texas

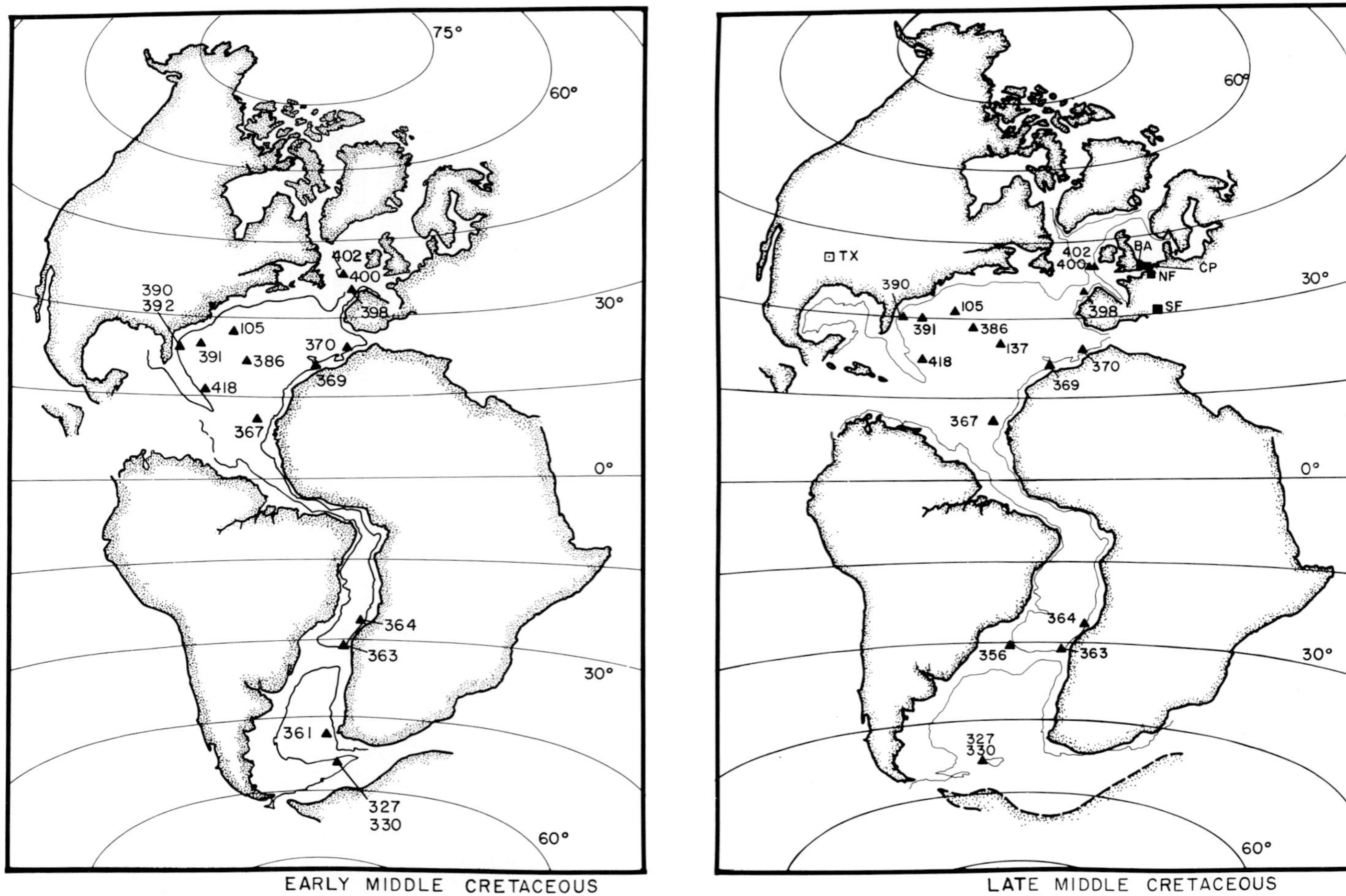

FIG. 1*A*, *B*.—Locations of DSDP-IPOD Sites where Middle Cretaceous sediments were recovered, indicated by dots, and site numbers of selected land sections marked by squares with the following abbreviations: BA = Barrington Cement Quarry, near Cambridge, lower Chalk, CP = Cope Point, Folkstone, Gault, NF = Northern France, Cenomanian (after Manivit, in Amedro, Damotte and others, 1978), SF = Southwestern France. For more information on sample locations see Tables 1 and 2. *A*, Sample locations for the early Middle Cretaceous (Zones NC 5 through NC 9). *B*. Sample locations for the late Middle Cretaceous (Zones NC 10 through NC 11).

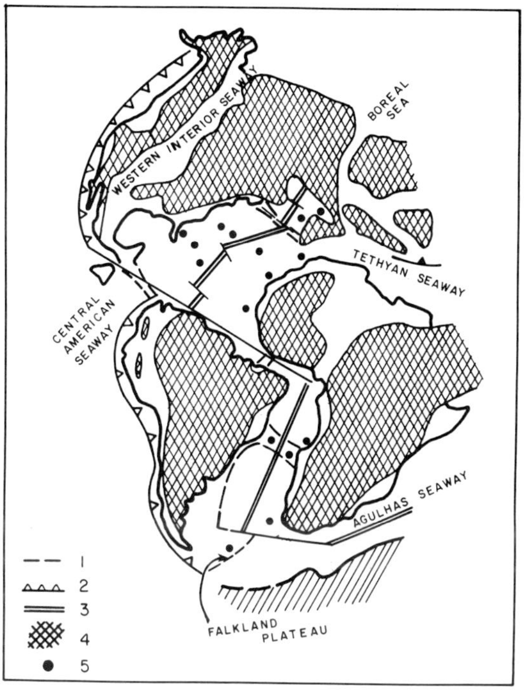

FIG. 2.—Paleogeographic reconstruction of the Middle Cretaceous continents surrounding the Atlantic Ocean. Explanation of symbols: 1. Shallow aseismic ridges and marginal plateaus. 2. Active continental margins 3. Ridge crests 4. Dry land 5. DSDP sites. (Based on various sources, including Smith and Briden, 1973; Gignoux, 1960; Thenius, 1977; Price, 1976; and Sclater and others, 1977.)

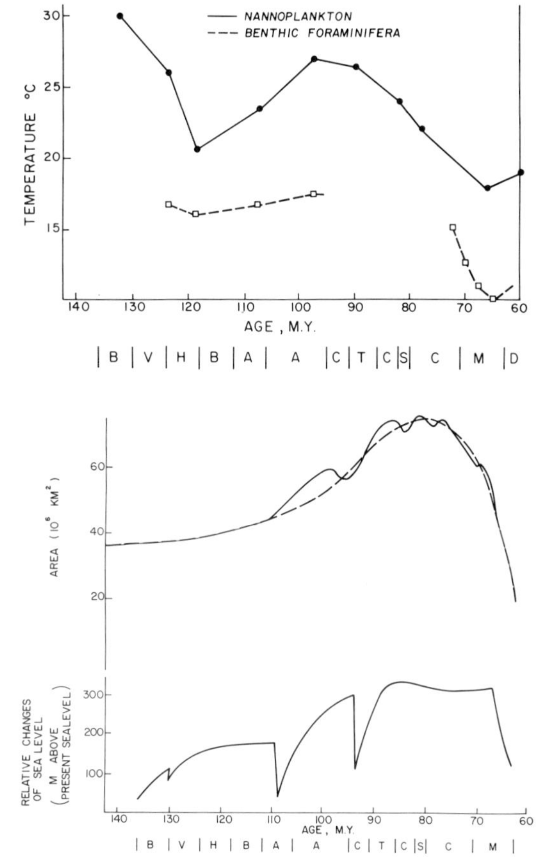

FIG. 3.—Cretaceous paleotemperatures and extent of epicontinental seas. *A,* Cretaceous temperature estimates based on oxygen-isotope ratios in fine fractions of deep-sea sediments, mostly nannoplankton remains (solid line) and benthic foraminifera (dashed line); the former reflects surface water temperature, the latter, bottom water temperature. Increased salinity of bottom waters would increase the differences between the two curves and might be partially responsible for it. Notice a steady increase in surface-water temperature during the Middle Cretaceous accompanied by a much smaller increase in bottom water temperature. (Based mostly on Douglas and Savin, 1975; Savin, 1977). *B,* Extent of epicontinental seas and sea level stand. Upper Curve: Extent of epicontinental seas, modified from Douglas and others (1973). Lower Curve: Relative sea level stand above present-day sea level, modified from Vail and others (1978).

mentological and paleontological data into a fairly good spatial framework. During the Mid-Cretaceous the North Atlantic was already a large but much more enclosed ocean basin about half the width of the present-day ocean, with a maximum depth of 5.5km in its deepest basins. The Mid-Atlantic Ridge divided the North Atlantic into two major basins: the western or *North American Basin* and the eastern or *West European-African Basin.* The shallow *Newfoundland - J-Anomaly Ridge* partly isolated the northern part of the North Atlantic from the southern part (Fig. 3). Deep connections to the Arctic Ocean were lacking during all of the Middle Cretaceous, and the connection to the South Atlantic did not open until the late Middle Cretaceous. Connections with the *Tethys Seaway* were fairly wide and probably quite deep. The connection with the Pacific across Central America was shallow and intermittently restricted exchange of deep and intermediate waters between the Atlantic and Pacific Oceans (Berggren and Hollister, 1974, 1977). Thus the North Atlantic Ocean had good surface water connections with the Tethys and moderate ones with the Pacific, but was a relatively isolated ocean basin with respect to deepwater exchange.

During the early part of the Middle Cretaceous the South Atlantic was a narrow mediterranean-type sea that had no connections to the North Atlantic nor the Tethys across Africa. A shallow barrier less than one kilometer deep, the *Walvis-Rio*

Grande Ridge, divided the South Atlantic into two basins. The southern boundary of the South Atlantic consisted mainly of the *Falkland Plateau,* which was no deeper than half a kilometer and restricted deep-water exchange with the Indopacific until the later part of the Middle Cretaceous (about 100 m.y.BP). At that time, seafloor spreading caused the eastern tip of the Falkland Plateau to clear the southern tip of the African continent, and a deep connection with the Indian Ocean, the *Agulhas Seaway,* formed (McCoy and Zimmerman, 1977).

CLIMATE AND CIRCULATION PATTERNS

Information pertinent to the reconstruction of the Middle Cretaceous climate is derived from paleobiogeographical, stratigraphic and oxygen-isotope studies (see reviews by Berggren and Hollister, 1974, 1977; Savin, 1977). Temperature gradients were much lower during the Cretaceous than today, perhaps as low as half the present temperature spread between the equator and the poles (Fig. 3A). Precipitation patterns are more difficult to estimate. There are good indications from palynological and organic-geochemical studies (McLachlan and Pieterse, 1978; Morgan, 1978; Tissot and others, 1979) that Middle Cretaceous climates in South Africa and Angola were fairly cool and dry, which is to be expected because greater continentality was caused by the closer proximity of the southern land masses. Extensive evaporite deposition in coastal lagoons in the Gulf of Mexico-Caribbean and Atlas regions and along the Brazilian and Angolan margin of the South Atlantic also points to a somewhat arid climate in the subtropical belt (Fig. 5). Occurrences of lignites of Mid-Cretaceous age along the East Coast of North America and Europe, and considerable input of detrital organic carbon along the eastern margin of the North Atlantic are indicative of lush vegetation and humid climates at mid-latitudes (30° and 45° N). No definitive evidence for Mesozoic glaciations have been found so far on our globe (Schwarzbach, 1974), and we can consider the Cretaceous an *acryogenic* time with ice-free poles. This has important consequences for atmospheric and oceanic circulation during the Middle Cretaceous. Lower temperature gradients would lead to a more sluggish atmospheric circulation resulting in more sluggish surface currents. Meridional transport of sensible heat by the oceans was larger during the Mid-Cretaceous than today. This resulted in a slight temperature decrease at low latitudes and a temperature increase at mid and high latitudes (Budyho, 1977). Precipitation would be high at mid to high latitudes under these conditions.

The lack of polar ice caps did not allow the formation of cold saline bottom waters by freezing of seawater on polar ice shelves. Thus, deep and bottom circulation was haline rather than thermohaline, and it was driven by excess evaporation at low and mid latitudes and in marginal seas (Chamberlin, 1906; Berger, 1970; Roth, 1978). Extensive Mid-Cretaceous evaporite deposits occur in marginal and mediterranean seas on both margins of the North and South Atlantic Ocean (Lotze, 1957; Institute of Petroleum, 1965; Evans, 1978; Arthur, 1979). Major sources for high salinity deep and bottom waters in the North Atlantic were the Yucatan-Gulf Coast-Florida evaporite basins and possibly other shallow carbonate banks along the eastern margin of North America. Brine flow from the shallow basins of the Atlas region and possibly from other parts of the Tethys margins are less certain sources. Injection of high-salinity waters from the South Atlantic (Thierstein and Berger, 1978) could have provided a small contribution to bottom waters only during the Late Albian to Cenomanian when a permanent moderately deep connection between the two ocean basins was established (Reyment and Tait, 1972; Kennedy and Cooper, 1975; Premoli-Silva and Boersma, 1977; Caron, 1978; Scheibnerova, 1978; Asmus and Ponte, 1973; Ponte and Asmus, 1978; Wiedmann and Neugebauer, 1978; Förster, 1978). During the early part

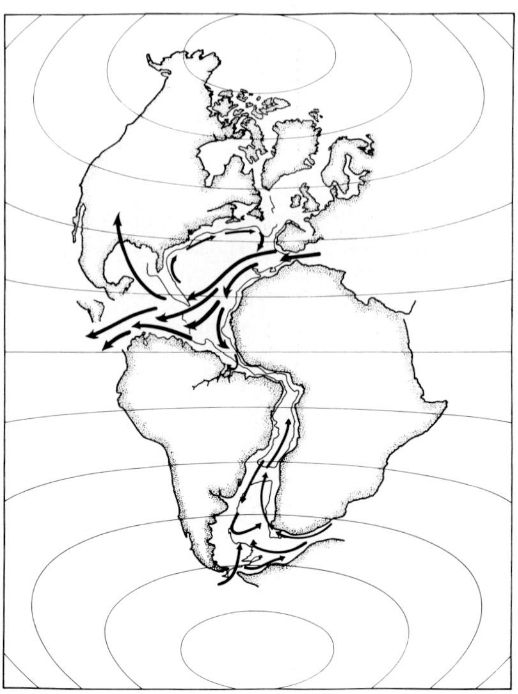

FIG. 4.—Inferred surface-water circulation during the Middle Cretaceous.

of the Mid-Cretaceous, *i.e.,* from the Barremian to the Middle Albian, brines from marginal basins and excess evaporation over carbonate banks and shelves was the main source of North Atlantic deep and bottom waters.

Increased salinity in deep and bottom waters was the major factor resulting in increased stability of the water column in the North Atlantic south of the Newfoundland - J-Anomaly Ridge. To the north of this ridge excess river inflow and high precipitation lead to a dilution of surface waters (Ryan and Cita, 1977; Arthur, 1979).

The *South Atlantic* was an isolated narrow mediterranean sea during the Early and Middle Cretaceous. Deep circulation of the South Atlantic is quite straightforward (McCoy and Zimmerman, 1977; Arthur and Natland, 1979). High evaporation rates in the northern end of the South Atlantic produced saline dense waters that sank to the bottom of the Angola-Brazil Basin; it was then forced to cross the shallow Walvis-Rio Grande Ridge, resulting in slight upwelling with increased mixing of deep and surface water, and finally descended into the Cape Basin. Bottom waters in both the southern and northern basins of the South Atlantic were quite stagnant, at least until deepwater connections with the Indopacific were established about 100 m.y.BP, and a wide passage to the North Atlantic had opened. Brine formation in the Angola-Brazil Basin ceased because this basin had become less isolated by Late Albian time. Some upwelling also occurred over the Falkland Plateau during the early Middle Cretaceous resulting in monospecific nannoplankton blooms (Wise and Wind, 1977; Noel and Manivit, 1978). Inferred circulation patterns for the Atlantic are shown on Figures 4 and 5.

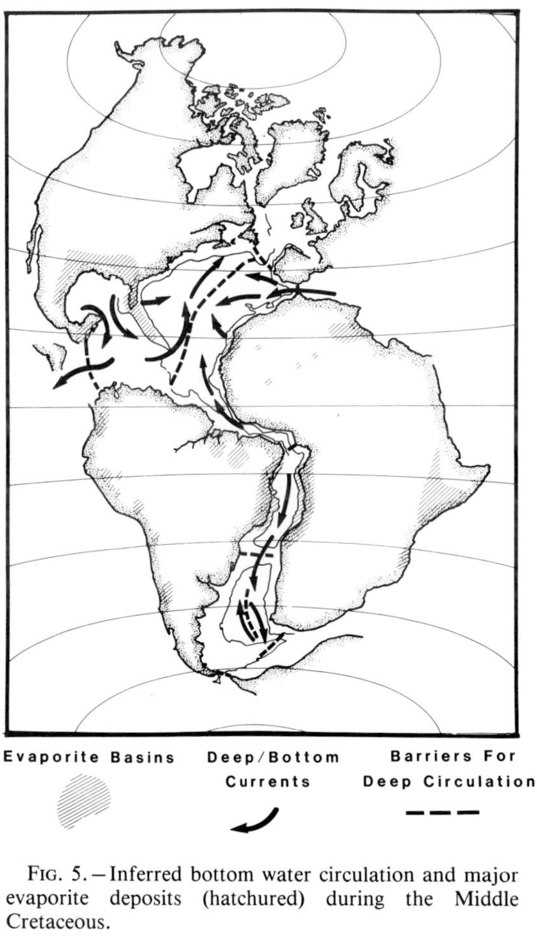

FIG. 5.—Inferred bottom water circulation and major evaporite deposits (hatchured) during the Middle Cretaceous.

According to W. W. Hay (oral presentation, AAPG-SEPM meeting, Houston, 1979) rapid deposition of evaporites in the South Atlantic during the Aptian could have reduced surface-water salinities by as much as 4–5‰ in the global oceans. Thus *paleogeographic isolation* (especially poor deepwater connections), *brine inflow* from marginal seas and *reduced surface-water salinities* contributed to great stability of the water column. *Low dissolved-oxygen concentration* in the much warmer deep and bottom waters in combination with slow mixing resulted in widespread stagnation of deep waters. The distribution of the major facies patterns discussed in the following section was greatly influenced by climate and oceanic circulation.

SEDIMENTARY FACIES PATTERNS IN THE ATLANTIC DURING THE MIDDLE CRETACEOUS

The methods of plate stratigraphy (Berger and Winterer, 1974) are used to determine the proper spatial and temporal framework of sedimentation in the Atlantic Ocean during the Mesozoic. North Atlantic sites were backtracked and the major facies patterns are shown for the Mesozoic of the Atlantic on Figures 6A and 6B. Somewhat more detailed facies diagrams of the major drilled sections are shown on Fig. 7. The distribution of the sites in a depth versus latitude plot (Fig. 8) shows how meager our coverage is, but all available sections with sufficiently well-preserved nannofossils were used, and only future drilling can help to increase the available information.

From the backtracking and facies distribution patterns we can deduce the following simplified history of sedimentation. During the Late Jurassic, pelagic limestones are the predominant facies in the North Atlantic (Jansa and others, 1977). The calcite compensation depth (CCD) was at about five kilometers depth at that time. During the Middle Cretaceous deepsea carbonates were mostly restricted to shallow regions (*e.g.,* Sites 390 and 392). Sites close to the crest of the Mid-Atlantic Ridge received calcium carbonate but at a fairly low rate resulting in marls rather than pure biogenous deepsea carbon-

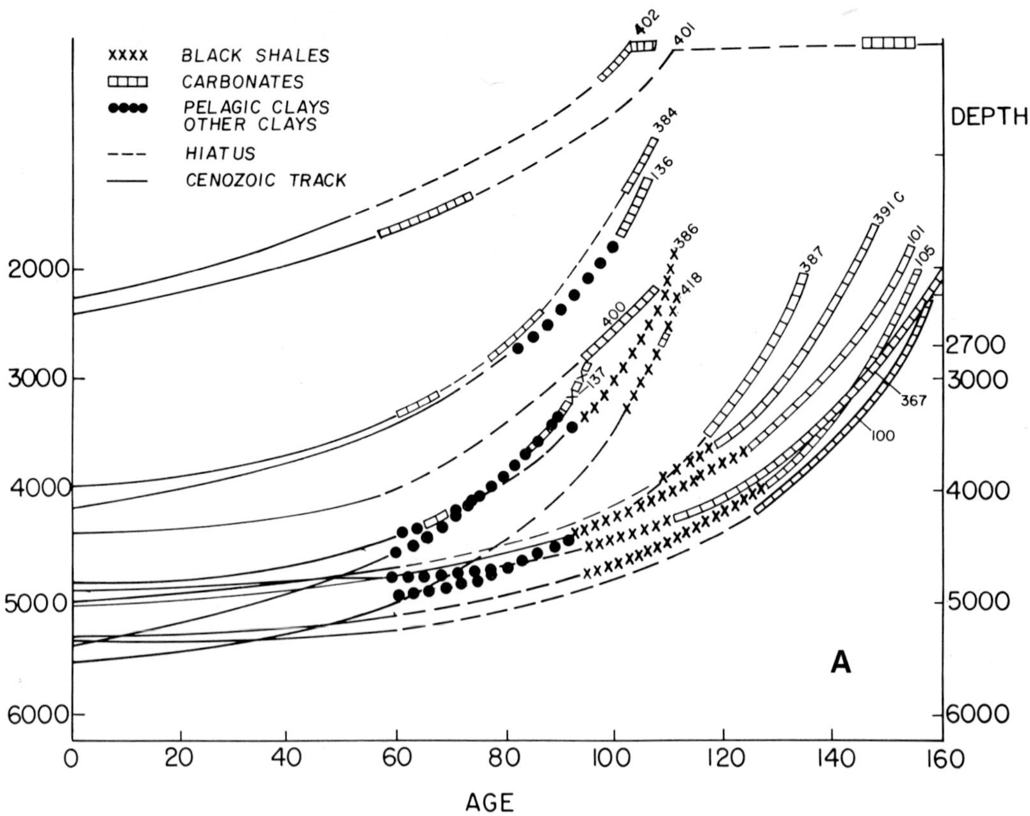

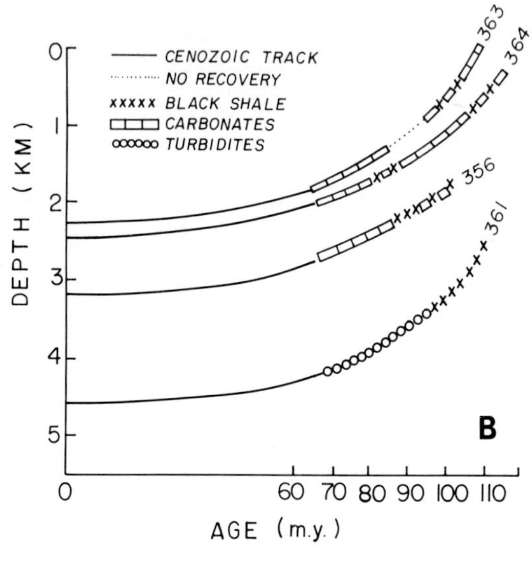

FIG. 6.—Paleodepth and facies reconstructions for the Mesozoic of the Atlantic: *A*, North Atlantic DSDP-IPOD sites. *B*, South Atlantic DSDP-IPOD sites.

ates. Black shales and mudstones with redeposited carbonates are the predominant facies in the deep western basin. Sediments enriched in organic carbon extend from the deep basin-floors to regions as shallow as a few hundred meters along the continental margin and up to the ridge crest (Sites 386 and 137) in the North Atlantic. Thierstein (1979) summarized the distribution of calcium carbonate and organic carbon in Mesozoic oceans based on calcium carbonate and organic carbon determinations stored at the DSDP data bank. The major change in carbonate deposition is generally interpreted as an indication of a shallowing of the CCD by about two kilometers around 115 m.y.BP (Fig. 9).

This rapid shallowing of the CCD by over two kilometers is difficult to explain. Rapid extraction of carbonates during evaporite formation (Arthur, 1979) and increased deposition of widespread shallow-water carbonates during low sea-level stands (see Fig. 3B) at the beginning of the Mid-Cretaceous are possible explanations. Widespread shallow-water rudist limestones in the Tethyan realm indicate that the latter process might have been important. Radiometric time-scales for the Mid-Cretaceous are too uncertain to demonstrate any

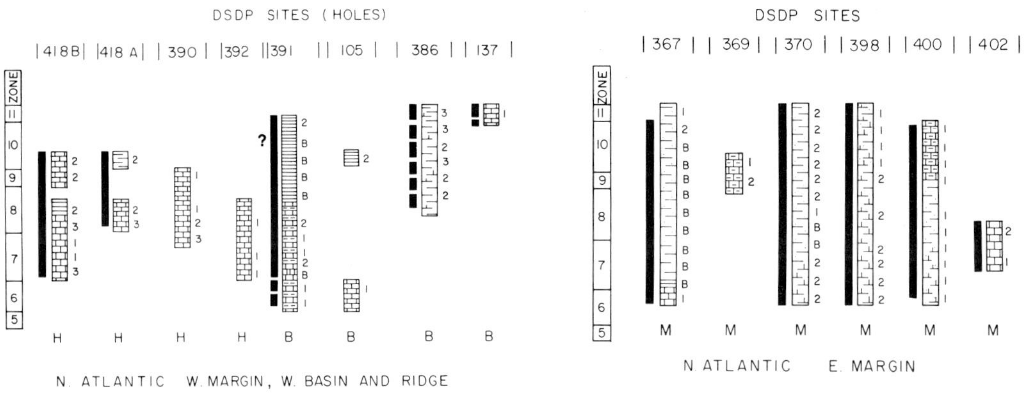

basin-basin fractionation between the major oceans. Also, too much of the Tethyan seafloor has been destroyed by subduction and collision or is too poorly known to make reliable estimates of the overall carbonate budget of the Lower and Middle Cretaceous oceans. It is also difficult to determine actual flux rates rather than relative abundances of organic carbon and carbonate and to fully evaluate the effects of dissolution and dilution by terrigenous input.

Sedimentation of pelagic carbonate and organic carbon are intimately linked in the Mid-Cretaceous as during other time periods. There is in general a negative correlation between organic carbon and calcium carbonate content of the sediments of Middle Cretaceous age (Thierstein, 1979). This is contrary to the conditions in the present-day oceans where a positive correlation is observed (Heath and others, 1977). This relationship indicates that dissolution of carbonate is largely due to carbon dioxide produced by the catabolism of organisms using organic carbon as an energy source. Breakdown of organic carbon took place both in the water column and in the sediments as indicated by the molds of coccoliths in "black shales" in the Cape Basin (Noel and Melguen, 1978; Noel and Manivit, 1978). Oxidation of organic matter in the water column would use up much of the available oxygen and would be responsible for the anoxic conditions in the warm and stagnant bottom waters. Dissolution of carbonates would in turn increase the relative proportion of organic carbon and detrital matter incorporated in the sediments.

The exact position of the boundary between oxygen-rich and anoxic waters plays an important part in the carbonate chemistry of the sediments (Berger, 1979). If sulphate reduction takes place at the sediment-water interface alkalinity may increase sufficiently to preserve calcium carbonate; if sulphate reduction occurs within the interstitial waters of the sediments below the interface, the alkalinity increase comes too late and much carbonate has already dissolved. Locally abundant pyrite and large fluctuation in the preservation of calcite in "black shale" sequences are an indication that fluctuations in the upper boundary of the sulphate reduction zone may have been important in the Mid-Cretaceous.

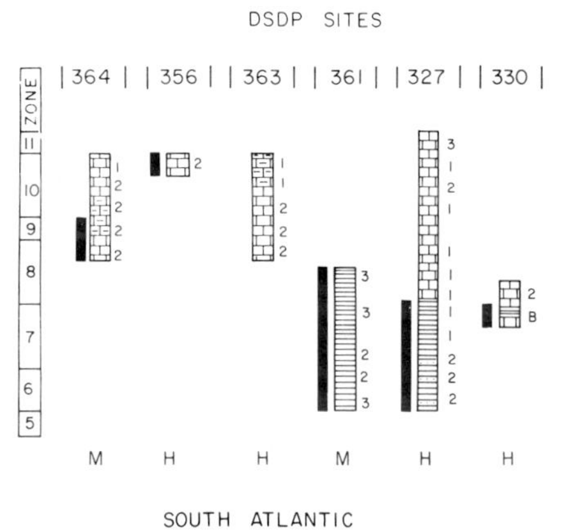

FIG. 7.—Lithofacies diagrams of Middle Cretaceous deep-sea sections recovered by DSDP-IPOD that were used in this study plotted against nannoplankton zones. Black bars to the left of the lithofacies columns indicate intervals rich in organic carbon. Numbers to the right of each column denote degree of etching of coccoliths in the best-preserved assemblages (1 = slightly etched; 2 = moderately etched; 3 = strongly etched; B = nannoliths absent). The letters below the lithofacies column indicate the site location with respect to bathymetry of the sea floor (H = rises, plateaus and marginal platforms isolated from much terrigenous input; B = deep oceanic basins; M = continental margin). Zones refer to NC zones and are explained on Fig. 11.

Speculations on the origin of black shales have been summarized in recent reviews by Fischer and Arthur (1977), Arthur and Natland (1979), Arthur (1979), Berger (1979) and Tissot and others (1979).

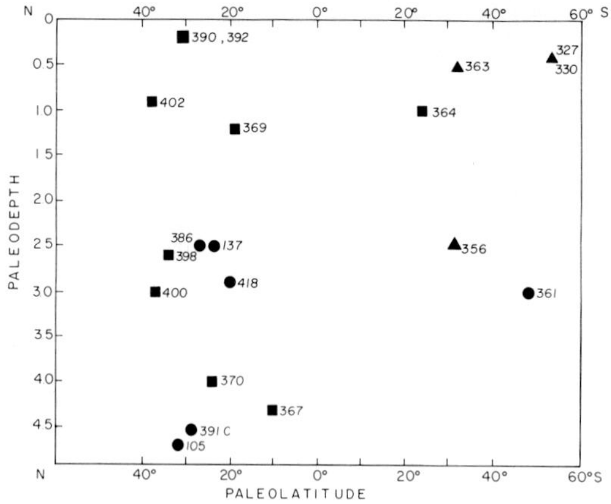

FIG. 8.—DSDP-IPOD sites used in this study plotted in a paleodepth-paleolatitude framework at 100 m.y.BP. (Symbols used: square = continental margin sites; circles = open ocean sites; triangles = sites on plateaus and shallow ridges.)

Earlier studies of the Mid-Cretaceous anoxia were mostly concerned with finding single mechanisms for the preservation of organic matter: either (1) more extensive oxygen-minimum zone (Schlanger and Jenkyns, 1976), (2) basinwide stagnation due to a fresh-water lid (Ryan and Cita, 1977), or (3) saline brine inflow (Roth, 1976, 1978; Thierstein and Berger, 1978). Many environmental conditions result in the preservation of black shales (Fig. 10). Recent studies of the organic geochemistry and palynology of Mid-Cretaceous shales have made it possible to estimate the relative contribution of detrital organic matter derived from terrestrial sources and marine (phytoplankton-derived) organic matter (Tissot and others, 1979; Habib, 1979). These studies show that most of the organic carbon in the sediments from the western North Atlantic is of terrestrial origin. Along the eastern margin of the Atlantic and in the Brazil-Angola Basin considerable amounts of marine organic matter have been preserved in Mid-Cretaceous sediments. Increased productivity due to upwelling as well as rapid burial were responsible for the increased preservation of marine organic matter. This corroborates studies of nannoplankton assemblages as shown in subsequent paragraphs on nannoplankton biogeography.

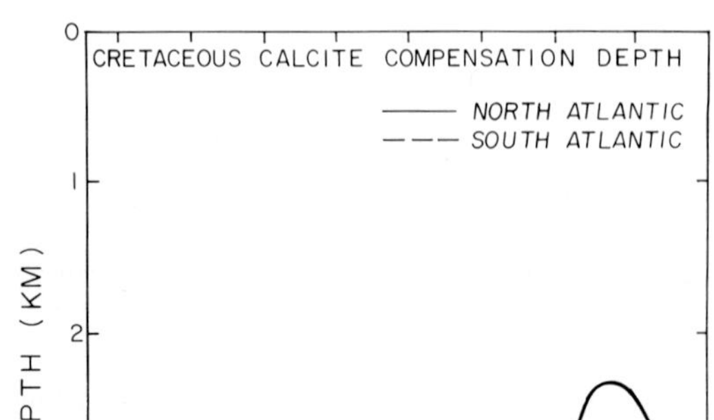

FIG. 9.—Generalized calcite compensation depths are here defined as the boundary between deep-sea carbonates and clays and are obtained by stacking age-depth facies traces of DSDP-IPOD sections. (After van Andel and others, 1977; Thierstein and Okada, 1979; and our observations.)

Cyclic alternation of oxidized and reduced sediments are reflected by changes in color (black, dark gray to greenish sediments vs. light gray to reddish sediments) and also by the presence of intensive burrowing in some beds and lamination in others (Dean and others, 1978; Gardner and others, 1978). Thus, dissolved-oxygen concentrations in bottom waters were fluctuating.

Dilution of carbonate and organic carbon were important along the European margin of the Atlantic and off North Africa. Dissolution of the carbonates resulted in a relative increase in organic carbon in some of the "black shales" (Tucholke and Mountain, 1977). Thus, the amount of calcium carbonate and organic carbon preserved in the sediments is determined by: (1) surface productivity and calcite/organic carbon ratio of the plankton, (2) dilution by terrigenous silicates, (3) influx of detrital organic carbon, (4) oxidation of organic carbon in the water column and at the sediment-water interface, (5) and dissolution of calcium carbonate at the sediment-water interface or within the sediment.

Plankton productivity was probably fairly low in the open Atlantic during the Mid-Cretaceous. Only at the eastern margin of the North Atlantic was plankton production high enough to preserve appreciable amounts of marine organic carbon. In the western North Atlantic, organic matter preserved in sediments is largely detrital. Decreased oxidation of organic matter in the oxygen-starved deep waters and at the sediment-water interface suffice to explain the occurrence of the organic carbon-rich deep basinal sediments. In continental margin settings, large detrital carbon input by rivers and rapid burial of organic matter played an important part in the origin of organic carbon-rich sediments (Fig. 10).

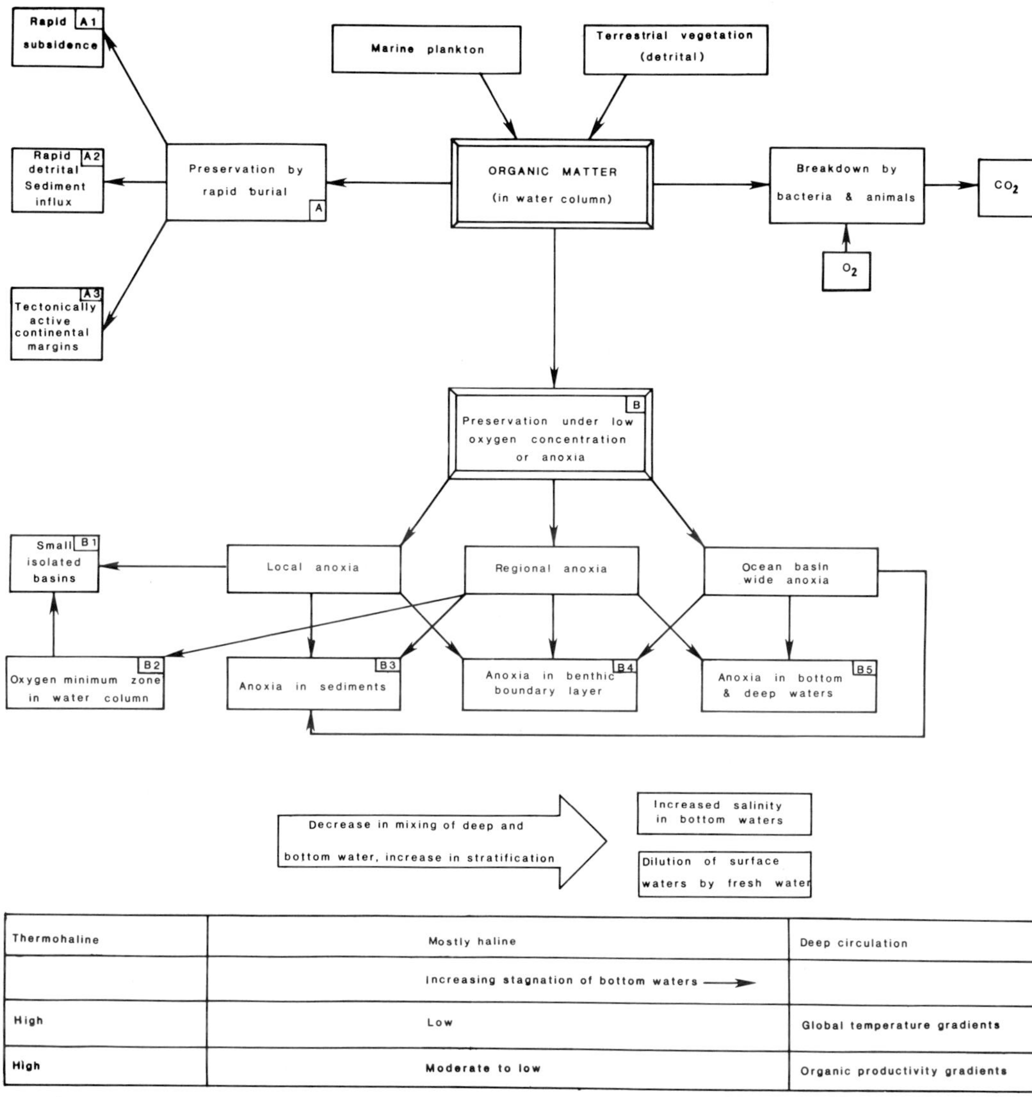

FIG. 10.—Sources and major pathways resulting in the preservation of organic matter in the marine environment. *A,* Rapid burial pathway: incorporation of organic carbon into the sediment before it has time to oxidize. This pathway is most important on continental margins and on abyssal plains with turbidity-current sedimentation. Conditions A2 and A3 were at least partly responsible for the preservation of organic matter at the eastern North Atlantic Sites 400, 402, 398, 370, 369, 367, and at Site 361. *B,* Anoxia pathway: preservation due to lack of oxygen in water column or interstitial waters within sediment. Modern examples of condition B1 are found in California Borderland basins and the Cariaco Trench. Oxygen minimum zones (B2) are important in the East Pacific and Indian Oceans today but probably require a strong pycnocline and cold deep waters to fully develop. Conditions B3, B4 and B5 require low dissolved-oxygen in deep and bottom waters and sluggish circulation. They are the most likely conditions resulting in black-shale formation in the Mid-Cretaceous Atlantic (see text).

Sediments at shallow sites in the western Atlantic (Sites 390 and 392) received mostly calcareous plankton remains with very little clay and were deposited under oxygenated conditions. Extensive turbidites transported carbonates into the deep Blake Basin (Site 391) resulting in an unusually carbonate-rich section for such a deep lower Middle Cretaceous site.

Sedimentation patterns in the South Atlantic have been summarized by van Andel and others (1977). Evaporites formed in the Angola-Brazil Basin during the Aptian. Dark laminated calcareous oozes, chalks, limestones, marls and claystones were deposited close to the continental margin in the Angola Basin and on the Walvis-Rio Grande Ridge. They commonly overlie carbonates

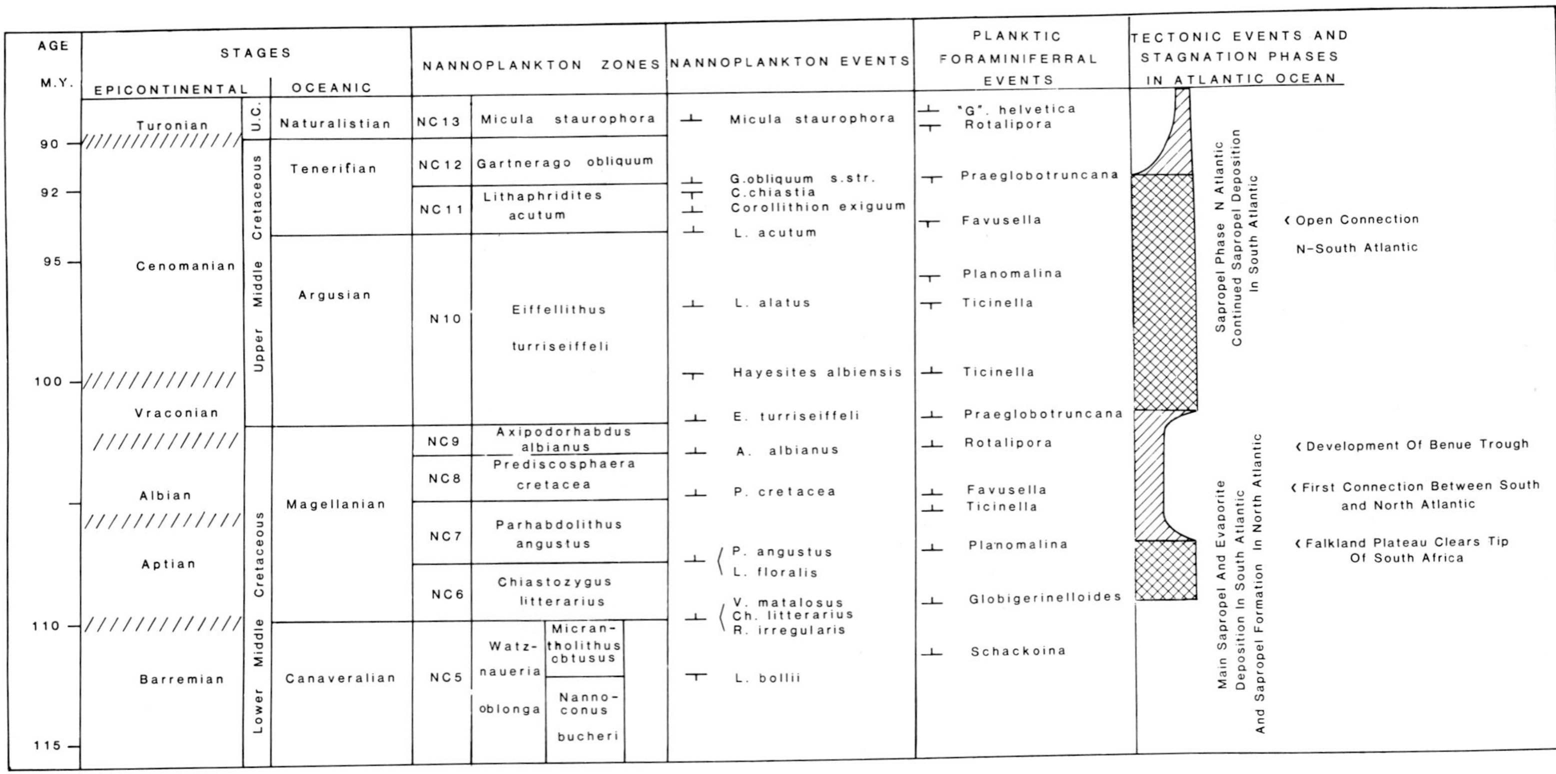

FIG. 11.—Middle Cretaceous biogeochronology and important oceanographic and tectonic events. Oceanic stages, NC-zones, correlation with radiometric time scale and classical land and oceanic stages after Roth (1978); planktic foraminiferal events after Sigal (1977).

indicative of shallowwater origin. Black shales and turbidites characterize the deeper Cape Basin (Site 361). Much terrestrial detritus and organic matter were transported to the deep Cape Basin at that time (Davey, 1978; Natland, 1978). Sedimentation on the Falkland Plateau was characterized by organic-rich shales and marls of early Middle Cretaceous age overlain by chalks deposited under oxygenated conditions, indicating a shallowing of the upper boundary of the anoxic deep waters with time, due to changes in the deep-water circulation related to the opening of the Agulhas Seaway.

BIOGEOCHRONOLOGY OF THE MIDDLE CRETACEOUS

One of the most frustrating aspects of Mesozoic paleoceanography is the lack of a reliable and stable time scale. Nannoplankton have provided the major means of dating and correlating oceanic sections of Lower and Middle Cretaceous age due to a dearth of foraminifera in many of the intervals recovered. The nannoplankton zonations for the Cretaceous have become quite consistent (Thierstein, 1976; Roth, 1978; Perch-Nielsen, 1979). The letter and number abbreviation system of nannofossil zones used in this paper is based on Roth (1978) and is summarized on Figure 11. The symbol "NC" stands for "nannofossil Cretaceous" and the numbering starts at the base of the Lower Cretaceous. If two zones are combined or cannot be separated because of the absence of the marker species, the numbers are separated by a slash (*e.g.*, NC8/9). We include Zones NC5 through NC12 in the Middle Cretaceous, an interval that includes the *Canaveralian, Magellanian, Argusian* and *Tenerifian* Oceanic Stages and the *Barremian, Aptian, Albian* and *Cenomanian* Land Stages. Although radiometric age boundaries for that time interval are uncertain, we assume about 115 m.y.BP for the lower boundary and about 90 m.y.BP for the upper boundary of the Middle Cretaceous (or "Mid-Cretaceous" to use a more informal term). Biostratigraphic resolution of nannoplankton stratigraphies is variable in the Middle Cretaceous. It is quite good for the earlier part of the Mid-Cretaceous (Zones NC5 through NC9), but leaves much to be desired for the upper part of the Middle Cretaceous (Zones NC10 and NC11). Although additional nannoplankton first- and last-appearance surfaces have been determined for the late Middle Cretaceous time interval, the markers proposed so far (*e.g.*, Manivit and others, 1977; Perch-Nielsen, 1979) are rare and have spotty occurrences. A combined biostratigraphy using nannoplankton and planktic foraminiferal events increases resolution considerably, since planktic foraminifera allow for a more detailed subdivision of the upper Middle Cretaceous (Fig. 11).

The establishment of an absolute chronology of the Middle Cretaceous is difficult. Various Cretaceous time scales have been proposed recently (van Hinte, 1976; Thierstein, 1976; Sigal, 1977; Roth, 1978). Differences in absolute ages of identical stratigraphic boundaries range from one to seven million years. We use the scale compiled by Roth (1978) with a short Cenomanian and a long Albian, which seems to be supported by radiometric dates of igneous and volcanogenic rocks (Obradovich and Cobban, 1975, 1977) rather than by glauconite dates that are frequently unreliable. Correlation of the calcareous plankton stratigraphic scale with land stages, based on sections with only poorly to moderately well-preserved plankton assemblages, is fraught with problems. This led Roth (1978) to propose oceanic stages which are also largely coincident with major events in the oceanic sedimentation history during the Middle Cretaceous. The Canaveralian-Magellanian boundary marks the transition from predominantly carbonate sedimentation in the Atlantic to the deposition of black shales. Anoxic sediments are most widespread during the Magellanian and Argusian and locally during the Tenerifian which marks the transition to oxic conditions over most of the oceans with only minor phases of anoxic deposition in the South Atlantic during later times. The biogeochronologic framework used in this study is summarized on Figure 11.

METHODS OF STUDY OF NANNOFOSSILS

A total of 313 samples were studied to determine nannoplankton distribution and preservation. The majority of the samples are from cores recovered by

TABLE 1.—DSDP-IPOD CORES USED IN THIS STUDY

Site	Paleolatitude[1]	Paleodepth[1]	Paleo-environment[2]
402	37.5°N	0.8	M
400	37 °N	3.0	M
398	34 °N	2.6	M
105	32 °N	4.7	O
390	31 °N	0.2	M
392	30.8°N	0.2	M
391	29 °N	4.5	O
386	27 °N	2.5	O
370	24 °N	4.0	M
137	22 °N	2.5	O
418,417	20 °N	2.9	O
369	19 °N	1.2	M
367	10 °N	4.3	M
364	24 °S	1.0	M
356	31 °S	2.5	H
363	32 °S	0.5	H
361	48 °S	3.0	O
327,330	53 °S	0.4	H

[1] At 100 m.y.BP
[2] M = Continental margin
O = Oceanic sites, deep basins to mid-ocean ridges
H = Shallow rises and plateaus with small detrital input.

the Deep Sea Drilling Project; 26 land samples were used for comparison (see Tables 1 and 2 for sample locations and paleolatitude, paleodepth of all sample locations). Sample preparation was kept simple, and any steps that might produce a bias in nannolith assemblages (*e.g.*, size fractionation, ultrasonication) have been avoided. A small amount of sediment was dispersed in a drop of water and spread on a microscope coverglass which was dried and later mounted with synthetic mounting medium for light microscope study. Age determination for all samples was done by the senior author using the zonation in Roth (1978) to assure a consistent set of criteria. About a quarter of the samples proved to be so poor that they were not used for more detailed investigations. Visual estimates of overall preservation were made using the scale proposed by Roth and Thierstein (1972) and further discussed in Roth (1973). Both visual preservation estimates and more quantitative dissolution indices were later used to select the 99 best-preserved samples for biogeographic studies. At least 300 specimens were counted in each sample using the light microscope and selecting random traverses. During our light-microscope studies 145 species were observed, of which 122 occurred in abundances of over one percent. Only 20 to 30 species occurred consistently and fairly abundantly. Morphological similarity was used as the main criterion for combining some species into larger groups for our subsequent biogeographical studies. Consistent co-occurrence of particular taxa was also used as a supporting factor in the consolidation of the unwieldingly large original species list. Table 3 shows the species combinations that are used in this paper.

TABLE 2.—LAND SAMPLES

Southeastern France
(see Thierstein, 1973 for details)

Samples	Age	Locality
TH 70/92	NC 10	Col de
TH 70/69	NC 10	Palleul
TH 70/60	NC 8	(Hautes-Alpes)
TH 70/188	NC 7	Leches en Dois (Drôme)
TH 69/356	NC 6	La Bédoule, Station de Cassis (Bouche-du-Rhône)
TH 70/162	NC 6 (base)	La Charce III
TH 70/156	NC 5 (top)	(Drôme)

England
(Samples collected by J. P. Beckmann, ETH, Zurich)

Gault Section, Copt Point, Folkstone, SE England
Gault 309, Bed XII, Upper Gault, Age: NC 10
Gault 306, Bed XI, Upper Gault, Age: NC 8
Gault 301, Bed III, Lower Gault, Age: NC 8

Barrington Cement Quarry, 15 miles SSW of Cambridge, England
BAR 323, Chalk, about 1 meter above Totteshoe Stone

Texas
(Samples provided by Merton Hill; see Hill, 1976, for more details)

Texas 6309, Duck Creek Formation, Grayson County, Age: NC 8
Texas 6259, Grayson Bluff, Denton County, Age: NC 11

TABLE 3.—SPECIES GROUPS USED IN THIS PAPER

Group	Species Included
Braarudosphaera sp.	all species of *Braarudosphaera* and *Micrantholithus*
Nannoconus sp.	all species of Nannoconus
Parhabdolithus splendens	*P. splendens, P. asper* and *P. swinnertonii*
Broinsonia sp.	*B. enormis, B. lata* and *B. signata*
Biscutum constans	*B. constans, B. gartneri*
Cretarhabdus	all species of *Cretarhabdus* and *Retecapsa*

PRESERVATION OF NANNOFOSSILS

Observation by light microscope showed that most coccolith assemblages (with the exception of samples from the lowermost Middle Cretaceous limestone) were more severely affected by dissolution than by overgrowth. In an effort to determine susceptibility of nannofossils to preservational changes, methods used by Roth and Berger (1975) for Quaternary coccolith preservation studies were adopted. All species were ranked according to their susceptibility to dissolution by using the following pairing analysis. It was observed that the better the preservation of the nannofossil assemblages the higher the species diversity in a sample. Uncertainty of paleodepth and the influence of proximity of the continents for some of the sites made the use of paleodepth impractical as a measure of preservation. Each species in one sample was compared with the same species in all neighboring samples that were within 20 degrees of latitude and belonged to the same age interval (either NC8 through 11 or NC5 through 7). If the species increased when going from the more highly diverse assemblage (better preserved) to an assemblage of lower diversity (more poorly preserved) it received a score of two, if it remained the same the score was one, and if it decreased the score was zero. A dissolution ranking was obtained by adding up all the normalized scores. As expected, judging from simple visual observations of Middle Cretaceous samples, *Watznaueria barnesae* is the species with the highest rank, *i.e.,* the most resistant one. Rare species that occurred only in very few samples received scores that are considered too low. Therefore Tables 4 and 5 show only the 15 most resistant species, which includes all of the common ones.

TABLE 4.—DISSOLUTION RANKING (ZONES NC 7 TO NC 5)

Rank	Species Name
1	*Watznaueria barnesae*
2	*Watznaueria ovata*
3	*Parhabdolithus splendens*
4	*Cyclagelosphaera margereli*
5	*Cretarhabdus* sp.
6	*Biscutum constans*
7	*Discorhabdus ignotus*
8	*Zygodiscus elegans*
9	*Lithraphidites carniolensis*
10	*Nannoconus* sp.
11	*Vagalapilla stradneri*
12	*Zygodiscus diplogrammus*
13	*Watznaueria supracretacea*
14	*Zygodiscus erectus*
15	*Stephanolithion laffitei*

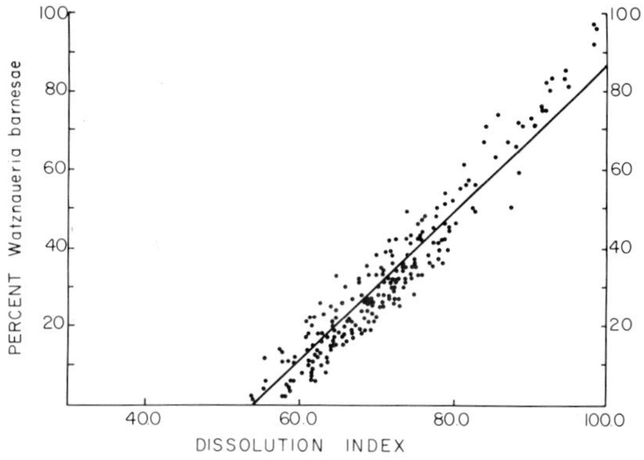

FIG. 12.—Dissolution index versus relative abundance of *Watznaueria barnesae*. The dissolution index is calculated as follows:

$$S = \Sigma_{pi} \bullet Ri/\Sigma_{pi},$$

where p_i is the proportion of species i, R_i is the rank of species i.

Using these dissolution rankings, a dissolution index was calculated similar to the one used in Roth and Berger (1975). The dissolution index is the sum of all species proportions multiplied with their respective rank divided by the sum of all species proportions. A plot of the dissolution index versus the abundance of *Watznaueria barnesae* shows an almost perfect straight-line correlation (Fig. 12). This shows that the dissolution index as calculated does not provide a more accurate measure of preservation than a visual estimate or simply the relative abundance of *Watznaueria barnesae* in a sample. Samples that contain more than 40 percent of this species are strongly affected by dissolution and were eliminated from our sample sets for biogeographic studies. Different ranking methods and dissolution index calculations that should eliminate some of the problems like spotty occurrences are in progress.

TABLE 5.—DISSOLUTION RANKING (ZONES NC 8 TO NC 11)

Rank	Species Name
1	*Watznaueria barnesae*
2	*Cretarhabdus* sp.
3	*Zygodiscus diplogrammus*
4	*Zygodiscus elegans*
5	*Parhabdolithus splendens*
6	*Watznaueria supracretacea*
7	*Biscutum constans*
8	*Manivitella pemmatoidea*
9	*Lithraphidites carniolensis*
10	*Zygodiscus erectus*
11	*Eiffellithus turriseiffelli*
12	*Prediscosphaera cretacea*
13	*Discorhabdus ignotus*
14	*Vagalapilla stradneri*
15	*Parhabdolithus angustus*

BIOGEOGRAPHY OF MIDDLE CRETACEOUS CALCAREOUS NANNOFOSSILS

Previous Work

The Middle Cretaceous calcareous nannoplankton was studied taxonomically by Manivit (1965, 1971) and Stover (1966). Gartner (1968) and Bukry (1969) did important taxonomic work on the Cretaceous calcareous nannoplankton using the electron microscope. Čepek and Hay (1969) published the first detailed zonation of the upper part of the Middle Cretaceous and the Late Cretaceous based on sections in the Gulf Coast and western interior of the U.S.A. Earlier investigations of DSDP core material were also largely biostratigraphic and taxonomic in nature (see Bukry and Bramlette, 1969; Wilcoxon, 1972; Worsley, 1971; Roth and Thierstein, 1972; Roth, 1973; Wise and Wind, 1977; Perch-Nielsen, 1977; Proto Decima and others, 1978; Čepek, 1978; Okada and Thierstein, 1979). Observation on preservational aspects and some remarks on nannoplankton biostratigraphy were added to some of the above biostratigraphic reports. Thierstein (1971, 1973, 1974) investigated the Lower and Middle Cretaceous stratotype sections of southwestern France and established a nannofossil zonation for the Middle Cretaceous that has global applicability and is used by the majority of nannopaleontologists. He also laid the groundwork for a sound Lower and Middle Cretaceous nannofossil taxonomy by using the light and scanning electron microscope to look at identical specimens (Thierstein and others, 1972). The late Maurice Black provided us with a monograph of the Gault clay with plates of astounding beauty; it was to be his last contribution to nannoplankton taxonomy, and it remained unfinished. It contains a wealth of information on the ultrastructure of coccoliths of Middle Cretaceous age (Black, 1972, 1973, 1975). Coccoliths from the classical Speeton Clay section (Berriasian to Barremian) of Northern England, first investigated by Black (1971), received a recent investigation of coccoliths resulting in a detailed range chart as well as some taxonomic revisions by Taylor (1978). She also showed that many of the low latitude markers used by Thierstein (1971, 1973) are absent from the Speeton section and that some of the assemblages have a boreal aspect with *Nannoconus abundans,* a species that so far has not yet been observed at low latitudes nor in the Southern hemisphere. A recent review of Cretaceous nannofossil biostratigraphy in the tethyan and boreal realm by Perch-Nielsen (1979) also contains some observations on the biogeography and ecology of nannofossils. Occurrences of *Nannoconus* in sedimentary rocks from high latitudes in the Southern Hemisphere are quite rare (Proto-Decima, 1974; Proto-Decima and others, 1979; Wise and Wind, 1977; Noel and Melguen, 1978), and the species are most abundant in the Tethys region and along the margins of the tropical Atlantic Ocean.

Coccolith studies in samples from the Cenomanian type section and other classical Middle and Upper Cretaceous sections by Verbeek (1977) and by Manivit and co-workers (Amedro, Bidar, and others, 1978) have greatly contributed to a better knowledge on the correlation of ammonite, foraminiferal and nannoplankton zonations for the late Middle Cretaceous.

Thierstein (1974) discovered the first Middle to Late Cretaceous nannoplankton species truly indicative of high latitudes, namely *Seribiscutum primitivum.* Wise and Wind (1977) found it also in great abundance in cores from the Falkland Plateau, and they identified a few other high-latitude species, namely *Sollasites falklandensis* and *Octocyclus magnus.* Manivit (in Amadro, Damotte and others, 1978) extended the northern biogeographic range of *Seribiscutum primitivum* from southern England to northern France. Thierstein (1976) summarized Mesozoic nannoplankton stratigraphy and provided some information of Late Cretaceous nannofossil biogeography.

Berger and Roth (1975) published a plot of diversity of coccoliths in the Mesozoic and also the biogeographic map of the Cretaceous nannolith genus *Nannoconus.* Roth (1976, 1978 and 1979) discussed the diversity changes of nannoplankton in the Cretaceous and explained the great increase in species diversity at the beginning of the Middle Cretaceous by low fertility of surface water due to denitrification in poorly oxygenated deep waters; these in turn were caused by brine flow from marginal lagoons and hypersaline seas into the deep ocean basins, resulting in a highly stable stratification of the oceans. This change in diversity of coccoliths and its coincidence with widespread black shales in the ocean prompted us to undertake this investigation. Coccoliths have proven to be excellent water-mass indicators in the Recent and Cenozoic as shown by McIntyre and coworkers (see papers in Cline and Hays, 1976; Haq and others, 1977).

Figures 1, 2 and 8 clearly show that the sample coverage available for this study is limited. The majority of samples used in this part of our study are moderately to well preserved. However, even the best preserved assemblages are not pristine (Fig. 7), and thus nannoplankton taphocoenoses must be distinctly different from the biocoenoses. Therefore only large-scale patterns of the more abundant species can be discussed here.

Relative abundances of species that showed pronounced trends in their biogeographic distribution patterns are shown on map reconstructions adapted largely from Sclater and others (1977). The reconstruction at 95 m.y.BP was used for Zones NC10 and NC11 (92–102 m.y.). There were not enough samples that belonged to Zone NC11 (92–94 m.y.) to make it worthwhile to use separate plots for that time interval. A reconstruction of the continents at 110 m.y.BP was used for Zones NC9 through NC5. The intervals of Zone NC5 (110–115 m.y.) were combined with values from Zone NC6 (108–110 m.y.) on the plots, because sample coverage in the lower Middle Cretaceous is very poor. Only well-preserved samples were used for these plots (Figs. 13–22). In order to detect possible latitudinal trends in species abundances and to see whether there are any strong dissolution effects on relative abundances of those species that showed interesting paleogeographic patterns, their abundances were also plotted on a paleolatitude-paleodepth diagram. As in Figure 8, a paleodepth at 100 m.y.BP was used for all time periods. Errors in paleodepth determinations using backtracking are at least 200 meters or probably even more for many sites; therefore, using the paleodepth at 100 m.y.BP for all sites at all time periods on Figures 19–22 seems to be within the limit of error. In order to detect east-west gradients and distance from the shore, the following letter designations are used on Table 1 to distinguish deep basin sites (0), sites on the continental margin (M), and sites on a plateau or rise (H).

BIOGEOGRAPHIC PATTERNS OF THE MAJOR SPECIES

Watznaueria barnesae (Figs. 13, 19)

During the time period of NC10/11 *W. barnesae* shows low concentrations along the African-western European margin in spite of the moderate degree of etching, at least in the samples from Site

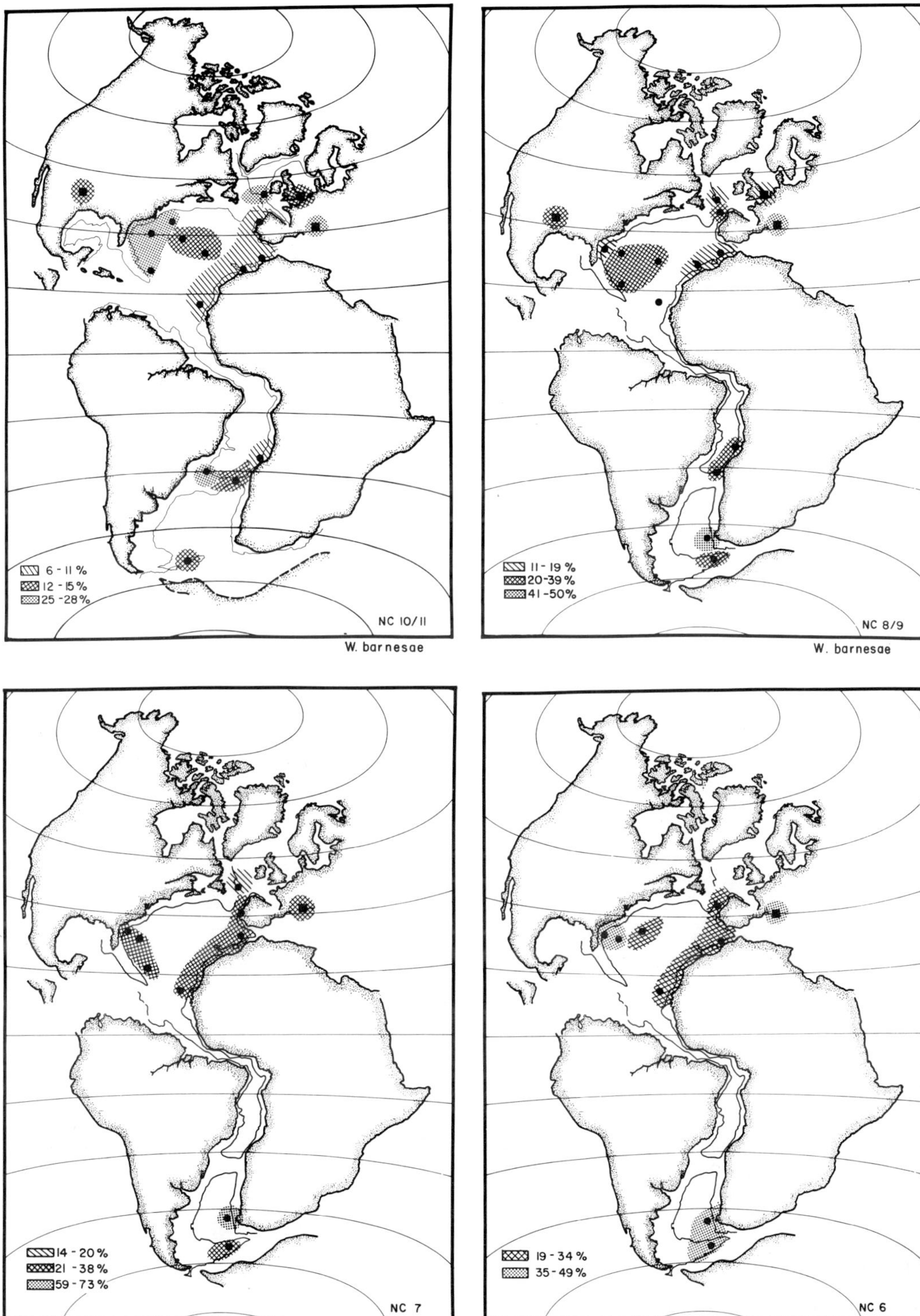

FIG. 13.—Biogeographic distribution of *Watznaueria barnesae* in the Atlantic Ocean. Plotted are the relative abundances in the best-preserved samples. *A,* Zones NC 10 to NC 11. *B,* Zones NC 8 to NC 9. *C,* Zone NC 7. *D,* Zone NC 6, including the top of NC 5.

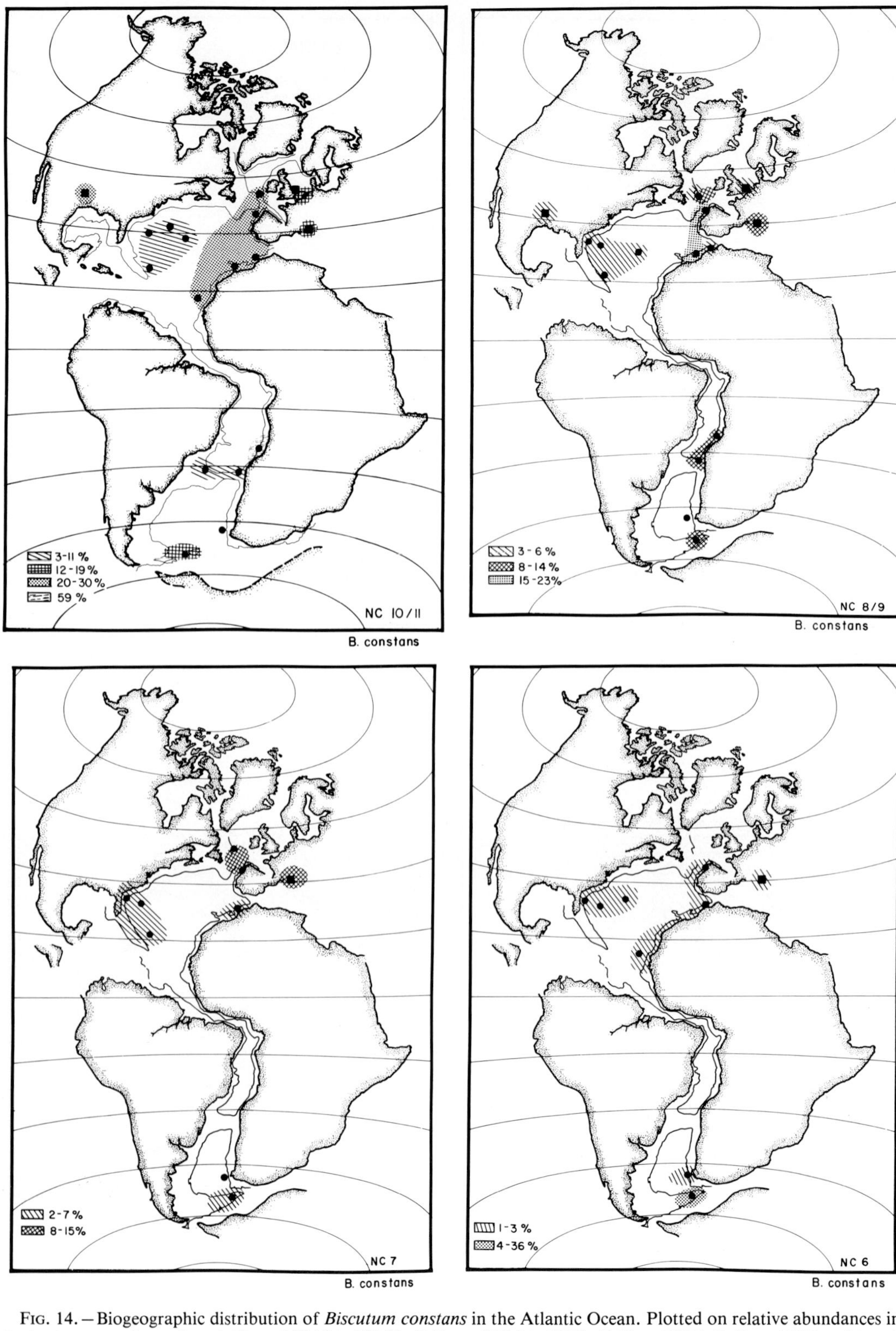

FIG. 14.—Biogeographic distribution of *Biscutum constans* in the Atlantic Ocean. Plotted on relative abundances in the best-preserved samples. *A,* Zones NC 10 to NC 11. *B,* Zones NC 8 to NC 9. *C,* Zone NC 7. *D,* Zone NC 6 and top of Zone NC 5.

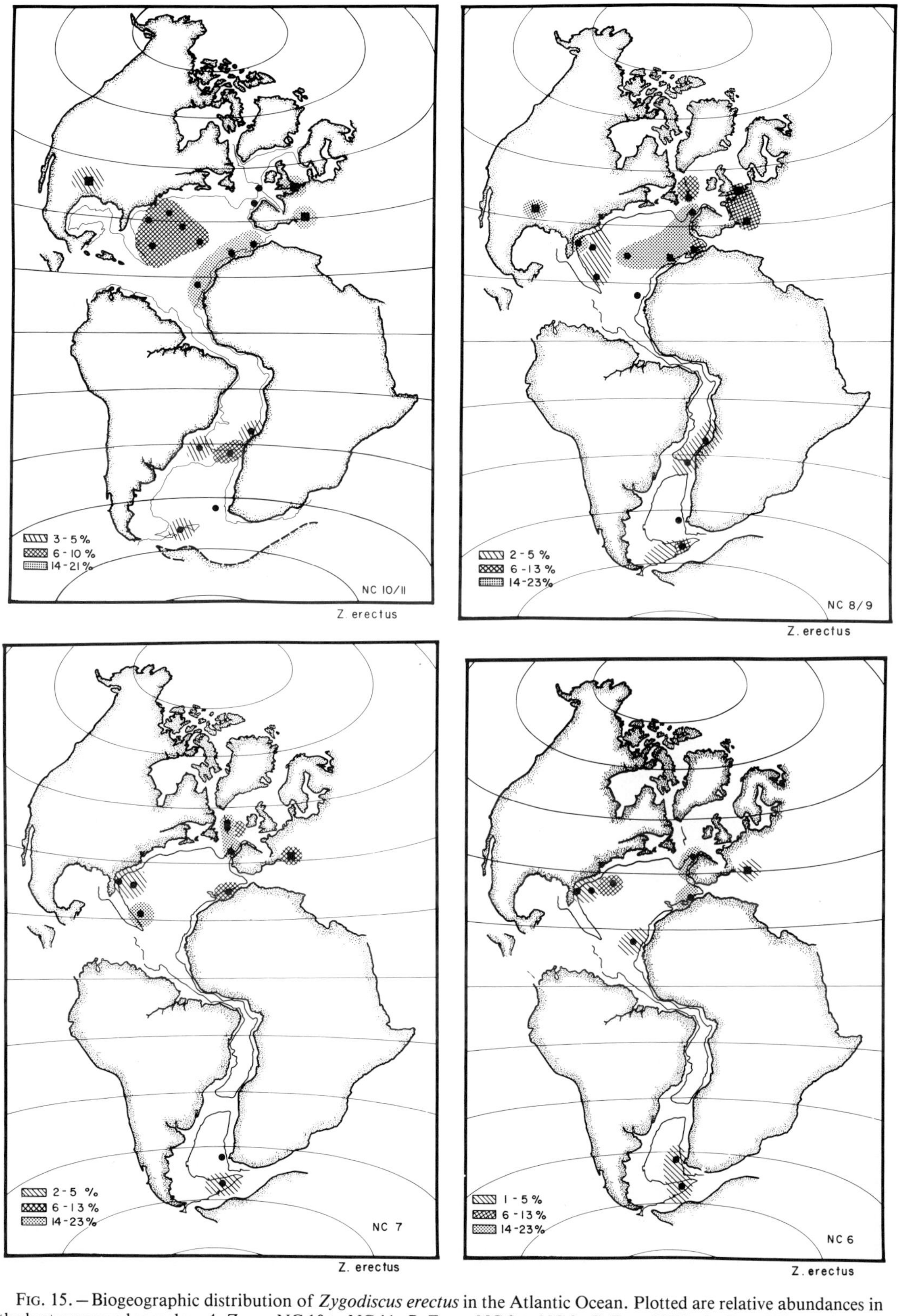

FIG. 15.—Biogeographic distribution of *Zygodiscus erectus* in the Atlantic Ocean. Plotted are relative abundances in the best preserved samples. *A,* Zones NC 10 to NC 11. *B,* Zones NC 8 to NC 9. *C,* Zone NC 7. *D,* Zone NC 6 and top of Zone NC 5.

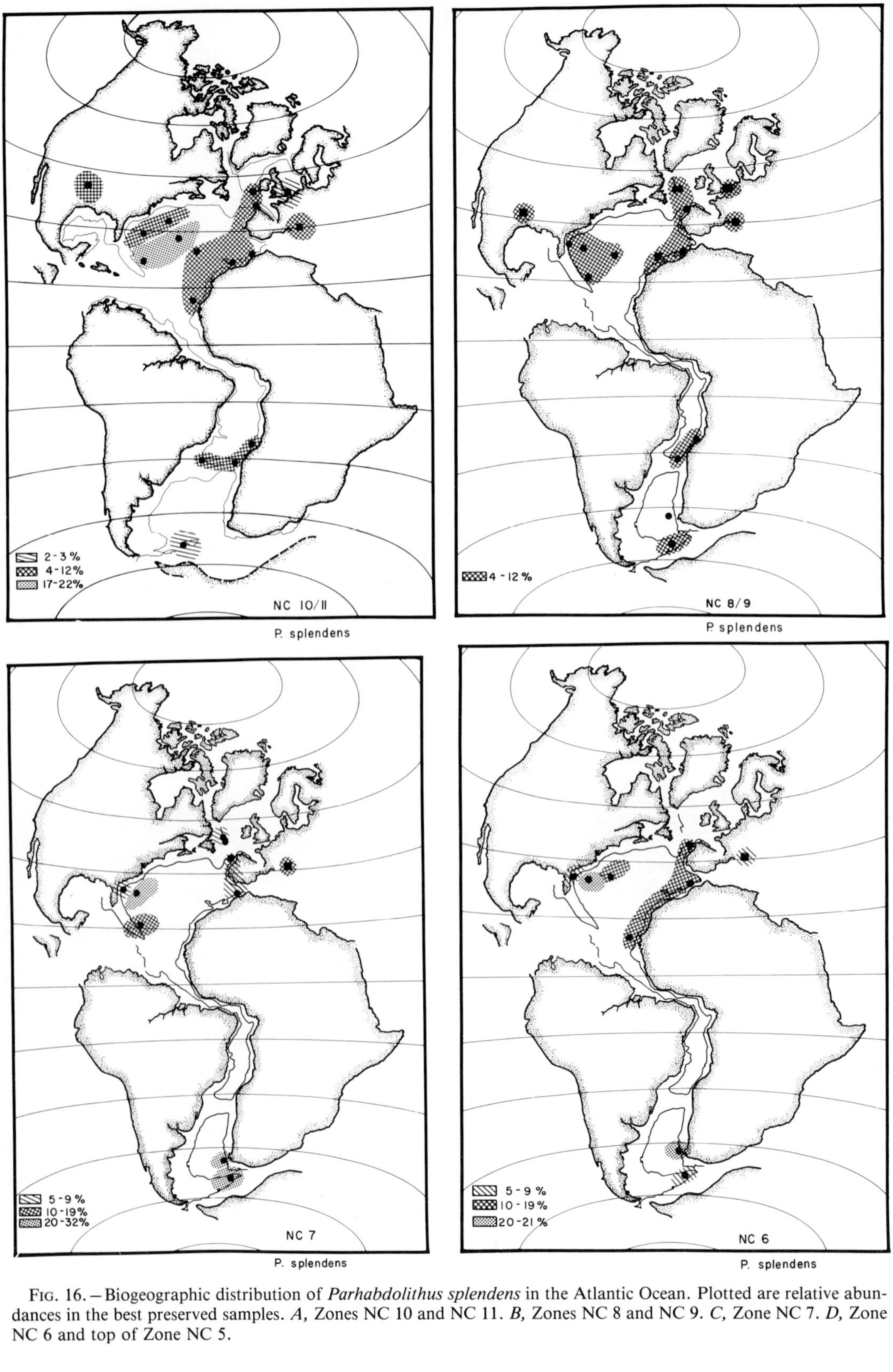

FIG. 16.—Biogeographic distribution of *Parhabdolithus splendens* in the Atlantic Ocean. Plotted are relative abundances in the best preserved samples. *A,* Zones NC 10 and NC 11. *B,* Zones NC 8 and NC 9. *C,* Zone NC 7. *D,* Zone NC 6 and top of Zone NC 5.

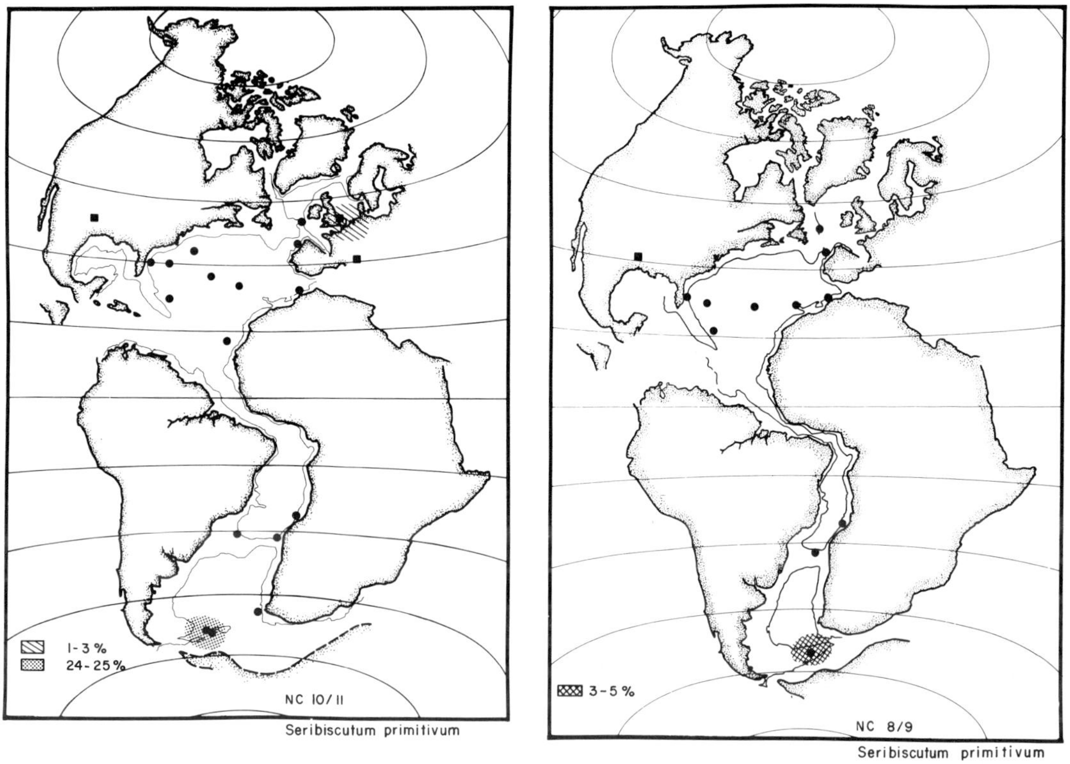

FIG. 17.—Biogeographic distribution of *Seribiscutum primitivum* in the Atlantic Ocean. Plotted are relative abundances in the best preserved samples. *A,* Zones NC 10 and NC 11. *B,* Zones NC 8 and NC 9.

370. Of the North Atlantic continental margin sites, Site 400 shows high *W. barnesae* values due to poor preservation. There is an increase in the relative abundance of this species with depth, but no latitudinal gradient was discovered. *W. barnesae* is an oceanic species that is less abundant in well-preserved assemblages from continental margins unless it has been enriched by dissolution. Nannofossil assemblages of moderate to poor preservation from epicontinental deposits (*e.g.,* from the Vocontian trough of south-eastern France) often contain abundant *W. barnesae.* Distribution in the South Atlantic also shows low values along the continental margin with higher values on the Walvis-Rio Grande Rise and the Falkland Plateau. This general trend continues for the time period of Zones NC8 and NC9. Distribution patterns are less distinctive during the early part of the Middle Cretaceous (Zones NC7 and NC5/6) with the lowest abundance of this species still along the continental margin and the greatest concentrations in deep basins or in samples that have undergone dissolution and diagenesis.

Biscutum constans (Figs. 14, 20)

This species is most abundant along continental margins and in epicontinental deposits, during the time period of Zones NC7 through NC11. The low abundance of this species at Site 370 during Zone NC7 time is due to considerable dissolution. During the earlier part of the Middle Cretaceous (NC5/6) *B. constans* is relatively rare and shows a relatively even distribution except for a very high concentration on the Falkland Plateau. Rapidly fluctuating conditions existed there during the earlier part of the Middle Cretaceous, and almost monospecific blooms and large fluctuations in the nannoplankton assemblages occur (see also Wise and Wind, 1977; Noël and Manivit, 1978). Depth-latitude plots do not show a distinctive latitudinal or depth trend but only increased abundances along continental margins.

Zygodiscus erectus (Figs. 15, 21)

This species is generally the most abundant species of the genus *Zygodiscus* except at high latitudes where *Z.* cf. *spiralis* or *Z. diplogrammus* are more abundant. The overall biogeographic distribution patterns of *Z. erectus* are similar to the pattern displayed by *Biscutum constans:* highest concentrations along the continental margins and in epicontinental basins, and lowest concentrations in the western and eastern basin of the North Atlantic. In the South Atlantic this species is most abundant at

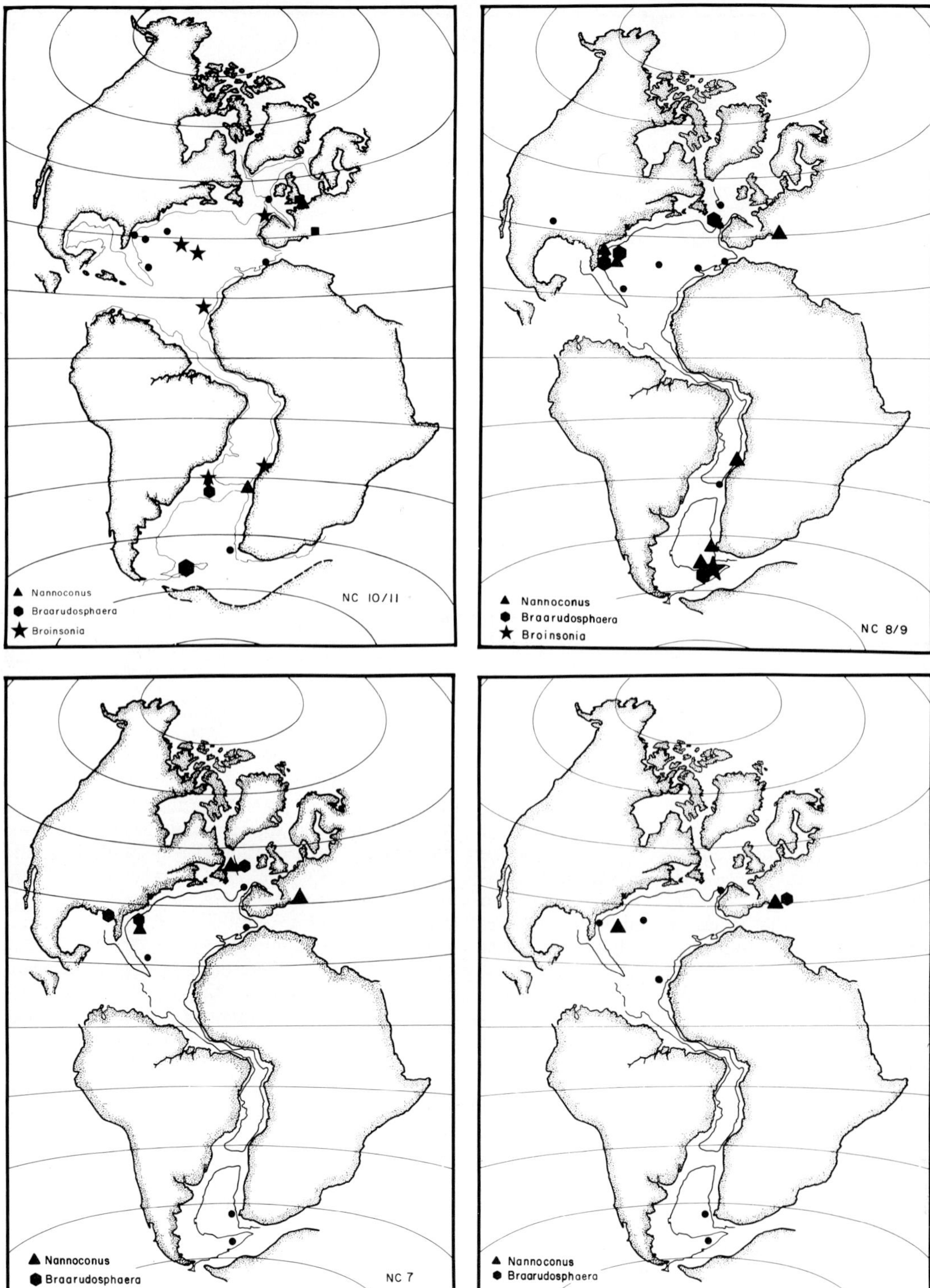

FIG. 18.—Biogeographic distribution of *Broinsonia* sp., *Nannoconus* sp. and *Braarudosphaera*. Relative abundances are always low, only a few percent of the total assemblage in well preserved samples; thus, relative abundances are not presented. Note that *Braarudosphaera* includes *Micrantholithus*. *A,* Zones NC 10 and NC 11. *B,* Zones NC 8 and NC 9. *C,* Zone NC 7. *D,* Zone NC 6 and top of Zone NC 5.

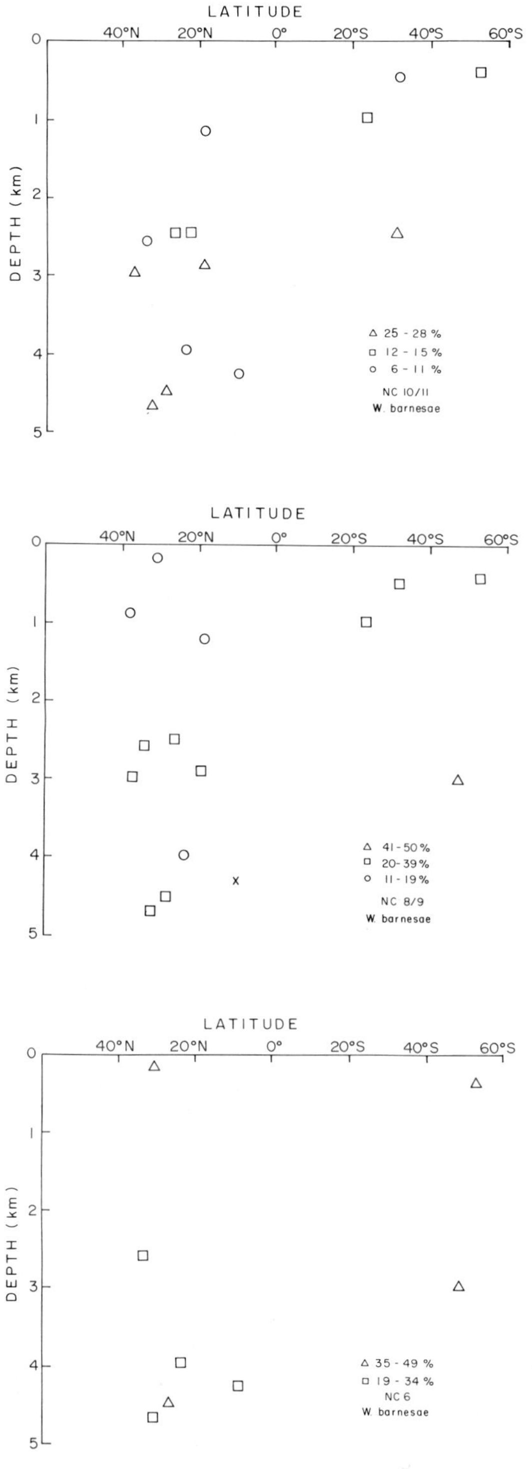

FIG. 19.—Relative abundance of *Watznaueria barnesae* in a paleolatitude-paleodepth framework. See Figure 9 for site indentifications. *A,* Zones NC 10 to NC 11. *B,* Zones NC 8 to NC 9. *C,* Zone NC 6, including the top of NC 5.

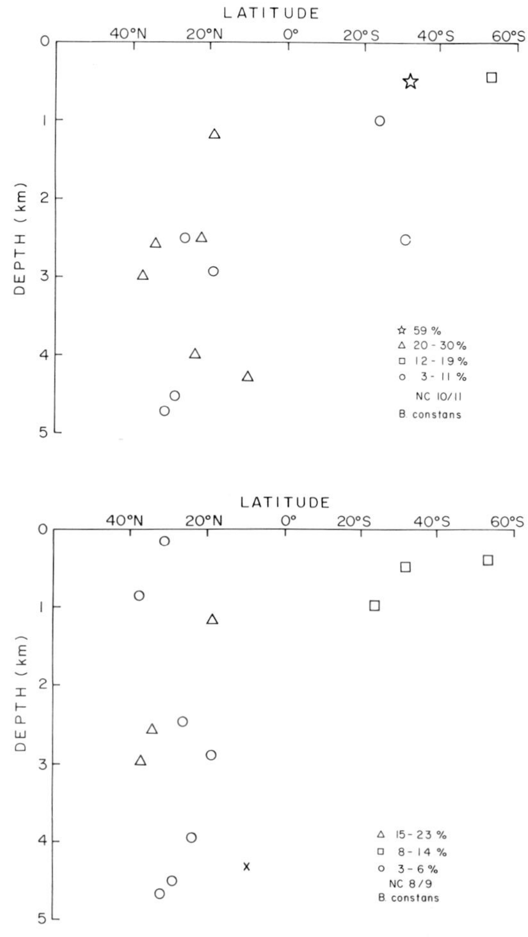

FIG. 20.—Relative abundance of *Biscutum constans* in a paleolatitude-paleodepth framework. *A,* Zones NC 10 and NC 11. *B,* Zones NC 8 to NC 9.

Site 363, with greatly fluctuating abundances, perhaps due to upwelling as deep Angola Basin-water rose to the surface when crossing the Walvis-Rio Grande Ridge. Neither a distinctive latitudinal nor a depth effect is apparent in the distribution of this species. However, the preference for continental margin settings, especially in the North Atlantic, and for plateau and rises with possible upwelling, seems quite pronounced during the whole Middle Cretaceous time period. Dissolution tends to reduce the relative abundance of *Z. erectus* (*e.g.,* at Site 370 during NC9 through NC7 time).

Parhabdolithus splendens and *P. asper* (Figs. 16, 22)

These two species are very similar, and counts for the two species were combined. *P. splendens* is somewhat more abundant in the upper Middle Cretaceous and *P. asper* in the lower Middle Cretaceous. The two species combined show the high-

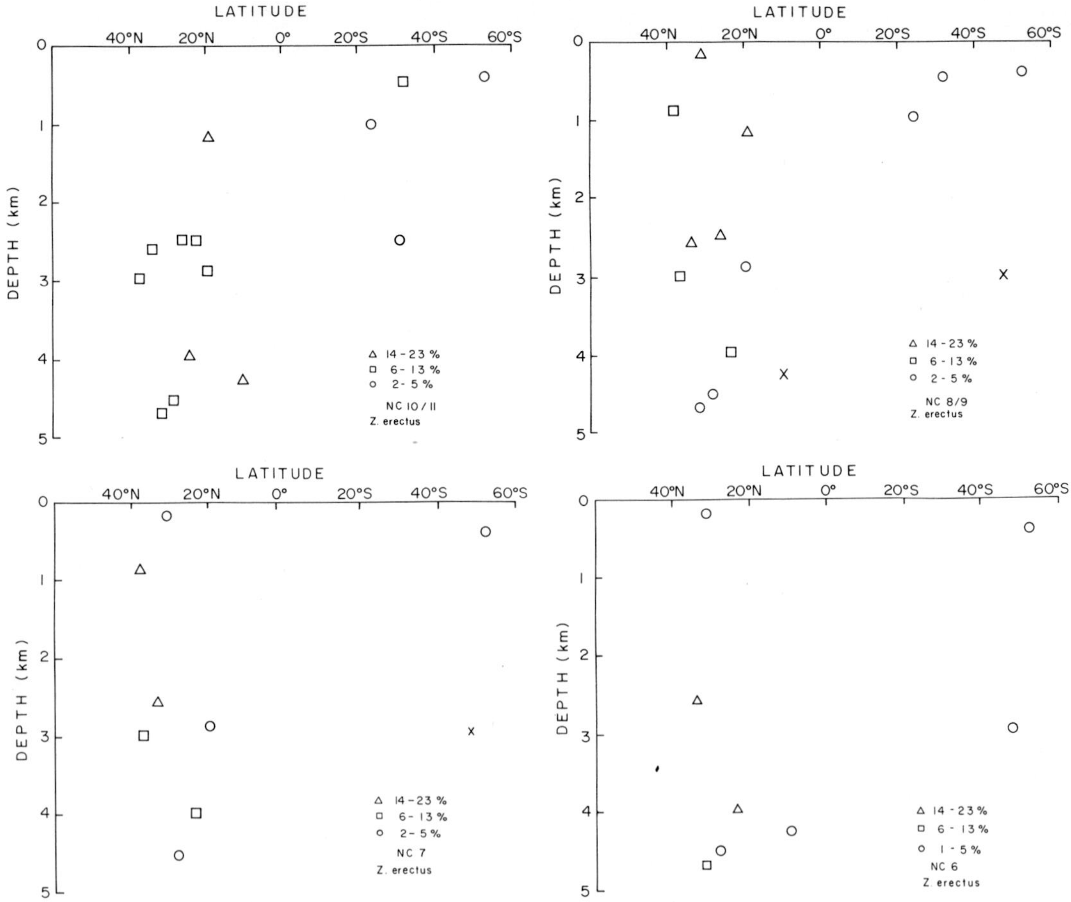

FIG. 21.—Relative abundance of *Zygodiscus erectus* in a paleodepth-paleolatitude framework. *A*, Zones NC 10 and NC 11. *B*, Zones NC 8 and NC 9. *C*, Zone NC 7. *D*, Zone NC 6 and top of NC 5.

est relative abundances in the oceanic samples of the North Atlantic basins during time periods NC10/11, NC7 and NC5/6. During NC8/9 time the concentration is quite uniform, which is difficult to explain. Perhaps it is caused by increased dissolution at most sites. Bowdler (1978) showed that the average dissolution index is greatest for time slice NC8/9 (see also Fig. 7, etching scale). This is reflected in less distinctive paleogeographic patterns for easily dissolved nannoplankton species for this time period. The depth-latitude plots for this species (Fig. 22) also show that neither depth nor latitude is an important factor controlling the distribution of this group of species and that oceanic sites contain the largest concentration.

Seribiscutum primitivum (Fig. 17)

This species first occurs in Zone NC8/9 on the Falkland Plateau. During the next younger time period (NC10/11) rare occurrences were also observed in Southeastern England (Gault Clay) and in Normandie (reported by Manivit, in Amedro, Damotte and others, 1978). All occurrences in the Atlantic are from shallow locations. *S. primitivum* is an excellent high-latitude indicator in the Middle Cretaceous.

Broinsonia sp. (Fig. 18)

The genus *Broinsonia sp.* is never very common in Middle Cretaceous sediments (1–2%, rarely up to 5%), and it is present consistently only in the upper Middle Cretaceous (Zones NC10 and NC11). Occurrences of over 1% of *Broinsonia* sp. are restricted to the eastern basin and continental margin of the North Atlantic and to the South Atlantic and epicontinental seas. Susceptibility to dissolution does not seem to be the only factor controlling this species. We consider it a form that is more common in waters of higher fertility and unstable "nearshore" environments and epicontinental seas.

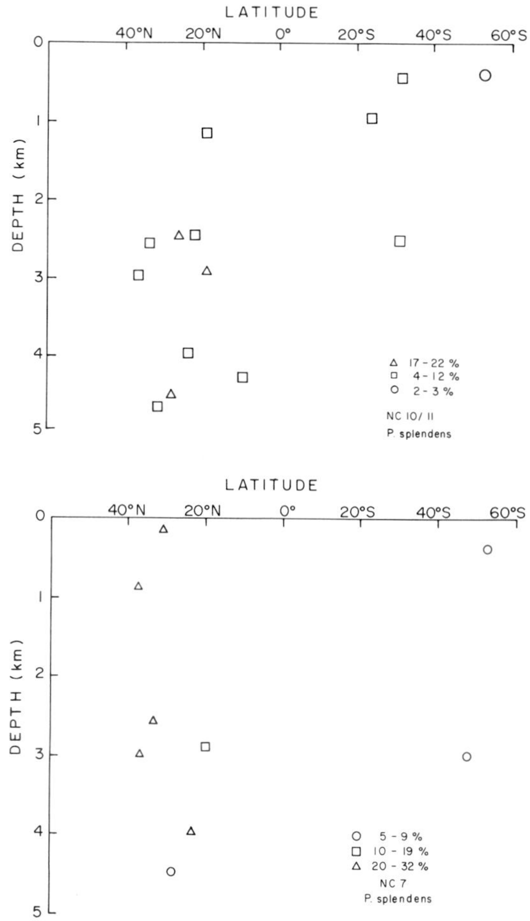

FIG. 22.—Relative abundance of *Parhabdolithus splendens* in a paleodepth-paleolatitude framework. *A,* NC 10 and NC 11. *B,* NC 7.

Braarudosphaera sp. (Fig. 18)

The braarudosphaerids (including *Micrantholithus*) are limited in their occurrence to the continental margin sites and to Site 391 where one finds ample evidence of sediment redeposition from shallower sources (see Benson, Sheridan and others, 1978). This group is a fairly good indicator of high fertility and other unstable conditions (see discussion in Roth, 1974 for Cenozoic to Recent occurrences of *Braarudosphaera*).

Nannoconus (Fig. 18)

Nannoconus is an indicator of continental margin, shallow plateau and epicontinental sea conditions. Preservation does play an important role in the distribution of this species as shown by Noël and Melguen (1978) and by Noël and Manivit (1978); dissolution within organic-rich sediments tends to destroy the coherence of the *Nannoconus* ultrastructure so that during usual preparation methods, where sediments are disaggregated, nannoconids disintegrate into calcite plates that are not recognizable as elements of *Nannoconus.* If pieces of the whole rock are viewed in the scanning electron microscope, the presence of *Nannoconus* becomes apparent. This might be one of the reasons why *Nannoconus* always seems most abundant in limestone samples. Nannoconids in Middle Cretaceous Site 361 were possibly transported from shallower depths as seems to be the case for at least some of the *Nannoconus* remains at Site 391.

Other Species with Less Distinctive Distribution Patterns

The other two common species of *Zygodiscus (Z. elegans* and *Z. diplogrammus)* also show a preference for continental margins and epicontinental seas (Bowdler, 1978). *Z. diplogrammus* can make up as much as 25% of the total assemblage in samples from the Gault Clay and might be a high latitude indicator, but further studies are needed to verify this trend in other parts of the world. Samples from the Falkland Plateau contain other species that are restricted to high latitudes such as *Sollasites falkladensis* and *Octocyclus magnus* (also described from the Gault Clay). High abundances of *Zygodiscus* cf. *spiralis* (forms illustrated by Hill, 1976) are also characteristic of austral assemblages. *Prediscosphaera cretacea* is very common in some Gault samples but otherwise shows no clear biogeographic patterns. *Cretarhabdus* sp. and *Retecapsa* sp. do not show clear biogeographic patterns and seem to be quite cosmopolitan with wide environmental tolerances.

CHARACTERISTIC ASSEMBLAGES OF MIDDLE CRETACEOUS NANNOPLANKTON

High-Latitude Assemblages

The assemblages from the lower Middle Cretaceous of the Falkland Plateau show extreme variation in species composition from sample to sample and include monospecific assemblages of *Corolithion* sp. and *Zygodiscus elegans* (Noel and Manivit, 1978); these are probably indicative of periodic blooms during times of particularly strong upwelling, when deep waters were forced to pass over the Falkland Plateau barrier. Other environmental instabilities could have caused such blooms. Late Middle Cretaceous assemblages from the same region consist of equally large concentrations of *Watznaueria barnesae* and *Seribiscutum primitivum,* with less abundant *Biscutum constans* and *Zygodiscus* sp.; all other species are rare. Samples from the boreal region (Gault Clay of Southern England) are also quite variable in composition indicating fairly unstable environmental conditions. *Seribiscutum primitivum* is present but about five to ten times less abundant in the Gault Clay than on the Falkland Plateau. Thus all the high-latitude assemblages we have studied from the Atlantic Ocean and surrounding inland seas are quite variable in composition. It is not known at the present time whether this is typical for high-latitude nannoplankton assemblages during the Middle Cretaceous or just due to the particular local environmental conditions at the two sites studied by us.

Continental-Margin Assemblages

Middle Cretaceous continental-margin assemblages in the North Atlantic are characterized by relatively rare *Watznaueria barnesae* and common *Biscutum constans, Zygodiscus erectus, Z. elegans* and *Z. diplogrammus*. In some samples *B. constans* is more abundant than *Zygodiscus;* in others the latter is more abundant than the former. Continental margins with high detrital input (eastern North Atlantic margin) generally do not contain common *Nannoconus,* a species that seems more abundant on shallow plateaus with low terrigenous input and in epicontinental seas.

The genera *Broinsonia, Braarudosphaera* and *Micrantholithus* have similar preferences for continental margin regions and epicontinental seas. *Prediscosphaera columnata* is locally abundant in continental margin samples (*e.g.,* at Site 167, Zone NC10–11; Gault) but less abundant at other continental margin sites. Its distribution pattern in space and time is quite "patchy".

Oceanic Assemblages

Coccolith and nannolith assemblages from sites further removed from continental margins show abundant *Watznaueria barnesae* and *Parhabdolithus splendens* and *P. asper*. The blade-shaped and long-ranging species *Lithraphidites carniolensis* might also be more common in oceanic samples, but the distribution patterns are quite patchy and were thus not plotted here. In the South Atlantic, Sites 356 and 363 on the Walvis-Rio Grande Ridge show more oceanic aspects, whereas Site 364 on the continental margin off Angola shows more continental margin aspects with somewhat more fluctuating temporal occurrence patterns of the more important species. This is perhaps due to the more variable circulation patterns in the eastern Angola Basin during the Middle Cretaceous.

Multivariate statistical approaches (Q-mode cluster analysis) by Bowdler (1978) showed that the separation of high- and low-latitude assemblages and continental margin and oceanic assemblages is possible. Time slices used were somewhat long (NC5/6, NC7/8 and NC9/11) so that temporal and spatial changes in assemblages could not be clearly separated. More refined statistical studies using shorter time slices are in progress.

PALEOCEANOGRAPHIC IMPLICATIONS OF NANNOPLANKTON STUDIES

Preservation

Nannofossil preservation is a more sensitive indicator of calcite dissolution than total calcium carbonate content of sediments because it is not affected by dilution with detrital material. Nannoliths show most pronounced etchings (E2 to E3) or have been totally dissolved in the Magellanian, especially in Zones NC7 and NC8/9 in the western North Atlantic at Sites 418 and 390, at Site 370 off Morocco and at the deep Cape Basin Site 361. Downslope redeposition of carbonates of Zone NC7 to NC8 age at Site 391, indicated by the presence of braarudosphaerids, nannoconids and other neritic forms, has resulted in better preserved nannolith assemblages at this site than at other deep-basin locations. Poor preservation of nannoliths is characteristic for the black shale sections of Argusian age at deep-basin Sites 391 and 105 and at the more shallow Site 386 in the western Atlantic. The sediments, rich in marine organic carbon at Site 367, show indications of upwelling and contain many barren intervals in the Magellanian and Argusian (Zones NC7 through NC10). Nannolith preservation at South Atlantic Sites 364, 356 and 363 is moderate in Zones NC8/9 and the lower part of Zone NC10, but generally improves toward the top of Zone NC10. It is not possible to infer nannofossil production in surface waters from the preservation of coccoliths because even if nannolith production is high, high dissolution rates can lead to large-scale destruction of coccoliths. Preservational changes are mostly a proxy record of increased dissolved carbon dioxide in bottom and interstitial waters.

Overall preservation of coccoliths in the Atlantic is poorest in the Upper Magellanian (Zone NC8/9, about 102–105 m.y.BP) when the proportion of organic-rich sediments is highest and the average calcium carbonate is lowest in the sections studied. River input of organic carbon and detritus and brine inflow from marginal evaporite basins were high during this regressive phase (Fig. 3B). Intensive dissolution of the carbonate during transfer, on the seafloor and within the sediments, during oxidation of organic matter and concurrent production of carbon dioxide resulted in low carbonate contents and poorest preservation of nannoliths. During the late Argusian (about 94–98 m.y.BP) more vigorous deep circulation and increased temperature gradients shown by the differentiation of a high latitude nannoflora and by oxygen isotope studies (Fig. 3A) resulted in a somewhat reduced stability of the water column and thus in slightly better mixing of deep waters; therefore less catabolically produced CO_2 accumulated in deep waters and more calcium carbonate with more delicate nannoliths were preserved in the sediments. The number of Tenerifian (90–94 m.y.BP) sections is too small to determine large-scale preservation patterns. Deep currents were certainly more pronounced in the Tenerifian and Naturalistian (86–90 m.y.BP) when the number of hiatuses in deep-sea sections is much greater and detrital organic carbon input was low due to covering of low-lying shelves by a global transgression of unmatched extent. By the beginning of the Upper Cretaceous (about 90 m.y.BP) meridional deep circulation had increased due to the opening of deep passages connecting the Arctic directly with the North Atlantic, the North Atlantic with the South Atlantic and the South Atlantic with the Indian Ocean, Tethys and Pacific Oceans.

More quantitative studies of nannofossil preservation, relating the composition of coccolith assemblages to organic carbon content of the sediments, distance from the continent, and paleodepths, are in progress. They promise to give a more detailed picture of the carbonate chemistry of Mid-Cretaceous deep-sea sediments.

Biogeography

Well preserved Mid-Cretaceous nannofossil assemblages provide an excellent proxy record of temperature and nutrient concentration in the surface waters of the Atlantic ocean. Although it is not possible to determine absolute temperatures, the nannofloral distribution patterns give a measure of temperature and fertility gradients.

It has generally been assumed that temperature gradients between low and high latitudes were smaller during the Mid-Cretaceous than today. This was borne out by our nannoplankton studies. Cold-water species show a bipolar distribution but are restricted to latitudes greater than 40°. High-latitude assemblages characterized by common *Seribiscutum primitivum,* as well as great variability in the composition of the nannofossil assemblages, are restricted to the southernmost region of the South Atlantic and the boreal seaway of England and northern France. The rare nannofossil species *Sollasites falklandensis* has only been found at southern high-latitude Sites 327 and 330 on the Falkland Plateau. A rare occurrence of *Seribiscutum primitivum* in the Upper Cretaceous of the Crimea (Thierstein, 1976) probably indicates a more southerly extent of the boreal assemblages on the Russian platform. Further studies of land sections from that region are needed to clarify epicontinental biogeographic patterns of nannofossil distribution. An increase in the temperature gradient during the latter part of the Middle Cretaceous, although suspected, could not be demonstrated with certainty using our present information. We speculate that opening of wider connections with the Arctic and the Indopacific to the south increased meridional circulation and resulted in more latitudinal biogeographic patterns in the Late Cretaceous (Thierstein, 1976, and in this volume; Wind, 1979). Upper Cenomanian and Turonian sections need to be added to our study. These intervals, however, do not contain abundant nannoplankton remains in the cores recovered from the Atlantic so far, making it difficult to determine the paleoceanographic conditions at the close of the Middle Cretaceous when black shale deposition was largely replaced by deposition of oxipelagites.

A wide tropical belt that extends from about 40° North to 40° South latitude is characterized by only minor latitudinal but strong east-west nannofloral gradients. Assemblages with relatively rare *Watznaueria barnesae* and abundant *Zygodiscus erectus* and *Biscutum constans* occur in well preserved samples from DSDP sites along the European and African continental margins, in the Angola Basin and on the Walvis-Rio Grande Rise. None of the nannofossil species are restricted to the eastern-margin assemblages, but the relative abundance of the most common nannofossil species differs significantly from oceanic deep basin assemblages. We interpret these continental-margin assemblages as indicative of increased nutrient concentration caused by upwelling along the eastern margin of the North and South Atlantic Oceans.

Nannoplankton assemblages from oceanic sites in the North Atlantic (Sites 105, 391C, 386, 418 and 137) generally contain more abundant *Watznaueria barnesae* and *Parhabdolithus splendens* and less abundant *Zygodiscus* spp. and *Biscutum constans* than "upwelling" assemblages. The nannoplankton assemblages from Site 391C in the Blake Basin contains redeposited sediments from shallower parts of the continental slope which explains the presence of *Braarudosphaera* in some of the samples. The nannoplankton assemblages from the western margin of the North Atlantic (Blake Nose Sites 390 and 392) differ from eastern-margin assemblages in the reduced abundance of *Zygodiscus erectus* and in the presence of *Nannoconus,* and resemble assemblages from the epicontinental seaway of Texas.

Nannofossil assemblages from the Walvis-Rio Grande Rise are somewhat less well-preserved but show greater similarities with the assemblages from the eastern North Atlantic margins than with deep basin assemblages. We infer high nutrient concentrations in surface waters along the eastern margin of the South Atlantic and upward mixing of nutrient-rich deep waters as they were forced over the shallow Walvis-Rio Grande Rise. Assemblages from the Falkland Plateau show abundant "upwelling indicators" together with high-latitude species.

Bloom assemblages composed almost exclusively of one species were observed in sediments from the Falkland Plateau and in the Angola Basin. They indicate rapidly changing conditions, such as periodic upwelling, that were favorable to only a few nannofossil species such as *Biscutum constans, Zygodiscus elegans* and *Corollithion.* Assemblages from the epicontinental seas of southeastern England also show great changes in the relative abundance of *Zygodiscus diplogrammus* and other "neritic" forms. Such fluctuations are indicative of unstable oceanic conditions in waters overlying the shelter of Northern Europe.

Well-preserved assemblages from the Gulf Coast region (Hill, 1976) contain assemblages that are most similar to the best preserved oceanic assemblages, especially the ones from Site 137. This was clearly shown by preliminary results of multivariate statistical analyses now in progress. The more common occurrence of *Nannoconus* in assemblages from the Gulf Coast than in oceanic assemblages indicates that it is possible to distinguish epicontinental-sea assemblages from well-preserved oceanic assemblages. Nannofossil assemblages from Southeastern France show a strong "neritic" aspect. *Nannoconus* and *Braarudosphaera* are more common than in oceanic basin and margin assemblages. Much poorer preservation is largely responsible for a secondary enrichment in *Watznaueria barnesae* relative to the more delicate species in many of these diagenetically altered assemblages. This makes direct comparison of these assemblages with assemblages from oceanic sites, and with better preserved epicontinental assemblages, difficult.

The large-scale patterns of nannofossil distribution in Mid-Cretaceous sediments allows us to test some of the hypotheses proposed for the formation of widespread black shales in the Atlantic. The presence of nannolith assemblages indicative of high nutrient concentrations caused by upwelling along the continental margins of the Iberic Peninsula and North and West Africa agrees with the presence of higher concentration of organic matter of marine origin in the eastern than in the western North Atlantic (Tissot and others, 1979) and also with the more pelagic palynofacies in that region (Habib, 1979). Epicontinental-sea assemblages at low latitudes (Texas) are more similar to eastern Atlantic oceanic assemblages than to the assemblages from the western margin (Blake Nose). This indicates higher nutrient concentration, probably due to river input on the Gulf Coast, than on the Blake Nose, which was separated from the continent by a wide shelf. The restriction of high-latitude assemblages to latitudes greater than 40° supports the hypothesis of equitable climates with small temperature gradients during the Mid-Cretaceous. However, considerable fluctuations of relative species abundances in samples from epicontinental seas of southeastern England, the Angola Basin and the Falkland Plateau shows that periodic fluctuations in nutrient supply and ocean mixing prevailed during the Mid-Cretaceous. Cyclic changes in sedimentation in the North and South Atlantic (Gardner and others, 1978; Arthur and Natland, 1979) also show fluctuation in water chemistry and sediment supply by rivers from continents. Variability in the preservation of nannofossils, alteration of sediment color and episodic burrowing in many Mid-Cretaceous sediments indicate cyclic changes in the stability of the Mid-Cretaceous water column, deepwater ventilation and dissolved carbon dioxide concentration which in turn resulted in changes of organic productivity in the surface waters and preservation of organic matter and calcite in the sediments. Although the number of oceanic deep-basin sites is relatively limited, we feel that there is convincing evidence for low oxygen concentrations reaching all the way to the bottom of the deep basins of the eastern Atlantic and the South Atlantic during most of the Mid-Cretaceous. In the eastern North Atlantic sediment facies indicate that at times only the sediments were anoxic and the depth of the boundary between reducing and oxidizing conditions fluctuated within the sediments which led to periodic oxidation of the surface sediments and burrowing. The western Atlantic geochemical studies of organic matter (Tissot and others, 1979) indicate more oxic conditions because most of the more labile marine organic matter in the sediments was destroyed by oxidation. Poor preservation of nannoliths over much of the section indicates that severe dissolution seems to have accompanied the oxidation of organic matter and resulted in a relative increase in the organic carbon content of the sediments.

In the South Atlantic we find evaporitic conditions in the Angola Basin and stagnant conditions in the Cape Basin during the early part of the Mid-Cretaceous. During the later part of the Mid-Cretaceous the Angola Basin and Falkland Plateau was overlain by oxygenated waters (Arthur and Natland, 1979). Stagnation in the South Atlantic was more stable than in the North Atlantic. Oxidized sediments do not occur in black shale cycles, but some layers are burrowed. This is probably due to the small size of the South Atlantic basins, the great topographic isolation and the large supply of brine from the northern end of the South Atlantic and from numerous marginal basins. The North Atlantic was already much wider and had better connections with the other oceans during the Mid-Cretaceous.

The notion of complete stagnation of deep and bottom water over a period of millions of years during the Mid-Cretaceous is thus a gross generalization and certainly not applicable to the North Atlantic. The Mid-Cretaceous oceans did have low dissolved-oxygen concentrations, and given sufficient topographic isolation they tended to become anoxic over much wider regions than during earlier or later geologic time periods. Surface-water dilution by river runoff and brine flux from marginal basins to deep basins contributed to stagnation in the topographically isolated Atlantic Ocean. Freshening of surface waters was never or very rarely strong enough to totally exclude coccolithophorids even in the North Atlantic basin to the north of the Newfoundland - J-Anomaly ridge. Thus, we feel that dilution of surface waters never exceeded a few parts per thousand and thus these waters never became brackish at the sites where nannoplankton samples were studied. Therefore, brine influx is the preferred mechanism for increasing the stability of the water column during the Mid-Cretaceous. The worldwide transgression at the end of the Mid-Cretaceous resulted in a destruction of evaporite basins and lead to a reduction of detrital organic carbon input. Stability of the water column and oxygen consumption due to oxidation of organic matter decreased. Deep-water connections within the Atlantic and between the Atlantic and the Indopacific improved and latitudinal temperature gradients increased. All these changes resulted in better ventilation of Atlantic deep basins, better preservation of deep-sea carbonates during the Late Cretaceous and more latitudinally controlled biogeographic distribution patterns of nannofossils during the Late Cretaceous. The replacement of a black-shale facies by multicolored clays and chalks in the Atlantic and in many surrounding epicontinental seas is most striking and reflects a major paleoceanographic change.

How did the "Mid-Cretaceous event" affect the evolution of marine organisms, especially neritic organisms? We observe a diversification among the calcareous nannoplankton in the Middle Cretaceous that cannot be completely explained by better preservation and an increase in the number of suitable sedimentary sections (Roth, 1978). Planktic foraminifera and ammonites also show significant diversification during the Mid-Cretaceous (Tappan, 1971; Kennedy 1977). Major extinctions did not occur during the Middle Cretaceous but there was a reduction in the occurrence of rudist reefs, especially rudist atolls, probably because of a rapid sea-level change.

There was definitely not a major catastrophy in the evolution of the biosphere in the Middle Cretaceous. Changes in salinity and oxygen content affected deep waters more than shallow waters and possible changes in surface-water salinity were slow enough that surface-water plankton could adapt to such changes. The general stagnation and low fertility of surface water resulted in diversification rather than major extinctions. The diverse Late Cretaceous nannoplankton assemblages evolved during the Mid-Cretaceous. They were perhaps too specialized and adapted to very narrow oceanographic parameters. Relatively small changes in environmental conditions resulted in major extinctions at the end of the Cretaceous.

ACKNOWLEDGMENTS

This research has been supported by National Science Foundation Grant OCE 76-21561. Oceanic sediment samples provided by the Deep Sea Drilling Project were recovered through the assistance of the U.S. National Science Foundation. We wish to thank Drs. J. P. Beckmann, M. Hill and H. R. Thierstein for providing land samples for this study. Discussions with W. H. Berger, R. Schroeder and H. R. Thierstein have been stimulating. Ms. Carleen Nutter provided assistance with computer programming. We are also grateful to all the marine scientists who have contributed to the success of the drilling operation in the world's oceans and to the administrators and politicians who had the foresight to support the Deep Sea Drilling Project that has made paleoceanography a reality.

REFERENCES CITED

NOTE: All references to Deep Sea Drilling Project Initial Reports have been abbreviated; for complete citations see *Appendix I* at the end of this volume, or the individual Initial Reports as published.

Amedro, F. Bidar, A., Damotte, R., Manivit, H., Robazynski, F., and Sornay, J., 1978, Echelles biostratigraphiques dans le Turonien du Cap Blac-Nez (pas-de-Callais, F.): Bull. Inf. Géol. Bass. Paris, v. 15, n. 2, p. 3–20.

Amedro, F., Damotte, R., Manivit, H., Robaszyndki, F., and Sornay, J., 1978, Echelles biostratigraphiques dans le Cénomanien du Boulonnais (macro-micro-nanno fossiles): Géologie Mediterranéenne, v. 5, n. 1, p. 5–18.

Arthur, M. A., 1979, Paleoceanographic events—recognition, resolution, and reconsideration: Rev. Geophys. Space Physics, v. 17, n. 7, p. 1474–1494.

———, and Natland, J. H., 1979, Carbonaceous sediments in the North Atlantic and South Atlantic: the role of salinity in stable stratification of early Cretaceous basins, *in* Talwani, M., Hay, W. W., and Ryan, W. B. F., eds., Results of deep drilling in the Atlantic Ocean, Maurice Ewing Series 3: Washington, D.C., Amer. Geophy. Union, p. 375–401.

Asmus, H. E., and Ponte, R. C., 1973, The Brazilian marginal basins, *in* Nairn, A. E. M., and Stehli, B. G., eds., The ocean basins and margins, the South Atlantic: New York, Plenum Press, v. 1, p. 87–133.

Benson, W. E., Sheridan, R. E., and others, 1978: Init. Repts. DSDP, v. 44, p. 1005.

Berger, W. H., 1970, Biogenous deep-sea sediments: fractionation by deep-sea circulation: Geol. Soc. America Bull., v. 81, p. 1385–1402.

———, 1979, Impact of deep-sea drilling on paleoceanography, *in* Talwani, M., Hay, W. W., and Ryan, W. B. F., eds., Results of deep drilling in the Atlantic Ocean, Maurice Ewing Series 3: Washington D.C., Amer. Geophy. Union, p. 297–314.

———, and Roth, P. H., 1975, Oceanic micropaleontology: progress and prospect: Rev. Geophys. Space Physics, v. 13, n. 3 p. 561–585, p. 624–635.

———, and Winterer, E. L., 1974, Plate stratigraphy and the fluctuating carbonate line: Internat. Assoc. Sedimentol. Spec. Pub. v. 1, p. 11–48.

Berggren, W. A., and Hollister, C. D., 1974, Paleogeography, paleobiography and the history of circulation in the Atlantic Ocean, *in* Hay, W. W., ed., Studies in paleo-oceanography: Symposium of Soc. Econ. Paleontologists Mineralogists Spec. Pub. no. 20, p. 526–586.

———, and ———, 1977, Plate tectonics and paleocirculation-commotion in the ocean: Tectonophysics, v. 38, p. 11–48.

Black, M., 1971, Coccoliths of the Speeton Clay and Sutterby Marl: Proc. Yorkshire Geol. Soc., v. 38.

———, 1972, British lower Cretaceous coccoliths, I, Gault Clay: Paleontogr. Soc. Monogr., pt. 1, p. 1–48.

———, 1973, British lower Cretaceous coccoliths, I, Gault Clay: Paleontogr. Soc. Monogr., pt. 2, p. 49–112.

———, 1975, British lower Cretaceous coccoliths, I, Gault Clay: Paleontogr. Soc. Monogr., pt. 3, p. 113–142.

Bowdler, J. L., 1978, Mid-Cretaceous calcareous nannoplankton paleobiogeography and paleoceanography of the Atlantic Ocean: Unpubl. Masters Thesis, Univ. of Utah, 182 p.

Budyko, M. I., 1977, Climatic changes: Washington D.C. Amer. Geophys. Union, 261 p.

Bukry, D., 1969, Upper Cretaceous coccoliths from Texas and Europe: Univ. Kansas Paleontol. Contr. Art. 52 (Protista 2), p. 1–79.

———, and Bramlette, M. N., 1969, Summary of Coccolith biostratiphy: Init. Repts. DSDP, v. I, p. 621–623.

Caron, M., 1978, Cretaceous planktonic foraminifers from DSDP Leg 40, southeastern Atlantic: Init. Repts. DSDP, v. 40, p. 651–678.

Čepek, P., 1978, Mesozoic calcareous nannoplankton of the eastern North Atlantic, Leg 41: Init. Repts. DSDP, v. 41, p.667–687.

———, and Hay, W. W., 1969, Calcareous nannoplankton and biostratigraphic subdivision of the upper Cretaceous: Tans. Gulf Coast Assoc. Geol. Soc., v. 19, p. 323–336.

Chamberlin, T. C., 1906, On a possible reversal of deep sea circulation and its influence on geologic climate: Jour. Geology, v. 14, p. 363–373.

Cline, R., and Hays, J. D., eds., 1976, Investigations of Late Quaternary paleoceanography and paleoclimatology: Geol. Soc. America Mem., v. 145.

Davey, R. T., 1978, Marine Cretaceous palynology of Site 361, DSDP Leg 41, off southwestern Africa: Init. Repts. DSDP, v. 40, p. 883–913.

Dean, W. E., Gardner, J. V., Jansa, L. F., Cepek, P., and Seibold, E., 1978, Cyclic sedimentation along the continental margin of northwest Africa: Init. Repts. DSDP, v. 41, p. 965–989.

Douglas, R. G., and Savin, S. M., 1975, Oxygen and carbon isotope analyses of Tertiary and Cretaceous microfossils from Shatsky Rise and other sites in the North Pacific: Init. Repts. DSDP, v. 32, p. 509–520.

———, Moullade, M., and Nairn, A. E. M., 1973, Causes and consequences of continental drift in the South Atlantic, *in* Tarling, D. H. and Runcorn, S. K., eds., Implications of continental drift to the earth sciences: New York, Academic Press, p. 517–537.

Evans, R., 1978, Origin and significance of evaporites in basins around Atlantic margin: Amer. Assoc. Petroleum Geologists Bull., v. 62, p. 223–234.

Fischer, A. G., and Arthur, M. A., 1977, Secular variations in the pelagic realm: Soc. Econ. Paleontologists Mineralogists Spec. Pub., no. 25, p. 19–50.

Förster, R., 1978, Evidence for an open seaway between northern and southern proto-Atlantic in Albian times: Nature, v. 272, p. 158–159.

Gardner, J. V., Dean, W. E., and Jansa, L. R., 1978, Sediments recovered from the northwest African continental margin, Leg 41, Deep Sea Drilling Project: Init. Repts. DSDP, v. 41, p. 1121–1134.

Gartner, S., 1968, Coccoliths and related calcareous nannofossils from Upper Cretaceous deposits of Texas and Arkansas: Univ. Kansas Paleontol. Contr. Art., v. 48 (Protista 1), p. 1–56.

Gignoux, M., 1960, Géologie Stratigraphique, 5th ed.: Paris, Masson, 759 p.

Habib, D., 1979, Sedimentary origin of North Atlantic Cretaceous palynofacies: *in* Talwani, M., Hay, W. W., and Ryan, W. G. F., eds., Results of deep drilling in the Atlantic Ocean, Maurice Ewing Series 3: Washington, D.C., American Geophysical Union, p. 420–437.

Haq, B. U., Premoli-Silva, I., and Lohmann, G. P., 1977, Calcareous plankton paleobiographic evidence for major climatic fluctuations in the early Cenozoic Atlantic Ocean: Jour. Geophys. Res., v. 82, p. 3861–3876.

Heath, G. R., Moore, T. C., and Dauphin, J. P., 1977, Organic carbon in deep-sea sediments: *in* Anderson, N. R., and Malahoff, A., eds., The fate of fossil fuel CO_2: New York and London, Plenum Press, 749 p.

Hill, M. E., 1976, Lower Cretaceous nannofossils from Texas and Oklahoma: Paleontographica Abt. B., Band 156, p. 103–179.

Institute of Petroleum, 1965, Salt basins around Africa: Amsterdam, Elsevier Pub. Co., 122 p.

JANSA, L. F., GARDNER, J. W., AND DEAN, W. E., 1977, Mesozoic sequences of the central North Atlantic: Init. Repts. DSDP, p. 991-1031.

KENNEDY, W. J., 1977, Ammonite evolution, *in* Hallam, A., ed., Patterns of evolution, as illustrated by the fossil record: Amsterdam, Elsevier Pub. Co., p. 251-304.

———, AND COOPER, M., 1975, Cretaceous ammonite distributions and the opening of the South Atlantic: Geol. Soc. London Quart. Jour., v. 131, p. 283-288.

LOTZE, R., 1957, Steinsalz und Kalisalze.I. Teil. Allgemein-geologischer Teil: Berlin, Gebrüder Bornträger, 465 p.

MANIVIT, H., 1965, Nannofossiles calcaires de l'Albo-Aptien: Rev. Micropal. 8/3, p. 189-201.

———, 1971, Les nannofossiles calcaires du Crétacé français de l'Aptien au Danien. Essai de biozonation appuyée sur les stratotypes: Fac. Sci. d'Orsay, Thèse doctorat d'état, 187p.

———, PERCH-NIELSEN, K., PRINS, B., AND VEEBECK, J. W., 1977, Mid-Cretaceous calcareous nannofossil biostratigraphy: Proc. Koninklijke Nederlandse Akademie van Wetenschappen, Amsterdam, Series B., v. 80, n. 3 p. 169–181.

MCCOY, F. M., AND ZIMMERMAN, H. B., 1977, A history of sediment lithofacies in the south Atlantic Ocean: Init. Repts. DSDP, v. 39, p. 1047-1079.

MCLACHLAN, I. R., AND PIETERSE, E., 1978, Preliminary palynological results: Site 361, Leg 40, Deep Sea Drilling Project: Init. Repts. DSDP, v. 40, p. 857-881.

MORGAN, R., 1978, Albian to Senonian palynology of Site 364, Angola Basin: Init. Repts. DSDP, v. 40, p. 915–951.

NATLAND, J. H., 1978, Composition, provenance, and diagenesis of Cretaceous clastic sediments drilled on the Atlantic continental rise off southern Africa, DSDP Site 361 — implications for the early circulation of the South Atlantic: Init. Repts. DSDP, v. 40, p. 1025-1061.

NOËL, D., AND MANIVIT, M., 1978, Nannofaciès de "black shales" aptiennes et albiennes d'Atlantique sud (Legs 36 et 40), Intérêt sédimentologique: Bull. Soc. Géol. France (7), v. 20, n. 4, p. 491–502.

———, AND MELGUEN, M., 1978, Nannofacies of Cape Basin and Walvis Ridge sediments Lower Cretaceous to Pliocene (Leg 40): Init. Repts. DSDP, v. 40, p. 487–524.

OBRADOVICH, H. D., AND COBBAN, W. A., 1975, A time-scale for the late Cretaceous of the western interior of North America, *in* Caldwell, W. G. E., ed., The Cretaceous System in western interior of North America: Geol. Assoc. Canada, Spec. Paper 13, p. 31–54.

———, AND ———, 1977, Speculations on the age span of the Albian stage (early Cretaceous): Geol. Soc. America Abstracts with Programs, p. 1117–1118.

OKADA, H., AND THIERSTEIN, H. R., 1979, Calcareous nannoplankton — Leg 43, Deep Sea Drilling Project: Init. Repts. DSDP, v. 43, p. 507–573.

PERCH-NIELSEN, F., 1977, Albian to Pleistocene calcareous nannofossils from the western South Atlantic, DSDP Leg 39: Init. Repts. DSDP, v. 39, p. 699–823.

———, 1979, Calcareous nannofossils from the Cretaceous between the North Sea and the Mediterranean, *in* Aspekte der Kreide Europas. IUGS Series A, n. 6, Stuttgart, Germany, p. 223–273.

PONTE, F. C., AND ASMUS, H. E., 1978, Geological framework of the Brazilian continental margin: Geol. Rundschau., v. 67, p. 201–235.

PREMOLI-SILVA, I., AND BOERSMA, A., 1977, Cretaceous planktonic foraminifers-DSDP Leg 39 South Atlantic: Init. Repts DSDP, v. 39, p. 615–641.

PRICE, R. J., 1976, Paleoenvironmental interpretations in the Albian of western and southern Europe, as shown by the distribution of selected foraminifera: First Int. Symp. on Benthic Foraminifera of Continental Margins, Maritime Sediments Sp. Pub. 1, p. 625–648.

PROTO-DECIMA, F., 1974, Leg 27 calcareous nannoplankton: Init. Repts. DSDP, v. 27, p. 589–621.

———, MEDIZZA, F., AND TODESCO, L., 1978, Southeastern Atlantic Leg 40 calcareous nannofossils: Init. Repts. DSDP, v. 40, p. 571–634.

REYMENT, R. A., AND TAIT, E. A., 1972, Biostratigraphical dating of the early history of the South Atlantic Ocean: Phil. Trans. Royal Soc. London, B264, p. 55–95.

ROTH, P. H., 1973, Calcareous nannofossils — Leg 17, Deep Sea Drilling Project: Init. Repts. DSDP, v. 17, p. 695–793.

———, 1974, Calcareous nannofossils from the northwestern Indian Ocean, Leg 24, Deep Sea Drilling Project: Init. Repts. DSDP, v. 24, p. 969–994.

———, 1976, Cretaceous calcareous nannofossil diversity and paleoceanography: Geol. Soc. America Abstract with Programs, v. 8, n. 6, p. 1078.

———, 1978, Cretaceous nannoplankton biostratigraphy and paleoceanography of the northwestern Atlantic: Init. Repts. DSDP, v. 44, p. 731–759.

———, 1979, Cretaceous calcareous nannoplankton diversity and paleoceanography: IV Int. Palynol. Conf., Lucknow (1976–77), v. 2, p. 22–33.

———, AND BERGER, W. H., 1975, Distribution and dissolution of coccoliths in the south and central Pacific: dissolution of Deep-Sea carbonates: Cushman Foundation for Foraminiferal Research Spec. Pub. No. 13, p. 87–113.

———, AND THIERSTEIN, H., 1972, Calcareous nannoplankton of Leg 14 of the Deep Sea Drilling Proj.: Init. Repts. DSDP, v. 14, p. 421–486.

RYAN, W. B. F., AND CITA, M. B., 1977, Ignorance concerning episodes of ocean-wide stagnation: Marine Geology, v. 23, p. 197–215.

SAVIN, S. M., 1977, The history of the earth's surface temperature during the past 100 million years: Ann. Rev. Earth Planet. Sci., v. 5, p. 319–355.

SCHEIBNEROVA, V., 1978, Aptian and Albian benthic foraminifers of Leg 40, Sites 363 and 364, South Atlantic: Init. Repts. DSDP, v. 40, p. 741–757.

SCHLANGER, S. O., AND JENKYNS, H. C., 1976, Cretaceous oceanic anoxic events—causes and consequences: Geologie en Mijnbouw, v. 55 (3–4), p. 179–184.

SCHWARZBACH, M., 1974, Das Klima der Vorzeit, 3rd Ed.: Stuttgart, Ferdinand Enke Verlag.

SCLATER, J., HELLINGER, E., AND TAPSCOTT, E., 1977, The paleobathymetry of the Atlantic Ocean from the Jurassic to the present: Jour. Geology, v. 85, n.5, p. 509–552.

SIGAL, J., 1977, Essai de zonation du Crétacé méditerranéen a l'aide des foraminifères plactoniques: Géologie Mediterranéenne, v. 4, n.2, p. 99–108.

SMITH, A. G., AND BRIDEN, J. C., 1977, Mesozoic and Cenozoic paleocontinental maps: Cambridge, Cambridge Univ. Press, 64 p.

———, AND DREWRY, G. E., 1973, Phanerozoic world maps: Sp. Pap. Palaeont., v. 12, p. 1–42.

STOVER, L. E., 1966, Cretaceous coccoliths and associated nannofossils from France and the Netherlands: Micropaleontology, v. 12, n. 2, p. 133–167.

TAPPAN, H., 1971, Microplankton, ecological succession and evolution: Proc. North American Paleontological Convention Part H, p. 1058–1103.

TAYLOR, R., 1978, The distribution of calcareous nannofossils in the Speeton Clay (Lower Cretaceous) of Yorkshire: Proc. Yorkshire Geol. Soc., v. 42, pt. 2, n. 10, p. 195–209.

THENIUS, E., 1977, Meere und Länder im Wechsel der Zeiter: Verständliche Wissenschaft, v. 114: Berlin, Springer Verlag, 200 p.

THIERSTEIN, H. R., 1971, Tentative lower Cretaceous calcareous nannoplankton zonation: Ecol. Geol. Helv., 64 v. 3, p. 459–488.

———, 1973, Lower Cretaceous calcareous nannoplankton biostratigraphy: Abhandlungen der Geologischen Bundsanstalt, Band 24, p. 1–52.

———, 1974, Calcareous nannoplankton: Init. Repts. DSDP, v. 26, p. 619–667.

———, 1976, Mesozoic calcareous nannoplankton biostratigraphy of marine sediments: Marine Micropaleontology, v. 1, p. 325–362.

———, 1979, Paleoceanographic implications of organic carbon and carbonates in Mesozoic deep-sea sediments, *in* Talwani, M., Hay, W. W., and Ryan, W. G. F., eds., Deep drilling in the Atlantic Ocean, Maurice Ewing Series 3: Washington, D.C., Amer. Geophys. Union, p. 249–174.

———, AND BERGER, W. H., 1978, Injection events in ocean history: Nature, v. 276, p. 461–466.

———, AND OKADA, H., 1979, The Cretaceous/Teritary boundary event in the North Atlantic: Init. Repts. DSDP, v. 43, p. 601–616.

———, FRANZ, H. E., AND ROTH, P. H., 1972, Scanning electron and light microscopy of the same small object: Micropaleontology, v. 17, n. 4, p. 501–502.

TISSOT, B., DEROO, G., AND HERBIN, J. P., 1979, Organic matter in Cretaceous sediments of the North Atlantic: contribution to sedimentology and paleoceanography, *in* Talwani, M., Hay, W. W., and Ryan, W. B. F., eds., Deep drilling in the Atlantic Ocean, Maurice Ewing Series 3: Washington, D.C., Amer. Geophys. Union, p. 362–375.

TUCHOLKE, B. E., AND MOUNTAIN, G. S., 1977, Depositional patterns and depositional environment of Cretaceous black clays in the western North Atlantic: Geol. Soc. America Abstracts with Programs, v. 9, n. 7, p. 1207.

VAIL, P. R., MITCHUM, R. M., JR., AND THOMPSON, W., III, 1978, Seismic stratigraphy and global changes of sea level, Part 4, *in* Payton, C. E., ed., Seismic stratigraphy—applications to hydrocarbon exploration: Amer. Assoc. Petroleum Geologists Mem., v. 26, p. 83–97.

VAN ANDEL, T. H., THIEDE, J., SCLATER, J. G., AND HAY, W. W., 1977, Depositional history of the South Atlantic Ocean during the last 25 million years: Jour. Geology, v. 85, p. 651–698.

VAN HINTE, J. E., 1976, A Cretaceous time scale: Amer. Assoc. Petroleum Geologists Bull., v. 60, n. 4, p. 498–516.

VERBEEK, J. W., 1977, Calcareous nannoplankton biostratigraphy of middle and upper Cretaceous deposits in Tunisia, southern Spain and France: Utrecht Micropaleo. Bull., n. 16, p. 1–156.

WIEDMANN, J., AND NEUGEBAUER, J., 1978, Lower Cretaceous ammonites from the South Atlantic Leg 40 (DSDP), their stratigraphic value and sedimentologic properties: Init. Repts. DSDP, v. 40 supplement, p. 709–755.

WILCOXON, J. A., 1972, Upper Jurassic-Lower Cretaceous calcareous nannoplankton from the western North Atlantic Basin: Init. Repts. DSDP, v. 11, p. 427–457.

WIND, F. H., 1979, Maestrichtian-Campanian nannofloral provinces of the South Atlantic and Indian Oceans, *in* Talwani, M., Hay, W. W., and Ryan, W. G. F., eds., Results of deep drilling in the Atlantic Ocean, Maurice Ewing Series 3: Washington, D.C., Amer. Geophys. Union, p. 124–137.

WISE, S. W., AND WIND, F. H., 1977, Mesozoic and Cenozoic calcareous nannofossils recovered by DSDP Leg 36 drilling on Falkland Plateau, S. W. Atlantic sector of the Southern Ocean: Init. Repts. DSDP, v. 36, p. 269–491.

WORSLEY, T. R., 1971, Calcareous nannofossil zonation of Upper Jurassic and Lower Cretaceous sediments from the western Atlantic: Proc. Second Planktonic Conf., Rome, v. 2, p. 1301–1322.

SEPM SPECIAL PUBLICATION NO. 32, P. 547–560, OCTOBER 1981

THE ENVIRONMENT OF DEPOSITION OF BLACK SHALES IN THE EARLY CRETACEOUS: AN ONGOING CONTROVERSY

HELMUT WEISSERT*
Department of Geological Sciences, University of Southern California, Los Angeles, CA 90007 U.S.A.

ABSTRACT

During the last 12 years of the Deep Sea Drilling Project, numerous Early Cretaceous black shale horizons have been discovered in all the major ocean basins. Similar organic-carbon rich sediments had been described earlier from various land outcrops in the Alps, in the Carribbean region and in the southern Andes. New and refined methods in sedimentology, geochemistry and geophysics have been applied in the analysis of these sediments, which can be classified into four regional groups: (a) central North Atlantic, (b) northern North Atlantic, (c) South Atlantic and Indian Oceans, (d) Pacific Ocean. The black shales of Barremian-Albian age in the central North Atlantic and Tethys were deposited in a deep basinal environment under low oxygen or anoxic conditions. The average content of organic matter in the often-laminated black shales is 2%. Terrigenous organic matter dominates in the organic spectrum of the western North Atlantic, while marine-derived material dominates in the eastern North Atlantic. The black shales in the northern North Atlantic, deposited on a passive continental margin in a depth of a few hundred to a few thousand meters, are parts of turbiditic sequences. The organic matter is of terrigenous origin. These deposits were not necessarily connected with anoxic conditions. The Aptian-Albian black shales in the restricted basins of the South Atlantic and the Indian Ocean were deposited under anoxic conditions in a water depth of a few hundred to a few thousand meters. The organic matter has a mixed terrigenous and marine origin. The rare occurrences of black shales in the Pacific are restricted to submarine rises. The finding of black shales in all major oceans encouraged authors to construct models explaining these deposits as a result of global oceanic events. Detailed geochemical and sedimentological investigations revealed shortcomings in the early global scenarios, and a series of alternative spatially restricted models were proposed. Models explaining the formation of black shales in pelagic environments, such as the central North Atlantic, the Tethys and the South Atlantic-Indian Ocean, emphasize the importance of the regional paleotectonic and paleoceanographic situation controlling the black shale facies patterns. Anoxic bottom-water conditions were established during periods of stratified watermasses. Models explaining the black shales in the northern North Atlantic underline the importance of the continental climate controlling the input rate of terrestrial organic matter into marine environments. In the future, global models have to integrate these regional models; the exciting problem of possible feedback mechanisms between paleoclimate and paleoceanography has to be solved.

INTRODUCTION

About 15 years ago a new period in the study of earth sciences began with the development of plate tectonics and sea-floor spreading concepts. The theory of plate tectonics facilitated the reconstruction of the paleogeographic position of lithospheric plates through time and therefore allowed the unraveling of earth history on a global scale. When the Deep Sea Drilling Project began 12 years ago, its major objective was to test the sea-floor spreading hypothesis (*e.g.,* Berger, 1979). With the early DSDP results the hypothesis gained wide acceptance, and scientists then focused their interests on various other problems related to deep-sea sediments; one of the most prominent problems is the analysis of the geochemical and sedimentary history of the oceans. The newly developed plate tectonics theory provided the geometric basis for such reconstruction, but the possibility of unraveling oceanic history was finally opened up by a combination of data from the Deep Sea Drilling Project and the application of new or refined methods in geochemical, sedimentological and geophysical analysis to the recovered sediments.

On one hand the marine record reflects the continuous evolution of the oceans and mirrors the stability of the equilibrium conditions existing among the various global geochemical reservoirs; on the other hand, several drastic changes in ocean history are indicated by a significant change in the faunal and sediment pattern within the sedimentary record (Fischer and Arthur, 1977; Berger, 1979). The biotic crises at the Cretaceous-Tertiary boundary and at the end of the Eocene are among the most prominent of these oceanic events (for a review see Thierstein, 1979; Arthur, *in press*). Another time of oceanic instability seems to have occurred during the Early and Middle Cretaceous, when a series of organic-carbon rich horizons were deposited globally in various oceanic environments. Two main intervals with black shale deposits have been recognized. A late Barremian through Albian age is assigned to the first black shale period; the second

* Geologisches Institut, ETH-Z CH-8092 Zurich, Switzerland

one is of Cenomanian-Turonian age (Schlanger and Jenkyns, 1976; Fischer and Arthur, 1977; Arthur and Schlanger, 1979). The recovery of carbonaceous sediments within these two time envelopes in numerous deep-sea cores from the major oceans, and the rediscovery of black shales of similar facies and age in many exposures on land, stimulated an ongoing controversy among paleoceanographers regarding the mode of formation of these beds. A group of "universalists" proposes that the black shales record a series of oceanic catastrophic events of global extent; a second group, critical of simplified global scenarios, supports multidepositional mechanisms for the formation of the carbonaceous horizons, which were not necessarily coupled with one another on a global scale.

The purpose of this review is to summarize this controversy on the paleoceanography of the Cretaceous. I will concentrate my analysis on studies of black shale facies of Hauterivian up to Albian age (for a discussion on the black shale facies at the Cenomanian-Turonian boundary, see Arthur, 1979; Arthur and Schlanger, 1979). The expression "black shale" is used as a strictly descriptive term for a marly or siliceous sediment with an organic carbon content exceeding 1% and up to 30%.

FROM A REGIONAL TO A GLOBAL PROBLEM

During the first three years of the Deep Sea Drilling Project black shale horizons alternating with coccolith limestones and variegated marls of Valanginian to Albian age were recovered from several sites in the eastern and western North Atlantic (Ewing and Hollister, 1972; Hollister, Ewing, and others, 1972; Hayes, Pimm, and others, 1972). An equable Early Cretaceous climate, causing a sluggish paleocirculation, and the tectonic configuration of the North Atlantic favored cyclic development of anoxic conditions in deep water and preservation of carbonaceous beds in open-marine environments, according to models by Lancelot and others (1972), Berger and von Rad (1972) and Saunders and others (1973). Luterbacher (1972), Bernoulli (1972) and Bernoulli and Jenkyns (1974) emphasized the facies similarities between the Early and Middle Cretaceous Atlantic and Tethys Oceans. Zeil (1956), Bitterli (1965), Fuganti (1961) and Scandone (1972) had previously described black shale occurrences from various pelagic sequences of Barremian up to Cenomanian age in Alpine térranes. Similarly Saunders and others (1973) recognized affinities between Cretaceous black shales from the Caribbean and a black shale facies from Venezuela.

The first descriptions of black shales outside the Tethys-North Atlantic realm were noted by the scientific parties of the DSDP Legs 25, 26 and 27 (Davies, Luyendyk and others, 1974; Veevers, Heirtzler and others, 1974; Simpson, Schlich and others, 1974). Both the Neocomian black shales from the Mozambique Ridge and the Hauterivian-Albian horizons in the growing Indian Ocean were again related to periods of anoxia in narrow, tectonically restricted basins (*i.e.,* Girdley and others, 1974).

The black shale problem, initially only of regional importance, gained significance after Hauterivian to Aptian carbonaceous sediments were drilled in the Pacific Ocean during DSDP Legs 32 and 33 (Larson, Moberly and others, 1975; Schlanger, Jackson and others, 1976). Intrigued by the finding of a thin organic-rich bed within an Aptian sequence of volcanoclastic sediments on the Manahiki Plateau, Jackson and Schlanger (1976, p. 925) suggested that this bed might be correlated with similar beds in the North Atlantic and raised "...the interesting possibility that this Early Cretaceous carbonaceous event may have been a very widespread, perhaps a worldwide oceanographic phenomenon." The existence of a widespread black shale facies of Early Cretaceous age in the Pacific Ocean has been confirmed during DSDP Legs 61 and 62 (Jenkyns and Schlanger, *in press;* Dean and others, 1979; Thiede and others, 1979).

Organic rich sediments of Oxfordian to Aptian age on the Falkland Plateau (DSDP Leg 36) and of Aptian-Albian age in the Cape Basin and in the Angola Basin (Leg 40) supported the hypothesis that during the Early Cretaceous carbonaceous sediments were deposited in all the major oceans (Barker, Dalziel and others, 1977; Bolli, Ryan and others, 1978).

Paleoceanographers had therefore recognized the global character of the black shale problem by the time DSDP Leg 41, 43, 44, 47 and 48 drilled black shale horizons from various locations in the central and northern North Atlantic (Lancelot, Seibold and others, 1978; Tucholke, Vogt and others, 1979; Benson, Sheridan and others, 1978; Montadert, Roberts and others, 1979; Sibouet, Ryan and others, 1979). However, numerous geochemical and sedimentological studies of these black shales revealed that there were problems in explaining their worldwide occurrence by using a single global scenario (*i.e.,* Roberts and Montadert, 1979a,b).

A MULTIDISCIPLINARY APPROACH

The transition from the recognition of a black shale facies in numerous DSDP cores and in land sections to a paleoenvironmental reconstruction of the Early Cretaceous involves the application of a wide variety of new and refined methods in the fields of micropaleontology, sedimentology, geochemistry and geophysics.

A paleotectonic framework is available from models constructed on the basis of the seafloor spreading theory (*i.e.,* Smith and Briden, 1977). Of primary interest is the bathymetry of the Early Cretaceous oceans. The most widely accepted method for the reconstruction of paleodepths is based on the rate of seafloor subsidence as the crust cools, moving away from the spreading center (Sclater and others, 1971). By backtracking a drilling site along a best-fit subsidence curve the depth of the black shale deposition can be determined (*e.g.*, Berger and von Rad, 1972). A primary question in comparing black shale deposits from different paleoenvironments concerns the source of buried organic matter. Elemental analysis of kerogen, determinations of lipid compounds, the measurement of $^{13}C/^{12}C$ ratios in organic matter, and palynological studies all prove to be useful tools in identifying the various sources of organic matter (*i.e.*, Didyk and others, 1978; Tissot and others, 1979; Habib, 1979b; Arthur, 1979). Sedimentological, mineralogical and micropaleontological data give information on the depositional processes and the paleoceanographic patterns. Supporting data are obtained from geophysical studies and stable isotope analysis of the carbonate phase. The analysis of the anisotropy of magnetic minerals in the black shales of the Bay of Biscay, for example, indicated the current-transport pattern during the deposition of the black shales (Hailwood and Sayre, 1979). The stable-carbon isotope composition of the carbonate phase in pelagic sediments has proved to be a promising tool for the reconstruction of the circulation and fertility in ancient oceans (Arthur, 1979; Scholle and Arthur, 1980; Weissert and others, 1979).

In the following summary the numerous data collected on Early Cretaceous organic-rich black shale sequences are classified into four regional groups (Figs. 1–3):

1) central North Atlantic and western Tethys Oceans
2) northern North Atlantic Ocean
3) South Atlantic and Indian Oceans
4) Pacific Ocean

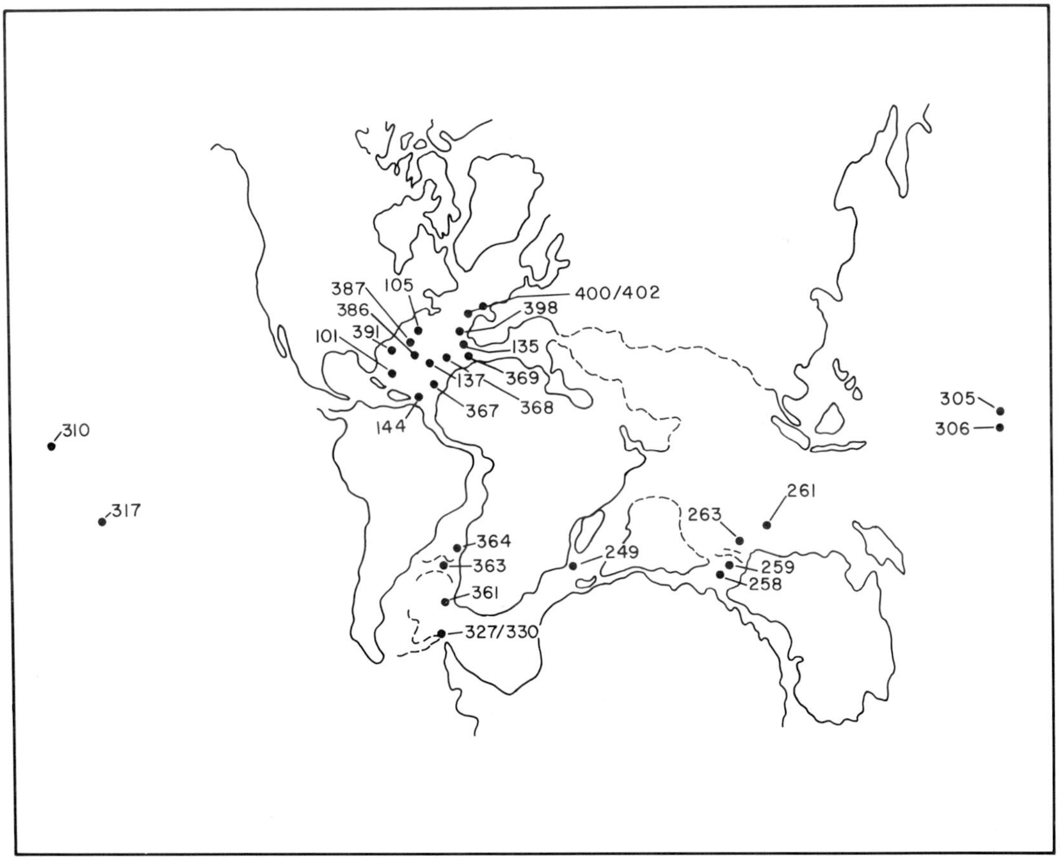

FIG. 1.—Locations of DSDP sites, where Early Cretaceous black shales were recovered (paleotectonic map after Smith and Briden, 1977).

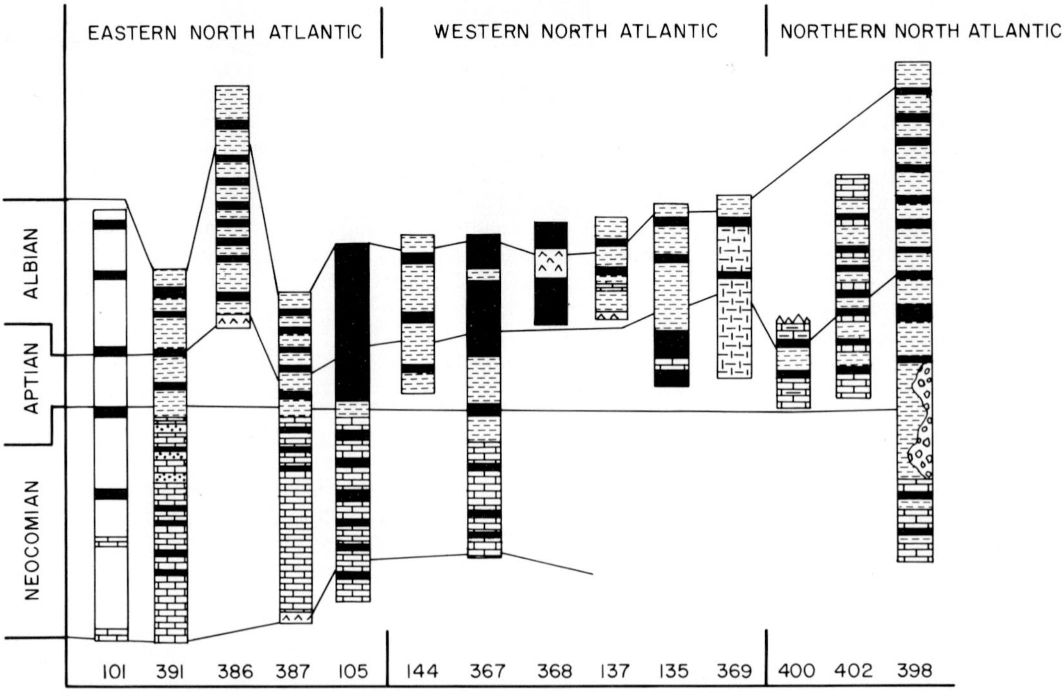

FIG. 2.—Early Cretaceous black-shale facies in DSDP cores from the North Atlantic Ocean (legend, see Fig. 3).

1) Central North Atlantic and Western Tethys Oceans

In the western North Atlantic, the Lower Cretaceous black shales were deposited in a deep-basinal environment within a depth range of 4000–5000 m (Tucholke and Vogt, 1979; Sheridan and others, 1978). The Mid-Cretaceous paleodepth for the sites drilled in the eastern North Atlantic was estimated as 3000 m or more by Berger and von Rad (1972) and Lancelot and Seibold (1978).

A notable feature of the Early Cretaceous black shale sections in the North Atlantic is the rhythmical alternation of carbonaceous sediments with limestone, marlstone and mudstone (Arthur, 1979). The Neocomian cycles generally consist of black shale-limestone couplets, while the Aptian-Albian organic-carbon rich sediments are intercalated in a sequence of variegated marls with a low content of $CaCO_3$, indicating a shallowing of the Aptian-Albian CCD (Lancelot and others, 1972; Berger and von Rad, 1972; Jansa and others, 1978; Tucholke and Vogt, 1979).

Similar black shale-limestone and black shale-variegated marl couplets, described from various regions in the southwestern Tethys Ocean, were probably deposited in depths of 2000–4000 m (Bosellini and Winterer, 1976; Hsü, 1976). The Neocomian black shales in the Southern Alps (Italy) are restricted to the basinal environments; they are missing on the adjacent submarine plateaus (Weissert and others, 1979). During the Albian, carbonaceous sediments were deposited both in the basins and on the submarine plateaus (Schlanger and Jenkyns, 1976).

The black shales in the central North Atlantic are often characterized by fine laminations which are generated by the alternation of light more calcareous and dark more carbonaceous sediment (Lancelot and Seibold, 1978; Dean and others, 1978; McCave, 1979). The general absence of bioturbation within these deposits may indicate a hostile environment for the bottom-dwelling fauna. Luterbacher (1972) emphasized the poor benthic foraminiferal assemblage in black shales from both the western North Atlantic and the Tethys Oceans. The black shales are composed of more than 50% clay minerals (illite, montmorillonite). Quartz, pyrite, zeolites and calcite are the most common among the remaining components (Lancelot and Seibold, 1978; Tucholke and Vogt, 1979; Dean and others, 1978). Clinoptilolite, interpreted as an alteration product of volcanic ash (Zemmels and others, 1972), is common in siliceous black shales, and may also form by direct transformation of radiolarians (Lancelot and others, 1979).

The average content of organic matter varies between 1% and 4% (Arthur and Natland, 1979), although values of up to 30% have been reported (*i.e.,* Jansa and others, 1978). Plant fragments and

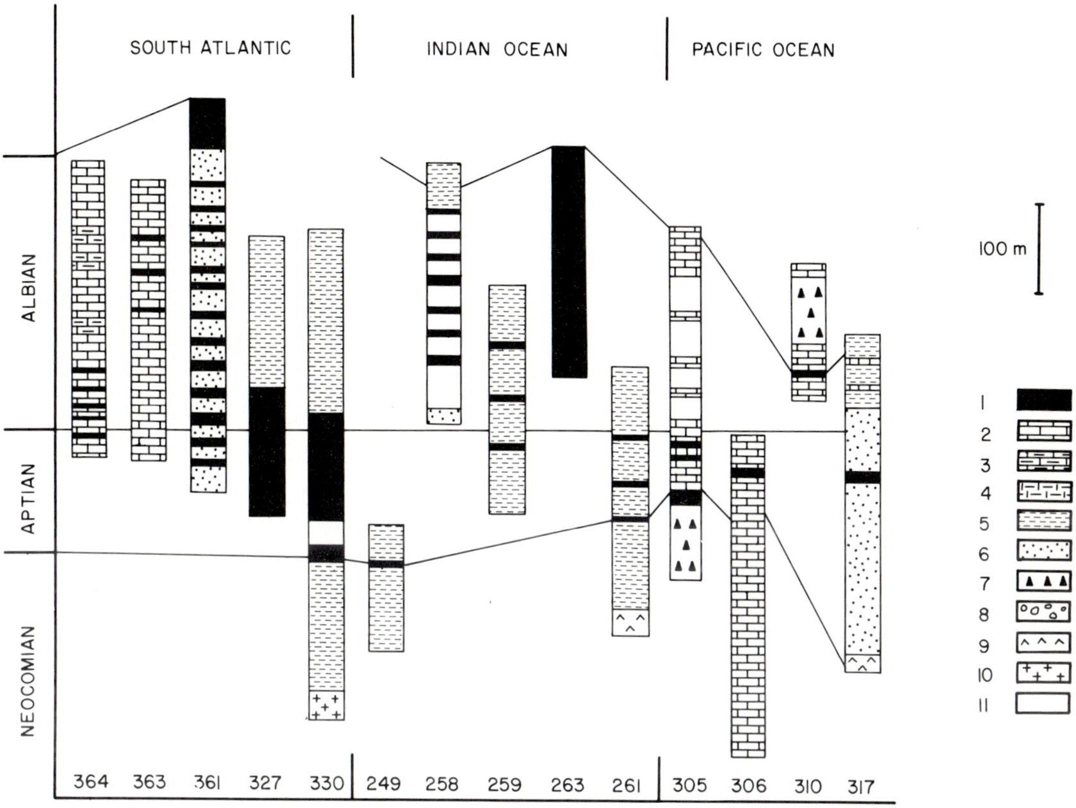

FIG. 3.—Early Cretaceous black shale facies in DSDP cores from the South Atlantic, the Indian Ocean and the Pacific Ocean. Legend: 1 Carbonaceous claystone, marl or limestone; 2 limestone; 3 marly limestone; 4 nanno ooze and nanno clay; 5 variegated marl, claystone; 6 sandstone; 7 chert; 8 conglomerate; 9 basalt, diabase; 10 granitic rocks; 11 sediments not recovered.

fish debris were reported from black shales off the coast of Morocco and from the Southern Alps (Dean and others, 1978; Weissert, 1979). Geochemical and palynological analyses indicate the origin of the organic matter. H/C versus O/C values in kerogen and the ^{13}C composition of the organic matter at the DSDP sites 386, 387, 391 and 105 in the western North Atlantic give evidence for a significant terrigenous source (Stuermer and Simoneit, 1978; Deroo and others, 1978b; Erdman and Schorno, 1979a; Cardoso and others, 1978; Dow, 1978; Kendrick, 1979; Simoneit, 1979; Tissot and others, 1979; Habib, 1979b). The black shales of Cenomanian age show an increasing amount of organic matter derived from marine organisms (*i.e.*, Kendrick, 1979). The degree of preservation of marine organic matter was dependent on the intensity of anoxia in the bottom water. During oxygenated conditions only the more resistant terrestrial compounds were preserved. One horizon, only, in the Albian sequence at Site 387 with a high percentage of marine-derived organic matter and a low pristane to phytane ratio ($Pr/Ph < 1$) indicated a very low oxygen content in the bottom waters (Simoneit, 1979). Marine organisms were the main contributor to the organic matter preserved in the black shale horizons in the eastern North Atlantic (Tissot and others, 1979; Deroo and others, 1978a; Summerhayes, 1980).

2) Northern North Atlantic

The Cretaceous black shales recovered at the DSDP sites 400A and 402 in the Bay of Biscay were deposited on a passive continental margin. The Lower Cretaceous Paleodepth at Site 400A was about 2000 m, while a depth of only 200 m is estimated for Site 402, where the black shales were part of a prograding depositional sequence on a carbonate platform (Roberts and Montadert, 1979a,b). At Site 398 off the coast of Portugal, the carbonaceous sediments of Hauterivian up to Cenomanian age were deposited in a graben structure at the southern margin of the Vigo Bank (Sibouet and Ryan, 1979). The depth of deposition ranged between 2500 m and 4000 m (Sibouet and Ryan, 1979; Rehault and Mauffret, 1979).

The Aptian-Albian sequence at Site 400A consists of a rhythmically bedded sequence, each of which starts with a bed of calcareous mudstone (de Graciansky and others, 1979). The top few centimeters of these beds are intensely bioturbated. Black, carbonaceous claystones with mottling overlie the mudstones. The black claystones contain up to 3% organic matter. Silty laminated claystones, interpreted as the autochthonous sediment, terminate the cycle. A rather high content of organic matter (up to 2%) in the laminated clays is interpreted as a result of partial reworking of the underlying sediment (de Graciansky and others, 1979). The hydrogen-poor compounds of the kerogen fraction of the organic matter indicate a terrestrial source for the organic carbon (Deroo and others, 1979a). Melieres (1979) analyzed the mineral fraction of the black claystones. The finding of the mineral assemblage illite-quartz-smectite supported the hypothesis of deGraciansky and others (1979) that the mudstone-black shale rhythms were deposited by turbidity currents. A well developed magnetic lineation, detected through anisotropy studies on magnetic minerals, originated from current activity during the deposition of the black shales (Hailwood and Sayre, 1979).

At Site 398 turbiditic sandstones and siltstones alternate with dark carbonaceous shales of Barremian-Aptian age. The Upper Aptian-Albian sequence consists of dark claystones with interbedded slightly burrowed mudstones (Arthur, 1979; Rehault and Mauffret, 1979; de Graciansky and Chenet, 1979). Graciansky and Chenet interpret the Barremian-Aptian sequence as a part of a large submarine fan. A low primary biogenic productivity and significant dilution by terrestrial material are explanations for the small number of microfossils in these sediments (Arthur, 1979). Fine laminations, the absence of bioturbation, and fluctuations in the bottom fauna are evidence for periodic anoxic conditions during the deposition of the black shales. The content of organic matter does not exceed 2% in the Barremian-Albian shales; in a Cenomanian bed values of up to 9% were measured (Deroo and others, 1979a). The organic matter in the Barremian-Albian sediments is entirely of terrestrial origin; marine organic matter contributes to the carbonaceous sediments in the Late Albian-Cenomanian (Cornford, 1979; Deroo and others, 1979a; Doerenkamp and Robert, 1979; Erdman and Schorno, 1979b; Simoneit and Mazurek, 1979; Habib, 1979a, b). The terrestrial organic matter in the Hauterivian-Aptian black shales was derived from Cretaceous land plants. Recycled organic matter dominates in the Late Aptian-Albian (Cornford, 1979).

3) South Atlantic and Indian Oceans

In the Early Cretaceous, the Falkland Plateau formed a shallow sill which separated the Antarctic-Indian Ocean from the juvenile South Atlantic (Barker, Dalziel and others, 1977). Two restricted basins, the Cape Basin in the south and the Angola Basin in the north, were separated by the Walvis-Sao Paulo Ridge (Arthur and Natland, 1979). During the Early Cretaceous the two basins had entirely different bathymetric conditions. The Cape Basin had a paleodepth of 1000–3000 m (Thiede and van Andel, 1977; Melguen, 1978; Arthur and Natland, 1979), whereas the Angola Basin was probably more shallow. Albian ammonites suggest inner shelf to epicontinental environments (Matsumoto and others, 1978). This paleontological evidence and the fact that the black shale facies is transitional to an evaporite facies imply that the Early Cretaceous carbonaceous sediments were deposited in a depth of only a few hundred meters (Arthur and Natland, 1979). No marine connection existed between the Angola Basin and the North Atlantic in Aptian-Albian times (Förster, 1978). The DSDP Site 249 on the Mozambique Ridge had a paleodepth of less than 200 m in the Early Cretaceous, while the Aptian-Albian dark shales in the eastern Indian Ocean at Sites 258, 259 and 261 were deposited in a deepsea environment (Kidd and Davies, 1977). At Site 263, close to the Australian continent, kaolinitic black clays have been deposited in shallow water of an embryonic oceanic basin (Veevers and Johnstone, 1974).

The Aptian-Albian black shales of the Angola Basin alternate with dolomitic limestones. The unusually high content of Na_2O in the dolomite minerals reflects a hypersaline environment during the deposition of the black shale-dolomitic limestone cycles (Matsumoto and others, 1978). The content of organic matter in the black shales varies around 6% (Arthur and Natland, 1979). Based on ^{13}C isotope data of the organic matter and lipid composition Simoneit (1978) concluded that both terrigenous and marine sources contributed to the deposition of these carbonaceous beds. The dominance of *Classopollis* in the palynomorph assemblage and the low pollen diversity suggest that the Albian climate around the Angola Basin was warm and semiarid (Morgan, 1978).

In the Cape Basin, at Site 361, the black shales alternate with carbonaceous sandstones (Kagami, 1978; Natland, 1978). Some of the shales are minutely laminated. The mineral assemblage illite-kaolinite-phillipsite, with little associated montmorillonite and mixed-layer clays, is an indicator for highly saline bottom water in the Aptian Cape Basin (Natland, 1978). The low carbonate content may reflect a low marine productivity or strong diagenetic dissolution (Melguen, 1978). The organic matter, which generally does not exceed 3% of the bulk sediment, has again a mixed marine and terrestrial origin (Foresman, 1978). McLachlan and Pieterse (1978) report a dominance of *Classopollis* in the pollen assemblage which is an indicator for a dry climate prevailing in South Africa in Aptian times. A benthic microfauna survived during the deposition of the black shales, but due to diagenetic dissolution of calcite only the chitinous linings of the tests are preserved (McLachlan and Pieterse, 1978).

The Jurassic and Lower Cretaceous carbonaceous sediments drilled on the Falkland Plateau alternate with horizons of porcellanite and limestone. The black shales, mainly composed of illite and montmorillonite, contain about 3% organic matter, which is mainly of terrigenous origin (Thompson, 1977). A comparable euxinic facies was described by Dalziel and others (1974) from the southernmost part of South America. Thompson (1977) suggests a direct relationship between the black shales found on the Falkland Plateau, the carbonaceous sediments recovered at Site 249 on the Mozambique Ridge, and Sites 258, 259, 261 and 263 in the then proto-Indian Ocean. Again the existence of an undisturbed parallel lamination and abundant finely divided pyrite within the black shales are evidence for an anoxic depositional environment which was unfavorable for a bottom-dwelling fauna (Simpson, Schlich, and others, 1974). The destruction of barriers during the ongoing fragmentation of Gondwanaland seems to coincide with a change from anoxic to oxygenated conditions in the above described environments (Thompson, 1977; Girdley and others, 1974).

4) Pacific Ocean

The few DSDP cores containing black shale facies in the Pacific are limited to submarine rises and plateaus; their paleodepth ranges between 2000 and 3000 m (Schlanger and Jenkyns, 1976). The carbonaceous sediments at Site 305 on the Shatsky Rise alternate with radiolarian limestones and cherts (Larson, Moberly, and others, 1975). A thin layer of organic rich sediment on the Manahiki Plateau is intercalated in a volcanoclastic sequence (Jenkyns, 1976; Jackson and Schlanger, 1976). Several sections drilled in the central Pacific indicate that the black shale facies was missing in the Early Cretaceous basinal environments (Sites 164, 166, 307, 310; Winterer, Ewing, and others, 1973; Larson, Moberly, and others, 1975).

GLOBAL MODELS VERSUS REGIONAL MODELS

Stratigraphic resolution in the Early Cretaceous is not sufficient to demonstrate synchroneity of the individual black shale deposits on a worldwide scale, and the sedimentology, organic geochemistry and mineralogy of the black shales vary greatly from one environment to the next. Nevertheless it is tempting to explain the formation of these Cretaceous carbonaceous sediments with a single global model. During the Early Cretaceous, large deltas in eastern North America, northwestern Europe, northwestern Africa and South Africa were outpouring into the North and South Atlantic oceans (*i.e.,* Ryan and Cita, 1977; von Rad and Arthur, 1979). The climate was warm and uniform and the temperature gradient between polar and equatorial regions was significantly smaller than today (Luyendyk and others, 1972; Savin, 1977). A reconstruction of the arrangement of the various lithospheric plates 100 million years ago reveals a large Pacific Ocean, a circumequatorial Tethys Ocean, and a series of small and still young oceans formed by the breakup of Gondwanaland and the separation of the Eurasian from the American continent (Luyendyk, 1974; Sclater and others, 1977; Smith and Briden, 1977). This particular configuration of ocean basins may have had a significant influence on the Cretaceous paleocirculation pattern. A large sill, the Panama Barrier, separated the North Atlantic from the Pacific, and probably only a surface circulation existed between the Atlantic and the Pacific Oceans (Saunders and others, 1973). The bottom circulation also may have been restricted in the Early Cretaceous Tethys realm, due to orogenic activity in the area of the present eastern Mediterranean (Smith and Briden, 1977). The North Atlantic and the South Atlantic were separated by a land barrier until late Albian (Berggren and Hollister, 1977; Förster, 1978). The South Atlantic, split into the Cape and Angola basins by the Walvis-Rio Grande Ridge, had a surface water connection with the Indian Ocean over the Falkland Plateau (Scheibnerova, 1978; Beckmann, 1978). The Wallaby Plateaus west of the Australian continent may have hindered bottom circulation between the Pacific and the Indian Ocean (Ryan and Cita, 1977).

Considering the general paleotectonic and paleoclimatic conditions during the Early Cretaceous, several models which explain the development of the black shale facies on a global scale can be constructed. Fischer and Arthur (1977), studying secular variations in the pelagic realm during the Mesozoic and Cenozoic, concluded that the Aptian-Albian coincided with a "polytaxic" period in earth history. Polytaxic periods are characterized by seas warmer than the present and by a sluggish marine circulation due to a uniform climate. The oxygen-minimum zone is intensified and expanded, leading to widespread deposition of organic-carbon rich sediments. The increased burial of organic matter in pelagic sediments of Early Cretaceous age disturbed the carbon-isotope budget of the oceans. Very positive $\delta^{13}C$ values measured in pelagic sediments of Aptian-Albian and of Cenomanian-Turonian age (Fischer and Arthur, 1977; Scholle and Arthur, 1980) may be construed as evidence for a disturbed oceanic circulation pattern during these time intervals (*e.g.,* Weissert and others, 1979).

According to Schlanger and Jenkyns (1976) the development of shallow epicontinental seas during Aptian-Albian and Cenomanian-Turonian times may have caused an increase in the production of organic carbon. An enlarged organic productivity as well as a sluggish circulation caused the expansion of the Mid-Cretaceous oxygen-minimum layer in restricted oceans and open seas. A slow renewal of bottom water may even have induced anoxic conditions in deepwater environments of small oceans. Supporting evidence for proposing a periodic expansion of the oxygen-minimum layer is given by the distribution of black shale in the Pacific, where it is restricted to submarine plateaus, and by the occurrence of carbonaceous sediments in shelf environments of the northwestern Tethys realm and in North America (Schlanger and Jenkyns, 1976; Byers and Larson, 1979). But it is incompatible with the findings of black shales restricted to basinal settings in the Atlantic and southwestern Tethys oceans (Sheridan and Enos, 1979; Weissert and others, 1979).

Thierstein and Berger (1978) and Thierstein (1979) postulate that during the Aptian the stability of the water column in the Atlantic and Pacific oceans was increased by periodic injections of brines into deep water, originating from the South Atlantic. The South Atlantic, filled with heavy brines, produced bottom waters which were dense enough to reinforce the haline stratification and oxygen-poor bottom waters in an already salinity-stratified Atlantic and Pacific oceans. The cyclicity of the black shale deposits is explained by a pulsing supply of saline bottom water, caused by possible climate fluctuations in the area of the South Atlantic.

Highly saline waters originating from the North American-Caribbean region and from the South Atlantic spilled periodically over shallow sills into the North Atlantic, Tethys and Indian Oceans and formed dense saline bottom water according to a model presented by Roth (1978). The resulting density stratification prevented a regular overturn of the water masses. Anoxic conditions were established in the bottom water as a consequence of stagnation and perhaps also because oxygen was sequestered as gypsum in the evaporite basins around the South Atlantic. Roth assumes that the oceanic productivity around 108 million years ago dropped because of a reduced nitrogen supply during times of extensive bacterial denitrification processes in anoxic water.

The periods of anoxia in the eastern Mediterranean during the Quaternary were taken as an analogue to the Cretaceous anoxic events in the global model by Ryan and Cita (1977). Increased freshwater influx during transitions from glacials to interglacials induced a stable stratification of the Mediterranean water by forming a low salinity lens over normal marine water (*i.e.,* Kidd and others, 1978). In the Cretaceous, large amounts of low density water may have been pouring into the Atlantic Ocean through huge deltas at times of increased rainfall. The resulting freshwater lens extended over the Atlantic and Pacific oceans. The depletion of oxygen in the stagnant deep water was caused by the decomposition of organic matter. Ryan and Cita suggest that the cyclicity of the black shale deposits indicates periodic overturn of the water masses. This mechanism may have been driven by heating of the bottom water through geothermal processes (Ryan and Cita, 1977; Natland, 1978).

Ryan and Cita (1977) emphasized the fact that the sedimentology and especially the organic composition of the various black shale deposits varies greatly through space and time. These differences have been taken into account to a certain degree by construction of the first global models (*e.g.,* Schlanger and Jenkyns, 1976). Later detailed geochemical and sedimentological investigations of black shale deposits from different paleogeographic environments revealed various shortcomings of the early global scenarios (Roberts and Montadert, 1979a,b). Subsequently, a number of alternative models explaining the formation of the black shales on a regional scale has been presented. In models based on data from pelagic realms such as the central North Atlantic, the Tethys or the South Atlantic, oceanographic parameters such as the circulation pattern or the biogenic productivity are selected as the main factors controlling the degree of anoxia in the deepwater environment. Models constructed for environments close to continents, as for the northern North Atlantic, emphasize the importance of climatically induced fluctuations in the input of terrestrial organic matter into marine settings. In the following section, regional models are presented which explain the formation of carbonaceous sediments in the South Atlantic, the central North Atlantic, the Tethys and the northern North Atlantic.

As a model for the formation of the black shale facies in the South Atlantic, stratified watermasses leading to periodic conditions of anoxia in the deep water has gained wide acceptance. The tectonically restricted Angola Basin was situated in a warm and dry low-latitude environment during Aptian times (Arthur and Natland, 1979). Dense hypersaline surface water and lagoonal water flowed into the deepest parts of the basin and caused the development of a stratified water mass (Bolli, Ryan and others, 1978; Arthur and Natland, 1979). The long residence-time of the bottom water facilitated the depletion of its oxygen content, and both marine and terrigenous organic matter were preserved in this environment (Melguen, 1978; Natland, 1978; Arthur and Natland, 1979). The adjacent Cape Basin was less restricted than the Angola Basin. The black shales at Site 361 were deposited in the deepest part of the basin according to Natland (1978). This supports the hypothesis that a periodic density stratification caused the formation of the anoxic deep water in which terrestrial and marine organic matter was not completely decomposed (McCoy and Zimmerman, 1977; Melguen, 1978; Natland, 1978; Arthur and Natland, 1979). Highly saline water originating from the Angola Basin spilled periodically over the Walvis-Rio Grande Ridge into the Cape Basin. In Albian time, anoxic conditions ceased as soon as the sill at the Aghulas fracture zone was deepened and the deep waters were oxygenated. High-latitude Indian Ocean water could replace the geothermally heated Cape Basin bottom water (Natland, 1978). The salinity-controlled stratification of the Aptian-Albian South Atlantic extended to the Falkland Plateau and possibly further into the Mozambique Graben and into the embryonic Indian Ocean (Thompson, 1977). Once the restricting sill at the northern end of the Mozambique Graben was removed during the Albian, oxygenated conditions started to dominate in these ocean basins (Girdley and others, 1974).

Anoxic bottom water in a stratified ocean has been proposed by various authors as the cause for the formation of black shale facies in the Barremian to Albian North Atlantic and Tethys oceans (Ewing and others, 1969; Lancelot and others, 1972; Saunders and others, 1973; Berger and von Rad, 1972; Bernoulli and Jenkyns, 1974; Lancelot and Seibold, 1978; Tucholke and Vogt, 1979; Arthur and Natland, 1979; Bourbon, 1978; Weissert and others, 1979). Saunders and others (1973) and Tucholke and Vogt (1979) suggest that tectonic barriers in the Caribbean region and in the Tethys blocked the deepwater circulation. Sluggish oceanic circulation conditions, prevalent in the Early Cretaceous, favored the development of an anoxic bottomwater mass under these circumstances. A humid climate in Aptian-Albian times supported an intensified growth of land plants on the European and North American continents (Sheridan and others, 1978). Detritus from this plant material was washed into the North Atlantic Basin through extensive deltas and contributed to the black shale deposits (Schlanger and Jenkyns, 1976; Ryan and Cita, 1977; Einsele and von Rad, 1979; Habib, 1979a,b). The palynological spectra from Barremian black shales from the Southern Alps (Italy) reveal a drastic change from a warm and dry climate to a cooler, more humid one (Weissert and others, 1979). During the Neocomian, a high evaporation-rate in the summer and an annual cooling of the surface waters may have made possible the formation of Tethys deep water in a way similar to that by which Mediterranean deep water is produced today (Lacombe and Tchernia, 1972). A more humid climate in the Barremian caused a replacement of the thermohaline circulation by a density stratification. Consequently the deepest waters of the western Tethys Ocean periodically became anoxic (Weissert and others, 1979).

The diversity in the composition of the organic matter in the western and eastern central North Atlantic induced Tissot and others (1979) to modify the models presented for a widespread zone of anoxic conditions in the bottom water of the North Atlantic. The dominance of terrestrial organic matter in the western basin is explained by turbidite transported land-plant material into generally oxygenated basins. Particle size-distribution studies in the black shales of the western North Atlantic by Pastouret and others (1979) support this model. Prevailing oxygenated conditions explain the low percentage of marine-derived organic matter in these black shales. In the eastern central North Atlantic, where both a marine and a terrestrial source for the organic matter have been determined, an extensive layer of anoxic bottom water existed during the deposition of the organic-rich sediments. Tissot and others (1979) argued that the Mid-Atlantic Ridge and the narrow connection between Tethys and Atlantic oceans in Cretaceous time hindered bottom circulation in the eastern basin. Deepwater exchange between the western Atlantic basins and the Pacific Ocean was not restricted according to this model. The model of Tissot and others (1979) contrasts with the explanation of Dean and others (1978) who suggest that the black shales in the eastern central North Atlantic were formed during periods of high productivity and that bottom waters were not essentially anoxic. A high productivity, due to upwelling along the west coast of North Africa, combined with anoxic conditions in the eastern Atlantic deep water may explain the high percentage of marine-derived organic matter found in various black shale deposits from the eastern North Atlantic (Einsele and Wiedmann, 1975; Arthur and Natland, 1979).

In comparison with other occurrences from the Atlantic Ocean, the black shales at Sites 398, 400 and 402, drilled along the continental margin of the northern North Atlantic, show significant differences in their sedimentological and geochemical composition. In the models proposed by Roberts and Montadert (1970a, b), Arthur (1979) and Habib (1979a, b) the importance of the terrigenous source of the organic matter is emphasized. Roberts and Montadert (1979a, b) concluded from sedimentological data that turbidity currents transported the organic matter into the Early Cretaceous Bay of Biscay, which was well oxygenated. The black shales at Site 398 were deposited under anoxic conditions during the Barremian-Aptian (Simoneit and Mazurek, 1979). In Arthur's model (1979), land-derived organic matter was periodically transported into this anoxic basin. The surface productivity was low, and the increasing percentage of preserved marine organic matter in Upper Albian time coincides with a decrease of terrigenous material in the bulk sediment. Intermittent anoxic conditions prevailed up to the Cenomanian along the coastal margin of Portugal.

SUMMARY: AN ONGOING CONTROVERSY

With the formulation of scenarios to explain the origin of the black shales recovered during the Deep Sea Drilling Project, a stimulating controversy on the paleoenvironmental conditions during the Early Cretaceous was initiated. Intrigued by the existence of Early Cretaceous black shales in all the major oceans, various authors proposed that the deposition of organic-carbon rich sediments was caused by a series of global oceanic events.

Because the stratigraphic resolution did not allow a worldwide correlation of individual black shale horizons, the Barremian-Albian and the Cenomanian-Turonian were proposed as the two main time-envelopes within which the oceanic anoxic events took place (Schlanger and Jenkyns, 1976). A periodic expansion of the oxygen-minimum zone or a repeated density stratification due to injection of high or low salinity waters into the world oceans were presented as possible mechanisms leading to the deposition of black shales in oxygen-depleted environments. Significant compositional differences among the recovered organic-carbon rich sediments, revealed by detailed sedimentological and geochemical analyses, induced various authors to propose a series of regional models as an alternative to the global scenarios. In these models, local tectonic and climatic conditions were emphasized as controlling factors for the black-shale facies pattern. Models based on studies of black shales in pelagic environments, such as the central North Atlantic, the Tethys or the South Atlantic-Indian ocean, accentuate the importance of oceanographic events causing the deposition of black shales in an anoxic environment. In models developed for black shale deposits in continental-margin environments like the northern North Atlantic, the importance of regional climatic conditions influencing the rate of redeposition of terrestrial organic matter into marine settings is underlined.

The proposed local effects causing the deposition of the Early Cretaceous carbonaceous sediments may only mask a series of global oceanic events. Indications that the Aptian-Albian and also the Cenomanian-Turonian represent periods of worldwide oceanic instability are given by the widespread occurrence of the carbonaceous deposits and by the carbon-isotope stratigraphy. Heavy carbon-isotope values in carbonates of Aptian-Albian and Cenomanian-Turonian ages suggest a worldwide disturbance of the carbon-isotope budget, caused by burial of excessive organic matter in the pelagic sediments. These indications of a global oceanic crisis during the Early and Middle Cretaceous, as well as the indications of a regionally controlled deposition of the black shales, will have to be integrated in a refined global scenario. Geologists are confronted with the exciting problem of unraveling feedback mechanisms between the paleoclimate, the paleoceanography and the specific paleotectonic situation in the Early Cretaceous. Did widespread anoxic conditions during the Cretaceous cause an imbalance in the global carbon budget which in turn affected the chemical composition of the atmosphere, as proposed by Ryan and Cita (1977)? Or was it a series of regional climatic or oceanographic events that triggered periods of global oceanic instability (*e.g.,* Arthur and Kelts, 1979)? In seeking an answer to these questions we may also learn more about the degree of perturbation that can be placed upon oceans and atmosphere without severely disturbing climatic and oceanic stability.

ACKNOWLEDGEMENTS

I am grateful to R. G. Douglas and D. S. Gorsline (University of Southern California) and K. Kelts and J. McKenzie (ETH, Zurich) who critically read earlier drafts of this manuscript. Their suggestions, as well as the helpful comments of M. A. Arthur (USGS, Denver) and S. O. Schlanger (University of Hawaii), who reviewed this article, led to improvements in the final manuscript.

REFERENCES CITED

NOTE: All references to Deep Sea Drilling Project Initial Reports have been abbreviated; for complete citations see *Appendix I* at the end of this volume, or the individual Initial Reports as published.

ARTHUR, M. A., 1979, North Atlantic Cretaceous black shales: the record at Site 398 and a brief comparison with other occurrences: Init. Repts. DSDP, v. 47, p. 719–751.

———, *in press,* Paleoceanographic events—recognition, resolution and reconsideration: Rev. Geophys. Space Physics.

———, AND KELTS, K., 1979, Evaporites, black shales and perturbations of ocean chemistry and fertility: Geol. Soc. America Ann. Mtg. Abstracts with Programs.

———, AND NATLAND, J. H., 1979, Carbonaceous sediments in the North and South Atlantic: the role of salinity in stable stratification of early Cretaceous basins, *in* Talwani, M., Hay, W. W., and Ryan, W. B., eds., Deep drilling results in the Atlantic Ocean: continental margins and paleoenvironments: Proceedings of the Second Maurice Ewing Symposium (1978), vol. 3, p. 375–401.

———, and Schlanger, S. O., 1979, Cretaceous "oceanic anoxic events" as causal factors in development of reef-reservoired giant oil fields: Amer. Assoc. Petroleum Geologists Bull., v. 63, p. 870–885.

Barker, P. F., Dalziel, I. W., and others, 1977: Init. Repts. DSDP, v. 36.

Beckmann, J. P., 1978, Late Cretaceous smaller benthonic foraminifers from Sites 363 and 364, DSDP Leg 40, southeast Atlantic Ocean: Init. Repts. DSDP, v. 40, p. 759–782.

Benson, W. E., Sheridan, R. E., and others, 1978: Init. Repts. DSDP, v. 44.

Berger, W. H., 1979, Impact of deep sea drilling on paleoceanography, *in* Talwani, M., Hay, W. W., and Ryan, W. B., eds., Deep drilling results in the Atlantic Ocean, Proceedings of the Second Maurice Ewing Symposium (1978), p. 297–314.

———, and von Rad, U., 1972, Cretaceous and Cenozoic sediments from the Atlantic Ocean: Init. Repts. DSDP, v. 14, p. 787–953.

Berggren, W. A., and Hollister, C. D., 1977, Plate tectonics and paleocirculation—commotion in the ocean: Tectonophysics, v. 38, p. 11–48.

Bernoulli, D., 1972, North Atlantic and Mediterranean Mesozoic facies: a comparison: Init. Repts. DSDP, v. 11, p. 801–822.

———, and Jenkyns, H. C., 1974, Alpine, Mediterranean and Central Atlantic Mesozoic facies in relation to the early evolution of the Tethys: Soc. Econ. Paleontologists Mineralogists Spec. Pub., 19, p. 129–160.

Bitterli, P., 1965, Bituminous intercalations in the Cretaceous of the Breggia River, S. Switzerland: Bull. Ver. Schweiz. Petrol. Geol. u. Ing., v. 31, p. 179–185.

Bolli, H. M., Ryan, W. B., and others, 1978: Init. Repts. DSDP, v. 40.

Bosellini, A., and Winterer, E., 1976, Pelagic limestone and radiolarite: a genetic model: Geology, v. 3, p. 279–282.

Bourbon, M., 1978, Mesozoic evolution of western North Atlantic and North Tethyan margins: a comparison: Init. Repts. DSDP, v. 44, p. 949–970.

Byers, C. W., and Larson, D. W., 1979, Paleoenvironments of Mowry shale (Lower Cretaceous), western and central Wyoming: Amer. Assoc. Petroleum Geologists Bull., v. 63, p. 345–375.

Cardoso, J. N., Wardroper, A. M., Watts, C. D., Barnes, P. J., Maxwell, J. R., Eglinton, G., Mound, D. G., and Speers, G. C., 1978, Preliminary organic geochemical analyses, Site 391, Leg 44 of the DSDP: Init. Repts. DSDP, v. 44, p. 617–623.

Cornford, C. H., 1979, Organic petrography of Lower Cretaceous shales at DSDP Leg 47B, Site 398, Vigo Seamount, eastern North Atlantic: Init. Repts. DSDP, v. 47, p. 523–528.

Dalziel, I. W., and others, 1974, South extremity of Andes: geology of Isla de los Estados, Argentina, Tierra del Fuego: Amer. Assoc. Petroleum Geologists Bull., v. 58, p. 2602–2512.

Davies, T. A., Luyendyk, B. P., and others, 1974, Init. Repts. DSDP, v. 26.

Dean, W. E., Gardner, J. V., Jansa, L. F., Cepek, P., and Seibold, E., 1978, Cyclic sedimentation along the continental margin of northwest Africa: Init. Repts. DSDP, v. 41, p. 965–989.

———, Thiede, J., and Claypool, G. E., 1979, Origin of organic-rich limestones of Mid-Cretaceous age, Mid-Pacific mountains and southern Hess Rise: Geol. Soc. America Abstracts with Programs.

de Graciansky, P. C., and Chenet, P. Y., 1979, Sedimentological study of Cores 138 to 56 (Upper Hauterivian to Middle Cenomanian): an attempt at reconstruction of paleoenvironments: Init. Repts. DSDP, v. 47, p. 403–418.

———, Auffret, G. A., Dupeusie, P., Montadert, L. and Mueller, C., 1979, Interpretation of depositional environments of the Aptian/Albian black shales of the northern margin of the Bay of Biscay (DSDP Sites 400 and 402): Init. Repts. DSDP, v. 48, p. 877–907.

Deroo, G., Herbin, J. P., Roucache, J., Tissot, B., Albrecht, P., and Schaeffle, J., 1978a, Organic geochemistry of some Cretaceous black shales from Sites 367 and 368, Leg 41, eastern North Atlantic: Init. Repts. DSDP, v. 41, p. 865–873.

———, ———, ———, ———, ———, and Dastillung, M., 1978b, Organic geochemistry of some Cretaceous claystones from Site 391, Leg 44, western North Atlantic: Init. Repts. DSDP, v. 44, p. 593–598.

———, ———, ———, and ———, 1979a, Organic geochemistry of Cretaceous mudstones and marly limestones from DSDP sites 400 and 402, Leg 48, eastern North Atlantic: Init. Repts. DSDP, v. 48, p. 921–930.

———, ———, ———, and ———, 1979b, Organic geochemistry of Cretaceous shales from DSDP Site 398, Leg 47B, eastern North Atlantic: Init. Repts. DSDP, v. 47, p. 513–522.

Didyk, B. M., Simoneit, B. R., Brassell, S. C., and Eglinton, G., 1978, Organic geochemical indicators of paleoenvironmental conditions of sedimentation: Nature, v. 272, p. 216–222.

Doerenkamp, A., and Robert, P., 1979, Optical study of organic matter from some samples of Cretaceous age, Leg 47B, Hole 398D: Init. Repts. DSDP, v. 47, p. 529–532.

Dow, W. G., 1978, Geochemical analysis of samples from Holes 391A and 391C, Leg 44, Blake-Bahama Basin: Init. Repts. DSDP, v. 44, p. 625–634.

Edgar, N. T., Saunders, J. B., and others, 1973: Init. Repts. DSDP, v. 15.

Einsele, G., and von Rad, U., 1979, Facies and paleoenvironment of Lower Cretaceous sediments at DSDP Site 397 and in the Aaiun Basin: Init. Repts. DSDP, v. 47, p. 559–577.

———, and Wiedmann, J., 1975, Faunal and sedimentological evidence for upwelling in the Upper Cretaceous coastal basin of Tarfaya, Morocco: IXth Int. Cong. on Sedimentology, Nice, Theme 1, p. 67–72.

Erdman, J. G., and Schorno, K. S., 1979a, Geochemistry of carbon: Deep Sea Drilling Project Leg 43: Init. Repts. DSDP, v. 43, p. 651–655.

———, AND ———, 1979b, Geochemistry of carbon: Deep Sea Drilling Project Legs 47A and 47B: Init. Repts. DSDP, v. 47, p. 553–560.

EWING, J. I., AND HOLLISTER, C. D., 1972, Regional aspects of deep sea drilling in the western North Atlantic: Init. Repts. DSDP, v. 11, p. 951–973.

EWING, M., AND OTHERS, 1969, Init. Repts. DSDP, v. 1.

FISCHER, A. G., AND ARTHUR, M. A., 1977, Secular variations in the pelagic realm: Soc. Econ. Paleontologists Mineralogists Spec. Pub. 25, p. 19–50.

FORESMAN, J. B., 1978, Organic geochemistry, DSDP Leg 40: continental rise of southwest Africa: Init. Repts. DSDP, v. 40, p. 557–567.

FÖRSTER, R., 1978, Evidence for an open seaway between northern and southern proto-Atlantic in Albian times: Nature, v. 272, p. 158–159.

FUGANTI, A., 1961, Ricerche geologiche e glacimentologiche sugli "scisti bituminosi uraniferi" de Mollaro (Val di Non, Trentino): Studie Trentini Sci. Atl., v. 44/1, p. 73–110.

GIRDLEY, W. A., LECLAIRE, L., MOORE, C., VALLIER, T. L., AND WHITE, S. M., 1974, Lithologic summary, Leg 25, DSDP: Init. Repts. DSDP, v. 25, p. 725–742.

HABIB, D., 1979a, Sedimentology of palynomorphs and palynodebris in Cretaceous carbonaceous facies south of Vigo Seamount: Init. Repts. DSDP, v. 47, p. 451–474.

———, 1979b, Sedimentary origin of North Atlantic Cretaceous palynofacies, *in* Talwani, M., Hay, W. W., and Ryan, W. B., eds., Deep drilling results in the Atlantic Ocean: continental margins and paleoenvironments, Maurice Ewing Series 3, p. 420–437.

HAILWOOD, E. A., AND SAYRE, W. O., 1979, Magnetic anisotropy and sediment transport directions in North Atlantic Early Cretaceous black shales and Eocene mudstones cored on DSDP Leg 48: Init. Repts. DSDP, v. 48, p. 909–948.

HAYES, D. E., PIMM, A. C., AND OTHERS, 1972: Init. Repts. DSDP, v. 1.

HEDBERG, H. D., 1937, Stratigraphy of the Rio Querecual section of northeastern Venezuela: Geol. Soc. America Bull., v. 48, p. 1971–2024.

HOLLISTER, C. D., EWING, J. I., AND OTHERS, 1972: Init. Repts. DSDP, v. 11.

HSÜ, K. J., 1976, Paleoceanography of the Mesozoic alpine Tethys: Geol. Soc. America Spec. Paper 170, 44 p.

JACKSON, E. D., AND SCHLANGER, S. O., 1976, Regional synthesis, Line Islands chain, Tuamotu Island chain and Manahiki Plateau, central Pacific Ocean: Init. Repts. DSDP, v. 33, p. 915–927.

JANSA, L., GARDNER, J. V., AND DEAN, W. E., 1978, Mesozoic sequences of the central North Atlantic: Init. Repts. DSDP, v. 41, p. 991–1032.

JENKYNS, H. C., 1976, Sediments and sedimentary history, Manahiki Plateau, southern Pacific Ocean: Init. Repts. DSDP, v. 33, p. 873–890.

———, AND SCHLANGER, S. O., *in press,* Significance of plant remains in redeposited Aptian sediments, Hole 462A, Nauru Basin: Init. Repts. DSDP, v. 61.

KAGAMI, H., 1978, Sedimentary features of Cape Basin and Angola Basin sediments: Init. Repts. DSDP, v. 40, p. 525–540.

KENDRICK, J. W., 1979, Geochemical studies of black clays from Leg 43, Deep Sea Drilling Project: Init. Repts. DSDP, v. 43, p. 633–642.

KIDD, R. B., AND DAVIES, T. A., 1977, Indian Ocean sediment distribution since the Late Jurassic: Marine Geology, v. 26, p. 49–70.

———, CITA, M. B., AND RYAN, W. B., 1978, Stratigraphy of eastern Mediterranean sapropels recovered during DSDP Leg 42A and their paleoenvironment significance: Init. Repts. DSDP, v. 42, p. 421–443.

LACOMBE, H., AND TCHERNIA, P., 1972, Caracteres hydrologiques et circulation des eaux de Mediterranee, *in* Stanley, D., ed., The Mediterranean Sea. A natural laboratory: Stroudsbourg, Pa., Dowden, Hutchinson & Ross, p. 25–36.

LANCELOT, Y., HATHAWAY, J. C., AND HOLLISTER, D. C., 1972, Lithology of sediments from the western North Atlantic, Leg 11, DSDP: Init. Repts. DSDP, v. 11, p. 901–949.

———, AND SEIBOLD, E., 1978, The evolution of the central northeastern Atlantic—summary of results of DSDP Leg 41: Init. Repts. DSDP, v. 41, p. 1215–1245.

———, ———, AND OTHERS, 1978: Init. Repts. DSDP, v. 41.

———, SLITER, W. V., AND WESTBERG, J., 1979, Evidence of direct transformation of radiolarians into zeolites in Mid-Cretaceous deep-sea clays: Geol. Soc. America Ann. Mtg. Abstracts with Programs.

LARSON, R. L., MOBERLY, R., AND OTHERS, 1975, Init. Repts. DSDP, v. 32.

LUTERBACHER, H. P., 1972, Foraminifera from the Lower Cretaceous and Upper Jurassic of the northwestern Atlantic: Init. Repts. DSDP, v. 11, p. 561–594.

LUYENDYK, B. P., 1974, Gondwanaland dispersal and the early formation of the Indian Ocean: Init. Repts. DSDP, v. 26, p. 945–952.

———, FORSYTH, D., AND PHILLIPS, J. D., 1972, Experimental approach to the paleocirculation of the oceanic surface waters: Geol. Soc. America Bull., v. 83, p. 2649–2664.

MATSUMOTO, R., UTADA, M., AND KAGAMI, H., 1978, Sedimentary petrology of DSDP cores from Sites 362 and 363, the Walvis Ridge, and Site 364, the Angola Basin, drilled on Leg 40: Init. Repts. DSDP, v. 40, p. 469–485.

MCCAVE, I. N., 1979, Depositional features of organic-rich black and green mudstones at DSDP Sites 386 and 387, western North Atlantic: Init. Repts. DSDP, v. 43, p. 395–406.

McCoy, F. W., and Zimmerman, H. B., 1977, A history of lithofacies in the South Atlantic Ocean: Init. Repts. DSDP, v. 39, p. 1047–1079.

McLachlan, I., and Pieterse, E., 1978, Preliminary palynological results, Site 361, Leg 40, DSDP: Init. Repts. DSDP, v. 40, p. 857–881.

Melguen, M., 1978, Facies evolution, carbonate dissolution cycles in sediments from the eastern South Atlantic (DSDP Leg 40) since the Early Cretaceous: Init. Repts. DSDP, v. 40, p. 981–1024.

Melieres, F., 1979, Minerology and geochemistry of selected Albian sediments from the Bay of Biscay, Deep Sea Drilling Project, Leg 48: Init. Repts. DSDP, v. 48, p. 855–875.

Montadert, L., Roberts, D. G., and others, 1979: Init. Repts. DSDP, v. 48.

Morgan, R., 1978, Albian to Senonian palynology of Site 364, Angola Basin: Init. Repts. DSDP, v. 40, p. 915–951.

Natland, J., 1978, Composition, provenance, and diagenesis of Cretaceous clastic sediments drilled on the Atlantic continental rise off southern Africa, DSDP Site 361: Init. Repts. DSDP, v. 40, p. 1025–1062.

Pastouret, L., Auffret, G. A., and Chambley, H., 1979, Microfacies of some sediments from the western North Atlantic: paleoceanographic implications: Init. Repts. DSDP, v. 44, p. 477–493.

Rehault, J. P., and Mauffret, A., 1979, Relationships between tectonics and sedimentation around the northwest Iberian margin: Init. Repts. DSDP, v. 47, p. 663–682.

Roberts, D. G., and Montadert, L., 1979a, Evolution of passive rifted margins—perspective and retrospective of DSDP Leg 48: Init. Repts. DSDP, v. 48, p. 1143–1153.

———, and ———, 1979b, Margin paleoenvironments of the northeast Atlantic: Init. Repts. DSDP, v. 48, p. 1099–1118.

Roth, P. H., 1978, Cretaceous nannoplankton biostratigraphy and oceanography of the northwestern Atlantic Ocean: Init. Repts. DSDP, v. 44, p. 731–759.

Ryan, W. B., and Cita, M. B., 1977, Ignorance concerning episodes of oceanwide stagnation: Marine Geology, v. 23, p. 197–215.

Saunders, J. B., Edgar, N. T., Donnelly, T. W., and Hay, W. W., 1973, Cruise synthesis, Leg 15: Init. Repts. DSDP, v. 15, p. 1077–1113.

Savin, S. M., 1977, The history of the earth's surface temperature during the last 100 million years: *in* Donath, F. A., Stehli, F. G., and Wetherill, G. W., eds., Annual Review Earth Planet. Sci., v. 5, p. 319–355.

Scandone, P., 1972, Studi di geologia lucana: carta dei terreni della serie calcareo-silico-marnosa: Soc. Naturalisti Napoli Bull., v. 81, p. 225–330.

Scheibnerova, V., 1978, Aptian and Albian benthic foraminifers of Leg 40, Sites 363 and 364, southern Atlantic: Init. Repts. DSDP, v. 40, p. 741–757.

Schlanger, S. O., Jackson, E. D., and others, 1976: Init. Repts. DSDP, v. 33.

———, and Jenkyns, H. C., 1976, Cretaceous oceanic anoxic events: causes and consequences: Geologie en Mijnbouw, v. 55, p. 179–184.

Scholle, P. A., and Arthur, M. A., 1980, Carbon isotopic fluctuations in pelagic limestones: potential stratigraphic and petroleum exploration tool: Amer. Assoc. Petroleum Geologists Bull., v. 54, p. 67–87.

Sclater, J. G., Anderson, R. N., and Bell, M. L., 1971, Elevation of ridges and evolution of the central eastern Pacific: Jour. Geophys. Res., v. 76, p. 7888–7915.

———, Hellinger, S., and Tapscott, C., 1977, The paleobathymetry of the Atlantic from the Jurassic to the Present: Jour. Geology, v. 85, p. 509–552.

Sheridan, R. E., and Enos, P., 1979, Stratigraphic evolution of the Blake Plateau after a decade of scientific drilling *in* Talwani, M., Hay, W. W., Ryan, W. B., eds., Deep drilling results in the Atlantic Ocean: continental margins and paleoenvironments: Proceedings Second Maurice Ewing Symposium, p. 187–203.

———, ———, Gradstein, F., and Benson, W. E., 1978, Mesozoic and Cenozoic environments: Init. Repts. DSDP, v. 44, p. 971–980.

Sibouet, J. C., Ryan, W. B., and others, 1979: Init. Repts. DSDP, v. 47.

———, and ———, 1979, Site 398: evolution of the west Iberian passive continental margin in the framework of the early evolution of the North Atlantic Ocean: Init. Repts. DSDP, v. 47, p. 761–775.

Simoneit, B. R., 1978, Lipid analysis of sediments from Site 364 in the Angola Basin, DSDP Leg 40: Init. Repts. DSDP, v. 40, p. 659–662.

———, 1979, Organic geochemistry of shales from the northwestern proto-Atlantic, DSDP Leg 43: Init. Repts. DSDP, v. 43, p. 643–650.

———, and Mazurek, M. A., 1979, Lipid geochemistry of Cretaceous sediments from Vigo Seamount, DSDP/IPOD Leg 47B: Init. Repts. DSDP, v. 47, p. 565–570.

Simpson, E. W., Schlich, R., and others, 1974, Init. Repts. DSDP, v. 25.

Smith, A. G., and Briden, J. C., 1977, Mesozoic and Cenozoic paleocontinental maps: Cambridge University Press, 63 p.

Stuermer, D. H., and Simoneit, B. R., 1978, Varying sources for the lipids and humic substances of Site 391, Blake-Bahama Basin, DSDP Leg 44: Init. Repts. DSDP, v. 44, p. 587–591.

Summerhayes, C. P., 1980, Organic facies of Mid-Cretaceous black shales in the deep North Atlantic: Abstract, Amer. Assoc. Petroleum Geologists. Annual meeting, Denver, 1980.

Thiede, J., and van Andel, T. H., 1977, The paleoenvironment of anaerobic sediments in the Late Mesozoic South Atlantic Ocean: Earth Planet. Sci. Letters, v. 33, p. 301–309.

———, AND OTHERS, 1979, Middle Cretaceous oxygen-deficient paleoenvironments in mid-Pacific mountains and on Hess-Rise, central North Pacific Ocean: Abstract, Amer. Assoc. Petroleum Geologists Bull., v. 63, p. 538–539.

THIERSTEIN, H. R., 1979, Paleoceanographic implications of organic carbon and carbonate distribution in Mesozoic deep-sea sediments, *in* Talwani, M., Hay, W. W., and Ryan, W. B. F., eds., Deep drilling results in the Atlantic Ocean: continental margins and paleoenvironment: Proceedings Second Maurice Ewing Symposium, p. 249–274.

———, AND BERGER, W. H., 1978, Injection events in earth history: Nature, v. 276, p. 461–464.

THOMPSON, R. W., 1977, Mesozoic sedimentation on the eastern Falkland Plateau: Init. Repts. DSDP, v. 36, p. 877–891.

TISSOT, B., DEROO, G., AND HERBIN, J. P., 1979, Organic matter in Cretaceous sediments of the North Atlantic: contribution to sedimentology and paleogeography, *in* Talwani, M., Hay, W. W., and Ryan, W. B., eds., Deep drilling results in the Atlantic Ocean: continental margins and paleoenvironments, Maurice Ewing Series 3, p. 362–374.

TUCHOLKE, B. R., AND VOGT, P. R., 1979, Western North Atlantic: sedimentary evolution and aspects of tectonic history: Init. Repts. DSDP, v. 43, p. 791–826.

———, ———, AND OTHERS, 1979: Init. Repts. DSDP, v. 43.

VEEVERS, J. J., AND JOHNSTONE, M. H., 1974, Comparative stratigraphy and structure of the western Australian margin and the adjacent ocean floor: Init. Repts. DSDP, v. 27, p. 571–585.

———, HEIRTZLER, J. R., AND OTHERS, 1974, Init. Repts. DSDP, v. 27.

VON RAD, U., AND ARTHUR, M. A., 1979, Geodynamic, sedimentary and volcanic evolution of Cape Bojador continental margin (NW-Africa), *in* Talwani, M., Hay, W. W., and Ryan, W. B., eds., Deep drilling results in the Atlantic Ocean: continental margins and paleoenvironments: Proceedings Second Maurice Ewing Symposium, p. 187–203.

WEISSERT, H., 1979, Die Palaeoozeanographic der südwestlichen Tethys in der Unterkreide: Mitteilungen aus dem Geol. Inst. der ETH und der Univ. Zuerich, nr. 226, 174 p.

———, MCKENZIE, J., AND HOCHULI, P., 1979, Cyclic anoxic events in the Early Cretaceous Tethys Ocean: Geology, v. 7, p. 147–151.

WINTERER, E. L., EWING, J. I., AND OTHERS, 1973: Init. Repts. DSDP, v. 17.

ZEIL, W., 1956, Zur Kenntnis der hoeheren Unterkreide in den Bayrischen Alpen: N. Jb. Geol. und Palaeont. Abh., v. 103, p. 375–412.

ZEMMELS, I., COOK, H. E., AND HATHAWAY, J. C., 1972, X-ray mineralogical studies, Leg 11: Init. Repts. DSDP, v. 11, p. 729–789.

Appendix

Chronological Listing of all DSDP Legs Published to Date by the U.S. Government Printing Office

Note: the following is an example of the reference style currently used by DSDP in the "blue book" volumes:

Karig, D.E., 1975. Basin genesis in the Philippine Sea. *In* Karig, D.E., Ingle, J.C., Jr., et al, *Init. Repts. DSDP,* 31: Washington (U.S. Govt. Printing Office), 857–879.

Leg #'s	Port to Port Dates	Ports	Sites	Co-Chiefs	Date of Publ.
1	Aug–Sep, 1968	Orange, Texas to Hoboken, New Jersey	1–7	Maurice Ewing J. Lamar Worzel	Jun, 1969
2	Oct–Nov, 1968	Hoboken, New Jersey to Dakar, Senegal	8–12	Melvin N.A. Peterson N. Terence Edgar	Feb, 1970
3	Dec, 1968–Jan, 1969	Dakar, Senegal to Rio de Janeiro, Brazil	13–22	Arthur E. Maxwell Richard P. Von Herzon	Apr, 1970
4	Feb–Mar, 1969	Rio de Janeiro, Brazil to San Cristobal, Panama	23–31	Richard G. Bader Robert D. Gerard	Sep, 1970
5	Apr–Jun, 1969	San Diego, California to Honolulu, Hawaii	32–43	Dean A. McManus Robert E. Burns	Oct, 1970
6	Jun–Aug, 1969	Honolulu, Hawaii to Apra, Guam	44–60	Alfred G. Fischer Bruce C. Heezen	Feb, 1971
7 Parts 1 & 2	Aug–Sep, 1969	Apra, Guam to Honolulu, Hawaii	61–67	Edward L. Winterer William R. Riedel	Aug, 1971
8	Oct–Dec, 1969	Honolulu, Hawaii to Papeete, Tahiti	68–75	Joshua I. Tracey, Jr. George H. Sutton	Nov, 1971
9	Dec, 1969–Jan, 1970	Papeete, Tahiti to Balboa, Panama	76–84	James D. Hays Harry E. Cook, III	Jan, 1972
10	Feb–Apr, 1970	Galveston, Texas to Miami, Florida	85–97	J. Lamar Worzel William Bryant	Mar, 1973
11	Apr–Jun, 1970	Miami, Florida to Hoboken, New Jersey	98–108	Charles D. Hollister John I. Ewing	Jun, 1972
12	Jun–Aug, 1970	Boston, Massachusetts to Lisbon, Portugal	109–119	Anthony S. Laughton William A. Berggren	Aug, 1972
13 Parts 1 & 2	Aug–Oct, 1970	Lisbon, Portugal to Lisbon, Portugal	120–134	William B.F. Ryan Kenneth J. Hsü	Jan, 1973
14	Oct–Dec, 1970	Lisbon, Portugal to San Juan, Puerto Rico	135–144	Dennis E. Hayes Anthony C. Pimm	Sep, 1972
15	Dec, 1970–Feb, 1971	San Juan, Puerto Rico to Cristobal, Panama	146–154	N. Terence Edgar John B. Saunders	Dec, 1973
16	Feb–Mar, 1971	San Cristobal, Panama to Honolulu, Hawaii	155–163	Tjeerd J. van Andel G. Ross Heath	Apr, 1973

17	Apr–May, 1971	Honolulu, Hawaii to Honolulu, Hawaii	164–171	Edward L. Winterer John I. Ewing	Aug, 1973
18	May–Jul, 1971	Honolulu, Hawaii to Kodiak, Alaska	172–182	LaVerne D. Kulm Roland von Huene	Jun, 1973
19	Jul–Sep, 1971	Kodiak, Alaska to Yokohama, Japan	183–193	Joe S. Creager David W. Scholl	Jun, 1973
20	Sep–Nov, 1971	Yokohama, Japan to Suva, Fiji	194–202	Bruce C. Heezen Ian D. MacGregor	Nov, 1973
21	Nov, 1971–Jan, 1972	Suva, Fiji to Darwin, Australia	203–210	Robert E. Burns James E. Andrews	Oct, 1973
22	Jan–Mar, 1972	Darwin, Australia to Colombo, Ceylon	211–218	Christopher C. von der Borch John G. Sclater	Feb, 1974
23	Mar–May, 1972	Colombo, Ceylon to Djibouti, F.T.A.I.	219–230	Arabian Sea Robert B. Whitmarsh Oscar E. Weser Red Sea Robert B. Whitmarsh David A. Ross	Apr, 1974
24	May–Jun, 1972	Djibouti, F.T.A.I. to Port Louis, Mauritius	231–238	Robert L. Fisher Elizabeth T. Bunce	Oct, 1974
25	Jun–Aug, 1972	Port Louis, Mauritius to Durban, South Africa	239–249	E.S.W. Simpson Roland Schlich	May, 1974
26	Sep–Oct, 1972	Durban, South Africa to Fremantle, Australia	250–258	Thomas A. Davies Bruce P. Luyendyk	Dec, 1974
27	Nov–Dec, 1972	Fremantle, Australia to Fremantle, Australia	259–263	John J. Veevers James R. Heirtzler	Sep, 1974
28	Dec, 1972–Feb, 1973	Fremantle, Australia to Christchurch, New Zealand	264–274	Dennis E. Hayes Lawrence A. Frakes	Jul, 1975
29	Mar–Apr, 1973	Lyttleton, New Zealand to Wellington, New Zealand	275–284	James P. Kennett Robert E. Houtz	Jan, 1975
30	Apr–Jun, 1973	Wellington, New Zealand to Apra, Guam	285–289	James E. Andrews Gordon Packham	Dec, 1975
31	Jun–Aug, 1973	Apra, Guam to Hakodate, Japan	290–302	Daniel E. Karig James C. Ingle	May, 1975
32	Aug–Oct, 1973	Hakodate, Japan to Honolulu, Hawaii	303–313	Roger L. Larson Ralph Moberly	Aug, 1975
33	Nov–Dec, 1973	Honolulu, Hawaii to Papeete, Tahiti	314–318	Seymour O. Schlanger Everett D. Jackson	Feb, 1976
34	Dec, 1973–Feb, 1974	Papeete, Tahiti to	319–321	Robert S. Yeats	Apr, 1976

APPENDIX I
(CONTINUED)

Leg #'s	Port to Port Dates	Ports	Sites	Co-Chiefs	Date of Publ.
		Callao, Peru		Stanley R. Hart	
35	Feb–Mar, 1974	Callao, Peru Ushuaia, Argentina	322–325	Charles D. Hollister Campbell Craddock	Aug, 1976
36	Apr–May, 1974	Ushuaia, Argentina to Rio de Janeiro, Brazil	326–331	Peter Barker Ian W.D. Dalziel	Jan, 1977
37	May–Jul, 1974	Rio de Janeiro, Brazil to Dublin Ireland	332–335	Fabrizio Aumento William G. Melson	May, 1977
38	Aug–Sep, 1974	Dublin, Ireland to Amsterdam, The Netherlands	336–352	Manik Talwani Gleb Udinstev	Dec, 1976
39	Oct–Dec, 1974	Amsterdam, The Netherlands to Cape Town, South Africa	353–359	Katharina Perch-Nielsen Peter R. Supko	Aug, 1977
40	Dec, 1974–Feb, 1975	Cape Town, South Africa to Abidjan, Ivory Coast	360–365	Hans M. Bolli William B.F. Ryan	Jul, 1978
41	Feb–Apr, 1975	Abidjan, Ivory Coast to Malaga, Spain	366–370	Yves Lancelot Eugen Seibold	Jan, 1978
Supplement to volumes 38, 39, 40, 41			336–370		Sep, 1978
42 Part 1	Apr–May, 1975	Malaga, Spain to Istanbul, Turkey	371–378	Kenneth J. Hsü Lucien Montadert	Apr, 1978
42 Part 2	May–Jun, 1975	Istanbul, Turkey to Istanbul, Turkey	379–381	David A. Ross Yuri P. Neprochnov	Apr, 1978
43	Jun–Aug, 1975	Istanbul, Turkey to Norfolk, Virginia	382–387	Brian E. Tucholke Peter R. Vogt	Jun, 1979
44	Aug–Sep, 1975	Norfolk, Virginia to Norfolk, Virginia	388–394	William E. Benson Robert E. Sheridan	Nov, 1978
45	Nov, 1975–Jan, 1976	San Juan, Puerto Rico to San Juan, Puerto Rico	395–396	William G. Melson Philip D. Rabinowitz	Jan, 1979
46	Jan–Mar, 1976	San Juan, Puerto Rico to Las Palmas, Canary Islands	396A–396B	Leonid Dmitriev James Heirtzler	Jan, 1979
47 Part 1	Mar–Apr, 1976	Las Palmas, Canary Islands to Vigo, Spain	397	Ulrich von Rad William B.F. Ryan	Sep, 1979
47 Part 2	Apr–May, 1976	Vigo, Spain to Brest, France	398	Jean-Claude Sibuet William B.F. Ryan	Nov, 1979
48	May–Jul, 1976	Brest, France to Aberdeen, Scotland	399–406	Lucien Montadert David G. Roberts	Aug, 1979

49	Jul–Sep, 1976	Aberdeen, Scotland to Funchal, Madeira Islands	407–414	Bruce P. Luyendyk Joe R. Cann	Mar, 1979
50	Sep–Nov, 1976	Funchal, Madeira Islands to Funchal, Madeira Islands	415–416	Yves Lancelot Edward L. Winterer	Jun, 1980
51, 52, 53 Parts 1 & 2	Nov, 1976–Apr, 1977	San Juan, Puerto Rico to San Juan, Puerto Rico	417–418	Leg 51 (Nov, 1976–Jan, 1977) Thomas W. Donnelly Jean Francheteau Leg 52 (Jan–Mar, 1977) Wilfred B. Bryan Paul T. Robinson Leg 53 (Mar–Apr, 1977) Martin F.J. Flower Matthew Salisburg	Feb, 1980
54	May–Jun, 1977	Cristobal, Panama Canal Zone to Long Beach, California	419–429	Bruce R. Rosendahl Roger Hekinian	Dec, 1980
55	Jul–Sep, 1977	Honolulu, Hawaii to Yokohama, Japan	430–433	E. Dale Jackson (deceased) Itaru Koizumi	Sep, 1980
56, 57 Part 1	Sep–Dec, 1977	Yokohama, Japan to Yokohama, Japan	434–441	Leg 56 (Sep–Oct, 1977) Marcus Langseth Hakuyu Okada Leg 57 (Oct–Dec, 1977) Roland von Huene Noriyuki Nasu	Nov, 1980
58	Dec, 1977–Jan, 1978	Yokohama, Japan to Okinawa, Japan	442–446	George deVries Klein Kazuo Kobayashi	Aug, 1980
59	Feb–Mar, 1978	Naha, Okinawa to Apra, Guam	447–451	Loren Kroenke Robert Scott	Jan, 1981

WORLD 1:40,0

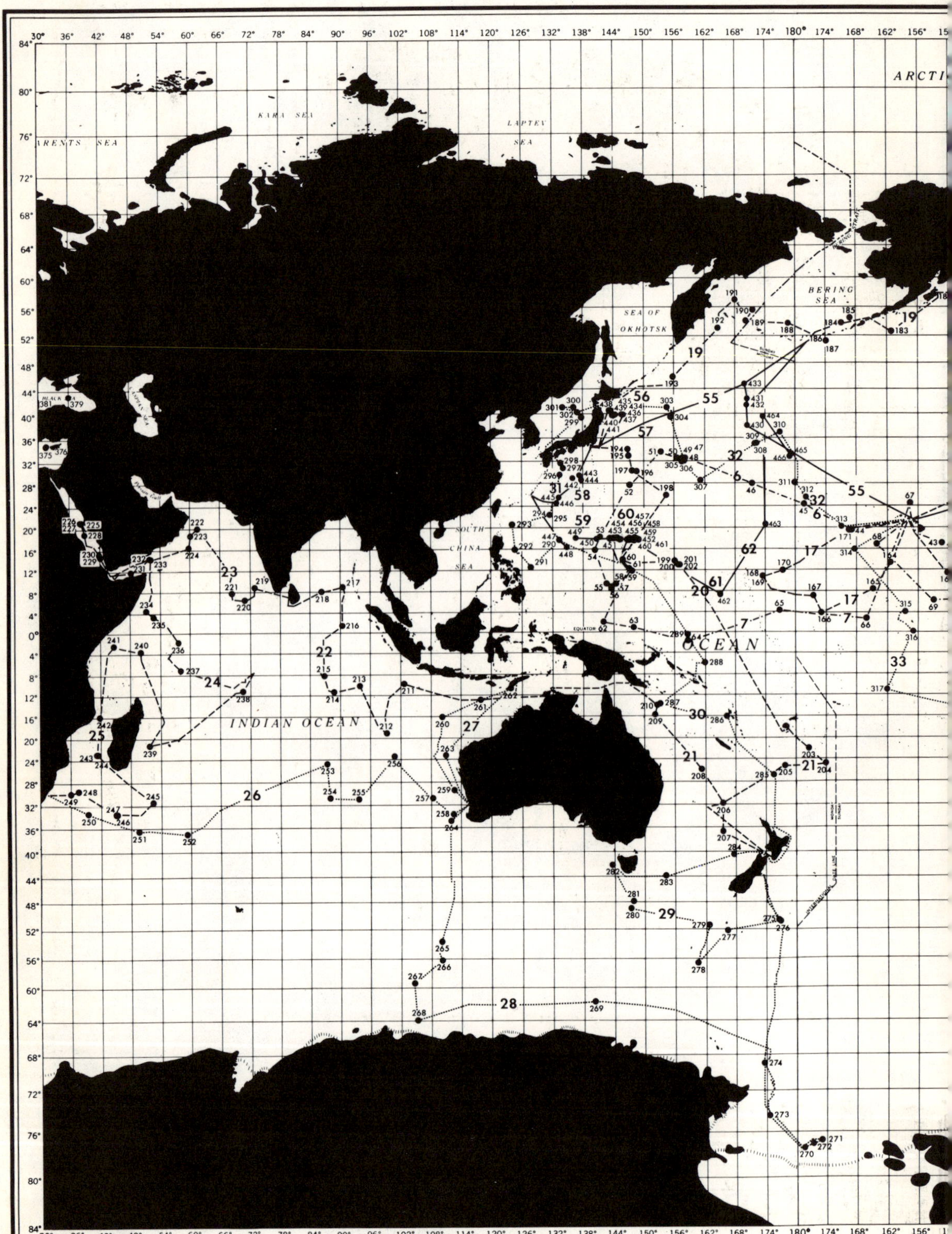

ARCTI
KARA SEA
LAPTEV SEA
BARENTS SEA
BERING SEA
SEA OF OKHOTSK
SOUTH CHINA SEA
INDIAN OCEAN
OCEAN
BLACK SEA
EQUATOR